S0-EDJ-956

Underwater Robotics

Science, Design & Fabrication

Underwater Robotics
Science, Design & Fabrication

Steven W. Moore, Ph.D.
California State University Monterey Bay

Harry Bohm

Vickie Jensen
Westcoast Words

Marine Advanced Technology Education (MATE) Center
Monterey, CA

Nola Johnston
Design and original illustrations

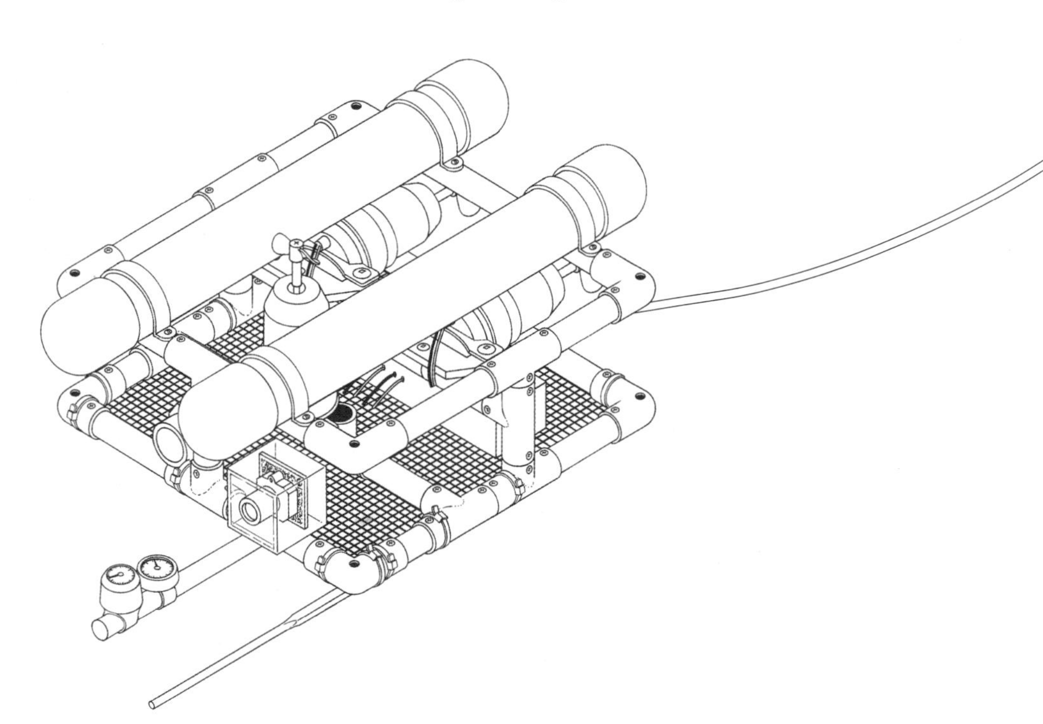

Authors: Steven W. Moore (PhD), Harry Bohm, Vickie Jensen

Editor: Vickie Jensen
Design and Illustration: Nola Johnston, Nola Johnston Graphic Design & Illustration
Textbook Project Coordinator: Jill Zande

Copy Editing: Ann Harmer, Bluff Hollow Editing
Chapter Consolidation: Caroline Brown
Indexing: Patricia Buchanan, Buchanan Indexing & Editing
Cover photo: © Patrick J. Endris/AlaskaPhotoGraphics.com
Back cover photos: Chess-playing ROVs, Scott Fraser, Long Beach City College, Electrical Department; ROV-building students and ROV in ocean, Dr. Steven W. Moore

Printed and bound in Hong Kong

Marine Advanced Technology Education (MATE) Center
980 Fremont Street
Monterey Peninsula College
Monterey, CA 93940
www.marinetech.org

Publisher's Cataloging-in-Publication data

Moore, Steven W.

Underwater robotics : science, design & fabrication / Dr. Steven W. Moore, Harry Bohm, Vickie Jensen ; Vickie Jensen, editor ; Jill Zande, textbook project coordinator.

p. cm.

Includes index.

ISBN 978-0-9841737-0-9

1. Remote submersibles--Design and construction. 2. Remote submersibles--Handbooks, manuals, etc. 3. Robotics. 4. Underwater exploration. 5. Underwater navigation. I. Bohm, Harry. II. Jensen, Vickie, 1946-. III. Underwater robotics : science, design, and fabrication. IV. Title.

TC1662 .M66 2010

623.82/05—dc22 2010926065

TABLE OF CONTENTS

PREFACE BY DREW MICHEL

Life member and fellow in the Marine Technology Society (MTS), chair of the MTS ROV Committee since 1992, co-chair of the annual Underwater Intervention Conference, senior member of the Institute of Electrical and Electronic Engineers (IEEE). Drew Michel has worked for over four decades in the undersea industry.

Underwater Robotics: Science, Design & Fabrication is the text and reference book that underwater robotics educators have been waiting for. Best of all, it lives up to expectations, with an amazing collection of technical information, stories, and photos in one convenient place.

The primary purpose of this book is to introduce newcomers to the very complex world of subsea technology, but it is also a fascinating look back at the development of subsea inventions for work and exploration. This text explains where we are today, explores possibilities for the future, and encourages individuals to consider careers in this field. Equally as important, the book provides readers with a sense of perspective—a look back at where we've traveled on this long subsea journey and the interesting characters who led the way. It starts with understanding the consequences of human exposure to pressure, moves through the courageous era of sitting in cramped submersibles, and arrives at today's reality of telepresence in the undersea world.

Underwater Robotics: Science, Design & Fabrication is an invaluable tool for young minds working on starter projects and provides the material to help them move to advanced options and continue to learn. It is also a reference book for those working in the industry and will be a valuable addition to the collections of those who have had long careers in the undersea arena. Picture a young student learning how to solder for the first time. Then see that student a year later, wiring the control system on his or her team's entry in the MATE International ROV Competition. Then, fast-forward 10 years to that same person, now an ROV supervisor, browsing through a well-worn copy of the book while sitting in the ROV control room of a major offshore energy project or scientific expedition. This book has the potential to inspire and enable those dreams.

Finally, a word about those who made this book happen. There are people in this world who go to work every day to earn enough money to be able to live and do the things they enjoy. Then there are others whose work is their love. The educators I have had the good fortune to deal with in my 44 years in this industry fit that latter category. The people at the MATE Center and many others who worked long and hard to make this book a reality certainly are part of that fraternity. They deserve high praise for their perseverance and dedication.

PREFACE BY DON WALSH, PhD

Captain USN (ret), oceanographer, marine policy specialist, former dean and professor (University of Southern California), Honorary President of the Explorers Club, and member of U.S. National Academy of Engineering. In 1960, Don Walsh and Jacques Piccard piloted the U.S. Navy bathyscaph Trieste *to the deepest point in the world's oceans, a feat never repeated.*

Underwater Robotics: Science, Design & Fabrication is most welcome in the world of ocean engineering. It is a well-organized survey of all major aspects of underwater engineering, and it leads students, educators, industry professionals, technology enthusiasts, and other interested readers through each subject area in an intelligent and engaging way.

Ocean engineering is a relatively new field of study. It is also an interdisciplinary field, in which a basic knowledge of mathematics, earth sciences, physics, computer technology, business, and the engineering arts and sciences is required, even for entry-level and support staff. While there are many excellent academic programs throughout the world that train ocean engineers and offer degrees in oceanography, the area of technical education for professionals who support ocean engineering work is still in the early stages of development. Qualified professionals are in high demand, yet few comprehensive training programs or learning materials are available—until now.

This book fills a real need for a broad introduction to oceanographic work. The stories, technical information, scientific knowledge, and projects provide an excellent "tour of the horizon" of this fascinating, wide-ranging interdisciplinary field. Some of the essential concepts and principles that apply to underwater technology can be daunting to those who are new to this work; the authors are to be congratulated for providing easy access to complicated material.

Underwater Robotics: Science, Design & Fabrication is produced by the Marine Advanced Technology Education (MATE) Center at Monterey Peninsula College in California. Supported by the National Science Foundation since 1997, the MATE Center works with schools and colleges nationwide to raise awareness of ocean science, technology, and engineering fields. MATE also strives to provide education and professional development for individuals working in these fields, and to recruit and train new technical professionals. This book is a significant contribution to that mission.

INTRODUCTION

There is a story behind every good book, and this one is no exception. Our story begins in 1999 with the MATE Center's discovery of the book *Build Your Own Underwater Robot and Other Wet Projects*, co-authored by Harry Bohm and Vickie Jensen. It was perfect timing, as our marine workforce studies indicated that the skills needed to design, build, and operate underwater vehicles (and other "wet" technologies) were in high demand. We immediately contacted Harry about expanding the book into a series of modules for a new college-level course on underwater robots at Monterey Peninsula College. Harry wrote those modules, keeping one step ahead of instructor Frank Barrows; both finished up amazed at the enthusiasm and learning going on in that classroom. In the summer of 2000, Harry delivered MATE's first summer institute focused on ROVs. The teachers who participated were also hooked. We knew that we were on to something!

Later that year, MATE asked Vickie and Harry to edit the modules into a full-scale textbook. As that book started to take shape, so did MATE's very first ROV competition. MATE called upon Harry's knowledge and expertise in the development of that program as well. From engineering the underwater props to creating the stories that put the mission tasks in context, Harry helped to build the competition from the ground up—just as he helped so many of the students to build their first ROVs.

As the competition gained momentum over the next few years, so did the textbook. In 2003, Dr. Steve Moore joined the writing team. His content expertise and classroom teaching experience were perfect complements to Harry's ability to think outside of the box and turn everyday materials into incredible inventions and Vickie's editorial prowess, organizational skills, and ability to channel creativity into words on a page.

Building underwater robots often takes more time, more prototypes, more money, and more effort than ever envisioned at the start of the project. It turns out that the same is true for underwater robot textbooks! The process of this writing project has taken us on an incredible journey of underwater robotics education, one that saw the modules MATE commissioned for our very first college course grow into this 769-page textbook. It also involved our summer faculty development institutes and MATE's annual competition program that engages thousands of teachers and students from around the world.

Underwater Robotics: Science, Design & Fabrication wouldn't have been possible without Harry setting the vision and Steve and Vickie contributing their knowledge and expertise to carry it out. The MATE Center is forever grateful to these individuals and the contributions that they have made to our mission.

And this is where our story ends—or, rather, begins with you and the book that you hold in your hands. We look forward to seeing how you write the next chapters.

—Jill Zande, MATE Associate Director and textbook project coordinator

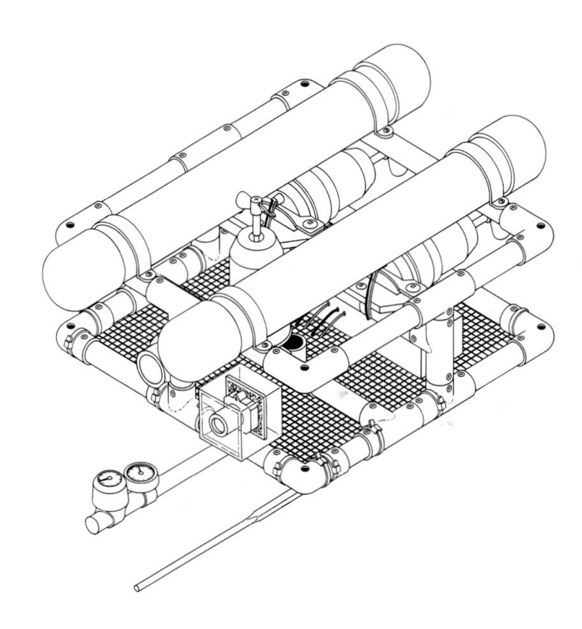

HOW TO USE THIS BOOK

Underwater Robotics: Science, Design & Fabrication welcomes you to the exciting world of underwater vehicle technology! This book provides an engaging blend of real-life stories, up-to-date technical information, and pragmatic "how-to" advice. It also guides you through the process of designing, building, and diving your own small custom-made, remotely-controlled underwater craft. You can use the *SeaMATE* ROV to carry out various underwater missions, from basic exploration to challenging search and recovery adventures.

The book is written primarily for high school and college students who want to have fun while developing marketable technical and teamwork skills. It is also a handy reference for educators, hobbyists, and others interested in marine technology, electronics, robotics, and similar topics. As an introductory book, this text assumes no prior experience with robotics or underwater vehicles; however, it does presume a working knowledge of high school algebra and geometry.

This textbook is structured to be highly flexible. If you are new to underwater vehicle design and construction (as many readers will be), it's probably best to read the chapters in order, since each builds to some extent on the previous chapters. On the other hand, if you already have experience with much of the material covered in this book, you may find it more useful to use this text as a reference and jump directly to the chapter you need at the moment. The choice is up to you.

How the Text is Organized

Underwater Robotics: Science, Design & Fabrication is not a conventional textbook, but then subsea technology is not a conventional academic discipline—it's interdisciplinary! So this book covers physics, marine biology, oceanography, math, chemistry, construction, tool use, design strategies, trouble-shooting, project planning, and teamwork—all subjects that will help you create more successful underwater vehicles.

Chapters 1–2 provide an introductory overview of manned and unmanned underwater craft (from earliest times to modern day) and offer pragmatic strategies for designing underwater robots.

Chapters 3–10 introduce the opportunities and challenges of working under water, then focus on specific technical solutions, ranging from structure and materials to power systems and payloads. Each of these chapters introduces relevant science and engineering concepts, as well as their pragmatic application to underwater robotics. There are ample visual and textual examples, from basic home-built robotic vehicles to complex commercial craft.

Chapter 11 is a different kind of technical chapter, one that moves beyond vehicle design and concentrates on the real world of subsea work and the logistics of preparing for and carrying out simple missions.

Chapter 12 provides you with an opportunity for guided hands-on learning, with detailed plans and instructions for assembling a "build-as-you-learn" ROV called *SeaMATE.* Readers can either construct this shallow-diving project in stages, coordinating it with their progress through each of the technical chapters, or they can tackle it independent of the other chapters. Chapter 12 also discusses options for upgrading *SeaMATE* or for building even more sophisticated ROVs or AUVs.

Textual Aids and Information

- Each of the chapters begins with a section called *Stories from Real Life* that highlights a real-world event related to the main focus of that chapter.
- All chapters also include an outline, a list of learning outcomes, the main text of the chapter, and a summary.
- The text features more than 500 photographs and illustrations. This wealth of color photographs helps readers connect with actual examples of subsea robotics. The straightforward illustrations are designed to expand and explain textual concepts.
- The authentic examples in this text help bridge the gap between theory and the excitement of on-the-job experience. While many of these focus on North America, the authors stress the worldwide application of underwater technology and exploration. Teachers and students are encouraged to supplement this text with other, locally relevant examples.
- Most measurements are given in metric units, such as meters and newtons; however, imperial equivalents are usually provided to assist those readers who may be more familiar with imperial (English) units, such as feet and pounds.
- The glossary contains definitions of hundreds of key terms. These are emboldened when first presented or explained in the text.
- A set of appendices provides helpful tables of information, useful facts, equations, and suggestions for sourcing parts, materials, and additional resources.
- A complete index helps locate page references to relevant topics as well as types and properties of vehicles, plus names of people, institutions, and individual craft.

UPDATES AND ORDERS

Book Orders

To get pricing and shipping information and to order individual copies of the textbook online, contact Westcoast Words at the address below. You are also welcome to call them for information regarding bulk orders, educational or wholesale discounts, and shipping costs.

Westcoast Words
Email: info@westcoastwords.com
Phone: 604-731-5565
www.westcoastwords.com

Updates and Comments

In the interest of keeping future editions of this book as accurate and current as possible, please forward any comments, corrections, or updates to the MATE Center.

Marine Advanced Technology Education (MATE) Center
Monterey Peninsula College
980 Fremont Street
Monterey, CA 93940
Phone: 831-645-1393
Email: info@marinetech.org
www.marinetech.org

DISCLAIMERS

Views Expressed

Any opinions, findings, conclusions, or recommendations expressed in this material are those of the authors and the Marine Advanced Technology Education (MATE) Center and do not necessarily reflect the views of the National Science Foundation or the Marine Technology Society's ROV Committee.

Safety Alert

Throughout the text, the authors have made a good-faith effort to alert readers to potential hazards and have stressed the importance of safety and common sense. However, it is not possible to anticipate and discuss every conceivable situation that might arise during the construction, testing, or operation of an underwater vehicle. Ultimately, it is up to each individual to make sure that he or she is aware of potential hazards and is operating within his or her safe limits.

Therefore, the authors and the Marine Advanced Technology Education (MATE) Center can accept no liability for accidents or injuries incurred during or in association with the construction, testing, or operation of any of the projects suggested herein. In particular, note that this book does NOT provide instruction in the design, construction, testing, operation, or maintenance of submersibles that carry living passengers. Likewise it does not cover boating and water safety skills or the safe use of high voltage power sources.

Endorsements

Today there are thousands of excellent parts and materials that are designed for or can be adapted to use in underwater robotics projects. The textbook has endeavored to give a representative sampling, however it is not possible to include mention of every product manufacturer or distributor. Note that where there is mention of a commercial product or citation of a trade name it does not constitute an official endorsement of that product by the authors or the Marine Advanced Technology Education (MATE) Center. Readers are always encouraged to conduct their own product research. *Appendix V: How to Find Parts* provides tips and suggestions for locating parts and materials suitable for a variety of underwater projects.

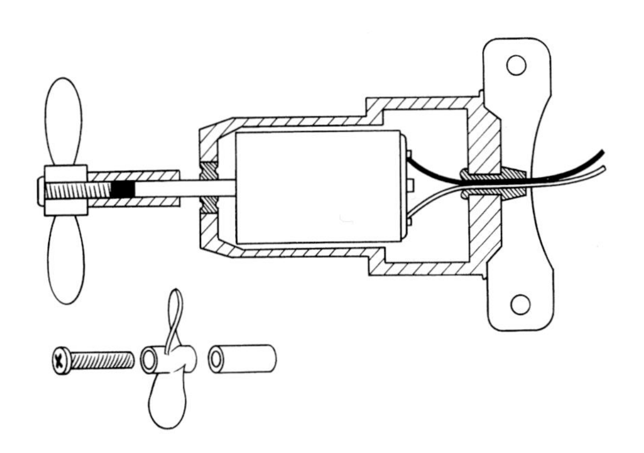

ACKNOWLEDGEMENTS

MATE salutes the many individuals, companies, organizations, universities and schools who have been extremely generous in sharing expertise and resources. Your efforts have made a significant contribution to the scope and quality of this book. While it is not possible to name everyone who helped, MATE would specifically like to acknowledge those that follow. We also salute readers like you, who dare to dream.

Heartfelt thanks to you all,

—Jill Zande, MATE Associate Director and textbook project coordinator

Funding

This textbook is based on work supported by the National Science Foundation under Awards DUE-0703197, DUE-0832284 and DUE-0302905, and printed with the assistance of the Marine Technology Society's ROV Committee. The MATE Center is extremely grateful for this on-going support.

MATE Leadership and Advisors

- Monterey Peninsula College, especially Michael Gilmartin, Dean of Instruction, Occupational and Economic Development
- MATE Center Director Deidre Sullivan
- Members of MATE's Textbook Advisory Committee
    - Frank Barrows, instructor (ret.) in Submersible and Automotive Technology, Monterey Peninsula College
    - Kevin R. Hardy, Director of Engineering, Deep Sea Power and Light formerly with Scripps Institution of Oceanography/UCSD
    - Drew Michel, MTS ROV Committee chair, CEO and Principal Consultant, ROV Technologies, Inc.
    - Ronald W. Roehmholdt, Director of Exhibits, Naval Undersea Museum
    - Al Trice, General Manager, ISE (International Submarine Engineering)
- MATE's National Visiting Committee (past and present members)
    - Dr. Lynne Carter, NOAA RISA Program, Southern Climate Impacts Planning Program
    - Dr. Thomas R. Consi, University of Wisconsin, Milwaukee
    - Elizabeth Corbin, President of Marine Technology Society
    - Dr. Sylvia Earle, Deep Ocean Exploration and Research
    - Dr. Robert Ford, Neo-Tech South, Inc.
    - Bruce Gilman, past President of the Marine Technology Society
    - Jim McFarlane, International Seabed Authority
    - Justin Manley, Liquid Robotics
    - Drew Michel, MTS ROV Committee chair, CEO and Principal Consultant, ROV Technologies, Inc.
    - John Peterson, Oceaneering International, Inc.
    - Jill Singer, State University of New York, Buffalo
    - Fritz Stahr, University of Washington
    - Jerry Streeter, past President of the Marine Technology Society
    - Dr. Don Walsh, Captain USN (ret), Honorary President of the Explorers Club, U.S. National Academy of Engineering member

Technical and Editorial Reviewers

- Mark Atherton, Kongsberg Mesotech
- John Bergman, CyVect Corporation
- Knute Brekke, Deputy Chief Pilot, ROV *Ventana*, Monterey Bay Aquarium Research Institute
- Ted Brockett
- Dr. Tom Consi, Shaw Associate Scientist, University of Wisconsin-Milwaukee, Great Lakes WATER Institute
- Nicole Crane, former MATE Center Director
- Dr. James P. Delgado, Executive Director, Institute of Nautical Archaeology
- Lee-Ann Ennis, marine educator, formerly with Vancouver Public Aquarium
- Tomiko Johnston, engineer, OceanWorks International
- Bill Kirkwood, Associate Director of Engineering, Monterey Bay Aquarium Research Institute
- Dr. Paul Kraeutner, Underwater Research Lab, Simon Fraser University
- Guy Immega, CEO (ret.), Kinetic Sciences Inc.
- Glenn McDonald, Woods Hole Oceanographic Institution
- Dr. Tom Murphree, Research Professor, Naval Postgraduate School
- Phil Nuytten, President, Nuytco Research Ltd.
- Scott Olson, Project Engineer, Perry Slingsby Systems
- Dr. John Pollock, President, Underwater Archaeological Society of British Columbia
- Dr. Jay Powell
- Nels Powell, Jensen Technologies
- Peter A. Robson and Associates
- Mary Schendlinger, editorial consultant
- Peter Thain, along with Beckie Thain, Lydia Burnett, Virginia Davis, Aubrey Senyard of the White Rock South Surrey Home Educators
- Steve Thone
- Bob Wernli, First Centurion Enterprises
- Li Xeuman (Lisa Li), School of Engineering, Simon Fraser University
- Students from the very first Spring 2000 Introduction to Submersible Technology class at Monterey Peninsula College
- 2001 MATE Summer Institute class: Bah Aahl, Joe Beydler, Ike Coffman, Ted Davis, Bill Falls, Sharon Flanagan, Kelly Kiefer, Barry Kilch, Steve Moore, Erica Moulton, Susan Phillips, Tom Rebold, Gary Scott, Joe Shewmaker, Bill Speed, Joel Spencer, Edd Spidell, Steve Struble, Annemarie Sullivan, Sandra Taylor, Susan Teel, Norman Thomas, Morrie Walworth, Joseph Zawodny

Authors' Families

The authors greatly appreciate their extremely patient and supportive spouses and families.

PHOTO AND ILLUSTRATION ACKNOWLEDGEMENTS

MATE sincerely appreciates the individuals, companies, organizations, schools, and universities who gave permission to use photographs and images. This textbook is significantly better because of their generosity!

As we sorted through thousands of photographs for this project, we tried to provide correct credit and caption references. If we've inadvertently erred, please contact us so that we can make the correction in future printings.

Principle Photographers

Randall Fox and Steve Van Meter/VideoRay LLC deserve special thanks for their extensive photo documentation of numerous MATE regional and international competitions. The authors have also given freely from their personal photo collections and produced a number of new images to illustrate or clarify concepts presented in the text. All of these photographs make a significant contribution to the educational impact of this textbook.

Key Image Contributors/Consultants

- Scott Fraser, Long Beach City College, Electrical Dept.
- Dr. Daniel Jones, SERPENT Project Coordinator, National Oceanography Centre (Southampton, UK)
- Glenn McDonald, WHOI
- Drew Michel, Marine Technology Society's ROV Committee
- Duane Thompson, MBARI
- Steve Thone

Illustrations

The Naval Undersea Museum, Keyport, Washington, kindly made a number of their illustrations available for use throughout the book. Authors Harry Bohm and Dr. Steven W. Moore designed the majority of the technical illustrations, which were rendered expertly by Nola Johnston. Westcoast Words also shared illustrations from *Build Your Own Underwater Robot and Other Wet Projects* and *Build Your Own Programmable LEGO® Submersible.*

Companies

- Alstrom; Barbara Gardner, Wes Gerriets
- American Underwater Search & Survey, Ltd.; John P. Fish
- Atlantis Submarines; Janet Griffiths, Dereck Hamada
- Bluefin Robotics Corp.; Deanna Abraham
- BlueView Technologies, Inc.; Rick Elento
- Comsub; Rod McLean for HYCO
- CSIP; Simon Gilligan
- Deepsea Power & Light; Kevin Hardy
- Deep Sea Systems International, Inc. (Division of Oceaneering International, Inc.); Chris Nicholson
- Deep Marine Technology Inc.; Lisa Robinson
- Divulgação Petrobras/Agência
- EdgeTech Marine; R.J. Jablonski
- Electropaedia website http://www.mpoweruk.com/performance.htm; Barry Woodbank
- Flexible Engineered Solutions Ltd.; Ian Latimer, Vikki Mason
- GRI Simulations Inc.; Stephen Dodd, Cory Sheppard
- IFREMER, France
- Imagenex Technology Corp; Disa Wilhelmsen
- Infotainment USA, Inc.; Mark Ward
- International Submarine Engineering (ISE) Ltd.; Jim McFarlane, Linda Mackay, Al Trice
- Inuktun Services, Ltd.; Roy Coles, Colin Dobell, Shiho Uzawa
- J.W. Fishers Mfg.; Chris Coombs, Jack Fisher
- John Wiley and Sons, Inc., Permissions Dept.; Judith Spreitzer
- Kongsberg Mesotech, Mark Atherton
- Kongsberg Maritime Ltd.; Lisbeth Ramde
- Kraft Telerobotics
- L-3 Communications Klein Associates, Inc.; Debbie Durgin
- Nuytco Research Ltd.; Phil Nuytten
- OceanWorks International; Jim English, Lyn Garbay, Adam Goldbach, Derek White
- Oceaneering International, Inc.; Peg Newman
- Parallax, Inc.; Lauren Davis
- Perry Slingsby Systems, Ltd.
- Power-Sonic Corporation
- Remote Presence, Ireland; Karl Bredendieck
- Rinspeed Inc.; Frank M. Rinderknecht
- Saab Seaeye Ltd.; James Douglas
- Saipem America; Jennifer Williams
- Schilling Robotics; Meagan Anderson, Jason Stanley
- SeaBotix, Inc.; Jesse Rodocker
- SeaTools, B.V.; Jan Frumau, Berdien Kaper
- SMD Ltd. (Newcastle upon Tyne); Victoria Bosi, Mike Jones
- Sonsub (Division of Saipem Group); Robert Corkren
- Southwest Research Institute (SWRI); Deborah Deffenbaugh, Matt James, B.K. Miller, Jesse Ramon
- Stone Aerospace; Dr. Bill Stone
- Sub Aviator Systems Ltd.; Alfred McLaren
- SubConn Inc.; Michael Stewart
- Submertec Ltd.; Peter Sharphouse
- Teledyne Impulse; Heather Butler, Raymond Hom
- Transocean (Houston); Guy Cantwell, Dr. Ian Hudson
- Tritech; Paul Hudson
- VideoRay LLC; Brian Luzzi
- Wikimedia Commons
- Wikipedia Commons

Institutions and Schools

- AUVSI Foundation; Daryl Davidson, Executive Director
- California Institute of Technology, Jet Propulsion Laboratory
- California State University Monterey Bay, Seafloor Mapping Lab; Pat J. Iampietro, Dr. Rikk Kvitek
- Canadian Scientific Submersible Facility, Yvonne Baier
- Carbonear Collegiate
- Cornell University Autonomous Underwater Vehicle Team
- Deutsches Bundasarchiv (German Federal Archives)
- Flower Mound High School
- Friends of the Hunley; Kellen Correia, and cpgphoto.com
- Hawaii Undersea Research Laboratory; Bernard Greeson, Terry Kerby, Rachel Orange
- Harbor Branch Oceanographic Institution; Geoff Oldfather, Jan Petri
- Institute of Nautical Archaeology; Dr. James P. Delgado, Chastity Hedlund
- Japan Agency for Marine-Earth Science and Technology (JAMSTEC); Toshihiko Chiba, Yukako Kawaguchi, Eiichi Kikawa, Shozo Tashiro
- Long Beach City College
- Marine Advanced Technology Education (MATE) Center
- Marine Institute, Memorial University of Newfoundland
- Marine Technology Society, ROV Committee
- Massachusetts Institute of Technology (MIT) Project Orca (see MIT below)
- Massachusetts Institute of Technology (MIT) Sea Grant College Program; Nancy Adams, Dr. Chryssostomos Chryssostomidis, Franz Hover, John Leonard, Henrik Schmidt, Brandy Moran Wilbur
- Monterey Bay Aquarium Research Institute (MBARI); Lisa Borok, David Fierstein, Kim Fulton-Bennett, Bill Kirkwood, Debbie Meyer, Todd Walsh
- NASA
- National Science Foundation
- NOAA
- NOAA/AOML; Rick Lumpkin
- NOAA Aquarius Reef Base; Otto Rutten
- NOAA National Ocean Service, Pacific Islands Navigation; Lt Jeffrey Taylor
- Oceanexplorer NOAA
- Rancho Santa Fe School; David Warner
- San Diego iBotics Student Engineering Society; Paul Wisecaver
- Scripps Institution of Oceanography, Marine Physical Laboratory; Dr. M. Dale Stokes
- Scripps Institution of Oceanography, UC San Diego; Rob Monroe
- SeaPerch Project; Susan Giver Nelsen
- SeaTech 4-H Club; Lee McNeil, Lee Thieman
- Simon Fraser University, Underwater Research Lab
- Society of Naval Architects and Marine Engineers (SNAME); Susan Nelson
- SONIA Team (http://www.sonia.estmtl.ca); Kevin Larose
- Southampton Oceanography Centre, U.K.
- U.S. Geological Survey, Coastal and Marine Geology Program; Patrick Barnard and Daniel M. Hanes
- United States Naval Historical Center; Donald M. McPherson, Kent Weekly
- University of Bath; Nic Delves-Broughton, Vicky Just, Dr. William Megill
- University of California San Diego, Argo Program; Megan Scanderbeg
- University of Oslo, National History Museum; Hans Arne Nakrem
- University of Washington, Applied Physics Laboratory; Russ Light, Brian Rasmussen
- Vancouver Maritime Museum
- Woods Hole Oceanographic Institution; Andy Bowen, Jayne Doucette, Erika Fitzpatrick, Dave Fratatoni, Christopher Griner, Dr. Susan Humphris, Tom Kleindinst, Glenn McDonald, Stephanie Murphy, Joanne Tromp

Other Individuals

(not listed above)

- James Bellingham, MBARI
- Michael Broxton, MIT
- Dick Buist
- Dr. Thomas R. Consi
- Scott Coté, Coté Consulting
- Dr. Susan Crawford, Naval Undersea Museum Foundation (Keyport, WA)
- Gail Drake, Battlefield High School and Northern Virginia Community College
- Patrick J. Endris, AlaskaPhotoGraphics.com
- Jonathan Gero, Harvard University graduate student, now with Space Science and Engineering Center, U. of Wisconsin-Madison
- Al Harvey, the Slide Farm
- Jay Henson
- Jeremy Hertzberg, Monterey Peninsula College
- Jim Hutchinson, Historical Diving Society U.K.
- Jaochim Jakobsen, Rebikoff-Niggeler Foundation
- Ian MacDonald, Texas A&M
- Neil McDaniel
- Dr. Angela Leimkuhler Moran,United States Naval Academy
- Scott Olson, Perry Slingsby Systems
- Dylan Owens
- Michael Paris
- Toby Ratcliffe, Naval Surface Warfare Center, Carderock Division
- Donnie Reid

- Christina M. Shaw, U.S. Navy Journalist 2nd Class
- Anne Smrcina, Stellwagen Bank National Marine Sanctuary
- Robert Spencer
- John Andrew Terschak
- Peter Thain
- Dr. L.T. Wang, Professor, Dept of Mechanical Engineering, National Taiwan University of Science and Technology
- Dr. Fred Watson, California State University Monterey Bay
- John F. Williams, U.S. Navy Chief Journalist
- Michael Derek Wood, SubOceanic Sciences Canada Ltd.
- Kyle Worcester-Moore
- Jill Zande

Chapter 1

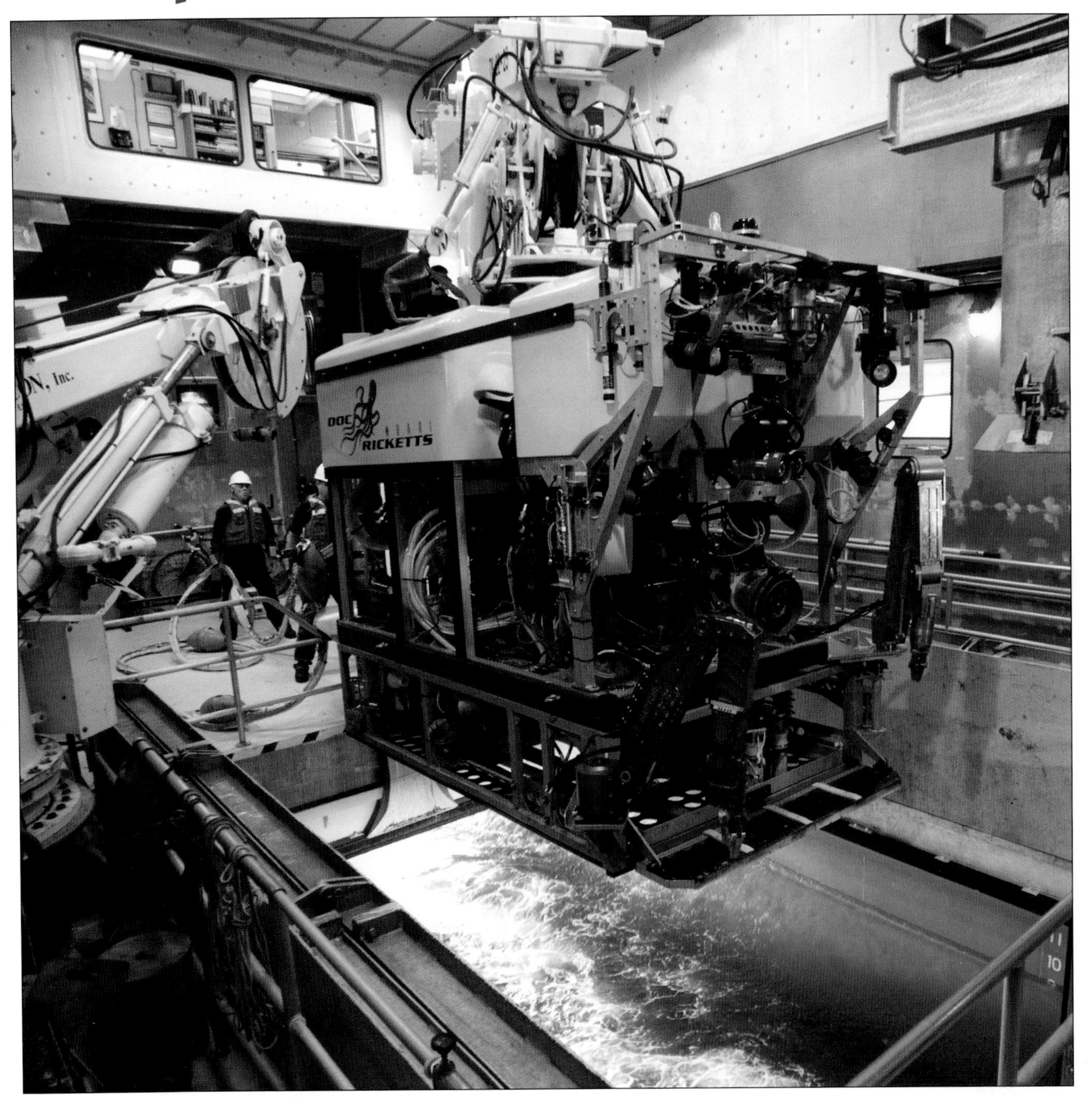

Underwater Vehicles

Chapter 1: **Underwater Vehicles**

Stories From Real Life: Cousteau and the Underwater World

Chapter Outline

1. **Introduction**
2. **Types of Underwater Vehicles**
   - 2.1. Submarines
   - 2.2. Submersibles
   - 2.3. Atmospheric Diving Suits
   - 2.4. Remotely Operated Vehicles (ROVs)
   - 2.5. Autonomous Underwater Vehicles (AUVs)
   - 2.6. Hybrid Underwater Vehicles
3. **A Brief History of Undersea Technology**
   - 3.1. Diving Bells
   - 3.2. Diving Gear
   - 3.3. Submarines and Related Technologies
   - 3.4. Submersibles for Deep Sea Research
   - 3.5. Robotic Undersea Vehicles
4. **Modern-Day Uses of Underwater Vehicles and Technology**
   - 4.1. National Defense
   - 4.2. Resource Extraction
   - 4.3. Science
   - 4.4. Telecommunications
   - 4.5. Construction, Inspection, and Maintenance
   - 4.6. Search and Recovery
   - 4.7. Archaeology
   - 4.8. Recreation and Entertainment
   - 4.9. Education
5. **A Detailed Look at a Work Class ROV**
   - 5.1. Power Source
   - 5.2. Control Room
   - 5.3. Operations Crew
   - 5.4. Deck Cable, Umbilical, and Tether
   - 5.5. Launch and Recovery System (LARS)
   - 5.6. The ROV
6. **Chapter Summary**

Chapter Learning Outcomes

- Describe the common types and uses of modern underwater vehicles.
- Identify motivating factors and key historic events in the evolution of underwater vehicles.
- Describe some of the major challenges confronted by developers of early underwater vehicles and describe how they were overcome.
- Name and describe the major subsystems of a modern work class ROV.

Figure 1.1.cover: MBARI's New* Doc Ricketts *ROV

The ROV Doc Ricketts *is the newest underwater research vehicle for the Monterey Bay Aquarium Research Institute (MBARI). It is shown here on a 2009 test dive, with launch through the moonpool of the research vessel* Western Flyer.

The Doc Ricketts *replaced the ROV* Tiburon, *which was built by MBARI staff in 1996. The new ROV features more sophisticated control systems and can handle much heavier loads. The* Doc Ricketts *was named after local marine biologist Ed Ricketts, who was made famous by John Steinbeck's book,* Cannery Row.

Image courtesy of Todd Walsh © 2009 Monterey Bay Aquarium Research Institute

STORIES FROM REAL LIFE: Cousteau and the Underwater World

In the 1950s and '60s, the world's attention was captivated by rockets, satellites, and the conquest of space. Suddenly, newspapers and TV broadcasts were full of images, scientific terminology, and even firsthand accounts of this previously unknown realm. The Space Race of the '60s was possible because of huge technological advances, particularly in data transmission, miniaturization, lightweight materials, and remote control of rockets.

These same significant advances were quickly adapted to another equally remarkable and little-known frontier—that of the ocean depths. The abyss was suddenly accessible. The star of this new era of oceanography was a Frenchman, Jacques-Yves Cousteau.

From an early age, Jacques Cousteau had two main preoccupations: water and photography. These passions would shape the course of his life. A sickly child, he pored over tales of the sea, reveling in the daring exploits of pearl divers, pirates, and smugglers. Against the advice of his family doctor, Cousteau learned to swim and tried out some of the underwater exploits he had read about. He quickly discovered the difference between the facts of diving and the fiction of diving storybooks.

Cousteau spent many an hour watching ocean-going ships and wondering why they didn't sink under their heavy loads. He was a boy who loved to tinker, and used his allowance to buy one of the first movie cameras available in France. Cousteau frequently took the camera apart and put it back together again. He was fascinated by the hardware, the chemicals, and the process of developing the film and making his own movies. When he was only 11 years old, his father brought home a blueprint for a marine crane. The boy used the drawings to build a 4-foot model of the crane. His proud papa showed it to an engineer, who pointed out that the boy had added his own improvement to the original design.

When it came time to choose a career, Cousteau opted to become a naval officer and documented his year-long working cruise with the same trusty movie camera. At the age of 26, a serious automobile accident ended his training as a navy flier and nearly cost him his life. Doctors wanted to amputate his left arm, but Cousteau refused and fought to recuperate. After eight months of painful therapy, he succeeded in wiggling one finger. He was on the mend! The French navy sent the young officer to recuperate on the Mediterranean coast, where Cousteau met Philippe Tailliez, another navy lieutenant, and Frédéric Dumas, a champion spear fisherman. The three became partners in adventure.

Tailliez encouraged Cousteau to swim in the sea, to strengthen his arm, and to take up the new sport of "goggling," the precursor to scuba diving. Cousteau vividly recalled the magic of his first clear view of the underwater world: "Sometimes we are lucky enough to know that our lives have been changed. It happened to me that summer's day, when my eyes opened to the world beneath the surface of the sea."

Scuba Gear

In the early 1940s, as World War II raged in Europe, Cousteau and his first wife, Simone, along with Dumas and Tailliez, cobbled together underwater diving masks, snorkels, and rubberized suits. Cousteau was determined to photograph under water, but couldn't find any easy way to do so. No quick solution presented itself, so he shot his first underwater movie without any breathing apparatus and by sealing the camera in a large glass jar!

These early frustrations pushed Cousteau to experiment with rebreathing apparatus. A French naval captain, Yves le Prieur, had already developed a type of diving gear that consisted of a cylinder of compressed air on the diver's back, connected by a hose to his face mask. But without any way to coordinate air pressure with water pressure, a diver was still restricted to shallow water. Using forged papers, Cousteau traveled to Paris in 1942, where he met Émile Gagnan in the Paris office of L'Air Liquide. The two men began joint experiments, adapting a type of regulator—one that allowed wartime automobiles to run on cooking gas instead of scarce gasoline—to one that would feed compressed air to a diver at his slightest intake of breath. The first test of their device in January 1943 was not completely successful, but they corrected the problems and patented the device as the Aqua-Lung. Their invention and the modifications that followed would change the world of diving and our perception of the planet.

Underwater Salvage and Documentation

The French navy agreed to form an Undersea Research Unit after the war, with Tailliez as commandant. Cousteau and his colleagues attempted increasingly deeper dives, working with more sophisticated underwater camera equipment. Cousteau dreamed of a research ship and helped arrange for a benefactor to purchase an old British minesweeper in 1950. The *Calypso* was refitted to become a floating research ship, complete with an underwater observation chamber. For years, Cousteau and a group from the French Navy Group for Undersea Study and Research worked to recover Greek and Roman treasures, among other projects. Determined to share the wonder of these undersea exploits with the world, Cousteau published *The Silent World* in 1953. The book was followed by a feature-length film of the same name and resulted in international acclaim. For many people, it was their first glimpse of the underwater world.

The Diving Saucer

As wonderful as scuba diving was, Cousteau dreamed of exploring even greater depths. But military submarines and deep diving bathyscaphs were simply too clumsy and too big. Instead, he envisioned "a radically new submarine, something small, agile." With financial backing from the National Geographic Society, L'Air Liquide, and the EDO Foundation, Cousteau began working with engineer Jean Mollard and chemist André Laban. He admonished them, "Throw the classic idea of submarines out the window and start with what we need."

The story goes that Cousteau picked up two soup plates and pressed them together, rim to rim, saying that was the shape he wanted for a sturdy diving craft. It should be highly maneuverable (this was more important than speed), able to carry one or two people, and light enough to be carried on board *Calypso*. Over the next six years, they worked on a design that featured a power plant and auxiliary systems outside the hull, freeing up more space for the humans inside. Crew should be comfortable, lying on their bellies on a mattress, instead of awkwardly kneeling, as in a bathyscaph. Émile Gagnan was the engineer responsible for the water-jet propulsion system, the ballast system, and a number of other unusual systems on this revolutionary submersible.

The small, yellow, turtle-shaped craft was only 2 meters (approx. 6 ft 5 in) in diameter and 1.5 meters (approx. 5 ft) high. Its pressure hull was 2 centimeters (approx. 3/4 in) of steel with a fiberglass covering, or fairing, around it that extended the saucer shape. *Hull No. I* had two windows, a top hatch, external lights, viewing ports, a hydraulic claw for collecting samples, and sophisticated cameras. It was built to descend to 350 meters (approx. 1,150 ft) for at least three hours. Best of all, it would be nearly as maneuverable as an underwater diver. Cousteau dubbed the innovative craft *La Soucoupe Plongeante*, or *Diving Saucer*.

The project suffered a significant setback during the very first unmanned test dive, when a cable retrieving the submersible snapped in rough seas. The diving saucer disappeared, finally settling on the bottom, 990 meters (approx. 3,250 ft) below. The only consolation for the inventors, when they picked up the sub's distant echo on sonar, was that the hull had survived a plunge to more than three times its proposed depth. Cousteau refused to give up. Over the next 18 months, technicians at the French Department of Underwater Research worked to complete *Hull No. 2*, or *DS-2*.

The final version resembled no previous underwater vehicle—it featured no propellers, rudders, or dive planes. Instead, it relied on patented hydrojets, mounted in the bow of the saucer that allowed it to spin on its axis and make extremely tight turns. Basic propulsion came from a jet pump on the stern that shot water streams out through flexible plastic pipes and produced speeds of up to 12 knots. A pilot could clamp either pipe to allow gradual turns. Tilting the sub up or down came from pumping mercury ballast between fore and aft trim tanks, another patented system. Unlike most of the deep water submersibles that would follow, this strange underwater craft could climb and dive at near-vertical angles, as well as bank.

Ballast was two 244.5-N (approx. 55-lb) pig iron weights and a 2,000-N (approx. 450-lb) emergency weight that hung underneath the hull. Occupants could fine-tune buoyancy by adding seawater to a small 45-liter (approx. 12-gal) tank situated between them. Best of all, there were 10 viewports that allowed occupants to use the cameras and offset strobe lights that Dr. Harold Edgerton of Massachusetts Institute of Technology (MIT) had perfected.

The crew likened the strange underwater craft to a comic-book version of a flying saucer. Unfortunately, the first underwater trial of the second hull (*DS-2*) in 1959 was nearly as disastrous as that of the first, when the copper battery compartments short-circuited and caught fire. However, Cousteau's vision paid off when the *Diving Saucer* and its crew finally filmed underwater life at a depth of 300 meters (nearly 1,000 ft). The unusual pictures dazzled scientists, the general public, and even Cousteau himself. Writing from inside the diving saucer, he noted, "We have no idea what the gray blobs are. We are probably the first creatures to see such a thing alive." Over the next three decades, this unusual underwater diving saucer took geologists and biologists on more than a thousand dives. Some of these missions ventured deeper than 300 meters—over three times greater than the limits of scuba diving—to explore the bizarre life forms that existed in absolute darkness and under tremendous pressure. This revolutionary vehicle heavily influenced the design of later submersibles, such as *Deep Quest* and *Deepstar*.

Underwater Habitats

Cousteau felt that if diving deep was good, then staying under water longer would be even better. So in 1962, he set up a manned habitat on the ocean floor. His *Continental Shelf Station No. 1* (*Conshelf 1*) in the Mediterranean allowed two scientists to stay there for a week, using scuba gear for daily work sessions outside the habitat. *Conshelf 2*, in the Red Sea, became home to five marine researchers for month-long sessions. In 1964 Cousteau set up *Starfish House*, a more ambitious habitat in the Sudanese Sea. For a time, Cousteau's diving saucer even operated from the undersea garage of *Starfish House*, to service the underwater habitat.

Cousteau and his son Philippe continued to work aboard the *Calypso*. Beginning in the late 1960s, they hosted a series of TV specials highlighting their diving adventures throughout the world's oceans. Many years later, *Calypso* was accidentally rammed and sunk in Singapore. Cousteau died a year later, in 1997. Certainly, his life and lifetime accomplishments were built around his passions—water and photography. Along the way, his inventions and documentary work introduced the rest of the world to the wonders of life under water.

1. Introduction

Kangaroos don't live under water. They can't. Their anatomy, physiology, and behavior are adapted for life on dry land, where they inhabit a "sea" of air, not water. We humans are also specialized for life on land. Although both people and kangaroos can swim, neither swims as quickly as it runs or hops. More importantly, because both species use lungs instead of gills to breathe, neither can survive long under water—at least not without some additional air source.

Ah, but this is where humans and kangaroos differ. Unlike the roos, we humans are constantly inventing new technologies that extend the capabilities of our organic bodies. These technologies enable us to visit, explore, and even exploit places we could not otherwise go. We build airplanes that allow us to fly miles above the earth's surface. We engineer space suits that enable us to walk around on the moon. And we create submersible vehicles that transport us to and from the dark and mysterious depths of the ocean.

When it's not possible (or wise) to go to these places in person, our clever inventions allow us to explore these locations in other ways. For example, we can deploy sensors to record and/or transmit information about hostile environments, such as the inside of an active volcano. We can enjoy various degrees of **telepresence** (a virtual sense of being there) through remotely operated robotic vehicles that move at our command and send back images by means of onboard cameras. It's even possible to have autonomous (self-controlled) robots accomplish work for us at remote or inhospitable locations, such as under the polar ice caps.

This book celebrates human ingenuity and invites you to become an active participant in the development of new technologies for exploring otherwise inaccessible or dangerous locations. It is a book for humans. It focuses particularly on the creative processes of designing and building small, remotely controlled robots and similar vehicles for exploring and working under water. Although submarines, submersibles, and other human-occupied vehicles are included throughout the text as sources for ideas and inspiration, this book does not provide instructions for building human-occupied vehicles; it sticks to the design of small remotely controlled machines or fully self-contained robots for underwater use. In case water isn't your thing, it is worth noting that almost everything you'll learn in this book can also be applied to the design and construction of robotic vehicles and sensor systems for exploring *any* remote or inhospitable environment, whether under water, on land, under ground, in the air, or even on other planets!

This first chapter is intended to whet your appetite for the challenge and adventure of robotic underwater vehicle design. It describes common types of underwater vehicles; it highlights some of the remarkable inventions that led to today's sophisticated underwater craft; and it introduces you to the basic anatomy of a large, modern-day remotely operated underwater vehicle. This introductory overview sets the stage for subsequent chapters, each of which addresses a different set of challenges and offers solutions that will help you design and build your own successful underwater robots.

Certainly Chapter 1 covers a staggering amount of information, but it's all fascinating, meaty stuff. And in reality, it only scratches the surface of the exciting history, evolution, and recent developments of underwater technology. Enjoy!

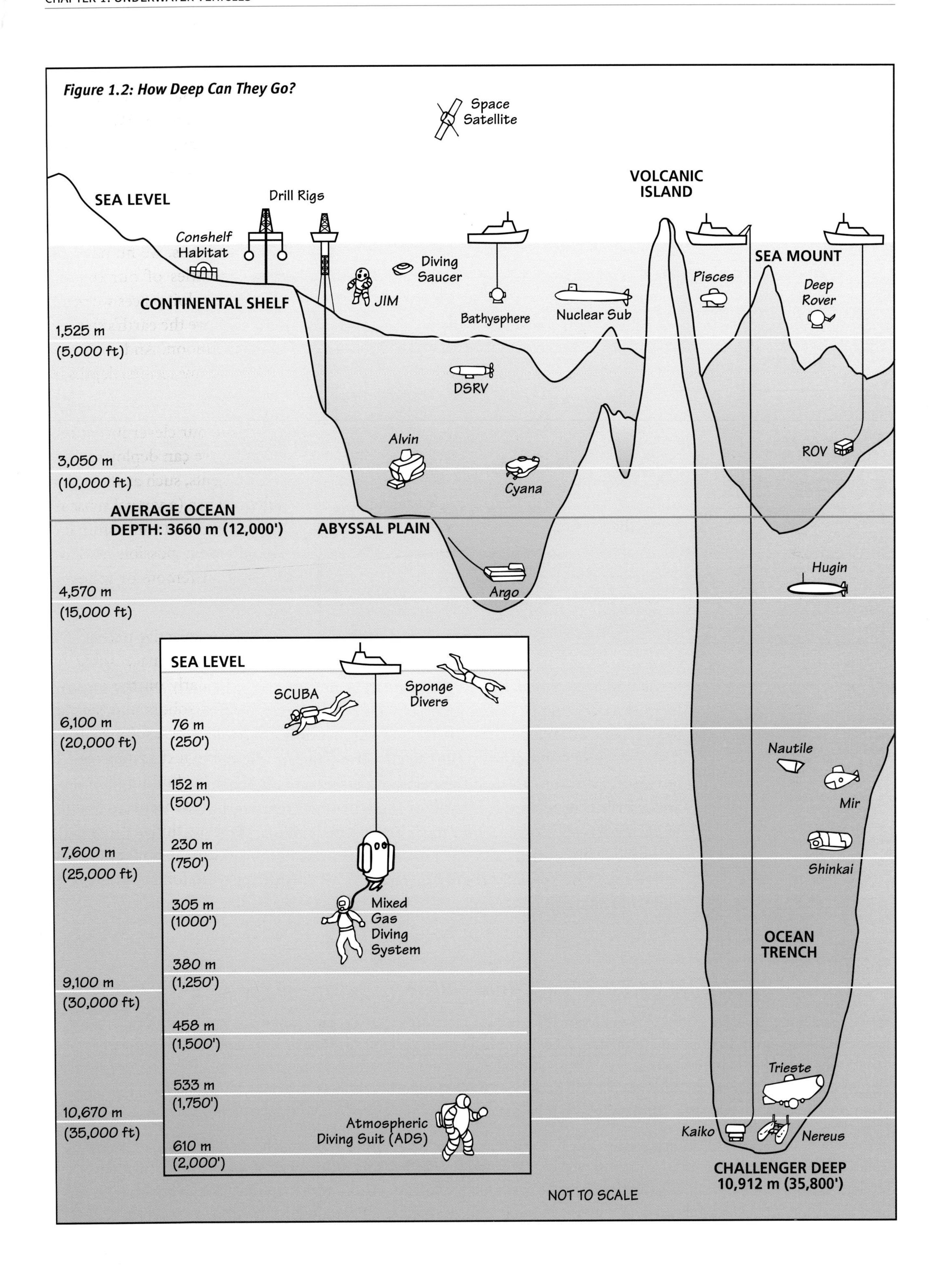

Figure 1.2: How Deep Can They Go?

2. Types of Underwater Vehicles

Underwater craft come in a wide variety of types and sizes and are used for many different purposes. (Figure 1.3.) Before proceeding much farther, it's useful to get a sense of what some of the major types of underwater craft are and how they are related.

Figure 1.3: The Range of Modern Underwater Vehicles and Structures

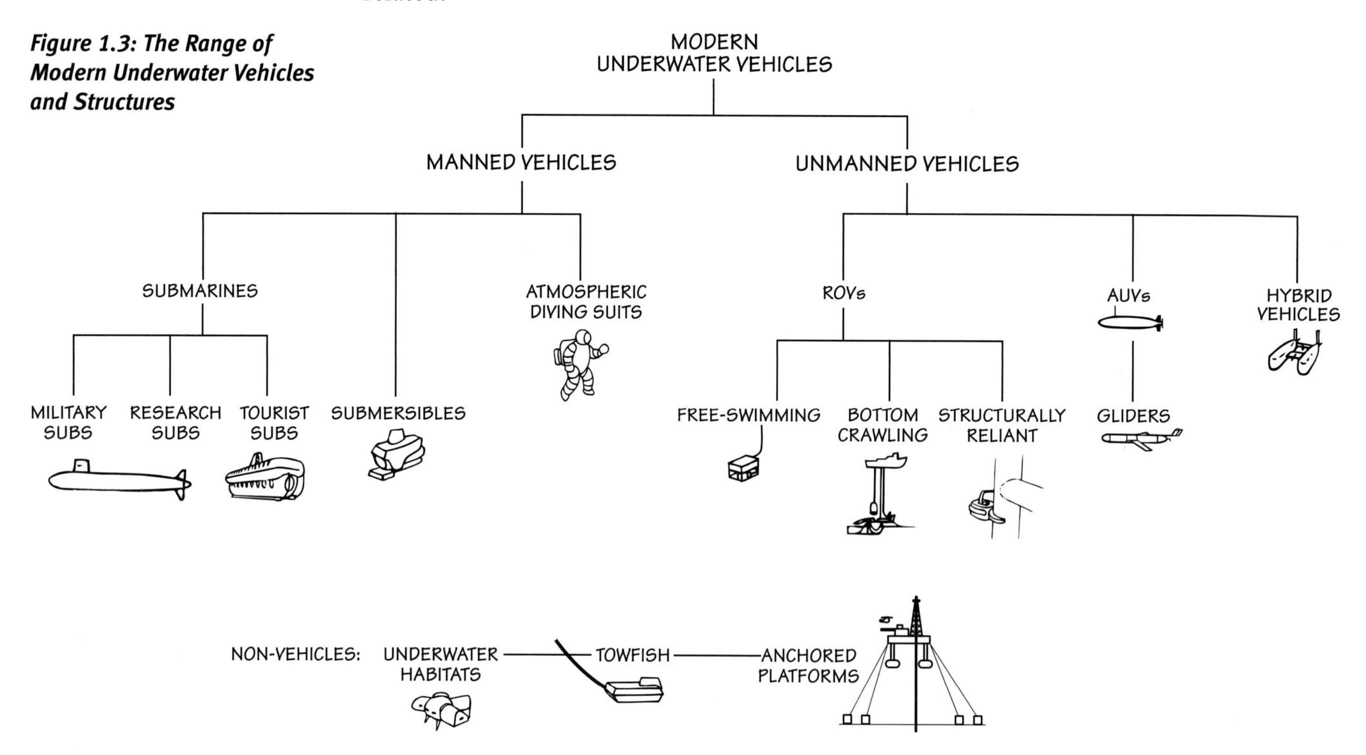

Some underwater vehicles carry human crew members or passengers and are known, for historical reasons, as **manned vehicles** (though women frequently "man" them, too). Examples include **submarines** and **submersibles.** Sections 2.1 and 2.2 (below) describe these vehicles in more detail. **Atmospheric diving suits (ADSs)**, which resemble space suits in many respects, are essentially one-person submersibles; they are described in Section 2.3. In all of these manned vehicles, human lives are at risk, so the design and construction must be extremely robust. The vessels are fitted with extensive safety systems and are subject to rigorous periodic inspections and regular maintenance by specially trained personnel. All of this makes them very costly to build, operate, and maintain.

TECH NOTE: WHAT'S A ROBOT AND WHAT'S NOT?

Strictly speaking, the term "robot" is generally reserved for machines that not only move but also exhibit the ability to sense and respond to their environments, all without direct human control. They usually contain computers or other programmable electronics that give them a rudimentary form of intelligence or at least the ability to make simple decisions and choose different actions based on sensor input. Sometimes they look human in form, but more often they are built in whatever form best suits their purpose.

In this book, the term "robot" is used broadly to include not only fully autonomous machines, like AUVs and conventional robots, but also remotely controlled mobile machines, like ROVs. This is not much of a stretch, since most modern ROVs include robotic capabilities to assist the human pilot. For example, many ROVs have depth and heading autopilots, which rely on computers to monitor vehicle sensors and automatically adjust propulsion systems to hold the vehicle in place. This frees the pilot to focus on other tasks.

An increasingly popular alternative is the use of **unmanned vehicles.** These are partially or fully robotic machines. They are usually operated by remote control or are programmed to operate entirely on their own without guidance from a human pilot. (See *Tech Note: What's a Robot and What's Not?*) Unmanned vehicles include **remotely operated vehicles** (**ROVs**), **autonomous underwater vehicles** (**AUVs**), and **hybrid underwater vehicles.** Each of these is described in more detail in Sections 2.4 through 2.6 below.

2.1. Submarines

Image by U.S. Navy Journalist 2nd Class Christina M. Shaw courtesy of Wikimedia Commons

Figure 1.4: Modern Military Submarines

Each of the Virginia *class (SSN-774) of U.S. submarines is powered by a S9G reactor that provides unlimited range. Pump-jet propulsors deliver quieter operation.*

Among the innovations incorporated in this class are extendable photonic masts in which three electronic imaging cameras replace the prisms and lenses of the old optical periscopes. Images are relayed via fiber optics to the control room.

Perhaps the best known of the manned (human-occupied) underwater vehicles are the modern military submarines, such as the Russian *Typhoon* class, the British *Vanguard* class, and the U.S. *Ohio* or *Virginia* class submarines. These are very large vessels, some nearly as long as two football fields (172 meters or approx. 564 feet) and weighing up to 48,000 tons when fully loaded and submerged. Most derive their power from nuclear reactors or diesel engines and can cover vast distances under water. Diesel-powered subs must periodically come to the surface, so they can run their engines to recharge their batteries, which they use for power while under water. Nuclear-powered subs can remain completely submerged for months at a time. They carry a sizeable crew and are used for military surveillance and stealthy bases from which to launch weapons, including nuclear missiles, that can be directed against either land-based targets or enemy fleets.

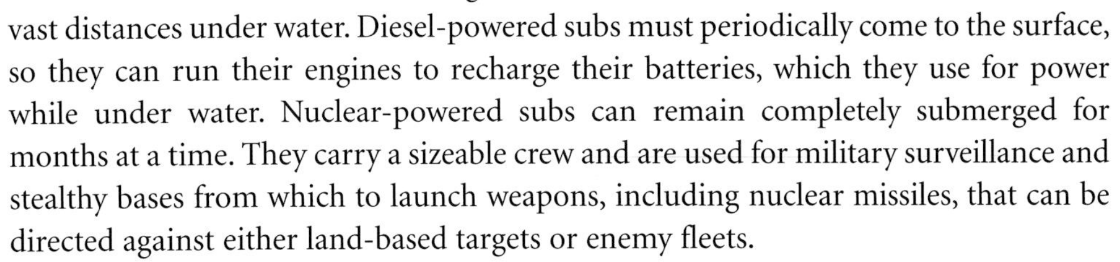

Not all modern submarines are so massive—or so deliberately stealthy. A few smaller ones conduct scientific research. More commonly, tourist subs make multiple dives per day, carrying groups of 25 to 50 passengers for a fish-eye view of underwater sights that few would see otherwise.

***Figure 1.5: Tourist Submarine* Atlantis**

The Atlantis *tourist submarines have introduced more than 11 million passengers to the underwater world at dive sites around the globe.*

Images courtesy of Atlantis Submarines

2.2. Submersibles

Submersibles are smaller-sized manned vehicles, commonly carrying two to three scientists and crew. Like the *Pisces V* in Figure 1.6, they may be fitted with a variety of sensors, robotic arms, and highly specialized gear to facilitate exploration and scientific research or other missions. While submarines operate independently, submersibles generally require the services of a nearby support vessel, or "mother ship," for launch, recovery, and maintenance.

Image courtesy of Hawaii Undersea Research Lab

Figure 1.6: **Pisces V**

The Hawaii Undersea Research Lab (HURL) utilizes two three-person submersibles for scientific research. Both Pisces V *and* IV *were built by International Hydrodynamics, have a maximum operating depth of 2,000 meters (approx. 6,500 ft), and typically undertake 6- to 10-hour daylight dives.*

Although unmanned vehicles (described below) are increasingly taking on missions that were once only carried out by submersibles, these versatile manned craft are still widely used for scientific applications. *Alvin* is probably the most legendary submersible of all, having taken more than 12,000 people on more than 4,000 research dives since its launch in 1964. The original 16-ton vehicle enabled scientists to document the existence of such unusual geological formations as "black smokers," the hydrothermal vents found on mid-ocean ridges. It was also involved in exploring the wreckage of the *Titanic.*

2.3. Atmospheric Diving Suits

The smallest manned vehicle (if you can even call it a vehicle) is an **Atmospheric Diving Suit**, or **ADS**—basically a one-person submersible. Most ADSs resemble an astronaut's space suit, and some are equipped with propulsion units. Unlike the soft suit worn by a scuba diver or the hard-hat rig of a commercial diver, an atmospheric diving suit has a rigid exterior that functions somewhat like a suit-of-armor hull to protect the occupant from exposure to extreme pressures. Complex articulated pressure joints allow the diver to move and work as deep as 600 meters (approx. 2,000 ft). Another significant advantage is that the diver works at surface air pressure inside the suit, so doesn't have to decompress when coming back to the surface, even after working in deep water. Still few in number, most modern ADSs are used for military purposes or undersea commercial work. (Figures 1.7 and 1.9 show ADS systems equipped with thruster packs for independent movement.)

2.4. Remotely Operated Vehicles (ROVs)

The most common unmanned vehicles are remotely operated vehicles, or ROVs. These robot-like machines allow their operators to pilot them from a relatively safe, dry, and comfortable place, usually on a ship or platform located at the surface of the water above the ROV. The pilot directs the ROV to perform its underwater work, usually communicating with the vehicle by means of a cable (called a **tether** or **umbilical**); this all-important tether also transmits electrical power to the robot and allows data, including audio, video, still images, navigational information, and other sensor readings to be sent back to the pilot.

Remotely operated vehicles range in size from small units that can be carried easily by one person up to large units the size of a small garage and weighing many tons.

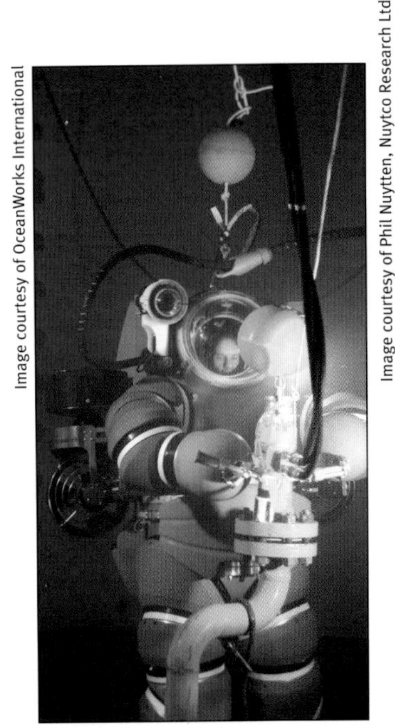

Image courtesy of OceanWorks International

Figure 1.7: **HardSuit Quantum *ADS***

Since the JIM *suit took its place in history as the first true working ADS, many versions have evolved, such as OceanWorks'* HardSuit Quantum *shown here.*

Image courtesy of Phil Nuytten, Nuytco Research Ltd.

Figure 1.8: The* JIM *Suit

Prior to the '70s, what we now call an "atmospheric diving suit" was generally called an armored diving suit, armored diving dress, or articulated diving suit. With the invention of the JIM *suit in the late 1960s, ADS came into common usage.*

The JIM *suit was the third generation of atmospheric diving suits developed by British inventor Joseph S. Peress;* Tritonia *was an earlier prototype. The final version was built by Underwater Marine Electronics Ltd. Later, Mike Humphrey and Mike Barrow fitted the* JIM *suit with limb joints; it was used until the 1980s. The* JIM *suit name relates to test-diver Jim Jarret.*

Image courtesy of Phil Nuytten, Nuytco Research Ltd.

Figure 1.9: The Propelled Atmospheric Diving Suit

Free-flying, more maneuverable ADS systems were first designed in the mid-1970s. Both the WASP and the Newtsuit/Hardsuit, shown here, can be equipped with a propulsion unit for independent movement.

Smaller ones may be used for scientific research, education, or inspection of difficult or dangerous liquid environments, such as the inside of pipelines and dams or the water-filled interior of a nuclear reactor. Drew Michel, chairman of the Marine Technology Society's ROV Committee, estimates that more than a thousand small ROVs are at work in a wide variety of sectors around the world. The commercial ROV industry is dominated by large **work class ROV systems**, with approximately 430 deployed worldwide. The bulk of these are owned by six major commercial operators. Most of these larger ROVs are equipped with two **manipulators** (robotic arms) as well as an interchangeable **tool sled** containing an assortment of tools for inspecting and maintaining the underwater components of oil rigs or underwater pipelines.

Regardless of size, ROVs generally fall into one of three sub-categories:

- tethered free-swimming ROVs
- bottom-crawling ROVs
- structurally reliant ROVs

2.4.1. Tethered Free-swimming ROVs

The tethered free-swimming ROVs category includes the majority of remotely operated vehicles and will be the focus of most of this book. Available in a wide variety of depth ratings and sizes, most free-swimming ROVs share the following characteristics:

- They are equipped with video/television cameras, so the pilot can see where the ROV is and where it's going.
- They are highly maneuverable, able to move up, down, right, left, forward, and backward.
- Most operate in mid-water or very near the bottom.
- Although some of these ROVs are powered by on-board batteries, the majority are powered from the surface through their tether/umbilical.
- Most free-swimming vehicles have their weight adjusted so that they just barely float, then use vertical thrusters (motors with a propeller attached) to get to depth.
- Aside from the smallest free-swimming ROVs, most are heavy enough and complex enough to require a surface control console, generator, winch, and crane to launch, operate, and recover them.

Tethered free-swimming ROVs are often further classified according to their cost and the heft of the work they can perform (light work, work, or heavy work systems).

TECH NOTE: THE RIGHT SIZE FOR THE JOB

Today's ROVs vary greatly in size and work. The largest, such as the *Triton XLS* or *Millennium Plus*, both at about 4,100 kilograms (approx. 9,000 lbs), are heavy-duty, commercial work class ROVs that are used to facilitate oil and gas explorations or to lay subsea telecommunications cable; a few assist with underwater construction. There are also large mission-specific ROVs for scientific research and subsea cable laying/maintenance. Any sizeable ROV employed at a deep water site is a powerful instrument, but requires a substantial support ship or oil-rig platform to support its mission.

Image courtesy of Dr. Daniel Jones, SERPENT Project, National Oceanography Centre, Southampton

Figure 1.10: Sizeable ROVs (top)

Since its commissioning trials in 2006, the UK Natural Environment Research Council's 6,500 meter (approx. 21,325 ft) depth-rated ROV ISIS *has tackled a range of diverse scientific missions, from researching marine life and how glaciers impact the ocean floor in Antarctica to investigating mud volcanoes off the coast of Portugal. It's shown here returning from its first trials dive in the Bahamas.*

Image courtesy of Brian Luzzi, VideoRay LLC

Figure 1.11: Eyeball ROVs (bottom)

At the other end of the size spectrum are less expensive, lightweight ***observation, "eyeball,"*** *or* ***micro-ROVs,*** *all of which are basically steerable underwater cameras. Their small size allows them to record subsea life at various depths, inspect nuclear reactor facilities, discover hazards, or conduct simple search missions. Some can even be hand-launched and recovered from a boat or dock. Eyeball ROVs are extremely useful and economical; not surprisingly, huge numbers of them are in operation worldwide.*

VideoRay is one of many companies producing excellent observation ROVs. In this photo, U.S. Park Service Submerged Resource Center diver Brett Seymour finished up a dive on Pearl Harbor wreckage using a VideoRay Pro 4, displayed in the foreground, and a VideoRay 3D High Definition ROV behind it.

TECH NOTE: TOW SLEDS

Towed mid-water and towed bottom craft are essentially underwater sensor platforms. Some people classify them as a type of ROV, but others argue that they shouldn't be, since these towed vehicles have no independent propulsion system. Regardless, they are exceedingly useful underwater devices. **Tow sleds**, as they are commonly called, range in size and may carry a wide range of sensors including cameras, seafloor mapping equipment, and other oceanographic gear. Depth is controlled by angling dive planes and/or by lengthening or shortening the towline. The principal advantage of tow sleds over other underwater vehicles is that they are relatively inexpensive to build.

Image courtesy of Woods Hole Oceanographic Institution

Figure 1.12: The* ARGO *Tow Sled

Although often dubbed a "dope on a rope," a tow sled is an underwater sensor platform that can provide a tremendous amount of information from the cameras and sonar sensors it carries. Working with a French-American team led by engineer Jean-Louis Michel, geologist Bob Ballard used ARGO*, equipped with large lights and cameras, in his search for the* Titanic*. By running* ARGO *close to the bottom in a long search-grid tow over a selected area, the wreckage of the famous ship was finally located on September 1, 1985. The original* ARGO *has since been upgraded to* ARGO II.

2.4.2. Crawling ROVs

Some ROVs are designed to crawl along the bottom or through pipes using caterpillar treads, legs, or suction cups. Most bottom-crawling ROVs used in commercial operations are heavily weighted ROVs with caterpillar treads like those of a bulldozer. These machines carry water jets, plows, or other tools for trenching and burying cable or pipeline in very deep water or for excavating the seabed. During the '70s, they were also used for mining manganese nodules (a potential ore source) on the ocean floor, but the extraction process proved to be cost prohibitive.

image courtesy of SMD Ltd., Newcastle upon Tyne

image courtesy of Inuktun Services Ltd.

image courtesy of SeaBotix Inc.

Image courtesy of Perry Slingsby Systems, Ltd.

Figure 1.13: Giant Trencher (top left)

Based in the UK, Soil Machine Dynamics (SMD) produces a variety of cable and pipe trencher systems. Its Ultra Trencher 1 (UT1) *is the size of a small house!*

Figure 1.14: Micro Pipe Inspector (top right)

Inuktun's series of tiny Versatrax vehicles, such as the Versatrax 150 *shown here, permit inspection of pipes in a range of pipe diameters. The vehicles can be configured for round pipe or flat surface work.*

Figure 1.15: ROV + Crawler (bottom left)

Attaching a crawler skid turns the LBV 150 SE-5² *into a versatile, stable inspection platform that can drive upside down over a ship's hull to inspect for corrosion, fouling, or other problems.*

Figure 1.16: Heavy-Duty Plows and Trenchers (bottom right)

Perry Slingsby Systems delivers a full line of specialized submarine cable plows and trenching/cable maintenance vehicles, such as the Triton T750 *pictured here with skid configuration.*

Considerably smaller vehicles with multiple caterpillar treads or wheels are designed to fit inside pipes and crawl along the length of the pipe with cameras or other sensors to inspect the interior for corrosion or other damage. Some small, free-swimming ROVs, such as SeaBotix's *LBV 150 SE-5*[2], can be fitted with a crawler skid attachment that enables it to switch from flying to driving on hard surfaces. Alternatively, the military and some universities have developed walking robots that use legs instead of treads or wheels to crawl along the bottom. Most of these are specialized for a variety of nearshore missions, such as detecting mines or sniffing out the source of toxic chemical leaks.

2.4.3. Structurally Reliant ROVs

As the name suggests, structurally reliant ROVs are attached to various types of underwater structures, such as ships' hulls or the pylons that support oil rig platforms. They are used primarily for cleaning and/or inspecting the structure with which they are associated. Most have no thrusters, moving instead by means of pulleys, cables, wheels, tracks, or various hydraulic mechanisms. (Note that some people do not classify these machines as ROVs, since they are not free to swim or crawl around independent of the structure.)

2.5. Autonomous Underwater Vehicles (AUVs)

"Autonomous" describes the independent nature of these tetherless underwater robots—an AUV carries its own power supply and has no physical link to the surface. Also, no pilot directly controls this type of robot. Instead, a pre-programmed computer or some sort of electronic circuit guides the vehicle on its mission. Onboard sensors connected to the computer bring in navigational information such as depth, speed, direction, and the location of potential obstacles. Additional systems may collect and store oceanographic data or other information for subsequent review by humans after the mission.

Image courtesy of International Submarine Engineering (ISE) Ltd. and IFREMER, France

Figure 1.17 (left):* Explorer *AUV

International Submarine Engineering (ISE) Ltd. has produced a series of Explorer *AUVs for IFREMER of Toulon, France. The autonomous vehicles can be fitted with various scientific devices for geo-physical, physical, and fisheries applications.*

Image courtesy of Dr. M. Dale Stokes, Marine Physical Laboratory, Scripps Institution of Oceanography

Figure 1.18 (right):* Fetch *AUV

The Fetch *AUV surveying in Bonaire, Dutch Antilles. The survey mission was part of a NOAA Office of Ocean Exploration expedition "Protecting a Shifting Baseline: Shallow to Deep Reefs on Bonaire."*

AUVs may return to the surface periodically during a mission to establish radio communication with a ship, shore facility, or satellites. Such a communication link can be used to retrieve data from the AUV, let the AUV update its position information, or provide the AUV with revised mission instructions. In some cases, AUVs can park themselves in underwater docking stations that are cabled to shore, thus enabling the AUV to recharge its batteries and exchange information with shore facilities through the cable.

Since they are tetherless, AUVs are usually easier to launch and recover than tethered ROVs. Most don't require large support ships and many are able to work in virtually inaccessible places, such as under the ice in polar regions. But these advantages are offset by some limitations, too. AUVs require very sophisticated systems for navigation and obstacle avoidance, so they tend to be expensive. They can store a limited amount of energy and are therefore restricted in the amount of work they can do before returning home to recharge batteries or otherwise replenish energy stores. Because of the physical properties of water that limit options for rapid, long-range data transmission, AUVs cannot transmit live video images nor easily receive complex commands while under water the way a tethered ROV can. And without such communication, they are entirely on their own in a very big ocean and can easily be lost without a trace if they become entangled, disabled, or simply fail to navigate accurately.

In spite of these challenges, AUVs offer tantalizing possibilities. Certainly, AUV design is a very active area of research, with new ideas and new systems being tested every day. The sizeable *Hugin* has already proven it can follow a pre-programmed track with high precision and efficiency, to create the detailed seabed maps necessary for deep water

TECH NOTE: GLIDERS

Gliders are a brilliant concept in AUV design, and they are revolutionizing the science of oceanography. A typical AUV glider looks like a torpedo with a pair of tiny wings and a tail rudder, but it has no propeller. When launched, the glider is "heavier" than water and sinks, gliding forward slowly as it does so, just as an airplane glider moves forward as it "sinks" through the air. And, just like an airplane glider, a glider-style AUV can cover vast horizontal distances without using any propellers, jets, or other engine power at all.

However, there's an exciting difference between these AUVs and an airplane glider: when the AUV nears the bottom of its descent, it doesn't have to land. Instead, a motor inside switches on briefly and drives an internal piston to expel some water from a chamber inside the vehicle. This reduces the effective weight of the glider until it starts to float, at which point it begins a long, slow *uphill* glide back to the surface. Once there, the vehicle receives position data from **GPS** satellites, so it knows exactly where it is, then relays that position information and any oceanographic data it has collected to the mission crew via satellite.

Image courtesy of Dave Fratantoni, Woods Hole Oceanographic Institution

Figure 1.19: Glider Power

This unique, thermally powered glider, developed by researchers and engineers at WHOI and Webb Research Corporation, uses heat energy from the water for its propulsive force. In this photo, David Sutherland, an MIT/WHOI Joint Program student in the Physical Oceanography Department, records the glider's ascent path during a test dive.

Then it can dive and repeat the process. The motor moves the piston in the opposite direction, taking in water to make the glider heavy again, and the vehicle resumes a new *downward* glide into the depths, using motion sensors and its rudder to control its direction. Thus, a glider is able to do a sequence of glides, repeating a down-and-up sawtooth pattern.

Since the onboard motors do not actually propel the vehicle, but instead operate only rarely and briefly (specifically, to change the piston position at the beginning of each long glide, and as needed to fine-tune the rudder position), these vehicles require very little stored energy. They can remain at sea for months and cover thousands of kilometers on a single battery charge. Newer designs are exploring ways to extend the duration and range of glider missions even more by extracting energy from thermal gradients or other features of the ocean environment. Small fleets of these gliders are now being deployed to collect oceanographic data from various locations and depths throughout the world.

oil and gas explorations. And AUVs that actually perform heavy physical tasks are under development. Other remarkable AUV developments include underwater gliders (See *Tech Note: Gliders*) and fleets of micro-AUVs that communicate with each other to accomplish complex tasks by working as a coordinated team.

Image courtesy of Woods Hole Oceanographic Institution and National Science Foundation

Figure 1.20: **Nereus**

The hybrid underwater vehicle Nereus *can operate tethered or untethered to a mother ship. Whether functioning as AUV (without tether) or ROV (with tether, manipulator arm, and toolsled),* Nereus *is participating in routine scientific investigation of ocean depths worldwide, including in the Mariana Trench near Guam, the deepest part of the world's oceans.*

Woods Hole Oceanographic Institution offered teams participating in the MATE's 2006 International ROV Competition the opportunity to name its new hybrid vehicle. The winning entry came from Monterey High School's ROV team. The students chose the name Nereus because that Greek mythological god was capable of changing forms, just as the hybrid craft does.

2.6. Hybrid Underwater Vehicles

The term "hybrid" generally suggests combining features of two or more normally distinct things, such as fruit trees that bear different varieties of apples or cars that can run on either gasoline or battery power. In the case case of unmanned underwater vehicles, it often refers to vehicles that combine the advantages of ROVs and AUVs in various ways to provide the best of both worlds. These are sometimes called hybrid remotely operated vehicles (HROVs).

Admittedly there are many different interpretations as to what "hybrid" means in the underwater world. One example is a vehicle that can switch between being an ROV and an AUV—it carries power and sophisticated navigational systems on board so it can operate independently as an AUV, but also allows for the connection of a thin, flexible **fiber-optic communication cable** for direct pilot control and live video feedback as an ROV. The ultra deep-diving *Nereus*, operated by Woods Hole Oceanographic Institution, exemplifies this type of hybrid. It can utilize AUV mode for surveying large areas of the deep ocean, then be reconfigured as an ROV for an in-depth visit to interesting-looking areas identified in the surveys.

TECH NOTE: IT'S CALLED A WHAT?

It would seem fairly easy to distinguish between the two major categories of underwater craft—they either carry humans or they don't, so they're manned or unmanned. But terminology does change with the times. For example, many use a new acronym for manned craft—HOV, for Human-Occupied Vehicles.

On the unmanned side of things, the Marine Technology Society prefers the umbrella term Unmanned Underwater Systems (UUS) for both ROVs and AUVs, while the U.S. Navy uses Unmanned Undersea Vehicles (UUV). Theoretically, this Navy term includes both tethered and tetherless robots; however, in actual practice, it's used pretty synonymously with AUVs. When it comes to hybrid underwater vehicles, some companies use the term HUV (for Hybrid Underwater Vehicle); others use ARV, a logical blend of AUV/ROV; still others use HROV for hybrid ROVs. Such "terminology wars" are common in science and industry.

Sometimes a vehicle name evolves, as was the case with the term "ROV." The original acronym for these new unmanned subs was RCV for Remotely Controlled Vehicle. However, RCV was also the trademark name for a specific robot manufactured by Hydro Products, so at a 1977 meeting of the RCV subcommittee of the Marine Technology Society's Submersible Committee, the members present agreed that from that day on, ROV would be the generic term—and it stuck.

Another interpretation of hybrid refers to ROVs, AUVs and other underwater devices working cooperatively as part of a greater system. For example, the hybrid AUV/ROV system *Swimmer*, by Cybernetix, uses an AUV to transport an ROV from an oil rig to an undersea docking station located on the bottom where work needs to be done. From there, the ROV operates on a relatively short tether, with power and communication provided to it through the docking station, which is permanently cabled to the rig. Thus, the ROV does not need a long tether descending all the way down from the surface through open water, and there's no need for the added expense of a separate surface support vessel to launch, tend, and recover the ROV.

A particularly exciting development in undersea vehicle technology is the coordinated use of several underwater robots working together as a team. ROVs, AUVs, drifters, surface vessels, aircraft, satellites, or other data-gathering technologies may work in collaboration to record data from several locations simultaneously, thereby providing a more integrated view of how various ocean processes interact across time and space. In some cases, large semi-autonomous vehicles are being used as undersea platforms from which to launch smaller craft.

3. A Brief History of Undersea Technology

Humans have been standing on shorelines or on boats and casting spears, traps, fishing nets, fishooks, or other inventions into the sea to pull food *up* from the depths for millennia. While these certainly qualify as "undersea technologies," the focus of this book is on a different kind of undersea technology—inventions that allow people to go down *into* the water, to look around and work in this mysterious, often beautiful, resource-rich world that is otherwise inaccessible to our air-breathing bodies.

It's often assumed that underwater technologies which allow humans to look around and work below the water's surface are all fairly recent inventions. Not so. For example, plausible designs for underwater breathing apparatus can be traced back more than 2,300 years. A number of different inventors were building and piloting small submarines well over 200 years ago. And unmanned robotic vehicles that provide humans with virtual eyes and hands beneath the waves were already working in the depths more than half a century ago.

These early inventors and the strange machinery they created still capture the imagination today. They inspire. They encourage us to dream and build. They also provide important lessons from their successes—and their failures. In fact, today's sophisticated subsea vehicles owe their existence to a long and colorful history of invention marked by one impressive breakthrough after another.

As you'll see, most of these advances were motivated by one (or a combination) of three broad goals:

- **Profit**: The expectation (or at least the hope) of making a living by harvesting marine life, recovering sunken treasure, tapping into offshore oil or gas reserves, or selling new undersea inventions to others has frequently inspired people to push the limits of existing subsea technology. Many individuals and corporations have taken great personal or financial risks to exploit undersea jackpots that nobody else had been able to reach before. Sometimes the profit motive has been strictly financial; other times it has includes fame or enhanced reputation.
- **Discovery**: Humans are explorers by nature. Having an immense expanse of wave-shrouded mystery literally lapping at people's toes has made undersea

exploration an irresistible activity since ancient times. These days, such exploration often takes the form of scientific research, whether it's mapping the ocean floor, cataloging new deep-sea species, or locating historic shipwrecks. It's been estimated that only one one-thousandth—possibly one ten-thousandth—of the seabed has been explored in any detail. Indeed, the surfaces of the moon and Mars have both been mapped with greater clarity than our own ocean floor, so much remains to be explored and discovered.

- **Military Advantage**: Warfare has always been a significant impetus to the development of new technologies, and undersea technology is no exception. In the earliest days of submarine development, the mere presence of an untried invention sometimes caused an enemy to withdraw its fleet. In more recent times, navies around the world have built or bought highly sophisticated submarines for the purposes of defense, offense, and undetected reconnaissance. Now these expensive manned vehicles are being supplemented (and in some cases replaced) with deep and shallow water robots.

While it is useful conceptually to think in terms of these three broad goals, it's important to remember that most advances in subsea technology can trace their roots to a combination of two or all three of these motivating factors. For example, many undersea inventions were (and still are) developed by individuals or companies hoping to make a profit by selling their designs to the military. And many undersea scientific discoveries have been made using technology originally developed for naval use. Thus, the history of underwater technology is a fascinating, complex, and interwoven one.

Another way to sort out your thinking about the history of undersea technology—and the one chosen to organize this section of the chapter—is to follow separately each of four interrelated threads in the evolution of subsea technology:

- The evolution of diving equipment, such as scuba gear, which has extended the capabilities of human divers
- The evolution of military submarines and related technologies
- The evolution of submersibles used for deep sea research
- The evolution of robotic technologies like ROVs and AUVs, which allow humans to work deep under water without going there physically

As with the motivations for advances in undersea technology, these threads are not isolated from one another. Each in its own way has influenced events in the other sectors.

3.1. Diving Bells

The very earliest records of people using simple equipment to help them work *under* water are drawings or written accounts of free-diving in the ocean to harvest shellfish, sponges, pearls, or other biological resources. The best "technology" (if it can even be called that) available to assist some of those divers included heavy rocks (held by the divers to help them descend rapidly to the bottom), ropes (used by those on a boat above to haul divers back to the surface quickly), simple goggles made from thin tortoiseshell or other materials (for clearer underwater vision), and knives, baskets, or other tools (to cut and carry the harvest). While undoubtedly helpful, these basic technologies did not solve the major limitation of free-diving—the fact that humans can hold their breath for a few minutes, at most, especially while actively working!

It might seem a simple matter to breathe through a long straw or other tube, but it turns out this won't work unless the diver is right at the surface. As a person descends, increasing water pressure squeezes the diver's chest, making it impossible to inhale. That's why you won't find extra long snorkels at any dive shop.

By the fourth century B.C., and possibly earlier, people had figured out a way to get around this problem and increase underwater endurance. Aristotle wrote about men (presumably Greek sponge divers) breathing under water by using the air trapped inside a large, heavy cauldron that had been lowered into the water upside down. You can see how this would work by inverting a drinking glass and shoving it under water to trap air inside. In this case, water pressure pressurizes the trapped air, too, making it possible for the diver to inhale. Although the supply of oxygen would soon run out and force the divers back to the surface, this simple technology allowed them to stay down and work much longer than was possible on a single breath.

The inverted cauldron was a predecessor to what later came to be known as a **diving bell**, so named because large church bells were sometimes adapted for this purpose. Suspended by a strong rope or chain, these bells were heavy enough that they could be depended on to stay upright under water, rather than tipping unexpectedly and releasing the trapped air. During the 1600s to 1800s, open-bottom diving bells became well known as a number of different styles were designed specifically for salvaging treasure from sunken ships. (See Figure 1.21.)

***Figure 1.21: Early Diver Working on the* Vasa**

The pocket of air trapped in the upper part of a diving bell allowed divers to breathe at 30 meters (approx. 100 ft) while working in bitterly cold water to salvage valuables from the sunken Swedish warship Vasa *(also written* Wasa*) in 1664.*

Using an assortment of tools, divers first loosened the ship's highly decorated bronze cannons from their carriages, then slung them through the gunports, and finally brought 50 of them to the surface.

During this same period, an English wool merchant named John Lethbridge decided to try a different approach to salvaging sunken treasure. In 1715, he invented a strong, person-sized, watertight, wooden barrel, which he called his "diving engine." It was equipped with a glass viewing port and two holes through which his arms could extend outside the barrel. Leather coverings wrapped tightly at the biceps kept the chamber watertight while permitting him to conduct salvage work using a variety of tools. Lethbridge communicated with surface tenders by a series of pull signals on a line secured to his wrist. Every few minutes, the diving chamber was hauled back to the surface, where assistants removed forward and aft plugs and used bellows to pump fresh air into the chamber. Then the plugs were replaced and the diver was lowered back into the water to continue his salvage work. Unlike a diving bell, Lethbridge's diving engine provided some protection for the diver from direct exposure to cold water and the intense underwater pressure. The system worked well, and Lethbridge became quite wealthy after salvaging treasure from a number of wrecks.

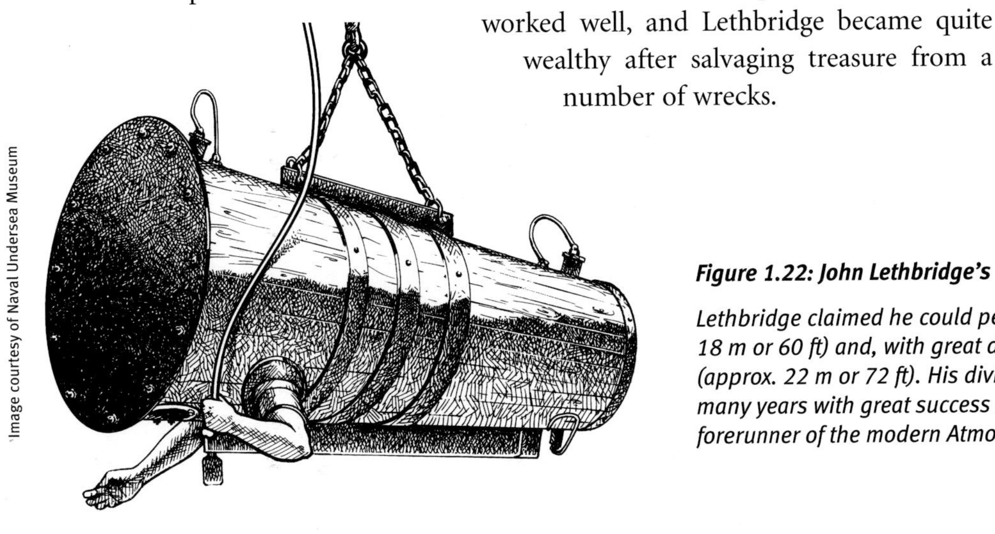

Image courtesy of Naval Undersea Museum

Figure 1.22: John Lethbridge's "Diving Engine"

Lethbridge claimed he could perform work at 10 fathoms (approx. 18 m or 60 ft) and, with great difficulty, achieved 12 fathoms (approx. 22 m or 72 ft). His diving engine was apparently used for many years with great success and can be considered the forerunner of the modern Atmospheric Diving Suit.

Figure 1.23: Hard-Hat Diving

The Siebe Gorman diving helmet was an integral part of standard diving dress, such as that worn by this WWII British hard-hat diver descending the ladder from the HMS Phoebe.

Image courtesy of Vickie Jensen

In spite of Lethbridge's success, diving bells continued to be a mainstay in underwater salvage. They were also used extensively in underwater construction projects, such as building foundations for bridge pylons. A major breakthrough in diving bell technology came when, in 1789, John Smeaton implemented an idea proposed 100 years earlier by Denis Papin. He used a pump to force fresh air from the surface down through a long hose into a diving bell. Not only did this continually replenish the diver's air supply, thereby giving him a limitless supply of oxygen, but it also eliminated another problem with diving bells—as a bell is lowered deeper into the water, increasing water pressure normally compresses the air inside, forcing it into a smaller and smaller volume in the top of the bell. By pumping pressurized air into the bell, the extra water could be forced out the bottom of the bell, restoring the original air volume inside and giving the divers more room to breathe at greater depths.

3.2. Diving Gear

Still, diving bells had some serious limitations. One was that they were terribly heavy and dangled straight down from the ship or dock. Divers had to remain in (or very near) the bell, so they were essentially tied to one spot on the seafloor and could not easily move about to explore the vicinity. That was all to change with the invention of the **hard-hat diving helmet**.

Image courtesy of Jim Hutchison, The Historical Diving Society

3.2.1. Hard-Hat Diving

The hard-hat diving helmet didn't look like a diving bell, although its function was somewhat similar. Essentially it was the equivalent of wearable air since the helmet trapped a bubble of air around the diver's head, and that air could be continuously replenished by Smeaton's air pump. However, unlike a large, heavy diving bell that couldn't be moved easily, the helmet was small enough to be carried on the diver's shoulders, making it portable. It had windows, so the diver could walk around on the seafloor and see what he was doing while wearing it. This new helmet had its beginning in 1823, when two English brothers and working divers, John and Charles Deane, applied for a patent on a smoke helmet they had used for fire rescue. Working with German inventor Augustus Siebe, they adapted the helmet for use in underwater salvage, a move that proved to be a major milestone in the development of modern diving apparatus. In 1827, Siebe and Gorman manufactured their first underwater helmet, complete with viewport and hose connections for surface-supplied air. Siebe continued modifications, adding a breastplate to which the helmet attached and a spring-loaded air exhaust valve. The design of this helmet and its associated suit, or "standard diving dress" as the equipment came to be called, was so sound and functional that it would serve as the principal feature of hard-hat diving for more than 150 years.

3.2.2. Rebreathers and Scuba Gear

Strangely, some technologies used under water had their origins under *ground.* In the late 1800's, resourceful inventors, such as Henry Fleuss, were working on the development of breathing devices to rescue workers from mines that had flooded or filled with toxic gas. Since this was prior to the days of extremely high-pressure gas cylinders, these devices had to recycle a person's air supply to maintain a compact size and weight, hence the name **rebreather**. During each breath, any oxygen consumed was replenished from a small oxygen canister, and exhaled carbon dioxide was chemically removed.

These rebreathers found early application not only in mine rescue, but also in firefighting, sport spearfishing, and the rescue of sailors trapped in submarines. Although rebreathers continue to be used today for specialized diving applications, they are complex devices, and they can be deadly if not maintained and adjusted just right.

In 1943, during World War II, French navy diver Jacques Cousteau, who had experienced some nearly fatal accidents on military rebreathers, teamed up with French Canadian engineer Émil Gagnan to invent a simpler, safer breathing system for divers. The pair modified a car engine valve to create a pressure regulator capable of delivering air from a high-pressure air cylinder to a diver at just the right pressure needed to balance the water pressure squeezing in on his chest.

Unlike a rebreather, their "open-circuit" system released the diver's exhaled breath into the water, rather than recycling it. Though less space- and weight-efficient than a rebreather and less suitable for very deep, extended, or stealthy dives, their "Aqua-Lung" (as they called it) retained many of the advantages of a rebreather. For example, it was self-contained and allowed divers to swim around freely, unencumbered by the long hoses and heavy weights typical of a hard-hat diving suit. Their simple and safe design caught on quickly and became a huge commercial success. Today's **scuba** gear (for **s**elf-**c**ontained **u**nderwater **b**reathing **a**pparatus) is a direct descendent of the Cousteau/Gagnan Aqua-Lung and relies on the same basic idea of using a pressure-regulating valve to feed air from a high-pressure cylinder to a diver at the same pressure as the surrounding water.

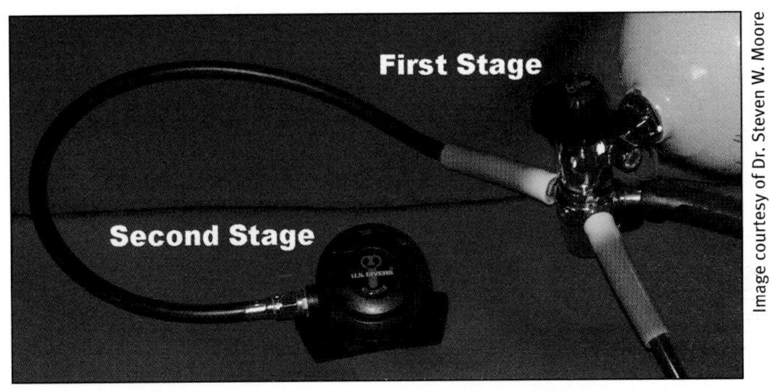

Image courtesy of Dr. Steven W. Moore

Figure 1.24: Modern Recreational Scuba Gear

Most recreational scuba divers today use the single-hose, two-stage regulator system illustrated here. The first stage, attached directly to the tank valve, reduces very high pressure air from inside the tank (typically about 3000 psi) to intermediate pressure in the hose (usually about 150 psi above the water pressure). The second stage, which fits in the diver's mouth, reduces the hose pressure to safe and effective breathing pressure by fine-tuning pressure in the mouthpiece to closely match the water pressure surrounding the diver's lungs.

Cousteau and Gagnan's invention revolutionized shallow water diving, but like earlier diving bells and hard-hat diving suits, the Aqua-Lung made underwater breathing possible only by feeding pressurized air to the diver's lungs, thereby enabling the diver to inhale against the external water pressure. Unfortunately, breathing pressurized air can cause problems. For example, air is mostly nitrogen, and beyond a certain depth, the increasing nitrogen pressure in a diver's air supply starts affecting the diver's brain in much the same way as a narcotic drug. The effect, called **nitrogen narcosis**, can cause divers to become careless, confused, or disoriented in deep water and may lead to fatal mistakes. It might seem better to fill a diver's tank with pure oxygen to get rid of the nitrogen, but oxygen becomes even more toxic than nitrogen at high pressures, causing sudden convulsions and death. One solution, widely used in technical diving, involves the use of specialized breathing mixtures containing carefully controlled proportions of helium, argon, or other gases (along with some oxygen), a technique called **mixed-gas diving**.

Another problem with breathing pressurized air for extended periods of time is that nitrogen dissolves under pressure into the diver's tissues. If the diver ascends too quickly, the dissolved nitrogen can form bubbles in the diver's tissues, just as carbon dioxide bubbles can form suddenly in a soda bottle when you release pressure by unscrewing the cap. The result can be a painful, debilitating, and sometimes fatal condition known as **caisson disease**, **decompression sickness** (**DCS**), or "**the bends**." To prevent this, a diver breathing compressed air must ascend toward the surface very slowly, stopping at intervals (called **decompression stops**) to let dissolved nitrogen diffuse gradually out of the tissues instead of forming dangerous bubbles.

For recreational divers, avoiding the bends is simple: follow prescribed dive tables or dive computer guidelines that keep dives short and shallow enough to avoid this problem. But for many military and commercial dive objectives, this simple solution is too limiting, because it may leave a diver without enough time to complete a task before needing to begin a return to the surface. In addition to mixed-gas diving (which can help reduce nitrogen bubbles), two other strategies were developed to enable divers to work longer and deeper without getting bent. The first was called **saturation diving**, a technique which enabled divers to work for extended bottom times in 30-90 meters (approx. 100-300 ft) of water and decompress only at the end of the job. (See *Historic Highlight: Saturation Diving and the World's First Aquanaut, and ADS.*)

The second was the atmospheric diving suit (introduced in Section 2.3 above). These rigid-walled diving suits resist external water pressure while providing the diver inside with air at regular atmospheric pressure. By allowing the diver to breathe air at normal sea-surface pressure, the need for decompression is avoided altogether.

HISTORIC HIGHLIGHT: SATURATION DIVING AND THE WORLD'S FIRST AQUANAUT, AND ADS

Robert Stenuit became the world's first aquanaut (a diver who remains on the seafloor for 24 hours or more) in 1962, utilizing Edwin Link's *Man-in-the-Sea* aluminum cylinder. This historic dive of 60 meters (approx. 200 ft) confirmed earlier theories by Captain George Bond of the U.S. Navy that living tissue would saturate to a limited extent at depth.

This meant that after sufficient duration, a diver's tissue gases would reach equilibrium at a particular water pressure, and that even increasing the time spent at that depth would require no corresponding increase in decompression time when the diver headed back to the surface. (Decompression refers to an ascent with periodic stops that allow for the nitrogen buildup in a diver's blood and tissues to safely diffuse, thereby avoiding decompression sickness.)

Today, **saturation diving** techniques regularly allow divers to work for extended bottom times in 30–90 meters (approx.100–300 ft) of water and decompress only at the end of the job. With mixed gases and modifications on standard lightweight helmets and hot-water suits, it is even possible for divers to work at depths of 300 meters (approx. 1,000 ft), using a diving-bell system for decompression.

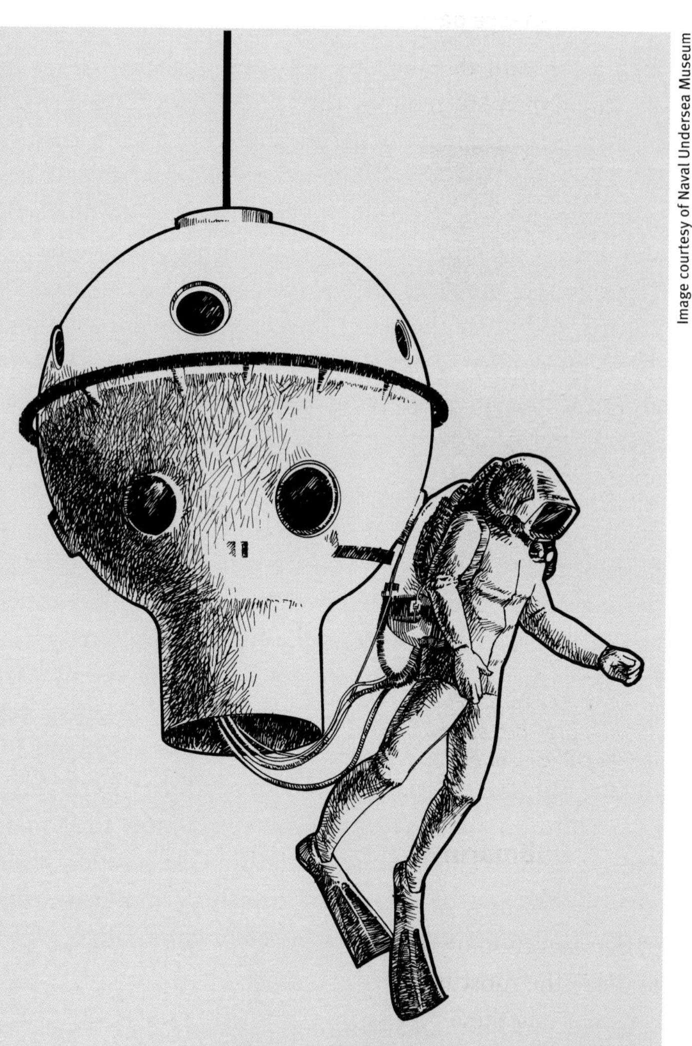

Image courtesy of Naval Undersea Museum

Figure 1.25: Pushing Diving's Depth Limits

Robert Stenuit and Edwin Link's Man-in-the-Sea *aluminum cylinder.*

HISTORIC HIGHLIGHT: UNDERSEA HABITATS

Undersea habitats—subsea working and living quarters anchored to the seafloor—are not underwater vehicles, because they are not designed to move around during normal use. However, their development certainly impacted the evolution of subsea research vehicles. Specifically, these underwater habitats were significant in generating information about how humans interact when they spend long periods in confined spaces and how they are affected by breathing various gases at depth over prolonged periods of time. These habitats also provided scientists with the opportunity to conduct long-term experiments under water without having to continually resurface.

During the 1960s and '70s, countries around the world experimented with a variety of habitats, including these highlights:

- Early in the 1960s, Captain George Bond of the U.S. Navy conducted a series of tests called "Project Genesis" to establish that a diver's body tissue becomes completely saturated with pressurized breathing gases within 24 hours at a given depth. These human saturation experiments were combined with underwater living projects carried out by the U.S. Navy in its habitats *Sealab I* and *II*, and later *Sealab III*.
- Jacques Cousteau also built on Bond's studies when he designed three *Conshelf* habitats, which had several men living and working under water for up to a month in the Mediterranean and Red Seas.
- During 1969–70, the waters of the Caribbean Sea became home to more than 50 scientists living and working in the *Tektite* habitat experiment. A unique feature of *Tektite II* was an all-female aquanaut team whose performance paved the way for women to be included in undersea missions.
- Of the more than 50 underwater research facilities, *Hydrolab* was one of the longest lasting, accommodating more than 500 scientists between 1966 and 1984.

Most habitats were not considered cost-effective and disappeared once companies and organizations found cheaper and simpler ways of working at depth. Today, the only operational underwater laboratory is *Aquarius*. Deployed at Conch Reef in the Florida Keys National Marine Sanctuary waters, this underwater habitat offers scientists the opportunity for long-term on-site research. *Aquarius* is supported by the National Oceanic and Atmospheric Administration (NOAA) and operated by their Undersea Research Center at the University of North Carolina Wilmington.

Image courtesy of Mark Ward of Infotainment USA, Inc., and NOAA

Figure 1.26:* Aquarius *Underwater Habitat

Aquarius *is the only operative undersea laboratory dedicated to marine science. First deployed in 1986, it is now located off the coast of Key Largo, Florida, near spectacular coral reefs some 20 meters (approx. 66 ft) beneath the surface. The underwater habitat provides scientists with a life-support system that allows them to live and conduct research under water during missions lasting a week or more.*

3.3. Submarines and Related Technologies

Although visions of wealth and discovery have stimulated many developments in undersea technology, there is no doubt that visions of military victory have fueled (and funded) the most impressive undersea technology advances the world has ever seen. Chief among these have been developments in submarines and related technologies.

3.3.1. Early Submarines

The concept of a submarine—literally a boat that can carry humans under the water—fascinated early inventors from many countries for many centuries. There are sketches

Image courtesy of Naval Undersea Museum

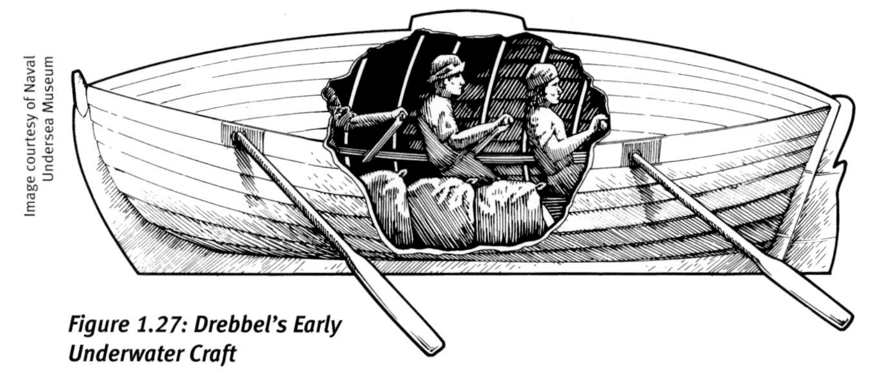

Figure 1.27: Drebbel's Early Underwater Craft

by Italian inventor Guido da Vigevano from the late 1200s and early 1300s of what appear to be designs for submarines. And in 1578, an Englishman named William Bourne published his idea of a "propelled diving boat" with leather-covered compartments that could be made smaller for diving or larger for surfacing. (How these size changes could make a sub dive or surface are explained in *Chapter 6: Buoyancy, Stability, and Ballast.*) In 1620, a Dutch court engineer by the name of Cornelius Van Drebbel is reputed to have built the first actual submarine. He borrowed from Bourne's idea to design a wooden submarine that submerged when crew members filled goatskins with water, then resurfaced when the skins were emptied. Drebbel's craft was covered with a greased leather skin and was propelled by 12 oarsmen breathing air from tubes to the surface.

Figure 1.28: The **Turtle**

Bushnell's design for the diminutive Turtle *included a screw for penetrating the hulls of enemy ships, piston pumps to empty ballast tanks, a safety device in the form of weights that could be dropped in case of emergency, a depth indicator, and a rudimentary form of conning tower.*

Submarine development received a boost during the American Revolution, which began in 1776, just as the Industrial Revolution was gathering steam in Britain. (See *Historic Highlight: The Industrial Revolution.*) At that time, an American patriot by the name of David Bushnell was intrigued by the idea of a small one-man submarine that could travel under water. He was also fascinated by the fact that a gunpowder charge exploded with greater effect under water than in air. He figured that by combining these two ideas, he could produce a significant stealth weapon—a machine that could move under water and attach a charge of gunpowder to the hull of unsuspecting enemy ships. The American inventor produced a small hand-powered wooden submarine, dubbed the *Turtle*, which attempted such an attack on the British warship HMS *Eagle* in New York Harbor. While the sub's simple weaponry was unable to penetrate the ship's hull, Bushnell is credited with establishing the submarine as a potentially formidable weapon of marine warfare.

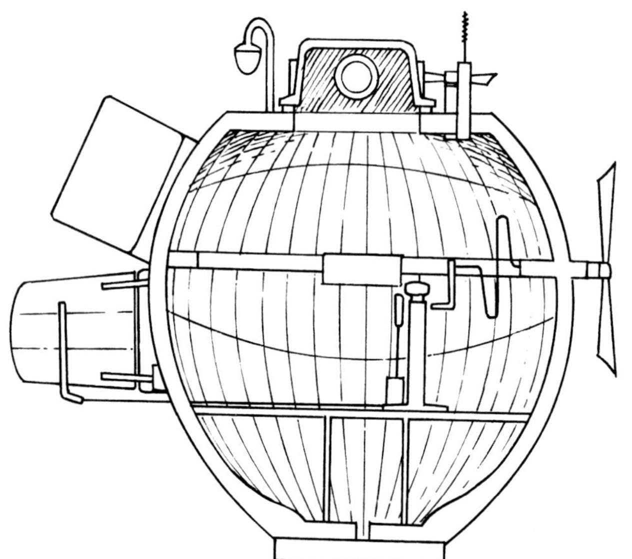

In 1864, during the U.S. Civil War, another human-powered submarine, the Confederate *H. L. Hunley*, succeeded in sinking a Union ship, the USS *Housatonic*, by means of an explosive charge mounted on a long pole (called a spar) attached to the front of the sub. Although the *Hunley*'s crew was killed in the effort, the vessel claimed the honor of launching the first successful submarine attack on another vessel. The South eventually lost the Civil War, but word of the successful submarine attack spread quickly and certainly impacted the submarine's future in navies around the world.

3.3.2. Moving Beyond Human Power

A major limitation of these early submarines was their total reliance on human power for propulsion. During the Industrial Revolution of the late 1700s and early 1800s, steam engines proved their value and quickly came to be the dominant power source for factories, locomotives, and many ships. (See *Historic Highlight: The Industrial Revolution.*) Not surprisingly, they were also tried as power sources for a few submarines such as the *Ictineo II* designed and built by Narcís Monturiol i Estarriol in the 1860s, the *Plunger* designed by John Holland around the turn of the century, and a fleet of

HISTORIC HIGHLIGHT: THE INDUSTRIAL REVOLUTION

While hard-hat diving was evolving, other events that would have major effects on undersea technology were unfolding around the world as part of the **Industrial Revolution**, which lasted from about 1760 to 1870. This revolution was marked by rapid expansion in agricultural production and industrial manufacture, as factory machinery and mass-production replaced at-home handwork and cottage industries. Beginning in Britain and then spreading throughout Europe, North America, and eventually the world, the Industrial Revolution marked a major turning point in human social history.

One of the greatest changes associated with the Industrial Revolution was the development of effective **steam engines** to produce mechanical power. Such power had previously been supplied (usually in much smaller quantities) by windmills, waterwheels, animals, or humans. Steam engines, like other inventions and scientific discoveries associated with the Industrial Revolution, not only changed the way of doing things, but sped up the rate of change itself—not unlike the impact of the computer on today's world.

Since they promised much greater speed and range than was possible with human power, steam engines were tried in a few submarines. However, there were significant technical difficulties so conventional steam engines never caught on as submarine power sources. What did catch on was the machine-oriented mindset of the Industrial Revolution, particularly the idea of replacing human power with mechanized engines of some sort, would ultimately revolutionize undersea propulsion. Today's nuclear submarines are, in a sense, steam-powered, since the reactor heat is used to produce steam, which spins turbines to power the sub.

Those old steam engines left another legacy, too. A branch of mathematics called "control theory," which originated with attempts to control the speed of early steam engines, paved the way for the modern computerized control systems used to control contemporary undersea craft. All in all, the Industrial Revolution inspired massive investment in the development of new materials and new technologies, much of which ultimately advanced underwater vehicle design.

British submarines called *K-boats* used in World War I. However, the insatiable appetite of most steam engines for air (which was hard to obtain under water), the tremendous amount of heat radiated by their boilers (which roasted crew members confined inside an enclosed hull), and the fact that tons of water sloshing around in the boilers played havoc with submarine stability, ultimately doomed this approach to failure.

Then, around the turn of the twentieth century, an Irish patriot named John Holland, who was working in the U.S., developed an innovative system for powering submarines. His design married electrical storage batteries and electric motors with internal combustion engines (similar in concept to today's car and truck engines). The batteries supplied power to electric motors that propelled the sub under water, but on the surface, the engines could propel the sub and recharge the batteries for the next dive. This was a pivotal step in achieving submarine power sources that were practical, powerful, reliable, and long-lasting. As a result of this breakthrough, Holland is considered the true father of the modern submarine. (See *Stories from Real Life, Chapter 7.)*

Image 63092, courtesy U.S. Naval Historical Center

Figure 1.29: John Holland

Inventor John Holland is shown in the conning tower of his submarine Holland, *circa 1898–1899.*

Like other early private entrepreneurs, John Holland sold designs and submarines to many countries, but his goal was to sell one to the U.S. Navy. He finally did so on April 11, 1900. The U.S. Navy purchased the *Holland VI* and extended a contract for six more that became the *Adder* Class. In 1902, the British overcame their skepticism of submarines and contracted for five new Holland boats that featured higher speeds, more torpedo tubes, and a new British invention—the **periscope**. However, general acceptance of submarines was extremely slow in coming for many naval officers; in the words attributed to British Admiral A. K. Wilson, any such stealthy underwater conveyance was "unfair, underhand, and damned un-English."

HISTORIC HIGHLIGHT: THE ACCEPTANCE OF SUBMARINES FOR WARFARE

During the 1800s, inventors around the world worked at designing and selling submarines, touting them as an advantage in warfare. Early on, many leading naval powers felt that subs were for weaker countries that could not support large fleets of real warships. However, military attitudes changed, and eventually every major European naval power began building submarines. Some of the best known are as follows:

- In 1801, the American inventor Robert Fulton built the first practical hand-powered submarine in France, hoping to attack British warships during the French Revolution. The 6.4-meter (approx. 21-ft) *Nautilus* had a bullet-shaped all-metal hull over an iron frame. To submerge, valves were opened to flood the sub's tanks; to ascend, compressed air was used to force the water out of the tanks. Horizontal planes at the stern provided depth control, much like the diving planes on modern submarines. The *Nautilus* featured a collapsible mast and sail for surface mobility and a hand-driven propeller for underwater travel. Frustrated by a lack of interest by European powers, Fulton returned to America to build the country's first steamboats.

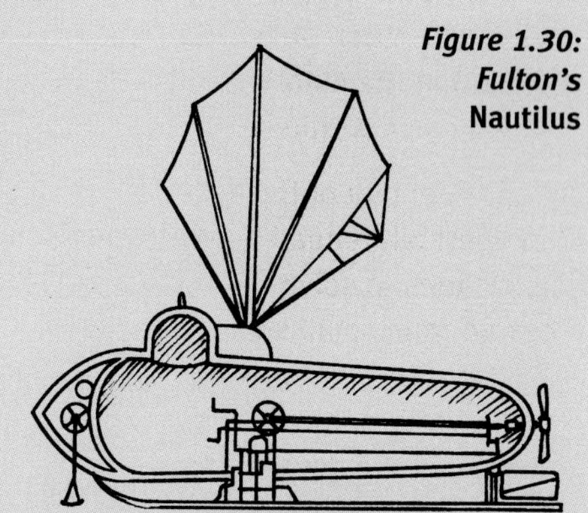

Figure 1.30: Fulton's **Nautilus**

- A number of inventors tried building submarines for both the Confederate and Union forces during the U.S. Civil War (1861–65). Brutus De Villeroi, a French inventor, first built the *Propeller* in 1862, followed by a green 14-meter (approx. 47-ft) version, dubbed the *Alligator.* This early Union sub was propelled by oars but was later refitted with a hand-cranked screw propeller; it featured an air purification system as well as an **airlock,** designed to allow a diver to leave the submerged submarine to place an explosive charge on an enemy ship. Neither sub was tested in battle. Scovel S. Meriam designed the human-powered *Intelligent Whale*, but it wasn't completed until 1866, after the Civil War ended.
- Early Confederate-built submarines included the 10-meter (approx. 33-ft) *Pioneer* in 1862 and the *American Diver* (later *Pioneer II*) in 1863. The South's historic submarine success came in 1864 with the human-powered *H. L. Hunley*, which sank the union ship USS *Housatonic* with an explosive device attached to the end of a pole. Although claiming a significant place in submarine history, the *Hunley*'s crew was killed in the effort.

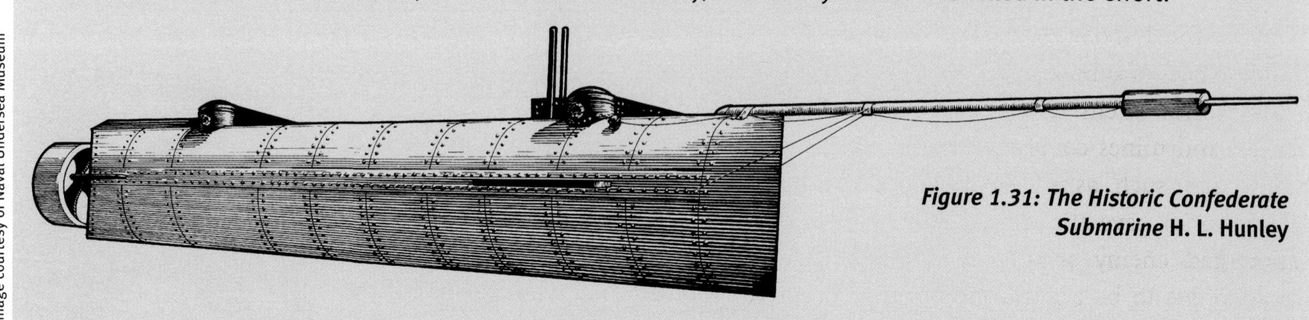

Image courtesy of Naval Undersea Museum

Figure 1.31: The Historic Confederate Submarine **H. L. Hunley**

- During the mid-1880s, Swedish designer and machine gun inventor Thorsten Nordenfelt worked with George Garrett, the British inventor who had already produced the *Resurgam,* to come up with the *Nordenfelt I*. This steam-driven submarine featured a torpedo held in an external tube, with two Nordenfelt machine guns mounted on deck for backup. However, subsequent subs based on this design were considered highly unstable.
- In 1888, the *Gymnote* ("Electric Eel") became the first submarine accepted by a major naval power and served as the basis for France's subsequent submarine fleet. This sub is also notable in that it was operated by an electric motor that generated 55 horsepower, with current supplied by 564 accumulators, or storage batteries. Its major drawback was that it had no internal means to refresh the accumulators at sea.
- In 1890, this sub was succeeded by the 49-meter (approx. 160-ft) *Gustave Zédé* and in the mid-1890s by the more practical *NARVAL* that had a steam engine for running on the surface and an electric motor for running submerged. Most importantly, the steam engine could recharge the sub's batteries at the surface by running the electric motors in reverse.

HISTORIC HIGHLIGHT: TORPEDOES

The term **torpedo** originally referred to a variety of underwater explosives, including passive floating mines. In the early 1800s, attempts were made to convert these stationary torpedoes into mobile weapons by mounting them on the ends of long poles, or "spars," affixed to the front of small attack boats or by using a submerged submarine to drag them on a long rope into contact with the hull of an enemy ship. The dragged torpedo didn't progress beyond the demonstration phase, but spar torpedoes were used by several navies and succeeded in sinking a number of ships during the American Civil War. Unfortunately, the effective use of these weapons required the small attacking vessel to come within close proximity of its target, making it vulnerable to enemy fire.

The development of the self-propelled torpedo in the mid 1860s overcame this problem. Giovanni Luppis (Austrian) and Robert Whitehead (English) devised the earliest version of such an "auto-motive" torpedo, although it would take decades to solve issues of range, accuracy, and effective delivery "platforms." Initially, torpedoes were launched from small ships and later from submarines and even airplanes. Not surprisingly, the development of submarines as an effective weapon of war was directly linked to the use and refinement of self-propelled torpedoes. It's also worth noting that some consider the invention of the self-propelled torpedo to be a pivotal event leading ultimately to the development of today's unmanned robotic vehicles.

3.3.3. World Wars I and II

Submarines improved with time and by World War I (1914–1919) had become a major threat to navies around the globe. Shortly after the declaration of war, one of Britain's smallest subs, the *E9*, sank the German cruiser *Hela*, thus becoming the first submarine to sink an enemy ship without sinking herself! At the beginning of the war, Britain had 78 of these smaller, fast subs. Germany, however, concentrated on a different strategy: a type of long-range submarine, soon to be known around the world as the **U-boat** for *Unterseeboot*. At first these subs were considered only for defense purposes, but later in the war, the German U-boat went on the offensive, becoming a menace so great that submarines overshadowed all other enemy threats.

In 1915, the U-boat sinking of the neutral passenger liner *Lusitania* brought the consequences of submarine warfare to newspaper headlines and to the attention of the entire world. Intensive retaliatory research on depth charges, aero bombs, and mines commenced in earnest. Other types of undersea technology, such as the **hydrophone** (essentially an underwater microphone) gave the Allies the ability to detect the presence of submerged enemy subs. Underwater acoustics, including **sonar**, would prove to be a hugely important field, both for military and civilian purposes. Today it remains one of the most active areas of undersea technology research and development. (See *Tech Note: Sonar.*)

Image courtesy of Deutsches Bundesarchiv (German Federal Archive) and Wikimedia Commons

Figure 1.32: WWII German U-Boat, Type II

Although Germany's subs were plagued by mechanical breakdowns, their effective torpedoes sank nearly one of every four ships entering the war zone during April 1917. This was an especially devastating blow for Britain, who relied on fleets of merchant ships to supply food and carry coal exports. U-boat successes were cut only when Allied ships began to sail in convoys with escort vessels that could sink the subs. WWI served as a wake-up call regarding the tremendous potential of submarines in combat and ultimately motivated an enormous investment in submarine research, including new devices for submarine communication, detection, and rescue.

World War II began when Britain and France declared war on Germany on September 3, 1939. In terms of the role that submarines and torpedoes played between the British and German navies, WWII was essentially a replay of WWI, but with more advanced

TECH NOTE: SONAR

Sonar (for "SOund Navigation And Ranging") encompasses a range of technologies that use sound waves to measure underwater distances, help vessels navigate in real time, measure water currents, detect and visualize submerged objects, or even map undersea mountains and valleys. Since sound travels farther than light under water, sonar can provide a form of long-range, remote sensing that's especially useful for navigating or finding objects in dark or murky water.

Passive sonar systems simply listen for things like submarine propellers, which make their own underwater noises. **Active sonar** systems send out a series of sound pulses, then listen for the echoes that bounce off underwater objects. Since the approximate speed of sound through water is known, active sonar equipment can use the time delay between when a pulse is sent out and when the echo returns to calculate the distance to the object that reflected the sound pulse. More sophisticated sonar systems keep track of the direction from which echoes arrive and can use that information to build up virtual maps or images of objects and terrain under water. (See Figure 1.33.)

Various types of sonar instruments can be mounted on ships, ROVs, AUVs, and towsleds. Sonar units can also function as independent units that are towed behind a vessel. In this mode, they're commonly referred to as *towfish* (not to be confused with towsleds described in *Tech Note: Towsleds* above). (See further discussion of sonar in *Chapter 3: Working in Water, Chapter 9: Control and Navigation, Chapter 10: Hydraulics and Payloads*, and *Chapter 11: Operations*.)

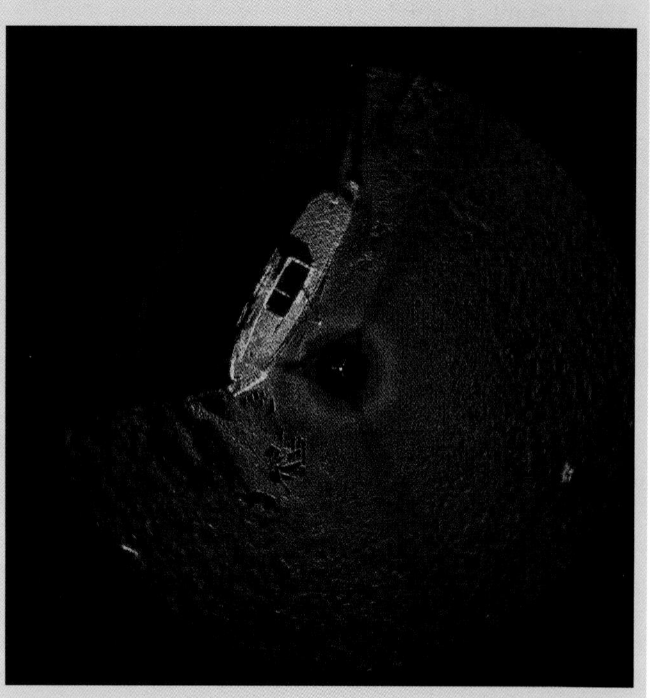

Image courtesy of Mark Atherton, Klongsberg Mesotech

Figure 1.33: The Advantage of Seeing Underwater

This image from a scanning sonar device shows a clear view of a shipwreck in Lake Michigan. The image is produced by the reflection of sound off hard and soft surfaces.

submarine technology on both sides. Britain's large merchant marine fleet still proved to be easy pickings for the German U-boats; in 1942, ships were being sunk at more than twice their replacement rate, with new U-boats joining the fleet at the rate of about one per day. During WWII, submarines still had limited underwater speed, range, and endurance, so they usually traveled on the surface at night and submerged only when necessary.

On the other side of the planet, Japanese submarine designers were moving in new directions from their German mentors, developing their own variety of cruising, coastal, and midget submarines, as well as submarine aircraft carriers and the highly effective "Long Lance" torpedo. In 1941, the Japanese navy used torpedo-equipped aircraft to attack the U.S. naval base in Pearl Harbor, thus bringing the United States into the war. Japan also used five two-man midget subs (called *Ko-Hyoteki*) in this attack. With the U.S. battle fleet nearly destroyed at Pearl Harbor, the United States immediately dispatched its submarines, but most had problems with faulty torpedoes. Once these issues were corrected, U.S. subs became considerably more effective in the Pacific.

HISTORIC HIGHLIGHT: UNDERWATER RESEARCH EFFORTS

Many countries have launched oceanographic expeditions over the years. Beginning in the 1870s, the Germans explored the South Atlantic and Indian Oceans, the Dutch investigated the Indian Archipelago, the Scandinavians surveyed the Arctic Ocean, while the Germans and Belgians explored the Antarctic. Many oceanographic institutions were also set up around this time: the Oceanographic Museum at Monaco, the University of Oslo Biological Station, the University of Bergen Geophysical Institute, etc.

Wealthy benefactors in North America also became interested in the subsea discoveries being made. The Scripps newspaper dynasty financed what eventually became known as the Scripps Institution of Oceanography in La Jolla, California. Rockefeller money endowed the Woods Hole Oceanographic Institution (WHOI) in Woods Hole, Massachusetts. The Johnson family founded Harbor Branch Oceanographic Institution in Ft. Pierce, Florida, and Packard family wealth later funded the Monterey Bay Aquarium in Monterey, California, and the affiliated Monterey Bay Aquarium Research Institute in nearby Moss Landing. These institutions have been joined by hundreds of public, private, and corporate groups, as well as government, military, and academic organizations.

Other underwater research centers around the globe include the Japanese Marine Science and Technology Center (JAMSTEC), the Institut Français de Recherche pour l'Exploitation de la Mer (IFREMER), the USSR's Shirshov Institute of Oceanology, and the Institute of Marine Technology Problems (IMTP), as well as the European consortium MAST (Marine Science and Technology), which includes Portugal, Italy, Denmark, and Belgium, and the U.K.'s National Oceanography Centre, Southampton (formerly Southampton Oceanography Centre). Some, such as the Institute of Nautical Archaeology (INA) at Texas A&M University and Bodrum, Turkey, have specialized interests.

3.3.4. The Cold War

The period following World War II, up until the 1990s, came to be known as the **Cold War**. Although there was no formal declaration of war and no open fighting, it was a period of deep mistrust between major world powers, particularly the United States and the Soviet Union (USSR). These suspicions meant that reconnaissance became a priority early in the Cold War. In the air, orbiting spy satellites and planes provided a bird's-eye view of enemy territory and activities as recorded in detailed photographs. On the ocean floor, the U.S. Navy set up a network of global hydrophone microphones (eventually known as **SOSUS** for **SO**und **SU**rvellance **S**ystem) that used sound waves to monitor Soviet ships and submarines over great distances.

The fierce competitiveness between the United States and the Soviet Union came into sharp focus on October 4, 1957, when the Soviets launched into space the world's first artificial satellite, called *Sputnik*. It was followed by *Sputnik II* a month later. Caught off guard, the U.S. began an intense catch-up effort. When the Soviet Union successfully accomplished the first manned orbital flight in 1961, President John F. Kennedy countered with a declaration that the United States would be the first country to put a man on the moon. The space race was on! Ultimately, this race spun off a huge number of technologies that found their way into undersea robots, from miniaturized electronic components and computers to sophisticated communication protocols and advanced materials.

During the several decades of the Cold War, the greatest cause for worldwide concern was the very real possibility of a nuclear war. Both the United States and the Soviet Union believed that the best deterrent to an enemy offense was to make sure their own arsenal of nuclear weapons was the biggest. The result was a perpetually escalating arms race featuring ever-more-advanced weapons, weapons delivery systems, and technologies for stealth, reconnaissance, and espionage. Significant among these were nuclear-powered submarines carrying nuclear missiles and highly sophisticated sonar systems.

The United States launched its first nuclear-powered sub, the USS *Nautilus*, in 1954. It had a submerged cruising speed of more than 20 knots, effectively unlimited range, and the ability to stay submerged for four months. By 1962, the Soviet navy had responded by adding some 23 nuclear-powered subs to its fleet. This fierce rivalry spawned fleets of large, fast, and incredibly stealthy submarines that could—and still do—move virtually undetected to any corner of the planet, armed with nuclear warheads on long-range guided ballistic missiles.

Other U.S. reconnaissance efforts included the development of the 105-meter (approx. 350-ft) *Halibut*, a submarine armed with lights, cameras, and an array of sensors to spy on enemy equipment in the depths, or the 41-meter (approx. 135-ft) Deep Submergence Vessel *NR-1*, the first deep-submergence nuclear-powered vessel for ocean search and retrieval missions. The Cold War finally ended with the collapse of the Soviet Union in the 1990s. However, the various technologies developed during this period of intense political rivalry continue to be utilized by countries around the world.

3.4. Submersibles for Deep Sea Research

While the development of submarines was undergoing significant acceleration throughout World Wars I and II and the Cold War, similar advances in submersible technology were taking place, starting as early as the 1930s.

In the world of undersea vehicles, form follows function. And when the function is to explore as deep as possible, as with submersibles, you use the most pressure-resistant shape that exists—a small sphere.

Deep sea exploration got its start in 1930, when a biologist by the name of Dr. William Beebe teamed up with Otis Barton, a young geologist-engineer, to design a strong, hollow, spherically shaped craft they could use to observe living marine life in the ocean's mid-water world, a previously inaccessible depth. (See *Historic Highlight: Setting Records with a Bathysphere.*) While the **bathysphere** they devised was a far cry from today's highly maneuverable submersibles, the cramped steel ball that hung from a steel cable suspended from a ship did allow them to descend in a series of record-making deep water dives off Bermuda. The deepest, in 1934, was to a depth of over 900 meters (approx. 3,000 feet), far deeper than any subs of the day could go. During these dives they recorded astounding life forms, including bizarrely shaped fish and many different bioluminescent creatures. They continued making bathysphere dives and observations until the onset of World War II.

HISTORIC HIGHLIGHT: SETTING RECORDS WITH A BATHYSPHERE

Dr. William Beebe and Otis Barton designed their record-setting bathysphere (from the Greek for "deep sphere") to observe mid-ocean marine life. There were no controls for maneuvering; the small steel sphere simply hung from a steel cable, which also supplied electricity. The bathysphere was 1.45 meters (approx. 4.75 ft) in diameter, with walls about 4 centimeters (approx. 1.5 in) thick. Its air supply consisted of two oxygen cylinders and chemicals for removing carbon dioxide.

Image courtesy of Naval Undersea Museum

Figure 1.34: First Exploration of the Mid-Water World

The exploits of Beebe and Barton caught the attention of a Swiss physicist by the name of Auguste Piccard. He realized that any metal sphere hanging by a single steel cable was extremely vulnerable. Instead he proposed a **bathyscaph** (also written bathyscaphe), or "deep boat"— basically an independent diving craft with its own buoyancy control system. Piccard built a series of maneuverable, pressure-resistant spheres, but it was his not-very-maneuverable *Trieste* that set the ultimate diving record in 1960 when it carried his son Jacques Piccard and U.S. Navy Lieutenant Don Walsh to what is believed to be the deepest point in the entire ocean. (See *Stories From Real Life* in Chapter 4 and *Historic Highlight: Journeys to the Ultimate Depth.*)

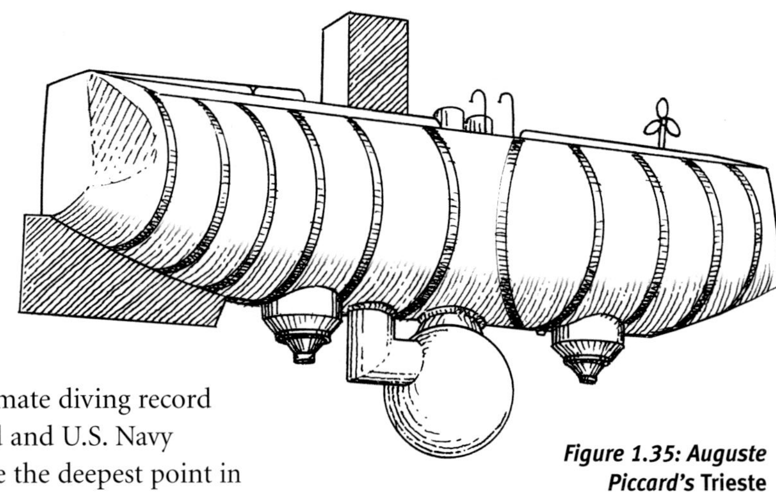

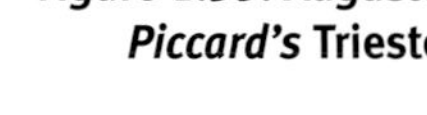

***Figure 1.35: Auguste Piccard's* Trieste**

HISTORIC HIGHLIGHT: JOURNEYS TO THE ULTIMATE DEPTH

In 1960, Jacques Piccard and U.S. Navy Lieutenant Don Walsh set a remarkable world record when they became the first (and so far only) human beings to visit the deepest point in the ocean. Using the innovative submersible *Trieste*, designed by Jacques' father August Piccard, they descended into the Challenger Deep in the Mariana Trench. Measurements gathered by their instruments at the bottom were used to calculate a depth of 10,916 m (about 35,814 feet).

Although their record-breaking journey has not yet been repeated by any other human beings, the Challenger Deep has been visited since then by two unmanned robotic vehicles. In December of 1995, the ROV *Kaiko* reached the bottom where its more advanced (and presumably more accurate) instrumentation measured the depth as 10,911 meters (35,798 ft). *Kaiko* returned several times over the next three years before being lost at sea when its tether broke on another deep dive.

In 2009, the hybrid AUV/ROV *Nereus* visited the area, measured a depth of 10,902 meters (35,768 ft), and retrieved samples. Minor discrepancies in depth measurements aside, it's clear that the Challenger Deep is aptly named and that humans can do phenomenal things with technology!

In 1964, the United States Navy built its manned deep submergence vehicle (DSV) *Alvin.* This submersible was based on an assessment of *Trieste's* amazing but complex deep-dive missions. The 7-meter (approx. 23-ft) *Alvin* could carry up to three people and was certified to dive at depths up to 1,828 meters (approx. 6,000 ft). Woods Hole Oceanographic Institution eventually took over operations of the craft, becoming a world leader in piloted deep sea exploration and later unmanned studies. Over the years, *Alvin* has been constantly modified and improved. (See *Stories From Real Life* in Chapter 11.)

Figure 1.36:* Alvin, *Submersible Workhorse

Image courtesy of Woods Hole Oceanographic Institution

Having logged more than 4,000 dives, this oceanographic workhorse has an impressive list of credits, including the first sampling of **hydrothermal vents** and **chemosynthetic communities** (which derive their energy from inorganic chemical reactions, rather than from food or photosynthesis) and expeditions to the *Titanic.*

Plans are under way to construct a new deeper-diving submersible to replace *Alvin.* When completed, the new

Image courtesy of Vickie Jensen

***Figure 1.37: The Research Submersible* Ben Franklin**

In 1969, six aquanauts aboard the 15-meter (approx. 50-ft) research submersible Ben Franklin *undertook a major undersea drift mission, traveling for 30 days with the Gulf Stream. They traveled some 1,444 nautical miles at depths ranging from 240 to 480 meters (approx. 800 to 1,600 ft) to observe, sense, and record previously unexplored inner space.*

vehicle will be capable of diving to 6,500 meters (approx. 21,325 ft) and will provide increased battery capacity, more viewports with overlapping fields of view between the pilot and scientists, improved maneuverability, and better lighting and video capacity. The new vehicle will be constructed in two stages. Stage 1 will integrate the new 6,500-meter personnel sphere with modified and enhanced *Alvin* components. Stage 2 will upgrade the remaining components to 6,500-meter diving depth.

Alvin was not the only manned submersible developed during the '60s:

- A number of research-oriented submersibles, such as the *Aluminaut, Deep Diver, Trieste II, Deepstar,* and *Ben Franklin,* were built in the United States.
- Outside of North America, France's *Cyana* and *Nautile,* Russia's *Mir I* and *Mir II,* and JAMSTEC's *Shinkai 6500* joined the ranks of notable research submersibles.
- Development of early commercial submersibles included the Canadian *Pisces* and the Vickers-Slingsby *LR2.*
- The *Asherah,* built for the University of Pennsylvania's Institute of Nautical Archaeology, was the first submersible for archaeological research.

These free-swimming submersibles, many with large viewports, powerful thrusters for moving and steering, and manipulator arms for collecting specimens, were a significant improvement over Beebe and Barton's original dangling bathysphere.

3.5. Robotic Undersea Vehicles

Because it could move by itself, the the self-propelled torpedo, developed in the 1860s, is sometimes regarded as the first undersea robot, albeit a primitive one by today's standards. (See *Historic Highlight: Torpedoes* earlier in this chapter.) Remotely controlled and "intelligent" autonomous underwater vehicles would not appear until much later.

Fast forward a hundred years to the 1970s and you can follow the development of three broad categories of underwater robots:

1. Tethered ROVs led the way, proving themselves in the commercial world of the offshore oil and gas industry.
2. Tetherless AUVs got a later start, but continuing technological advances are pushing more commercial viability, particularly for survey work.
3. Hybrid vehicles may offer the best of both robots, with vehicles able to operate in tethered or tetherless mode, depending on the mission.

The development of each of these types of vehicles is explored in greater detail below.

3.5.1. Early ROVs

During the first decade of the Cold War, a French nautical archaeologist, diver, underwater photographer, and prolific inventor named Dimitri Rebikoff created a diver propulsion vehicle he called the *Torpille.* It was essentially a torpedo that could tow a diver. In 1953, he modified this vehicle by adding a TV camera and remotely controlled steering (via cable/tether) to search for shipwrecks in deep water. This device, named *POODLE,* is generally regarded as the first true ROV and represents a

Images courtesy of Rebikoff-Niggeler Foundation

Figure 1.38: ***Dimitri Rebikoff and*** **POODLE**

major milestone in the history of underwater robotics. Using *POODLE*, he successfully located a number of wrecks.

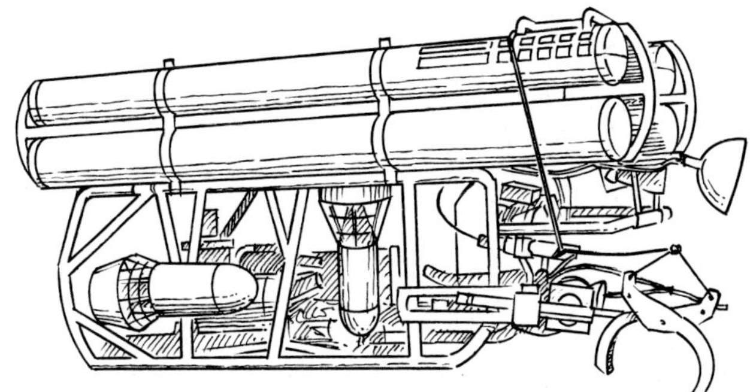

Figure 1.39: **CURV**

About the same time, the British Royal Navy and the U.S. Navy began experimenting with the use of remotely controlled underwater robots to locate and recover expensive weapons that had been lost during military practice exercises. For example, in the 1950s, the Royal Navy used an ROV called *Cutlet* to retrieve lost torpedoes and mines, and the U.S. Navy used a series of vehicles called **C**able-controlled **U**ndersea **R**ecovery **V**ehicles (**CURV**s) for the same purpose.

Occasionally, these specialized robots found more exciting work, like recovering real nuclear bombs! In 1966, newspaper headlines around the world focused on the importance of subsea vehicles when a hydrogen bomb was lost off the coast of Spain during a U.S. Air Force B-52 refueling operation. Submersibles such as the *Aluminaut*, *Alvin*, and *Perry Cubmarine* were tasked with the job of finding and retrieving the bomb. *Alvin* did locate it but didn't have the capacity to make the underwater recovery, so the U.S. Navy's remotely controlled CURV robot was called in and completed the mission. In 1973, *CURV III* made the news when it helped rescue the trapped crew of *Pisces III*. (See *Stories From Real Life* in Chapter 3.)

While none of these early robotic vehicles were particularly capable compared with those of today, they all served to advance the technology to an operational state and to increase awareness of the great potential of undersea vehicles.

3.5.2. Oil Boom of the '70s

In the 1950s and 60's, many subsea companies got their start and many types of subsea equipment were developed in response to Cold War military research and plump military budgets. But eventually those budgets were cut worldwide, and subsea robotic vehicles suddenly had to pay their own way. As it turns out, they proved extremely useful in the challenging and dangerous environment of the offshore oil industry, so when steadily growing demand for energy fueled an **oil boom** in the 1970s, commercial development of unmanned underwater vehicles took off.

Developments during this time included:

- Machinery and techniques to locate fossil fuel resources offshore
- Methods, vessels, and equipment for drilling wells and maintaining offshore production facilities
- Techniques and machinery for laying pipelines and/or using ships to transport fuel products

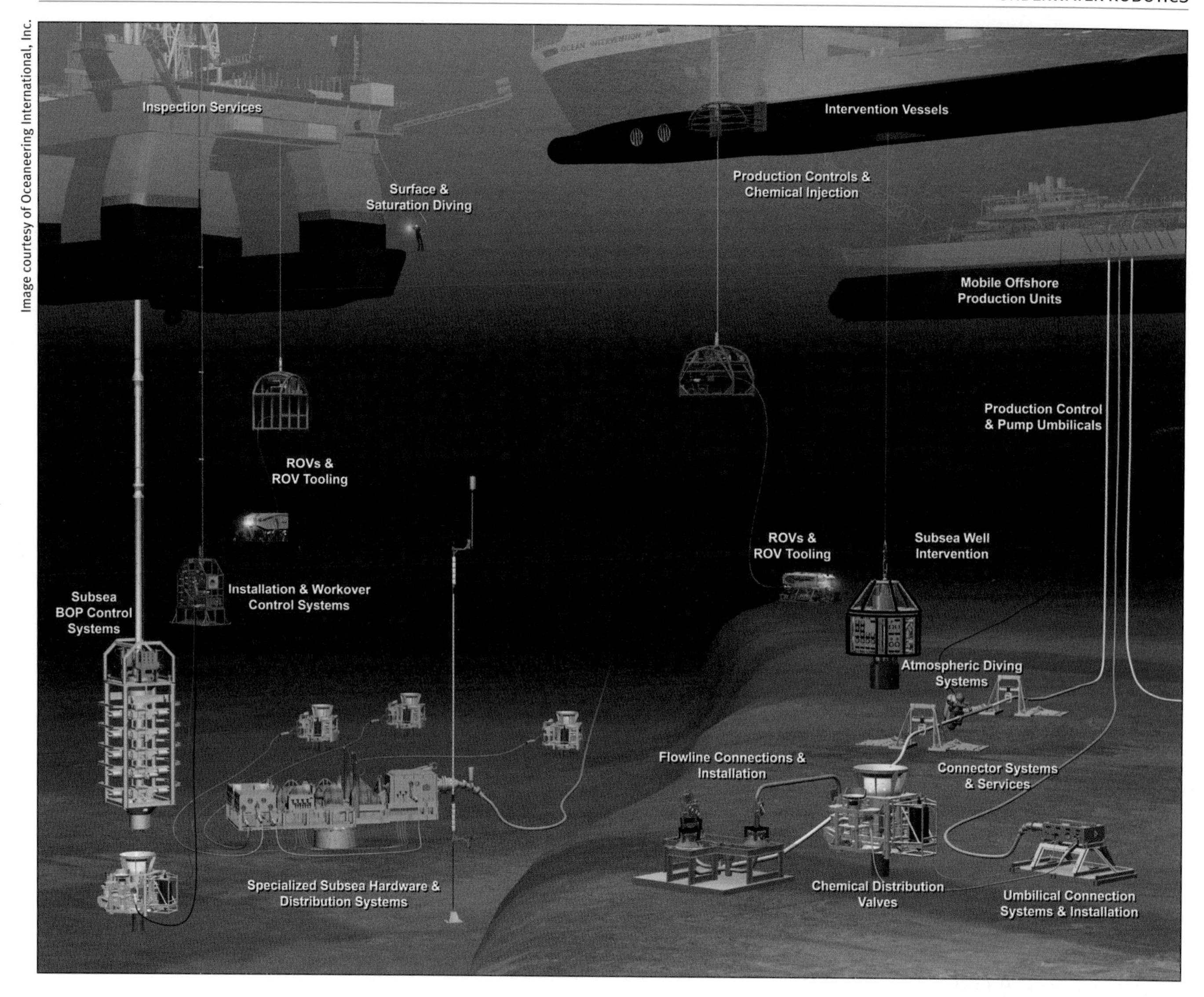

Figure 1.40: Offshore Production Facilities

This diagram shows the variety of underwater technologies, equipment, and techniques used in the offshore oil/gas industry.

Today, the oil boom has broadened somewhat into a search for alternative energy sources, but the world's reliance on petroleum resources continues to expand. For subsea robotics, this translates into ongoing support from the oil and gas industry that includes the quest to develop increasingly sophisticated, deeper-working, heavy-duty vehicles that are efficient, reliable, and cost-effective.

3.5.3. Expansion of Commercial ROV Applications

According to the Marine Technology Society, there was only one industrial manufacturer of ROVs at the beginning of the 1970s. And the only real academic presence in ROV research and development was the Heriot-Watt University of Edinburgh, with its *ANGUS* line of vehicles. But interest in ROVs, particularly among commercial firms, was growing quickly. By the end of 1982, the French company Société ECA had produced more than 200 of its small mine neutralization vehicle, the popular *PAP-104*, for navies around the world. By the mid-'80s, some 27 companies had developed more than 500 ROVs, the vast majority of them for use in the offshore petroleum industry.

Designers pushed underwater robots to do increasingly difficult tasks. The result was a surge in the development of ROVs, ranging from those weighing less than 45 kilograms (approx. 100 lbs) to those the size of a mini-van and weighing more than 4,540 kilograms (approx. 10,000 lbs). Whatever their size, ROVs have earned their stripes in the commercial world, proving that they can accomplish more work and at significantly greater depths than had been possible previously.

The next stage in ROV development was the emergence of **Low-Cost ROVs (LCROVs)**, which offered easily portable vehicles at prices even academic institutions could afford. These smaller vehicles cost less than $50,000, are typically all-electric, and operate above 300 meters (approx. 1,000 ft). Those with a price tag around $10,000 are tagged **VLCROV—Very Low-Cost ROV**. More recent developments have the power source carried on board and a thin fiber-optic tether to relay commands and data to and from the ROV. A new class of electric ROVs has emerged; some of these, such as the *Quest*, manufactured by Schilling Robotics and owned and operated by Canyon Offshore, can operate at 3,050 meters (approx. 10,000 ft) with much less power than was previously needed to work at this depth.

Image courtesy of ROV Committee of the Marine Technology Society

***Figure 1.41: The* RCV 150**

Developed in the 1970s, the RCV 150 *was an early tethered robot developed by Hydro Products. RCV stood for Remotely Controlled Vehicle. In this 1980 photo, an* RCV 150 *is handing a wrench to a diver while an* RCV 225 *looks on.*

HISTORIC HIGHLIGHT: SUBSEA OIL AND GAS RESOURCES

Advances in technology directly impacted the ability to access and harvest subsea resources. For example, the first offshore oil-drilling platforms in the world were essentially nearshore wooden pile rigs (actually dock extensions) built off the southern California coast in the late 1860s. The first truly offshore oil production facility was a wooden structure installed out of sight of land south of Morgan City, Louisiana, by Kerr McGee in 1947.

Not surprisingly, oil and gas harvesting has expanded further offshore to keep up with today's energy demands. Shell's *Perdido* project is an excellent example of how the oil and gas industry is confronting the challenge of the ultra deep water frontier. Billed as a "technological tour de force" the spar-type platform will be able to boast a number of world records—the deepest oil development, the deepest drilling and production platform, and the deepest subsea well.

The *Perdido* spar platform is located 400 kilometers (approx. 250 mi) south of Galveston and moored in 2,380 meters (approx. 9,600 ft) of water. When completed in 2010, it will be almost as tall as the Eiffel Tower and will gather, process, and export oil from 22 direct verticle access wells in three oil fields, all within a 45 kilometer (approx. 28 mi) radius. Shell operates *Perdido* on behalf of its partners BP and Chevron.

What's ahead? Deep water drilling techniques have continued to move forward, and now offshore oil platforms have become technologically sophisticated structures that are essentially self-contained factories on the ocean. Currently ExxonMobil estimates that it gets 25 percent of its total worldwide production from deep water. Fixed platforms may be replaced by more economic floating production systems—these floating structures can service multiple deep water wells and their associated seabed hardware.

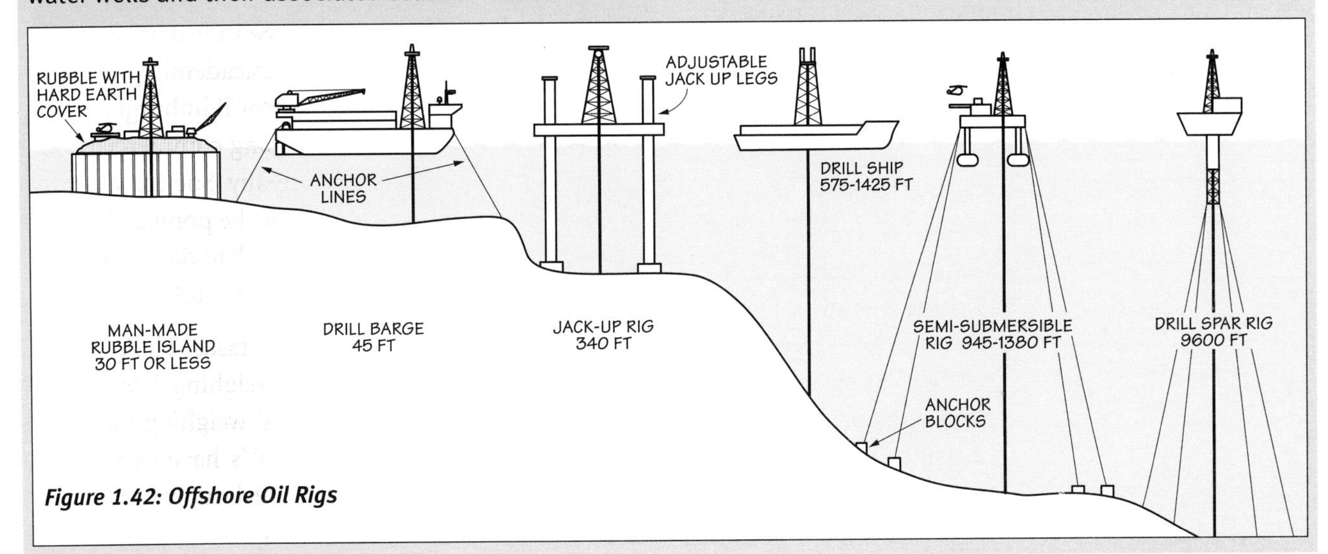

Figure 1.42: Offshore Oil Rigs

Image courtesy of Michael Paris

Figure 1.43:* MiniROVER *at Work on* City of Ainsworth *Shipwreck

Early major developers of LCROVs included Deep Ocean Engineering with its Phantom *vehicles and Benthos that marketed Chris Nicholson's* Mini-ROVER. *In this photo, diver Jacques Marc and* MiniROVER *inspect the remains of the* City of Ainsworth *wreck in Kootenay Lake.*

The world's insatiable need for energy—notably oil and gas—continues to drive ROV development. The Marine Technology Society reports that 95 percent of the 600 commercial work class ROVs in the world are engaged in oil and gas exploration/production or telecommunications underwater cable work. The remaining 5 percent are mostly used for treasure hunting and science. But the ROV success story applies to both ends of the size spectrum. As the availability and flexibility of smaller ROVs and LCROVs increased dramatically, these small but useful machines have become regular work tools for a wide variety of users.

3.5.4. The Development of AUVs

AUVs have had a somewhat slower developmental history than their tethered ROV cousins. Dimitri Rebikoff had already experimented with both ROV and AUV designs in the 1960s, so the idea of getting rid of a cumbersome tether was not new, but for a long time, AUVs were perceived as too high-tech and high-risk. More recently, ROVs have been tasked to work at greater and greater depths, a challenge that significantly multiplies the problems of tethers. These greater working depths have encouraged commercial interest in the development of AUVs.

By the 1970s, Massachusetts Institute of Technology (MIT) was working on early AUVs, as was the Soviet Union. The development of AUVs began in earnest during the 1980s, with many designed as test-bed vehicles for other technologies (e.g., control systems, sonar, propulsion units, etc.). Once the military recognized that AUVs had potential as stealth reconnaissance platforms, they allocated significant money to the development of large, sophisticated vessels. Internationally, the Soviet Union, Canada, Denmark, France, and the U.K. were launching AUVs, while academic organizations such as MIT Sea Grant, Florida Atlantic University, the University of Washington's Applied Physics Lab, and Woods Hole Oceanographic Institution were forging ahead in the U.S. In the late 1990s, the petroleum and telecommunications industries also became interested in the development of AUVs as tetherless robots developed a track record for excellent survey capability at lowered cost. Like ROVs, AUVs cover the size spectrum, from large behemoths to miniaturized, reconfigurable ones.

Image courtesy of Applied Physics Laboratory, University of Washington

***Figure 1.44:* SPURV 1**

The Applied Physics Laboratory at the University of Washington in Seattle developed one of the earliest operational AUVs. Their torpedo-sized Self-Propelled Underwater Research Vehicle, *or* SPURV 1, *was used successfully in mid-water research.*

3.5.5. The Development of Hybrids

Individual hybrid vehicles or systems of cooperative craft are one of the latest in subsea robotic developments and certainly reflect emerging technologies. Perhaps the most publicized hybrid systems, where two underwater vehicles worked cooperatively, were associated with the discovery of the *Titanic*—first was the testing of the *Argo/Jason* system in 1985, a mission that highlighted the capabilities of new camera and sonar systems when the ship was first discovered. A year later, a second system involved testing the *Jason Jr.* prototype, a small ROV that was flown from the manned submersible *Alvin* in order to film the interior of the historic wreck.

Image courtesy of Woods Hole Oceanographic Institution

Figure 1.45: ROV Cooperation

ROVs worked together to explore the wreck of the Titanic. *In the case of this photo, one ROV is filming the other.*

Some individual hybrid vehicles proved to be ineffective. For example, *Duplus* was a hybrid combination of ROV and manned submersible that could operate as an ROV but also submerge with a single operator and carry out a mission. It proved ineffective as an ROV because it had to be large enough to carry a human and all of the required life support and backup systems, making it too big and underpowered. Conversely, it was not particularly effective as an ADS because it carried all of the electronics and other things necessary for an ROV. What is proving to be workable are single craft that can switch between ROV and AUV (i.e., tethered and tetherless) modes.

Whether hybrids are individual vehicles or multiple craft with the capacity to interact or support each other, they are essentially robotic history in the making—but, as with ROVs and AUVs, they must prove themselves in terms of reliability and cost-effectiveness to have a significant impact.

3.5.6. Ocean Observing Systems

A particularly exciting development in undersea research and operations involves the coordinated use of multiple undersea robots, all equipped with sensors and functioning together as a team. Sometimes these robotic vehicles also work in collaboration with buoys, surface ships, aircraft, and satellites, to gather multi-dimensional data about some particular region in the ocean. These elaborate efforts and the technology behind them are often referred to as **Ocean Observing Systems (OOS)** if the data are made available for widespread distribution in real-time or near real-time.

By gathering information from several nearby locations simultaneously, these systems provide a more integrated view of how various ocean processes interact across time and space. They also have applications in monitoring ecosystem health, fisheries management, climate change, marine forecasting, pollution, and national security.

Image courtesy of David Fierstein © 2001 MBARI

Figure 1.46: Taking the Ocean's Pulse

Ocean observing systems are an exciting new development that utilizes an integrated system of multiple underwater robots and other subsea, surface, and air technologies to provide real-time data on the state and health of our oceans.

During August, 2000, the Monterey Bay Aquarium Research Institute (MBARI) conducted a large-scale, multi-disciplinary, multi-institute field experiment in Monterey Bay to assess the role of iron on phytoplankton blooms in Monterey Bay. Their MUSE *project involved three ships, two aircraft, two satellites, two AUVs, several drifters, nine moorings, six gliders, and a host of small boats, many of which are shown in this image by David Fierstein.*

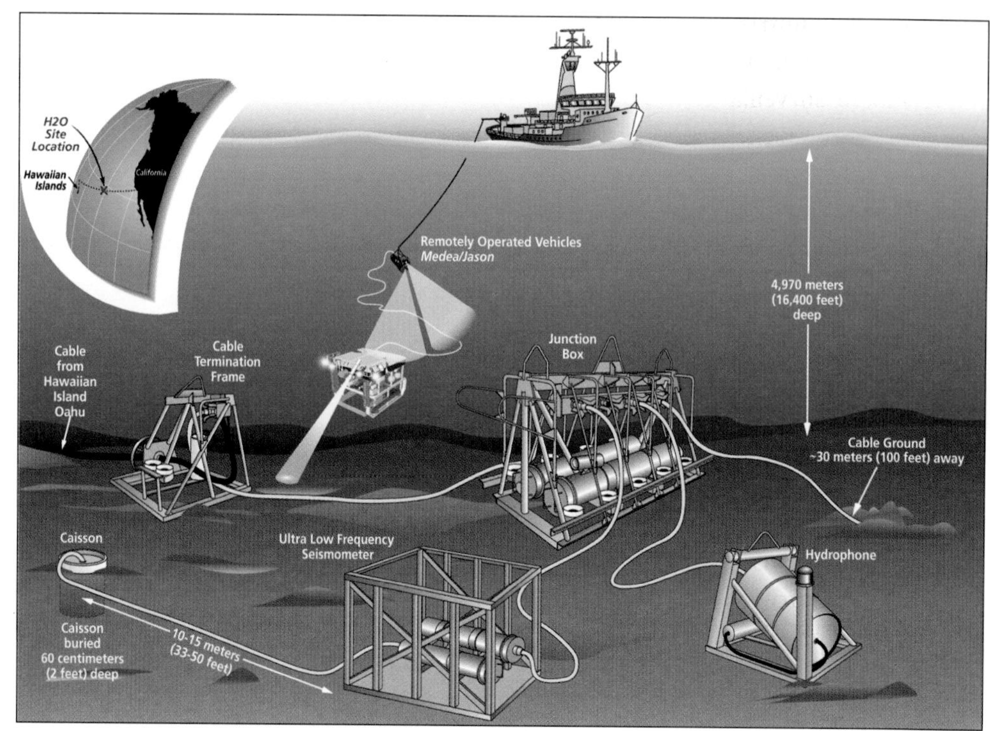

Image by Jayne Doucette, courtesy of Woods Hole Oceanographic Institution

Figure 1.47: Integrated Underwater Technologies

The Hawaii-2 Observatory (H2O) is a long-term, seafloor observatory to enable scientists to conduct ongoing seismic observations and other scientific tests. In setting up H2O, Jason *and* Medea *acted in concert to help scientists splice an abandoned subsea telephone cable to a junction box that now provides continuous power for up to six instruments. Since the initial installation, the* Jason/Medea *combo has successfully plugged and unplugged equipment for installation and servicing.*

4. Modern-Day Uses of Underwater Vehicles and Technology

All underwater vehicles are designed and built for a purpose—that is, they have some kind of work or mission to accomplish. That work may be purely observational in nature, as it is in survey, exploration, or inspection tasks, or it may require physical **intervention**, such as adjusting valves on an oil manifold, burying a telecommunications cable, or salvaging items from a wreck.

Images courtesy of Kongsberg Maritime

***Figure 1.48: Joint-Venture* Hugin**

Sometimes military groups partner with local companies to produce new equipment such as Hugin. *This AUV is powered by a unique aluminum oxygen fuel cell that Kongsberg Simrad developed in conjunction with the Norwegian Defense Establishment (FFI).*

4.1. National Defense

Most coastal countries rely on a variety of manned and unmanned underwater craft for military search, detection, and reconnaissance missions, underwater communications, subsea surveillance, and self-defense of ships, subs, and harbors. While the standard naval arsenal includes a variety of submarines, there remains a significant safety advantage in using unmanned versus manned equipment. Indeed, today's new military mandate for "battlefield dominance via unmanned systems" specifies an increasing emphasis on robotic systems on the ground, in the air, and at all depths in the seas.

The military sector also continues its development of various subsea technologies. Extremely accurate views of the ocean bottom are now possible, due to advances in acoustic and rapid-profiling instruments. Various types of sonar technologies are teamed with instruments and computers to analyze the character of the returning echoes, and with graphic recorders to generate images of bottom topography. These survey/mapping procedures are now widely used by a variety of other sectors, including resource extraction, construction, scientific exploration, and telecommunications.

Image courtesy of Oceaneering International

***Figure 1.49: Joint Venture* Remora**

The United States and Royal Australian navies partnered with OceanWorks International to fabricate various hybrid submarine rescue systems. The U.S. version, an advanced untethered Submarine Rescue Vehicle (SRV), is capable of 650-meter (approx. 2,100-ft) depths. The Australian Advanced Submarine Rescue Vehicle (ASRV) is aptly named Remora *for the small sucker fish that easily attaches itself to manta rays or other larger fish.*

Figure 1.50: Shallow-Water* REMUS *AUV

Hydroid's light-weight, low-cost REMUS *was designed by Woods Hole Oceanographic Institution. Available in various sizes and depth ratings,the AUV can accomplish a variety of worldwide missions that include environmental monitoring, mine countermeasures, and hydrographic surveys.*

Images courtesy of Kongsberg Maritime

Military interest has shifted from deep water to shallow water technologies. For example, a lobster-sized robot crustacean named *Robolobster* is designed to detect and dispose of mines hidden in coastal waters and shores. Hydroid's *REMUS* is another shallow water AUV that's already proven itself, working under extremely adverse conditions to detect and clear nearshore underwater mines in Iraq, among other missions.

4.2. Resource Extraction

The sea contains a wealth of useful resources, ranging from all types of seafood and offshore fossil fuel reserves to chemicals, including valuable minerals and pharmaceutical compounds. ROVs and AUVs play an ever-increasing role in locating and retrieving these resources. However, the biggest challenge may be to assess and manage resources so that they are sustainably harvested. The other test will be extracting resources while minimally impacting the surrounding environment and species.

Contemporary underwater vehicles already service the seafood, fossil fuel, and chemical/mining resource sectors. For example:

- Underwater equipment can detect schools of fish, provide seafloor profiles that help fishermen keep from snagging their nets, and track data on water characteristics that directly impact seafood.
- Prior to drilling offshore wells, the oil and gas industry maps potential sites with survey ships that either tow an array of sensors or employ robotic vehicles fitted with sophisticated mapping equipment. Survey work is big business. A North Sea shallow gas drill site survey typically costs $150,000–$250,000. As a result, AUV

survey systems are quickly gaining commercial acceptability because they can save time over similar ROV work, they cost less because of smaller support crews, and they offer higher quality survey data than conventional ship-based systems. At least one company is already working on a multi-vehicle approach that uses multiple micro-AUVs for search and data acquisition.

- Once wells are dug and an underwater gas/oil field is established, work class ROVs maintain wellheads and equipment that can be as much as two miles below fixed platforms. The recent development of floating production systems for even deeper wells may see the use of hybrid intervention systems, where an AUV routinely delivers an ROV to the appropriate subsea worksite.
- Underwater craft may help subsea mining ventures deal with two huge economic challenges: 1) the expense of locating sufficient resource concentrations and 2) the high cost of getting the valuable assets to the surface and to market.

Image courtesy of Divulgação Petrobras/Agência Brasil and Wikimedia Commons

Image courtesy of Dr. Daniel Jones, SERPENT Project, National Oceanography Centre, Southampton

Figure 1.51 (left): Oil Rig

This photo of Oil Platform P-51 was taken off the coast of Brazil in January 2009 and gives a sense of the scope and complexity of an oil rig. P-51 is the first 100% Brazilian oil platform and is estimated to produce about 180,000 barrels of oil and 6,000,000 cubic meters of gas per day, once it reaches full production.

Figure 1.52 (right): Oil Rig ROV

This underwater shot shows Oceaneering's Magnum 155 *ROV with tooling skid working next to the top of the Lowered Marine Riser Package module of the blow-out preventer at the Rosebank North Field, Faroe-Shetland Channel, U.K.*

4.3. Science

Scientists usually answer a scientific question by first proposing several *possible* answers, called hypotheses, then making observations or conducting experiments to help them figure out which of those possible answers is most likely to be correct. Sometimes a new technology enables scientists to do previously impossible observations and experiments, thus opening up whole new frontiers to scientific discovery. For example, the invention of the telescope revolutionized the understanding of the cosmos, and the invention of the microscope revolutionized the science of biology and medicine. Today, advances in underwater vehicle technology are ushering in a new age of deep sea (and deep lake) research by making it possible for scientists to observe and manipulate things they couldn't even imagine 10 or 20 years ago.

Here are just a few of the exciting fields within science that have been accelerated by the use of underwater vehicles and equipment:

- **Oceanography**: Oceanographers study the physics, chemistry, geology, and biology of the sea, answering questions such as, "What causes currents, tides, tsunamis, or undersea volcanoes?" or "Can we predict the next harmful explosion of algae growth that might produce enough toxin to make some seafood dangerous to eat?"
- **Pharmacology**: Chemicals that can be used as the starting points for the manufacture of new drugs to treat and cure human illness are being discovered in many species of marine organisms, including several deep sea species accessible only via deep-diving ROVs.

Image courtesy of Woods Hole Oceanographic Institution

Figure 1.53: The Scientific ROV

The Jason *ROV was built by WHOI's Deep Submergence Laboratory and first launched in 1988. A second generation* Jason *took to the water in 2002 as a sturdier, more advanced ROV. It carries sonar imagers, water samplers, a variety of cameras, and dual manipulators in order to carry out its scientific missions.*

- **Acoustics**: Scientists and engineers are studying how sound can be generated and heard under water and how it can be used to form images of underwater terrain and objects. This information is improving remote sensing tools used for underwater mapping and surveys. The push for better underwater vehicle positioning and navigation is fueling much of this research.

- **Microbiology**: By studying microorganisms living in deep sea hydrothermal vents and other seemingly inhospitable places, scientists are learning how life might have originated on earth and how it might exist now on other planets or elsewhere in space.

- **Ocean engineering**: Engineering improvements to offshore structures like oil rigs, underwater habitats, transoceanic cables, and even undersea vehicles themselves rely on scientific data collected by underwater vehicles about corrosion, wear, biological overgrowth, and other problems.

- **Marine biology**: ROVs, submersibles, and even AUVs are being used to find, follow, and study fish, jellyfish, squid, mammals, plants, and other organisms, as well as many aspects of marine ecology. This technology has been especially helpful in observing deep sea creatures going about their daily lives. Specialized equipment also allows the collection of jellyfish and other delicate specimens that would be squished if retrieved using traditional nets.

- **Limnology**: Limnology is the study of freshwater lakes and similar systems. Underwater vehicle technology is allowing unprecedented access to the bottom of the deepest lakes in the world.

- **Energy:** Underwater vehicles play a crucial role in scientific studies related to finding and using oil reserves, as well as alternative sources of energy, such as methane hydrates from the seafloor and undersea geothermal vents, not to mention mechanical energy from tides and waves.

Image courtesy of Deep Sea Systems International Inc.

Figure 1.54: Sampling the Underwater World

Deep Sea Systems International's Global Explorer *ROV has demonstrated excellent performance in geological and biological sampling as well as high resolution digital and HDTV imaging. Transportation costs are economical due to the system's low overall weight and modular components.*

Designers of scientific subsea vehicles face the challenge of creating or adapting craft for specific areas of scientific research. For example, a marine biologist might need a quieter vessel, so as not to disturb the organisms being studied. Other requirements could be vehicles with longer underwater capacity, more sophisticated observing and recording instruments, or greater agility. One such interesting example is *U-Tow*, developed for data sampling. It is specifically designed to carry a variety of oceanographic sensors while undulating through the water, ranging from just below the surface clear down to the seafloor.

4.4. Telecommunications

Today, a growing network of communication cables crisscrosses the bottom of the world's seas and oceans. Prior to laying subsea telecommunication cables, survey ships use AUVs or ROVs to carry out mapping runs and determine the most advantageous routes. Once these surveys are completed, cables are spooled out from supply vessels following those routes, while bottom-crawling ROVs handle the actual seabed installation. These procedures require very expensive surface support ships as well as custom-built equipment to lay and bury the actual telecommunication cable a meter deep. The cost of these massive projects is enormous—but not when compared to the losses incurred when cables are not well buried and protected.

TECH NOTE: THE SERPENT TEAM

A unique academic-industrial partnership called SERPENT (for **S**cientific and **E**nvironmental **R**OV **P**artnership using **E**xisting i**N**dustrial **T**echnology) is encouraging collaboration between members of the offshore oil and gas industry and scientists from academic institutions.

SERPENT came about in recognition of the scientific potential of industrial ROVs. The partnership works with the oil and gas sector and ROV companies to place scientists on board drilling platforms or drill ships, providing them with access to ROVs and, most importantly, to the video data collected by the vehicles. Even when scientists are not on board, ROV operators are encouraged to take still images of interesting or unusual creatures that can be sent to scientists for evaluation. Images and information are then shared with students and teachers around the world, via the project's website.

SERPENT is headquartered at the National Oceanography Centre at Southampton University in England. The project is a win-win partnership for the research community and offshore industry, providing scientists with a unique opportunity to assemble a comprehensive picture of life in the deep sea, while allowing oil and gas companies to demonstrate that environmental consciousness and petroleum exploration/extraction are not mutually exclusive.

4.5. Construction, Inspection, and Maintenance

Projects such as bridges, offshore drilling and production facilities, dams, piers, and cable and pipeline operations all require some sort of underwater intervention work. The first step is generally a comprehensive site survey by ROVs or AUVs, followed by core and/or sediment sampling. The actual construction work may involve placing concrete for pylons and coffer dams, welding and cutting steel, conducting tests and inspections, or laying pipe or cable. In shallow water the visibility is usually too bad to use an ROV, so divers handle those jobs. However, robots replace divers for deep water work. In cable lay operations, a trencher ROV can operate from deep water to the shore.

Image courtesy of Saab Seaeye Ltd.

Figure 1.55: Falcon in the Bore Hole

Small eyeball ROVs, such as the Seaeye Falcon, make inspection of reservoirs, dams, tunnels, piping, etc., relatively effortless, efficient, and economical.

Inspection and maintenance requirements necessitate underwater work, even after construction is finished. For example, offshore oil fields have formations and equipment that need ongoing maintenance. Piers and bridges require structural inspections. Reservoirs, tanks, dams, and nuclear reactors also must have periodic scrutiny—using robots for inspections often means that the structures can be kept in near-continuous operation, with minimal downtime. ROVs have taken over much routine underwater inspection and maintenance work, particularly at greater depths.

Image courtesy of JW Fishers Mfg.

Figure 1.56: Icy Inspection

The Fisher SeaLion *ROV was put to the test documenting a 10,000 gallon per minute water leak from a high altitude reservoir in Utah. Due to extremely poor visibility from algae growth and summer runoff, divers could not find the source of the leak.*

Personnel from Adventure, Depth and Technology (ADT) and Advanced Diving Systems (ADS) worked for nine days in near zero temps, with all equipment transported in by snowmobiles and sleds. The ROV located the first leak within hours, and during the remaining days, it successfully tracked and pinpointed several more faults.

4.6. Search and Recovery

While stories of recovered sunken treasure grab the headlines, the fact is that most search and recovery work is undertaken to locate and recover bodies, downed aircraft, sunken boats, valuable equipment, stolen property, or evidence of criminal activities. Underwater search techniques and equipment are commonly used by civil groups, police departments, Coast Guard/Navy crews, or search and rescue teams.

Locating the target generally involves some type of ROV, ranging from a simple observation robot to one equipped with multiple sensors and cameras. Other searches may involve a towed platform equipped with advanced sonar scanning equipment and/or high-definition cameras or video equipment. Recovery may call for more sophisticated equipment and techniques. (See *Historic Highlight: The Development of Side Scan Sonar* and *Chapter 9: Control and Navigation* and *Chapter 10: Hydraulics and Payloads* for more information on side scan sonar.)

Image courtesy of JW Fishers Mfg.

Figure 1.57: Searching for Evidence

Underwater metal detectors (magnetometers), carried by divers or attached to ROVs, boats, or other platforms, are one of many important tools for location and recovery of evidence—in this case a gun and knife as murder weapons. Aquatech of Santa Rosa, CA provided the expertise and equipment to help the Sonoma County Sheriff's Department search for the weapons. The equipment included a JW Fisher Pulse 8X metal detector with a boat-deployed 8 by 48 inch oval coil (pictured in the inset).

The search area was described as "a tidal river with a foot and half of water at low tide and two feet of silt." The mag pinpointed a handgun in less than two hours—but it turned out to be the wrong one! Returning to the scene, both the murder weapons were recovered and the suspects charged.

4.7. Archaeology

Underwater archaeology is a rapidly growing field that benefits from the whole range of scientific underwater techniques and tools developed to study shallow and deep water sites. While the actual search for an historic site is often a tedious procedure involving towed sonar equipment and other sensors, it remains a compelling adventure. However, locating a shipwreck is just the first step; ideally the discovery is followed by careful assessment, identification, and documentation. Recovering items or even raising the shipwreck also relies on underwater vehicles and more extensive topside support craft.

Underwater archaeologist George Bass was the first to use a submersible and employ stereo photography in mapping ancient shipwrecks. Today, sonar-equipped submersibles, ROVs, and AUVs are commonly used to map, sample, and excavate deep water archaeological sites. Other specialized sonar technology can even reveal structures *beneath* seafloor sediment.

HISTORIC HIGHLIGHT: THE DEVELOPMENT OF SIDE SCAN SONAR

In the 1950s, famous electrical engineer Harold "Doc" Edgerton and student Martin "Marty" Klein pioneered the development of side scan sonar at Massachusetts Institute of Technology. Their work made a significant impact on search operations. For the first time, it became possible to use sound to "see" the bottom in near real time and to scan an area larger than had ever been covered by prior techniques. Klein has continued to work on sonar technology, developing sonar systems for commercial applications.

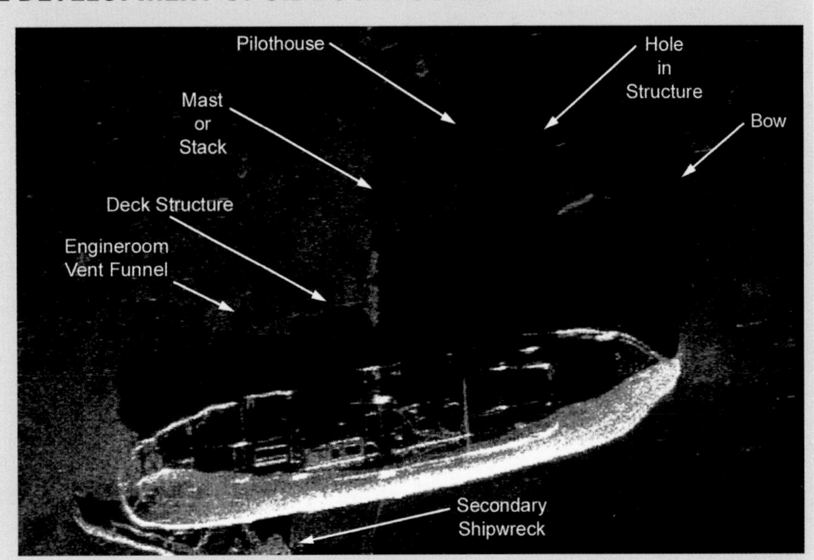

Figure 1.58: Side Scan Sonar Image of a Shipwreck

Image courtesy of Oceaneering International Inc.

4.8. Recreation and Entertainment

Images courtesy Saab Seaeye Ltd.

Figure 1.59: Lights, Camera, Action!

Sometimes an ROV can be just the answer to a film crew's woes. A Bollywood production filming an underwater shark attack was stopped cold. Their fake shark looked like the real thing but certainly didn't swim like one. The problem was solved by attaching a Seaeye Falcon *ROV to the belly of the monster prop. ROV pilots Steve Wilkinson and Nicolas Stroud of Marine Solutions delivered realistic swimming motions for the 400 kilogram shark in both pool and open ocean shots. The cameraman simply cropped the assist divers and the ROV out of the scenes, and the shark got its lunch.*

Today, millions of people enjoy the safe sport of scuba diving as a means of exploring the underwater world. Others prefer a drier approach, booking passage on one of the many tourist subs that allow passengers to view the wonders of reefs and wrecks in air-conditioned comfort. A few resorts even have their own rental fleets of mini-subs available for day excursions. There's also the growing popularity of micro-ROVs that can be used for remote underwater sightseeing from a dock or boat deck.

Sometimes underwater vehicles assist with making movies. They may provide a safety backup for underwater filming and stunts. They even get starring roles, such as the underwater work platform and manned submersibles in *The Abyss* or the long list of submarine movies (*The Hunt for Red October*, *Das Boot*, *Silent Service*, etc.). More commonly, underwater vehicles serve as camera platforms for shooting underwater scenes, notably in James Cameron's epic film *Titanic*. Production on this Academy

HISTORIC HIGHLIGHT: UNDERWATER TOURISM

The first actual tourist submarine was the 40-passenger Swiss-built *Auguste Piccard*, launched in 1964 for use at the Swiss National Exposition in Lausanne. During this event, the sub carried some 33,000 fascinated passengers to the bottom of Lake Geneva. A much more upscale version of the tourist sub involves high-profile submersible trips to deep water sites like the *Titanic*, at a cost of $50,000 a person. There are also two- or three-person submersible trips to observe the behavior of six-gill sharks or visit shipwrecks in polar regions. Actual tourist submarines such as the fleet of *Aquarius* subs offer a significantly more affordable under water experience at many destinations around the world.

Award winner began in 1995 when the filmmaker used ROVs to shoot footage of the actual RMS *Titanic* on the ocean floor. Cameron returned to the wreckage in 2001, executing a series of tandem dives on the *Mir 1* and *Mir 2* submersibles and shooting with the aid of improved lighting on the ROV *Medusa* and two mini-ROVs, dubbed *Jake* and *Elwood.* The result was even more spectacular underwater footage for his large-format documentary *Ghosts of the Abyss.*

4.9. Education

It may seem like a great leap to link dripping underwater vehicles and dry-land classroom education, but it's not as far-fetched as you might think. Increasing numbers of teachers and students are building shallow-diving ROVs such as *SeaMATE* and *Sea Perch.* This kind of hands-on education effectively demonstrates how scientific principles and theoretical concepts work in real life. It also generates tremendous classroom enthusiasm and motivation. There's educational support in the form of teacher-training workshops, on-line curriculum resources, student ROV/AUV design and building competitions, and materials such as this textbook provided and supported by a number of groups. These include the Marine Advanced Technology Education (MATE) Center, the MIT Sea Grant College Program, the Society of Naval Architects and Marine Engineers and the Office of Naval Research's SeaPerch Project, the Association for Unmanned Vehicle Systems International (AUVSI), the Naval Undersea Warfare Center Keyport and Naval Undersea Museum Foundation's ROV Challenge Program, and others.

Image courtesy of Vickie Jensen

Figure 1.60: MATE's International Student ROV Competition

Since its inception in 2002, the MATE Student ROV Competition has involved thousands of students, teachers, and mentors from around the world. The annual ROV competition is just one of the ways that the MATE Center fills its educational mandate.

In other classrooms, students are exploring the subsea world via cameras carried on board underwater craft that may be half a world away. The series of *JASON* Projects, involving Bob Ballard and Woods Hole Oceanographic Institution, led the way with on-line, subsea-based learning that provides captivating math and science instruction for students and professional development for teachers. This type of exciting outreach is being extended by other projects. For example, the California State Park System is using the *ROVing Otter* developed at California State University, Monterey Bay, to help students across the state explore the kelp forests of Point Lobos State Reserve via the internet. MIT graduate student Dylan Owens has developed *REX II*, a modified ROV, built in connection with the University of Hawaii's Hawaii Institute of Marine Biology. It hopes to bring the operator station to local classrooms with internet access so students can operate the vehicle themselves. (See Figure 1.64.)

If learning about subsea robotics provides an exciting gateway to a better understanding of the subsea world, it also offers a tremendous introduction to various career opportunities. All of the underwater robotics applications mentioned above require the creation or adaptation of vehicles and equipment. All offer interesting career opportunities in terms of research, design, construction, operation, maintenance, and sales of those subsea products.

Professional organizations in the maritime sector are a rich source of career information. These include the American Society of Naval Engineers (ASNE), the Association of Diving Contractors International (ADCI), the Association for Unmanned Vehicle Systems International (AUVSI), the Institute for Electrical and Electronics Engineers (IEEE), the Marine Technology Society (MTS), the Oceanic Engineering Society (OES), and the Society of Naval Architects and Marine Engineers (SNAME), among many others. Many of these organizations post career information, offer scholarships, sponsor meetings and events, provide volunteer opportunities, and publish journals or newsletters.

Government agencies such as NOAA also have education sections on their websites (check out "OceanAge" for interviews with various ocean scientists). Commercial companies who provide underwater equipment or services may also post job listings

Image courtesy of Dr. Angela Leimkuhler Moran, United States Naval Academy

Image courtesy of Dr. Susan Crawford, Naval Undersea Museum Foundation

Figure 1.61 (top left): The Sea Perch ROV and the SeaPerch Program

The Sea Perch Program was created originally by MIT Sea Grant College Program in 2003 to train teachers to build Sea Perch *ROVs with their students. Their vehicles were based on the original* Sea Perch *ROV designed by Harry Bohm for* Build Your Own Underwater Robot and Other Wet Projects. *MIT is still following its mandate to provide information and teaching materials.*

In 2007, the Office of Naval Research (ONR) and Society of Naval Architects and Marine Engineers (SNAME) set up a similar but separate SeaPerch Project. Its goal is to increase college enrollment in engineering and technical studies by getting middle and high school students in hands-on science such as building ROVs.

Figure 1.62 (top right): Naval Undersea Museum Foundation's Student Projects

Navy mentors work with students as part of a wide variety of student ROV projects sponsored by the Naval Undersea Museum Foundation at Keyport, WA.

on their websites. An on-line search may also turn up career information sites such as www.marinecareers.net. Similarly, one of the goals of this textbook is provide an introduction to subsea career options.

TECH NOTE: OCEAN CAREERS—NAVIGATING YOUR WAY TO A BETTER FUTURE

A wealth of career options awaits eager students of all ages interested in ocean technology. From maintaining aquarium exhibits to designing ship hulls or operating ROVs on offshore platforms, there are career opportunities in ocean science, exploration, industry, recreation, education, and the military. All of these areas are looking for qualified, motivated individuals.

Image courtesy of the MATE Center

Figure 1.63: Hands-On Learning

Terry Eddy served as an intern aboard the R/V Thompson *in 2003. The mission involved using the* Jason *ROV to explore seamounts in the Pacific Ocean.*

To help provide good, relevant, easily accessible career information, the MATE Center and its partners in the Centers for Ocean Science Excellence (COSEE) developed OceanCareers.com. This interactive website gathers and synthesizes the best available information about ocean-related careers, the knowledge and skills required to work in these careers, the educational institutions that provide the skills and training, professional societies that support workers in these career fields, and ocean-related jobs and internships.

Much of the information on required skills comes from the U.S. Department of Labor and other governmental occupational listings, as well as from the MATE Center's Knowledge and Skill Guidelines (KSGs). With the help of professionals from the field, MATE developed KSGs for several ocean occupations, including marine technician, hydrographic survey technician, and aquarist. The KSGs were then used to identify requirements, or skill competencies, that are common to two or more occupations. Examples include safety and seamanship, navigation, electronics, hydraulics, and oceanography. The competencies are a critical link between the workplace and the classroom because they help students see how what they are learning in the classroom can be applied in the real world. The MATE Center's website (www.marinetech.org) gives a full listing of their Knowledge and Skill Guidelines for ROV technicians.

There is no substitute for hands-on learning. Many schools and university programs with an oceanographic focus have coursework that is integrated with onboard work experience. Check out their websites. In the United States, other hands-on opportunities (volunteer or paid) are available through NOAA, the Monterey Bay Aquarium Research Institute (MBARI), Woods Hole Oceanographic Institution (WHOI), the U.S. Environmental Protection Agency (EPA), or the Environmental Careers Organization (ECO). Don't forget to check out your local aquarium, marine research station, or any business providing marine-related products and services to see if they could use a helping hand. Some professional maritime organizations offer internships or volunteer opportunities. All these options are excellent ways to build your resumé and explore career options.

The MATE Center's Technical Internship Program is a typical way to gain experience at sea through work/study ventures. It places college and university students in internship positions on board research vessels, including those that are part of the University-National Oceanographic Laboratory System (UNOLS) fleet. As interns, the students work alongside marine technicians and ships' crew, gaining valuable workplace experience with mission operations, as well as a knowledge of shipboard procedures and safety requirements.

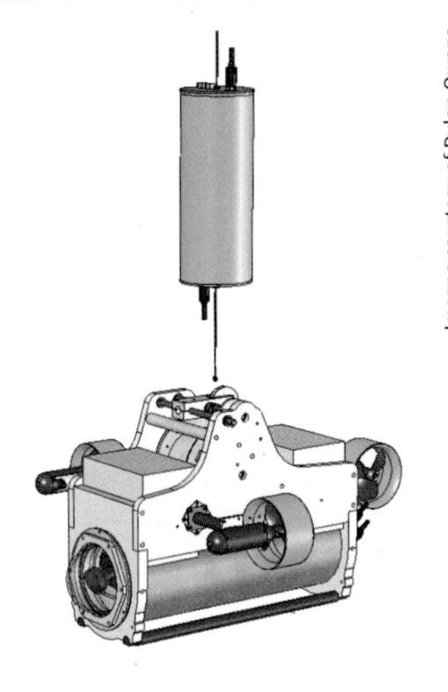

Images courtesy of Dylan Owens

Figure 1.64: Bringing the Underwater World Inside

Rex II *is similar to a conventional ROV, but the tether runs vertically to a float at the water's surface. The ROV hangs below the surface float and uses an onboard winch to reel the tether in or out, thereby moving up or down.*

A radio antenna sticks off the float into the air, sending video and remote control commands between the operator and the vehicle. Because the information is transmitted wirelessly, the vehicle can move around more easily and the operator does not need to be right nearby.

5. A Detailed Look at a Work Class ROV

Since ROVs are one of the most widely used robotic vehicles today and will likely be the first type of underwater vehicle you design and build, it's important to understand some basic ROV anatomy and to develop a general understanding of how the various parts of an ROV function. In this section, we take a look at a typical work class ROV, such as those used in the offshore oil and gas industry. These industrial workhorses are far bigger, more versatile—and more costly—than the first ROV any beginner is likely to build, but they provide a great overview of what an ROV is, illustrate a broad range of features and capabilities you might want to try replicating on a smaller scale, and introduce some of the issues and challenges associated with trying to operate an ROV under less than ideal "real world" conditions.

There are hundreds of work class ROVs in operation today around the world. The vast majority provide support for the offshore oil and gas industry or help to lay and maintain subsea telecommunications cables. These vehicles are highly specialized, costly tools that operate from a support platform such as a large ship or drilling platform. This large support platform is necessary, because the ROV itself comprises only about a third of a complex system of people and equipment that produces power, controls the vehicle, gets it into and out of the water, and accomplishes the subsea work that needs to get done.

The mechanical side of a work class ROV system consists of these major subsystems:

- a power source (generator) and power distribution unit (PDU)
- the control console (housed in the control shack)
- deck cable, umbilical, and tether
- the launch and recovery system (LARS), with crane and/or winch
- tether management system (TMS)
- the ROV itself

The technical chapters in this book (Chapters 3–11) expand upon concepts and terminology presented in this section. Specific technical chapter references are generally given at the end of the appropriate paragraph.

TECH NOTE: SIZE CLASSES OF COMMERCIAL ROVS

Heavy Work Class Vehicles and Ultra Heavy Duty (not electric)

This class handles the biggest equipment (tools, skids, etc.) and requires big ships for deep water operations.

- They service primarily oil and gas operations.
- Horsepower range is 200-300. Weight is over 4,500 kilograms (10,000 lbs), about the size of a minivan.
- Work depths of 3,000 meters (approx. 10,000 ft) are now common, with some able to handle depths to 6,000 meters (approx. 20,000 ft).
- Vehicle thru-frame lift (not swimming) capacities are generally to 5,000 kilograms (approx. 11,000 lbs).
- Most have two sophisticated and powerful manipulator arms, which function like human arms and hands to grasp things and operate tools.

Work Class ROVs

This class of ROV generally refers to electro-hydraulic vehicles with these characteristics:

- They handle drilling support, construction operations, pipeline or cable inspections, and a variety of other tasks.
- Horsepower range is 100-200; weight is 1,000-3,600 kilograms (approx. 2,200-8,000 lbs).
- Most carry moderate **payloads**, including two manipulators (robotic arms), and are capable of lifting and swimming with up to 680 kilograms (approx. 1500 lbs).
- Like their heavy duty cousins, all work class ROVs rely on powerful dual manipulators with multiple degrees of freedom (functions).

High-Capability Electric ROVs

This more recent development is a compact work class system that utilizes an all-electric propulsion system. The decreased size and weight of the system means a corresponding decrease in deck space and installation time. Although not as powerful as most of their hydraulic cousins, they perform many tasks at greatly reduced cost.

- The quietness of electric motors gives these ROVs a significant advantage for military and science missions.
- Horsepower is typically 20-100. Weight range is 1,000-2,200 kilograms (approx. 2,200-4,850 lbs).
- Depth capacity to 900 meters (approx. 3,000 ft) with a few to 6,100 meters (approx. 20,000 ft).
- Payload capabilities are in the range of 100–200 kilograms (approx. 220–440 lbs).
- A single manipulator arm is common, although the larger of these vehicles may have two.

Small (Electric) ROVs

Small ROVs are by far the most numerous of all robotic vehicles. Most function in some kind of flying eyeball capacity. Technological improvements have resulted in increased capabilities, performance, and working depths. Sometimes these portable robots are called "suitcase" ROVs because the entire unit can fit into one or two suitcase-like containers.

- They are used primarily for inspection and observation tasks, search and rescue, and scientific exploration.
- Horsepower range varies considerably, but often is less than 10.
- Most are considered "low cost" (hence LCROVs), with price tags ranging from about $10,000 to $100,000.
- Typically, they are all electric; most operate in water depths less than 300 meters (approx. 990 ft).
- Few have a manipulator unless it's a very basic gripper type.

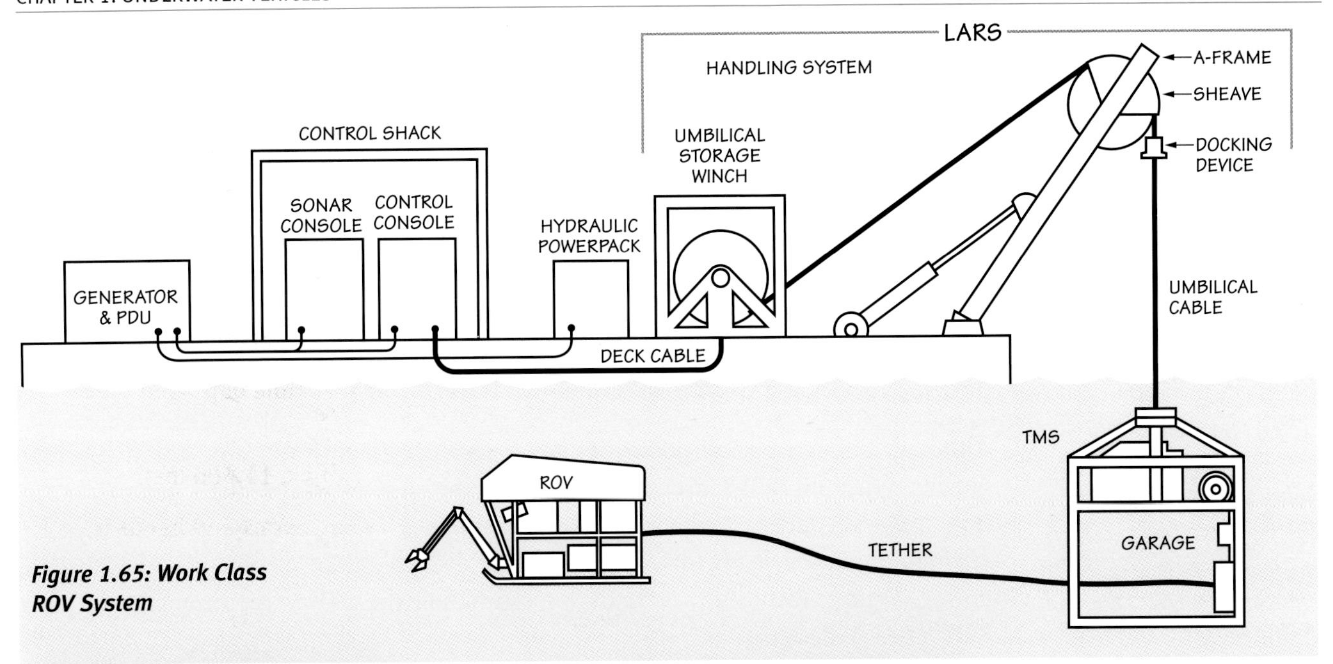

Figure 1.65: Work Class ROV System

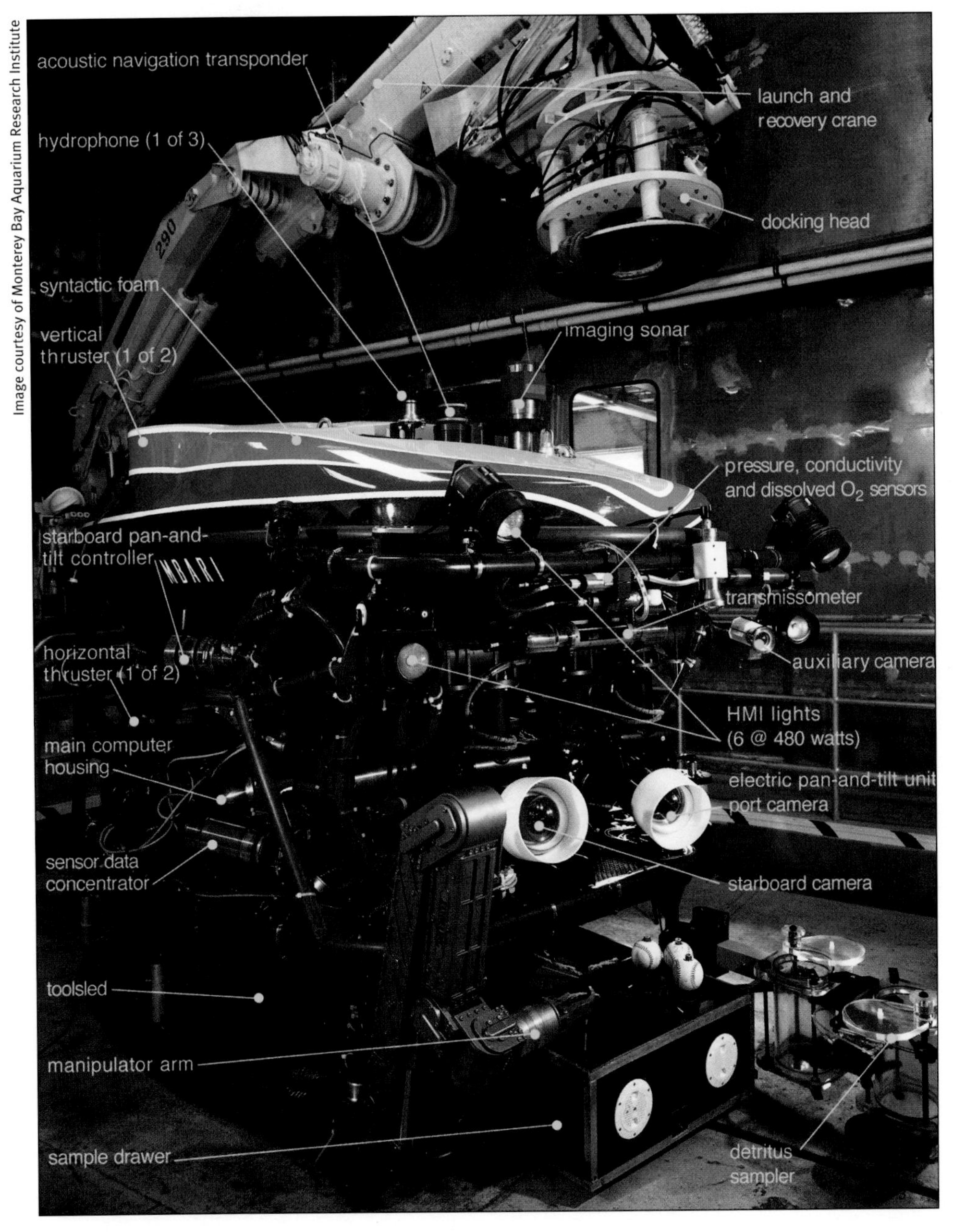

Image courtesy of Monterey Bay Aquarium Research Institute

Figure 1.66: The* Tiburon *Legacy

The Monterey Bay Aquarium Research Institute's Tiburon *ROV helped scientists explore Monterey Canyon and beyond until it was retired in 2008. This photo with labeled components helps readers to understand the complexity of a scientific ROV's "anatomy."*

5.1. Power Source

Different classes of ROVs have different power requirements, depending on the jobs they're expected to do. Most work class ROVs use a combination of electrical and hydraulic power while under water. The electricity is produced on the surface by a dedicated diesel or gasoline generator. High-voltage power from this generator is delivered by cable to a **power distribution unit** or **PDU**, which distributes this electrical power to the ROV (via the tether) and to related systems needed to operate the ROV, like the winch and the control console. On the ROV, a portion of the electrical power is converted to hydraulic power, which is then available to operate thrusters, manipulator arms, and tools.

Although this is a typical power system for a work class ROV, some are all-electric. Smaller types of ROVs use a variety of power sources, from small generators, to standard wall outlets, to batteries. (See *Chapter 8: Power Systems* and *Chapter 10: Hydraulics and Payloads* for more complete discussions of electrical and hydraulic power systems, including methods for estimating the power requirements of your ROV or AUV design.)

Figure 1.67: Control Room Interior

The Subsea 7 Clansman ROV control van onboard the transocean semi-submersible drilling rig Jack Bates.

Image courtesy of Dr. Daniel Jones, SERPENT Project, National Oceanography Centre, Southampton

Image courtesy Saab Seaeye Ltd.

Figure 1.68: Containerized Control Shacks

There's a significant cost and efficiency advantage to a dedicated ROV control room such as this portable container.

5.2. Control Room

The controls for a work class ROV are housed most often in a specially designed shipping container called the **control room**, control van, or control shack. Located on the support vessel or occasionally on a dock or platform, the control room houses a pilot's console, a sonar system (for tracking the location of the ROV), a monitoring and control panel for power systems, manipulator controls, and an audio/video recording system. The control shack also provides important humidity control for sensitive computer equipment, as well as protection and temperature regulation for the components and personnel inside. Most ships don't have a dedicated ROV control room, so a complete work class ROV system usually comes with a shipping container that has been converted to serve as a portable control shack. It can be moved easily from one ship to another, and set-up time for the ROV is minimized because the modular nature of the container allows operators to leave all the components of the control system in place during storage and transport.

Smaller "fly-away" systems are those where the underwater vehicle and necessary hardware and software, including controls, are transferable via commercial air freight. (*Chapter 11: Operations* provides more information.)

5.3. Operations Crew

Although a tiny inspection ROV can sometimes be launched, piloted, and recovered by one person, this is not so with a work class ROV. Safely managing one of these big beasts typically involves a team of about three people working 12-hour shifts. They spend much of their time in the control shack while the ROV is in the water.

- The **supervisor** is the shift leader, a job that requires an all-around seaman and offshore operations-savvy person (often someone with commercial diving experience), who also knows some hydraulics and electronics.
- The **pilot** not only flies the vehicle but also functions as the electronics technician, running and maintaining all electronics systems. They assist with manipulator tasks, sonar ops, photo and video recording, and may interface with survey people.
- The **ROV winch operator** runs the deck winch and maintains all mechanical systems.

Image courtesy of Anne Smrcina, Stellwagen Bank National Marine Sanctuary, NOAA

***Figure 1.69: Maintenance on the* Hela**

Crews working on scientific or commercial subsea ROVs and AUVs regularly split their duties between running/tracking the vehicle when it's in the water and maintaining/repairing/charging the vehicle when it's out of the water. Here, four pairs of hands work on the Hela *ROV from the National Undersea Research Center at the University of Connecticut (NURC-UConn). Deployment was from the R/V* Connecticut *during a 2003 mission to the wreck of the coastal steamship* Portland, *lost in 1898 during the great "Portland Gale" (a storm of the century). The wreck has been called "New England's* Titanic*" and is within the Stellwagen Bank National Marine Sanctuary (SBNMS). NURC-UConn is now the Northeast Undersea Research, Technology and Education Center (NURTEC) at the University of Connecticut.*

Depending on the nature of the work, other people may also be in the control shack—chief scientists, survey personnel, and project leaders, among others. (See also *Chapter 11: Operations.*)

It's not uncommon for crew members to rotate or share job responsibilities, so being flexible and having a wide variety of skills is important. In addition to piloting the ROV while it's under water, crew members also coordinate launch and recovery activities; this is always a critical time, because any mistakes can be costly or even deadly.

5.4. Deck Cable, Umbilical, and Tether

With smaller ROVs, you'll hear the terms "cable," "tether," and "umbilical" used interchangeably to refer to the power/control cable. In that context, they all mean the same thing. But in larger inspection class ROV systems and on all work class ROV systems, there are three distinctly different cables, each specialized for its own important function. (See Figure 1.70.)

- The **deck cable** connects the control room to the vehicle launch and recovery system (LARS), which typically includes some sort of crane, a winch, and other equipment for getting the ROV in and out of the water safely. (LARS components are described in more detail in the next section.) This cable serves as the first leg in a three-part power and communications link between the control room and the ROV. To perform that function, it carries copper wires that deliver electrical power and optical fibers that carry optically encoded pilot commands, sensor data, video signals, and other information. In addition to direct support for the ROV, the deck cable commonly contains a couple of added conductors for video and voice communication between the LARS winch operator and the control room. This cable may be laid on the deck of the ship (hence the name deck cable) or routed along an overhead structure or other protected area.
- The **umbilical** is a very long, heavy-duty, armored cable stored on the LARS' winch drum. This umbilical serves not only as the second leg in the ROV's communications and power link, but also as a lift wire able to support the weight of the ROV and associated equipment during launch and recovery. On large commercial work class ROV systems, the umbilical can be as large as 5 centimeters (approx. 2 in) in diameter and have two or three layers of steel armor. During operations, it spools out from the LARS winch down through the water to the depth where the ROV is working. In some cases, the umbilical connects directly to the ROV, but more often it ends at something called the "garage," a cage-like structure that holds and protects the ROV during launch and recovery.

- Extending from the garage to the ROV itself is a comparatively short and much more flexible cable called the **tether**, which forms the final leg in the three-part link from the control room to the ROV. The tether has the same core electrical wires and optical fibers as the umbilical, but is smaller in diameter and lighter, because in place of steel armor it has a Kevlar-strength layer and soft polyurethane jacket. The tether allows the ROV to move freely about in the vicinity of the garage, without having to drag thousands of meters of heavy, stiff umbilical around everywhere it goes. When the ROV returns to the garage, the tether is reeled in and stored by a motorized spool called the tether management system (TMS).

When a work class ROV is on deck, there's a truly impressive quantity of umbilical wrapped on the winch drum. The actual length and diameter of the umbilical depends on the type of ROV being used and its mission, but it's not uncommon to have many kilometers (i.e., several miles) of thick cable spooled neatly onto a large drum with the help of some pretty beefy machinery.

Figure 1.70: Launch Mechanisms

A strong fixed crane, as in the left hand photo, or A-frame, as in the right hand photo, are the basic requirements for any heavy-duty launch and recovery system.

Image courtesy of Schilling Robotics

Image courtesy of Dr. Daniel Jones, SERPENT Project, National Oceanography Centre, Southampton

5.5. Launch and Recovery Systems (LARS)

In the open ocean, waves and swell cannot be ignored. They rock even the largest ships and can crush an unprotected ROV against a ship's hull, mash it against the pylons of an oil platform, or snap its umbilical, sending it to a watery grave. As a result, the transition from air to sea and back again is often the most treacherous phase of an ROV mission. The term LARS describes the equipment that enables mid- to large-size ROVs to make this transition safely. The essential components of a typical LARS are a base frame, an A-frame (or crane), a winch, and a slip ring assembly.

The base frame is simply a rigid structure that serves as a foundation for the other parts of the LARS. The **A-frame** or crane is a strong overhead structure—sometimes shaped like a capital letter "A" (hence the name)—supporting a pulley over which the umbilical is run. This allows the reinforced umbilical to be used like a strong rope to lower the ROV into the water and to lift it back out again later. The A-frame may be fixed in place over an opening in the ship's deck, called a **moonpool**, or it may be hinged so that it can pivot in over the deck to pick up the ROV just before launch, then pivot (with the ROV) out over the side or stern of the ship to lower the ROV into the water.

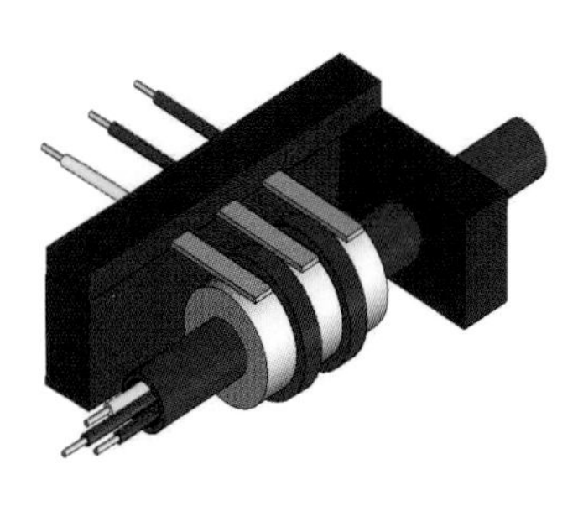

Figure 1.71: Slip Ring Assembly

A slip ring is a mechanism for transferring electrical power or data from a stationary platform (such as a ship's deck) to a rotating structure (such as the spool for an ROV's umbilical) without twisting the wires.

This diagram shows the basic idea for a simple power transfer slip ring. The rotating blue shaft and blue disks are made of some insulated material and support three brass rings, which are electrically isolated from each other. Inside the shaft, each of the three rings is connected to one of three wires running through the shaft, so as the shaft rotates, the wires inside rotate with it. Each brass ring is in contact with a spring-loaded strip of metal that maintains electrical contact as the ring rotates under it, allowing transfer of electricity between the wires mounted on the stationary block and those on the rotating machinery.

More (or fewer) wires can be accommodated by adding (or removing) brass rings and insulating spacers. Some types of slip rings have additional features for transmitting optical data between stationary and rotating optical fibers.

The **winch** is a large, powerful motorized spool that winds and unwinds the umbilical stored on it. It must be strong enough to lift the weight of the ROV, any TMS (or cage), and assorted tooling packages. In some cases, the LARS winch is **heave-compensated**, which means it's fitted with a special mechanism to adjust for the rise and fall of the ship in large waves. This minimizes the effect of ship movement during launch and recovery, usually the most dangerous time in any ROV operation.

The **slip ring assembly** is a specialized electromechanical connector that transfers power and data from the copper wires and optical fibers on the fixed winch junction box (where the LARS end of the deck cable is attached) to the rotating umbilical junction box on the winch drum (where the upper end of the umbilical is attached). Without this slip ring assembly, the deck cable would get twisted and damaged as the drum rotated.

It is common for ROVs deployed from semi-submersible drilling rigs and some vessels to use a **capture frame** that guides the ROV down a pair of wires through the air-sea interface. This type of arrangement is sometimes called a **cursor system**.

Figure 1.72: A Capture Frame/Cursor LARS

ROV inside garage

Capture frame supports the ROV and garage, particularly during launch or recovery in heavy weather conditions.

During launch, the garage/ROV safely descends to the base of the capture frame. At mission depth the ROV then exits the garage. This procedure is reversed for recovery.

Images courtesy of Oceaneering International, Inc.

Regardless of the launch and recovery system used, the objective is always to keep the ROV from being damaged during the crucial transition from the ship's deck to beneath the wave action. (*Chapter 11: Operations* provides general operational considerations, including launch and recovery procedures.)

Obviously, a small ROV operating in a backyard swimming pool isn't going to need a huge and complex LARS to manage miles of armored umbilical, but you might be surprised how many problems arise from tether-related issues even in these small craft. Indeed, tether design and management are among the most frequently underestimated challenges that confront novice designers, builders, and operators of ROVs. *Chapter 8: Power Systems* discusses some of the tether-related issues you'll need to consider as you design or operate small ROVs. *Chapter 11: Operations* provides practical tether management suggestions for use with smaller, shallow-diving ROVs. *Chapter 12: SeaMATE* includes plans for constructing a tether for a shallow-diving ROV and discusses options for building deeper-diving vehicles.

5.6. The ROV

A work class ROV consists of several basic subsystems that are found on just about any ROV, big or small. These include a frame; flotation; thrusters; lights and cameras; an assortment of navigational sensors and other instruments; various grippers, manipulator arms, or other tools; and the electronic and/or hydraulic systems needed to power and control many of these systems.

The frame is usually an open, box-shaped structure made from welded aluminum struts that have been treated to prevent corrosion from saltwater. Like the frame of an automobile or bicycle, it serves as a rigid foundation to which all of the other parts of the vehicle can be attached. Underwater vehicle frames are discussed in *Chapter 4: Structure and Materials.*

The frame and most of the other components on an ROV are heavy and would tend to sink like a rock in water if it were not for a large block of relatively lightweight flotation material called **syntactic foam** mounted on top of the frame. This flotation is designed to almost exactly balance the sinking tendency of the other parts of the ROV, thereby allowing the ROV to float effortlessly in mid-water, like the underwater equivalent of a blimp. This has a number of advantages. For example, it conserves power that would otherwise be needed to lift the ROV, and it improves visibility near the bottom by avoiding the clouds of sediment that would be stirred up if a helicopter-like "down draft" was needed to lift the vehicle off the sea floor.

The thrusters are motorized propellers used to generate forces to propel the vehicle through the water. A typical work class ROV will have a half-dozen or more thrusters oriented in different directions to give the craft great maneuverability in the water. *Chapter 7: Moving and Maneuvering* describes thrusters, thruster placement, and related issues.

Lights and video cameras provide the ROV pilot with a way to see what's happening around the ROV from the control room far above, and these images can be recorded for later review. In conjunction with the cameras, navigational sensors—including sonar systems, depth sensors, and compass—help the pilot know where the ROV is, where submerged objects of interest are, and which direction the ROV is facing. Other sensors can collect a variety of data for scientific or other purposes. *Chapter 9: Control and Navigation* explores a wide variety of cameras and electronic sensors used for navigation or other purposes.

TECH NOTE: TETHER MANAGEMENT SYSTEMS (TMS)

The term **TMS** is used to describe a method of efficiently handling tether deployment, particularly for deep-working ROVs. The TMS can be housed in one of two basic arrangements—either a garage (sometimes called a cage or launcher) arrangement or a top-hat assembly. In the **garage tether management system**, the ROV is housed in a cage-like frame that includes a motorized tether reel and slip ring assembly. When the garage reaches the worksite location, the retaining clamps securing the ROV are released and the robot is piloted out of the cage, with a lightweight tether spooling out behind it. At the end of a job, the tether retracts onto the spool as the ROV is flown back into the garage; then both robot and garage are raised to the surface for recovery. The alternative **top-hat tether management system** has the ROV attached beneath a frame that houses the motor-driven tether reel and associated slip ring assembly. Just as on the LARS drum, this slip ring assembly allows the transition from fixed to rotating optical fibers and copper conductors.

Either garage or top-hat TMS system allows the tether to spool out (and back in) as the ROV goes about its work. An underwater camera mounted on the TMS permits the operator in the control van to monitor the progress of the tether. The advantage of such a system is that the ROV is decoupled from any surface wave action that might be yanking on the umbilical and is relatively unencumbered in performing its mission, since it's free of the greater drag of the bulkier umbilical. Another advantage is that both garage and top-hat arrangements are heavily weighted so they hang straight down during deployment, keeping the ROV as stable and protected as possible in rough seas.

Effective tether management is crucial because the tether/umbilical is one of the ROV's greatest advantages—and limitations. It effectively transmits power, control signals, and data, but it creates drag, which can severely impede the motion of the ROV. The longer the tether and the greater the currents, the greater this problem becomes for large, deep-diving craft. Small ROVs operating in shallow water with debris or seaweed may encounter tether entanglement issues. (*Chapter 11: Operations* discusses tether management, particularly for smaller ROVs, in much greater detail.)

Image courtesy of Perry Slingsby Systems

Figure 1.73: Tether Management Systems: Top-Hat System and Garage

An effective tether management system such as the top-hat system on Perry Slingsby's Triton XLS *or the garage housing for Oceaneering's* Millennium *reduces tether drag and the chance of entanglement while ensuring maximum transmission of power, control signals, and data.*

Image courtesy of Oceaneering International Inc.

Most work class ROVs are equipped with robotic manipulator arms. The ROV operators can use these to grab onto things, pick up things, or to use tools that saw, weld, rotate valves, and so on. The mission requirements determine which interchangeable tool tray is attached to the bottom of the ROV's frame. Each modular unit carries specific tools for specialized tasks like digging, drilling, cleaning, inspecting, etc. A high degree of modularity enhances the functionality of an ROV and makes switching the tool packs easier. (Tool trays and other specialized **payload** equipment are discussed in greater detail in *Chapter 10: Hydraulics and Payloads.*) Although sophisticated manipulators and tool trays like those found on work class ROVs are rarely found on smaller, low-cost ROVs, some of these smaller vehicles have simple hooks, scoops, or grippers that can be used to pick up small items or perform other simple tasks.

A work class ROV will also need some advanced electronic and hydraulic systems to power and control the thrusters, cameras, lights, sensors, manipulator arms, and other devices on the ROV. *Chapter 8: Power Systems*, *Chapter 9: Control and Navigation*, and *Chapter 10: Hydraulics and Payloads* cover these topics in much more detail. Since electrical circuits and water are generally incompatible, all of the control electronics are usually housed in waterproof, air- or oil-filled containers called electronics "canisters." *Chapter 5: Pressure Hulls and Canisters* describes why and how to build such containers.

Admittedly, work class ROV systems such as the one described here are considerably bigger and more complex than a simple observation ROV, but all ROVs share similar challenges and can borrow from each other's solutions, regardless of size. The project plans and ideas in *Chapter 12: SeaMATE* guide you through the process of building your own small ROV. Even though *SeaMATE* is much smaller and simpler than a work class ROV, you'll find that it faces many of the same challenges and requires many of the same basic subsystems as a work class ROV weighing and costing substantially more!

6. Chapter Summary

This chapter has introduced you to the fascinating history and tremendous diversity of underwater vehicles—from the first attempts to build manned craft to the most recent robotic developments.

Early inventors had to deal with huge challenges as they learned first hand about the effects of water pressure on human bodies and vehicle hulls, and as they adapted machinery to replace human-powered propulsion. Events such as World Wars I and II, as well as the undeclared Cold War, spurred the development of subsea technology and gave rise to such impressive underwater achievements as today's massive nuclear submarines. The Space Race of the 1960s and the Oil Boom of the 1970s gradually shifted the emphasis from man-in-the-sea to robot-in-the-sea.

The current spectrum of underwater robots, both remotely controlled and autonomous, includes numerous commercially viable ROVs as well as many more experimental ROVs, AUVs, and hybrid vehicles. These range in size from heavy-duty work class ROVs that routinely perform missions on deep gas and oil installations to tiny eyeball robots that can inspect the interiors of pipes in nuclear reactors. Today's underwater vehicles are employed in a diverse array of jobs. The majority work in the underwater oil/gas sector, but many others collect scientific data, perform search and recovery missions, survey routes for subsea cables, or introduce the undersea world to tourists and students alike.

Regardless of size and complexity, all underwater vehicles must cope with a similar set of challenges brought about by the physical realities of working under water. And to a surprisingly large extent, they all require similar approaches and subsystems to overcome those challenges. In the remainder of this book, you'll be learning what you need to know to build small ROVs or AUVs and in the process learning a lot about how to anticipate and overcome the challenges of conducting *any* underwater mission with *any* size or type of vehicle. Have fun, and prepare to get wet!

Chapter 2

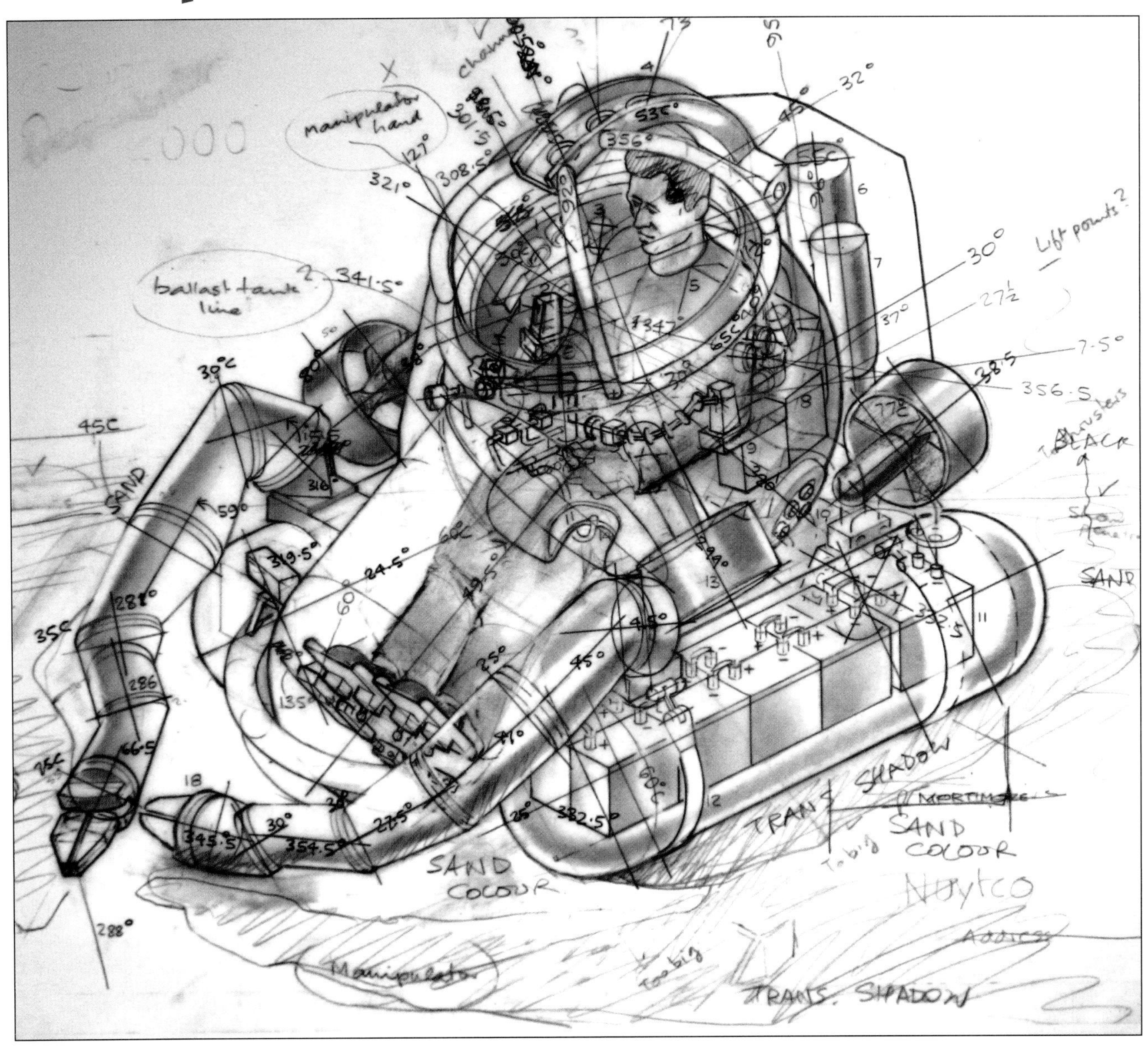

Design Toolkit

Chapter 2: **Design Toolkit**

Stories From Real Life: Quest for a Ship of Gold

Chapter Outline

1. **Introduction**
2. **Common Design Challenges**
   - 2.1. Strategies for Successful Projects and Missions
   - 2.2. Strategy A: Keep Your Eye on the Mission
   - 2.3. Strategy B: Build an Effective Team
   - 2.4. Strategy C: Be Proactive about Project Planning, Management, and Safety
   - 2.5. Strategy D: Think in Terms of Systems
   - 2.6. Strategy E: Use the Design Spiral
   - 2.7. Strategy F: Research, Research, Research
   - 2.8. Strategy G: Build and Test Prototypes
3. **A Design Methodology for Underwater Vehicles**
   - 3.1. Stage 1: Accepting the Mission
   - 3.2. Stage 2: Crafting a Mission Statement
   - 3.3. Stage 3: Identifying the Mission Tasks
   - 3.4. Stage 4: Establishing Performance Requirements
   - 3.5. Stage 5: Identifying Constraints
   - 3.6. Stage 6: Listing Vehicle Systems
   - 3.7. Stage 7: Generating the Concept Design
   - 3.8. Stage 8: Fabricating the Vehicle
   - 3.9. Stage 9: Conducting Sea Trials
   - 3.10. Stage 10: Carrying Out Operations
   - 3.11. Stage 11: Evaluating Ops and Writing a Report
4. **Chapter Summary**

Chapter Learning Outcomes

- Describe the major challenges when designing and building an underwater vehicle, regardless of size.
- Describe specific strategies to address common design and building challenges (for example, how to be proactive about team safety), as well as how to keep a project on schedule and within budget.
- Explain the nature and importance of various design strategies, for example:

  1) being proactive about project management

  2) using the design spiral approach
- Explain how the various design methodology stages, (for example, identifying constraints) help define an underwater vehicle before it ever gets built.

***Figure 2.1.cover: Concept Drawing of* Deep Worker**

Systematically working through the steps in a design methodology will help you arrive at a realistic concept drawing of your vehicle, such as this one of the Deep Worker *submersible.*

A concept drawing, along with parts lists, cost estimates, and other information you've collected, allows you to assess the feasibility of building it.

Image courtesy of Phil Nuytten, Nuytco Research, Ltd.

STORIES FROM REAL LIFE: Quest for a Ship of Gold

From earliest times, men have gambled physical and financial resources in the quest to retrieve valuable cargo from shipwrecks. Early practical equipment adapted for salvage operations included Edmund Halley's patented diving bell in 1691 and John Smeaton's mechanical air compressor in 1788. Even so, divers were limited to shallow-water wrecks.

As technology evolved, so did mechanized retrieval strategies. Working from an anchored platform, a steam-shovel clambucket could bite into and scoop up chunks of historic cargo. The development of atmospheric diving suits, pressure hoses, and suction pumps allowed a more directed salvage. However, treasure lost at great depths remained elusive. This Story From Real Life, *based on Gary Kinder's book* Ship of Gold in the Deep Blue Sea *and on-line accounts, details Tommy Thompson's determination to do what had never been done before—salvage a deep-water shipwreck.*

Gold! The word alone was enough to ignite the dreams of thousands of men and women when the precious metal was discovered in 1848 at Sutter's Mill. Some 85,000 gold seekers tried any means possible to get to northern California, with the majority coming from the eastern United States. San Francisco exploded from a sleepy settlement of 459 to a major seaport with thousands of masts in the harbor.

Most prospectors would return home with little but disappointment to show for months, even years, of desperately hard work. A few of the lucky ones headed back with saddlebags or packsacks full of nuggets and gold dust. Whether rich or poor, most traveled back to their East Coast homes by sea, choosing either the dangerous route around the tip of South America or the quicker, safer Panama route, which involved traveling by ship from California to Panama, then crossing the Isthmus by dugout canoe, mule, or foot, and finally boarding a different ship for the final leg to the East Coast. From 1849 to 1869, prospectors opting for the Panama route traveled home with some $711 million in gold.

Not all of them made it. In September 1857, the Atlantic sidewheel steamer *Central America* left Panama, bound for New York carrying $1,595,497.13 worth of consigned gold, along with an unknown quantity of unregistered gold dust and nuggets, gold coins from the new San Francisco Mint, and gold bars, some the size of building bricks. Days out of Havana, the ship encountered a storm that quickly escalated into a fierce hurricane. Despite a desperate bailing effort over several days, the vessel sank. Most of her 500 passengers—along with 21 tons of gold—were lost at sea.

Well over a century later, a young maverick engineer by the name of Tommy Thompson became fascinated by the "deep ocean," as the vast unknown and unexplored depths below 200 meters (approx. 650 ft) were called. He reasoned that the possibility of locating a deep-water treasure wreck would encourage the money and the minds necessary to design equipment capable of working far below the surface. Others had found and recovered sunken treasure before, but by the late 1980s, none had done so at the phenomenal depth of 2,600 meters (approx. 8,500 ft).

If anyone could pull off the feat, Tommy Thompson was the man. As a young boy, he'd always asked questions, collected parts, and built things—or taken them apart to see how they worked. As he got older, Thompson knew he wanted to be an inventor. But there was no college offering that specialization, so he chose Ohio State University, home to one of the largest engineering schools in the world. Determined to become an ocean engineer, Thompson bucked the odds and became the only student out of 8,000 in the College of Engineering to go to sea. Luckily, the dean of the School of Mechanical Engineering was himself a marine engineer interested in the deep ocean. He encouraged Thompson to take chances, to look and think beyond the conventional solutions everyone else was trying. After graduation, Thompson continued to work on numerous inventions and projects.

Eventually, Tommy Thompson joined Battelle Memorial Institute, a privately funded organization involved in contract research for both government and private industry. Most importantly, it was a place where a real inventor could flourish. His assigned mission was to figure out an overall approach to mining the deep ocean. This topic soon proved the wisdom of his college dean who said, it's relatively easy going down to great depths, but the ability to work in the deep ocean is quite another matter. Thompson eventually came to the conclusion that radical new technologies could be developed, but the cost to build and operate the equipment would be far greater than the value of the recovered minerals.

Fortunately, much of the technological knowledge gained about deep-ocean mining provided insight on deep-water shipwreck recovery. In characteristic style, Thompson attacked the idea of finding such a treasure ship with research, research, and more research. He broke the problem down into several parts, first determining if any feasible deep-water wrecks were worth salvaging. This involved researching historic records and ships' logs for probable course and cargo, and getting to know the historic archives of several libraries, all the while building a network of contacts.

This methodical research process deleted wrecks that were not feasible targets. Slowly, the sidewheeler *Central America* rose to the top of Thompson's list for several important reasons: 1) there were survivors' reports that could provide data to help locate the ship; 2) the wreck was in deep water, so new technology would be crucial to finding and recovering it; 3) the considerable depth should have kept the ship undamaged by storms, tides, or other natural occurrences; and 4) the ship carried enough documented treasure to interest investors in sponsoring a search and recovery expedition.

Thompson began serious research in 1983, working with Bob Evans, a geologist and longtime associate. They compiled extensive information from passenger and crew accounts and took this info to Dr. Lawrence Stone, an expert on search theory. Plotting their data and creating thousands of computerized models of possible sinking scenarios, he arrived at a search area of 2,250 square kilometers (approx. 868 square miles).

Thompson's quest was more than a simple treasure hunt—his vision was a scientific pursuit that was logical, one that would allow him to learn along the way and, in the process, push back the frontier of deep-water exploration. Furthermore, he wanted to explore and document the finding of a valuable deep-water wreck before recovering the valuable and historic cargo.

Once the *Central America* became a feasible target on paper and a probable search area was established, the next technological challenge was actually locating the target in the depths. Thompson felt that the sonar prototype *SeaMARC I*, developed by the University of Hawaii and Raytheon Company, had the potential to search large areas of deep ocean. *SeaMARC* was the first of its kind—a deep ocean, near-bottom, towed bathymetric sonar system. The sonar prototype fit Thompson's goal to obtain reliable, repeatable results for target acquisition data, in spite of weather conditions and mechanical problems. The biggest barrier, however, was not so much adapting *SeaMARC* technology to more efficient side scan sonar, but its million-dollar price tag. Dealing with that kind of money necessitated setting up Thompson's high-risk venture as a viable business and beginning the process of capturing investor interest. Eventually, the project became legally viable as the Columbus-America Discovery Group.

Now to find the ship. The first physical stage of the mission was a grueling 40-day sonar search aboard an old Louisiana mud boat. It towed a modified *SeaMARC* side scanning sonar, sweeping the target area grid in three-mile swaths. Analyzing those sonar images took place during the winter of 1986–87. Thompson also began designing a robotic machine that could access a deep-water wreck and its cargo. Although the military had used ROVs for years, the oil and gas industry had adopted them only in the 1970s.

Ten such robots existed, and at that time, none had the capacity for the kind of work Thompson needed.

He envisioned a glorified underwater Swiss army knife with an array of saws, grabbers, drills, blowers, cameras, and lights that could be remotely operated from a surface ship. Important considerations included how deep such a piece of equipment would have to work, how far from a harbor, typical sea states, what the target was made of, and what size pieces would have to be retrieved. Stability would be an important issue, since any robot equipped with a robotic arm would shift its center of stability when retrieving a load. Thompson was also adamant that the design had to be able to survive working at sea—that meant it had to be dumb, simple, and flexible.

The resultant ROV wasn't pretty, but it was workable. In 1987, Thompson went back to the target area with *Nemo* as the centerpiece of the expedition. If the ROV could handle complex, heavy-duty tasks in the deep ocean's harsh environment, it would be the first "working presence" more than 2.4 kilometers (approx. 1.5 mi) below the surface. But first, the group needed to determine whether or not the most probable sonar target was actually the *Central America*, a task further complicated by the presence of other competitors in the area. In order to protect the site legally, Thompson was advised to retrieve some item from the still-unidentified wreck, to prove that Columbus-America had accessed the site and was legally in control of it. Fortunately, *Nemo* managed to retrieve a single lump of coal that won the group a federal court injunction to claim the site. It was the first such claim ever made on a shipwreck site accessed only by a robot—a precedent in international maritime law on the basis of telepresence.

During the winter of 1987–88, the Columbus-America group purchased and retrofitted an old Canadian icebreaker, the *Arctic Discoverer*, but plans to definitively prove the wreck was the *Central America* were delayed as the refurbished ship waited for equipment. Meanwhile, Bob Evans had reanalyzed some other targets from the 1986 sonar survey, turning up an intriguing anomaly on the bottom at another, closer site. When the operation finally left the dock, the decision was made to stop at this new locale, test out equipment, and compare the two sites. *Nemo* went over the side on Sunday, September 11. A weary crew stared at hours of watery images projected on 12 video monitors. Then suddenly, an unbelievable image scrolled by—the rusting sidewheel of a ship, lying in the mud. It was, indeed, the *Central America*!

If the discovery of this deep-water wreck was an engineering success, it was the lure of its treasure that financed the expedition. Now the second part of the mission kicked in—recovery of the cargo. By mid-October, the weather was getting snotty, and the crew was tired of weeks at sea. Just days before they were due to leave the site, the photographer-videographer spotted a new color on some images—gold! The precise nature of *Nemo*'s dives meant it could pinpoint the locale where those photos had been taken. As the ROV flew over to that location again and the lights were adjusted to reveal true colors, the monitors were suddenly filled with brilliant images—ingots stacked like loaves of bread and towers of historic gold coins!

Using the latest search technologies and an innovative ROV, Thompson found the *Central America* and its prized cargo. But more than that, he approached the problem methodologically, researching the literature, investigating the lessons learned by others, thinking creatively, and using his network of contacts to connect with experts in the field. He laid out the performance requirements of the underwater technologies that he would need to locate and uncover the shipwreck and used them to guide his approach either to find what already existed or to design and build his own. He didn't do it alone, of course, but rather assembled a team of people who contributed their own strengths and expertise to the project. It wasn't always smooth sailing; often Thompson's team found themselves back at the drawing board. But they went about their rethinking and re-engineering with an open mind to each others' ideas and always with their mission in mind—to find a ship of gold in the deep blue sea.

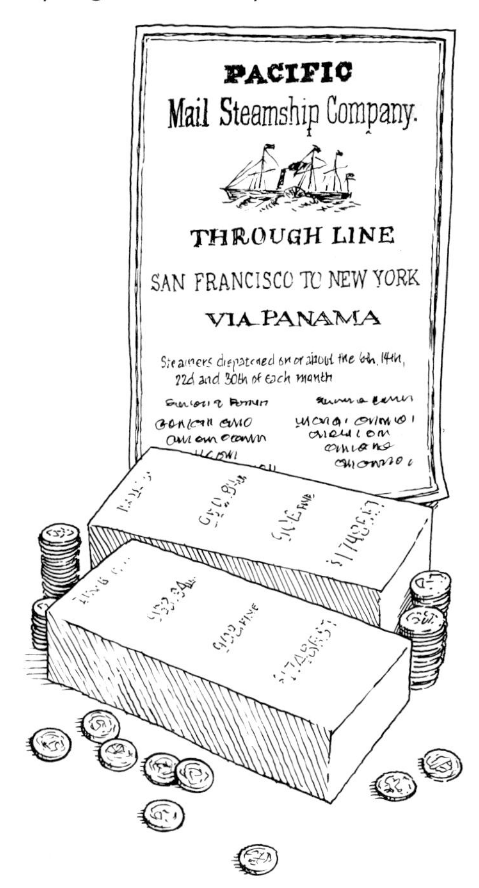

Image courtesy of Steve Van Meter, VideoRay LLC

Figure 2.2: The Challenge of Building a Team and an ROV

Participating in a competition encourages students to develop both technical and teamwork skills.

1. Introduction

People get into ROV and AUV projects for many different reasons; furthermore, they come from a variety of backgrounds and experiences. Some are students building an ROV for a competition. Some are teachers exploring new ways to make their classes more fun and more valuable for their students. Some are technology enthusiasts who want to try something new. Regardless of motivation, anyone undertaking an underwater vehicle design project will encounter similar issues and challenges.

The first half of this chapter outlines the major challenges you are likely to face when undertaking an underwater vehicle project. It also provides a "toolkit" of pragmatic strategies you can rely on to overcome those challenges and keep your project on track. The latter half of the chapter is devoted to a detailed case study of a real ROV project successfully completed by a group of community college students. This example not only outlines the sequence of stages necessary to complete an underwater vehicle project, but also illustrates application of the strategies introduced earlier in the chapter.

This chapter provides a way to organize your thinking, planning, designing, and construction when building an underwater robot. It provides the theory you'll put into practice in *Chapter 12: SeaMATE*, which takes you step by step through building an actual shallow-diving ROV. In between these two important chapters—one on design and one with actual fabrication instructions—are a series of technical chapters that provide greater detail about important issues of buoyancy, power, control, etc., as they relate to underwater craft. Note that this textbook is designed for individual flexibility—you may choose to read it straight through, then begin building *SeaMATE* or another vehicle. Or you may choose to build various ROV systems as you study particular technical chapters. For example, you can build the frame while studying *Chapter 4: Structure and Materials* or work on thrusters while reading *Chapter 7: Moving and Maneuvering*. The choice is up to you. Once you have completed *SeaMATE*, you should have enough learning and confidence under your belt to begin more advanced projects—which brings you full circle back to the planning and organizational theory of this chapter. That's the way it is with underwater robots—and underwater robot textbooks—everything is interrelated!

2. Common Design Challenges

The issues and obstacles you are most likely to encounter when designing and building an underwater vehicle usually fall into one of the following categories:

1. You're not clear exactly how or where to start.
2. It's difficult to translate the general goals of the project into the precise vehicle specifications you need in order to select and order suitable parts.
3. You discover, sometimes partway into the project, that some of the technical challenges of the project are beyond your prior training and experience.
4. The project has (or soon will) run up against one or more constraints, such as limited resources (e.g., time, budget, access to tools, and facilities), rules and regulations required by competition organizers or other "clients," or obstacles imposed by the physical realities of a harsh underwater working environment.
5. There may be personality conflicts or other challenges in working with teammates, vendors, or other people who are directly or indirectly involved in your project.
6. You discover that fixing a problem in one place creates new problems in other parts of the design—everything seems to affect everything else!

2.1. Strategies for Successful Projects and Missions

There really is no secret to creating a functioning underwater vehicle. It's just a matter of knowing how best to approach the project. The next section of this chapter provides a tangible set of strategies you can use to address the challenges listed above and keep your project moving to completion on time and within budget. They will increase your fun and learning while decreasing frustration and setbacks.

The strategies presented here are time-tested. They work. They are also general enough that they can be applied to any technical design, regardless of whether you're planning a science fair project, a competition-level vehicle, or a high-tech, high-budget submersible.

The strategies are tagged with letters (for example, Strategy A, Strategy B) for convenient reference; however, they do not need to be followed in any particular order. Use each of them whenever it makes sense to do so, and feel free to use some or all of them simultaneously. Build on what you learn from your mistakes.

2.2. Strategy A: Keep Your Eye on the Mission

Underwater vehicles are generally designed and built for a reason. That reason might include exploration and discovery, search and recovery, repair and maintenance, or a host of other possibilities. The specific purpose of each dive or closely related set of dives is often described in terms of a **mission**—it's what you hope to to accomplish during the dive(s). Examples of a mission could include recovering a valuable object lost at sea, inspecting the interior of a large pipe for cracks, or repairing part of an offshore oil rig. Designing and building even a simple ROV is a significant endeavor, and it takes a worthy mission or two to justify investing all of that time, effort, and money.

A design project should revolve around the very reason for building the equipment in the first place. That might sound completely obvious, but it's surprisingly easy to get lost in the daily details of a large project and gradually forget to think about why you started the project and where you're headed. Once that happens, it's easy to wander off course and waste valuable time pursuing directions that don't contribute to mission success.

When you keep your eye on the mission, all of the brainstorming, planning, fund-raising, inventing, and testing you do are shaped by the mission and contribute directly to its success. Make sure you schedule time every now and then to step back from the building chaos and remind everyone about the mission. This gives you an opportunity to reassess whether things are still heading in the right direction. Later in this chapter, you'll learn more about how to craft a clear mission statement that can help guide you through the design and construction process.

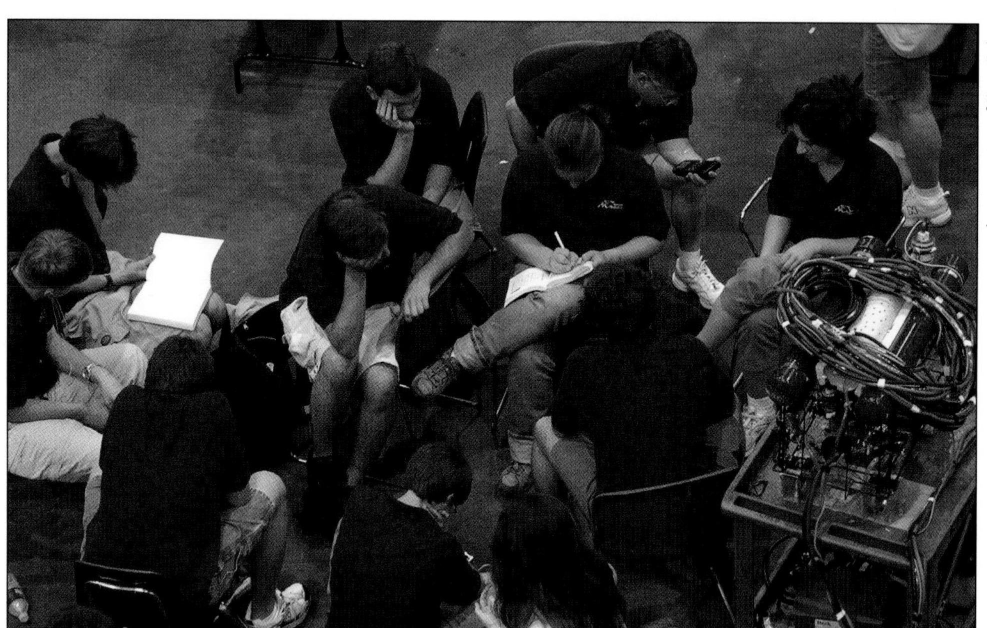

Image courtesy of Harry Bohm

Figure 2.3: Teamwork

When it comes to building ROVs, often many heads are better than one.

2.3. Strategy B: Build an Effective Team

Project success begins with assembling an effective project team. The team may consist of only one person, but more commonly it will be a group of people working together to complete the project. Even a one-person show will need to enlist the services of others, such as parts vendors or technical advisors. Thus every project involves some degree of teamwork or collaboration among many people.

Sometimes the team membership is beyond your control. This might be the case, for example, in an ROV team made up of students selected by a teacher. In other cases, you may need to choose who is on the team. Either way, the team will be most effective if you can identify and emphasize complementary skill sets among your team members. A three-person team consisting of one person who has experience soldering, another who has experience cutting and gluing PVC pipe, and a third who has experience with different types of boat propellers is likely to have an advantage over a team with three people all of whom are experienced only in soldering, even if they are excellent at soldering.

TECH NOTE: CONFLICT RESOLUTION

Tensions and disagreements occur in almost any group, especially when things are not proceeding according to plan. Resolving these issues is essential. Energy put into arguments, frustration, and anger is energy that is not going into productive vehicle design and construction. Even more importantly, working on projects like this should be fun! If the team isn't having fun, enthusiasm will fizzle. The pace and quality of the work will be compromised. And that can lead to a host of unwanted technical, management, and emotional issues. While there is no single best way to tackle thorny conflict situations, consider these guidelines if things start to get strained between team members:

- Try to let an "ugly" moment pass. Avoid the "boil-over reaction." Remember that everyone starts to unravel when the hours are long, when low on food or sleep, and when under pressure. Take a deep breath. Maybe even take a break.
- Find ways to renew the enthusiasm the team had at the beginning of the project—this can range from a regular pizza night and team shirts to watching a movie featuring underwater vehicles or inviting an expert to meet with the group.

Although it is helpful to have experts on your team, it is not essential as long as team members are willing to learn new things. What is important is having team members who can work effectively together and who know how to obtain the expert information they need from books, websites, knowledgeable people, and other sources. *Strategy F: Research, Research, Research* offers some specific suggestions on how to gather the information you'll need.

2.4 Strategy C: Be Proactive about Project Planning, Management, and Safety

Once hooked on the idea of an underwater robot, it's tempting to hotfoot it to the hardware store and buy materials. Enthusiasm is high. It's time to get building!

Or is it?

Great projects don't just happen; they are the result of careful planning and coordination. Your projects will be more successful, more efficient in the use of time, energy, and money, less frustrating, and more fun, if you can resist that primal impulse to rush off and build something immediately. Instead, invest some of that initial enthusiasm in planning the project and setting up an effective project management system. Admittedly, this methodical approach may not sound as exciting at first, but it works and leads to greater fun, excitement, and success in the long run.

There are several aspects to planning and managing a project well. Each of them is described below.

2.4.1. Identify a Project Manager

A project without a project manager is like a boat without a rudder. The project manager directs the team's efforts and is ultimately responsible for successful completion of all aspects of the project. Team members look to the project manager for guidance on such critical issues as milestones, budget, staffing, procurement, and coordinated effort—the result is a final product that comes together on the shop floor and during sea trials. Sometimes a teacher or boss automatically fills this role. In other cases, the role falls to a team member with the drive to move the project through to completion. Needless to say, it is a challenging and crucial job—one that requires attention to details and organizational skill.

2.4.2. Make Sure the Project is Clearly Defined

It is critical that the project manager and all team members have a shared understanding of what the vehicle needs to accomplish (i.e., its mission), how it will do that, and (roughly) how much time, money, and other resources are available for the project. Without this shared understanding, team members may unintentionally work toward different, potentially conflicting, goals. This wastes valuable time, money, or other resources. One of the first things a good project manager often does is help the team work together to develop a clear list of project goals, expectations, and constraints. This list may evolve over time as new information becomes available; that's ok, as long as everyone on the team is made aware of any changes.

2.4.3. Establish and Maintain Effective Communication Channels

Clear and effective communication among members of the team, contractors, suppliers, and advisors is critical to the success of every project. Discuss needs, technical problems, priorities, scheduling, and budget limitations. Use progress meetings, emails, telephone calls, letters, and informal talks. Make sure everyone in the

- Drop negativity and criticism. Eliminate phrases like "should have," "impossible," "can't," "stupid."
- Evaluate and critique ideas, not people. If you disagree with somebody's idea, say you disagree with the idea and explain why, but don't berate the person.
- When there's a serious difference of opinion, look for a win-win solution. Rather than pushing for a personal victory, find a way for all sides in a conflict to feel they've been heard and included.
- Don't gossip about other team members, even in confidence. This causes deep divisions in the team.
- Strive for a positive attitude. (Your glass should always be half full rather than half empty.)
- Remember that mistakes are part of the learning process and are not "bad." Most mistakes can be corrected. In fact, often people learn more from their mistakes than from their successes.
- Be tenacious when facing technical setbacks. This is an opportunity to improve things. Analyze the problem by rebuilding it one step at a time. Remember that there are always solutions. Look for outside-the-box ones if necessary.

TECH NOTE: THE SHOEBOX FILES

A shoebox (or a series of them) is a great way to keep track of a small project's expenditures. First get a box with a lid, label it "Receipts," and put it in a convenient location. Then make sure that everyone buying project supplies gets receipts and puts them in the shoebox. The trick is to do this immediately.

When the time comes to make an accounting of expenditures, simply empty the box, organize the receipts by date, enter the amounts in your accounting journal or computer spreadsheet, and total them. Separate boxes can also be used as a convenient way for storing correspondence, product brochures, and other useful documents.

group knows what's going on. Be clear about meeting times and locations, as well as what needs to get accomplished at each session. If team members cannot make a meeting, they should let someone know ahead of time and find out what they missed.

2.4.4. Break the Project into Smaller Sub-Projects

Every project, no matter how big, can be broken down into a series of smaller, more manageable sub-projects. Among other advantages, this helps team morale by converting a potentially overwhelming project into a collection of relatively doable tasks. It also makes it easier to distribute the workload among team members, because different people can focus on different sub-projects. *Strategy D: Think in Terms of Systems* provides a particularly effective method of subdividing a large project into smaller tasks that can easily be distributed to sub-groups of the team.

2.4.5. Set Realistic Goals, Milestones, and Deadlines

While the overall project objective is to see the vehicle hit the water and run the mission, there are many intermediate goals that must be completed along the way in order to do that. These include:

- completing the concept design
- obtaining all essential parts and tools for construction
- completing fabrication of the vehicle
- conducting sea trials

All of these intermediate goals show progress toward the ultimate mission objective and deadline. Developing a timeline for achieving these intermediate targets, often called milestones, is what scheduling is all about. Milestones with deadlines are among the best motivators for completing a job on time, but the goals and deadlines must be realistic.

When defining goals or milestones, focus on tangible products or "deliverables" rather than on processes, and get specific about dates. This helps the team assess whether a goal was met—if so, then they've made good progress. Nebulous statements like, "Work on thrusters during March," are problematic, because it's not clear how much work needs to be done to meet this goal, nor whether there will be anything useful to show for that work when it is done. A more useful milestone is, "Complete three working thrusters by March 31."

When you set goals and deadlines, it's important that they be *realistic*. Unfortunately, this is easier said than done, particularly if you have limited experience with similar-sized projects. Discuss the calendar with other team members, check that supplies and parts are available, and assess how to deal with other circumstances or events (such as schedule conflicts among team members, exams, etc.) that will impact the project schedule.

Recognize, as you do this, that unanticipated complications invariably arise and make projects take longer than originally expected. For example, a propeller you order may be out of stock, so either you'll need to wait for the back-ordered part to arrive or redesign your vehicle so it can use a different propeller. Both options take extra time. Be sure to pad your schedule with ample time to absorb a few of these unanticipated delays. Most beginners underestimate the amount of time needed to complete a project as complex as an underwater vehicle. One common recommendation is to estimate how long it should take to complete the project if everything goes according to plan, then budget triple that amount of time! So if you think it will take two months, budget at least six months. If you have only two months available, cut the size and complexity of your project to one-third of your original plan before you even begin.

If you reach a milestone ahead of schedule, that's great. You're ahead of the game and will have extra time to absorb unexpected delays as you work toward subsequent milestones. If you fall behind, revise the project schedule or devise a catch-up strategy. If a project falls *way* behind schedule and deadlines can't be extended, it may mean that the project was too ambitious in the first place and needs to be scaled back to something more manageable. If that's the case, the sooner the team recognizes the problem and makes appropriate changes in their list of goals and expectations, the better.

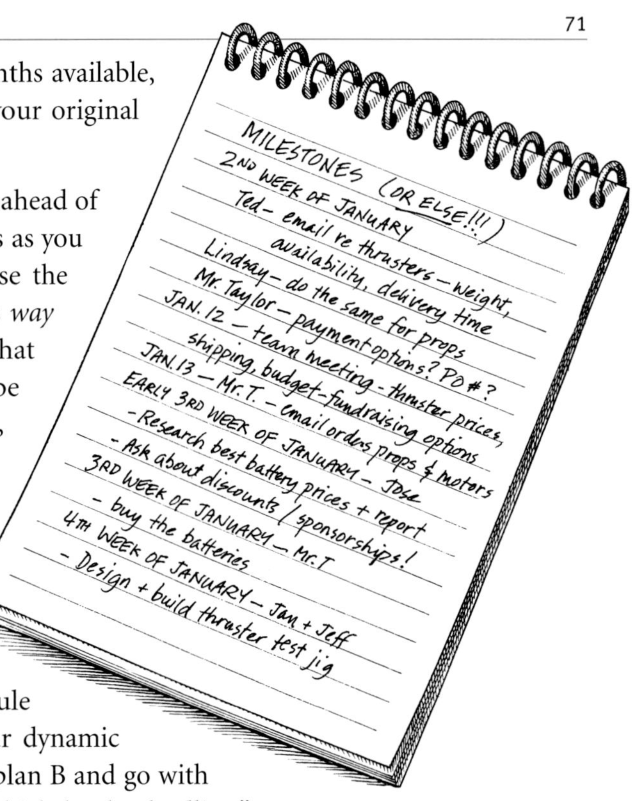

Figure 2.4: Sample Schedule

Another excellent way of coping with uncertainty is to build "decision points" into the schedule, whereby the scope of the project is automatically adjusted based on progress made by specific dates. For example, your schedule might include a decision point that says, "If we can't get our dynamic ballast system working well by February 1, then we switch to plan B and go with a simpler static ballast system to make sure we complete the vehicle by the deadline."

Once you've gone to the trouble to make a schedule, don't forget to use it! Post it on the wall or on a team website. Make sure everyone on the team checks it frequently, so the team stays on track and completes the vehicle by the deadline. Adjust or update the schedule and/or expectations as needed, to keep the project moving toward completion while maintaining a positive working atmosphere.

2.4.6. Maintain Careful Records

Certain records are essential to the effective management of a project. For most small ventures, you can get by with only a project notebook, calendar, "shoebox file," shopping list, and your to-do lists. (See *Tech Note: The Shoebox Files.*)

A project notebook can be just a small pad or blank booklet to record lists, ideas, notes, contact info, prices, and first-hand data on experiments. Keep it with you all the time. Most scientists and engineers have a well-used notebook to collect project data. This information can be referred to later when you need to prepare more formal documentation, such as a project report. Here are suggestions for a basic project notebook:

- Use numbered pages and add the date to each entry.
- Write in waterproof pen.
- Don't erase any notes or mistakes. Simply strike them out with a single line. Sometimes what seems like a mistake or a waste of space turns out to be something important needed later.
- Write down all meeting notes, design ideas, and technical or experimental results in your notebook. Again, you never know which piece of that information you're going to need to refer back to later, so capture it while it's fresh in your mind.
- Include notes on conversations with anyone connected to the project.
- Include raw data and sketches necessary to produce a working drawing, manual, or project report.

Figure 2.5: Sample Project Notebook Page

SEA MATE PROJECT
MARCH 21, 2010

THRUSTER ASSEMBLY NOTES

— HAVE OBTAINED THREE THRUSTERS – JOHNSON 500

– HAVE PURCHASED THREE DIFFERENT PROPS (1", 1½", 2")
2 BLADE NYLON PROPS

– PROPS WERE NOT THE TYPE SPECIFIED SO DID A TEST TO SEE HOW POWER-EFFICIENT THEY WERE – TEST RESULTS BELOW FOR FORWARD THRUST ON FULL.

PROP TYPE	SPEED	SCALE READING	AMPS
2 BLADE 1"	AHEAD FULL	5 OZ.	2 A
2 BLADE 1½"	AHEAD FULL	10 OZ.	2.5 A
2 BLADE 2"	AHEAD FULL	15 OZ.	4.0 A

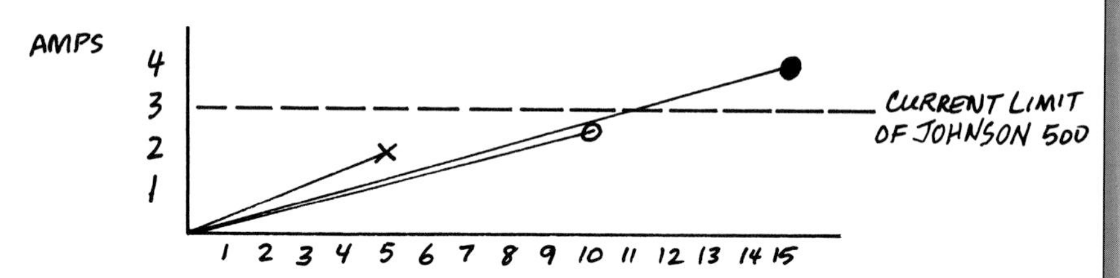

x = 2 BLADE 1" PROP
o = 2 BLADE 1½" PROP
● = 2 BLADE 2" PROP

— FROM THE TEST RESULTS THE ~~2~~ 1½" PROP IS BEST — PURCHASE 2 MORE 1½" PROPS — RETURN OTHERS
— ASSEMBLED ONE THRUSTER WITH 1½" PROP
— DRILLED HOLES FOR MOUNTING THRUSTER ON (VERTICAL) FRAME
— MOUNT THE HORIZONTAL THRUSTERS ON FRAME WITHOUT PROPS

- Don't forget contact numbers and addresses of everyone associated with the project. It can be helpful to designate a section at the end of the notebook for this contact info, so it's all in one easy-to-find place.
- Add a list of suppliers and product information.

Keeping lists sounds almost too simple, but it's one of the most basic project management tools. Lists keep things moving and save time by reminding team members of what's still to be accomplished. These lists should be kept in a centralized location where all team members can read or update them. Lists can be recorded in the project notebook(s), but it often helps to post them on a whiteboard or other prominent place in the project work area, too. Some particularly helpful lists are:

- **Shopping lists for needed parts or supplies**: Team members heading for the store should take a copy of the list with them and cross off items from the master list as soon as they've been purchased, to avoid duplication.

- **To-do lists that identify remaining tasks**: The tasks should be prioritized frequently, so that the most urgent and important items are clearly identified on the lists and completed first. Revise these lists as the project progresses. Tick off the items that have been completed and add new items as they come up.
- **Lists of contact information for sources of parts, supplies, or information**: It is often helpful to separate this information into two lists: one for people, vendors, books, and other resources your team has already used and may need to contact again, and another for prospective sources of information, parts, or supplies that may or may not prove useful.

An effective filing system for receipts, correspondence, product brochures, and other documents that are not part of the data in your notebook is key to finding information quickly, so it's important to establish one from the very beginning. While it seems easy enough to keep track of costs and purchases at the start of a project, you can soon become deluged by receipts, notes, research results, and product information unless there is some sort of organizational scheme. The filing system need not be fancy. Even a series of labeled shoeboxes can do the trick. (See *Tech Note: The Shoebox Files.*) The trick is to get a procedure going right away and make sure everyone involved knows what info (receipts, notes, etc.) must be kept and where it should go. A good filing system also lessens the chances of misplacing critical information, which can cost time and money.

2.4.7. Develop and Track the Project Budget

Money is almost always a limiting factor in any project, and some competitions have very specific budget limitations. So it's important to know at the very beginning of the project how much money the team will have available for parts or other project needs and what fraction of the available funds will be allocated to each of those needs. Failure to plan can result in too much money being spent on some parts of the vehicle, leaving insufficient funds to purchase other critical parts.

In developing the plan, it's necessary to do some research to find out how much various parts are likely to cost. In the initial stages of planning, rough estimates of cost are usually adequate. Remember that the advertised price is not likely to be the actual cost to your budget. Sales taxes and shipping costs will increase the true cost, while educational discounts, donations, or other factors may decrease it.

Figure 2.6: Sample Budget Spreadsheet

This sample budget spreadsheet was part of a technical report submitted by St. George's School in Vancouver, B.C. for the 2007 MATE International ROV Competition.

3.4 Budget/Expense Sheet:

Items	Source	Cost	Donations
Brushless motors (4)	eBay	$200.00	
Brushless controllers (4)	eBay	$120.00	
Brushed controllers (2)	eBay	$60.00	
Rule 1100gph (1.156L/s) bilge pumps (4)	Wal-Mart	$120.00	
Sevyler SBM props (4)	Sevyler Canada		
Graupner props (2)	Hobby Shop	$6.00	$32.00
Traxx props (2)	Hobby Shop	$6.00	
Basic Atom 28 and Mini Atom Bot Board (2)	Lynxmotion	$110.00	
RS-485 (SN75176) transceivers (2)	HVW Technologies	$24.00	
Parallax servo controller	HVW Technologies	$45.00	
Igus Cat5e Cable 50' (15.24meters)	Igus		
Power cord 16 gauge 50' (15.24meters)	Home Depot	$15.00	$250.00
Miscellaneous wire	McMaster	$15.00	
2.75" (6.985cm) clear cast acrylic tubes (4)	McMaster	$76.00	
4" (10.16cm) clear cast acrylic tubes (2)	McMaster	$48.00	
Cord grips (12)	McMaster	$48.00	
Acetal 4" (10.16cm) diameter			
6" (182.88 cm) length	McMaster	$35.00	
6mm diameter 318 stainless steel rod	McMaster	$21.00	
PVC pipe and fittings	Home Depot	$80.00	
Cameras (3)	Supercircuits	$60.00	
Video baluns (3)	eBay	$30.00	
Saitek joystick	Fry's	$39.00	
Electric solenoid	Electronic Goldmine	$3.00	
Scrap metal and plastic pieces	Mr. Kay's basement	$0.00	
Misc. items (screws, connectors, o-rings etc.)	Various sources	$200.00	
Total		$1,361.00	$282.00
Grand Total including donations		**$1,643.00**	
Budget (funding from Saint's Auxiliary)		**$2000.00**	
Surplus		**$639.00**	

All costs are in USD and rounded up to the nearest dollar.
Items purchased online include shipping and handling costs.

In addition, remember that you will likely need to buy extra tools, parts, and materials that do *not* end up in the final vehicle. For example, you might need only 12 feet of PVC pipe for your vehicle frame, but if that pipe only comes in 10-foot sections at the store, and the store can't or won't cut shorter lengths for you, then you'll need to budget enough money to buy 20 feet (two 10-foot sections) of pipe.

Likewise, you may need to buy extra parts for testing and prototype development. For example, you might know that your vehicle design will require four thrusters made from bilge pump motors, but you might not know which model of bilge pump will produce enough thrust. In that case, it would be a good idea to budget enough money to buy and test a few promising candidates. If you budgeted enough for six thrusters, you could start by buying one each of three *different* models (call them A, B, and C) and testing them to see which of the three works best for your purposes. Then buy three more of the winning model (say it's B) to complete your set of four thruster motors. That's a total of six bilge pumps purchased as part of the design and construction process, even though only four of them will end up in the final vehicle. If you don't build these extra expenses into your budget planning up front, they'll sneak up on you later and force you to cut corners on critical tools, parts, or supplies, ultimately jeopardizing the mission.

Before any money gets spent, it's important to establish a formal purchasing process, including who can make purchases and how expenditures will be handled (refunds only with receipts, purchase orders, etc.) Once the project is under way, you'll want to record *all* expenditures accurately and subtract them right away from the total amount of funding still available for the project. That way your team (or at least the project manager) will know each day exactly how much money is still available for new purchases.

The calculations required to keep track of how much money is still available can be done with pencil and paper, calculator, or computer spreadsheet program. Any method will work, though spreadsheets offer a number of advantages for those who have access to them and know how to use them.

Whatever method you use, keep it organized and accessible, because all of the budget information and other records you are keeping will be crucial in compiling a final project report. (See *Section 3.11. Stage 11: Evaluating Ops and Writing a Final Report.*)

2.4.8. Make Safety a Top Priority

Image courtesy of Steve Van Meter, VideoRay LLC

Figure 2.7: Safety First!

Safety glasses for eye protection are a minimum safety precaution when working with power tools or soldering. Ear protection may be needed around loud tools and machinery. In addition, you should avoid loose hair, loose clothing, or jewelry that might get caught and pull you into moving parts such as power saw blades or ROV propellers. Metal jewelry can contact electrical systems, causing severe burns or electrocution.

When working around water, a brightly-colored personal flotation device (PFD) and electrical safety devices such as GFIs and fuses are standard safety precautions. In cramped work areas or areas where people or machines are working overhead, it's a good idea to wear a hardhat, in case somebody drops a wrench on your noggin or you stand up too fast under a low steel beam. Also, keeping ropes, extension cords, and long tethers neatly coiled and stowed out of the way reduces the likelihood of someone tripping and falling.

Good project planning and management includes attention to safety issues. Underwater vehicle projects are potentially dangerous, because they bring water, electricity, and power tools into close proximity. Slipping on a wet pool deck, electric shock, electrical fires, drowning, and severe or fatal injuries from power tools are all valid concerns and should never be overlooked or casually dismissed. You don't need to be paranoid about safety issues, but you will want to make sure that you and other team members are aware of all significant hazards associated with the project and have taken reasonable steps to reduce the likelihood of an accident. You also need to pay extra attention to what you're doing when danger is near.

A project manager who is really on the ball will have the team establish and follow some basic safety policies and emergency response procedures and will also ensure appropriate safety training for all team members. This needs to happen during the early stages of the project, before it's needed, not after somebody gets hurt! For example, prior to beginning actual construction of the vehicle, the team might develop and agree to follow a rule that nobody works alone with power tools. That way there will always be someone who can summon help in case of a serious injury that leaves the victim unconscious or bleeding profusely. The team might also ask a team member or neighbor who's already handy with power tools to demonstrate their safe use for team members who have little or no prior power tool experience. Another sensible safety rule requires that everyone wear safety glasses to protect their eyes while cutting, drilling, or soldering.

It's always a good idea to make sure current emergency phone numbers are posted in workshops, around swimming pools or docks, and in project notebooks. That way you won't need to waste time looking up the right number to call during a real emergency.

2.5. Strategy D: Think in Terms of Systems

An underwater vehicle may have hundreds or even thousands of parts, all interacting in complicated ways. The normal human brain cannot think effectively about all of those parts and interactions simultaneously, so it's important to find some way to simplify and organize your thinking about these machines without overlooking any important details.

One effective way to do this—and one that comes very naturally to most people—is to think hierarchically in terms of systems composed of subsystems. A **system** is a collection of parts (and/or processes) that work together to do (or be) something the individual parts cannot do (or be) alone. A **subsystem** is a system that functions as one of the parts within another system. A **supersystem** is a system composed of other systems.

Hiding in the definition of a system are three features common to all systems:

1. parts
2. interactions between those parts
3. new characteristics or functions that result from the interactions among the parts and *do not exist without those interactions*

The new characteristics are called **emergent properties**. Emergent properties are what distinguish a system from a random pile of stuff and what make the systems concept such a powerful one for simplifying design and analysis of complex machines or processes.

A bicycle is a good example of a system. First it has parts such as its frame, wheels, handlebar, seat, chain, pedals, etc. Second, those parts interact—for example, the chain helps transmit mechanical force from pedals to the rear wheel. Third, emergent properties arise from the interactions. The most significant of these emergent properties is that the bicycle can transport a rider from one location to another, which is something none of the bicycle's parts can do by itself.

A bicycle wheel is a good example of a subsystem. It's clearly a part of the bicycle system, but at the same time, it's an entire system in its own right, with hub, spokes, rim, tube, and tire interacting to produce a structure that can support the weight of the bicycle and rider while rolling smoothly over pavement. If you focus on the wheel as a system, instead of as a subsystem, then the bicycle becomes a supersystem, because

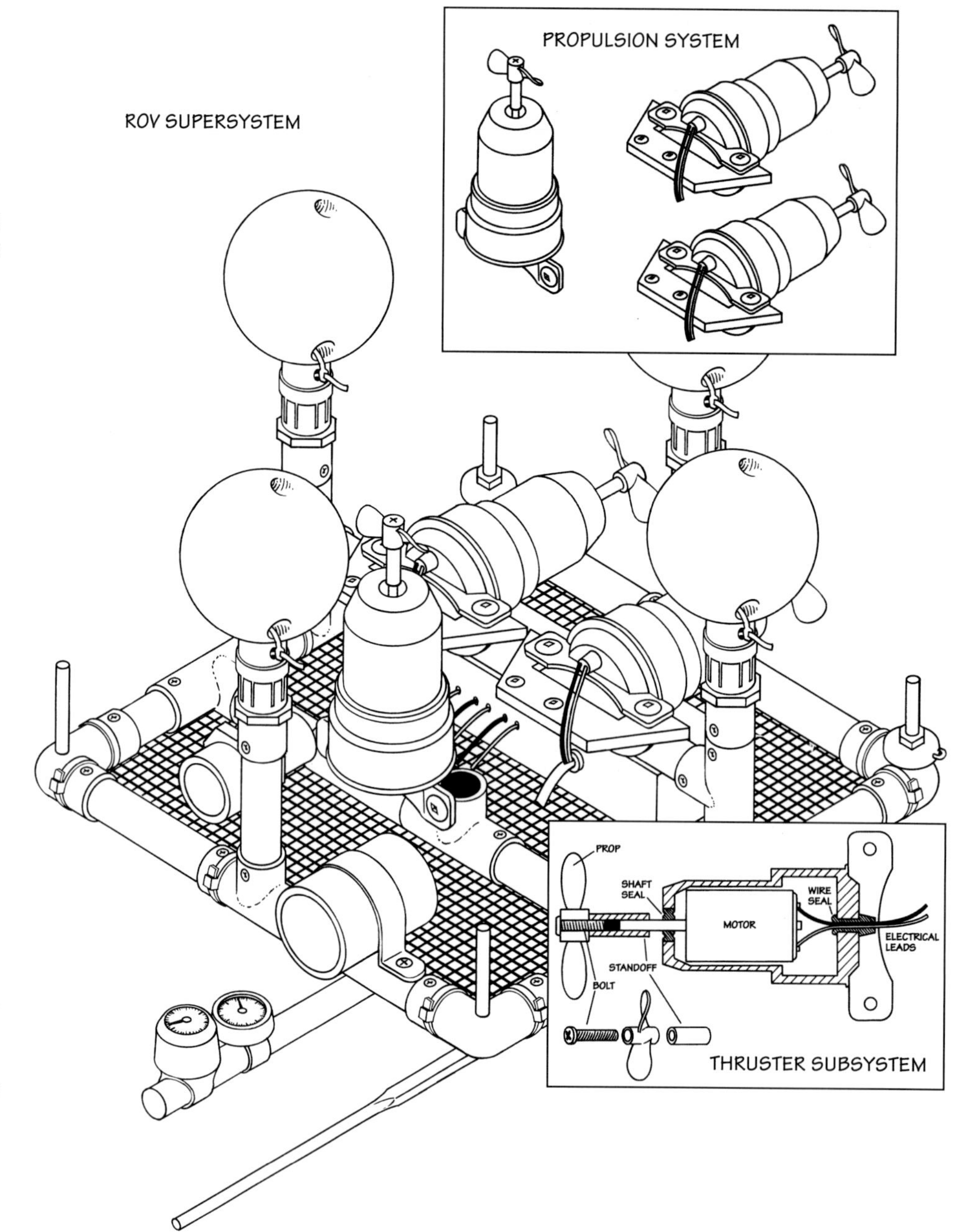

Figure 2.8: Supersystem, Systems, and Subsystems

Regardless of size or complexity, you can think of any ROV as a supersystem that's composed of other systems. Propulsion components would be just one of the vehicle's systems, and a single thruster would be a subsystem of the propulsion system.

the wheel is part of the bicycle. As you can see, the terms system, subsystem, and supersystem are relative terms, not absolute ones. Any system can be viewed as any one of the three. It all depends on your perspective and your objective.

Thinking of an ROV or AUV as a supersystem composed of systems offers at least three distinct benefits, as detailed in the following paragraphs.

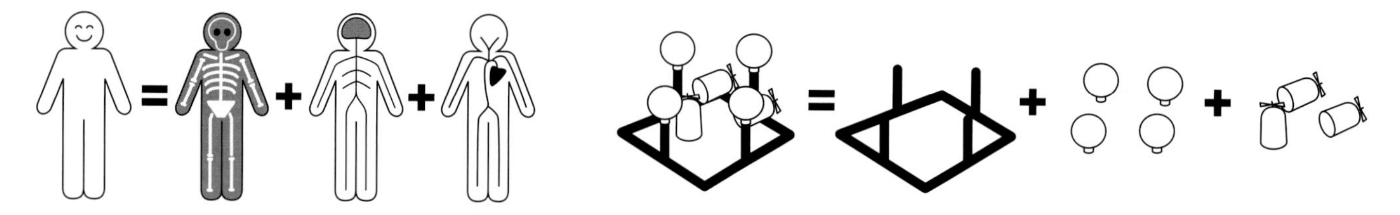

Figure 2.9: Human and Underwater Vehicle Supersystems and Systems

An adult human body is a supersystem made up of some 100 trillion cells, 206 bones, 600 muscles, and 22 internal organs. These bits of flesh and bone provide components for the body's various systems—skeletal, muscular, nervous, circulatory, respiratory, and digestive.

Similarly, every ROV is a supersystem robot that can be divided into various systems—structure, ballast, propulsion, power, control, navigation/sensors, payload, and tether.

2.5.1. System Benefits

There are at least three benefits of a systems approach—simplification, sensible division of labor, and better compatibility with other systems.

Benefit #1: Simplification Thinking in terms of systems, subsystems, and supersystems allows you to organize an overwhelming number of parts and processes into a well-defined hierarchical structure. Then if you have a problem or issue at a particular moment, you can largely ignore everything else and focus on the appropriate section of structure with its particular level of detail.

So instead of thinking about an ROV as a thousand screws, wires, bits of plastic, and other pieces, you simplify your approach and see it as a relatively small number of interacting systems, each defined by its relevant emergent properties (Table 2.1).

If you are designing or trying to repair a particular subsystem (say, a thruster), then you can shift your attention to that level, viewing the thruster now as *the* system with its own set of subsystems (motor, propeller, waterproof housing, etc.), meanwhile ignoring the tether, camera, and other parts of the ROV.

Table 2.1: Examples of Systems Commonly Found on Simple ROVs

System	*Relevant Emergent Property*
Structure/Frame	*Holds all of the ROV parts in a functional spatial arrangement. May also provide some protection from collisions or pressure. May also be streamlined to reduce drag.*
Ballast	*Adjusts tendency of ROV to float or sink.*
Propulsion (thrusters)	*Provides propulsive forces needed to move ROV through water.*
Power (battery or other)	*Provides power to the ROV and associated equipment.*
Control	*Allows pilot to control the thrusters, tools, cameras, lights, or other subsystems on the vehicle.*
Navigation Sensors (camera, etc.)	*Provide pilot with information about where the ROV is and what it's doing.*
Payload	*Generally consists of a gripper or manipulator for picking up things, a set of tools for working on an underwater maintenance project, or a variety of other systems with whatever capabilities are called for by the mission.*
Tether	*Transmits data and (usually) power between the ROV and the surface.*

Benefit #2: Sensible Division of Labor Thinking of your vehicle as an integrated group of systems and subsystems provides an easy and sensible way to distribute responsibility for different parts of the vehicle to different parts of the team. In this approach, each group of people becomes responsible for delivering a different system with a specific set of emergent properties. For example, the thruster group might be assigned the task of producing four thrusters, each of which can generate two pounds of thrust. The tether group might be responsible for producing a tether 30 meters (approx. 100 ft) long that can carry electrical power to each thruster and return one live video signal to the pilot. Because these systems are part of a larger supersystem (the ROV), there will be interaction and interdependence between them (for example, the tether will need to be able to carry enough power to run all four thrusters at full speed). Aside from these compatibility issues, the two groups can work largely independently, even on different schedules, coming up with whatever design delivers the required emergent properties.

If mission success is the primary goal and time is limited, it makes sense to assign team members to systems where their particular skills are needed most. People with strong mechanical skills can design and build the vehicle frame, manipulator arm, and other mechanical parts. The electronics geeks can handle the sensors, control systems, and

TECH NOTE: ELEMENTS OF GOOD DESIGN

Three features contribute to an effective overall design: functionality, serviceability, and aesthetics.

Functionality: Fundamentally, a vehicle structure has to perform its job well. It must be sturdy enough to withstand water pressure at mission depth as well as the knocks it will get during launch and recovery. It might also need to be streamlined to reduce entanglement or drag. The vehicle also needs to be tough in an electrical sense. Sloppy wiring, substandard electronic components, or poor soldering will cause no end of troubles.

Serviceability: A large part of a vehicle's robustness comes from its serviceability. If it's designed so that repairs and maintenance can be easily done, that will help keep the vehicle at work in the water instead of on the workbench for repairs.

Aesthetics and Workmanship: Beauty and functionality go hand in hand. For example, sloppy wiring or poor soldering not only look ugly but will cause no end of trouble.

electrical power distribution. On the other hand, if the focus of the project is education and if learning is more important than even mission success, it may make more sense to assign people to the system with which they are *least* experienced.

Benefit #3: Better Compatibility with Other Systems in Less Time As long as communication channels remain open, having different team members work on different systems simultaneously not only saves time, but it makes it easier to spot and correct incompatibilities between different system designs early in the process. This means your team won't have to backtrack as far or as often. Imagine the frustration of having the whole team finish designing and building a tether before starting work on the thrusters, only to discover that the thrusters will need more power than the just-finished tether can deliver. Ouch!

Before moving on, take a brief look at the typical systems of a small ROV. Most of these are shared by other underwater vehicles, including human-occupied submersibles and AUVs. Each of these systems will be revisited in much more detail in the subsequent technical chapters.

2.5.2. Common ROV Systems

In Chapter 1, you were introduced to the typical systems that make up a large work class ROV. These same systems are found on most smaller ROVs, too. Here's a brief review of these systems.

Structure

The structural system serves as your vehicle's skeleton and sometimes its skin, too. In addition to providing the craft with its overall shape, it also has to provide these basic features:

- pressure resistance
- strength
- lifting points for launching and recovery
- corrosion resistance

If possible, the structure should also be aesthetically pleasing to the eye. Cost, material properties, and ease of fabrication are all considerations when selecting appropriate materials. (See *Chapter 4: Structure and Materials.*)

Ballast

The **ballast system** adjusts the vehicle's natural tendency to drift upward or downward through the water column without any active propulsion. The ballast system also controls the vehicle's ability to remain upright, as opposed to leaning to one side or even turning upside down. Most ROVs and many other vehicles use a simple **static ballast system** of fixed weights and floats to make the vehicle about neutrally buoyant, so it hangs effortlessly in the water, just as a blimp "hangs" in the air. Vehicles with static ballast systems use thrusters to move up and down through the water. Other vehicles (e.g., most submarines) employ a more complicated **dynamic ballast system**, for example, using compressed gases, pistons, or other methods to force water in or out of special chambers on the vehicle. As water comes in, the vehicle gets heavier and sinks. As water goes out, the vehicle gets lighter and rises.

Designing an underwater vehicle's ballast system requires that you pay careful attention to the weight and location of all vehicle components, as well as the size and location of any components that are in contact with water. In *Chapter 6: Buoyancy,*

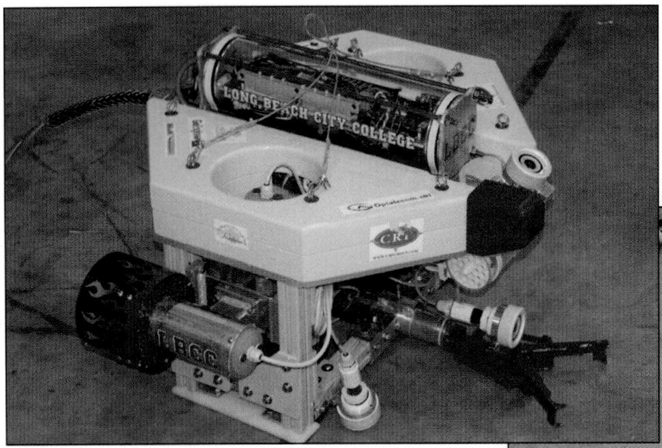

Images courtesy of Steve Van Meter, VideoRay LLC

Figure 2.10: Creative Options for Ballast and Buoyancy

Student ROVs utilize highly creative variations for ballast and buoyancy, ranging from foam blocks to beverage cans.

Stability, and Ballast, you will learn how to use weight statement tables to keep track of these details and use them to adjust the buoyancy of your vehicle.

Propulsion

This system makes the vehicle move. This happens by transforming electrical energy into motion, typically by means of thruster units (a motor and propeller combination). (See *Chapter 7: Moving and Maneuvering.*) The key factors to consider in any propulsion system are:

- thruster power consumption
- number of thrusters
- location of thrusters
- thrust (force) produced

The propulsion system is intimately interwoven with the power distribution system, since a vehicle's velocity and endurance are dependent on how much energy is available, how quickly it can be delivered, and how efficiently it gets transformed into motion.

Power

All underwater robots use electrical energy. The function of the power distribution system is to channel that energy to the various systems of the vehicle that need it: thrusters, lights, camera, manipulator, etc. Note that the source of electricity varies from one type of vehicle to another. For example, most ROVs channel energy through an umbilical that connects the vehicle to a topside power source, whereas AUVs use energy stored on board the vehicle. Regardless of how the power is supplied, the power distribution system performs essentially the same function in all these vehicles. From a design standpoint, the thruster motors and any bright lights are usually the most power-hungry components on the vehicle.

Control

It's relatively simple to transform energy into physical movement so that an underwater vehicle moves through the water. Controlling that movement is a bit more of a challenge. The control system takes energy and directs it to accomplish a set of desired actions. Thrusters, manipulators, and pan-and-tilt mechanisms for cameras are all examples of components that require some type of control, whether through simple or complex methods.

TECH NOTE: ELECTRICAL WIRING—THE VEHICLE'S NERVOUS SYSTEM

Electrical wiring is an integral part of the power distribution system. You can think of wiring as the nervous system of a vehicle, since it links every electrical component to its source of energy. (See *Chapter 8: Power Systems*.) Here are some key considerations when designing the electrical system for your vehicle:

- Bundled wire conductors take up more room inside pressure cans, junction boxes, control panels, and on the frame than you might imagine. Failure to allocate enough space for wiring can cause a lot of difficulties during construction.
- Try to minimize the number and size of the conductors in a tether, to reduce bulk, stiffness, and mass.
- Selecting the appropriate type of conductors is key—be sure to match the correct gauge, flexibility, and jacket insulating materials with the wiring's intended purpose.

Most control systems on underwater vehicles employ some sort of **feedback**, which means that the result of some process gets fed back around to become part of the input to that process. In this case, it means that cameras, compasses, depth gauges, or other sensors monitor the vehicle's actions and feed that information back to the pilot or some automatic control system to improve control. Operating an underwater vehicle without feedback would be like driving a car while wearing a blindfold and earplugs. (See *Chapter 9: Control and Navigation* for a more detailed explanation of feedback.)

Navigation/Sensors

Among the most important sensors used in control are those that provide feedback about the vehicle's location. These usually include a camera (with lights), compass, and depth gauge; sometimes they include more advanced sensors, such as echosounder altimeters and other sonar systems. The information from these navigational sensors is critical, since the vehicle is usually out of sight and working in a murky, dark world where it's very easy to get lost or disoriented. Navigational sensors are so important that they are often regarded collectively as their own system. (See *Chapter 9: Control and Navigation.*)

TECH NOTE: SENSOR LOCATION

An important design consideration regarding sensors is where to place them on the vehicle. For example, on *SeaMATE*, the video camera has to be positioned to relay effectively what is ahead of the vehicle. At the same time, it also needs to "see" the depth gauge and compass in order to read their dials and transmit that information back to the surface. So the gauge and compass are usually placed in the upper or lower portion of the camera's field of view.

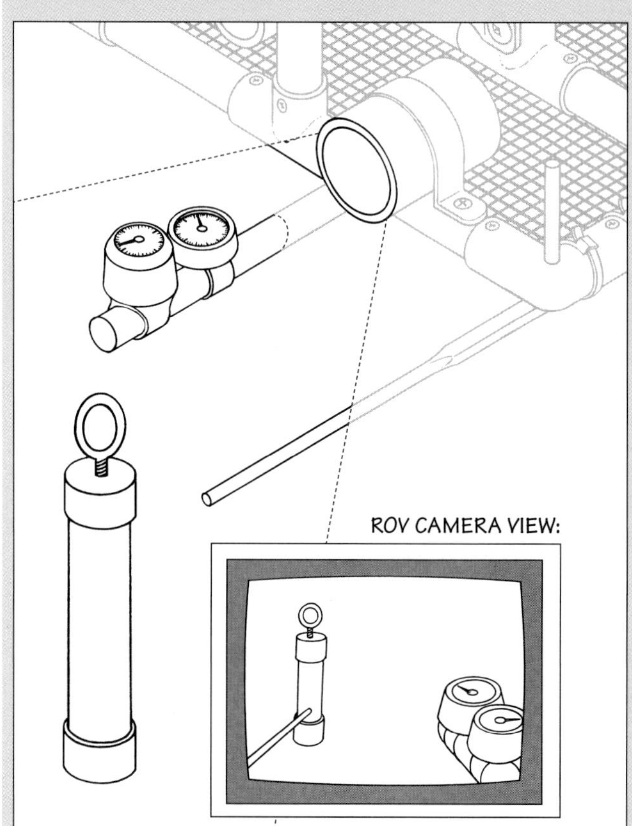

Figure 2.11: Optimum Sensor Position

Some sensors are sensitive to interference from various sources—for example, sonar may be sensitive to noise from thrusters, a compass may react to magnetic fields from a thruster motor, etc. Another common sensor interference problem on ROVs occurs when the vehicle is near the bottom and the thruster wash kicks up silt, thus obscuring the camera. This can be hard to avoid, especially when the bottom is particularly silty. (See *Chapter 9: Control and Navigation*.)

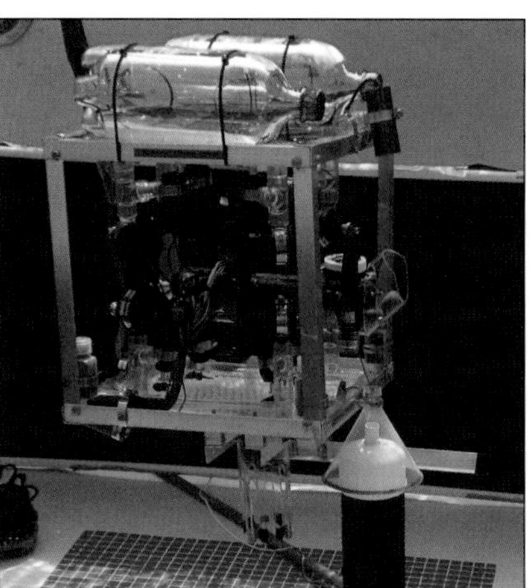

Figure 2.12: Payload Variations

Some missions can be handled with a familiar manipulator; other scenarios inspire mission-specific payloads, as evidenced by this variety of student-built ROVs.

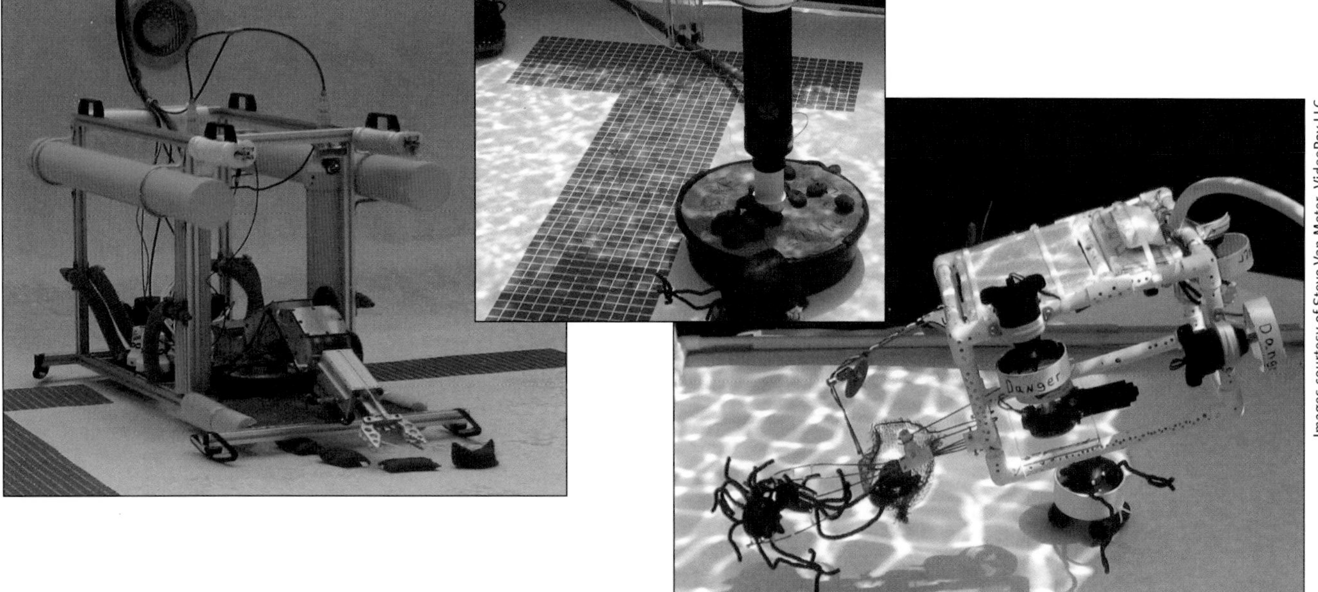

Images courtesy of Steve Van Meter, VideoRay LLC

Payload

With the exception of flying eyeball ROVs that earn their keep by providing underwater observation, underwater vehicles aren't much good if they can't do any work. So payloads are aptly named—they are the various tools or devices that a submersible, an ROV, or an AUV carry to accomplish its mission—and often to produce revenue. Payloads vary as each mission varies. They can be as simple as the fixed pick-up probe on *SeaMATE* or as complex as swap-out tool trays that may include work tools, manipulators, and/or scientific data-collecting instrumentation.

Devising payloads is not only challenging, but fun. For example, student-built ROVs have picked up stuff from the pool bottom using payloads as diverse as chopsticks, dustpans, BBQ tongs, waterproofed toy robot arms, and sucking devices adapted from vacuum cleaners.

The temptation is always to add just one more device to an underwater vehicle's payload—but keep in mind that each additional component impacts other systems. For example, manipulators and tool trays generally need complex control mechanisms to operate them. Heavy-duty tools can significantly impact the physical size and mass of the vehicle. Or a new component might have major power needs in addition to those of the basic thrusters and lights. (See *Chapter 10: Hydraulics and Payloads.*)

Tether

The tether connects an ROV to the surface. It usually contains wires to supply electrical power, and wires or fiber-optic cables to transmit information (commands, video images, and sensor data) back and forth between the ROV and the pilot on the surface.

The power wires are usually considered part of the power distribution system, and the signal wires or optical fibers are usually considered part of the navigation and control systems, but the tether still warrants consideration as its own system because there are a number of issues, including length, strength, durability, drag, buoyancy, flexibility, and storage/deployment/retrieval that are distinctly tether-specific.

2.6. Strategy E: Use the Design Spiral

You'll quickly learn that one of the greatest challenges in designing, building, and maintaining an underwater vehicle (or any other complex system) is that interactions between systems within the supersystem complicate design and repair. Often it seems as though fixing or improving one part of the system creates a whole new set of problems in other parts of the system, leading to tremendous delays and frustration.

For example, you might discover that your ROV is underpowered, so you decide to use bigger thrusters. Fine, but by moving to bigger thrusters, you have also increased vehicle weight and placed additional demands on the power system. To compensate, you must change the ballast and increase the capacities of the power supply and thruster control circuitry. All of that costs a lot more money, which means you won't be able to afford that nice camera upgrade you were hoping for. Just remember that in systems, everything affects everything else.

The **design spiral** is a strategy for confronting this type of problem throughout the design phase of a project. It takes advantage of the natural hierarchical structure of systems by allowing you to focus on the right system level at the right time, yet acknowledges that there are interactions within and among levels that must be taken into account.

Figure 2.13: Design Spiral

In the design spiral, each system gets revisited again and again, each time with more precise knowledge of the other systems. Ultimately the process converges on the final vehicle design.

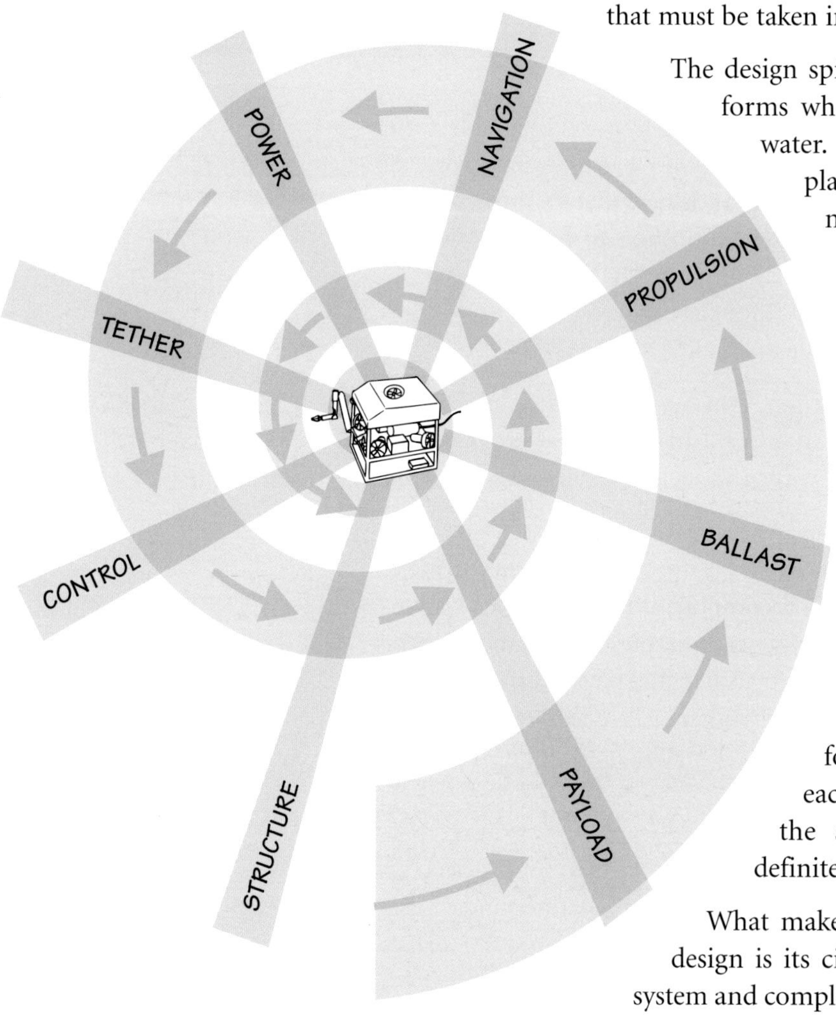

The design spiral is like the vortex of swirling water that forms when you pull the plug on a bathtub full of water. It spirals round and round, passing near places it has visited before, but always a little narrower and more focused than on previous passes.

The vortex is wide at the top, where it starts. Similarly, at the beginning of any project, the options are wide open. There is usually an undisciplined mass of ideas and data as well as a wide range of possible options for solving tricky technical problems. But as soon as you start making firm decisions about any particular system, the options begin to narrow, just like the vortex. Deciding on a thruster type determines the power necessary to propel the vehicle, and that narrows the list of options available for controlling that power. As you address each round of technical solutions in your design, the spiral tightens and plans become more definite.

What makes the design spiral so powerful in systems design is its circular nature. Instead of starting with one system and completing its design before moving on to the next system, you keep circling around, revisiting systems over and over again throughout the design process. Each time around, you come armed with additional information about the evolution of the other system designs. This ensures a high level of compatibility among all the systems in the vehicle—each one is designed with good knowledge of all the others.

Thus, the design spiral allows you to grow the different systems in parallel, letting the evolution of each inform the evolution of all the others. If different groups are working on different systems, each group can be applying the design spiral concept to its own system, since each system is composed of subsystems! It's like having smaller secondary vortices whirling around within the larger one.

But a note of caution here. The design spiral doesn't just happen spontaneously. It requires a conscious effort on the part of the design team, or at least the project manager. Somebody needs to make sure that information is flowing regularly among the different groups working on the various systems. This can happen through frequent team meetings, where each group presents an update followed by discussion. Or it can happen through a project manager who regularly makes the rounds among the groups and relays crucial information about what's happening with each of the other systems.

The ultimate goal of tightening the design spiral and resolving all of the major system issues is to achieve a **concept design**. This is the end result of the planning phase—a specific road map for what you intend to create, in this case a working underwater vehicle. The concept design will include a set of detailed drawings, functional specifications, and perhaps even a mock-up of what you intend to build—but to reach that point, it's necessary to work through a **design methodology**, or series of design stages, as described later in this chapter.

2.7. Strategy F: Research, Research, Research

Knowledge is power. The more you can learn about the techniques and technologies available to you, the better your vehicles will be, and the more likely that your missions will succeed.

A couple of hours invested each week in reading books at the local library, surfing the web, or talking with people knowledgeable about underwater vehicles (or related topics) can save your team many weeks of hard work, not to mention money. That's because you can learn from other people's past successes and failures much more quickly than you can recreate those successes and failures yourself. Don't overlook the incredible gold mine of project reports from various competitions that are posted on their websites. Not only can you compare the variety of cameras, thrusters, control systems, etc. that other teams elected to use, but you can learn from their mistakes, which are also detailed in project reports. A team may decide to put one or two team members in charge of research, or all team members may share the responsibility (and benefits) equally.

The most productive teams usually have one or more members who are good at **networking** and have—or are willing to develop—a long and varied list of people they can contact for information on a wide variety of topics (See *Tech Note: Networking*). Through networking, the local librarian, hobby store owner, or a retired engineer can extend the capabilities of your team. Family members and schoolteachers or staff people who facilitate an intensive project can also be an important part of your extended team network.

TECH NOTE: NETWORKING

Networking refers to the process of meeting and forming mutually beneficial relationships with other people, usually through being introduced by people you already know. Having an extended network of such contacts can help you quickly access information or other resources you need. It can also help you spread the word about projects you are working on.

Whenever possible, take advantage of opportunities to meet and get to know new people who are interested in the kinds of things you are interested in. They may be able to provide you with information, publicity, parts, or other resources. They may be able to help you find jobs. Perhaps most importantly, they may be able and willing to introduce you to still more people, allowing your network to grow in size and effectiveness.

There are two levels of research. The first is just to go out and survey the range of possibilities to get ideas. The second is to pin down details about exact parts, prices, etc. During this latter stage, it's important to gather all the information you will need for the design. This includes not only model numbers, price, and availability, but physical dimensions, weight, type of materials, power requirements (if any), etc. The exact information you need to look for during this second, more detailed phase of research will become clear as you read through the technical chapters later in this book.

Sometimes the information you gather through your research will open up completely new avenues and possibilities. Other times it will warn you that a path you were planning to take would be a mistake. At this early stage, keep your plans and thinking flexible, and do as much up-front research as you can. It's much easier to make changes early in the design process than it is to do so after you've already purchased and assembled a large fraction of your vehicle. (You will find out more about this amazing process of research in *Section 3.7.2.* of this chapter.)

2.8. Strategy G: Build and Test Prototypes

A **prototype** is a relatively simple draft version of a mechanism, machine, circuit, device, or other system that you can cobble together much more quickly and at much less expense than the real thing. The purpose of building a prototype is to test the basic feasibility of a design idea or to learn other things about how well it will work *before* you go to all the time and trouble to build the real thing. A prototype is a wise investment anytime you can build it much faster and much less expensively than the final version. A prototype is especially important if you're planning to mass-produce a bunch of similar items that have never been tested before. Make one first as your prototype and test it thoroughly before you bother to crank out the rest of them.

A vehicle design project may involve dozens of prototypes of different systems and their subsystems, etc. Once you start building and testing prototypes regularly, you'll quickly discover how valuable they can be. They often reveal that something you've *perceived* as a big technical obstacle and have spent weeks worrying about how to solve is, in fact, a non-issue, while another situation you had completely overlooked is the one that is actually going to stop your vehicle dead in its tracks.

One example of a prototype might involve a control system circuit. Before going to the trouble, expense, and time required to design and manufacture a fancy circuit board for a newly designed control circuit, an experienced engineer might solder together a crude version of the circuit, just to see if the basic circuit design works. If it does, then she can proceed with plans to make the fancy board. If it doesn't, she's just avoided wasting a lot of time, effort, and money. Likewise, before you machine a bunch of fancy metal parts for a gripper, you might try mocking up a model of the gripper jaws with cardboard or Plexiglas, just to make sure the basic mechanism you have in mind will actually open and close the gripper jaws as expected.

Figure 2.14: From CAD to Reality

The CAD version of the student-built Sonia *AUV is quite similar to the finished vehicle.*

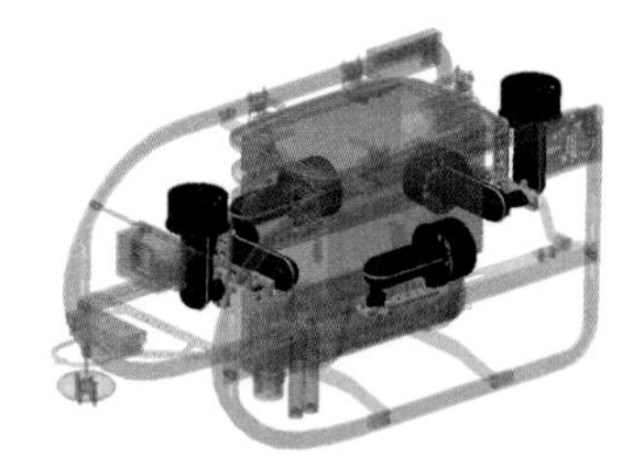

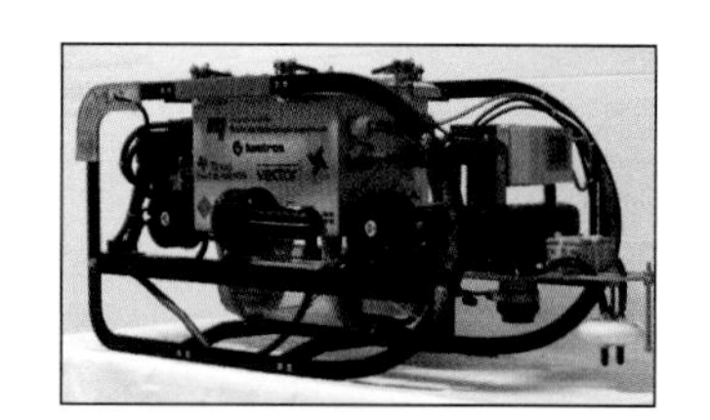

Images courtesy of Team SONIA, www.sonia.etsmtl.ca

3. A Design Methodology for Underwater Vehicles

A **design methodology** is a time-proven sequence of stages and procedures followed to plan, define, design, build, test, and implement a specific type of system. The final segment of this chapter takes you, stage by stage, through a design methodology for underwater vehicles that will transform your ROV or AUV project from a nebulous idea in the back of your mind to a completed working vehicle that's in the water, doing the mission it was designed to do.

The methodology presented includes 10 stages:

Stage 1: Accepting the Mission

Stage 2: Crafting a Mission Statement

Stage 3: Identifying the Mission Tasks

Stage 4: Establishing Performance Requirements

Stage 5: Identifying Constraints

Stage 6: Listing Vehicle Systems

Stage 7: Generating the Concept Design

Stage 8: Fabricating the Vehicle

Stage 9: Conducting Sea Trials

Stage 10: Carrying out Operations

Stage 11: Evaluating Ops and Writing a Report

The first four stages are really about clarifying exactly what it is you are trying to accomplish with your design. Though many beginners are tempted to skip these stages because they don't sound like as much fun as drilling, sawing, and soldering, such clarification is absolutely essential. Without it, your design efforts will be unfocused and will result in an uncoordinated, poorly optimized design that may or may not be able to accomplish the tasks you want it to do.

Stage 5 helps you complete a realistic assessment of your available time, money, etc.

Stages 6 and 7 translate your project goals into a feasible design that is sufficiently detailed for you to evaluate whether or not it's consistent with your constraints. If not, this is where you decide whether to revise the design or abandon the project. (Most teams confronted with this decision simply revise the design, because they're having too much fun to abandon the project!)

Stages 8 and 9 are where all that preparation really pays off—the saws and drills start humming, sparks start flying, and a real-life ROV or AUV begins to materialize on the workbench. Sea trials confirm which parts of your design are effective and which ones need to go back to the shop for some design revisions or adjustments.

Stage 10 is where your completed vehicle finally goes to work, accomplishing the underwater mission(s) for which it was designed and built! Finally, after ops are complete, there's the important Stage 11 in which you evaluate the vehicle's performance, suggest improvements, and summarize everything in a glorious project report.

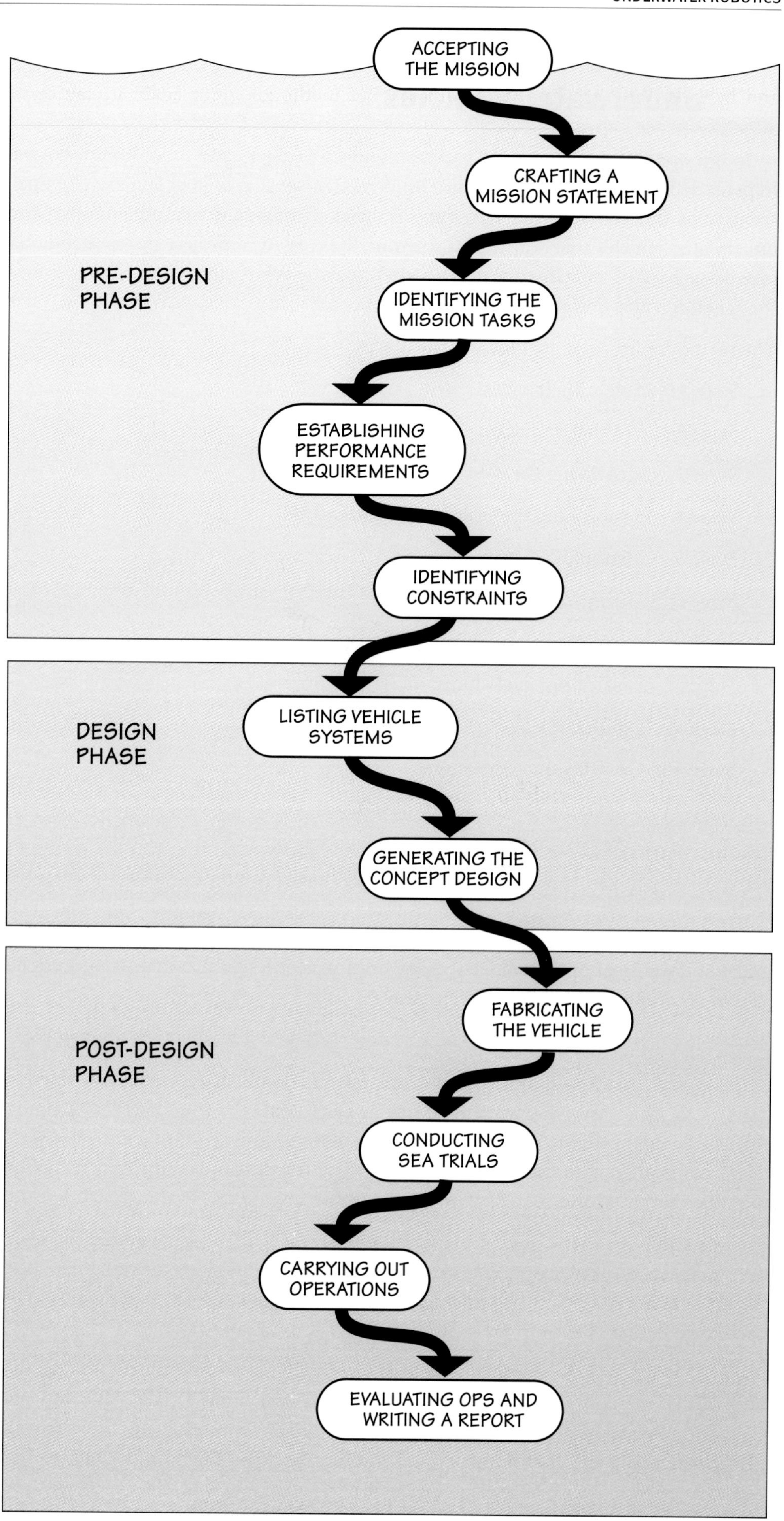

Figure 2.15: Design Stages

Sometimes it helps to think of the parts of this methodology as being divided into three basic phases: pre-design, design, and post-design.

Certainly there are many ways to attack a design challenge, but this methodology works quite well for small-scale underwater vehicles or robots, including ROVs, AUVs, and hybrids. Whether you follow this specific methodology or adapt it, pay close attention to the salient points in each stage, for they are the keys to figuring out a successful design.

3.1. Stage 1: Accepting the Mission

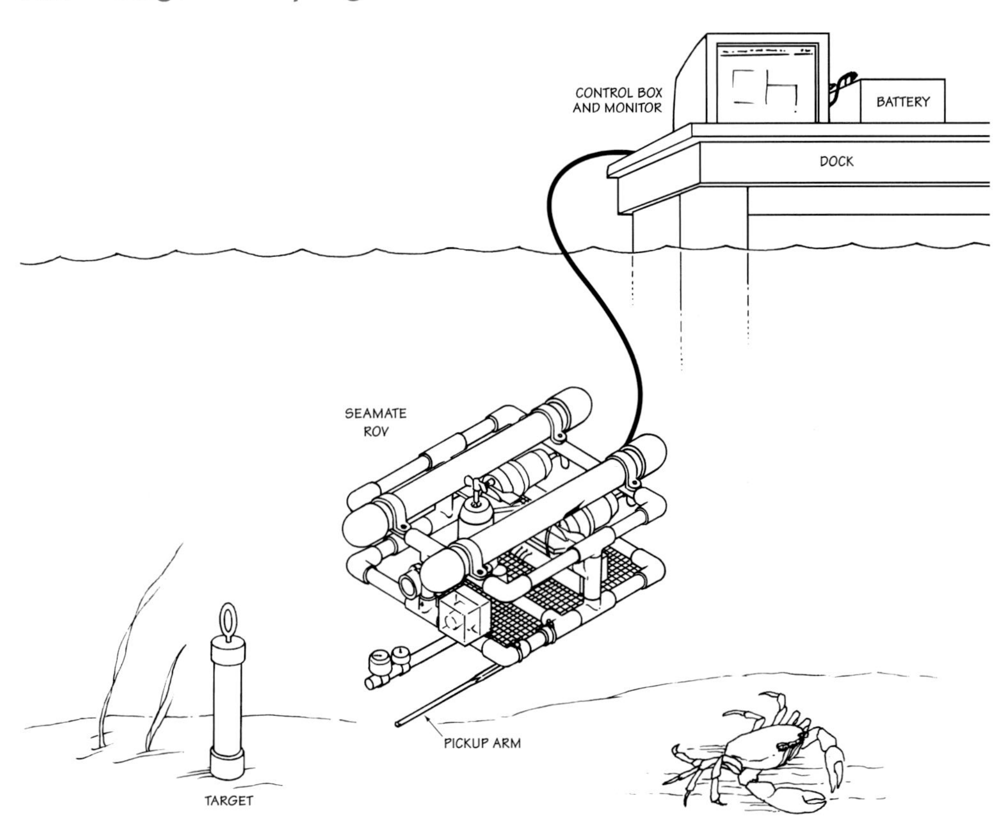

Figure 2.16: Bozonium Recovery

NEWS FLASH!

KSBW News Channel 8 in Monterey has just been informed of a pending emergency situation on the Monterey Peninsula. At approximately 3:30 p.m. this afternoon, the commercial fishing vessel Lucky Dog *was about 10 miles off Moss Landing, just completing her day's run of bottom net trawling. While retrieving her nets, a crew member spotted an unidentified container with four small cylinders inside. Captain Iam Chaos of the vessel notified the U.S. Coast Guard by radio. They identified the cylinders as those containing a highly dangerous nuclear fusion bozonium crystal, stolen three weeks ago from a nearby military test facility. As the vessel* Lucky Dog *was making fast at Pier 70 in Monterey Harbor, the captain's young son opened the box and inadvertently dropped the four cylinders. Fortunately, they fell overboard—if the cylinders had landed on the pier, none of the crew would have survived to tell the story. But now four highly explosive cylinders are loose at the bottom of the harbor. Each is about the size of an empty paper towel roll—not very heavy, but volatile. The Coast Guard and local military experts have declared the situation to be hazardous. Navy and civilian teams are on site with ROV equipment for retrieval, but so far all units and support vessels are too large to work in the small confines of the harbor. So the MATE team at Monterey Peninsula College has been requested to build a special-purpose ROV to handle the emergency. Project manager Will Doit said that the team will begin work immediately to develop a mini-ROV that can handle the mission. Stand by for further news and developments as they happen.*

TECH NOTE: ASKING THE RIGHT QUESTIONS

Designing underwater vehicles is all about asking the right questions, which means those that can be answered in a technically useful way. Here are some examples (presented in no particular order):

Why: Why is it needed? Why will it work? Why is it not working?

What: What do you need? What does it do? What are the options? What is the problem?

Where: Where can it be found? Where does it operate?

When: When is it needed? When does it get built? When does it get tested?

Who: Who will build it? Who will test it? Who will pay for it?

How: How can it be done? How much will it cost? How many are needed?

Which: Which option or solution works best?

Is: Is the design meeting the mission requirements? Is the project on budget and on schedule?

This rescue scenario was much more than an amusing story to the students in Frank Barrows' Introduction to Submersible Technology course offered by the MATE Center at Monterey Peninsula College. This was the mission scenario that would determine more than 80 percent of their final grade!

In reality, of course, there was no dangerous bozonium. The scenario was fictitious, but the mission—to retrieve four PVC pipe cylinders (the simulated bozonium canisters) from the bottom of Monterey Harbor—was real enough. The very first thing the students in Frank's class had to do was to decide whether or not to accept the mission. This is the first decision any person or group must make when faced with a design opportunity.

In some cases, like the example above, the mission opportunity may be presented to you by someone else. In other cases, you may invent the mission yourself. Perhaps you lost an important set of keys off the end of the dock and are contemplating building a simple ROV to retrieve them. In a sense, you are presenting yourself with a mission. Regardless of who comes up with the mission, you still have to decide whether or not to pursue it. For example, if that particular set of keys contains a very hard-to-replace opener for a specialized laboratory, it may be well worth the trouble to attempt a retrieval.

All of the brainstorming, planning, fund-raising, inventing, and testing associated with an underwater vehicle project revolve around the mission. Before committing to a mission, you must do a simple cost-benefit analysis, weighing what benefits you will get out of the mission against your best estimates of what will be required in terms of the time, effort, money, tools, facilities, and expertise to design and build a vehicle capable of completing the mission. In most cases, you will need to do some research and ask some questions to find out more about what the project entails. You must be honest with yourself about your level of interest and motivation, the resources available to you for the project, and the amount of time you can devote to the project without unduly straining personal relationships and other commitments you already have in your life. Keep in mind that a project of this type will probably take three times longer to complete than your first estimates. On the plus side, if you do accept the mission, you'll learn new things, meet new people, and push yourself like never before!

In the case of Frank's students, 80 percent of their grade was riding on successful completion of the mission. That single fact outweighed all other factors, so of course, they accepted the mission. What they finally completed was a small eyeball ROV, made from hardware-store technology and equipped with a simple manipulator that would retrieve the canisters from the bottom. The robotic vehicle they designed and built is very similar to the *SeaMATE* ROV plans in this book.

3.2. Stage 2: Crafting a Mission Statement

The imaginary mission scenario above is packed with information—some of it relevant, some unimportant. But none of it gives enough of the specifics you need to actually design and build an ROV. So how do you transform a scenario such as this into useful information for a vehicle design? You start by crafting a **mission statement.** A mission statement summarizes—ideally in one or two clear, objective sentences—what the vehicle must do to successfully complete its mission. In subsequent stages, this mission statement will be used as the starting point for a more detailed analysis of the steps required to complete the mission and the vehicle capabilities needed to complete those steps.

In Frank's scenario, the MATE team at Monterey Peninsula College has been asked to build a special-purpose ROV to handle an emergency down at the harbor. It might seem logical that the mission is therefore to build an ROV that can save Monterey from death by bozonium! Not so fast. That might be a good mission for the *students* in Frank's class, but it's not a useful mission statement for design purposes. What the students need to help them design their ROV is a mission statement for the ROV (not for the students). In this example, a mission statement for the ROV that could be used for design might read:

Mission Statement:

Locate and recover four simulated bozonium canisters (sections of PVC pipe) from the bottom of Monterey Harbor by the end of the semester (May 25).

Once the mission statement is accurate and complete, you can move on to the third stage—figuring out precisely what jobs the vehicle has to do under water to successfully complete its mission. This means assessing the mission tasks.

3.3. Stage 3: Identifying the Mission Tasks

Once the mission statement is defined, the next step is to think about the step-by-step series of specific actions the vehicle must perform to accomplish the mission. These actions are called the **mission tasks**.

Frank's students might break their mission down into the following sequence of mission tasks:

Mission Task 1: Dive to the bottom of the Monterey Harbor.

Mission Task 2: Search the bottom for one of the cylinders.

Mission Task 3: Once a cylinder is located, grab it.

Mission Task 4: Surface with the cylinder.

Mission Task 5: Return to the dock with the cylinder.

Ongoing: Repeat these mission tasks until all four cylinders are recovered.

Once the mission statement has been translated into a series of mission tasks, it becomes much easier to start thinking about what specific capabilities the vehicle must have and to start brainstorming about how you might provide your vehicle with those capabilities. For example:

Task 1: Dive to the bottom. The vehicle needs to be able to descend vertically. For it to do so, some sort of ballasting mechanism or vertical thruster will be required.

Task 2: Search the bottom. Searching in the murky water of Monterey Harbor will work best if a systematic search pattern, such as a search grid, is used. To follow a search grid, the ROV must be able to maintain a specified course for a certain distance and be able to turn onto another course once that run is completed. This task requires Frank's students to use multiple horizontal thrusters to effect these precise maneuvers. It also necessitates two sensors—one (a compass) to maintain a straight course and the other (a video camera) to "see" what's in front of the vehicle.

Task 3: Grab the cylinder. This task demands some means of retrieving the cylinder, perhaps a pick-up probe, scoop, or mechanical gripper attached to the ROV.

Task 4: Surface with the cylinder. The ROV must have the ability to lift the weight of the cylinder to the surface, using either powerful vertical thrusters, a variable ballast tank, or both.

Murphy's Law states, "Anything that can go wrong will go wrong." Though obviously intended as a joke, it's surprising how often Murphy's Law rears its ugly head. Most experienced engineers can recite this law in their sleep, and they make it a habit to design their products to withstand (realistic) worst-case scenarios.

With ROVs and AUVs, there's plenty that can go wrong. You may wish to design accordingly, particularly where safety might be an issue. Start by making sure your mission requirements have identified all the environmental factors that are likely to interfere with a successful mission.

TECH NOTE: A VEHICLE'S FOOTPRINT

The three-dimensional excursion area in which a vehicle operates is often referred to as its "footprint." Each underwater robot's operational footprint is dependent on its thrust capacity, vehicle speed, and tether/umbilical length and drag, but it can be highly impacted by environmental conditions such as weather, currents, temperature variations, plant life, artificial structures, debris, etc. Furthermore, these "wild card" factors may vary in intensity during the course of a single mission.

While designers make their plans based on a basic, or standard, vehicle footprint, they must also take into account various factors that can suddenly and quite dramatically affect the vehicle's area of operation.

Task 5: Return to the dock with the cylinder. This task requires the pilot to maneuver the ROV alongside the dock while still holding onto the cylinder. This is easily done by having the tether handler verbally relay steering instructions to the pilot. Or the pilot can simply watch the ROV at the surface to effect a successful docking. Once the ROV is alongside the dock, the cylinder can be retrieved by hand (since it's not really bozonium).

At this stage, it's not necessary to spend a lot of time working out the details of the specific hardware needed to accomplish these tasks. Just keep those great ideas in the back of your mind, or better yet, jot them down in your project notebook.

3.4. Stage 4: Establishing Performance Requirements

Once you've generated the mission tasks and some general ideas about how to accomplish them, it's time to start getting serious about what it's going to take, in a technical sense, to pull off a successful mission. This is where you start putting some numbers on things and imagine realistic worst-case conditions that may have to be overcome during the dives. This is done by specifying the requirements for a successful mission in the real (not ideal) world.

Finding those bozonium canisters in the harbor seemed to Frank's students like a straightforward mission, but they soon found out that even a simple mission is fraught with vexing details. At the beginning of the project, students jumped right into assembling an underwater camera and even began building a frame. But they had forgotten a number of factors that sent them back to the drawing board. Many of these involved environmental conditions, such as the currents that ran around the pier pilings or the mud bottom that their thrusters would stir up, thus obscuring the camera. There was also an ocean surge that would drag a lightweight ROV back and forth, making it difficult to maintain a straight course while searching for a canister and to hold a steady position next to a canister while trying to pick it up. (See *Tech Note: Murphy's Law.*)

When confronted with these and other problems, Frank hinted that the students would be wise to look at all the logistical, operational, and environmental factors that would impact their design. Engineers refer to these factors, and the capabilities the project must have to deal effectively with them, as the **performance requirements**. The performance requirements should be elaborated *before* the actual vehicle designing begins, because they directly impact the vehicle's ability to complete the mission.

The performance requirements can be developed by answering a set of questions about environmental and logistical conditions and issues. Answering these questions can help you discover and order technical information that's directly relevant to your vehicle design. Some sample questions are listed below, grouped into categories that correspond to common problem areas. Some straddle multiple categories.

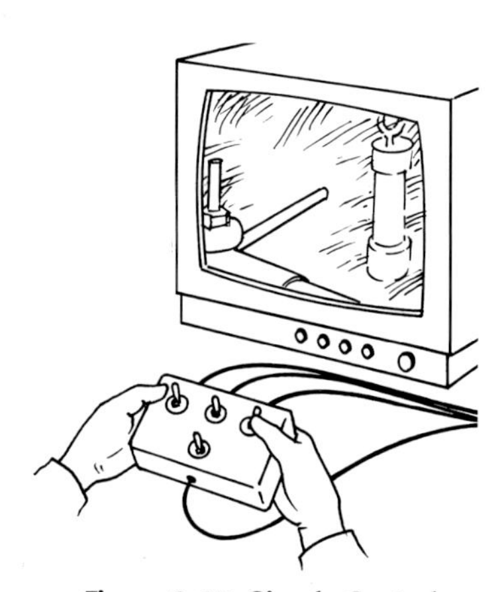

Figure 2.17: Simple Control System for Recovery of Bozonium

3.4.1. Environmental Issues

- Will the vehicle operate in saltwater or freshwater?
- How deep will the vehicle need to go?
- What is the pressure at the maximum anticipated depth?
- What is the anticipated temperature range of the water?
- Are there surf or surge conditions to consider? If so, what are they?
- What is the anticipated visibility (i.e., how far can the cameras see) under water?

Image courtesy of Harry Bohm

Figure 2.18: Environmental Issues Impact Vehicle Design

Operating an underwater vehicle in dark, icy conditions places special demands on the vehicle as well as its crew!

- What is the maximum velocity of water currents?
- Are the currents constant, or do they vary in speed and direction with waves, tidal cycles, or other factors?
- If the vehicle will be operating near or at bottom, what is the seabed like?
- Is the vehicle operating in an ecologically sensitive area? If so, are there protected species, habitats, or formal restrictions to consider?
- Are there natural or artificial structures, coral, fishing gear, or other obstacles that might block or entangle the vehicle?
- Are there sea lions or other large, curious animals present that might attempt to bite, pull, or otherwise play with (and possibly damage) the vehicle?
- Is there kelp, seagrass, or other vegetation in the dive area that might get tangled in propellers?
- Are there spaces between boulders or among the branches of sunken trees where the vehicle might get wedged and trapped?

3.4.2. Velocity/Maneuverability Issues

- Does the vehicle have a tether, and how will the tether affect vehicle maneuverability?
- How quickly does the vehicle need to move vertically (meters per second or feet per second)?
- How fast does the vehicle have to travel in a horizontal direction (meters per second or feet per second)?
- How quickly does the vehicle need to turn or rotate?
- In what directions does the vehicle have to maneuver?
- How far does the vehicle need to travel under water?
- Does the vehicle ever need to move directly sideways ("crabwalk")?
- Does the vehicle ever need to turn on its side, flip upside down, or do other unusual maneuvers?

3.4.3. Structural Issues

- What are the major forces the structure might have to endure? Possibilities include water pressure, lifting forces, thruster forces, currents, impacts with boulders or other submerged structures, and impacts with boats or docks during launch/recovery.
- Are there any vehicle weight requirements that might affect the choice of materials for the structure?
- Does the vehicle structure have to be made from corrosion-resistant materials?
- Does the structure need to be streamlined? The answer will depend on vehicle velocity, power, and water currents.

3.4.4. Control Issues

- Will the thruster motors need speed control, or are simple forward, off, and reverse adequate?
- How many thrusters need to be controlled?
- What else needs to be controlled on the vehicle? Are there manipulator arms, video lights, camera pan/tilt/zoom mechanisms, saws, suction-samplers, or other tools associated with the payload?
- Are special sensors required to provide the feedback necessary for effective control of any devices on the vehicle?
- Do any of the control requirements demand large amounts of power that might require special power amplifier circuits?
- Will anything be controlled by some means (hydraulic, pneumatic, etc.) other than electricity?

3.4.5. Energy/Power Issues

- Is the source of energy for the vehicle located on board the vehicle or supplied externally via the tether?
- What is the power source? Options include generators, batteries, and fuel cells.
- Which vehicle components consume the most energy? How much do they need?
- What is the maximum power (watts) that needs to be supplied to the vehicle?
- Are there any limitations to the weight, volume, or cost of the power source?
- If the vehicle has a tether, how long does it have to be? To what extent does this length interfere with power transmission over the tether?
- Do the vehicle motors, motor control circuits, lights, or other systems generate enough heat to damage parts of the vehicle? If so, does the vehicle have mechanisms for dissipating this excess heat?

3.4.6. Navigation Issues

- How will the vehicle navigate in the water?
- If the vehicle is an ROV, how will the *pilot* know where it is and where it's going? If the vehicle is an AUV, how will the *vehicle* know where it is/where it's going?
- Are special sensors required to provide information about vehicle location, direction, and obstacles for navigation purposes?

- How precise does the navigation need to be?
- What conditions in the operations area (low light, silt, etc.) could adversely affect navigation (related to environmental issues)?

3.4.7. Sensor Issues

- What external conditions (light, depth, direction, temperature, etc.) need to be monitored by sensors?
- What internal vehicle conditions (battery charge, motor temperature, leaks, etc.) need to be monitored by sensors?
- What types of sensor(s) are required to measure these things?
- How precise and reliable do the various sensor measurements need to be? This affects cost of sensors selected.
- How many sensors do you need on the vehicle?
- How will these sensors be powered?
- How will data from these sensors be monitored and used?

Figure 2.19:* Mir *Launch and Recovery

Many launch-and-recovery procedures for specialized craft, such as the Mir *submersibles, require a team of people as well as specialized lifting mechanisms.*

Images courtesy of James P. Delgado

3.4.8. Launch-and-Recovery Issues

- Will the vehicle be launched from a ship, small boat, dock, rocky coastline, or sandy beach?
- Will the vehicle be launched by hand, cranes, or other mechanisms?
- How many people will be needed to launch the vehicle safely?
- What is the maximum permissible weight of the vehicle (in air)?
- Will the vehicle need strong lifting points on the frame?
- How big will the waves, winds, and currents be during vehicle launch and recovery? Will tidal cycles affect this?
- What are the maximum allowable outside dimensions (length, width, height)?

Image courtesy of The Marine Institute, Memorial University of Newfoundland

Figure 2.20: Simple Launch and Recovery

Even vehicles that can be launched and recovered easily by hand will benefit from well thought out procedure and team work.

3.4.9. Operational Issues

- How long does the vehicle have to remain submerged in order to complete the mission tasks?

- How long does the vehicle have to be on deck for servicing between dives?
- How long can the vehicle operate under water before returning to the surface for servicing?
- How long is the transit time from the shore base to the dive site?
- What sort of transport is required to move the vehicle system from the shore base to the dive site?
- What sort of watercraft is required to provide on-site logistical and launch and recovery support (LARS) for the vehicle system?
- How many people are required to launch, pilot, and recover the vehicle? What about servicing it?

3.4.10. Safety Issues

- Who is piloting the vehicle?
- Who is launching and recovering the vehicle?
- Who is servicing it in the field?
- Do the operators need special training or certifications?
- What hazards are present in the mission?
- How will you free the vehicle if it becomes snagged or entangled on the bottom?
- Are there any regulatory certifications or special testing required?
- What design features are needed to minimize hazards?
- Are all the materials and power sources safe to use in a wet environment?
- Do the materials or power sources used in the vehicle pose any special hazards to plant or animal life under normal operating conditions? If the vehicle is lost at sea or runs aground on rocks and breaks open, will it release anything toxic?
- What procedures are required to maintain the vehicle in a safe operating condition?
- Are there special hazards associated with launch and recovery?
- Are there special hazards associated with vehicle repairs done in the field (i.e., on a boat, dock, poolside patio, or beach, rather than back at the workshop).

These questions don't require detailed answers yet. It's just important to be thinking about them and answering as many of them as you can as soon as you can. For now, just enter any answers you have in your notebook or some type of data table. Any gaps in the data can be addressed during later stages as you tackle more in-depth research about very detailed technical issues.

Here are some examples of a few of the many performance requirements Frank's students identified after doing a little research:

- The ROV must operate in seawater to a depth of about 3 meters (approx. 10 ft). (This is the depth of the water in the harbor under the dock where the bozonium canisters were dropped.)
- There is some current (up to 10 centimeters/second) in the protected harbor waters but no strong surge, and there is plenty of time to recover the canisters, so vehicle speed is not a crucial parameter as long as it's fast enough to outrun the current.

HISTORIC HIGHLIGHT: CREATIVE THINKING

Figure 2.21: Simon Lake's **Argonaut Jr.**

When designing underwater vehicles, there are always difficult technical challenges, often made more so by very modest budgets.

Simon Lake is a prime example of an early inventor who refused to let a lack of money get in the way of his determination to explore under water. His first sub, the 4.2-meter (approx. 14-ft) *Argonaut Jr.*, was built in 1894 with a total budget of $15 for the prototype. Since he couldn't afford steel, the ungainly looking craft was built of thin yellow pine planking, with canvas sandwiched between the layers, and painted with coal tar to seal the seams. Although the sub was essentially a box on wheels, it boasted an airlock and diver's compartment that provided access to the seafloor. When on the bottom, the *Argonaut Jr.* was propelled by a man turning a crank with foot-powered bicycle pedals.

With such a modest budget for the prototype, Lake purchased soda-water fountain tanks from a bankrupt drugstore to use as the sub's compressed air reservoir. The compressed air pump was a recycled plumber's hand pump. He built his own diver's helmet and suit from assorted bits and pieces. Despite its humble beginnings, Lake's *Argonaut Jr.* was successful and attracted the attention of the press and several wealthy investors. As a result, Lake was able to proceed with a full-sized bottom crawler—the *Argonaut.*

- The vehicle must be able to grab and lift a PVC cylinder that is about 5 centimeters (approx. 2 in) in diameter, 30 centimeters (approx. 12 in) long, and has an effective weight in water of about 4 ounces.
- The dive can occur during daylight hours and is in shallow water, so lights probably are not needed.
- The harbor has a soft, muddy bottom, so the ROV thrusters must be positioned to avoid stirring up the silt on the bottom; otherwise, visibility may be reduced to near zero.
- There are wharf pilings nearby. The ROV must avoid getting its tether tangled around the pilings.
- The ROV team will need to get permission from the harbormaster before going down to the dock.
- There is no electrical power available at the site, so all power will need to be provided by portable batteries.

3.5. Stage 5: Identifying Constraints

A **constraint** is something that limits or restricts the design in some way. Constraints can come in many forms, from physics to finances. For example, Frank's students had a budget of $250 to build their ROV. This turned out to be a huge constraint, one that forced them to shop carefully for the least expensive parts that would do the job and to rummage through the junk-parts bin to cut costs even further. It also motivated them to ask local businesses to donate some parts.

Sometimes a technological development or other event changes a constraint. For example, 15 years ago, a video camera cost hundreds of dollars, so that price often proved to be a huge impediment for a simple ROV built in a classroom or home workshop. Today the cost of such a camera has dropped dramatically, so camera cost is now a much less severe constraint.

TECH NOTE: DESIGN TRADE-OFFS

All designs end up being exercises in compromise—engineers call these **"design trade-offs."** Compromise is all about the reality that it's never possible to have it all—just try getting the ideal hydrodynamic shape, top energy efficiency, and state-of-the-art components, all while staying within the confines of a realistic budget. Because systems always affect other systems, the solution is to trade off one design feature for another.

For example, going to the required depth may require a strong (and heavy) hull that is inconsistent with the requirement for hand-launching the vehicle. So you must decide if it is better to trade ease of launch and recovery for greater depth, or vice versa. Or perhaps you want to use super-deluxe thrusters, but find their cost would eat most of the project budget and force you to cut back on the number of navigational sensors, so the trade-off is the superior performance of these expensive thrusters versus better navigation.

The key is to address the major design trade-offs before starting to build, since it costs more (in time, money, effort, and team morale) to change something during fabrication than when the vehicle is still on paper.

Even though constraints are often viewed as limitations, try not to think of them in a negative sense. In fact, they can help narrow down an overwhelming number of possible designs to a manageable few. These restrictions are also invariably educational, and they provide some of the challenges that make ROV-building fun. So think of a constraint as a reality check on your ideas, one that forces you to work with what you have, rather than what you want. This is a good design skill to cultivate. Many technological innovations have come about because of limitations that forced the designer to get creative and consider other alternatives. The most important point is this—don't ignore constraints! Taking time to consider fully their implications will save a considerable amount of wasted effort and money.

The following paragraphs highlight some common categories of constraints that you should consider when evaluating the constraints confronting your project.

3.5.1. Human Resources

People are the most important resource a project needs, since their creativity, expertise, and experience drive the design. Difficulty in finding such experts can act as a constraint. Just remember that not everyone on your team will be skilled, and that's fine—in fact, the most important qualifications are the ability to get along, to be eager to learn, to be excited about building underwater vehicles, and to be fully committed to that goal. Frank's team of 15 students was inexperienced with ROV-building, but they were dedicated and eager to learn, so they were successful in the end.

3.5.2. Physics

The laws of physics govern the natural world—in the air, on land, and under water. So these laws and the properties of water constrain almost every aspect of an underwater vehicle in one way or another. For example, water pressure requires that the structure of any underwater vehicle be strong enough to resist the tremendous forces caused by that pressure. And the physics of electricity demand that wires be big enough to carry the required electrical current without overheating, melting, or starting a fire on the vehicle. Most of the technical chapters that follow in this book are devoted to these laws of physics, the constraints they place on underwater vehicle design, and proven solutions for working within those constraints.

3.5.3. Availability of Workspace, Tools, Test Facilities, and Transportation

Access to workshop and test facilities can also place a constraint on your project. Large, sophisticated ROVs require lots of space with welders, machine tools, and a crane; otherwise, the ROV won't get built no matter how good the design is on paper. So look carefully at the requirements of your intended project and match those needs with the facilities you have available. This advance planning should take into account the availability of dedicated workshop space, tools such as drill presses and soldering irons, use of test tanks and electronic test gear, and some means of transporting your vehicle and the project team to and from the operations area. (See *Chapter 12: SeaMATE* for specifics on setting up a workbench and test tank.)

3.5.4. Availability of Components, Materials, and Supplies

It's common to start a project with the idea of using a specific component for your vehicle, only to find that the part is no longer made, is out of stock, or has been replaced by a new model with slightly different specifications. Since lack of availability can constrain a design, always check:

- Is the part still being made? If it has been replaced with a newer version, have the specs changed in ways that would affect your design?

- Is the part in stock?
- Does the part have to be custom fabricated?
- How long will it take to have the part delivered?
- What will it cost to purchase and deliver?
- Will there be a future cost increase that makes the part uneconomical?

If the part is not available, then ask the supplier for other options or search for a substitute. This is not always a bad thing—sometimes a new part actually improves on your original idea.

3.5.5. Time

Time can present a serious roadblock. For example, if Frank's students had come up with a glorious design for their ROV but then determined that it would take three semesters to complete, that would have been a serious constraint because they had a one-semester time limit. In this case, the problem would not have been with their design *per se*, but with how that design conflicted with the reality of their deadline. Because of this, they would have needed to rethink their design.

3.5.6. Budget

Money, or rather the lack of it, is often the biggest constraint for any project. This is true for prominent companies producing cutting-edge subsea technology products as well as for the amateur builder. How much money you can bring into a project, and how soon you can bring it in, drives what can be purchased for parts, what talent can

TECH NOTE: IN-KIND CONTRIBUTIONS

Don't be put off by a small budget. Many practical innovations occur when money is in short supply. For example, effective networking can result in getting others to contribute hardware and/or services. Donations like this that do not involve direct gifts of money are called **in-kind contributions.** Many teams participating in competitions solicit thrusters, propellers, electronics, PVC pipe, cameras, and other parts for their vehicles from manufacturers, thus stretching their budgets. Often these in-kind contributions amount to hundreds or even thousands of dollars in equipment and materials.

However, don't expect something for nothing. In exchange for their gifts, most companies (really, the people in the companies who made the decision to donate) will want to know that they have done something that makes a real difference to someone's project. They'll want to see that you're seriously committed, so develop a solid plan first, then do what it takes to give them the confidence to invest in you. That will require making some phone calls or personal contacts to introduce your group and its mission. Most companies will also expect (or at least hope for) some free advertising, so be sure to acknowledge their support whenever and however you can. Tell people about the donation. Put a sticker with the company logo on your ROV. If your team has a website, be sure to include your thanks to the sponsors in a prominent position, and let the companies know that you are doing so.

Some companies will want to publicize their generosity, because doing good things for the community helps them attract and retain customers and stockholders. If asked, be sure to provide the company with photos and team member interviews for this purpose. By doing so, you not only strengthen your relationship with a helpful company, but you also get some excellent publicity for your team through the company's publications and presentations. Such positive exposure raises the visibility of your group and may ultimately lead to additional support from other individuals or companies. It will also benefit future student teams.

Finally, don't forget about the time-honored practice of hosting a bake sale or car wash to raise seed money and awareness about your project. Keep a list of sponsors throughout the project, and when the project is done, always remember to go back and thank all the people and companies for their help and support.

TECH NOTE: *DEEP JEEP*'S MINIMAL BUDGET

Many designers and technicians have amazing stories about the challenge of creating "a safe something out of next to nothing." Will Forman's experience of designing and building *Deep Jeep*, a two-man submersible capable of operating at 600 meters (approx. 2,000 ft), is a classic example. Back in 1964, *Deep Jeep* was one of the U.S. Navy's first American-built deep sea research vehicles. It was a bare-bones project. As it neared completion, George Bond, a renowned U.S. Navy doctor and commander, agreed to review the submersible's environmental system.

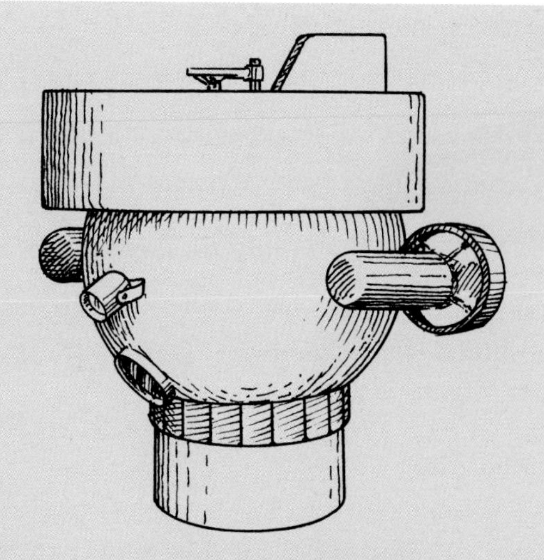

Figure 2.22: **Deep Jeep**

Dr. Bond went over every detail of the submersible's equipment, but was particularly astonished at the extremely basic system of assessing and cleaning air inside the capsule.

- A scrubber assembly consisted of four round lab sieves, commonly used for sifting and grading sand. The bottom two trays held soda lime for absorbing carbon dioxide. These assembled trays were topped off by a 12 volt DC motor and blower that had seen former service as a defroster in Forman's family station wagon.
- Carbon dioxide measurement was compliments of a World War II surplus Navy submarine unit, called a Dwyer, which cost $39.
- The submersible's oxygen indicator was an inexpensive mine safety apparatus.
- Auguste Piccard had passed on a tip about using a simple thermometer in the soda lime scrubber trays where elevated temperature would indicate whether the scrubber's chemical reactions were still taking place to reduce the carbon dioxide level.
- An aircraft altimeter, zeroed before diving, indicated pressure inside the hull and provided a backup indication of oxygen content: excess pressure (negative altitude) was the sign of excess oxygen, and vice versa.

Dr. Bond spent two days inspecting the equipment and analyzing data test results, then gave his official ok. The entire system had been constructed for some $200. Bond remarked that he had just come from a review of the *Moray* project, where the continuous read-out oxygen analyzer alone had cost $50,000!

be attracted to the team, what test and construction facilities can be used, what tools can be purchased, and the scope of vehicle operations that can be conducted. Good money management (see *Strategy C* earlier in this chapter) will help keep track of what monetary resources are available. Networking, scrounging, and in-kind donations can also help to stretch a small budget. (See *Tech Note: In-Kind Contributions.*)

3.5.7. Safety

Safety is an important consideration and potential constraint for many aspects of an underwater vehicle's design and its operation. Admittedly, very few life-threatening situations arise with a small battery-powered ROV like *SeaMATE*, but there are still some safety issues that may affect design. For example, you should always have a fuse in line with the main electrical power source to reduce the chance of a fire if some circuitry develops a short. And when working from a pool deck, marina dock, or small boat, it's always possible that someone could fall into the water or strain their back when launching or recovering your vehicle, particularly if you didn't think carefully about the placement of handles, lift points, or other mechanisms that make it easier to get the vehicle in and out of the water safely.

Getting in the habit of thinking about potential safety hazards and taking active steps to avoid them is important, not only during vehicle design, but also during construc-

Image courtesy of Steve Van Meter, VideoRay LLC

Figure 2.23: Safe Launch and Recovery Procedure

tion, maintenance, operations, launch, and recovery. Developing these habits early and using them consistently will help prepare you for bigger projects later, where safe working habits can quite literally mean the difference between life and death. For companies, universities, institutes, and agencies that have people working with subsea vehicles, safety is always a *huge* concern. Accidents occur all too easily and can result in needless injury or death, medical expenses, negative publicity, project delays, and lawsuits. If you ever decide to apply for a job in subsea technology or any other potentially hazardous field, good safety habits developed early and used routinely will go a long way toward helping you get and keep that dream job.

3.6. Stage 6: Listing Vehicle Systems

This is a quick stage in which you decide, at least tentatively, which systems your vehicle will need and maybe start thinking about what each one will contain and how it will interact with other systems in your vehicle.

You'll recall that *Strategy D: Think in Terms of Systems,* presented earlier in this chapter, introduced the flexible concept of a supersystem (e.g., your entire vehicle), a system (e.g., its thrusters), and a subsystem (e.g., the bilge pumps, props, shafts, couplings, etc.). Some of the systems commonly found on small ROVs include: Structure, Ballast, Propulsion, Power, Control, Navigation/Sensors, Payload, and Tether. The purpose of these systems is also introduced in *Strategy D,* and each is covered in much greater detail in the various technical chapters that follow.

Most vehicles will require most or all of these systems. A review of your mission tasks and mission requirements should make it fairly clear which of these systems you need for your particular mission.

Once you have your list of required systems, you are ready to proceed to the next stage—generating the concept design. As you begin that process, you may find it helpful to let different team members focus on different systems, but remember that there must be good exchange of information between the groups to ensure that each system will work well with the others.

3.7. Stage 7: Generating the Concept Design

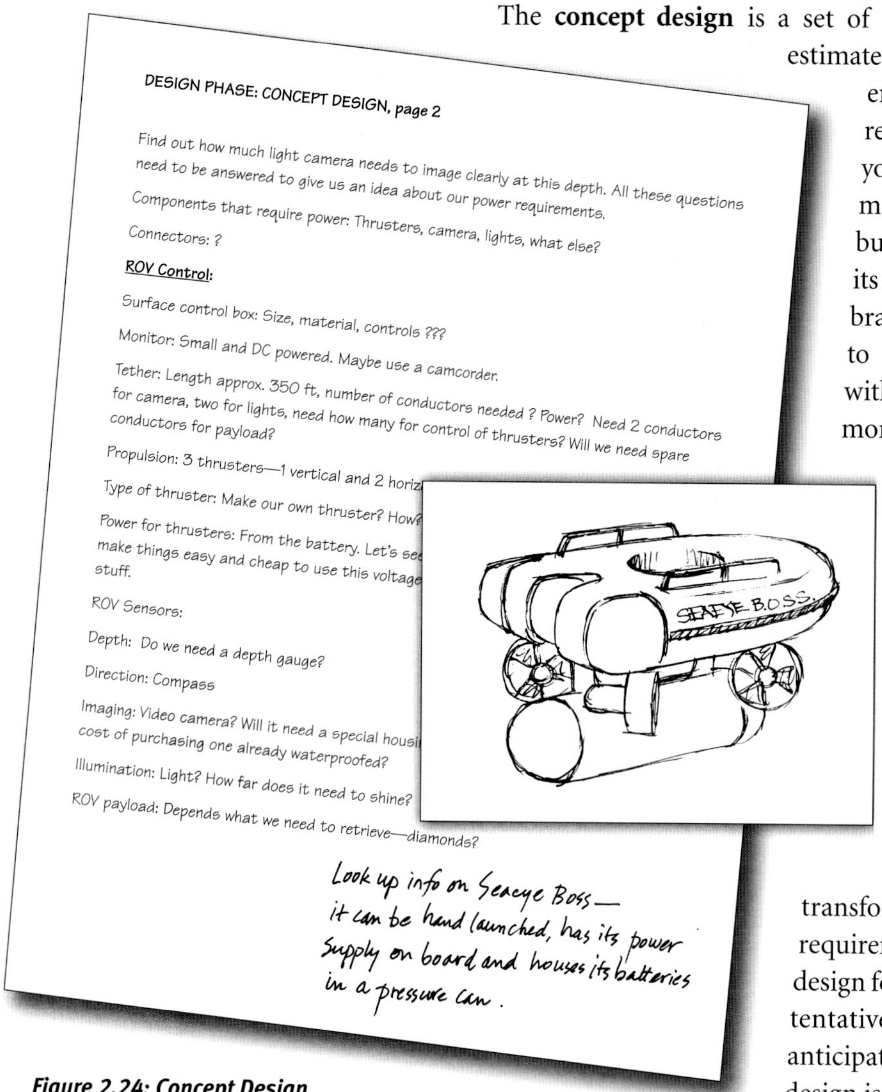

DESIGN PHASE: CONCEPT DESIGN, page 2

Find out how much light camera needs to image clearly at this depth. All these questions need to be answered to give us an idea about our power requirements.

Components that require power: Thrusters, camera, lights, what else?

Connectors: ?

ROV Control:

Surface control box: Size, material, controls ???

Monitor: Small and DC powered. Maybe use a camcorder.

Tether: Length approx. 350 ft, number of conductors needed ? Power? Need 2 conductors for camera, two for lights, need how many for control of thrusters? Will we need spare conductors for payload?

Propulsion: 3 thrusters—1 vertical and 2 horiz

Type of thruster: Make our own thruster? How

Power for thrusters: From the battery. Let's se make things easy and cheap to use this voltage stuff.

ROV Sensors:

Depth: Do we need a depth gauge?

Direction: Compass

Imaging: Video camera? Will it need a special housi cost of purchasing one already waterproofed?

Illumination: Light? How far does it need to shine?

ROV payload: Depends what we need to retrieve—diamonds?

SEAEYE B.O.S.S.

Look up info on Seaeye Boss— it can be hand launched, has its power supply on board and houses its batteries in a pressure can.

Figure 2.24: Concept Design

The **concept design** is a set of diagrams, instructions, parts lists, cost estimates, and other information that provide enough information for you to get a realistic sense of what it will take to build your vehicle. Developing a concept design may not seem as exciting as actually building a vehicle, but don't underestimate its importance. It's an extremely challenging brain game as well as the most effective way to evaluate ideas and arrive at solutions without expending a great deal of time or money.

Unlike the previous stage, which you can usually complete in just a few minutes, generating the concept design is a major undertaking. In your project planning, you should budget time for this, because it will require several days or weeks of research and team meetings to explore the options, make decisions, and write up the results.

Generating a concept design transforms your lists of mission tasks, mission requirements, and constraints into a rough design for your vehicle, complete with diagrams, tentative parts lists, estimated costs, and anticipated time to completion. The concept design is what you will evaluate for feasibility, to see if you can actually build the vehicle within the available time and budget. If not, you'll either need to revise the concept design, increase the available resources, or abandon the project.

During the concept design stage you will see, for the first time, a reasonably clear picture of what your proposed vehicle will look like when it's done. It is also during this stage that you will see the design spiral (*Strategy E*) come into play and reveal its power as a design tool.

The process of generating the concept design can be broken down into four smaller steps, two of which are repeated numerous times:

1. Entering the design spiral
2. Exploring the options*
3. Narrowing the options*
4. Completing the concept design

* *These two alternate and are done multiple times for each vehicle system.*

The overall process is summarized in Figure 2.25. For the purposes of the illustration, assume a simplified, hypothetical vehicle with only these four systems: propulsion, power, navigation, and structure. Refer to that figure as you read the text explanations of each step below. Of course, a real vehicle will have more systems (tether, payload, etc.), but the basic process is exactly the same, regardless of the number and type of systems.

In the figure, arrows trace the conceptual path you'll follow through the design spiral to arrive at a final concept design. In this example, that path enters the spiral (*Step #1*) at the propulsion system, then proceeds to the power system, then to the navigation system, and finally to the structural system before returning to the propulsion system to begin a second "lap" around the spiral.

Each time the path encounters a system, it enters through a box labeled with a plus (+) sign. This represents the process of exploring options (*Step 2*) for that system, which generally *increases* the range of possibilities you are considering. Immediately thereafter, your path enters a box labeled with a minus (-) sign. This represents the process of narrowing down the options (*Step 3*) by eliminating those that are not realistic. This generally *decreases* the number of options you are considering for that system.

Each time you complete a lap, you know more than you did before about the options you are still considering for each system. Since every one of those systems affects every other one, this additional information helps you further narrow the list of viable options for each system as you come to it. This general reduction in the number of realistic options with each lap of the spiral is represented by the shrinking size of the squares as you move toward the center of the spiral.

The process of spiraling continues (here for three laps, but in reality for as many laps as are needed) until you arrive at a single, tentative, "best" option for each system. Because the selection of the best option for each system is based on knowledge of all the other systems, the designs are likely to be compatible, so you can combine them (*Step 4*) to form the overall concept design for the vehicle.

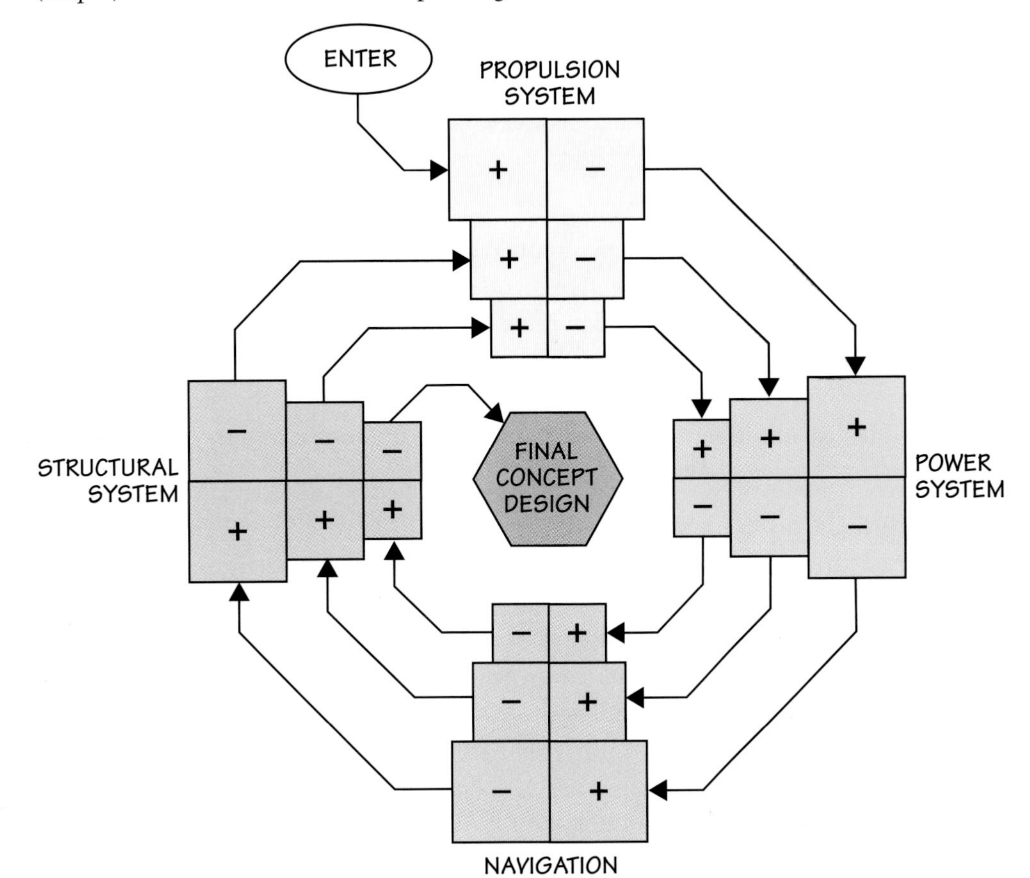

Figure 2.25: Generating the Concept Design

As explained in the text, arriving at a final concept design is a process of visiting and revising various systems, each time with more information which thereby allows you to make informed choices. Once all system options are selected, each system should be complete and compatible. The result is a final concept design of your vehicle.

In reality, there's no need to follow this spiral pathway precisely. The spiral is only a metaphor. It's perfectly okay to bounce around among the systems, modifying the design of each whenever you learn something that gives you an idea about how to improve it or make it more compatible with another system. The main point is that you want to "grow" all systems in parallel, letting the evolution of each system's design inform the evolution of the others. To look at it another way, the various groups working on each system should talk to each other frequently, exchanging information so no group's design ends up being incompatible with another's.

Regardless of whether you interpret the spiral literally, visiting each system exactly once per lap and always in the same order, or whether you wander casually among the systems, you'll follow the same four steps: 1) entering the design spiral at some point, 2) alternately expanding and 3) narrowing the options for each system multiple times, and ultimately 4) combining the best designs for each system into the final concept design for the vehicle. These four steps are described in more detail with examples below.

3.7.1. Stage 7, Step 1: Entering the Design Spiral

One of the first difficulties you'll encounter when developing a concept design is figuring out where to start. After all, your vehicle is a complex system with multiple systems. Is it best to start with the propulsion system? The power system? Some other system?

Rest assured that it doesn't matter too much where you enter the design spiral, so don't lose sleep over this decision and don't spend much time on it either. Remember, the power of the design spiral is that you'll be revisiting each system multiple times, refining its details based on what you've learned about the other systems during previous "laps" around the spiral. By the time you complete a couple turns, so much information will have been shared among all the systems that it won't really matter where you started.

Still, you must start somewhere. It often saves time to let the vehicle's mission guide your entry point. Go back to your mission task list (*Stage 3*) and ask, "What is the primary mission task the vehicle has to perform?" Start at the system most directly involved in performing that task.

Frank's class decided their prime mission task was retrieving the bozonium canisters. To do so, they would need an effective grabber device to pick up the canisters. As a result, they entered the design spiral at the payload system. The most crucial decision the students had to make was deciding if the payload tool would be a simple probe or an elaborate manipulator. This decision would affect the type of thrusters needed, the amount of ballast and flotation required, the size of the frame, and the complexity of the control and power systems.

Not surprisingly, it's common for team members to have different opinions about where to start the design, even when all agree on the nature of the mission. A typical example is a team of naval engineers designing a torpedo. Basically, they are crafting a high-speed AUV with a one-way mission—to move stealthily through the water, find, hit, and destroy a target. To do that, the torpedo has to get from its launch tube to the target as quickly as possible. The mission calls for speed, in which case the propulsion system is an obvious entry point. However, other designers might argue that speed is important, but hitting the target is trickier to do. For them, the prime mission task is guidance, so they would choose to work on the navigation system first. Others might argue the explosive payload is the most important system; after all, there's no point in getting to the target quickly if the warhead fails to explode. The point is that there can be many valid opinions as to what the prime mission task is and many (equally valid) places to enter the design spiral.

If your team can agree on a logical entry point following a modest amount of discussion, that's good. If not, don't let this non-critical decision delay progress. Flip a coin or draw straws to select where you'll enter the design spiral, then move on! Remember, you'll get a chance to visit *every* system during the first lap around the design spiral, and again during every lap thereafter.

TECH NOTE: CHOOSING OPTIONS

As you progress deeper and deeper into the design spiral, notice that there are generally at least two ways to accomplish each thing. While it's good to have options, making the best choice can be difficult. In the *SeaMATE* camera example above, there are two methods for waterproofing the inexpensive video camera option—a pressure housing made from PVC pipe fittings or encapsulation. A third option involves a more expensive off-the-shelf underwater camera. To make the best choice, ask the question, "Which option conforms most closely to the design constraints?"

Here are the answers:

- Option A: PVC housing conforms to all constraints except robustness. There is a danger of flooding the camera.
- Option B: Encapsulation conforms to all constraints—there is a deeper depth rating and no danger of leakage. But with no experience with encapsulating procedures and without knowing the best compound to use, you would need to research and practice encapsulation techniques before choosing this option.
- Option C: The OTS underwater camera has a higher price, but this might be offset by reducing other project costs.

Obviously, you still haven't made an informed choice as to which option to choose. There are still some unknowns about the encapsulation procedures and the price of the OTS underwater camera, but further testing and research should sort that out.

Let's say you tried various potting compounds and found that encapsulating is really not difficult to do. And after an internet search, you couldn't find an off-the-shelf underwater camera in your price range. So now you are down to Options A and B. Option A requires some machining to construct the camera housing, but with no access to a lathe, this option is effectively eliminated. That leaves Option B. Having already tested out encapsulation techniques, you feel confident in building a robust underwater camera—and Option B also conforms most closely to the important constraint of low cost.

Just remember to make these decisions systematically, one subsystem at a time, one option at a time. As you come up with solutions to the issues in each subsystem, the design spiral tightens up, and you approach the completed concept design for your vehicle.

Keep in mind that this process will take you farther and faster once you have read and understood the technical chapters. You may want to read those chapters before you start your project, then return to this chapter for a refresher on how to progress through the various stages of your project.

3.7.2. Stage 7, Step 2: Exploring the Options

Suppose that you decide to enter the design spiral at the propulsion system as shown in Figure 2.25. The first thing you'll want to do is gather information about the range of possible ways you might propel your vehicle through the water. The goal when exploring options, particularly during the first lap around the spiral, is to make sure you don't overlook any design ideas that might be the ideal solution for your propulsion needs. Therefore, you'll want to cast a wide net and keep an open mind. Several approaches can be used in parallel to gather the information you need.

It's always productive to start with a **brainstorming** session where you and your teammates get together and offer suggestions for possible versions of each system. The idea behind brainstorming is to develop an extensive list of options to consider as a starting point. Brainstorming is most productive if suggestions are not critiqued during the session because that may inhibit the free flow of ideas. Just list the ideas as they come up and plan to evaluate them later. For each idea, you may want to develop a list of related keywords that could be used to search for additional information about that idea. The results of a brainstorming session on propulsion might include dozens of different ideas. A hypothetical list might include the following suggestions:

- waterproofed model rocket engines
- fish-like fins or flippers
- propellers
- glider-like wings
- Caterpillar tracks or wheels
- trained dolphins towing the vehicle
- pressurized gas
- water jets

Some of these ideas might seem ridiculous, but that's okay. The completely unworkable ones will get filtered out naturally as part of *Step 3: Narrowing the Options*. For now, a long list of diverse and creative ideas is more likely to produce a winning ticket than a short list restricted immediately to "sensible" ideas.

Once you have brainstormed a list of possibilities, it's time to do some research on any that look even remotely promising. Sure, your team might not have any trained dolphins this week, but don't be too quick to scrap that idea; perhaps it makes sense to consider pulling the vehicle along using a boat or other conveyance? Or maybe some research on how dolphins swim so efficiently can provide insights you could use to design a more efficient propulsion system for your ROV.

One of the best places to begin such research is in books, trade journal (magazine) articles, project reports, and website entries. Read up on basic background material and learn what other people have done with similar kinds of projects. This is commonly known as a **literature review**. It's valuable because it provides insight into the types of problems others have encountered when tackling similar projects and how they've solved those problems. Building on the research and activities of others can often save considerable time and money and will usually turn up some great ideas your team may not have thought of on their own.

Throughout the design spiral, but particularly during that first critical lap, it's important to keep an open mind and to search broadly rather than narrowly. For example, even if you know you're going to build an ROV, don't limit your search to only ROV projects. Look at AUV projects, terrestrial robot projects, biological examples, and any other avenues you can think of that might provide ideas you could use or adapt for your vehicle.

The fact that you are reading these words means that you have already started your literature review! And you will learn more information that will be helpful in identifying options and constraints for each vehicle system as you continue into the technical chapters. But don't limit yourself to just this textbook or only to written sources. You can extend your "literature" review beyond the printed word by contacting knowledgeable people by email or by phone. Keep in mind that you may be dealing with busy people who can't get back to you right away or who may be away at sea, so remember to be polite and patient. *Appendix VI: How to Find Additional Information* lists some great ideas for finding information on subsea technology and related subjects.

An effective literature review will invariably lead you to ideas for specific products or product categories about which you'll want to gather more detailed information, including prices, features, and availability, before you purchase anything. Gathering this information is known as **product research**. For example, one ROV team's website may suggest a particular brand and model of video camera that they used for a simple navigation system, but you'll want to review that camera as well as others with similar specifications as you try to decide which is best for your project. You may find a newer, better camera that wasn't even available when that team built their project. Or the website might have introduced you to a type of electrical connecter you've never heard of before, inspiring you to investigate those (and related) connectors in greater detail to see what opportunities they might provide for your project.

Image courtesy of Dr. Steven W. Moore

Figure 2.26: Product Research

It's fun and informative to search catalogs and websites for information about parts, materials, or other products you might use in your underwater project. Remember to search widely, since parts not originally intended for underwater use might be adapted or spark new design ideas in your mind.

Hopefully, you've already browsed the aisles of the local hardware stores, electronics shops, and marine suppliers. This is another great form of product research and, more importantly, a source of ideas and inspiration for adapting off-the-shelf items to your vehicle's needs. While running laps in the design spiral, it's good to head back to those stores again, each time bringing a more focused list of items to investigate for pricing info and for testing out ideas.

Other sources of product information include the websites of companies that produce and sell metal and plastic materials, small robots, electronic components, hobby supplies, and subsea equipment. Try and get hard-copy catalogs, as well—admittedly most product info is available on-line, but there still is something magical about actually leafing through catalog pages.

If your team gets involved in a formal robotic vehicle competition or you get to observe one, take advantage of the opportunity to see how other teams have configured their vehicles and solved problems—that's one of the true advantages of these nail-biting events and an excellent form of research. Many heads are always better than one, and you're bound to come away with new ideas and enthusiasm for your *next* vehicle.

Whichever resources you use, remember to record all potentially useful specification data for anything that looks promising while you have the resource information in front of you. A critical aspect of research is keeping track of this flood of new information, so use a project notebook and some sort of filing system to record it. (See *Strategy C: Be Proactive about Project Planning, Management, and Safety.*) You don't need to write down everything about every possible product, but do be sure to note anything that might help you make a final decision, such as size, weight, materials, mounting options, availability, and, of course, price. That may seem like a lot of work, but it's not nearly as hard (nor as frustrating) as trying to go back and track down some piece of critical information you know you've seen before, but didn't bother to record at the time.

3.7.3. Stage 7, Step 3: Narrowing the Options

As you conduct your literature review and product research, you'll find some doors opening and beckoning you to explore further, while others slam shut. For example, you might learn about a new way to waterproof cameras that you had never thought about before, opening a door to far more camera options than you originally thought possible. On the very same day, you might learn that one of the propellers you were thinking of using for propulsion is no longer available, thus closing that door.

Generally speaking, literature reviews and project research open more doors than they close, so at some point, you'll need to get serious about boiling down a broad list of possibilities to a short, specific list of things you're actually going to buy or build. To some extent, you can do this as you go along, essentially performing Step 2 and 3 simultaneously. After all, there's no need to spend time researching more sonar system options if you discover early in your product research that even the least expensive sonar systems cost ten times more than your entire project budget. However, be careful about narrowing the search too early or you may miss some great opportunities.

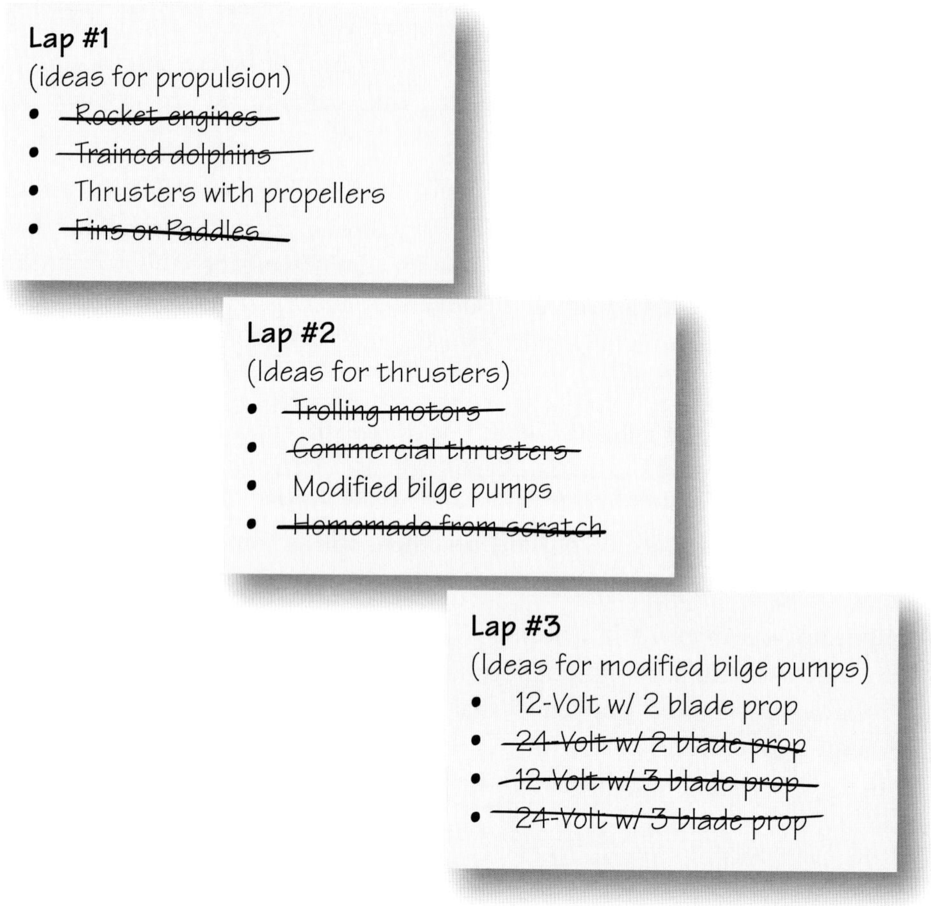

Lap #1
(ideas for propulsion)
- ~~Rocket engines~~
- ~~Trained dolphins~~
- Thrusters with propellers
- ~~Fins or Paddles~~

Lap #2
(Ideas for thrusters)
- ~~Trolling motors~~
- ~~Commercial thrusters~~
- Modified bilge pumps
- ~~Homemade from scratch~~

Lap #3
(Ideas for modified bilge pumps)
- 12-Volt w/ 2 blade prop
- ~~24-Volt w/ 2 blade prop~~
- ~~12-Volt w/ 3 blade prop~~
- ~~24-Volt w/ 3 blade prop~~

Figure 2.27: Homing in on a Concept Design

The process of arriving at an optimal design for each system is an iterative one—that means it's one you do again and again, generally improving as you go. In the first lap around the design spiral, think big, exploring a broad range of ideas, then narrow that list to one (or two) most-promising options. During the next lap, research the range of possibilities for the chosen option before narrowing that second list to one or two top choices. Continue the process in subsequent laps, each time getting more focused and specific, until you arrive at a suitably precise concept design.

The checklist that follows presents various considerations you can use to help narrow your array of options to realistic ones. Not surprisingly, some of these topics may remind you of those discussed earlier in *Stage 5: Identifying Constraints.* This is where those constraints come into play, helping you to evaluate and refine your list of options.

Price: Many otherwise desirable or promising options will have price tags that exceed your total project budget. These must be removed from consideration, unless you can get them donated or can find a way to increase your budget. Other items might fit within the total project budget, but be so expensive that they would seriously undermine funding for other systems. Decisions about these items must be made carefully by the team in the context of the entire project, weighing the pros and cons of spending money on each system at the expense of others. For example, you might discover that if you go with the cheaper of two workable camera options, you can afford an upgrade to more powerful thrusters, which would allow your vehicle to work in stronger currents. As always, let mission needs drive the decision.

Availability: Some items may be out of stock or not available in your area. You might have the option of waiting, or having it shipped from farther away, or you might not. If not, scratch those items off your list.

Relevance to mission success: Some items or item features you uncover in your research may be very cool and highly desirable (emergency flare launchers, day-glo paint, etc.), but if they contribute little or nothing to mission success, you should be wary of spending team resources on them. Don't underestimate the value of such things for boosting fun and team morale, but keep your eye on the target and make sure you don't expend resources on non-essential items or features at the expense of something critical to the mission.

A good example might involve a decision about whether or not to buy a nice color video camera for your ROV or go with a less expensive black-and-white model. If the mission requires you to identify fish species based on subtle color differences, then the ability to see color is critical to mission success. On the other hand, if your mission is to locate a brick on the bottom of a swimming pool, the black-and-white camera will suffice.

Assembly Time: In some cases you may have the option of buying something or building it yourself from scratch. For example, you might be able to buy a thruster or build one yourself. While the latter approach can sometimes save you money, it can also be extremely time-consuming, particularly if it's something you've never built before. If time is a limiting factor (and it usually is), the build-it-yourself option may not be a realistic one.

TECH NOTE: COMPARING OPTIONS

One task for Frank's class was to locate the bozonium canisters on the bottom of Monterey harbor. A sophisticated, high-resolution sonar system could probably do this, but it would exceed the available budget and would very likely be beyond the technical expertise of the team members to install and use properly. Therefore sonar would not be a *realistic* option for completing this task.

On the other hand, a modestly-priced video camera probably would be a realistic option, if it could be made to work under water. Research on this front might have uncovered the following three video camera options:

1. An inexpensive video camera enclosed in a homemade pressure canister.
2. An inexpensive video camera encapsulated within a block of clear epoxy.
3. A modestly priced commercial off-the-shelf (COTS) underwater video camera.

Although each of these options might be realistic, they are not all the same. Option 1 would take lots of time to build and might leak, but it would be a great learning experience. Option 2 would probably be quick and easy, but it would permanently entomb the camera, making it impossible to adjust or repair later. Option 3 would also be quick and easy (if there were no delays in ordering and shipping), and it might be more robust and reliable than a homemade version, but it would probably cost more and leave less money available to spend on other systems.

Once you've made an accurate assessment of the various options, you'll also want to consider how each of them will impact other systems. When that's completed, you can then proceed to make a sensible choice, based on your team's abilities and desires as well as your project's mission and budget.

Required expertise: Some otherwise enticing options may simply be beyond the technical capabilities of your team. It's good to challenge yourself, because that's how you learn new skills, but don't overdo it. You should probably wait a project or two before pursuing the nuclear-powered, hyper-turbo drive system that came up during the brainstorming session.

Compatibility: One of the biggest constraints that will narrow your options is compatibility with other vehicle systems and components. Here is where the design spiral really comes into play. For example, the promising camera options you've discovered might include 5-volt, 12-volt, and 18-volt models, but the only viable thruster motor option you found might be a 12-volt model. If you want your entire vehicle to operate from a single supply voltage, then the thruster option will, in fact, constrain your camera options.

Prototype test results: Alas, few things ever work exactly as expected, and some don't work at all. Therefore, whenever possible, it's a good idea to test design ideas to make sure they will work for your vehicle.

If you or someone you know already dabbles in similar projects and has a garage full of spare parts, you may be able to cobble together prototypes to test ideas without having to buy anything. If not, you might be wise to invest some of your project money and time in purchasing and testing items before you commit to using them in your design.

One team in a competition forgot this proviso and built a new type of thruster design that they presumed would be superior to the standard prop-and-bilge pump motor arrangement. They were so sure it would work better that they built *eight* of them. To their dismay, these new thrusters didn't seem to be putting out much thrust when it came time for sea trials. So the team finally did some testing to compare the force of the new thrusters to that of the standard ones. Surprisingly, the standard units outperformed the new ones! If the team members had done this test on a prototype thruster beforehand, they would have saved themselves a hundred hours building eight thrusters that just didn't measure up—a hard lesson learned. (To their credit, the team doggedly put in the hours to fix the problem and ended up winning first place!)

3.7.4. Stage 7, Step 4: Completing the Concept Design

After several laps around the design spiral, you will have uncovered and considered a wide variety of possible designs for each system, and you will have narrowed the list in each case to a much smaller set of realistic alternatives.

The final round of narrowing the options consists of identifying the "best" option from among the available alternatives for each system. Of course, this decision cannot be made for any one system without simultaneously considering its impact on all the others.

Congratulations! By the time you reach this final stage of completing the concept design, all the major design issues have been addressed, each of the systems is now nearly complete in terms of specific details, and you're fairly confident that the various systems should interlink without major problems. In other words, the design spiral should be converging on a final design, and you should be getting a good sense of what the vehicle is going to look like, how big it's going to be, and what it's going to be capable of doing.

Take a minute to look at what you've accomplished so far! You will have made informed decisions about the details of each of the various systems, with much of that information coming from reading the technical chapters of this book and your other research. So you know the specifications, prices, availability, and delivery times for your major vehicle components as well as for a set of reasonable alternatives. You have also researched or tested the feasibility of any innovative ideas your team may have generated. In other words, you're a heck of a lot smarter and further ahead with your vehicle before you've even started to build it! This is a huge accomplishment.

All that remains to finish off the concept design is to complete the following tasks, which are detailed below this list:

- produce a description of your vehicle's major technical features—these are called functional specifications
- generate concept drawings of what the vehicle will look like
- estimate a project budget
- estimate the time to complete the project
- evaluate whether it is feasible to proceed to fabrication

Functional Specifications: "Specs," or more completely, **functional specifications**, provide a synopsis of the main technical features and performance capabilities of the vehicle. Essentially they are the "boiled down" answers to the mission requirements issues that you got after going through the design methodology procedure. See *Tech Note: Functional Specifications for SeaMATE* for an example of a typical list of specifications. Don't worry if you don't know what some of these things are yet or don't know how to figure them out. You will by the time you finish reading the technical chapters.

Concept Design Drawings: Aside from keeping written notes in your notebook, you've probably been making diagrams. Now is the time to synthesize those sketches and notes to make a **concept drawing**. A concept drawing should be accurate in terms of what the vehicle design looks like, but it is not necessarily meant to specify where each nut and bolt should be. Those kinds of detailed drawings are done once the concept design has been approved for fabrication.

Estimated Project Budget: In order to decide if a design is feasible to build, it's essential to know its projected cost. If you've been recording the prices of each major component, you can easily devise a simple spreadsheet that lists all these vehicle items; you'll also want to include any other major expenses related to fabrication (such as sea trials, transportation, and shop costs). This is just a preliminary estimate, so don't worry if you can't account for everything—just make a reasonable guess so you can come up with a total. The bottom line should reveal if this vehicle falls within your financial resources—in which case you can proceed! If not, revisit the design spiral to see where it is possible to reduce vehicle costs.

Estimated Completion Time: One of the toughest project tasks is estimating the time to complete a project.

Figure 2.28: Composite Drawing Showing Perspective, Side, and Top Views

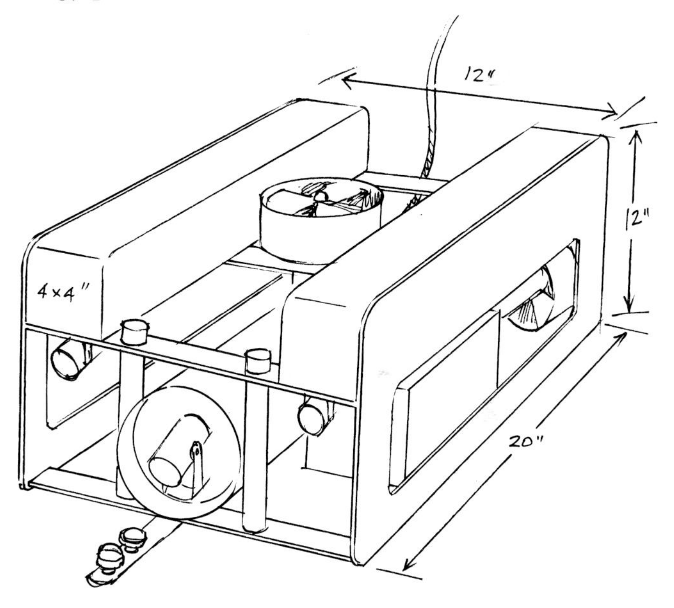

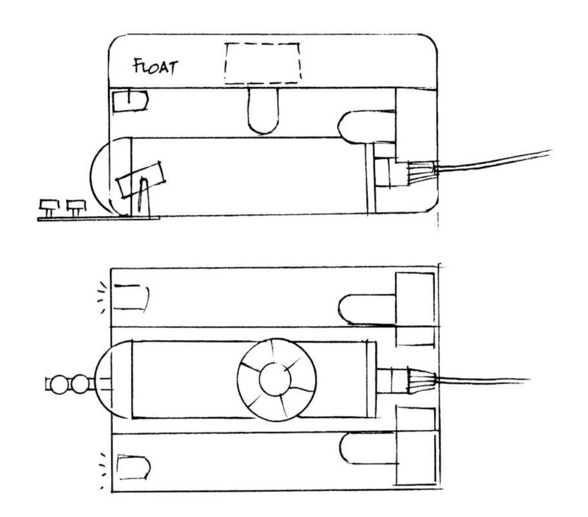

TECH NOTE: FUNCTIONAL SPECIFICATIONS FOR *SeaMATE*

Note that the order of presentation of these functional specs is arbitrary. Imperial measurements are presented first, sometimes followed by metric equivalents.

Operational Depth: 33 ft (approx. 10 m)

Collapse Depth: unknown

Operating Environment: fresh or saltwater, pool, marina, lake, or slow river with ROV launched from a fixed or floating platform (shore, boat, wharf, etc.)

Dimensions: (approx.) 12 inches (30 cm) long by 12 inches (30 cm) wide by 10 inches (25 cm) high

Displacement: (amount of water displaced by submerged vehicle): (approx.) 1 pound of freshwater

Weight in Air: (approx.) 4 pounds (17.6 N)

Floats: two sections 15 inches long x 1 ½ inches (approx. 38 cm x 3.8 cm) Schedule 40 or Class 160 PVC pipe with glued end caps

Ballast Weights: coated lead or steel weights

Power: 12 to 14 V DC, 8 amps from power source

Endurance: dependent on the capacity of surface power source

Propulsion: three Johnson 500 gallons-per-hour bilge pump motor cartridges, spinning #3003 Dumas two-bladed props

Velocity: 1.6 feet per second (approx. 0.5 m/s) in forward direction

Maneuvering: up, down, forward, reverse, turn right and left, spin right and left

Control: three miniature DPDT (double pole double throw) switches hardwired via tether directly to each of the three thrusters. Switches control only full on and off, forward and reverse rotation of each thruster (no variable speed). One miniature SPST (single pole single throw) switch for ON-OFF control of camera and light.

Structure: 1/2–inch Schedule 40 PVC pipe and and fittings, with mounting brackets for thrusters, camera, and optional compass/depth gauge

Sensors: black-and-white CCD (charge-coupled device) video camera in a waterproof housing and one 12 V reflector light in waterproof housing with viewing on surface video monitor. Other sensor options: compass and depth gauge viewed by camera

Tether: 40 feet (approx. 12 m) in length, consisting of three pairs of thruster conductors, 1 video coaxial cable, 1 pair conductor for light(s) or spare

Navigation: by sight, using camera, compass, and depth gauge

Payload: one 6-8 inch (15 – 20 cm) plastic chopstick or metal rod to function as a probe for retrieving small objects

(See *Strategy C: Be Proactive about Project Planning, Management and Safety* earlier in this chapter.) Usually any first estimate of timing is wildly optimistic and doesn't allow for delivery delays, fabrication problems, sea trials, personal circumstances, and the like. At a minimum, we suggest doubling your first time projection to get a realistic estimate. To be even more conservative, triple it. If this estimated completion date is past the mission deadline, then revisit the design spiral to see how to reduce fabrication time or simply schedule more working hours. Once you've reached a workable estimate, fill in the calendar with clear milestones and realistic deadlines to ensure the vehicle will be completed by the project deadline.

Final Evaluation: After completing this design procedure and producing a great concept vehicle, it may seem strange to think that there might be a question as to whether or not you are ready to build it. In fact, some large projects get to this stage and the decision is made to pull the plug—the design concept does not appear realistic for the mission, budget, or time constraints. That's why it's important to evaluate your work before taking the next step of committing money and time for fabrication. So evaluate your concept design using these key questions:

- Have you included all the systems necessary to accomplish the mission?
- Has the research on these systems been comprehensive?
- Has there been good communication among team members so ensure all the systems will work together comprehensively?
- Is the design feasible in terms of time and budget?

Negative answers to any of these queries should encourage revision of the weak aspects of the design. The real trick is to make these changes within your budget and time frame, rather than getting ensnared by the revision process. Sometimes it may be better to accept what you have, even if it is not perfect, as long as the vehicle design is capable of accomplishing the mission.

3.8. Stage 8: Fabricating the Vehicle

The **fabrication** stage is the really exciting one, where those concept drawings finally come to life on the workbench or shop floor. Let's take a closer look at these distinct steps in fabrication: detail design, procurement, construction, and troubleshooting.

3.8.1. Stage 8, Step 1: Detail Design

Once the concept design is approved, the next major step is to take those concept drawings and draft a set of detailed technical drawings that show exactly where each nut and bolt, so to speak, is to go. These are the drawings you'll use on your workbench to build and assemble each of the vehicle components. There are several basic types of these technical drawings—those that show how a part is made are called **fabrication drawings**, while those that show how parts are put together are called **assembly drawings**. Electrical plans are called **schematics**.

Simple vehicle projects don't require a full set of technical drawings prior to construction. For example, *SeaMATE* was assembled on the workbench simply by laying out all the pipe fittings, thrusters, and camera components. The frame layout came together using only a ruler and pencil and by dry-fitting the pipe pieces together. This process is called **mocking up**. The advantage of mocking up is that you see exactly how everything is going to come together without doing a lot of calculations. You simply "see as you go" and trial-fit in place as you build. The disadvantage with mocking up is that it usually takes longer to correct an error if you make a mistake. If you're designing on paper or using a CAD (Computer Assisted Drawing) program, such as SolidWorks®, changes are as easy as erasing a line or undoing a computer command. Note that if you're going to have something fabricated at an outside machine shop, they generally will need detailed technical drawings in order to make the parts, so it might be helpful to have both—a limited set of technical drawings and a scaled mock-up.

Producing a materials and components list is an important part of your detailed design. This list specifies every part and type of material you are going to need. Remember to list tools that will have to be purchased.

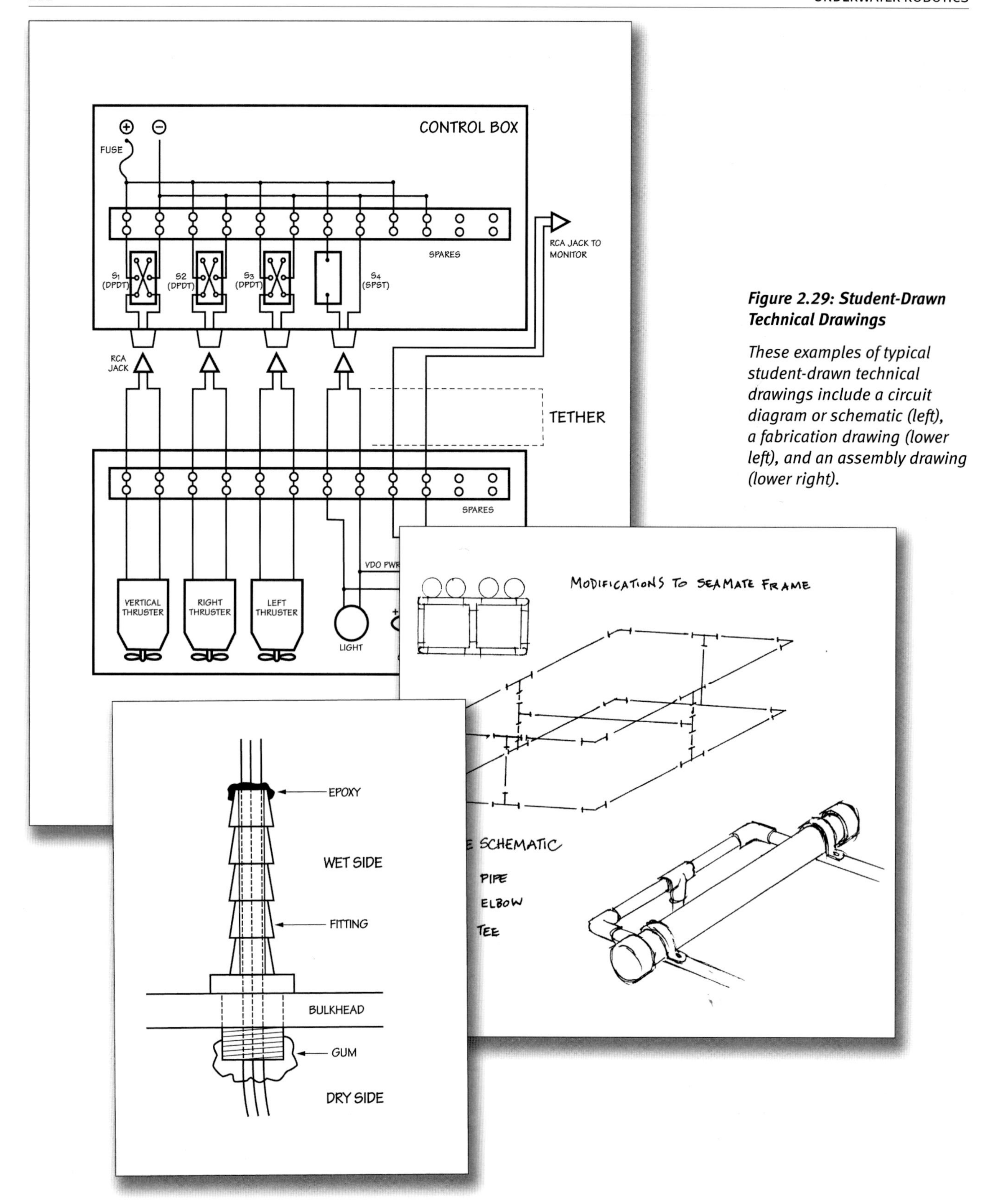

Figure 2.29: Student-Drawn Technical Drawings

These examples of typical student-drawn technical drawings include a circuit diagram or schematic (left), a fabrication drawing (lower left), and an assembly drawing (lower right).

3.8.2. Stage 8, Step 2: Procurement

When you've completed the detailed technical drawings and generated a materials/ components list, it's finally time to order stuff. As well, remember to assess your workspace and purchase the tools needed to assemble the vehicle. This purchasing process is called **procurement.** In addition to buying the stuff, a huge part of project management is tracking what is bought and recording how much it costs. (See *Strategy C: Be Proactive about Project Planning, Management, and Safety* earlier in this chapter).

It's important to do this type of record keeping, because it's very easy to run into cost overruns at this stage, especially if your initial estimates for component prices have changed or if an item is no longer available. In such cases, it may be necessary to revisit the design spiral to modify the design or find an alternative solution.

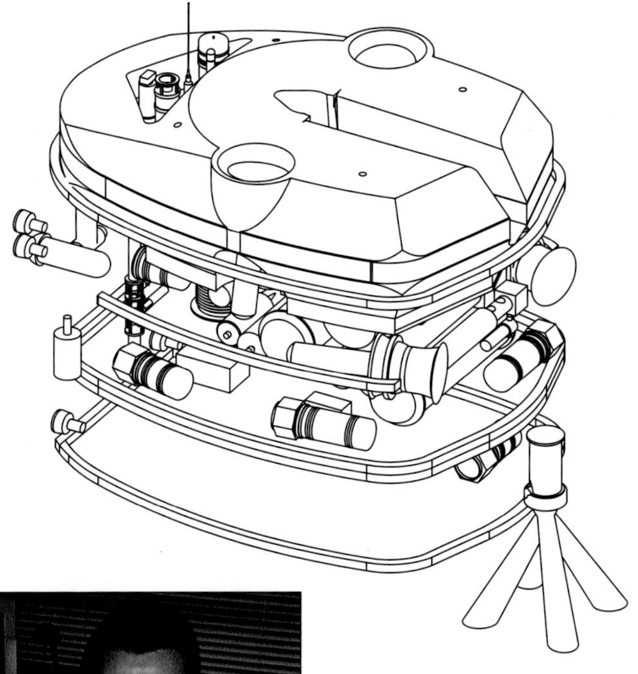

3.8.3. Stage 8, Step 3: Construction

Finally, the moment has come. The plans are finished, the parts have arrived, and your workspace is set up. Now it's time to start building!

Some teams do things in order and one at a time—they assemble the frame, then the wiring, then the tether, and so on. Although it takes longer, the advantage is that each system assembly dovetails nicely into the next. Any errors you make are only in one place and are easily caught.

Others prefer to use a systems approach, delegating jobs to various team members, then piecing together the completed systems. The advantage here is time savings. The downside is that things can get complicated when the group working on one system encounters problems and is forced to modify the design in a way that renders it incompatible with one or more other systems. Good communication may alert the other groups in time for them to make appropriate adjustments, but if they've already finished the affected part of their system, they'll have to undo some of their handiwork. For example, if the person constructing the manipulator is forced to make some changes in the dimensions of the mounting bracket at the point where it would connect to the frame, and the frame guy has already glued the structure together according to the original dimensions, the manipulator may not fit onto the frame—and suddenly there's a big problem.

Figure 2.30: Three Versions of Tiburon

MBARI's Tiburon *in three states: an early CAD drawing, Bill Kirkwood holding a LEGO model, and the real ROV.*

Images courtesy Bill Kirkwood (top) and Vickie Jensen

Such fabrication changes, or revisions, are inevitable. To minimize issues, team members should build their assembly precisely to the specs whenever they can, so that their system dovetails with the others. If changes are necessary, be sure to tell everyone else affected in the group via the project manager. A multi-pronged assembly approach can be effective only if the project manager oversees that each part is made accurately and according to plans, and that any necessary revisions are brought promptly to the attention of the whole team.

Vehicle construction often takes a lot longer than anticipated. Sometimes there are technical setbacks. Sometimes parts arrive later than expected. Sometimes component assembly turns out to be more complicated than planned. Any number of things can—and will—slow the process down. In order to get the vehicle finished on time, start early, adjust the schedule, adapt the budget, get more people involved, and put in more late nights at the workbench. The end result should be worth it.

Image courtesy of Dr. Susan Crawford, Naval Undersea Museum Foundation

Figure 2.31: Fabrication—At Last!

Most ROV builders see fabrication as the most exciting stage. This is where advance research and planning pays off, helping to ensure that the process of building the vehicle is as stress-free as possible.

3.8.4. Stage 8, Step 4: Troubleshooting

At every opportunity during construction, you should test the parts and systems you are working on to **troubleshoot** (i.e., diagnose and fix) any problems. It's much easier to fix little problems as you go than it is to wait until you have accumulated dozens of problems, some of which may be buried deep inside a fully assembled vehicle. Testing as you go also makes it easier to pinpoint the source of a problem. For example, if early testing showed that part A worked and part B worked, but they don't work now that they've been connected together, then the problem is probably related to the connection between them, rather than something specifically in part A or part B. If you didn't test A or B before hooking them together, you won't have a clue whether the problem was in part A, part B, or the connection.

Troubleshooting is like doing the work of a detective or scientist. First you try to come up with some possible explanations for the problem. Then you devise tests or experiments to determine which of your guesses is correct. For example, if a thruster doesn't work, the problem could be that the battery is dead, or that a wire is broken, or that the motor is burned out, or that the propeller shaft is jammed, or that it isn't receiving the proper commands to turn on. Swapping in a fresh battery tests the possibility that the battery is dead. Trying a different set of wires tests the idea that the wires are broken, and so on. This type of systematic troubleshooting approach can quickly pin down the source of any trouble.

3.9. Stage 9: Conducting Sea Trials

Finally, the last bolt has been tightened and the final wire soldered. You've completed your preliminary troubleshooting and everything seems to be working fine—at least on the workbench. But now you need to conduct systematic reliability and endurance tests in the water under conditions similar to the real mission conditions. These in-water tests are called **sea trials**. The point of these trials is to reveal any remaining

TECH NOTE: BACK TO THE BASICS

"Never put a screen door in a submarine." Ron Roehmholdt, Naval Undersea Museum, Keyport, Washington, USA.

As an undersea technology educator, Ron Roehmholdt often uses off-hand humor to challenge his students' perceptions of what is and isn't important in solving technical problems. He knows that when faced with a frustrating problem, it's easy to forget the most obvious or simple solutions. For example, if you turn on a TV or computer and nothing happens, it's easy to assume the problem must be something serious. However, sometimes the solution can be as simple as discovering that the device is unplugged or that the circuit breaker has tripped and cut off the power.

Finding the solution to most problems should begin with simply going back to the basics. And that is what Ron Roehmholdt is pointing out with his screen door analogy. When you build an underwater vehicle, always remember the fundamental fact that water is wet, so don't design or build stuff on a vehicle that will act like a screen door and let water in.

Image courtesy of Steve Van Meter, VideoRay LLC

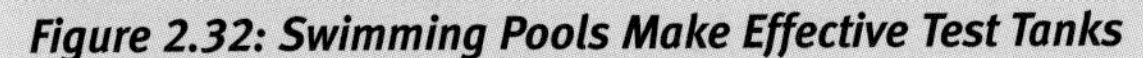
Figure 2.32: Swimming Pools Make Effective Test Tanks

TECH NOTE: SWIMMING POOLS

Swimming pools and even hot tubs are popular test facilities for small ROV projects. However, be aware that this type of water usually contains chlorine-based disinfectants. While not as corrosive or electrically conductive as seawater, chlorinated water can still cause shorting and corrosion problems. Less corrosive options include a bathtub, a wading pool filled with tapwater, or a freshwater lake. Remember also that the chemical disinfectants in pools or hot tubs will slowly attack the molecular structure of most plastics, so it's a good idea to give your vehicle a rinse in tapwater after testing it in a pool or hot tub. (Tapwater also contains disinfectants, but at much lower concentrations.)

deficiencies in your vehicle that need to be corrected before it attempts the mission. (See *Chapter 12: SeaMATE* for a more detailed explanation of sea trial activities.)

If possible, begin sea trials in a freshwater environment, such as a large test tank or lake. That way, if there is a leak, there's less chance of major damage from a shorted circuit. (Saltwater is highly conductive and corrosive, so most electrical components become worthless once wetted with saltwater.) Final trials can then be conducted in saltwater if that is where the vehicle has to function.

Sea trials often reveal minor problems that send you back to the workshop. Larger problems may send you back to the drawing board before you can go back to the workshop, and that can take a lot of time—another reason to be conservative when planning your project schedule. Once sea trials have been completed successfully, your ROV should be ready to perform the mission for which it was designed and built.

3.10. Stage 10: Carrying Out Operations

This is the thrilling (sometimes also nerve-racking) stage where you put your vehicle to work and conduct its mission. *Chapter 11: Operations* will describe in detail how missions are run, so we will not dwell on those details here. Specifically, Chapter 11 will explain how you can use simple observation-class ROVs to conduct shallow-water inspections or surveys and even salvage sunken objects in shallow bodies of water. For some team members, the operations stage marks the end of a long journey, albeit a rewarding one. For many, however, the end of one project is just a stepping stone to the beginning of the next project, usually a more advanced one with even greater challenges and more exciting opportunities! (See *What's Next? Going Beyond SeaMATE* near the end of Chapter 12.)

3.11. Stage 11: Evaluating Ops and Writing a Report

Most design projects conclude with a final **project report** or similar documentation. Teachers, contest organizers, and employers frequently *require* a formal written report or an oral presentation about each project, including a summary of the budget and a detailed technical description of the vehicle and its performance capabilities. Even if

Figure 2.33: Sample Project Reports

Reading project reports from other teams is an excellent way to get ideas for manipulators, cameras (including types, housing options, pan-and-tilt mechanisms), thrusters, etc., as part of your research.

Also check out the "troubleshooting" and "lessons learned" sections of these reports to avoid costly mistakes, thereby saving both time and money.

you're just doing the project for fun and nobody is *requiring* you to submit a final report, you will probably want to write one for your own (or others') future reference. Otherwise you might have to repeat time-intensive research or reinvent wheels you've already invented once.

A typical project report for a small undersea vehicle consists of a well-organized summary, roughly 10–20 pages long (including figures and tables), of key information about the project and the vehicle. Such reports usually include most or all of the information listed in the *Tech Note: What Belongs in a Final Project Report*, though not necessarily in exactly that order.

The party requesting the report may mandate particular content and may specify the format and sequence of sections. If so, follow their instructions precisely. If not, any well-organized, logical sequence is likely to be acceptable. If you'd like to look at some sample reports for ideas, the MATE website archives project reports from past ROV competitions (www.marinetech.org/rov_competition/report_examples.php).

These reports illustrate a variety of report styles and many include valuable information about "lessons learned the hard way" that you might find helpful as you embark on your own ROV or AUV design projects.

Keep in mind that the content and emphasis of your final report may vary somewhat, depending on the intended audience. For example, an industry client may be most interested in mission results and suggestions for equipment modifications. On the other hand, instructors may be most concerned with the details of how the group approached troubleshooting during design/construction and what they learned from the experience. If you're writing the report for your own future reference, you'll probably want to be sure to record data that will save you time on your next project. Examples of such information might include component numbers or vendors for parts that took a long time to find but that ended up working very well. You'll also want to note solutions to frustrating mistakes that you don't want to have to repeat and recover from all over again!

It might seem sensible to wait until your project is all finished before you start writing up your report, but this is a mistake. Your final report will be much quicker and easier to write, much more accurate, and much more impressive if you take notes as you go. That way the report evolves along with the project. For example, you may decide that you want an underwater photo of your ROV on the cover of your report. If you've started thinking about your report early on and have the idea ahead of time, you can arrange to have somebody in the water with a waterproof camera on the day of your first sea trials. But if you wait until after the project is finished to start thinking about the cover and didn't happen to have a camera in the water the day of sea trials, you'll either have to give up on your dream cover shot or waste unnecessary time and effort trying to "recreate" the sea trial scenario all over again. Likewise if you have a budget page and vehicle specifications sheet half written on a computer, you can drop information straight into those pages as it becomes available and while it's still fresh in your mind, rather than having to search back through months of notes looking for that particular needle in the haystack.

TECH NOTE: WHAT BELONGS IN A FINAL PROJECT REPORT?

Project reports vary in their content and organization, but they commonly include the following types of information:

- Most reports begin with a cover and/or title page. This may feature an informative photograph or diagram of the vehicle as well as the name of the vehicle and the team or organization that designed and built it. The title page typically specifies the date of the report.
- Longer reports will usually include a table of contents; however, this may not be needed for short reports of a few pages or less.
- The main body of the report will usually begin with an introductory statement about the origin and purpose of the project. This statement may include a broad goal (e.g., to learn about ROV construction) and/or a more focused statement about the specific mission(s) for which the vehicle was designed (e.g., to recover bozonium canisters from the bottom of Monterey Harbor.)
- Somewhere in the report, there is usually a brief summary of all design options that were *seriously* considered or explored, plus an explanation for why the final design was chosen over the other options.
- A major portion of the report will usually be dedicated to a detailed description of the final design, including parts lists, technical diagrams, and photos. This may be augmented by brief descriptions of any major challenges encountered and any special methods used for construction, testing, or troubleshooting.
- Most reports will include a one- or two-page "Vehicle Specifications" section. This is basically a list or table summarizing the vehicle's key characteristics, including its physical dimensions, weight (in air), maximum working depth, maximum speed, maximum thrust, power requirements, and any other details that would help prospective users quickly understand the requirements and capabilities of the vehicle without having to read the whole report. (See *Tech Note: Functional Specifications for SeaMATE.*) Sometimes the specifications are embedded in the technical description section of the report, and sometimes they are attached to the end of the report as a separate appendix.
- Most reports will include a budget summary. This does not need to be super-detailed, but should provide an overall sense of the cost of the vehicle and of each major vehicle system.
- Some reports will include a summary of basic operations and the results of any missions. Many conclude the main body of the text by reflecting back on the goals and purpose of the project and assessing whether or not they were achieved.
- Most reports will also include a list of suggestions for future improvements. No vehicle is ever perfect. What seems to work just fine in sea trials suddenly turns cranky when the pressure's on. Expect that things will come up that you never expected and learn from them. Take notes on your vehicle's performance under a variety of conditions and document ideas for improvements or modifications.
- Near the very end of the report, it's common to include a bibliography or similar list of key references (books, websites, expert advisors) that were used to obtain information vital to the project's success. This list should be formatted in one of several acceptable standard citation formats.
- It is customary and considerate to include an acknowledgements section—either at the very beginning or very end of the report—thanking any people, companies, or organizations that contributed significant time, expertise, parts, money, or other resources to your project. You don't need to thank every company that sold you parts at regular prices, because in a sense you've already thanked them with your money, but you should thank people or companies that *volunteered* their time or expertise, *donated* parts, gave you special *discounts*, fed your team pizza at 3:00 a.m. the night before the contest to help you finish on time, offered consistent moral support, or otherwise went "above and beyond the call of duty" to benefit your project in some special way.

While documenting as you go, it's a good idea to have all team members contribute information to the report. Certainly, each team member with expertise in a particular area should have primary responsibility for that portion of the report; however, it's also a good idea to get input from everyone else, too. For example, when summarizing vehicle performance or making recommendations for improvement, it's likely that the person handling the vehicle's tether will have different insights than the person operating the pilot's controls.

Always think broadly about what might be useful in a final report. Don't limit yourself to the vehicle alone or to the ideas that worked. Take notes on problems getting the ROV to the pier, issues with the launch and recovery system, team dynamics under stress, schedule conflicts, or other factors that ultimately impacted the vehicle design or the mission. These issues may turn out to be useful material for a final report. Be sure to record what didn't work in addition to what did work. Also, record unexpected operations conditions or thoughts about the parts you wished you had but didn't. All of that can be useful information when it comes to reporting on lessons learned.

4. Chapter Summary

Most people think the creation of an ROV or AUV takes place in the workshop. However, if you're smart about it, most of the creative work will actually take place beforehand, through a careful planning and design process. If these critical processes are done well, the construction and testing will go much more smoothly. Furthermore, you'll use your time, money, and other resources far more efficiently, and your vehicle will have a considerably greater chance of completing the underwater missions for which it was designed.

The first half of this chapter provides a number of specific strategies you can use to overcome common obstacles encountered when planning and building a vehicle or similar type of project. For example, one strategy is to remain focused on the mission, so you don't waste time, energy, or money working on things that don't help accomplish your goals. Another is to be proactive about project planning, management, and safety. Yet another is to use the design spiral as a way of coping with the difficult challenge of coordinating the designs of a large number of vehicle systems that must work together seamlessly. Feel free to craft strategies that work most effectively for your project.

The second half of the chapter outlines the value of following a step-by-step design methodology to move systematically through the process of planning, designing, building, and testing your vehicle. This design methodology is only an outline of the process—the additional details you will need to design and build a functional vehicle will come to light as you proceed through the technical chapters that begin with Chapter 3. Each of the technical chapters will explore one or two vehicle systems in detail, giving you the knowledge necessary to research, purchase, and build each system and to integrate those systems into one capable underwater vehicle. The last chapter in this book, Chapter 12, is the culmination of all this learning and provides a step-by-step procedure for building your own ROV. One of greatest advantages of using a design methodology is that it acts like a template that you can use for any vehicle design. So once you've mastered the basics, have fabricated your own vehicle, and are ready for something new, come back to *Chapter 2: Design Toolkit* to begin planning and researching your next challenge!

Chapter 3

Working in Water

Chapter 3: **Working in Water**

Stories From Real Life: *Pisces III*

Chapter Outline

1. **Introduction**
2. **Physical Properties of Water**
    - 2.1. Chemical and Electrical Properties of Water
    - 2.2. Mechanical Properties of Water
    - 2.3. Acoustic Properties of Water
    - 2.4. Optical Properties of Water
    - 2.5. Thermal Properties of Water
3. **Water Movements**
    - 3.1. Currents
    - 3.2. Tides
    - 3.3. Waves, Swell, and Surge
    - 3.4. Tsunamis
4. **Plant and Animal Life in Water**
    - 4.1. Biofouling
    - 4.2. Entanglement
    - 4.3. Large Animals
5. **The Water Column**
    - 5.1. Light
    - 5.2. Temperature
    - 5.3. Salinity
    - 5.4. Density
    - 5.5. Plant and Animal Life
6. **Chapter Summary**

Chapter Learning Outcomes

- Describe ways that the physical properties of water differ from those of air.
- Explain how each of those differences presents challenges and/or opportunities for those designing or using underwater vehicles.
- Describe how physical and biological characteristics of water vary with depth in the ocean.

Figure 3.1.cover: Phantom ROV Begins a Mission

The moment any vehicle slips under water to begin a mission, it is subjected to a range of physical properties that can differ significantly from those of air.

Image courtesy of Neil McDaniel

STORIES FROM REAL LIFE: *Pisces III*

The complexity of working under water is brought home most forcibly when something goes wrong. Disastrously wrong. During a routine recovery operation in the summer of 1973, the submersible Pisces III *flooded, snapped its towline, and sank to the bottom with two crew members. With time running out, rescue personnel had to draw on their collective expertise, training, and resourcefulness in order to turn a nearly fatal underwater accident into a triumphant rescue.*

In August 1973, the two-person submersible *Pisces III* and its mother ship *Vickers Voyager* were hard at work far out on the Atlantic, some 240 kilometers (approx. 150 mi) off southern Ireland. A team of pilots and divers was burying a section of transatlantic armored telephone cable. Their work routine was a familiar one: dive, bury the cable, return to the surface, fix the submersible, debrief, eat, and then sleep while *Pisces*' batteries recharged.

One of the dive teams consisted of pilot Roger Mallinson, the most experienced engineer and maintenance technician of the submersible crew, paired with Roger Chapman, as observer. On August 25, they were about to complete their fourth dive. A technical hitch the night before had robbed both men of sleep. Now, having spent nine hours in the cramped, humid sphere, they were eager to get back to the mother ship.

Pisces III rose to the surface, bobbing up and down in the waves. A diver clambered aboard, attaching a towline in order to haul the submersible to the stern of *Voyager* for recovery. Inside the tiny submersible, the two men could practically smell the bacon and eggs awaiting them on board. Suddenly, a water alarm sounded. They weren't overly concerned—condensation in the aft sphere had triggered such an alert before—but then the alarm sounded again.

Mallinson and Chapman heard loud, garbled voices over the VHF radio. They felt the submersible tip sharply backward and watched as the needle on the depth gauge began dropping. At 175 feet (approx. 53 m), their sudden descent was halted abruptly when the towline, still attached to the stern of the submersible, reached its limit and snapped taut.

Now a new nightmare began as *Pisces III* and her crew were jerked up and down, following the motion of the mother ship tossing on surface waves. Trapped inside the capsule, the two men dodged flying equipment and fought seasickness. Both realized that somehow the aft sphere had become flooded with seawater, making the submersible over a ton heavier than she should be. At that weight, they knew the towline couldn't stand the strain.

In the chaos, Roger Mallinson managed to release the drop weight, lightening *Pisces III* by 400 pounds (approx. 181 kg). But seconds later, the towline snapped anyway, and the submersible headed back down. He called out the depths over the underwater telephone " . . . 300 feet . . . 350 feet . . . 400 feet." Hitting bottom was certain, so the two crew members first disconnected the large sonar, dangling by its electrical cable. Then automatically, they began to go through safety procedures, switching off electrical power to prevent a fire, and moving anything soft to the back of the sphere to cushion the impact.

In the dark, they watched the luminous depth gauge as the submersible continued its stern-first drop—1,200 feet . . . 1,300 feet . . . 1,400 feet. *Pisces III* hit the seabed at exactly 1,575 feet (approx. 480 m). Mallinson and Chapman were now trapped on the Atlantic seabed, far deeper than anyone had ever survived a submarine accident before.

At first, neither man dared move in the total darkness. Then slowly, they shifted position, and Mallinson reached up to try the main 120-volt breaker. The small dome lights came on! The inside of the sphere was a shambles, but they had power, and the *Pisces III* was still watertight—there was hope!

Aware that any rescue attempt would take some time, the two men searched for life-support supplies in the chaos, and repositioned equipment to conserve vital power and oxygen. The air scrubber worked, as did the underwater telephone, their only communication link with the surface. The men relayed vital details about their physical status, the remaining amount of oxygen, and the state of their CO_2 canisters to the mother ship. Then, there was nothing else to do but shiver in the damp chill and wait. Fortunately, Roger Mallinson had followed proper pre-dive procedures the night before, so they had two full canisters of oxygen that would last 30 hours—perhaps as long as three days if they kept physical movement to an absolute minimum.

On the surface, a tremendous rescue operation swung into action at several points around the globe. At Vickers Oceanics' head office in Barrow-in-Furness, England, plans and timing for the rescue operation were carefully mapped out, based on data supplied from the stranded submersible. Vickers routinely operated two *Pisces*-class submersibles, so the company sent an urgent call to *Pisces II*, working in the North Sea; this second sub and mother ship were ordered back to port immediately. Vickers also asked for assistance from International Hydrodynamics Ltd. (HYCO), operating *Pisces IV* and *V* on the east and west coasts of Canada. The Royal Naval Submarine Base in Scotland immediately deployed subs to the area.

On site, two surface buoys marked the general position of the downed submersible. However, exact positioning could be ascertained only by telephone communication with the two men, so the Vickers *Voyager* could not leave the site to pick up experts or rescue vehicles until a suitable relief ship could be found in the area, to maintain the vital communication link. As the winds freshened, it took all of the captain's piloting and seamanship skills to keep the mother ship in position.

Detailed plans were put in place to fly the other *Pisces* craft, experts, and crews to Cork, Scotland, the nearest port from which rescue craft could be deployed. These people came from a variety of countries, but they shared a common language and expertise relating to submersible operations. Requests for equipment were quickly filled, and supplies were loaded onto ships and planes. At last, the Royal Fleet Auxiliary Vessel *Sir Tristram* arrived on site as relief vessel, so the Vickers mother ship could head for Cork to pick up personnel, underwater vehicles, and supplies. Even at full speed, *Voyager* would not be back on site for 27 hours, plus loading time. As the hours ticked by, word came that *CURV III*, the U.S. remotely controlled underwater recovery vehicle, was also bound for Cork, and the Canadian cable ship *John Cabot* would stand by to transfer *CURV* from shore to ship.

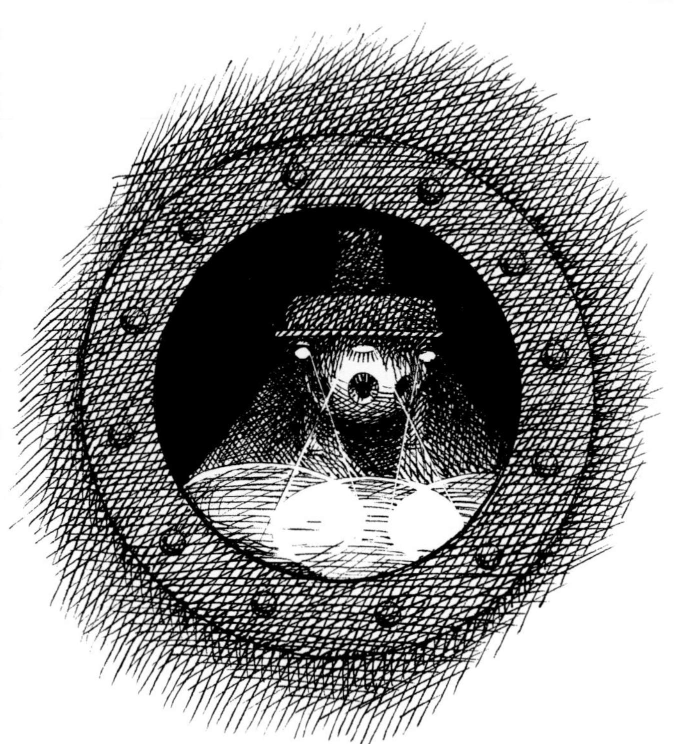

Meanwhile, Mallinson and Chapman tried to keep calm and still in the damp cold; their only exertion was limited to periodically maintaining radio communication and scrubbing CO_2 from the foul atmosphere of the tiny sphere. Ironically, the action of shaking the chemical canisters, so that what little remained of the absorbent compound could react with the CO_2, made the men breathe all the more heavily.

The plan that evolved was for a rescue submersible carrying a lift line to locate *Pisces III* and attach the line so that the ship could winch *Pisces* back to the surface. If the weather was cooperative, the procedure might take only a few hours. Alas, the weather worsened as *Voyager* arrived back on scene, carrying both *Pisces II* and *Pisces V*.

Only 24 hours' worth of oxygen remained for the men trapped below when the seas calmed enough to launch *Pisces II* and Vickers' two most experienced pilots. The rescue submersible headed down. But before reaching the bottom, the buoyant rescue line that had been lashed to the submersible bent its metal manipulator arm, rendering *Pisces II* useless as a rescuer. Careful not to convey any sense of alarm to Mallinson and Chapman, the *Pisces* crew continued to try and locate the downed craft. However, the rescue submersible developed a slight leak in one of its penetrators when it reached the bottom. There was no choice but to abort the mission and return to the surface.

While technicians concentrated on repairing the damaged manipulator and leaking penetrator on *Pisces II*, crews launched *Pisces V* in rough seas. It searched in vain for six long hours, unable to locate its sister submersible, which lay hidden from the sonar in a slight hollow. Recalled to the surface, technicians discovered that the gyro compass on *Pisces V* was not giving a true reading, but there was no time for repairs. Far below, the two trapped men coped with the severe headaches and confused thinking caused by high levels of CO_2.

The *Pisces V* was towed to an approximate position of the stricken submersible and descended again, clutching the line and snap hook in its manipulator. This time, the rescue sub's sonar spotted the downed craft, and within 20 minutes, the crew of *Pisces V* had the snap hook attached to the lifting point! Then somehow, as the rescue submersible carefully backed up, the hook fell away. Its pilot watched in horror as the buoyant line began pulling the hook upward. Acting immediately, he swung his craft around and just managed to grab the end of the hook with his manipulator before it disappeared from view. This time, however, he could only attach it to the propeller guard. Although there was now a definite link to the surface, the crew was bitterly disappointed; the propeller guard was too flimsy to haul the stranded submersible to the surface.

Pisces V was ordered to stand by. On the surface, technicians continued through a third night of rescue operations, working without sleep to repair the manipulator on *Pisces II*. The *John Cabot* arrived on scene with *CURV III*. *Pisces II* was launched in heavy weather, only to be recovered immediately when the submersible's water alarm sounded. Again, weary technicians went back to work. Time was running out.

With only hours to spare, *Pisces II* was launched again in rough water conditions. This time, its crew was able to attach a specially designed toggle to the downed craft. But one line wouldn't be enough to lift *Pisces III*, so *CURV* was launched. Fortunately, operations went exactly as planned, and the robot successfully placed a second toggle on the aft sphere of the stricken submersible. Now, with two strong lines and the *John Cabot*'s powerful lifting gear, the operation could begin, but the big swells generated by the continuing bad weather would make the maneuver extremely difficult.

Inside the *Pisces III* sphere, rising levels of CO_2 had kept the men comatose for hours, but now that the lift was beginning, fear and excitement took over. Then the mad pendulum motion started again. For the next 75 minutes, in an atmosphere fouled by excrement, urine, and vomit, there was absolutely nothing Chapman and Mallinson could do but hang on. At 350 feet (approx. 106 m), the lifting line became entangled with *CURV*'s tether. Exhausted crews scrambled to straighten them out. With only 20 minutes of oxygen remaining, divers battled rough seas to attach more lift lines. Finally, *Pisces III* broke clear of the water. The hatch was opened, and eager hands reached down to help the two men climb out of the capsule into fresh air.

Fortunately, this was a narrow escape and not a disaster. To this day, no one knows how the hatch was ripped off *Pisces III*, causing the aft sphere to flood and the vessel to sink. But every mariner knows that the sea is full of surprises—incidents occur that defy all odds, and at any time, crew and equipment can be pushed to extreme limits. Sometimes vehicle structure and human experience, training, and resourcefulness are not enough to effect a successful rescue operation.

Fortunately, sometimes they are.

1. Introduction

Your intuition about how the world works has been honed through years of living in air, not water. This air-based intuition serves you well on land, but it can backfire under water, where the laws of physics *seem* to be different. For example, if you drop a log on land, it falls *down*, threatening your toes. But if you drop the same log under water, it floats *up*, threatening your chin. In actuality, the laws of physics are exactly the same in both environments; however, the way those laws manifest themselves can be surprisingly different. That's because water and air are very different substances with very different properties.

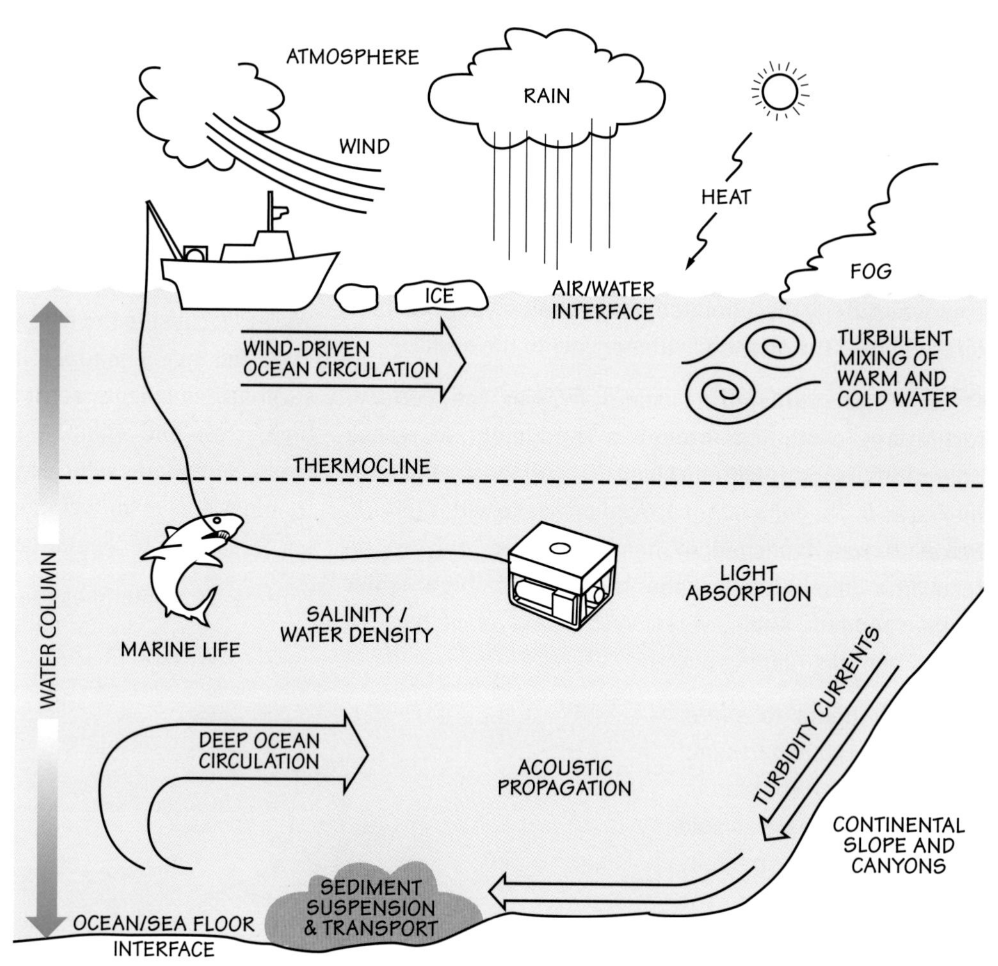

Figure 3.2: The Challenge of Working in Water

The design of any underwater craft must take into account a wide variety of environmental factors in the atmosphere, air-to-water interface, water column, and seafloor. Often the operation of underwater vehicles is influenced by several of these factors simultaneously.

Here are some examples of the possibly counterintuitive aspects of working in water:

- Metals are usually very strong and reliable building materials on land, but they can corrode quickly and crumble under water if not properly protected.
- An object that is immensely heavy and almost impossible to move on land may seem to weigh nothing and be very easy to move under water.
- Pressure changes associated with changing altitude in air are subtle and rarely noticed. In water, however, pressure changes are hundreds of times greater than in air, so they cannot be ignored at any depth without risking serious consequences.
- Radio signals provide excellent opportunities for communication on land (e.g., TV, radio, cell phones) and for wireless remote control of robots and other devices across the room or across the solar systems. But they cannot effectively

penetrate water so are nearly useless for underwater vehicle communication and control.

- Vehicles operating in water can quickly accumulate thick growths of algae, barnacles, and other living organisms, which can interfere with vehicle performance. This isn't a problem on land—cars may get muddy, but they don't require mowing!
- Electrical control systems that operate safely in air can short-circuit, malfunction, and even create a deadly electrocution hazard when used in or near water.

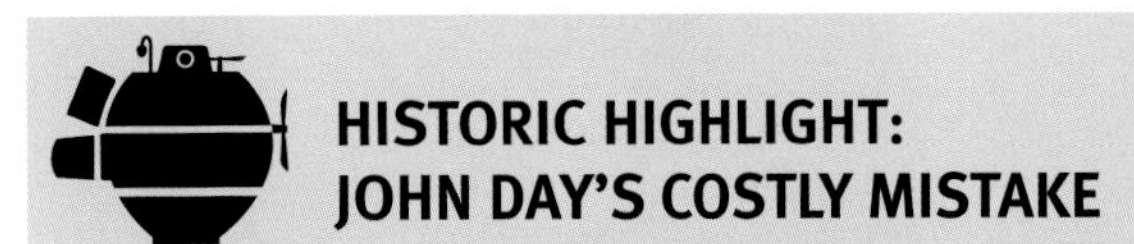

HISTORIC HIGHLIGHT: JOHN DAY'S COSTLY MISTAKE

Many devices for underwater exploration failed because their inventors did not understand the physical properties of water and how those properties affect underwater vehicles. John Day, a British wheelwright, is one such tragic example. In the 1770s, he converted a small boat into a diving machine by installing a watertight cabin. At the time, underwater feats were a popular form of commercial entertainment, and Day succeeded in submerging in his vehicle to 30 feet.

Encouraged by this initial success, Day connected with a wealthy gambler who promised him a greater financial return if he could stay submerged at a deeper depth without harm. Day converted a 50-ton sloop into a second underwater boat and proposed to remain submerged at 130 feet for 12 hours. The betting began and on the appointed day, the necessary ballast was added until Day and his "submarine" sank. Alas, he never surfaced again, thus becoming the first recorded submarine casualty. In all likelihood, John Day died because he did not adequately understand and appreciate how quickly water pressure increases with depth and neglected to make his cabin strong enough to resist the intense pressure at 130 feet.

This chapter introduces the properties of water (both physical and biological) that you need to know to build successful underwater vehicles. Some of these properties may be obvious to you, and some may not. But history has shown over and over again that misunderstanding or ignoring any of them can result in problems ranging from missed opportunities and poor vehicle performance to tragic loss of human life.

This chapter begins with an overview of the physical properties of water that affect vehicle design—chemical, electrical, and acoustic properties. Next, the chapter discusses how waves, currents, and tides can affect underwater vehicles. A brief section on plant and animal life highlights additional things for ROV and AUV designers to consider. The chapter concludes with a look at how the various physical and biological properties of water change with depth.

The technical chapters that follow after this one expand on the introductory technical information in this chapter and will give you a deeper understanding of the distinctive benefits and challenges of working under water. They will also offer practical solutions to the most vexing challenges you are likely to encounter.

2. Physical Properties of Water

Working on an ROV is a lot like working on a car. The tools and techniques are similar. Yet the final products are designed to function in very different environments, so they look and perform very differently. For someone who is used to working on machines that operate in air, water provides exciting new challenges and opportunities, most of which can be traced, either directly or indirectly, to differences between the physical properties of air and water. These include the following differences:

- chemical properties, such as salinity
- electrical properties, such as electrical conductivity
- mechanical properties, such as density, viscosity, and pressure
- acoustic (sound-related) properties
- optical (light-related) properties
- thermal (temperature-related) properties

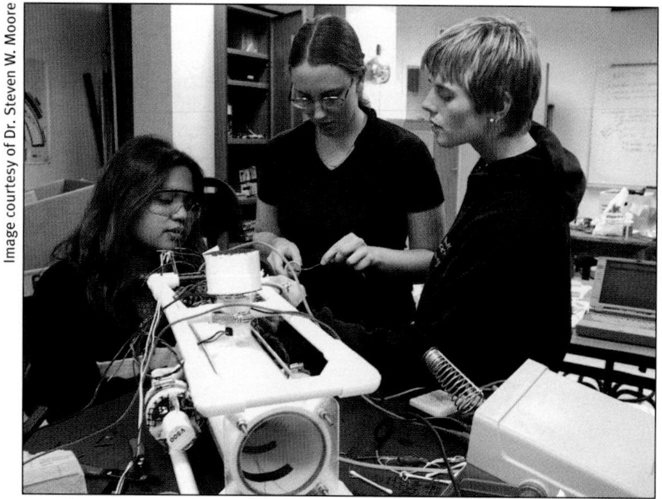

Figure 3.3: Science and Fabrication

Students at California State University, Monterey Bay, assembling the BOB-II *ROV, which they and their classmates designed. The students were able to make good decisions about vehicle shape and materials because they understood the physical properties of water and how those properties would affect their vehicle.*

2.1. Chemical and Electrical Properties of Water

Water is a simple but remarkable compound. Each water molecule is composed of one oxygen atom covalently bonded to two hydrogen atoms, hence its chemical formula: H_2O. The angle formed by the bonds (see Figure 3.4) and the strong affinity of the oxygen atom for electrons give the water molecule an asymmetric electrical charge distribution that accounts for many of water's unusual and important properties. Among these are its outstanding ability to dissolve salts, its tremendous capacity to absorb or release heat without much change in temperature, and its extremely unusual ability to freeze into a solid phase (ice) that floats on top of its liquid phase. If it weren't for these qualities and many other special properties of water, the earth would be a very different place, and life as we know it would not exist.

Water's ability to dissolve salts has important implications for underwater vehicle design. When a salt compound dissolves in water, the salt molecules break apart into electrically charged atoms (or groups of atoms) called ions. Because they tend to fit into the spaces between the water molecules, adding mass without adding much volume, these dissolved ions increase water's density. Thus, saltwater is heavier than an equivalent volume of freshwater. This can have a surprisingly strong effect on whether an underwater vehicle floats or sinks. The ions also increase water's **electrical conductivity**. This increased capacity to conduct electrical current can short out electrical equipment that comes in contact with the water, block radio signals, and accelerate corrosion of metals.

TECH NOTE: WHY IS THE OCEAN SALTY?

The earth's crust is composed of various minerals, including a variety of different kinds of salts. As freshwater from rain or melting ice flows over and through the crust, it dissolves tiny amounts of salt and carries that salt to the ocean. Once it's there, evaporation returns some of the water to the atmosphere, but leaves the salt behind. That evaporated water eventually forms clouds, turns to rain or snow again, and cycles back through the earth's crust, picking up another tiny load of salt and carrying it to the sea. Over millions of years, this constant input of tiny amounts of salt, coupled with the concentrating power of evaporation, builds up high levels of salt in the ocean. A similar process has increased the saltiness of landlocked lakes such as the Great Salt Lake of Utah.

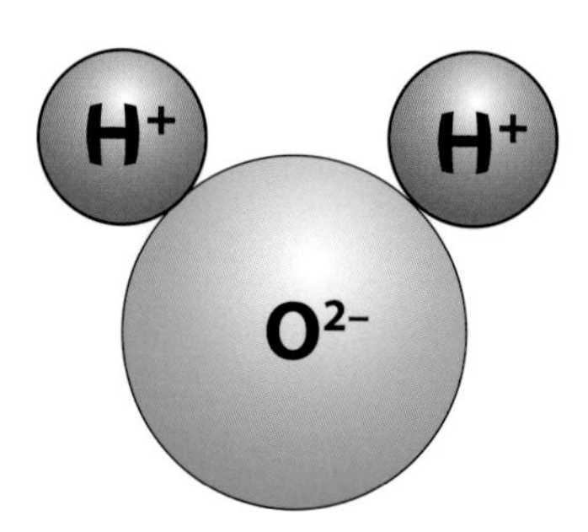

Figure 3.4: Water Molecule

Many of water's unique physical properties can be traced to the arrangement of atoms in each water molecule. Hydrogen (H) atoms in this molecule have a slight positive charge, and the oxygen (O) atom has a slight negative charge, so the unbalanced arrangement gives the molecule a lopsided electrical charge distribution.

2.1.1. Salinity

The term **salinity** is used to quantify the concentration of salts dissolved in water and is often expressed in parts per thousand (ppt or ‰) by mass or weight. The ocean has an average salinity of about 35‰. In other words, 1,000 kilograms (approx. 2,200 lbs) of seawater contains about 35 kilograms (approx. 75 lbs) of dissolved salts. (See Figure 3.8.) Note that 35‰ (parts per thousand) is the same as 3.5% (parts per hundred, or percent). The exact value varies somewhat, depending on local conditions, such as the amount of freshwater input from rain or rivers and local rates of evaporation, but the salinity of the open ocean is typically between 33‰ and 37‰. Pure freshwater (e.g.,

TECH NOTE: CTDs

One of the most important instruments in oceanographic research is the **Conductivity, Temperature, and Depth Sensor**, or **CTD**. This instrument records **electrical conductivity**, temperature, and water pressure at various depths. The pressure readings are used to determine the depths at which each of the other readings were taken. The temperature and conductivity readings can be combined to calculate the salinity. Thus, the CTD data can be used to calculate and plot how temperature and salinity vary with depth. (See *Tech Note: Practical Salinity Units*.) CTDs may be lowered on a cable from a ship or carried on board an underwater vehicle, such as an AUV. (See *Tech Note:* Argo *Floats*.)

Image courtesy of Harvard graduate student Jonathan Gero

Figure 3.5: CTD Deployment

Releasing a Niskin carousel/CTD profiler from the SSV Robert Seamans *during a scientific cruise run by the SEA Education Association.*

TECH NOTE: *ARGO* FLOATS

An ***Argo* float** is essentially a free-drifting, autonomous CTD that records measurements of salinity, temperature, and depth. (See *Tech Note: CTDs*.) Deployed from ships or aircraft, each float sinks to a preset depth of some 1,000–2,000 meters (approx. 3,300–6,600 ft) and drifts with deep currents. Then at regular intervals, the float rises, taking vertical temperature and salinity measurements as it ascends. At the surface, the float transmits its position along with new stored data to a relay satellite before sinking back to depth.

Three thousand *Argo* floats have been deployed around the world in a massive international cooperative effort that involved ships from several countries, including New Zealand, Japan, Germany, Spain, and the U.S. The objective of this ongoing deployment is to provide a quantitative description of the changing state of the upper ocean and the patterns of ocean climate variability. All collected data are relayed and made publicly available on the internet within hours after collection. Currently, all floats are distributed over the global oceans at an average 3-degree spacing, providing 100,000 temperature/salinity profiles and velocity measurements per year. Check out the *Argo* website at www.argo.ucsd.edu for distribution maps, photos of deployment, and continuously updated information.

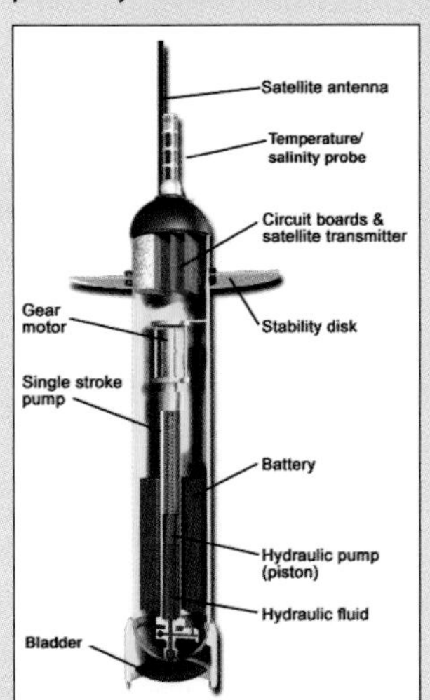

Cross section schematic courtesy of Southampton Oceanography Centre, UK

Figure 3.6: Cross Section of an* Argo *float

Image courtesy of website www.argo.ucsd.edu and the *Argo* Program

Figure 3.7: Float Recovery

The Provor, *a type of* Argo *float, awaits recovery by the Japanese Coastguard vessel* Takuyo.

TECH NOTE: PRACTICAL SALINITY UNITS

Historically, scientists measured salinity by lowering a sample bottle over the side of a ship, bringing the bottle into the lab, then chemically analyzing the concentration of ions in the water. This was difficult and time-consuming work. Fortunately, salinity can be estimated with reasonable accuracy from conductivity and temperature, both of which can be measured much more quickly and easily than true salinity. One complicating factor is that the exact relationship between conductivity, temperature, and salinity depends on the relative proportions of different ions present in the solution, and this can vary from one body of water to the next. To get around this problem, the **practical salinity unit,** or **psu**, was developed. This unit is defined as part of the Practical Salinity Scale of 1978 (PSS-78), which is based on how the **conductivity** of the sample solution compares to the conductivity of a standard, precisely defined solution of potassium chloride. Since a practical salinity unit is a ratio of two conductivities, it is a dimensionless unit. However, the standard is defined to facilitate compatibility between the two systems for describing salinity: ocean water with a true salinity of 35‰ will have a practical salinity of about 35 psu.

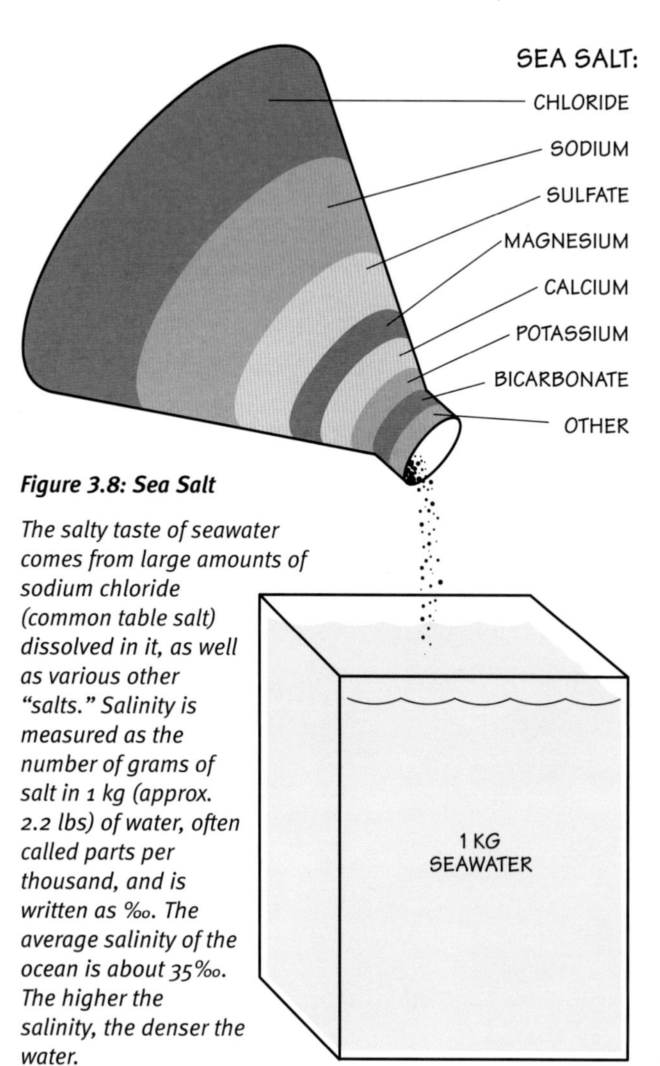

Figure 3.8: Sea Salt

The salty taste of seawater comes from large amounts of sodium chloride (common table salt) dissolved in it, as well as various other "salts." Salinity is measured as the number of grams of salt in 1 kg (approx. 2.2 lbs) of water, often called parts per thousand, and is written as ‰. The average salinity of the ocean is about 35‰. The higher the salinity, the denser the water.

distilled or deionized water) has a salinity of 0‰. Although trace amounts of salt are found in all streams, lakes, and other natural bodies of freshwater, these natural freshwater sources, as well as tapwater and swimming pool water, typically have salinities less than 1‰ and can usually be regarded as having a salinity of 0‰. Desert lakes with no outflow other than evaporation can concentrate salts to very high salinities. For example, the Dead Sea between Israel and Jordan has a salinity of about 300‰.

Most underwater vehicles operate either in freshwater or seawater, and this book will focus on those two environments. For the purpose of sample calculations, salinity values of 0‰ for freshwater and 35‰ for seawater will be assumed throughout this textbook, unless stated otherwise.

2.1.2. Electrical Conductivity

The **electrical conductivity** of a material (whether solid, liquid, or gas) is a measure of how easily electrical current can flow through that material when a voltage is applied across the material. Electrical current is simply a flow of charged particles from one location to another. In most electrical and electronic circuits, the charged particles that flow are electrons moving through metal or semiconductors. However, in water, electrical current is carried predominantly by dissolved ions.

Materials that have high electrical conductivity are called **conductors**. Those with low electrical conductivity are called **insulators**. Seawater is a good conductor because of all the dissolved ions it contains. Dry air is a good insulator. Freshwater is somewhere in between. A practical consequence of this is that saltwater (and to a lesser extent, freshwater) can cause short-circuiting of electrical equipment. Since most underwater vehicles use electrical circuits for everything from video camera signals to navigation and control of propulsion systems, designers of underwater vehicles must take steps to protect electronics from exposure to water, including water vapor that may condense on the

SAFETY NOTE: ELECTRICAL SAFETY NEAR WATER

Electricity and water can be a dangerous—even fatal—combination. In most cases, the simplest way to avoid trouble is to avoid working with electricity near water. But with underwater vehicles, most of which are powered by electricity, that's nearly impossible. Therefore, it is critical for people who design, operate, or work near underwater vehicles to pay particular attention to electrical safety.

Basic electrical safety requires a good understanding of how electricity works. This book doesn't cover electricity in detail until *Chapter 8: Power Systems*. In the meantime, play it extra safe. You absolutely need to know that the regular 110–120 volts AC electrical power common in household and school wiring in the U.S. and Canada can (and does) kill hundreds of people every year. Electricity is invisible and travels near the speed of light, so there's no way you can see it coming, and your quick reflexes couldn't save you, even if you did see it coming. So here are the safety basics:

- Never work on live, high-voltage electrical circuits, wet or dry.
- Never handle or work on vehicles, vehicle circuits, vehicle controls, television monitors, or other devices that draw power from outlets, generators, extension cords, or similar power sources around water without first verifying that the circuit is protected by a Ground Fault Circuit Interrupter (called a GFCI or more commonly a GFI). (See *Chapter 8: Power Systems* for a more detailed explanation of GFIs.)
- If you don't know where the GFI is or how to make sure it's working properly, check with someone who knows this information and knows how to test the GFI.
- If you can't confirm the presence and proper function of a GFI, don't take chances.
- Know that while a GFI greatly improves electrical safety around water, it is not a total guarantee of safety; there are still ways to get electrocuted by a GFI-protected circuit, even in dry situations and even when the GFI is working properly.

inside of vehicles during missions. Another consequence of water's conductivity that designers *must* be aware of is that water can create a serious electrocution hazard. (See *Safety Note: Electrical Safety Near Water.*)

Simple devices are available for measuring the electrical conductivity of water. In general, they work by calculating how easy it is to force a certain amount of electrical current through a specific amount of water. When combined with temperature measurements, these conductivity measurements can be used to estimate the salinity of seawater. (See *Tech Note: Practical Salinity Units.*)

2.1.3. Corrosion

The term **corrosion** applies both to an electrochemical process that eats away metals and to the end product of that process. Rust is a familiar example of corrosion. Metal corrosion requires the removal of electrons from the metal. In air, this usually happens very slowly if at all, since air is a good electrical insulator, but in water—particularly salty water with lots of dissolved ions to help carry electrical current away from the surface of the metal—severe corrosion can happen very quickly. Corrosion rates will be accelerated even more if electrical voltages from batteries or other power sources are present to actively drive electrons from one place to another.

Corrosion is generally undesirable. Mild corrosion can discolor or roughen metal surfaces, causing excessive friction in moving parts or leakage around gaskets and seals. More severe corrosion can eat through metal hulls and/or seriously weaken metal beams or other critical structural members, potentially causing the flooding or collapse of a vehicle. Since parts of underwater vehicles are commonly made of metal, designers and maintenance crews must pay particular attention to the corrosive

TECH NOTE: SYSTEMS OF MEASUREMENT

Quantification—or measurement—is one of the foundations of science and engineering. Over time, the measurements regularly used to specify length, time, mass, speed, weight, and other quantities have become standardized into two systems of measurement:

- the Système International d'Unites (SI), commonly called the **metric system**
- the British Engineering Units (BE), commonly referred to as the **imperial system**

Both systems are commonly used in subsea engineering, so it's important for students of underwater vehicle design and operation to be comfortable working in either system and to be able to convert rapidly and accurately between the two. For that reason, we use both systems in this book. Common measurement conversions used in underwater vehicle work are presented in *Appendix I: Dimensions, Units, and Conversion Factors*. (The *Historic Highlight: Unit Mix-up Crashes NASA Mars Mission* is an example of why it's particularly important to pay attention to units and unit conversions.)

properties of water, to methods for reducing the rate of corrosion, and to proper inspection techniques. (See *Chapter 4: Structure and Materials* for more information on corrosion.)

2.1.4. Radio Signal Attenuation

Radio signals are a form of electromagnetic radiation. They travel easily and almost instantaneously through air and the vacuum of space and are used for many forms of wireless information transfer above water, including radio and television broadcasts, cell phones, walkie-talkies, wireless computer networks, and even communication with distant space probes exploring other planets in our solar system. Under water, however, the energy of radio waves is quickly absorbed by the water molecules and dissolved ions, thereby weakening or "attenuating" the signals. Typical low-cost, unlicensed radio signals, such as those used in radio-controlled toys, cannot penetrate seawater at all, and they penetrate even freshwater for only about 1 meter (approx. 3 ft). This is unfortunate, as these signals would otherwise provide an ideal communication channel for control of underwater vehicles and receipt of underwater video images.

There are some radio-controlled model submarines, and there have been successful experimental radio-controlled ROVs, such as *PURL I* built by Simon Fraser University, but these operate well only in freshwater and most of them must remain very near the surface. Model submarines, for example, typically lose contact with the surface controller if the sub dives deeper than about 1.5 meters (approx. 5 ft) in freshwater and won't work at all under saltwater. Modern military submarines can receive coded instructions by very low frequency (VLF) and extremely low frequency (ELF) radio waves, even in seawater, but extremely large specialized antennas and other equipment are needed to send these transmissions. As well, the sub still needs to be within a few meters of the surface to receive the signals, and communications are limited to very low data rates. It takes several minutes to transmit a short text sentence, and it is impractical to send more complicated information, such as sounds or images.

2.2. Mechanical Properties of Water

Before discussing the mechanical properties of water, it's important to define and understand four basic physics terms used to describe some of those properties:

- force
- weight
- mass
- gravity

In the process of trying to learn these scientific terms here, it may be necessary to *unlearn* (at least temporarily) other definitions you may have used or heard for these words. It's not that the more familiar definitions are wrong, but when these four terms are used in an engineering or physics context, as they are in this book, they have very specific meanings that may be more precise, or even completely different, than their meanings in everyday speech.

2.2.1. Force

In everyday speech, the word "**force**" has many different meanings. However, in a physics or engineering context, force almost always refers to a physical *push* or *pull* that can be measured and described completely in terms of two attributes: its *magnitude* (i.e., how strong the push or pull is) and its *direction* (i.e., whether the push or pull is up, down, north, south, diagonal, etc.) Examples of force include the tension in a rope and the attraction or repulsion between two magnets.

There are many acceptable units for measuring the magnitude of a force. The two used most commonly throughout this book are the **newton** and the **pound**. Newtons are metric units abbreviated with a capital **N** and roughly equal to the weight of one small apple. Pounds are imperial units abbreviated **lbs** (but see *Tech Note: Sloppy Units*).

Figure 3.9: Weight on Different Worlds

An astronaut weighs one-sixth as much on the moon as he does on the earth.

2.2.2. Weight

Weight is a familiar example of a force. Fortunately, the everyday meaning of the term "weight" is exactly the same as its precise physics/engineering meaning. If something has a lot of weight, it is difficult to lift and is described as "heavy." If something has only a tiny amount of weight, it is easy to lift and is described as "light." Regardless of whether an object is heavy or light, the reason it has weight on the earth is because the gravitational pull of the earth is attracting it—actually pulling it "down" toward the center of the earth. Since weight is a force, it can be measured in newtons or pounds (or any other valid unit of force).

HISTORIC HIGHLIGHT: UNIT MIX-UP CRASHES NASA MARS MISSION

Working with two different measurement systems can be confusing. On September 3, 1999, a $125 million orbiter that was supposed to move into orbit around Mars that day evidently went off course, fell into the red planet's atmosphere, and burned up. The source of the problem turned out to be a course adjustment error caused by a simple mix-up between imperial and metric units. This happened, in part, because NASA straddles the world of science, which relies primarily on metric units, and the world of American industry, which relies primarily on imperial units. The required course correction was transmitted in imperial units, but the orbiter was expecting metric units. NASA has since announced that it will use only metric units to plan and run critical missions in the future.

Like NASA, underwater vehicle designers straddle both the metric and imperial world. Someday underwater vehicle designers in the U.S. and Britain may move to an all-metric philosophy, as NASA has done, but for now, it's important to know both systems—and to know them well enough to detect and correct the type of error that brought down the Mars orbiter.

2.2.3. Mass

In casual conversation, the terms mass and weight are often used interchangeably, but they are not the same thing, and confusing them can lead to problems. **Mass** (sometimes called **inertia**) quantifies an object's intrinsic resistance to acceleration. It is *not* a force. An object with a large mass has a lot of inertia and will be harder to accelerate than an object with a small mass. Once it's moving, the more massive object with its greater inertia will be harder to stop or turn than the smaller one.

Compared to weight, mass is a more reliable measure of how much matter is in an object, because it doesn't depend on the strength of the local gravitational field. When

TECH NOTE: CALCULATING WEIGHT ON EUROPA

It may seem strange that one of Jupiter's moons, Europa, may be an exciting prospect for underwater vehicle designers. But there is good evidence that Europa may be hiding a large ocean of liquid water beneath its frozen exterior. This ocean could harbor extraterrestrial life and begs to be explored by robotic vehicles.

To design a successful extraterrestrial mission to Europa, you must be able to determine precisely the weight your vehicle will have on that moon. Since Equation 3.1 is universal, it can be used to calculate the weight of *any* object on *any* planet or moon, as long as you know the mass of your vehicle, the mass of the other planet or moon, and the radius of that planet or moon. The mass and radius of Europa are known quantities that can be looked up easily.

Many research groups are planning robotic missions to Europa and other such places. Some of these missions have already been launched, and more are likely in the years ahead. When it comes to landing an exciting job designing robots to explore distant worlds, those who understand how to apply Newton's Law of Gravity will have an advantage over those who do not.

an astronaut travels to the moon, his weight drops to about 1/6th of its magnitude on Earth (because the moon has 1/6th the gravity), but his mass does not change. (See Figure 3.9.) For this reason, scientists use mass more often than weight to describe how much physical matter is present in an object.

The mass unit used most commonly in this book is a metric unit called the **kilogram**, abbreviated **kg**. For a discussion of imperial mass units, see the *Tech Note: Sloppy Units.*

2.2.4. Gravity

Gravity is that familiar, yet mysterious, phenomenon that keeps the planets in orbit around the sun and holds you securely on the earth's surface. Sir Isaac Newton determined that a gravitational force of attraction exists between any two objects that have mass, even if one or both of the objects are small. Moreover, he developed a formula that can be used to calculate the magnitude of this force of attraction. This formula is known as **Newton's Law of Gravity**:

Equation 3.1

$$F = \frac{m_1 m_2 G}{d_2}$$

In this equation, F is the force of attraction, $m1$ and $m2$ are the masses of the two objects, G is the Universal Gravitational Constant (a number you can look up on the web or in almost any physics textbook), and d is the distance separating the centers of the two objects. This equation is simply a shorthand way of saying,

> *The force of gravitational attraction between two objects equals the mass of the first object times the mass of the second object times the Universal Gravitational Constant divided by the square of the distance between the centers of the two objects.*

(Note: If the correspondence between the equation for Newton's Law of Gravity and the preceding sentence isn't clear to you, now would be a good time to review standard algebra notation. You can also check out *Appendix II: Useful Constants, Formulas, and Equations* at the back of this text.)

Here on the surface of the earth, where $m2$ (Earth's mass) and d (Earth's radius) are both constants, we can combine them with G to form a single constant value for the earth, g (lower case):

Equation 3.2

$$g = \frac{m_2 G}{d_2}$$

Once this is done, Newton's gravity equation simplifies to:

Equation 3.3

$$F = mg$$

The constant g is known as the **earth's gravitational acceleration**. In the metric system, $g = 9.8\ m/s^2$. In the imperial system, $g = 32.2\ ft/s^2$. Note that these values of g work only on the earth. (See *Tech Note: Calculating Weight on Europa.*)

Equation 3.3 makes it easy to relate the mass of an object to its weight on or near the earth's surface and vice versa, because F now represents the object's weight. To convert mass to weight, multiply by g. To convert weight to mass, divide by g.

Don't be misled by the name "gravitational *acceleration*." When an object is dropped near the earth's surface, it does initially accelerate downward at a rate equal to g; however, Equation 3.3 and its constant g apply equally well to objects that are *not* accelerating. For example, to convert your body mass in kilograms to your weight in newtons, multiply your mass by *g*, even if you're sitting in a chair and not going anywhere.

Also, do not fall into the common misconception that the earth pulls stronger on you than you pull on it. That's simply not true. Newton's famous 3rd law of motion states that every force is matched by an equal and opposite force. If you weigh 120 pounds and fall out of an airplane, the earth will be pulling down on you with a force of 120 pounds, and you will also be pulling *up* on the earth with a force of 120 pounds. Therefore you and the earth will fall toward each other! However, since you have much less mass (inertia) than the earth does, the 120-pound force of mutual attraction will have a much bigger effect on your motion than it does on the earth's motion, and you will end up accelerating toward the earth much faster than it accelerates toward you.

TECH NOTE: SLOPPY UNITS

Since *g* is a constant here on the earth, an object's weight and mass are always proportional to one another. For example, if a 20-kilogram (44-pound) boy grows into an 80-kilogram (176-pound) man (a four-fold increase in mass), his weight will also have increased four-fold. That consistent proportionality has allowed people to get rather sloppy about the distinction between weight and mass. For example, bathroom scales commonly report a person's "weight" both in terms of pounds (an imperial system unit for *force*) and kilograms (a metric unit for *mass*), but what a scale actually measures is the force pressing down on it. Since pounds are a legitimate unit for force, the pound readings will be accurate anywhere (even on the moon, where they will correctly show the person as having only one-sixth as many pounds as on the earth). But the kilogram reading on these scales is calibrated based on the earth's gravitational acceleration, so on the moon the same scale would lie about the person's mass, implying that he had lost most of his body mass when, in fact, he had lost none.

Actually, the confusion gets worse. When it comes to weight versus mass, the term "pound" is itself an ambiguous unit, as are its common abbreviations (lb and lbs). It is better to be explicit about "pounds-force" (lbf) or "pounds-mass" (lbm), since both exist in the imperial system and since they are *not* the same thing (though they are numerically equivalent on the earth's surface). An alternative imperial unit for mass is an obscure, little-known unit called the "slug." One slug equals about 32.2 lbm (anywhere) and weighs about 32.2 lbf (on earth).

Once you have an understanding of gravity and the earth's gravitational acceleration, you can revisit the distinction between weight and mass at a more pragmatic level. There are three important reasons for understanding the difference between weight and mass and knowing how to convert between them:

1. Knowing the distinction between mass and weight will help avoid confusion later and provide a more accurate understanding of how things work. This is particularly important when immersing objects in water, where flotation can partially or completely offset weight, but has no effect on mass.

2. Confusing weight and mass frequently results in careless (and possibly costly or even dangerous) calculation errors. Equation 3.3 shows that such confusion will often lead to numerical answers that are in error by a factor of *g* (9.8 m/s^2 in the metric system or 32 ft/s^2 in the imperial system), because that is the factor used to convert between mass and weight.

3. Understanding the weight/mass distinction and being able to calculate each in places with different gravitational accelerations is particularly relevant now that some ROVs and AUVs are being designed for operation on other moons or planets, where the gravitational acceleration is different than on the earth! (See *Tech Note: Calculating Weight on Europa.*)

HISTORIC HIGHLIGHT: NICE ROUND NUMBERS

Pure freshwater has a density of *exactly* 1,000 kilograms per cubic meter. This nice round number is not just a coincidence. Water is such an important chemical on the earth that its physical properties were used originally to define and link together many of the standard units in the metric system.

In this case, the "gram" was originally defined as the mass of one cubic centimeter of freshwater. Since there are, by definition, *exactly* 1,000,000 cubic centimeters in a cubic meter and *exactly* 1,000 grams in one kilogram, this ends up giving you exactly 1,000 kilograms per cubic meter of freshwater.

Originally, this standard was based on water's density at the temperature of an ice-water mixture (which, incidentally, was also used to define zero on the centigrade temperature scale). More recently, the definition was adjusted and is now based on the density of water at the temperature where water has its greatest density, which is about 4° Celsius (approx. 39° F).

Armed with an understanding of the technical meanings of the terms force, weight, mass, and gravity, you are now ready to learn about some important mechanical properties of water.

2.2.5. Density and Specific Gravity

Everyday experience makes it obvious that a truckload of lumber weighs much more than a handful of small rocks. In that respect, the lumber seems *heavier* than the rocks. But if that truckload of lumber and those few small rocks both get dumped into a lake, the lumber will float, whereas the rocks will sink. In the water, the lumber seems somehow *lighter* than the rocks. Clearly it's important to be careful about what we mean by "heavier" and "lighter."

When it comes to fluids in general, and floating and sinking in particular—all of which are critical subjects in underwater vehicle design—the concept of **density** turns out to be more useful for comparisons than the concept of weight. Density in this context is defined as mass per volume:

Equation 3.4

$$Density = \frac{mass}{volume}$$

The density of any uniform material, including water, can be determined by taking a sample of the material and then dividing the sample's mass by its volume.

In metric, densities are often expressed in kilograms per cubic meter (kg/m^3). Freshwater, for example, has a density of exactly 1,000 kg/m^3. This means that one cubic meter of water has a mass of 1,000 kilograms. (See *Historic Highlight: Nice Round Numbers*.) In the imperial system, the density of freshwater is 62.4 pounds per cubic foot.

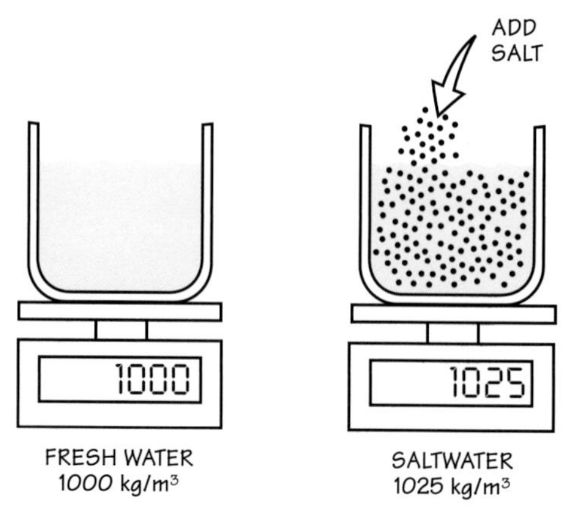

Figure 3.10: Density of Water

At 4° Celsius, freshwater has a density of 1,000 kilograms per cubic meter (approx. 62.4 lbs/ft^3). Add the extra salt found in seawater and you are adding more mass, which causes it to become more dense. Saltwater has an average density of 1,025 kilograms per cubic meter (approx. 64 lbs/ft^3).

The density formula can be extended to describe the density of an object, but if the object is made up of two or more different materials or has spaces filled with air, water, oil, or other materials, then the density calculated by this formula will be an *average* density for the object. Things that have an average density less than that of water (e.g., most logs) will float, whereas things with an average density higher than that of water (e.g., most rocks) will sink.

Specific gravity is a common alternative way of expressing the density of a material or object. It is particularly useful for those who are concerned about things floating and sinking. Specific gravity expresses the density of the material or object as a fraction of the density of pure water:

Equation 3.5

$$SG_{material} = \frac{Density_{material}}{Density_{freshwater}}$$

For example, granite has a density of about 2,700 kg/m^3, so its specific gravity is:

$$(2{,}700\ \text{kg/m}^3)/(1{,}000\ \text{kg/m}^3) = 2.7$$

Note that there are no units listed for specific gravity. That is because density appears in both the numerator and denominator, so the units cancel out. Note also that the specific gravity of freshwater is going to be exactly 1, because any quantity divided by itself is always 1. Table 3.1 lists the density and specific gravity for granite, freshwater, seawater, and several other materials.

Table 3.1: Density and Specific Gravity of Various Materials

MATERIAL	DENSITY (kg/m^3)	SPECIFIC GRAVITY
Air (sea level, room temp, dry)	1.3	0.0013
Wood (balsa, dry)	170	0.17
Petroleum oil	880	0.88
Wood (live oak, dry)	950	0.95
Freshwater	1,000	1
Seawater (average)	1,025	1.025
Rock (granite)	2,750	2.75
Steel (average of several alloys)	7,800	7.8
Lead	11,300	11.3
Gold	19,300	19.3
Platinum	21,500	21.5

Specific gravity comparisons are especially useful in determining whether a particular material (or object) will float or sink in a particular fluid. If the material has a specific gravity less than that of the fluid, it will float on that fluid. (Example from Table 3.1: Dry balsa wood will float on petroleum oil.) If the material has a specific gravity greater than that of the fluid, it will sink in that fluid. (Example from Table 3.1: Dry live oak wood will sink in petroleum oil.) To determine whether something will float or sink in freshwater, simply check to see if the specific gravity of the material is less than or greater than 1. To determine whether something will float or sink in seawater, it's only slightly more complicated; just compare the specific gravity of the material to that of seawater, which is about 1.025.

The specific gravity figures in Table 3.1 agree nicely with practical experience: granite rocks sink, but dry wood floats. Thus, density and specific gravity resolve the lumber-and-rocks paradox presented at the beginning of this density section. *Chapter 6: Buoyancy, Stability, and Ballast* discusses the important role of density and specific gravity calculations in the design of underwater vehicles and provides a number of sample calculations. *Chapter 4: Structure and Materials* and *Appendix IV: Material Properties* list the densities and specific gravities for a number of materials used commonly to build underwater vehicles.

2.2.6. Hydrostatic Pressure

Pressure is defined as the amount of force per area. Whenever an object is immersed in a fluid on the earth (or anywhere else there is a gravitational field or some form of acceleration), it experiences pressure caused by the weight of the fluid above pressing down on it. In water, this pressure is called **hydrostatic pressure**.

The deeper the object is immersed, the taller (and therefore heavier) the column of fluid the object must support, and the greater the pressure acting on the object. As you can see from Table 3.1, the density of water is about 800 times greater than that of air,

SAFETY NOTE: DON'T RISK A LIFE!

Remember, this book does *not* provide enough information for you to design vehicles that can safely transport people or pets under water, even at shallow depths.

Dangerous pressures and pressure-related forces that can crush improperly designed vehicles and their occupants—even a meter or two (approx. 3–6 ft) under water—are only two of many potentially fatal hazards. Drowning, fire, electrocution, and toxic fumes are a few other hazards, and there are many more.

Fortunately, this book does provide enough information for you to have plenty of fun and adventure designing and building fairly sophisticated *unmanned* underwater vehicles, including ocean-going ROVs and AUVs.

If you dream of building submarines or submersibles for human crew or other passengers, that's great. Learning to build ROVs and AUVs with this book will provide you with a good foundation for more advanced study, but you will definitely need to follow up with additional training and certifications beyond what this book can provide.

so pressure increases about 800 times faster on an object descending through water than it does on an object descending at the same rate through air. For example, if you descend vertically 10 meters (about 33 ft, or three floors in a typical building) through air near sea level, the air pressure acting on your body will increase by about 0.1%. This tiny increase normally goes unnoticed. In contrast, if you descend from the sea's surface to a depth of 10 meters, the pressure acting on your body will double—that's a 100% increase! If not equalized properly, this pressure increase would probably cause severe ear pain and rupture your eardrum.

The normal air pressure acting on you at sea level is called, appropriately, one **atmosphere** of pressure and is equal to about 100 kilopascals (15 psi). Kilopascals are a metric unit for pressure; psi is a common imperial unit for pressure. (See *Chapter 5: Pressure Hulls and Canisters* for more information.) As a general rule, hydrostatic pressure increases by about 1 atmosphere for every 10 meters of additional depth. At a depth of 60 meters (approx. 200 ft), for example, pressure has increased by 6 atmospheres, going from 1 atmosphere at the surface to a total of 7 atmospheres (700 kPa or 105 psi) at 60 meters depth.

For underwater vehicle designers, the most important consequences of hydrostatic pressure changes are the mechanical forces associated with pressure *differences* between the inside and outside of submerged containers. The hull of a submerged submarine, for example, has high-pressure water pressing in from the outside and relatively low-pressure air (usually at about 1 atmosphere of pressure) pressing out from the inside. This causes a pressure imbalance, which produces an inward force on the submarine's hull. Because water pressure increases so quickly with depth, these pressure differences and associated forces can become enormous as the submarine dives, and they can crush the hull if the sub goes too deep.

To get a sense of the magnitude of this problem, consider the hydrostatic force pressing inward on a submarine hatch with a diameter of 63.5 centimeters (25 in) and a surface area of 0.32 square meters (approx. 3.4 ft^2) when the sub is operating at a depth of 100 meters (approx. 330 ft). Outside the sub, the water pressure at this depth is about 1,100 kilopascals. This external pressure is partially offset by the sub's internal atmospheric

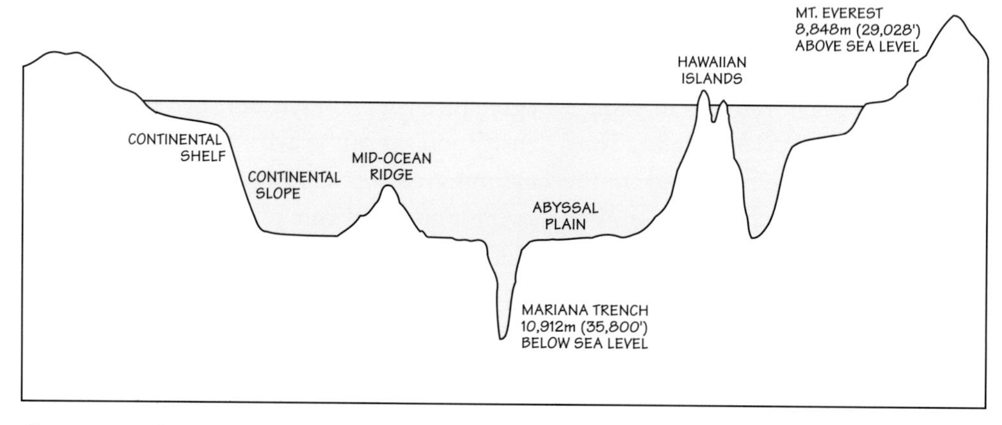

Figure 3.11: The Ocean's Features

The first scientific charts of the ocean floor issued in the late 1950s depicted the wide range of seafloor features—mountains, ridges, seamounts, trenches, continental shelves, and abyssal plains.

Of particular interest were the great average depths of the ocean, as well as the extreme depths, including the Mariana Trench and the Mid-Ocean Ridge, a range of globe-encircling underwater mountains. In the 1970s, this ridge was explored extensively by the French-American Mid-Ocean Undersea Study, using the French bathyscaph Archimède *and deep submersible* Cyana, *along with the U.S. submersible* Alvin.

TECH NOTE: THE PRESSURE CRUSH

On deep-diving operations, such as tours aboard *Mir* submersibles that take scientists and tourists some 3,800 meters (approx. 2.4 mi) down to view the remains of the *Titanic*, ordinary disposable foam cups are placed in a mesh bag attached to the submersible before its descent. When the vehicle returns to the surface, these cups have become miniature versions of themselves—exotic souvenirs bearing testimony to the intense pressure exerted at full ocean depth.

Image courtesy of Harry Bohm

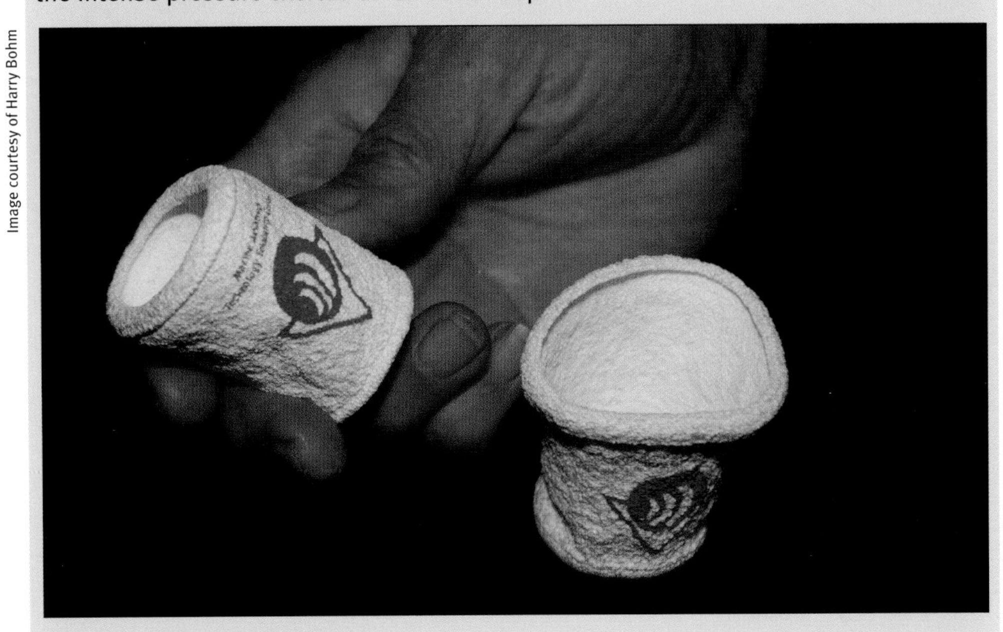

Figure 3.12: Pressure Souvenirs

These disposable coffee cups were significantly scrunched when taken to a depth of 1,250 meters (approx. 4,100 ft) on board the ROV Ventana, *owned and operated by the Monterey Bay Aquarium Research Institute.*

pressure of about 100 kilopascals, leaving a 1,000-kilopascal pressure difference that must be supported by the hatch. When you multiply this pressure difference by the surface area of a 63.5-centimeter diameter hatch, you end up with an inward force on the hatch of about 320 kilonewtons (approx. 72,000 pounds)!

If that much force is acting on a small hatch door, imagine the total force that must be acting on the entire length of a large submarine at that depth. Over the years, pressure-induced forces have crushed many submarines that went too deep, usually killing all crew members aboard. (See *Safety Note: Don't Risk a Life!*) Keep in mind that the hypothetical submarine in this example is only 100 meters (approx. 330 ft) deep. The ocean averages about 4,000 meters (approx. 2.5 mi) deep, and the deepest part of the ocean is roughly 11,000 meters (approx. 6.8 mi) deep.

Pressure is an important factor even for small, unmanned vehicles operating at swimming pool depths. Just a meter or two under water, increased pressure acting on the hulls or pressure canisters of vehicles can force water to enter through tiny existing cracks. It can also deform the structures enough to break seals and flood the interior of the vehicle. Water pressure can also compress squishy foam floats, reducing a vehicle's ability to return to the surface of the pool after a dive. More than one team at an ROV competition has learned this lesson the hard way.

Chapter 5: Pressure Hulls and Canisters will give you the necessary tools to calculate the hydrostatic pressure and pressure-related forces your vehicles will encounter at any

TECH NOTE: HULL COMPRESSION

Submarines certainly creak and groan under the strains imposed on them by pressure. World War II submariners often played a cruel trick on new crew members by tying a string across the inside diameter of the pressure hull and pulling it taut. As the greenhorns commenced their first deep test dive toward collapse depth, a nerve-racking experience even for old hands, the string would sag noticeably as the pressure hull shrunk inward under the relentless water pressure.

Modern submersible pressure hulls also compress under pressure. Harbor Branch Oceanographic Institution's *Johnson-Sea-Link* submersible has a 5 1/4-inch thick Plexiglas™ sphere that routinely contracts some 3/8 inch when the submersible is at its working depth of 3,000 feet (approx. 915 m).

depth and will teach you how to build unmanned vehicles that can withstand those pressure-related forces. Until you get to that chapter, remember that pressure increases very rapidly as a vehicle descends through water and that this pressure increase can cause all sorts of problems, from leaking seals to complete vehicle collapse.

TECH NOTE: USING WATER PRESSURE TO ADVANTAGE

Whenever possible, clever inventors and engineers try to make underwater pressure work for them, rather than against them. For example, underwater viewing ports are often designed with a conical shape, so that increasing water pressure wedges the port ever more tightly into the opening, thereby creating an excellent seal.

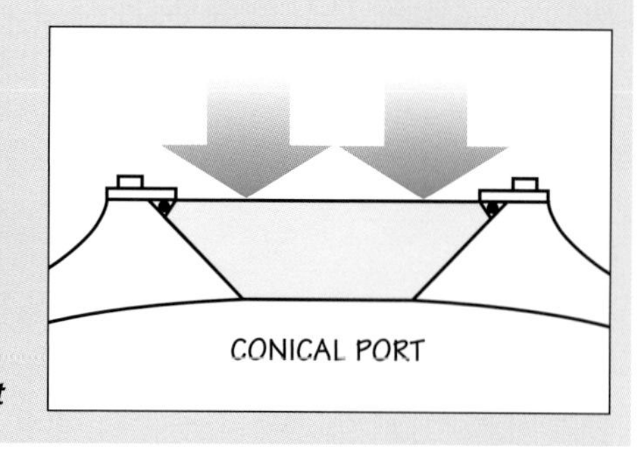

Figure 3.13: A Conical Viewport

2.2.7. Buoyancy

The term **buoyancy** has two related yet distinctly different meanings when describing the interactions between objects and fluids:

- In one case, buoyancy is the upward force exerted by the fluid on an object immersed in it.
- In the other case, buoyancy is the overall tendency of the object to float in the fluid, and that depends both on the (first definition of) buoyancy and on the weight of the object.

To avoid confusion, this text refers to the first type of buoyancy as **buoyant force**, and reserves the term "buoyancy" for the second definition. Be warned, however, that this explicit distinction is not often made elsewhere—usually the context makes it clear which definition is being used.

Remember that pressure always increases with depth. An important side effect of this fact is that the pressure pushing up on the bottom of a submerged object is always greater than the pressure pushing down on the top of it. The net effect is that some or all of an object's weight is always supported by the fluid in which it is immersed. (See Figure 3.14.) The same buoyant force effect is also present in air, but usually it is so small that it goes unnoticed. Blimps and hot air balloons *do* obtain significant lift from the buoyancy provided by air pressure, but only because their volumes are so enormous. In water, however, the effects of buoyant forces are quite dramatic, for both small and large vehicles. That's why a tiny cork and an aircraft carrier made of tens of thousands of tons of steel can both float easily in water without sinking.

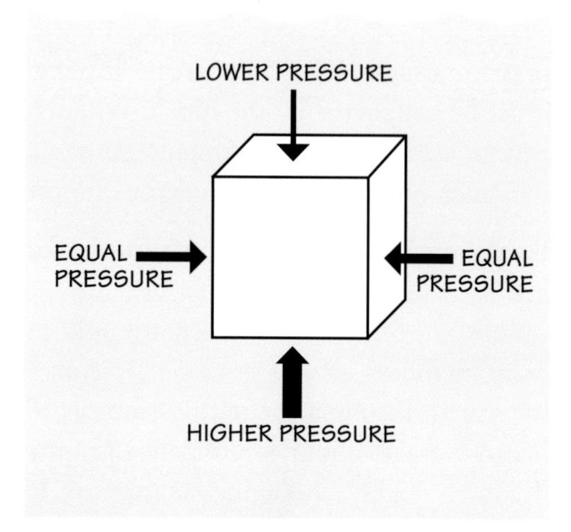

Figure 3.14: Buoyant Force

Water pressure is greater on the lower (deeper) side of an object and less on the upper (less deep) side. This results in a net upward force called buoyancy. The pressures on the right and left (and front and back) sides are the same and cancel each other out.

Most underwater vehicles are designed so that their buoyant force is about equal to their weight. Balancing weight with buoyancy in this way allows the vehicle to hover effortlessly (a condition known as **neutral buoyancy**) in mid-water and makes it possible to control up and down movements of the vehicle with gentle effort from vertical thrusters. Near neutral buoyancy is achieved by making the vehicle's average density approximately equal to the density of water. Although simple in concept, controlling buoyancy can be difficult in practice, particularly if air or other compressible gases are used as part of the buoyancy control system. These buoyancy topics, including Archimedes' Principle, are covered in detail in *Chapter 6: Buoyancy, Stability, and Ballast.*

HISTORIC HIGHLIGHT: DIFFERENT WATER, DIFFERENT DENSITY

Seawater is slightly denser than freshwater—an important fact to remember in the subsea world. The first pilot of the Reynolds *Aluminaut* experienced the difference between freshwater and saltwater density in 1964, in a near disaster. When this 76-ton aluminum submersible underwent sea trials in Long Island Sound, it was ballasted (weighted) for saltwater density. However, when crossing the mouth of the Connecticut River, the *Aluminaut* began to sink rapidly. Why? The buoyant force provided by the freshwater was less than that provided by the saltwater, hence less able to support the weight of the craft. Fortunately the pilot was alert and quickly blew water out of the ballast tanks, dropped weights, and applied full upward power to the vertical thruster. Through his fast action, he managed to prevent a dangerous meeting of submersible and seafloor.

Figure 3.15: **Aluminaut**

Image courtesy of Woods Hole Oceanographic Institution

2.2.8. Compressibility

In general, gases have greater **compressibility** (i.e., they are "squishier") than liquids. For example, if you increase the pressure acting on a cubic meter of air by a factor of 100, the air volume will be reduced to 1/100th of what it was before. If you do the same thing with a cubic meter of water, its decrease in volume, though present, will be so small that you can't even see it. This is a hugely important difference for the underwater vehicle designer. (See Figure 3.16.)

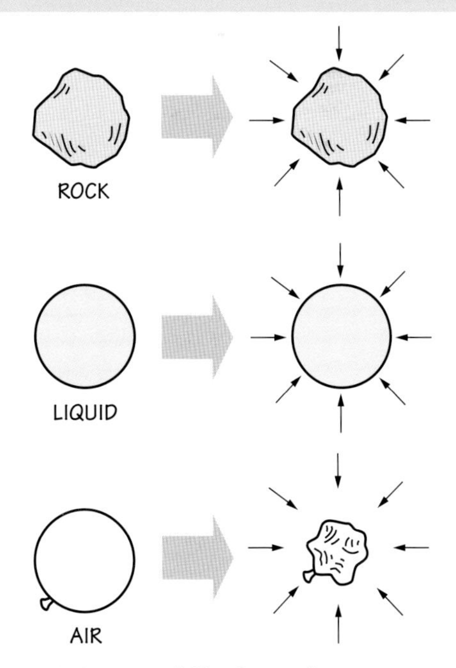

Figure 3.16: Compressibility Comparison

Solids compress the least amount under pressure and liquids compress only very slightly. However, gases (including air) compress easily. Hence, hollow-pressure hulls must be designed to resist water pressure's compressive forces.

There are two places where this difference really comes into play. First, gases exposed to the ambient water pressure surrounding a vehicle will shrink in volume as the vehicle goes deeper and expand as it rises. This can greatly complicate control of vehicle buoyancy, as described in *Chapter 6: Buoyancy, Stability, and Ballast.*

In the second instance, hollow compartments such as electronics canisters can sometimes be filled with fluids instead of gases to help them support the crushing hydrostatic pressures encountered at depth. (See *Historic Highlight: Fighting Pressure with Pressure.*) This strategy is discussed in more detail in *Chapter 5: Pressure Hulls and Canisters.*

HISTORIC HIGHLIGHT: FIGHTING PRESSURE WITH PRESSURE

Compared to gases, solids and liquids are essentially incompressible, so they can resist tremendous compressive pressures without much change in volume. The designers of *Trieste* took advantage of this fact to create a unique flotation system for their submersible. They modeled their vehicle after a dirigible, which suspends a small cabin for people beneath a large hull filled with a lighter-than-air gas, such as helium. *Trieste* featured a small pressure-resistant sphere for her two crew members, suspended beneath a 50-foot long x 12-foot diameter (approx. 50 m x 3.5 m) hull filled with lighter-than-water gasoline. If the hull had been filled with air, it would have been much more buoyant, but it would never have survived *Trieste*'s dive to the bottom of the deepest part of the ocean, where pressures are sufficient to compress air to less than 1/1000th of its surface volume. With incompressible gasoline supporting the inside of the hull, the water pressure could not compress the hull far enough to cause damage. Modern ROVs and AUVs use a different strategy for flotation, but incompressible fluids are still used to fill thruster housings, electronics canisters, and many other containers that must survive extremely high pressures.

Photo #NH 96799, courtesy of U.S. Naval Historical Center

***Figure 3.17:* Trieste**

Trieste's unusual shape provided safe accommodation for crew and room for sufficient gasoline flotation.

2.2.9. Viscosity

The thickness or "goopiness" of a liquid or gas is referred to as its **viscosity** (more precisely, its "dynamic viscosity"). Honey and molasses are examples of fluids with high viscosity. The dynamic viscosity of fluids tends to be strongly dependent on temperature—the colder the temperature, the greater the viscosity. At typical room temperature, freshwater has a viscosity about 50 times greater than that of air. This can affect the movement of underwater vehicles (see next section on drag) and can impact underwater **visibility** by allowing water to suspend particles in it more easily than in air. (See *Section 2.4: The Optical Properties of Water.*)

2.2.10. Drag

When an underwater vehicle (or any other object) moves through water, the fluid exerts a force that tries to resist the movement of the object through the fluid to some extent. Technically, the term "fluid" includes both gases and liquids, so this applies to vehicle movement through both air and water. This resistance to relative motion is called **drag**. Drag comes from two sources: **pressure drag** and **viscous drag**. When an object moves through fluid, both sources of drag are always present, but one usually dominates over the other.

For very tiny objects, such as single-celled animals, viscous drag is the dominant source of drag. But for larger, faster-moving objects like underwater vehicles, pressure drag usually dominates and directly affects vehicle motion. Pressure drag is caused by fluid inertia, which makes it hard for the vehicle to accelerate water out of its way as it tries to pass through the fluid. The fluid inertia also reduces the fluid's ability to flow instantly back into the space left behind the passing vehicle, so it creates a sort of "suction" in the back. Together, these effects create a net force pushing on the front and pulling on the back, thereby resisting the vehicle's motion. Water is about 800 times denser than air, so pressure drag is a significant factor with big consequences in underwater vehicle design. Any underwater vehicle designer is wise to remember these drag issues:

- Given equal speed requirements, pushing a vehicle through water typically requires much more power than pushing it through air.
- Maximum practical speeds are much less in water than in air.
- Vehicles coast to a stop more quickly in water than they do in air.

See *Chapter 7: Moving and Maneuvering* for more information on how pressure drag impacts underwater vehicles.

The second source of drag is called **viscous drag** or **skin friction drag**. Viscous drag is caused by the friction of water molecules sliding past each other and past the surface of the object as the object moves through the fluid. This type of drag is most significant for very small, slow-moving particles such as silt grains, single-celled organisms, or other microscopic objects. Although viscous drag is not as significant as pressure drag in resisting typical underwater vehicle movements, it does help water support large amounts of suspended particulate matter. That can impact underwater visibility for vehicles, thereby affecting navigation, as described later in this chapter.

2.3. Acoustic Properties of Water

Sound is the propagation of mechanical vibrations through a solid, liquid, or gas. The study of sound is called acoustics, and the sub-specialty that examines sound propagation through water is called **hydroacoustics**. Water's acoustic properties are directly related to its mechanical properties and could be included under that heading, but hydroacoustics plays an important enough role in underwater vehicle operations to warrant its own section.

HISTORIC HIGHLIGHT: EXPERIMENTING WITH UNDERWATER SOUND

Leonardo da Vinci (1452–1519) appears to have made experiments in the transmission of underwater sound. He may have used an underwater listening tube as a receiver and a bell that clanged under water as a transmitter. When he placed the bell some distance away from the listening tube, he noticed a delay between the time the bell was rung under water and the time he heard the sound with the listening tube, so he knew underwater sound transmission was not instantaneous.

But he also noted that the sound coming through the water reached his ears sooner than the sound traveling the same distance through air. Thus, he observed that sound travels more quickly through water than through air.

HISTORIC HIGHLIGHT: UNDERWATER ACOUSTICS

In 1826, Daniel Colladon first measured the speed of sound in water at Lake Geneva, Switzerland, laying the foundation for underwater acoustics as a research discipline. His experiment involved positioning two boats 16 kilometers (approx. 9.9 mi) apart. Then someone on the first boat hit an underwater bell with a hammer while simultaneously igniting flash powder. Using a stopwatch, he compared the time it took for the receiving boat to "hear" the bell to the time it took to see the bright flash. Of course, the flash was seen instantaneously, having traveled at the speed of light. Colladon determined the speed of sound under water to within .21 percent of the currently accepted value of 1,438 meters/second (approx. 4,718 ft/sec).

TECH NOTE: SOUND IN AN "INCOMPRESSIBLE" FLUID?

Sound travels through air, water, and other media as waves of pressure, which alternately compress and expand the spaces between the molecules. In a gas, this makes sense. Gases are "squishy," so it's easy to visualize how the gas can be compressed and expanded. On the other hand, solids and liquids are often described (even in this book) as "incompressible," so how can sound propagate through them? The reality is that *everything* is compressible to some extent, even water or solid rock. Sound propagation is all about energy transfer through the medium. If the medium is nearly incompressible, as water is, then a microscopic bit of compression can still store and propagate plenty of energy.

Image courtesy of Patrick L. Barnard and Daniel M. Hanes of United States Geological Survey, Coastal and Marine Geology Program, and Rikk G. Kvitek and Pat J. Iampietro of Seafloor Mapping Lab, Division of Science and Environmental Policy, CSUMB

Figure 3.18: Giant Underwater Sand Waves Off San Francisco's Golden Gate Bridge

The U.S. Geological Survey collaborated with the Seafloor Mapping Lab at California State University, Monterey Bay, to generate these high-resolution images of an enormous sand-wave field hiding under water 30 to 100 meters (approx. 100 to 325 ft) deep near the mouth of San Francisco Bay. A research boat used sonar equipment that bounced sound off the seafloor and timed the echoes to measure water depth as the boat crisscrossed the area. These depth measurements were combined with precise position data from GPS satellites to create these images. Color-coding highlights the depth.

2.3.1. Sonar

As it turns out, water propagates sound quite well, and this fact is exploited extensively for underwater vehicle operations. Sound provides an effective means of communicating information and acoustically visualizing the environment in a world where the radio signals and cameras (that can do these things on land) don't work very well. Most significant among the uses of underwater sound is with a device called sonar. (See Chapter 1 *Tech Note: Sonar.*) A simple example of sonar use can be found in an **echosounder** (also called an **altimeter** or **fathometer**) used on many ROVs and AUVs to measure their distance above the seafloor. These devices transmit a pulse of sound directed downward toward the seafloor. After sending the pulse, the altimeter listens for the return echo and times how long it takes for the sound to make the round-trip to the seafloor and back. It then multiplies this time by the known speed of sound through water (roughly 1,500 m/s, approx. 5,000 ft/s) to calculate the round-trip distance to the seafloor. Then it divides the round-trip distance by two to get the one-way distance.

Other sonar devices, such as sector scanning sonar and multibeam sonar, use the same basic idea, but they keep track of both the distance and the direction to underwater objects. This information can be used to construct sound-based "images" of the surroundings, including terrain, sunken ships, submarines, fish schools, and other objects. (See Figure 3.18.) In water, sonar can almost always "see" farther than cameras, and it works even in completely dark or muddy water, where light-based vision systems are useless, so it's an extremely valuable tool in underwater work.

An array of sound transmitters and receivers anchored to the bottom or attached to the underside of a ship can use sound to pinpoint the location of an ROV or AUV while it's working. In this approach, sound propagation time is used to determine the distance from the vehicle to each of the points in the array, and that information is used to calculate the exact location of the vehicle.

In addition to using sonar for seabed visualization, sound can be used to directly communicate information to and from an underwater vehicle. Unfortunately, there are limitations as to how quickly information can be sent in this way. Sound is adequate for sending text commands, small data amounts, or simple voice messages, but it is not adequate for sending live video images or other signals that require the transmission of large amounts of data in a short amount of time. (For more sonar specifics, check out *Chapter 9: Control and Navigation*, *Chapter 10: Hydraulics and Payloads*, and *Chapter 11: Operations.*)

2.4. Optical Properties of Water

2.4.1. Absorption

When light shines through water, some of the light energy is absorbed by the water molecules. This **absorption** happens even in perfectly pure, clean water. The farther the light travels through the water, the more light gets absorbed. For example, as sunlight descends through the ocean from the surface, it gets filtered out even in the clear waters of a tropical sea. Three meters (approx. 10 ft) down, the effect is hardly noticeable, but by 300 meters (approx. 990 ft) down, it's getting pretty dark.

An interesting aspect of this light absorption is that different colors are absorbed at different rates. Among the colors in the visible light spectrum, red is absorbed most quickly, with orange, yellow, green, and blue following in that order. (See Figure 3.19.) Ten meters (approx. 30 ft) under the surface, for example, the underwater scene has taken on a decidedly bluish or greenish color. A red fish or a red sponge will look dark brown at that depth, because most of the red light that would bring out those colors is gone. Going still deeper, say to 30 meters (approx. 100 ft), you enter a world of deep blues and essentially no other visible colors. Going still deeper, even the blues gradually fade away until all hints of sunlight from the surface are gone, and the only natural light present is specks of light from bioluminescent animals.

If you want to see true colors more than a couple of meters below the surface, or see at all at night or in very deep water, your vehicle needs its own light source, such as headlights or video lights. These can cut through the darkness and restore natural colors, bringing out the brilliant reds, oranges, and yellows of some sponges, corals, and other marine organisms.

Figure 3.19: Light Penetration with Depth

Different colors penetrate water to different depths. Reds and oranges get filtered out most quickly, while blues penetrate the farthest, which is why photos taken at depths of about 30 meters (approx. 100 ft) are almost pure blue in color.

Image courtesy of Neil McDaniel

Figure 3.20: Color Change with Depth

All objects in the background of this underwater photo appear blue or black, because at the depth where this picture was taken, the optical properties of water have filtered out all other colors.

The bright yellows of this crinoid, a marine animal, are visible only because bright artificial lights on the camera illuminated the foreground with white light.

2.4.2. Turbidity and Backscatter

Even with lights, **visibility** can be severely limited under water. Natural water bodies are rarely as clear as the tropical seas described above. Most natural water has microscopic bits of clay, plankton, or other material suspended in it, giving it a cloudy appearance. In extreme cases (like some rivers flowing through easily eroded land), the water may be so full of particulate matter that it is essentially opaque, looking more like brown paint than water. **Turbidity** is the technical term for cloudiness caused by suspended particles (usually microscopic but sometimes sand-sized).

Due to its density and viscosity, water has a greater capacity than air does to keep particulate matter suspended for long periods. To see this yourself, put a pinch of flour into each of two jars—one full of air and one full of water—and then cap the jars and shake them up. After you stop shaking, you'll see that the flour settles very quickly to the bottom of the jar with air, but it remains suspended for a long time in the water-filled jar. In natural bodies of water, the particles causing turbidity are usually bits of clay, silt, sand, microscopic plants and animals, or other organic debris.

Underwater craft that must navigate and work in dark or turbid waters carry powerful lights to illuminate the depths, but even these can push back the gloom for only a short distance. If the placement of lights is too close to the camera lens, light gets reflected back from the suspended particles and straight into the lens, an effect called **backscatter**. The resultant images are filled with bright specks, making them look as if they were taken in a heavy snowstorm or blizzard. Fortunately, there are other helpful technologies, such as infrared cameras (which are less sensitive to backscatter and can see a bit farther in murky water) and sonar imaging, which uses sound waves instead of light waves to "see" objects. Sonar can penetrate extremely dark or murky waters.

TECH NOTE: A SIMPLE WAY TO MEASURE VISIBILITY

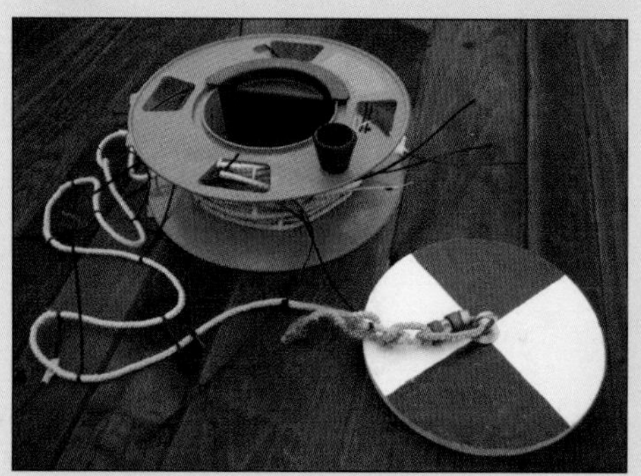

Image courtesy of Dr. Steven W. Moore and Dr. Fred Watson

Figure 3.21: Secchi Disk

How do you measure visibility in water? Scientists often use a simple yet effective device called a **Secchi disk**. You can easily make one for yourself. It typically consists of a metal or weighted plastic disk, 20 centimeters in diameter, with alternating white and black quadrants. (See Figure 3.21.) The center of the disk is attached to a rope that has been marked off in regular increments, say every 10 centimeters. The rope is used to lower the disk slowly into the water (with the white/black pattern facing up) until it just disappears from view. Then it's pulled back up a bit until the disk just barely becomes visible again. At that point, the marks on the rope can be used to determine how far the disk is beneath the water's surface.

The result, known as the **Secchi depth**, provides a standardized measure of the attenuation of light in the water column. The method is somewhat subjective, because it depends on lighting conditions and other factors. More precise methods are available, if needed, but the simplicity and low-cost of the Secchi disk method makes it extremely popular for most routine measurements of water clarity. For best results, the method should be used on the shady side of a boat or dock within about 2 hours of noon.

Pilots of vehicles operating near soft bottoms of sand or mud must be careful not to stir up so much sediment with the thrusters that they can no longer see anything through the camera. In extremely turbid environments, gritty sediment can actually damage vehicles by wearing away seals around propeller shafts or by otherwise interfering with moving parts.

Attenuation refers to the overall decrease in light intensity with distance as light passes through water (or another substance). Attenuation in natural bodies of water results from a combination of absorption (by water molecules) and scattering (by particles). Visibility in these waters can range from 0 to about 300 meters (approx. 990 ft), with poor visibility (tens of meters or more). (See *Tech Note: A Simple Way to Measure Visibility.*)

Image courtesy of Dr. Steven W. Moore

Figure 3.22: Refraction

The light rays traveling from the submerged part of this pencil to the camera must first pass through water then through air, thereby arriving from a different angle than those coming directly through the air. As a result of this refraction, the viewer gets the impression that the pencil is split in two. Note that the underwater portion of the pencil also appears to have a larger diameter compared to the part of the pencil that's in air. Such magnification is a common consequence of refraction in air/water vision systems, though the magnification factor and other details depend on the shape of the air/glass/water interface.

2.4.3. Refraction

One more optical difference between water and air is worth mentioning. Light travels at different speeds through different transparent materials, such as air, water, glass, or plastic. The speed of light through a material is related to that material's refractive index. When light rays pass between two materials having different refractive indices, the light rays bend. This bending, called optical **refraction**, can be observed easily by putting a pencil into a clear glass of water and noticing how the pencil appears to break and/or bend where it enters the water.

Camera lenses and eyeglasses rely on refraction to form clear images by bending light in carefully controlled ways as it passes from air to glass and back to air again. However, since the shape of these lenses is based on an air-glass interface, they do not work well under water. Even though the lenses of our eyes are not made of glass, they work on the same principle and require an air-eyeball interface to work properly. That's why everything appears blurry under water unless you wear a face mask or goggles to trap some air in front of the lenses of your eyes.

Refraction also has important optical consequences for underwater vehicles. One consequence of refraction is that the light rays get bent so that a person or camera looking out into the water sees a narrower field of view, as if looking through a telephoto lens. This makes underwater objects look larger and closer than they really are when viewed through a flat viewport. This is something an ROV pilot needs to be aware of. Flat ports also create greater color separation under water, causing some images to lose their sharpness.

Figure 3.23: Refraction Effects

When viewed through a flat viewport, bent light rays can make underwater critters seem larger, closer, and more ferocious than they really are!

2.5. Thermal Properties of Water

Water has some unusual thermal properties that can influence underwater vehicle design. These include its large heat capacity and thermal conductivity.

2.5.1. Heat Capacity

For one thing, water has an unusually high **heat capacity**, meaning it can absorb or release a lot of heat energy without large changes in its temperature. This capacity enables the ocean to stabilize temperatures on planet Earth and helps keep the temperature of the planet within a range compatible with life. It also means that the temperature of any given part of the ocean does not change very much over time. This makes life much easier for the underwater vehicle designer, since so many things are strongly dependent on temperature, from the output of electrical sensors to the strength and stiffness of structural materials.

Still, water does have different temperatures at different times and places, and this can have additional implications for vehicle design. For example, water temperature affects water density. Warmer waters are less dense and therefore more buoyant than colder waters (except within 4° Celsius of freezing, where the reverse is true). This causes some portions of ocean water to sink while others rise, and that plays an important role in driving ocean currents. Temperature-dependent water density also affects the buoyant forces acting on submerged vehicles. Finally, temperature affects the mechanical properties of some plastics and other building materials. For example, PVC pipe is a great, inexpensive building material for warm-water applications, but may become too brittle to use reliably in waters that are near freezing.

2.5.2. Thermal Conductivity

Water also has a much higher **thermal conductivity** than air. This means that cold water can draw heat out of a warm object much faster than cold air can. Likewise, warm water can pump heat into a cold object much faster than warm air can. That is why you can be comfortable in a room at typical "room temperature" but will quickly become chilled in a swimming pool or pond with water at that same temperature. Because of this high thermal conductivity, a vehicle's hull will quickly come to the same temperature of the surrounding water. High-powered vehicles can use this property to their advantage by allowing the water to act as a natural heat sink, drawing away excess heat before it can cause problems with the vehicle. Unfortunately, this can result in other problems. For example, a camera port can become cooled to the point that water condenses on the inside of the port, causing it to cloud up and ruin the camera view.

3. Water Movements

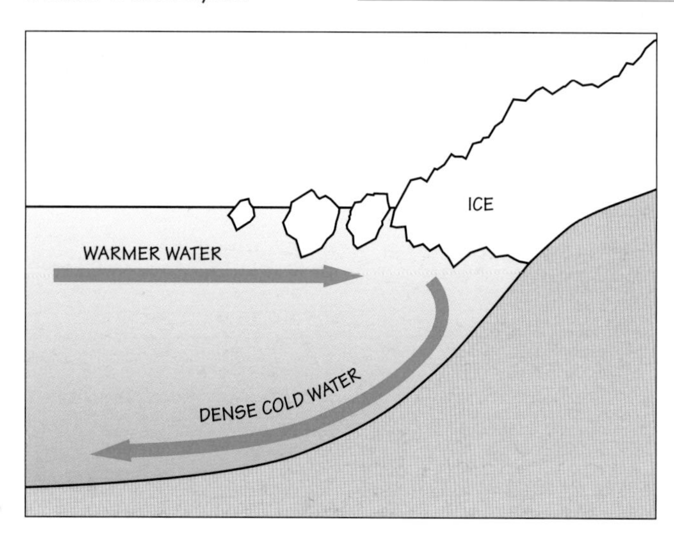

Figure 3.24: Thermal Movement and Major Deep Water Currents

Major deep water currents occur when cold, dense water from the polar regions sinks and spreads outward toward the equator. Warmer, less dense water flows in to replace it closer to the surface.

Like air, water in the natural world is almost always in motion. Even in seemingly calm ponds or lakes, water is usually flowing, though sometimes those flows are imperceptibly slow. In other places, water motion is more obvious—sometimes even violent—as when waves crash against a rocky shore. In rivers or in the ocean, water motion is a major factor that must always be considered when designing or operating underwater vehicles.

3.1. Currents

More or less steady flows of water moving from one location to another are called **currents**. They may last

HISTORIC HIGHLIGHT: ATTEMPTS AT MAPPING CURRENTS

In early times, hardy men attempted underwater salvage and construction, but their understanding of other aspects of the ocean was extremely limited. A prime example is the **Gulf Stream current** that originates in the Gulf of Mexico and flows all the way north to Iceland. The existence of this mysterious ocean highway was noted by the Spanish explorer Ponce de Leon in 1513, but it wasn't mapped until the 1750s, when Benjamin Franklin became intrigued while serving as Deputy Postmaster General of the American colonies. He pieced together scraps of information about the Gulf Stream from sea captains and whalers and plotted the path of this massive current. Savvy American ship captains used his chart to gain swifter transit from North America to Europe, then returned home by an alternate route.

A century and a half later, Prince Rainier of Monaco discovered that the Gulf Stream current splits as it gets to eastern Canada, with one branch heading northeast toward Ireland and Great Britain and the other curving south past Spain and Africa. Avidly committed to oceanographic research but lacking today's scientific instrumentation, this early oceanographer relied on a simple technique to make his discovery: he launched a flotilla of glass bottles, wooden barrels, and copper balls, all stuffed with requests for information, written in 10 languages. Prince Rainier founded the Oceanographic Institute in Paris in 1906, and the Oceanographic Museum in Monaco in 1910.

A somewhat similar, if unintentional, mapping of ocean currents occurred in May 1990, when a storm-tossed cargo ship lost a container of Nike running shoes in the northeastern Pacific. Over the next several years, the 60,000 shoes ended up providing useful data abut surface currents as they washed ashore on beaches around the world.

Figure 3.25: Global Ocean Currents

This global ocean currents schematic is derived from NOAA's Global Drifter Program data. Additional information is available at http://www.aoml.noaa.gov/phod/dac/gdp.html.

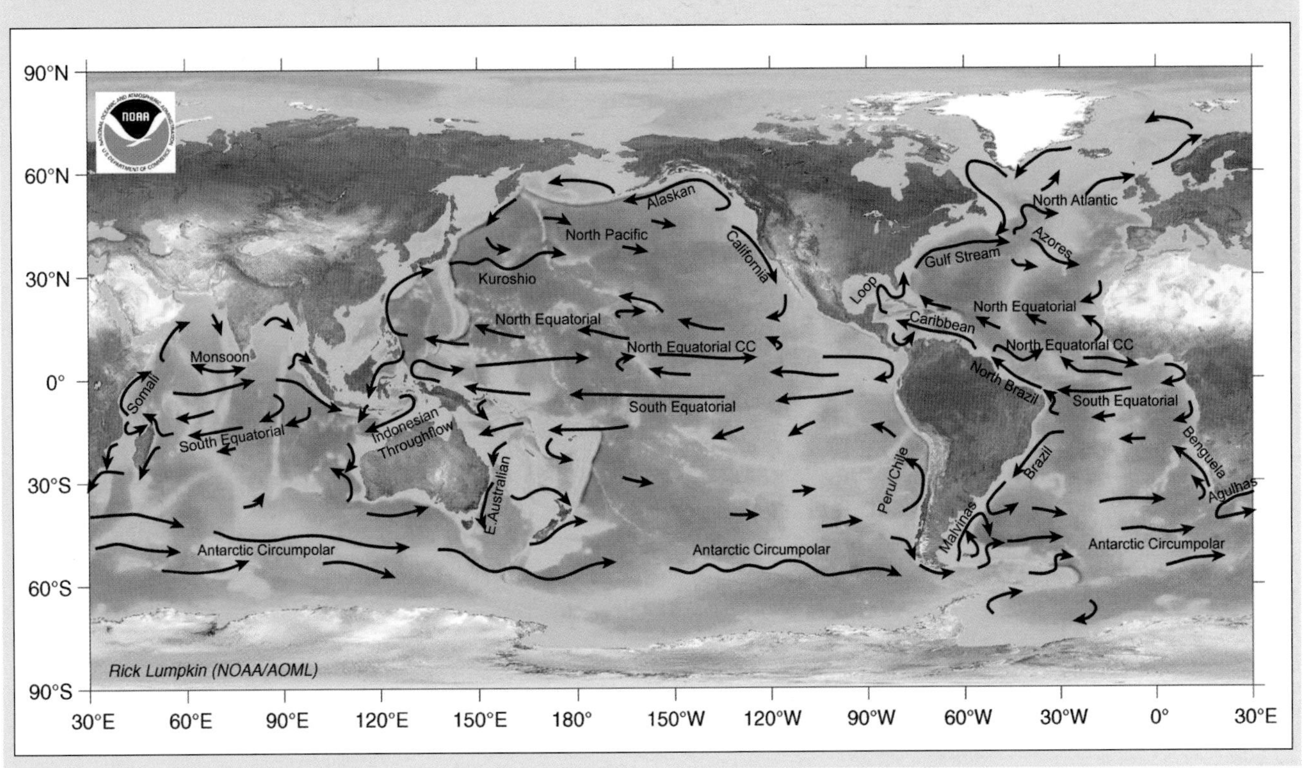

from a few minutes to thousands of years, depending on where they are located and what drives them. They may be caused by gravity (as when water flows downhill in a river or when hot, less dense water from a deep sea thermal vent rises through the colder, denser water surrounding it), by wind (as when wind friction moves water across the surface of a lake), or by complex interactions among wind, temperature, salinity, gravity, topography, and the earth's rotation (such as the giant currents that circulate throughout the world's ocean.) (See Figure 3.25.) There are many types of currents:

- **Long-shore currents** flow parallel to shore; they are often part of much larger ocean current systems.
- What appears to be a one-directional current on a small scale is often part of a large mass of rotating water, like an **eddy**. In the ocean, very large eddies (sometimes thousands of miles in diameter) are called **gyres**.
- **Rip tides** are misnamed in that they are actually strong currents (not tides) that flow offshore from places where breaking waves pile water up along a coastline. Rip tides are a threat to unwary swimmers and those launching or working with ROVs in such areas.
- In some places, silt or other material mixed with water can cause water density to increase, resulting in the rapid sinking of such water. The resulting **turbidity currents** can be very violent and powerful, scouring rock to form underwater canyons.
- Another type of current is a **tidal current** (see next paragraph).

3.2. Tides

Tides are changes in the height of the ocean surface caused by interactions between the gravitational attraction of the sun and moon, the mass of the ocean water, the rotation of the earth, and undersea topography. Tides typically cause sea level to rise and fall once or twice a day, depending on location. The difference between water levels at high tide and low tide varies, depending on the phase of the moon, the season, and the location. This tidal change tends to be less near the equator and more at higher latitudes. For example, in the tropics, tidal exchange is often less than 1 meter (approx. 3 ft), whereas in Nova Scotia's Bay of Fundy, the high and low water levels vary as much as 16 meters (approx. 52 ft)! In shallow areas, the tidal changes in water depth can temporarily strand boats. **Tidal currents** are created as bays, river mouths, lagoons, estuaries, shallow coastal areas, and other partially obstructed portions of coastal areas fill or drain in response to tidal ocean height fluctuations. These tidal currents may be mild or very strong—and potentially dangerous. For example, directly under the Golden Gate Bridge, at the mouth of San Francisco Bay, tidal currents can exceed 2 meters per second (over 5 mi/hr), which is fast enough to pile up huge underwater sand waves (see Figure 3.18) and much faster than most small underwater vehicles or human swimmers can move in water.

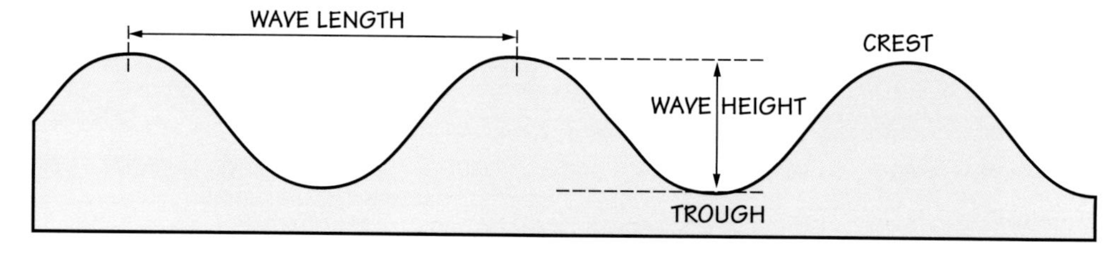

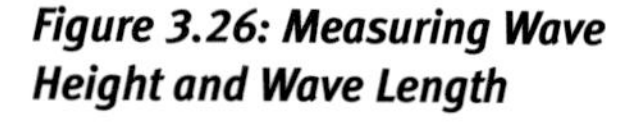
Figure 3.26: Measuring Wave Height and Wave Length

The height of a wave is measured from its trough (lowest point) to its crest (highest point). Wave amplitude is ½ the wave height. Wave length is the horizontal distance between two crests.

3.3. Waves, Swell, and Surge

Waves are another common form of water movement. Waves are produced when wind blows over the surface of water. The wind pushes on tiny ripples, gradually increasing their size. The longer and faster the wind blows, the larger the waves can become. During large storms in the open ocean, waves can be more than 15 meters (approx. 50 ft) high. There are credible recordings of rogue or "freak" waves even taller than that. Books and movies such as *The Perfect Storm* feature these freak waves.

Far from the storms where big ocean waves originate, the waves may spread out into a smooth, undulating form known as **swell**. This is the classic water motion that causes seagoing ships to slowly rise and fall at sea. Large swell or large wind-driven waves can make the launch and recovery of underwater vehicles from support ships technically challenging and very dangerous.

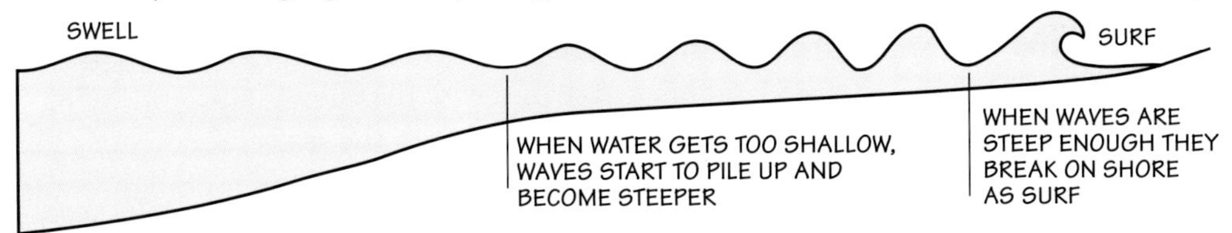

Figure 3.27: Wave Formation

As waves approach the shallow water off the coast, they slow down and bunch up. Wave length gets shorter, but wave height increases. If the waves get steep enough, they form surf waves, called breakers, which crash on the shore. Wind direction and type of coast affect three main types of surf waves: spilling, plunging, and surging. Surfers love spilling breakers!

As a diver or ROV descends toward the seafloor in shallow areas, the swell at the surface transforms into surge near the bottom. **Surge** is a back-and-forth motion of water parallel to the seafloor. It can be quite strong and can make maneuvering ROVs in shallow water very challenging. It can also smash ROVs into submerged rocks.

Another change occurs when open ocean swell approaches shallow water. The lower portions slow down due to friction with the bottom, causing the upper parts of the swell to outrun the lower parts. As a result, the swell piles up and forms a **breaking wave**, sometimes called **surf**. Near shore, large breaking waves can capsize or crush boats and other vehicles.

TECH NOTE: WAVE MOVEMENT

Waves appear to move forward, but the actual particles of water in a wave move mainly in a circle, returning to nearly their original position, over and over again. Any object, such as a bottle, floating in the water follows the circular path of these water particles as the energy passes through them. If the object moves horizontally, it is primarily because of the forces of wind, tide, or currents acting on it, rather than wave action.

Figure 3.28

3.4. Tsunamis

On rare occasions, an underwater earthquake or landslide can create a sudden shift in sea level that creates a very dangerous wave or series of waves known as a **tsunami** (sometimes called a tidal wave).These can come ashore with little warning, pushing deadly walls of churning debris in their paths, and wreaking devastation on low-lying seashore communities. Sometimes ocean water recedes shortly before the waves hit, attracting curious onlookers to the beach and increasing the casualties. On December 26, 2004, a 9.2-magnitude earthquake off the island of Sumatra, Indonesia, triggered a series of deadly tsunamis. As the sea rushed in over the top of coastal towns and villages, it killed approximately 230,000 people throughout many parts of Indonesia, Sri Lanka, India, and Thailand. Some victims were killed as far away as South Africa, 8,000 kilometers (approx. 5,000 mi) from the epicenter of the earthquake.

Tsunamis are rare events and highly unlikely to affect your ROV or AUV operations, but if you are at the seashore and feel an earthquake or observe an unusual retreat of water resembling a sudden very low tide, it would be prudent to move quickly to higher ground.

4. Plant and Animal Life in Water

The total volume of the ocean provides 300 times more space for life than terrestrial and freshwater environments combined. That's impressive, when you consider how much life grows, walks, or crawls on the surface of the earth or flies in the atmosphere. Every cubic centimeter of the sea supports life! Most life in the sea is microscopic, but some of it, including the great whales, is huge. One important reason for building and piloting underwater vehicles is to explore, observe, and study this amazing **biodiversity**.

Unfortunately, the same plants and animals that are interesting to study can also create problems for those who design, operate, or maintain underwater vehicles.

4.1. Biofouling

Vehicles that stay in the water for more than a few days usually become covered in growths of algae, barnacles, sponges, or other organisms. This **biofouling** is difficult to prevent and can increase drag on the vehicle, jam moving parts, or obscure the view through otherwise clear camera ports. The anti-fouling paints and other coatings used to discourage the growth of marine organisms on ship hulls can also be applied to underwater vehicles to reduce this problem. There are even some clear coatings for discouraging growth on optical surfaces, like the periscope lenses on submarines.

Figure 3.29: Biofouling

Just as vehicles can have issues with biofouling, any structure that stays underwater will attract marine growth. In this photo, a Sonsub ROV inspects drill rig legs covered in algae, barnacles, sponges, or other organisms.

Image courtesy of Sonsub/Saipem

4.2. Entanglement

An even greater threat to underwater vehicles, particularly small ones used near shore, is getting tangled in something growing on, or otherwise attached to, the bottom. For example, the vehicle or its tether can become entangled and hopelessly trapped in kelp, some of which resembles strong, thick, underwater vines. Vehicles or their tethers can also get caught in the branches of dead, submerged trees.

In areas where fishing is popular, the presence of fish or shellfish usually leads to the presence of submerged and often abandoned fishing gear, including nets, lines, hooks, and traps. These can present navigational hazards and can severely entangle an underwater vehicle or its tether.

Even if the plants or fishing gear are not big or strong enough to trap an ROV on the bottom, some vegetation (such as surfgrasses) or monofilament line can get caught in thruster propellers and prevent them from turning.

4.3. Large Animals

In rare cases, large marine animals can cause problems, too. Sharks, which have exquisite sensitivity to electricity and are normally attracted to tiny electrical signals produced by the muscles of their prey, will sometimes bite electrical cables (including ROV tethers) and electric motors! Sea lions have been known to play tug-of-war games with small ROVs. This is one of many reasons why you should never let a tether get wrapped around your ankle. There are even cases of octopuses "attacking" (or at least interfering with the operation of) ROVs. (See Figure 3.30.)

Image courtesy of Michael Derek Wood, SubOceanic Sciences Canada Ltd. and SAAB Seaeye Ltd.

Figure 3.30: Octopus versus ROV

ROV pilot Mike Wood captured this remarkable video image when a giant Pacific octopus attacked his Seaeye Falcon ROV.

The underwater robot was working off Vancouver Island to locate and recover receivers used to track Pacific offshore salmon migration.

5. The Water Column

Temperature, salinity, and other water properties usually vary with depth in any natural water body. To help describe and think about this variation, scientists frequently imagine a tall column of water extending from the surface all the way down to the bottom of the ocean, lake, or whatever. It's almost as if they insert a big clear plastic tube down through the water and focus their attention on the water contained within the tube. This concept, called a **water column**, helps them visualize, think about, and describe the way different properties of water change with depth. Sometimes one or more of these properties change smoothly and gradually with depth; other times they remain relatively constant for a distance, but then transition quickly to a new value, forming distinct layers like those in a tall, skinny cake. (See Figure 3.31.)

5.1. Light

Light is one of several properties that change with depth in the water column. During the day, surface waters are illuminated by the sun. In clear water (e.g., in tropical ocean waters or clear lakes), this light can penetrate over 100 meters (approx. 325 ft) deep and generally makes it easy to see things at shallow depths. In nutrient-rich waters, however, sunlight often supports prolific growth of photosynthetic algae (called **phytoplankton**), turning the surface waters a murky pea-soup green. This source of turbidity greatly reduces light penetration to deeper layers. Non-living sediment from river runoff or other sources of turbidity can also limit light penetration.

In a water body, the upper layer where light penetrates far enough to support significant photosynthetic activity is called the **euphotic zone**. (See *Section 5.5: Plant*

HISTORIC HIGHLIGHT: PROBING THE DEPTHS

How deep is the ocean? To answer this question, early scientists initially relies on the observations of explorers such as Sir James Clark Ross, who accompanied the ships *Erebus* and *Terror* into the Antarctic Ocean in the 1840s.

Ross used a sounding line 4 miles (about 6.4 km) long to take the first depth measurements of the deep sea. Even though the ocean depths could be measured, it was still assumed they were devoid of life. Occasionally, reports surfaced of strange sea creatures that fishing vessels had hooked or caught in their nets, but such specimens were often damaged or dead, if not rotten, by the time scientists got to see them.

Then in 1872, just as readers were thrilling to Jules Verne's *Twenty Thousand Leagues Under the Sea*, HMS *Challenger* left England on a four-year expedition to the major oceans of the world. It was one of the earliest systematic scientific exploration voyages, one that brought back abundant information, measurements, and samples of marine plants and animals.

and Animal Life below.) Extending from the lower limit of the euphotic zone all the way to the bottom of the lake or ocean is the **aphotic zone**. This is a dark region, both day and night. The only natural light here is provided by the eerie points of light generated by bioluminescent organisms. Visual exploration at these depths requires artificial lighting.

5.2. Temperature

In non-freezing conditions, the water temperature of surface waters tends to be warmer than deeper waters because sunlight warms the surface layer. Warm water then remains trapped at the surface, because it is less dense and "floats" on top of the colder, denser water layers below. In the ocean, surface temperatures range from over 30°C (86°F) in the tropics to less than 1.8°C (approx. 35°F) in polar regions.

Wind and wave action can stir things up a bit, mixing the generally warmer surface layer with cooler layers below, but these surface influences extend only a limited distance beneath the surface. In very deep bodies of water, such as the ocean, the result is a relatively warm **mixed layer** near the surface and a cold, calm, unmixed layer of water below it. The thickness and temperature of the mixed layer at the surface varies considerably, depending on sunlight intensity, wind speed, season, and other factors.

The warm and cold layers are separated by a region in which the temperature changes quickly with depth. Such a layer of rapid temperature change is known as a **thermocline**. In tropical seas, the thermocline beneath the mixed layer is usually distinct and located fairly deep (on the order of 100 meters [approx. 325 ft] or more), but at higher latitudes, it approaches the surface and becomes less pronounced, coming all the way to the surface (and therefore disappearing entirely) before reaching polar waters.

Beneath the thermocline associated with the mixed layer, water temperature tends to decrease gradually with depth, although some thermoclines are found in deeper water. Regardless of latitude, below 1,000 meters (approx. 3,300 ft) the ocean is a more or less uniform 4° Celsius (approx 39°F); this is the temperature at which water has its highest density and sinks beneath all other water. Thus, equatorial deep water temperatures at the equator are within a few degrees of polar deep water temperatures.

5.3. Salinity

Since salty water is denser than freshwater, you might expect the saltiest ocean water to sink and accumulate at the bottom of the sea. To some extent, you'd be right, since the deepest regions typically have higher salinity than mid-water regions. But the highest salinity can occur at the surface, where sun and wind can cause high evaporation rates that concentrate salts. This phenomenon is most pronounced at mid-latitudes. Near the equator and poles, the freshwater input of rain or ice melt dilutes salts faster than evaporation concentrates them.

As mentioned earlier, salinity is often measured with a CTD instrument. This instrument can be lowered through the water column to produce a graph of salinity versus depth. This type of graph is called a **salinity profile**. A layer of water in which salinity changes rapidly with depth is called a **halocline**.

5.4. Density

Water density depends on temperature, salinity, and pressure. For example, salty water is generally denser than freshwater. In general, denser waters sink beneath less dense waters (though strong wind-driven currents or other factors can upset this balance). A layer of water in which density changes rapidly with depth is called a **pycnocline**.

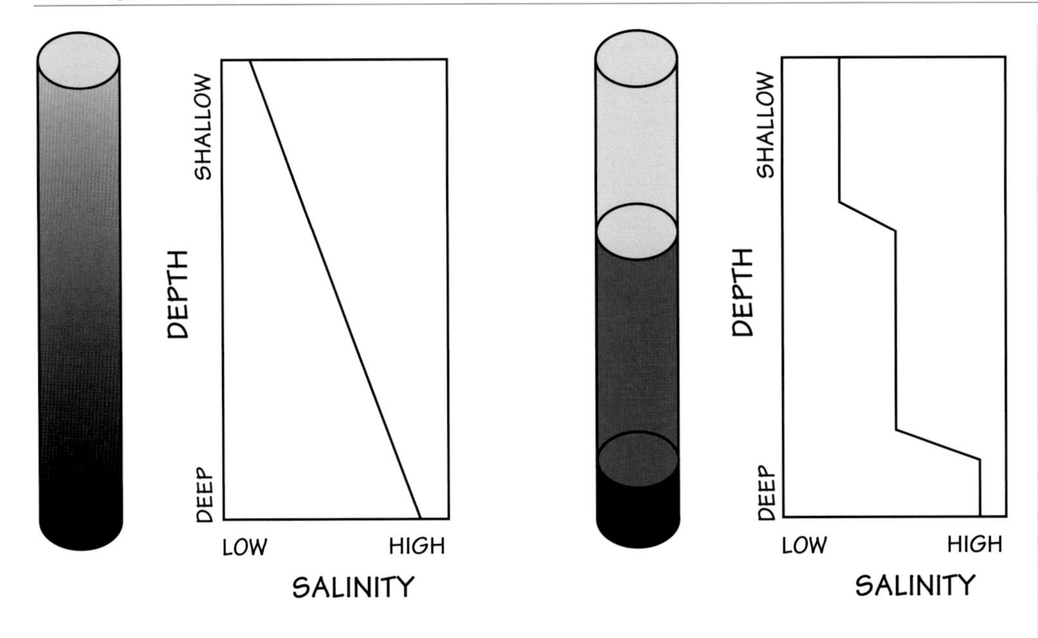

Figure 3.31: Patterns of Change within the Water Column

In this hypothetical example, pictures and graphs are used to illustrate two of many possible patterns in the distribution of salt concentration within a body of water. The picture and graph on the left represent the water column in a body of water where salinity increases gradually from low salt concentrations in shallow water to high salt concentrations in deep water. The picture and graph on the right depict a different water body (or perhaps the same water body at a different time of year) in which the salinity is stratified, forming distinct layers.

In both cases, the shallow waters have low salinity, and the deep waters have high salinity, but in the case on the right, the transition between low and high salinities occurs in sudden jumps, rather than gradually. Scientists call these abrupt transitions in salinity over relatively short vertical distances ***haloclines****. They separate the water column into layers, within which there is relatively little change in salinity with depth. The resulting graph resembles a series of stairsteps. Similar sudden changes in temperature occur and are called* ***thermoclines****. Similar sudden changes in water density also occur and are called* ***pycnoclines****.*

5.5. Plant and Animal Life

Life occurs throughout the water column, from sunlit surface waters to the deepest, darkest depths of the ocean. As a rule, each species inhabits a particular depth range, with some being found near the surface, others on or near the bottom, and still others in between. Some species move freely up and down, while others are restricted to very narrow depth ranges. Some species even undergo daily vertical migrations. For example, the enigmatic chambered nautilus, some species of fish, and many zooplankton hang out in the deeper, darker waters during the day (presumably to avoid detection by visual predators), then move up into the euphotic zone at night to feed on the abundant phytoplankton and other life found at those shallower depths.

The most distinct layering of life within the water column occurs within the euphotic zone. Here the quantity and quality of light change rapidly with depth. This produces a correspondingly rapid change in the distribution of photosynthetic organisms that depend on that light and in the non-photosynthetic organisms that feed on the photosynthetic ones.

In shallow water a few meters deep or less, there is usually abundant light for photosynthesis, so a rich assortment of sea grasses, algae, and phytoplankton is usually present. For example, shallow, rocky areas in temperate oceans often support lush kelp

HISTORIC HIGHLIGHT: *ALVIN*'S COMMUNICATION PROBLEM WITH THERMOCLINES

Sometimes a thermocline causes problems between a mother ship and a manned submersible, because it can reflect sound like a mirror reflects light, preventing the sound from crossing the thermocline.

During *Alvin*'s first 2,000-meter (approx. 6,500-ft) dive, the pilots realized that the mother ship *Lulu* was not receiving their acoustic communications; meanwhile, the crew aboard the mother ship radioed the U.S. Coast Guard for help in locating the submersible.

When *Alvin* finally rose through the thermocline that was preventing their acoustic signals from getting through, they discovered that *Lulu* had steamed off several miles in the wrong direction, searching for them. There was nothing to do but bob about on the surface and wait for dusk, when they could release flares to alert Coast Guard search planes to their whereabouts.

HISTORIC HIGHLIGHT: *TRIESTE'S* ENCOUNTER WITH A PYCNOCLINE

Pycnoclines (regions of the water column in which water density changes suddenly) can interfere with a submersible's descent or ascent because of their effect on vehicle buoyancy.

This occurred during the *Trieste* mission when the craft got stuck in mid-water, unable to pass through a temperature-related change in water density. The crew had to jettison some of its lighter-than-water gasoline to descend.

Another option for most modern vehicles is to increase downward thrust and attempt to power through the pycnocline. Of course, a submersible or robotic craft may face the opposite challenge when it attempts to return to the surface.

forests; in tropical seas, extensive coral reefs are built by coral animals with the help of photosynthetic algae living inside their tissues.

As depth increases, the light fades, photosynthetic rates drop, and photosynthetic organisms thin out. Most, if not all, photosynthetic organisms are gone by 100 meters (approx. 325 ft) down. Without the oxygen input from photosynthesis, oxygen levels drop. At even lower depths, life is still present (even abundant), but it's composed almost exclusively of predators and scavengers. Some of these, like sharks, fit the typical image of a predator—swift, powerful bodies with big teeth. Others are more subtle about their dining habits. For example, delicate, graceful jellyfish feed by capturing minute prey in deadly tentacles, which are laced with poisonous stinging cells. On the bottom, deep sea corals that look more like beautiful plants than predators use the same type of stinging cells to subdue unwary prey.

If you were to take an ROV on a dive from the surface to the bottom of the sea, you might see the following sights along the way. At the surface, brilliant rays of sunshine filter down through the water. There might be a slightly cloudy, greenish cast to the water, caused by microscopic phytoplankton there. You might also see some of the larger zooplankton or fish feeding in the area. The water-oxygen sensor on your ROV would reveal relatively high levels of oxygen near the surface.

As you descend, the surface light fades into blackness, making it necessary to switch on the ROV's lights. In mid-water, well below the euphotic zone, but still far above the bottom, these lights would reveal a sight that might surprise you as it did the first people who saw and studied it—**marine snow**. (See Figure 3.34.) These white specks visible in the lights are actually a rain of tiny dead bodies, animal waste, shed skins, sloughed-off mucous, and other organic debris that falls continually from the biologically productive euphotic zone above. Illuminated in the bright ROV lights, the bits of debris look white against the dark background and resemble snowflakes in a car's headlights. There are living things here, too. Jellyfish, tiny crustaceans, and other organisms are feeding and reproducing here among the bits of marine snow.

Image courtesy of SeaBotix

Figure 3.32: The Euphotic Zone

The ambient daylight and luxurious kelp growth in this photo indicate that the ROV is working in the euphotic zone.

TECH NOTE: NEW DISCOVERIES IN THE WATER COLUMN

Oceanographic researchers are still discovering new life forms and even new life processes in the deep sea. Among the most amazing of these is the existence of hydrothermal vent communities and the **chemosynthesis** which fuels them.

Chemosynthesis is the process by which bacteria found at these vent sites use chemicals as an energy source to create food. These specialized bacteria are found within the tissues of some vent organisms, such as the giant, blood-red tubeworms and supersized mussels seen in the photo, where they deliver nutrients directly to their hosts. This energy pathway is similar to the process of **photosynthesis** in which plants use sunlight to make simple sugars. Other vent organisms, such as crabs and snails, feed off free-living bacteria or scavenge, devouring the dead or living tubeworms and mussels.

In addition to investigating hydrothermal vent communities, ROVs and deep submersibles are allowing biologists and biological oceanographers to observe delicate living animals that were previously unknown or known only from dead, mangled specimens brought up in fishing nets or bottom trawls. Among these are ctenophores (comb jellies), like *Beroe*, a transparent mid-water creature whose eight bands of cilia break white light into pulsating rainbows of color.

These discoveries remind us of how little is known, and how much is yet to be discovered!

Image courtesy of Woods Hole Oceanographic Institution

Figure 3.33: New Scientific Knowledge

A robotic arm facilitates temperature sampling in a tube worm colony.

At the bottom, you might discover a vast plain of fine sediment—a thick accumulation of old marine snow. This is nutrient-rich muck, and it's crawling with worms, sea stars, sea cucumbers, and other bizarre denizens of the deep who dine on the nutrients trapped within. Crabs, rays, and others are here, too, predators feeding on the scavengers.

Such a journey down through the water column highlights not only how the chemical and physical properties of water vary with depth, but also how that variation is reflected in the distribution of plants and animals that inhabit it.

Image courtesy of Neil McDaniel

Figure 3.34: Marine Snow

The white specks visible in this image are marine snow—particles of suspended organic matter that appear similar to slowly falling snow when illuminated by vehicle lights. Marine snow is a testament to the prolific biological activity happening far above in the water column.

Marine snow can interfere with photography, but in open water, where there are no other landmarks to focus on, it can be useful to ROV pilots. By watching the movement of the white specks on the screen, the pilots can tell if their vehicles are ascending, descending, or turning.

6. Chapter Summary

This chapter is the first of the technical chapters in this textbook (Chapters 3-11). It summarizes some of the important properties of water and of natural water bodies as they relate to underwater vehicle design. Among these are the corrosive and electrically conductive salinity of ocean water, the extreme mechanical pressures and forces encountered at depth, the advantages of using sound instead of light or radio waves for communication and "vision" under water, the problem of biofouling, and the vertical distribution of various water properties and life forms in the water column.

It's important to know about these properties because they place enormous constraints on vehicle design. However, they also provide some opportunities not available to designers of land-based vehicles.

Subsequent chapters will explore many of these properties in greater technical detail and will describe methods for confronting the challenges and taking advantage of the opportunities provided by working in water.

Chapter 4

Structure and Materials

Chapter 4: **Structure and Materials**

Stories From Real Life: *Trieste*

Chapter Outline

***Figure 4.1.cover: Exacting Structure for* Nereus**

Working at full ocean depth requires a structure that is strong, lightweight, and able to handle the intense pressure of 15,751 psi, or more than 1000 times the atmospheric pressure at sea level. This photo of the ROV/AUV Nereus *under construction shows WHOI senior welder Geoff Ekblaw tackling the critical weld quality of the vehicle's aluminum structure. For the highly successful* Nereus *project, he combined more than 30 years of welding experience with specialized equipment.*

Image courtesy of Tom Kleindinst, Woods Hole Oceanographic Institution

Chapter Learning Outcomes

- Describe the typical shape and purpose of each of the following functional subsystems of an underwater vehicle's structure: frame, pressure hull (or canisters), and fairing.
- Select appropriate building materials (shapes and substances) to use for the structure of a small ROV or AUV and describe methods for fabricating and assembling the required parts.
- Explain the causes and consequences of metal corrosion and discuss methods to control it.

STORIES FROM REAL LIFE: *Trieste*

Auguste Piccard was determined to explore the earth's extremes—its height and its depth. To that end, the Swiss physicist, inventor, and explorer designed a number of vehicles, from hot air balloons to bathyscaphs such as the Trieste. *The success of these craft to reach record-setting heights or depths depended on Piccard's understanding of the laws of physics, as well as his creative application of these principles to the design of new craft.*

In the early 1900s, scientists were convinced that all life existed in the top 150 meters (approx. 500 ft) of the oceans. The deep abyss was considered a dark, sterile void. The first two deep sea explorers to challenge this mindset were Otis Barton and Dr. William Beebe. On August 15, 1934, the two men pretzeled themselves into the bathysphere that Barton had developed, constructed, and funded. The cramped steel sphere was then bolted shut and lowered by means of a cable to a record depth of 922.9 meters (approx. 3,028 ft) near Bermuda. Barton and Beebe became the first deep sea explorers. However, it was not an easy victory—even reputable scientists made fun of the bathysphere and questioned Beebe's identification of phosphorescent fish.

But if the greatest ocean depth was to be conquered, it would take a far different kind of craft than Beebe and Barton's bathysphere—one with a structure strong enough to withstand much greater pressure; one that didn't simply dangle from a cable; one that could be brought back to the surface more easily; one that afforded greater viewing and comfort; and above all, one that creatively conformed to the laws of physics that govern all undersea craft.

Auguste Piccard had given considerable thought to the idea of such a "deep boat," or bathyscaph. Initially Piccard, a Swiss physics professor, had chosen to study cosmic rays and had built a stratospheric balloon, the Belgian *Fonds National de la Recherche Scientific (FNRS)*, from which he could conduct scientific experiments. In 1931, he and collaborator Paul Kipfer made their first pioneering balloon ascent, shooting up roughly 14 kilometers (approx. 9 miles) into the stratosphere in just 28 minutes. The pressurized aluminum gondola that Piccard designed allowed him and other colleagues ultimately to ascend some 22 kilometers (approx. 14 mi) without having to wear pressure suits.

Back on land, Piccard continued to explore the earth's extremes, but by the mid-1930s, his interest shifted to the ocean depths. He planned to modify his pressurized balloon gondola idea and create a steel craft capable of withstanding the extreme pressures of the abyss. Belgium agreed to finance construction plans for the *FNRS-2*, as the underwater craft was to be known, but work was halted when Europe became embroiled in World War II. At the end of the war, skyrocketing costs and the devalued Belgian franc cut some of the vehicle's desirable features. Finally, in November 1948, the *FNRS-2* completed its first successful unmanned dive. The bathyscaph went to 1,500 meters (approx. 4,500 ft), but would require repairs, modifica-

tions, and more tests to go any deeper. Eventually, the French Navy agreed to take over the *FNRS-2*. In 1954 the rebuilt and renamed *FNRS-3* set a manned depth record of 4,050 meters (approx. 13,300 ft), reaching the bottom of the Atlantic Ocean 160 miles off Dakar.

Meanwhile, Piccard's son Jacques was finishing his studies in Trieste, Italy. There the young man met a professor who offered to help the Piccards raise money to build a new vessel. The only stipulation was that it be named after his beloved city of Trieste. The deal was made. Auguste Piccard worked on the design, while Jacques Piccard traveled through northern and central Italy to move construction of the new bathyscaph forward.

Trieste's unusual structure was the result of two primary requirements—the need to provide and regulate adequate buoyancy and the need to protect humans at greater depths. The solution for the vessel's two crew members was a pressure-resistant globe nestled underneath the massive flotation tanks. The sphere was made of hard-forged steel alloy, which allowed deeper dives. The original capsule was rated to a depth of about 6,100 m (approx. 20,000 ft), enough to explore some 98 percent of the ocean. It had an air weight of 89 kN (10 tons) and walls 8.9 centimeters (approx. 3.5 in) thick. Additional ballast weight came from 80.2 kN (9 tons) of iron shot. To counteract the heavy nature of the bathyscaph's protective steel chamber, Piccard attached a large 15-meter (approx. 50-ft) steel float for buoyancy and filled it with gasoline to keep it from being crushed. *Trieste* was built a decade before the development of flotation materials such as syntatic foam, so gasoline, which is less dense than water, was the cheapest and most easily available crush-proof flotation medium for a vessel of this size. Basically, the bulk of *Trieste* was a glorified gasoline tank holding roughly 85,000 liters (approx. 22,500 gal) of "lighter-than-water" gasoline, enough to counteract the enormous weight of the vehicle and keep it buoyant.

Air-filled tanks at both ends of the large flotation structure added extra buoyancy at the surface and were flooded for descent. For fine-tuning buoyancy or when *Trieste* encountered thermoclines, the pilot could release up to 12.5 kN (1.4 tons) of gasoline from a central storage compartment. To ascend, the pilot released the iron pellets stored in two large hoppers on either side of the cabin. An electromagnet system fused the BB-sized shot into a solid mass, but when the pilot turned off the current, they became pellets again and trickled out of the hopper. The maximum descent rate achieved was 3 ft/sec; the ascent rate was 5 ft/sec. Because gasoline is more compressible than seawater, the submersible would naturally accelerate

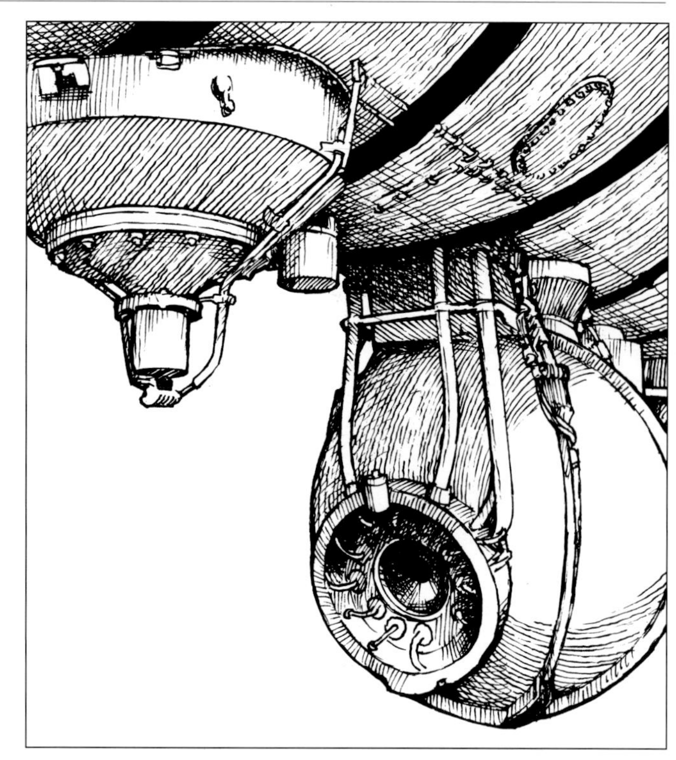

during descent and ascent. So the pilot had to monitor speed continually and make adjustments. For example, if *Trieste* ascended too fast, the gasoline would get so cold from the rapid expansion that the control valves could seize.

Despite the novel design, *Trieste* was not without problems. Hindered by a modest budget and the need to conserve battery power, *Trieste* was outfitted only with small mercury-vapor and incandescent lamps, barely adequate in the darkness of the deep. And while the bathyscaph was technically an untethered vessel, in fact, she had very little mobility. Her oil-bathed electric motors each delivered only 1.5 kw of power (2 horsepower), giving *Trieste* a maximum horizontal speed of 1 knot (1.85 km/hr). It was more elevator than submarine!

It was inevitable that competition would develop among the countries engaged in building bathyscaphs. Despite the lack of any public challenge, the race was on to see who would go first and who would go deepest. The French were the first to break Barton's existing record; then *Trieste* descended to 3,167 meters (approx. 10,390 ft). The contest see-sawed back and forth, but by the late 1950s, both bathyscaphs had descended more than 3 kilometers (approx. 2 miles). In 1956, Jacques Piccard talked to U.S. scientists and the Office of Naval Research (ONR), who agreed to conduct a series of dives to evaluate the bathysphere's potential. At the same time, the Soviet Union launched *Sputnik*, the world's first orbiting satellite. The space race was just about to heat up—and so was the contest in the deep frontier. The U.S. Navy put forward an unexpected offer to buy *Trieste*; the Piccards accepted, with one proviso, that Jacques Piccard could go on any dives presenting special problems.

In the back of everyone's mind was a descent into Challenger Deep—the Mount Everest of the ocean, located in the Mariana Trench near Guam. The floor of this greatest of all ocean trenches lay nearly 11 kilometers (approx. 7 miles) down. The French were already talking about plans for a "super bathyscaph." Not to be outdone, Jacques Piccard began negotiations for a new pressure sphere for *Trieste*, this time with a depth rating of 10,970 meters (approx. 36,000 ft). To balance this new heavier sphere, the float was filled with 128,700 liters (approx. 34,000 gallons) of gasoline. When completed, the 12.5-centimeter (approx. 5-in) thick sphere was calculated to survive water pressure of just under 124,000 kPa (18,000 psi).

Originally, the U.S. expedition called for three dives into Challenger Deep, giving each one of the five principals involved in *Trieste*'s diving program a chance to make history: Lieutenant Don Walsh, naval officer in charge; Andreas Rechnitzer, head of the Deep Submergence Program at the Naval Electronics Laboratory and scientist in charge; geologist Robert Dietz; physicist Kenneth Mackenzie; and *Trieste*'s most experienced pilot, Jacques Piccard. Then the program was cut back to only one dive, with Walsh and Rechnitzer selected. When Piccard found out, he reminded naval officials of his option to go on any dive presenting "special problems." The final choice for the ultimate depth dive became Piccard and Walsh.

Towing the bathysphere 320 kilometers (approx. 200 miles) to the dive site was no easy task, and on the day of the dive, 7.5-meter (approx. 25-ft) seas assaulted the small rubber boat trying to get Piccard and Walsh on board the *Trieste*. Finally, they scrambled down the entrance tunnel that led through the flotation tank and into their fortified chamber. Descent began at 8:23 a.m. They had dropped only 102 meters (approx. 335 ft) when the bathysphere literally bounced off a cold thermocline and stopped its descent. Piccard couldn't wait for the mass of gasoline to cool, so he jettisoned some of it, tapping into *Trieste*'s reserve buoyancy and its margin of safety.

At 9,875 meters (approx. 32,400 ft), a strong explosion shook the bathysphere. The men realized that something outside their tiny steel cabin had broken, but they had no idea what. As the descent continued, the intense cold of the depths added to their tension. Finally, at 12:56 p.m., the sounder indicated the bottom was near, and Piccard began releasing metal shot to slow their descent. As *Trieste* touched down, he looked out the forward viewport and saw a flat fish swim slowly away. There *was* life in the depths, where pressure was nearly 8 tons per square inch! Walsh was able to find the cause of the explosion: a large plastic window in the chamber outside the pressure capsule had cracked. With luck, it would hold.

The descent to the deepest part of the ocean had taken 4 hours and 48 minutes. Walsh and Piccard reached over to shake hands. They had made it! Their instruments showed a depth of 11,512 meters (approx. 37,700 ft). Later corrections applied to this number suggested the actual depth was closer to 10,916 meters (approx. 35,814 ft) under water. Either way, they were sitting on the bottom in the deepest part of the ocean and were as deep as any human had ever been.

Unfortunately, after only 20 minutes on the bottom, it was time to begin the long ascent. Piccard dropped more ballast and the slow rise began. The historic bathysphere reached the surface at 4:56 p.m. on January 23, 1960. They had won the race to the depths. Far more importantly, the carefully thought out structure and continual testing of this unusual deep ocean craft had allowed them to write a new chapter in subsea history. Piccard was presented with the Navy's Distinguished Public Service Award and Walsh the Legion of Merit.

In subsequent years, as new and improved equipment became available, *Trieste* would undergo several modifications, and a renaming to *Trieste II*. But what did not change was the bathyscaph's record-setting descent to the deepest ocean depth—a feat made possible at that time by the vessel's carefully designed and fabricated structure. To date, no one has repeated this remarkable achievement in a human-occupied vehicle.

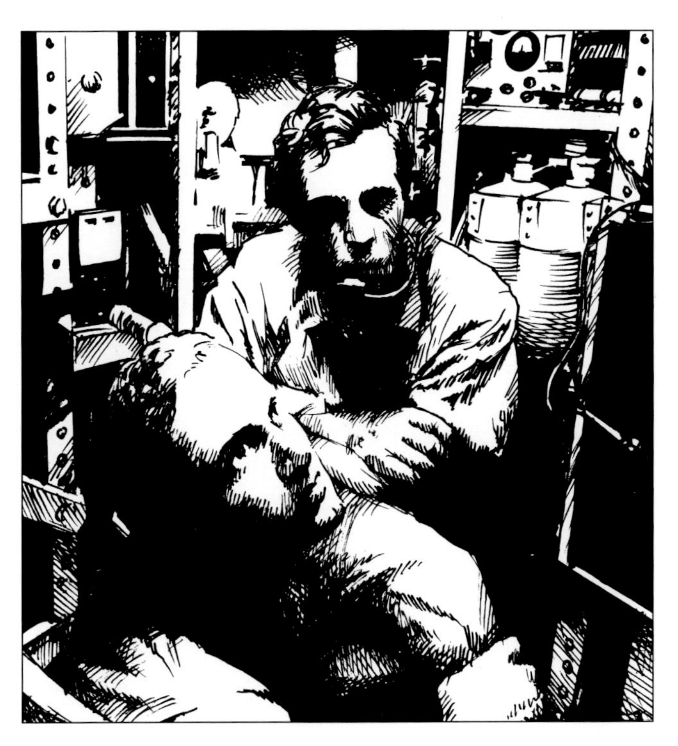

1. Introduction

Image courtesy of Hans Arne Nakrem; © Natural History Museum, University of Oslo, Norway

Figure 4.2: Model of an Armor-Plated Placoderm (genus: Bothriolepis)

Some 400 million years ago, the seas were ruled by the earth's first vertebrate super-predators. These were a group of armor-plated, shark-like fish called placoderms. A placoderm's skeleton supported its muscles and internal organs, while providing armor to protect its delicate internal organs. It also gave the big fish a streamlined shape, so it could move swiftly through the water to attack its prey.

Like a placoderm's skeleton, the **structure** of an underwater vehicle performs the following critical functions:

1. It provides physical support, attachment points, and mechanical protection for all other parts of the vehicle, including cameras, flotation, and thrusters.
2. It shields certain subsystems (such as electronic circuits) and any air-breathing passengers (such as human crew members) from water and water pressure.
3. It defines the vehicle's overall shape, thereby influencing both its performance and its appearance.

This chapter begins with an overview of three basic structural subsystems, each of which performs one of the functions described above. It then discusses some of the performance criteria and other considerations that separate successful structural designs from problematic ones. The hallmarks of effective structures include high strength and stiffness for their weight, compatibility with an aqueous environment, and ease of assembly, maintenance, and repair.

This chapter explores how shape and substance impact structural performance. It also provides recommendations for good shapes and good materials (mostly certain metals and plastics). One of the interesting challenges for vehicle designers is to blend technical skill with artistic talent—the result is a vehicle form, like a good architectural structure, that is both functional and pleasing to the eye.

Finally, the chapter concludes with an examination of metal corrosion and its prevention, as well as some suggestions on how you can get started building your own underwater vehicle frame, pressure canisters, and fairing.

1.1. The Technical Chapters

This chapter is one of the early technical chapters in this book. You may be wondering why these chapters begin with vehicle structure and materials, instead of some other subsystem. After all, you can't design your structure until you know what it's going to be supporting and protecting. How, for example, can you design a frame to support a thruster if you don't even know how big and heavy that thruster is going to be or how much thrust it will generate? Of course, that argument works in the other direction, too—you can't design your thruster until you know the size and shape of the structure it needs to push through the water. The design spiral (introduced in *Chapter 2: Design*

Toolkit) serves as a reminder that this chicken-and-egg problem permeates the design of complex systems with multiple interacting subsystems and requires that the separate designs of the vehicle's many subsystems evolve in parallel.

The goal of this first technical chapter, then, is *not* to arrive at a final structural design for your vehicle, but rather to open your eyes to structural issues, options, and constraints that may impact the design of your final structure as well as the design of other parts of your vehicle. This chapter will survey the different types of structural building materials available and explore methods for modifying and joining those materials to make completed structures. It will also look at how the shape and size of structural elements affect overall structural performance. Ultimately, of course, the material in this chapter will help you put the finishing touches on your vehicle's structure, but only after your other vehicle subsystems have started coming into focus.

In addition to helping you get an early start on thinking about the sizes, shapes, and materials to use for *all* of your vehicle subsystems (not just the structure), this chapter provides a solid introduction to some basic chemistry and physics. This foundation will serve as a good stepping stone into some of the more challenging topics presented in later technical chapters.

2. Structural Overview

An underwater vehicle's structure is divided conceptually (and often physically) into three functional subsystems, each of which performs a different role (Figure 4.3).

1. The **frame** (or **framework**) serves as the primary skeleton of the vehicle. It is usually made of interconnected beams, struts, plates, or other load-bearing members. The frame defines the vehicle's overall shape, and it provides mechanical support and attachment points for weights, floats, thrusters, cameras, lights, and other vehicle components. The frame often extends out and around delicate parts of the vehicle to protect them from impacts, in much the same way that your ribcage protects your vital internal organs. This chapter is concerned primarily with this skeletal framework for underwater vehicles.

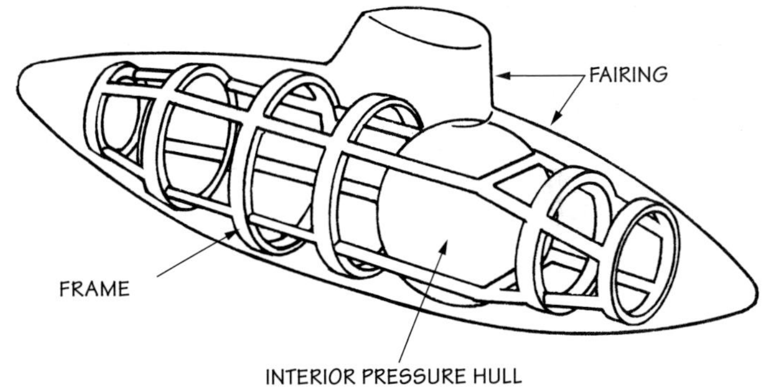

Figure 4.3: Structural Subsystems

This simplified drawing of an underwater vehicle's structure illustrates how it may involve several functional subsystems, each of which performs a different role.

2. The **pressure hull**, and its smaller counterparts called **pressure canisters** ("cans") or **pressure housings**, are strong, rigid, watertight containers, usually containing air at normal atmospheric pressure, that are designed to protect their contents from water and extreme pressure. A large vehicle such as a submarine may have one enormous pressure hull to enclose the entire crew and all electrical equipment, whereas a small vehicle like an ROV is more likely to have several small pressure canisters, each one enclosing a different camera, light, navigational sensor, motor, or other electronic device. Nearly all pressure hulls and canisters are spherical or cylindrical in shape because these geometries are particularly resistant to pressure damage. *Chapter 5: Pressure Hulls and Canisters* is devoted entirely to the design and construction of these crucial structural components.

3. The **fairing** is a skin or shell that covers all or part of the frame of many (but not all) vehicles to give them a smoother, more streamlined shape. A fairing helps the vehicle slide more easily through the water, thereby improving speed and energy

efficiency. Fairings are used commonly on vehicles that must travel quickly or far, but they are not often used on ROVs or other craft that travel slowly and cover relatively small distances on each mission. Some military submarines have thick steel fairings, which double as armor. *Chapter 7: Moving and Maneuvering* discusses fairings and their contributions to vehicle efficiency in greater detail.

In many vehicles, these three functional subsystems (frame, pressure hull/canisters, and fairing) are physically distinct units, and they are often made of completely different materials. For example, in a submersible, the frame might be fabricated from aluminum beams welded together, the pressure hull might consist of an acrylic sphere, and the fairing might be made of glass-reinforced plastic (fiberglass).

In other vehicles, two or all three of these functions may be combined into a single structural unit. For example, in a military submarine one large streamlined pressure hull made from steel or titanium might serve all three functions.

In ROVs, which typically lack a fairing, it's common to have a box-shaped steel, aluminum, or plastic frame supporting an assortment of pressure canisters and other equipment made from a variety of metals and plastics. AUVs often do have a plastic or fiberglass fairing and may have a slender, streamlined pressure hull because they rely on self-contained battery power and must be very efficient.

Figure 4.4: Variations in Structure and Materials

These student-built ROVs illustrate the diversity of structural framework and fabrication materials available. Generally, the choices are determined by mission, budget, imagination, and skill level of the fabrication team. Looking at other underwater robots, talking with their builders, and researching on-line are all excellent ways to explore workable options.

Image courtesy of Steve Van Meter, VideoRay LLC

Image courtesy of Dr. Susan Crawford, Naval Undersea Museum Foundation

Image courtesy of Harry Bohm

3. Structural Performance Criteria

At a fundamental level, the primary role of a vehicle's structure is to manage all of the mechanical forces (i.e., the pushes and pulls) the vehicle will encounter during its lifetime.

In order to do so, the frame portion of the structure has these requirements:

- It must support the (usually heavy) weight of the entire vehicle and all its parts (including the frame itself), both in and out of the water.
- It must effectively withstand and distribute forces that get concentrated at specific points in the frame. These include lifting points that are used whenever the vehicle is being lifted in or out of the water, as well as feet and tie-down points used to stabilize and secure the vehicle for transport over rough seas or bumpy roads.
- It has to withstand hydrodynamic forces associated with currents, swell, waves, and vehicle propulsion without damage and without undue flexing or other deformation that might adversely impact the function of other vehicle systems.
- It must be able to position and securely attach cameras, sensors, thrusters, and other components so that they are effective, protected, and convenient to access for adjustments, repairs, or maintenance.
- It must support any forces associated with tools or other payloads, including not only the weight of the payload itself, but also any forces transferred to the frame by tools as they lift, push, pull, twist, or cut while working.
- It needs to protect the vehicle from collisions with hard objects, such as submerged rocks or timbers, as well as from the inevitable bumps and scrapes associated with launch, recovery, and transport, particularly in rough conditions.
- It should protect (or at least not damage) the umbilical during launch, operation, recovery, and transport of a tethered vehicle.

***Figure 4.5: Mission-Driven Vehicle Structure for* Robolobster**

Prof. Joseph Ayers gave his biomimetic underwater robot, Robolobster, *a familiar skeletal shape constructed out of industrial-strength plastic. Like real lobsters, this unusual underwater robot uses its antenna to detect minute changes in seawater, allowing it to locate and destroy mines in the shallows. Multiple legs permit movement in any direction while the claws and tail provide stability in turbulent water.*

Image courtesy of Robert Spencer

***Figure 4.6: Mission-Driven Vehicle Structure for* Nereus**

The weight-saving, buoyancy-producing ceramics technology of Nereus' *structure allows it to reach the greatest depths of the world's oceans. The robotic vehicle can function as an ROV with a micro-thin, fiber-optic tether or switch into free-swimming, autonomous mode.*

Image courtesy of Woods Hole Oceanographic Institution

***Figure 4.7: Mission-Driven Vehicle Structure for* Super Aviator**

The extremely streamlined structure of the new Super Aviator *allows the two-man submersible to be flown like an airplane to depths of 500 meters. Owned and operated by Sub Aviator Systems Ltd., the wide-ranging mini-sub can be fully configured for undersea exploration and scientific-data collection.*

Image courtesy of Alfred McLaren, Sub Aviator Systems Ltd.

The pressure canisters or pressure hull portions of the structure have these requirements:

- They must resist being crushed by water pressure forces, which can be enormous if the vehicle is operating at great depths.
- They have to maintain watertight seals against the forces of water pressure and, if required, provide a way to access/service interior components.
- They may have to allow wires and/or piping to penetrate into the canister while still remaining watertight.
- They should provide some way of being securely mounted to the frame.

The fairing portion of the structure has these requirements:

- It should provide a streamlined shape that minimizes so-called "drag forces" that resist the motion of a vehicle through water.
- It needs to smooth the surface of vehicles working in kelp or other places where the vehicle might get snagged and caught. In this sense, the fairing works to reduce or eliminate forces that could otherwise trap the vehicle. This is especially important for small, low-powered vehicles that are not powerful enough to break free of such entanglements.

Before you can design an adequate structure for your vehicle, you must have a realistic sense of the nature and magnitude of *all* the mechanical forces each piece of the vehicle's structure will need to withstand across the full range of expected operating, launch, recovery, storage, and transport conditions, and you must keep these forces in mind throughout the structural design and construction process. Of course, the magnitude of these forces may be seriously impacted by the vehicle's mission. For example, a commercial ROV servicing oil rigs at 1,000-meter (approx. 3,300-ft) depths will have different structural needs than a student-built vehicle competing in a swimming pool. Each of these underwater crafts has a structure, but the materials and design may vary significantly because of differing mission requirements.

3.1. Strength and Stiffness

Strength and stiffness are arguably the most critical aspects of a structure's mechanical performance. In simplified terms, the **strength** of a structure is a measure of how much force the structure can withstand before breaking or suffering other permanent

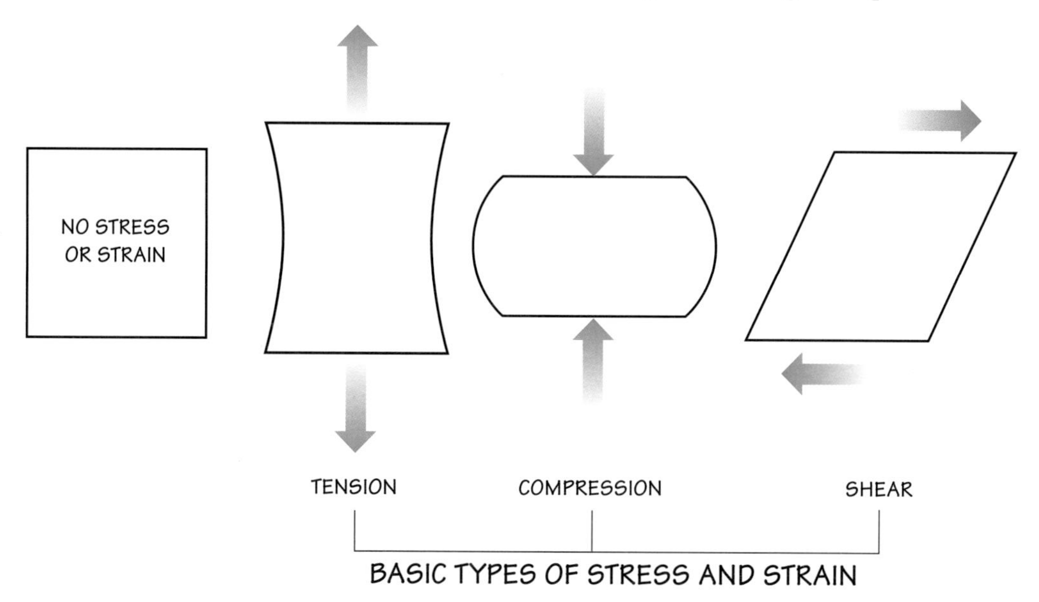

Figure 4.8: Stress and Strain

*Forces applied to an object are often quantified in terms of a **stress**, or force per area. The deformations they cause are generally quantified in terms of a **strain**, or fractional change in shape.*

These can ultimately be broken down into combinations of just three basic types. Tension is associated with a pair of pulling forces that tend to stretch an object. Compression is associated with a pair of pushing forces that tend to squash an object. Shear is associated with a pair of forces that act in opposite directions but not in line with one another, so that one end of the object tends to slide past the other end.

damage. The **stiffness** of a structure is a measure of how much force it takes to bend or deform the structure by a specified amount. It is possible for structures to be strong but not stiff (for example, a rope) or stiff but not strong (for example, a piece of toast). Underwater vehicle structures usually need to be both strong and stiff at the same time, and they need to stay that way when wet. (Forget the toast.) For more precise definitions of strength and stiffness, see *Tech Note: Measuring Strength* and *Tech Note: Measuring Stiffness.*

Most vehicle structures are composed of many separate parts that have been joined together. The strength and stiffness of the overall structure depends on at least three things:

1. the strength and stiffness of each part
2. how those parts are arranged into triangles, rectangles, or other shapes within the structure
3. how those parts are welded, glued, bolted, or otherwise joined together

Thus, for example, metal beams arranged in triangles (which are an inherently stiff geometry) and welded at the corners will produce a stronger and stiffer frame than rubber beams arranged in squares (which can flop over into a parallelogram shape) with their ends joined like hinges.

TECH NOTE: DUCTILE AND ELASTIC MATERIALS

Once deformed, some materials, like putty and copper, will stay deformed; these are called **ductile materials.**

Others, such as rubber or spring steel, will bounce back to their original shape; these are called **elastic materials.**

TECH NOTE: MEASURING STRENGTH

Strength is a critical property for any structural part, so it's important that you understand precisely what the term means. It's also important to recognize that strength is defined differently for parts than it is for materials.

In the case of a *part*, such as a bolt, strength refers to how much force the part can withstand before breaking or sustaining permanent damage. This is usually measured by watching a spring scale or other force-measuring device while it applies a gradually increasing force until the part fails. The maximum force supported (immediately prior to failure) is that part's strength. However, the result depends on how the force is applied and how you define "failure." For example, many parts have a different strength when stretched (called tensile strength) than when compressed (called compressive strength), and some give different strength results depending on how quickly the force is applied. Likewise, a part may become permanently deformed before it actually breaks. The maximum strength measured before permanent deformation is called the yield strength. The maximum strength measured before breaking is called the ultimate strength. From a practical standpoint, the way force is applied during the measurement test, and the definition of failure, should correspond as closely as possible to the anticipated forces the part will encounter while doing its job in your vehicle's structure and the type of failure that would represent a problem for your vehicle.

Defining the strength of a *material*, such as aluminum, is a bit more challenging. The goal is to quantify the strength of the material in some way that is independent of the size and shape of the part or sample used for the measurement. To do this, engineers select samples of material with simple geometries (commonly a cube or long rod) and take advantage of the fact that the amount of force a part can support is roughly proportional to its cross-sectional area. To measure the ultimate tensile strength of aluminum, for example, a long aluminum rod is stretched (while measuring the force applied) until it breaks. This gives the ultimate tensile strength of the aluminum rod as a part. Then that maximum force is divided by the cross-sectional area of the rod to get the strength of aluminum as a material. Unlike the strength of a part, which is measured in units of force, the strength of a material is measured in units of force per area, which is called a stress. Stress units are just like pressure units. Thus, the strength of a material may be expressed in units of pressure, such as pounds per square inch (psi), millions of newtons per square meter (megapascals, or MPa), or any other valid units for pressure.

TECH NOTE: MEASURING STIFFNESS

Stiffness, like strength, is a critical property for a structural part, so it's important that you understand precisely what the term means. It's also important to recognize that stiffness, like strength, is defined differently for parts than it is for materials.

The stiffness of a part is usually defined as the force required to produce a certain (usually very small) amount of change in the part's shape. A stiff object will bend or deform less than a flexible object when both are subjected to the same force. Of course, it's possible to make a stiff object bend farther than a flexible object if you apply a much larger force to the stiff object; that's why the ratio between force and deformation is so important in the definition of stiffness. The tensile stiffness of a length of rope, for example, can be measured by hanging a known weight from the rope and measuring how far the rope stretches. If a 10-newton weight causes a 0.02-meter increase in length, the stiffness is 10 newtons/0.02 meters = 500 newtons/meter. As with strength, the stiffness of a part usually depends on how you apply the force used to measure it. It also depends on how you measure the deformation or bending of the part.

The stiffness of a material is measured differently. First, the force applied is divided by the cross-sectional area supporting the force to convert the force into stress. If the rope in the previous example had a cross-sectional area of 1 cm^2, the stress would have been 10 newtons/cm^2. The change in length (or other dimension) is expressed as a percentage or fraction of the original length to convert it to a dimensionless number called a strain, so if the sample of rope was originally 2 meters long, the strain would have been 0.02 meters/2 meters = 0.01. (Note that the meters cancel out, so strain is dimensionless; it has no units.) The stiffness of the material is then defined as the stress per strain. Since strain is dimensionless, the units for material stiffness end up being the same as the units of stress, which are like pressure. The tensile stiffness of the rope material would have been (10 newtons/cm^2)/0.01 = 1000 newtons/cm2 = 10 million newtons per square meter = 10 MPa. Note that the stiffness of a material is thus expressed in units of stress or pressure (force per area).

Image courtesy of Vickie Jensen

Figure 4.9: Truss-Style Bridge

A bridge frame based on triangles can be very strong and stiff for its weight. This vehicle bridge incorporates triangles, X-braced rectangles, and beams that are themselves trusses made of lots of little triangles. Other examples of trusses that provide high strength and stiffness include tower cranes, radio towers, and that famous Paris landmark, the Eiffel Tower.

The strength and stiffness of each part in turn depend on the shape and size of the individual part as well as the type of material (metal, plastic, etc.) from which the part is made. Thus, a large steel bolt is usually stronger than a small steel bolt, and a steel bolt is usually stronger than a plastic bolt of the same size. (Take another look at the Tech Notes to see how strength and stiffness are measured *differently* for parts than they are for materials.)

As you design your vehicle's structure, pay careful attention to the size, shape, material composition, arrangement, and connections among the various parts to make sure that the final structure meets your mission's strength and stiffness requirements. Suggestions for particularly good shapes and materials to use for strong and stiff structures are provided later in this chapter.

3.2. Weight

Weight turns out to be another important factor in the mechanical performance of structures. Less weight is usually better than more weight. Even though the effects of buoyancy can easily offset weight under water, every vehicle must be hauled out of the water from time to time and transported over land or sea by another vehicle. Heavier underwater vehicles require bigger, more complex, and more expensive equipment to launch, recover, and transport them. Even when the vehicle is small and light enough to launch by hand, lighter is better. A smaller, lighter vehicle is easier to carry and store, and it's much safer to launch and recover by hand than a heavier one. It's also worth pointing out that smaller, lighter vehicles are often less expensive, because they require less material to build.

Unfortunately, keeping weight down can be challenging, because adding weight is often the most obvious way to meet other design goals. For example, if you need to make a structure stronger or stiffer, you can almost always do so by adding more structural elements and/or making the existing elements thicker—but this adds weight.

There are three things you can do to reduce vehicle weight while preserving adequate strength and stiffness:

1. Reduce the overall *size* of the vehicle as much as possible.
2. Take advantage of *substances* that offer high strength and stiffness for their mass density.
3. Take advantage of structural *shapes* that offer high strength and stiffness for their weight.

Image courtesy of Dr. Steven W. Moore

Figure 4.10: Ease of Hand Launch

Ranger Kevin Brady demonstrates the ease with which a light-weight vehicle can be launched. Small size also facilitates recovery, transport, and storage.

The first of these approaches—reducing vehicle size—reduces weight for a couple of reasons. One reason is that a smaller vehicle displaces less water, so it is less buoyant and therefore needs less weight to submerge. (This topic will be explored in much greater detail in *Chapter 6: Buoyancy, Stability, and Ballast.*) Another reason is that large vehicle structures are so heavy that they need additional framing just to support their structure! In that respect, smaller vehicles use structural materials more efficiently. Additional advantages of smaller vehicle size include less resistance to movement through the water and improved ability to maneuver among obstacles (like wharf pilings) or explore confined spaces (like the interior of a sunken ship). One good way to reduce vehicle size is to minimize wasted space inside pressure canisters.

The second approach is to build your structure out of *substances* that offer a lot of strength and stiffness for their mass density compared to other materials. Such materials are said to have a good **strength-to-weight ratio** or good **stiffness-to-**

weight ratio. Metals are popular structural materials, in part because they have excellent strength-to-weight ratios and stiffness-to-weight ratios compared to most other building materials. Even though metals are heavier (more precisely, denser), they are so much stronger and stiffer than the alternative materials, such as wood or plastic, that you don't need as many pounds of metal to get the same job done. Paradoxically, then, the overall vehicle structure can be made lighter by using a "heavier" material. (See *Table 4.1: Properties of Some Metals and Plastics Used in Underwater Vehicle Structures.*) Plastics, in general, are not as strong or as stiff as metals, but they also are not as dense. Their strength-to-weight and stiffness-to-weight ratios can be competitive with those of metals in applications where extreme strength and stiffness are not required. Substances with good strength-to-weight and stiffness-to-weight ratios that also meet other criteria for use in underwater vehicle structures are described in a separate section later in this chapter.

A third strategy is to choose structural *shapes* that offer high strength and stiffness for their size and weight. This important topic is covered later in this chapter.

3.3. Compatibility with the Underwater Environment

Another critical performance consideration is the structure's ability to maintain its strength and stiffness under water. Most parts of an underwater vehicle's structure will come into direct contact with freshwater and/or saltwater for extended periods of time. Materials that quickly dissolve or become mushy in water are clearly unsuitable as structural materials, but even materials that seem waterproof for a while can gradually be weakened through prolonged contact with water.

Image courtesy of Vickie Jensen

Figure 4.11: Corrosion and Material Breakdown

Without careful attention to materials selection and compatibility, and without vigilant maintenance, the marine environment will exact a harsh toll on metals and plastics alike.

One reason for this is that water accelerates a number of chemical reactions that gradually eat away many otherwise ideal structural materials, most notably metals. This process is called **corrosion**. If not prevented, or at least managed adequately, corrosion can weaken structural members over time and eventually lead to catastrophic failure.

Some materials are much more susceptible to corrosion than others. Corrosion of metals is of particular concern for underwater vehicle designers, so is discussed in greater detail later in this chapter. Plastics are generally immune to corrosion in seawater and freshwater, but you should know that their molecular structure can be attacked by the chlorine compounds used to disinfect swimming pools and hot tubs. Most plastics are also susceptible to gradual damage by the ultraviolet radiation present in sunlight. Some plastics can soften and deform while sitting on the dock during a hot, sunny day, while others may become brittle and fracture easily in cold water. Some plastics gradually absorb water and swell just enough to jam precision moving parts, cause leakage around watertight seals, or even crack the structural framework.

Another reason that some materials do not work as well under water as they do on land is that the biological environments are different. Plant- and animal-derived materials, such as wood, leather, and cloth made from natural fibers, wear out more quickly under water than they do on land; that's because these nutrient-rich organic materials rot more quickly in water or get eaten, burrowed into, or otherwise damaged by marine organisms such as shipworms.

Concerns about biological compatibility can run in the other direction, too. It's important to avoid using structural materials or other vehicle components that are likely to release toxic chemicals into natural bodies of water; such chemicals might harm plants or animals living there and might eventually make their way into human drinking water supplies.

As you design your structure, remember to choose materials that are compatible with the chemical and biological environment in which your vehicle will operate. A list of many suitable materials will be provided later in this chapter.

HISTORIC HIGHLIGHT: EARLY MATERIALS FOR UNDERWATER WORK

Inventors in each era are confined by the materials and technology available to them. Despite the limitations of their day, many came up with creative ways to adapt available materials to underwater work.

Animal Products: In earlier times, divers in the Persian Gulf used goggles made from tortoise shells, while divers in the Mediterranean breathed under water by means of animal skins or bladders filled with air. Animal bladders also served as ballast tanks in early submersible vessels. Leather was used for many undersea applications because it was cheap and water-resistant. Diving suits, helmets, hoses, straps, and seals were all made from leather until well into the nineteenth century, when vulcanized rubber and rubber/cloth laminates came into being.

Wood: For centuries, wood has been used to construct ships, boats, and underwater devices. Plentiful and cheap, wood was easily worked and could be made watertight. Early diving bells were simply heavily weighted wooden casks that were turned upside down to trap pockets of air. Wood was effective for shallow-water explorations and was more common than metal, which was expensive and difficult to work at that time. As a result, some of the first submarines were made of wood. Today, synthetic and metal materials are stronger, cheaper, and more versatile than wood, so its use in undersea vehicles has declined significantly.

Metals: By the early 1800s, the invention of the steam engine facilitated production of cast iron and then wrought iron, which replaced wood for submarine hulls. By the late 1800s, steel processing had been upgraded so that refined steel became a better option than iron. Lead has always been commonly used for weights.

Fiber and Cloth: Undersea use of various fibers was limited mainly to ropes and lines until the development of heavy cotton, linen, and hemp fabrics. Early helmets and diving suits were made of heavy canvas backed with leather; later models employed rubber for water resistance. Woolen underwear was every early diver's choice for insulation beneath the waves. With the development of rayon in 1884 and nylon in 1938, synthetic fibers began to have an impact on undersea equipment. Today, ropes are made of nylon and Kevlar (both are plastics), while suits of rubber and synthetic materials have replaced natural-fiber diving suits. Modern composites such as glass, carbon, and boron fibers mixed with resin are the choice for modern undersea equipment because they are both lightweight and strong.

Rubber: True rubber is a natural secretion from a tropical tree, but it was not considered useful until Charles Goodyear heated it with sulfur to produce a strong and durable material called vulcanized rubber. As soon as the technology allowed, this new type of rubber replaced leather hoses and diving suit linings. Now it is used in the manufacture of gloves, goggles, masks, fins, and nose plugs, as well as for hull seals and cable casings. Most rubber today is a synthetic petroleum product. Its elastic and insulating properties make it an ideal undersea material, although most kinds are very susceptible to attack by the sun's rays.

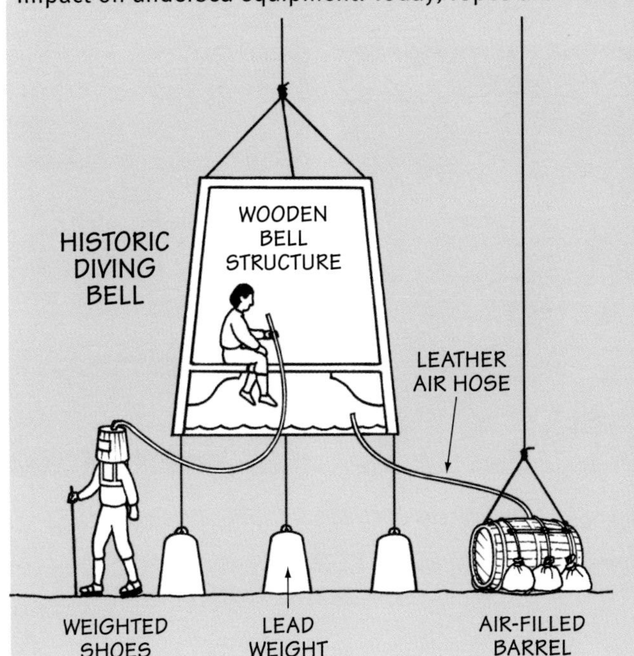

Figure 4.12: Early Materials Used for Underwater Structures

4. Other Structural Design Considerations

In addition to the critical performance criteria discussed above, you will want to consider a number of other important factors as you design your vehicle's structure.

4.1. Ease of Construction, Maintenance, and Repair

Throughout the structural design process, it helps to visualize the entire sequence of steps needed to build your frame, pressure canisters, and fairing (if required), as well as how you will perform any routine maintenance or repair on the structure. This will minimize the chance of problems down the road.

Simpler is usually better. That's because simple designs are generally easier to build and modify and also because less complicated designs usually have fewer things that can go wrong with them. So plan carefully to make the construction process as straightforward as possible. One way to simplify the design and construction processes is to use parts you buy ready-made at a store. Such parts are called **COTS** for **Commercial Off-The-Shelf**. These components are contrasted with **DIY** which stands for **Do-It-Yourself** and refers to parts, projects, or systems that you build yourself from scratch. (See Figures 4.13 and 4.14.) Each approach has its advantages. DIY offers more opportunities for mistakes and is therefore invariably more educational. COTS is usually much faster, more reliable, and sometimes less expensive. COTS parts are also more likely to be compatible with standard fasteners, making them relatively easy to connect together or to disconnect for repair or replacement. Note that at some level, every DIY project relies on COTS parts or materials. For example, most people who bake their own bread "from scratch," don't grow the grain and grind the flour. Likewise, students building DIY parts for ROVs rarely make their nuts and bolts or manufacture their own transistors.

Even if you get most of your parts ready-made from the hardware store or other source, you will probably need to custom-make or modify at least a few of them. Life will be easier if you select materials that are easy and cost-effective to cut, drill, bend, cast, or otherwise form into the specialized parts you need. Make sure materials and the parts you make from them are easy to glue, weld, bolt, or in some way join together, so that smaller pieces of material can be connected to make larger parts of the structure.

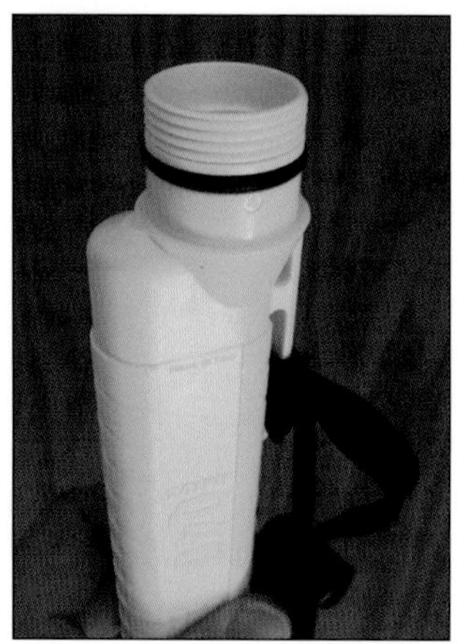

Images courtesy of Dr. Steven W. Moore

Figure 4.13: DIY Versus COTS Lights

These two images compare DIY underwater pressure canisters being assembled from plastic PVC pipe (left) with a COTS dive light housing that's easy to "repurpose" into a small electronics canister for an ROV (right).

Figure 4.14: DIY Versus COTS Motor Control Circuits

These two images show a DIY motor control circuit (left) compared with a COTS motor control circuit (right).

Keep in mind that the shape of the structure, the materials used to build it, and the order in which parts are assembled can all have a profound impact on how easily the final structure can be put together, maintained, and repaired. A structural element that is located in an inaccessible position on the vehicle may be difficult to inspect for corrosion and even more difficult to repair or replace if corrosion is found. A welded frame may be more rigid and reliable than one held together by bolts, but a frame held together by bolts may allow easier replacement of a damaged section. Here again, the use of mass-produced COTS parts can be advantageous, in case replacement of a part of the structure becomes necessary.

4.2. Cost and Availability of Parts and Raw Materials

No matter how well a structural component meets your other design requirements, it won't do your project any good if you can't find it or afford it.

Some parts, including nuts, bolts, and other fasteners, may already be readily available in their final form at an affordable price. Others may need to be custom-made in the shop by cutting, drilling, bending, or otherwise modifying a piece of generic building material, such as a plastic pipe or a piece of sheet metal. Still others might need to be created by casting or otherwise forming solid parts out of initially liquid substances such as molten metals or epoxy resins.

Avoid going too far down a particular design pathway without checking to make sure that all the required parts (or at least any materials and tools you need to make the parts) are available and affordable. Keep in mind that any money or other resources

TECH NOTE: SAFETY FACTOR

A passenger elevator suspended from cables will commonly have four or more strong cables supporting it, even though any one of the cables is plenty strong enough to support the fully loaded elevator all by itself. This deliberate over-design provides an extra "margin of safety" that will keep passengers safe even if one of the cables weakens or breaks between elevator inspections. The margin of safety can be quantified by specifying a number called the **safety factor**, which is the ratio between what's provided and what's needed. For example, if each cable can hold 1.25 times the weight of the fully loaded elevator all by itself, an elevator with only one cable would have a safety factor of 1.25 (sometimes written as 1.25X). But since there are actually four cables in this elevator, the total safety factor is 4 x 1.25 = 5X. This 5X safety factor means the elevator is designed to support five times as much weight as it normally needs to support.

The term "safety factor" applies even when no human safety is at stake, and it can apply to subsystems other than vehicle structure. For example, if your unmanned ROV mission calls for a dive time of 30 minutes in a lake "just to see what's down there," you might want to design a battery system that can provide power for at least 45 minutes. This 1.5X safety factor might save the day if your mission ends up taking a few minutes longer than expected, or if some unanticipated currents or other challenges cause your vehicle to consume electrical power a little more quickly than usual.

Safety factors can range from none at all to quite large, depending on how serious it would be to have something fail and how difficult or expensive it would be to build in a larger safety factor. In the elevator example above, a 5X safety factor is large but reasonable, because human lives are at stake. In that case, it's easy to justify the slight added cost for the extra cables. In the ROV example, the more modest 1.5X factor is also reasonable, because it is fairly easy to do and not very expensive, yet it provides an adequate margin of safety in this non-critical application. In some cases, limits on what is technically or financially possible force safety factors to be less than ideal. Such is often the case when pushing the limits of a new technology—for example, to go where no robot has gone before. In such cases, you must weigh the high risk of losing your robot due to inadequate safety factors against the potential benefits of a successful mission.

TECH NOTE: STYLE MEETS ENGINEERING

Structure not only serves an important functional role, but it also has a dramatic effect on the overall appearance of the vehicle. Because of this, structural design provides an excellent opportunity to blend your artistic talent with your engineering skill. Making a good-looking hull that is also functional is a designer's dream and usually a little harder to accomplish than it may seem.

you spend on your vehicle's structure are resources that are *not* available to spend on other vehicle systems or operations. Therefore, even if you *can* afford something, ask yourself if it would be wiser to save that money for something else that's more important. That said, a vehicle's structure is no place to skimp, since a structural failure can destroy a vehicle and doom the mission. Spend what's necessary to produce a good, reliable structure (see *Tech Note: Safety Factor*), but be wary of spending any more than that, at least until you're absolutely sure all other needs have been met.

4.3. Safety

A vehicle's structure impacts safety in many ways, so safety must be a priority consideration throughout the structural design process.

As previously stated, this is an introductory book and does *not* cover structural systems (or any other vehicle system) in enough detail for you to design vehicles that are safe for living passengers or crew. In such vehicles, structure-related dangers include the very real possibility of drowning (if the structure leaks) or being crushed in a violent implosion (if the structure collapses under the enormous forces imposed by water pressure). History has shown repeatedly that even professionally designed underwater vehicles sometimes experience structural failure, and if people are on board, the consequences are frequently fatal. For this reason, pressure hulls, frames, and materials used in all manned submersibles are required to meet rigid safety standards set by various agencies such as the American Bureau of Shipping, Lloyds Registry, and the American Society of Mechanical Engineers. Moreover, only individuals certified by the appropriate agencies are permitted to work on the structural components of manned underwater vehicles.

But safety concerns aren't limited to manned vehicles. Even small ROVs and AUVs can and do injure people. For example, a small vehicle being transported on a rocking boat can roll around on the deck or slide off a table or cradle, potentially injuring someone nearby. This risk is greatly increased if the designers of the vehicle forgot to think about adding tie-down points or other structural elements for stabilizing the craft during transport. Any sharp edges or sharp points on the structure further increase the risk of injury to people working on or near the vehicle.

Awareness of potential hazards, coupled with common sense and a proactive attitude about safety, can go a long way toward making your structural designs reasonably safe.

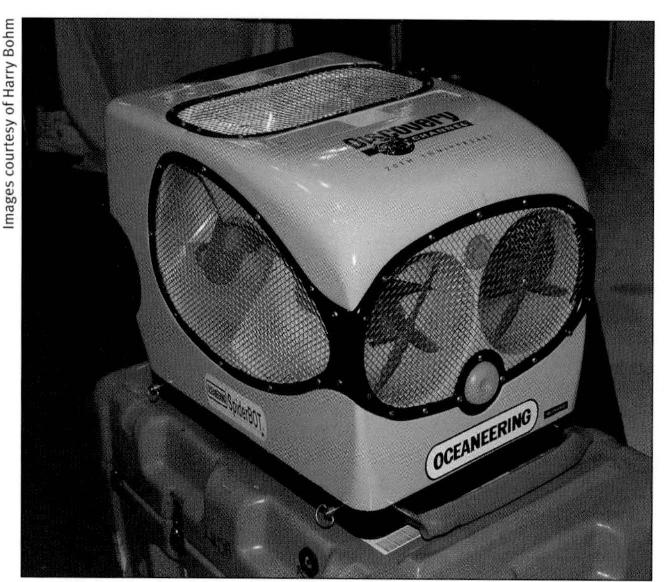

Images courtesy of Harry Bohm

Figure 4.15: Style and Function

The smooth finish and rounded corners of Oceaneering's compact SpiderBOT ROV allow it to function in dangerous areas such as shipwrecks; the structure also provides good protection for the propellers.

4.4. Aesthetics

Mission success usually depends more on performance than on looks; nonetheless, there's something uniquely satisfying about designing a vehicle that excels at both. Surprisingly enough, the two objectives often go hand in hand, because most humans instinctively find functional forms aesthetically pleasing. For example, graceful, streamlined vehicles, which offer lower drag and therefore better performance, are generally considered more attractive than chunky, lumpy ones.

A good-looking or strikingly unusual vehicle can garner more attention from potential sponsors (and maybe your future employers), boost team morale, and provide evidence that the team has well-rounded skills that encompass both art and engineering. Since the vehicle's structure is its most visible subsystem, it makes sense to invest some time and energy in thinking about how to make it look as good as possible, as well as be consistent with excellent performance and successful completion of the mission.

Different structural materials have different colors, textures, and finishes. If you can find two or three different materials that meet all other design constraints, you might as well go with the one(s) you find most attractive.

5. Good Structural Shapes

Workable structural shapes are generally those that provide high levels of strength and stiffness with a minimum amount of material and weight, yet are easy to obtain or build.

5.1. Good Shapes for Pressure Hulls and Canisters

Cylinders and spheres are particularly effective at resisting water pressure and are therefore the most common shapes used for pressure hulls and pressure canisters. These shapes and their structural importance are discussed in much greater detail in *Chapter 5: Pressure Hulls and Canisters*, so they will not be described further here.

5.2. Good Shapes for Frames

Frames are most commonly (and most easily) made by joining a bunch of straight beams or rods. Selecting an effective shape for a frame is therefore a matter of selecting beams or rods with good structural properties and connecting them together into polygons that also have good structural properties. In this context, "good structural

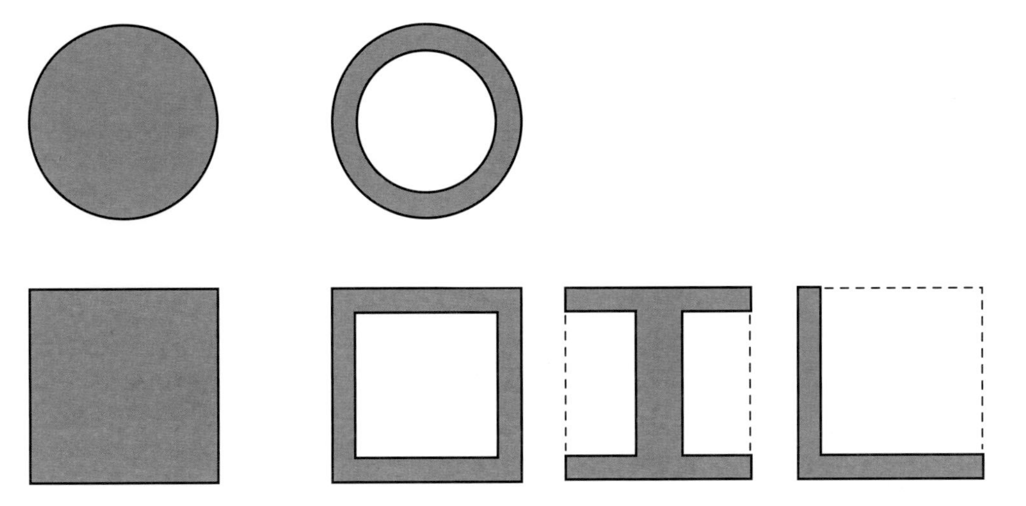

Figure 4.16: Strong but Lightweight Beams

When it comes time to select beams for a vehicle frame, remember that a hollow pipe or cylinder is nearly as stiff as a solid, round rod of the same length and diameter, but is much lighter (and usually much less expensive).

Likewise, square tubing, I-beam, and L-shaped right angles are all nearly as stiff as solid square bar stock of similar size, but less heavy and less costly. These thin-walled shapes have the added advantage that they are easier to cut and drill than their solid counterparts.

properties" means high strength and stiffness with minimal material and minimum weight. Ease of fabrication is another important consideration.

While the simplest shape for a beam is usually a solid rectangular or circular cross-section, this is not usually the best shape for a frame element. It turns out that most of the strength and stiffness in these supports is provided by material near the outer surface, because that's the part of the beam that gets stretched or compressed the most when something attempts to bend the beam. This means that you can get rid of a lot of material near the center of the beam (thereby eliminating a lot of weight) without decreasing the strength and stiffness very much. A hollow pipe, for example, is almost as stiff as a solid cylinder of the same diameter, but not nearly so heavy.

For this reason, hollow pipes, rigid square "tubing," and beams with cross-sections that are shaped like the letters "I" (called "I-beams"), "L" (called "angles" or "right angles"), or a square-shaped "U" (called "channel") are popular starting materials for building frames. (See Figure 4.16.) These thin-walled shapes have the added advantage that they are easier to cut and drill. They can also be joined easily by drilling a hole through two pieces and inserting a bolt or rivet through the holes.

It's important to note that both the length and thickness of a beam have a large effect on its mechanical properties. In the case of a long rod or beam, diameter or thickness influences stiffness, but not as strongly as length does. For example, doubling a solid, round rod's diameter makes it about *four* times stiffer, but doubling its length makes it about *eight* times more flexible. The upshot of this is that a round rod twice as long must have *more than* twice the diameter (about 2.8 times greater, actually) to maintain the same stiffness. Since the volume (and hence the weight) of the rod is proportional to the length multiplied by the diameter squared, a rod twice as long needs to be about 16 times as heavy to maintain the same stiffness! This is another argument for keeping your vehicle dimensions as small as possible.

TECH NOTE: STIFFENING A RECTANGLE

Rectangles and squares made by connecting four beams at their ends are inherently less stiff than triangles. This is particularly true if the joints can pivot, as they can if joined by a single bolt or rivet that loosens. The reason is that rectangular assemblies can fold into parallelograms without changing the length of any side, whereas triangles cannot change shape without also changing the length of at least one side—something beams won't allow.

Rectangle A was unsupported and has collapsed. Rectangle B is reinforced with a diagonal brace, which effectively turns it into two triangles and makes it very stiff. For this rectangle to collapse, the diagonal brace would have to get shorter or longer, and that isn't going to happen easily. By using a diagonal brace, you can get by with one simple bolt or rivet to join each corner (no need for multiple bolts or welding at each corner); this adds lots of stiffness without adding much additional weight. Replacing the brace with two thin cables in an "X" pattern, as in rectangle C, provides an even lighter option for stiffening a rectangle. You need two cables, since cables cannot resist compressive forces. One cable prevents collapse to the right, while the other prevents collapse to the left. In some cases, the central part of the rectangle must be left open (for example, to avoid blocking the view of a camera). In these cases diagonal braces and/or cables across the center of the rectangle may not be an option. An alternative that is ok, but not quite as stiff, is to use one or more corner braces, as in rectangle D.

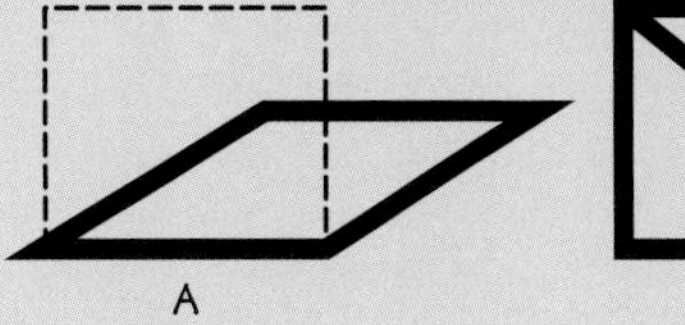

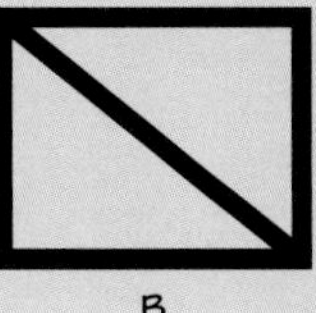

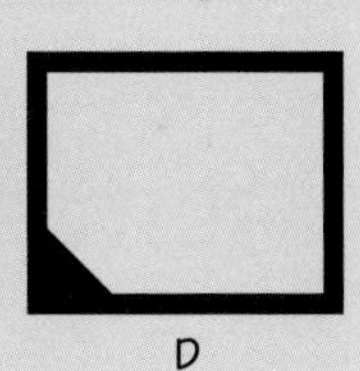

Figure 4.17: Options for Stiffening a Rectangular Shape

When it comes time to join your beams together into a framework, you need to decide how to do the joining (welding, gluing, bolting, riveting, etc.) and how to arrange the beams in the final frame. Gluing or welding, if done well, can produce strong, rigid joins between beams, but either procedure requires some skill, and such joins are difficult or impossible to undo without cutting the beams. Bolts and rivets, on the other hand, don't hold quite as well, but they can be secure enough for most purposes and are much easier to undo if a section of frame needs to be adjusted or replaced.

As for the spatial arrangement of beams within the overall framework, it's good to remember that *triangles* are inherently stiffer as structural arrangements than rectangles or any other polygons. This is the principle upon which trusses are based. A truss is an open framework of triangles and triangle-reinforced rectangles designed to support loads that are heavy compared to the weight of the structure itself. You can find examples of their use in skyscrapers, bridges, tower cranes, radio towers, and many other structures. (See Figure 4.9.)

Image courtesy of Peter Thain

Figure 4.18: Optimal Fairing Shape

The Bissell Missile, *a student-built ROV, features a sleek fairing creatively recycled from the housing of a hand vacuum cleaner.*

5.3. Good Shapes for Fairings

For optimal fairing shape, just look at any fish that spends a lot of time actively swimming through its watery environment. Over millions of years, their body forms have evolved for speed and energy-efficiency. Although there is tremendous variation in the details from species to species, a common theme is evident. Almost all fish (not to mention marine mammals, airplanes, and even some race cars) are basically shaped like teardrops. They have fairly smooth surfaces that are blunt and rounded at the front, but tapered to a point or thin edge at the back. This tapering is particularly important, because as water separates and flows around opposite sides of the fish (or vehicle), the tapered rear end brings it gently back together, minimizing the turbulence that robs a fish (or vehicle) of speed and efficiency.

6. Good Structural Materials

In addition to choosing the size and shape of each structural part used in your vehicle, you'll need to select the material or substance from which the part is made. Over the years, a number of materials—mostly metals and plastics—have proven themselves worthy of serious consideration in underwater structural design. These materials meet the demanding strength-to-weight, stiffness-to-weight, and water compatibility constraints for use in underwater vehicle structures. Some of these structural materials and a few others also serve non-structural roles within underwater vehicles.

TECH NOTE: WHAT IS A "MATERIAL?"

The term "material" is somewhat ambiguous and potentially confusing in discussions of structural materials or building materials. On one hand, the word may refer to a *substance* (such as metal, wood, plastic, or glass) from which objects can be made. On the other hand, it may refer to an *object* (such as a metal beam, wooden board, plastic pipe, or glass dome) already made of a particular substance, that has been produced in some standard shape for sale to people who will probably modify it and/or combine it with other objects to make useful things. If you pay close attention to the context in which the word "material" is used, its meaning will usually be clear.

Images courtesy of Duane Thompson, Monterey Bay Aquarium Research Institute

Figure 4.19: Options for Structure Materials

The structural framework of these similarly sized student-built ROVs demonstrates three different types of materials: squared aluminum extrusion, PVC piping, and sheet plastic.

Keep in mind as you read through this section that all potential materials have both advantages and disadvantages. Which materials are best for your particular project will depend on the details of your mission, what you can find in the local hardware store, your fabrication skills, your budget, and other factors.

6.1. Metals

Compared to other structural materials, metals are chemically simple, sometimes consisting of a single type of atom (one element). More often, they consist of a mixture of two or more elemental metals and possibly some non-metal elements, like carbon. These metal mixtures are called **alloys**. Each alloy has unique properties determined by the different elements used and the ratios in which they are mixed. The material properties of many alloys may be further modified by heating or "working" (repeatedly bending and squishing) the alloy. Metals that occur as elements (but are often sold commercially as alloys with improved properties) include iron, nickel, copper, gold, silver, titanium, and aluminum. Familiar alloys that always have two or more elements include steel, stainless steel, brass, and bronze.

Metals are popular structural materials for all sorts of things (not just underwater vehicles), because most are hard, stiff, wear-resistant materials that offer excellent strength-to-weight ratios, yet they can also be cut, drilled, bent, welded, cast, and otherwise fabricated into a wide variety of useful structural components, including beams, plates, bolts, rivets, and even wire rope or cable. In large underwater vehicles like submarines, metals (particularly steel alloys) are the dominant structural materials used. Metals also play important roles (structural or otherwise) in most smaller vehicles.

One serious drawback of metals—especially for underwater work—is that metals as a group tend to be much more susceptible to corrosion than other structural materials. Fortunately, some of the alloys have good corrosion resistance. And there are ways to reduce corrosion in other alloys to the point where the beneficial properties of the metal outweigh the disadvantages presented by corrosion. The causes of metal corrosion and ways to prevent it are discussed later in this chapter.

Here are brief descriptions of some structural metals that every underwater vehicle designer should know about.

6.1.1. Steel

The term "steel" is actually a general term referring to a large group of different alloys in which iron is the dominant elemental metal. Additives include things like carbon, manganese, or silicon. The various steel alloys differ in their hardness, stiffness, corrosion resistance, and other properties. Some are easier to machine than others. Even though steel is dense (See Table 4.1), it makes for lighter vehicles because it offers a good strength-to-weight ratio; you don't need as many pounds of steel to achieve a particular level of strength as you would pounds of most alternatives. This advantage, coupled with steel's relatively low cost, abundance, and ease of machining and welding, have made steel alloys the structural material of choice for most large underwater vehicles, such as military submarines, as well as surface ships, oil platforms, and other large marine structures.

The main drawback to steel is that most of its alloys rust easily. Fortunately, these alloys can be protected from corrosion with paint and many other types of anti-corrosion coatings. So steel is often more cost-effective than other materials, even after you factor in the added cost for maintenance.

Another drawback is that most steel alloys distort magnetic fields and can become magnetized, so they may interfere with compass readings used in underwater navigation. This issue can be solved by using a modern **gyro compass,** which does not rely on a magnetic field to give direction. These gyro compasses are regularly used in deep-diving submersibles and submarines. Another solution is to use non-magnetic materials such as plastic or stainless steel.

6.1.2. Stainless Steel

Stainless steel (or just "stainless" for short, sometimes abbreviated "SS") is a specialized subset of steel alloys in which the iron is mixed with chromium (and sometimes other materials) to make it far more resistant to corrosion. This makes some stainless alloys (like Type 316) ideal for many undersea applications. Stainless is used in hydraulic fittings, pneumatic fittings, electrical fittings, and all kinds of fasteners (nuts, bolts, screws, etc.). The frames of some ROVs, such as the Phantom 350 from Deep Ocean Engineering, are made from tubular stainless.

Small pressure cans are often made from stainless steel, but larger pressure hulls and cans are not, because stainless steel is expensive and more difficult to machine than

HISTORIC HIGHLIGHT: THE WORLD'S FIRST ALUMINUM SUBMARINE

Launched in 1964, the *Aluminaut* was the world's first aluminum submarine. It was built for Reynolds Metals Company by the Electric Boat Division of General Dynamics and was operated by Reynolds Submarine Services Corporation until 1970. The 15.5-meter (approx. 51-ft) sub had an operating crew of three and could accommodate three to four scientists. It weighed about 80 tons and its design depth was about 4,570 meters (approx. 15,000 ft). Submerged endurance was 32 hours at three knots, with water ballast and steel shot used for buoyancy control. The aluminum sub had four viewports and was equipped with sonar, manipulators, and about 26.7 kN (approx. 6,000 lbs) of scientific payload for various types of oceanographic and salvage missions.

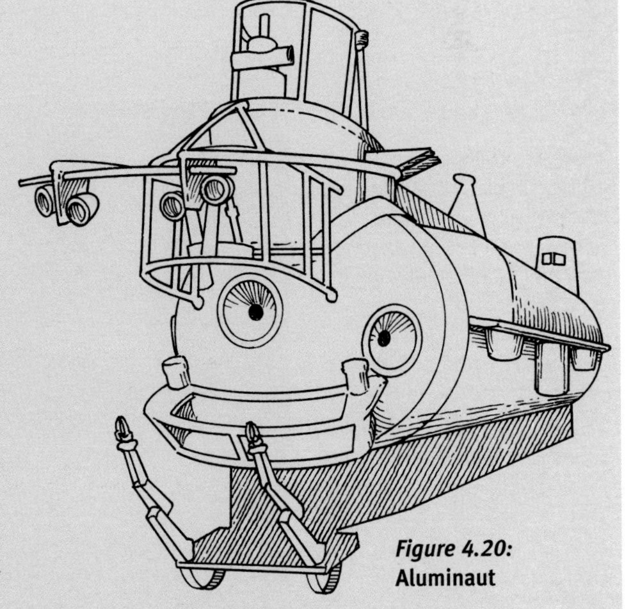

Figure 4.20: **Aluminaut**

Notable missions included assisting with the recovery of the hydrogen bomb off Palomares, Spain, and the 17-hour salvage of the research submersible *Alvin*. In addition to surveys for the U.S. Naval Oceanographic Office, *Aluminaut* made movies with Jacques Cousteau and Ivan Tor Studios. The historic aluminum submarine is now on permanent display at the Science Museum of Virginia, in Richmond.

other steel alloys. It also requires special welding techniques and is prone to stress cracking, unlike the regular steel and aluminum normally used in large structures. While the strength-to-weight ratio of stainless steel is similar to that of other steels, it is simply not cost-effective for large structures.

6.1.3. Aluminum

Aluminum is an elemental metal, but the types you can buy are generally aluminum alloys. Aluminum alloys have been used extensively for pressure hulls and canisters. Aluminum alloys 6061 and 6063 are particularly popular for underwater vehicles because of their very good corrosion resistance, good strength-to-weight ratio (which exceeds that of carbon steel), and wide availability. Though stiff, aluminum is a relatively soft metal and is easier than steel to cut and drill, particularly if you are using hand tools. It is also reasonably affordable. For mid-ocean-depth pressure hulls, the trade-off is often whether to use aluminum or titanium. Usually aluminum is chosen because there are no special tooling requirements for any machine shop and because welding aluminum is now a routine trade skill. While not as susceptible to corrosion as steel, aluminum does corrode in seawater, so some anti-corrosion methods are usually necessary.

6.1.4. Titanium

Titanium alloys are used in some undersea structures (as well as in aircraft frames, human joint replacements, and other specialized applications) because they have excellent resistance to saltwater corrosion, a very high strength-to-weight ratio, and high resistance to fracture. Some titanium alloys have strength-to-weight ratios more than three times that of common steel alloys. Because the Soviet Union had a reliable supply of titanium, it was used to construct pressure hulls for their nuclear submarine fleet. This gave these vessels a distinct depth advantage over steel-hulled subs. Titanium hulls have also been used in deep-diving submersibles such as the *Nautile*, *Alvin*, and *Shinkai 6500*. Unfortunately, titanium is *much* more expensive than steel. The price has dropped somewhat in recent years, and titanium is now being used more often in small undersea vehicles for pressure canisters, manipulator arms, and other parts; however, its cost will have to drop considerably lower to make it a viable choice for

***Figure 4.21:* Shinkai 6500**

The deep submergence vehicle Shinkai 6500 *can dive the deepest of any current manned submersible. Its thick titanium hull allows two pilots and a researcher to descend to 6,500 meters (21,000 ft)! Completed in 1990,* Shinkai *is owned and operated by JAMSTEC (Japanese Agency for Marine Earth Science and Technology).*

Image courtesy of Japan Agency for Marine-Earth Science and Technology (JAMSTEC)

most school groups or others on a tight budget. Another disadvantage is that welding titanium requires special techniques that raise fabrication expenses. For vehicles that will be diving no deeper than a few hundred meters, less expensive alternatives, such as stainless steel or aluminum, should provide adequate structural support.

6.1.5. Brass

Brass is a shiny, yellowish-colored metal alloy made by combining copper and zinc. It is hard and strong, yet can be machined with shop tools and is corrosion-resistant. It is used for plumbing fittings and for screws or bolts employed in corrosive environments. Because of its combination of strength, corrosion-resistance, and attractive finish, it is often used for railings and similar fixtures on boats. **Naval brass** is an alloy specifically formulated for high corrosion-resistance in seawater.

6.1.6. Bronze

Bronze is a brownish-red metal alloy similar to brass, but it is made by combining copper with tin (or sometimes other metals) instead of zinc. Like brass, bronze is fairly hard, strong, and corrosion-resistant and is often used for valves and other plumbing fittings.

Some important properties of these structural metals and plastics are summarized in Table 4.1.

Table 4.1: Properties of Some Metals and Plastics Used in Underwater Vehicle Structures

Material	*Strength (Ultimate tensile strength in MPa)*	*Stiffness (Elastic modulus in GPa)*	*Density (kg/m³)*	*Corrosion Resistance*	*Cost Comparison (Approximate retail price for a solid round rod, 1-inch diameter, 36-inch length as of March 2009, in U.S. dollars)*	
					Per rod	*Per kg*
Carbon steel (medium)	*500*	*200*	*7860*	*poor*	*$20*	*$6*
Stainless steel (316)	*550*	*195*	*8000*	*good to excellent*	*$50*	*$13*
Aluminum (6061)	*310*	*70*	*2700*	*good*	*$15*	*$12*
Titanium (Ti-6Al-4V)	*1000*	*110*	*4430*	*excellent*	*$500*	*$244*
Naval brass (485)	*430*	*100*	*8440*	*fair to excellent*	*$75*	*$19*
Bronze (316, tempered)	*450*	*115*	*8860*	*good to excellent*	*$125*	*$30*
PVC (extruded)	*46*	*1.9*	*1360*	*excellent*	*$6*	*$10*
ABS (molded)	*40*	*2.2*	*1050*	*excellent*	*$17*	*$35*
Polycarbonate (extruded)	*67*	*2.4*	*1200*	*excellent*	*$26*	*$47*
Acetal (cast)	*60*	*3.1*	*1420*	*excellent*	*$15*	*$22*
Acrylic (cast)	*80*	*3.3*	*1200*	*excellent*	*$14*	*$25*

Note: These are approximate, averaged values intended for general comparison purposes only. The physical properties for any given piece of material may vary considerably from those listed, depending on exact alloy composition, temperature, environmental exposure, and other factors.

Do not rely on these generic numbers for precise or critical design decisions. Likewise, prices among vendors are affected by the quantity of material purchased and fluctuate with market conditions.

Sources: www.onlinemetals.com, www.matweb.com, and www.mcmaster.com, accessed March 2009.

6.2. Plastics

Plastics are organic polymers consisting of interlocking chains of carbon-based molecules. They are familiar materials, used in everything from beverage bottles to playground equipment, synthetic textiles, and the dashboards of most cars. Because of their low cost, easy availability, excellent corrosion resistance, and ease of fabrication, plastics are also the dominant structural materials used in fabricating most small, low-budget underwater vehicles.

Low-cost plastics are readily available in a variety of shapes at local hardware stores and on-line distributors. They have moderate strength-to-weight ratios and are easy to drill, tap, saw, sand, and otherwise fabricate into parts, even with hand tools. Many can be easily melted or glued together. As a rule, plastics are much more corrosion-resistant than metals, though they can be damaged over time by exposure to sunlight and chlorine compounds used in swimming pools. Polymers have adequate strength-to-weight ratios to make them useful for shallow-water applications (say, less than 100–200-meter/approx. 325–650-ft. depths), but their limited strength and stiffness make them less suitable than metals for large and/or deep-diving vehicles, which must support large forces. Though waterproof, some plastics can absorb minute quantities of water when immersed and can swell just enough to cause problems where precision fit is critical.

Images courtesy of Vickie Jensen

Figure 4.22: PVC Pipe and Connection Options

TECH NOTE: PVC PIPE GRADES

Schedule 40, 80, and 120 Pipe and Fittings

PVC pipe comes in several grades, called schedules, with Schedule 40 and Schedule 80 the most widely used. Schedule 40 is commonly found in the plumbing section of hardware stores and is rated for typical household water pressures. Schedule 80 pipe (less common in hardware stores, but easy to order on the internet), is thicker and able to withstand higher pressures. Both Schedule 40 and Schedule 80 PVC pipe and fittings are generally stocked up to about 4- or 6-inch diameters, and larger sizes can be special-ordered.

Class 160 and 200 Pipe

These pipe-sizing standards apply to thin-wall pipe used in applications such as irrigation systems. Typically, Class 160 pipe has a 1- to 2-inch nominal inside diameter and Class 200 pipe has a ¾-inch and ½-inch nominal inside diameter.

You can build the *SeaMATE* frame using either Class 200 ½-inch pipe or Schedule 40 ½-inch pipe. Note that there are no Class 200-sized pipe fittings, but Schedule 40 PVC pipe fittings are compatible for Class 200 pipe.

The listing that follows is a synopsis of just a few of the many types of plastics available for underwater vehicle constructions. There are many others in common use, so before settling on any polymer, remember to research a potential material's properties to ensure it is capable of providing the structural integrity your vehicle will need to complete its mission. Most of these materials come in ready-to-assemble shapes that require no machining or welding to fabricate into a frame. However, you can also get them in standard plates, bars, and other shapes from which you can machine your own parts.

6.2.1. PVC (Polyvinyl Chloride)

PVC, or polyvinyl chloride, is a material manufactured by using a polymerization process that combines vinyl chloride with other chemicals to produce a plastic that can be melted, then molded or extruded into various shapes and products such as pipe and pipe fittings. (The familiar white plastic water pipes and fittings found in hardware stores are made from PVC.) This plastic comes in several standard colors, including white, gray, bluish, lavender and (almost) clear. PVC, like other plastics, is not as dense as metal.

This type of plastic is very popular as a raw material for shallow-water ROV and AUV projects because of its very low cost, wide availability in hardware stores, outstanding corrosion resistance, and ease of fabrication. It can be cut and drilled with no trouble using standard hand or power woodworking tools. It can be glued or melted together. Note that PVC's low melting point can be a disadvantage, since this type of plastic tends to melt easily when worked with power tools, such as a lathe; however, if lathes are run with carbon bits and at slow speeds, PVC can be machined to accuracies (tolerances) of 1/1000 of an inch (about half the width of a human hair)—good enough for reliable O-ring seals that can keep high-pressure water from leaking into pressure canisters. (Use of O-rings is explained in *Chapter 5: Pressure Hulls and Canisters.*)

In large vehicles, PVC is not often used for major structural members because it is brittle, particularly at the low temperatures found in many deep water bodies, so it can shatter upon impact. However, it finds other uses on these larger/deeper vehicles as attachment brackets, small component structures, insulating surfaces, and parts for various sensors and instruments.

6.2.2. ABS (Acrylonitrile-Butadiene-Styrene)

This is a tough and impact-resistant plastic. If struck, it will dent and sometimes break, but rarely shatter like PVC, even at low temperatures. In many parts of the United States it is available at low cost in hardware stores in the form of black plastic sewer pipes and fittings, while in other parts of the country it may be unavailable, depending on local building codes. ABS pipes often have bubbles mixed into the plastic. This makes the pipes lightweight; however, the foamy internal texture can make it difficult to machine smooth surfaces for O-rings or other purposes. ABS and PVC are both acceptable as structural materials for most shallow water, non-critical ROV and AUV applications.

6.2.3. Polycarbonate (commonly known by the trade name Lexan®)

Polycarbonate is a tough, impact-resistant plastic that comes in a limited selection of colors, including almost clear. It can be cast into complex shapes and is frequently used for underwater camera housings. It is not as "crystal clear" as acrylic (described below), so it is rarely used for camera ports or viewing ports; however, it is more resistant to scratching and shattering than acrylic, and thin sheets can be made clear enough to be useful as shatterproof windows in homes and businesses. It tends to be more expensive than some alternatives.

6.2.4. Acetal (commonly known by the trade name Delrin®)

This plastic, often black, dark gray, or opaque white in color, is strong and impact-resistant. It has a self-lubricating, slippery finish, much like Teflon, and is sometimes used for low-friction bearings, gears, and other mechanical parts in corrosive environments where steel ball bearings or other steel parts would rust. Like PVC, it is less dense than metals and easy to fabricate. It can be heat-welded. The material is often used for instrumentation housings and pressure canister end caps. Acetal has low moisture absorption, so does not swell in contact with water. However, is it a bit "rubbery," so is prone to flexing. For example, if used for a long propeller shaft, acetyl might require extra bearings for support along its length.

6.2.5. Acrylic (commonly known by the trade name Plexiglas®)

Acrylic (a methacrylate polymer) has excellent corrosion resistance and is easy to fabricate; it can be machined (though it is somewhat brittle and can shatter unless special drill bits and saw blades made for plastic are used), glued, extruded, and cast into molds. It is sold in a variety of shapes (both standard and unusual). It is also sold in a veritable rainbow of different colors (both opaque and transparent), probably because it is popular for making backlit advertising signs. Like other plastics, acrylic has low density (approaching that of water) and a good strength-to-weight ratio.

One of acrylic's most distinctive and useful properties is that it can be made optically clear and shaped into smooth plates or domes. Consequently, it is often used for camera ports and view ports. Sometimes acrylic is put to use for bigger tasks. For example, the main pressure hulls for the submersibles *Deep Rover* and *Deep Rover II* are made from two hemispherical acrylic shells. The sphere on each of the *Johnson-Sea-Link* submersibles is now made as two hemispheres that are then bonded together. The older spheres were actually made up of 12 individual pentagons glued together.

Clear pressure canisters can also be made from cast acrylic pipe. While good at resisting uniform compressive loads, such as those produced by water pressure, acrylic is not good for lifting forces or serious impacts, so it is not generally used as a frame material. Avoid using extruded acrylic tubing for pressure cans if the vehicle will be diving below 10 meters (approx. 33 ft), since the walls of extruded materials are weaker than cast versions and not uniform in thickness.

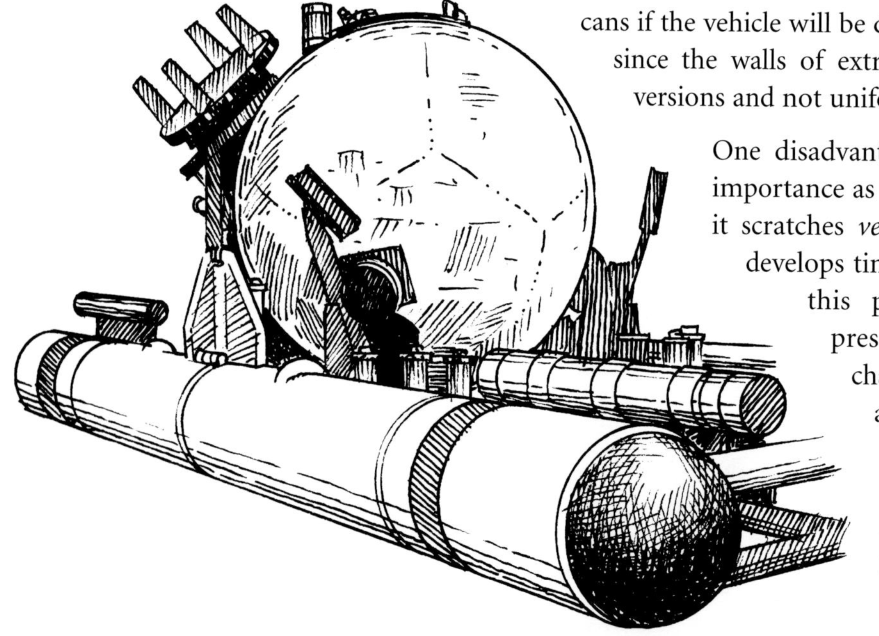

***Figure 4.23:* Makakai**

The Makakai *was one of the first submersibles to be built using five-sided pieces of acrylic for its pressure hull. The vintage submersible is stored at the Naval Undersea Museum in Keyport, Washington.*

One disadvantage of acrylic, particularly given its importance as a material for optical windows, is that it scratches *very* easily. Another problem is that it develops tiny cracks (called "**crazing**") over time; this process is accelerated by repeated pressurization cycles and temperature changes. Fortunately, the aging process in acrylic is predictable, and the crazing provides an early warning of impending failure, so acrylic parts rarely fail catastrophically before being replaced.

6.3. Other Structural Materials

In addition to metals and plastics, other types of materials have proven useful in underwater vehicle structures.

6.3.1. Glass

Glass has long been used for viewports in ships and for submersibles, but it wasn't until the early 1960s that experiments were conducted using glass for full-sized pressure hulls and dome ports. In spite of some success, the advantages and cost-effectiveness of acrylic soon relegated glass to a secondary role in pressure vessel design. Glass does have clarity and raw glass is inexpensive, but the cost of fabricating the mating surfaces for glass-to-steel flanges (a rim used for attachment) is high. Another major disadvantage is that glass can shatter easily upon impact.

In a spherical shape, glass has a strong ability to resist water pressure, an advantage that has long been demonstrated by fishermen using glass spheres for flotation on their trawling nets. (See *Historic Highlight: Glass Sphere History.*) See *Section 4.6.4. Glass Spheres* in *Chapter 5: Pressure Hulls and Canisters* for modern applications and manufacturers of glass pressure spheres.

HISTORIC HIGHLIGHT: GLASS SPHERE HISTORY

For centuries, fishermen have used glass spheres as floats for their nets; some Japanese commercial fisheries still use them to support their trawl nets, which can be submerged at 100 meters (approx. 325 ft.) or more. In the 1950s and '60s, commercial divers often made cheap but effective work lights from conical, sealed-beam, glass car headlights that could easily resist pressure up to 30 meters (approx. 100 ft).

Before the common use of acrylics, Will Forman, the American submersible inventor of *Deep Jeep*, did a lot of groundbreaking research into the use of glass as a material for pressure hulls and viewports. He found that borosilicate glass (a tough heat-resistant type of glass that contains a minimum of 5 percent boric oxide) had an extraordinary ability to resist pressure, especially if it was spherical. Today, hollow borosilicate glass spheres are made for the subsea industry, with the typical depth rating for a 12-inch diameter sphere being around 7,000 meters (approx. 23,000 ft) and for 17-inch spheres around 6,700 meters (approx. 22,000 ft). These glass spheres are commonly used for deep water flotation on oceanographic instrument buoys, as instrument housings, and even as pressure cans for vehicle electronics.

6.3.2. Glass-Reinforced Plastics (GRP)

Glass-Reinforced Plastics (GRP) are commonly known by the trade name Fiberglas® or the generic equivalent "fiberglass." They are one example of a class of increasingly popular materials known as composites. **Composites** consist of strands or particles of one substance embedded inside another substance, providing added strength and toughness, just as rebar does in concrete. Many of the structural and exterior materials used in newer cars, bicycles, planes, boats, and other machines are composites made from glass, carbon, or Kevlar fibers embedded in a plastic matrix of polyester or epoxy. As a group, composites tend to have very high strength-to-weight ratios (often much higher than those of metals), excellent corrosion resistance, and toughness—which means they cannot be easily cracked, torn, punctured, or broken.

Fiberglass is lightweight, inexpensive, and easy to fabricate without specialized equipment, though many of the chemicals used are toxic and require well-ventilated work spaces, rubber gloves, and filtered breathing masks.

GRP can be made into a variety of shapes—including thin, streamlined shells—and is often used for fairings in both aquatic and terrestrial vehicles. Almost all underwater vehicles have some type of GRP involved in their structure. For example, fiberglass struts and screens are used in many small ROV frames, and fiberglass fairings are used in many AUVs. Smaller pressure canisters are being made using similar GRP technology, but larger pressure hulls from this type of material are still in the experimental stage.

HISTORIC HIGHLIGHT: THE WONDER OF FIBERGLASS

Fiberglass is a popular term for extremely fine glass fibers that serve as a reinforcing agent in any of several polymers ("plastics"). This composite material provides additional strength. Fiberglass is technically referred to as fiber-reinforced polymer (FRP) or glass-reinforced plastic (GRP).

Although there were earlier experiments with creating fine glass fibers, the fiberglass commonly known today was invented for insulation purposes in 1938 by Russell Games Slayter. Owens-Corning successfully marketed the product as Fiberglas®.

Today, fiberglass applications are widely varied: building and thermal insulation materials, reinforced or strengthened materials, heat- and corrosion-resistant fabrics, and even entire boat frames and hulls. It is also being used for underwater vehicle structures, including frames and fairings.

Image courtesy of Dick Buist

Figure 4.24: Shark Trike

The fairing on this cycle helps reduce wind resistance without adding a lot of weight. (It may also help clear people and pesky sea lions out of the way.) Because it is easy to obtain, easy to form into smooth shapes, stiff, lightweight, and waterproof, fiberglass makes a great material for underwater vehicle fairings, too.

HISTORIC HIGHLIGHT: IMPROVISING MATERIALS

While it's always great to be able to buy the exotic materials you want, sometimes the budget simply won't stretch that far. So creative designers and inventors may be forced to come up with ways to use or adapt existing equipment or materials. Several designers of underwater habitats in the 1960s and '70s were great at improvising. For example, the shell of South Africa's only underwater habitat was constructed from two obsolete molasses crystallizer vessels donated by a local sugar company. The Soviet Union's *Kitjesch* habitat was a modified railroad tank car with holes cut for ports. Czechoslovakian builders created *Klobouk* by turning an industrial cauldron upside down, and the U.S. *Suny-Lab* was fashioned from a used cement mixer! By improvising, you'll be following in the footsteps of some of the greatest minds in the history of underwater invention, many of whom were working on underfunded projects.

6.3.3. Ceramics

Ceramic materials are a promising option for undersea use since they are corrosion-resistant, waterproof, thermally conductive, non-magnetic, and electrically insulating. They permit good radiation of acoustic energy (which is important for many underwater sensors), have a high strength-to-weight ratio, and huge compressive strength.

Although brittle in nature, current research efforts have developed lightweight, pressure-resistant ceramic vessels, or housings, that have travelled to the very deepest parts of the ocean—and back! (See Figure 4.25.)

Image courtesy of Tom Kleindinst, Woods Hole Oceanographic Institution

Figure 4.25: Glued to Their Work

Christopher Griner (left) and Glenn McDonald use a special epoxy to glue together sections of ceramic tubing, or "housings," for the hybrid vehicle Nereus. *These pressure vessels provide safe shelter (including maintaining pressures at a surface-like one atmosphere) for the batteries and electronics that will run the vehicle. Titanium endcaps allow wires into and out of the ceramic housings.*

"Ceramics, like your coffee cup, are brittle in nature and have only fair tensile strength, but they have a huge compressive strength," said McDonald. The ceramics on Nereus *can resist the pressure of deep water (as much as 18,000 psi while trench pressures are around 16,500 psi). "This ceramic pressure housing can have a thin wall and actually provide flotation, whereas a titanium vessel built for this depth would have a thick wall and not provide any flotation." In 2009,* Nereus *made several successful dives to the deepest parts of the ocean.*

6.3.4. Rubber

Although rubber might not seem like a good structural material because of its soft, flexible nature, it serves at least two important structural roles. First, rubber O-rings, which look like little rubber donuts, are by far the most common method for making watertight seals in pressure canisters, as they are simple and inexpensive, yet provide reliable watertight seals, even under extremely high pressure differentials. Second, rubber "crash bars" are used as bumpers on some vehicles to protect them from impacts.

Table 4.2 summarizes common uses of some of the materials described in this chapter.

Table 4.2: Structural Roles of Some Materials Used Commonly in Underwater Vehicle Design

Materials		Frame	Pressure Hull (large housing)	Pressure Canister (small housing)
Metals	steel	yes	yes	not often
	stainless steel	yes	no	yes
	aluminum	yes	yes	yes
	titanium	yes	yes	yes
Plastics	acrylic	no	yes	small cans only
	PVC	small vehicles	no	small cans only
	Delrin	small vehicles	no	small cans only
Composites	GRP	yes (fairings)	experimental	yes
	syntactic foam	flotation	no	no
Other	glass	no	experimental	yes
	ceramic	no	experimental	experimental
	rubber	bumpers	O-ring seals	O-ring seals

7. Other Useful Materials

While the preceding lists have focused on materials of structural importance, it's important to recognize that material properties—and by extension, material choices—can be important in other vehicle systems, as well. A few particularly useful materials that have not yet been covered above are mentioned here to round out the list of materials used commonly in underwater vehicles.

7.1. Syntactic Foam

This is a low-density composite material that consists of hollow glass microspheres embedded in a polymer matrix. Commonly used for crush-proof flotation and stability in underwater vehicles, syntactic foam is described in more detail in *Chapter 6: Buoyancy, Stability, and Ballast.*

7.2. Non-Structural Metals

A few metals play important *non-structural* roles in underwater vehicle design. For example, lead, which is about 11 times denser than water, is important for vehicle ballast and stability, as described in *Chapter 6: Buoyancy, Stability, and Ballast.* Copper wires and gold-plated connectors are important in electronic systems used for vehicle control and navigation, which are discussed in *Chapter 9: Control and Navigation* and extensively in *Chapter 12: SeaMATE.*

7.3. Teflon

Teflon™ (the same material used in non-stick frying pans) is actually the brand name for a synthetic fluoropolymer with numerous applications. This hard, slippery, plastic-like substance can be used where low friction is critical, as in seals around rotating propeller shafts.

7.4. Potting Compounds

One effective way of waterproofing some electrical circuits and other systems is to embed them in epoxy resin, polyurethane rubber, or some other material that starts out as a liquid mixture, then solidifies into plastic- or rubber-like material, sealing the circuit safely inside. The use of potting compounds is explored a bit more in *Chapter 5: Pressure Hulls and Canisters* and in *Chapter 12:* SeaMATE.

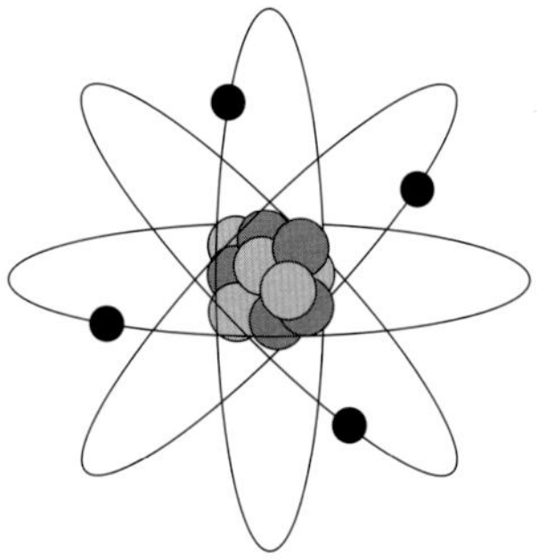

Figure 4.26: Corrosion Happens at the Level of Individual Atoms

An atom consists of the central nucleus, made up of protons and neutrons, surrounded by clouds of orbiting electrons.

8. Metal Corrosion

Corrosion is the gradual deterioration or eating away of a material, usually a metal, through chemical action. Early stages of corrosion often appear as a faint discoloration on the material's surface. More advanced stages are characterized by significant loss of material and weakening of structural members, sometimes to the point of failure (Figure 4.27).

Although corrosion can affect just about any material, most of the concern about corrosion is related to metals. Corrosion of steel is of particular concern because of its importance as a structural material in everything from skyscrapers to paper clips. The reasons for steel's popularity include its relatively low cost and excellent strength-to-weight ratio. Even at sea, where corrosion can be particularly severe, steel is the dominant load-bearing material in large structures, including offshore oil rigs, ships,

Image courtesy of James P. Delgado

Figure 4.27: Corrosion Claiming Historic Sub

Corrosion is slowly but surely claiming the remains of the little-known, long-forgotten U.S. Civil War–era submarine Sub Marine Explorer. *After serving with the Union naval forces, engineer Julius Kroehl built his 36-foot craft, complete with lockout dive chambers so that a diver could disarm torpedoes or set charges.*

Kroehl's submarine was financed by the Pacific Pearl Company, who felt it might also be adapted to the pearl trade. With the war winding down, the U.S. Navy declined to purchase the innovative sub, so Kroehl and his invention went to work harvesting pearls in the deeper waters off the coast of Panama. At some point after 1869, the submarine was abandoned. One hundred and forty years later, the historic hull lies rusting in the surf of a forgotten island off Panama.

and submarines. In smaller underwater structures, other metals such as aluminum often play a larger role than steel because of their greater corrosion resistance. Just remember that no metal, not even gold, is completely immune to corrosion.

Corrosion is a complex process that depends on many interacting factors, and scientists are still unraveling its mysteries. However, the basics of corrosion are understood well enough and provide a powerful foundation for controlling corrosion in underwater vehicles.

To understand corrosion, think small—very small—because corrosion happens at the level of individual atoms (Figure 4.26). The diameter of an atom is about one-*millionth* the diameter of the period at the end of this sentence. If you are feeling a bit rusty about terms like atom, ion, molecule, proton, electron, positive/negative charges, ionic bonds, and covalent bonds, you may want to review these high-school chemistry concepts before tackling this section of the chapter.

In a chunk of pure metal or metal alloy, the number of electrons in each atom typically equals the number of protons, so the positive and negative charges balance out, and the atom has no net charge (i.e., it is neither positive nor negative). But the outermost electrons of metal atoms are only loosely bound to the nucleus. Corrosion of metals happens when something steals these loosely held outer electrons, leaving the metal atom with a net positive charge. This process of electron removal—whether from a metal atom or any other kind of atom—is called **oxidation**, because oxygen atoms are often (though not always) the electron thieves. Oxidation of metal atoms causes metal corrosion.

Once created, the positively charged metal ions usually do one of two things, depending on whether the metal is in dry air or in contact with water. In dry air, the positively charged metal ions will combine with the now negatively charged oxygen atoms to form a metal-oxide coating on the surface of the metal. Rust is a familiar example of a metal oxide. It forms when oxygen steals electrons from—then forms an ionic bond with—iron atoms in the steel alloy. The rate of oxidation in *dry* air is usually slow, in part because the layer of metal oxide formed during the reaction often acts as an insulating barrier that blocks further contact between oxygen and the bare metal. In many metals, such as aluminum and stainless steel, this oxide layer stops the oxidation reaction almost as soon as it starts, leaving the metal looking as if it hadn't corroded at all.

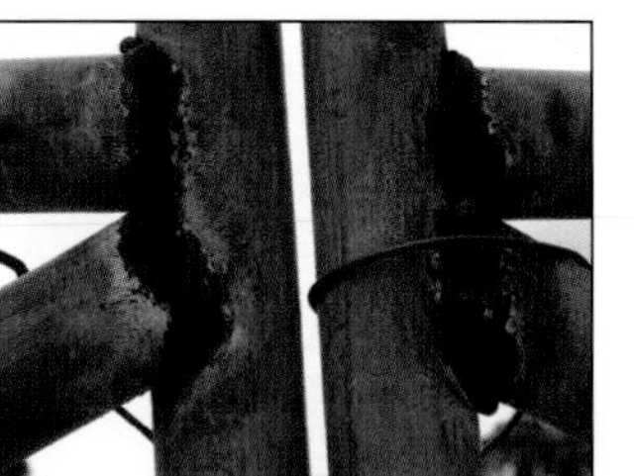
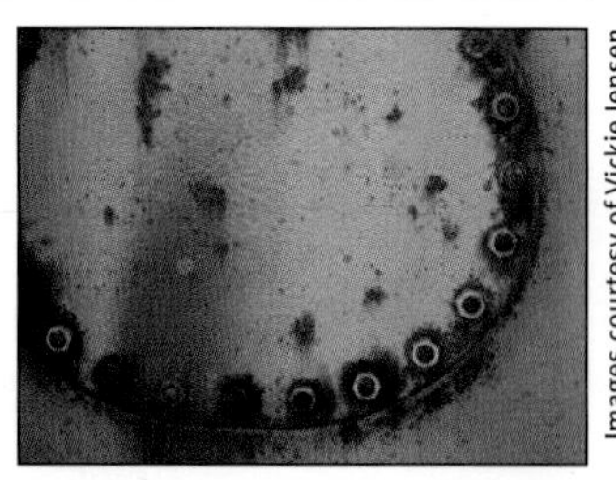

Figure 4.28: Metal Corrodes Quickly Near Water

Images courtesy of Vickie Jensen

When water is present, however, the situation is very different. The metal ions created in these oxidation reactions dissolve readily in the water and are carried away. Eventually this leaves pits or other gaps in the metal surface. As these metal ions wash away, fresh metal is exposed to attack by more oxygen or other electron-hungry chemicals dissolved in the water. Thus, in water, corrosion rates are greatly accelerated. Saltwater is particularly corrosive, because the dissolved salt ions transport electrical charges well, thus facilitating the theft of electrons from the metal.

Interestingly, a metal object will often corrode much more quickly when exposed to saltwater spray or a thin film of saltwater than when completely submerged in it. The reason is that these conditions combine the high oxygen concentrations of air with the electrical-charge–transporting properties of saltwater. This is why giving your ROV or AUV a good freshwater rinse after it comes out of the ocean and drying it promptly can go a long way toward minimizing corrosion damage.

Two related processes are notorious for *accelerating* rates of metal corrosion in boats and underwater vehicles. These are **galvanic corrosion** and **electrolytic corrosion**.

8.1. Galvanic Corrosion

The rate at which a metal corrodes in water can be affected by contact with other metals. Different metals have different affinities for electrons. That means some metals are more willing to give up their electrons than others. If two different metals are brought into electrical contact under water, the more electronegative one (i.e., the one least willing to give up electrons) will try to steal electrons from the less electronegative one. The more electronegative one will therefore corrode more slowly than it normally would (sometimes not at all). Meanwhile, the less electronegative metal will corrode more quickly than it would by itself. This accelerated rate of corrosion, which results from contact between dissimilar metals in the presence of water, is called **galvanic corrosion**. In these reactions, the metal that gives up its electrons to the other metal is called the **anode**, and the one that steals them is called the **cathode**.

TECH NOTE: GALVANIC CORROSION AND MOISTURE

Galvanic corrosion is associated with wet (or at least damp) environments, including humid air. Why isn't it also a problem in dry places? Consider the example of steel and aluminum. Steel is more electronegative than aluminum, so steel will try to grab electrons away from aluminum whenever they are in electrical contact. In dry air, this tug-of-war doesn't get very far—as soon as electrons start to move toward the electronegative metal, that piece of metal starts to develop a net negative charge, and the other piece starts to develop a net positive charge. Since electrons are attracted to positive charges, this effect starts pulling the electrons right back where they came from, and nothing ends up going anywhere.

If moisture is present, however, positively charged ions present in the water will give the extra electrons on the steel a place to go, so they will hop off the steel, and the reverse voltage needed to stop the corrosive flow of electrons will not build up like it did in dry air.

If a boat, ROV, or other vehicle has parts made of different metals connected electrically (either by direct contact or by wires), and if those parts are both touching the water, then galvanic corrosion can occur. For example, a steel outboard motor mounted on a boat with an aluminum hull could cause corrosion of the aluminum hull over time, particularly if the engine was left in the water all day every day. The effect can even operate across multiple boats in a marina. If two boats—say one with an aluminum hull and the other with a steel hull—are touching each other, or if they are electrically connected through the marina's electrical grounding system (as they should be for safety purposes whenever using electrical power supplied by the marina), then one hull could corrode the other.

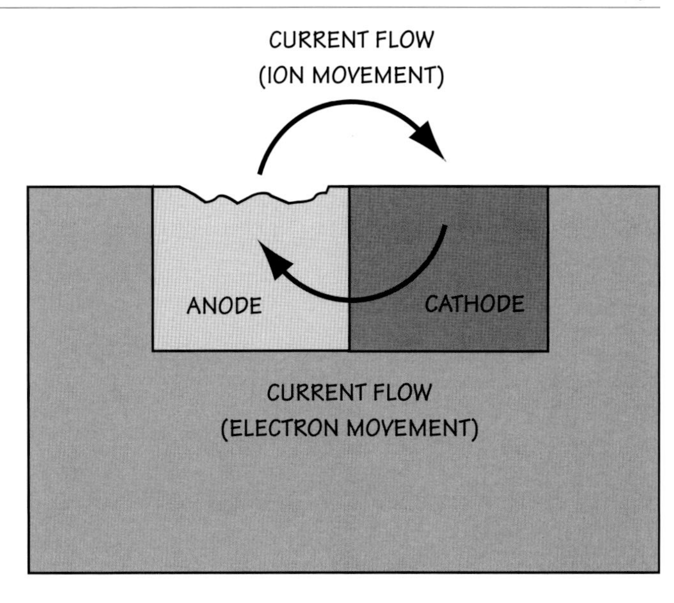

Figure 4.29: Corrosion Circuitry

This simple diagram illustrates how electrical charges must move in a circular pathway or "circuit" for corrosion to proceed. Anything that interrupts this cyclic flow can be used to control corrosion.

(Note: The actual direction of electron movement is from anode to cathode, opposite the direction indicated by the current flow arrow in this diagram. For an explanation of this seeming contradiction, see Tech Note: Which Way Does Current Really Flow? *in* Chapter 8: Power Systems.*)*

8.2. Electrolytic Corrosion

Since corrosion involves the movement of electrons or other charged particles from one place to another, it is fundamentally an electrical process. Therefore, it can be influenced by voltage, which acts like an electrical "pressure" trying to push electrical charges from one place to another. In galvanic corrosion (described above), the voltages that push the charged particles along are generated by intrinsic differences between the electron affinities of dissimilar metals and are generally small (typically about 2 volts or less). But in **electrolytic corrosion** (commonly, though not as precisely, referred to as "**electrolysis**" among mariners) the voltages are supplied by an external power source, such as a battery or electrical outlet, and the voltages involved are generally greater (commonly tens or even hundreds of volts), so corrosion can occur much more rapidly.

Boats, ROVs, and similar vehicles usually have plenty of electrical systems on board for communication, navigation, and other purposes. It's all too easy to inadvertently create or augment corrosive processes by accidentally letting stray electrical current from the batteries or electrical power sources flow between metal parts and the surrounding water. This can happen through faulty wiring or through moisture that gets splashed, spilled, or condensed onto electrical contacts and allows electrical current to "leak" over damp surfaces to places where it isn't supposed to be.

It's important to recognize that electrolytic and galvanic corrosion are fundamentally the same process—oxidation of metals at the atomic level in response to voltage differences that move electrical charges from one place to another. They differ only in the source of the voltage that drives the process. In galvanic corrosion, the voltage comes from differences in the electron-attracting properties of the metals themselves, which create the equivalent of a small battery. In electrolytic corrosion, the voltage is applied by some external electrical power source, such as a regular battery or AC power main.

8.3. How to Control Corrosion

There are a number of ways to reduce or prevent metal corrosion. All of them work by slowing or stopping the removal of electrons from atoms in the metal that you're trying to protect.

Electrical charges must move in a circular pathway for corrosion to proceed. The removal of electrons from the metal being corroded is just one part of that cycle, but the corrosive process can be slowed or halted by blocking *any* part of the cycle. Each of the prevention methods described below does this in one or more ways. Note that some practical corrosion control strategies may employ several, if not all, of these protective measures:

1. **Use Corrosion-Resistant Materials**: The best way to avoid metal corrosion is to avoid using metals, if possible. Use plastic or other corrosion-resistant materials instead. If you must use metals, keep in mind that some metals and alloys are naturally more corrosion-resistant than others. Unfortunately, many of the most corrosion-resistant metals have their own drawbacks. Some are not strong or stiff enough to make good structural materials. Others are simply too expensive. Gold is a great example of a very corrosion-resistant metal that is almost never used as a structural material because of its high cost. However, certain alloys, such as Aluminum 6061 and Stainless Steel 316, have been formulated specifically to resist corrosion while preserving other useful properties at moderate cost. As a result, they make excellent materials for building underwater vehicle structures. Note that corrosion-resistant alloys will still corrode, albeit slowly, even in the best of circumstances, and they can corrode much more quickly in the presence of galvanic or electrolytic corrosion processes. You should also be aware that these corrosion-resistant metals can contribute to galvanic corrosion of other metals they touch.

2. **Paint Exposed Metal**: Sometimes limited budgets, engineering requirements, or other factors force you to build structures out of a metal, such as steel, that is not corrosion-resistant. Does this mean the vehicle's structure is doomed to a short saltwater career? Not necessarily. One of the simplest and most common methods of corrosion protection in these cases (and sometimes even with corrosion-resistant metals) is to coat the metal surface completely with some electrically insulating material that acts as a barrier to oxygen, water, other corrosive chemicals, and/or electrical currents. Paint is the most familiar example. In fact, numerous boat, ship, and submarine hulls are annually sandblasted and repainted with protective types of paint. And many a sailor remembers hours spent chipping rust off various deck fittings before repainting them. Other coatings, such as plastic or rubber, can be used instead of paint.

Unfortunately, these protective coatings are not infallible. Tiny scratches and other imperfections eventually expose bits of the underlying metal to corrosive attack. When this happens, electrical currents can actually become concentrated at the gaps, intensifying corrosion rates at those sites. If not stopped by some other means, this intense, localized corrosion often loosens the protective coating on adjacent areas of the metal, thus allowing the corrosion to spread rapidly *under* the coating. Because of this, badly peeling paint is a common sight on corroded steel or aluminum surfaces near the ocean.

Image courtesy of Vickie Jensen

Figure 4.30: A Freshly Painted Hull

Cleaning hulls and repainting them with protective types of paint are part of regular maintenance procedures for marine craft.

3. **Anodize Metal Surfaces**: This is really a variation on the paint theme, but the coating is applied, ironically, through a corrosive process. Aluminum, for example, is often treated through a carefully controlled electrolytic corrosion process that grows a nice thick, uniform layer of aluminum oxide on the surface. This is called **anodization.** The oxide layer that forms is extremely hard and serves as an effective barrier to scratches and uncontrolled corrosion. The oxide layer can even be dyed red, green,

purple, or other colors during the anodizing process, resulting in beautifully colored, corrosion-resistant aluminum!

Image courtesy of Vickie Jensen

Figure 4.31: New "Zincs"

Sacrificial anodes, commonly called "zincs," further protect a freshly painted steel hull.

4. **Use Sacrificial Anodes**: This approach, like anodizing, is a clever example of corrosion used to fight corrosion. Recall that during galvanic corrosion, one metal (the cathode) avoids oxidation by stealing electrons from the other metal (the anode). Corrosion of the cathode is slowed or halted, while corrosion of the anode accelerates. Suppose you have a steel-hulled vehicle that needs to be protected from corrosion. You can do so by making it the cathode in a galvanic reaction. To do this, you simply add a less electronegative metal, such as zinc. If a piece of zinc is attached to the steel hull, the zinc will act as a **sacrificial anode**, allowing itself to get eaten away relatively quickly while it protects the steel cathode by donating electrons. Zinc is used so commonly for this purpose that these sacrificial anodes are often called simply "**zincs.**" Since the protected hull is acting as a cathode, this technique is one example of something called **cathodic protection**. (See *Tech Note: Other Examples of Cathodic Protection.*) These zincs eventually need to be replaced, of course, but it's a lot easier (and cheaper) to replace a little block of zinc every now and then than it is to repair or replace a massive, badly corroded steel hull. As long as the zinc block is inspected regularly and replaced before it disappears, this technique provides a good degree of protection for the hull. If a good coat of well-maintained paint also protects the steel hull, the little zincs won't have a lot of work to do anyway and will last a longer time. Inspecting the condition of zincs on a boat hull is a great task for a small ROV like *SeaMATE.*

5. **Pay Attention to Galvanic Compatibility**: Sometimes structural failure can be traced to a metal component that was unintentionally serving as a sacrificial anode. This is particularly likely to happen when fasteners made from one type of metal are used to connect structural elements made from another type of metal. For example, if the metal bolts connecting the metal beams in your frame donate electrons to that frame, the bolts will dissolve away and the frame will eventually fall apart! Table 4.3 summarizes the compatibility of base metals (e.g., beams) with metal fasteners (e.g., bolts) for subsea applications.

TECH NOTE: OTHER EXAMPLES OF CATHODIC PROTECTION

Cathodic protection refers to protecting a metal from corrosion by making it the cathode in an electrochemical circuit. The zincs described in the main text of this chapter provide cathodic protection to the main hull or motor casing by acting as sacrificial anodes. Galvanized nails (steel nails coated with a layer of zinc) are another example of cathodic protection used in a common household item. The zinc coating on the steel nail acts as a sacrificial anode, protecting the nail from rust.

Yet another example can be found in underground pipelines. When steel pipelines run through moist soil, corrosion can be a big problem. Instead of using zincs or other sacrificial anodes, it's common to paint the pipe (thereby blocking most corrosion) and to attach the pipe to the negative terminal of an electrical power supply with the other terminal of the supply grounded. If the paint on the pipe gets scratched and something attempts to steal electrons from the pipe, the negative power supply terminal protects the steel pipe by providing a steady stream of electrons, just like a sacrificial anode. Unlike a sacrificial anode, however, the power supply doesn't get used up and doesn't need to be replaced.

Table 4.3: Galvanic Compatibility Chart

	FASTENER			
BASE METAL	**Aluminum**	**Carbon Steel**	**Silicon Bronze**	**316 Stainless Steel**
Aluminum	*Neutral*	*Compatible*[1]	*Not Compatible*	*Not Compatible*[2]
Steel and Cast Iron	*Not Compatible*	*Neutral*	*Compatible*	*Compatible*
Copper	*Not Compatible*	*Not Compatible*	*Compatible*	*Compatible*
316 Stainless Steel	*Not Compatible*	*Not Compatible*	*Not Compatible*	*Neutral*[3]

Source: abbreviated and excerpted from a chart supplied by the International Nickel Company, Inc. (INCO)

[1] *This combination is generally compatible, but some enlargement of the bolt hole may occur over time.*

[2] *Base metal tends to be cathodic because its larger surface area increases the electrical potential of the metal.*

[3] *There may be some corrosion under the head of the bolt over time.*

TECH NOTE: TEST TANK AND SWIMMING POOL CHEMISTRY

Even though it's not considered saltwater, the water in most swimming pools, test tanks, and hot tubs should be considered a corrosive environment. That's because these relatively small bodies of water generally contain disinfectants such as chlorine or bromine that are routinely added to reduce harmful bacteria. (Chemical levels are monitored with a test kit that indicates when more chemicals are required to maintain the correct ratio of chemicals to water.) Though these chemicals keep the bacterial count down, they also play havoc with materials like rubber and various metals. For example, when rubber is exposed to disinfecting chemicals, it is degraded and eventually loses its elasticity. If the pool or tank is located outside, the combination of sun and chemicals will accelerate this process. Plastics that are regularly immersed may become brittle over time. Aluminum may begin to corrode. The key to preventing deterioration of any of these materials is to rinse them thoroughly with freshwater after removing them from the tank. And always store gear out of the sunlight.

Another potential problem in tanks and pools is the **pH** (a unit of measurement for acidity and alkalinity) of the water. The pH value of solutions can range from 1 to 14. The number represents the negative exponent of the hydrogen ion concentration in the water. Distilled water has a neutral pH of 7. A pH value over 7 means the solution is basic, or relatively high in alkalinity. A number lower than 7 means the water is acidic. Strong acids, such as battery acid and the acid in the human stomach, have a pH level of 2 to 4. Strong bases, such as lye, have a pH of 10 or more. A well-balanced pH level for a test tank or pool is close to neutral, somewhere between 6 and 8.

Image courtesy of Randall Fox

A tank with water that is too acidic will encourage corrosion of metals like aluminum and carbon steels. Even poor-quality stainless steel will pit in acidic solutions. You can keep your pool or tank correctly pH balanced by regularly monitoring the water, using a pH testing kit (often part of the chlorine test kit). Standard chemicals such as soda ash (a base to neutralize acidity) or acid (to neutralize alkalinity) can be purchased from swimming pool or hot tub suppliers. As mentioned above, always rinse your gear with freshwater after using it in a pool or tank, even if it was only in for a quick test.

Figure 4.32: Corrosive Environments

The water in pools and test tanks is not as innocent as it appears.

6. **Insulate Metal-to-Metal Contact**: Since electrons must flow from one metal to the other for galvanic or electrolytic corrosion to occur, the simple expedient of inserting rubber, plastic, or some other insulating material between the fastener and the base metal can dramatically reduce corrosion rates when marginally compatible metal combinations are used. Even something as simple as wrapping Teflon tape around stainless steel bolts to keep them from touching aluminum frame pieces can help.

7. **Eliminate Stray Electrical Currents**: A battery or other DC electrical power source supplies electrons to circuits from its negative terminal and removes them from circuits at its positive terminal. Normally, the battery-powered flow of electrical current circulates within the intended circuit, but sometimes wires or water in the wrong place or a poorly designed circuit allows some electrical current to take an alternate path through the metal frame material. If the current flow is oriented so that hull voltage is more negative than the water voltage, electrons may flow out of the frame material and into the water (causing metal corrosion at the same time) on their way to the positive terminal of the battery. Even AC circuits (in which the electrical current alternates direction of travel many times each second) can accelerate corrosion, because the corrosion process is not perfectly reversible. As the designer of an underwater vehicle that will probably include electrical circuits, you must be careful to avoid unintentionally causing electrolytic corrosion. You can do this by making sure that the water is never (intentionally or unintentionally) a current-carrying part of your electrical circuits.

9. Fabrication and Assembly

When it comes time to actually start building your vehicle's structure, you'll need to do three things:

1. Obtain the tools, parts, fasteners, and raw materials needed to complete the job.
2. Cut, drill, or otherwise modify raw materials or other parts to fabricate any custom parts that are needed.
3. Connect all the parts together into a final structure(s).

These are not necessarily sequential steps. For example, you may not know exactly where you need to drill that hole until after you've started assembling your frame and can see how the pieces line up. Or you may get a better idea for a raw material to use as a starting point after you uncover some unanticipated problems with the material you originally purchased. That's ok; it's all part of the process. Good planning can reduce, but not eliminate, this type of learn-as-you-go construction.

To give you a flavor of the construction process, as well as a tangible staring point for vehicle structure and other subsystems, a specific example of the fabrication and assembly process is provided by *Chapter 12: SeaMATE*. It guides you in meticulous detail through the construction of a complete ROV capable of diving in fresh or saltwater to a depth of 10 meters (approx. 33 ft), where the absolute pressure is twice what it is at the surface. The material is organized to parallel the technical chapters, so you can (if you want to) start building the frame of that vehicle right now, based on what you've learned in this chapter. Alternatively, you can wait until you've read all the technical sections, then use *Chapter 12: SeaMATE* to tie it all together through a real ROV project. It's up to you. One nice thing about that chapter is that it includes specific recommendations about the workspace, tools, safety equipment, parts, and raw materials you'll need for the project. *Appendix V: How to Find Parts* suggests techniques for finding that special widget you are looking for.

The chapter you're reading now, like the other technical chapters, focuses instead on a specific subsystem—in this case, vehicle structure. The background information presented here not only provides a useful foundation for the *SeaMATE* project, but also gives you ideas and facts you can use to go beyond *SeaMATE* or to venture off in a completely different direction, if you choose to do so.

9.1. Obtaining Tools

Any serious structural construction project for an underwater vehicle will require tools for cutting, drilling, smoothing, or otherwise fabricating parts, but unless you're building a deep-diving vehicle, they need not be fancy or expensive. Power tools can be helpful, but are not required. A standard set of the usual hand tools found in many people's homes or garages—screwdrivers, wrenches, saws, a drill with drill bits, pliers, tape measures, and a handful of other basic tools—should suffice. *Chapter 12: SeaMATE* provides a complete list of essential tools for simple underwater vehicle projects.

If you don't own these tools, chances are you can borrow them from a friend or neighbor. Make sure you take good care of them and return them promptly, so your friends will still be your friends. If you damage a tool, get it repaired or replaced in a timely manner.

If you want to buy tools for your own workshop, you can find an impressive selection at most hardware stores and home improvement centers. Alternatively, you can order tools on-line. Some distributors sell all kinds of tools, but it is also common for tool companies to specialize either in woodworking tools (most of which also work well for plastic) or metalworking tools. Other companies sell the raw materials and tools needed for particular types of projects. For example, a distributor catering to people who build and fly radio-controlled model airplanes as a hobby would likely sell balsa wood, plastic, paints, tiny engines, propellers, and an assortment of tools useful for turning these raw materials into finished flying machines.

In many (but not all) cases, you can do a job more quickly, easily, and precisely with power tools than with hand tools. For some fabrication needs, particularly those associated with deep-diving vehicles, they may be essential. If you plan to buy power tools, be warned that you usually get what you pay for, and good-quality power tools can be quite expensive, costing hundreds or even thousands of dollars for something like a high-quality milling machine or lathe. Power tools—particularly saws—can also be extremely dangerous if used improperly and should never be used without proper training and supervision. If you do use power tools, make sure you work in the presence of another person present who can summon medical help and administer first aid if that should ever become necessary.

If you do decide to invest in a few power tools, consider cordless tools. They are usually a bit more expensive than their corded plug-in counterparts, but they are more convenient to use and present much less of an electrical shock hazard if you use them to work on your vehicle near water. A cordless drill-driver is probably the most useful general-purpose power tool.

9.2. Obtaining COTS Parts

Whenever possible, you should strive to build as much of your vehicle as you can out of **commercial off-the-shelf (COTS)** parts. Unlike custom parts, which usually take a long time to produce, commercial parts are usually readily available and in sufficient quantities. They also come in standard sizes that are likely to connect more easily to other COTS parts. Because they are mass-produced, they will probably have consistent sizes, shapes, and other properties. That means you can easily replace a lost or damaged part quickly and simply by going back to the store where you got the first one. Mass production also tends to drive the price down, so you can often buy a commercially available component for less than it would cost you to make your own from raw materials. Also, the COTS product will usually have been improved over several generations by the manufacturer, so the quality and reliability may be much better than you are likely to achieve on your first try, anyway. Among the most useful of commercial parts, from a structural perspective, are various fasteners (nuts, bolts, screws, rivets, clips, etc.), brackets, and fittings.

Don't worry about missing out on a chance to do some creative fabrication—no matter how much of your vehicle you make out of COTS parts, there will be plenty of times you'll need to adjust or modify some of those parts to get everything to fit together and work properly.

The best place to obtain commercial off-the-shelf parts depends entirely on the type of components you are looking for. Retail outlets specializing in parts for do-it-yourself underwater vehicle projects are currently few and far between. That may change someday, but for the time being, you'll need to be creative in envisioning how COTS parts designed originally for *other* kinds of projects can be repurposed to your vehicular needs.

Hardware stores and home improvement centers are a great place to start. Just wander up and down the aisles, looking for anything that might meet your vehicle's structural needs. It's helpful to go with a rough draft of your design plus an open mind. That way you'll have some idea of what you're looking for, but you'll also be able to modify your design on the fly to take advantage of whatever is available in your local store.

Some sections of a hardware store are bound to be more productive than others. For structural parts, which must come into contact with water, focus on aisles that feature products used with water, like the plumbing department, the gardening area, and the stainless-steel section of the nuts and bolts department.

Kitchen stores, outdoor recreation outlets, and other suppliers that sell plastic products for use around water can also be productive places to look for items that might do double duty as structural parts. For example, a sturdy plastic nalgene bottle with a watertight screwtop lid might make a reasonable pressure canister for shallow dives. Other potentially fruitful places include hobby shops, boating supply stores, and specialty outlets that cater to robotics enthusiasts. All of these types of stores exist in on-line versions, too.

9.3. Obtaining Raw Materials for Custom Parts

If a COTS part for a particular job is not available, you may need to make a custom part. Sometimes it's possible to do this by making some minor modification(s) to an existing commercial part. For example, you might bend a bracket slightly and file away one corner of it for a better fit on your vehicle frame.

If a sufficiently similar COTS part is not available, however, you may need to start from scratch. Fortunately, most of the metals, plastics, and other materials used to construct parts for underwater vehicle structures come in a variety of standard shapes and sizes. These pre-shaped raw materials can be purchased from an on-line

Figure 4.33: Standard Shapes of Materials

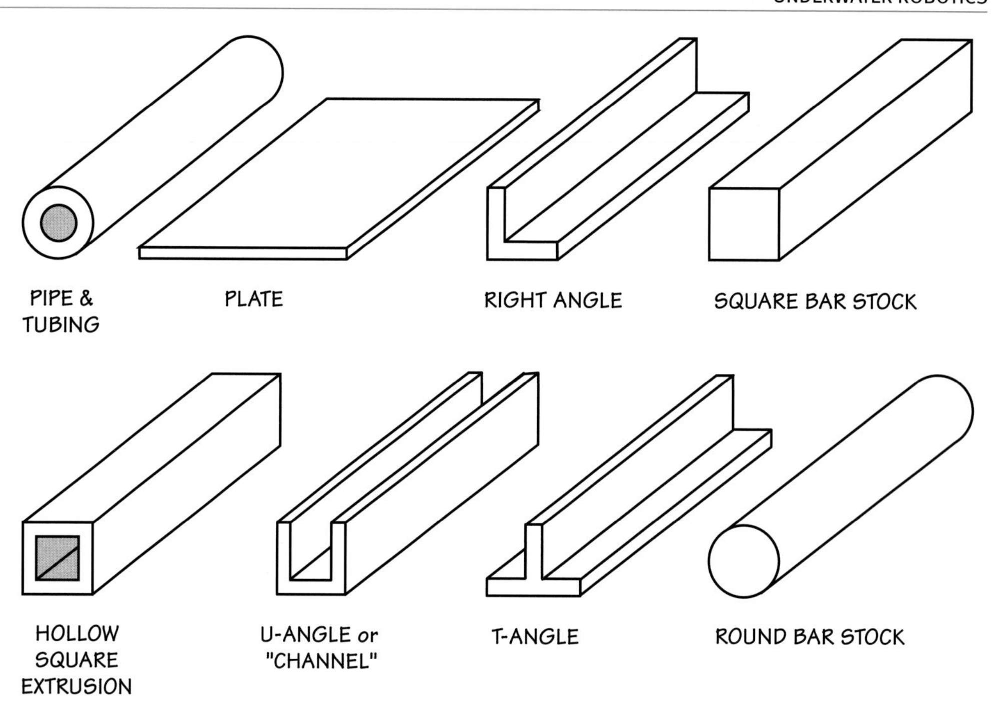

distributor, a local hardware store, or other supplier, then modified by cutting, bending, drilling, joining, or other means to make the parts needed for your vehicle's final structure.

Large on-line distributors of these raw materials usually have more extensive and/or specialized inventories, but local retail stores are more likely to have a friendly employee who can help you find what you need, and they may have better prices for common items, particularly when you factor in the cost of shipping. If you're shopping on-line, you can probably find more or less exactly what you're looking for, as long as you know what it's called. If you're shopping in a local hardware store, you may need to think more broadly about what you're looking for. For example, that frame tube you need may be in the bathroom aisle, disguised as a shower curtain rod.

You may also be able to find suppliers who use these materials in massive quantities and are willing to donate their scraps or sell them to you at very low prices. For example, a company building aluminum boats might have a two-foot long "scrap" piece of aluminum headed for the recycle bin that you could use as the main backbone of your ROV frame. A sign painter might have pieces of colorful acrylic plastic and be willing to sell them by the pound for a very reasonable price or maybe even give them away.

Here are common terms for some standard shapes in which raw metal and plastic (and some other) building materials are sold. You can combine these terms with the name of the material you want in keyword searches to find lists of on-line distributors of things like "acrylic plates" or "aluminum extrusions." You can also usually find them simply by going straight to a supplier of various plastics or metals and looking through the list of shapes and sizes they offer for each type of material.

- **Blocks**: This term specifies rectangular chunks of solid material. In a block of material, the length, width, and thickness are similar enough that no dimension is more than a few times greater than any other. Blocks are very stiff, but also very heavy, so they are rarely used in underwater vehicles except as a starting point from which to machine other parts. Blocks of solid metals and plastics are uncommon in hardware stores, but they may be more readily available in many industrial supply houses and through on-line vendors.

- **Plates and Sheets:** These terms refer to flat layers of material, usually metal or plastic. Plates and sheets are most commonly sold in rectangular sections, but sometimes they are sold as discs or other shapes. A layer that's too thick to bend easily is often called a "plate," whereas a thin, flexible layer is more likely to be called a "sheet;" however, there is no formal dividing line between the two terms. The thickness of metal sheets is often expressed in terms of gauge thickness: 18, 16, 14, or 12 gauge, for example, with smaller numbers corresponding (counter-intuitively) to thicker sheets. Plate thicknesses are generally expressed in millimeters or inches, with 1/8", 1/4", 3/8", 1/2", 3/4", 1", and 2" being common imperial thicknesses for metals and plastics. Only the thinner versions are available in hardware stores. Thicker ones can be ordered easily from on-line suppliers.

Image courtesy of Dr. Steven W. Moore

Figure 4.34: Joining Angles

Extruded angles make good frame materials because they have good stiffness-to-weight ratios and can be joined easily by drilling a hole through the flat sides, then joining them with a nut and bolt (as shown) or a rivet. Nuts and bolts frequently work loose if there is any rotation around the joint, so use triangular beam arrangements to prevent rotation. Don't forget to use lock washers, too.

- **Bar Stock:** This refers to a long rod of solid material that may be round, square, rectangular, or hexagonal in cross-section. Other shapes have better strength and stiffness-to-weight ratios, so bar stock is not widely used for structural members in underwater vehicle frames. Nonetheless, these shapes can be very useful for some other purposes. For example, round rods can be used as propeller shafts, and rectangular stock can be cut into smaller blocks that are ideal for machining into specialized parts.
- **Extrusions:** Extrusions are long pieces of material, similar to bar stock, but usually with more complex cross-sectional shapes. Extrusions are made by forcing molten material, often aluminum or thermoplastic, through a specially shaped opening, which determines the shape of the cross-section. Extrusions come in several standard shapes (many of which resemble letters of the alphabet in cross-section), as well as a bewildering variety of custom shapes for specialized purposes. Some standard extrusions include:
    - L—called "angles" or "right angles"
    - squared-off U—sometimes called "channels" or "U-angles"
    - T—called "T angles" or simply "Ts"
    - E—which is commonly used with aluminum for the tracks on sliding patio or shower doors
    - hollow squares—called "square tubing," though it is not flexible like most other forms of tubing
    - tubes with airfoil-shaped cross-sections
- **Pipe:** Pipe refers to a rigid, hollow cylinder of long length (compared to the diameter). Pipes are among the most useful building materials for underwater vehicle frames and pressure canisters. Like an extruded angle, a cylindrical pipe is generally a better choice than a solid rod for making a stiff yet lightweight frame. That's because a hollow pipe is almost as stiff as a solid cylinder of the same diameter, but not nearly so heavy. The cylindrical shape of a pipe is inherently good at resisting pressure, so pipes are also a logical choice for constructing effective pressure canisters. (The use of cylindrical shapes for pressure housings will be discussed in much greater detail in the next chapter.) In general, pipe can be cut, bent (but not sharply), machined, and welded or glued. Pipe sizes may be specified in terms of inside diameter, outside diameter, wall thickness, and length. Wall thickness may be specified indirectly in terms of the pressure rating of the pipe. For example, Schedule 80 PVC pipe is rated for higher pressures and has thicker walls than Schedule 40 PVC pipe.

- **Tubing**: This term usually specifies long, flexible, hollow cylinders, like a garden hose. (Note that there are exceptions, such as hollow square "tubing" extrusions made from aluminum, which are not flexible.) Usually made of plastic or rubber, flexible tubing is manufactured by different techniques to permit high-tolerance dimensions and flexibility, or ease of bending. Tubing is used for small-radius bending installations. Because of its great flexibility, tubing is not generally used for structural purposes, but rather for flow systems of gas, oil, and other fluids.
- **Castings**: Castings are shapes formed by pouring molten metal or a liquid polymer (e.g., plastic) into a mold, then allowing it to harden. Castings can be made into complex shapes, such as clear acrylic domes used for camera viewports. Repeated castings can be produced from the same mold, making it a cost-effective approach when many identical pieces are needed. In terms of tolerances (precise dimensions), castings are considered rough in relation to objects machined with a lathe or milling device. However, castings can still be quite smooth and accurate, and they are often machined later to tolerance.

9.4. Suggestions for Your First Vehicle Structure

The range of possible structural designs, including all the potential shapes and materials that can be used, is truly enormous—and potentially overwhelming. Sometimes it helps to know what others have tried and found successful, just as a starting point.

The following suggestions are just that—suggestions. Feel free to do something slightly or completely different. But know that these approaches have been used successfully by many groups of people building their first simple, inexpensive underwater vehicle.

For a vehicle frame, consider either of the two following options:

- **PVC pipe and fittings**: The plumbing section of hardware stores will commonly stock white plastic PVC water pipe up to 2 inches or more in diameter. These pipes are an especially popular material for small ROV frames, because they are widely available, inexpensive, corrosion-resistant, and easy to cut, drill, and join in a home or school shop, even if only hand tools are available. Sections of pipe can be joined together easily by means of pipe fittings, which come in a wide range of configurations for connecting different numbers and sizes of pipe sections at various angles. PVC pipes can be glued into smooth "slip" fittings (using PVC pipe primer and PVC cement), threaded and screwed into threaded fittings, or connected by drilling holes through the fitting and pipe and joining them with nuts and bolts.
- **Extruded aluminum angles**: Extrusions that have an L-shaped cross-section are useful for building vehicle frames, because they are stiff for their weight and easily joined by bolting or riveting through holes drilled through the walls of the material. They can be ordered on-line in corrosion-resistant types 6061 and 6063 from companies that sell aluminum and other metals. They can also usually be found along with other extruded aluminum products in hardware stores. Look in the aisle with other metal rods, and if you can't find it there, try the bathroom section, where it may be found among the shower and shower door repair supplies. These angles come in a variety of sizes. The bigger ones are stiffer and stronger, but also much heavier. Choose the smallest size that seems to be stiff

enough for your needs. Look carefully to see if the inside corner of the angle has been filleted (rounded). While this feature makes the angle stiffer and stronger, it can also get in the way of joining pieces together securely. Remember to think about galvanic compatibility when choosing fasteners for frames made of aluminum. Many common fasteners, including stainless steel, can steal electrons from aluminum, causing rapid corrosion around thc holes through which the bolts pass. Welding is an option that avoids this issue, if you have the skill and proper equipment.

For a pressure canister:

- Read the next chapter, which covers pressure hulls and pressure canisters in greater detail and offers some pragmatic suggestions on how to design and construct them.

For a fairing:

- Glass-reinforced plastic (GFP), often called fiberglass, is an ideal material from which to make a sturdy, smooth, streamlined, teardrop-shaped fairing. Remember that it's toxic and needs to be applied carefully. There are plastics stores that specialize in raw materials, tools, and safety equipment for doing fiberglass work.

Image courtesy of Randall Fox

Figure 4.35: PVC Construction

You might also consider something as simple as stretching some Lycra fabric or mounting some thin sheets of stiff plastic over a frame that has a streamlined shape.

Materials and tools for non-structural uses are described in the relevant chapters. For example, electronic materials and devices are described in *Chapter 9: Control and Navigation.*

10. Chapter Summary

This chapter has introduced the three functional subsystems of an underwater vehicle's structure: the frame, pressure hull (or pressure canisters), and fairing. Collectively, these subsystems manage the mechanical forces acting on the vehicle and its parts, so mechanical properties such as strength and stiffness are important performance characteristics for structures.

The strength and stiffness of a structure depend on the materials (substances) used to make the parts, the size and shape of each part, the arrangement of parts, and the method used to join the parts together.

Corrosion can be a serious threat to the strength and stiffness of structures, particularly those made of metal. The chapter includes an introduction to corrosion, as well as various methods that can be used to control it, including the use of corrosion-resistant materials, painting, cathodic protection, use of compatible fasteners, and avoidance of stray electrical currents.

Certain metals and plastics are the most common materials used for building underwater vehicle structures, primarily because they offer good compromises between mechanical performance, compatibility with the aqueous environment, ease of fabrication, availability, and cost. The chapter closes with a discussion of these materials, as well as time-tested structural options for simple, shallow-diving craft.

Chapter 5

Pressure Hulls and Canisters

Chapter 5: **Pressure Hulls and Canisters**

Stories From Real Life: *Squalus* and *Thresher*

Chapter Outline

1. **Introduction**
2. **Pressure**
   - 2.1. Atmospheric Pressure
   - 2.2. Pressure Differentials
   - 2.3. Gauge Pressure Versus Absolute Pressure
   - 2.4. Pressure Units
   - 2.5. Devices for Measuring Pressure (and Depth)
   - 2.6. Calculating Hydrostatic Pressures Under Water
   - 2.7. Calculating Hydrostatic Pressures on Other Worlds
3. **Pressure-Related Forces on Submerged Objects**
4. **Basic Principles of Pressure Hull Design**
   - 4.1. Size
   - 4.2. Shape
   - 4.3. Materials
   - 4.4. Using Pressure to Advantage
   - 4.5. Choosing Canister Size and Single or Multiple Cans
   - 4.6. Pressure Canister Options
5. **Calculating Pressure-Related Forces on Spheres and Cylinders**
6. **Constructing Leak-Proof Openings**
   - 6.1. O-Rings
   - 6.2. Pressure Hull Penetrators
   - 6.3. Pressure Can Access
7. **Pressure-Compensation Techniques**
   - 7.1. Oil Compensation
   - 7.2. Gas Compensation
8. **Encapsulation (Potting)**
9. **Adding a Card Cage**
10. **Chapter Summary**

Chapter Learning Outcomes

- Calculate the magnitude of the hydrostatic pressure-related forces acting on various parts of an underwater vehicle at any depth, in either freshwater or saltwater.
- Recommend effective shapes, sizes, and materials for pressure-resistant and leak-resistant hulls and canisters.
- Describe specific techniques for getting rotating propeller shafts, camera images, or wires through the walls of these containers.
- Describe relatively low-cost and easy-to-build yet effective designs for pressure canisters that can be used for small, unmanned vehicles diving to maximum depths of about 100 meters (approx. 325 ft).

***Figure 5.1.cover: Hydrostatic-Testing* Deep Worker 2000**

Southwest Research Institute prepares Deep Marine Technology's Deep Worker 2000 *submersible for a hydrostatic pressure test. SwRI operates ocean simulation chambers with diameters up to 90 inches and pressures to 30,000 psi.*

Image courtesy Southwest Research Institute and Deep Marine Technology, Inc.

STORIES FROM REAL LIFE: *Squalus* and *Thresher*

As navies around the world began acquiring submarines in the early 1900s, the accepted wisdom was that no crew could survive when a sub went down. But Charles "Swede" Momsen, a lieutenant commander in the U.S. Navy, refused to believe that. He proposed adapting a diving bell that could be lowered to a stricken sub and mated with its escape hatch so as to rescue any survivors. Momsen's proposal was ignored until the tragic ramming of the U.S. submarine S-4 in 1927. Trapped crew members could only tap frantic messages to divers on the outside of the hull until all succumbed.

As a result of this tragedy, Momsen was given the green light to develop his idea. He spent two years testing and adapting a diving chamber before being transferred to another naval department to concentrate on training submariners to use his individual breathing apparatus for submarine escape. Lt. Commander Allan McCann took over successful completion of the project. Despite its severe limitations, the McCann rescue chamber would prove its worth when the U.S. submarine Squalus *sank in 1939 and thirty-three men were rescued from the stricken sub. Two and a half decades later, the U.S. nuclear attack sub* Thresher *sank to a much greater crush depth and imploded. This time, there were no survivors. The long, frustrating search for the wreckage—and for answers—once again focused attention on the need for committed resources, technology, and deep-ocean vehicles.*

Squalus

It was a Tuesday just like any other on the Atlantic coast. But this particular Tuesday, May 23, 1939, would change the lives of many, to say nothing of altering the future of submarine rescue around the world.

The U.S. Navy's newest fleet-type submarine, *Squalus*, was completing a series of test dives. The admiral in charge of the Portsmouth Navy Yard felt confident that subs like the *Squalus* and its sister ship *Sculpin* would prove far superior to the German U-boats. If the United States got drawn into the looming conflict in Europe, he was sure these new fleet subs would give a good account of themselves. The state-of-the-art *Squalus* was 94.5 meters (approx. 310 ft) long and armed with eight torpedoes. Its crew of 59 represented 28 states; nearly half of the men were married.

This particular test dive would begin with *Squalus* riding with its ballast tanks high and dry, allowing it to travel at top speed on the surface. Then the sub would initiate a timed emergency battle descent, attempting to reach periscope depth at 15 meters (50 ft) in just 60 seconds. It had missed that goal by five seconds in an earlier run-through.

The captain ordered *Squalus* rigged for diving. The klaxon, or dive horn, sounded throughout the sub, signaling the first diving alert. Operators of the bow and stern planes

Squalus *was built in the 1930s as part of the vanguard of the Fleet Class, a new type of submarine constructed for the U.S. Navy. On May 23, 1939, while undergoing diving trials, it sunk in 73 meters (approx. 240 ft) of water. After 113 days,* Squalus *was raised. The entire electrical system was replaced and the sub was recommissioned on May 15, 1940, as the* Sailfish. *It went on patrol in the South Pacific during World War II.*

were directed to angle them to hard dive. The lights on the control panel all shifted to green, indicating the sub was secure. The final dive alert sounded, and *Squalus* started its slide down into the cold Atlantic. All thoughts were on the speed of the dive, as the depth indicator moved more quickly. At the target depth of 15 meters, the captain and civilian test superintendent from the shipyard were both pleased; the dive time was now only a fraction over the 60-second goal.

Then suddenly, they heard a message over the battle phone that completely contradicted the indicators on the control board: "Sir! The engine rooms! They're flooding!"

Tons of seawater were blasting violently into the aft compartments of the sub. For a brief moment, the sub's bow rose sharply, and those inside thought that *Squalus* was heading to the surface. Then they sensed the submarine sliding backward and deeper. Less than five minutes after starting its dive, the sub touched down on the bottom, helpless in 73 meters (approx. 240 ft) of frigid water, with no electricity, no heat, and limited air.

No one on the surface was aware of a problem until *Squalus* failed to reappear at the expected time. Its sister sub was immediately dispatched to search the area. As word spread, worried wives and family members began to gather on shore. The site of the stricken sub was located, and official calls went out to submarine rescue vessels such as the *Falcon* and to key personnel—specifically, Charles "Swede" Momsen.

And on that particular Tuesday in 1939, Momsen probably knew as much about submarine rescue and escape as anyone. Prior to heading up an experimental deep-sea diving unit, he had spent 14 years inventing or working on any number of devices that might help save a trapped submariner. They ranged from smoke bombs, telephone marker buoys, and new deep-sea diving techniques, to escape hatches, artificial lungs, and the McCann rescue chamber, a great pear-shaped device designed to rescue crew trapped in a sunken sub. But none of these devices had ever been tested in an actual submarine disaster.

When *Squalus* went down, Momsen had most recently been working on a series of deep-diving tests in which nontoxic helium replaced nitrogen in a diver's normal oxygen/nitrogen mix. This enabled a diver to go well past 100 meters (approx. 325 ft), the accepted working limit of the day. His team of rescue divers would need all of this expertise, along with the McCann rescue chamber.

Working in near-impossible conditions, the divers managed three successful trips to the stricken sub, bringing up groups of survivors in the rescue chamber. On the fourth and final trip with the last of the submariners, the wire haul-up cable jammed, and the cable began to unravel. Rather than risk severing the cable completely, the rescue personnel sent the McCann chamber—and those inside—back to the bottom.

With time running out, divers worked in vain in the numbing cold and depth to attach a new retrieving cable. The only chance for the last group of survivors still aboard the sub was a daring attempt to carefully blow ballast and control the rescue chamber's ascent to the surface. It was a gamble they would have to take. And it worked! When the McCann chamber finally broke the surface, a deckhand on the support ship managed to secure a clamp below the broken section of cable, and the last of the surviving submariners were hauled on board.

Although 26 crew members died immediately in the sub's ill-fated test dive, 33 survivors were brought back to the surface over a period of 39 hours. Headlines in the *New York Times* announced, "Man won a victory from the sea early this morning." Momsen's divers had proven that crew could be rescued from a stricken submarine.

In further dives, the Navy was able to verify that there were no more survivors, but the mystery remained. What had caused *Squalus* to go down? Had the high-induction valve failed to close, even though the control board registered that it had, causing the sudden flooding of the aft engine rooms? Had a crewman made a terrible mistake? Or was the sub a victim of sabotage? The only way to answer those questions was to raise the *Squalus* and examine it in dry dock. That task would be one of the toughest, deepest jobs Momsen and his experimental mixed-gas divers would undertake. Wildly optimistic staff officers believed it would take only three weeks to raise the damaged submarine. In fact, the job was finally accomplished after 640 dives over over 113 days. Ultimately, the cause of the *Squalus* tragedy was found to be failure of the main induction valve. That mystery was solved only after mounting the greatest undersea rescue and deepest salvage operation of its day.

Thresher

Twenty-four years later, another tragic sinking took over U.S. headlines. On April 9, 1963, the USS *Thresher* cruised out of Portsmouth Harbor in the early dawn, with 129 personnel on board. The powerful nuclear attack sub was known for stealth more than speed. *Thresher* had been overhauled earlier in the year and sent back on patrol, despite troubling results from recent tests of its silver-brazed pipe joints. Accompanying *Thresher* on its sea trials was the submarine rescue ship *Skylark*, equipped with a McCann rescue chamber that could be used on a stricken sub at depths up to 255 meters (approx. 836 ft). Alas, by the 1960s, modern nuclear subs were spending most of their time deeper than that.

Thresher's thick hull meant it could dive deeper than any previous sub. Its secret test depth, very deep for its day, was probably around 300–396 meters (approx. 990–1,300 ft), and the sub would have to come close to this depth rating before being cleared to return to duty. *Thresher*'s skipper called for crew to "Rig for deep submergence." As the sub slipped into deeper water, *Skylark* received a message that *Thresher* was "Proceeding to test depth." What happened next is uncertain, but experts believe the disaster may have begun with a silver-brazed pipe that broke at test depth; this led to a series of events that automatically shut down the sub's nuclear reactor. Unable to restart its reactors immediately, *Thresher*'s commander could not keep the submarine from sinking deeper and deeper—to crush depth. Moments later, the Navy's secret underwater listening system, designed to track Soviet subs, picked up a powerful implosion that signaled the end of *Thresher* and its crew.

The tragic loss shocked the American public. To make matters worse, the submarine seemed to have vanished, and no one knew exactly what had happened. *Thresher* had been the first of a series of new attack submarines the Navy had planned to build, so finding the wreckage and identifying the cause of the disaster were crucial. But the search only led to more frustration. The bathyscaph *Trieste* was called in; it made five dives without success; fortunately, a second round of dives finally located *Thresher*'s remains, now a pile of twisted wreckage resembling an "automobile junkyard." Fortunately, enough debris remained to determine that one of the suspect brazed joints probably broke, and the resultant spray of high-pressure water could have shorted out an electrical panel, causing a shutdown of the reactor and a sequence of further problems culminating in the sub's sinking.

The loss of a nuclear submarine and its crew, along with the long, frustrating search to locate the sub pointed out two issues—the U.S. Navy needed to instigate better submarine safety procedures and needed to improve its deep-ocean capacity. The answer to the first problem was the implementation of the SUBSAFE program. The second involved convening a panel of 58 experts who issued a top-secret report a year later that resulted in the formation of the Deep Submergence Project Office with the funding and the official sanction to issue proposals for new research, new technologies, and new submersibles.

This deep-sea initiative flourished in a climate of Cold War paranoia, quickly widening its scope beyond the perimeters of rescue. If the U.S. could launch spacecraft, it could also explore the ocean's depths. A swarm of new vehicles soon appeared, including Westinghouse's *Deepstar 4000*. Grumman launched the *Ben Franklin*. The U.S. Navy's *Turtle* and *Sea Cliff*, as well as the University of Hawaii's *Makali'i*, were built by General Dynamics. *Trieste II* replaced the older bathyscaphe, and the nuclear-powered research sub *NR-1*, capable of reaching 900 meters (approx. 2,950 ft), was launched. And in 1964, before any of these vehicles was completed, Reynolds Metals Company launched the hefty 76-ton *Aluminaut*, and Woods Hole Oceanographic Institution launched its much smaller 14-ton submersible *Alvin*. All these vessels were proof that a new era of deep-sea exploration was finally under way.

1. Introduction

Image courtesy of U.S. Naval Historical Center and Donald M. McPherson

Figure 5.2: Implosion!

The U.S. submarine F-4 *is pictured in drydock at Honolulu, Hawaii, on September 1, 1915 after being raised from over 100 meters (approx. 330 ft) and towed into port. Note the large implosion hole in the sub's port side and the salvage pontoons used to support her during the final lift.*

Water pressure is the single greatest physical threat facing any underwater vehicle, whether it's a low-budget plastic ROV operating in a wading pool or a state-of-the-art nuclear attack submarine operating in the depths of the ocean. Though silent and invisible, this enemy is lurking everywhere under water. Pressure hulls and pressure canisters are vehicle structures designed specifically to resist this threat. They play a critical role in shielding sensitive systems, including human occupants and electronics, from flooding and vehicle collapse.

In *Chapter 3: Working in Water*, you learned about many of the challenges associated with building underwater systems, including the high pressures and pressure-related forces caused by the weight of water. You may recall that any object submerged in water must support the weight of all the water directly above it. This weight creates a pressure that pushes in on all surfaces of the vehicle. If the vehicle's structure can't handle that pressure, it will be damaged and possibly leak or even collapse. Tragically, a number of submarine crews have perished when their subs dove too deep and ended up beneath more water than the hull could support. (See *Stories From Real Life: Squalus and Thresher* in this chapter.)

This chapter builds on the general, qualitative introduction in Chapter 3 and takes a more in-depth look at pressure issues. It gets more *quantitative*, giving you the numbers and equations necessary to calculate the pressures at any depth (in either freshwater or saltwater) and to calculate the forces those pressures will impose on submerged vehicles. As the numbers will show, these underwater pressures and the forces they impart can be enormous.

After giving you an appreciation of the magnitude of this pressure problem, the chapter will discuss what to do about it by examining tried-and-true techniques for building pressure hulls and pressure canisters. This is critical stuff. It can make or break the success of any underwater mission, regardless of whether the action takes place in a shallow wading pool or in the deepest regions of the sea. Pressure-related issues—and how to deal with them—are among the most essential things for underwater vehicle designers to understand thoroughly.

Many of the pressure hulls and canisters used in commercial or military underwater vehicles require specialized materials, tools, or skills and are beyond the budget and technical capabilities of the average school group or ROV/AUV enthusiast. Although some of these materials and techniques are mentioned in this introductory book because they are interesting, educational, or inspirational, the emphasis here is on practical information you can use to achieve adequate pressure protection for unmanned vehicles down to depths of about 100 meters (approx. 325 ft) while working on a shoestring budget.

2. Pressure

Technically, both liquids and gases are considered fluids, and both can be under pressure. When you push on a fluid, it tries to flow out of the way. If, however, the fluid is confined so it can't move out of the way, it gets *pressurized* instead. For example, when inflating a bicycle tire or automobile tire with a pump or air compressor, you are pressurizing the air inside. On a molecular level, this means more air molecules are being crammed into the confined space in the tire than would normally be there. This increases the number of molecules bouncing off the inside of the tire each second. Since each collision imparts a tiny outward force, the increased number of collisions per second increases the overall outward force, thus inflating and stiffening the tire.

A tire pressure gauge is one way to measure this pressure. In the U.S., tire gauges usually display the pressure in units of **pounds per square inch**, abbreviated **psi**. So a bicycle tire inflated to 60 psi has every square inch of the tire being pushed outward with a force of 60 pounds.

Note that pressure and force are *not* the same thing. A force acts only in one direction at any given time. A pressure, on the other hand, pushes equally hard in all directions at once. The units are different, too. For example, a *pound* is an imperial unit of force, but a *pound per square inch* is an imperial unit of pressure. Pressures are quantified by describing how hard they push against a surface of a particular size.

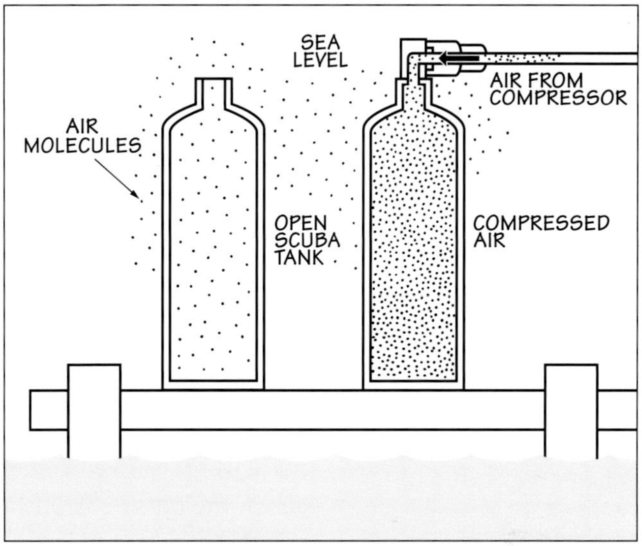

Figure 5.3: Compressing Air

Air molecules have a lot of space between them, so air can be compressed by forcing the molecules closer together. The air inside an open scuba tank has the same density as the surrounding air, but when the air is compressed into the tank, its density and pressure increase.

Although tire pumps and air compressors are one way to create pressure, they are not the only way, nor even the most common. Anything that attempts to squash a liquid or gas into a confined volume will create **fluid pressure**. The most common source of fluid pressure in nature is the weight of the fluid itself. In a glass of water, for example, the weight of the water near the top of the glass presses down on the water in the bottom of the glass, slightly pressurizing that deeper layer of water. In deep bodies of water, such as the ocean, the pressures at the bottom due to the weight of all the water above can be downright frightening.

2.1. Atmospheric Pressure

Strange though it may seem, this chapter about underwater hull design begins with a brief discussion of above-water air pressure. That's because an understanding of atmospheric pressure is helpful for understanding some basic aspects of water pressure, including some of the units used to measure water pressure and the difference between "absolute" and "gauge" pressures.

In the water-glass example above, the weight of the water in the upper portion of the glass caused pressure in the lower portion. This happens in the atmosphere, too. The "sea" of air in which we live is over 30 kilometers (approx. 20 miles) deep. When there's that much of something above us—even something as "light" as air—the weight adds up to a substantial amount. At sea level, the air pressure caused by the weight of the atmosphere is a surprising 14.7 psi. This particular amount of pressure has a special name—it's appropriately called **one "atmosphere" of pressure**, abbreviated one "**atm**." You'll also hear the term **ambient pressure** which simply means the surrounding pressure.

Memorize this (approximate) relationship, given here in imperial units, later in metric units. You'll need it often in this and later chapters.

1 atm = 14.7 psi

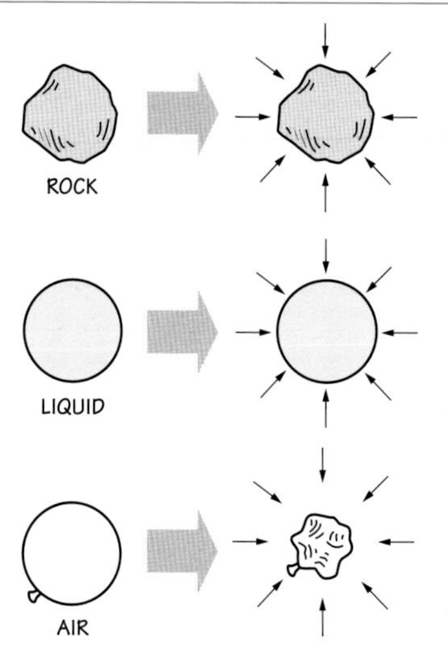

Figure 5.4: Compressibility

Some kinds of matter are more compressible than others. Solids compress the least amount under pressure; liquids compress only a little; gases (including air) compress easily. Pressure hulls and canisters are designed to resist the compressive forces caused by water pressure.

Think about the implications of this for a moment. Right now, as you read these words, *every square inch* of your body is being squashed by the surrounding air with a force of roughly 15 pounds (slightly less, if you're at high altitude). If you add that up over all the square inches of skin on your body, you'll discover that there are literally tons of force trying to squash you right now! If you jump into a pool, a lake, or the ocean and dive beneath the surface, the weight of the water above you adds even more pressure squeezing in on your body. Even if you submerge only a couple of meters (approx. 6 ft), you can feel this increased pressure on your eardrums.

How do humans survive such crushing forces? The answer is primarily the fact that your body is composed mostly of solids and liquids, which readily resist uniform compression, even under extreme pressures. Air cannot resist pressure without getting compressed, but the few air spaces in your body—lungs, sinus, middle ear, for example—are all connected through tubes and passageways to the outside of your body, so pressurized air gets in and pushes outward from the inside just as hard as the air on the outside is pushing inward. Thus, the pressures are in balance throughout your body, and you don't feel them. There are some situations, such as scuba diving or high-altitude flying, where internal and external pressures can get out of balance. The consequences can range from mild discomfort to severe pain or even death, depending on the magnitude of the pressure difference. (See *Tech Note: Clearing the Ears.*)

2.2. Pressure Differentials

You already know that extreme pressures can crush a submarine, even a high-tech sub made of steel or titanium. So how can fish, octopi, and other marine animals with comparatively soft, fragile bodies survive and thrive at the very bottom of the sea without the slightest hint of any harm from the intense pressures they endure? Clearly, it's not high pressures, *per se*, that cause trouble. It's something else.

To understand what's going on here, think about an empty plastic water bottle or soda bottle with its cap off. The empty bottle isn't really empty. There is air inside the bottle, and the air inside the bottle is at the same pressure as the air outside the bottle. Since

TECH NOTE: WHAT IF THERE WAS NO ATMOSPHERIC PRESSURE?

Admittedly, it's a bit scary to visualize tons of force pressing in on your body, the way the atmosphere does. You might even be wondering if there's some way to remove all that pressure, so your body wouldn't have to deal with it. Well, there is. It's called a vacuum pump, and it's a common piece of laboratory equipment. Don't be too quick to suck all the air out of your body, though. Not only do you need air to breathe (for the oxygen in it), but you actually *need* that pressure pressing in on your body to survive. Without it, there is nothing to keep liquids—including your blood and other body fluids—from vaporizing.

When you heat water on a stove, it boils because the vapor pressure of the hot water exceeds the pressure of the atmosphere, allowing bubbles of water to form and expand against the external pressure. In a vacuum, however, water will boil even without heat, because the vapor pressure of cold water is still more than zero. Astronauts, who have only the thin shell of a space suit separating them from the vacuum of space during a spacewalk, are all too aware of this. If they get a bad leak in their suit, the air pressure that keeps their body liquids liquid will escape, their blood will boil (without heating), and they will die a rapid, painful death.

the air inside is pressing out on the walls of the bottle exactly as hard as the air outside is pressing in, the thin plastic wall of the bottle is supported and does not collapse. But if you reduce the internal pressure by hooking the bottle up to a vacuum pump and sucking out the air inside, the bottle will collapse inward. This is the principle at work when a child sucks all the air out of a juice box with a straw and the walls of the juice box are drawn inward. The walls collapse because the interior pressure becomes less than the exterior air pressure. This difference between pressure areas, or zones, is called a **pressure differential**.

ATMOSPHERIC PRESSURE
AIR EVACUATED BY PUMP
VACUUM
VACUUM PUMP

Figure 5.5: The Effect of Reducing Internal Pressure

The opposite happens if the internal pressure is greater than the external pressure. If you hook up a tire pump and force air into the water bottle, its sides will bulge outward. Or if you fill a balloon and keep on blowing air in, it will eventually pop. Here again, a pressure differential is at work. In the early days of high-altitude jet liners, the fuselage airframes of these aircraft were dangerously stressed because near sea-level air pressure was maintained inside for the comfort of passengers, and this inside pressure was much greater than the lower high-altitude atmospheric pressure outside. Aeronautical engineers have since designed airframes to withstand this air pressure differential. In addition, the internal air pressures on many planes are now lowered a bit during flights to help further reduce the pressure differential.

Fish and octopi living at the bottom of the sea don't get squashed, and they don't explode, because they don't try to maintain any pressure differentials across their body walls. Their soft innards are at exactly the same pressure as the surrounding water. They're like that soda bottle with its cap off. A submarine is like the soda bottle with

TECH NOTE: CLEARING THE EARS

All divers have to deal with the problem of "clearing the ears." A descending diver often feels ear discomfort or pain that gets worse with increasing depth. This pain is caused by a pressure differential across the eardrum. The water pressure acting on the outside surface of the eardrum is greater than the air pressure acting on the inside surface, so the eardrum gets pushed inward and distended. This hurts. If the diver continues to descend without equalizing pressure, the pain may become intense, and eventually the pressure differential may cause the eardrum to burst.

A person feeling discomfort in the ears while diving should immediately stop descending or even ascend a little, then "clear the ears" (equalize the pressure) by closing the mouth, pinching the nose, and blowing gently. Through this procedure, air is forced from the throat via the eustachian tube into the middle-ear region, increasing the pressure on the inside surface of the eardrum. This procedure is also known as the **Valsalva maneuver**. When the discomfort eases, the pressure has been equalized. Once the outer water pressure and air pressure inside the middle ear are approximately equal, it is safe to continue the descent, provided the diver continues to clear the ears periodically on the way down.

The opposite problem, called a "reverse ear block," can occur on ascent, but this is uncommon, as air seems to escape from the middle ear more easily than it enters. One reason people should avoid diving while sick with a cold is that their eustachian tubes are often inflamed and clogged with mucus, making it difficult or impossible to clear the ears.

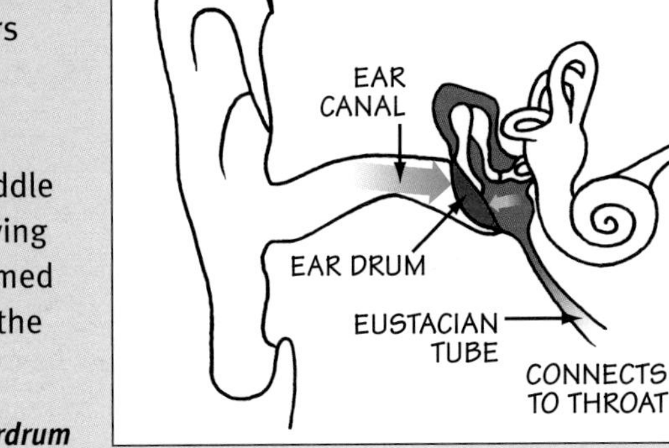

Figure 5.6: Pressure Differential Across the Eardrum

some of the air pumped out of it—the pressure inside is not enough to balance the pressure outside, so the hull has to be strong enough to make up the difference.

The key message here is that pressure *differentials*, rather than pressures, present some of the greatest challenges in subsea design and operations. Even a small pressure differential can cause leaks or structural damage. If you know you'll have a large pressure differential—perhaps because the inside of your vehicle will be at atmospheric pressure while the outside is exposed to extreme water pressure—then you know you'll need to build a very strong hull. On the other hand, if you can figure out some clever way to reduce the pressure differential (for example, by using the oil-compensation technique described later in this chapter to match the internal pressure to the external pressure), then you may be able to get by with a simpler, lighter, and less expensive hull—like an octopus does.

2.3. Gauge Pressure Versus Absolute Pressure

At this point, you need to be warned of a potential pitfall when talking about pressure values. Pressure-measuring devices always display on some type of meter the *difference* between the pressure being measured and some reference pressure. Two different reference pressures are in common use. One is a vacuum (zero pressure), and the other is whatever ambient pressure happens to be outside the meter (normally something very close to 1 atm of air pressure). Devices that use the vacuum reference are said to report **absolute pressure** readings (**psia**), whereas those that use the local ambient pressure reference are said to report **gauge pressure** readings (**psig**). The two types of readings typically differ by about 1 atm (14.7 psi), so you need to be careful to keep track of which reference is used!

TECH NOTE: CONFUSING TERMINOLOGY

It can be easy to confuse "gauge pressure" (a pressure measured with respect to atmospheric pressure) with "pressure gauge" (a device used to measure pressure). To make matters even worse, you may encounter gauge pressures written as "pressure (gauge)," which means gauge pressure.

Another common example of sloppy and potentially confusing word (mis-)usage is the substitution of "pounds" for psi. For example, someone might ask you to inflate a tire to "40 pounds," or a scuba diver might come back from a dive saying there's still "1,000 pounds" of air left in his tank. In these cases, what they really mean is 40 psi or 1,000 psi. If the scuba diver actually had 1,000 pounds of air in his tank, he wouldn't be able to lift it!

For example, a typical tire pressure gauge, which reports gauge pressure, will read zero for a completely flat tire, even though that tire still has about 1 atm of air pressure inside. (The air leaked out, but it wasn't sucked out, so it stopped leaving the tire as soon as the internal pressure matched the external pressure.) On the other hand, a **barometer**, which is used to measure the pressure of the atmosphere itself, reports pressure values in absolute terms, so if attached to the completely flat tire, it would reveal the roughly 1 atm of pressure still in there.

In underwater vehicle design, both absolute and gauge pressures are used frequently, so it's essential to understand the difference between them and know how to convert readily from one to the other. It's not difficult. Just remember that absolute pressures are the "true" or "real" pressures, and that gauge pressure readings (which are the more common ones) have 1 atmosphere subtracted from the absolute pressure. This gives a convenient zero reading at the surface of the ocean.

A quick look over Table 5.1 will illustrate how this works.

Table 5.1: Absolute and Gauge Pressure Comparison

	PRESSURE READING	
	Absolute	*Gauge (referenced to 1 atm)*
Vacuum (e.g., outer space)	*0*	*-1 atm*
Sea level	*1 atm*	*0*
10 meters deep in seawater	*2 atm*	*1 atm*
20 meters deep in seawater	*3 atm*	*2 atm*
30 meters deep in seawater	*4 atm*	*3 atm*

TECH NOTE: GAUGE PRESSURES AT HIGH ALTITUDES

For most practical applications, particularly anywhere near sea level, you can safely assume that absolute pressure and gauge pressure differ by 1 atm. However, at high altitudes, where ambient air pressure is slightly less than 1 atm, the two pressure readings will differ by less than 1 atm. Fortunately, this effect is minor and of little practical significance for underwater vehicle design and operation. It can usually be ignored, even for work in high-altitude lakes. (These lower atmospheric pressures do, however, produce a significantly increased risk for scuba divers, who should receive special training before diving in high-altitude lakes.)

2.4. Pressure Units

So far, this chapter has presented only psi and atm as units for pressure measurements; however, many other units are in frequent use. You should know about the common ones, since you'll probably encounter them sooner or later. Remember that all units used for pressure must directly or indirectly refer to some standard amount of *force per area.* Get to know these units:

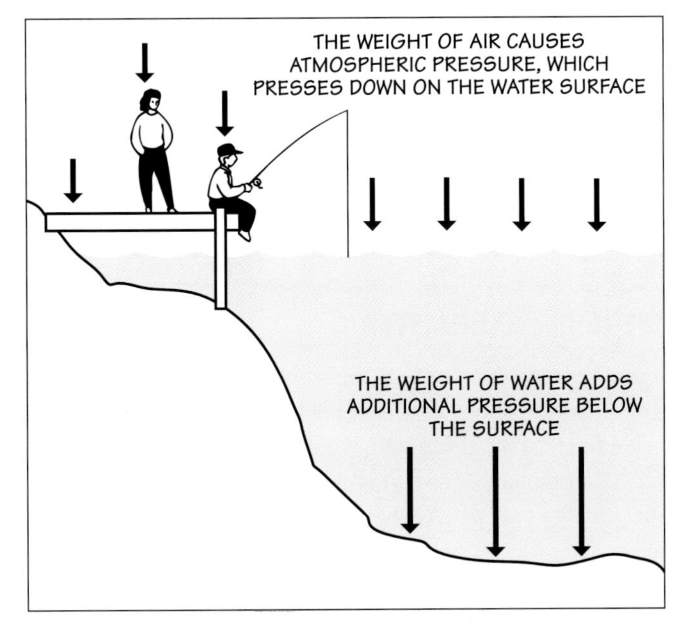

Figure 5.7: The Combined Effect of Air Weight and Water Weight

The weight of the earth's atmosphere presses down on the surface of the ocean and other bodies of water. Thus, both air weight and water weight contribute to the total pressure present beneath the water's surface.

Some pressure gauges include the 1 atm of additional pressure contributed by the weight of air in their pressure readings, but most do not.

If your pressure gauge reads 1 atm (or 14.7 psi or some other equivalent of 1 atm) at the surface of the ocean, it is including the contribution of air weight and is therefore measuring absolute pressure. If it reads 0 atm (or 0 psi or 0 anything else), it is ignoring the contribution of air weight and is therefore measuring gauge pressure.

- **psi:** You've already been introduced to "psi," which stands for one pound (of force) per one square inch (of area).
- **atm:** You have also been introduced to the atmosphere, or atm. One atm is formally defined as the average atmospheric pressure at sea level when the air temperature is 20 degrees C (68 degrees F). This is a case where the force and area are indirectly defined by referring to a real-world pressure.
- **Pa**: In the metric system, where forces are commonly measured in newtons (N) and areas are commonly measured in square meters (m^2), a natural unit for pressure is the N/m^2. This unit is also known as a Pascal (Pa) in honor of Blaise Pascal (1623–1662), a famous French mathematician, physicist, and religious philosopher who contributed a great deal to our understanding of fluids and statistics during his short lifetime. One kilopascal (kPa) is 1,000 Pa.

- **bar:** The bar is an alternative metric unit for measuring fluid pressure. It is defined as 100,000 Pa, or 100 kPa. Outside of North America, pressure gauges such as those used on scuba tanks often use bars to express pressures. It turns out that 1 atm = 101,300 Pa = 1.013 bar, so a bar is nearly (but not exactly) equal to 1 atm. Smaller pressures may be expressed in millibars: 1 **mbar** = 1/1000 of a bar.

- **inches or mm of mercury (in. Hg, mm Hg):** Mercury is a dense, shiny, silver-colored metal that is liquid at room temperature. Barometric pressures used for weather forecasting have traditionally been based on how far air pressure could push a column of mercury up into a vacuum-filled glass tube, so it's natural to express these pressures in millimeters of mercury (metric) or inches of mercury (imperial). (See *Tech Note: The Mercury Barometer.*) Since the chemical symbol for mercury is Hg, these units are abbreviated **mm Hg** and **in. Hg**, respectively. One atm (the pressure unit) is approximately 760 mm Hg or 29.9 in. Hg. Actual atmospheric pressure at sea level on any given day may be slightly higher or lower than these values, depending on the weather.

The following equations (some are approximate, others exact) summarize these units and can be used to calculate conversion factors among them:

In imperial units:

1 atm = 14.7 psi = 30 in. Hg

In metric units:

1 Pa = 1N/m2
1 bar = 100,000 Pa
1 atm = 101,300 Pa = 1.013 bar = 760 mm Hg.

TECH NOTE: THE MERCURY BAROMETER

Although 1 atmosphere describes a constant unit of pressure, it's important to remember that this unit is defined in terms of an *average* atmospheric pressure under very specific conditions. Pressure in the real atmosphere (known as **barometric pressure**) is constantly fluctuating slightly above or below this average due to changes in temperature, weather, and other factors. Sensitive pressure-measuring instruments, called barometers, can track barometric pressure changes to help predict changes in the weather.

The mercury barometer is one of the simplest and oldest devices invented to measure barometric pressure. It consists of a tube with a vacuum inside that is inverted in a small pool of mercury. The tube is graduated along its length in inches or millimeters, the zero point being the average level of the mercury in the container. The units of measurement are millimeters of mercury (mm Hg) or inches of mercury (in. Hg).

Because the tube has essentially zero pressure inside it, the atmospheric pressure acting on the pool of mercury forces it upward into the tube where the pressure, or trend of pressure, is measured accordingly. When the atmospheric pressure drops, the level of mercury in the tube drops. When the atmospheric pressure rises, the mercury rises. Weather forecasters often use the expressions "the mercury is dropping" or "there is a rise in mercury" to describe a change in air pressure and weather. Mercury rising generally indicates good weather, whereas mercury dropping means a storm is brewing.

Figure 5.8: Mercury Barometer

2.5. Devices for Measuring Pressure (and Depth)

Over the centuries, people have developed many different methods to measure pressure and have adapted some of these for measuring depth. Here are some in common use:

Manometers: These devices are very simple in design and easy to make in a home or school shop. The basic idea consists of a U-shaped, clear glass or plastic tube partly filled with colored water, mercury, or some other fluid that is visible through the wall of the tube. If the pressures on the two ends of the tube are equal, the level of the fluid on each side will be equal, but if the pressures are different, the fluid will be forced farther up one side of the tube than the other. Provided the density of the fluid is known, the difference in fluid height between the two sides of the tube can be used to calculate the pressure difference with great precision. This extremely sensitive technique is great for measuring tiny pressure differences (a few percent of one atm or less), but is not used for large pressure differences, such as those often encountered in subsea work.

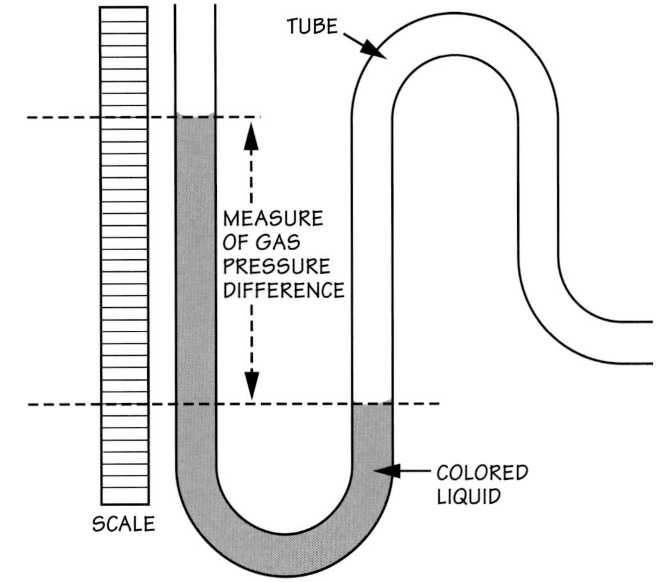

Figure 5.9: Manometer

Barometers: A barometer is a variation on the manometer theme. Though they often look very different, the physical principle behind the operation of the two devices is similar. Barometers are used commonly to measure absolute atmospheric pressure for the purpose of forecasting weather. For a summary of how a barometer works, see *Tech Note: The Mercury Barometer.*

Dial Pressure Gauges: These are the familiar gauges used for measuring the pressure of air in tires and of liquids or gases in pipes or tanks. On the front of the gauge there's often a round dial with a needle that moves in response to pressure and points to a number representing the amount of pressure. There's also a hollow stem that can be attached to a tire valve or pipe fitting, or just left open to the outside air/water, so that the pressure to be measured can enter the gauge. Inside the gauge, the needle is connected mechanically to either a pliable diaphragm that flexes when pressure is applied to one side of it or to a curved tube (called a Bourdon tube) that straightens when pressure is applied to the inside of the tube. Dial pressure gauges used in undersea applications are commonly calibrated in units of psi, kPa, or bars.

Integrated (Electronic) Pressure Sensors: Modern pressure sensors often use solid-state sensors and integrated circuits to produce an electrical signal that varies with pressure. These signals cannot be read directly by a human operator; however, they are easily interfaced with other electronic devices and are therefore ideal for digital displays, computer-based recording, automated control systems, long-distance pressure signal transmission, and many other applications.

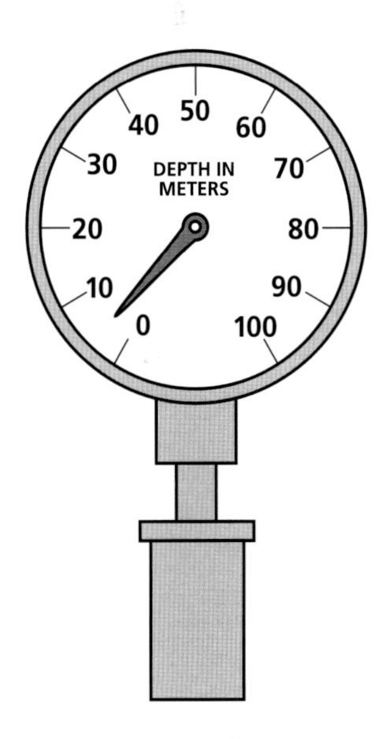

Figure 5.10: Depth Gauge

A pressure gauge can be calibrated to read in meters or feet, thereby converting it to a depth gauge.

Depth Gauges: Devices for measuring pressure can also be used to measure depth. Incompressible liquids, including water, exhibit a straightforward, linear relationship between depth and pressure, so a pressure gauge can be converted into a depth gauge simply by modifying the units displayed. Older depth gauges used by submersibles and scuba divers were essentially corrosion-resistant dial pressure gauges with the units on the dial face calibrated in feet, meters, or some other depth unit instead of pressure units. Many newer depth gauges are based on electronic pressure sensors, but they do the conversion from pressure to depth electronically and display the result as a depth reading.

HISTORIC HIGHLIGHT: WRONG CONVERSION FACTOR

To convert accurately from pressure to depth, you must know the density of the fluid involved. The importance of using the correct density was demonstrated dramatically when Jacques Piccard and Don Walsh made their record descent to the Challenger Deep in 1960. Prior to that time, depth soundings had shown that the deepest place in the ocean was about 10,880 meters (approx. 35,700 ft) beneath the surface. So as the submersible *Trieste* descended toward the bottom, the two men were puzzled to see the depth gauge in the submersible pass 10,972 meters (approx. 36,000 ft), nearly 100 meters *below* the expected bottom!

When their sounder finally indicated bottom approaching, Piccard began releasing steel shot to slow their descent. At touchdown, the depth gauge read 11,521 meters (approx. 37,800 ft). What was going on? The mystery was solved when someone realized that the depth gauge had been calibrated in Switzerland—using *freshwater* density. *Trieste*'s descent had actually been to 10,910 meters (approx. 35,800 ft).

2.6. Calculating Hydrostatic Pressures Under Water

Vehicles operating under water may encounter two different sources of water pressure. One—the focus of this chapter—is called **hydrostatic pressure**. Hydrostatic pressure is pressure that exists even in a *static* situation, where water is not moving. The most common source of hydrostatic pressure is the weight of water itself. In contrast, **hydrodynamic pressure** is pressure associated with water *motion*. When a vehicle moves through water or has water currents moving past it, pressures increase on the upstream side of the vehicle as water gets shoved up against it and decrease on the downstream side as water fills in behind the vehicle.

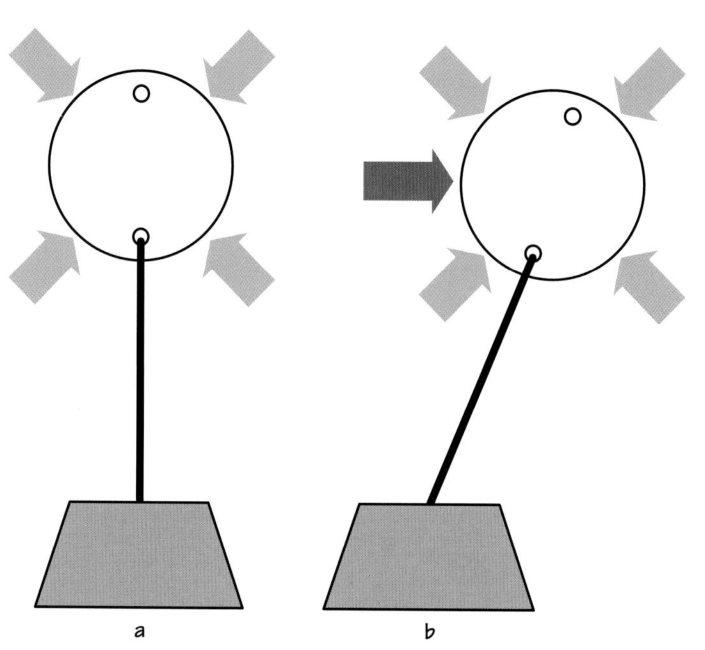

Figure 5.11: Two Sources of Water Pressure

In calm water (a), hydrostatic pressure (represented here by blue arrows) presses in on all sides of a submerged object, such as this submerged float, but there is no hydrodynamic pressure. In flowing water (b), the hydrostatic pressure continues to press in on all sides, but now dynamic pressure (dark grey arrow) is added to the mix. In this example, the hydrodynamic pressure forces the float to one side. In small underwater vehicle design, hydrostatic pressures are important because they can crush a vehicle or cause leaks. Hydrodynamic pressures are important because they can impede vehicle movement and make vehicles difficult to control in currents.

For most underwater vehicles operating at modest speeds, hydrostatic pressures are many times larger (often thousands of times larger) than hydrodynamic pressures and pose the greatest threat to vehicle survival. The primary role of pressure hulls and canisters is to deal with these potentially dangerous hydrostatic pressures. Hydrostatic pressure will therefore continue to be the focus of this chapter. Dynamic pressure issues, which can impact vehicle speed, fuel efficiency, and other aspects of performance, will be addressed in later chapters.

To design effective pressure hulls and canisters, it's essential that you be able to estimate the approximate hydrostatic pressure at different depths, in both freshwater and seawater. A good place to start is to memorize the following two relationships for water bodies on planet Earth:

- **In freshwater, pressure increases by 1 atm every 34 feet of depth.**
- **In saltwater, pressure increases by 1 atm every 33 feet of depth.**

These relationships are not exact; they are rounded to the nearest whole foot. Nonetheless they are accurate enough for typical underwater vehicle design purposes.

The slight difference between the fresh- and saltwater values is due to the slightly greater density of seawater as compared to freshwater. Note that 33 feet is almost exactly 10 meters. Thus, many people find it easier to remember the metric version:

- **In saltwater, pressure increases by 1 atm every 10 meters of depth.**

Unless you need extreme precision, this super-simple metric relationship is close enough to use for most freshwater calculations, too.

Note also that all these relationships can be applied at any depth. Pressure in seawater increases by 1 atmosphere every 10 meters whether you are going from 0 to 10 meters,

TECH NOTE: FEEL THE PRESSURE

To feel hydrostatic pressure directly, try this simple exercise. Fill a tall plastic garbage can with water to a depth of about 60 cm (approx. 2 ft). Then slide your hand and arm into a large, clean plastic trash bag all the way up to your shoulder. Now reach your bag-covered arm straight down into the water as far as you can without letting any water run into the bag. You will feel the water (hydrostatic pressure) pressing in on your hand and arm through the plastic bag. Try spreading out your fingers while your hand is submerged.

Now move your hand up and down slowly through the water with your fingers still outstretched. Notice how the strength of the "squeeze" increases as your hand moves deeper?

TECH NOTE: THE DIRECTION OF HYDROSTATIC PRESSURE

If you were to place a stack of heavy bricks on your head while standing on a hard floor, you would certainly feel the increased weight pressing down on your head. You might even notice the floor pressing up harder on the soles of your feet as your feet get mashed against the floor, but your sides wouldn't experience any increased force. It may seem odd, then, that hydrostatic pressure, which comes from the weight of water above, is felt as an all-around squeeze, rather than a strictly vertical force sandwiching the top and bottom of a submerged object. The difference arises because water (unlike bricks) can flow to redistribute forces.

You can see this effect when squeezing a water balloon (Figure 5.12). The water balloon will bulge out sideways, even if you're only pressing straight down on it. If you imagine a submerged object as being similar to an object surrounded by water balloons, you can see how some of the balloons in contact with the sides of the object would be under pressure from the weight of balloons above them and would bulge out sideways, thereby pressing in on the sides of the object. While balloons may help to visualize what's going on by conceptually dividing the water into separate chunks, it's the ability of water to flow and redistribute forces, not the balloons themselves, that translates water's vertical weight into sideways forces. Thus, an object immersed in water experiences pressure on its sides as well as its top and bottom.

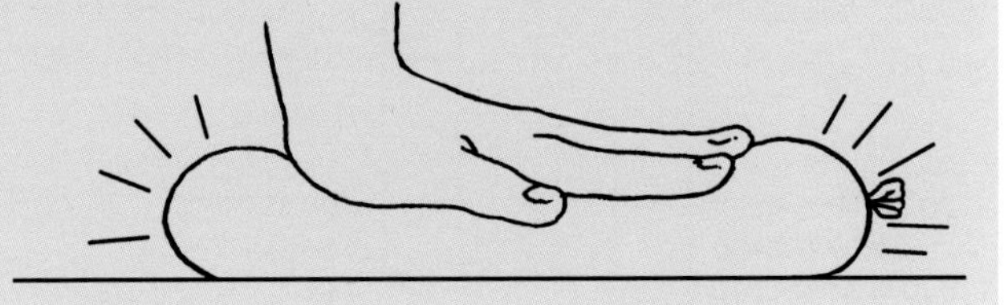

Figure 5.12: Hydrostatic Pressure Affects All Sides

Figure 5.13: Pressure Increase with Depth

In seawater, pressure increases at the rate of approximately 1 atm (14.7 psi) for every 10 meters (approx. 33 feet) of depth. Absolute pressure values include the contribution of atmospheric pressure to the total underwater pressure and are always 1 atm (14.7 psi) higher than the corresponding gauge pressure. Gauge pressure is zero at the surface and increases in direct proportion to water depth.

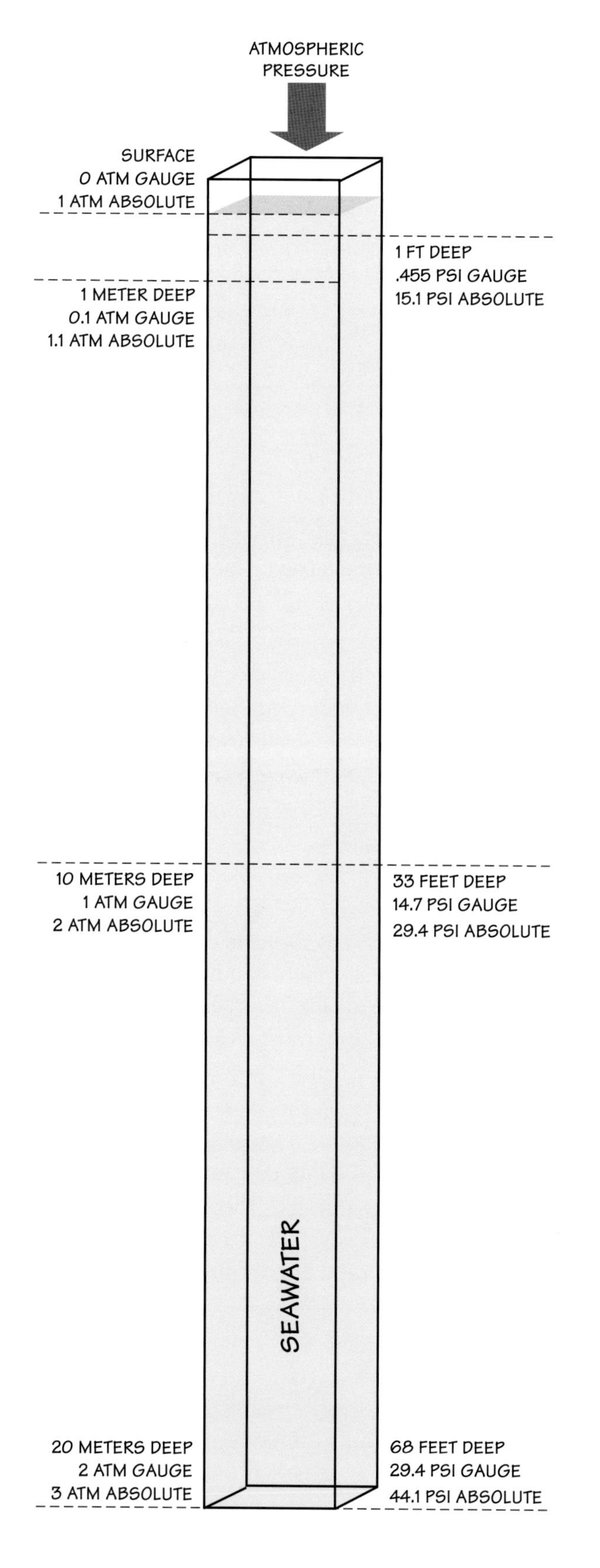

from 10 to 20 meters, or from 1,000 to 1,010 meters. This is a convenient side effect of the fact that water is largely incompressible, so equal *changes* in depth will result in equal *changes* in pressure, no matter what depth you are working at. This makes underwater pressure calculations much easier than pressure calculations in air, where the relationship between altitude change and air pressure change varies, depending on the elevation!

Finally, don't forget that absolute pressure is generally 1 atmosphere higher than gauge pressure:

Equation 5.1 $$psia = psig + 1\ atm$$

With these easy-to-remember relationships and some basic algebra, you can quickly find the gauge pressure and absolute pressure at any depth in either freshwater or saltwater.

For example, to find the absolute pressure (call it A) at a depth of 150 feet in saltwater, you can begin by figuring out the increase in water pressure from the surface down to 150 feet, then adding in the extra 1 atm of pressure from the atmosphere at the surface. The first part is equivalent to finding the gauge pressure (call it G) at 150 feet. One way to do this is to set up a pair of proportional ratios and solve for the gauge pressure:

$$(1\text{ atm}/33\text{ feet}) = (G\text{ atm}/150\text{ feet})$$

Rearranging to solve for G, you get:

$$G = 150\text{ feet} \times (1\text{ atm}/33\text{ feet}) = 4.55\text{ atm}$$

To convert this to absolute pressure, simply add 1 atm:

$$A = G + 1\text{ atm} = 4.55\text{ atm} + 1\text{ atm} = 5.55\text{ atm}$$

Once you have the answer needed in atm, you can easily convert to any other common units, provided you know how many of those units equal 1 atm.

For example, to convert to psi, we use the fact that 1 atm = 14.7 psi:

$$5.55\text{ atm} \times 14.7\text{ psi/atm} = 81.6\text{ psi}$$

Remember that these simple relationships are only approximate, though in most cases they'll produce answers that are accurate to within a few percent.

Keep in mind, though, that these relationships are based on several key assumptions, among them that water is incompressible and that the water is located on planet Earth, with its surface open to the atmosphere at an altitude near sea level. For most practical applications—particularly in the local lake, pool, or shallow ocean dive site—these assumptions are very good approximations. However, there are unique situations where these assumptions are over-simplified or even completely wrong. In those cases, the cavalier use of these relationships can get you into trouble. One place where the assumptions can become problematic is in extremely deep water, such as full ocean depth. (See *Tech Note: Compressibility of Water.*) An example of a place where the assumptions are just plain wrong and guaranteed to lead to major errors is discussed in the next section on pressure calculations for the seas of Jupiter's moon Europa.

TECH NOTE: COMPRESSIBILITY OF WATER

For most purposes, scientists consider liquids to be incompressible, but there are many instances when even small changes in the density of a liquid need to be considered. Oceanographers measure the density of ocean water very precisely, for example, so that they can predict ocean phenomena related to currents and marine life.

Similarly, deep sea vehicle engineers must account for the increase in water density to the full ocean depth of 10,700 meters (approx. 35,000 ft) when designing pressure hulls and ballast systems for vehicles that will enter these depths. The water at full ocean depth is subjected to such intense pressure that its density is increased by about 6 percent compared to the average density of water at the surface.

TECH NOTE: FRESHWATER PRESSURE EXERCISE

Try to verify that freshwater pressure increases by 1 atm for every 34 feet of depth, as stated above. To do this, first calculate the weight of a column of freshwater 1 foot wide x 1 foot long x 34 feet high. If you recall that freshwater weighs 62.4 pounds per cubic foot, you should get 2,121.6 pounds. Since all of this weight is supported by 1 square foot of floor area at the bottom of the column, you've just calculated the pressure (in pounds per square foot) at a depth of 34 feet of freshwater.

To convert this to psi, you'll need to calculate the number of square inches in one square foot, then divide the pounds per square foot by the number of square inches per square foot. The answer should be 14.7 psi. This is one atmosphere, so you have verified that pressure increases by 1 atmosphere for every 34 feet of depth in freshwater. (Note: If you got about 177 psi, instead of 14.7 psi, your answer is off by a factor of 12. This is a common mistake. Go back and double-check your calculations for the number of square inches per square foot—it is *not* the same as the number of inches in a foot.)

If you want even more practice, verify the corresponding relationship given for saltwater, where pressure increases by 1 atmosphere for every 33 (not 34) feet of depth. For this calculation, you need to know that the density of seawater is about 64 pounds per cubic foot.

2.7. Calculating Hydrostatic Pressures on Other Worlds

The method described above works well enough for most practical underwater vehicle situations here on planet Earth, but it relies on several assumptions that would not be valid in the exciting underwater vehicle missions now being planned for other worlds. In these cases, you must adopt a more systematic, step-by-logical-step approach, using basic physics.

Consider Jupiter's second moon, **Europa**, as an example. Many scientists think Europa has an ice-covered, salty ocean that could potentially harbor alien life forms. That ocean is thought to be about 100 km (approx. 60 miles) deep!

Some day in the not-too-distant future, an AUV from Earth may explore the bottom of this ocean for signs of life. If you were to become one of the engineers helping to design such a craft, you'd certainly need to estimate the maximum water pressure your

Image courtesy of Jet Propulsion Laboratory, California Institute of Technology, and NASA

Figure 5.14: Europa

This image of Europa was taken in 1979 by Voyager 1 *from a distance of about 2 million kilometers (1.2 million miles). The bright areas are probably ice deposits while the darkened areas may be the rocky surface or areas with a more patchy distribution of ice.*

The most unusual features are the long linear structures crisscrossing the surface which are over a thousand kilometers long and 200 – 300 hundred kilometers wide. They resemble cracks in Arctic ice and are thought to be evidence that Europa's surface is ice on top of an ocean of liquid water.

vehicle would need to withstand while exploring the bottom of Europa's ocean. Here's how you could do so:

On a scientific mission like this, you'd almost certainly use metric units. Lacking information to the contrary, you could start by assuming the saltwater there is similar in density to that of Earth's ocean, or about 1,025 kg/m^3. (Note that it's essential to use mass density here, not weight density, because you are no longer working in Earth's gravity.)

Recall that weight equals mass multiplied by gravitational acceleration. So you could convert this water mass density to a water weight (in newtons) per unit volume on Europa by using the fact that Europa's gravity is known to be only about 13.5 percent that of Earth's, or about 1.32 m/s^2. Thus,

1025 kg/m3 x 1.32 m/s2 = 1353 N/m3

As on Earth, the pressure at depth would be equal to the weight of a column of fluid—extending vertically between that depth and the surface—divided by the cross-sectional area of that column. For the sake of simplicity, imagine a fluid column with a 1 square meter cross-section; you can see that the pressure would increase by about 1,350 N/m^2 (i.e., 1,350 Pa) for every meter of depth. So at the bottom of a 100 km (=100,000 m) deep sea, there would be a pressure of 135,000,000 Pa.

Converting this to more familiar units, note that 1 (Earth) atm = 101,300 Pa, so this is a pressure of 135,000,000 Pa /101,300 Pa per atm = 1,333 atm. Converting this to psi, you get: 1,333 atm x 14.7 psi/atm = about 19,600 psi.

For comparison, the pressure at the deepest part of Earth's ocean (a mere 11 kilometers down, but in a stronger gravitational field) is roughly 1,100 atm, or about 16,200 psi.

One complicating factor, of course, is that Europa's ocean appears to be covered by a thick layer of ice. However, the density of ice is not vastly different from that of liquid water, so your answer wouldn't be far off if you simply measured your vehicle's "depth" from the top of the ice layer rather than the top of the liquid layer. Keep in mind, after all, that these are very rough estimates, so don't get too hung up about things that change your answer by only a few percent.

Another complicating factor is that you might not know what the atmospheric pressure is at the top of the ice layer, so it would be difficult to estimate the absolute pressure, rather than the gauge pressure. However, this is not likely to present much of a problem for the following reason. Since Europa's gravity is much less than that of Earth, its atmosphere is probably thinner and has less pressure—maybe a few psi at most. Compared to the nearly 20,000 psi you estimated for the bottom of Europa's sea, the contribution of atmospheric pressure to the total can probably be ignored safely.

3. Pressure-Related Forces on Submerged Objects

There's an old adage that says, "It's not the fall that hurts; it's the sudden stop at the bottom." In a similar sense, it isn't the pressures, or even the pressure differentials, that crush an underwater vehicle; it is something else that's closely related. In this case, it is the forces that pressure differentials impose on a vehicle's structure.

Recall that a force is a physical push or pull. Whenever a surface separates two regions of different pressure, that surface experiences a force trying to push it toward the region of lower pressure. For example, a submarine hull, which separates the interior of the sub from much higher pressures outside, experiences a force trying to push the hull into the interior of the sub. The magnitude of the force is proportional to the pressure differential (which is why pressure differentials are important), but it is also proportional to the size (specifically, the surface area) of the surface separating the two pressure regions.

In mathematical terms, this means that the force the surface must support is equal to the difference in pressure from one side of the surface to the other, multiplied by the area of the surface:

Equation 5.2 ***Force = Pressure differential x Surface area***

To get valid numbers from this formula, the units used for area must be the same as those used in the denominator of the pressure units. For example, you can multiply square inches by psi (pounds per square inch) to get pounds of force, but it does not work to multiply square meters by psi directly. You must first convert square meters to square inches if you want your force expressed in pounds, or convert psi to Pa, which are newtons per square meter, if you want force expressed in newtons.

To see how Equation 5.2 works, let's review the example of the submarine hatch from *Chapter 3: Working in Water.* In that example, a submarine with a hatch 25 inches in diameter is working at a depth of 100 meters (approx. 325 ft). A submarine hatch is circular in shape, and a circle's area is πr^2, so a 25-inch diameter hatch has a surface area of 491 square inches. In seawater, you now know that pressure increases by 1 atm for every 10 meters of depth, so at 100 meters, the pressure has increased 10 atm above what it was at the surface, or 10 x 14.7 psi = 147 psi. The absolute pressure is 1 atm higher, or 147 + 14.7 psi, or about 162 psi. Of course, that extra atm is also present inside the sub pressing outward, so the differential pressure acting across the hatch is simply the gauge pressure, 147 psi. Multiplying 147 psi by 491 square inches yields a total force of 72,177 pounds.

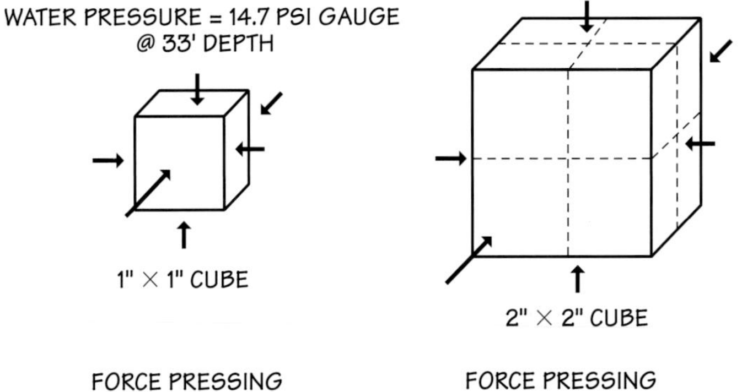

Figure 5.15: Force and Surface Area

Forces due to pressure are proportional to the surface area on which they act.

Now think about the mind-boggling forces that must be supported by the entire hull of a large military submarine working 100 meters (approx. 325 feet) beneath the ocean's surface. Such a sub might present 6,000 square feet of surface area (864,000 square inches) to the water above. At 100 meters depth, the gauge pressure is 10 atmospheres, or 147 psi. That produces a total force of 864,000 square inches x 147 pounds per square inch = more than 127 *million* pounds of force! And a sub at a depth of 100 meters is barely scratching the surface of the ocean, which averages almost 4,000 meters (approx. 2.5 mi) deep!

Such submarine examples are impressive, but what about the pressures and forces encountered by a small ROV in a school swimming pool? Are they big enough to be a problem? Suppose you are in the U.S. (where imperial units are commonly used) and want to build a little ROV for use in the deep end of the local swimming pool. Presume the pool is 12 feet deep. Let's say you're on a very tight budget, so the plan is to use an empty half-gallon waxed cardboard milk carton as a waterproof container to house the electronics.

After installing the electronics and other components, you close the container, seal it shut, and prepare the ROV for its maiden voyage. Note that you sealed the milk carton while sitting on the pool deck where the absolute pressure is 1 atm, so the pressure inside the carton is also 1 atm, or 14.7 psi. Then you proceed to send your ROV to the bottom of the 12-foot-deep pool where the absolute pressure is (and you should verify this to practice your calculations) about 19.9 psi. The differential pressure between the inside and outside walls of your milk carton is therefore 19.9 psi - 14.7 psi = 5.2 psi.

That doesn't seem like a very big pressure differential, but suddenly your ROV quits moving and refuses to return to the surface under its own power. When you haul it up by the tether, you discover that your seal has cracked loose and the electronics inside the carton are sopping wet. What happened? Some simple force calculations provide the answer: A typical half-gallon milk carton has a rectangular shape with sides that are about 4 inches wide and 8 inches tall. Each side therefore has a surface area of about 4 x 8 = 32 square inches. Multiply the pressure differential (5.2 pounds per square inch) by 32 square inches and you get 166.4 pounds! Obviously, there is no way the thin walls of a waxed cardboard milk carton will support 166 pounds without deforming quite a bit. This deformation evidently cracked the seal loose and let water flood the carton. Clearly, pressure hull design, whether for the deep sea or the backyard pool, is not a trivial matter.

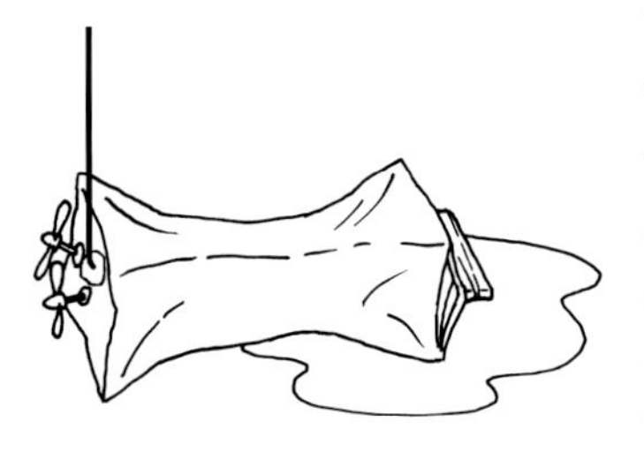

Figure 5.16: Swimming Pool Pressure Differentials

Milk cartons make lousy ROV housings, even for swimming pool depths.

4. Basic Principles of Pressure Hull Design

Subsea designers work with four primary imperatives when designing a pressure-resistant submersible vessel:

1. Minimize the size of all pressure hulls and canisters to reduce surface area.
2. Use pressure-resistant shapes.
3. Use materials that have a high strength-to-weight ratio in order to minimize vehicle mass. (Note: This option is often constrained by the project budget.)
4. Design the structure so that the pressures encountered work with you, rather than against you, whenever possible.

4.1. Size

If it's possible to find a way to reduce the forces acting on the hull, you can generally get by with less expensive hull materials. As discussed above, the net force acting on a submerged surface is equal to the differential pressure multiplied by the area of the surface. So you can lessen these forces by doing either or both of the following:

- Reduce the differential pressures to which the hull is exposed.
- Reduce the surface area of the hull.

The hull or canisters of most vehicles enclose a 1 atm environment, so the differential pressure to focus on is the gauge pressure of the water at the maximum mission depth. This pressure can, of course, be reduced by not diving as deep, but that option may not be consistent with the mission goals.

That leaves surface area as the remaining option. In general, smaller objects have less surface area, so the smaller you can make a pressure hull, the more pressure-resistant it can be. A smaller hull has the added benefit that it requires less total material. This can free up money in the materials budget, sometimes enough to afford a better grade of hull material or to support other project costs.

4.2. Shape

Just as roadway vehicles have wheels and airplanes have wings, underwater vehicles have characteristic shapes related to their function. Most obvious among these are spheres, cylinders, and combinations of these two simple geometric shapes. This is not just a fashion statement; it has to do with the excellent pressure-resisting properties of these shapes.

A sphere is the most pressure-resistant shape of all. Its surface is curved outward everywhere so any attempt to flatten or collapse one section of the sphere causes an outward push of adjacent sections. In water, pressure is uniform over the entire surface of the sphere, so it counteracts any bulge. This makes a sphere self-supporting. As a result, it is highly resistant to collapse, even under extreme pressure.

Conversely, a cube is a lousy shape for resisting pressure, because bowing one face inward does not require another face to bulge outward. Figure 5.17 shows a cross-section of a cube subjected to water pressure; notice how the sides of the first cube can all be forced inward simultaneously. They do not reinforce each other; consequently they buckle. The second cube has internal reinforcing that joins the sides with braces, so an inward push on one face tends to transfer the force and create an outward push on adjacent faces. Thus, the sides now do a better job of supporting each other. Note that by adding reinforcing at the corners, the load-bearing parts of the cube are beginning to approach the profile of a sphere!

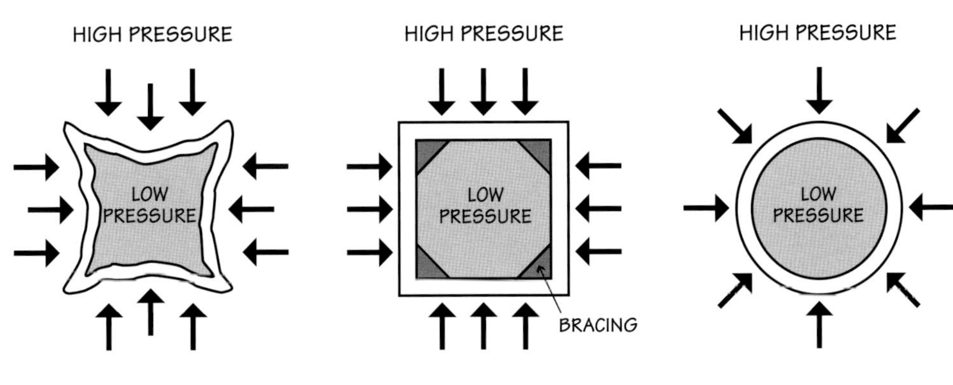

A cube or rectangular box is a poor pressure-resistant shape.

Bracing the inside of a box provides better support and helps to keep it from collapsing when water pressure acts on it.

A sphere is one of the most pressure-resistant shapes.

Figure 5.17: Pressure and Shape

Though not quite so pressure-resistant as a sphere, the roundness of a cylinder makes it largely self-supporting. It is much better at resisting collapse under pressure than a rectangular box. It does have to be made somewhat stronger—that is, thicker and heavier—than a sphere of the same diameter to resist pressure at the same depth. It may also require internal stiffeners, which add even more weight and may get in the way of equipment that has to be positioned inside. Despite these drawbacks, the cylinder shape is widely used, as it is easier to manufacture than a sphere and allows for much more efficient use of internal space than a sphere does. Some underwater vehicles use a configuration that attempts to blend the advantages of both shapes—the strength of the sphere and the efficient internal space of the cylinder—by capping each end with a hemisphere. (See Figure 5.18.)

Figure 5.18: The Benefits of Combining Shapes

A cylinder with two hemispheres gains the advantages of internal space and increased pressure resistance.

TECH NOTE: FINITE ELEMENT ANALYSIS (FEA)

A computer technique known as **finite element analysis**, or FEA, can be used to predict the effects of extreme pressure, extreme heat, or other factors on parts early in the design process. With FEA, potential problems can be identified while those parts are still on the drawing board. This advanced technique was used to design pressure-resistant ceramic/titanium housings for the HROV *Nereus*, which dove nearly 7 miles deep, where external pressures reach 16,500 psi!

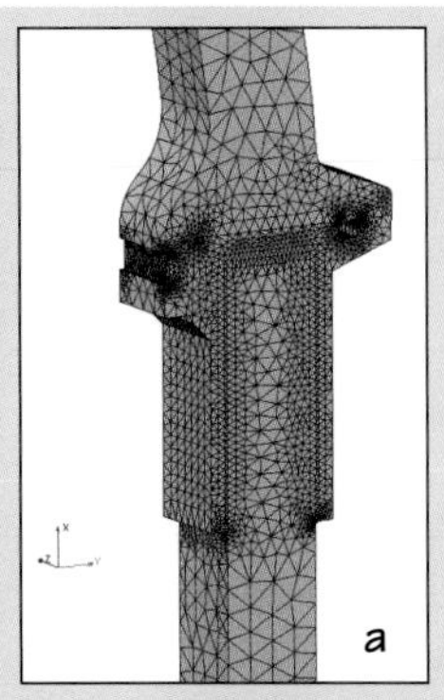

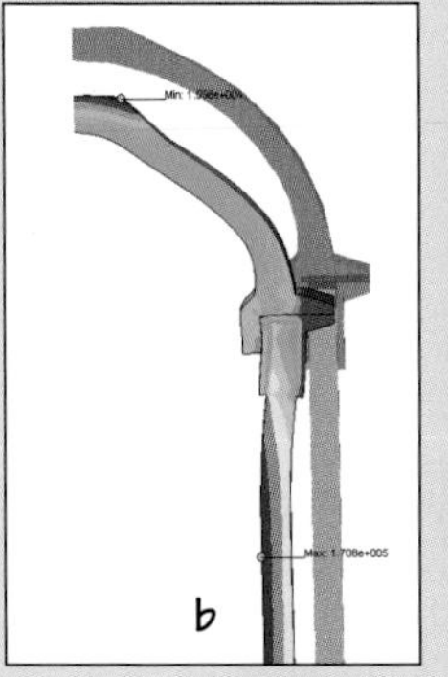

Images courtesy of Glenn McDonald, Woods Hole Oceanographic Institution

Figure 5.19: Finite Element Analysis (FEA)

a) This closeup cross-section shows how a portion of the titanium cap fits over the edge of the ceramic cylinder and illustrates how the FEA software breaks each part of the model into small triangles for analysis.

b) This diagram compares the housing's normal shape at the surface (gray) with the exaggerated deformed shape the FEA program predicted the housing would have at depth (colored). The color-coding indicated where the greatest stress or deformation would occur. If this had exceeded allowable limits, the shapes or materials used in the design could have been modified to resolve the problem before the parts were actually made.

c) The finished product, with white ceramic cylinder and titanium endcaps.

4.3. Materials

In an ideal world, all pressure-resistant containers would be constructed of materials that are extremely lightweight (more precisely, low-density), incredibly strong, totally corrosion-resistant, and maybe even beautiful. Alas, very few materials come close to meeting all these ideals at once, and those that do (such as titanium) tend to be very expensive and difficult to machine. Generally, such specialized materials are used only where the mission requires it and is important enough to justify the extra cost.

As with other aspects of engineering design, wise trade-offs and compromises are the name of the game. Typically, you start with your mission and knowing how deep the vehicle will need to dive. That lets you determine the maximum pressures the vehicle will encounter. Next you look at vehicle size and anticipated pressure-related forces. Then you take a hard look at the budget to determine how much is available to spend on hull materials and fabrication. With luck, you'll be able to identify one or more materials that can be used to make a hull that's able to withstand the pressures at your mission depth and that also fits within your budget. If not, it's time to scale back mission goals—or start fund-raising.

High-strength steel is the traditional material of choice for pressure hulls on larger vehicles, such as submarines. Pressure hulls on smaller deep-diving submersibles tend to be made of aluminum, titanium, or acrylic. Pressure canisters on ROVs and AUVs are made from a wider variety of materials, including different metals and plastics. In these smaller-pressure canisters, the cost of the material is less of an issue than in a large submarine, so there is more emphasis on corrosion-resistance and ease of machining. Steel is rarely used for these smaller craft.

For the very budget-conscious builder of small ROVs or AUVs with modest missions in a swimming pool, lake, or shallow ocean environment (where pressures aren't too high), several inexpensive and easily machined materials that will work can usually be obtained from the local hardware store or by mail order. A time-honored favorite

Images by Todd Walsh, courtesy of Monterey Bay Aquarium Research Institute

***Figure 5.20: Titanium Hull for* Environmental Sample Processor (ESP)**

Monterey Bay Aquarium Research Institute's ESP *is a self-contained robotic laboratory that collects and tests samples of seawater for different types of genetic material. Molecular biologist Chris Scholin and his team have recently adapted* ESP *to work in deep water, a complex challenge that included adding a titanium pressure housing that could handle the immense pressure at mission depth. Engineer Scott Jensen inspected the upper half of* ESP*'s pressure housing prior to assembly. Once he was assured the mating surface was clean and smooth, the team of MBARI engineers aligned the two halves, then tightened the connecting bolts.*

among school groups is PVC water pipe. (See *Chapter 12: SeaMATE.*) This material comes naturally in cylindrical shapes, which are very pressure-resistant. It is highly corrosion-resistant. It is widely available in a range of standard sizes. It can be machined easily with hand or power tools. And it has ready-made fittings that allow you to combine different sections easily. Best of all, it's very inexpensive compared to most alternatives. Other polymers or aluminum extrusions are also good options, as described in *Chapter 4: Structure and Materials.*

4.4. Using Pressure to Advantage

The smart designer tries to use pressure differentials to advantage. For instance, a submersible hatch is designed to open outward rather than inward, so that water pressure provides the force to seal it. (See Figure 5.35.) All undersea pressure-resistant structures are designed with this principle in mind: whenever possible and safe, let fluid pressure do the work of keeping the vessel watertight. For example, submersible hatches are designed so that water pressure forces them closed rather than open. (See Section 6.2.3 later in this chapter.)

4.5. Choosing Canister Size and Single or Multiple Cans

Of course, the *size* of the pressure canister is determined by what you have to put inside it. And the *number* of cans depends on whether you're going to try and put all the electronics and batteries in one large can or use multiple smaller cans—for example, one for the electronics, one or two for batteries, possibly a smaller housing for an externally mounted camera, and a couple of cans for lights.

The question of whether to use a larger, single pressure canister or several smaller ones is one you'll want to look at from several angles before making a decision. A single can has these advantages:

- A single can is able to contain all the electronics and batteries, thus minimizing the number of external electrical connectors needed.
- A single can simplifies construction, generally requiring less complex framework and brackets.
- A single can is often more cost-effective.
- A single can may have less drag than multiple cans, thereby providing better energy efficiency.
- A single can may (or may not) be easier to maintain.

Multiple cans have these advantages:

- Multiple pressure canisters may offer more flexibility in terms of the placement of components. For example, batteries could be housed in a different can than the electronics, and both canisters could be electrically connected using electrical connector cables (also called whips).
- Multiple cans present more options for distributing weight to various locations on the frame; this may help with trim, not blocking thrusters, etc.
- If a problem or leak develops in one canister, it generally does not impact components in the other canisters.
- For vehicles with on-board power (batteries), that particular canister can be designed for easy access and change out, with an extra power pack ready to go.

Regardless of the size or number of canisters you opt to use, each will require mounting brackets. And the endcap(s) may have one or several penetrators to accommodate electrical wiring, pressure relief valves, fill ports, etc.

Choosing the number and size of canisters depends on the dimensions and weight of what's going inside. There's seldom a perfect choice; usually it's a case of trade-offs—you get this advantage but lose that one. Just make a selection, keeping in mind your vehicle's mission statement, power requirements, and your research on the component size, weight, capacity, and power specs of various subsystems—sensors, camera(s), lights, etc. Of course, your decision directly impacts frame shape, materials, bracketing, and thruster location/orientation, so there may be some adjustments that pop up. But as you close in on these choices, it means you can begin to work up a weight statement table in order to estimate your vehicle's total effective weight. (Chapter 6, Section 3.1 details making a weight statement table.) Then you'll be able to calculate the amount of flotation and ballast required. That's how you start moving along in the design spiral!

Figure 5.21: Ensuring Battery Fit

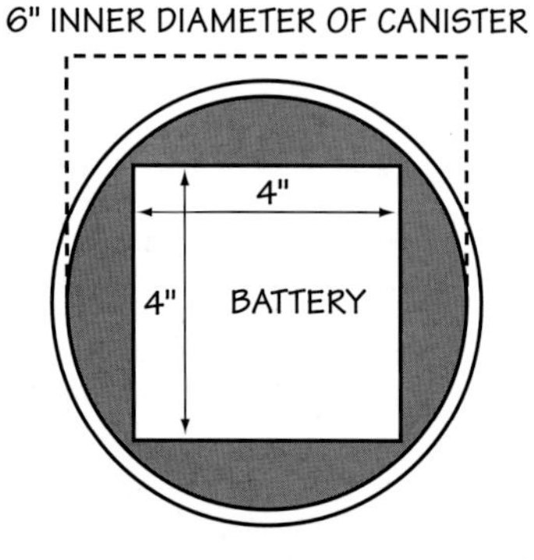

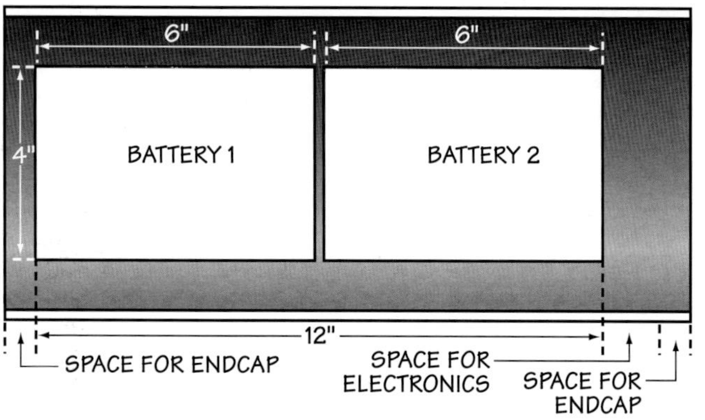

Greater mission depths will require single or multiple canisters to be outfitted with a pressure relief valve to equalize any pressure buildup inside the can. Undue pressure could either prevent you from opening it (due to low inside pressure) or cause a forceful ejection of the endcap when opened (due to high internal pressure). In addition to the pressure relief valve, installing a fill port will allow you to vacuum-test and nitrogen-purge the can. An oil-compensated canister will also need a fill port to pour in oil and vent any gases that batteries might produce. (See *Safety Note: Seal Screws and Pressure Relief Valves* and *Tech Note: Pressure Can Vacuum Test* in this chapter for an explanation of pressure can safety features.)

All these decisions are interconnected. For example, if you ever opt to design a small AUV or hybrid ROV that carries its power source (likely batteries) on board, the decision about the optimum number and size of canisters probably hinges on the type, number, and requirements of the thrusters you're considering. These thruster specs will allow you to draw up a power budget (also called a power usage table), as explained in *Chapter 8: Power Systems*, and this power budget then helps you decide which batteries can meet your specifications.

Next, you figure out the dimension and configuration of your battery pack. (See Figure 5.21.) This data will help you determine which size canister(s) will work best. To check the fit, you'll need to make a scaled sketch of the internal diameter of the battery canister of choice. (Use a CAD program or even the drawing function of a word processor/spreadsheet program.) Then, using manufacturer's dimensions, draw the chosen batteries inside this circle to see if they will fit in that space and if there will still be room for the other components. If so, you've figured out one canister size! Fortunately, if the standard battery packs don't fit your canister dimensions, it's often possible to custom order battery packs in non-standard configurations that will work.

4.6. Pressure Canister Options

Here are some types of pressure cans that might prove useful for small ROV fabrication, particularly those for deeper-diving missions.

4.6.1. Cylindrical Pressure Cans

These tubular-shaped pressure canisters are designed to withstand external water pressure while preserving a constant, lower internal air pressure. They can be made from aluminum, steel, titanium, ceramics, or plastics, and they come in various sizes and pressure ratings. For a deeper-diving vehicle, you'll want to carefully estimate, test, or secure accurate technical design and manufacturing data regarding the cans you select. Manufacturers of custom-made cans should be able to provide this information. However, the cost for a custom pressure housing can be quite high. If you include an optically polished dome that can handle greater depths, expect a hefty add-on fee.

Figure 5.22: Titanium Housing for MOBB _Seismometer_

Technicians at the Monterey Bay Aquarium Research Institute install the Monterey Ocean-Bottom Broadband (MOBB) *seismometer in a cylindrical titanium pressure housing before deploying it at a depth of about 900 meters off Monterey Bay.*

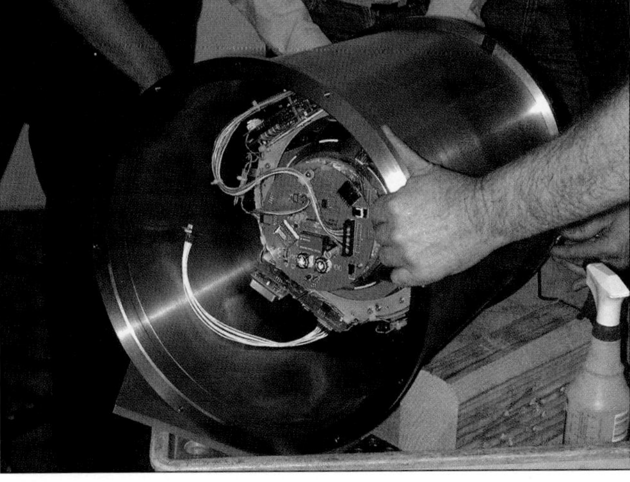

A cheaper alternative would be to use standard off-the-shelf plastic canisters from manufacturers like PREVCO, which offers stock sizes rated to 100 meters; they're very affordable and have flat, optically clear, acrylic endcaps at both ends. These commercially available containers come in various sizes with optional types of endcap closure methods: through bolted (TB), stress rod, (SR), clamp block (CB), threaded tube (TD), threaded endcap (TE), and threaded collar (TC). If you choose this option, remember to order a pressure relief valve and fill port. Table 5.2 provides sample data for a PREVCO threaded-collar can.

Table 5.2: Threaded Collar (TC) Closure Data Table

MODEL NUMBER	Depth Rating (meters)	Material	Inside Packaging Diameter (1) (inches)	Inside Packaging Length (inches)	Outside Diameter (inches)	Outside Length (inches)	Approx. Weight in Air (lb)	Approx. Weight in Water (lb)
P2013-5-TC	130	Plastic	2.05	5.00	3.70	7.85	2.5	0.0
P3.5011-7.9-TC	110	Plastic	3.50	7.90	5.05	10.70	4.0	-3.0
P4.501SS-8.6-TC	100	Plastic	4.50	8.60	6.25	12.10	6.4	-5.0
P6.501SS-7.7-TC	100	Plastic	6.50	7.70	8.75	12.40	11.5	-12.0

1. Packaging dimensions are the maximum size for internal hardware.
Source: PREVCO Subsea Housings

Note that commercially available canisters can be limiting in terms of space and internal arrangement options, especially if you prefer using square, non-spillable lead-acid batteries. So another option to explore is using made-to-order canisters that are supplied in factory-specified sizes. You could order a 15-centimeter diameter canister, which the factory can supply in varying lengths and endcap styles. This might provide more options for internal configuration of batteries and electronics.

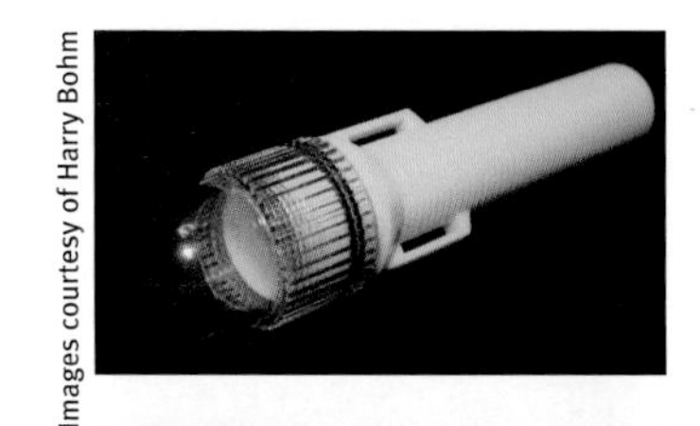

Images courtesy of Harry Bohm

Figure 5.23: Economical Dive Lights

You can modify some dive lights rated to 100 meters or more for use as pressure cans on an ROV.

4.6.2. Dive Lights

Some housings for dive lights are rated to 100 meters or more, and it is possible to modify these for use as pressure cans on an ROV. For example, you might use a series of dive lights to house various subsystems—two for the batteries, one for the control electronics, one for the camera, and two for flotation. Of course, multiple cans require more electrical connectors. Compare the cost of these flashlight housings to the cost of off-the-shelf pressure cans to see if there is any savings in using them.

4.6.3. Oil- or Gas-Compensated Housings

As noted in Sections 7.1 and 7.2 below, it is possible to use oil or gas to adjust the internal pressure of a pressure can so that it is the same as or slightly greater than the external water pressure at various depths. However, this technique adds a certain level of technical complexity. Check to see if any oil-compensated or gas-compensated housings are commercially available for smaller underwater robots. If not, you can create an oil-compensated housing yourself pretty easily. Or you can tackle the complexity of a gas-compensated system.

4.6.4. Glass Spheres

One of the most interesting but quiet advances in underwater technology involves the use of a very traditional material—glass. A glass sphere is strong and gets stronger as it is compressed, which is good news for deep-diving vehicles. The underwater technology company Teledyne Benthos has patented VacuSealed® glass spheres (in various sizes and with optional protective plastic shells) for deep water flotation and for pressure canisters that can be used to house cameras, navigational systems, or other electronics. The positive buoyancy factor of glass spheres is another helpful factor. You may want to consider glass spheres as potential pressure cans for a deeper-diving or hybrid ROV. However, this tempting, if unusual, approach often complicates the structural design more so than by using a conventional cylindrical can shape. Also, it may be more difficult to take the sphere apart to service batteries or other components. As well, a glass sphere is more fragile and far heavier than a plastic cylindrical can rated for 100 meters.

TECH NOTE: NEW USES FOR GLASS

Traditionally, glass spheres have been used for flotation, as detailed in *Chapter 4: Structure and Materials*, but in the 1990s, Dr. James Bellingham of MIT developed the AUV *Odyssey* using two 30-centimeter (approx. 12-inch) diameter glass spheres as pressure cans for the vehicle's electronics and batteries. The glass spheres were then covered with a molded plastic external fairing. Undersea bulkhead connectors seated on the top of the spheres allowed cabling to connect to external components, such as the propulsion motor, dive plane actuators, and sensors.

Kevin Hardy is another glass innovator. While a senior designer at the Scripps Institution of Oceanography, he devised a low-cost digital camera probe housed in a single glass sphere to photograph deep-sea fish at a depth of 7 kilometers (approx. 4.5 mi).

Image courtesy of James Bellingham

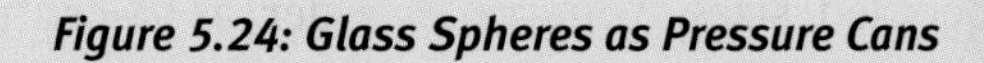

Figure 5.24: Glass Spheres as Pressure Cans

The AUV Odyssey *has voyaged around the world. The technology was licensed by Bluefin Robotics.*

5. Calculating Pressure-Related Forces on Spheres and Cylinders

Image courtesy of Vickie Jensen

Figure 5.25: Pressure on a Pilot's Viewport Dome

The pilot's viewport in this tourist sub operating to a depth of 100 feet must be able to withstand over 50,000 pounds of force pressing in on it!

Most of the sample pressure calculations in this chapter have used circles, cubes, or other flat shapes. That's because these shapes are easy to think about while learning the basics of pressures and pressure-related forces. However, you will probably use spheres and cylinders, rather than cubes or other flat-sided shapes, as the usual pressure-resisting shapes for your underwater vehicle hulls. This section of the chapter shows how easy it is to estimate the pressure-related forces acting on submerged spheres and cylinders, even though they have curved surfaces.

One interesting property of pressure is that the net force it exerts on a surface is always perpendicular to the surface at every point. In the case of a curved surface, this means that the direction of the force changes as you move to different locations on the surface, as shown by the sphere in Figure 5.17 that compares the pressure resistance of a cube to that of a sphere.

You might reasonably expect that this makes the calculation of overall net force quite complicated. In one sense it does, since you need calculus to work through the addition of all those little forces acting in all those different directions. However, if you do the calculus, a surprisingly simple result emerges. The overall force acting on a half-sphere has exactly the same magnitude and direction as the net force acting on a flat circular area of the same diameter.

Look at this example: Calculate the approximate force (in pounds) pressing in on a hemispherical acrylic dome used as a viewport on a tourist submarine, if that dome has a diameter of 18 inches and is at a depth of 100 feet.

To do so, all you need to do is calculate the surface area of a circle with the same diameter (9-inch radius) and multiply that area by the (gauge) water pressure 100 feet down.

The area of a circle is given by πr^2, so we have: $3.14 \times (9)^2 = 3.14 \times 81 = 254$ square inches.

One hundred feet is about 99 feet, which corresponds to 3 atm (gauge), according to the memorized relationship between depth and pressure in saltwater. One atm is 14.7 psi, so the total pressure pushing in on the dome is: 3 atm x 14.7 psi/atm x 254 square inches = about 11,200 pounds!

For the well-being of the tourists on board, it had better be a thick, strong dome.

A similar relationship holds for cylinders. You can mentally split a cylinder in half lengthwise and calculate the net pressure on that half, as if the cylinder were actually a rectangle having the length of the cylinder and a width equal to the diameter of the cylinder.

6. Constructing Leak-Proof Openings

One of the greatest challenges in pressure hull and canister design is how to keep water out while still allowing wires, equipment, or even rotating propeller shafts to go in or out. In other words, how can you waterproof a hull or canister that has to be full of holes? And once it's waterproofed and pressure-proofed, how can you open it and reseal it again and again, to change out parts, batteries, or crew members (in human-occupied vehicles)?

This section discusses materials, devices, and techniques for doing just that. Some terms, such as hatches or viewports, generally apply to larger hulls. Others, such as endcaps, generally refer to the closures on smaller canisters. Typically, a canister is an open cylinder, fabricated from a pipe, with one end sealed with a removable **endcap**. The other end can be capped with either a welded plate or another removable endcap. Regardless of their size, these openings must all deal with the challenges of water and pressure—and they use similar techniques for doing so.

6.1. O-Rings

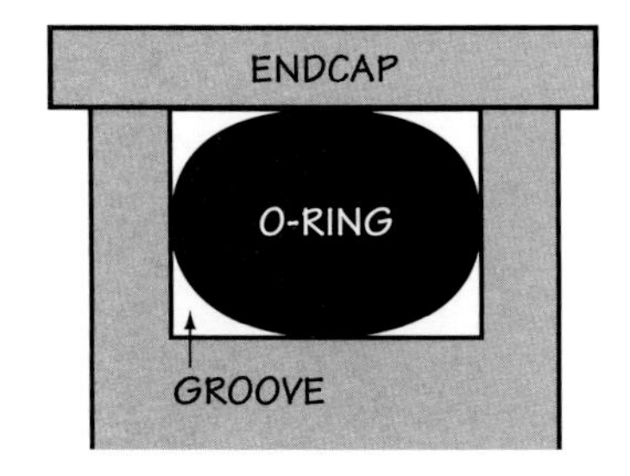

Figure 5.26: O-Ring in Position

This cutaway view shows a black O-ring properly sandwiched in an endcap groove.

An **O-ring** is a doughnut-shaped piece of rubber or other elastic material that is usually placed in a groove and sandwiched between smooth surfaces to create a watertight seal. (See Figure 5.26.) If used properly, O-rings can seal tightly against very high pressures. As such, they are a common and critical component of hatches on pressure hulls, lids on pressure canisters, and many other parts of undersea vehicles. In fact, they account for over 90 percent of all watertight seals used in underwater technology. Not surprisingly, you'll hear them referred to again and again in design and construction discussions of how to keep water out of underwater craft. No doubt you'll end up using O-rings in your own underwater vehicle designs.

O-rings function in one of two capacities. Most often they are used as a **static seal** in which the O-ring and the sealing surfaces do not move in relation to each other during normal operation. A typical example would be an endcap seal (see Figure 5.27). They can also be used as a **dynamic seal** in which the O-ring forms a watertight seal between moving parts, such as a rotary shaft or piston. However, other types of seals, such as cup seals, are usually a better choice for moving parts. Nonetheless, O-rings are a good place to start learning how to seal openings when large pressure differences are involved.

O-rings are commonly made from neoprene, but other materials such as Viton, silicone rubber, Teflon, and Buna-N are also used. The hardness of the ring material is important: static seals require a softer material; dynamic seals need a harder material. Since O-rings need to move slightly during pressure changes, they should be lubricated with a very thin layer of silicone grease.

As a simple device for keeping water out, the O-ring has many advantages:

- It's inexpensive.
- It's removable and replaceable.
- It fits conveniently into a machined groove or "seat."
- It seals by pressure from the sealing surface, so is squished into the groove, filling it completely and effectively. As the pressure differential increases, so does the effectiveness of the O-ring seal.

However, to work effectively, the O-ring groove must be the correct dimension—too big and the O-ring will not compress enough to seal; too small and the O-ring will deform and extrude. A design rule of thumb is that the O-ring should be squeezed to about 90 percent of its original cross-sectional diameter. Thus, the O-ring groove must allow some extra space for the ring to bulge into. If the groove is exactly the same size as the O-ring, then the ring may extrude and fail.

Image courtesy of Harry Bohm

Figure 5.27: Black O-Rings Positioned in Endcap Grooves

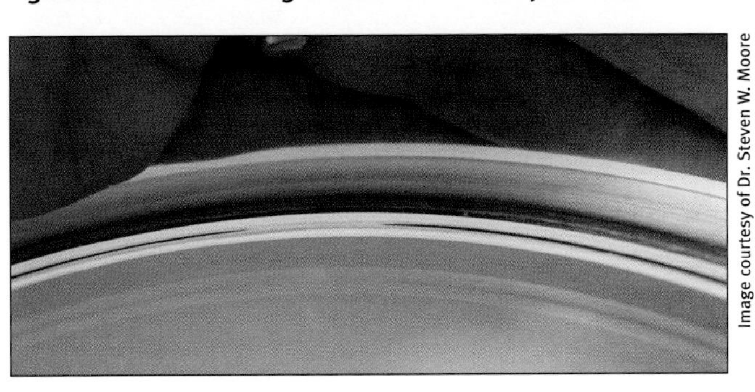

Image courtesy of Dr. Steven W. Moore

Figure 5.28: Poor O-Ring Seal

One advantage to using acrylic or other clear materials for pressure housings is that clear materials allow you to see gaps in O-ring seals that might otherwise be difficult to detect.

The visible gap in the thin black line reveals a place where the black O-ring is not pressed firmly against the acrylic end-cap of this PVC housing. This discovery allowed the designer to modify the design and create a reliable seal before submerging any electronics.

To keep things perfectly watertight, the sealing surfaces must be machined to within 1/1000th of an inch tolerance (approx. 0.025 mm). The surfaces must be smooth, with no irregularities, and sharp corners in the groove must be "broken" (a machinist's term for rounding the sharp edges). In dynamic applications, where the O-ring is sealing a moving piston or rotating shaft, the surfaces are often polished until they are extremely smooth, to minimize wear on the O-rings. In some dynamic sealing jobs, O-rings are not the best option. Other types of seals, such as rubber or Teflon cup seals (also called U-cup seals), may be more appropriate.

U-cup seals are another mechanism for helping keep water out. They're used to seal the motor shaft in trolling motors and in the bilge pump thrusters for the *SeaMATE* ROV. (See *Chapter 12 Tech Note: Bilge Pump Seals.*) U-cup seals are fine for shallower-diving vehicles, but O-rings are the better choice for deeper-diving vehicles.

TECH NOTE: HANDLING O-RINGS

O-rings are basic equipment, but they require special care to be effective.

Removing an O-ring from a Piston Seal Endcap: Push on the sides, then up with your fingers. This deforms the circular shape and lifts the O-ring out of the groove. You can then grab the extruded section and slip it off the endcap. To remove a very small O-ring, you may need to use a bamboo skewer or a blunted dental tool. Just be careful not to damage the O-ring or the smooth groove surface while removing it.

Figure 5.29: Techniques for Removing O-Rings

Images courtesy of Harry Bohm

Cleaning and Inspecting for Flaws: A dirty O-ring will fail, so be sure to regularly clean each one. Do so with 70 percent isopropyl alcohol and lint-free tissue. Then run your fingers around the ring—it's easier to feel flaws and nicks than it is to see them. If there's any doubt about the integrity of an O-ring, discard it.

Cleaning the Groove: After cleaning the ring itself, clean the groove with 70 percent isopropyl alcohol and a lint-free tissue or cotton swab. Check for nicks and grit. Lubricate the groove with silicone oil or grease. Note: Do not use any petroleum-based lubricants on an O-ring. Dow Corning #200 silicone oil is recommended for use in pressure can seals.

Replacing the Seal: Before mating an O-ring with the sealing surface, make sure the surface is clean and lubricated with a thin coat of silicone oil or grease. Then stretch the O-ring over the endcap and place it in the groove. Roll the ring with your fingers to get any twists out of it and to seat it firmly. When setting the sealing surface against the O-ring, do so evenly and with gentle pressure. Never force an O-ring seal to fit. If the sealing surface does not mate up easily, then stop, remove the endcap, and determine the cause of the problem.

6.2. Pressure Hull Penetrators

As mentioned at the beginning of this section, pressure hulls and canisters need openings so that things like wires and cables can get in and out of them. These openings are blocked with viewports, hatches, and penetrators, depending on their function and size. Naturally, installing any of these in a watertight structure involves some challenges.

First of all, making a hole in a pressure can or hull automatically weakens the structure. So, depending on the material and the size of the hole, the area must be reinforced in some way to compensate for this weakness. For pressure hulls, the opening is commonly reinforced with a thickened section, called a **boss**. (See Section 6.2.2.) In the case of smaller pressure canisters, the endcap is usually made thicker than that required for a canister without penetrations.

Secondly, most openings must be easy to open and close, yet maintain a consistent and reliable watertight seal when closed.

6.2.1. Bulkhead Penetrators

A bulkhead is a nautical term for the partitions or walls separating compartments in a ship, but even the wall of a pressure hull or a canister endcap can be considered a bulkhead. A **bulkhead penetrator** (often simply called a **penetrator**) refers to a special type of pipe, tubing, or electrical fitting that penetrates the wall of the pressure housing or hull. It prevents pressurized water from coming into the hull, while allowing fluids, gases, or electrical cables to be brought from outside the hull into the interior or vice versa. (See Figure 5.32.)

Because bulkhead penetrators must be removed for maintenance, they should not be welded in place to the hull or endcap, and epoxy should not be used to seal threads or

TECH NOTE: ELECTRICAL CONNECTORS

Electrical **connectors** are a critical part of the circuitry that delivers (and controls) electrical power going to a vehicle's thrusters, lights, and other systems. Such connectors come in a genuinely overwhelming number of styles, each with a variety of common and not-so-common names plus a bewildering array of configuration options. A recent search for the word "connector" on the website of just one popular electronics distributor returned over 360,000 items!

Most connectors, like the familiar plugs that connect household appliances to wall outlets, the little jacks that you plug in to recharge digital cameras and cell phones, or the ethernet cables used to connect computers in a wired network, are designed for use in dry conditions only. Although you can sometimes get away with using such connectors for a short time in freshwater, they will soon corrode, and in saltwater they will short out immediately upon contact with the water. This presents a serious challenge for those building underwater systems. A convenient, though sometimes expensive, solution is to find and use electrical connectors manufactured specifically for underwater use.

Professional designers of underwater systems take great care in selecting the appropriate subsea connectors to match the power and control system they are developing. For example, they must contain the right number of wires, and those wires must be capable of handling the anticipated voltage and current safely. Even when you limit yourself to underwater connectors, the variety available can be bit of a minefield for the amateur, particularly since the selection of connectors from various companies is not necessarily standardized. As a small underwater vehicle designer, you should familiarize yourself with the basic elements of an underwater connector and some of the common options. A good way to do this is to visit the websites of several manufacturers of subsea connectors and read the descriptions, purposes, advantages, and disadvantages of the different types of connectors they offer.

Subsea connectors are broadly referred to by series. For example SubConn Inc. classifies their connector products as: Standard Circular Series, Micro Series, Low-Profile Series, Metal Shell Series, Power Series, Special Series, Penetrator Series, and so on. Teledyne Impulse uses Wet Pluggable, Miniature High Density, Rubber Molded etc. Each series is optimized for a different purpose. These designations will differ from company to company but the styles of connectors are often similar. In fact, a series from one company may be (but isn't always) interchangeable with a similar connector series from another firm.

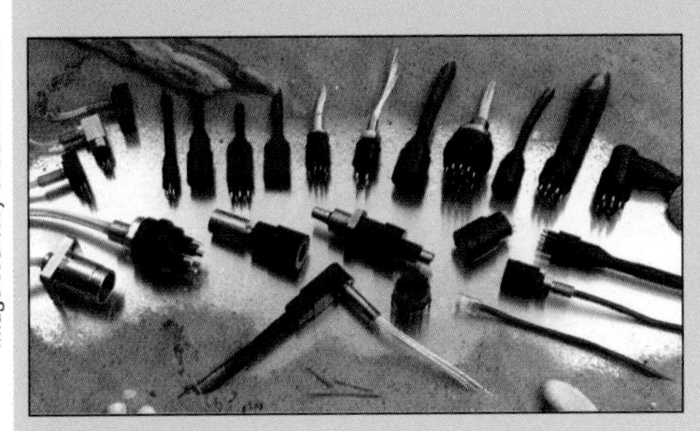

It's important to recognize that most underwater connectors are not "wet mateable." In other words, even though a connection made in the air will remain watertight after being submerged, you can't plug and unplug them under water. All connection and disconnection must be done while the connectors are completely dry.

Figure 5.30: Various Electrical Connectors

This photo shows the wide range of connector and penetrator styles available for underwater vehicles.

Figure 5.31: Electrical and Hydraulic Penetrators

Note the variety of penetrators used on a typical work class ROV such as the Millenium Plus model.

Image courtesy of Oceaneering International and Drew Michel

Figure 5.32: Bulkhead Penetrator

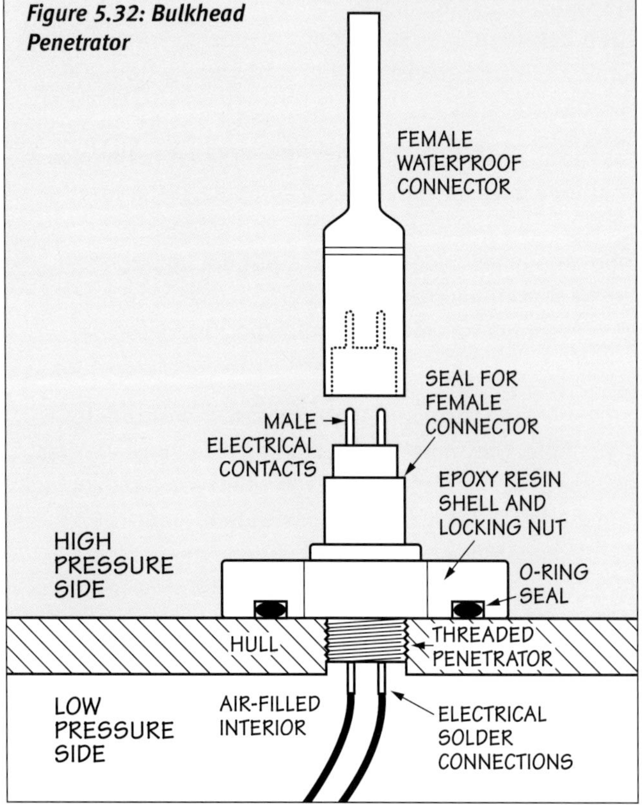

FEMALE WATERPROOF CONNECTOR

SEAL FOR FEMALE CONNECTOR

MALE ELECTRICAL CONTACTS

EPOXY RESIN SHELL AND LOCKING NUT

HIGH PRESSURE SIDE

O-RING SEAL

HULL

THREADED PENETRATOR

LOW PRESSURE SIDE

AIR-FILLED INTERIOR

ELECTRICAL SOLDER CONNECTIONS

sealing surfaces. In a manned vessel, all bulkhead penetrators through which liquids and gases are brought into the interior must have a valve, called a sea cock, on the low-pressure side of the penetration in order to shut down a potential leak into the pressure hull. In unmanned systems, this valve is not necessary.

Whether the bulkhead fitting is used for liquids, air, or cables, the basic external, or shell, design is similar. However, internal construction may vary, particularly for the electrical bulkhead penetrator. This is because it has to make a conductive connection from the cable outside the bulkhead to a cable inside.

6.2.2. Penetrator Boss

This is a circular piece of thick plate welded around the hull opening, to structurally reinforce the area around the hole and to strengthen the hull. The boss face and inside hole are machined so that the face acts as a sealing surface for an O-ring on the penetrator. For larger viewports and hatches, this boss is called a **flange**.

Image courtesy of Harry Bohm

Figure 5.33: Electrical Bulkhead Penetrator

This typical electrical bulkhead penetrator is being fitted into a piston seal type endcap.

6.2.3. Hatches

People and gear pass in and out of pressure hulls through openings called **hatches**. A hatch is mounted on a machined flange and swings on a spring-loaded hinge. An O-ring makes the seal. Contrary to popular belief, the inner hatch locking mechanisms, called "dogs," do nothing to keep water from coming in. Technically, you wouldn't have to tighten a hatch, because the water pressure presses in against it upon descent, sealing it ever tighter. Dogs are there to prevent accidental opening at the surface, as happened when the submersible *Pisces III* sank off the coast of Ireland in 1973. (See *Stories from Real Life* in Chapter 3.) Dog latches are also necessary if for some reason the internal pressure is higher than the surface air pressure. Without the dogs, the hatch would pop open as soon as the vehicle surfaced, and the craft could be down-flooded by waves.

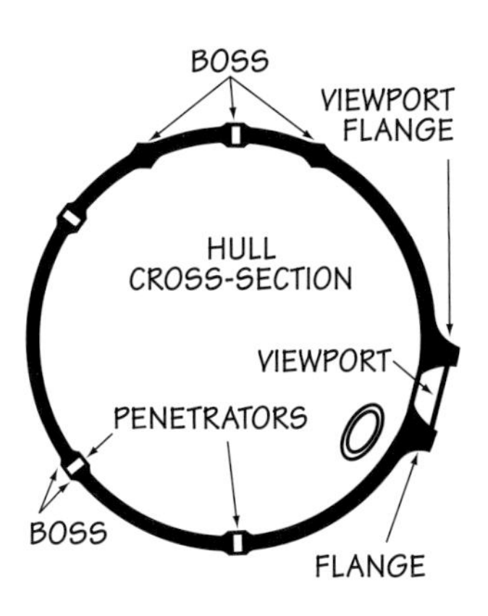

Figure 5.34: Typical Submersible Pressure Hull

This simplified diagram of a typical submersible pressure hull shows the flanges and bosses that reinforce the penetrations.

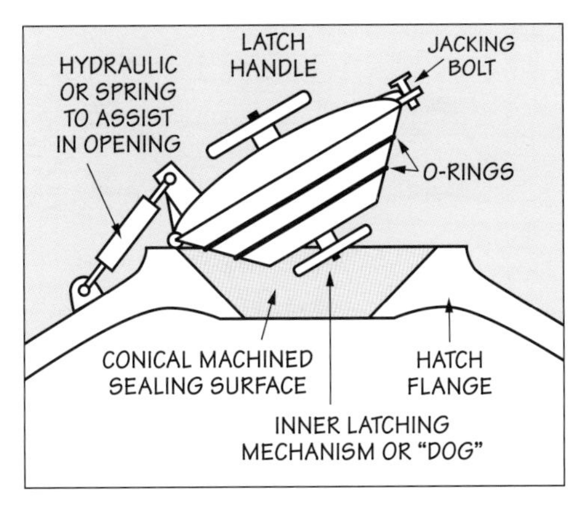

Figure 5.35: The Plug Hatch

There are two basic types of hatches: the **seat hatch** and the **plug hatch**. The first simply relies on water pressure to effectively seat the hatch O-ring—hence the name seat hatch. A plug hatch has an emergency jacking bolt so that outside crew can pry open the hatch, breaking the seal in case of lower pressure inside the sub. Such a procedure would be necessary only if the crew were incapacitated and could not open the hatch from inside.

6.2.4. Viewports

A viewport is basically a window inside a submersible hull or a clear protective area in front of a deep water camera. Adding a viewport results in two problems: 1) cutting any hole in a pressure hull weakens it, and 2) using different materials for the viewport and for the hull creates various issues that the designer must deal with. For example, different materials expand or contract at different rates, potentially breaking seals. Fortunately, designers have come up with several unique shapes for effective underwater windows. (See *Tech Note: Types of Viewports.*)

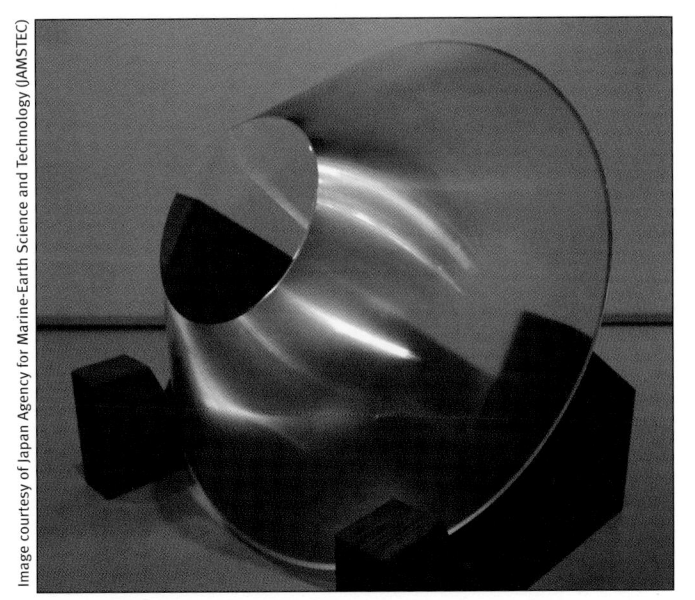

Image courtesy of Japan Agency for Marine-Earth Science and Technology (JAMSTEC)

***Figure 5.36: Viewport for* Shinkai 6500**

The Shinkai 6500 *submersible has three viewports fabricated from highly transparent methacrylate resin. The total thickness of 14 centimeters and specific conical shape allow each viewport to withstand a pressure of around 680 atmospheres at a depth of 6,500 meters.*

TECH NOTE: TYPES OF VIEWPORTS

There are three basic shapes for viewports: flat, conical, and spherical (domed). Of course, each has its advantages and disadvantages.

Flat Ports: These flat windows are easy to fabricate and install. They offer the least strength, so are used in shallow-water applications. Like windows on land, they are often made of glass, but are small in diameter and can be quite thick.

Conical Ports: These cone-shaped ports are widely used down to full ocean depth. The window is seated in the conical flange with an O-ring seal, an arrangement that allows more than 60 percent of the viewport to be supported against pressure by the flange, affording a large viewing area and minimizing the area subjected to direct sea pressure. Typically, sizes are 15–30 centimeters (approx. 6–12 in) in diameter on the outside face and 7–10 centimeters (approx. 3–4 in) on the inside face. August Piccard used conical viewports when designing *Trieste*.

Spherical (Dome) Ports: This type of viewport has the largest viewing area and best resistance to water pressure for its area. These ports can be as small as 5 centimeters (approx. 2 in) in diameter for underwater camera housings and as large as the huge 1.8-meter (approx. 6-ft) diameter domes in shallow-water tourist submarines. For submersibles diving to 330 meters (approx. 1,000 ft), the spherical ports are normally 60–90 centimeters (approx. 2–3 ft) in diameter.

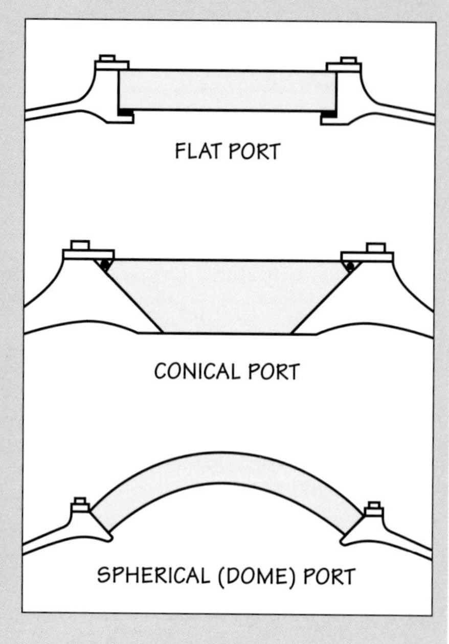

Figure 5.37: Types of Viewports

6.3. Pressure Can Access

Pressure canisters are usually designed to carry electrical and electronic components. Because these cans are smaller than pressure hulls—as small as 2.5 centimeters (approx. 1 in) in diameter and 5 centimeters (approx. 2 in) long—they can be made from lighter materials, such as PVC, Delrin, GRP, titanium, stainless steel, and aluminum. Their smaller size also makes them easy to fabricate in a small machine shop. Admittedly, there is no designated maximum size for a canister, so the distinction between canister and pressure hull can get somewhat blurry.

The dimensions of the pipe are determined by the components inside, and the thickness of the pipe wall is determined by the depth at which the can will operate. The endcap has to be equally robust, but it must also be easy to remove and reinstall for canister access and servicing.

The three most common styles of pressure canisters, categorized by their method of sealing, are as follows:

6.3.1. Piston Seal

In a **piston seal** can, part of the endcap fits down inside the end of the cylindrical can, like a piston in a cylinder. An O-ring sandwiched between the "piston" and the inside of the can creates a waterproof seal. Piston seals work well because external water pressure compresses the diameter of the can slightly, squeezing the O-ring. As with other O-ring seals, the internal and piston seal

END CAP WITH SEAL
CABLE PENETRATOR
END CAP WITH PENETRATOR
SCREW SEAL
PRESSURE CAN
MACHINED INTERNAL SURFACE FOR O-RING SEAL FACE
O-RING
MACHINED SURFACE FOR FACE SEALING WITH PENETRATOR O-RING

Figure 5.38. The Piston Seal Can

surfaces must be machined to 1/1000th of an inch tolerance and must be smooth and free of sharp edges. Most O-ring failures occur because they are nicked on insertion and/or because dirt or hair gets caught in the O-ring groove.

6.3.2. Flange/Face

A **flange** or **face seal** can is relatively easy to fabricate without a lathe or other expensive tools and therefore makes an ideal choice for small-scale, low-budget projects. In this type of can, a flat flange is glued or welded to the end of the can. An O-ring or flat rubber gasket is then sandwiched between this flange and a flat plate serving as the endcap. Retaining bolts clamp the end plate to the flange, squeezing the O-ring or gasket. The retaining bolts and water pressure all function to squeeze the O-ring against the flange-sealing surface. Because the seal surface is easy to maintain and remove, it is the workhorse can for many ROVs. These cans can be 1 atm inside, but are also easily adapted for oil-filled or air-compensated use. (See Section 7.)

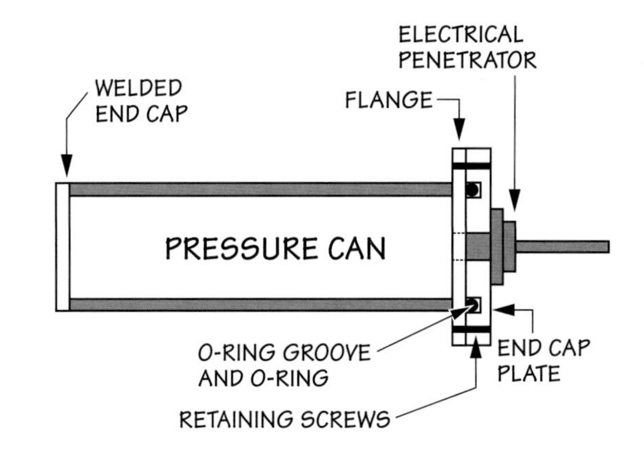

Figure 5.39. Flange or Face Seal Can

6.3.3. Jam Jar Variation

"Jam jar" refers to common glass jars used for home-canning, where such a jar is sealed by means of a gasket and a lid that screws down onto the mouth of the jar. In the underwater version, a threaded endcap is screwed down onto an O-ring located in a shallow hollow at the base of the threads on the can. The endcap has a groove that fits snugly around the O-ring when tightened. This effective shallow-water seal is commonly found on waterproof flashlights.

Upon surfacing, any internal pressure is relieved by slowly turning the endcap, so there is no need for a seal screw. (See *Safety Note: Seal Screws and Pressure Relief Valves.*) However, if the internal pressure increases enough, it can exert sufficient force so that the endcap seizes up, making it very difficult to unscrew. That's why an endcap should just be hand-tightened—never forced tight. See Figure 5.40 for a deep water variation that's a cross between a jam jar and piston seal can.

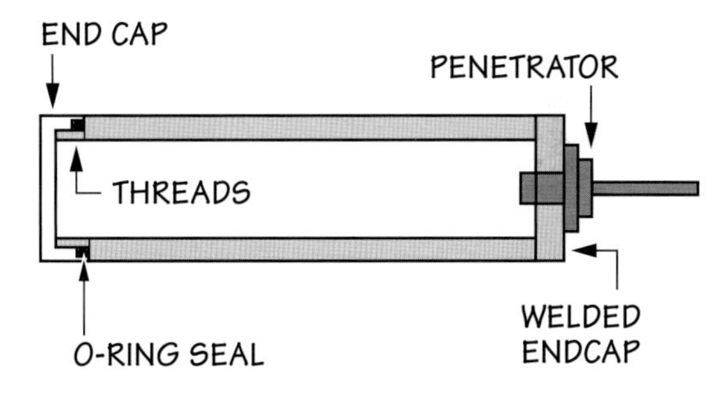

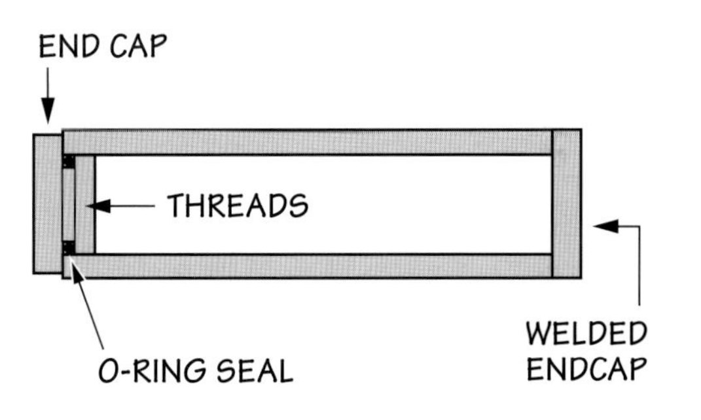

Figure 5.40: Jam Jar Variation

There are several different ways to create jam jar housings. Two popular variations are shown here. The upper one has a lid and O-ring arrangement common to underwater flashlights and other shallow use. The lower variation is better at resisting extremely high external pressures.

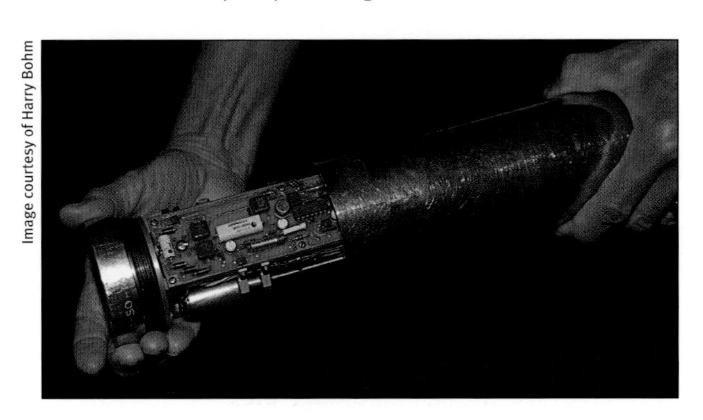

Image courtesy of Harry Bohm

Figure 5.41: Camera with Jam Jar Endcap

SAFETY NOTE: SEAL SCREWS AND PRESSURE RELIEF VALVES

Sometimes pressure inside a pressure housing builds up to a high level, creating a dangerous situation in which the endcap can blast out and strike someone when the canister is opened. This pressure buildup can happen when a lot of water has leaked into a can, compressing the air inside into a relatively small fraction of its original volume. It can also happen when batteries inside give off hydrogen gas, a situation sometimes referred to as battery off-gassing. Even the heat from the electronic instruments inside can warm up the air sufficiently to make it expand and pressurize. Because it's not possible to tell whether a can is pressurized merely by looking at the outside, it's a good idea to utilize a seal screw or a pressure relief valve.

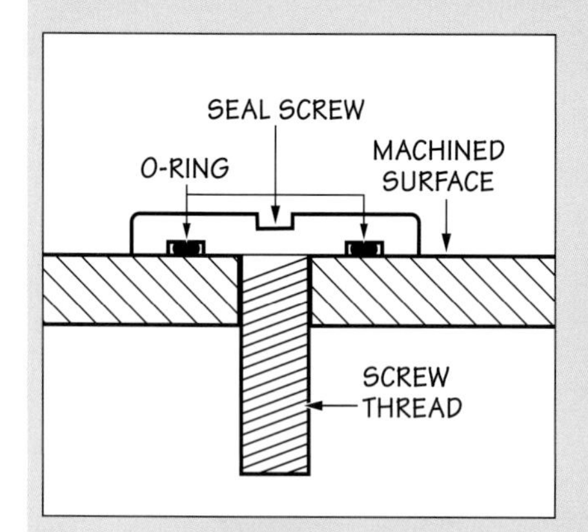

Figure 5.42: Seal Screw Cross Section

A **seal screw** is basically a simple machine screw that has been fitted with an O-ring. It's used to plug a tiny hole that has been drilled and threaded through the endcap or housing. The head of the screw seats against an O-ring or a gasket when the screw is tightened, sealing the hole to prevent water entry. After a vehicle is recovered from the water, the seal screws are loosened, allowing any internal pressure to equalize with the outside air in a slow, controlled manner. After equalization, the endcaps can be removed safely.

To prevent mishaps, follow these suggestions when using seal screws:

- Always make sure seal screws are tight before submerging a can.
- Always turn the seal screw(s) to relieve the pressure inside a can before removing the endcap(s).

A **pressure relief valve** is essentially an automatic seal screw. It works because the tension of a calibrated spring keeps a poppet pushed tight against an opening (the seat) in the bottom of the valve. The seat is opened, or "cracked," when the pressure inside the canister is greater than the combined force of the ambient pressure and the calibrated spring. This forces the poppet off the seat, allowing gas inside to vent. Once the interior pressure is the same as the force exerted by the calibrated spring, the poppet gets pushed back against the seat. This relief, or crack, pressure is adjustable by the user, usually at a pressure differential of 1–5 psi higher than the external or ambient pressure surrounding the canister. Follow the manufacturer's instructions for making this adjustment.

Figure 5.43: Pressure Relief Valve Cross Section

When the gas pressure inside the pressure canister is higher than ambient pressure, it forces the poppet upward against the force of the calibrated spring, "cracking" the pressure relief valve. Gas is then released to the outside. When pressure inside is reduced, the combined force of the spring and the ambient pressure pushes the poppet back against the seat, closing the relief valve.

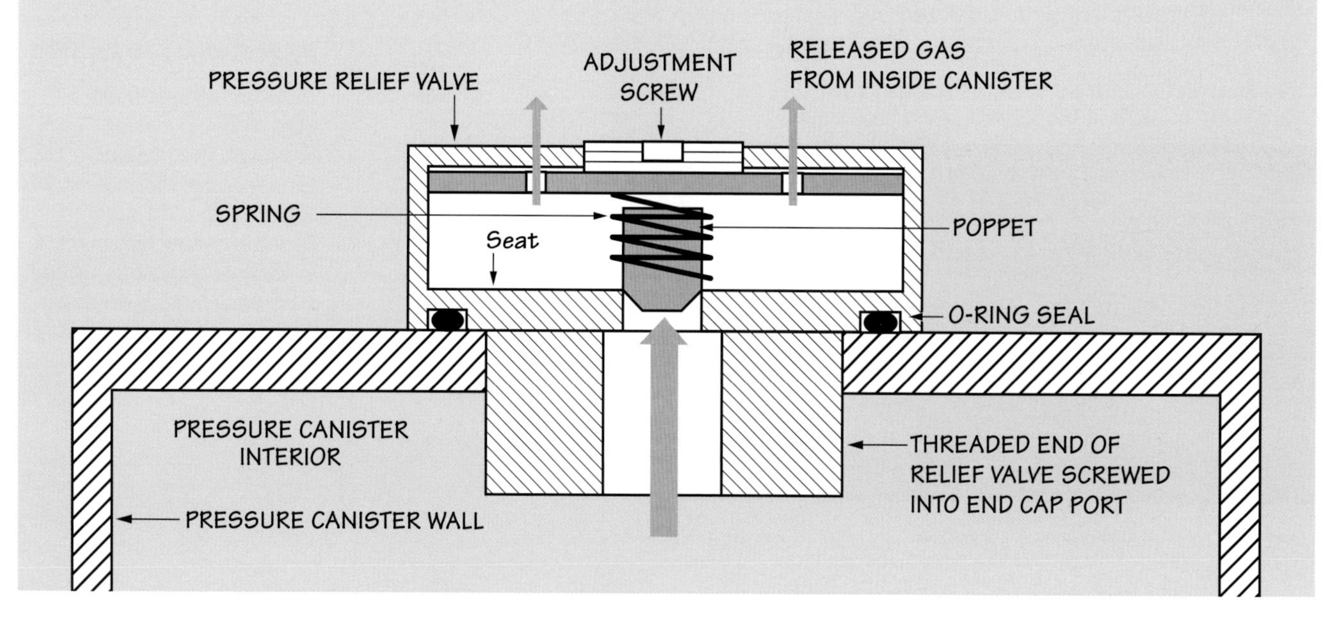

TECH NOTE: PRESSURE CAN VACUUM TEST

One of the most reliable ways to check for leaks in a pressure canister is to suck all the air out of the canister and see if any can leak back in. This is usually done by attaching a vacuum pump to the pressure canister you want to test, as shown in Figure 5.44.

The connection to the canister is usually made through a special gas-tight fill port fitting. Note that a **fill port** is a threaded port through which you can fill the canister with some type of liquid or gas. With valve A open and valve B closed, the vacuum pump removes almost all the air from the pressure canister. Then valve A is closed to "seal off" the pressure canister from the rest of the system except for the vacuum gauge, which displays the pressure inside the canister. If that gauge reading is near zero atmospheres absolute and remains constant for many hours, that's a good indication that there are no leaks in the canister.

On the other hand, if the gauge shows increasing pressure over time, this indicates that air is leaking into the pressure canister. Following the test, valves A and B are opened to let air flow back into the canister. This restores the internal pressure to 1 atmosphere, so the test system can be disconnected from the canister.

Figure 5.44: Vacuum Test Setup

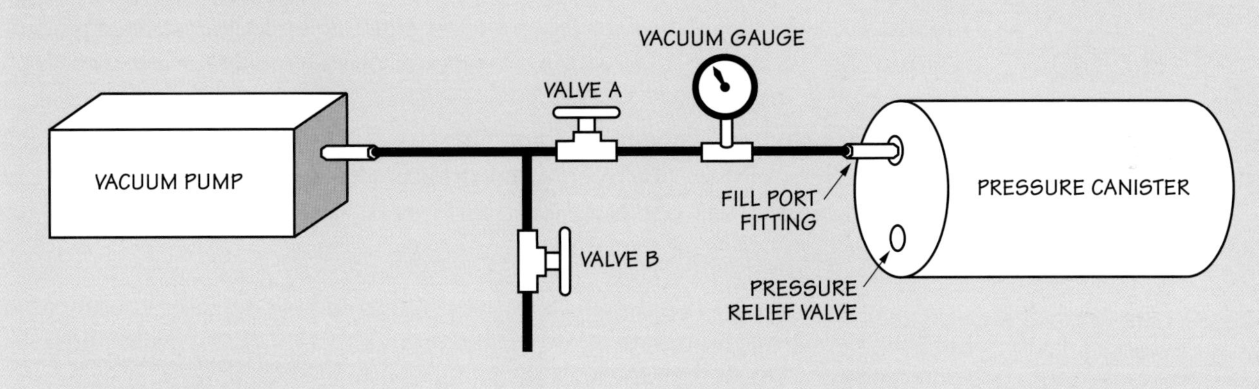

TECH NOTE: PREVENTING CONDENSATION IN A PRESSURE CAN

If you pour a glass of cold milk and leave it sitting on the countertop for a few minutes, the outside of the glass will become damp. This happens because normally-invisible water vapor in the air condenses into water droplets on the cold glass surface. Similarly, if you immerse an air-filled pressure canister in cold water, the same process may cause water to condense on the inner surfaces of the canister when those surfaces cool. The resulting dampness can fog-up camera view ports, promote corrosion, and even short out electronic circuits. There are three common strategies for reducing this problem:

- Minimize the total amount of empty (i.e., air-filled) space inside the canister, since this will limit the total number of water molecules available for condensation.
- Put some type of desiccant product inside the canister to absorb water vapor before it can condense. This must be done in accordance with the desiccant manufacturer's instructions, because desiccant exposed to too much air for too long will saturate with water and be useless for absorbing more.
- Replace the air (and water vapor) in the canister with a dry, inert gas, such as nitrogen. This process is called **nitrogen-purging**. To do this, you need access to a tank of nitrogen with a pressure or flow-rate regulator, as well as a fill port and seal screw in your canister. By using an appropriate hose and fitting you can add nitrogen gas to the canister through the fill port. This flushes most of the air and water vapor out through the open seal screw. Immediately after flushing, you must close the seal screw and the fill port to trap the dry nitrogen inside and prevent air from re-entering.

7. Pressure-Compensation Techniques

For canisters and housings that do not require an internal environment of air at 1 atm pressure, additional options are available to resist outside pressure and reduce the threat of leaks. These include non-conductive liquid compensation (commonly oil) and gas compensation (commonly air). These techniques are so effective that they free you from many of the usual constraints on canister size, shape, strength, and materials.

7.1. Oil Compensation

Some electronics, motors, and batteries can operate under pressure in an oil-filled housing with no air. Since oil, like other liquids, is virtually incompressible, it protects the housing from being crushed and allows simpler, lighter housings to survive and function well, even at full ocean depth. Oil has the added advantages of lubricating motor bearings and helping to cool motors or electronics. However, the oil does transfer ambient pressure to whatever's inside, so it works only if the components have no enclosed air spaces in their structures and can function under the expected pressure. Oil may also be incompatible with some types of plastic, rubber, or other materials. Mineral oil is the typical choice, but other types of specially formulated oil for high-voltage electrical circuits are commonly used in commercial/military systems. Generally, these oils are not recommended for amateur use. Oil-compensated cans commonly protect wet-cell batteries, transformers, junction boxes, and deep sea thrusters.

Because oil density changes slightly with temperature and pressure, the housing needs a bladder-compensation chamber or flexible (e.g., rubber) wall to equalize inner and outer pressure. The walls of the housing do not have to be thick to resist pressure; they

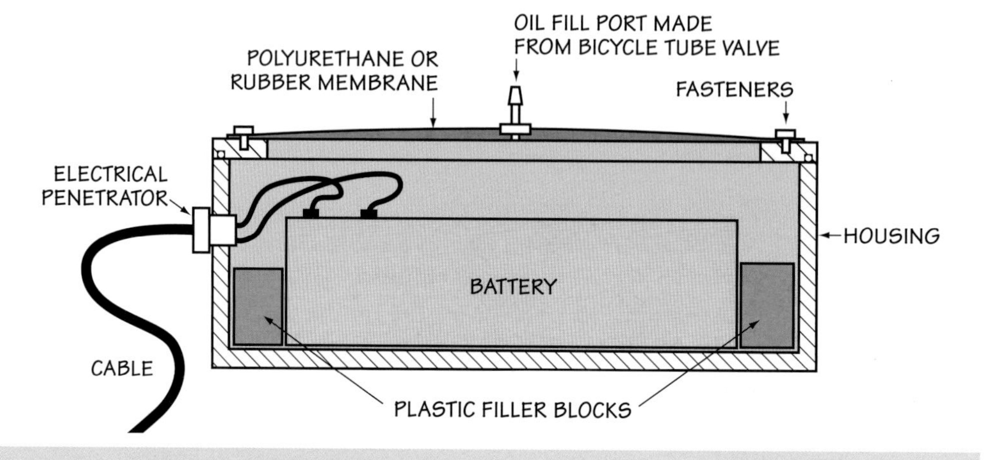

Figure 5.45: Oil-Filled Housing

TECH NOTE: MAKING AN OIL-FILLED HOUSING

Rugged watertight equipment cases, such as those made by Pelican or Underwater Kinetics, can make good oil-filled housings if modified with a compensation membrane or piston to balance inner and outer pressure. This option is best suited for sealed lead-acid batteries that are mounted externally. (Chapter 8 has detailed info on batteries.)

You can also oil-fill bilge pump motors so they can take deeper depths if you add a compensation bladder or piston. However, brushed motors will shed carbon and contaminate the oil over time, making it conductive, reducing the efficiency of the motor, and necessitating frequent oil changes. That's why oil-filled thruster housings are more suitable for brushless motors. (See *Chapter 7: Moving and Maneuvering* for info on brushed and brushless motors.)

only need to be strong enough to contain the interior components when the can is lifted into the air. Since there is liquid pressure both inside and outside the canister, a pressure-resistant shape is not needed. The housing can even be cube-shaped.

7.2. Gas Compensation

Gas-compensated (or gas-comped) canisters are similar to oil-filled housings, but they use a gas instead of a liquid. A compressed gas, such as dry, pressurized air or nitrogen, is pumped into the can or released as needed, so that the interior is kept at a pressure that matches or slightly exceeds that of the ambient water pressure. The system is controlled by sensors and valves that measure the pressure differential between the interior of the can and the water surrounding it. If the internal pressure is set slightly higher than the external pressure, then the gas will leak out if there is a tiny leak. Canisters that use pressurized air to resist exterior water pressure are often referred to as "air-comp systems."

TECH NOTE: AIR-COMPED ROV

Images courtesy of Dr. Steven W. Moore

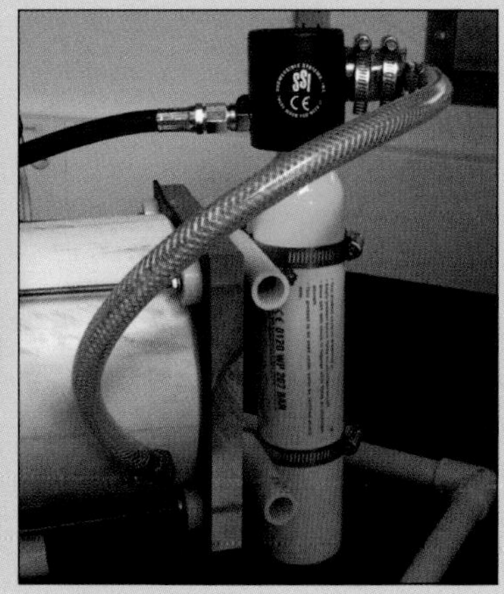

Figure 5.46: Air-Comped Thrusters

The *ROVing Otter* is a small ROV designed for dives to depths of 100 meters (approx. 325 ft). It uses air-compensated thrusters to reduce the probability and severity of leaks, particularly around the propeller shafts. The compressed air and precise pressure regulation (including the overpressure relief essential for safety) are provided by a Spare Air™ unit that is normally sold as an emergency air supply for scuba divers.

As the ROV descends, the ambient water pressure increases and becomes higher than the air pressure inside the ROV. This pressure imbalance opens a valve inside the scuba regulator, which allows high-pressure air to flow from inside the scuba cylinder into the interior of the ROV until the pressures are once again balanced. Then the valve closes. When the ROV ascends, the ambient water pressure outside drops, making the air pressure inside higher than the water pressure outside. This opens a different valve in the regulator, allowing the excess air pressure inside the ROV to escape, maintaining the balance between inside and outside pressures.

The ROV has one large central pressure canister for its camera and electronics, plus four thruster housings and two video light housings. All seven housings and the Spare Air™ regulator are interconnected by clear plastic hoses and brass hose barb fittings. These connections provide a watertight passage for air and for the wires that carry electrical power to the motors and lights.

A big advantage of having all housings connected in this way is that a single air unit can compensate all of them at once. A big disadvantage is that a leak in any one hose or housing (or accidentally using up all the air in the small scuba tank) can result in *everything* getting flooded. At one point, *ROVing Otter* learned this the hard way, but the team's quick response salvaged most of the essential parts (Figure 5.47). *ROVing Otter* lived to dive again!

SAFETY NOTE: AIR-COMPED ROV

Because of the serious risk of explosion and/or arterial gas embolism associated with high-pressure compressed air, air-comped systems should be designed, built and used only under the close supervision of an adult who has a good working knowledge of compressed gas physics and who is trained and experienced in the inspection and maintenance of scuba cylinders and scuba regulators. While air compensation has its advantages, oil compensation is a much simpler, safer, and less expensive way to waterproof your thrusters and should probably be tried first.

TECH NOTE: WHAT TO DO IF YOUR VEHICLE FLOODS

Despite all your precautions, leaks are a common occurrence, particularly during the testing stage of any new design. To minimize damage to the vehicle, it's important to notice if your vehicle begins acting a little heavier than usual and to know what to do if your vehicle gets water inside.

If your vehicle floods, act immediately, especially if the vehicle is in saltwater:

- Disconnect any external electrical power *if you can do so safely.*
- Open the vehicle and pour out the water.
- If there are internal batteries, immediately disconnect them, because applied voltages greatly accelerate the corrosion process.
- Rinse all affected parts under tapwater as soon as possible—seconds count!
- After rinsing the affected components, gently blot them dry with cloth or paper towels.
- Complete the drying process by warming the affected circuits in sunshine, under a hair dryer (on low setting and with care to avoid melting anything), or in an oven set to its lowest setting.

The good news is that a very quick response to a leak can often save many electronic components.

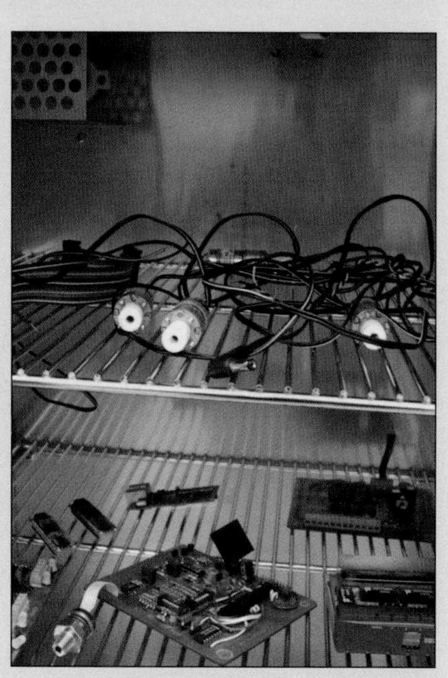

Images courtesy of Dr. Steven W. Moore

Figure 5.47: Resuscitating Drowned Electronics

Though it might seem counter-intuitive, pouring freshwater over electronics really is the best thing to do following a seawater disaster. In 2006, failure of ROVing Otter*'s air compensation system during an ocean dive flooded the main electronics canister, completely immersing nearly $1000 worth of components in corrosive saltwater.*

Quick disconnection of the battery power followed by thoroughly rinsing and drying of the vehicle's guts saved every single piece of electronics, including a network camera, six motors, a battery, a small computer, numerous electronic sensors, two lasers, two motor controllers, and a computer networking router! Following an initial tap water rinse in the sink, everything was disassembled and soaked for 30 minutes in distilled water. Then each part was dried carefully with a paper towel. Final drying took several hours in a low-heat laboratory drying oven.

Like the oil-comped system, the canister walls of a gas-comped system can be thinner, and the can itself larger and of any shape because the inner gas pressure supports the housing's walls. The disadvantage is that the gas-compensation system adds technical complexity. It also places limits on how many times the ROV can dive and surface during a mission. That's because the compensating gas is stored on board the vehicle in a pressure container such as a small emergency back-up scuba tank and pressure regulator such as Spare Air™. These get used up as the vehicle changes depth, since air must be injected into the can during descent, and gas must be vented from the canister during ascent. For most robotic vehicles, it is usually simpler just to use a 1 atm pressure canister. Still, there are some vehicles in which air-compensated canisters are a viable solution.

8. Encapsulation (Potting)

Simply put, **encapsulation**, also known as **potting**, is a method of waterproofing components by embedding them in a solid or rubbery material, thereby often eliminating the need for any housing whatsoever. For example, some electrical and electronic circuits can operate when encased in solid epoxy. To do this, the circuit is placed in a cup or other simple mold with all the wire leads extending up and out of the mold, away from the circuit board. Then liquid epoxy or rubber is poured into the mold until it completely covers the circuit and allowed to set. This technique works well as a waterproofing technique, but only if the circuits have a very low heat output. (Epoxies are good thermal insulators, so heat-producing circuits can overheat if encapsulated.) It's important to remember that epoxy encapsulation is permanent—once the epoxy is poured, the circuit cannot be changed or repaired.

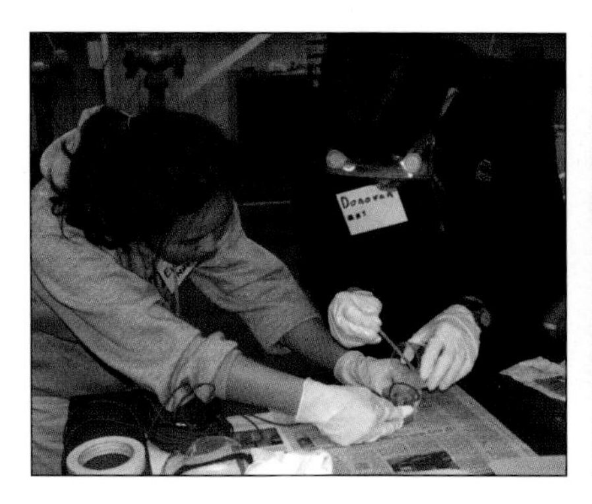

Images courtesy of Randall Fox

Figure 5.48: Potting a Camera and Camera in Position on ROV

TECH NOTE: ENCAPSULATED HOUSINGS AS A DEEPER-DIVING OPTION?

The *SeaMATE* ROV project demonstrates how a tiny camera can be encapsulated to make it pressure-proof. So couldn't this procedure work to protect other electronic components in a deeper-diving ROV? The answer is yes, but . . . The "but" of the answer has to do with the expense of the electronics, ease in servicing, and heat dissipation.

You can get away with potting an inexpensive camera, because its low cost means it is literally cheaper to replace the camera than fix it if there's a failure. But motor controllers, radio-controlled components, and microprocessors are all expensive. And if they fail, the encapsulation material surrounding the multitude of tiny wires and printed circuit boards must be removed and cleaned before you can troubleshoot. This is a major inconvenience, especially if there is only a simple loose wire somewhere in the circuit or a reset switch that needs to be triggered. Furthermore, if polymers such as epoxy or urethane are used for encapsulation, it's impossible to remove the cured material, forcing you to replace the whole expensive unit.

Remember, too, that the encapsulation material typically insulates the components it surrounds. Thus, if you encapsulate a motor controller that generates a lot of excess heat during operation, the heat cannot escape, and the motor controller will quickly burn out.

A non-permanent alternative to epoxy commonly used for shallow-water applications is to encapsulate electric circuits in beeswax or toilet flange sealing wax. Both resist water well, and the circuit can be removed from the wax for repairs by gently warming the wax until it melts again. To encapsulate a circuit in wax, the melted wax is poured over a circuit that has been mounted in a sealed open-top project box. Once the wax is cooled, any voids can be filled with more wax. When all is filled and cooled, the project box lid is closed; then the electronics protected inside the box are ready to use under water.

While encapsulation protects components from contact with water, it provides little or no protection from pressure. Therefore, the encapsulated components must be capable of normal operation when exposed to the pressures found at the maximum dive depth of the vehicle. See *Chapter 12: SeaMATE* for further discussion of encapsulation techniques.

9. Adding a Card Cage

As you begin to explore more sophisticated and complex vehicles, you'll also want to look at ways to simplify accessing and servicing the parts and subsystems inside your pressure housings. One technique for electrical cans is to add a frame inside the can that securely supports electrical components. This frame, called a **card cage,** makes it easy to remove and replace the whole internal component assembly during servicing. There aren't any one-size-fits-all card cages, so you'll have to design one based on the type of pressure housing you decide to use and the types of components you need to mount inside the can. Card cages can be fabricated from plastic plate or aluminum sheet metal held together with stainless steel fasteners.

Here are suggested requirements for your card cage design:

- It should facilitate ease of servicing for battery change outs, etc.
- It should support the necessary terminal strips and fuse holder locations.
- It can facilitate the optimum location of bulkhead connector(s) on the endcap, any two-piece "pin and socket" interconnections (often called Molex connectors), and the number and type of wires necessary.

Figure 5.49: Card Cage Examples

A card cage is often essential to support the electronics and wiring inside a pressure canister. The University of British Columbia's AUV Gavia *has a robust digital imaging camera card cage as part of the endcap that inserts into the sleek nose cone module (pressure housing).*

The card cage for the ROVing Otter *is not attached to the endcap but rather slides into the pressure canister.*

Images courtesy of Harry Bohm (left) and Dr. Steven W. Moore (right)

- It should include an integrated endcap-to-card cage assembly so that the card cage can be removed easily or inserted together with the endcap. Fastening or gluing a card cage to the inside of the pressure can is never recommended since it complicates servicing.

Fortunately, when you're at the concept vehicle stage, it's not necessary generally to work out a detailed card cage design. However, if the fit of the components inside the card cage looks like it might be overly tight, you will want to show enough detail to indicate how to optimize components so they will go into the pressure canister you have chosen. This initial design can be refined into a detailed drawing if you progress toward fabricating your concept vehicle. What you will need, however, is an estimate of the weight of the card cage materials to help with buoyancy calculations, as explained next in *Chapter 6: Buoyancy, Stability, and Ballast.*

10. Chapter Summary

This chapter has explored causes and solutions for the greatest single threat facing any underwater vehicle, regardless of its size or complexity—hydrostatic pressure. This pressure can force water into a vehicle through tiny holes or cracks and can cause large enough forces to crush a vehicle outright.

Hydrostatic pressure is caused by the weight of water. It can be measured as either gauge or absolute pressure; the latter includes the added pressure caused by the weight of the atmosphere pressing down on the surface of the water. In this chapter, you have learned how to calculate the gauge and absolute hydrostatic pressure at any depth in either fresh or saltwater. You have also learned how to calculate the forces caused by pressure differentials across the surface of pressure hulls and canisters of different sizes. These forces can be impressive, even in a shallow swimming pool, and are the main reason that constructing crush-proof and leak-proof pressure housings is such an exciting challenge.

The solutions presented here, from basic pressure hull shapes like spheres and cylinders, to the almost magical sealing powers of O-rings and pressure-compensated designs, are all time-tested techniques used in fabricating underwater craft. Some of the solutions, such as strong steel hulls and welded bosses, are applicable primarily to large vehicles. Others, such as the use of PVC pipe sections with flange-type endcaps for pressure canisters, are ideally suited for smaller, low-budget vehicles.

Truly, the inventors of early submarines and submersibles would be astounded at the array of tools, equipment, and knowledge available to today's designer and builder. But while the technical details of designing and building underwater craft have changed dramatically, fluid pressure and the challenges it presents remain the same. Understanding them is still essential to any successful underwater vehicle design.

Chapter 6

Buoyancy, Stability, and Ballast

Chapter 6: **Buoyancy, Stability, and Ballast**

Stories From Real Life: *Ben Franklin*

Chapter Outline

Chapter Learning Outcomes

- Explain why things in water sink, float, or tip over and how ballast systems can be used to control these processes.
- Describe the difference between positive, negative, and neutral buoyancy and explain why the designers of most underwater vehicles strive for near-neutral buoyancy.
- Know why and how to use a simple weight statement table to design a vehicle with the desired degree of buoyancy.
- Describe how some active ballast systems function and explain why most ROVs and AUVs rely on the simpler static ballast approach.

Figure 6.1.cover:* UT-1 *Ultra Trencher

Soil Machine Dynamics (SMD) has built the world's largest underwater robot for CTC Marine Projects, a UK contractor. The Ultra Trencher UT-1 *is the size of a small house and weighs in at 60 tons. Huge blocks of syntactic foam (yellow) at the top of the frame provide flotation to offset its weight and keep the vehicle upright.*

This gigantic trencher has the capacity to bury oil and gas pipelines up to 1 meter in diameter in tough soils, working at a depth of 1.5 kilometers.

Image courtesy of SMD Ltd., Newcastle upon Tyne

STORIES FROM REAL LIFE: *Ben Franklin*

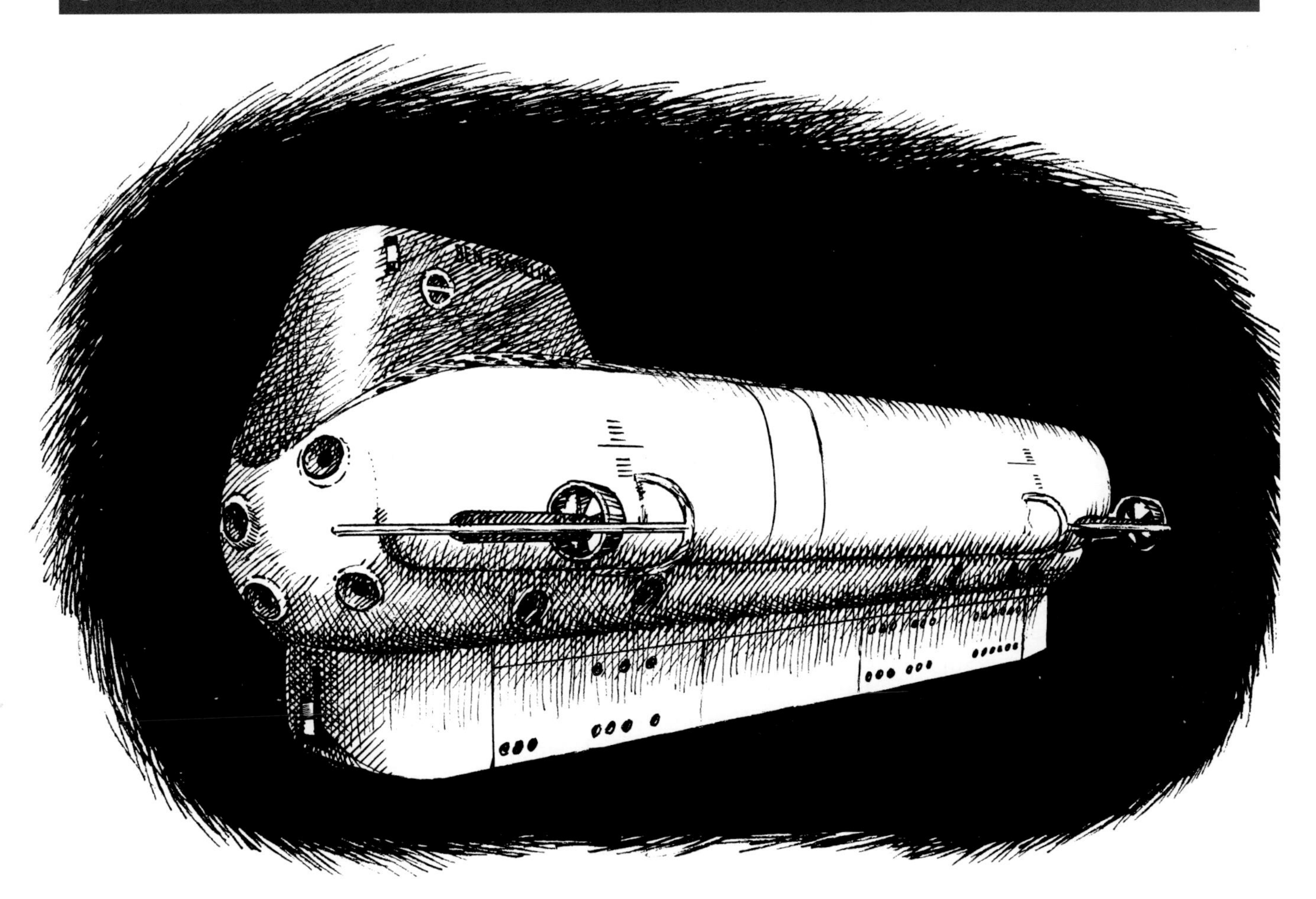

In 1969, the mesoscaph Ben Franklin *transported six scientists in a 30-day "drift dive" in the Gulf Stream current. It was the first time in history such a continuous long-term observation had ever been attempted below the surface.*

The key to this success was a combination of modern materials and an unusual design that centered on the concept of neutral buoyancy, allowing Ben Franklin *to turn off its motors and drift easily at a consistent depth in a fast-moving river of ocean water.*

By 1953, the bathyscaph *Trieste* had begun setting depth records, but the father/son design team of Auguste and Jacques Piccard was acutely aware of the tremendous effort and cost that each such mission entailed. They began to muse on the possibilities for designing a different kind of underwater vehicle, one that could still carry out research at respectable depths in the mid-ocean region. This type of submersible would not be dedicated to deep dives, so Auguste christened it a mesoscaph, meaning a vessel intended for intermediate depths.

Building such a submersible out of modern materials would mean the hull could be lighter; hence, the vessel wouldn't need such a large float as that used by *Trieste*. To make such a lighter-weight craft submerge, it could be equipped with a propeller in the vertical axis, somewhat like a helicopter, but pushing down instead of lifting up. This would also give the advantage of total security—if the motor or propeller broke down, the vessel would automatically come back to the surface. Jacques Piccard began making sketches and experimental models, but soon realized that using two spheres instead of one provided greater stability, buoyancy, space, and range.

However, the Piccards' first actual mesoscaph was not produced for research purposes, and it certainly didn't operate in the mid-ocean region. Instead, it was a tourist vehicle that became the focal piece of the Swiss-based World Exposition of 1964. Aptly named the *Auguste Piccard*, it took more than 20,000 passengers to the bottom of Lake Geneva over the course of the Exposition—all without incident.

A year later, Jacques Piccard connected with Grumman Aircraft Engineering Corporation (later Grumman Aerospace Corporation). Like other major American aircraft manufacturers, Grumman had opened up departments of submarine research as part of an overall plan to diversify. The company was particularly interested in Piccard's idea for a mesoscaph to explore the massive Gulf Stream current. Grumman hired Piccard and drew up a contract for a vessel that would be built in Switzerland, so as to benefit from previous mesoscaph experience, but launched in the U.S.

There were many reasons for choosing to design and build an underwater vehicle specifically to study the Gulf Stream. This amazing mass of water, including the Florida Current that flows from the Gulf of Mexico to Cape Hatteras, is greater than all the rivers of the world combined. The Gulf Stream is warmed by the tropical sun in the Gulf of Mexico, its waters moving perpetually around the tip of Florida and up the east coast of North America toward the cooler north, where the heat dissipates. There the cooler waters sink and begin a return journey. This Gulf Stream cycle has a tremendous effect on the weather patterns of the Western Hemisphere.

Early Spanish shipmasters of the mid-1500s knew enough about the Gulf Stream current to take advantage of it when sailing east. But it was Benjamin Franklin, acting as deputy postmaster general of the United States, who had the massive current stream put on sailing charts of the Atlantic used by mail packet ships sailing between England and the colonies. And it was Franklin who became the first scientist to make studies of the Gulf Stream. Appropriately, Piccard's newest mesoscaph was eventually named the *Ben Franklin*.

No one had ever attempted an underwater drift mission. Simply put, the challenge was to drift as a consistent underwater depth and with a minimum of noise so as to take acoustical measurements and observe sea life.

Jacques Piccard began working with Grumman on a 15-meter (approx. 49-ft) submersible capable of taking six crew members and scientists on a 30-day drift at various depths, all within the Gulf Stream along the east coast of the United States. The scientific goal was to collect a variety of measurements and conduct experiments involving temperature, salinity, oxygen content, density, sound velocity, and other conditions—all while traveling in the Gulf Stream. Another goal was to monitor the psychological and physical stresses that crew would undergo during a 30-day confinement. Grumman rightly guessed that this data would prove helpful in obtaining a *Skylab* contract for its space division.

The success of the mission depended on designing a rigid pressure hull that would not compress significantly, even when subjected to water pressure at a depth of 600 meters (approx. 2,000 ft), but that could be ballasted to be neutrally buoyant. Such a hull would enable crew to keep the mesoscaph at a consistent depth in the Gulf Stream without vertical drift. In other words, when the submersible got itself positioned and stabilized at its average drift depth of 200 meters (approx. 650 ft), it could easily stay there without needing to use power. This buoyancy system worked relatively well.

The cylindrical steel hull was constructed in six segments, not unlike thick slices of baloney, and then welded together. The outside diameter, like that of the *Auguste Piccard*, was 3.15 meters (10.4 ft). There were 29 viewports and 13 thickened penetrator areas on the pressure hull. The American Bureau of Shipping (ABS) oversaw practically every phase of construction, in the process establishing basic rules for certification of other non-military submarines.

Piccard's new mesoscaph had four 25 hp motors outside the hull at all four "corners." Each was equipped with a four-bladed prop and mounted in a steerable pod. Cruising speed was 2.5 knots (4 knots maximum). Power came from 230 kN (approx. 52,000 lbs) of lead acid batteries, accounting for 20 percent of the total weight of the submersible. These were also housed outside the pressure hull, with oil insulating the batteries from the sea. Each battery included a system that allowed the oil to refill any

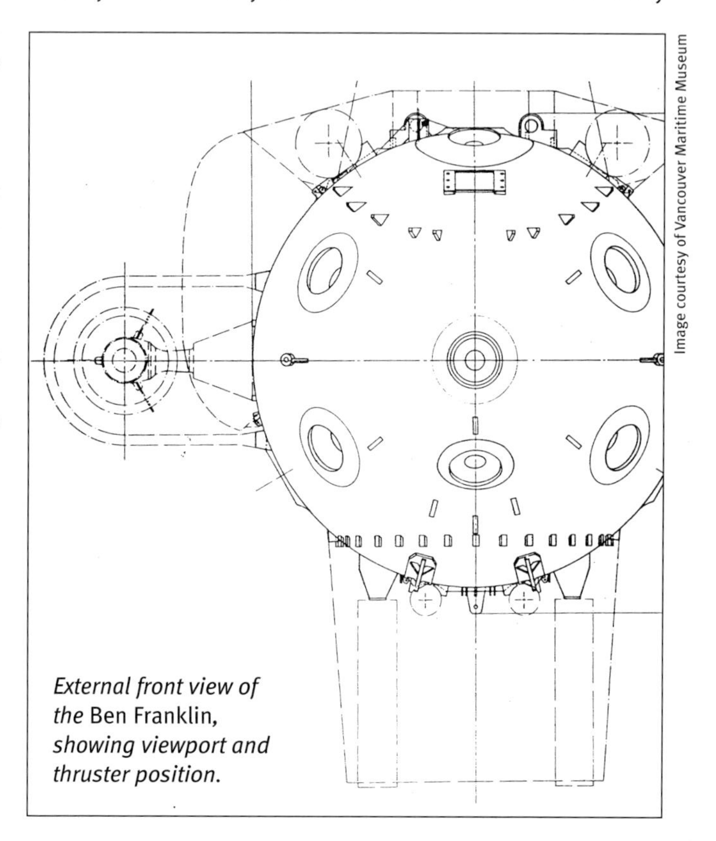

External front view of the Ben Franklin, *showing viewport and thruster position.*

Image courtesy of Vancouver Maritime Museum

space freed by the compression of the gas from the batteries, particularly at the beginning of a dive. Placing the batteries in the lowest part of the submersible played a decisive role in stabilizing the craft.

Four fiberglass tanks located outside of the pressure hull could give 213.3 kN (approx. 48,000 lb) of buoyancy when blown free of water. Two pressure-resistant tanks in the keel provided an additional 30.2 kN (approx. 6,800 lb) of variable buoyancy for depth keeping. For emergency ballast, there was 53.4 kN (approx. 12,000 lb) of droppable iron shot. A heavy guide chain hung 9 meters (approx. 30 ft) below the *Ben Franklin*, another idea Piccard borrowed from hot-air ballooning and used in the design of his submersible *Trieste*. In this case, the chain links always touched bottom first. Freeing the mesoscaph of the weight of even a few links was enough to bring it into equilibrium and stop the descent.

Was the relatively simple design of this highly sophisticated research vessel a success? Certainly, the premise of neutral buoyancy was brilliantly effective. And even though the craft never became an enduring undersea workhorse, the mission yielded valuable data that NASA used in designing the first space station, *Skylab*, in the early '70s. For the 30-day drift, NASA monitored the civilian crew continuously, photographing them every two minutes as they worked, conducted research, wrote in their journals, and coped with confinement in a less than ideal space. The crew de-stressed listening to music, particularly the newly released Beatles' album *Yellow Submarine*, but life aboard their own sub was often a significant challenge. Without modern insulating materials, the cold, damp, metal hull was never comfortable. Despite hundreds of pounds of silica gel to absorb moisture, humidity remained between 72 and 100 percent. Continuous monitoring for bacterial contamination was essential. When the hot water tanks did not perform as planned, the men had to forgo showers for the mission! Not surprisingly, NASA collected an impressive amount of realistic data from *Ben Franklin*'s crew of six.

While no sensational marine discoveries were made on the drift voyage, the quantity of information collected added significantly to knowledge of the Gulf Stream and of the ocean in general. All previous expeditions had been conducted from the surface; even other research subs generally went down for a only few hours. *Ben Franklin* represented a new method of research and observation in the ocean—the first time that continuous long-term mission had been carried out below the surface. Careful attention to buoyancy was critical to the success of this unusual mission.

Internal view of the front end of the Ben Franklin

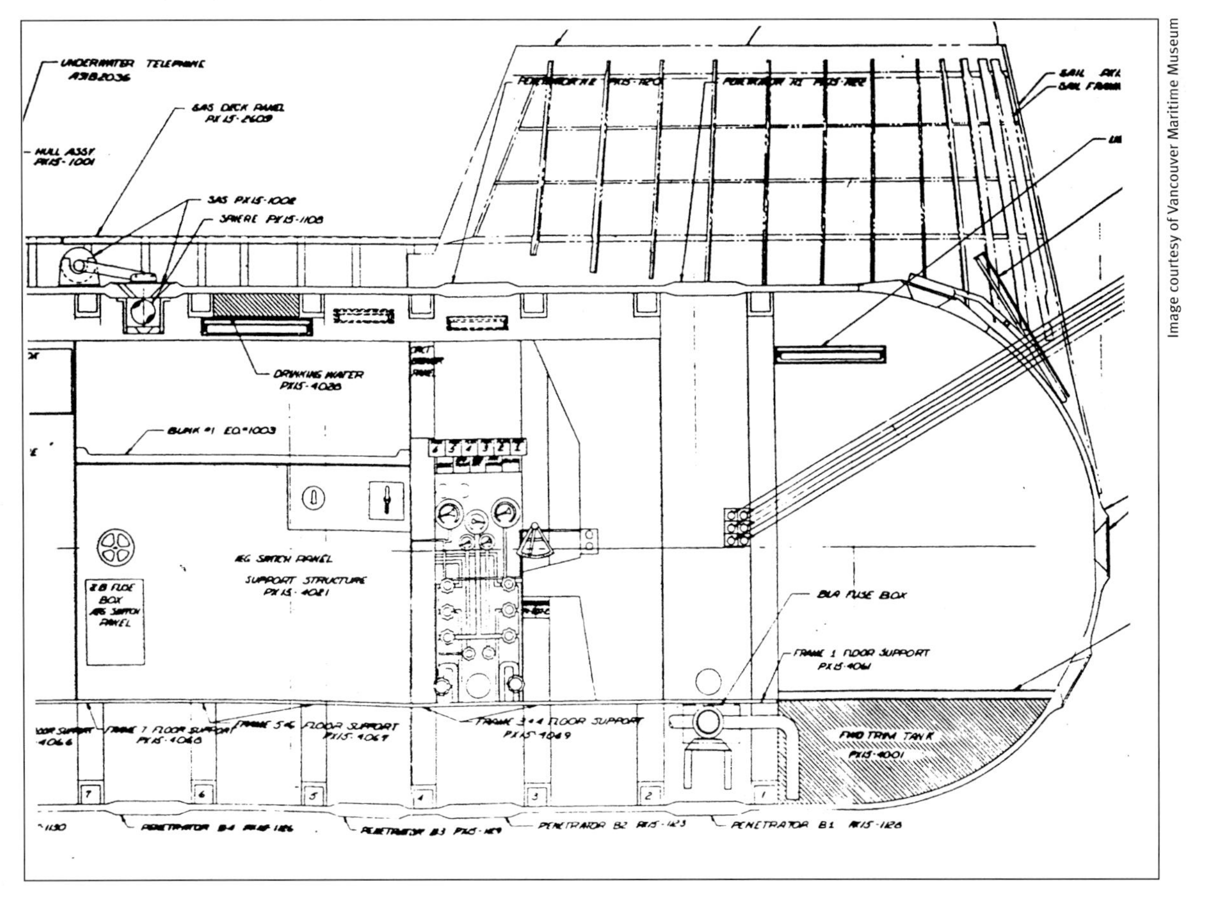

Image courtesy of Vancouver Maritime Museum

1. Introduction

Image courtesy of NASA, Johnson Space Center

Figure 6.2: The Challenge of Achieving Stability and Trim

Helicopters and submarines enjoy a freedom that cars and bicycles do not—in addition to being able to travel forward and backward, or right and left, they can move straight up and straight down! But there is an important difference between how helicopters and submarines "hover" in mid-air or mid-water. Helicopters use engine power and rotors to generate enough thrust to overcome the vehicle's weight and support it in the air. This consumes energy rapidly and severely limits the time a helicopter can stay aloft without refueling.

Underwater vehicles could use this same power-hungry approach to hovering, but they don't. They don't need to. Water has a much higher density than air, so submarines, along with whales and other underwater objects, can use **buoyancy** to offset their weight completely. This allows them to hover without expending any energy whatsoever. This chapter explores the physics of sinking and floating and explains how you can use that knowledge to design underwater vehicles that do neither—until directed to do so.

This chapter also examines the related issues of vehicle **trim** (i.e., orientation) and **stability**. It looks at the physical factors that determine whether an underwater vehicle sits upright, leans to one side, or flips completely upside down. It also discusses why some vehicles are more resistant to tipping and are better able to reorient themselves into an upright position when knocked over by a wave or other outside force. The tendency to remain upright (or at least to remain oriented in a predictable and controllable direction) is called stability, and it depends directly on the placement of flotation and weights (**ballast**) on the vehicle.

It's important to invest some time in learning the physical principles described here. Once you understand them, you can have them work *for* you to improve vehicle performance, rather than having them work *against* you. Armed with the knowledge gained in this chapter, you will be able to build more stable vehicles.

2. Why Things Float or Sink

Many objects float when you put them in water. Even objects that sink seem to weigh less in water than they do in air. This first section of the chapter reviews why this is so and details how to calculate whether your ROV or AUV will float or sink, even before it's put in the water. This is an important skill, because it allows you to predict problems and make required changes early in the design and construction process. Otherwise you might spend hundreds or thousands of hours building your vehicle, only to discover that it floats like a cork and can't dive, or sinks like a rock and can't return to the surface.

TECH NOTE: BASIC STABILITY

A simple demonstration using an empty plastic water or soda bottle will illustrate some of the forces that affect stability. Choose a plastic bottle that is smoothly round (no ridges or ribs), uniformly cylindrical, and without any paper label. Empty the bottle completely, then cap it tightly. Put a mark on the bottom end, as in the illustration. Then place the bottle in a sink or large pan of water and give it a rolling spin. Do this several times. You'll see that the bottle rolls easily in the water and also that the mark doesn't necessarily end up in the same place twice.

Now tape a large bolt or nail to one side of the bottle and spin the bottle again. Notice that it takes more effort to get the bottle to roll and that it spins fewer times before settling. Notice also that the mark you made on the bottle always ends up in the same position with each spin and that the bottle always ends up with the weight (nail or bolt) in the water. If you try and balance the bottle so that the weight stays out of the water, it's just not possible. The weight always moves downward. Because the weighted bottle always returns to its original position after a spin (i.e., some movement from an external force), it is said to be **stable**.

This simple stability demonstration has big implications for any submerged vehicle. If your craft is not stable, it may not remain upright when subjected to any external force such as waves, winds, or current.

So how do you get an underwater vehicle to be stable when it is submerged?

- Place the heaviest weights (ballast) low down on the vehicle's structure.
- Place most of the flotation high on the vehicle structure, well above the weights.

Figure 6.3: Stability Experiment

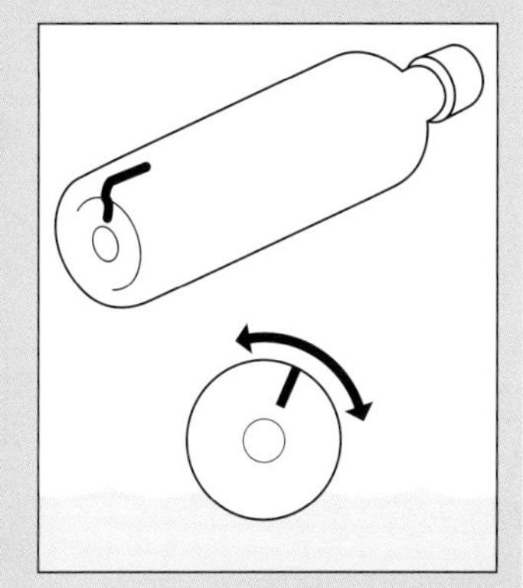

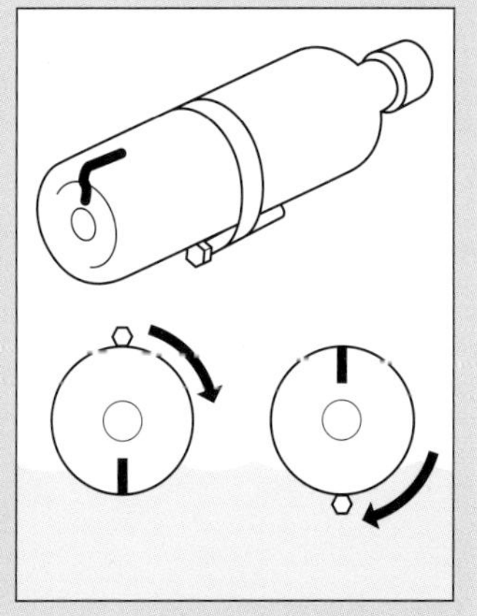

2.1. Buoyant Force

When comparing some of the properties of water and air in *Chapter 3: Working in Water* and in *Chapter 5: Pressure Hulls and Canisters*, you learned that water is about 800 times denser than air. That means pressure increases about 800 times more quickly when you descend through water than when you descend through air. Thus, even a tiny change in depth can produce a significant change in water pressure.

When pressure presses against the surface of an object, it creates a force. The direction of that force is perpendicular to the surface. The size, or magnitude, of the force is equal to the pressure multiplied by the area of the surface. Since pressure increases with depth, the pressure-induced force pushing up on the bottom of a submerged object is always greater than the pressure-induced force pushing down on it from above. The combined effect is a *net upward force* called the **buoyant force**. This buoyant force is always present for any submerged object, though it may or may not be strong enough to make the object float. (More on this later in the section on positive, negative, and neutral buoyancy.)

Image courtesy of Harry Bohm

Figure 6.4: Murphy's Law at Work

More than one ROV design team has had to rescue their sinking vehicle.

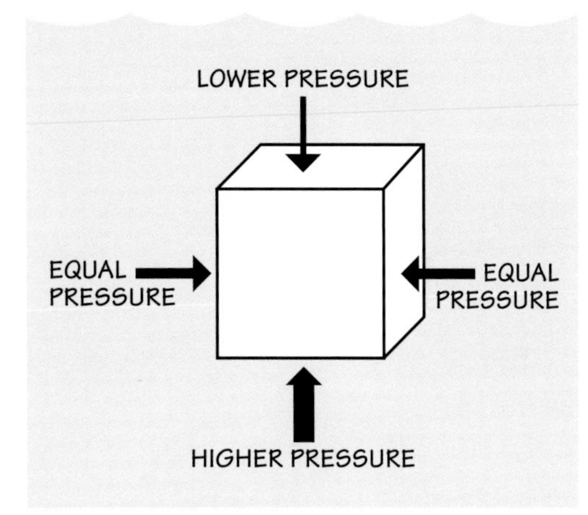

Figure 6.5: Buoyant Force

Water pressure is greater on the lower (deeper) side of an object and less on the upper side. This results in a net upward force called the buoyant force. The pressures on opposite sides are the same and cancel out.

It's easy to think about how this works if you imagine a cube immersed in water. Each of the six faces of the cube experiences a force equal to the area of that face multiplied by the average pressure of the water acting on that face. The symmetry of the cube guarantees that the force pushing on the left side of the cube is exactly balanced by the force pushing on the right side, so there is no net force trying to push the cube left or right. Likewise, the forces on the front and back of the cube cancel, so there is no net force trying to move the cube forward or backward. However, because of the change of pressure with depth, the forces on the top and bottom do *not* cancel. The bottom surface of the cube is located in slightly deeper water (and therefore exposed to greater pressure) than the top surface of the cube. Since the faces of the cube have equal area, the greater pressure acting on the bottom surface will overpower the pressure acting on the top of the cube and try to lift the cube toward the surface. This upward force (force from below minus force from above) is the buoyant force. Again, whether this buoyant force will succeed in lifting the cube depends on other factors, which will be described shortly.

Although the mathematical details get more involved for objects with more complicated shapes than a cube, the end result is the same: any submerged object experiences an upward buoyant force. This does not necessarily mean that the object will float, only that it will seem to weigh less in water than in air, because at least some of its weight will be supported by buoyant force.

2.2. Archimedes' Principle

It's one thing to determine that water pressure creates an upward buoyant force on a submerged object. It's another thing to determine precisely how strong that force is.

One approach to determining the magnitude of the overall force involves mathematically combining all of the smaller forces generated by the pressures acting on each bit of each surface of the object. This approach is straightforward for simple cases like the cube example above, but gets more involved and usually requires calculus for more complicated object shapes, such as those with curved surfaces. Fortunately, an ancient Greek mathematician and scientist named Archimedes found an easier way. He made some careful measurements and discovered the following general rule:

> **The magnitude of the buoyant force acting on an object is equal to the weight of the fluid displaced (i.e., pushed out of the way) by the object.**

This simple yet useful relationship is known as **Archimedes' Principle**. It is useful because it allows you to predict the buoyant force on an object even before it's placed

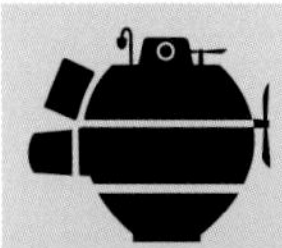

HISTORIC HIGHLIGHT: ARCHIMEDES' BATH

Rumor has it that Archimedes (287–212 BC) was inspired to investigate buoyancy—and ultimately to formulate his famous principle—while taking a bath one day. The story goes that when he stepped into a full-to-the-brim bathtub, he noticed that the water overflowed the tub. He also noticed that his body felt lighter. Based on these two observations and additional experiments, he came up with an explanation of why things float that is now commonly known as Archimedes' Principle.

Figure 6.6: Inspiration Strikes!

in water. This gives you the opportunity to detect and correct any major buoyancy problems while your vehicle is still on the drawing board. That's almost always easier, faster, and cheaper than doing it after you've already built the vehicle.

To apply Archimedes' Principle, all you need to know is the density of water and the volume of water that the vehicle will eventually displace. For a completely submerged vehicle, the submerged volume is simply the total volume of the vehicle, which can usually be estimated from some basic geometric formulas. Once you know that volume, it's a simple matter to calculate how much that volume of water weighs.

For example, Archimedes' Principle tells us that the buoyant force acting on a completely submerged basketball will be equal to the weight of a basketball-sized sphere of water. A basketball has a diameter of about 9 inches and weights less than 1.5 pounds. The volume of a sphere is determined with this equation:

Equation 6.1

$$V = \frac{4}{3}\pi r^3$$

For a basketball, that works out to a volume of about 381 cubic inches, or 0.22 cubic feet. Multiplying this by 62.4 pounds per cubic foot for freshwater yields a buoyant force of about 13.8 pounds in freshwater. This is much greater than the weight of a basketball and explains why pushing a basketball to the bottom of a swimming pool is easier said than done.

2.2.1. Why Boats Float

Archimedes' Principle explains why boats float (and, as you'll see later, also explains why they sink). To understand why boats float, think about a crane lowering a boat into a lake. When the boat is hanging from the crane, completely clear of the water, all of the boat's weight is supported by the crane. Now imagine that the crane begins slowly lowering the boat into the water. As the hull of the boat first touches the lake, it begins to displace water. As it does so, Archimedes' Principle tells us that the boat begins to experience a buoyant force acting upward on it. At first, this buoyant force is small, because the amount of displaced water is small, so the crane still has to support most of the weight of the boat.

As more of the boat goes into the water, more water is displaced and the buoyant force grows larger. As this happens, more of the boat's weight is supported by the water and less is supported by the crane. Eventually, a point is reached at which the boat has displaced just enough water to create a buoyant force that *exactly* counteracts the weight of the boat. At this point, the cable on the crane goes slack, because it is no longer supporting any of the boat's weight—the boat is floating. So how much does the displaced water weigh? Based on Archimedes' Principle, we know it weighs exactly as much as the boat!

Using similar logic, you can see that any object floating on the surface will have sunk into the water just far enough to displace a quantity of water that weighs exactly as much as that object.

2.3. Positive, Negative, and Neutral Buoyancy

The boat in the previous example shows how Archimedes' Principle can be applied to a surface vessel; however, the law's usefulness is not limited to objects floating at the surface. It applies equally well to objects that are submerged, including underwater vehicles.

TECH NOTE: BUOYANCY, BUOYANT FORCE, AND EFFECTIVE WEIGHT

It's easy to confuse "buoyancy," "buoyant force," and "effective weight." These are related concepts, but they are not the same. The **buoyant force** is an always upward force acting on a partly or completely submerged object. The buoyant force may or may not be large enough to make the object float. **Buoyancy**, on the other hand, refers to the object's overall tendency to float and results from competition between two forces: the object's weight, which tries to pull it down, and the object's buoyant force, which tries to pull it up. As explained in the main text, buoyancy can be positive (if the object floats), negative (if the object sinks), or neutral (zero, if the buoyant force and weight exactly balance one another).

Effective weight is the opposite of buoyancy. Like buoyancy, it can be positive, negative, or neutral, but the algebraic sign is reversed. An object that sinks has a positive effective weight (i.e., it still seems to weigh something in water), but one that floats has a negative effective weight. In the latter case, the object's actual weight is still a positive number and pulls downward, but because the object's buoyant force is larger than its actual weight, the object tends to "fall" upward—hence the concept of a negative effective weight. Effective weight is also known by several other names, including wet-weight, submerged weight, and in-water weight.

Whether a submerged object sinks downward toward the bottom or floats upward toward the surface is determined by the outcome of a tug-of-war between two opposing forces. One force is the object's weight, which always tries to pull the object down. The other is the object's buoyant force, which always tries to lift it toward the surface. Based on Archimedes' Principle, it's possible to make the following three statements about the tug-of-war and its outcome:

- If an object weighs less than the weight of the water it displaces, its buoyant force will overpower its weight, and the object will float to the surface, eventually coming to rest with part of the object above the water line (like a floating boat). For example, a block of wood weighing one pound floats because it would displace *more* than one pound of water if fully submerged. Objects that float are said to have **positive buoyancy** or to be **positively buoyant** or simply to be **positive**.
- If an object weighs more than the weight of the water it displaces, the object's weight will overpower its buoyant force, and the object will sink. For example, a chunk of lead weighing one pound sinks because it displaces *less* than one pound

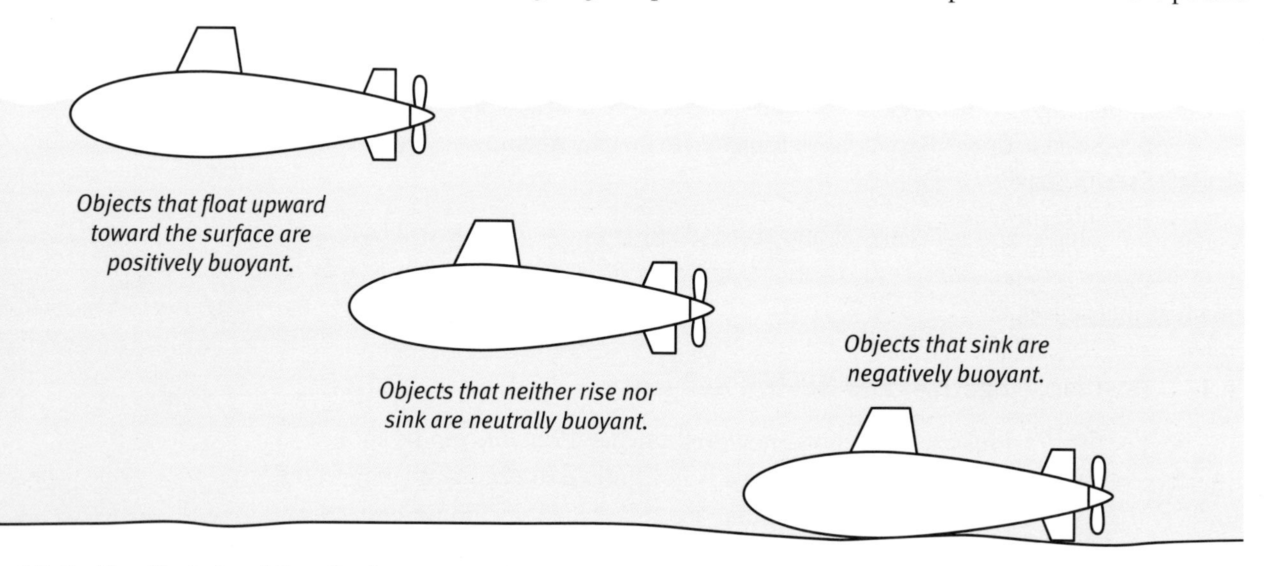

Figure 6.7: Positive, Neutral, and Negative Buoyancy

of water. Objects that sink are said to have **negative buoyancy**, or to be **negatively buoyant**, or simply to be **negative**.

- If an object weighs *exactly* the same as the weight of the water it displaces, the tug-of-war between weight and buoyancy will end in a tie, and the object will hang motionless in mid-water, neither sinking nor floating. This is how a submersible can hang in mid-water without expending energy. In this case, the object is said to have **neutral buoyancy**, or to be **neutrally buoyant**, or simply to be **neutral**. (It's worth pointing out here that achieving *perfect* neutral buoyancy is a little like balancing a marble on the edge of a sharp knife—it's not going to happen. Nonetheless, you can get pretty close to neutral buoyancy, and this idealized situation remains a very important and useful concept in underwater vehicle design.)

2.3.1. Why Boats Sink

So what happens when a boat sinks? Here again, Archimedes' Principle provides the answer. Boats most commonly sink when they take on excess water. Sometimes this occurs because they run into something and get a hole punched in their hull. Other

TECH NOTE: PREDICTING POSITIVE, NEGATIVE, AND NEUTRAL BUOYANCY

There are several different, though equivalent, ways to predict whether an object will sink or float *before* it goes into the water. All of the ones listed below are based on Archimedes' Principle.

1. Compare Weight: As stated earlier, Archimedes' Principle compares the weight of the object to the weight of the water the object displaces. If the object weighs less than the water it displaces, the object is positively buoyant and floats. If the object weighs more than the water it displaces, the object is negatively buoyant and sinks.

2. Compare Density: Compare the object's average density to the density of the water. Technically, density is mass per volume, though you can also use weight per volume for this comparison. If an object's density is greater than that of water, the object will sink. If it's less, the object will float. So a chunk of steel sinks in freshwater, because it has a density of about 7,800 kilograms per cubic meter, which is greater than the 1,000 kilograms per cubic meter of freshwater. For objects that have non-uniform density (such as a steel pressure canister, which has steel on the outside and air on the inside), you can use the average density. To get the average density, divide the object's total mass by its total volume.

3. Compare Specific Gravity: Another method is to look at the specific gravity of the object. **Specific gravity** is defined as the density of the object or material divided by the density of freshwater. (For objects with non-uniform density, you can use average density to calculate specific gravity). By definition, freshwater has a specific gravity of 1.000. Anything with a specific gravity greater than 1.000 is denser than freshwater and will sink in it. Anything with a specific gravity less than 1.000 is less dense than freshwater and will float in it. Seawater has a specific gravity of about 1.025, so any object with a specific gravity of more than 1.025 will sink in seawater, and anything with specific gravity less than 1.025 will float.

Use whichever version of these comparisons—weight, density, or specific gravity—is easiest; they are all based on Archimedes' Principle and are saying exactly the same thing in slightly different ways. With just a little arithmetic, it's possible to convert from any one of the approaches to any one of the others.

TECH NOTE: WHY ARE UNDERWATER CRAFT SO HEAVY?

An underwater vehicle is much heavier than a boat of the same size. Archimedes' Principle helps explain why this is so. Most underwater vehicles are designed to be almost neutrally buoyant when submerged. To achieve this, their weight must equal the weight of the water they displace.

Recall that freshwater has a density of 1,000 kilograms per cubic meter (equivalent to 62.4 lbs/ft^3, and seawater is even denser. So a small ROV or AUV that displaces only one cubic foot of water must already weigh over 60 pounds.

While it may be possible for one or two strong people to launch and recover a 60-pound vehicle by hand under very calm, stable conditions, larger vehicles or rougher conditions necessitate a crane or other form of assistance to keep people safe. As you design your vehicle, keep in mind that you'll need to keep its displacement volume very small if you plan to hand-launch it. If it's bigger than about a half cubic foot in volume, you should plan on using some extra equipment to launch and recover it safely.

times it can happen when a wave swamps the boat, filling it with water. Still other times, it can happen when the boat capsizes, thereby allowing water to flow inside the overturned vessel. In all these cases, an excessive amount of water enters a space from which it was previously excluded (displaced) by the hull.

In light of what you have just learned about positive, negative, and neutral buoyancy, there are two equally accurate ways to explain why this additional water causes a boat to sink. Both explanations are based on Archimedes' Principle and a transition from positive to negative buoyancy. This transition is caused by a reversal in the relative magnitudes of the buoyant and weight forces.

Explanation #1: Decreased displacement volume decreases the buoyant force. Water is now occupying a space from which it had been previously excluded. In this respect, the volume of water being displaced by the hull has decreased, as if the boat's hull had shrunk to a smaller size. The reduced displacement volume means a reduced buoyant force. If that buoyant force decreases until it is no longer sufficient to support the boat's weight, the boat becomes negatively buoyant and sinks.

Explanation #2: Increased weight. The alternative approach is to assume that the displacement volume has *not* changed (i.e., the hull is still displacing the original amount of water), but rather that you've just added a whole lot of very heavy cargo (in the form of water) to the boat. If the total weight builds up to the point where it exceeds the buoyant force provided by hull displacement, the boat becomes negatively buoyant and sinks.

These are not two different processes, just two different ways of thinking about the same process. You can use whichever explanation makes the most sense to you, but don't mix and match—if you think of sinking as caused by reduced displacement volume, don't count the extra water weight inside the boat. If you count the extra water weight, don't include the reduced displacement volume.

3. Designing for Optimal Buoyancy

Figure 6.8: A Lesson on Buoyancy

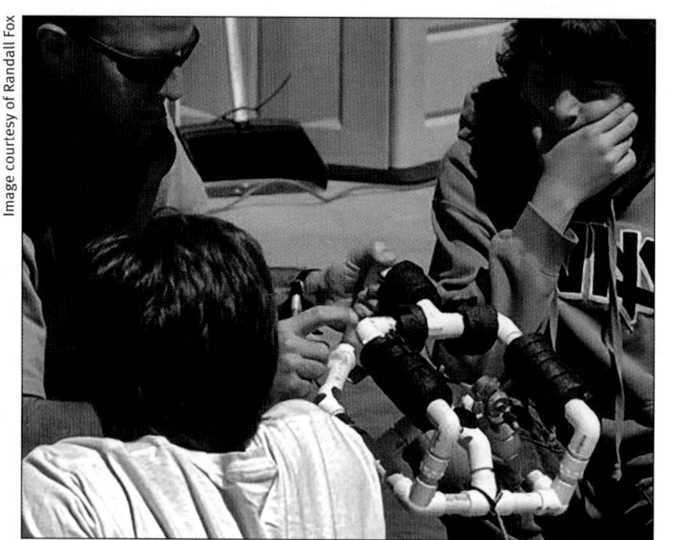

Image courtesy of Randall Fox

Different types of vehicles demand different types of buoyancy. A regular ship or boat needs to maintain strong positive buoyancy at all times. On the other hand, a bottom-crawling vehicle used to dig trenches and bury communications cables needs to be negatively buoyant for traction on the bottom.

Mid-water vehicles, such as submarines, submersibles, and most ROVs and AUVs, are best served by near-neutral buoyancy. Neutral buoyancy allows such a vehicle to hover at any desired depth with minimal expenditure of energy, and when it needs to move vertically, the necessary movements require only small forces (much less than the vehicle's weight). This reduces the size, complexity, and cost of the systems needed for vertical movement.

While it's always possible to achieve your buoyancy goals by adding enough weight or flotation to your vehicle after it's done, strapping large bunches of weights or floats to the outside of a finished vehicle is not normally recommended. Such additions add to the total size, weight and drag of the vehicle, reducing its performance and efficiency.

A much better solution is to design a vehicle for optimal buoyancy from the start, so the need for external floats or weights will be eliminated (or at least minimized). This extra up-front design effort will be repaid later in the form of a vehicle that is smaller, lighter, sleeker, faster, more efficient, and more convenient to operate than it otherwise would have been.

As mentioned earlier, perfect neutral buoyancy is a hypothetical condition that cannot be precisely achieved in practice. Given that, is it better to be a little bit positive or a little bit negative, or does it not matter at all? It turns out that it does matter. For a variety of reasons, it's almost always better to make a "neutrally buoyant" vehicle slightly positive in buoyancy.

One reason is that this strategy makes it easier to recover the vehicle in the event of a propulsion system failure because the vehicle will drift back to the surface on its own. For safety reasons, this is essential in manned submersibles, but it's also a good idea in unmanned vehicles.

A second reason is that it's easier to make adjustments to a vehicle or its payload while the vessel is floating passively on the surface than while it's resting on the seafloor (or the bottom of the pool).

A third reason to opt for slightly positive buoyancy is that it's easier and more effective to add weight than to add flotation when the vehicle's buoyancy needs to be trimmed. (See *Tech Note: Is It Better to Add Floats or Weights?*)

In summary, as you design your underwater vehicle, you will probably want to aim for slightly positive buoyancy. For a small, low-cost, unmanned vehicle, a net buoyant

TECH NOTE: IS IT BETTER TO ADD FLOATS OR WEIGHTS?

When designing a vehicle, it's good to recognize early in the planning process that your vehicle will probably need to have its buoyancy trimmed (adjusted) when it first enters the water. That raises an interesting question: Are you better off designing a slightly positive vehicle and adding weights to trim it, or designing a slightly negative vehicle and adding floats? It turns out it's generally easier and more effective to add weight than to add flotation.

There are at least four reasons for this. First, solid metal weights are easy to find or fabricate in a wide variety of shapes and sizes, whereas floats come in a comparatively limited range of sizes, shapes, and types. Second, solid chunks of metal don't compress or leak the way many floats can, so you don't have to worry about changes in buoyancy with depth. Third, unlike floats, which must displace water to work, weights can be placed inside a vehicle's pressure hull or canisters, where they do not add to the displacement volume or drag and do not increase the risk of entanglement. Fourth, metal weights pack more buoyancy-adjusting power into a smaller package than floats do.

An example may help to clarify that last point. Suppose your vehicle's buoyancy is off by 5 newtons (roughly 1 pound). If the vehicle is too heavy, you'll need to add enough flotation to displace 5 newtons of additional water, plus a bit more to compensate for the weight of the float itself. That translates into more than 500 cubic centimeters of displacement. If the float is a sphere, it will need to be about 10 centimeters in diameter to displace this volume of water. If, on the other hand, the vehicle is 5 newtons too buoyant, you'll need to add 5 newtons of net weight. Lead weights have a specific gravity of about 11.4, so you can achieve this added weight using only $500/11.4 = 44$ cubic centimeters of lead. A spherical lead fishing weight only 2.2 centimeters in diameter would do the trick. This is much smaller than the corresponding 10-centimeter float size.

In short, if you have the choice, design your vehicle to start out somewhat more positive than you really want it to be, then trim it back down to ideal buoyancy with metal weights. If the weights will be in contact with water, make sure they are lead or another corrosion-resistant metal.

force equal in magnitude to about 4 percent of the total vehicle weight is a reasonable design goal, with the understanding that you'll probably add weights later to trim that down to 1–2 percent of vehicle weight. Thus, for example, a vehicle weighing 50 pounds should be designed to displace about 52 pounds of water, then have its buoyancy fine-tuned with about 1 pound of lead (or other metal) weights.

3.1. The Weight Statement Table

When the moment arrives to slide your vehicle into the water for its maiden voyage, two quantities are going to take center stage. The first is the overall weight of the vehicle. This will determine how easy it is to lift the vehicle in and out of the water and whether you will need additional people and/or equipment to launch and recover the vehicle safely. The second is the overall buoyancy, which will determine whether (and how strongly) your vehicle floats or sinks before you power up those thrusters.

To determine the overall weight and buoyancy before the vehicle goes into the water (and ideally before it even leaves the drawing board), you'll need to know the weight and buoyancy of each of its components. It's usually straightforward to measure the weight of each part. Depending on its size, you can use postal scales, spring (fish) scales, kitchen scales, laboratory scales, or bathroom scales.

Determining the buoyancy of a part is sometimes more complicated. Although there are ways to measure buoyancy directly, it's usually easier to calculate it from related measurements of displacement volume, displacement weight, or buoyant force.

The **weight statement table** is a design tool you can use to record these important quantities. (See Table 6.1 for a sample weight statement table.) The table is particularly helpful in predicting your vehicle's overall buoyancy. As you now know, an underwater vehicle floats, sinks, or hovers, depending on the outcome of a tug-of-war between its weight and its buoyant force. The weight statement table helps you keep track of the tug-of-war score as you add, remove, or modify vehicle parts in your design. If you are striving for neutral buoyancy, you can use the weight statement table to make sure this tug-of-war ends in a tie. If you want a vehicle that has a buoyancy of +1 pound, you can use the table to make sure the buoyant force is one pound bigger than the weight.

A weight statement table typically includes one row for each item and columns for important characteristics of each item, including description, weight, displacement, and related quantities. (See sample table below.) You can create a weight statement table with pencil and paper, use a computer spreadsheet program, or come up with some other format of your own. The key is to have a system that lets you record and update the relevant information in an organized way as your design evolves.

Most groups find that a computer spreadsheet is the most convenient and powerful way to build a weight statement table, because you can embed formulas in the spreadsheet. If done correctly, these formulas will automatically recalculate total weights, displacement volumes, and buoyant forces as you add or modify parts in your design. If you don't know how to use a spreadsheet yet, this is a good time to ask someone to show you how. You will find spreadsheet skills useful for a wide variety of projects beyond your underwater vehicle project.

3.1.1. Building the Table

To create a weight statement table, start by making a row for each part in your vehicle. For example, you might have one row for the main pressure canister, another for each thruster, and still another for a gripper arm. You will also need several columns in the table. One column should contain the name or description of the part. You will also need to include a column for weight. Each part contributes weight, even if it's buoyant, so you'll need to enter its weight and include that amount in the total downward force acting on the vehicle. Include another column for displacement. If a part displaces water when the vehicle is submerged (as some, but not all, parts do), then you'll need to record this displacement volume and add it to the total displacement volume of the vehicle. Objects that are completely enclosed within a pressure canister (for example, any on-board circuitry) will contribute to vehicle weight, but not to buoyant force. Objects that touch water (including the pressure canisters enclosing other objects) will contribute to both weight and buoyant force.

Eventually, you'll need to convert the displacement volumes to buoyant forces. You can do this individually for each item, or you can wait and do it just once to convert the total displacement volume to the total buoyant force. Most people prefer to do it in each row, so they can see the direct contribution of each part to both weight and total buoyant force. In fact, most people take it a step further and add a column that shows

Table 6.1: Sample Weight Statement Table for a Small Thruster

Project:	*Vertical Thruster for King High School ROV*
Team Members:	*Shana, Jay, Kelli, Mike*
Planned Dive Site:	*Fisherman's Wharf*
Water Density at Dive Site:	*Seawater = 1025 kg/m³ = 1.025 g/cc*
Last updated:	*2 March 2010*

Part	***Weight of Part (gf)***	***Approx. Shape and Dimensions of Part***	***Volume of Displaced Water (cc)***	***Weight of Displaced Water (gf)***	***Buoyancy (gf)***
Motor Housing	*145*	*Cylinder. See note #1 below*	*172*	*176*	*+ 31*
Motor (Beetle B16)	*65*	*Not important here, because it doesn't displace water.*	*None*	*0*	*– 65*
Propeller	*6*	*Complex shape. See note #2 below*	*6*	*6 See note #2*	*0*
Shaft	*5*	*Cylinder. Diam = 0.4 cm, Length = 3 cm*	*Assumed negligible*	*Assumed negligible*	*0*
TOTAL	*221*	*—*	*178*	*182*	*– 39*

Notes:

1. Thruster motor housing is made from a section of 1" Sch 80 PVC pipe with endcaps and is roughly cylindrical in shape. The cylinder's outside dimensions are approximately 1.5 inches diameter by 6 inches long. Converting to metric, the cylinder is approximately 3.8 cm diameter by 15.2 cm long. Using the formula for the volume of a cylinder, the displacement volume for this cylinder is 172 cc.

2. This is a Robbe 60 mm diameter 3-blade plastic prop. Because of its complex shape, the propeller's buoyant force was measured directly by suspending it in a bowl of freshwater on a kitchen scale. The mass of seawater displaced would be only slightly higher (6 cc x 1.025 g / cc = 6.15 g), so it was rounded to 6 g.

3. 1 gram force (gf) is the weight of 1 gram of mass on Earth. See Chapter 3, Sections 2.2.2 and 2.2.3 for a discussion of the difference between weight and mass.

the buoyancy (buoyant force minus weight) for each part. If the part's buoyancy is positive, it helps the vehicle float; if negative, it helps the vehicle to sink.

Be very explicit and careful with units in a weight statement table. People commonly get into trouble here because they get sloppy with units and unit conversions. To reduce this problem, it helps to adopt a standard unit for weight and another for displacement volume, then convert everything to those units before entering it in the table. For small vehicle projects, it's almost always easiest to work in metric units, using **grams force** for weight and cubic centimeters for volume. (One gram force, gf, is the weight of a 1-gram mass in Earth gravity.) Avoid the temptation to ignore units while performing calculations and unit conversions, because attention to the units will often reveal otherwise invisible mistakes, like multiplying when you should have divided.

The sample table above (Table 6.1) illustrates one way in which a weight statement table might be organized. This particular one is for illustration purposes only and does not cover an entire vehicle—just a single homemade thruster consisting of a motor, a propeller, and a cylindrical pressure canister used to house the motor. Nonetheless, it contains all the key components of an effective weight statement table. The table for a complete vehicle would normally include many more rows (and more footnotes), since the entire vehicle would include many more parts.

Here are some things to notice about this table:

1. **Title and Header Information**: First, the top of the table contains basic information needed to identify what the table is for and who is involved in the project. It also contains a water density entry to help you convert from displacement volume to displacement weight lower in the table. The "last updated" date is important, because you'll probably keep modifying the table as you add or change parts in your vehicle design, and you'll need to know which version of the table is your most recent one. If you use a spreadsheet program to create and update your table, you may also want to encode the revision date in the file name each time you update the table.

2. **Units of Force**: Weight and buoyant force in this table are expressed in grams force (gf). One gf represents the weight (on Earth) of 1 gram of mass (about 0.0098 newtons). For a variety of reasons, using gf in these weight statement tables for small underwater vehicles turns out to be much more convenient than using other common units of force. Note that most kitchen scales, postal scales, and other small scales that claim to be reporting mass or weight in grams are, in

TECH NOTE: ADVANCED TIPS FOR USING WEIGHT STATEMENT TABLES

To fill in the table for an entire vehicle, it's critical to list *all* the parts of your vehicle, since every item contributes to the total weight of the vehicle. It's ok to lump parts together, if it's more convenient to do so. For example, you could use one line for a gripper arm, even though the arm itself is made of dozens of smaller components, such as wires, a motor, bearings, a pressure canister, and jaws. The only caveat here is to be careful with certain parts, such as a thruster motor shaft, where only part of the shaft might be in contact with water. In such a case, it's permissible (in fact, ideal) to mentally split the part into an in-water portion and an out-of-water portion, then include those on separate lines of the table as if they were physically separate parts. If you're treating the whole thruster (including the shaft) as one lumped part, you don't need to worry about this level of detail. Parts with very small weights *and* very small displacement volumes (for example, decals from your team's sponsors) can safely be ignored.

fact, reporting only weight and only in gf, even though the readout may say "g" instead of "gf."

3. The **Part column** contains short, descriptive names to help you keep track of parts, to make sure you've included all relevant parts in the table.

4. The **Weight of Part (gf) column** is where you enter the weight of each part. Note that this column heading, like some others in the table, explicitly specifies the units for all entries entered in the column. This is a good idea, because it reduces clutter in the table and, more importantly, reduces the chance of calculation errors due to mixed-up units. Of course, this means that you'll need to express every part's weight in grams-force before entering it in the table. If you have a scale that measures weight directly in "grams" (which are really grams-force with those scales), that's easy. If not, you'll have to convert from other units, such as ounces or pounds. Every single part should have an entry here, even if that entry simply says "negligible" for very tiny parts, because every part contributes weight to the vehicle.

5. The **Approximate Shape and Dimension column** is used to summarize information needed to calculate the displacement volume of the part. If the part does not get wet, this information is not really needed, and you can just say so. In many cases, the detailed shape and dimensions won't fit in the table cell easily; in these cases, you can refer the reader to a numbered "note" in the notes section at the bottom of the table, where you have room to explain in more detail the dimensions and methods used to calculate the displacement volume.

6. The **Volume of Displaced Water (cc) column** is the volume of water displaced in cubic centimeters (cc or ml). This volume can be calculated from the part's shape using geometric formulas, or measured directly using one of the techniques described in *Tech Note: Determining the Displacement of Odd-Shaped Parts.* If you use formulas, be especially careful when converting areas or volumes between different units. For example, there are 2.54 cm in one inch, but $(2.54)^2$ = about 6.45 *square* centimeters in one *square* inch and $(2.54)^3$ = about 16.39 *cubic* centimeters in one *cubic* inch. If the part does not come into contact with water, then it does not displace water, so you can just write "None."

7. The **Weight of Displaced Water (gf) column** is used to record the weight of the water displaced by the part. If the part does not contact water, this value will be zero. If the vehicle will operate in freshwater, the number in this column will match the number in the volume column, because freshwater has a weight density of exactly 1 gf per cc. If the vehicle will operate in seawater, where the density of the fluid is slightly higher (1.025 g per cc), the exact number will be slightly higher; however, due to rounding, this difference may not be apparent for parts with low volumes, as in the case of the propeller in the sample table.

Figure 6.9: Calculating Volume and Buoyancy for Different Geometric Shapes

Calculating the volume of objects immersed in water is a crucial step in calculating net buoyancy. ROV components come in a variety of shapes and sizes. The formulas given here are for figuring out the volume of solid or sealed objects of regular shapes. To calculate buoyancy, multiply volume x weight density of water. (For more geometric formulas, refer to Appendix II: Useful Constants, Formulas, and Equations.*)*

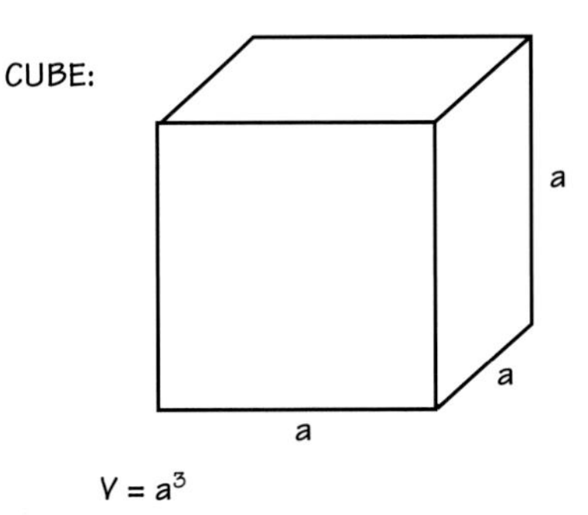

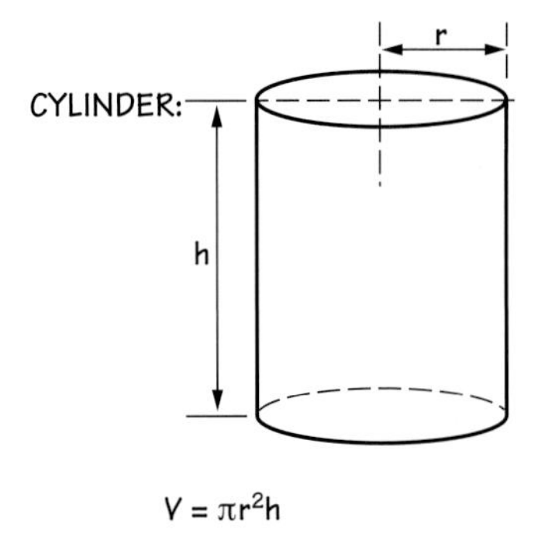

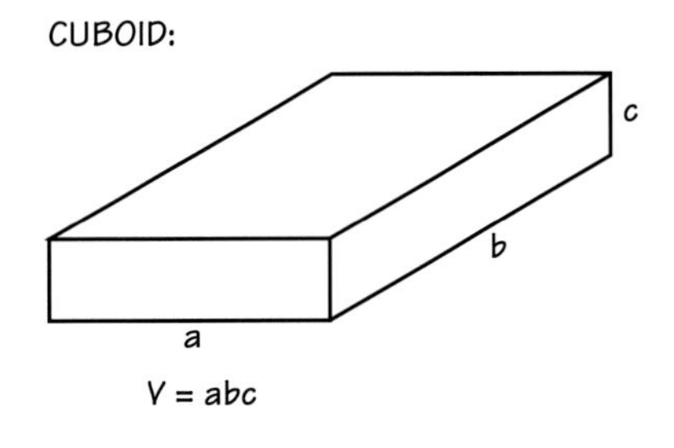

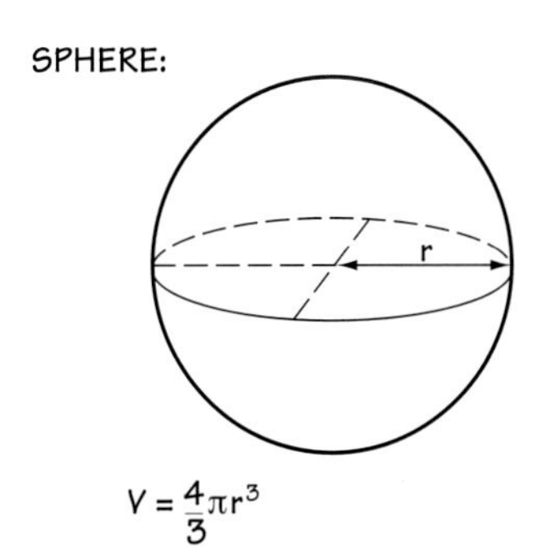

8. The **Buoyancy (gf) column** is where you subtract the weight of the part from the weight of the displaced water to see which force wins the weight-versus-buoyant force tug of war. This calculation gives you the overall buoyancy of the part. (Remember, buoyancy is not the same as buoyant force; see *Tech Note: Buoyancy, Buoyant Force, and Effective Weight* in Section 2.3.1.) If this buoyancy is positive, the part floats by itself and will help to float the vehicle. If the buoyancy is negative, the part will help to sink the vehicle.

9. The **TOTAL row** near the bottom of the table keeps track of the sum of the values in each of the other columns. Two of these totals—weight and buoyancy—are especially important, particularly when the weight statement table covers the entire vehicle instead of just the one thruster illustrated in this sample table. The total weight of all the parts combined tells you how heavy the vehicle will be out of the water. You want to keep this number small, if possible, to make it easier to transport, launch, and recover the vehicle. Remember that 1,000 gf is equivalent to 2.2 pounds. The other important total value, particularly in the context of this chapter, is the total buoyancy. This number tells you whether the vehicle will float or sink by itself. You probably want your ROV or AUV to have slightly positive buoyancy. As described earlier, you probably want to make this number positive and equal in magnitude to about 4 percent of the total vehicle weight prior to any last minute fine-tuning of buoyancy. For example, if your vehicle's total weight is 10,000 gf, you want the total buoyancy to be somewhere around +400 gf.

To complete the table accurately, you will need to perform a variety of measurements and calculations. See the Tech Notes in this section of the chapter for some helpful suggestions and examples.

A common question that arises in the course of filling out a weight statement table is: "How accurate and precise do I need to be in my measurements and calculations?" Do you need to measure every tiny nut and bolt to the nearest gram? Do you need to worry if your "cylindrical" motor housing isn't a perfect cylinder, but instead has some bumps or ridges on it that might throw off calculations based on the cylinder volume formula? The answer to these questions depends on how accurate and precise you

TECH NOTE: WEIGHING PARTS

One challenge in completing a weight statement table for a small ROV or AUV is that you may need accurate weights for a wide range of objects, from things as light as a single nut or bolt to as heavy as a bucket full of water or more. To cover this range of weights, you will probably want at least two different-sized scales.

Small postal scales designed to weigh letters and the triple-beam balances designed for use in many science classrooms work well for small objects like nuts and bolts. Flat-topped, digital kitchen scales provide another excellent option. These are designed for measuring the ingredients in food recipes and combine modest capacity (up to about 5 kg or 11 pounds) with good precision (often measuring down to 1 gram, which is much less than 1 ounce) at an affordable price. Some even include buttons to switch between imperial and metric units and "tare" (zeroed) buttons for automatically subtracting the weight of a container used to hold fluids or small parts. If you'll be weighing wet parts, make sure whatever scale you use is stainless steel, plastic, or otherwise water-resistant, or cover it with a light sheet of plastic.

For objects weighing several pounds or more, a standard bathroom scale works well. These won't weigh individual small things accurately; however, there is a trick you can use to get precise weights of small things on a larger scale, *if* you have very large numbers of identical small things. For example, if 100 identical bolts weigh about 2 pounds on your bathroom scale, then it's safe to assume that each bolt weighs about 2/100 = 0.02 pounds.

TECH NOTE: DETERMINING THE DISPLACEMENT OF ODD-SHAPED PARTS

Some parts, such as propellers, have complex or irregular shapes that do not lend themselves to easy geometric calculations of volume. Fortunately, there are other ways to determine displacement volume that work with both simple and complex geometries.

One straightforward approach is to submerge the part in a container of water and directly measure the volume of water displaced. If the part is small, the best way to do this is to slide it into a half-filled measuring cup, beaker, or graduated cylinder that has markings to indicate the precise water level. As long as the part submerges completely and the water does not overflow, you can subtract the original water level from the final water level to get the apparent change in water volume (which corresponds to the volume of the part). If the part is relatively large (say, the size of your fist or greater), you can immerse it in a completely full bucket of water, catch the overflow, and measure the volume of the water that overflowed. For accurate results, you must 1) fill the bucket to overflowing in order to make sure it is *completely* full before you submerge the part, 2) move slowly to avoid sloshing excess water out of the bucket, 3) catch every drop of overflowing water, and 4) avoid submerging your fingers or other devices used to push the part under water—otherwise you might inadvertently count their volume as belonging to the part.

Here again, once you know the displacement (volume), you can multiply it by the weight density of water to obtain the displacement weight, which is numerically equal to the upward buoyant force.

When estimating displacement volumes using these immersion techniques, be extremely careful about things like hollow PVC pipe that can trap variable amounts of air. If this happens, the displacement volumes in these pipes can vary between dives or even within a dive as air escapes, shifts around, or compresses. To avoid this problem, many ROV and AUV builders drill holes in hollow frame members to allow any trapped air to escape before a dive. In a weight statement table, you would need to remember that the hollow inside of such a frame member is full of water and therefore does not count as displacement volume; however, the pipe wall material itself would count.

TECH NOTE: DETERMINING THE DISPLACEMENT OF PARTS WITH SIMPLE GEOMETRY

Some parts have shapes that can be closely approximated by combinations of cubes, cylinders, spheres, or other simple geometric forms. In these cases, it's often easiest to measure a few key dimensions of the part, then use simple geometric formulas to calculate the displacement volume. Volume formulas for some common shapes are given in Figure 6.9 and *Appendix II: Useful Constants, Formulas, and Equations*.

As an example, consider the challenge of estimating the displacement of a PVC pipe frame that's allowed to fill with water, as is typical of such frames. In this case, the plastic walls of the pipe displace water, but the hollow interior of the pipe does not. To estimate the displacement of a straight section of the frame, you can first calculate the displacement volume of that section as if the pipe were a solid plastic cylinder, then subtract the volume that's occupied by the cylindrical plug of water inside the real pipe.

Once you know the displacement (volume), you can multiply it by the weight density of water to obtain the displacement weight, which is numerically equal to the upward buoyant force.

Figure 6.10: Calculating Displacement Volume of Hollow Pipe Walls

need your answer to be. The more carefully and accurately you make your measurements, the more closely your actual vehicle weight and buoyancy will match your predictions. That said, it's nearly impossible to get an exact match, so don't worry about it excessively. If you do a reasonably careful job with your measurements and calculations and make sure you don't leave out any large or heavy parts, your final vehicle design should be a lot closer to its ideal weight and buoyancy than if you had not bothered to do a weight statement table at all! And as you'll see soon, any minor errors in your predictions can be corrected later when you go through a processes called "trimming" your vehicle's buoyancy.

You can begin a weight statement table as soon as you like, but it becomes essential to have one and use it as you define your concept design. To avoid unpleasant surprises, be sure to update the table as changes are made throughout all stages of design and construction.

Feel free to customize a weight statement table to suit the needs of your particular project. The important thing is to make sure it keeps track of the weight and displacement volume of each part, so you can add them all up. Remember, the table should

TECH NOTE: ONE WAY OF MEASURING BUOYANT FORCE DIRECTLY

Although you can calculate an object's buoyant force by multiplying its displacement by the weight density of water, it's also possible to measure that force more directly, without having to first determine the displacement volume. These measurements are best done with a digital kitchen scale or other scale that can handle fairly heavy weights, yet provides good fine-scale resolution of weight.

One method for measuring buoyant force simply measures the weight of water displaced by the object. To do this, start by filling a bucket completely with water and weighing it carefully on the scale. Write down this combined weight of bucket and water. Then move the bucket to a sink, refill if any water spilled, and immerse the object completely in the water. Let the overflowing water go down the drain. If the object floats, you'll need to push it completely under the water surface, but don't submerge your fingers or otherwise displace much more water than the object itself. Also, make sure the object is not touching the sides of the container while you are pushing on it. You may need to use some toothpicks or other little sticks with negligible volume to push the object under. As you remove the object, the water level in the bucket will drop. Move the bucket back to the scale (being careful not to spill any of the water) and re-weigh it. The original weight of the full bucket minus the new weight of the partly full bucket equals the weight of the missing water that overflowed from the bucket. Since this is the water displaced by the object, Archimedes' Principle tells you that this number is also the upward buoyant force acting on the object when it's submerged.

Figure 6.11: Calculating Displacement via the Immersion Method

also record the density of water at the planned dive site, so you can accurately convert the displaced water volume to a displaced water weight for comparison with the total vehicle weight.

3.2. Adjusting Vehicle Buoyancy

Once the vehicle is built and in the water, you will probably discover that it's slightly more or slightly less buoyant than you predicted from your weight statement table. If you've included every part and if you've been thorough and careful with your measurements and calculations, the difference between the observed and predicted buoyancy should be slight. However, the two will almost never match exactly. Fortunately, you can fine-tune a vehicle's buoyancy by adding weights (which increase vehicle weight without dramatically increasing displacement) or by adding flotation (which increases vehicle displacement without dramatically increasing weight). This process is called adjusting the buoyancy. With small ROVs and AUVs, the usual goal is to adjust these vehicles to be just slightly positive, so they float at the surface but can be driven downward easily when it's time to dive.

TECH NOTE: ANOTHER WAY OF MEASURING BUOYANT FORCE DIRECTLY

A slightly different way to measure buoyant force does essentially the same thing, but without any overflowing water. Fill a bowl or beaker (one large enough to contain the part you want to measure) about 2/3 full with water and put it on the scale. Make sure the scale can comfortably handle that much weight with room to spare. Write down the combined weight of the container and water. Next, completely submerge the part in the water *without letting it touch the sides or bottom of the container*. If an item can sink in water, you should tie or tape a thin thread to it and suspend it in the water by the thread. If the part normally floats, you'll need to poke it under and hold it there with a thin object, such as a needle or toothpick. Regardless of whether you suspend it or push it under, the part must be *completely* submerged and must *not* touch the bottom or sides of the container. While the item is under water, record the new weight, which should be higher than the first weight you wrote down. The increase in measured weight is the buoyant force acting on the object. For accurate results, you must make sure that no water overflows and that the combined weight of the container + water + buoyant force does not exceed the limit of the scale. This approach is illustrated in Figure 6.12.

If you are able to measure the buoyant force acting on the object directly, you don't really need to calculate the displacement volume, but if you want to (to fill all the columns in your weight statement table), you can use the density of water and the fact that the buoyant force equals the displacement weight to calculate the displacement volume. For example, if the original weight of the water and container is 287 grams (gf, in this case), and the weight with the submerged part in the water is 292 grams, the buoyant force is 292 grams – 287 grams = 5 grams. That means the displacement volume of the part must be 5 cubic centimeters.

Image courtesy of Dr. Steven W. Moore

Figure 6.12: Measuring the Buoyant Force of an Oddly Shaped Part

This photo shows a simple, direct way to measure the buoyant force of an individual part, particularly an oddly shaped one such as this propeller. The red wire in this photo is supporting all of the weight of the propeller except that portion of the weight supported by its buoyant force. That buoyant force, the weight of the water, and the weight of the beaker are all transferred directly to the kitchen scale. If the scale was "tared" (zeroed) just before the propeller was submerged, then the number on the display is the buoyant force acting upward on the propeller. In this case, the display shows it is equal to the weight of a 6-gram mass.

If your scale does not offer this tare feature, you can just record the weight of the water and beaker before submerging the propeller, then subtract that recorded weight from the total weight measured while the propeller is submerged. For accurate results, remember that the propeller must be completely submerged, but must not touch the sides or bottom of the container. Also, the volume of wire or string submerged should be small enough to displace a negligible weight of water.

Image courtesy of Steve Van Meter, VideoRay LLC

Figure 6.13: Adjusting the Buoyancy

You will need to re-adjust the buoyancy whenever you modify the vehicle (by changing a payload, for example) and whenever you move it from freshwater to saltwater or vice versa.

As mentioned earlier, buoyancy adjustments are most easily and effectively made by adding small weights to a vehicle that is a bit too positively buoyant. If possible, add weights *inside* pressure canisters or hulls. Not only are weights more effective in this location (because they don't displace any additional water and therefore don't add anything to the vehicle's upward buoyant force), but they also don't change the size or shape of the vehicle, so they don't increase drag.

A design team that's on the ball will anticipate this need to adjust the amount of weight at the dive site and will build in features that make it possible to add or subtract weight quickly, easily, and reliably (you don't want the weights falling off mid-dive), even while the vehicle is in the water. One way to do this is to mount some vertical posts—often called trim tubes or posts—on the frame. Then you can add or subtract stainless steel washers to add or subtract weight. If you use bolts for the posts, you can use wingnuts to lock the washers in place.

4. Why Things in Water Tip or Flip

Image courtesy of Randall Fox

Figure 6.14: "She's listing a bit, Captain!"

Many students on ROV teams work hard to get their vehicle's buoyancy adjusted just right on the drawing board. Later they work hard to trim their finished vehicle perfectly while holding it at pool's edge. Alas, as soon as they let go of it, they watch (with understandable dismay) as their vehicle rolls over on its side or flips completely upside-down.

It turns out that you not only have to add the right *amount* of weight or flotation, but you also need to put it in the right *place.* That's because the distribution of weight within the volume of water displaced by the vehicle determines whether or not a vehicle sits upright. The location of the weights and floats also has a profound effect on how resistant the vehicle is to being tipped over by waves, currents, tether drag, or other forces.

Fast-moving vehicles such as submarines and torpedoes can use **dive planes** (essentially underwater wings or rudders) to generate forces for stabilizing the vehicle, but these work only while the vehicle is moving fairly quickly through the water. Very slow-moving or stationary vehicles, including most ROVs and AUVs, can't rely on dive planes for stability. Since this book is focused on the design of small ROVs and AUVs, coverage here is limited to stabilization methods that work for slow-moving vehicles.

Figure 6.15: Positioning Weights for Effective Vehicle Trim

Weight placed in the wrong position tips the vehicle down. Weight shifted toward the centerline levels the vehicle.

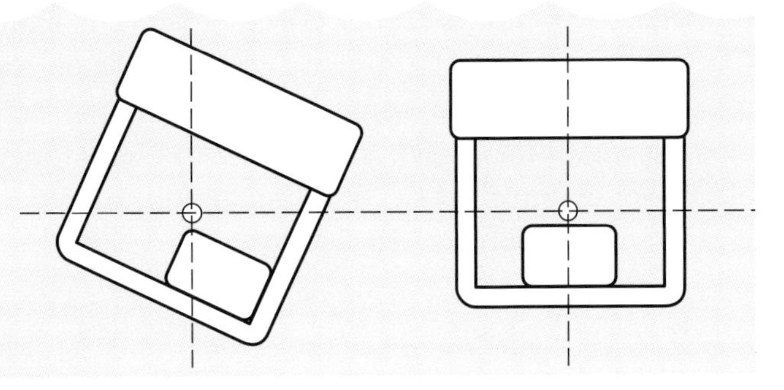

4.1. Preliminary Concepts

Three concepts are central to understanding how to get a vehicle to sit upright and how to keep it that way when it is stationary or moving slowly relative to the water around it. These concepts are:

- center of gravity
- center of buoyancy
- torque

4.1.1. Center of Gravity (CG)

The overall weight of a vehicle is the sum of the weights of each of its parts—every nut and bolt, pipe, battery, wire, motor, camera, and so on. Thus, the effect of gravity on the vehicle is, in reality, distributed across the entire vehicle in a very complicated way. Fortunately, it turns out that for many types of analyses, including those involving buoyancy and stability, the net effect of all of the separate little weights acting together is exactly equivalent to having the entire weight of the vehicle concentrated at one very special point called the **center of gravity**, or **CG**. The CG is essentially the average location of all of the vehicle's mass. Sometimes the CG is called the **center of mass**.

Every object has exactly one CG. Normally the CG is buried somewhere deep inside the object where you can't see it, but that doesn't diminish its usefulness as a concept. The beauty of the CG idea is that you need to focus on only one point, instead of millions of points, when trying to analyze the effects of weight on vehicle performance.

TECH NOTE: FINDING THE CG

Although it's always possible to calculate the location of the center of gravity for any object or collection of objects, the mathematics can become quite involved, so in practice, it's often easier to find the center of gravity empirically. One way to do this for relatively light objects is to hang the object in air (not water) from a string. Do this twice, each time using a different suspension point on the object. Because the object will always hang with its center of gravity directly beneath the point where it is attached to the string, a line drawn straight down from that attachment point will always pass through the CG. By using two different attachment points, you can get two different lines through the CG. The intersection of the lines will mark the location of the CG. See Figure 6.17, where this method is illustrated for a sheet of plastic cut into the shape of the letter *L*. (Note that this happens to be one of those odd cases where the CG is located outside the object.)

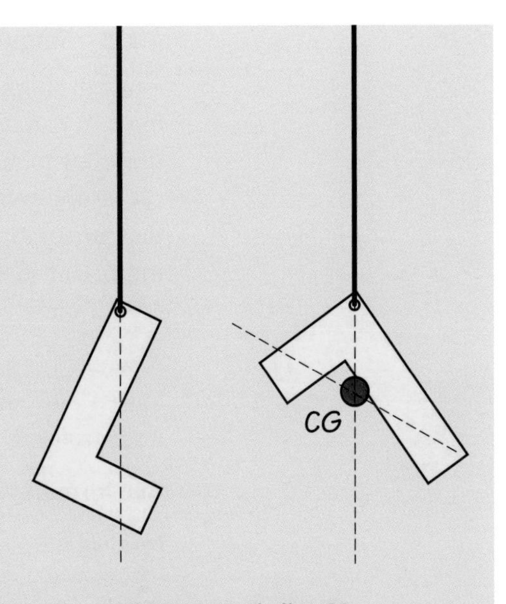

Figure 6.16: Finding CG

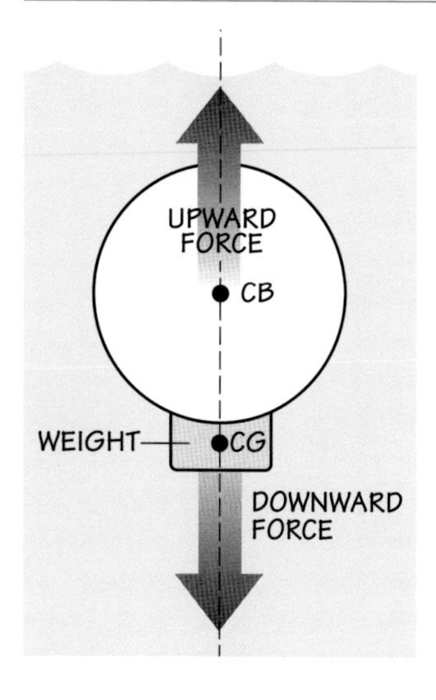

Figure 6.17: CB and CG

The center of buoyancy (CB) is located at the geometric center of the portion of the vessel that is under water. The buoyant force acts upward through this point.

The center of gravity (CG) is located at the point where all the effects of gravity acting on the vessel appear to concentrate. Total vessel weight acts downward through the CG.

In a perfectly symmetrical object with a perfectly uniform distribution of mass, the CG will be at the exact geometric center of the object. More commonly, the CG is off-center, due to a non-symmetrical object shape and/or non-uniform distribution of mass within the object. In some cases, the CG may even be located outside the object. For example, the CG of a donut is located in the center of its hole, where there is no yummy donut material whatsoever.

The exact location of the CG can be determined theoretically, using calculus or similar methods, for any object of any size or shape, but this isn't typically necessary. In most cases, you don't need to know exactly where the CG is, as long as you can estimate roughly where it is and know how to move it around if necessary. (More on this below.) If you do need to find the CG precisely, there is a simple technique you can use to find it empirically. (See *Tech Note: Finding the CG.*)

4.1.2. Center of Buoyancy (CB)

Buoyancy results from an infinite number of separate little pressure-related forces distributed in a complicated way over the entire vehicle surface, much as weight is distributed over all of the vehicle's mass. However, as with weight, we can often proceed as if all of those effects are concentrated at a single point. In this case, that point is called the **center of buoyancy**, or **CB**. An object's center of buoyancy is located where its center of gravity would be if the object had a perfectly uniform distribution of mass, which is almost never the case. An even better way to think about it is to visualize the exact shape of water displaced by the object. Then the CB of the object is the center of gravity of that object-shaped blob of water.

Like the CG concept, the CB concept makes calculations easier by allowing you to treat a complicated, distributed set of millions of forces as if they were a single force concentrated at a single point.

On extremely rare occasions, the CG and CB for an object are located at exactly the same place. (See *Tech Note: Zero BG Vehicles.*) This usually happens only if the object has a uniform distribution of mass, like a solid chunk of metal. For most objects, it works out that the CG and CB are in different locations. Whenever they are separated, they can influence vehicle orientation through something called torque.

4.1.3. Torque

You can think of **torque** as a twisting force, though technically speaking, it is not a force. When using a screwdriver to put a screw into a wall, you are applying torque to the screw to make it turn. For underwater vehicles, torque is the physical principle that causes the vehicle to rotate into a particular orientation in the water, then maintain that orientation. So understanding how and why torque is generated is especially important in designing vehicles that will orient (and re-orient) themselves properly, even if buffeted by waves or other insults.

A wrench is a tool designed specifically to help generate high torque to tighten or loosen nuts and bolts, so this familiar tool can help you understand torque and how it's calculated.

The formula for torque is:

Equation 6.2

$$\tau = FL$$

In this equation, the Greek letter tau (τ) represents torque. It is equal to the product of F (the applied force) and L (a distance called the **moment arm**, or **lever arm**).

Relating this equation to the wrench example, F is how hard you are pushing or pulling on the wrench handle. L, the moment arm, is defined as the *shortest distance* from the pivot point (in this case, the bolt) to the line of action of the force. (See Figure 6.18.) This line of action passes through the place where your hand is pushing or pulling on the wrench handle and runs in the direction that you are pushing or pulling. If you are pushing or pulling perpendicular to the wrench handle, the moment arm is simply the distance between your hand and the bolt (i.e., roughly the length of the handle). If you are pushing or pulling at some different angle, the moment arm will be shorter (again, see Figure 6.18).

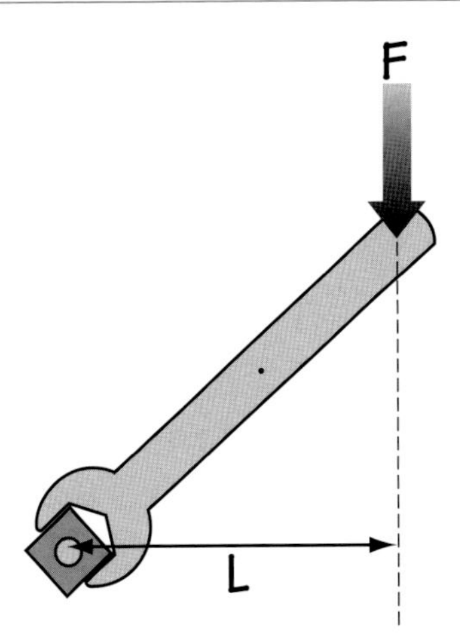

Figure 6.18: Understanding Torque

The torque produced by a wrench is the product of the force (F = heavy, single-headed arrow) applied by a person's hand to the wrench handle multiplied by the length of the moment arm (L = small, double-headed arrow), which is defined as the shortest distance from the pivot point to the line along which the force acts. It's important to remember that the moment arm is always measured perpendicular to the line of action of the force.

In keeping with the formula, torque is usually measured in units that have dimensions of force times length, such as newton-meters (metric) or foot-pounds (imperial). For example, pushing with a force of 20 newtons perpendicular to the end of a wrench handle that is half a meter long generates a torque of 20 x 1/2 = 10 newton-meters.

Note that it is critical to measure the length of the moment arm along a line *perpendicular* to the direction, or line of action, of the force. Forgetting to do so and assuming that the moment arm is always the length of the wrench handle is a common mistake. You can understand why this point is so important if you see that pushing straight toward the bolt along the length of a wrench's handle creates no torque, even if the handle of the wrench is long and the force applied is great. In such a case, the line of action of the force passes right through the pivot point, so the length of the lever arm (i.e., perpendicular distance between pivot point and line of action of force) is zero. That makes the product of force and moment arm zero, so the overall torque is zero.

In practice, Equation 6.2 is seldom used to calculate a precise torque value for a small underwater vehicle. Rather, its usefulness is more conceptual—it helps you think about how force and moment arms interact to produce torque, thereby revealing what your options are if you need to change the amount of torque in a given situation. If, for example, you discover you need about twice as much torque as you have for a particular purpose, this formula tells you that you can get it in a variety of different ways. You could double the force. You could double the moment arm (by using a pipe to extend the wrench handle, for example). You could even use half as much force with four times the moment arm length, since ½ x 4 = 2. So the formula tells you that you have options, and it tells you what they are.

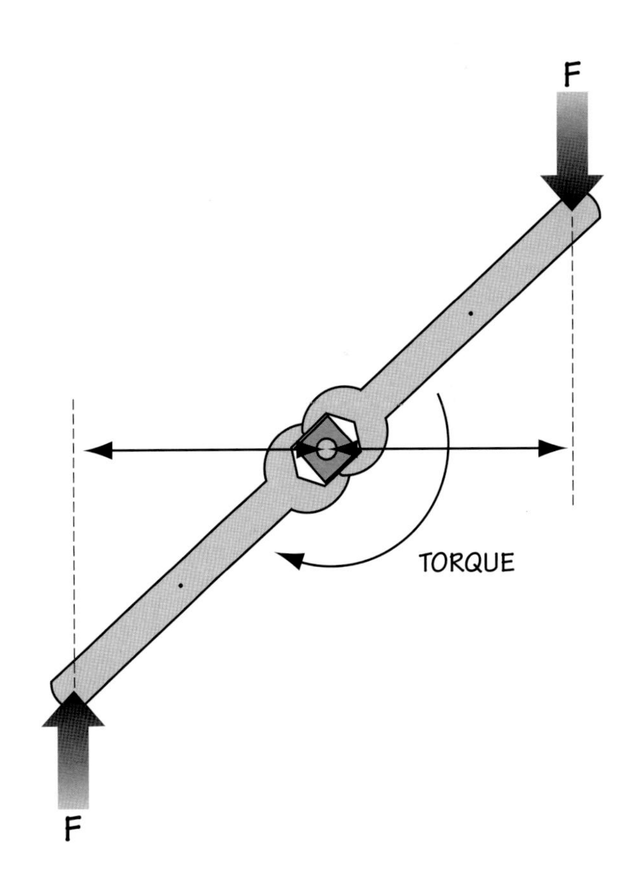

Figure 6.19: A Couple

Two wrenches used as shown produce a couple. Equal forces (F) are applied in opposite directions along (dashed) lines that are parallel to each other. The tendency of each wrench in this example to move the bolt up or down is cancelled by the influence of the other wrench, so the bolt experiences no net tendency to move out of its present location; however, the torques generated by the two wrenches work cooperatively, doubling the total torque on the bolt.

TECH NOTE: CALCULATING TORQUE PRODUCED BY A COUPLE

There are two ways to calculate the torque produced by a couple, as you can see by thinking about the double-wrench example in Figure 6.19. There are two wrenches, and each contributes half the overall torque, so one way to calculate the total torque is to calculate the torque produced by one wrench and double it. The other is to take the force acting on one wrench and multiply it by the distance separating the two lines of force action (i.e., multiply one force by twice the length of one moment arm). Both methods will give you the same result.

4.1.4. Couples

The one-wrench method of generating torque is great for turning something like a bolt, which can spin on one axis but is otherwise anchored solidly in a wall or other structure. But an ROV hovering in mid-water is not anchored. It is free to scoot sideways (or up or down), as well as to rotate if you push on it with a wrench (or with a thruster), so you need to think a bit differently when you want to generate or control the torques that rotate an ROV or other free-floating underwater vehicle to achieve stability.

In these cases, it is more useful to think in terms of **couples**. In the world of torque, the term "couple" refers to a pair of *equally strong* forces that act in *opposite directions* along *parallel lines.* (Figure 6.19) In a couple, the tendency of each applied force to move the object laterally (rather than to rotate it) is cancelled out by the presence of the opposing force. At the same time, the torques produced by each force add together. Thus, a couple offers twice the tendency to rotate, yet does so without causing any tendency to move away from the center of rotation. This can be very useful under water, when you want a vehicle to turn without moving sideways or up and down at the same time.

As you'll see in the next section, the matched weight and buoyancy of a submerged, neutrally buoyant vehicle are a good example of a couple—the two forces are equal in magnitude, but opposite in direction. The couple associated with these two forces is the primary source of torque involved in passive underwater vehicle orientation and stability.

4.2. How CG and CB Determine Vehicle Orientation

Any time an underwater vehicle is submerged, it will orient itself so that its CB is directly above its CG. To understand why this is so, you need to recognize that weight and buoyancy form a natural couple that can exert a torque on the vehicle, changing its orientation in the water. Once you recognize this, you can place your floats and weights in places that help stabilize, rather than destabilize, your vehicle.

4.2.1. Weight and Buoyancy as a Torque-Generating Couple

Figure 6.20 is a modified version of Figure 6.19. It represents a submerged, neutrally buoyant vehicle oriented so that its center of buoyancy (CB) is located above and to the right of its center of gravity (CG). This is *not* a stable situation and would not persist for long, but it might exist for a brief time immediately after a vehicle was tipped sideways by a yank on the tether or other force.

Think of the pair of wrenches as a portion of the rigid frame of the vehicle. Note that the generic forces (F) present in Figure 6.19 have been replaced by two specific forces: the vehicle's buoyant force and its weight, which act at the center of buoyancy (CB) and the center of gravity (CG), respectively. Since the vehicle is neutrally buoyant, these two forces must be equal in magnitude, even though they are opposite in direction. Thus, the weight and the buoyant force form a *couple.*

The buoyant force pulls up on the right, while the weight pulls down on the left. The vehicle neither rises nor sinks, because the upward pull of the buoyant force is counteracted by the downward pull of the weight. However, the torques generated by the two forces do *not* cancel. Each generates a torque that tends to rotate the vehicle counterclockwise. Thus, the weight and buoyant forces combine to generate a single large torque that tries to rotate the vehicle counterclockwise, as indicated by the word "torque" in the figure. Since the object is suspended passively in the water and not

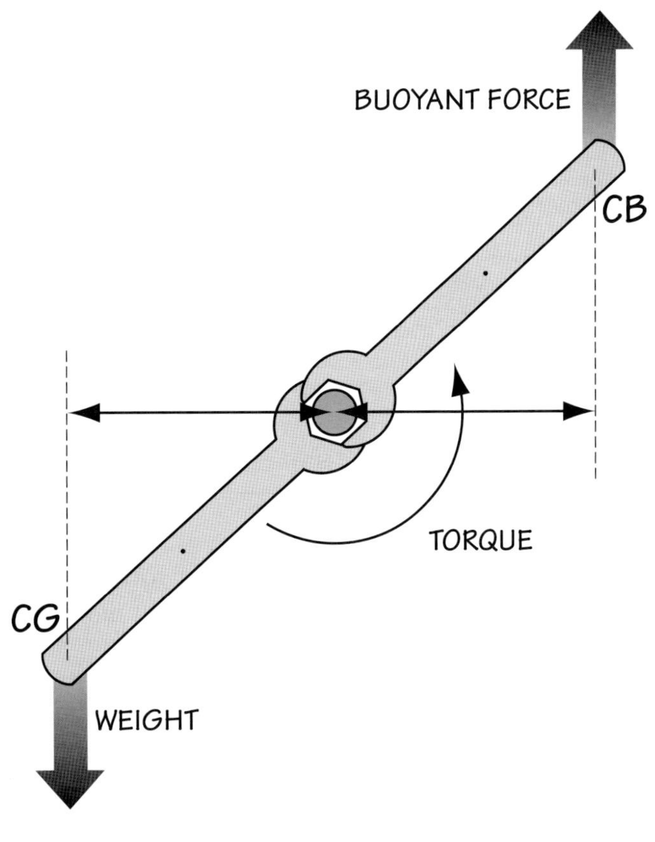

Figure 6.20: Weight and Buoyancy Form a Couple

The weight and buoyant force in a neutrally buoyant submerged vehicle form a couple. The torque produced by this couple always tends to rotate the vehicle frame until the center of buoyancy (CB) lies directly above the center of gravity (CG). Refer to the main text for a more detailed analysis of this figure.

anchored to anything solid, it has no way to resist this torque and will rotate freely until the CB is directly above the CG.

Once the CB is directly above the CG, the lines of action for each force pass directly through the center of rotation. That means the moment arms become zero, so there is no longer any torque. However, as soon as anything tips the vehicle away from this orientation, the lines of action associated with the two forces separate again, the moment arms reappear, and a torque is produced to counteract the tipping.

Note that if the vehicle were tipped to the left (counterclockwise) instead of the right, the resulting torque would be in the opposite direction (clockwise) and would again return the vehicle to a position where CB was directly above CG. This is what makes a submerged vehicle stable in its orientation. In other words, this is why CG always hangs directly beneath the CB. The trick to ensuring that your vehicle design will be stable (always returning to an upright position) is to make sure that CB is directly above CG when the vehicle is upright. In practice, you do this by adjusting where CB and CG are, relative to the frame of the vehicle.

When a vehicle is designed properly so that the torque produced by interactions between the buoyant force and the weight tends to return the vehicle to an upright orientation, the torque is called a **righting torque**, because it "rights" the vehicle by restoring it to an upright orientation.

In a real vehicle, CB tends to be located near the geometric center of the vehicle, because it's always at the center of the vehicle-shaped blob of water that's been displaced. Since floats displace water, they extend the

Figure 6.21: Center of Gravity and Center of Buoyancy

To understand how the Center of Gravity (CG) and the Center of Buoyancy (CB) of an ROV can be located in different places, imagine a perfectly cube-shaped ROV made from eight smaller cubes arranged as shown.

Now imagine that the four purple cubes on the bottom layer are filled with lead or another high-density material while the four yellow cubes on the top layer are filled with air or styrofoam or some other low-density material. Only the heavy cubes on the lower, purple layer would contribute significantly to the mass of the ROV, so the CG (represented by the black marble) would be located near the geometric center of that purple layer.

On the other hand, all eight of the small cubes would displace water, regardless of what they were made of, so the CB (represented by the white marble) would be located right at the geometric center of the large cube.

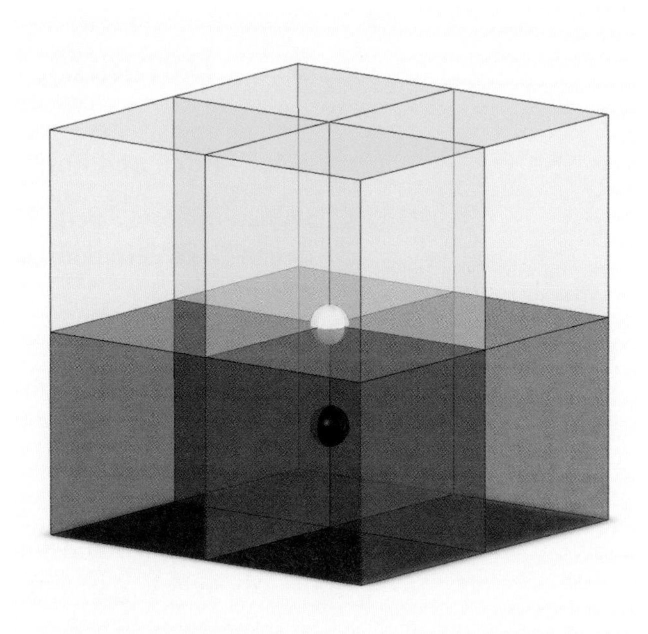

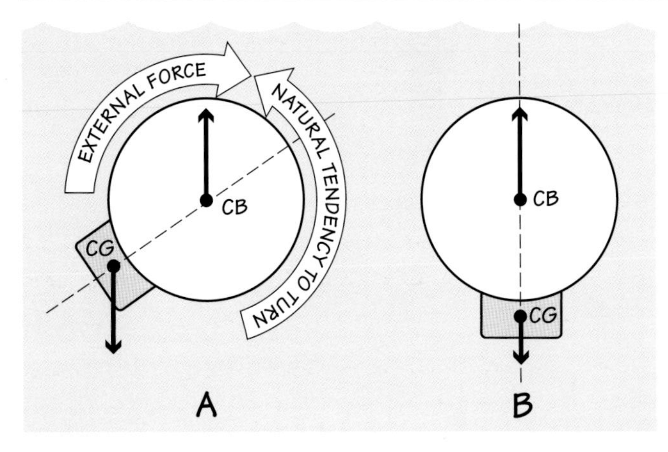

Figure 6.22: CB, CG, and Stability

A vehicle will naturally tend to right itself with the center of buoyancy (CB) above the center of gravity (CG). If an external force tips the vehicle over, the misalignment of the buoyant force and the weight (the forces associated with CB and CG) will generate a torque that helps to rotate the vehicle back to its upright position.

geometric outline of the vehicle and tend to shift CB in their direction. Thus, you can use careful placement of floats to adjust the location of CB. The CG, on the other hand, tends to be shifted away from the geometric center toward where the majority of the weight is concentrated. You can adjust the location of CG by adjusting where you put the densest, heaviest weights on the vehicle. Therefore, you can control how your vehicle orients itself in the water by using floats and weights to control where the CG and CB are, relative to the frame of the vehicle.

4.2.2. Place Floating Parts High, Sinking Parts Low

Generally, when building an underwater vehicle you want to position parts that float (i.e., those that are positively buoyant) above parts that sink (i.e., those that are negatively buoyant). You can see which are which by looking at the Buoyancy column in your weight statement table.

If, instead, you were to start with the floating parts on the bottom of the vehicle and the sinking parts on the top, then the CG would start out above the CB. That would be an unstable situation, because as soon as your vehicle submerged in the water, the righting torque described in the previous section would would kick into action. This would flip the vehicle 180 degrees, until the CB was on top. At that point, your vehicle would be hanging upside down in the water! The "righting" would have gone all wrong. Similarly, if you start with the floating parts on the right (starboard) side and the sinking parts on the left (port) side, then when your vehicle is released in the water, the physics you just learned will force your vehicle over onto its side, so that the starboard side actually faces up and the port side hangs down.

In practical terms, the need to keep floats high and weights low suggests the wisdom of placing heavy components, such as motors and batteries, as low as possible on your vehicle and placing buoyant components, such as floats or air-filled parts of pressure canisters, high on your vehicle. This ensures the righting torque you've created actually will right the vehicle.

5. Trimming a Vehicle's Orientation

5.1. Pitch and Roll

Deviations from a perfectly upright, level orientation are a common problem in ROV design. Such deviations are often described in terms of pitch and roll angles (Figure 6.23 below):

- **Pitch** is the angle of the forward or backward lean of the vehicle. A +5 degree pitch means the front of the vehicle is tilted *up* 5 degrees above horizontal; a -10 degree pitch means the nose is tilted *down* at an angle of 10 degrees below the horizontal.

- **Roll** is the angle of sideways leaning to the right (starboard) or left (port). For example, if you define roll as positive to the right, then a +3 degree roll means the vehicle is leaning 3 degrees to the right; a -3 degree roll means it's leaning 3 degrees to the left.

5.2. Trimming Pitch and Roll

When you put a vehicle in the water for the first time (or after adding or removing a payload), you probably will need to adjust its overall buoyancy, as well as its pitch and roll. As described earlier, buoyancy is trimmed by adding or subtracting floats and/or weights. Pitch and roll are adjusted by rearranging where those floats and/or weights are located on the vehicle. The overall process of **trimming a vehicle** often includes the process of adjusting buoyancy, as well as pitch and roll. When all three are adjusted properly, the vehicle is said to be "in trim."

Many underwater vehicles strive for zero pitch and zero roll most of the time. However, some ROVs are trimmed with a slight upward pitch, so that they don't plow up mud when moving forward along the bottom. Other specialized craft may need to alter their pitch or roll to perform their missions. For example, inspection ROVs must be able to orient themselves at all sorts of odd angles, to view different surfaces inside pipes or other situations. Similarly, rescue submarines can control their pitch and roll

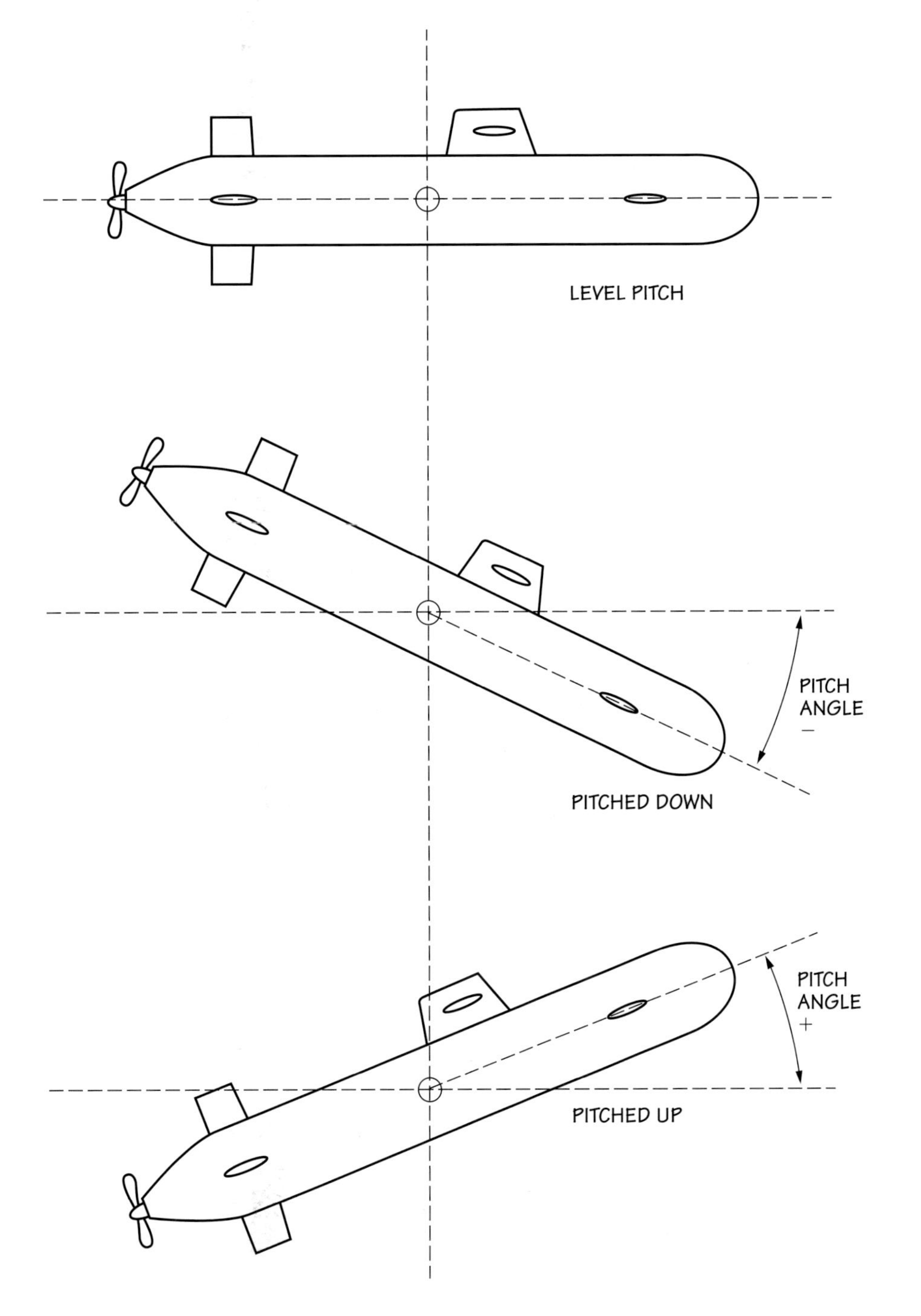

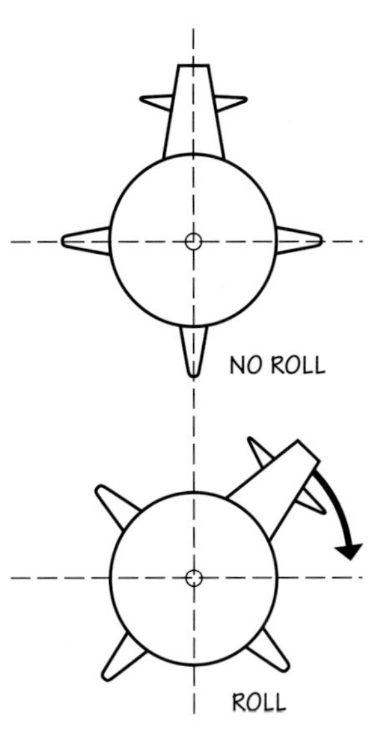

Figure 6.23: Pitch and Roll

Pitch refers to tipping nose up or nose down. Roll refers to rotating left or right. The speed with which a vessel rolls, whether at the surface or when submerged, is a direct indication of its stability.

When BG is small, the vessel has a slow roll and a weak tendency to return to an upright position. Such a craft is said to be tender.

When BG is large, the roll is quicker and the vessel has a stronger tendency to return to a stable position. Such a vessel is said to be stiff.

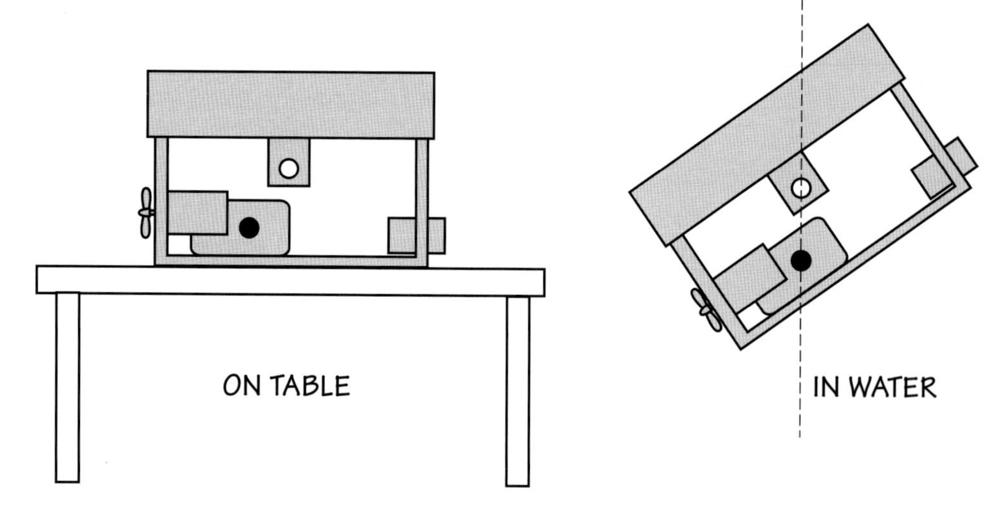

Figure 6.24: Figuring Out Trim

This pair of figures illustrates a common problem that plagues many ROV-building teams. The figure at left represents an ROV, oriented upright, sitting on a table where the design team has been assembling the vehicle. The placement of motors, batteries, and other weight is such that the CG (dark dot) is NOT under the CB (light dot). Unfortunately, this fact is not obvious when the vehicle is sitting on the table, because the table prevents the ROV from tilting.

The figure on the right shows what happens when the vehicle is submerged in water. The torque generated by the couple associated with the weight acting through CG and the buoyant force acting through CB rotates the vehicle until the CG hangs directly beneath the CB. This causes the vehicle to hang tilted in the water. If the CG had been directly under the CB while the ROV was sitting level on the tabletop, this tilting would not have happened in the water.

over a wide range, to facilitate mating of the rescue hatch with the hatch of a downed submarine, which may be oriented at an unusual angle when the damaged sub is resting on the seafloor.

You've already learned that when trimming a vehicle's buoyancy, it's the *amount* of weight and flotation that matters. However, when trimming a vehicle's pitch and roll, it's *where* you put the floats and weights that matters, because that's what determines where the CB is, relative to the CG. The challenge is therefore figuring out where the floats and weights should go.

To determine where to put the weights or floats, you need to visualize where the CG and CB of your vehicle are and determine how they need to be moved to improve the orientation of your vehicle. Remember that CG will always hang directly below CB in a submerged vehicle, for the reasons explained in Section 4.2.1. above.

As a general rule, the CG will move in the same direction as the weights. So if you want to shift your CG forward to tilt the vehicle front downward, then you'll need to move some of the vehicle's weight closer to the front of the vehicle. (See *Historic Highlight: German U-Boats and Shifting Human Weights.*)

Similarly, the CB will "follow" the floats, so moving floats forward will tend to tilt the front of the vehicle upward. If you get more change than you want, simply move the weight or float closer to the vehicle's CG or CB. As you make these adjustments, remember that you should generally be keeping weights low on the vehicle and floats high on the vehicle.

In practice, achieving good trim is done in two steps. The first step takes place during the design phase, when care is taken to make sure that the weight and buoyant force

HISTORIC HIGHLIGHT: GERMAN U-BOATS AND SHIFTING HUMAN WEIGHTS

In a submerged submarine, any shift in weight fore or aft will affect the sub's pitch (i.e., whether its nose points up or down). During World War II, German U-boat captains used this characteristic to advantage during emergency dives. When the alarm sounded, all crewmen with no duty stations in the control room or engine room immediately ran to the bow compartment. Moving the crew's weight to the bow shifted CG forward, thereby tipping the sub's nose down and facilitating a rapid dive.

Some modern "glider-style" AUVs rely on the same principle; an electric motor under computer control shifts a weight forward to tilt the glider nose down for diving or backward to point the glider nose up for ascending.

will balance, and to make sure that the position of batteries, motors, and other heavy parts will place the CG more or less directly under the CB. Unfortunately, it is usually difficult to judge the precise location of either point, but with experience, you can get better at estimating these locations.

The second step takes place during initial in-water tests. During these tests, you note any excessive floating, sinking, or tilting of the vehicle in the water, and add, remove, or shift weights, as needed, to correct any problems. Although some trial-and-error is usually involved, a design team that knows what they're doing will never need to resort to random guessing. The direction and degree of tilt can tell a lot about what changes need to be made and where they need to be made. (But this is true only if you understand the relationships between weights, floats, CB, CG, and vehicle orientation, as described in this chapter.)

The savvy design team anticipates the second step and designs in some easy method to adjust the number, size, and location of weights and floats. For example, they might make the exact location of the batteries adjustable forward or backward, to help trim the pitch of the vehicle. Or they might add posts or moveable weights, as shown in Figure 6.25.

Image courtesy of Dr. Steven W. Moore

Figure 6.25: Adjustable Ballast Weights

This ROV ballast system, built by Stanley Moore, has a threaded stainless steel post mounted on each of the four corners of its frame. Washers can be stacked on these posts and held in place with a nylon nut. This system lets you add weight to the ROV in small increments, allowing precise trimming of buoyancy, pitch, and roll.

The overall positive or negative buoyancy of the craft is controlled by adjusting the total number of washers used.

Once overall buoyancy is adjusted properly, the static pitch and roll of the craft can be trimmed by changing how those washers are distributed among the four posts. For example, to tilt the nose down more, some of the washers are moved from the rear posts to the front posts. To lean the ROV further to the right, some of the washers are moved from the posts on the port side to the posts on the starboard side.

Note that buoyancy and orientation are interrelated, because they both depend on weights and floats. If you want to change buoyancy without changing orientation, you'll need to add your floats directly above the CB or add your weights directly below the CG. That way, the lever arms associated with these new forces will be zero, so they will create no torque and will not affect vehicle orientation.

If you want to change vehicle orientation without changing buoyancy, you'll need to keep the weights and floats in balance, but relocate them to shift CB and/or CG as needed, to achieve the desired degree of tipping in one direction or another. You can do this either by moving existing weights and floats on a properly buoyant vehicle to new locations on the vehicle, or by adding matched pairs of weights and floats that have no net effect on buoyancy. For example, to tip a vehicle's nose up without changing vehicle buoyancy, you can slide existing floats forward and/or move existing weights backward, or you could add a new float to the front of the vehicle while adding a new matched weight to the back.

If you need to change both buoyancy and orientation at the same time, you can do so by adding a float or weight off-center. For example, if you want to make your vehicle less buoyant, *and* lower its nose, *and* lower the starboard side of the vehicle all at the same time, you may be able to accomplish all three goals by adding a single weight to the front right corner of the vehicle.

6. Stability

In the context of underwater vehicle pitch and roll, stability refers to the tendency of a vehicle to return to an upright position—that is, to "right" itself—whenever it is accidentally tipped or flipped over by some disturbance, such as a powerful push from one of its thrusters, a yank on its tether, shifting payloads, or the swirling water currents caused by a crashing wave. A vehicle that will consistently return to an upright position in spite of such disturbances is said to be "stable."

For the reasons discussed in the previous section, any completely submerged vehicle with its CB above its CG is going to be stable, as long as nothing alters the position of CB or CG relative to the vehicle's frame. However, some submerged vehicles may be more resistant to tipping than others; such vehicles tend to return more quickly to an upright position when tipped over. In most cases, higher levels of stability are better, and this section describes how to achieve that.

6.1. Vehicle Stability Under Water

The strength with which a vehicle attempts to right itself depends on how quickly the righting torque grows as the vehicle is tipped farther and farther from its upright position. Remember that the force couple associated with the righting torque is the one caused by the interaction of the vehicle's weight and buoyant force. Remember also that the torque produced by a force couple is equal to the magnitude of one of the forces multiplied by the distance separating the lines of action of the forces (i.e., twice the moment arm for each force). For a given degree of vehicle tipping, then, you can increase the righting torque by increasing the magnitude of the forces (more weights and floats) or by increasing the distance between CB and CG, so as to increase the moment arm. (See Figures 6.26-6.28.)

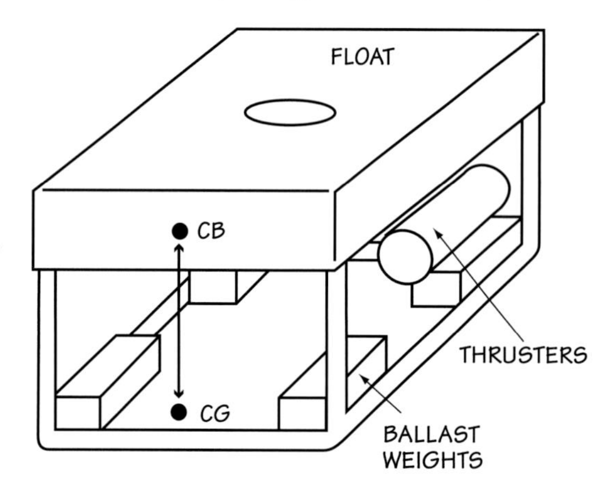

Figure 6.26: BG

The physical distance between the center of buoyancy (CB) and the center of gravity (CG) is called BG. The larger the BG, the more stable the undersea vehicle, whether at the surface or under water.

In completely submerged vehicles, the magnitude of these forces is generally set by the size of the vehicle. That's because vehicle size determines how much water is displaced and therefore what the buoyant force will be, and vehicle weight needs to be about the same as the buoyant force for neutral buoyancy. Therefore, from an underwater vehicle design standpoint, the distance between CB and CG becomes the most important determinant of how stable the vehicle is. This distance is known as **BG**.

To gain maximum stability for your underwater vehicle, you need a large BG. In other words, you need to separate CB and CG as far as possible. As you know by now, to do this, position the dense parts of your vehicle (batteries, motors, etc.) as low as possible on the vehicle and put floats and other low-density parts as high as possible. Note that there are some specialized missions that call for zero BG craft. (See *Tech Note: Zero BG Vehicles.*)

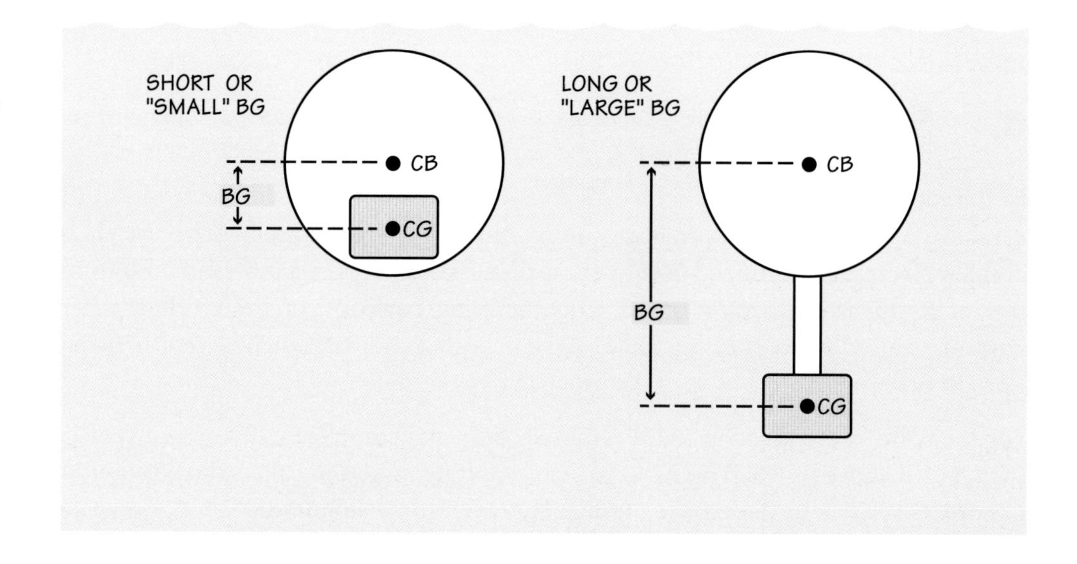

Figure 6.27: Increasing Stability

A small BG makes for a less stable vehicle. A large BG increases stability.

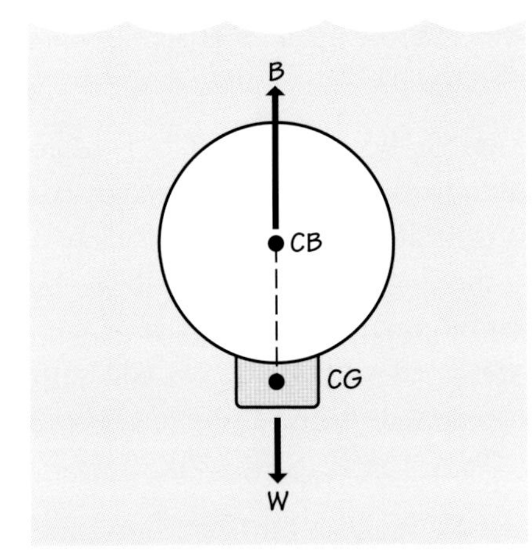

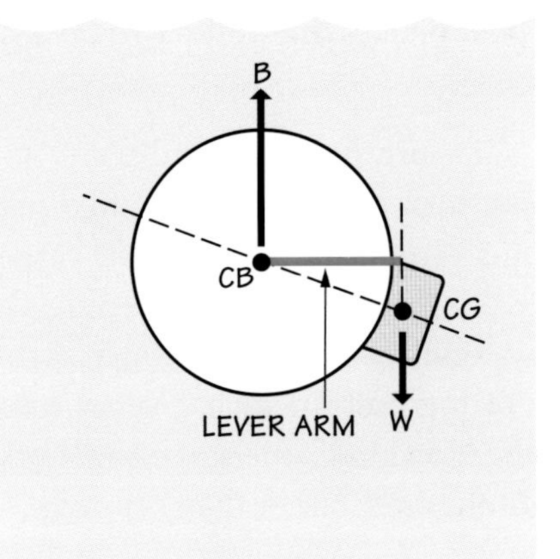

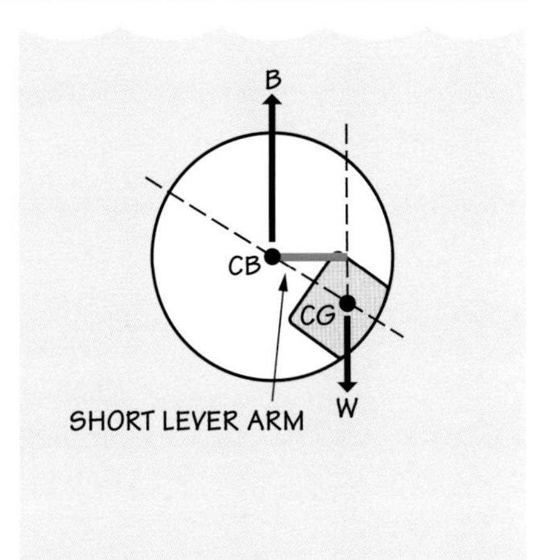

Figure 6.28: Stability Under Water

If the CB and CG are vertically in line with each other, there is no moment arm and therefore no torque, so the vehicle tends to stay oriented that way.

If the vessel is tilted to one side or the other, a moment arm is produced. This moment arm, in concert with the weight (W) and buoyant force (B), creates a torque that acts to right the vehicle, bringing it back into an upright position.

If the CG is located closer to the CB, the moment arm is shorter and hence less effective. Less torque is produced so it takes longer to bring the vessel back to an upright position.

TECH NOTE: ZERO BG VEHICLES

Some undersea vehicles are purposely designed to have a **zero BG**. Such a vehicle can change direction and attitude very easily, because it has no preference for any particular orientation; however, it is also difficult to control. Operators need computers and special thruster arrangements to keep the craft oriented in the desired direction. In fact, without computer-assisted attitude control, it is almost impossible for an ROV pilot to manually fly such a vehicle.

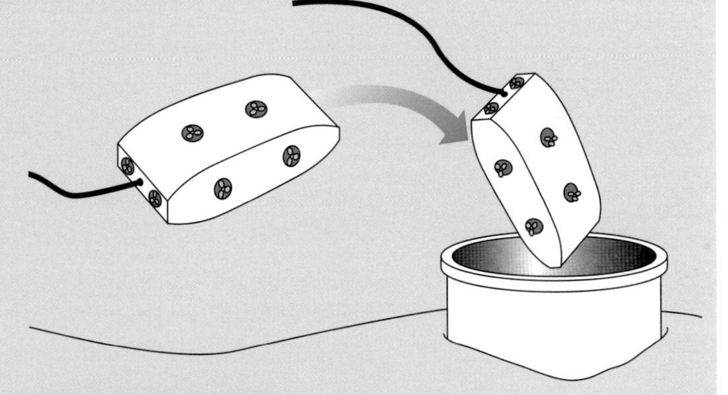

Figure 6.29: Zero BG ROV

A zero BG underwater vehicle is very maneuverable, but the pilot needs the assistance of a sophisticated thruster controller to fly it.

The *SUTEC* ROVs from the early 1990s have a near zero BG design. This allows them to actually hover and to move in any direction in any attitude, even pointing downward vertically. This is handy for examining the interior of vertical pipes or for moving around shipwrecks. Torpedoes also have a small BG to facilitate maneuvering. However, the majority of undersea craft have a large BG. Certainly, no undersea vehicles have a CG *above* the CB. This condition would not work physically—the vehicle would just flip.

6.2. Stability on the Surface

Strange though it may seem, making any vessel stable on the surface can actually be *more* difficult than making it stable under water. Even so, stability on the surface is no guarantee of stability under water.

When a vehicle is completely submerged, the shape and position of the displaced volume of water does not change relative to the vehicle frame, so the CB does not change position. Likewise, the CG is generally fixed in location relative to the frame. Since the CG will always try to hang directly below the CB, the fixed locations of CG and CB result in a predictable, stable situation.

When a vehicle is on the surface, however, the CB can shift around as the vehicle pitches and rolls. This happens because the hull may displace very different-shaped volumes of water located in very different locations relative to the vehicle's frame and

CG as the boat pitches and rolls. This greatly complicates the analysis of stability for surface vessels.

Boats and ships are particularly difficult to keep stable, because their CG is often located *above* their CB—something that just wouldn't work at all under water. For example, on a cargo ship, much of the cargo mass is located high on the vessel; as a result, the CG is often above the waterline. On the other hand, the CB must always be below the waterline, because it's defined as the center of gravity of the displaced water. The shape of the hull on such a vessel must be designed so that the CB will shift rapidly to the "downhill" side as the ship begins to tip, so that the resulting torque will tend to right the craft rather than capsize it.

A submarine must be stable both on the surface, where it acts a bit like a boat, and under water. The submarine's long, cylindrical shape makes it an interesting craft to study in order to understand the relationship between surface stability and underwater stability. Subs are designed so that most of the heavy weights, such as the lead-acid batteries, are located in the lowermost sections, while the flotation is located in the uppermost section. The submarine designer does this to make the submarine stable under water.

On the surface, a long, lean submarine has buoyant forces acting all along the length of the hull. It is like a canoe—very difficult to tip forward or backward (though it can roll fairly easily from side to side). If a wave or something else tries to tip a sub (or canoe) forward, for example, the bow gets pushed down into the water, displacing extra water there. This generates a sudden increase in the buoyant force acting upward under the bow. The increased buoyant force has a huge lever arm (almost half the length of the sub), so it creates a correspondingly huge torque that very effectively resists any further forward tipping of the bow. The same process likewise prevents a backward tip, so the sub (or canoe) has a great deal of longitudinal (lengthwise) stability. Of course, a canoe or sub has considerably less resistance to tipping to the side, because its width is so much smaller than its length.

Once the sub submerges, the location of CB and the buoyant forces don't change much, so a (motionless) sub's stability is then entirely dependent on BG, which usually offers much shorter lever arms and correspondingly lower torques. Thus, the vessel's stability—particularly its longitudinal stability—is greatly reduced under water. A moving sub can use other tricks, such as dive planes that act like airplane wings or rudders, to provide extra stability while it's moving through water, but these won't help much when it's stationary.

Lack of knowledge about lowered longitudinal stability of submerged vehicles resulted in numerous highly unstable submarine designs in the early years. Those built in the

HISTORIC HIGHLIGHT: *LE PLONGEUR*

An early example of lowered longitudinal stability was the 4-meter (approx. 13-ft) French sub, *Le Plongeur*, launched in 1863. On the surface, the vessel behaved well enough, but as ballast was added and longitudinal stability decreased, the vessel surged out of control. Like a playful dolphin, it plummeted steeply downward until the correcting gear took effect and it swung to the surface, where another "correction" would instigate a steep dive.

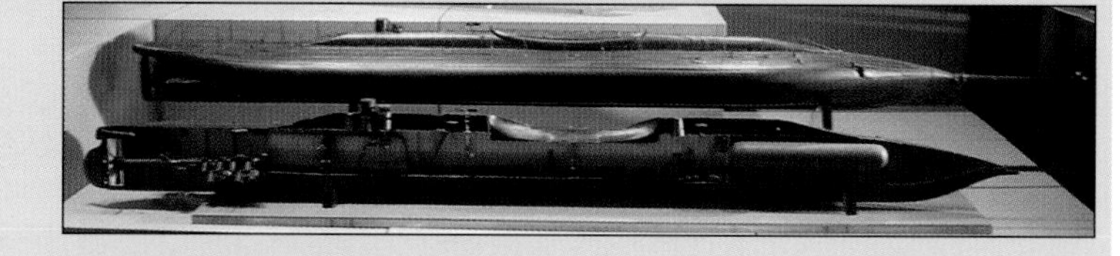

Figure 6.30: Model of* Le Plongeur *at the Musée de la Marine, Paris

Image courtesy of Wikimedia Commons

late 1890s and early 1900s were particularly notorious for losing control of trim and surfacing unexpectedly or making fatal uncontrolled dives because they could not compensate for diminished longitudinal stability once submerged.

6.3. Shifting Weights and Loss of Stability

Sometimes a vehicle that initially seems stable suddenly becomes unstable—or, more precisely, it shifts suddenly and unexpectedly to a new, but unwanted, stable orientation. The capsizing of a ship is a bad situation but a good example, since an upside-down ship is usually stable in that new position!

Anything that can shift the position of the CB or the CG in an underwater vehicle can potentially cause this type of problem and must be evaluated carefully to make sure it can't result in disaster. A group of college students piloting their just-finished ROV in a kelp forest off the coast of California learned this the hard way. They discovered that a heavy battery inside an ROV needs to be anchored in place, so it can't move. The students had placed a large battery in the center of their ROV, but it was not anchored well and managed to slide forward just a little bit when the ROV was tipped forward by a wave. Unfortunately, this small shift in battery position was enough to shift the CG of the ROV forward, which caused the ROV to tip even farther forward. The increased angle of the ROV sent the battery sliding quickly toward the front of the vehicle, which it struck like a battering ram, breaking the watertight seal on the front port. Within a few seconds, the ROV filled with seawater and went straight to the bottom. The vehicle and all of its internal electronics were a complete loss. The moral of this story is to make sure that all weights and floats on your vehicle are securely

HISTORIC HIGHLIGHT: FREE SURFACE EFFECT

In 1885, the Swedish industrialist Thorsten Nordenfelt encountered free surface effect problems with his steam-powered *Resurgam*. The sub's boiler was never completely full or completely empty, so any inclination (tipping) caused the water in the boilers to shift in the direction of the tilt, with disastrous results when the vehicle lost stability, surfacing and diving uncontrollably.

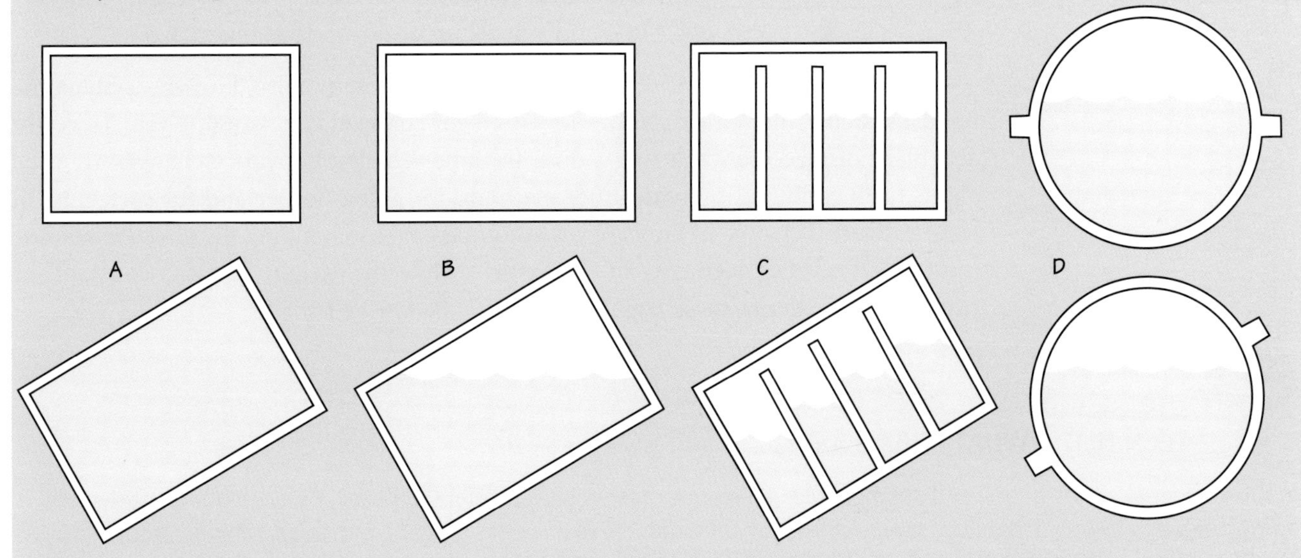

Figure 6.31: Free Surface Effect

A full tank has no free surface, even when inclined (A). When a partially full tank is inclined, all the liquid settles to the lower corner, shifting CG toward that corner and increasing the tilt even further (B). When baffles are added, the liquid does not collect in one place; thus, less weight is shifted when the tank is inclined (C). A spherical tank has lower free surface effects (D).

anchored to the frame, so they can't move around and inadvertently shift the location of the CB or CG.

A less obvious (but no less serious) source of this type of problem is fluid-filled compartments in a vehicle. If the vehicle tips a bit in one direction, fuel or other fluid in tanks can flow toward the downhill side of the tank, adding weight to that side and shifting the CG of the vehicle in that direction. This drives even more fluid in that direction, resulting in a vicious cycle that quickly upsets the stability of the vehicle. This dangerous phenomenon, known as the **free surface effect**, has caused the demise of many ships and submarines over the years. Modern vessels incorporate spherical tanks, baffles, or other features to minimize free surface effects. (See *Historic Highlight: Free Surface Effect.*)

7. Ballast Systems

Ballasting describes the process of adding, removing, and arranging weights (or sometimes flotation) on an undersea vehicle to adjust its overall buoyancy and trim. Ships are ballasted to be positively buoyant, so they'll float on the surface, but most underwater vehicles are ballasted for something much closer to neutral buoyancy. That way they can dive and return to the surface easily. Historically, a wide range of ballast ideas were tried. Some worked, and some didn't. Today, underwater craft use well thought-out ballast systems to submerge and surface in a controlled manner.

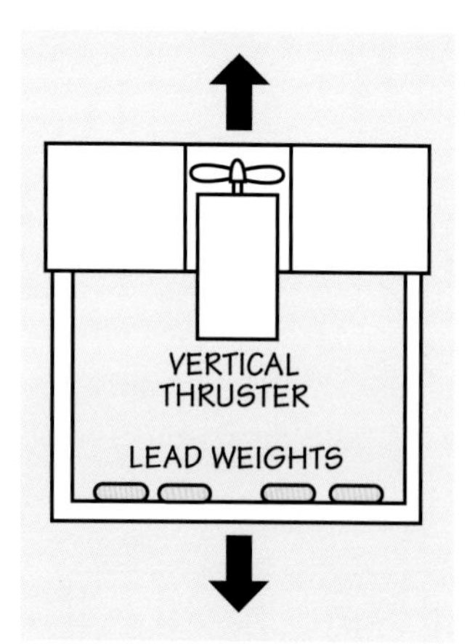

Figure 6.32: A Static Ballast System

An ROV is ballasted so that it is very close to neutral buoyancy. A vertical thruster is used for depth control. Using this type of static ballast system, a sophisticated ROV can often hover at fixed depths, varying up or down by only several centimeters.

There are two basic types of ballast systems—static ballast systems and active ballast systems.

In a **static ballast system**, the buoyancy of the vehicle is pre-set at an appropriate value and left unchanged throughout each dive. This type of ballast system is common in ROVs and some AUVs, where the system is usually designed to make the vehicle just slightly positive at all times. Diving and surfacing are then accomplished with thrusters, which can easily push the almost neutrally buoyant vehicle up or down. (See Figure 6.32.)

In an **active ballast system** (also known as **dynamic ballast system**), the vehicle's buoyancy is actively changed (by changing vehicle weight and/or its displacement) during a dive. In these vehicles, buoyancy changes are used alone, or in combination with other methods, for going deeper and for surfacing. In most cases, this change in buoyancy is accomplished by changing the effective volume (and therefore displacement) of the vehicle while the weight remains constant, or nearly so. (See the *Historic Highlight: Historic Ballast Systems.*)

TECH NOTE: WHAT IS BALLAST?

In the context of ships, boats, and submersible vehicles, ballast generally refers to heavy, dense material added low on the vessel to improve stability. It may consist of solid objects like chunks of steel, batteries, or even rocks. Alternatively, it may take the form of a liquid (usually water) held in tanks. Although ballast is usually placed low on a vehicle, some active ballast tanks used on submarines are placed high on the vehicle. That's because these tanks actually function more like adjustable floats; when the sub needs to surface, compressed air replaces ballast water in the tanks, making them positively buoyant and helping to lift the submarine toward the surface.

Air is used in many ballast systems, both static and active. Therefore, to understand the details of how static and active ballast systems work, it's first necessary to learn how air behaves under pressure.

HISTORIC HIGHLIGHT: HISTORIC BALLAST SYSTEMS

In 1578, the Englishman William Bourne published a description of his propelled diving boat in his book *Inventions and Devices*. His idea was to construct a submarine whose weight remained constant but whose size (hence displacement) could be made smaller for diving and larger for surfacing. He planned to accomplish this by utilizing leather-covered ballast-tank bulkheads that could be compressed or enlarged by means of large screws. It was the first time that Archimedes' Principle was applied to submarine design.

In 1620, Cornelius Van Drebbel, a Dutchman and court engineer, borrowed on Bourne's idea of flexible ballast, but his human-powered invention was designed to use goat skins that were filled with water for submerging and emptied for surfacing.

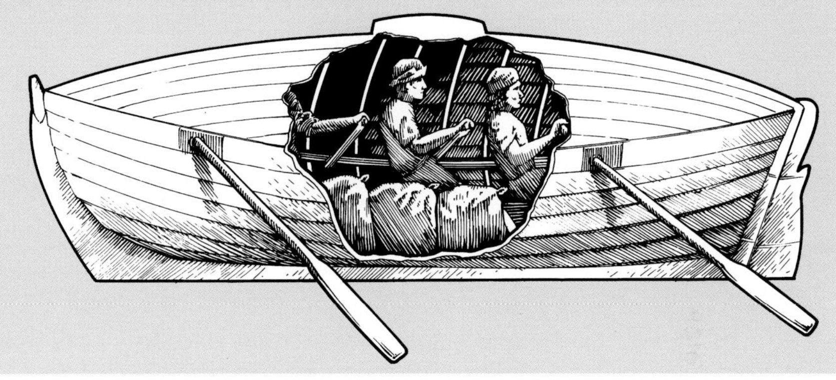

Figure 6.33: A Static Ballast System

Image courtesy Naval Undersea Museum

7.1. Air Under Pressure

All materials compress (that is, get reduced to a smaller volume) under pressure. However, some compress a lot, even under modest pressure, whereas others do so hardly at all, even under extreme pressure. Air and other gases compress relatively easily. Water and other liquids are comparatively difficult to compress and, for simplicity, are often regarded as incompressible.

Here's an example of how striking the difference is: Imagine exactly 1 liter of air and 1 liter of water (1 liter is about 1 quart), each collected in a separate plastic bag at the seashore, where the atmospheric pressure is 1 atm absolute. Now imagine attaching both bags to a deep-diving ROV that dives with the bags to a depth of 1,000 meters (approx. 3,300 ft), where the absolute pressure is about 100 atm. At that pressure, the air will be compressed to a volume of only 0.01 liters, or 1 percent of its original volume. (In imperial units, that means the volume of air that originally filled a quart-sized bag has been compressed to a volume of about 2 teaspoons!) The water, on the other hand, will have been compressed only a little—to a volume of about 0.995 liters (still nearly a whole quart). In other words, the air volume will have been reduced by 99 percent, whereas the water volume will have been reduced by only about 0.5 percent.

For this reason, it's almost always ok to ignore the compressibility of water, even at depths of hundreds of meters, but it is *not* advisable to ignore the compressibility of air, even in a swimming pool.

To calculate changes in air volume with pressure, start by noting that air, like most gases, behaves very much like an ideal gas—that is to say, it conforms very closely to the following algebraic equation, which is known as the **Ideal Gas Law**:

Equation 6.3 $$PV = nRT$$

In this equation,

P = absolute pressure of the gas

V = volume of the gas

n = number of gas molecules measured in units called "moles"

R = universal gas constant

T = Kelvin (absolute) temperature

Fortunately, you don't need to remember all these details for simple underwater vehicle design projects. As it turns out, in most simple ballasting situations, there's a fixed number of air molecules enclosed within some space, and that air is held at a nearly constant temperature, by virtue of being immersed in a large body of water. Under these circumstances, n, R, and T are constants, so the ideal gas law simplifies to:

Equation 6.4 $$PV = constant$$

In other words, pressure times volume equals a constant. You don't even need to worry about what the value of that constant is. You just need to know that if pressure goes up, volume goes down by the same proportion, and vice versa. This important relationship is known as **Boyle's Law**. Boyle's Law says, for example, that if the external pressure acting on an air-filled balloon is increased by seven-fold, the air volume in the balloon will be compressed to $1/7^{th}$ of its original volume. It also says that if the air volume is forced to remain constant by being enclosed inside a rigid vehicle hull or canister, then the air pressure inside that hull will also remain constant, regardless of what the ambient water pressure is doing outside the hull. (See *Tech Note: The Effect of Depth on Air.*)

TECH NOTE: BOYLE'S LAW AND SCUBA DIVERS

All scuba divers are required to understand Boyle's Law as part of their scuba training and certification. For a scuba diver, failure to understand this law can be fatal. That's because a device called a scuba regulator delivers compressed air to a diver's lungs at a pressure that is closely matched to the ambient (surrounding) water pressure. This pressure-matching function of the scuba regulator is what makes scuba diving possible. Without it, a diver could not breathe, even with an air tube leading all the way to the surface, because the external water pressure at depth would compress the diver's lungs with overwhelming force, making it impossible to inhale.

Unfortunately, the compressed air that makes it possible to breathe at depth also presents a serious danger if the diver holds his or her breath while ascending. Boyle's Law tells us why. As the diver ascends, the ambient water pressure surrounding the diver decreases, allowing the pressurized air inside the soft, extensible lungs to expand. For example, if a diver takes a full breath of air at a depth of 33 feet (where the absolute pressure is 2 atm), then tries to surface while holding that breath, the air in the lungs will try to expand to twice the volume of the lungs. Human lungs can withstand a maximum pressure differential of only about 1 or 2 psi (less than 1/6 of one atm), so if the diver does not release that extra air volume by exhaling, it can easily result in a lung rupture, which can inject a stream of air bubbles into the diver's bloodstream. Once injected, the bubbles can clog small blood vessels, cutting off the blood supply to parts of the brain and other vital organs, causing severe injury or even death.

This problem can even occur while a diver is sitting on the bottom in shallow water, if large waves are passing overhead. Imagine that a wave crest passes over just as a diver inhales. When the wave trough passes over a few seconds later, the diver's depth and ambient water pressure will be lower than when the diver inhaled—so much so that the air in the diver's lungs could have expanded beyond the safe lung volume. Because of these effects, there are two extremely important rules new scuba divers learn: 1) Never hold your breath while scuba diving. 2) Blow as you go (i.e., blow bubbles as you ascend to the surface)!

TECH NOTE: THE EFFECT OF DEPTH ON AIR

Figure 6.34

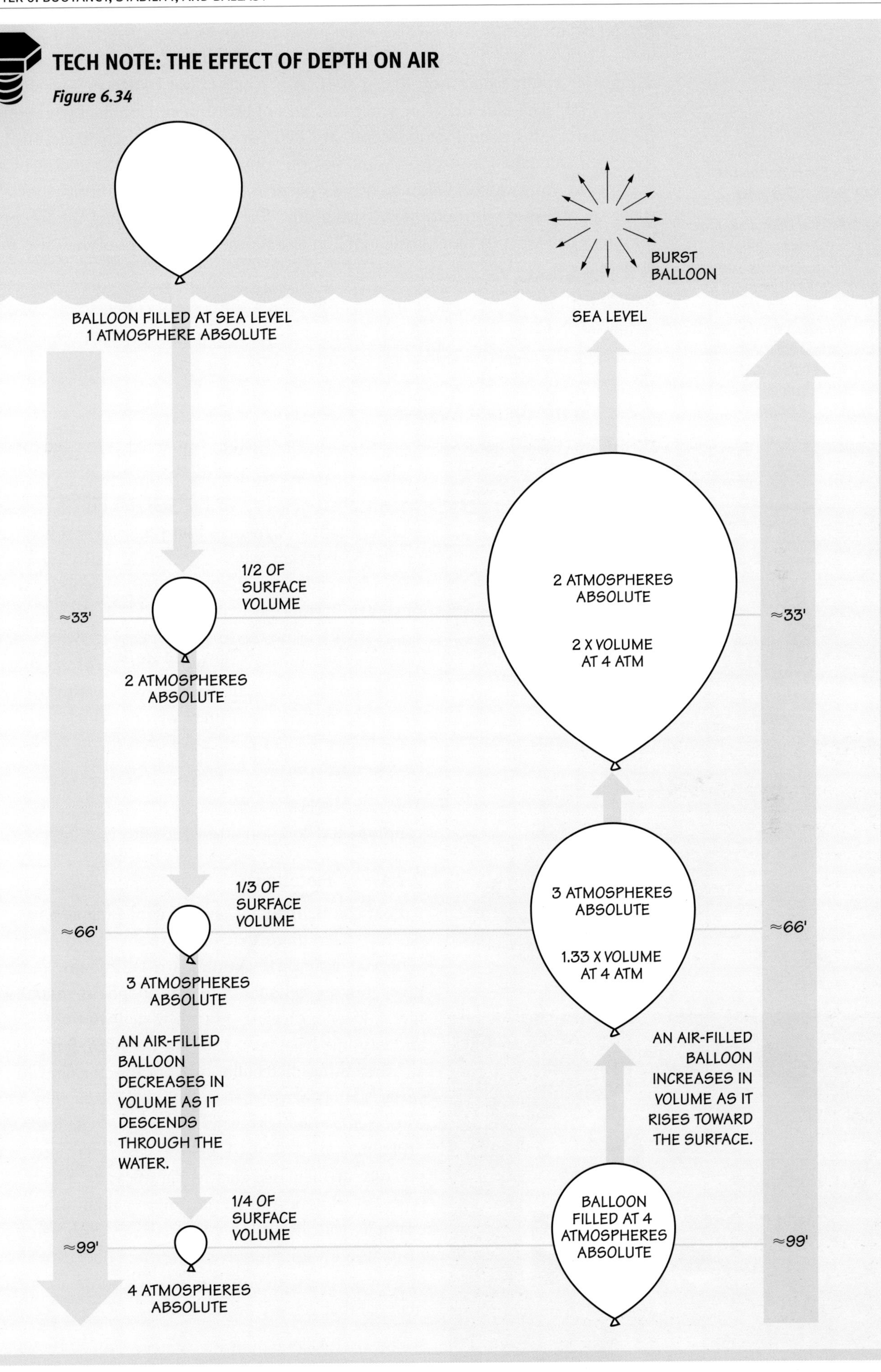

It's important to note that while water pressure increases in a nearly linear fashion with depth—1 atm for every 10 meters (approx. 33 ft) of seawater, regardless of depth—air volume does not. If you start with 1 cubic foot (0.028 m^3) of air at the surface and take it down 33 feet, you gain 1 atm of pressure and lose half your volume. If you start with 1 cubic foot at a depth of 1,000 feet and take it down 33 feet (to 1,033 feet), you still gain 1 atm of pressure, but this time you lose only 3 percent of your volume. As you can see, volume changes occur much more dramatically in shallow water than they do in deep water. Ironically, this means that using air to control buoyancy can be even more challenging in a shallow swimming pool than it is in the deep sea!

Armed with this basic understanding of how air volumes can change under water in response to depth-related pressure changes, you are ready to learn about static and active ballast systems used to control vehicle buoyancy.

7.2. Static Ballast Systems

Most ROVs use a static ballast system coupled with vertical thrusters to achieve depth control. In such a system, the buoyancy of the vehicle remains constant and is usually designed so that the vehicle is just barely positively buoyant when no vertical thrust is applied. That way, it takes very little thruster power to move it up or down. Furthermore, if power is lost, the vehicle will float slowly back to the surface for easy recovery. For this static ballast approach to work, the thrusters must be able to produce enough downward thrust to overcome the slight positive buoyancy of the vehicle. A simple static ballast system is effective even for ROVs working at extreme depths.

The static ballast system on an ROV includes two main components: flotation and weights.

7.2.1. Flotation

A float can be made of anything, as long as its average density is less than that of the water surrounding the vehicle. Common examples are stiff foam, pressure-proof hollow cylinders, or empty spheres made from metal, plastic, or glass. Even low-density fluids like oil and gasoline have been used for flotation. Soft, squishy air-filled things like balloons work fine for floats on the surface, but because of their compressibility, they do not work well on diving vehicles. In fact, they can lead to real problems, as described below. The increased water pressure at increased depths significantly affects some types of flotation materials more than others. For example, you may be able to get away with using pipe insulation for flotation in a swimming pool. But as soon as you go deeper, you'll need to choose a flotation material that is less compressible.

To be useful, any buoyant material must be able to provide at least enough buoyant force to overcome its

Figure 6.35: Flotation for Static Ballast Systems

Regardless of their size, most ROVs use a static ballast system that relies on some type of buoyant material for flotation. This is true in the case of student-built ROVs, as in the first photo, or for the heavy work class ROVs, like the Sonsub Innovator shown in the second photo.

Image courtesy of 2008 MATE International Competition Dive Support Team

Image courtesy of Sonsub Saipem

Image courtesy of Randall Fox

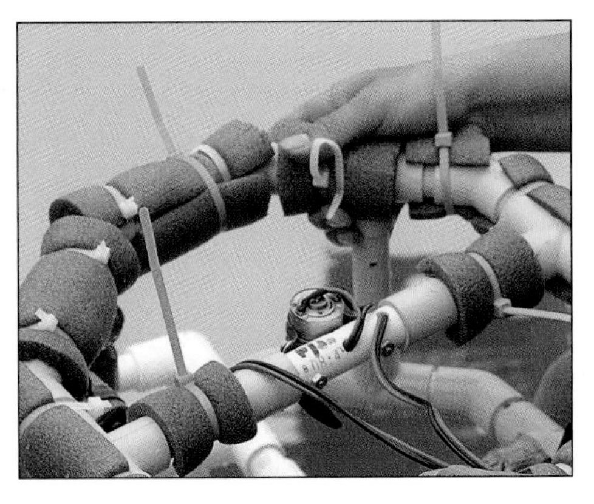

Center and right: Image courtesy of Duane Thompson, Monterey Bay Aquarium Research Institute

Figures 6.36: Flotation and Depth

Flotation options such as the soft foams used to insulate water pipes or plastic pop bottles may be easy to come by and use on basic PVC ROVs, however their buoyancy capacity is only useful in very limited swimming pool depths. Flotation made out of PVC piping offers greater depth capacity.

own weight. This is relatively simple to achieve on the surface, but it becomes more challenging with increasing depth. As depth increases, pressure increases, so the walls of the float must be made thicker (and therefore heavier) to keep the float from being crushed. At some depth, the walls must be so thick that they weigh more than the amount of water the float displaces—at that point the *float* becomes negatively buoyant and turns into a ballast *weight*! Materials like steel or titanium, which have much higher strength-to-weight ratios than plastic, allow you to go deeper, but they do not eliminate this problem entirely.

That's not the only complication. Regardless of how it is constructed, a float used in a static ballast system must be able to maintain a nearly constant displacement in spite of increasing pressure as the ROV dives to its deepest operating depth. If water pressure can squeeze the flotation material to a much smaller size during a dive, like a balloon filled with air, then the amount of water the float displaces will decrease, and so will its ability to provide an adequate buoyant force. In fact, the ROV may lose so much of its buoyancy that it will begin to sink uncontrollably and be unable to return to the surface! Many an otherwise excellent ROV has been stuck helplessly on the bottom of a swimming pool during an ROV competition because the designers used water-pipe insulation or other compressible foam for flotation without understanding the consequences.

Foam: Today's very deep-diving ROVs generally use syntactic foam for flotation. Syntactic foam is rigid foam made by embedding millions of microscopic hollow glass balls (called "microspheres") in epoxy. This composite material was developed in the 1970s. Each sphere acts as a microscopic float. Because each microsphere is so small, it has very little surface area for pressure to push on and experiences only a tiny compressive force, even at very high pressures. It can therefore survive without collapsing, even at full ocean depth! Embedding millions of these microspheres in epoxy results in a relatively strong, low-density material that can be molded, cut, or otherwise shaped and sized to meet a wide range of flotation needs. While this composite material is ideal for advanced commercial, scientific, and military ROVs, syntactic foam is toxic to make and cut, and it tends to be very expensive, so it is usually not an option for low-budget, student-built ROVs.

For flotation on shallow-diving ROVs, school groups and hobbyists frequently use common, inexpensive foams such as styrofoam, the neoprene rubber used in wetsuits, or the foam insulation designed to insulate water pipes. Of these three, styrofoam may provide the most constant buoyancy, because it is usually stiff enough to resist significant compression, at least at shallow depths. (Note that Styrofoam™ is a trademarked name for polystyrene thermal insulation, but you'll often hear it used generically. People also use the term styrofoam for expanded polystyrene foam, used for disposable coffee cups.)

Soft, rubbery foams compress much more easily and will lose significant buoyancy, even at the bottom of a swimming pool. In deeper water, this can lead to serious problems. As foam compression reduces buoyancy, the vehicle may begin sinking, which compresses the foam even more and promotes even faster sinking. Unless the pilot reacts quickly, the craft may lose so much buoyancy that the vertical thrusters cannot reverse the sinking. The vehicle then sinks faster and faster as the foam becomes more and more compressed, ultimately plunging uncontrollably into the depths (or at least to the limits of its tether).

It's important to be sure that any foam used for flotation is a *closed-cell* foam. This means that the individual bubbles in the foam are not connected to each other. The air spaces in open-cell foams will quickly soak up water like a sponge, making that foam less and less effective as flotation. You can tell if a foam is closed-cell by trying to blow air through it. If you can, it means there are passages between the bubbles, so it's open-cell. If you can't, it's probably closed-cell foam. However, even closed-cell foams tend to break down and lose some buoyancy over time.

Plastic Pipe: Sealed sections of air-filled PVC pipe also make great low-cost floats that do not compress enough to affect buoyancy at swimming pool depths. For safety reasons, this type of float should *not* be used for diving deeper than a few tens of meters, unless an over-pressure relief valve is installed, because a tiny leak can pressurize the interior of the pipe during a dive, causing it to explode upon returning to the surface.

Ceramic and Plastic Spheres: Ceramics are now being used for deep sea flotation. For example, DeepSea Power and Light produces hollow 90-millimeter (approx. 3.5-in) diameter ceramic spheres, called SeaSpheres™, that are strong, lightweight, made of high-purity alumina (aluminum oxide—Al_2O_3), and rated for extreme depths. Six-inch trawl floats commonly used by commercial fishing boats are similar to ceramic spheres in shape; these robust, low-cost floats are made from plastic, have a depth rating of 400 meters (approx. 1,300 ft), and are manufactured with holes for easy mounting options.

7.2.2. Weights

The weight used to ballast a vehicle usually takes the form of several separate pieces of metal that are arranged on the lower part of the ROV frame. Weights can be added or removed until the vehicle just barely floats at the surface. They can also be rearranged on the frame to adjust pitch and roll. Once the craft is trimmed properly, you'll want to use some method of securing these weights so they don't shift or fall off.

Lead is an ideal weight for static ballast systems because of its high density, easy availability, and relatively low cost. Lead has a low melting point, so it can be conveniently poured into molds to make complex shapes. It is easily worked with machine shop tools and does not corrode much in seawater. The disadvantages of lead are that it has some environmental toxicity and is toxic to people if ingested. For this reason, you should always wash your hands before eating if you have been handling lead. Steel and iron pellets (or "pigs") are also good low-cost weights that are environmentally safe; their downside is that they rust quickly in seawater.

If space is available inside pressure canisters positioned near the bottom of the ROV, it is slightly more effective to place weights inside these canisters than on the outside. As noted earlier, this is because weights inside add weight without adding displacement. In contrast, weights on the outside have some fraction of their weight offset by their own buoyant force. For very dense materials like lead, which is about 11 times denser than water, the difference between putting the weight inside and outside is small, so it isn't worth heroic measures to squeeze the lead inside. But if you've got empty space low on the inside of a canister and your lead weight will fit in there easily, why not take advantage of the slightly greater effectiveness?

HISTORIC HIGHLIGHT: EMERGENCY DROP WEIGHTS

Human-occupied submersibles are designed with fail-safe systems that allow them to jettison battery pods, thrusters, or other heavy equipment so as to get their occupants to the surface quickly. For example, heavy battery pods can be held in place by elecromagnets. In the event of a complete power failure, the magnets would turn off, dropping the batteries and giving the craft emergency buoyancy.

Image by KoS, courtesy Wikimedia Commons

The French bathyscaph *FNRS-3* gave a dramatic demonstration of this safety feature during a record-setting dive in 1954. After a lengthy three-hour descent, the vessel touched down on the seafloor. Shortly afterward, the exterior lights went out, and the startled occupants felt two tremors. A fuse had blown, cutting power. The tremors that the crew felt were the two heavy batteries hitting the seafloor. Now some 1,134 kilograms (approx. 2,500 lbs) lighter, the bathyscaph demonstrated the power of positive buoyancy, shooting back to the surface in less that half the time it had taken to descend!

Figure 6.37: **FNRS-3**

7.3. Active Ballast Systems

As mentioned earlier in this chapter, active (dynamic) ballast systems are ones that actively change the buoyancy of a vehicle during the course of a dive. Because of their complexity, they are generally found only on sophisticated vehicles, such as submersibles and submarines. However, simple versions also find application in some AUVs and specialty underwater craft.

In principle, changing the buoyancy of a vehicle during a dive can be done by modifying the vehicle's weight and/or volume (displacement). Active ballast systems can be used alone for vertical position control, but more often they are used in combination with other methods of depth control, including dive planes and propulsive forces.

Changing a vehicle's weight is sometimes done with **drop weights**. For example, a positively buoyant submersible can be equipped with detachable weights heavy enough to pull the submersible (gently) to the bottom, thereby conserving thruster power for other purposes; when the vehicle is ready to return to the surface, it simply releases those weights.

Vehicle weight can also be changed by pumping water in or out of a rigid-walled, pressure-proof chamber called a **hard ballast tank**. Such tanks can be positioned inside

HISTORIC HIGHLIGHT: *TRIESTE*'S UNIQUE BALLAST SYSTEM

Trieste is the only manned submersible to reach the very bottom of the ocean. Its historic journey to the Challenger Deep occurred in 1960. At that time, syntactic foam had not been invented and soft ballast tanks would not work to achieve buoyancy at the extreme pressures encountered at 10,910 meters (approx. 36,000 ft), so the designers came up with a unique ballast system that used gasoline for flotation. Gasoline is nearly incompressible and less dense than water, so it floats. Thus, the bulk of *Trieste* was basically a glorified gasoline tank that provided buoyancy. Iron pellets served as drop weights. These pellets were carried in hoppers with a magnetic ballast release.

To start its historic dive, *Trieste* released some air used for flotation at the surface, free-flooding the conning tower. During the descent, gasoline had to be vented to allow *Trieste* to sink through the thermoclines and into the denser water below. When the submersible was ready to ascend, the iron pellets were released from the hoppers, thereby decreasing weight.

Jacques Piccard added an interesting ballast feature by dangling a heavy chain below the submersible. Its purpose was to keep the vehicle a specified distance off the ocean floor without expending extra energy and control. As the bathyscaph approached the bottom, the chain touched down. Each link resting on the ocean floor reduced *Trieste*'s weight. Once a number of links lay on the bottom, the weight difference was enough to allow the submersible to achieve near-neutral buoyancy. Piccard had done his calculations so carefully that the bathyscaph hovered over the bottom without having to resort to gasoline and pellets to maintain trim! For years, a similar dragging chain method has been used when towing pipelines to location.

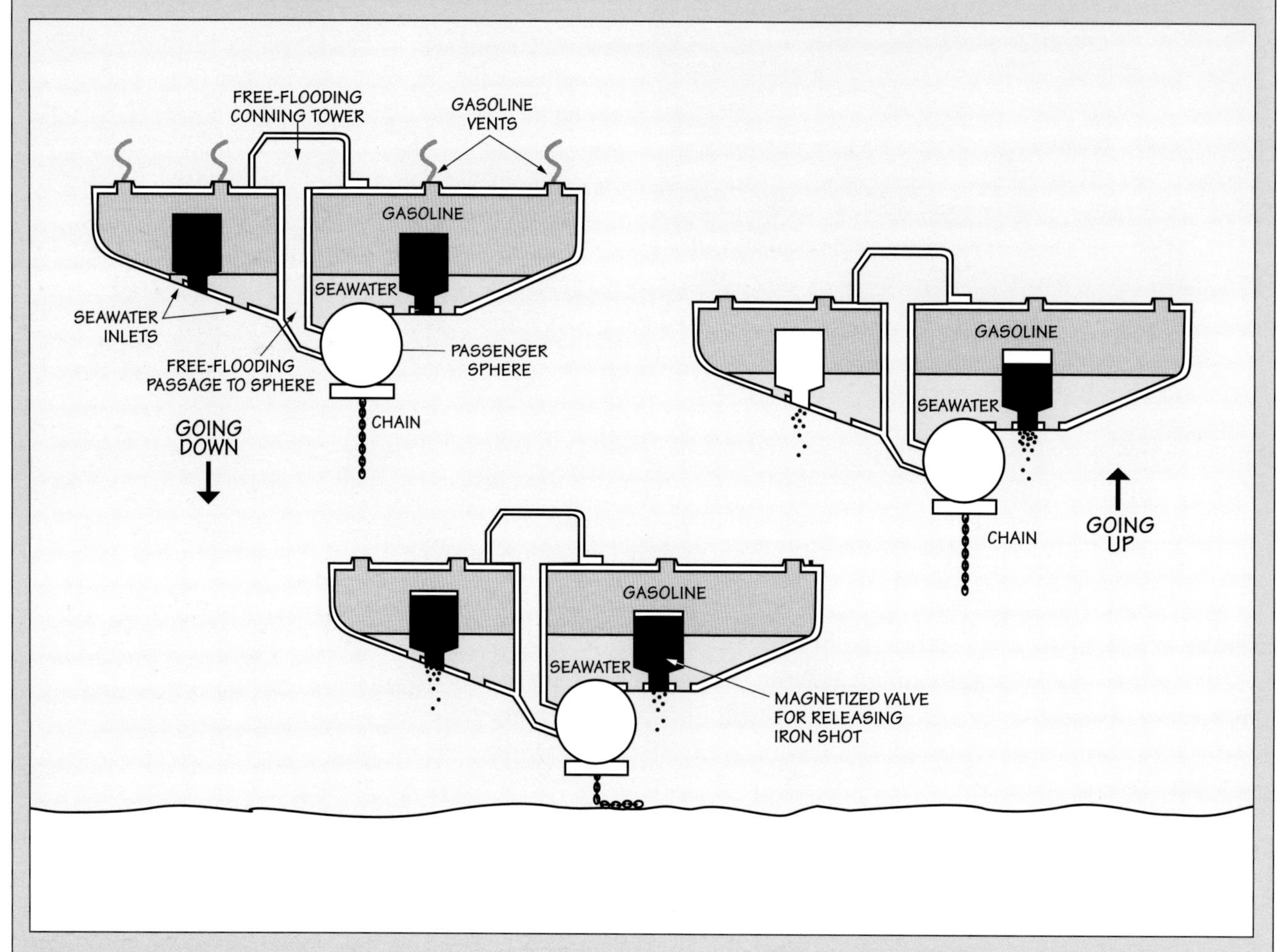

Figure 6.38:* Trieste's *Ballast System

In 1960, Jacques Piccard and Don Walsh were on board Trieste *as it descended to the deepest part of the earth's oceans—the Challenger Deep.* Trieste *employed a ballast system that juggled gasoline, sea water, iron pellets, and a heavy chain to adjust buoyancy.*

or outside the pressure hull. They are built to resist pressure at depths greater than the operating depth of the submersible and are usually emptied by a high-pressure pump.

Changes in displacement are generally made by using pressurized air and a series of control valves to replace water with air (or vice versa) inside **soft ballast tanks,** which are open to the surrounding water at ambient pressure through holes on their bottom surface. One set of valves lets pressurized air into the top of the tanks, forcing water out the bottom. This effectively replaces water with air and makes the tanks more buoyant. If the vehicle needs to dive, another set of valves allows the air to escape from the top of the soft ballast tanks, letting water refill the tanks.

The deeper a vehicle operates, the greater the quantity of compressed gas required to blow the same volume of tank. Submersibles often operate at greater depths than submarines, but stowing large quantities of compressed air is not feasible. So submersibles utilize smaller-volume soft ballast tanks, usually carrying only enough high-pressure air for just one or two blows of the tanks.

Some vehicles (primarily submersibles and AUVs) use an oil-filled reservoir in a tank for fine buoyancy control. This serves as a substitute for soft ballast. Oil is pumped into a flexible bladder that expands on the outside of the pressure hull. This increases the submersible's displacement without adding weight, causing it to gain buoyancy. Conversely, the displacement decreases when the oil is pumped back into an internal reservoir tank, so the vehicle gets less buoyant.

Pairs of **trim tanks** can be used to adjust the pitch of a vehicle without changing its buoyancy. One tank is located near the front of the vehicle and one near the back. Liquid is pumped from the forward tank to the aft tank, or vice versa, to shift weight from one end of the vehicle to the other, thereby adjusting pitch for efficient operation. An alternative to the trim tank is a shifting weight mechanism that can mechanically move a heavy weight (usually the battery pods) back and forth. Shifting the batteries has the same result as moving liquid in the trim tanks. International Hydrodynamics' early submersible *Taurus* had a mechanism that moved its batteries in this manner.

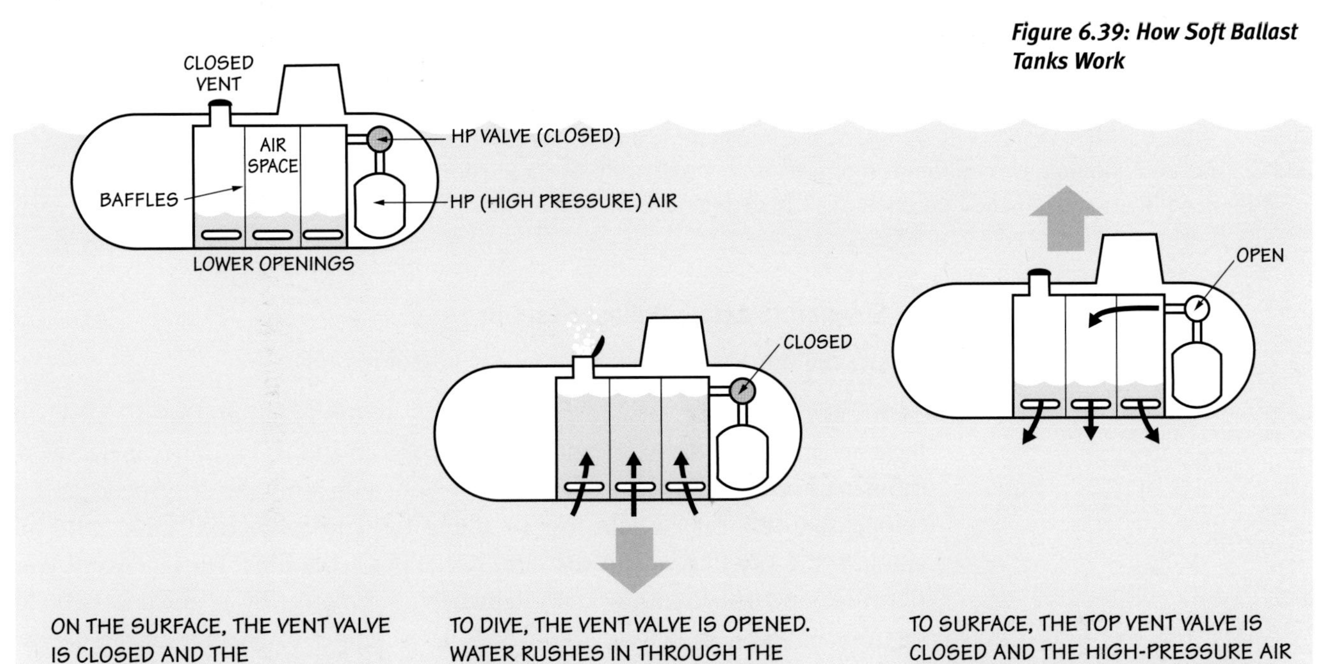

Figure 6.39: How Soft Ballast Tanks Work

The remainder of this section highlights some of the ballast systems used to control depth and trim in various types of underwater vehicles. It also includes a few examples of hybrid ballast systems, which combine both active and static ballast systems.

TECH NOTE: SOFT BALLAST TANK DEMO

To construct and operate a very crude soft ballast tank, take a small, clear plastic water bottle with its cap removed and push it top down into the water. You will notice that an air space is left inside. This is like a diving bell, in which the air trapped inside displaces the water.

Now take the bottle out of the water and punch a hole in the bottom of it with a nail. Insert a soda straw through the hole and seal the opening around the straw with chewing gum so that no water leaks in. Now push the bottle under water, straw end up, until half of the bottle is submerged. Notice how the water slowly rises up into the bottle. You can actually feel the air escaping through the end of the straw. Note that the water inside the bottle stops rising when it's reached the level of the water outside. If you push the bottle deeper until it is completely submerged, the water will rise inside until the bottle is filled completely. Remove the bottle and empty it.

Now push the bottle under water again, only this time keep your finger over the end of the straw so that the air is trapped inside. Notice that little or no water enters the bottle. If you release your finger, water starts rising in the bottle. Cover the hole again and the water stops rising. Now let the water rise until the bottle is about half full, then blow into the straw. If you blow hard enough, water is forced out through the bottom opening as the bottle is refilled with air. This demo illustrates how submarines can use valves and pressurized air to adjust the amount of water in a ballast tank to change buoyancy.

SAFETY NOTE: CAUTION REGARDING ACTIVE BALLAST SYSTEMS

It's fun and informative to study the various methods used to solve buoyancy and trim challenges in commercial, scientific, and military underwater vehicles. Doing so helps reinforce the basic physics principles presented in this chapter. It can also be inspiring, as many of these complicated active ballast systems serve as remarkable testaments to human ambition and ingenuity. Keep in mind, however, that the complexity of most of these systems is way beyond what is required for successful missions with small shallow-water ROVs or AUVs.

In fact, some of these systems—particularly the ones employing compressed air or other gases—can be extremely dangerous and should never be replicated, even on a small scale, without competent guidance from people experienced in the design and operation of such systems.

7.3.1. Submarine Active Ballast Systems

Submarines require active ballast systems for several reasons:

- Subs carry large and varying payloads, such as torpedoes and missiles. When these are launched, the weight in the sub is reduced and its buoyancy increases. A submarine must maintain its depth when launching a missile or torpedo, and it compensates for that weight loss by filling torpedo-room trim tanks with an amount of water that weighs the same as the projectile fired. During World War II, firing one torpedo immediately lightened a submarine by as much as 1,000 lb (approx. 454 kg), so it took a knowledgeable skipper to balance all the depth-keeping considerations of a submarine, especially when making a submerged daylight attack on a convoy. More than one submarine was lost in action after the sub fired a torpedo at periscope depth and its trim-compensation system then failed. Forced to surface, the submarine was a sitting duck for enemy fire.

- Modern subs have a large crew complement and may remain at sea for long periods, so large quantities of fuel, food, and water are consumed during a voyage. Trim ballast tanks compensate for this ongoing loss of weight.
- Damage to a submarine often results in flooding, so the vessel is divided into watertight compartments. That way, if one section floods, it can be sealed off to prevent the whole vessel from being flooded. Most submarines are built so that if partial flooding does occur, the soft ballast tanks can be blown dry to regain positive buoyancy.

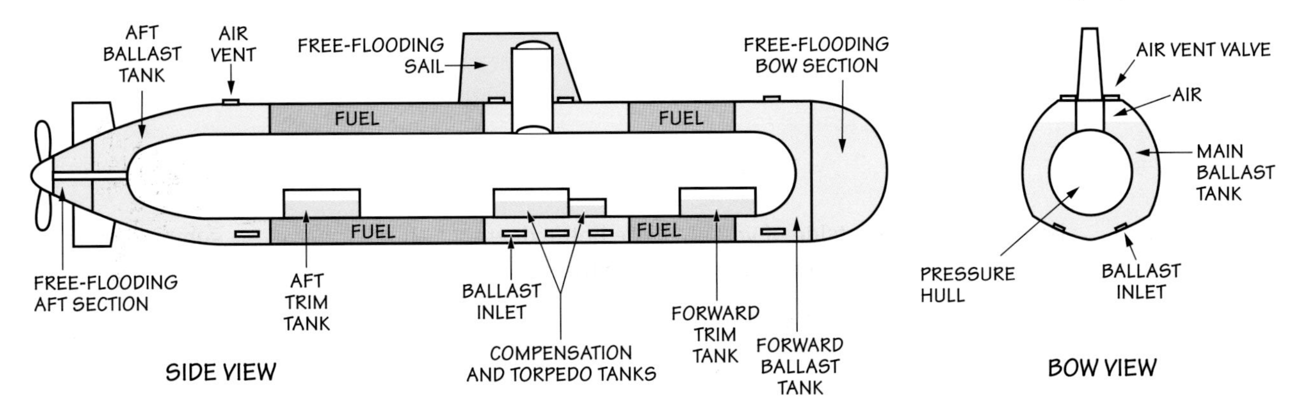

Figure 6.40: Active Ballast System on a Submarine

Simplified drawing of a submarine's active ballast system, showing the location and type of ballast tanks.

7.3.2. Submersible Active Ballast Systems

Unlike submarines, submersibles generally do not need to deal with large changes in weight. Therefore, most of the buoyant force in a typical modern submersible is provided through a static ballast system employing syntactic foam, as described earlier. However, some submersibles supplement their static ballast system with a small active ballast system for these reasons:

- Submersibles often require very precise control of depth to perform delicate tasks, such as collecting specimens. It doesn't make sense to use a lot of vertical thruster power in this situation, because that can stir up bottom sediments, destroying visibility and burning up battery power. Therefore, submersible pilots often use a small active ballast system instead of thrusters when they must achieve and hold perfect neutral buoyancy for long periods of time.
- Some submersibles employ a small soft ballast tank for gaining extra buoyancy after a dive. For example, the vehicle might need to compensate for the weight of heavy samples it has collected from the ocean floor.

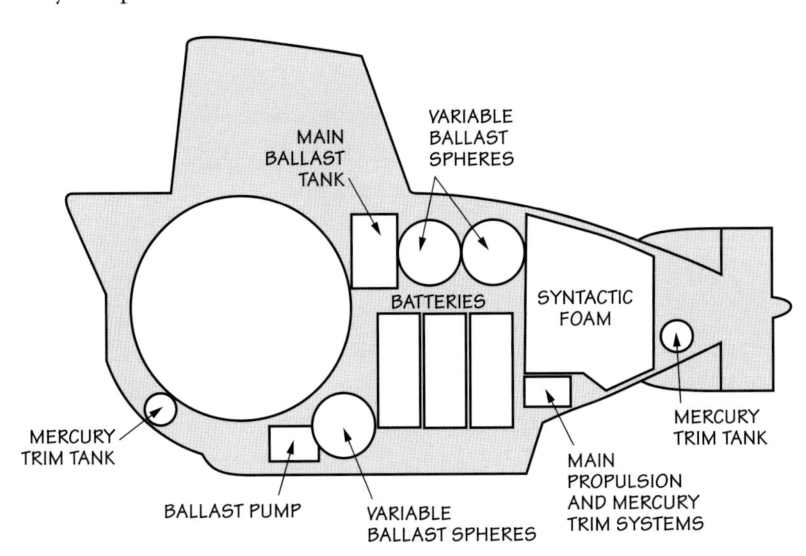

Figure 6.41: Active Ballast System on a Submersible

This simplified drawing shows the active ballast system on the submersible Alvin.

- A soft ballast tank is also a safety mechanism that provides extra buoyancy if an emergency occurs. At the surface, it provides reserve buoyancy, keeping the submersible higher out of the water.

HISTORIC HIGHLIGHT: HOLLAND'S BREAKTHROUGH WITH BUOYANCY AND STABILITY

In order to dive easily, ROVs, AUVs, submarines, and submersibles all need to achieve a buoyant state that is close to neutral buoyancy. Most early subs were not good at this. Essentially, they "pulled the plug" and sank, sometimes rather suddenly. John Holland took a new approach to the problem of longitudinal stability when he designed the *Fenian Ram*, launched in 1881. Prior to that, vehicles had been ballasted to neutral buoyancy in order to submerge. The *Fenian Ram* always maintained a slight positive buoyancy and submerged by diving rather than by sinking. Holland accomplished this by means of horizontal rudders (dive planes) fitted at the stern, along with the normal vertical rudder.

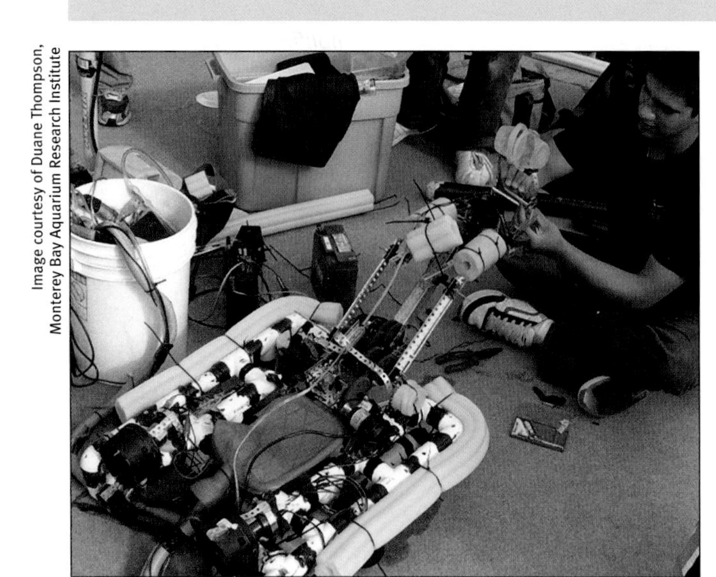

Image courtesy of Duane Thompson, Monterey Bay Aquarium Research Institute

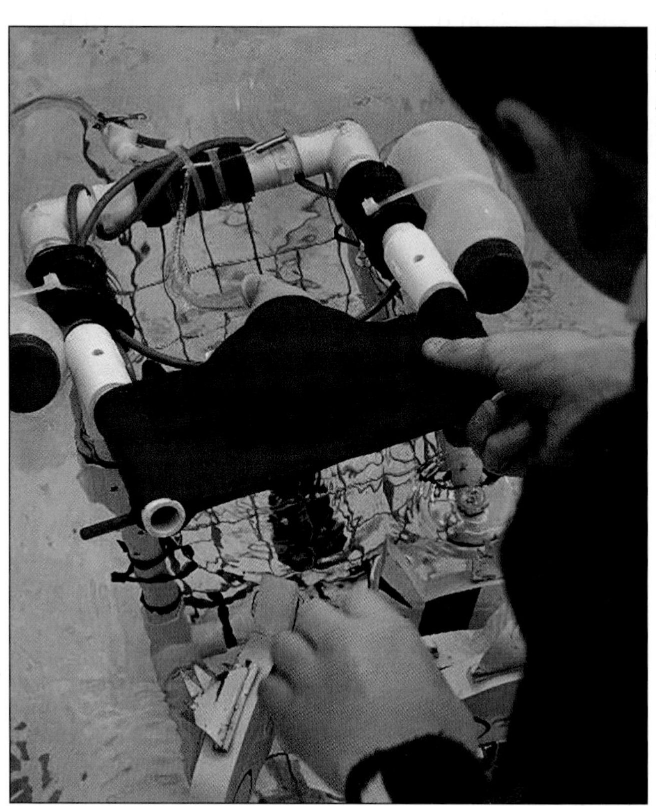

Image courtesy of Randall Fox

Figure 6.42: Active Ballast Systems for Small ROVs

It is possible to use materials, such as a hot water bottle or a balloon, to fashion innovative active ballast systems for hand-built, shallow-diving ROVs.

7.3.3. Active Ballast Systems for ROVs and AUVs

The ballast systems used for ROVs and AUVs are as varied as their diverse missions. Most depend primarily on a static ballast system, but some supplement that with an active ballast system. This is particularly true for experimental or hybrid vehicles with unusual propulsion systems or those designed for specialized tasks. Here are some examples:

- ROVs designed to retrieve heavy objects from the seafloor may use an active ballast system to help the vertical thrusters lift the load. Years ago, students at Florida Atlantic University built an ROV that used an air ballast tank to gain additional lift when the submersible was recovering large objects from the bottom. This ROV was ballasted conventionally with floats and weights to near neutral buoyancy at the surface and used a vertical thruster to submerge and hover. Once the object was secured, air was blown into the ballast tank to increase displacement, so the ROV could rise to the surface with the object.

- Some AUVs, notably **gliders**, follow a sawtooth pattern up and down through the water column as they collect oceanographic data. Some do this by carrying drop weights. The weights make them negatively buoyant and help them glide to the bottom. Once they've reached their intended depth, they can drop the weights, and glide "up" as they float toward the surface. Other gliders do a similar thing without drop weights by using motors and pistons to pump oil in or out of a bladder, thereby changing displacement and buoyancy. These gliders may also shift their trim to angle their nose down while descending or up while ascending.
- The *Slocum* class of gliders uses an innovative, experimental, chemical-based, "thermal engine" system to change displacement during its sawtooth movement pattern. This AUV contains a plumbing system with various fluids in it. One of these is a wax-like substance that is liquid at surface ocean temperatures, but freezes in the deep ocean. Pressure changes associated with the expansion and contraction of this fluid are stored and used to inflate or deflate a bladder filled with another fluid, thereby changing vehicle displacement. Although this type of glider uses batteries for communication and valve control, the ocean's natural thermoclines provide the ultimate energy for propulsion, so the glider can can operate for months.

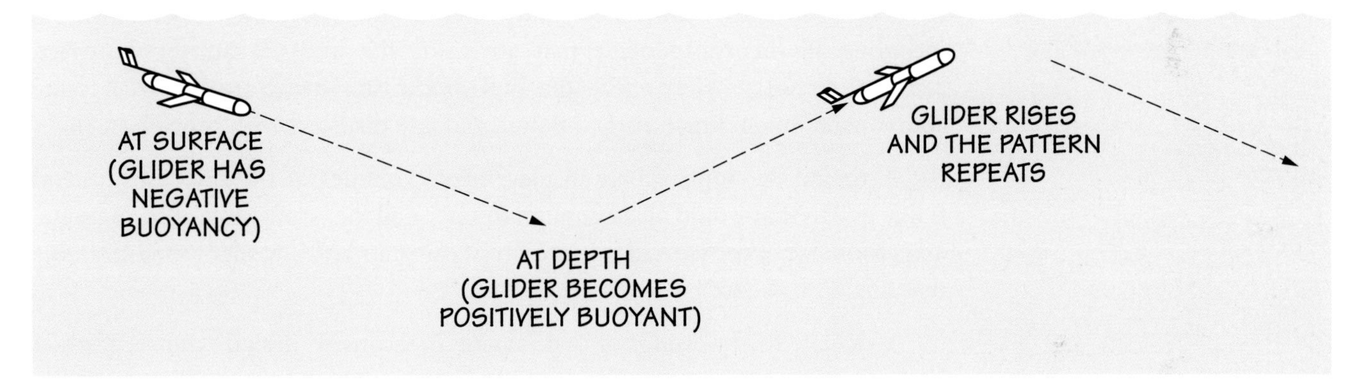

Figure 6.43: Glider Sawtooth Pattern

Simplified diagram of a glider-type vehicle with a modified soft ballast system that enables it to change displacement and produce a sawtooth movement pattern.

8. Practical Tips for Ballasting and Trimming a Small Underwater Vehicle

Studying the components of complex ballast systems does not negate the basics. Here's a short list of practical suggestions for adjusting the buoyancy and trim of a small, low-budget, underwater vehicle designed for shallow water (less than 10 meters).

- Keep it simple.
- Review and follow the two basic design suggestions discussed earlier in this chapter:

  (a) Strive to make your vehicle slightly positive in buoyancy.

  (b) For increased stability, place weights and dense components low on the vehicle; position less dense parts and floats high on the vehicle.
- Whenever possible, try to design a ballast system around inexpensive, readily available, waterproof materials.

  (a) For flotation, consider fishing floats, sealed sections of PVC pipe, sealed film canisters, or any other rigid container with air inside. If you use foam, use a closed-cell variety, and keep in mind that it may lose buoyancy at depth due to compression.

(b) For weights, lead fishing weights or scuba weights work well, but lead toxicity is a potential concern. Steel (for example, short sections of "rebar") also work well, but most steel will rust quickly if not shielded from water. Steel washers are easy to attach to a frame, and they offer good "fine-tuning" ability. Several smaller weights offer more flexibility than one big chunk but may be harder to attach.

- Design for adjustability.

  (a) Make sure your frame includes some easy, yet reliable, system for attaching various amounts of flotation and weight. Hooks are easy but risky, since they can allow weights or floats to slide off if the vehicle tips too far. Cable ties (also known as zip-ties or tie-wraps) work well. Washers can be stacked on bolts mounted to the frame and held in place with a wing nut.

  (b) Make sure your frame allows some flexibility for locating your floats and/or weights, so you can easily move the CG or CB to level off your vehicle.

- Use a weight statement table to keep track of the total weight and buoyant force acting on your vehicle.

  (a) Figure out the weight of *everything* you must include for a successful mission, including all the components that go inside the hull or canister (batteries, cameras, circuitry, wires, etc.), the hull or canister itself, any external frame components (including nuts and bolts), and any thrusters with propellers, etc.

  (b) Estimate the approximate displacement (volume) of the smallest hull and frame design that would allow enough room for all that stuff while allowing some extra room for easy access to the interior of canisters and any slight modifications that might be needed.

  (c) Calculate the buoyancy of that "smallest reasonable design" and see if it's at least as great as the weight. If not, plan on making the pressure canisters a little bigger or adding more floats to provide more buoyancy. When you do this, remember to adjust your weight statement table to account for the additional material required for the larger canisters or floats. Note also that some forms of flotation suitable for shallower depths are too compressible to use in deeper water.

  (d) Tweak all this, using the design-spiral approach, until you get something on paper that should definitely float, but has some easy way to add some weight later for trim.

Once you start actual construction (or want to evaluate design ideas you're considering), test, test, test, and test some more. Theory is great for guiding you close to a correct answer. However, you need to include all the relevant factors and avoid making any mathematical errors, or theory will guide you in the wrong direction. Just remember that no theory ever covers all aspects of reality, so always test each component of your ballast system as you go. This helps you develop a good understanding of how well it works (or doesn't work) and helps you avoid any unpleasant surprises later.

If you opt to take on a greater challenge, such as designing an observation craft that would operate at 100 meters instead of 10, remember that all the rules and considerations pertaining to buoyancy, stability, and ballast that you just studied in this chapter apply to deep-diving as well as shallow-diving vehicles.

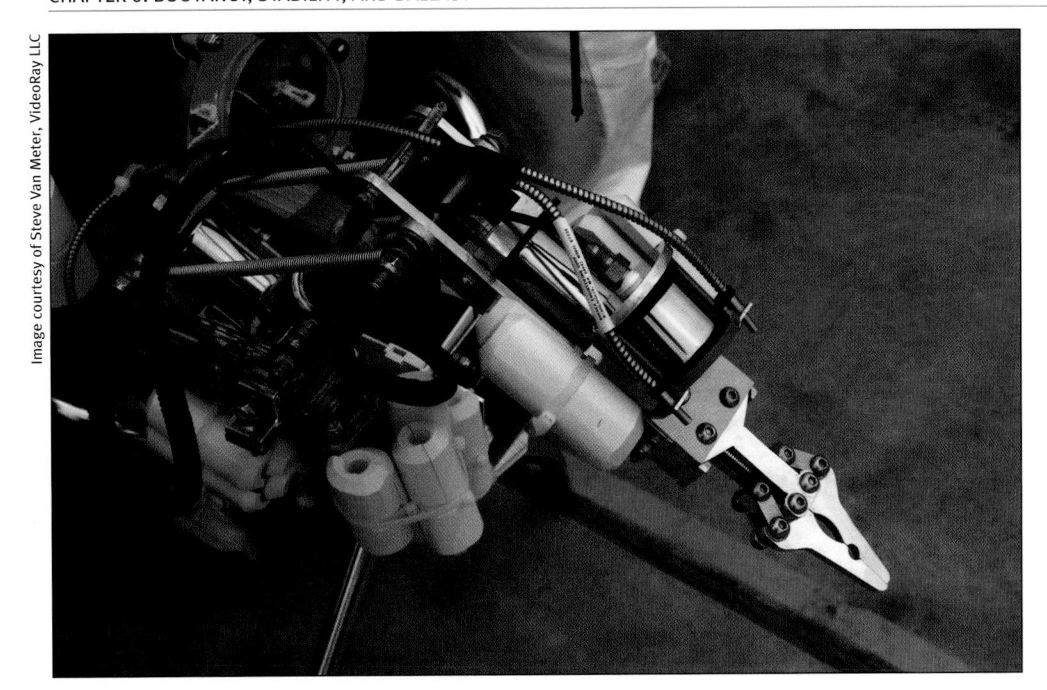

Image courtesy of Steve Van Meter, VideoRay LLC

Figure 6.44: Trimming a Manipulator

Manipulators are typically positioned low and forward on the vehicle structure. However, this can move the center of gravity forward, tipping the front of a vehicle downward. To counteract this trim issue, you may have to add flotation to the front of the vehicle, as shown in this photo, or on the manipulator itself.

9. Chapter Summary

This chapter has given you ways to predict and control (even during the design phase) the answers to each of these important questions:

Will your vehicle float, sink, or be neutrally buoyant in water?

Will your vehicle sit upright when submerged in water?

Can your vehicle be tipped over easily, and if it does get tipped over, how quickly can it return to an upright position?

The first question deals with buoyancy. You have learned that the buoyancy of a vessel is determined by a tug-of-war between the vessel's weight (which pulls down) and its buoyant force (which pushes up). According to Archimedes' Principle, the buoyant force, which always acts upward, is equal in magnitude to the weight of the water displaced by the vehicle. With the help of a weight-statement table, you can keep track of the weight, displacement, buoyant force, and buoyancy of each part of the vehicle and use that information to determine the overall weight and buoyancy of your vehicle.

A static ballast system can be designed to maintain a constant (usually slightly positive) level of buoyancy for the vehicle, so that vertical thrusters can move it up or down easily. Alternatively, an active ballast system can be employed to help the vehicle move up and down through the water column by causing controlled changes in vehicle buoyancy. Both static and active ballast systems need to be "trimmed" through adjustments in weight, displacement, or both, to provide just the right amount of buoyancy.

The second question deals with vehicle orientation. You've learned that a completely submerged vehicle orients itself with its center of gravity (CG) hanging directly beneath its center of buoyancy (CB) and that you can control the position of these points (and hence vehicle orientation) through careful placement of vehicle components, including weights and floats. If you want to actively change the orientation of a vehicle under water, you can do so by deliberately moving objects or fluids from one portion of the vehicle to another. Such movements are often used to trim the pitch and roll of a vehicle. And if you don't want orientation to change unexpectedly, you must be careful to guard against unplanned movements of objects or fluids inside the vehicle, such as those associated with the free surface effect or loose weights.

The third question deals with stability. You've learned that a stable vehicle is hard to tip over and will return to an upright position quickly if it does get tipped over. Under water, a vehicle with widely separated CG and CB is going to be more stable than one in which CG and CB are close together. In most (but not all) ROVs and AUVs, high stability is a desirable attribute. In practical terms, this means that you'll usually want to make sure air-filled canisters, floats, and other positively buoyant parts are located as near the top of your vehicle as possible and that batteries or other dense, heavy components are located as low as possible.

Chapter 7

Moving and Maneuvering

Chapter 7: **Moving and Maneuvering**

Stories From Real Life: Finally, the Practical Submarine

Chapter Outline

1. **Introduction**
2. **The Basics of Moving a Vehicle through Water**
   - 2.1. Newton's Laws
   - 2.2. Forces on Underwater Vehicles
   - 2.3. Combining Forces
3. **Estimating Thrust Requirements**
   - 3.1. Theoretical Approach
   - 3.2. Empirical Approach
4. **Producing Thrust**
   - 4.1. Electric Thrusters
5. **An Introduction to Electric Motors**
   - 5.1. Types of Electric Motors
   - 5.2. Anatomy and Function of a Brushed DC Motor
6. **An Introduction to Propellers**
   - 6.1. How a Propeller Works
   - 6.2. Types of Propellers
7. **Building Your Own Thrusters**
   - 7.1. Choosing a Motor
   - 7.2. Waterproofing a Motor
   - 7.3. Attaching a Thruster
   - 7.4. Selecting and Attaching a Propeller
8. **Thruster Placement**
   - 8.1. Analysis of a Common Three-Thruster Arrangement
   - 8.2. Analysis of Some Common Four-Thruster Arrangements
   - 8.3. Analysis of a More Elaborate Thruster Arrangement
   - 8.4. A Checklist for Thruster Placement
9. **Chapter Summary**

Figure 7.1. cover: Moving in a Fluid World

Moving through a three-dimensional fluid world presents unique challenges and opportunities. This ROV, a modified RCV 150, upgraded with fiber optics, high definition cameras, and enhanced hydraulics, is used by the Hawaii Undersea Research Laboratory to scout targets for full investigation and sampling using the human-occupied submersibles Pisces IV *and* V.

Image courtesy of Terry Kerby, Hawaii Undersea Research Laboratory

Chapter Learning Outcomes

- List and describe the physical forces affecting underwater vehicle motion.
- Determine how much thrust is required to propel your vehicle through the water at a particular speed.
- Describe how electric motors and propellers work and explain how to match a prop with its motor to maximize the output thrust and efficiency.
- Design an electric thruster capable of propelling a small vehicle at depths of up to 10 meters (33 ft).
- Describe various options for thruster placement and identify the simplest configuration that will meet your mission's requirements for maneuverability.

STORIES FROM REAL LIFE: Finally, the Practical Submarine

A number of flamboyant, over-confident submarine promoters flourished between the late 1700s and the mid-1800s. Many made extravagant promises for machines that either failed to live up to their designers' claims or never even got off the drawing board. John Holland and Simon Lake were a breed apart—they built submarines that worked.

Both these inventors were poor and struggled throughout their careers to raise enough money to fund their inventions. Holland was serious and scholarly, with some influential connections; Lake was a natural showman, beloved by the press and the public, but largely ignored by academics and military people. Despite their different styles, they turned out design after design, always building on what they learned from a previous model. Their subs were powered with internal combustion engines, state-of-the-art technology in the late 1800s. Both men eventually sold submarines to countries around the world. They were indeed key figures in early submarine history.

John Holland

Poor eyesight kept John Holland from joining the Merchant Marine as a young man. After a stint teaching school, he left Ireland and sailed to America as a steerage passenger. He had little in his pockets, other than drawings of a submarine and the conviction it could work. Holland's Irish patriotism drew him to the powerful Fenian Brotherhood, the American counterpart of the Irish Revolutionary Brotherhood. The Fenians seized on Holland's submarine idea as a way to strike at Britain. They began raising money, and in 1876, Holland demonstrated to the group a 75-centimeter (approx. 30-in) model of his submarine. Given the go-ahead, he then constructed a real vessel 4.4 meters (approx. 14.5 ft) long. Although it promptly sank upon launching, a week later it was afloat and working.

Called simply *Boat No. 1*, this first sub was powered by a 20-cylinder gasoline engine patented by George Brayton. Holland knew this new type of engine was the wave of the future for submarines. On a per-horsepower basis, it was lightweight. Also, unlike steam engines, it could start and stop instantaneously. The only major disadvantage was that it required a source of air when submerged. The use of electric cells (such as those in Claude Goubet's sub under construction in France) or stored compressed air (as had powered Simeon Bourgeois' boat a decade earlier) eliminated the need for air when submerged. But unlike Holland's sub, neither Goubet's nor Bourgeois' vessels could be recharged while under way.

Holland removed the engine and machinery from his first sub and began work on an improved 9.3-meter (approx. 31-ft) three-man version, dubbed the *Fenian Ram*. Powered by a stronger Brayton engine, this craft had a streamlined hull shape to optimize underwater speed, and a pneumatic gun that looked impressive but was inaccurate. Far more effective was the vessel's ramming power, which Holland estimated at about 50 tons (450 kN). Launched in 1881, the *Fenian Ram* could dive and surface with relative ease. The internal combustion engine still presented problems for propulsion when submerged, but at least this second sub could stay under water for two and a half hours.

Holland's early submarine designs all featured slight positive buoyancy. Unlike other subs of the day that simply sank in order to submerge, his craft tended to descend at an angle with enough speed to make dive rudders effective. His ballasting arrangements prevented free-surface surging, an effect that plagued many other inventors' subs when their ballast tanks began to empty and the remaining liquid shifted. The sudden movement played havoc with longitudinal stability. Unlike these rival craft, the *Fenian Ram* could run slightly below the surface, keeping level or diving by means of its dive rudders.

Holland's next project was a 4.9-meter (approx. 16-ft) replica of the *Fenian Ram*. Unfortunately, this *Boat No. 3* became the subject of a brotherhood dispute and eventually floundered while under tow. Disgusted, Holland severed all connections with the Fenians and tried to raise money by selling submarine plans to anyone, even the British, but to no avail. Finally, to make ends meet, he was forced to take a job at a dredging company for $4 a day. There he happened to meet one Elihu B. Frost, a company lawyer who joined him in the incorporation of the John P. Holland Torpedo Boat Company.

In 1888, Holland won a U.S. Navy submarine design contest, but before he could start work, the Navy diverted the prize money to the completion of a new cruiser. The contest was reopened in 1892, and Holland won it again. Finally, in 1895, Congress authorized money for John Holland to build a sub called the *Plunger*. The Navy immediately established a high-ranking supervisory committee, whose members knew little about building submarines. Holland grew so frustrated with their actions that in 1899, he began constructing his sixth submarine, the *Holland VI*, at his own expense. In the end, Holland's version worked, and the Navy's committee version didn't. Unfortunately, by now, Holland had been forced to sell most of his stock to finance the project and was broke.

In 1900, the Holland company was taken over by Isaac Rice, a German-American patent lawyer and battery manufacturer, who made it a subsidiary of his Electric Boat Company. He had supplied Holland with storage cells in the past and predicted a worldwide expansion of submarine production—and battery use. Rice was a savvy businessman and knew Holland's designs could be marketed. Their sleek, porpoise-like shape and minimal auxiliary electrical load permitted the vessels to move at a speed of 7.1 knots for more than three hours while submerged.

Suddenly, navies around the world were getting into submarines. France had taken an early commanding lead with a variety of designs. Several Holland submarines sold to Japan and to the Royal Navy, who started their submarine fleet with the delivery of five boats in 1902. Russia bought a Holland sub. The word spread—Holland boats traveled easily at the surface, dived quickly, could fire a torpedo when submerged, and returned to the surface, all predictably and on command. Germany was still skeptical and waited until 1906 to take delivery of a Krupp-Germania design that would form the basis of its first *U-boat* fleet.

On April 11, 1900, the U.S. Navy officially started its own submarine fleet, purchasing the *Holland VI* for $150,000. It was renamed the USS *Holland (SS-1)*. The sub was powered by a 45 hp gasoline engine, designed by Nicholas Otto, that provided surface propulsion and power for recharging batteries. A 50 hp electric motor provided power when submerged. A specialized engine-clutch arrangement allowed the engine to drive the propeller directly or recharge the batteries, with the motor serving as generator. In addition, the sub had a 200 psi (13,800 kPa) air system, a trim pump, air compressor, pneumatic steering, dive plane controls, and a periscope. Once the U.S. Navy became firmly committed to the idea of submarines, they bought successive Holland designs, each slightly larger and improved.

When John Holland's business agreement with Electric Boat expired, he applied on his own for new patents for submarine design improvements, but the company contested his actions in court and stopped him from doing any more work. Holland died a poor, broken man. Fortunately, his submarine legacy proved enduring, particularly his work in the effective use of hull shape, dive plane controls, and the internal combustion engine and batteries. The results of his legacy were truly workable submarines that were able to move and maneuver effectively.

Simon Lake

Flamboyant and a quarter-century younger than Holland, Simon Lake envisioned a different future for subsea craft—one that featured salvage, oyster gathering, underwater mining, and cable maintenance. He read Jules Verne's *Twenty Thousand Leagues Under the Sea* as a youth and began a lifetime of underwater experimentation. As a schoolboy, he ran tests to see how long he could stay beneath the capsized hull of a canvas canoe he had built. By age 14, he had sketched plans for a submersible that became the basis of his future designs. Just before turning 20, he read of John Holland winning a Navy contract and decided to build his own submersible.

Lake's first prototype sub was the 4.2-meter (approx. 14-ft) *Argonaut Jr.*, which he built of yellow pine planking and canvas in 1894. Essentially a box on wheels, the ungainly-looking craft was propelled when on the bottom by a man turning a crank with foot-powered pedals. The sub boasted wheels, an airlock, and a diver's compartment that provided access to the seafloor—all built on a meager budget of $15!

Despite its humble beginnings, *Argonaut Jr.* was successful and attracted the attention of wealthy investors. However, Lake saw only a bit of this money. Most of it was siphoned off by a promoter, but there was enough to build a full-size bottom crawler that took shape right next to John Holland's *Plunger*. The new 11-meter (approx. 36-ft) *Argonaut* was built of steel plate, this time in the shape of a more conventional boat. On the surface with ballast tanks empty, it easily rode the waves with a 30 hp gasoline engine turning a propeller. When the ballast tanks were filled, the new sub descended to the bottom, and the same engine was connected to two large, toothed driving wheels. A third wheel in the stern steered the craft. Initially, the engine sucked air through a canvas hose connected to a surface float; later, this was replaced by two long intake and exhaust masts. An even larger version, *Argonaut II*, had a hull that was some 6 meters (approx. 20 ft) longer and could carry up to 31 people. Unlike Holland's subs, all of the *Argonauts* were designed to operate either at the surface or on the bottom.

Lake was enthusiastic about his craft, and his exploits made the newspapers. He loved taking visitors for seafloor tours, and once on the bottom, was not above some thrilling theatrics, especially for those who were slightly claustrophobic. *Argonaut II* was the scene of a much publicized underwater picnic. Several dignitaries from Bridgeport, Connecticut (the scene of Lake's new operations), were invited aboard for an underwater journey in which they picked shellfish from the seafloor, made chowder, and passed the jug—all while submerged. This demonstration was effective enough to secure funding for Lake's next venture, the 19.5-meter (approx. 64-ft) *Protector*.

This new sub was a multi-purpose vessel that ran on a gasoline engine at the surface and on batteries when submerged. Lake hoped to sell *Protector* to the U.S. Navy, since its design included minesweeping and mine-laying capabilities and three torpedo tubes. But the Navy turned him down, so he offered it to Russia and Japan. Russia bought the sub in 1904, renamed it the *Osetr*, and ordered five more. Lake then built submarines for Austria and took out German patents, as well. In 1910, he built a more conventional submarine, the 48-meter (approx. 157-ft) *Seal*, and in the next year finally closed a deal with the U.S. Navy for the *Protector*. Lake sold other subs to the U.S. until 1925, when the Navy began building its own.

Simon Lake was well along in years when he met Louis Reynolds of Reynolds Metals Company and sold him on the idea of someday building aluminum submarines. Lake died in 1945 at the age of 78, but his aluminum sub concept eventually materialized in the *Aluminaut*.

Simon Lake and John Holland were both genuine visionaries who successfully married their ideas of underwater craft to new technologies, particularly the internal combustion engine. Despite the funding difficulties that plagued them throughout their careers, both men are true pioneers in the development of submarine propulsion systems.

1. Introduction

Image courtesy of Frank M. Rinderknecht and Rinspeed Inc.

Figure 7.2: The sQuba Car

Unlike a regular automobile that operates in a two-dimensional world, this unusual James Bond-like sports car and other underwater vehicles need to be able to maneuver in a three-dimensional fluid world.

Any vehicle, whether it's a tricycle, airplane, dump truck, or submersible, must be able to transport people, equipment, or other items from one location to another. Doing so under water requires special methods for propelling the vehicle in a controlled manner through a three-dimensional aquatic world.

Whether it's an underwater sports car suitable for a James Bond mission or a not-so-secretive ROV, your vehicle will have these requirements:

1. It must have some device(s) for generating the required propulsive forces.
2. It requires some way to supply those devices with the power they need to do their job.
3. It needs some way to control the strength and direction of the forces produced.

The next three closely related technical chapters cover these topics.

- *Chapter 7: Moving and Maneuvering* covers the basic physics behind vehicle movement. It also describes how you can build force-generating thrusters and discusses how to arrange those thrusters on your vehicle for effective maneuvering.
- *Chapter 8: Power Systems* addresses how to supply the power needed so that your vehicle's thrusters and other systems can do their jobs.
- *Chapter 9: Control and Navigation* discusses how you can regulate the power flowing to thrusters in precise ways to achieve good control over the location, speed, and direction of your vehicle.

2. The Basics of Moving a Vehicle through Water

The ancient Greek philosopher Aristotle (384–322 BC) pondered many fundamental questions about the natural world, including, "What makes something move?" When he observed a horse-drawn cart, he figured that the cart moved because the horse pulled it, and that it stopped because the horse stopped pulling it. Though not quite accurate, this Aristotelian notion of motion seems reasonable, even today, and it prevailed for 1,800 years.

Then, in the sixteenth century, an Italian astronomer and philosopher named Galileo (1564–1642) performed simple experiments that challenged Aristotle's ideas about motion. For example, some of Galileo's experiments showed that a ball could keep rolling for a long time, even if the initial push that got it rolling had ceased. This and other experiments led Galileo to conclude that forces were not required to keep an object moving, but that they were required to start an object's motion, and to stop or change that motion. Though somewhat counterintuitive, this insight revolutionized scientific understanding of how objects move. Scientists now recognize, for example, that the cart stops when the horse quits pulling because there is friction in the wheels that stops the cart.

2.1. Newton's Laws

In the late seventeenth century, Sir Isaac Newton (1642–1727), an English mathematician, scientist, and philosopher, built on Galileo's work and developed three mathematical equations that are simple in form, yet profoundly influential. They describe with remarkable precision how the movement of all objects, from the size of sand grains to entire galaxies, is controlled by basic mechanical forces. These relationships are known as **Newton's Laws of Motion**. Because of their precise, quantitative nature and awesome predictive power, Newton's laws form the foundation of most of modern science and engineering, and they are an ideal place to start any serious discussion of vehicle motion.

Describing Newton's equations as "simple in form" might be a bit misleading. It's true that the equations are elegantly simple-looking. For example, the second law (his most famous one) can be written as:

Equation 7.1

$$F = ma$$

This is just a mathematical way of saying that the net force (F) acting on an object equals the object's mass (m) multiplied by the object's acceleration (a). However, there's a lot of engine hiding under the hood of this little equation. For one thing, F represents not just one force, but the sum of *all* forces acting on the object. Moreover, this and his other two equations of motion are **vector** equations, not algebra equations. That means you need to keep track of not only the *sizes* of the forces and acceleration, but also the *directions* in which they act. Applying these equations correctly to predict general motions of a three-dimensional vehicle in a three-dimensional fluid world requires vector calculus, which is beyond the scope of this book.

Fortunately for your needs, the essential information necessary to begin thinking accurately about underwater vehicle motion can be extracted from each of Newton's three equations. Those "laws" are translated below into three corresponding—and much more intuitive—statements of practical significance for the underwater vehicle designer.

Newton's (Simplified) Laws of Motion for Vehicle Control

1. For a vehicle to maintain constant speed and direction, all of the forces acting on it must be in balance, so their effects cancel.
2. For a vehicle to change speed and/or direction, there must be an imbalance in the forces acting on it, so the net force is not zero.
3. To push a vehicle in one direction, you must push something else equally hard in the opposite direction.

One thing you can see immediately from these paraphrased laws of motion is that understanding the combinations of forces acting on a vehicle is essential for understanding how to make it move and how to control its motion.

2.2. Forces on Underwater Vehicles

Five types of force influence the motion of an underwater vehicle:

- drag
- weight
- buoyancy
- thrust
- lift

These are not the only types of force in the world, but these are the five force categories that directly impact vehicle movements in water. Each is described below. Recall that in this textbook forces are typically expressed in newtons (metric) or pounds (imperial).

2.2.1. Drag

The naturally occurring force that resists relative motion between an object and the fluid surrounding it is called **drag**. (See *Chapter 3: Working in Water* for an introduction to this topic.) Drag is usually undesirable, but also unavoidable. Vehicle designers must understand drag thoroughly, because drag limits a vehicle's maximum speed, plays an important role in vehicle steering, and consumes most of the energy used to power underwater vehicles.

Drag will be revisited in more detail later in this chapter, but for now you can think of it as something that acts a bit like a set of brakes that is perpetually on. Drag always tries to stop relative motion between the vehicle and the water around it. This makes the vehicle try to do whatever the water is doing. In calm water, such as a pond or swimming pool, drag will try to slow and stop a moving vehicle. In moving water, such as a flowing river or surging sea, drag will try to carry a vehicle along with the moving water, making it harder to control and potentially crashing it into submerged boulders or other obstacles.

Therefore, to move a vehicle through water or to maintain vehicle position in moving water, other forces are required to overcome drag. That's the job of an underwater vehicle's propulsion system—to generate the needed forces. Usually these forces come in the form of thrust from one or more propellers. The propulsion system must also be able to adjust the strength, direction, and timing of propulsive forces to produce the desired vehicle movements.

One very important aspect of drag is that the size of the drag effect depends on the speed at which the vehicle is moving through the water. The greater the difference between vehicle speed and water speed, the more aggressively drag "applies the brakes" to force the vehicle to move at the same speed and direction as the water. A vehicle that is not moving at all (relative to the water) experiences no drag.

2.2.2. Weight

Chapter 6: Buoyancy, Stability, and Ballast introduced **weight** and its relationship to buoyancy. You'll recall that weight is always a downward force caused by interaction between the vehicle's mass and gravity. Acting by itself, weight tends to pull a vehicle downward through the water.

2.2.3. Buoyancy

Chapter 6 also discussed **buoyancy**, an upward force acting on the vehicle, caused indirectly by the weight of water. The water weight results in a pressure gradient, with higher pressures deeper down in the water column. Water pressure pushing up from below a vehicle is therefore higher than the pressure pushing down from above, so the net effect is an upward buoyant force. Acting by itself, buoyancy tends to push a vehicle upward toward the surface.

2.2.4. Thrust

An actively generated, energy-requiring, propulsive force used specifically to cause or control vehicle motion through the water is called **thrust**. Rotating propellers are the most common method for generating thrust, but many other possibilities exist. (You'll see more on this in the next section.) In principle, thrust may be directed in any direction, though most vehicle designs employ a limited range of possible thrust directions. Thrust directed vertically can be used, in combination with weight and buoyancy, to control a vehicle's vertical position and vertical movements through the water. Thrust directed horizontally can be used to propel the vehicle forward or backward, execute right or left turns, or spin the vehicle in place. Strategies for orienting thrusters to produce a wide range of desirable vehicle motions will be discussed near the end of this chapter.

2.2.5. Lift

Lift is a little like drag, in that this force can exist only when there is relative motion between the vehicle and the water. Unlike drag, however, lift always acts *perpendicular to* (i.e., at right angles to) the direction of vehicle motion through the water rather than exactly opposite that direction. Although the term "lift" suggests an upward force, lift can also act downward or sideways.

Lift may be created intentionally by using fixed "wings" or adjustable control surfaces. Examples of control surfaces include the dive planes on a submarine, which can be angled up or down to help the sub surface or dive, and the rudder on a ship, which is used to steer the vessel left or right. Lift may also be produced unintentionally, because almost any surface on a vehicle can generate lift when the vehicle moves through the water. Like drag, lift becomes a more significant force as vehicle speed through the water increases; it plays an important role in the control of fast-moving vehicles like submarines and torpedoes, but is often negligible in slow-moving vehicles like ROVs.

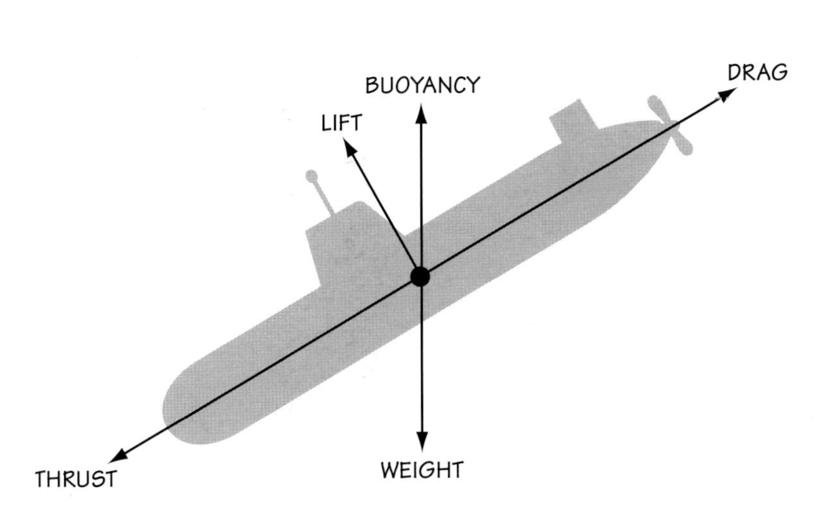

Figure 7.3: Forces Acting on a Moving Underwater Vehicle

Five basic forces act on an underwater vehicle as it propels itself through the water. In this diagram, each arrow represents one of the forces, as if a rope were tied to the vehicle and pulled in the direction the arrow is pointing.

***Thrust** produced by propellers or other methods pushes or pulls the vehicle in a particular direction (usually forward) through the water.*

***Drag** acts exactly opposite to the actual direction of vehicle motion through the water, usually opposing thrust.*

***Weight** always acts straight downward.*

***Buoyancy** always acts straight upward.*

***Lift** acts at right angles to the direction of vehicle motion and may be large or small, depending on vehicle speed and shape. Lift is insignificant in many ROV applications.*

TECH NOTE: LIFT

Some fast-moving vehicles, such as airplanes and torpedoes, maneuver with the help of wings, rudders, or similar control surfaces. These structures create a force called lift, which is oriented perpendicular to the direction the vehicle is moving through the air or water. The term "lift" originates from airplanes, in which upward force is produced by wings that lift the forward-moving plane into the air, but lift can be oriented in other directions, as well. For example, the spoiler on the back of a race car creates downward lift to press the rear wheels firmly against the roadway for better traction. The rudder on a ship uses sideways lift to push the rear of the ship to port or starboard, thereby steering the ship in a new direction. Submarines adjust the angle on short, wing-like structures called dive planes to produce either upward lift (which aids in surfacing) or downward lift (which aids in diving).

Unfortunately, wings, rudders, dive planes, and other such control surfaces provide significant amounts of lift only at high speeds. They are pretty much useless on very slow-moving vehicles, such as ROVs. The good news for people designing slow-moving ROVs is that this shortens the list from five significant forces to only four: weight, buoyancy, thrust, and drag.

2.3. Combining Forces

The force (F) in Newton's laws refers not to individual forces acting alone, but to the *net* force created when *all* the forces acting on a vehicle are considered simultaneously. Thus, underwater vehicle motion is always determined by a five-way interaction among the five types of force just described. The underwater vehicle designer who carelessly ignores any of these force categories invites disaster.

Keep in mind that these five force categories are just categories, not a complete list of forces. For example, if you have six thrusters on your vehicle, you may have six distinct thrust forces you need to think about for just the thrust category, in addition to any forces in the four other categories. One of the challenges facing your design team will be how to keep track of these multiple forces in a realistic yet manageable way during the design phase.

In some cases, it is possible conceptually to replace a large number of individual forces acting at different places on the vehicle with one equivalent force acting at a single point. This approach was illustrated in *Chapter 6: Buoyancy, Stability, and Ballast*, where you learned that the individual weights of each part of the vehicle can be combined into a single force (the total weight) acting at the vehicle's center of gravity (CG). Likewise, distributed buoyant forces can be combined into one force acting at a point called the center of buoyancy (CB).

In addition, you may be able to thoughtfully (as opposed to carelessly) ignore some of the five listed force categories altogether. For example, if you carefully design your vehicle for neutral buoyancy at all planned dive depths, then from that point on, the vehicle's weight and buoyancy will cancel, so you can discount their contributions to vehicle motion. In addition, if your vehicle moves slowly, you can probably disregard the effects of lift, because it will be too small to matter. (See *Tech Note: Lift*.) Keep in mind that these ignored forces are not gone. They just cancel out or are small enough that they don't contribute much to the overall net force controlling vehicle motion. If you employ an active ballast system (which changes buoyancy and/or weight, so those two forces no longer cancel) or build a fast vehicle (which will likely experience significant lift), then you can no longer overlook these force categories.

Most probably, you will choose to employ a static ballast system trimmed for neutral buoyancy. If so, you will be able to discount vehicle weight and buoyancy to a large extent when designing their propulsion system. And since most of the vehicles are likely to be slow ROVs or AUVs, lift can also be safely ignored. Therefore, the rest of this chapter will focus on interactions between the two remaining force categories that have a strong influence on the movements of these vehicles: drag and thrust.

3. Estimating Thrust Requirements

To estimate thrust requirements, you need a good sense of how drag affects underwater vehicle movement. Start by comparing the motion of a rocket ship in the vacuum of space (where drag is negligible) to that of a vehicle in water (where drag is very important).

When a rocket ship floating around in outer space fires its rocket engine, the engine produces thrust that creates a force imbalance. In accordance with Newton's second law, this force imbalance causes the rocket ship to accelerate. In other words, its velocity starts changing. If the rocket is initially motionless, it will begin to move. If the rocket engine continues to burn steadily for several minutes, acceleration will continue for several minutes. The speed of the rocket will be increasing steadily that entire time and may reach impressive values—several miles per second or more!

Newton's laws apply equally well in water; however, the observed result is very different. The reason for this difference is drag. When an underwater vehicle switches on one of its thrusters, that thruster creates a force imbalance, just like the rocket engine does. In accordance with Newton's second law, this causes the vehicle to begin accelerating, just as a rocket begins accelerating. However, for a vehicle in water, this acceleration period is *very* short-lived. As the underwater vehicle gains speed through the water, the force called drag appears and grows rapidly. Drag opposes vehicle motion, and before long (usually within a second or two), it has grown big enough to cancel out the effect of the thrust. The steady thrust is still there, but now it's exactly balanced by the drag, so these two forces cancel, and there is now zero *net* force. In accordance with Newton's first law, this force balance kills any further increase in speed, even though the thruster is still generating thrust. If thrust levels are later increased or decreased, the vehicle will speed up or slow down until drag again balances thrust.

Thus, in outer space, where drag is negligible, steady thrust results in steady acceleration and *ever-increasing speed.* But in water, where drag is significant, steady thrust results in only a very brief period of acceleration followed by *constant speed.* The constant speed achieved is determined by the balance between propulsive thrust and drag. Drag increases with speed and decreases with streamlining, so the balance point will be reached at higher speeds when higher values of thrust and/or more streamlined vehicle shapes are used.

Almost everybody seems concerned about building a faster vehicle, but faster is not always better, particularly when you consider all the associated costs. Faster vehicles have bigger problems with drag and are therefore usually more difficult and expensive to build. They also consume energy much more quickly, requiring bigger surface-supplied power systems and limiting dive time on any battery-powered dives. You'll save yourself lots of time, frustration, and money if you get realistic about the maximum speed you need and understand how thrust, drag, and speed interact.

To calculate how much thrust your propulsion system must produce, you'll need to estimate the maximum drag on your vehicle. Since drag is speed-dependent, you'll need to know the maximum required speed of your vehicle before you can determine the maximum drag and the thrust needed to overcome it. There are two basic ways to estimate the drag on a vehicle during the early stages of the design spiral. One is theoretical; the other is empirical. Each approach has its advantages.

3.1. Theoretical Approach

The theoretical approach is a mathematical method in which you use a formula and certain simplifying assumptions to estimate the drag on your vehicle at different speeds. The main advantage of this approach is that it requires only a pencil and paper (and maybe a calculator), so it is relatively fast and easy and it can be done even when your design is still on the drawing board. You don't need to have built anything yet. Its main disadvantage is that the simplifying assumptions used for this technique mean the results are never exactly the same as reality, even if you do the calculations correctly. And, of course, if the wrong formula is used, if the formula is applied incorrectly, or if mathematical mistakes are made during the calculations, the results can be way off.

When an object moves through water, two types of drag act on the object. One type is called **viscous drag** (also known as **skin friction drag**). It is caused by the natural tendency of fluid molecules to stick to each other and to objects in contact with the fluid. This type of drag is what makes it hard to stir honey or other viscous (i.e., thick and goopy) fluids. The other type of drag is called **pressure drag**. Pressure drag is caused by the inertia or mass of the fluid. As an object moves through a fluid, it must push the fluid out of the way, thereby creating a high-pressure region in front of the object. Meanwhile, a relatively low-pressure region forms immediately behind the object, because fluid inertia resists acceleration of the fluid back into the space behind the passing object. The pressure drop from front to back creates a net force retarding the object's motion through the fluid. The same is true if fluid flows past a stationary object, as shown in Figure 7.4. Note that a high pressure zone forms on the front side (upstream) of the object and a low pressure zone forms on the back side (downstream) of the object.

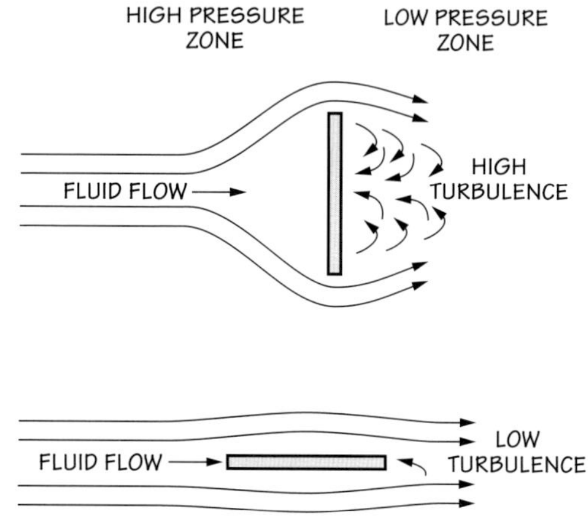

Figure 7.4: A Demonstration of Pressure Drag

When a flat plate is set perpendicular to the flow of fluid, it presents a huge frontal area that disrupts the stream, resulting in a high-pressure zone in front of the plate and a low-pressure zone immediately behind. This causes turbulence and results in pressure drag, a force that counteracts motion. If you set the same plate so that only its edge is facing the flow, turbulence and the pressure differences from front to back are minimized. This streamlining reduces the amount of pressure drag and allows the plate and water to slip past each other more easily.

The relative importance of viscous and pressure drag depends on several factors:

- the size and shape of the object
- the density and viscosity of the fluid
- the relative speed of the object and the fluid

For extremely small, slow-moving objects like bacteria and amoebas moving through water, viscous drag is the dominant type of drag, and pressure drag is negligible.

TECH NOTE: REYNOLDS NUMBER

Students interested in the quantitative details of comparing pressure drag with viscous drag will enjoy learning about a mathematical quantity dubbed the **Reynolds number**, often denoted "Re." Simply put, Re captures the relative importance of inertial and viscous forces in fluid flow situations. This dimensionless number can be calculated from the size of the object, the density and viscosity of the fluid, and the relative speed of the object and the fluid. As a rule of thumb, when the Reynolds number of a sphere or other blunt (non-streamlined) object is higher than 1, pressure drag is the dominant source of drag. When it's lower than 1, viscous drag is the dominant source.

A typical ROV cruising from one place to another has a Reynolds number in the tens of thousands or more, reflecting the fact that pressure drag can be thousands of times more important than viscous drag in these vehicles. Interestingly, modern streamlined hull designs are good enough that pressure drag on aircraft and submarines has been reduced to the point where viscous drag is now a significant component of overall drag in these vehicles. Certainly, drag reduction has become a productive area of research for improving their performance.

However, for larger, faster objects like underwater vehicles, pressure drag becomes much more significant and usually dominates. This is particularly true for "bluff bodied" (i.e., non-streamlined) vehicles, like ROVs. In these non-streamlined vehicles, pressure drag is usually thousands of times larger than viscous drag, so viscous drag can safely be ignored.

Whenever pressure drag is much larger than viscous drag, you can use the following pressure drag equation to get a rough estimate of the total drag:

Equation 7.2

$$D = \frac{1}{2}\rho C_d A U^2$$

The algebraic symbols used in this formula are defined below. Appropriate metric units are suggested. Certain other combinations of units can be used, but substitutions must be done carefully to avoid errors.

D = **Drag** (in newtons): Drag has already been described above.

ρ = **Fluid Density** (in kilograms per cubic meter): Since your vehicle will be operating in water, the fluid density to plug in here is the density of water. This varies somewhat with temperature and salinity, but for rough calculations, you can assume a value of about 1,000 kg/m^3.

Cd = **Drag Coefficient** (a dimensionless number with no units): This number captures the effect of vehicle shape on drag. Streamlined shapes have a lower drag coefficient than other shapes. (See Figure 7.5 for some examples.) That's why fast underwater vehicles, as well as airplanes, and even some cars, have fairings to make their shapes more streamlined. Determining an accurate drag coefficient using theoretical approaches is difficult, but for rough calculations, you can generally assume that an ROV or any other vehicle that is not particularly streamlined has a drag coefficient of about 1.

A = **Frontal Area** (in square meters): To determine the frontal area of your vehicle, imagine standing in front of it, making a life-sized photo image, then measuring the surface area contained within the vehicle outline. (See *Tech Note: Estimating Frontal Area.*) However, for an ROV with a complicated shape (including struts and wires and pressure canisters, for example), it can be difficult to get an accurate estimate of frontal area using this method. Fortunately, it is possible to get a conservative estimate

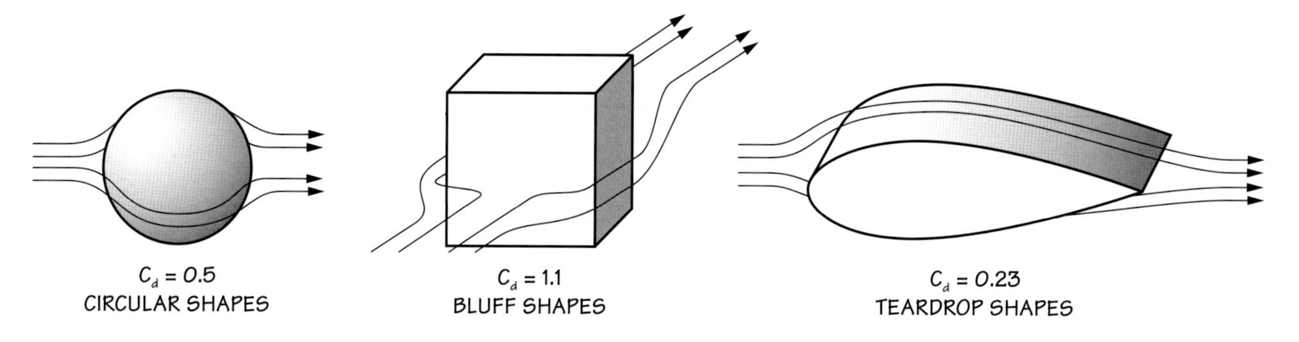

Figure 7.5: Approximate Drag Coefficients

The pressure drag equation (Equation 7.2) shows us that drag depends on several things, including the drag coefficient, C_d. The purpose of the drag coefficient in this equation is to capture the effect of object shape on drag. Objects with lower C_d values have more streamlined shapes and experience less drag than similar-sized objects with higher values of C_d under similar flow conditions.

(that is, one that overestimates the drag and therefore underestimates maximum vehicle speed) much more easily by simply using the frontal area of the smallest box your ROV would fit into.

U = Speed through Fluid (in meters per second): Use the maximum desired speed of your vehicle for this velocity in the pressure drag formula.

Note that speed is squared in the pressure drag formula. This means that drag increases rapidly with increasing speed. For example, you must quadruple your thrust just to double the speed. You would need 100 times as much thrust to go 10 times faster! Furthermore, increasing thrust is not trivial. It requires bigger motors or more of them. It demands a bigger power source and will typically drain batteries or fuel more quickly. Usually, all of this extra stuff makes for a bigger vehicle and therefore more drag. In other words, even if you increase your thrust to four times what it was before, you'll still get less than twice as much speed for all that time, money, and effort!

This is a good reason to avoid the temptation to go for extra speed, unless it's really needed. Moving quickly under water is both difficult and costly. You'll save time, money, and headaches if you can live with lower vehicle speeds. Always let the mission requirements determine the maximum speed required. If your vehicle needs to cover

TECH NOTE: ESTIMATING FRONTAL AREA

For theoretical drag calculations, you will need to estimate the frontal area of your vehicle. To do this, just picture the size and shape of the smallest hole you'd need to cut out of a piece of cardboard or sheet metal for your vehicle to slide through as it moves forward through the water. The surface area of the piece you'd have to remove equals the frontal area of your vehicle.

For example, in this figure of an idealized blue and purple "submarine" traveling in the direction of the red arrow, the frontal area of the sub is the same as the area of the keyhole-shaped piece of material removed to allow passage of the sub. Once you understand the concept of frontal area, you can usually estimate it pretty closely without actually having to cut out a real piece of cardboard. But if you want a precise answer, cut a hole just big enough for your vehicle, then lay the hole over some graph paper and count the squares you can see through the hole. Remember to be careful when converting units of area. For example, there are 100 centimeters in a meter, but 10,000 square centimeters in a square meter.

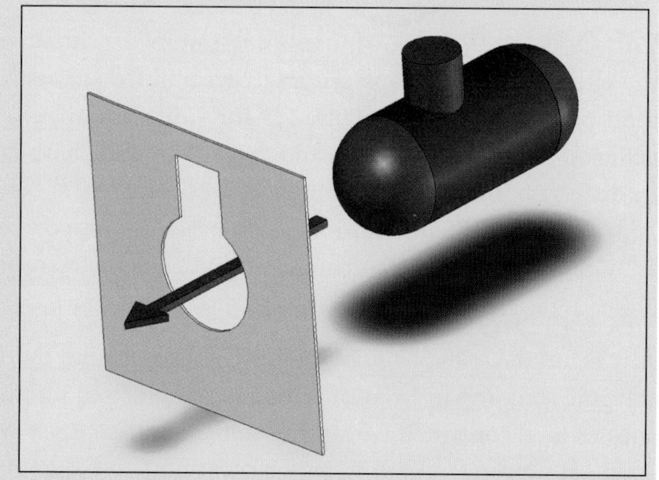

Figure 7.6: Calculating Frontal Area

large distances for a survey in limited time, outrun a challenger in an underwater robot race, or hold position against a fast-moving current, you may need high speed capability. But if you are exploring a lake bottom or recovering a sunken object in relatively calm water, speed may not be a critical factor.

3.2. Empirical Approach

The empirical approach provides an alternative to the theoretical approach for determining the drag on a vehicle. It works by directly measuring the force required to tow the vehicle through water at various speeds. The advantage of this approach is that it can produce very accurate drag data without requiring much in the way of calculations. However, it has some drawbacks not found in the theoretical approach. In particular, it requires more equipment, more time, and an early commitment to at least a tentative vehicle design. To use this method, you'll need to have:

- a nearly finished vehicle, or at least a physical model of your vehicle that closely resembles its eventual size and shape
- some spring scales, a long piece of string, a stopwatch, and a tape measure (or other tools for measuring forces and speeds)
- access to a swimming pool or other large, calm body of water where you can easily tow the vehicle in a straight line for at least 15 meters (approx. 50 ft)

The basic idea is to tow the vehicle through water at a constant speed while using a spring scale or some other device to measure the tension in the tow string. Tie one end of a long string to the ROV and the other end to the hook on the spring scale. Then hold the top of the spring scale and pull it along to tow the ROV through the water, as shown in Figure 7.8. The tension in the string (which equals the force on the ROV) can then be read directly from the spring scale. In practice, it's often easiest to do this with two or more people. One person can tow while focusing on the spring scale and trying to maintain very constant tension in the string, while the other person times how many seconds it takes the vehicle to move exactly 10 meters (or some other known distance) and calculates the vehicle's speed. By collecting speed-versus-tension data for a variety of different towing forces, plotting those data on a graph, and connecting the dots with a smooth line, you can create a graph that will allow you to estimate the force required to move the vehicle at the maximum speed required for its mission. (See Figure 7.9.)

Figure 7.8: Measuring Drag with a Spring Scale

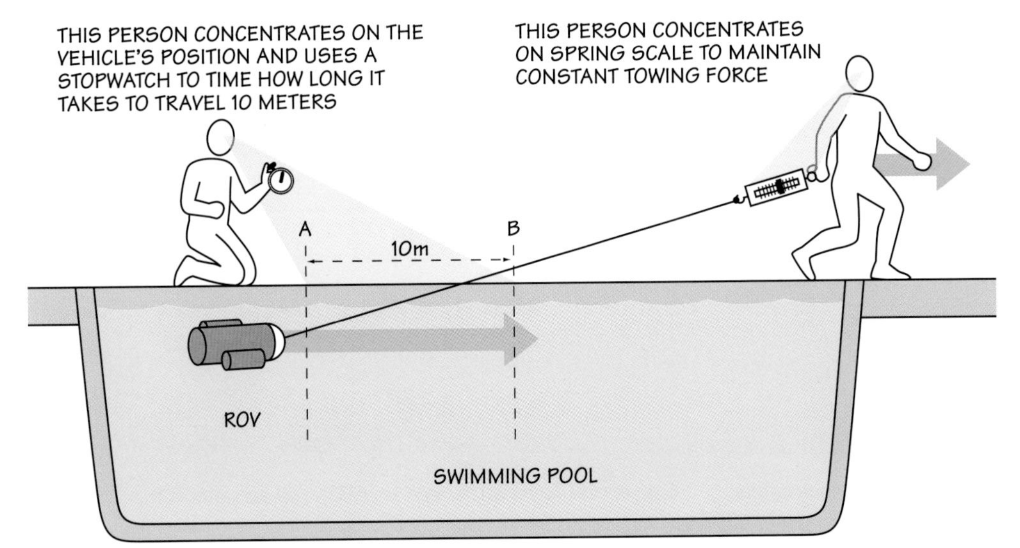

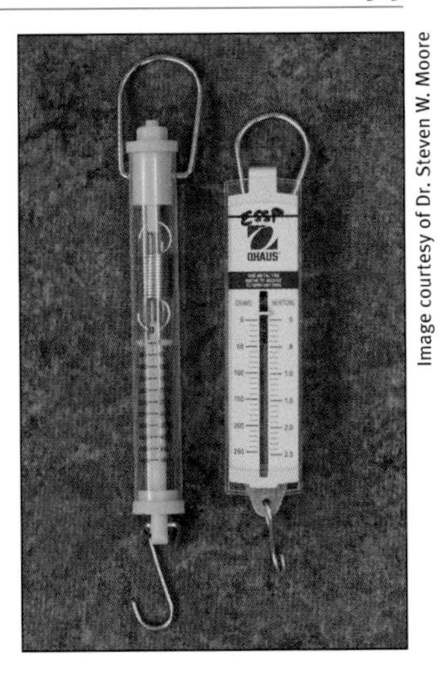

Image courtesy of Dr. Steven W. Moore

Figure 7.7: Measuring Drag with a Spring Scale

A spring scale is a simple device for measuring weight and other forces. When you hang a weight from the hook or otherwise place the scale in tension, it stretches the spring inside by a distance that's proportional to the weight. As the spring stretches, a marker attached to the spring slides along a ruler-like scale that's calibrated in force (or sometimes mass) units.

The best spring scales to use for drag measurements are inexpensive, lightweight versions similar to those shown, which you can get from bait shops, science education supply houses, and other sources. Spring scales come in a wide variety of force ranges optimized to weigh everything from baby birds to full grown elephants, so be sure you get ones designed for an appropriate force range.

For most small ROV or AUV projects, scales with maximum ranges of 10-50 newtons (2-10 lbs) are probably about right. It's good to keep two or three different ones around, each optimized for a different range. For example, you might want a 10 newton and a 50 newton scale. The 10 newton one will give more precise readings for small forces, but will max out long before the 50 newton scale does. Large and/or fast moving vehicles might require scales designed to measure more than 50 newtons.

Here are a few tricks that will help you get more accurate results:

- Use a string that's at least 5 meters long, so you're pulling mostly horizontally on the vehicle, rather than up at an angle. After all, you want to measure drag, not vehicle weight.
- Make sure the spring scale gives an accurate reading when pulling horizontally, and use one that's easy to read accurately while walking along slowly and holding it at arm's length over a pool. The spring scales used in some school classrooms and those used to weigh fish are good for this.
- Ballast the vehicle to make it a bit negatively buoyant, so it will remain just a little under the surface while it's being towed. Results will be inaccurate if the vehicle is breaking the surface.
- Give the person doing the towing about 5 meters or so to get the applied tension stabilized before starting to measure the speed. A good way to do this is to put coins or other markers at distances of 0, 5, 10, and 15 meters along the side of a pool or dock. Start the vehicle from the 0-meter mark, but don't start the stopwatch until the instant the front of the vehicle passes the 5-meter mark. Stop the timer 10 meters later, the instant the front of the vehicle passes the 15-meter mark. To calculate speed, just divide 10 meters by the number of seconds. You'll get a result in meters per second.
- Make sure the range of speeds you measure includes at least five distinctly different speeds, with at least two above the maximum speed you expect your vehicle to need under its own power and two below that maximum speed. You do not need to match the maximum speed closely during the tests, because you'll be able to estimate the corresponding thrust requirements from the graph, as long as you have some data points above and some below that value.
- Repeat the measurements for each tension level at least three times and calculate an *average* time for traveling 10 meters at each tension level.

Here's an example: Suppose you have determined that your mission requires a maximum vehicle speed (relative to the water) of about 1.5 feet per second. That's just a little less than 30 meters per minute, or about 20 seconds to go 10 meters. During your drag tests, you start with low string tensions, like 2 newtons, but discover this is too little, and eventually, you work your way up to tensions that pull the vehicle faster than the maximum speed required by the mission. You might end up with a data table that looks something like this:

Table 7.1: Empirical Data for Thrust Versus Speed Test

Target speed: 30 meters per minute = 10 meters in 20 seconds					
Towing tension used (in newtons)	*2*	*5*	*10*	*15*	*20*
Measured time (in seconds) required to move 10 meters:					
First trial	*48*	*30*	*22*	*19*	*16*
Second trial	*51*	*32*	*22*	*18*	*16*
Third trial	*47*	*31*	*24*	*18*	*17*
Average of 1st, 2nd, and 3rd trials	*48.67*	*31.00*	*22.67*	*18.33*	*16.33*
*Calculated speed in meters per second**	*0.205*	*0.323*	*0.441*	*0.545*	*0.612*
*Calculated speed in meters per min***	*12.3*	*19.4*	*26.5*	*32.7*	*36.7*

** Conversion formula: 10 meters/measured time in seconds ** Conversion formula: (speed in m/s) x (60 seconds/min)*
Calculated speeds have been rounded to three significant figures.

The tension in the string during each of these drag tests equals the drag on the vehicle at that tow speed and is the force your vehicle's propulsion system would need to produce to move your vehicle at the same speed. If you were to plot string tension as a function of tow speed using these data, you would get the following graph:

Figure 7.9: Graphing Tow Test Results

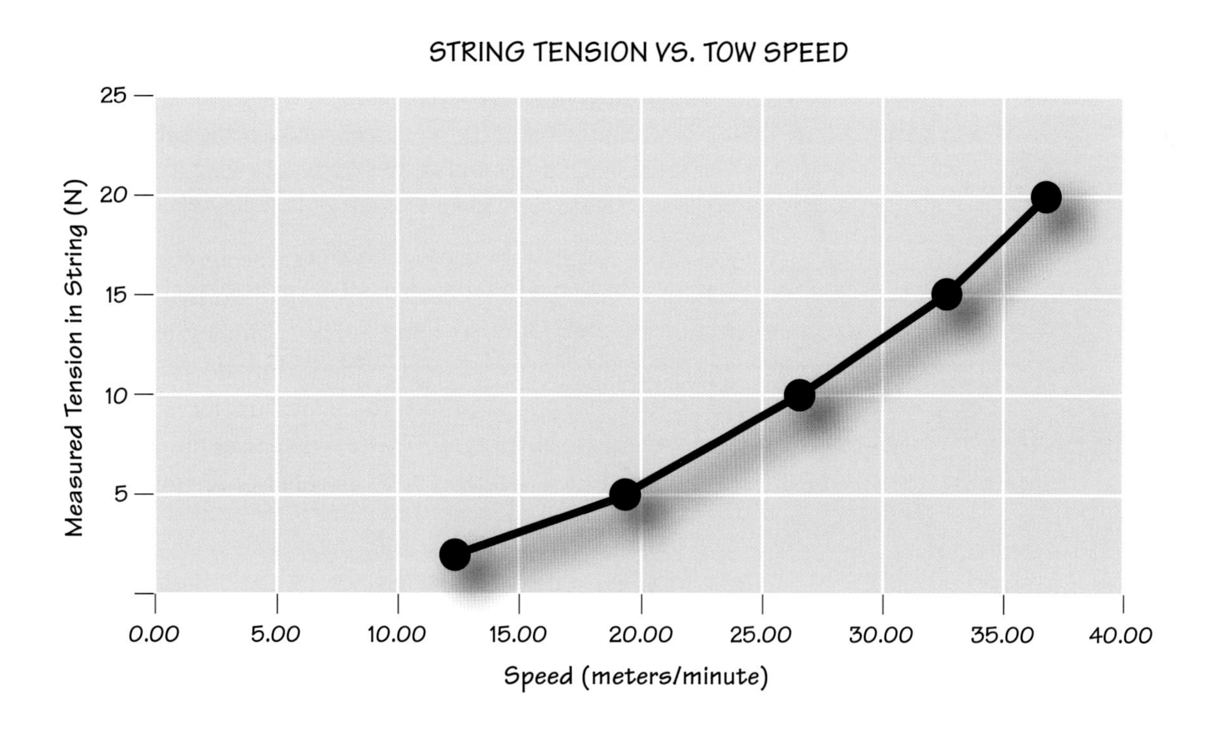

You can use this graph to eyeball approximately how much thrust your vehicle would need to generate to achieve the required mission speed. In this example, that's 30 meters per minute, so you would start by finding 30 meters/minute on the horizontal speed axis. Next, travel vertically up the graph from that speed until you hit the plotted line, then go sideways from there to the corresponding force value on the vertical axis. In this case, the graph tells us that it would take about 13 newtons of thrust to move this particular vehicle through the water at 30 meters/minute.

If you make careful measurements and cover a sufficiently wide range of towing forces, the plot you get should be a curved, parabola-shaped line showing that required thrust increases quickly with speed and is roughly proportional to the speed *squared*. For example, in the graph above, doubling the speed from 15 to 30 meters/min requires a two-squared (i.e., four-fold) increase in thrust from just over 3 newtons to about 12 or 13 newtons.

4. Producing Thrust

Once you have figured out how much thrust you need to overcome drag and meet the mission's speed requirements, your next job is to figure out how to produce that thrust.

Over the years, a wide range of innovative methods have been devised to propel vehicles through water. Some of these methods, such as paddlewheels and sails, work well only for surface vehicles. Others require some solid surface to walk or crawl on. The latter group includes vehicles that use robotic walking legs or bulldozer-like

TECH NOTE: REDUCING DRAG

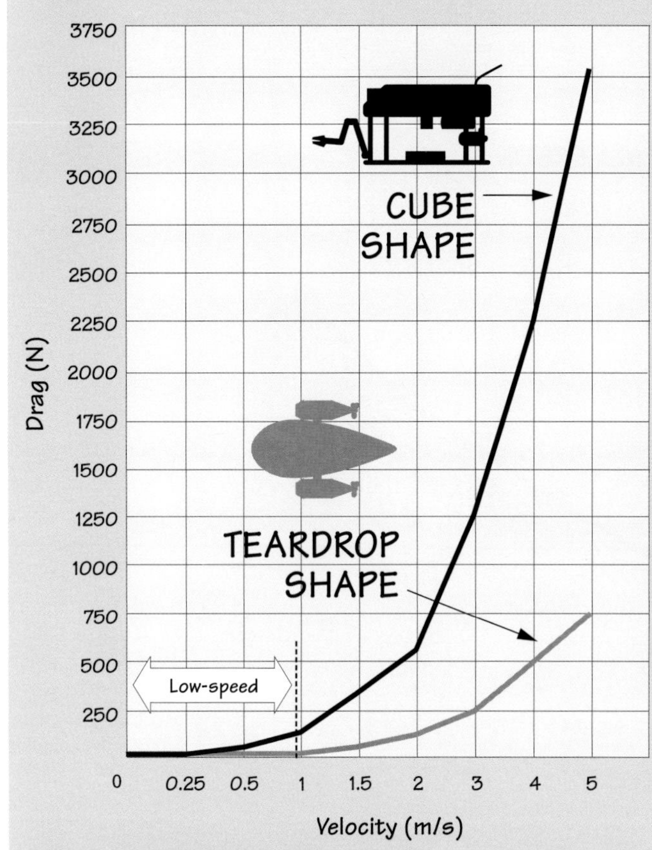

Figure 7.10: How Shape Affects Speed

This graph shows the drag acting on a large cube and equally large teardrop shape at various speeds. Note that drag increases very quickly with increasing speed, so drag is a big issue for fast-moving vehicles, but not for slow-moving ones. Second, drag at all speeds is lower for streamlined objects than for similarly sized non-streamlined objects.

One question facing a design team is whether or not to streamline the shape of their vehicle. All vehicles can reduce drag by using streamlined shapes for their hull and any protruding parts, and effective streamlining will always improve the speed and efficiency of a vehicle. However, streamlining is tricky and not always worth the effort. Designers of slow-moving vehicles in particular often don't bother, because drag is not as big an issue at low speeds as it is at high speeds.

You may have noticed that high drag shapes such as cuboids and cylinders are prevalent in ROV open-framed structures, but not in torpedoes and other fast-moving craft, where streamlined shapes predominate.

If you are building an ROV or similar slow-moving vehicle, you probably don't need to streamline it, but if you want to try, here are a couple things to keep in mind: First, the lowest drag shape resembles a long skinny teardrop shape (also called a foil), with the rounded end at the front and the tapered end at the back. A typical torpedo is a modified teardrop shape made by combining a hemisphere, cylinder, and cone, all of which are much easier to manufacture than a true teardrop shape. On an ROV, the drag associated with cylindrical frame members can be reduced considerably by making them tapered on the trailing edge, like an airplane wing. In some cases, the simple expedient of wrapping a rope in a spiral around a cylindrical frame member can disrupt the turbulence associated with pressure drag enough to help reduce overall drag on the vehicle.

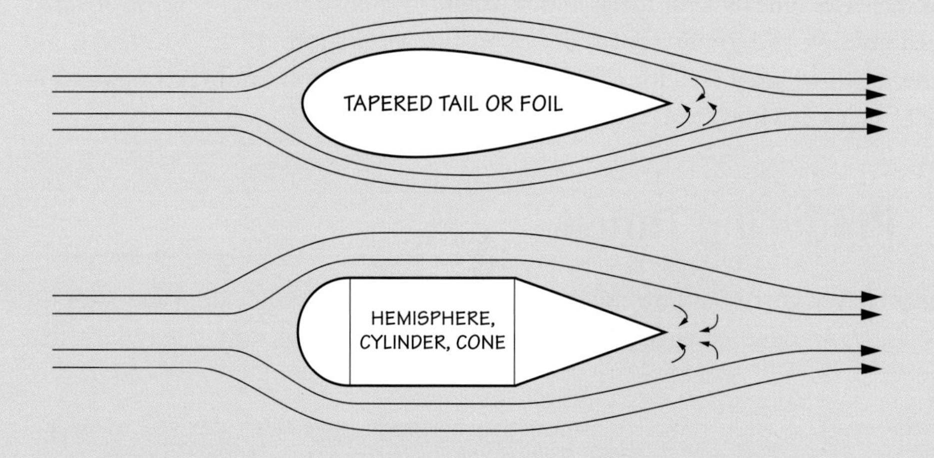

Figure 7.11: Drag-Reducing Shapes

If you really want to get serious about streamlining, your best bet may be to fabricate a teardrop-shaped hull or fairing out of fiberglass or a similar type of glass-reinforced plastic.

caterpillar treads. These are most common on negatively buoyant vehicles that move around on the bottom, but a few walking robots have suction cups on their feet so they can move around on the underside of ships to scrub marine growth off the hulls.

Most ROVs and AUVs generally rely on thrusters for locomotion since they spend the bulk of their time working in mid or deep water where paddlewheels and sails don't function and where there's nothing for tractor treads, wheels, or suction cups to grip or push against. A thruster is a device designed to generate thrust—that is, a pushing force—when there is nothing solid to push against.

A rocket engine is an example of a thruster used in outer space. It works by burning rocket fuel, which expands quickly, forcing gas molecules out the back of the rocket at extremely high velocities. These gas molecules have mass (not much, but some), so based on Newton's second law, the rocket engine must be applying a force to the molecules to accelerate them to such high speeds before ejecting them. Newton's third law tells us that while the engine is pushing the molecules out the back of the rocket, the molecules must be pushing the rocket engine forward with an equal but opposite force. This is the force propelling the engine forward, along with the rocket that's attached securely to it.

In water, it's a little easier. You don't need to carry a load of molecules to push against, because you can push against the water molecules already surrounding the vehicle. Even though water is not solid, it does have mass, so a force is required to accelerate it. Any device attached to the vehicle that can accelerate water backward must (according to Newton's second law) be applying a force to that water. And (according to Newton's third law) the water must push back with an equal but opposite force. This equal but opposite force can be used to move the vehicle.

TECH NOTE: BOTTOM CRAWLERS

Creating propulsive thrust has always been one of the most significant challenges for underwater vehicle inventors and designers. However, this task is often impacted by mission variables, such as whether the vehicle's mission is predominantly mid-water, for example, or moving along the seabed. Most early submarines were intended primarily for warfare maneuvers at or near the surface. However, in 1894, Simon Lake designed his prototype *Argonaut Jr.* and subsequent *Argonaut* bottom crawling submarines to be able to carry out commercial work such as salvage, underwater mining, and other diving operations.

Today, several high-tech, job-specific underwater vehicles such as jetters, trenchers, and ploughs carry out commercial jobs such as laying and/or burying cable and pipeline—just as Simon Lake predicted. Like *Argonaut Jr.* and *Argonaut*, these modern underwater vehicles move by utilizing propulsive force acting on some type of wheels, tracks, or treads which push against the solid surface of the seabottom.

Figure 7.12: Modern Bottom Crawlers

The Arthropod 600 is a product of the Netherlands-based company Seatools. Their pipeline trencher is designed to bury pipe in hard soils at water depths between 4 and 350 meters (approx. 13 and 1150 ft). The trencher is also equipped with a separate tool to handle removal of soft soils and sand. Its powerful tracks enable it to handle slopes up to 10° in any direction.

Image courtesy of Seatools B.V.

Many methods have been devised to generate thrust by pushing water, including:

- human hands or feet doing basic paddling and kicking motions
- paddles and oars
- propellers
- impeller-driven water jets
- reciprocating fish-like tails
- **magnetohydrodynamic drives** (MHDs), which use interactions between magnetic fields and electric currents to move seawater
- gas generators that expel gas produced by a chemical reaction into the water
- **Magnus effect drives** that rotate smooth cylinders to generate a propulsive pressure differential

Of all these devices, the familiar propeller found on the stern of ships and motorboats has proven the most popular and is by far the most common.

Every propulsion method, including the humble propeller, requires some source of energy. For example, human hands and feet can propel a vehicle only when the human body has been supplied with adequate chemical energy in the form of peanut butter and jelly sandwiches (or suitable alternatives). Once fed, those hands or feet can provide the power to row oars or turn cranks that drive propellers. In modern times, the energy needed to propel underwater vehicles usually comes in the form of electricity or electrically powered hydraulics. The electricity usually originates above water, where it is produced by power plants, diesel-powered generators, or other means. From the surface, it may be carried under water via a tether, or it may be stored in batteries, which are then carried on board the vehicle.

TECH NOTE: MHD AND THE *YAMATO-1*

The *Yamato-1* is an experimental ship fitted with a magnetohydrodynamic drive (MHD) and twin thrusters. MHD technology, also known as Caterpillar drive or M-drive, uses magnetic fields and electric current to push seawater past the vehicle, thereby generating thrust. Though still largely experimental, MHD has no moving parts and should therefore provide greater reliability and much lower noise levels than conventional propulsion systems.

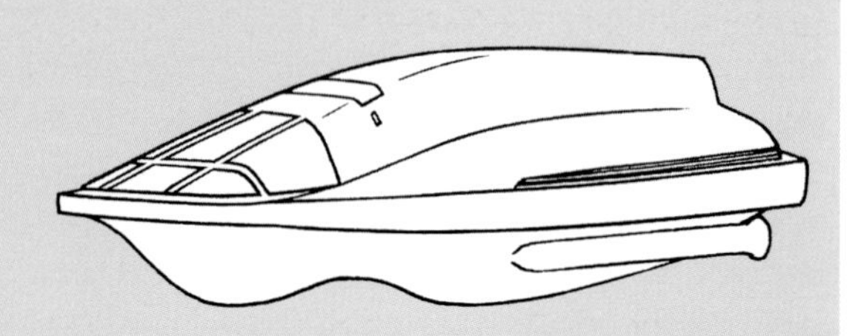

Figure 7.13: **Yamato-1**

4.1. Electric Thrusters

An electric thruster is simply a propeller attached to a waterproofed electric motor.

Although you may eventually wish to experiment with other propulsive methods, electric thrusters are the ideal place to start thinking about your propulsion system design, because of their relative safety and simplicity, as well as the low cost and wide availability of conventional propellers and small electric motors you can use for building them.

It is possible to buy some wonderful ready-made electric thrusters for small ROVs and similar vehicles. The LBV thrusters made by SeaBotix are just one example. Unfortunately, commercially available thrusters tend to be expensive and are outside the price range of typical school teams and other groups building small underwater vehicles. Most of these groups have no choice but to build their own less expensive thrusters.

The middle of this chapter therefore concentrates on providing basic information you need to build effective low-cost thrusters for your vehicle. It includes an introduction to electric motors and an introduction to propellers. It then discusses major things to consider as you select and assemble the parts for your thrusters.

5. An Introduction to Electric Motors

Electric motors convert electrical energy into rotational energy, which can be used to drive many things, including the propeller on an electric thruster. Unlike gasoline engines and diesel engines, which require oxygen to burn their fuel, electric motors do not require air to operate. This is a big advantage in underwater applications. They are also easier to start, run well over a wider range of speeds, and are generally much cleaner than fuel-burning engines. They can be safer, too, particularly if low-voltage motors are used. They come in a wide range of types and sizes, from huge industrial motors weighing many tons to microscopic nanotechnology motors smaller than a human hair. With a bit of resourcefulness, you can always find one just right for your application.

Image courtesy of Randall Fox

Figure 7.14: Even Simple Electric Motors Can Power Your ROV!

In a typical motor, the outside part is stationary and attached securely to something solid, such as a building foundation or vehicle frame. This stationary part of the motor is called the **stator**. The inside part of the motor rotates and is called the **rotor**. The **motor shaft**, which transmits rotational energy from the motor to other machinery, forms the core of the rotor. Bearings support the shaft and rotor within the stator.

All conventional electric motors control the flow of electric current through coils of wire to produce fluctuating magnetic fields, which spin the rotor within the stator. These coils are often called the motor **windings**. Pieces of iron or steel in the center of some or all of the coils are often used to concentrate and shape the magnetic fields for improved motor performance. This combination of a winding around a core of ferrous metal is often called an **electromagnet**.

5.1. Types of Electric Motors

When you start looking through catalogs or on-line sites for electric motors, the enormous variety of motors available can seem overwhelming unless you are armed with a basic understanding of how motors are classified and know which types of motors are likely to be useful for your project.

One of the most fundamental ways of classifying motors is by their power output. Most small ROVs and AUVs use small motors classified as **fractional horsepower motors**. This means that the maximum power output of the motor is less than one horsepower (746 watts).

Another way of classifying motors is based on the nature of the electrical power needed to run them. There are two categories: **AC motors** depend on *alternating current*, which oscillates rapidly back and forth between positive and negative voltages, typically 50 or 60 times each second. The electricity available in standard wall outlets is one example of alternating current. AC motors depend on the fluctuating nature of AC current to produce the fluctuating magnetic fields needed to run the motor. They are the usual motors in household appliances that plug into the wall, such as refrigerators, kitchen blenders, vacuum cleaners, and power tools with plug-in cords. **DC motors** run on *direct current*, which means they require a relatively steady (non-oscillating) input voltage, like that supplied by batteries. Unlike AC motors, DC motors (or the circuits that control them) must produce their own form of oscillating electric current to generate the fluctuating magnetic fields needed to make the motor turn. They use something called a **commutator** to do this. Commutators are explained later in this chapter. DC motors are common in battery-operated devices. (See also *Chapter 8: Power Sources* for more information on electric motors.)

Most small, low-cost underwater vehicles built in homes or schools are battery-operated and rely on DC motors. At least five different sub-categories of DC motors are of potential interest to designers of such vehicles:

- brushed DC motors
- brushless motors
- stepper motors
- servo motors
- gear motors

The first of these sub-categories is the **brushed DC motor**, or **BDC motor**. Brushed DC motors are the least expensive and most common of the small DC motors. These are the ones you'll find in toys and cordless power tools. In underwater vehicles, BDC motors are used most commonly for propulsion but can also be used for many other purposes. Brushed motors are so named because they contain electrically conductive pads or contacts, called "brushes." These are used in their commutators to transfer electrical current from the non-moving stator to the spinning rotor, which is where the windings that produce the oscillating electric fields are located. Details of how a brushed DC motor commutator works are explained later in this section of the chapter.

A **brushless motor,** or **BLDC motor**, on the other hand, uses permanent magnets instead of electromagnets on the rotor, so this type of motor does not need brushes to transfer current between the stator and rotor. This results in a simpler mechanical design and generally improves reliability. However, it also requires a more complex electronic control circuit, called a brushless motor controller, which creates the fluctuating electrical currents needed in the stator windings to make the motor run. In most respects, BLDC motors offer better performance than BDC motors, but they also tend to be significantly more expensive, in large part because of the sophisticated control electronics required.

Two other sub-categories of DC motors, **stepper motors** and **servo motors**, may also be of interest to people designing small underwater vehicles. These motors are designed for precision control of shaft rotation angle rather than raw turning speed or power, so they are not usually used for propulsion and are not described in this chapter; however, they are perfect for control of things like camera angle or water-sample volume, so they are included in *Chapter 9: Control and Navigation.*

Gear motors, a final variation on the motor theme, are worth mentioning in a chapter on propulsion. Gear motors can be large or small, AC or DC, brushed or brushless. They are simply an electric motor (of any type) with a set of gears attached to the motor shaft. This set of gears is usually used to increase the torque (twisting force) of the output shaft.

Here's why gear motors are particularly useful. Unlike gasoline engines, which deliver their maximum torque at fairly high RPM, most electric motors, regardless of type, deliver their maximum torque when completely stopped or running at very low speed. Unfortunately, at these low speeds, motors draw large amounts of current and most can quickly overheat. Therefore, electric motors are normally operated at high speeds, in spite of the reduced torque. For applications that require high speeds and don't need much torque, the motor shaft can be connected directly to the device it's driving. But for applications that require high torque at low speed—as many propulsion systems do—it may be necessary to use a gear motor to convert the high-speed, low-torque motor output into a low-speed, high-torque gearbox output. This allows the motor to operate at high RPM, where it is more efficient, while still providing high torque at the gearbox output. (For additional information on gear motors, see Section 7.1.3 below.)

5.2. Anatomy and Function of a Brushed DC Motor

As mentioned above, the brushed DC motor is one of the most abundant and least expensive types of motors available. And because many of them are designed for use with low-voltage batteries, they can be safer than AC motors designed to run on household current. Collectively, these factors make brushed DC motors very popular for school and home projects. Before you try to select a motor for your project, it is instructive to gain a basic understanding of how a brushed DC motor functions. This is true even if you ultimately decide to go with something other than a brushed DC motor, because knowledge of how a brushed DC motor works provides a great foundation for understanding how other electric motors work.

The rotation of a brushed DC motor, like that of all electric motors, is driven by magnets. If you've ever played with magnets, you know that a magnet can either attract or repel another magnet, depending on how you line up the north and south poles of the two. If you bring the north pole of one magnet close to the south pole of another magnet, the magnets will attract each other and try to stick together. Conversely, if you try to bring two north poles together or to bring two south poles together, the magnets will repel each other. Carefully timed patterns of attraction and repulsion between magnets on the rotor and magnets on the stator are what drive motor rotation.

In a typical brushed DC motor, the rotor magnets are *electromagnets*. Like a permanent magnet, an electromagnet has a north and south pole, but *unlike* a permanent magnet, an electromagnet's poles can be swapped almost instantly by reversing the direction of electric current flow through the magnet's coils.

All conventional electric motors reverse electric currents in order to reverse magnetic fields as the motor rotates. In AC motors, this results from natural fluctuations in the polarity of the voltage. In brushless motors, this results from control of the current's timing and direction by a solid state electronic circuit called a brushless motor controller. In brushed DC motors, this results from the operation of a mechanical commutator—an ingenious device that reverses the direction of current through each rotor electromagnet (and thereby reverses each magnet's north-south polarity) about every half-revolution.

The photograph in Figure 7.15 shows what the electromagnets and commutator in an actual brushed DC motor look like. The diagrams in Figure 7.16 explain how such a commutator controls current in the electromagnets to produce rotation of the motor.

TECH NOTE: ANATOMY OF A SIMPLE BRUSHED DC MOTOR

Dissecting an inexpensive toy motor is a great way to learn more about how a brushed DC motor works. In this example, the stator (non-rotating part) of the motor includes a metal casing (visible on the left), two crescent-shaped permanent magnets (one red, one green), and a plastic endcap (on the right) that supports two copper "brushes" connected to electrical terminals on the outside of the motor. The rotor (rotating part) of the motor consists of electromagnets and a commutator, both attached to the motor shaft. This motor's rotor has three electromagnets, each wrapped in a coil of blue-green wire, and a commutator made up of three separate copper plates arranged in a cylinder around the motor shaft. The plates are connected electrically to the coils of the electromagnets.

When the motor is properly assembled, with its endcap in place, the two brushes rest lightly on opposite sides of the commutator. When an appropriate DC voltage is applied to the motor terminals, current flows through the brushes to the electromagnet coils via the commutator, energizing the electromagnets and causing the motor to turn. The brushes maintain contact with the commutator, even while the motor is spinning. However, the commutator does more than just transfer electricity to and from the electromagnets on a spinning rotor. It is also an ingenious mechanism for switching the direction of electrical current flow through those electromagnets (hence switching their magnetic polarity) with just the right timing to ensure that interactions between the stator and rotor magnets cause the motor to turn. See Figure 7.16 for a more detailed explanation of how a commutator does its job.

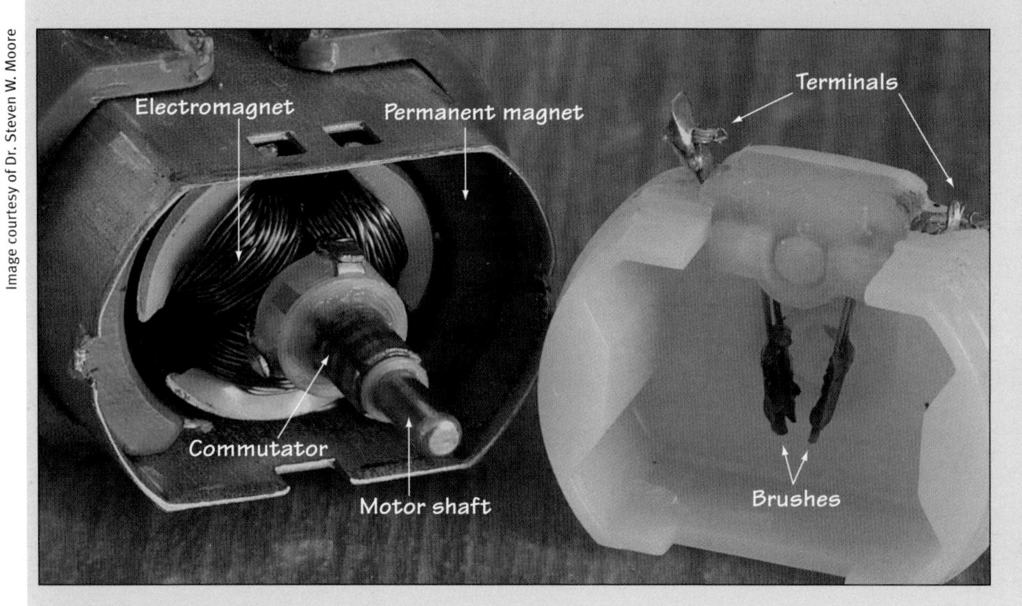

Image courtesy of Dr. Steven W. Moore

Figure 7.15: Looking Inside a Brushed DC Motor

TECH NOTE: HOW THE COMMUTATOR IN A BRUSHED MOTOR WORKS

Figure 7.16 below illustrates how the commutator in a brushed DC motor controls the polarity of electromagnets mounted on the rotor in order to cause rotation of the motor shaft. Keep in mind that these diagrams show only one electromagnet and two commutator plates; most actual motors have at least three electromagnets and at least three commutator plates. Some have many more.

In both diagrams, the letters "N" and "S" stand for the north and south magnetic poles, respectively. The small gray box on top of the red arrow in each diagram represents an electromagnet attached to the rotor. The two larger gray boxes on either side of the motor represent the permanent magnets of the stator. The yellow and blue portions of the diagram represent a pair of curved commutator plates and the wires connecting them to the electromagnet coils, which are not shown. The small purple rectangles represent the stationary brushes that glide over the rounded surfaces of the moving commutator plates, transferring electrical power from the battery to the electromagnet as the motor shaft turns.

In the diagram on the left, the electromagnet is pointed upward, like an analog clock hand pointed at 12 o'clock. In this orientation, current from the positive terminal of the battery flows into the electromagnet through the yellow side of the commutator and out through the blue side. In this hypothetical motor, a yellow-to-blue current flow direction turns the business end of the electromagnet into a north ("N") magnetic pole. Since opposite magnetic poles attract and like poles repel, when in the 12 o'clock position, the electromagnet is pushed away from the "N" stator magnet on the left and pulled toward the "S"stator magnet on the right. The electromagnet is attached firmly to the motor shaft, so this drives the motor shaft in a clockwise rotation, as indicated by the red arrow.

Momentum (and, in an actual motor, forces produced by additional electromagnets) carries the electromagnet all the way around to the 6 o'clock position (right diagram). As it makes this journey, an interesting thing happens. The yellow and blue commutator plates rotate with the shaft 180 degrees and end up swapping electrical roles. Now electric current from the positive terminal of the battery is entering the electromagnet through the blue side of the commutator instead of the yellow side and exiting through the yellow side instead of the blue side. This is the reverse of what was happening in the first diagram, and it reverses the flow of electrical current through the electromagnet's coils. This in turn changes the polarity of the electromagnet from "N" to "S" while the electromagnet is pointed downward. Now interactions with the stator magnets (which are permanent magnets and have not changed polarity) force the electromagnet from right to left, perpetuating the clockwise rotation of the motor shaft.

Momentum (and other electromagnets not shown) now carries this electromagnet back up to the 12 o'clock position, where the cycle begins anew. This cycle will repeat itself over and over, producing continuous motor-shaft rotation, as long as adequate electrical power is supplied to the motor terminals.

This simplified motor is helpful for understanding how a commutator promotes continuous rotation; however, it is oversimplified and would not work very well if actually built. For one thing, whenever the electromagnet passed through the 3 o'clock or 9 o'clock positions, each of the brushes would be in contact with both the yellow and blue commutator plates simultaneously, shorting out the battery. Even if this didn't happen, the motor would have a hard time starting whenever the electromagnet were sitting at the 3 o'clock or 9 o'clock position, since little or no torque is produced in that position (zero moment arm). Finally, if the motor ever did manage to run, the unbalanced weight of a single electromagnet spinning around off-center would cause violent shaking of the motor.

In a real motor, such as the one pictured in Figure 7.15, three (or more) electromagnets are distributed evenly around the circle to balance weight and to guarantee that at least one electromagnet is always in a good position for a strong "power stroke" to get the motor started and keep it running with plenty of oomph! Additionally, the presence of three or more commutator plates evenly spaced around the motor shaft ensures that the two brushes can never be touching the same plate at the same instant, thereby avoiding short circuits.

In brushless motors, the permanent magnets are on the rotor, and the electromagnets are on the stator. These motors use sensors or other techniques to determine the position of the motor shaft, so the control circuitry can correctly time reversals in the polarity of the stator's electromagnets.

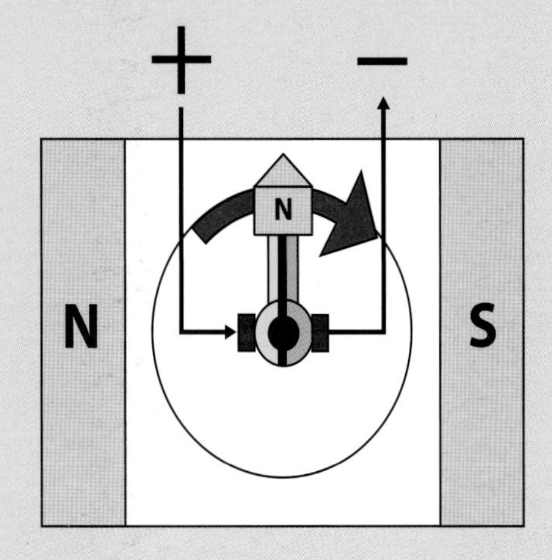

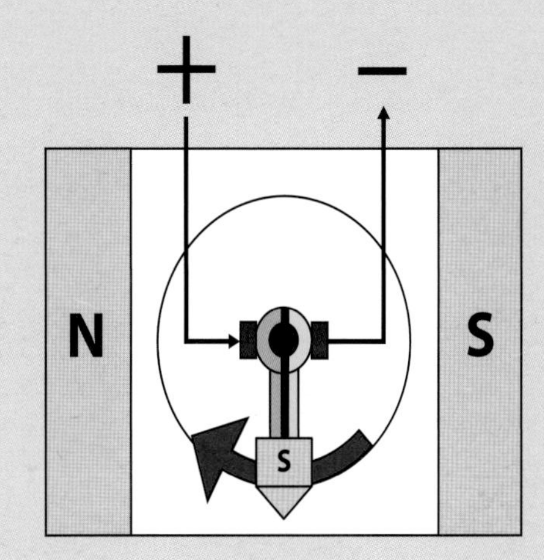

Figure 7.16: BDC Commutator in Operation

6. An Introduction to Propellers

Figure 7.17: A Standard Fixed Three-Blade Prop

Propellers (commonly called "props") are the most common device for converting the power of a rotating motor shaft into fluid motion for the purpose of producing thrust. Before exploring how these familiar devices work, it's helpful to learn the names for some of the parts associated with conventional props:

- **Shaft**: the long rod that connects the propeller to the motor or engine
- **Blade**: one of the several twisted fins or foils on the prop (It is the rotation of the blades through the water that produces the thrust that moves the vehicle.)
- **Hub**: the center solid disc or cone that fits over the end of the shaft and connects the blades to the shaft.
- **Keyway**: a matching set of lengthwise slots cut into the shaft and hub of larger propellers that fits a rectangular piece of metal, called a key, which acts to transfer torque from the shaft to the propeller.
- **Cotter pin**: a metal rod or pin that fits into a hole drilled sideways through both the hub and shaft (A cotter pin is commonly used on smaller outboard motor props, instead of a key. Like a key, it guarantees that the prop rotates whenever the shaft rotates. Note that a small hobby prop usually has no keyway or cotter pin. Instead, the propeller is just pressed tightly onto a smooth shaft or screwed onto the end of a threaded shaft.)
- **Blade face**: the high-pressure front side of the blade that normally pushes the water
- **Blade back**: the low-pressure back side of the blade
- **Blade root**: the point where the blade attaches to the hub
- **Leading edge**: the flatter edge of the blade, which cleaves or bites into the water when the prop is spinning in its normal forward direction
- **Trailing edge**: the rounder edge of the blade and the rear edge when the prop is spinning in its normal forward direction

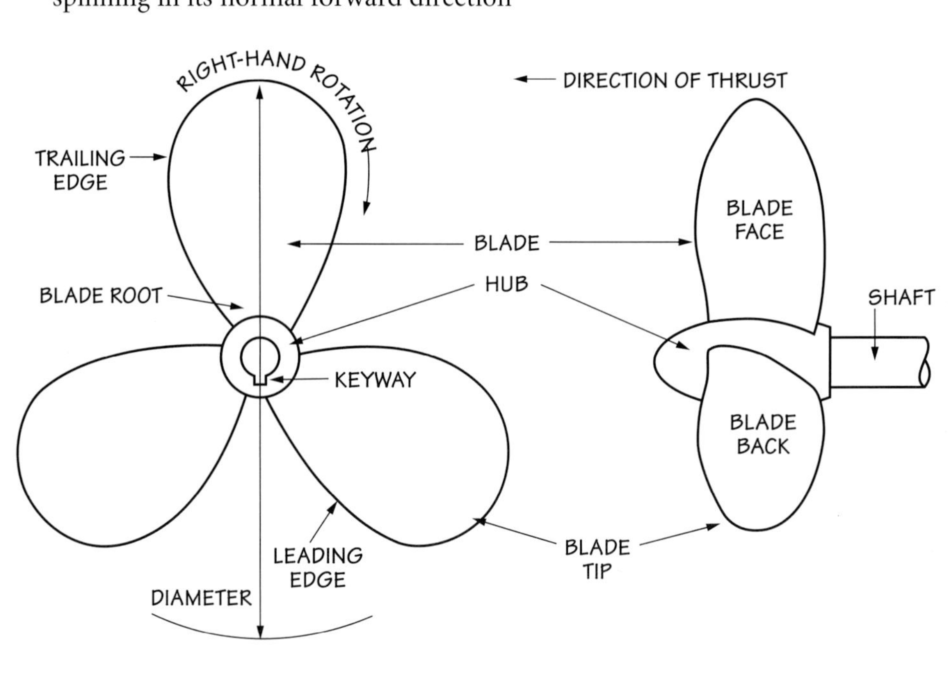

Figure 7.18: Parts of a Propeller

Propellers are used commonly to generate propulsive thrust in both water and air. The basic principle of operation is the same, even though they have different blade shapes optimized for use in water or air.

Today, propellers are particularly popular for both surface vessels and for underwater craft because of their relative low cost, ease of use, and high efficiency. They can be effective at high or low speeds and can be made in a wide range of sizes, from the huge propellers used on the *Titanic* to the small plastic props used on *SeaMATE.*

The popularity of props has also inspired a number of significant variations on the basic propeller concept, including:

- **Impeller-driven water jets**: Impellers are similar to propellers, but are contained within a conduit and optimized to accelerate water through an opening to create a powerful jet of water.
- **Kort nozzles:** These ducted props utilize specially shaped housing that is placed around propellers to improve their thrust and efficiency; more on these later in this chapter. (See Section 6.2.)
- **Hubless propellers:** These unique, solid-state, electric ring propellers have no center shaft (Figure 7.19).

Figure 7.19: Hubless Propeller

The hubless propeller is an innovative design that offers at least two major advantages over conventional propellers. First, the tapered propeller blades stick inward from a rotating ring, rather than outward from a rotating shaft, so it is nearly impossible to get fishing line, ropes, ROV tethers, or other gear tangled in the propeller. Second, there is no propeller shaft to seal. The propeller ring is rotated directly by magnetic fields passing directly through the walls of the duct. For bearings, the prop ring rides on a thin film of water between the prop and duct.

6.1. How a Propeller Works

When a propeller rotates, its blades slice through the water or air at an angle, pushing the propeller (and any vessel connected to it) through the medium, much as the angled threads on a wood screw pull the rotating screw through a piece of wood. Indeed, mariners often refer to props as "screws." In addition to being angled, most propeller blades also have an airfoil-shaped cross section, which acts like an airplane

HISTORIC HIGHLIGHT: THE EVOLUTION OF PROPELLERS

The invention of the steam engine in the late 1700s and early 1800s generated an amazing wave of technological change for vessels. Boats and ships traveling on the surface no longer had to rely on wind filling their sails in order to move. However, these first steam-powered vessels got their propulsive thrust from paddle wheels, not propellers. The technology necessary to manufacture the large, heavy props necessary for ocean-going ships simply wasn't available yet. In addition, little was known of how propellers worked or how to connect them effectively to a ship.

The reign of paddlewheels didn't last. In naval warfare, paddlewheels were easy targets for cannonballs, and their large size limited the space available to mount cannons for defense. Understandably, the military therefore explored alternatives that would keep the propulsion system hidden and protected beneath the water line. Modern propeller systems were born in the 1800s, when a Swedish inventor named John Ericsson finally came up with an effective propeller system design. He not only developed an improved propeller, but also made it practical by figuring out the complete system for transferring power from the steam engine to the propeller. In addition, he initiated careful scientific studies of how and why propellers work.

Simple, crude propellers were used as early as the 1700s for underwater vehicles. For example, the *Turtle*, built in 1776, relied on one crude hand-driven propeller for forward propulsion and another for diving.

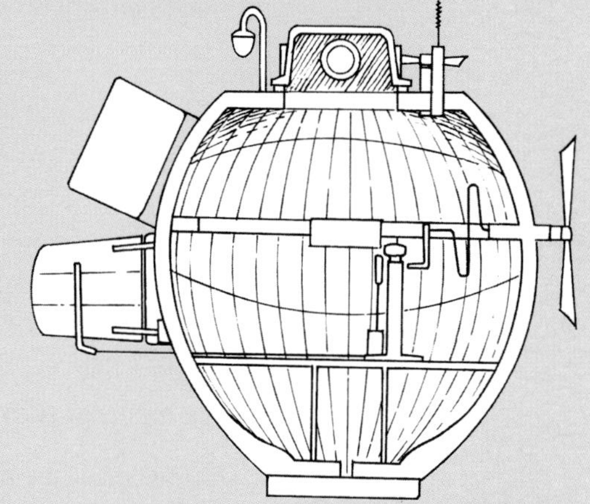

Figure 7.20: The* Turtle, *1776

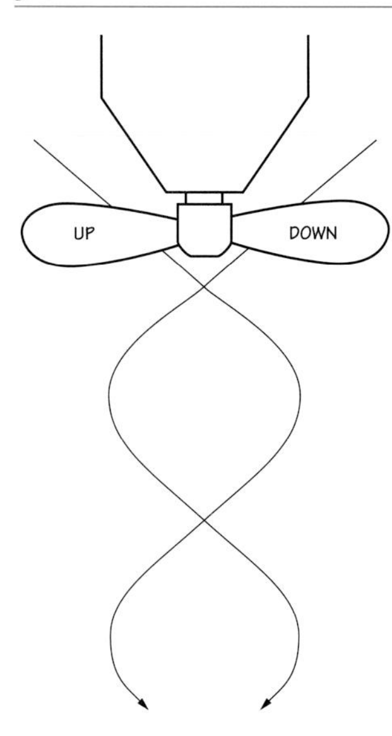

Figure 7.21: How a Propeller Works

As a propeller spins, its angled blades, coupled with the blades' airfoil-shaped cross sections, create a strong pressure differential that accelerates water backward through the prop. The force produced by the propeller to accelerate this mass of water (per Newton's second law of motion) is matched by an equal and opposite force (per Newton's third law of motion), which pushes the rotating propeller forward. Since the propeller is attached to the vessel, the vessel is also propelled forward.

wing to increase the pressure difference between the face and back side of the blade. This increases thrust. Most blades have a relatively straight and blunt leading edge and a rounded, tapered trailing edge. This streamlined shape (lower drag coefficient) improves efficiency. Propellers optimized for use in air tend to have long, skinny blades. Props optimized for use in water tend to have short, wide blades. The differences are directly related to differences in the density and other physical properties of air and water.

Propellers can generate either forward or backward thrust, but generally produce less thrust in reverse than in the forward direction. There are two reasons for the reduction in reverse thrust. First, to generate backward thrust, most propellers must reverse their direction of rotation. This means that the airfoil-shaped blades so perfect for running forward are now running backward through the water. An airfoil has a very low drag coefficient and generates good lift when running forward, but not so much when running backward. That's why you don't see a lot of airplanes flying around backward—it doesn't work very well. Second, there are usually structures, including the motor housing and associated mounting brackets, that block a significant portion of the water flow coming from a propeller when it's operating in the reverse direction.

6.2. Types of Propellers

A number of characteristics are used to classify propellers. Designers of small craft need to pay particular attention to the following ones:

- diameter
- pitch
- number of blades
- blade shape
- material
- handedness
- method of attaching prop hub to shaft
- ducted props or not

Diameter: The diameter of a propeller is the diameter of the circle swept out by the tips of the blades as the propeller rotates. Propeller diameters can vary from less than one centimeter to many meters. Larger props usually require larger motors, but also tend to produce more thrust.

Pitch: The pitch of a propeller is the distance the angle of the blades would move the propeller forward through the water after exactly one full revolution of the prop—if the prop did not slip. To understand **slip**, think of walking up a steep sand dune. As you climb, sand gives way under each step, causing your foot to slip backward. Thus, each step produces less progress than it would have if you had been walking up solid stairs instead of loose sand. The same thing happens to a propeller in water, because water, like loose sand, "gives" under pressure. Slip becomes more severe when a propeller is trying to move a big load like a large boat or submarine.

In general, props with "high pitch" (steep blade angles and relatively high pitch-to-diameter ratios) are designed to push vehicles along at high speeds, but they do so at the expense of good acceleration. These are the props you'll see on speedboats and remote-controlled racing boats. On the other hand, props with low pitch are designed to accelerate well and maintain a large pushing force at low vehicle speeds, but they can't move the vehicle very fast. Low-pitch props are the ones you'll find on tugboats. From a performance standpoint, you can think of high pitch and low pitch like high gear and low gear on a car or bicycle.

The most common and the least expensive props are **fixed-blade props**. These props have a constant pitch that cannot be changed. To change speed, the vehicle must change the rotational speed of the propeller shaft. To reverse direction, the vehicle must reverse the propeller's direction of rotation. Most surface and undersea craft use fixed-blade props for propulsion.

An alternative to the fixed-blade prop is the **variable-pitch prop.** These are designed so that the angle of the blades can be changed while the prop is rotating. This is done by adjusting mechanical or hydraulic linkages that run inside the propeller shaft and into the hub of the prop to rotate the blades. Vessels with variable-pitch props can change vessel speed and even direction without having to change the speed or direction of propeller-shaft rotation. For example, to go in reverse, the variable-pitch prop can be adjusted so the blades push the water forward instead of backward. To idle, the prop pitch can be adjusted so water is forced neither backward nor forward while the prop is rotating. The variable-pitch prop is very efficient, because it allows the engine to run at its optimal speed over a wide range of vessel speeds. However, the additional mechanical/hydraulic linkages make this option much more complicated and expensive than a fixed-blade prop. Variable-pitch props are rarely, if ever, used on small, low-cost underwater vehicles.

Number of Blades: Propellers designed for use in water can have anywhere from two to seven (rarely more) blades attached to the hub, but three or four are most common. Props with fewer blades are generally more efficient, since they require less power to achieve a given level of thrust; however, props with more blades can push or pull harder at low speeds.

Blade Shape: Blades come in an endless variety of shapes, but one important parameter is the ratio of length to width. Airplane propellers tend to have long, skinny blades, whereas boat propellers tend to have short, wide blades. Certainly, long, skinny blades can be used in water and are more efficient under light loads, but they tend to "stall" (lose thrust due to excessive production of turbulent eddies behind the blades) if pushed too hard. Short, wide blades are somewhat less efficient, but they can push much harder without stalling.

Material: Most small propellers are made of some kind of metal or plastic. As a general rule, metal props are more durable than plastic props, but plastic ones are lighter, more corrosion-resistant, and usually far less expensive. Plastic props are used on most small, hand-built ROVs built by school groups, particularly if the propeller is protected inside a shroud or duct where impact damage is not a serious concern.

Handedness: Some propellers come in both a **right-handed (RH)** version, where the prop rotates to the right (clockwise as viewed from the rear) when moving forward, and a **left-handed (LH)** version, where the prop rotates counterclockwise when moving forward. All props also have a slight tendency to move a vessel sideways while rotating; this phenomenon is known as **prop walk**. You can cancel out prop walk effects by combining an RH and LH prop on any boat or other vessel that has two propeller shafts side by side.

Method of Attaching Prop Hub to Shaft : It's important to get a good, tight fit when attaching the propeller hub to the shaft, because that's what transfers rotational energy

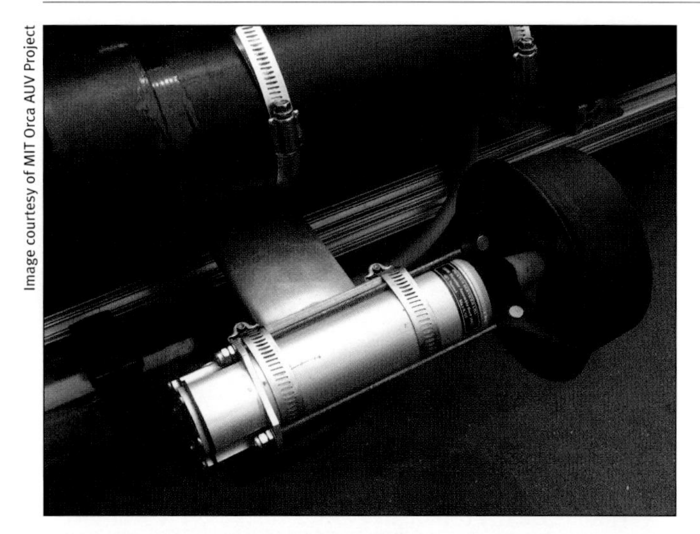

Image courtesy of MIT Orca AUV Project

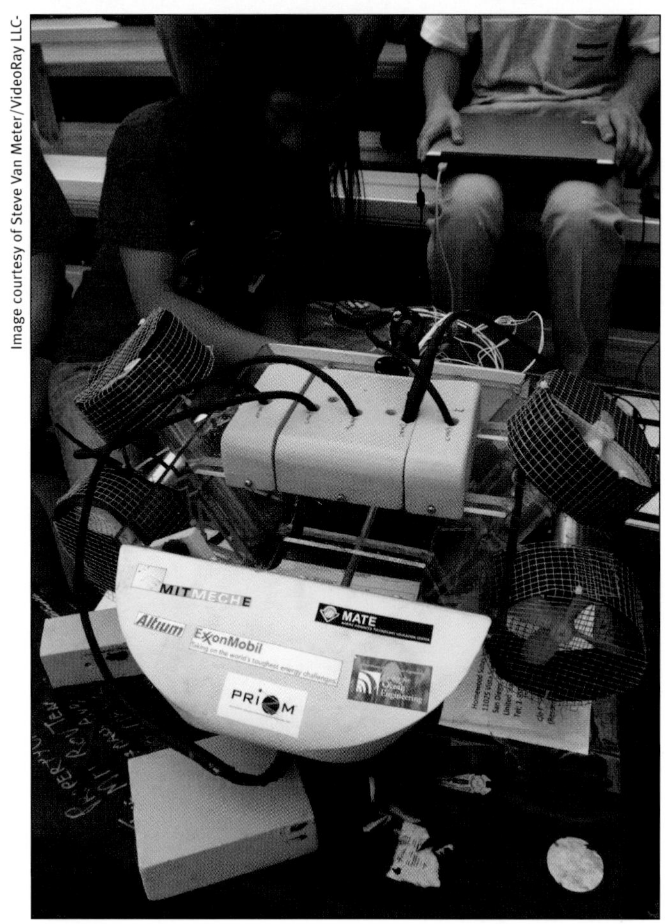

Image courtesy of Steve Van Meter/VideoRay LLC-

Figure 7.22: Ducted Props

A ducted prop may come as an integral part of a commercial thruster, such as this one manufactured by Inuktun shown in the top photo. MIT's Orca *AUV uses six of these thrusters. Another alternative is to create your own ducted props such those pictured in the lower photo of the MIT team's ROV for the 2008 MATE ROV Competition.*

from the motor to the prop. For small ROV/AUV designers, the main issue is how the hub attaches to the shaft. In the case of a threaded fitting, the hub and shaft must have matching thread types. And there must also be some way of keeping the prop from unscrewing itself and falling off. In other situations, the prop can be held on the shaft by a tight friction fit—snug enough so there's no slippage that would "waste" rotational energy. In still other cases, the prop can be locked onto the shaft by a cotter pin, bolt, or other device inserted through a hole drilled sideways through both the hub and shaft.

If the motor shaft and propeller are designed to fit together, the attachment challenge is often fairly straightforward. If not, you'll have to get creative and invent your own adapters to connect the two so that your prop spins effectively and drives your vehicle through the water.

Ducted Props or Not: Some propellers are designed to be mounted inside a short cylinder that helps guide fluid straight back through the prop, rather than letting it get thrown off sideways, which wastes energy. If done properly, such ducting can dramatically increase both thrust and efficiency. However, at high speeds, the added drag associated with the duct can make the whole affair less efficient than no duct at all. For this reason, ducted props are not normally found on fast vehicles.

A duct also provides some protection for the prop from collisions with submerged boulders or logs, and it helps protect fingers and other body parts from injury. Ducted props work best when there is very little space between the outer edge of the blades and the inside of the duct.

The thrust and efficiency of this arrangement can be increased further by using a carefully shaped duct called a **Kort nozzle**. It combines the physics of an airfoil with that of a tapered nozzle to achieve higher water velocities at the exit end of the duct. Making a good Kort nozzle is a fun challenge, because it works only if shaped just right.

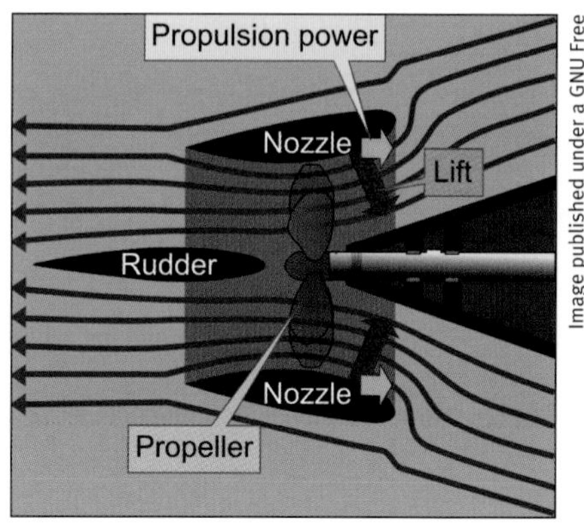

Image published under a GNU Free Documentation Licence, WikiMedia Commons

Figure 7.23: Kort Nozzles

Note the specific taper of the Kort nozzle, which serves to speed up the water as it exits the duct.

TECH NOTE: THINKING OUTSIDE THE PROPELLER

Researchers at the University of Bath have used biomimetics to come up with an alternative to propeller-driven propulsion. The *Gymnobot* has a long fin running underneath the AUV, and the undulations it produces create a wave which propels the robot through the water. The idea for the fin was borrowed from an Amazonian knifefish.

As Dr. William Megill, Lecturer in Biomimetics at the university explains, "This form of propulsion is potentially much more efficient than a conventional propeller and is easier to control in shallow water near the shore."

Postgraduate student Keri Collins developed *Gymnobot* as part of her PhD work. She hopes to continue studying the effect of vortices that fish create with their fins and tails. "It will be particularly interesting to see how well our machine will be able to mimic the fish's ability to improve thrust generation by changing the oscillation envelope of the fin from a constant amplitude to one which is flared toward the tail."

Image courtesy of Nic Delves-Broughton, University of Bath

Figure 7.24: **Gymnobot**

Postgraduate researchers Keri Collins (left) and Ryan Ladd from the University of Bath helped develop the Gymnobot, *an AUV based on the Amazonian knifefish.*

7. Building Your Own Thrusters

Building a thruster is not easy. However, commercially available thrusters are often beyond the financial means of school groups and other enthusiasts building small ROVs or AUVs. Unless you have substantial purchasing power or can get a set of professional thrusters donated to your project, building your own is likely your only option.

You can find a detailed parts list and step-by-step instructions for building a small but effective ROV thruster in *Chapter 12: SeaMATE.* If you have not built a thruster before, this is an ideal way to get some experience and increase your confidence.

If you are feeling more adventurous or have different mission requirements, you may want or need to design and build your own thrusters. The rest of this section outlines some basic approaches, highlights some key issues to consider, and offers suggestions to get you started in the right direction. Chapter 12 walks you through building thrusters for the *SeaMATE* ROV, but the details of other thruster designs are left up to your creativity and resourcefulness.

The process of designing a thruster echoes the process of designing an entire underwater vehicle or any other complex mechanical system. As such, it represents a great opportunity to apply the design spiral principles described in *Chapter 2: Design Toolkit.* Like your vehicle, each thruster is a system composed of subsystems: the motor, motor housing, propeller, and shaft. Choices you make about any of these subsystems

TECH NOTE: *SeaMATE* THRUSTERS

SeaMATE uses only three thrusters to produce its motion. Two thrusters are mounted horizontally about 150 to 200 millimeters (approx. 6 to 8 in) apart to provide forward, backward, and turning motions. The other one is mounted vertically to propel the vehicle up and down. *SeaMATE*'s thrusters are made from standard 500 gph (gallons per hour) bilge pumps manufactured by the Johnson Pump Company. They are waterproof and fully submersible.

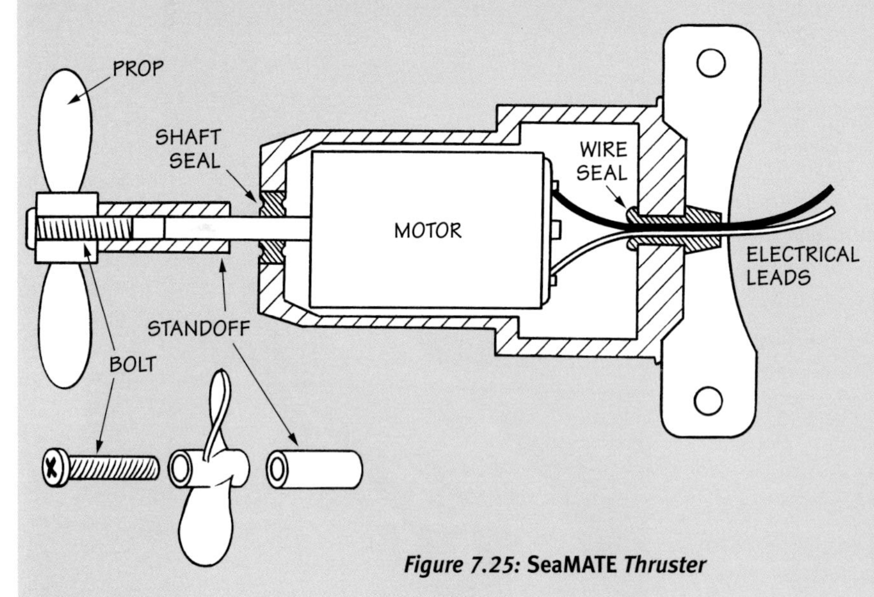

Figure 7.25: **SeaMATE** *Thruster*

Although bilge pumps were never designed for use as ROV thrusters, they have a surprising capability to withstand water pressure up to 20 meters (approx. 66 ft) deep. By attaching a small 1½-inch, two-bladed nylon propeller from Dumas (model #3003), they develop around 1 pound (4.4 N) of thrust each at 12.5 volts (measured at the motor), with an average current draw of around 3 amps. That's quite impressive for a small unit not even designed to act as a thruster. These bilge pumps are low-cost and perfect for building your own underwater robot at home or school.

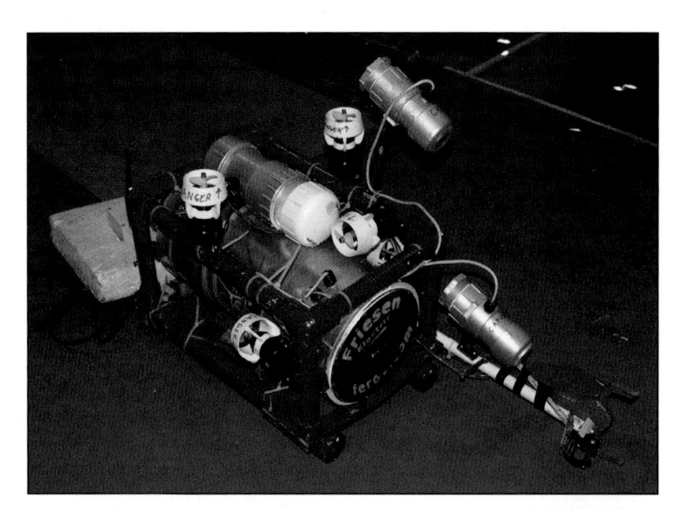

Top images courtesy of Steve Van Meter, VideoRay LLC

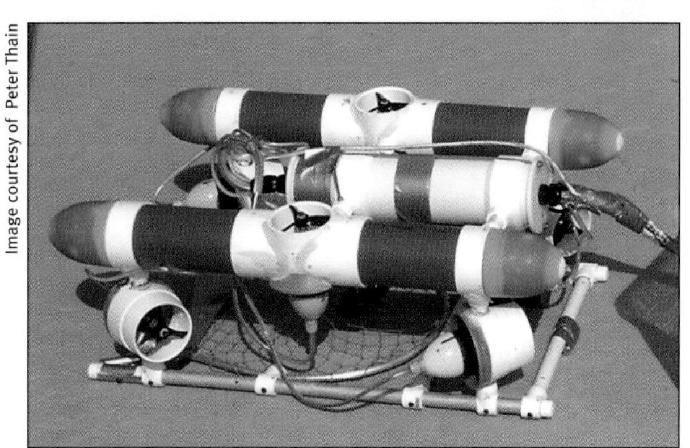

Image courtesy of Peter Thain

Figure 7.26: Do It Yourself Ducted Props

In addition to the challenge of building your own thrusters, you may also want to add DIY ducted props of one sort or another, as shown in these student-built ROVs.

will invariably impact the design of the other subsystems. Thrust capabilities must match the size and required speed of the vehicle; as well, each propulsive unit must be supplied with adequate power. Furthermore, the unit must be able to handle the water pressure at mission depth. Therefore, as you design your thrusters, you need to think about what motors to use, how you're going to waterproof them, what type of propeller you're going to use and how you'll attach it, how you will attach the thrusters to the vehicle, how they will affect the vehicle, and how you're going to supply them with power. It's best not to lock in the design of any one component without first considering all these interdependencies.

Image courtesy of Dr. Steven W. Moore

Figure 7.27: Lots of Motors to Choose From!

You have lots of choices when selecting motors to use for the thrusters on a small ROV or AUV. This photograph illustrates just a few of many popular options. The large black motor at upper left is a trolling motor with the prop removed. The smaller black motor next to it is a replaceable bilge pump "cartridge." Next in line is a silver-colored, brushed motor used in model racing boats. The motor at far right is one removed from the inside of a cordless drill/driver. The gold motor at lower left with the green circuit board attached is a brushless motor and speed controller used in model airplanes.

7.1. Choosing a Motor

While you can start anywhere in the thruster design spiral, it's convenient to start by thinking about what motor to use. Odds are good that you'll want to use some sort of fractional horsepower DC motor. But you'll need to pin things down further than that for each motor before you can go shopping.

- How powerful does each motor need to be?
- How much voltage and current can you realistically (and safely) provide for them?
- How much can you afford to pay (especially since you'll need at least three or four of these motors)?
- Do you want to use brushed or brushless motors?
- Do you need a gear motor?
- How are you going to waterproof the motors?
- How are you going to attach the propeller to the motor shaft?
- How important is it to be able to obtain the motors (and any necessary replacements) quickly and easily?

The information provided below will help you develop informed answers to most of these questions.

7.1.1. Electrical Specifications

Unless you have a compelling reason to do otherwise, start by looking for a motor that can run on 12 volts DC. This voltage is high enough to power some moderately strong motors without requiring huge electrical currents, yet low enough to be pretty safe around water, including saltwater. It is also readily available in the form of rechargeable, sealed lead-acid batteries which come in a variety of sizes and offer good value (in terms of energy storage capacity per unit price), compared to most other types of batteries. (See *Chapter 8: Power Systems* for battery specifics.)

Under optimal conditions, it's possible to get about 1 newton of thrust per 3 watts of electrical power supplied to a thruster motor. Watts are equal to volts multiplied by amps, so this translates into about 4 newtons (just under 1 pound) of thrust per amp of electrical current consumed by a 12 VDC motor. Professionally designed ROV thrusters and trolling motors come close to delivering this efficiency, but many homemade thrusters produce only about 1/3 to 1/2 of this optimal value. In imperial

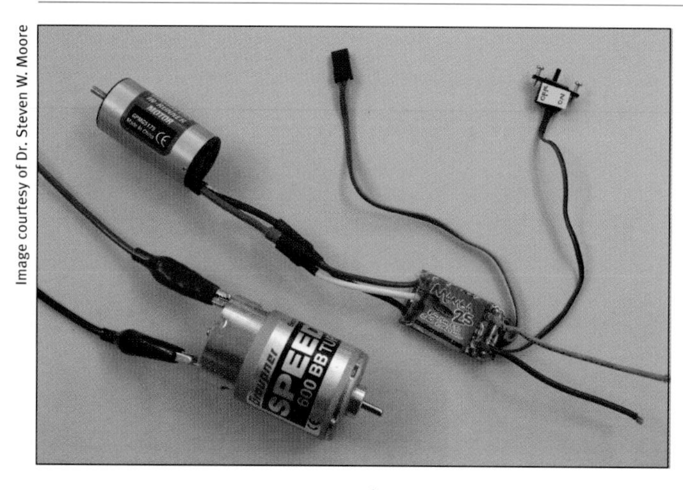

Image courtesy of Dr. Steven W. Moore

Figure 7.28: Comparison of Brushed and Brushless Motors

A brushed motor, like the silver-colored one shown at lower left, is very simple to hook up. It requires only two wires connecting the motor terminals to the battery terminals to make it run. You can add an ON/OFF switch, directional control, and speed control circuits if you want to, but they are optional.

On the other hand, a brushless motor (like the gold-colored one shown at upper left) requires a motor control circuit (small green circuit board) and associated control circuitry (not pictured) to run at all. The wiring is correspondingly more complex. Three power wires lead from the motor to the motor controller. Power from the battery is supplied to the motor controller through two wires (here, pink and black at lower right). Coded electronic control signals that tell the controller what speed (and sometimes what direction) to run the motor are sent via a 3-wire connector from a microcomputer or radio receiver unit, and there is often a 2-wire ON/OFF switch attached as well.

terms, you should be looking for 12 VDC motors that will draw about 3 amps or 36 watts for every pound of thrust you think you need. (Motor manufacturers typically specify voltage and either amps or watts, but not both.)

As recommended earlier, be realistic about your vehicle speed and thrust requirements. High thrust is not critical for a small ROV casually maneuvering in a swimming pool, lake, or harbor where currents are minimal and tether lengths are relatively short. Under these conditions, a couple pounds of thrust in total should be adequate. Since most small vehicles would have two horizontal thrusters, each would need to produce only about one pound of thrust, as *SeaMATE*'s thrusters do.

If you are planning to operate your vehicle in more exposed coastal areas and/or at greater depths where surge or other currents and additional tether drag will place greater demands on the propulsion system, you will probably need (or at least want) several times this amount of thrust. Commercial vehicles small enough to be carried, launched, and recovered safely by one person and designed to operate in shallow ocean settings commonly provide between 10 and 30 pounds of thrust. Most of these use supply voltages higher than 12 V to allow for lower electrical currents and greater efficiency through longer tethers, but these higher voltages also require additional safety considerations in the design of the vehicle and its power and control systems. Chapter 8 discusses these power-supply issues in greater detail.

7.1.2. Brushed Versus Brushless Motors

Once you've identified a reasonable wattage for your DC motor, you'll need to decide whether to use a brushed or brushless variety. Each has its advantages.

Brushed motors are the most popular for simple, low-cost underwater vehicles, primarily because they are affordable, easy to find, and very easy to use. However, brushless motors offer distinct advantages of their own and are the preferred motor type for many deep-diving and high-performance vehicles.

Table 7.2 summarizes significant differences between these two motor types. Some of these differences are explained in more detail below.

Table 7.2: Comparison of Brushed and Brushless DC Motors

	Brushed (BDC)	*Brushless (BLDC)*
Availability	*Widely available*	*More specialized*
External motor controller	*Optional*	*Required*
Price	*Less expensive**	*More expensive*
Performance	*Good*	*Better*
Sparkless commutator	*No*	*Yes*
Compatible with oil compensation	*Not recommended*	*Yes*
Good heat dissipation	*No*	*Yes*
Reliability	*Good*	*Excellent*

*Especially if used without a motor controller.

A brushed motor is simpler to use. It will run as soon as you connect it with two wires to an appropriate battery. That's because the commutator, which switches current flow through the windings, is built right into the motor. A simple and inexpensive switch circuit can be used to control on/off and forward/reverse operation of this type of motor. More elaborate control of speed and direction is possible with an optional **brushed motor controller** circuit, but this is not required. Thus, a brushed motor allows you to start out simply and inexpensively, yet offers the option of upgrading to more elaborate control later if you want to and have the budget to do so.

A brushless motor, on the other hand, will not run at all and may even be damaged if you hook it directly to a battery. With this type of motor, you must use a separate **brushless motor controller** circuit (sometimes called an **electronic speed control**, or ESC) and an R/C (remote control) console (plus radio receiver) or a microcontroller circuit to send appropriately coded speed (and in some models, direction) commands to the motor controller. All of this adds complexity and expense right up front. In exchange for the added expense and complexity of brushless motor setups, you get precision speed control (and sometimes direction control) with a high performance motor. Motor control options and techniques for both brushed and brushless motors are discussed in greater detail in *Chapter 9: Control and Navigation.*

There are lots of different brushless motor and controller options. The controllers and control consoles typically cost significantly more than the motor itself, so do some careful research on features and costs for all the required parts before shopping.

Although brushless motors require extra control circuitry, they offer some significant performance advantages. Many of these advantages center around the fact that brushless motors do a better job of dissipating heat. Most of the heat produced by a motor is generated in its windings. This heat must be removed as quickly as it is produced, or the motor will overheat. A brushed motor cannot shed heat very effectively, because its windings are located on the rotor, where they are thermally insulated from the outside of the motor by the layer of air between the rotor and stator. Ventilation holes and fans built into some brushed motors help, but these don't work as well inside an enclosed underwater pressure housing. This limited ability to shed excess heat limits the amount of power that can be delivered to (and hence by) a small brushed motor.

In contrast, the windings on a brushless motor dissipate excess heat easily, because they are attached directly to the outside surface of the motor. This means that a brushless motor can be pushed much harder than a brushed motor of the same physical size without overheating—a feature that makes them popular among remote-control model enthusiasts for powering high-performance model race cars, airplanes, and helicopters. If you attach the motor to the inside of a metal housing or use oil-compensation in a metal housing, the housing itself will help with cooling by conducting heat directly from the motor to the cool water outside the housing.

Speaking of oil-compensated housings, brushless motors work better in oil than brushed motors do. The reasons for this are explained later in the section on waterproofing your thruster motor.

Brushless motors offer another advantage to consider, particularly if your payload will contain sensitive scientific instruments, analog video cameras, or other electronic systems that depend on tiny electrical signals. The commutator on a brushed motor generates a continuous stream of sparks as the brushes make and break contact with the spinning commutator plates. These sparks create electromagnetic interference, which can impede the proper operation of some electronic devices, particularly sensors and communication equipment. The commutation for a brushless motor takes place within solid-state circuitry, so brushless motors do not make sparks when operating normally and are less likely to create electrical interference.

7.1.3. Gearing and Gear Motors

It takes a lot more torque to spin a prop in water than it does to spin it in air. This presents a problem for electric motors, which, as mentioned earlier, tend to be more efficient when generating high speed than when generating high torque. A prop that offers too much resistance can slow a motor way down and cause it to draw excessive electrical current. At best, this is inefficient. At worst, it can overheat and destroy the motor or its control circuitry and potentially cause a fire. (See Section 7.4.3 for specific info about how to optimally match a propeller to a motor.)

Fortunately, because the density of water is so much greater than that of air, propellers used in water don't need to be as large nor spin as fast as air-optimized propellers do in air to produce comparable thrust. You may find you can get by just fine with a small propeller that allows the motor to continue spinning at fairly high speeds.

Alternatively, you may be able to use a **gear motor** to drive a larger propeller. As mentioned earlier (Section 5.1), a gear motor is simply a motor combined with a set of gears (called a "gearbox") attached to the shaft. The gears convert the (high, but wimpy) motor shaft rotations into slower (but higher torque) gearbox shaft rotations. Thus, the output shaft of the gearbox can be used to drive a bigger propeller without overloading the motor.

An important spec for a gear motor is its **gear ratio**. The gear ratio quantifies how many revolutions the motor shaft must complete to produce one full revolution of the shaft coming out of the gearbox. For example, a gear motor with a gear ratio of 4:1 driving a propeller would take four motor revolutions to produce one revolution of the propeller shaft. Therefore, the propeller would spin only 1/4th the speed of the motor shaft, but with about four times as much torque.

If you think you might want to use a gear motor, there are several different styles to choose from, but **planetary gear motors** are ideal for many ROV and AUV applications. The gearbox on a planetary gear motor has several little gears revolving around a central gear, like little planets orbiting around the sun, hence the name. (See *Tech Note: Planetary Gears.*) These planetary gear motors can be used for thrusters as well as payload mechanisms, such as gripper jaws, that might benefit from high torque. They offer good gear reduction ratios in a compact, cylindrical package, and they have an output shaft that is parallel to and concentric with the axis of the motor housing. All of this makes it relatively easy to fit the motor and gear head into a cylindrical pressure housing with a propeller shaft emerging conveniently along the central axis of the housing.

Planetary gear sets can be stacked to produce gearboxes with a wide range of gear ratios, but be careful not to overdo the gear reduction. Unless you choose an unusually large propeller, you'll still need fairly high shaft speed (probably several hundred to a few thousand RPM) to produce useful thrust. If in doubt, start experimenting with gear reduction ratios of about 4 to 8, then branch out from there.

Here the design spiral paradox strikes again. You can't select the ideal gear motor (or even decide if you need one) until you know what propeller you're going to use, but you can't know what propeller to use until you have some idea of what kind of motor you'll be using to drive it. In practice, one effective strategy is to pick a motor based on what has worked well for others doing similar projects, then experimentally determine an optimal (or near optimal) propeller with the help of a **thruster test jig**, as described later in this section.

TECH NOTE: PLANETARY GEARS

Like many other gear heads used with electric motors, planetary gear heads convert high-speed, low-torque motor shaft rotations into low-speed, high-torque output shaft rotations, thereby allowing the motor to turn with more torque than it otherwise could. This is often useful when using electric motors to drive moderately large propellers in water.

Planetary gearheads have gears orbiting like little planets around a rotating shaft. As the central shaft rotates, it spins the "planets." This causes the planets to roll along the inner surface of the outer ring, which remains stationary. The axles for all the planets are anchored to a plate that turns more slowly (but with more torque) than the central driving shaft and gear. An output shaft can be attached to this rotating plate to produce slower, higher torque.

Noteworthy features of planetary gearboxes are that they are compact, can be stacked one after another to produce extremely high torque (watch your fingers!), and have an output shaft that lines up exactly with the motor shaft. (Most gearboxes don't do this.)

Planetary gear heads are remarkably ingenious mechanisms with strikingly beautiful symmetry, and they are truly fascinating to dissect, but don't try it unless you enjoy reassembling complex puzzles and have plenty of spare time on your hands.

Images courtesy of Dr. Steven W. Moore

Figure 7.29: A Planetary Gear Motor

These photos show (a) a small planetary gear motor, (b) the same motor with its gear head disassembled, and (c) a close up view of one set of planetary gears. This particular gear motor consists of a small, brushed DC motor (green label, wires protruding) connected to a cylindrical gearbox. The second photo shows what the inside of the gearbox looks like. This one has three layers of gears inside it. In each layer, there is a central drive gear surrounded by three planetary gears (third photo).

When the gearbox is assembled, the brass gear on the motor shaft becomes the drive gear for the three white plastic gears in the top layer. As the drive gear in each layer spins, it causes the planetary gears to roll along the inside surface of the gearbox housing. The shafts of these planetary gears are connected by a metal disc to the drive gear in the next layer down (or to the output shaft in the final layer). Since each planetary gear completes less than one "orbit" per full revolution of the drive gear, each layer turns more slowly (but with more torque) than the layer above it. In this particular gear motor, each layer reduces speed and increases torque by a factor of about four, so the three layers together produce almost 4 x 4 x 4 = 64 times as much torque as the motor can produce without the gearbox.

7.2. Waterproofing a Motor

Waterproofing a motor can be the most difficult aspect of designing and building an effective thruster. The main challenge is that you need to get a rotating propeller shaft through the wall of a pressure housing without letting any water inside, even though that water is trying to force its way in under high pressure. Whenever possible, consider letting the pros tackle this problem. One way to do this is to look for a suitable size and type of motor that comes already encased in a professionally made waterproof housing.

7.2.1. Trolling Motors

Commercial trolling motors are already waterproofed and usually designed to operate from 12 V batteries. However, they are generally much bigger, much more powerful, and much more expensive than bilge pump motors. While trolling motors do provide phenomenal thrust on the order of 240 N (54 lb), using 720 W of power to do so, they are not without their problems—notably current draw, size, weight, and depth rating. Their excellent thrust output is oh-so-tempting, but the larger trolling motor draws way more current than a smaller bilge pump motor. Furthermore, the motor is physically too large and too heavy to be used on most small deeper-diving ROVs. Additional weight would necessitate additional flotation. The larger size of these thrusters would require a much larger, beefier frame. Furthermore, trolling motors are rated to shallow depths only, so additional pressure-proofing measures are advisable for deeper missions. (See *Historic Highlight: Adapted Trolling Motors.*) For example, to bring the wires from the externally mounted thrusters into the pressure can, you would need to find some way to block water from entering the hollow tubing where the motor's power conductors exit the housing, perhaps by using a plug of epoxy.

Figure 7.30: Trolling Motors on a Student-Built ROV

Image courtesy Steve Van Meter, VideoRay LLC

For these reasons, trolling motors are overkill for most small vehicle projects; however, they can deliver a lot of thrust if you really need it (and if you can supply the large electrical currents these motors demand). One major advantage trolling motors have over bilge pump motors is that their shafts are already designed to accept

HISTORIC HIGHLIGHT: ADAPTED TROLLING MOTORS

Low-cost ROVs (LCROVs) and home-built submersibles have used trolling motors for thrusters since the 1970s. *SEACLOPS,* an LCROV built in the late 1980s, used electric trolling motors that were capable of operating at depths of 100 meters (approx. 325 ft) with no modification to the dual U-cup type shaft seals, other than replacing the standard prop with a cut-down 15-centimeter (approx. 6-in) model airplane propeller to reduce the load and hence its power consumption. While these vehicles successfully used trolling motors at these depths, there's no guarantee you would be so lucky. Remember that there's always additional risk anytime you push a component beyond its normal use or stated limitations.

Figure 7.31: Modified Trolling Motors

The original SEACLOPS *had three trolling motors. However there was no waterblocking between the trolling motor/thruster and the pressure canister, so if there was a leak in the thruster the whole pressure can would flood. A rebuild project to correct this problem involved waterblocking the hollow strut with epoxy. The photo on the right shows a trolling motor with the guard and prop removed.*

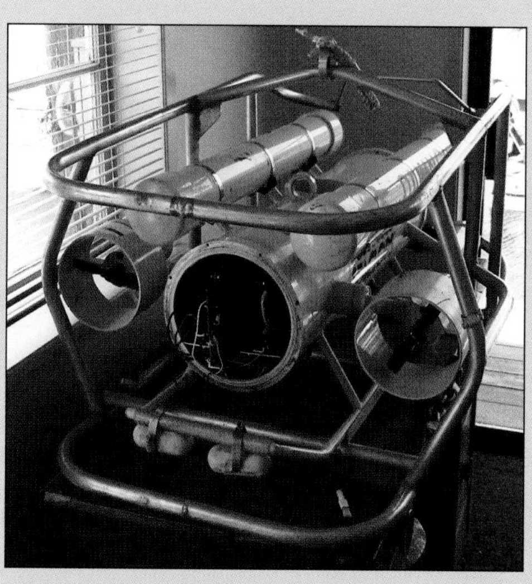

Images courtesy of Harry Bohm

a propeller. If you decide to go this route, get a propeller specifically made to go with the motor you have. Other props will most likely reduce performance and efficiency. If you plan to operate your vehicle in saltwater, make sure you get a trolling motor designed for ocean use.

7.2.2. Bilge Pump Motors

Bilge pump motors are used to pump water out of the bottom of boats. They operate from 12 V batteries and are already waterproofed, so it's not surprising that they are the most popular motors for small underwater vehicle projects. They are reasonably affordable, widely available in boating stores, and offer a good size and power output for small underwater craft operating in calm water. However, they must be modified to accept a propeller. Methods of attaching a propeller to a motor shaft are described later in this section.

Neither trolling motors nor bilge pump motors are intended to operate much deeper than about half a meter beneath the water's surface; however, these systems tend to be over-designed for reliability. Many people have successfully operated them at depths of 10 meters (approx. 33 ft) or more. You may be able to do this, too, but anytime equipment is pushed beyond the conditions for which it was designed, the risk of failure increases dramatically. It would be best to avoid going too deep, unless you can easily afford to replace the motors and anything else that might get damaged by a leak.

7.2.3. Small Commercial ROV Thrusters

If you're specifically interested in deeper-diving vehicles, say to 100 meters (approx. 325 ft), you might want to research commercial off-the-shelf thruster options. Currently only a few subsea companies, such as Inuktun (Figure 7.22), SeaBotix, and VideoRay provide small, low-voltage thrusters suitable for a deeper-diving project. For example, the BDT150 from SeaBotix has close to the right dimensions, weight, thrust, and power requirements needed for a small ROV. There's also a brushless motor version, the HPDC150. This efficient motor exerts dramatically more thrust for its size and power consumption, but is priced higher and needs a special controller supplied by the company. Both thrusters have subsea electrical connector options that would match up with standard subsea bulkhead connectors used to bring wiring in and out the pressure can. Your research may turn up other suitable options. Whatever your choice of thrusters, ensure that its pressure-proof connectors match the style of bulkhead connector you are using.

7.2.4. Build-Your-Own Thruster Housings

If you are willing and able to build your own waterproof motor housings, you open up access to other types of motors. These include the strong DC gear motors used in cordless power tools and the high-performance brushless motors used in model R/C (radio-controlled) race cars and airplanes. Before you go down this road, be sure to review *Chapter 5: Pressure Hulls and Canisters* for information on how to construct pressure-resistant, waterproof housings. Remember that you will need to provide watertight shaft seals for the propeller shaft, as well as some way to get electrical power to the motor. You may also want to provide a way to open the housing easily for repair or replacement of thruster parts. Note that adding this flexibility increases the complexity of the housing and may increase the risk of leaks if not done carefully.

Because it is difficult to seal a rotating shaft against external water pressure—particularly high pressure—you might consider making an oil- or air-compensated housing as described in *Chapter 5: Pressure Hulls and Canisters* whenever you are designing a vehicle for dives deeper than about 5 or 10 meters (approx. 16 to 33 ft). An oil-compensated housing is much easier and safer to make than an air-compensated housing, though it is also messier.

If you decide to go with an oil-compensated housing, it's best to use a brushless motor. (Don't forget that you'll need a matching brushless motor controller.) Brushless motors normally work fine in light mineral oil. The oil's viscosity will reduce speed and efficiency somewhat, but the slight loss in performance is usually well worth the

TECH NOTE: MACHINING PARTS FOR A CUSTOM THRUSTER HOUSING

If you have access to a metal-working lathe and know how to use it safely, you may be able to machine custom parts for your own thruster housings out of PVC, aluminum, or other stock materials.

The main challenge will be waterproofing the seal on the rotating motor shaft. As discussed in the latter part of *Chapter 5: Pressure Hulls and Canisters*, O-rings are one good way to make watertight seals around a shaft, but these require smooth surfaces manufactured with precise dimensions—something a metal-working lathe is good at producing.

This diagram illustrates one design for a thruster nose cone that uses O-rings for seals. The nose cone (yellow in diagram) is the most complex part of this particular housing. It holds two low-friction sleeve bearings (magenta) that support the propeller shaft (orange). A pair of small O-rings lubricated with grease forms a watertight seal between the rotating shaft and the non-rotating nose cone. The nose cone itself fits like a cork into the end of a long cylindrical motor housing (gray); note that only one end of this long cylinder is shown. A second pair of larger O-rings forms a watertight seal between the nose cone and the motor housing. A setscrew (green) passes through the motor housing and partway into the nose cone to keep the cone from sliding out of the motor housing during normal use. If motor inspection, repair, or replacement becomes necessary, the setscrew can be removed to allow the nose cone, shaft, and attached motor (not shown) to be pulled out of the motor housing. Note that the setscrew penetrates the motor housing *outside* the larger pair of O-rings so that water leaking along the screw's threads does not gain access to the interior of the motor housing.

The accompanying photograph shows the nose cone being machined from gray PVC plastic on a metal-working lathe. You can see the finished nose cone in Figure 7.36.

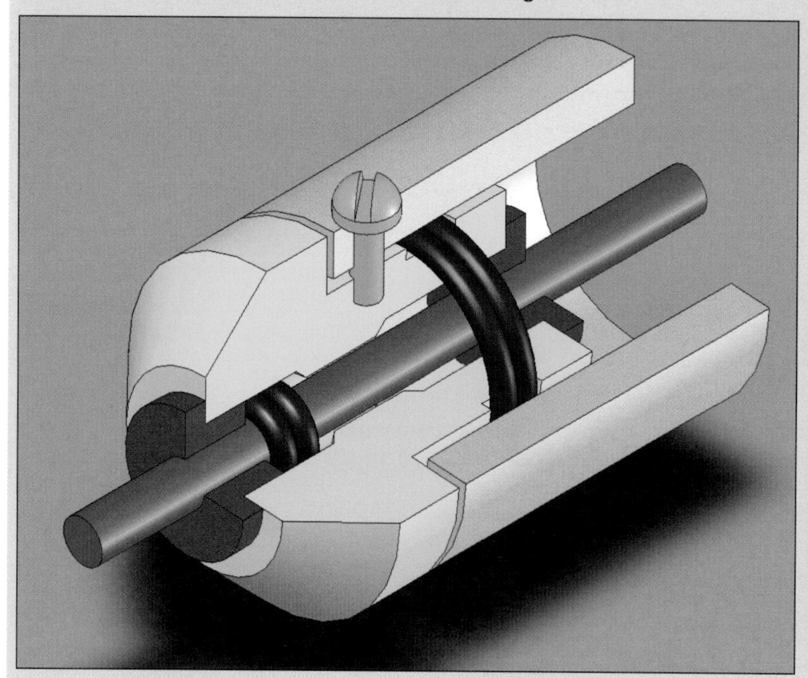

Diagram and photo courtesy of Kyle Worcester Moore and Dr. Steven W. Moore

Figure 7.32: Custom Thruster Housing

added protection against leakage, particularly at great depths. Thrusters designed professionally for full ocean depth routinely use brushless motors in oil-compensated housings.

If necessary, brushed motors can also be used in oil-compensated housings, but this is not recommended. As a brushed motor gains speed, its brushes ride up on a thin layer of oil, like car tires hydroplaning on a wet road. This disrupts electrical contact between the brushes and the commutator plates, momentarily cutting power to the motor. Without power, the motor quickly slows, the brushes re-establish contact with the commutator, and the motor revs up again. This process repeats indefinitely, resulting in jerky, inconsistent motor behavior and greatly reduced performance. In addition, carbon particles and bits of worn brush material eventually contaminate the oil, so it needs to be replaced frequently—a messy job. In spite of these difficulties, some groups have managed successful deep dives with inexpensive home-made vehicles using oil-filled bilge pump motors containing brushed motors.

One other option for waterproofing a motor takes advantage of something called **magnetic coupling**. This technique completely eliminates the need to pass a rotating propeller shaft through the motor housing and makes it much easier to keep the motor dry. Modern neodymium or "rare earth" permanent magnets provide extremely strong magnetic fields at close range and can be used to transmit mechanical forces and motions directly through the walls of a motor housing, thereby eliminating the need to pass a rotating shaft directly through the housing wall. This makes it much easier to make the motor housing watertight. For this to work, the housing must be made of a non-magnetic material, such as plastic, aluminum, or stainless steel, and the walls must be thin so the distance separating the magnets is small. Inside the housing, the motor spins a disk or cylinder with magnets mounted on it. The magnets pass just beneath the surface of the housing, so their magnetic field will extend outside the housing. On the outside, another set of magnets attached to the propeller follows the rotation of the magnets inside the housing, thereby transferring the rotational energy to the propeller.

Though elegantly simple in concept, magnetic coupling is not easy to do in practice. The method requires particular geometries and precise rotational alignment to get strong coupling between the magnets—otherwise, it will not work. Some submersible aquarium pumps already employ this technique to drive impellers, and they might be modified to drive propellers for an ROV.

7.3. Attaching a Thruster

As you design your thruster housing (or select an existing bilge pump or trolling motor housing), remember to think about how you are going to attach it to the vehicle. Otherwise, you may end up with a thruster that looks great and works even better, but can't easily be connected to your vehicle's frame.

The method you use to mount your thrusters need not be complex or sophisticated. For example, many groups simply glue or strap their thrusters to sections of PVC pipe on the vehicle frame. Others design and build their own brackets out of plastic or metal to keep the thrusters securely in place. Ideally, any method you use should have these characteristics:

- provide strong, rigid, waterproof and corrosion-resistant support that won't flex or vibrate excessively, even when the thruster is operating at full power under water

- minimize, as much as possible, blockage of the thruster's water jet in both the forward and reverse thrust directions, to preserve maximum thrust
- minimize drag created as the vehicle moves through the water
- be resistant to getting snagged or entangled by underwater obstacles
- allow some way to adjust the position and angle of the thruster, since minor changes in these parameters can have a significant effect on vehicle performance
- allow for easy removal and reattachment of the thruster to facilitate maintenance, repair, or replacement

If you are using ducted props, you may find it easiest to attach the motor housing only to the duct, then attach the duct to the vehicle frame.

Image courtesy of the MATE Center

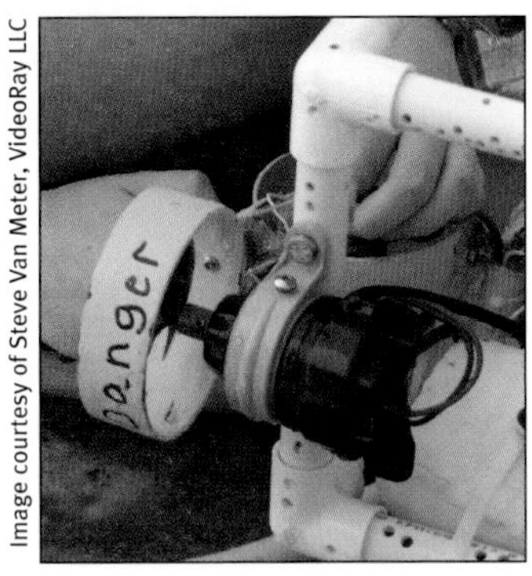

Image courtesy of Steve Van Meter, VideoRay LLC

Figure 7.33: Simple Thruster Attachments

There are a number of simple, yet effective, methods of attaching thrusters to a vehicle's frame. The first photo shows two cable ties. The second photo shows a bracket and screw arrangement which provides a secure way to attach a thruster to the frame. Note also that the prop is protected by ducting.

7.4. Selecting and Attaching a Propeller

Like motors, propellers come in thousands of different sizes, shapes, and styles, so it can be a daunting task to select just the right one for your vehicle's thrusters. The ideal prop should be effective, affordable, and easy to obtain.

As with motors, it helps to narrow the field by taking a serious look at what others have been using successfully. At the same time, you want to be open to other ideas, because you might discover something even better.

Most small ROVs built by school groups use plastic propellers in the 25–75 millimeter (approx. 1–3 in) diameter range. These are available from many hobby shops, either locally or on the internet. Modified model airplane propellers have also been used successfully by others.

Hopefully, you'll narrow your list to a few candidate props. Before you make a final decision and start mass-producing several thrusters for your vehicle, you'll want to obtain a sample of each potential prop to test with whatever motors you are considering. That's the best way to find out which motor+propeller combination delivers the best performance. To do that, you'll need to figure out some way to attach each propeller to each motor. If you're going to use a ducted prop, you'll also need to figure out how to make and attach an appropriate propeller duct. (See Figure 7.34.) Then you'll need to have some way of measuring how much thrust each propeller+motor combination produces under water and how much electrical current it consumes in the process. The sections that follow discuss each of these steps in turn.

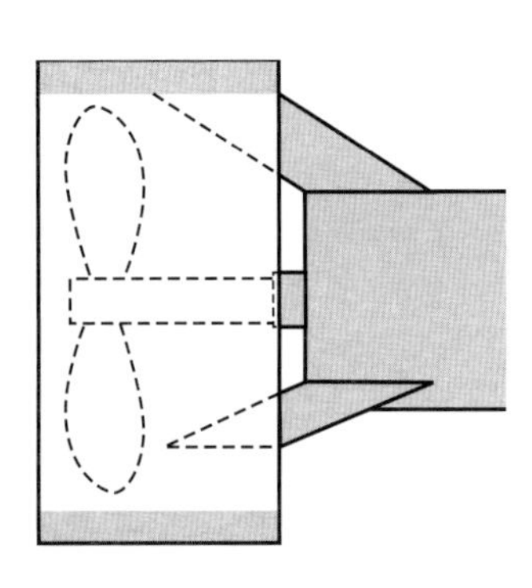

Figure 7.34: Possible Duct Mounting

Brackets shaped like rocket fins can be an effective way to mount a duct around a thruster prop.

7.4.1. Connecting the Prop to the Motor Shaft

Second only to waterproofing the motor, securing a prop to the motor shaft is one of the greatest challenges you are likely to encounter in making a thruster. This is true whether you are just trying to mount a propeller temporarily on your motor for preliminary testing or to mount it permanently. There are three basic problems you'll need to solve.

First, the motors and props usually come from completely different sources, so the shaft of the motor is almost always the wrong diameter or shape to fit into the hub of the propeller. That means you'll have to modify the shaft, modify the hub, or fabricate some type of adapter to connect them together. For example, sometimes the prop has a threaded fitting, in which case the motor shaft must have the matching thread type; as well, there must be some way of keeping the prop from unscrewing itself and falling off during a dive. In other cases, the prop is held on the shaft by a tight friction fit, which will work only if the shaft and hub are *exactly* the right sizes. In still other cases, the prop is locked onto the shaft by a cotter pin, bolt, or other device inserted through a hole drilled sideways through both the hub and shaft. This will work only if both the hub and shaft have matching holes that line up properly. Finding a way to get your props attached to your motors is likely to require considerable creativity on your part.

Second, even if you can manage a reliable connection between the motor and prop, it may be difficult to achieve the precise centering and alignment needed to avoid serious vibration problems. Shaft vibration can reduce efficiency, shake things enough to loosen or damage parts over time, and cause leakage around the propeller shaft seals.

Third, if you ever plan to disassemble your thruster for motor replacements or other maintenance or upgrades, you'll need to make sure that the method you use to couple prop to shaft can be undone and does not interfere with opening the motor housing.

If you don't need to disassemble things and if your prop already fits the shaft reasonably well, the simplest option is to use epoxy to glue the prop directly to the shaft. Sometimes a roll-pin or similar gizmo can be used as a makeshift adapter for a prop hub that is too large for the shaft. A roll-pin is basically a small metal tube with a slit along one side. Roll-pins come in lots of sizes so you can probably find one that will fit snugly over your motor shaft, effectively increasing the shaft's diameter. If you're lucky, the new diameter will fit perfectly into the prop hub. To form a durable connection, you'll want to glue both ends of the roll-pin in place with epoxy. Remember to clean all surfaces of oil or grease before trying to glue them and to be very careful to make sure everything is aligned perfectly while the epoxy hardens. It's worth the time and effort to build a little jig that will hold the prop in correct alignment with the motor while the glue hardens.

Images courtesy Dr. Steven W. Moore

Figure 7.35: Home-Made Shaft Coupler

In this photo, a lathe has been used to drill a perfectly centered hole in the end of a 3/8-inch aluminum round rod. This hole fits snugly over the output shaft of the bilge pump motor shown on the right. A second hole was drilled perpendicular to the first all the way through the rod, then a tap was used to cut threads in the hole for a small screw. When the motor shaft is inserted in the end of the rod, the screw can be tightened against the flat area cut on the end of the motor shaft, securing the coupler to the shaft. In the next step, the 3/8-inch rod will be cut to about a 1-inch length, and a second pair of holes will be drilled to accommodate the prop shaft and another setscrew.

Figure 7.36: Modified Commercial Shaft Coupler

The white cylindrical object in the center of this photo is a commercially available shaft coupler used to connect the 4-millimeter diameter output shaft of a small gear motor to a 3/16-inch propeller shaft. Neither shaft is visible in the photo, as both are hidden inside the coupler or thruster housing nose cone (the gray object with the double O-ring seal). This particular type of coupler has a spiral slit cut in it, making it slightly flexible, so it can bend while rotating. This helpful feature allows it to accommodate slight shaft misalignment while still transferring torque effectively. Recessed setscrews visible at either end of the coupler are used to tighten the coupler onto the shafts. In this instance, the designer was not able to find a coupler that matched both shaft dimensions exactly, so he purchased a 4 mm to 4 mm shaft coupler, then used a 3/16-inch drill bit to enlarge one end to fit the prop shaft. This type of creative problem-solving is essential to successful ROV and AUV design.

If you have a lathe, drill press, or milling machine, you may be able to make your own adapters to join the motor shaft to the propeller (Figure 7.35). If not, your best bet is to try to find a commercial **shaft coupler** with one end sized to fit the motor shaft and the other end sized to fit a shaft that mates with the propeller. (See Figure 7.36.)

7.4.2. Using a Ducted Prop

As described earlier, ducted props, including Kort nozzles, can significantly improve thruster performance if done correctly. On the other hand, they can reduce performance if not done correctly. The easiest way to make a duct is to take a hollow cylinder of appropriate diameter, such as a section of PVC pipe or part of a tin can, and mount it around a propeller. For best effect, you'll need to shape the duct to enhance water-flow velocity, but this is tricky to do well.

Ducted props work best with certain types of propellers (usually having clipped blade tips) and when there is very little space between the outside edge of the prop and the inside of the duct. That means the duct must be centered exactly on the prop's axis of rotation and held there rigidly. Otherwise, thruster vibrations or minor bumps against submerged objects might force the duct into contact with the rotating prop blades. Finding a way to hold the duct rigidly in place without blocking too much water flow through the thruster is an interesting design challenge.

Many groups find that a series of three slender, parallelogram-shaped ribs, arranged like fins on a rocket (Figure 7.34), provide a good compromise between the conflicting demands for rigid attachment and unobstructed water flow.

Image courtesy Dr. Steven W. Moore

Figure 7.37: Prop Choice

When matching the propeller to the motor, it's extremely important to test prop variations, including pitch, size, length, and number of blades.

7.4.3. Matching the Propeller to the Motor

To maximize the thrust you get out of your thruster while minimizing the amount of energy wasted, it's extremely important to "match" the propeller to the motor. This means choosing a propeller with the right diameter, number of blades, blade pitch, and other characteristics to transfer maximum power from the motor to the water. Figure 7.37 shows the range of plastic props you might consider.

If a propeller is too small or otherwise offers too little resistance, the motor turning it will spin too quickly and use most of its energy overcoming motor friction, rather than moving water. This situation is like riding a bicycle in a gear that's way too low—you waste lots of energy on moving your legs rapidly up and down, but don't make much forward progress. On the other hand, if the propeller is too large or otherwise offers too much resistance, the motor can get bogged down, draw excessive current, and overheat. At best, this will be an inefficient use of power. Worse, it could destroy the motor circuitry or even start a fire. Running a motor with a propeller that is too large is like riding a bicycle in a gear that's way too high. You are practically standing on the pedals and working up a great sweat, yet hardly moving. So finding the right prop for your motor is like finding the optimal gear ratio to use during a bike ride. The best choice allows you to move as quickly as possible without overloading the motor.

Although there are theoretical ways to calculate optimal propeller-motor combinations, these methods are complex and beyond the scope of this book. Fortunately, there's an easy and fun alternative. Presuming you've found an effective method of attaching various props to various motors, you can measure the performance of each prop+motor combination directly. All you need to do is get a large container of water (bucket, bathtub, ice chest, garbage can, swimming pool, or whatever) and rig up some mechanism that can hold the spinning propeller in the water while allowing you to measure how hard it's pushing (or pulling) and how much electrical power the motor is consuming.

For a very rough estimate of thrust production and motor efficiency in a *small low-voltage* thruster, you don't even need a test mechanism; you can just use your arm. Attach a prop to a motor, hook it up to a fused power supply, hold the motor in your hand with the propeller under water, and turn on the motor to see what happens. (First read *Safety Note: Thruster Testing Dangers.*) By observing how vigorously the water moves and by feeling how strongly the propeller pushes or pulls against your hand, you can quickly get a sense of whether one prop+motor combination produces a lot more thrust than another and whether or not a particular combination has problems overheating the motor, battery, or wires.

For more subtle comparisons, however, you'll want some way to obtain precise and repeatable measurements of thrust and electrical power. For that, you'll want to build a **thruster test jig** and find a suitable meter for measuring electrical voltage and current. You will also want to record the measurements you get during each test for later analysis and comparison. (See *Tech Note: Building a Simple Thruster Test Jig* and *Chapter 12 Tech Note: Propeller Matching* for two thruster test jig variations.)

It's possible for motors to overheat if overloaded with a propeller that's too hard to turn, so check the motor frequently during these tests to make sure it's not getting too hot, especially when it seems to be turning a lot slower than its unloaded speed. Be alert for the smell of melting wire insulation.

You should try several different propellers with each motor to see which one produces the most thrust. If possible, try different diameters, different numbers of blades, and different pitches. You may also wish to try several different motors. Be sure to take careful notes, so you don't have to repeat the entire process.

These tests will be most informative if you not only measure the thrust, but also measure the electrical power flowing to the motor during these tests. To do this, you'll need to monitor both the voltage and the electrical current. Be advised that while

TECH NOTE: OPTIMAL DC MOTOR SPEED

A motor that spins quickly but isn't attached to anything consumes energy but does no useful work. Likewise, a motor that tries hard but is not strong enough to turn its load consumes energy but does no useful work. You want to find a motor between these two extremes, one that can turn its shaft and apply torque at the same time, so as to convert some of its input energy into potentially useful output energy. An important goal of propeller-matching is to maximize the fraction of input power that gets converted to output power.

The power output of a motor is defined as the shaft torque times the shaft speed. For a typical permanent-magnet DC motor, this value reaches a maximum when the motor is working just hard enough to slow its shaft speed to half its maximum (unloaded) speed at the supplied voltage. A properly matched propeller will thus slow its motor to about half its unloaded speed. If you are musically inclined, you may be able to hear this halving of rpm as a one-octave drop in the tone of the spinning motor when a spinning well-matched prop is moved from air to water.

Assuming the propeller is well designed, it should do a good job of converting a large fraction of that output power into directed water movement and thrust. Thus, a well-designed propeller that slows a motor to half its unloaded speed should give you maximum thrust output per watt of electrical power delivered to the motor.

standard **multimeters** can easily handle the voltages you are likely to be using, most can't handle the number of amps (a unit of electrical current) used by even a small motor without blowing the multimeter's fuse. (See *Chapter 8: Power Systems* for a discussion of volts, amps, and fuses.)

Be sure you use a meter that can measure up to 10 amps for bilge pump motors and similar-sized motors; more for larger motors. Fortunately, some readily available and not-too-expensive meters are designed to handle up to a hundred amps or more. These include clamp meters, used commonly to measure current in automotive or home-power circuits. If you use one of these, make sure it can measure DC current, not just AC current.

To use a clamp meter, you open the jaws at one end and close them around a wire to measure the current flowing through that wire. It's important to measure the current in only one of the two wires carrying current to/from the motor, but it doesn't matter which one.

You will probably discover that larger propellers cause motors to draw larger currents. If you are planning a battery-powered mission where battery life is a limiting factor, it's worth noting that higher currents will drain a battery more quickly. A propeller that produces only half as much thrust, but draws only one-tenth as much electrical power while doing so, may be a better choice in that situation than the propeller producing the most thrust. For another example of how thruster test jig results can help you select an appropriate propeller, see the *Tech Note: Interpreting Sample Test Jig Results.*

SAFETY NOTE: THRUSTER TESTING DANGERS

Although thruster testing is fairly safe, there are some potential hazards you should be aware of. First, you are mixing electricity and water. Props spinning in shallow test tanks invariably splatter water all over the place, and unprotected motors and wires will get wet. Avoid doing these tests with more than about 24 volts in the vicinity, unless you take special precautions to avoid electric shock.

Second, the mechanical connection between the prop and motor is often temporary during these tests and may be unreliable. Wear eye protection in case a fast-spinning prop comes loose and flies at your face.

Third, it is not uncommon to discover that the prop you are testing is too big for your motor. When this happens, some overloaded motors can get very hot very quickly. Temperatures can get hot enough to burn skin, melt wire insulation in the motor circuit, or even start a fire. Batteries powering an overloaded motor can also get hot and may even explode.

To be safe, limit initial tests with any new prop+motor combination to a few seconds at first, then feel the motor, supply wires, and battery to make sure nothing is getting excessively hot before trying a longer test. Gradually increase the length of each test, checking temperatures again during and after each test, until you are confident that nothing is overheating.

If you find a prop+motor combination you think you want to use in your vehicle, give it a good long test in water lasting at least five minutes to make sure nothing overheats. Modest warming of a motor that has been running for awhile is perfectly normal and nothing to be worried about, but if anything smells like burning electronics, smokes, or gets too hot to hold comfortably in your fingers, shut down the test and scratch that prop+motor combo off your list.

TECH NOTE: BUILDING A SIMPLE THRUSTER TEST JIG

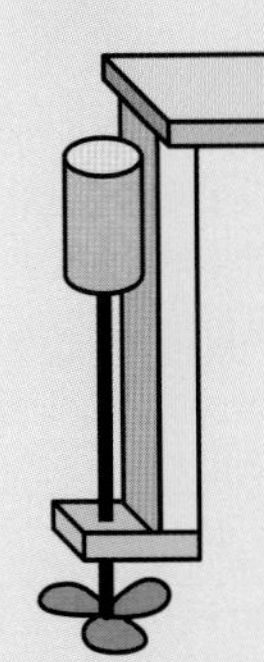

Figure 7.38: Simple Thruster Test Jig

This diagram and the accompanying photos illustrate one way you can build a simple thruster test jig suitable for use with any small electric motor—waterproofed or not. The functional heart of the jig is a long piece of flexible wooden molding (green in the diagram), with one end clamped to the edge of a table and the other end cantilevered out into thin air. The cantilevered end must be free to flex up and down like the end of a springy diving board. Ideally, the end should bend downward about one inch for every pound of weight placed on the end of the board. To make the board more flexible, extend it out farther from the edge of the table. To make it stiffer, shorten the cantilevered section.

Hanging straight down from the flexible end of the diving board is a stiff vertical piece of wood (yellow in the diagram) to which you securely tape or otherwise attach a motor for the tests. The vertical piece of wood—about 1 inch x 2 inches x 12 inches is a good size—should be nailed or glued to the underside of the molding *before* clamping the molding to the table.

If your motor is already waterproofed, you can use electrical tape, duct tape, or another water-tolerant tape to secure it near the bottom of the vertical board, which will be submerged in a bucket or other container of water. If the motor's not yet waterproofed, you'll need to mount it higher (as shown) so it doesn't get wet. In this case, you'll also need to use a shaft extension (probably custom-made) to transfer motor torque to the propeller located under water. To control vibration in a long shaft extension, you'll probably need to pass it through a hole that's drilled just slightly bigger than the shaft in a small piece of wood (blue in the diagram) or through an L-shaped bracket attached to the bottom of the vertical piece of wood.

This test jig relies on gravity to make force measurements; gravity acts vertically, so for accurate thrust data, it's important that the motor shaft be nearly vertical and that the end of the diving board be nearly horizontal. The weight of the vertical board, motor, shaft extension, propeller, wires, and other accessories will probably cause your diving board to sag enough that the "vertical" board won't be vertical anymore. To fix this problem, you'll need to slide a small board or other spacer between the table and molding, as shown in Figure 7.39, lower left photo. This will cause the diving board to leave the table surface at an upward angle, so that it will be horizontal at the tip. You may need to adjust the position of the spacer and/or clamp to get the end of the diving board properly leveled.

Once the diving board end is leveled, you will likely need to adjust the height of the water container. Ideally, the prop will be at least several inches beneath the water surface, yet far from the bottom and sides of the container. In the photo, a small bucket is resting on a stepping stool, to make the water level high enough to cover the prop; however, it would be better to use a larger, deeper container, such as a garbage can full of water sitting directly on the floor. The idea is to keep the prop deep under water and as far as practical from the walls and bottom of the container to more accurately simulate thrust in open water conditions. A prop too near the surface or too close to the bottom or sides of the container will produce inaccurate thrust data.

If you now connect power to the motor, the prop should spin and produce thrust. Swap the wires, if necessary, so that the thrust produced is upward, lifting the tip of the flexible diving board higher. (This is straightforward if you're using a DC brushed motor, but potentially a bit trickier if you're using a DC brushless motor and brushless motor controller.) Be careful—if you've got a strong thruster combination and a nice, flexible diving board, the prop can literally climb all the way out of the water and buck like a wild bronco! It's often best to hold onto the end of the diving board to stabilize it when you first turn on a motor and release it gradually only when you're convinced it's going to behave itself.

To measure how much upward thrust is created, start stacking weights in the small box taped to the diving board. For accurate results, the weights must be placed directly over the motor. Keep adding weight until the diving board is forced back down to its original position, where it was before you turned the power on. At that point, the weight you've added is exactly equal to the static or "bollard" thrust produced by that motor-propeller combination. You can use known weights, like laboratory weights, or simply use nuts and bolts, rocks, fishing weights, or other heavy objects during the tests, then move these informal weights onto a scale to weigh them after each test run.

For accurate thrust data, you'll need some way to determine when your diving board has returned *precisely* to its original height. In Figure 7.39, the top left photo shows a thin metal rod lying across the chair with some red beanbags to hold it in place. This rod lines up with (but does not touch) a black triangle drawn on some yellow tape around the vertical motor mount board, as you can see in the photo on the right.

This photo was taken before the motor was turned on. As soon as the motor was turned on, the thruster action lifted the assembly with the black and yellow arrow about two inches above the metal rod. Lead fishing weights were then added to the box while the motor was running until the triangle was forced back down into alignment with the metal rod. The weights were then removed from the box and placed on a kitchen scale to determine the force generated by the thruster.

If you want to measure how much thrust your propeller and motor combination can produce in reverse, you can swap the motor wires and use a spring scale to pull up on the end of the diving board until the black triangle lines up with the metal rod.

In the top left hand photo, you can see a black battery on the large cardboard box at the far left. The red wire going from the battery to the motor is passing through a large hole in the end of a device known as a clamp meter. Clamp meters are used to measure the amount of electrical current (in amps) flowing through a wire circuit. You can multiply this reading by the voltage of the battery to determine the watts of power being used by the motor during the thruster tests.

A modified version of this thruster test jig uses a rigid board instead of a flexible board, with a pivot point in the middle like a teeter-totter. Weights are added to the end opposite the thruster until the teeter-totter balances and the beam is perfectly horizontal. Then the motor is turned on (which makes the beam rotate about the pivot) and weights (or a spring scale) are applied to one end or the other of the beam (depending on whether the thruster is operating in forward or reverse), to restore the teeter-totter to a level position. The amount of weight or spring tension required equals the thrust produced by the thruster. (See *Tech Note: Propeller Matching* in *Chapter 12: SeaMATE* for another thruster test jig variation.)

Keep in mind that these designs are only examples to help you get started. They are not the only or even the best ways to build a thruster test jig. People have built jigs based on sliders from drawers, bicycle wheels, springs hung from the ceiling, parallelograms made from kids' toys, and lots of other mechanisms. Feel free to come up with any design that gives you accurate data. Thruster test jig design is one place where you can let your creativity run wild and have lots of fun converting old junk from the garage into fascinating contraptions for gathering real scientific data!

Images courtesy of Dr. Steven W. Moore

Figure 7.39: Test Jig Set-Up

TECH NOTE: INTERPRETING SAMPLE TEST JIG RESULTS

This hypothetical example illustrates how you might go about trying to determine which of three props is the best one to use with a particular DC motor rated at 12 volts and 3 amps (12 VDC/3A). To collect this information, use a thruster test jig to measure the thrust produced and one or more meters to measure voltage and electrical current delivered to the motor during each test. (See *Tech Note: Building a Simple Thruster Test Jig*.)

Note: In this hypothetical example, each prop was tested four times, each time with a different voltage applied to the motor (3, 6, 9, and 12 volts), but this isn't essential. Applying the maximum voltage you intend to supply to the motor (in this case it's 12 volts) would be enough to provide the data needed for a useful propeller comparison.

It is usually most convenient to record your data for each prop during the test in a table, like this one:

Test Prop #1: Red plastic two-blade, 1 inch diameter, right-handed (RH)

Voltage (V)	*Current (A)*	*Thrust (lb)*	*Power (W)* *P=V x I*
3 volts	*1*	*.25*	*3*
6 volts	*1.25*	*.5*	*7.5*
9 volts	*2.25*	*1.25*	*20.2*
12 volts	*3.0*	*2.25*	*36*

Note that voltage, current, and thrust are *measured* during the tests, but power (in the fourth column) is *calculated* from the voltage and current measurements.

Here are the data for the other two props:

Test Prop #2: Black plastic three-blade, 1 inch diameter, right-handed (RH)

Voltage (V)	*Current (A)*	*Thrust (lb)*	*Power (W)*
3 volts	*.6*	*.5*	*2.0*
6 volts	*2.3*	*1.75*	*14*
9 volts	*2.7*	*2.5*	*25*
12 volts	*3.0*	*2.75*	*36*

Test Prop #3: White plastic three-blade, 1.5 inch diameter, right-handed (RH)

Voltage (V)	*Current (A)*	*Thrust (lb)*	*Power (W)*
3 volts	*1.3*	*1.0*	*4.0*
6 volts	*2.6*	*2.25*	*16*
9 volts	*3.3*	*2.5*	*30*
12 volts	*3.5*	*3.0*	*40*

Plotting these results in graph form makes it easier to compare the data. Since you are particularly interested in how much thrust each propeller produces for a given power input, you would create a graph to display thrust (on the vertical axis) versus power (on the horizontal axis).

By studying the graph below, you can quickly spot some important results from these tests. First, notice that the line connecting the data points for prop #1 is always below the other two lines. This means the amount of thrust produced by prop #1 is lower than the thrust produced by either of the other two props for all values of input power. This tells you that prop #1 is less efficient at converting input power to thrust. Moreover, the maximum thrust produced by prop #1 is less than the maximum thrust produced by either of the other two props. Prop #1 might be a great prop for some *other* motor, but with this motor, it is a weak performer that wastes a lot of energy. It makes good sense to eliminate it from consideration and focus attention on props #2 and #3.

Prop #3 achieves a higher maximum thrust than prop #2. It also appears to be more efficient at the lower power input levels. Therefore, prop #3 might seem to be the obvious choice, but be careful. Notice that the highest thrust level required an input power of more than 40 watts. The motor is rated for only 12 volts and 3 amps, which corresponds to 36 watts. Something's amiss. A quick look back at the data table confirms that there's a problem brewing. At both 9 volts and 12 volts, prop #3 draws *more than* the rated 3 amps maximum current for this motor. In other words, if you supply the motor with 9 or 12 volts while this propeller is attached, you run the risk of overheating and damaging the motor.

Prop #2 performs *almost* as well as prop #3 in terms of both maximum thrust and efficiency, and it does so without ever exceeding the motor's maximum current rating. Equipment tends to wear out more quickly if operated beyond its rated limits, so if long-term motor reliability is important to the success of your missions, prop #2 is the sensible, conservative choice.

On the other hand, if your mission calls for mostly slow, deliberate, low-power maneuvering punctuated by rare and very brief demands for high thrust, prop #3 might be the better choice. It is more efficient than prop #2 at low speeds and probably wouldn't have time to overheat during an occasional brief burst of power.

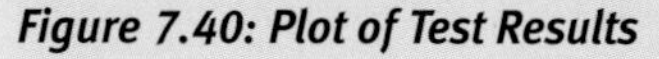

Figure 7.40: Plot of Test Results

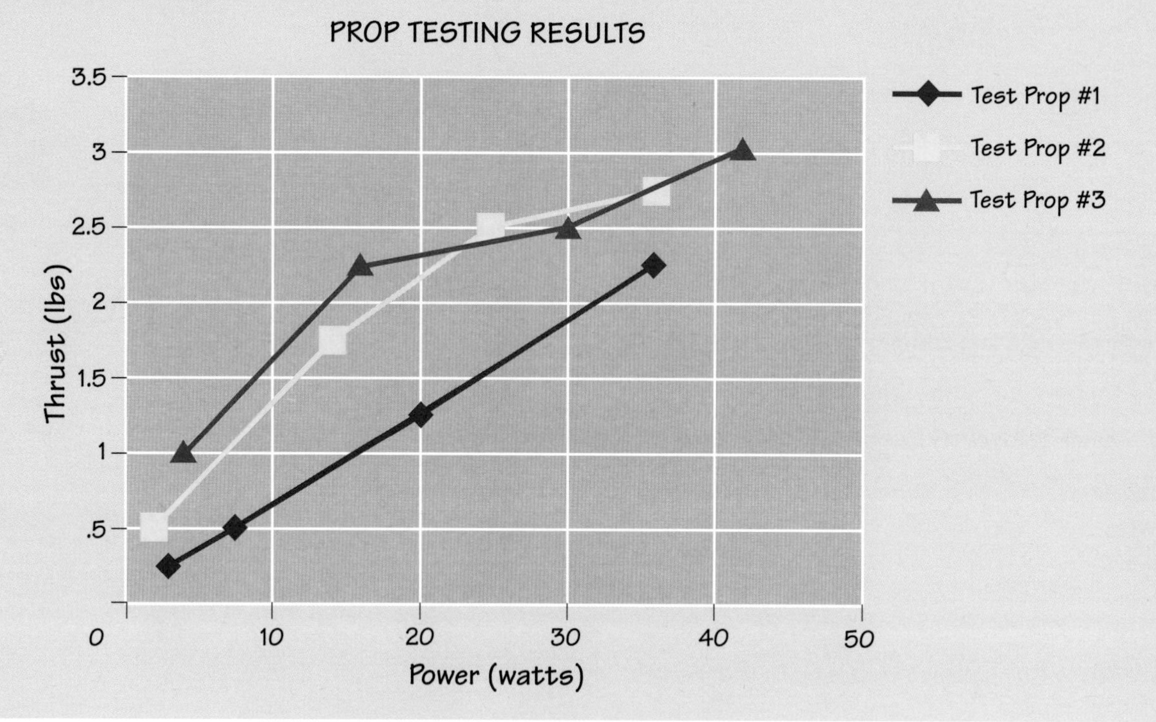

8. Thruster Placement

Once you've figured out how to buy or build good thrusters, you've solved one of the biggest challenges to creating a successful vehicle—but you aren't done yet. You still need to determine exactly how many thrusters are needed and where to put them on your vehicle. The number and placement of thrusters not only determine the range of possible maneuvers, but also strongly affect the performance and the efficiency of the vehicle.

The number of thrusters on an underwater vehicle can range from one (as on many torpedoes) to seven (as on some work class ROVs) or even more. Small ROVs typically use three or four. There are many options for positioning those thrusters on a vehicle; however, certain arrangements are more effective than others for particular purposes. Rather than provide an exhaustive survey of the nearly infinite range of possible thruster configurations, this chapter analyzes a few of the most common three- and four-thruster arrangements used for small underwater craft. These are by no means the only, or even the best, ways to arrange thrusters on your vehicle, but the analysis of these few arrangements should provide a good foundation for designing your own thruster placements to meet the requirements of your particular mission.

8.1. Analysis of a Common Three-Thruster Arrangement

The thruster arrangement shown in Figure 7.41 (or something very much like it) is common in vehicles having three thrusters. One thruster is oriented vertically and used to control vertical movements of a more or less neutrally buoyant vehicle. The other two thrusters are placed on either side of the vehicle and used to power horizontal movements. This pattern is simple, yet permits a surprising range of vehicle maneuvers.

8.1.1. Vertical Thruster Placement

Ideally, the vertical thruster is placed directly in line with the center of the vehicle, to minimize pitch- and roll-inducing torques that would be generated by an off-center thruster. Sometimes the thruster can be located in a vertical duct that passes directly through the center of the vehicle (Figure 7.41). Unfortunately, in most cases, the distribution of pressure canisters, payloads, or other vehicle components makes this ideal placement impractical.

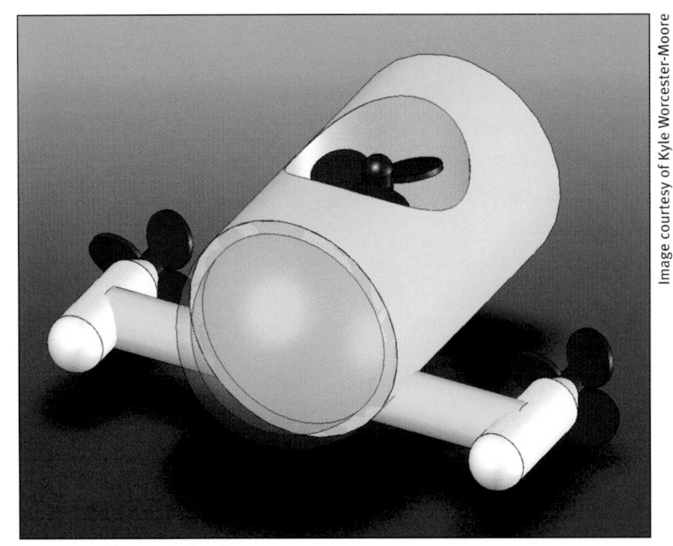

Image courtesy of Kyle Worcester-Moore

Figure 7.41: Common Arrangement of Three Thrusters

Vehicles with three thrusters commonly have one located centrally and oriented vertically, to move the vehicle up or down, while two horizontal thrusters, one on either side, drive horizontal movements, including forward, reverse, and turns.

In a vehicle with a central pressure canister that won't allow a vertical duct, one possible alternative is to place the vertical thruster directly above or below the center of the vehicle. However, this is not recommended, because the housing will block thrust in one of the two directions. For example, if the thruster is mounted above the vehicle, it can push the vehicle down (because nothing blocks the thruster's upward jet of water); however, it can't pull the vehicle up very well, because the downward jet of water blasts against the hull. This blocks the momentum of the water and pushes the vehicle down almost as hard as the thruster is pulling it up. In addition, this high placement of the thruster is likely to reduce vehicle stability, since thrusters are generally denser than water and need to be kept low on the vehicle frame. Locating the thruster directly beneath the center of the vehicle would improve stability, but isn't ideal, either. The thruster can push the vehicle up by jetting water downward, but it can't effectively pull the vehicle down, because the vehicle body blocks flow again. In addition, this lower position is particularly prone to stirring up sediment or becoming tangled in vegetation whenever the vehicle is operating near the bottom.

If the vehicle is very stable (lots of flotation placed high and lots of weight placed low), as many well-designed vehicles are, any pitch or roll induced by an off-center vertical thruster is likely to be small and may be within acceptable bounds. In that case, the vertical thruster can be placed almost anywhere, allowing flexibility to choose positions that do not obstruct water flow or interfere with other aspects of the mission.

Additional options for vertical thruster arrangements present themselves when a fourth thruster is added, as described later in this chapter.

8.1.2. Horizontal Thruster Placement

The two remaining thrusters in this three-thruster arrangement are oriented horizontally, one on either side of the vehicle as shown (Figure 7.41). This configuration works in much the same way as the caterpillar treads on either side of a bulldozer, and it can produce an impressive range of forward, backward, turning, and spinning motions.

Figures 7.42-7.45 diagram the range of motion that two horizontal thrusters can produce. Note that the dome on each of the figures represents the front of the vehicle, the small white arrows inscribed in the "thrusters" on either side of the vehicle indicate the direction of the *thrust* produced (which is opposite the direction of the water flow through the thruster), and the shaded arrows show the resulting movement of the whole vehicle.

Figure 7.42: Driving Forward or Backward

If both horizontal thrusters generate equal forward thrust (by pushing water backward), the vehicle will move forward. The more thrust they generate, the faster the vehicle will move. If both generate equal reverse thrust, the vehicle will move backward. (See Figure 7.42.)

If the thruster on the port (left) side generates forward thrust while the thruster on the starboard (right) side generates equally strong reverse thrust, the vehicle spins in place toward the right (clockwise as viewed from above). If the thrust pattern is reversed, the vehicle spins to the left (counterclockwise as viewed from above). (See Figure 7.43.)

Figure 7.43: Spinning Right or Left

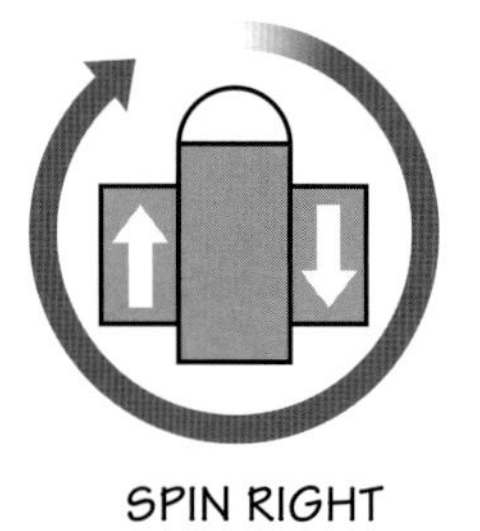

Here's what's happening: The two thrusters are acting as a force couple, generating a torque or moment about the center of the vehicle. The concept of a couple was introduced earlier in *Chapter 6: Buoyancy, Stability, and Ballast* in connection with the stability of a vehicle. This is like the example of two wrench handles pulled in opposite directions to rotate a bolt, only this time it is thrusters rather than hands that are generating the force, and a vehicle frame rather than wrench handles converting those forces into torque and rotation. Note that the forward and reverse thrusts cancel (so the vehicle does not move forward or backward), but the torques produced by those thrust vectors add together (both clockwise or both counterclockwise), spinning the vehicle in place. (See Figure 7.43.)

Figure 7.44: Veering Right or Left

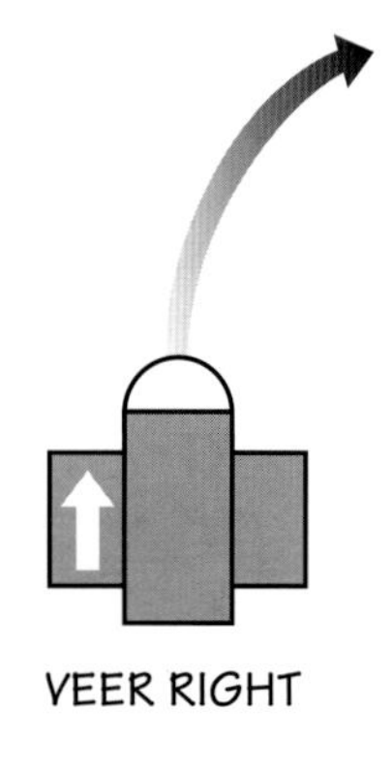

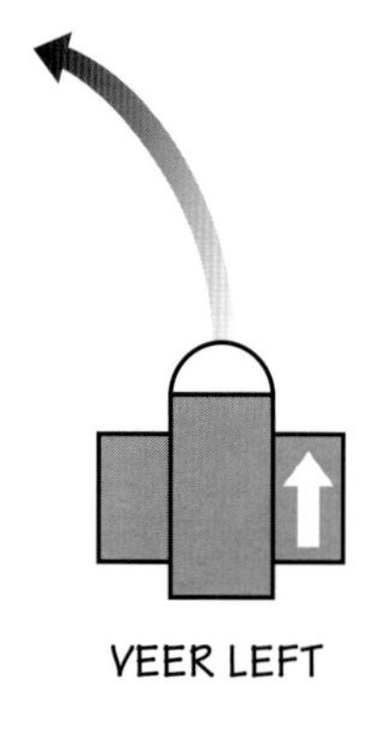

What happens if one horizontal thruster operates alone, without the other one doing anything? For example, what if the port thruster generates forward thrust while the starboard thruster is turned off? In this case, there is no reverse thrust to counteract the forward thrust, so the vehicle will move forward. However, it will not do so in a straight line. It will instead veer off to the right, making a broad, sweeping turn. If the starboard thruster generates forward thrust while the port thruster is off, the vehicle will veer to the left. (See Figure 7.44.)

These broad turns happen for two reasons, both of which depend on the fact that a line drawn through the thruster along the direction of the thrust does not pass through the center of the vehicle. The first reason has to do with the location of that line of action relative to the vehicle's center of gravity (CG). Anytime a force acts out of line with the CG of an object, that force will tend to both push and turn the object.

Once the vehicle is moving, a second reason for the turn comes into play. A vehicle moving through water generates drag. Although the drag is distributed across many parts of the vehicle, the combined effect of all those little drags is equivalent to one big drag acting more or less at the center of the vehicle. (The exact location depends on the details of the vehicle's shape.) The forward thrust on one side of the vehicle combined with drag pushing backward on the center of the vehicle creates a force couple and associated torque, not unlike that created when the two thrusters operate in opposite directions. This adds to the vehicle's tendency to turn as it moves forward. The drag does not exist unless the vehicle is moving forward, so the vehicle doesn't just turn in place, but moves forward as it is turning. Note that broad, sweeping turns can also be made in reverse by using only one thruster at a time in reverse.

Table 7.3 summarizes the wide range of basic movements that can be achieved with simple FORWARD-OFF-REVERSE control of two horizontal thrusters, one on each side of the vehicle.

Table 7.3: Basic Maneuvers with Two Horizontal Thrusters

Port (left) thruster action	Starboard (right) thruster action	Vehicle motion
Off	Off	Stop
Forward	Forward	Straight forward
Reverse	Reverse	Straight reverse
Forward	Reverse	Spin to right
Reverse	Forward	Spin to left
Forward	Off	Forward while turning to right
Off	Forward	Forward while turning to left
Reverse	Off	Reverse while turning right
Off	Reverse	Reverse while turning left

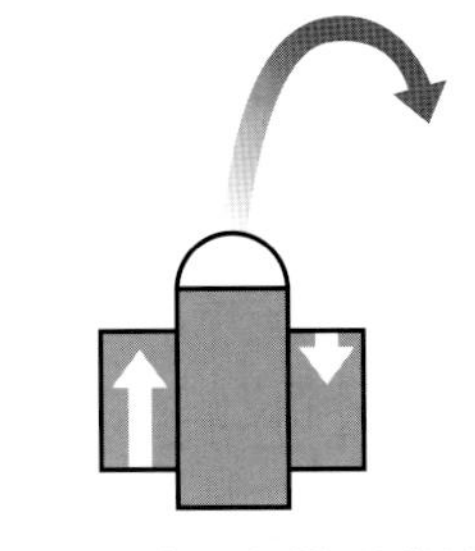

Figure 7.45: Turning Sharply

If you are able to control the speed of each thruster, then more subtle variations and combinations of these basic maneuvers become possible. For example, you can make a tighter turn to the right by starting with a regular "veer right" turn (port thruster full forward), then adding a bit of reverse thrust on the starboard side (Figure 7.45).

8.2. Analysis of Some Common Four-Thruster Arrangements

If a fourth thruster is available, even more options open up. At a minimum, the central vertical thruster of the three-thruster arrangement can be replaced with two vertical thrusters, one mounted on each side of the vehicle (not shown). Running both of these thrusters together in the same direction produces a net upward or downward force. You now have twice as many vertical thrusters for increased vertical thrust. Moreover, neither thruster requires that a duct be placed through the middle of the vehicle, and neither thruster has one of its thrust directions blocked by a central pressure hull or payload. That means strong vertical thrust can be generated in both the up and down directions. The roll tendency generated by each thruster cancels that produced by the other, so roll is not a problem with this arrangement, even for a vehicle with limited stability, as long as both thrusters are run in the same direction at more or less the same thrust level.

In spite of these advantages, this may not be the best way to use a fourth thruster, if you have one available. Other arrangements allow an even greater range of maneuvers. For example, the configuration shown in Figure 7.46 modifies the three-thruster arrangement shown earlier by adding a fourth thruster oriented sideways through

Figure 7.46: Four Thrusters in Modified Three-Thruster Placement

This four-thruster arrangement allows the vehicle to move straight sideways (to "crabwalk"), in addition to vertical movements, forward/backward movements, turns, and spins.

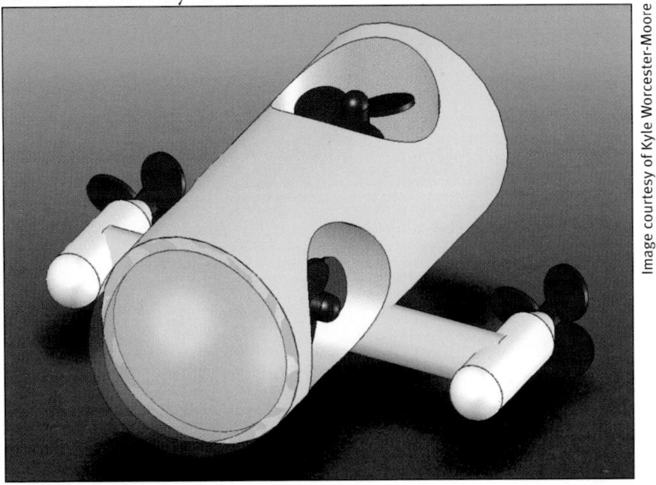

Image courtesy of Kyle Worcester-Moore

Figure 7.47: Alternate Four-Thruster Placement

This four-thruster arrangement allows the vehicle to move straight sideways, a motion called a "crabwalk," in addition to many other movements, but does so without requiring any thrusters to occupy a central position in the vehicle. It relies on thrust vector addition to produce vertical and crabwalk motions, as described below in the Tech Note: Vectored Thrusters.

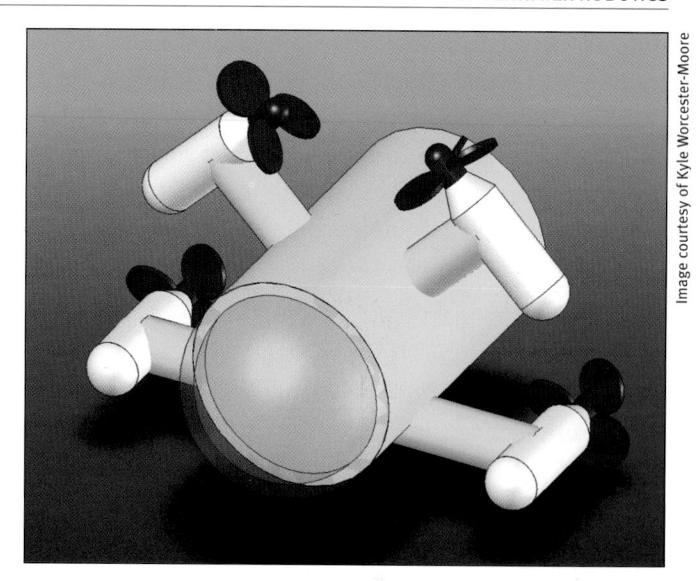

Image courtesy of Kyle Worcester-Moore

TECH NOTE: VECTORED THRUSTERS

These two diagrams illustrate how vectored thrusters interact to produce a wide range of vehicle motions. Each diagram shows the ROV from Figure 7.47 as it would appear when viewed from the stern. In the diagram on the left, both of the vectored thrusters are generating force diagonally upward and toward the midline, as represented by the heavy black arrows. Each of these thrust vectors is equivalent to the combination of a vertical force (blue arrow) and a horizontal force (red arrow). In the left hand diagram, the two red horizontal thrust components are equal in magnitude but opposite in direction, so they cancel, meaning the vehicle does not move sideways. However, the blue vertical thrust components are both directed upward and cooperate to propel the vehicle toward the surface. Note that if both of these thrusters now reversed direction, both red arrows would be directed outward (and would again cancel), and both blue arrows would be directed downward, so the vehicle would dive. Thus, this arrangement allows straight up and straight down motion.

In the diagram on the right, the port thruster is creating a propulsive force downward and to the left, while the starboard thruster is creating a force directed upward and to the left. In this case, it is the vertical (blue) thrust components that cancel, so the vehicle does not move vertically. Instead, this time it is the horizontal (red) components that cooperate, so the vehicle is pushed sideways in a crabwalk to the left. If both of these thrusters now reversed their directions, the blue arrows would again cancel, and the red arrows would cooperate in the opposite direction, so the vehicle would crabwalk to the right. It's worth noting that thrusters operating as shown in the rightmost diagram are also likely to roll an unstable vehicle; however, if the vehicle is very stable, the extent of rolling should be small and not present a serious problem.

If each thruster offers independent speed control, then the pair of vectored thrusters can generate even more movements than up, down, and sideways. For example, if the starboard thruster in the left diagram were throttled back to 50 percent of full speed, then both blue arrows would still cooperate and together provide 150 percent of one thruster's vertical thrust upward, but now the red arrow on the right would be only half as strong and would cancel only half of the red arrow on the left. Thus, a 50 percent rightward thrust would be superimposed on the otherwise vertical climb of the vehicle, resulting in an angled climb, mostly upward, but also somewhat to the right. By precisely controlling the relative speed and direction of these two vectored thrusters, the craft can actually be made to move in any direction within that vertical plane, making vectored thrusters extremely useful. More advanced ROVs often use several pairs of vectored thrusters for sophisticated maneuvering, as shown in Figures 7.49 and 7.50.

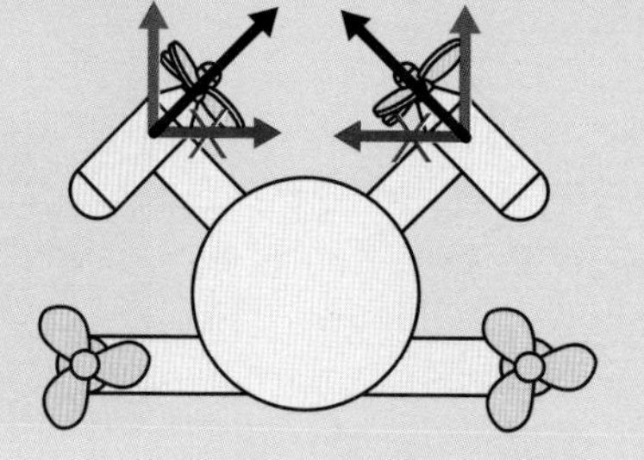

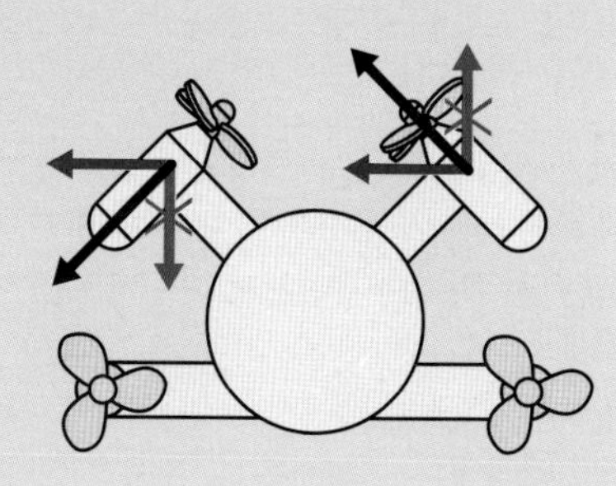

Figure 7.48: How Vectored Thrusters Work

the hull. With this new thruster configuration, the vehicle can move straight sideways (right or left) without first having to turn itself in the direction of travel. This type of sideways motion (sometimes known as a "crabwalk") is extremely helpful, if not essential, for many inspection and tool-use tasks, because it allows the vehicle to move laterally while cameras, payload tools, and other devices attached to the front of the vehicle remain pointed forward.

As discussed earlier, putting thruster ducts right through the middle of the vehicle can be problematic, so the design shown in Figure 7.46 may not be an option for you. Fortunately, another arrangement, based on something called vectored thrusters, avoids this through-hull problem completely, yet preserves all the maneuverability of the through-hull arrangement. Figure 7.47 shows an example using vectored thrusters.

In this arrangement, two thrusters have been mounted at an angle, so each points somewhat vertically and somewhat horizontally at the same time. In this configuration, thrust from the two thrusters can be made to interact in ways that produce a useful range of vertical and horizontal forces to move the vehicle up, down, sideways, or to any angle in between. (See *Tech Note: Vectored Thrusters.*)

Image courtesy of SAAB Seaeye Ltd.

Figure 7.49: Four Vectored Horizontal Thrusters

The Falcon ROV *by SAAB Seaeye features four vectored horizontal thrusters. The position and angle of these four thrusters is clearly visible in this photograph of the vehicle's underside, taken from below and behind the vehicle.*

8.3. Analysis of a More Elaborate Thruster Arrangement

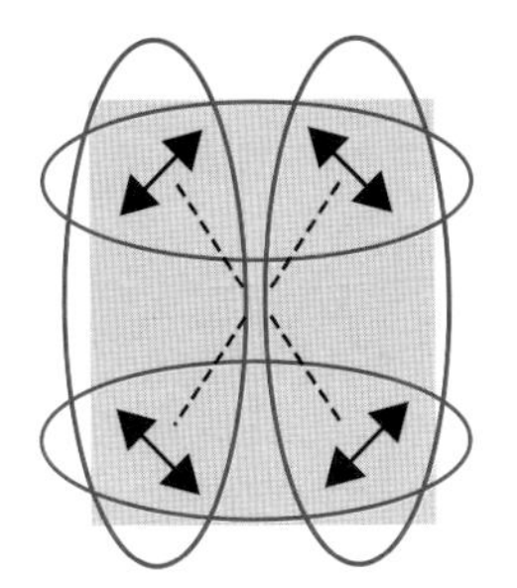
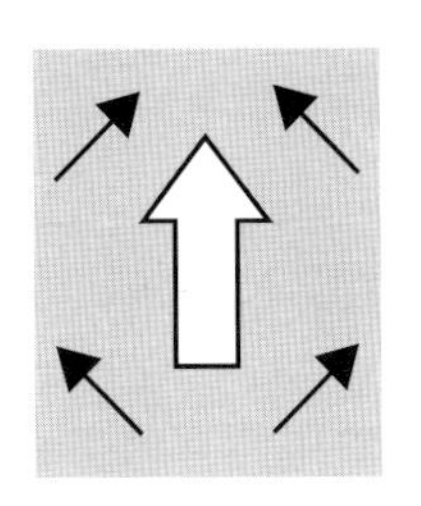
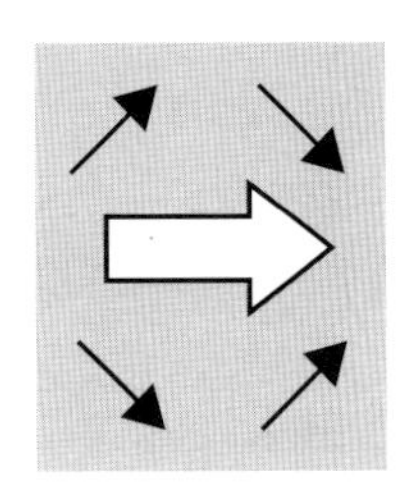

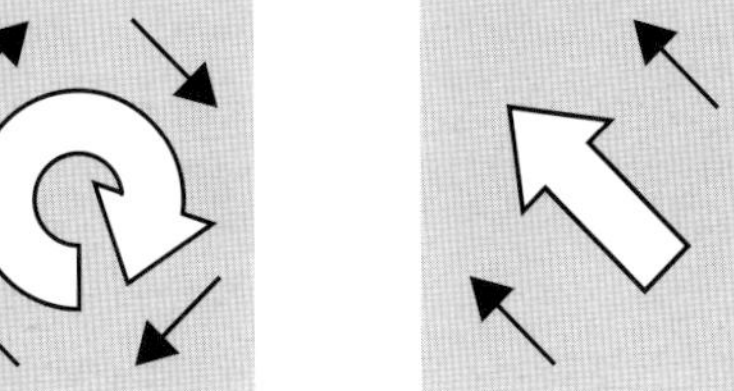

Figure 7.50: A Versatile Arrangement of Four Horizontal Thrusters

This diagram shows four horizontal thrusters as they are commonly arranged in an advanced ROV, as viewed from above. These four thrusters are placed near the corners of the ROV and oriented so that their thrust directions (small, solid arrows) each point roughly perpendicular to a line between the thruster and the center of the vehicle (dashed line). This creates four pairs of vectored thrusters (red ovals). Each vectored pair can push its edge of the vehicle in any direction.

By mixing different combinations of speed and direction of these four thrusters, a tremendous range of vehicle motions in the horizontal plane is possible. In the four simple examples shown, the larger arrows indicate the overall vehicle behavior in response to the combined effect of the individual thrusters.

Work class ROVs commonly have six or more thrusters arranged in what might at first seem like strange positions and angles. Strange, that is, until you recognize that they are configured quite sensibly to produce a number of interacting vectored thruster pairs. These provide extreme maneuverability and good control. They also provide some level of redundancy, so that basic maneuvers are still possible, even if one thruster fails.

For example, four horizontal thrusters are often arranged with one at each corner pointing diagonally rather than straight ahead or straight sideways. This can create *four pairs* of vectored thrusters (one pair in front, one at the back, and one on each side), which can be used alone or in combination to coordinate a truly impressive range of motion. (See Figure 7.50.) If the four horizontal thrusters in the corners have independent speed and direction control (as they usually do), they can be used to move the ROV in *any* horizontal direction, *with or without* simultaneously changing the direction the ROV is facing. The ROV can also do a controlled spin or turn while sliding sideways or even following a broad arc in any direction. This same thruster arrangement can be used to hold the ROV stationary in a current with the vehicle facing any direction, no matter which direction the current is coming from. As you can see, this is an extremely versatile thruster arrangement.

Taming this level of maneuverability to make it manageable and useful isn't easy. It usually requires sophisticated navigational instruments and custom computer software to translate intuitive pilot commands into appropriate combinations of thruster speed and direction. In commercial operations, it's often worth the extra effort and cost, because it gives the pilot enough control to get the job done reliably, even when the scenario requires precise positioning of cameras, tools, or other payloads in the presence of strong currents, tether drag, and other "real world" challenges.

A number of school groups have successfully designed, built, and operated low-cost vehicles using advanced thruster arrangements, including one or more pairs of vectored thrusters. For people with some prior ROV or AUV design experience, taking a design to this level can be a fun and rewarding challenge. The second half of *Chapter 9: Control and Navigation* introduces some of the more sophisticated electronics and programming skills needed to coordinate the action of multiple vectored thrusters for smooth control. However, if you are working on your first ROV or AUV design, you would do well to begin with a simpler design that does *not* rely on vectored thrusters.

8.4. A Checklist for Thruster Placement

Here is a checklist of issues to consider as you design your thruster placement. Some review issues have been discussed already; others bring up new points to consider.

TECH NOTE: MEASURING BOLLARD PULL

Bollard pull (also known as bollard thrust) describes how hard a vessel can pull against a stationary object. It's an important measure of performance for slow, but powerful craft such as tugboats, which must push and pull large ships, and ROVs, which must continually "tow" their long tethers through the water.

After you get your vehicle assembled and into the swimming pool for some test runs, it's fun (and informative) to measure its bollard pull. You can make this measurement easily using the same string and spring scale you may have used for your empirical drag measurements. Just tie a long string between your ROV and a spring scale, as shown in Figure 7.51. Then use the scale to measure how hard the ROV can pull on the string when you give it full throttle. If you have two horizontal thrusters running at 100% power, you might expect the vehicle's bollard pull to be twice the thrust of a single thruster (which you probably measured earlier with a thruster test jig); however, limitations in the electrical system and/or the presence of frame elements that partially block thruster output can siphon off energy and give you a bollard pull that's less than what you'd expect. If it's way less, there's probably a problem. See if you can figure out what's wrong and come up with a way to resolve the problem.

Image courtesy of Dr. Steven W. Moore

Figure 7.51: Students Measuring ROV's Bollard Pull

- **Obstacles to Flow**: Thrusters create force to move a vehicle by accelerating water in the opposite direction. Any part of the vehicle that blocks the stream of high-velocity water exiting a propeller will reduce the effective thrust—sometimes dramatically. If you plan to drive thrusters in both forward and reverse, try to leave clear pathways on both ends of the thruster. With most thrusters, the motor housing will block part of the reverse thrust, resulting in some thrust loss, but this is largely unavoidable in a conventional thruster. (The hubless propeller mentioned earlier largely avoids this problem, but entails much greater expense.) Thrust loss can be reduced by streamlining the motor housing and by giving the prop a longer shaft to move it a bit farther from the motor housing. This approach has its limits, though, because a longer prop shaft is more prone to bending and vibration.
- **Turning Moments**: It's easy to create turning moments (torques) that can cause a vehicle to pitch, roll, or yaw, even when you don't intend to do so. Anytime you have two thrusters producing thrust along different lines of action, even if those lines are parallel to each other, there is the potential to create a couple and a turning moment. In addition, anytime a single thruster acts along a line that does not pass through the center of the vehicle, the result is a turning moment. Remember that these turning moments are not always a bad thing—in fact, they're essential for making a vehicle turn when you want it to! The point is to be conscious of the turning moments associated with every thruster placement, so they won't catch you by surprise; otherwise, you may end up generating unexpected pitch, roll, or yaw.
- **Thrust Asymmetry:** As mentioned earlier in this chapter, propeller blade shapes are optimized for efficiency in one rotational direction (forward) and don't work as well in the other direction (reverse). In addition, the motor housing typically blocks a significant fraction of the already diminished reverse thrust produced by the propeller. Finally, many DC electric motors are likewise optimized for rotation in one direction, even though they can usually be driven in either direction. For all of these reasons, most thrusters produce stronger thrust in one direction than in the opposite direction. It can be helpful to orient your thrusters so that their most powerful and efficient direction is the direction needed most often and/or the direction needing the most thrust. For example, a flying eyeball ROV that's trimmed for slight positive buoyancy might want its strongest vertical thrust downward, to overcome the positive bias in the buoyancy. On the other hand, an ROV designed to lift heavy rock samples from the seafloor might want to orient vertical thrusters so their strongest thrust is upward, for heavy lifting.

TECH NOTE: THRUSTER PLACEMENT PROBLEMS WITH PITCH AND ROLL

Here's an example of how the improper placement of thrusters can generate a pitch and roll problem. Because thrusters are usually denser than water, most ROV thrusters are mounted low on the vehicle's frame, to lower CG and increase vehicle stability. That's fine, except that this placement causes the vehicle to pitch in a nose-up direction (essentially "popping a wheelie") whenever the vehicle moves forward and to pitch nose-down whenever the vehicle backs up. This can be very frustrating if, for example, you try to drive straight forward to pick up something with a manipulator arm, only to discover that the induced pitch lifts the arm up and over the target item. When you back up for a second attempt, the manipulator swings down hard, crashing into and possibly damaging the object and/or the arm. Advanced work class ROVs make use of additional thrusters and sophisticated computer algorithms to anticipate and counteract this pitch artifact, but this sophisticated level of control is not something to attempt on your first ROV project.

- **Prop Torque**: When a vehicle tries to spin a prop in one direction, the prop tries to spin the vehicle in the opposite direction. (Newton's third law strikes again!) Provided the ROV has a large BG, it can resist this torque easily, and only the prop rotates enough to notice. However, a slender, torpedo-shaped ROV or AUV with a small BG can roll quite far—maybe even flip over or corkscrew through the water—whenever a drive thruster is switched on. Some designs incorporate two counter-rotating props or special control surfaces to solve this problem.
- **Prop Walk**: In addition to creating strong thrust along the axis of the propeller (i.e., parallel to the prop shaft), a spinning prop can create a slight sideways force. This effect, called "prop walk," is most pronounced when the flow of water through the prop is not parallel to the prop shaft or when the prop is very near the water surface. The effect is often more severe with high-pitch props. Though this is a significant issue for some boats, propwalk is usually negligible in ROVs and AUVs. Using two thrusters in parallel, with a right-handed prop on one side and a left-handed prop on the other, should eliminate this problem.
- **Power Consumption**: While having more powerful thrusters or adding more thrusters can provide more total thrust and may increase the number of possible thruster configurations for greater maneuverability, the additional power demand comes with costs. It may require upgrading to a larger (usually more expensive) power supply and maybe a modified tether. Larger power supplies may require additional cooling systems. If batteries are supplying the power, a larger power drain also forces a choice between bigger batteries or shorter run-times. Be sure to carefully weigh the advantages of more thrust against the costs of more power.
- **Thruster Density**: Thrusters tend to be heavy and denser than water, so their placement can strongly influence the vehicle's center of gravity and therefore its stability. This argues for placing thrusters low on the vehicle whenever possible (but see next item).
- **Prop Damage and Reduced Visibility**: Mounting thrusters low on the vehicle helps with stability, but there are drawbacks to doing this, too, particularly with vehicles that operate on or very near the bottom in natural bodies of water. First, unducted propellers may be damaged if they strike rocks or other hard objects on the bottom. Second, the water currents generated by props can suck debris off the bottom and into the propellers, potentially entangling them in kelp, seagrasses, rope, fishing line, fishing nets, or other materials. Third, the high-velocity water expelled from the propellers can easily stir up sand or silt, severely reducing visibility and hampering navigation.
- **Thruster Spacing for Fast Turns**: How far a thruster is placed from the vehicle's center of rotation has a large impact on the speed of turns and is an important consideration if you may need to make fast turns (as in underwater robot battles).

TECH NOTE: ADJUSTABLE THRUSTER CONFIGURATIONS

Another interesting option is to experiment with systems for changing thruster placement (or its equivalent) on the fly. For example, a vertical thruster could be used to push a neutrally buoyant ROV under water, then that thruster could be rotated 90 degrees to push the ROV horizontally through the water, then rotated 90 degrees again to drive the ROV back to the surface. You might also want to experiment with adjustable nozzles or rudders to redirect the flow of water from pumps or propellers. The possibilities are nearly endless!

If the thruster is placed too close to the center of rotation, its moment arm is short and the torque generated will have trouble overcoming vehicle inertia and drag. If the thruster is placed too far from the center of rotation, vehicle rotation speed will be limited by how quickly the thruster can pull itself through the water. For small, low-budget vehicles, the optimal distance is best determined experimentally; therefore, some provision for adjusting that distance easily during early in-water trials should be built into the design of the vehicle.

- **Impacts of Thrusters on Nearby Circuits and Sensors**: Navigational compasses and many other electronic sensors can be fooled by the strong electromagnetic fields generated by electric motors and by the heavy electrical currents coursing through their supply wires. Therefore, great care should be taken to separate thrusters (and all other motors) and motor wiring from compasses and other components that may be sensitive to these electromagnetic fields. These fields typically decrease rapidly with distance, so a little separation goes a long way. As a general guideline, assume an inverse square law—that means that doubling the distance from 1 centimeter to 2 centimeters will reduce the disturbance by a factor of 2-squared, or 4. Increasing the distance from 1 centimeter to 10 centimeters (a factor of 10) will reduce the disturbance by a factor of 10-squared, or 100.

- **Cost**: Thrusters and their controllers can cost big money. If you choose to spend more money on these systems, then you will be left with less to spend on other parts of your vehicle. Additional thrusters also add weight, place greater demands on power circuits, and may cause other problems. While it's fun to fantasize about what could be done with 17 super-charged thrusters, it's also a good challenge to see how well you can maximize maneuverability and performance in a project that can afford only three or four thrusters.

- **Multiple Correct Answers**: Optimal thruster placement depends on a large number of interacting factors, including the vehicle's mission, environment, and budget. Accordingly, there is no one best answer for thruster arrangements. Combine what you have learned in this chapter with creative experimentation and additional research (including conversations with others who have built underwater vehicles) to find out what would work best for your particular mission and situation.

9. Chapter Summary

Underwater vehicle motion is determined by a five-way interaction among five types of force: drag, weight, buoyancy, thrust, and lift. Newton's laws of motion can be used to understand how these forces interact to cause and control underwater vehicle movements.

Most small ROVs and AUVs are designed to be neutrally buoyant (so weight and buoyant forces cancel) and slow-moving (so lift is negligible). Therefore, this chapter focuses on the remaining two force categories, specifically the production of *thrust* needed to overcome vehicle *drag*.

The amount of drag depends on vehicle size, shape, and speed and can be estimated theoretically or measured empirically to determine how much thrust your vehicle needs to accomplish its mission.

Although there are many ways to generate thrust, most small ROVs and AUVs rely on propulsive devices called thrusters made by attaching a propeller to a waterproofed electric motor. There are many types of electric motors and many types of propellers. Each has advantages and disadvantages, so the choice of appropriate motors and propellers involves careful consideration of many factors. Once some tentative selections have been made, a thruster test jig can be used to identify propeller/motor combinations that offer particularly good thrust and efficiency.

The number and placement of thrusters on an underwater vehicle are extremely important. They determine the range of possible maneuvers and strongly affect both the performance and the efficiency of the vehicle. Typical configurations range from simple three-thruster arrangements to complex arrangements with six or more thrusters arranged in vectored pairs for advanced maneuverability and control.

Armed with the technical knowledge from this chapter, you should be able to take your vehicle farther and faster, maybe even adding a twist or two. Of course, if you want to know how to power and control those twists, you'll have to move on to the next two chapters.

Chapter 8

Power Systems

Chapter 8: **Power Systems**

Stories From Real Life: The *Hunley* and Nuclear Subs

Chapter Outline

Figure 8.1.cover: Battery Powered Robots

This team has their power system figured out and is ready to launch. There's the robot (on the left of the photo), the power source (the battery on the right), and the power delivery mechanism (the tether in the middle).

Image courtesy of Randall Fox

Chapter Learning Outcomes

- Estimate the total amount of energy (in joules and watt-hours) and the maximum instantaneous power (in watts) that your vehicle will require to complete its mission.
- Explain why batteries are a good power source for small underwater vehicles, and select an appropriate battery (or combination of batteries) to supply the energy and power requirements of your vehicle.
- Describe options for placing batteries (on board the vehicle or on the surface), and explain how you would distribute electrical power from those batteries to the parts of your vehicle that need it.
- Describe the important roles of fuses and voltage regulators in power distribution systems.

STORIES FROM REAL LIFE: The *Hunley* and Nuclear Subs

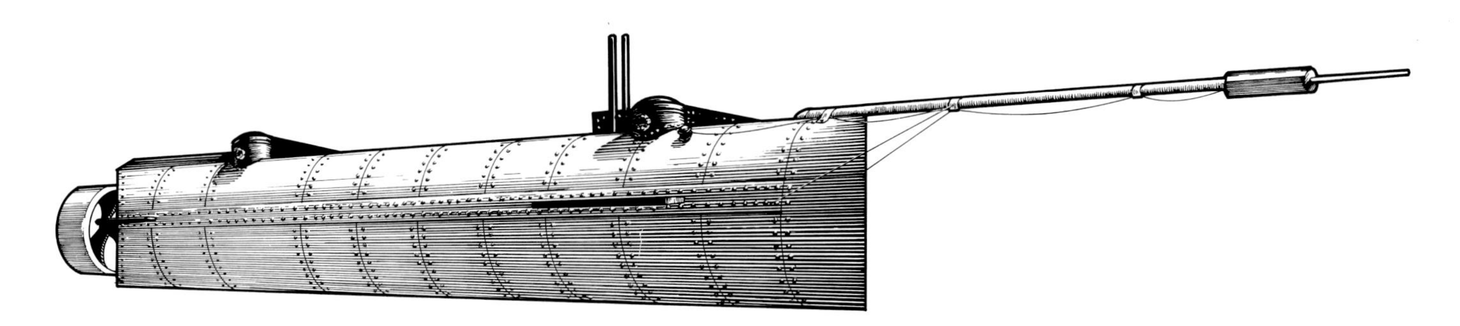

For very early submarines, the only effective driving force was human muscle power. Then as time and technology advanced, the power sources for subsea vehicles evolved from hand-cranked propellers to engines using steam, compressed air, gasoline, diesel, and finally nuclear fuel.

This amazing progression becomes all the more graphic when you picture eight sweaty men huddled shoulder-to-shoulder in a dark, dank hull, turning a crankshaft in order to propel the Confederate sub H.L. Hunley. *Fast forward almost a hundred years and compare that image with the first nuclear submarine, the USS* Nautilus, *which travelled under water for more than 100,000 miles, powered by a nuclear mass the size of a baseball.*

The quest for the ultimate secret weapon in war has often advanced the cause of naval vessel design. This was especially the case in the U.S. Civil War (1861–1865) which proved to be the crucible for a number of submarine inventions. Also known as the War Between the States, the South (Confederate side) was desperate for some way to break the North's (Union side) naval blockades of key southern ports; this tactic was effectively crippling the South's ability to receive war supplies from abroad. Lacking the funding to build a large naval fleet of its own, the South resorted to small, fast torpedo ships that rode extremely low in the water. By running awash at the surface, they offered a minimal target when attacking larger Union ships under cover of darkness. The stealthy craft attempted to sink or damage an enemy vessel by attaching an explosive spar torpedo to its hull, then detonating the device from a distance with a pull cord. Aptly enough, these small, ironclad *David* class vessels took their name from the David and Goliath story. While these Confederate boats were never all that successful, the idea of their stealthy, semi-submersible approach held huge appeal for naval designers. Some even dreamed of a boat that might submerge completely!

A number of inventors—some patriots, some opportunists, some both—attempted workable models powered by human propulsion for both the North and the South. Desperate for a strategic advantage in the war, each side considered submarine craft despite the terrible reputations some gained after the drowning deaths of the crews who powered them. None of these historic vessels were true submarines—in fact, they ran at the surface, submerging only when required by relying on crude dive planes and all the muscle power their straining crews could produce. As was the case with most of these early prototypes or craft, subsequent naval tests produced less reliable results than had been touted, so the projects either sank or were abandoned.

Of all these hand-powered submarines, the one that achieved the greatest historic fame was the *H.L. Hunley*. Launched in 1862, this Confederate submarine is best known for its successful attack on the Union's *Housatonic* on February 17, 1864. The *Hunley*'s crew of eight, under the command of Lieutenant George Dixon, used a hand crank to propel the sub at speeds reportedly up to 4.6 miles per hour (7.4 kph). Operating in the dark, the submarine was able to ram the enemy ship and attach a mine carried at the end of sub's bowsprit. Then it backed off and pulled the long detonation rope. When the 16-gun Union steam sloop blew up, the *Hunley* became the first submarine ever to sink an enemy ship during war.

Triumphant, the small sub surfaced only long enough to signal shore, then disappeared! None of the sub's crew survived to claim their fame. What actually happened to the *Hunley* that historic night became a mystery that motivated treasure hunters, mercenaries, and explorers for the next 137 years. The mystery remained unsolved until 1995, when author Clive Cussler organized and funded an underwater search that located the *Hunley*'s hull buried deep in the silt, not far from where the *Housatonic* sank.

Since then, the famous hull has been raised in a massive recovery operation, finally allowing for proper burial of her crew. It also gave scientists and marine historians the opportunity to discover a number of design innovations that indicate the *Hunley* was far more advanced than the crude sub many had supposed. For example, internal frames running inside the submarine provided lateral strength. The iron plates on the hull did not overlap, as previously thought; they were fastened together with backing plates and rivets. These rivets were driven from the inside, not the outside, and the protruding parts were then pounded down and smoothed, to minimize drag. These extra innovations were all done in order to streamline the hull. The propeller shaft and steering cables were wisely housed inside protective hollow tubes, which may also have contributed to watertightness. Other aspects of hull construction, as well as two rectangular iron ballast blocks in the rear of the sub, all indicate that the *Hunley*'s builders took a knowledgeable and creative approach to balance and design. What *Hunley* and other early submarines lacked was not ingenuity, but power.

During the Industrial Revolution, steam engines quickly revolutionized industrial machinery as well as travel by rail and surface ship. But adapting steam to power a suitable engine for submarines was still years away. The French came close with the lengthy *Le Plongeur*, powered by an 80 hp engine that ran on vast quantities of compressed air. Despite the sub's technological advances, it could not maintain underwater stability; instead it dived and then surfaced, repeatedly and uncontrollably, more like a playful dolphin than any war craft.

However, it was only a matter of time before engines were adapted—more or less successfully—to working under water. When small gas and steam engines proved unsatisfactory under water, inventors like John Holland turned to internal combustion motors for surface propulsion *and* for charging a sub's accumulator batteries. These batteries powered the electric motors that turned the sub's propellers when submerged. This reliable power system fostered the development of submarine fleets around the globe. Then in 1904, the French Navy's construction of the sub *Aigrette* saw more efficient diesel engines replacing gasoline engines. Diesel engines would remain the "standard" submarine power plant for the next fifty years.

The increased speed of these new power systems translated to increased stability and maneuverability for submarines. But there were still serious issues to be resolved—noise, fuel consumption, running at the surface in order to recharge batteries, mechanical breakdowns, marginal living conditions at sea, and torpedoes that routinely refused to do what they were supposed to.

Not surprisingly, World War I and particularly World War II served to advance submarine design. The German *U-boats* gained notoriety around the world for their effectiveness. During World War II, Nazi subs, hunting in "wolf packs," came close to altering the course of world history. Between the wars, countries around the world, including Russia and Japan, worked to build impressive fleets of submarines, but none could close the gap with Germany.

By the close of World War II in 1945, there were thousands of submarines around the globe. In terms of power, maneuverability, and reliability, all were a far cry from the early subs of the American Civil War. However, it can be argued that even these sophisticated submarines were really still surface vessels that could submerge for only limited periods of time. In fact, there was still one submarine to come—the nuclear-powered *Nautilus*—that would rewrite subsea history.

The early dream of subs as true stealth craft was finally actualized largely through the vision and bull-headedness of one man—Admiral Hyman George Rickover. Even as a young naval officer, Rickover was considered a bit of a fanatic; it was a trait that initially threatened to stunt his career. As World War II segued into the Cold War, Rickover took over representing the Navy in a joint military-civilian task force—their assignment was to find practical uses for atomic energy. Knowing full well the Navy's gluttonous use of power, Rickover quickly saw that controlled nuclear reactions could be used to produce a steady supply of electricity, a fact that could transform the Navy *and* submarine propulsion.

He felt that the fission process was the perfect propulsion scheme for a submarine—it consumed very small amounts of uranium and involved no combustion, hence required no air. Rickover pushed the idea of atomic power to anyone with political or military influence. After initial rejection, he was given command of a new nuclear department at the Bureau of Ships, and then became head of a Naval Reactors Branch. Now almost like a man possessed by his mission, Rickover pressed Westinghouse to develop a nuclear power plant small enough to fit into a submarine. Moreover, the company had to do it quickly, and it had to be perfect. The Electric Boat Company was tasked with its own challenge, building the first nuclear submarine—*Nautilus*.

The 320-foot (approx. 98-m) submarine was launched on January 21, 1954, within the eight-year time frame Rickover had promised. Suddenly, the United States had the most potent and deadly weapon afloat, one that significantly impacted naval tactics and strategies. Although the craft wasn't perfect—its cooling pumps were audible 10 miles away—the submarine quickly established new underwater performance records. *Nautilus* averaged 19.1 knots on one voyage, an unheard-of speed for conventional subs. And it became the first submarine to transit under the polar ice cap, traveling 1,830 miles (approx. 2,950 km). Rickover became a hero as subsequent nuclear-powered submarines rolled down the ways, replicating and then besting *Nautilus*' accomplishments.

Like *Hunley*, *Nautilus* deserved to capture the historic spotlight. The basis for its claim to fame was a revolutionary new propulsive force—nuclear power—which allowed a dramatic increase in range and operational flexibility. As a result, *Nautilus* also became the first submarine whose power system allowed it to inhabit—not just visit—the depths.

1. Introduction

Figure 8.2: Decoding the Mystery of Electrical Power

Image courtesy of Steve Van Meter, VideoRay LLC

When a car runs out of gas, it stops. When the battery in a portable radio dies, the music stops. When an animal runs out of food, it starves to death and stops living. These things stop because they have run out of something called **energy**. Machines, living things, and even non-living processes like wind and ocean currents all require some source of energy to keep going. Underwater vehicles are no exception.

To be useful, energy must not only be available, but also available at an adequate *rate*. Even a trainload of rice wouldn't keep you from starving if you were allowed to eat only one grain of that rice each day. You'd be getting a regular supply of energy, but not quickly enough to keep you alive. The rate at which energy is delivered, converted, or used is called **power**. Like every other machine, living thing, or physical process, underwater vehicles need to have their energy supplied at an adequate rate—they require adequate power.

This chapter covers the energy and power requirements of underwater vehicles and explores ways to supply enough of both, so that vehicles can complete their missions successfully.

This chapter also reviews some of the systems that have been used historically to supply underwater vehicles with power and discusses which of these approaches are best suited for powering small ROVs and AUVs. For a variety of reasons, almost all small underwater vehicles used today are powered by electricity, so this chapter includes an introduction to basic electricity concepts you'll need to know to design electrical power systems for these vehicles. Batteries are discussed in detail, because they are relatively safe, convenient, and economical power sources for hand-built ROV and AUV projects. Many additional electrical concepts, including how to regulate the amount of electrical power flowing to various vehicle subsystems, are covered later in *Chapter 9: Control and Navigation.*

The final portion of the present chapter details methods of transmitting and distributing electrical power to the various subsystems of a small underwater vehicle. This is no trivial matter, particularly if power must be delivered to the vehicle via a long tether, or if the vehicle's different subsystems require different voltages. The chapter concludes with summary descriptions of electrical power distribution systems for three different ROVs. These examples can be used as starting points for your own ROV or AUV power system designs.

HISTORIC HIGHLIGHT: EVOLUTION OF UNDERWATER POWER SYSTEMS

Over the years, underwater craft have been powered in a fascinating variety of ways.

The earliest vehicles, like the *Turtle* (1776), *Nautilus* (1800), *Hunley* (1860s), and *Sub Marine Explorer* (1865), were propelled by the muscles of their human crew. One or more people inside would paddle with oars or turn propellers, using cranks, pedals, or treadmills. However, the speed, range, and other capabilities of the earliest craft like *Turtle* and *Nautilus* were severely limited by humans' need for a steady supply of oxygen and by the limited power output they could generate, particularly when compared to the power needed to move a pressure hull that was large enough to enclose and protect them.

In the 1860s, inventors turned to the use of gear linkages to increase speed using the same amount of sweaty effort. Much like a modern 10-speed bicycle, these gears allowed more efficient and effective use of human muscle power. A classic example is the 11 meter- (approx. 36 ft) long *Sub Marine Explorer*, designed and built by Julius Kroehl between 1863 and 1866. Based on pre-construction plans and archaeological reconstruction, the early sub featured a flywheel that one of the crew members could crank every few seconds to generate a speed of 3 to 4 knots. The reliance on gears to amplify human effort was soon superseded as inventors turned to mechanical engines to power their submarines

Initially, inventors toyed with small gas and steam engines, but neither proved satisfactory. The answer was electric-powered submarines, and it came from four inventors working in four countries—Josiah Tuck with his *Peacemaker* in America; J. F. Waddington and his *Porpoise* in England; France's Claude Goubet, with his namesake, *Goubet I*; and the Spaniard Isaac Peral, whose design featured an electric motor powered by accumulator batteries. Their power system ideas inspired many subsequent submarine designs.

Then in the late nineteenth-century, the Irish-American inventor John P. Holland successfully introduced internal combustion engines to submarines, particularly the concept of using those types of engines to run on the surface and to charge batteries that would power the sub while submerged. The development of submarines accelerated as navies around the globe adopted and adapted Holland's designs or pursued their own. In 1904, with the French Navy's construction of the sub *Aigrette*, gasoline engines were replaced with more efficient diesel engines, introducing what would become and remain the "standard" submarine power plant.

Image #NH46756, USS *H-5*, courtesy of the Naval Historical Center

Figure 8.3: Engine Room of WWI Diesel Submarine, circa 1919

By the early to mid-1900s (notably during World War I and II), the submarine fleets of most nations ran on diesel engines and electric motors. Electric motors were relatively quiet and did not require air, so they were ideal for secret underwater military operations.

The motors drew their power from electric batteries. Unfortunately, batteries alone could power a sub for only a few hours before they needed to be recharged. Early on, the best way of recharging them was to use an electric generator powered by a diesel engine, but these engines required a steady supply of air. Consequently, the batteries could be recharged only while the sub was on the surface or at least close enough to the surface to use a snorkel for air. These engines also generated a lot of noise. Running a noisy engine while close to the surface made the sub easier to detect, limiting its wartime effectiveness as a stealth vehicle.

Because of these limitations, a great deal of military research went into inventing truly air-independent propulsion (AIP) systems for submarines, torpedoes, and other underwater vehicles. One approach was to use liquid oxygen (or other chemical oxidants), stored in tanks on board the vehicle, instead of using oxygen from the air to burn diesel fuel or other conventional fuels. The heat generated by these reactions could be used to drive pistons in a Sterling engine (a type of engine that requires heat but not air) or could be used to boil water to make pressurized steam, which could then be used to drive a steam turbine. Both of these systems could drive a propeller directly or drive a generator that recharged batteries.

In 1954, a new era in submarine propulsion began with the commissioning of the USS *Nautilus*, the world's first nuclear-powered submarine. Nuclear power provides an alternative way to produce enough heat to boil water and drive a turbine, but, unlike a conventional steam turbine, it does not require air or any other source of oxygen. Furthermore, it takes advantage of the phenomenal energy density of matter ($E = mc^2$), which packs a practically unlimited supply of energy into a very small volume of fuel. The power output of the reactors is sufficient to propel a sub at high speed almost indefinitely. With modern technology, power supplied by these nuclear reactors can even be used to convert seawater into a steady supply of oxygen and fresh drinking water for the crew while the sub is submerged and under way. This enables a modern nuclear sub to cruise around under water at respectable speeds without surfacing for many months at a time. Thus, a nuclear submarine can even cross under the Arctic ice and quickly move undetected into almost any corner of the world's oceans.

In recent years, there has been a growing emphasis on smaller unmanned vehicles for many underwater missions. For the most part, these ROVs and AUVs rely on electrical power; however, there is considerable variation in how that electrical power is supplied and how it is used. Work class ROVs often store their energy as diesel fuel in tanks on a ship or oil platform. This fuel is used to power a surface generator to make electricity that is then sent down the umbilical/tether to the ROV. Once at the ROV, the electrical current may be channeled to power motors, lights, cameras, etc. Or it may run a hydraulic pump that produces hydraulic power for hydraulic thruster motors, manipulator arms, or other mechanical systems.

Most AUVs are totally electric, carrying their own batteries for power. A few particularly innovative ones draw most of their power directly from the environment. For example, the *SLOCUM* glider AUV extracts propulsive energy from naturally occurring temperature gradients that exist throughout temperate and tropical seas. It is capable of traveling tens of thousands of kilometers without recharging. It does this by using battery power only for navigation and minor steering corrections, not for basic propulsion.

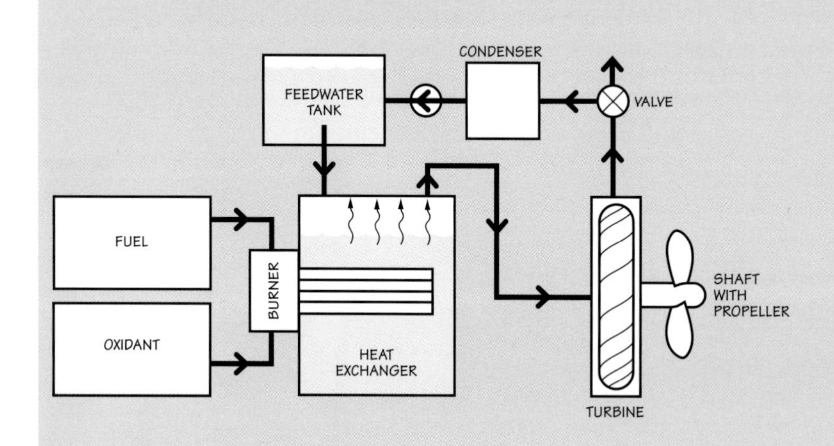

Figure 8.4: Simplified Schematic of a Steam Turbine

In this simplified schematic of a steam turbine, the fuel and oxidant are used by the burner to produce heat. This heat is transferred to the exchanger. Steam is produced, which is injected into the turbine at high pressure. This spins the turbine, which causes the shaft and propeller to rotate. Excess steam is either shunted directly overboard or recycled through the condenser. The condenser cools the steam and returns it to the feedwater tank as water.

2. Energy, Power, and Efficiency

2.1. Energy

Energy is that enigmatic, invisible *something* that runs the universe. It's the capacity to do things or to change things. Technically, energy is often defined as "the ability to do work," but this work isn't just any old work—it's defined in very precise physical or chemical terms for specific situations. For example, in mechanical situations, work can be equated with a force multiplied by a distance. So if you push a heavy desk across the floor, the work you have done is defined as the force with which you pushed horizontally on the desk multiplied by the distance you moved the desk across the floor. On the other hand, if you work up a sweat struggling to move a desk that's too heavy to budge, then technically you haven't done any work at all, because you covered no distance!

In spite of its abstract and sometimes counterintuitive nature, the concept of energy turns out to be a profoundly useful one. It helps scientists and engineers think about everything from the origins of life in the universe to the best way of fixing a clogged kitchen sink. In fact, energy is arguably the most important concept across all of the physical and life sciences, as well as in engineering and other disciplines where scientific knowledge is applied.

Energy sustains life, but energy can also end life. This chapter discusses energy sources, storage methods, and transmission methods that can potentially injure or even kill you if used improperly. Therefore, **pay extra attention to the Safety Notes throughout this chapter**. For example, school-based ROV and AUV projects, by their very nature, put young and often inexperienced learners at great risk of accidentally mixing electricity with water. If appropriate safety measures are taken, the risk of injury when this (inevitably) happens is essentially zero, so you can have fun and not worry. However, if basic electrical safety precautions are not learned and/or followed, then mixing electricity with water can result in death by electrocution.

TECH NOTE: MASS AS A FORM OF ENERGY

In everyday experience, mass and energy seem very different. However, in the year 1905, Albert Einstein published a now-famous paper in which he derived a mathematical result showing that mass can be equated with more conventional forms of energy. The equation he presented was (essentially): E = mc2. This tiny equation changed the world forever.

Though simple, it is mind-boggling in its implications. The “E” in the equation represents energy. The “m” represents mass. The “c” represents the speed of light, which is a huge number, since light is extremely fast (300,000,000 meters per second or about 186,000 miles per second). Squaring the speed of light therefore results in a humongous number, so what Einstein’s equation tells us is that even the tiniest speck of matter contains a frighteningly huge amount of energy.

In normal situations, mass does not readily change into other forms of energy, at least not on a noticeable scale. However, intentional conversion of mass to other forms of energy is possible with nuclear technology. This fact was made vividly and horrifyingly clear to the world—and changed the course of history forever—on August 6, 1945, when the United States converted less than 1 kilogram (approx. 2 lbs) of mass into heat energy in the sky above Hiroshima, Japan. The resulting atomic blast leveled most of this large city and killed an estimated 140,000 people. A second atomic bomb was dropped on the city of Nagasaki three days later. These first (and so far only) atomic weapons ever used against a human population brought a rapid end to World War II and marked the beginning of a long Cold War between the United States and the Soviet Union.

Today, conversion of mass into heat energy is used routinely for peaceful generation of electricity. In fact, more than 400 land-based nuclear power plants generate roughly 17 percent of the world’s electrical power. Smaller on-board reactors provide electrical power for many military submarines. However, the safe disposal of radioactive reactor waste and the continuing threat of nuclear weapon use remain serious concerns.

Image courtesy of Harry Bohm

Figure 8.5: Generating Nuclear Power

TECH NOTE: ENERGY CONVERSIONS

Energy conversions have been happening since the beginning of the universe and are still happening all around us all the time. This process is illustrated in the following example:

- For billions of years, matter (a form of energy) has been getting converted to solar energy (light and heat) through nuclear reactions happening deep inside the sun.
- Millions of years ago, some of this solar energy was converted by photosynthesis into chemical energy inside the tissues of ancient plants and marine algae, because those organisms used the solar energy to build the molecules they needed to grow new cells and tissues.
- That ancient chemical energy remains stored to this day and is super-concentrated in fossil fuels such as coal, oil, and natural gas, which are the chemical leftovers of those long-dead organisms.
- In a conventional power plant, those energy-rich fossil fuels are burned to convert the stored chemical energy into heat energy.
- That heat energy is then used to boil water to create energy in the form of a volume of pressurized steam.
- This pressure-related energy is converted to rotational mechanical energy when the steam is forced through turbines, which are like high-tech, high-speed windmills.
- The rotational mechanical energy of the turbines is then used to turn large generators, which convert the rotational energy into electrical energy inside the power plant.
- That electrical energy flows through wires from the power plant to homes and businesses, where it gets converted into a wide range of other useful forms. These include heat (e.g., toasters and ovens), radio waves (e.g., cell phones), sound (e.g., stereo systems), motion (e.g., kitchen blenders), chemical energy (e.g., battery chargers), and many others.
- Some of it even gets converted back into light (light bulbs), which is the same form the energy was in millions of years ago when it made its original journey from the sun to Earth.

Image courtesy of Harry Bohm

Figure 8.6: Solar Power

2.1.1. Forms of Energy

Energy can exist in many different forms. Common ones include:

- heat
- light, radio waves, and other forms of electromagnetic radiation
- energy stored in the chemical bonds that join atoms together to form molecules

- mechanical energy, like that associated with moving objects, pressurized gases, or compressed springs
- electrical energy used to run light bulbs, motors, and many other things
- mass (See *Tech Note: Mass as a Form of Energy*.)

2.1.2. Energy Conversion

Energy can be converted from one form to another. Indeed, energy conversion is a surprisingly common process. For example, when you eat lunch, the solar energy stored in the chemical bonds of your food is converted into energy that pumps your blood, pedals your bicycle up a steep hill, and powers your eyeballs and brain so you can read these pages. Ultimately, most of this energy is converted into the heat that warms your body. For some other examples of energy conversions, see the *Tech Note: Energy Conversions*.

2.1.3. Conservation of Energy

Although energy can be converted from one form to another, it can never be created or destroyed. This fact is one of the most fundamental laws of physics and is known as the **Law of Conservation of Energy**. This law says that the total amount of energy present *before* a particular process or event is the same as the total amount of energy present *during* and *after* that process or event.

For example, when you turn on a light bulb, some of the electrical energy going into the bulb gets converted into light and some gets converted into heat, but if you carefully measure the amount of light energy produced and the amount of heat energy produced, you'll find they add up to exactly the amount of electrical energy originally supplied to the bulb.

2.1.4. Quantifying Energy

The amount of energy available determines how much work you can do, so it's important to be able to quantify energy. Units are needed to give numerical amounts meaning. Unfortunately for beginners, there's a wide variety of different energy units in common use. Some (but not all) of them are introduced below.

TECH NOTE: CONSERVING ENERGY—IT'S NOT JUST A GOOD IDEA, IT'S THE LAW!

The Law of Conservation of Energy says that energy is *always* conserved (i.e., never created or destroyed). Why, then, do we need to carpool, insulate our homes, and turn off lights in unoccupied rooms to conserve energy? Good question. The apparent conflict is not actually a conflict at all, just a difference in perspective. The physical law treats all forms of energy as being equal. It says you can convert between different forms, but you'll always end up with the same amount of energy you had originally.

The advice you hear about carpooling and other forms of energy conservation is not wrong, but it's based on the fact that some forms of energy are more easily used for certain purposes than others. For example, if you have some gasoline, you can use the energy in that gas to propel your car. As you drive, the car's engine converts the chemical energy stored in the gasoline into other forms of energy, primarily motion and heat. When you apply the brakes to stop the car, the brakes convert the energy of motion into even more heat through friction. Thus, most of the energy originally present in the gasoline ultimately gets converted to heat energy, leaving behind a trail of slightly warmer air.

If you could gather up all that warm air and convert the heat energy back into mechanical energy to propel your car, there'd be no problem. But you can't. In that sense, the energy originally stored in the gasoline, while still present in a *technical* sense as heat, has disappeared in a *practical* sense. So when you "conserve energy" by carpooling, for example, you are helping to keep more energy available in an easily useable form.

The standard metric unit for energy is the **joule** (J). One joule is roughly the amount of energy it takes to lift an apple 1 meter off the floor. More precisely, it's the amount of energy required to move an object 1 meter using 1 newton of force applied in the same direction as the motion.

Sometimes this "force x distance" method of quantifying energy is captured explicitly by using metric units of **newton-meters** (Nm), which are precisely equivalent to joules (1 Nm = 1 J). You can also use imperial units of **foot-pounds** (ft-lbs), where 1 ft-lb is equivalent to about 1.356 joules.

Another common metric unit for energy is the **calorie** (lower case "c"), which is defined as the amount of heat energy it takes to raise the temperature of 1 cubic centimeter (1 milliliter) of pure liquid water by 1 degree centigrade. One joule is equivalent to about 4.184 calories. One thousand calories is called a **kilocalorie**. (See the *Tech Note: Dietary Calories* for more information about the kilocalorie.)

The amount of electrical energy purchased from a power company is often measured in **kilowatt-hours** (kWh), so you may see this unit on an electricity bill. One kWh = 3.6 million joules. A smaller version of this unit, used later in this chapter, is a watt-hour (Wh) (1 Wh = 1/1,000 kWh).

If you purchase energy from a power company in the form of natural gas (methane), the amount of natural gas energy you are billed for may be expressed in **BTU** (British Thermal Units). One BTU = about 1,055 joules.

See *Appendix I: Dimensions, Units, and Conversion Factors* for a list of energy conversion factors and other useful information.

TECH NOTE: DIETARY CALORIES

The "Calories" used to describe the amount of energy in foods are not the same as the official metric calories used to measure energy. Calories with the capital "C" are, instead, kilocalories, which are equal to 1,000 metric calories. In other words, they're equal to the amount of heat energy required to raise the temperature of an entire liter (1,000 cubic centimeters) of water by 1 degree centigrade.

By convention, regular calories are denoted by a lower case "c", whereas dietary calories are denoted by an upper case "C". Unfortunately, many people get rather sloppy about following this convention, so it's usually necessary to determine which kind of calorie is being used by context. If you're talking to a chemist, they're probably using calories (small "c"). If you're reading a food wrapper or talking to a dietician, they're probably using Calories (upper case "C").

2.2. Power

Power is one of those messy words that gets used in lots of different ways with lots of different meanings. You can hear "the powers that be," "he has no political power," or "she's a powerful speaker." But in this chapter and generally in engineering, the term power has a very specific meaning that relates to energy.

2.2.1. The Difference between Energy and Power

The terms energy and power are often used interchangeably in everyday language, but it's important to remember that they are *not* the same thing. Technically, energy refers to an *amount* of work that can be done, whereas power refers to *how quickly* that work can be done. In fact, power is defined as energy (converted between forms or used to do work) per unit time:

Equation 8.1

$$Power = \frac{Energy}{Time}$$

As a practical example of the difference between energy and power, consider a typical car that uses gasoline for fuel. The amount of *energy* available to the car depends on the amount of gas in the tank (since gas stores the energy), not on the size or type of engine. In contrast, the amount of power available to the car (that is, how quickly energy can be extracted from gasoline and used to accelerate the car) depends on the size and type of engine, not on the amount of gas in the tank.

When designing your vehicle power systems, it's important to think about both the total amount of energy and the maximum amount of power required for the mission.

2.2.2. Quantifying Power

Like energy, power can be measured and expressed in several different units. Here are the two most commonly encountered ones:

Watt: The metric unit for power is the **watt** (W). One watt = 1 joule/second. One **kilowatt** (kW) is 1,000 watts.

Horsepower: A common imperial unit for power is the **horsepower** (Hp). One horsepower is defined as the amount of mechanical power required to lift 33,000 pounds 1 foot in 1 minute. It was originally based on the amount of power that one strong draft horse could produce. 1 Hp = 746 W.

2.2.3. Power Transmission

Power often needs to be delivered from one physical location to another. For example, the power generated by a car engine needs to be transmitted to the wheels, and the electrical power produced in a hydroelectric dam needs to be transported to homes and businesses.

There are many different ways to deliver power. Mechanical power can be transferred through rotating shafts, gears, pulleys, sliding rods, levers, or other mechanisms. It can also be delivered by using pressurized liquids or gases to transmit force and movement through a series of pipes, tubes, valves, or pistons. When power is transferred in this way, liquids (e.g., hydraulic fluid, brake fluid, or water) are usually preferred over gases (e.g., air), because the compressibility of gases allows a significant fraction of the energy to be stored (rather than transmitted). This makes for "squishier" (i.e., less easily controlled) movement of pistons, and the compressed gases can create an explosion hazard if the pressures are high enough.

Electrical power is usually moved from one location to another through copper wires or other metal conductors. For reasons discussed later in this chapter, long-distance transmission of large quantities of electrical power is usually done using high-voltage alternating current (AC).

There are other ways of delivering power, including light beams and radio waves, but the amount of power transmitted by these means is generally too small to be of use for purposes other than illumination or communication (information transmission). A notable exception is the microwaves used in a microwave oven. Obviously, these are providing enough power to heat food.

2.3. Efficiency

Whenever energy is converted or used to perform work, the original form of energy is usually changed into *two or more* different forms at the same time. Unfortunately, only one of these forms is normally useful. When combined with the Law of Conservation of Energy, this tendency to produce multiple forms of energy during a conversion process results in some bad news. It means that you never get as much useful energy out of a machine or process as you put into it, because some fraction of the original energy always gets converted into one or more forms you don't need or want. The fraction of total energy put into a machine or process that comes out in the desired form is referred to as the **efficiency** of that machine or process.

Equation 8.2

$$\textit{Efficiency} = 100\% \times \frac{\textit{Energy Output (desired form)}}{\textit{Energy Input (total)}}$$

For example, an incandescent lightbulb works by using electrical power to heat a tiny wire (the filament) until it is hot enough to glow brightly. Thus, electrical energy is converted by the light bulb into both heat and light. When you turn on the light switch you are seeking light, not heat, but you inevitably get both. Any energy that's converted to an unwanted form is usually viewed as "lost" or "wasted" energy, although it's not really lost in the strict sense defined by the Law of Conservation of Energy.

Note that since energy and power are *proportional* to each other (energy = power/time), it's equivalent to say that efficiency is a fraction of the total *power* put into a machine or process that emerges in the desired form of *power*:

Equation 8.3

$$\textit{Efficiency} = 100\% \times \frac{\textit{Power Output (desired form)}}{\textit{Power Input (total)}}$$

As an example, consider the incandescent light bulb again. A 60-watt incandescent light bulb requires 60 watts of electrical power going in, but only about 6 watts of that power actually emerge in the form of visible light. The other 54 watts come out as heat. If light is what you want, the light bulb is said to be about 100% x (6/60) = 10% efficient. If, on the other hand, you are using the light bulb as a heater to keep your pet lizard warm, you are lucky. The bulb is 100% x (54/60) = 90% efficient as a heat generator. As you can see from this example, before calculating efficiency, you need to clarify what form of energy output you want. Table 8.1 lists energy conversion efficiencies for a variety of devices and processes.

Table 8.1: Efficiency of Some Common Energy Conversions

Device or Process	***Energy input***	***Energy output***	***Approximate efficiency***
Incandescent light	*Electricity*	*Light*	*10%*
Gasoline engine (automobile)	*Energy stored in chemical bonds of gasoline*	*Vehicle motion*	*20%*
Compact fluorescent light	*Electricity*	*Light*	*30%*
Power plant (coal-fired)	*Energy stored in chemical bonds of coal*	*Electricity*	*30%*
Small electric motor (operating under optimal load conditions)	*Electricity*	*Motion*	*70%*
Large industrial electric motor (many horsepower; operating under optimal load conditions)	*Electricity*	*Motion*	*95%*

Sources: Wikipedia for most values. U.S. Dept. of Energy for coal-fired power plant.

This definition of efficiency applies whether energy is being used to perform useful work, converted from one form to another, or is simply being transmitted from one location to another. For example, all conventional methods of power transmission lose some power along the way due to processes like friction in mechanical systems or resistive heating in electrical systems. These processes siphon off energy at some rate (i.e., they siphon off power) between the transmission source and destination. Thus, efficiency is an important concept whenever you are using, converting, or transporting power.

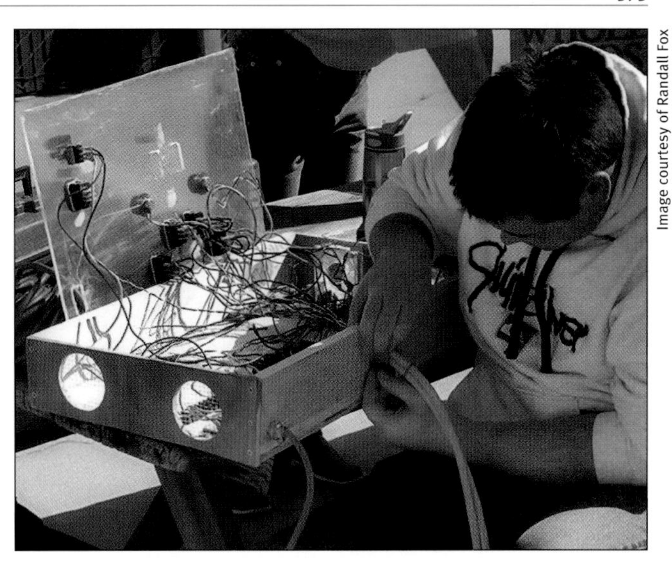

Image courtesy of Randall Fox

Figure 8.7: Double-Checking the Wiring on an ROV Power System

2.3.1. Design Implications

If you knew how much output power your ROV would need to perform its job and could somehow determine the efficiency of all the power transmission, conversion, and usage steps involved, you could theoretically calculate the amount of input power required to complete various tasks associated with your vehicle's mission. Unfortunately, conversion efficiencies vary widely (from less than 1% to nearly 100%) and change with conditions, so even published values are a rough guide at best. However, as a starting point, you can assume that most energy conversions in well-designed machines have an efficiency of between 10% and 40%. Some may be better, some may be worse, but on average, they'll probably be somewhere in that range.

This limited efficiency can create serious problems, particularly in systems that require several steps of energy conversion. By the time all the conversions are completed, there's often only a tiny fraction of the original energy actually left for useful work. For example, suppose you want to propel an ROV with a battery-powered thruster. At least three energy conversions are involved in this process:

- First, electrical power is converted to rotational power by a motor. However, a thruster motor gets warm and makes noise while running. Since heat and sound waves are both forms of energy, it's immediately apparent that some of the electrical power going into the motor is being converted into forms other than rotational power.
- Second, the output power of the motor is used to spin the propeller, but here some of the rotational power put into the motor shaft is lost through friction in the shaft seal and any bearings before it reaches the propeller.
- Third, the prop's rotational power must be converted to thrust, but in this case, a significant fraction of the prop's rotational power is lost due to drag on the blades as they slice through the water and to energy wasted by flinging water sideways instead of pushing it straight backward. (This is one reason ducted props can be more efficient.)

If, for the sake of this example, you assume that each of these three conversion steps is, on average, about 30% efficient (i.e., about 3/10 of the power makes it through each conversion step), then the overall efficiency is 0.3 x 0.3 x 0.3 = 0.027 = 2.7%. In other words, less than 3% of the power delivered by the battery to the motor actually contributes directly to propelling the ROV through the water. If you take the inverse of this figure, you discover that you need to supply the thrusters in this example with about 37 times as much power as you really expect to get out of them!

Don't bother memorizing this number, because this example is only hypothetical, and the real value for your vehicle is going to be heavily dependent on the design of the propulsion system and many other factors. What you should bother to remember is that you'll need to put *much* more power into your thrusters and other vehicle systems than you expect to get out of them. Once your vehicle is completed, you might want to measure the power going in and compare it to the power coming out for various systems on the vehicle. This will help refine your intuition about how much efficiency you can expect to get out of your own designs and will give you a benchmark to try to beat in your next design!

Another important lesson you can draw from the example above is the value of avoiding unnecessary energy conversions. For example, don't use electricity to run a motor to turn a generator to make electricity, unless you have some compelling reason to do so, because you'll never get as much electrical power out of this series of conversions as you put into it in the first place.

Before moving on from this discussion of efficiency, it's worth noting that anytime a system is converting only a small fraction of its input power into useful work, it must be converting a large fraction of its input power into something else. Usually, that something else is heat, and heat can spell trouble. Unless heat escapes as quickly as it's being produced, the temperature will rise. This can overheat parts, causing them to malfunction, melt, or even catch on fire. A designer needs to be aware of this potential threat and assess whether or not enough heat could be generated and trapped to cause problems. If so, it may be necessary to reduce heat production through improved efficiency or to design a cooling system or other mechanism for getting rid of this excess heat more quickly.

Vehicles operating under water have both an advantage and disadvantage when it comes to getting rid of excess heat. On the positive side, cool bodies of water are excellent **heat sinks** that can be used to draw away and absorb excess heat produced by a vehicle. On the negative side, the motors, electronics, lights, and other heat-producing innards of ROVs and AUVs are often confined in densely packed, airtight spaces with no access to cooling breezes. And in inexpensive vehicles, the pressure canisters are often made of thick-walled plastic cylinders that act like a layer of thermal insulation, slowing the loss of heat to the surrounding water. Metal canisters are much better at transferring excess heat out into the water.

Clearly, quite a number of factors conspire to make overheating difficult to predict and control. As a quick guide, though, if your motors, lights, or other systems get warm to the touch when operating on your workbench, you may need to worry about them getting too hot when placed inside the confines of a plastic pressure canister.

Poor efficiency wastes energy, reduces maximum power, and generates unwanted heat. Therefore, it's important whenever possible to maximize efficiency. Here's a summary of the most important messages from this discussion.

1. Minimize the number of energy conversion steps in any underwater craft.
2. Make sure each conversion step is as efficient as possible.
3. Check to make sure nothing is in danger of overheating.

If you heed these recommendations, you'll design vehicles that can move faster and accomplish more work over longer periods of time with smaller, safer, and less expensive power supplies (and with less likelihood of overheating) than if you ignore them.

3. Vehicle Power Choices: What's Realistic?

All underwater vehicles require some energy source (usually stored on the vehicle or at the other end of a tether) and some way to deliver that energy at appropriate rates to the various subsystems that need it. In many cases, the stored form of energy will need to be converted to other forms, either before or after distribution, in order for it to be useful, just as the chemical energy stored in gasoline must be converted to electrical energy before it can be used to power your car's headlights.

In most underwater vehicles, large or small, the greatest power demands come from the propulsion system, so that should be your primary focus when evaluating potential power system designs for your vehicle. But don't overlook the power needs of other systems, because they may place additional demands on power system design.

As described in the *Historic Highlight: Evolution of Underwater Power Systems* early in this chapter, underwater vehicles have been powered in the past by everything from human muscles to nuclear reactors and even temperature gradients in the ocean itself. Some of the approaches described there would work for small unmanned vehicles, but most would not. Therefore, it's important to have some criteria you can use to help you decide which power system options might work well for your project.

3.1. Criteria for Evaluating Power Systems

When evaluating potential power systems to decide which kind to use, there are a number of important factors to consider. As with most other aspects of vehicle design, the best choice always depends on the mission and is usually a tradeoff among a number of competing constraints. Careful consideration of the following questions will help lead you in the right direction:

- **Can the power system store (or extract from the environment) enough energy to complete the entire mission?** Are you planning a quick 10-minute trip to view the bottom of a swimming pool, or are you disassembling a huge decommissioned oil platform? The total amount of energy that needs to be stored and delivered for each of these missions is vastly different.

- **Can the power system meet the peak power demands of your vehicle?** Most missions involve short periods of high power demand interspersed among longer periods with low power demand. It's critical that your power system be able to deliver energy quickly enough to meet the peak demands, not just the average demand.

- **How much physical space will the power system require?** Some power systems are more space-efficient than others. It's usually best to use the most compact system that will meet your needs, because that results in a smaller, lighter, and more streamlined vehicle. Don't forget to include the entire power system—not just the engines or motors, but also any fuel tanks, batteries, or other energy storage compartments. Keep in mind that most pressure canisters are cylindrical, so a power system with a boxy shape may require a larger pressure canister to house it than a power system with a cylindrical shape, even if the two systems have identical volumes.

- **Can the system function well under water?** Air and other viable sources of oxygen are in short supply under water; therefore, many power systems that are ideal for cars, airplanes, and other machines operating in air are impractical for most underwater craft. Exposed electrical systems are generally not compatible with water, but well-designed pressure canisters and various encapsulation methods can be used to separate circuitry from direct contact with water.

- **How easy is it to obtain, install, use, maintain, and eventually retire the system?** Some power sources, like standard household batteries and electric motors, are easy to find, use, and dispose of properly (especially if there's a household hazardous waste program in your area). Others, such as nuclear reactors, are not. Power system simplicity is a particularly valuable attribute for small groups working on their first small vehicle design project(s).
- **How easily can energy be distributed from the source to the various systems that need it?** For example, stiff mechanical linkages, such as gears and drive shafts, can transmit power effectively, but electrical and hydraulic systems are much more forgiving when it comes to spatial dimensions and machine movements, so they are often much easier to use.
- **What forms of power are required by the various vehicle systems?** For reasons of efficiency and convenience, it is generally desirable to minimize the number of energy conversions required. Therefore, it is good to standardize on a single form of power, if possible. These days, this factor weighs heavily in favor of electrical power systems, since most modern-day vehicle systems (e.g., navigation systems, control systems, camera systems, and lighting systems) require electrical power, and all other systems (propulsion systems, for example) can be designed to use electrical power, too.
- **How safe is it?** Some power sources, such as low-voltage batteries, are relatively safe for beginners to use. Others, like nuclear reactors or even small gasoline engines, can be very dangerous and are best left to people who have ample experience with these technologies.
- **How much does it cost?** As always, cost is an important consideration, particularly for designers operating on a tight budget. In some cases, cost can be an overriding factor. When estimating costs, be sure to include the initial cost of purchasing and installing the system, as well as the recurring costs associated with system maintenance and routine replenishment of fuel, batteries, etc.

Image courtesy of the Cornell University Autonomous Underwater Vehicle team

Figure 8.8: CAD Drawing of Cornell University's* Triton *AUV and Battery Pod

The Cornell University Autonomous Underwater Vehicle team designs and builds AUVs to participate in the AUVSI and ONR's International Autonomous Underwater Vehicle (AUV) Competition.

The top CAD illustration of their Triton *AUV shows the battery pod. The bottom illustration shows its position low down on the frame, under the thruster.*

3.2. Electrical Power—A Logical Choice

When all of the above criteria are considered together, electrical power systems come out on top for most contemporary small ROV and AUV projects, whether they're built by beginners or professionals. That may change with time as power technology evolves, but for the near future, electrical power systems are an excellent choice.

Although electrical power systems must be isolated from water and require more space to store a given amount of energy than fossil or nuclear fuels do, they offer some distinct advantages that more than compensate for these few drawbacks. The advantages include:

- convenience
- simplicity
- air-independence
- low cost
- flexibility and scalability
- ease of maintenance

- ease of power distribution
- compatibility with a wide array of sensors, motors, lights, and other useful devices already designed to run on electricity
- safety, in the case of home-built or school-built projects, since low-voltage batteries can be used

There are essentially two methods used to deliver electrical power to a small underwater vehicle. One way is to draw the power from batteries placed directly on board the vehicle. Nearly all AUVs and a few specialized ROVs use this approach. The other is to deliver electrical power to the vehicle through wires bundled inside its tether. This is the most common method of supplying power to ROVs. The tether-supplied power can come from any surface source of electrical power, including batteries, portable generators, or standard electrical outlets. Each of these options will be discussed in more detail later in the chapter, but first it's important to master some basic electrical concepts and vocabulary.

4. An Introduction to Electricity and Electric Circuits

This section of the chapter provides an introduction to basic concepts about electricity and electric circuits, along with some related vocabulary. These are essential for understanding the electrical power systems used in small underwater vehicles. Some of these concepts are explored in greater detail later in this chapter. Others are developed further in *Chapter 9: Control and Navigation*, which explores more sophisticated electronic circuits used for sensors, navigation, control, and communication. Due to space constraints, this book can provide only a rudimentary introduction to electricity. Fortunately, many excellent books and websites are available for those who wish to learn more about this fascinating and useful subject.

Image courtesy of Randall Fox

Figure 8.9: Wiring a Simple Switch

Electricity is definitely weird stuff. You can't directly see it, smell it, or taste it, and you generally don't want to feel it! Not surprisingly, electricity is difficult to define in any simple way. Dictionary meanings are frequently incomprehensible and sometimes even conflicting. Don't worry, though; it turns out you don't need to define electricity precisely to be able to use it effectively. What you do need to do is develop a solid intuitive grasp of the key electricity concepts described below.

4.1. Charge

As you may recall from the discussion of corrosion chemistry in *Chapter 4: Structure and Materials*, **charge** is an electrical property associated with protons and electrons. Protons (usually found in an atom's nucleus) are said to have a positive (+) electric charge, whereas electrons (usually found orbiting the nucleus of an atom) are said to have a negative (-) electric charge. The term "electricity" refers to a host of related phenomena that happen when these charged particles move (or try to move) from one place to another. Often these movements happen (or try to happen) because opposite charges attract each other (so electrons and protons try to stick together) and identical charges repel (so two electrons or two protons attempt to move away from one another).

4.2. Current

Electrical **current** is simply the flow of charged particles from one location to another. These charged particles may be protons, electrons, or ions. Ions, you may recall, are atoms or groups of atoms in which the number of electrons does not equal the number of protons.

In most electric circuits, the charged particles that move and carry the current are electrons flowing through metal or other conductive materials. However, in water, particularly saltwater, electrical current is carried by ions such as Na^+ and Cl^-. By convention, electric current is defined as moving from (+) toward (-), but see *Tech Note: Which Way Does Electric Current Really Flow?*

In equations, electrical current is commonly represented by the letter "I" (sometimes "i"). The metric and imperial units for electrical current are the same: **amperes** (commonly called "**amps**" for short and abbreviated "A"), so a statement that says, "The current is 14 amps" could be written as: I = 14 A.

TECH NOTE: WHICH WAY DOES ELECTRIC CURRENT REALLY FLOW?

You may have heard some people say that electricity flows from (+) to (-) and others say it flows from (-) to (+). Who's correct? The answer: Both. (Well, sort of.)

By convention, electrical current is *formally defined* as flowing from more positive (+) voltages toward more negative (-) voltages. In that sense, electricity *always* flows from (+) to (-). Since opposite charges attract and identical charges repel, this formally defined current direction is the same as the direction that positively charged particles would flow while trying to run away from the positive terminal of a battery or other voltage source and run toward the negative terminal. It is also exactly *opposite* the direction that negatively charged particles would flow as they flee the negative terminal in search of the positive terminal.

If electric current were always carried by positively charged particles, the formal definition of electric current direction would make perfect sense all the time, and life would be simple. Unfortunately, electric current in most circuits is carried by electrons, which are negatively charged particles. Thus, the actual physical movement of particles in an electric circuit is normally opposite the formally defined direction of current flow. In this sense, you can understand why someone might say electricity flows from (-) to (+). Fortunately, a negative charge flowing backward is functionally equivalent to a positive charge flowing forward, so these seemingly contradictory ideas about electricity's direction are, strange as it may seem, in perfect agreement. Nonetheless, many beginning students of electricity perceive the conflict as real and find it confusing.

If you find it confusing, too, you can blame Benjamin Franklin for your headache. He knew about static electricity experiments in which glass and resin were attracted to each other after being rubbed together, and he correctly suspected that some invisible "electrical fluid" was flowing from one to the other. Unfortunately, electrons had not yet been discovered, and he guessed wrong about the direction things were moving. He assigned a (+) value to glass on the assumption that it had gained some of the mysterious electrical fluid. Many years later, his "electrical fluid" turned out to be electrons, and they were flowing from glass to resin, not resin to glass. Unfortunately, by the time this was figured out, the (+) and (-) designations that Franklin had assigned were too firmly entrenched to be changed. If he had originally assigned a (+) value to the resin, instead of to the glass, then electrons today would be defined as having a positive charge, protons would be negative, the labeling of battery terminals would be reversed, and the electrons flowing through circuits would indeed be moving in the same direction as the formally defined direction of current flow.

4.3. Voltage

Anytime you separate positive charges from negative charges, you produce an **electric field.** A charge dropped into the middle of this electric field will be pulled toward the side with the opposite type of charge and away from the side with the same type of charge. For example, an electron will be pulled toward the positive charges and away from the negative charges. The strength of this tendency for a charged particle to be pushed or pulled can be quantified in terms of a **voltage.** You can think of voltage as a kind of electrical pressure that can push electrical current through wires, just as water pressure can push water through pipes.

It's important to remember that voltage is all relative. It's like altitude in which how high you are depends on whether you're measuring from sea level or the bottom of your ladder. Thus, what matters in circuits is not the absolute voltage, but the *difference* in voltage between two points. It's this voltage difference that makes electrical current flow through a wire, just as it's a difference in altitude that makes water flow down a stream, or a difference in pressure between one end of a horizontal pipe and the other end that forces water through the pipe. Accordingly you should always speak of the voltage *across* a device (i.e., the voltage difference from one end of the device to the other). It is incorrect to talk about how much voltage is flowing *through* a device. Current flows through devices; voltage does not.

In equations, voltage is commonly represented by the letter "V" (sometimes it's written "E" or "e" for electro-motive force). Thankfully, the metric and imperial units for voltage are the same: Volts. Volts are abbreviated "V", so a statement that the voltage difference between two points is 6 volts could be written as: E = 6 V.

Be warned that "V" is used much more commonly than "E" for voltage, but this introductory text uses "E" in most cases, to avoid a potentially confusing notation such as: V = 6 V.

4.4. Resistance and Ohm's Law

Whenever electrical current flows through a wire, light bulb, or any other object, it encounters the electrical equivalent of friction, which converts some of the energy of the flowing electricity into heat and causes a drop in voltage from one end of the object to the other. This friction-like property in electric wires and other components is called **resistance.** It is usually denoted by the variable "R" and quantified in units of ohms, which have the symbol Ω (the Greek letter "omega").

The resistance of an object (such as a length of wire) is defined in terms of the amount of voltage it takes to force a given amount of current through the object. Thus:

Equation 8.4

$$R = \frac{E}{I}$$

This simple equation is one form of "**Ohm's Law.**" It can be rearranged to solve for any of the three variables, provided the values of the other two are known. For example, Ohm's Law is commonly written:

Equation 8.5

$$E = I \times R$$

This is arguably the most famous (and useful) of all equations in electronics. Later in this chapter, you'll see how to use it to calculate the voltage drop that occurs in any tether that's supplying electrical power from the surface to an ROV working in the depths. The triangle diagram in Figure 8.10 is an easy way to remember and use Ohm's Law.

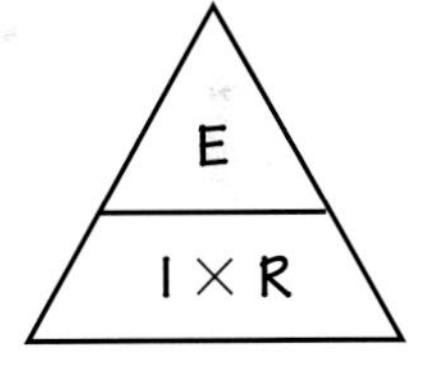

Figure 8.10: Ohm's Law Triangle

A simple way to remember Ohm's Law is to draw a triangle with E at the top, and I and R below. If you need the equation for voltage, just cover the E with your finger and you'll see the formula: I X R.

This works for current and resistance, as well. If you cover the R, you'll see that resistance is equal to E divided by I, and if you cover the I, you'll see that current is equal to E divided by R.

Simply put, if you know the value of two variables of Ohm's Law, then you can calculate the third.

4.5. Insulators, Conductors, and Semiconductors

Objects or materials that offer high resistance to electric current are called **insulators**. Familiar examples of good insulating materials include glass, most plastics, rubber, and dry air.

Objects that have low resistance conduct electricity easily and are called **conductors**. Familiar examples of good conductors include metals and saltwater. Wires are made out of metal (usually copper) because metal is a great conductor of electricity.

Some materials have electrical properties that fall in between those of insulators and conductors. Many of these materials can have their properties modified to make them more or less conductive, and this turns out to be extremely useful in the manufacture of many important electronic components, including transistors and integrated circuits. These materials with adjustable conductivity (and circuit devices made from them) are called **semiconductors**.

4.6. Circuits

Electric current typically flows in closed loops called **circuits**. A regular circuit will have at least three fundamental parts connected together to form the loop:

1. There will be some **voltage source** (such as a battery or a wall outlet), which provides the electrical "pressure" needed to push electrons or other charged particles around the loop.
2. There will be a **load**, which in this context refers to any device(s) and/or process(es) being powered by the circuit. For example, in a flashlight circuit, the

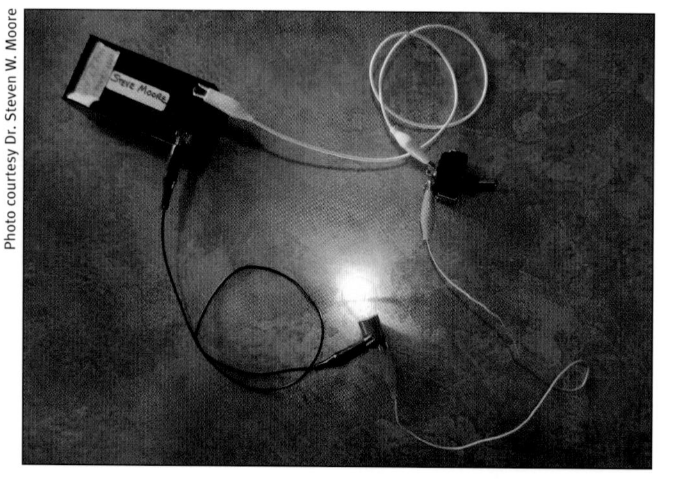

Photo courtesy Dr. Steven W. Moore

Figure 8.11: Simple Light Bulb Circuit

This photograph illustrates a simple circuit used to power a light bulb and to switch the light ON or OFF. The positive terminal of a battery (black rectangular box in upper left) is connected by a wire (white) to one terminal of a switch. The other terminal of the switch is connected to one terminal of a light bulb (the "load" for this circuit) by another wire (yellow). The remaining terminal of the light bulb is connected to the negative terminal of the battery by yet another wire (black), completing the circuit.

When the switch is flipped to the ON position (as shown), electric current flows around the circuit and lights the bulb. The light bulb can be turned off by interrupting the flow of current around the loop, either by flipping the switch to the OFF position or by disconnecting any of the wires.

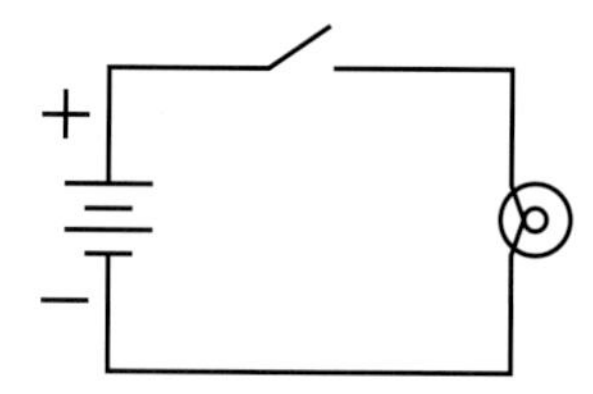

Figure 8.12: Schematic for Simple Light Bulb Circuit

This is a schematic diagram of a simple light bulb circuit, such as the one inside a flashlight or the one pictured in Figure 8.11. The stack of two long and two short horizontal lines at far left is a very common schematic symbol for a battery. The plus (+) sign shows which end is the positive terminal of the battery, and the minus (-) sign shows which end is the negative terminal. The circular curlicue symbol on the right side of the diagram is one of many standard symbols for a light bulb. The little ramp-like symbol at the top represents an electric switch. Other straight-line segments represent wires connecting the positive terminal of the battery to the switch, the switch to the light bulb, and the light bulb back to the negative terminal of the battery.

These wires and the devices they interconnect form a complete circuit around which current can flow whenever the switch is in the ON position. By convention, switches in schematic diagrams are normally shown in their "open," or OFF, position, which current cannot jump across, so the light is OFF. It is understood that the switch would need to be closed before the light bulb or other load would turn ON.

load would be the light bulb. If you were talking about a circuit that supplies power to an entire ROV, then the whole ROV would be considered the load.

3. Almost all circuits include one or more **switches** to turn the load ON and OFF. Each switch acts like a little conductive drawbridge that electrons or other charged particles must travel over to cross an insulating gap. When the switch is in the ON position, it's as if the bridge is down, so the electrons or other charge carriers can cross the bridge and continue on their way around the loop. When the switch is in the OFF position, it's as if the bridge is raised, stopping traffic. This blocks the flow of charged particles around the loop, creating a traffic jam that instantly backs up all the way around the loop, bringing current throughout the circuit to a standstill, and shutting off power to the load. (See Figure 8.12 and *Tech Note: Common Schematic Symbols.*)

Note: The term "circuit" is also used frequently in electronics to refer to just a portion of a complete circuit. This can be a bit confusing, since these partial "circuits" do not contain a complete loop around which electric current flows.

4.7. Schematic Diagrams

A schematic diagram (often just called a **schematic**) is a technical drawing used to show, in a clear and unambiguous manner, how various electronic components within a circuit are connected. Different symbols, which are more or less standardized (see *Tech Note: Common Schematic Symbols*), are used to represent different kinds of electric components, such as batteries and switches. Straight lines or combinations of straight line segments joined end-to-end are used to represent wires or other conductive pathways that can carry electric current from one component to the next.

Unlike mechanical drawings, which usually emphasize the precise shape, dimension, and placement of each part, a schematic diagram usually ignores these things. Instead, a schematic focuses on showing how electric current and electrically encoded information flow among the various components in a circuit. For example, the diagram in Figure 8.12 is a schematic diagram of the light bulb circuit pictured earlier in Figure 8.11. It shows in a clear and uncluttered way that the positive terminal of the battery is connected by a wire to a switch and that the other side of that switch is connected by another wire to a light bulb, and that a third wire connects the other side of the light bulb to the negative terminal of the battery; however, it does not specify how long the wires are, whether the wires travel in straight lines or are coiled, where the switch is located, or any other such location-specific geometric information.

Although there are some variations in schematic symbols and conventions, the differences are small enough that an experienced person can usually look at a schematic drawn by someone else and quickly figure out what the different components are and what the overall circuit is designed to do.

It is conventional (though not strictly required) to organize a schematic diagram so that higher voltages are near the top of the diagram and lower voltages are near the bottom of the diagram. This makes it easier to visualize what's happening, because you can think of the battery as "pumping" the charges up to a higher voltage, then letting them flow "downhill" through the circuit on their way to the lowest voltages. When information is flowing through a circuit, for example, data from an electronic depth sensor flowing to a digital display, it's conventional (but again not required) to organize the schematic so the information flow through the diagram is from left to

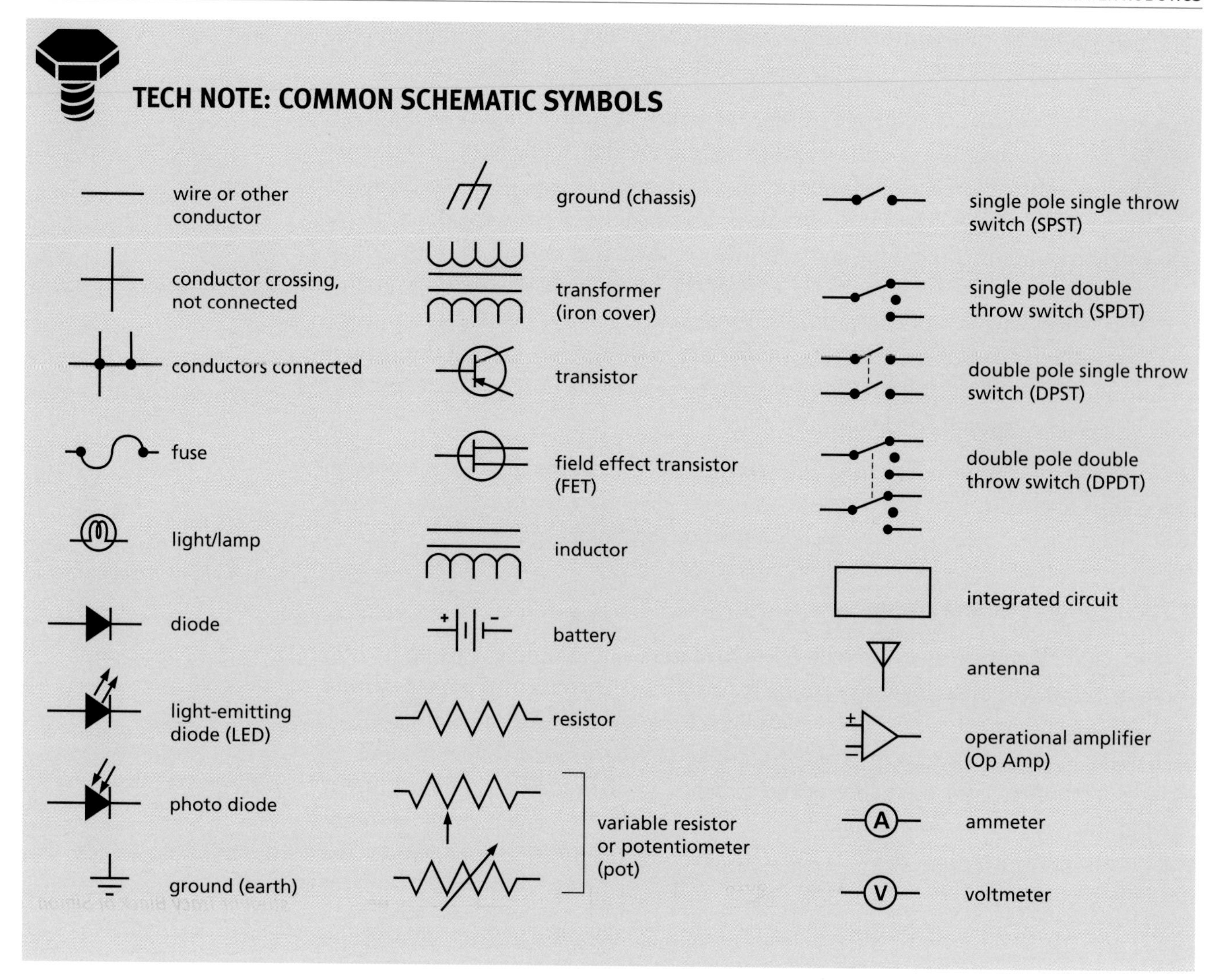

right, as if you were reading the words in a line of text. Thus, the sensors would commonly be on the left and data displays would be on the right.

You will learn more about schematic diagrams and schematic symbols as you progress through this chapter and the next.

4.8. Open Circuits and Short Circuits

An **open circuit** occurs when a gap forms somewhere in the circuit, preventing current from flowing around the loop. An open circuit may be intentional (as when a switch is flipped to the OFF position) or it can also be unintentional, as when a wire breaks or comes loose, or when a light bulb or some other component in the circuit burns out. Unintentional open circuits are a common cause of equipment problems.

A **short circuit**, or **short**, is a different type of error condition in which electric current finds some unauthorized "shortcut" through which it can bypass some or all of the usual load that it's supposed to flow through. Under normal conditions, the load provides resistance that limits the amount of current flowing through the circuit; however, when a short occurs, electricity can bypass the usual current-limiting parts of the load, making it way too easy for the battery (or other power source) to push current through the circuit. This can result in dangerously high current levels capable of melting wires, damaging components, and starting fires. In addition, shorts can give high voltages access to places where those high voltages are not normally present, and this can present a serious threat to human safety.

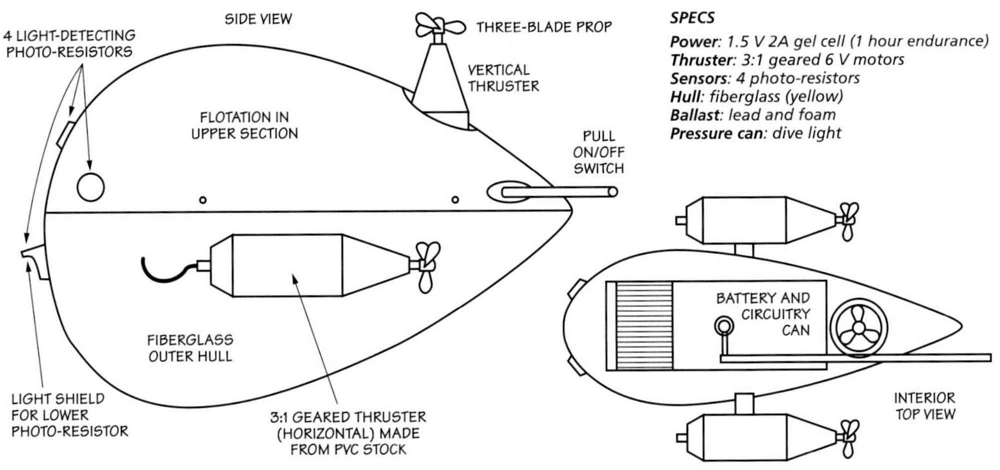

Figure 8.13: Light-Seeking AUV **Sunfish**

This small light-seeking AUV Sunfish *works by using each of its four photoelectric detectors to sense the intensity of light. Then the circuitry converts this signal to a current. Comparators (chips) compare the current produced by the sensors—the higher the current, the brighter the detected light. Then the circuitry switches on the thruster and moves the AUV toward the brightest light. The design is so successful that the tiny AUV rarely fails to find the light source, demonstrating the feasibility of using light in clear water to bring an AUV back to its retrieval point.*

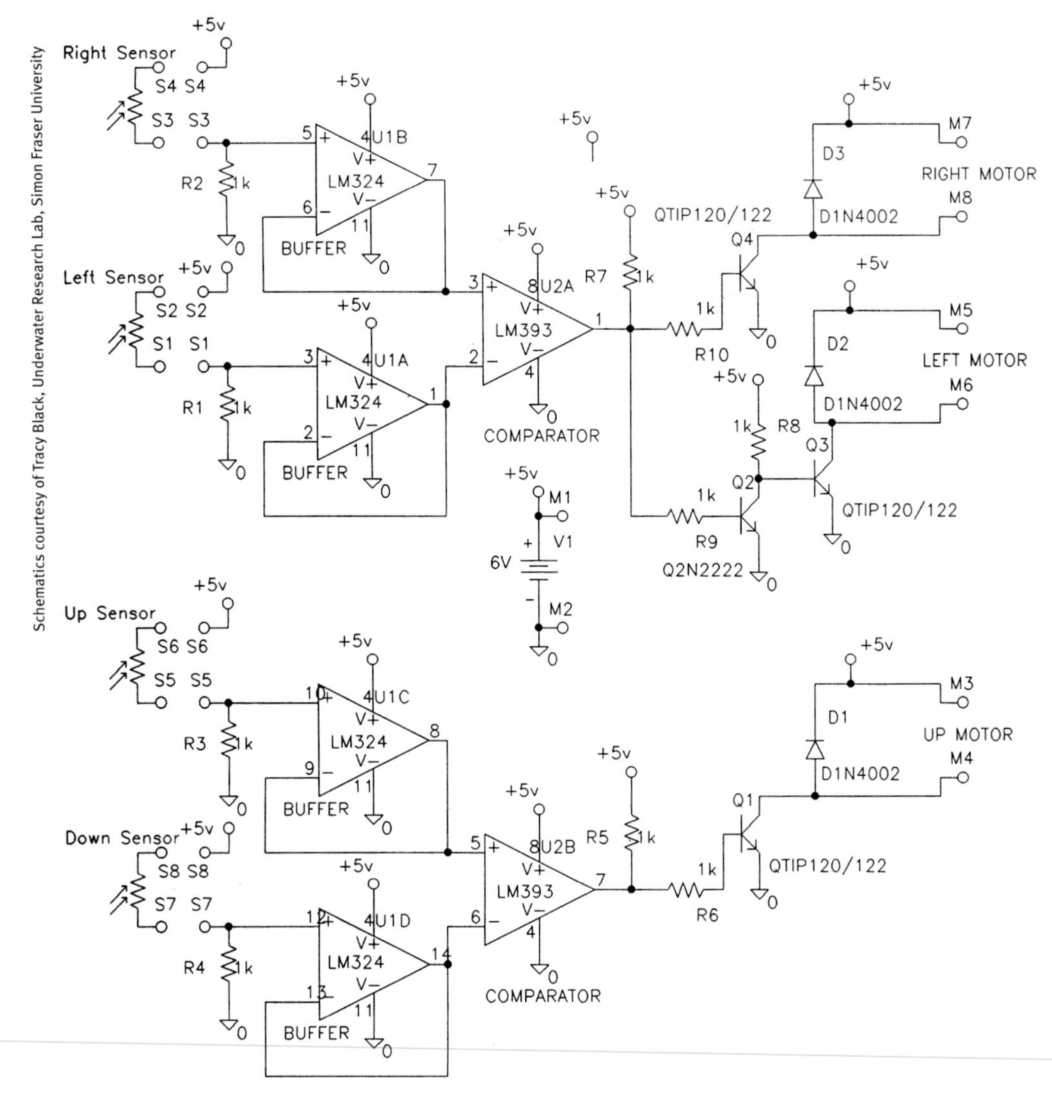

Schematics courtesy of Tracy Black, Underwater Research Lab, Simon Fraser University

Figure 8.14: Electrical Schematic for AUV **Sunfish**

This electrical circuitry was designed by engineering student Tracy Black of Simon Fraser University's Underwater Research Lab for the small light-seeking AUV Sunfish *shown in Figure 8.13.*

Fuses, circuit breakers, and similar devices are used to minimize the risk to circuits, people, and property when short circuits happen. These protective devices are described in more detail later in this chapter.

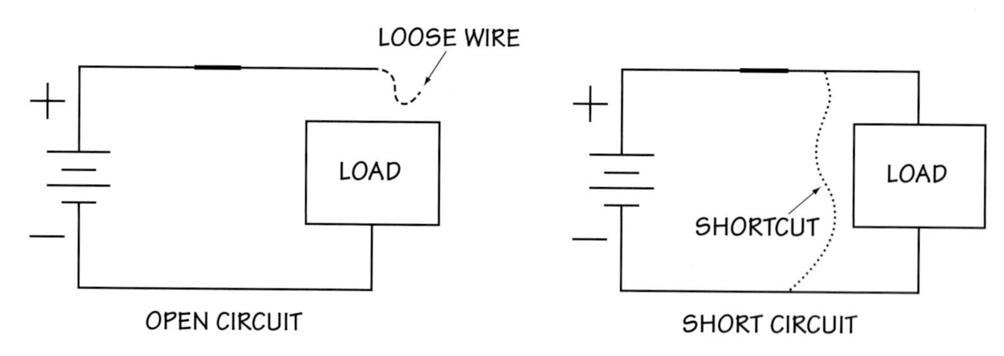

Figure 8.15: Open Circuits and Short Circuits

In these two schematic diagrams, dashed lines are used to illustrate examples of an unintended open circuit and an unintentional short circuit. In the open circuit example on the left, a wire has come loose, breaking the circuit and preventing current flow through the load (e.g., the light bulb or motor), even though the switch is in the closed (ON) position.

In the short circuit diagram on the right, a conductive pathway has accidentally formed where it should not be (perhaps because somebody left a metal tool rattling around inside the circuit housing or because two bare wires touched when they should not have), thereby creating a shortcut through which electric current can return to the battery without flowing through the normal load.

4.9. Ground

For convenience, one point in a circuit is typically defined to be at zero volts (just as sea level is commonly defined to be at an altitude of zero meters or feet). This point in a circuit is commonly called the **reference ground**, or simply **ground**. Once this reference ground has been identified, you can talk about *the* voltage at a point in the circuit as if it were some absolute thing. For example, if point Y has already been defined as the reference point, then you can say "point X is at 3 volts" instead of the much more cumbersome "point X is 3 volts above the voltage at point Y." Defining a reference ground just makes it easier to talk about voltages in a circuit.

The point in the circuit with the lowest voltage is frequently, but not always, chosen as the reference ground. Any other part of the circuit connected directly to the reference ground point will also be at (or very nearly at) the same voltage and may also be considered "ground." It's important to realize that reference grounds are not the only type of ground found in electrical systems and that one device or system may have two or more different grounds.

The term "ground" originally comes from the idea of "grounding" a circuit by connecting it, quite literally, to the soil on the earth's surface. Indeed, **grounding** is sometimes called **earthing**. This type of ground is called an **earth ground**. For safety reasons, building codes in most areas require that electrical appliances be "grounded" in this way. This is normally done through a three-pronged power cord, which has a "ground" wire on it, in addition to the two wires used to carry electrical power from the power plant to the device and back again. The ground pin connects metal parts of the appliance frame to a ground wire in the wall outlet, which ultimately connects to a metal rod or water pipe embedded deep in the soil under the building. In the event of a short in the appliance wiring, electricity that might otherwise present a serious electrocution hazard for someone touching the appliance flows harmlessly through the ground wires to Earth, ideally tripping a circuit breaker in the process for added protection against electrocution and electrical fires. (See circuit breaker explanation later in this chapter.)

Earth ground can be used as a reference ground, but it doesn't have to be. In fact, some electric circuits, such as the electric navigation systems in an airplane, would be difficult to connect to an earth ground!

Some equipment may use separate reference grounds for different sub-circuits, particularly if those circuits might interfere with one another. For example, in advanced underwater vehicles and many other complex electrical circuits, the circuits used for electric motors often have their ground wires separated from the ground wires of circuits used for sensors. This is done because motors usually generate lots of electrical spikes or other interference on the power and ground lines. These spikes can cause errors in the measurements of sensitive sensor circuits, so it's important to keep the motor and sensor circuits as separate as possible. Using separate grounds helps a lot.

Proper grounding is a complex subject unto itself. Readers interested in the details are invited to consult more advanced texts on electricity and electric circuits.

TECH NOTE: MORE CIRCUIT TERMINOLOGY

Electronic circuits are made up of many different types of components, often called devices, each of which is designed to control, regulate, or use electricity in some specific way. Here are a few of the more common and widespread ones, along with at least one important function for each type:

- **Wires** are conductive pathways used to route electricity to particular locations.
- **Switches** block or allow the flow of electrical current.
- **Resistors** can be used to limit current flow through a circuit.
- **Capacitors** store energy in the form of electric fields.
- **Inductors** store energy in the form of electric currents.
- **Diodes** act as one-way valves for electric current. Some give off light.
- **Transistors** are semiconductor devices used as amplifiers or switches.
- **Integrated circuits** (ICs) are complex, miniaturized circuits that may contain thousands or even millions of transistors and other components embedded inside a single tiny piece of semiconductor material. There are thousands of different types of ICs, each with a distinctive function. The microprocessors in computers are just one example of an IC manufactured for a specific purpose.

Chapter 9: Control and Navigation describes most of these components and many others in greater detail.

4.10. Power in Electric Circuits

The electrical power delivered to a device is equal to the current running through the device multiplied by the voltage difference across its input terminals:

Power (watts) = Current (amps) x Voltage (volts)

or, more succinctly:

Equation 8.6

$$P = E \times I$$

In a resistor, all that power is dissipated (i.e., released) as heat. All electronic devices act at least partly like resistors, even if they aren't designed to be resistors, so they also release heat when current is flowing through them. There are two major reasons that the designer of a power system for an underwater vehicle (or any other circuit for any other purpose) must be concerned about power:

1. The power source must be able to satisfy the combined power demands of all systems on the vehicle. To design a power system that can do this reliably, it's crucial that the designer know how much power each vehicle system requires and can estimate the peak power demand when all vehicle systems are running at full power at the same time.
2. As mentioned, all devices produce heat at some rate when electric current is flowing through them. If too much heat is produced too quickly, the device (or

something else near it) may overheat. And any overheated device may operate incorrectly, suffer permanent damage, or even start a fire. The technical specifications on most electric components specify the maximum power the device can handle without overheating, and designers need to pay close attention to that information.

4.11. AC and DC Electricity

In the context of electricity, the abbreviations "AC" and "DC" refer to **alternating current** and **direct current**, respectively.

AC power sources, AC appliances, and AC circuits produce or use electricity in which the currents in the circuit (and the voltages that drive those currents) alternate back and forth, usually between positive and negative values, and usually many times each second. For example, the voltage provided by a standard household wall outlet in the U.S. and Canada crosses back and forth between positive and negative values 60 times each second. If you plot a graph of voltage versus time for this power source, you'll see a pattern that looks like a sine wave (shown as the red line in Figure 8.16). Although many AC signals are sine waves, they don't have to be sinusoidal to qualify as alternating current. The tiny electrical impulses in your nervous system are an example of AC signals that are not sinusoidal.

AC voltages can change very quickly; those in microphone circuits often oscillate back and forth thousands of times each second, and those in radio circuits can be vibrating back and forth at frequencies ranging from thousands to millions or even billions of times each second.

DC power sources, DC appliances, and DC circuits, on the other hand, are characterized by voltages and currents that are more or less constant in time, like the voltage at the terminals of a freshly charged battery. Currents in DC devices may be switched ON or OFF, but when they are ON, they maintain fairly steady, constant values (as shown by the blue line in Figure 8.16).

It is common practice to convert between AC and DC forms of electricity. For example, the AC "adapters" you plug into a wall to recharge portable devices commonly convert the AC power from the wall outlet into DC power used to recharge batteries in the portable device. And some battery-powered devices, such as cell phones and radio

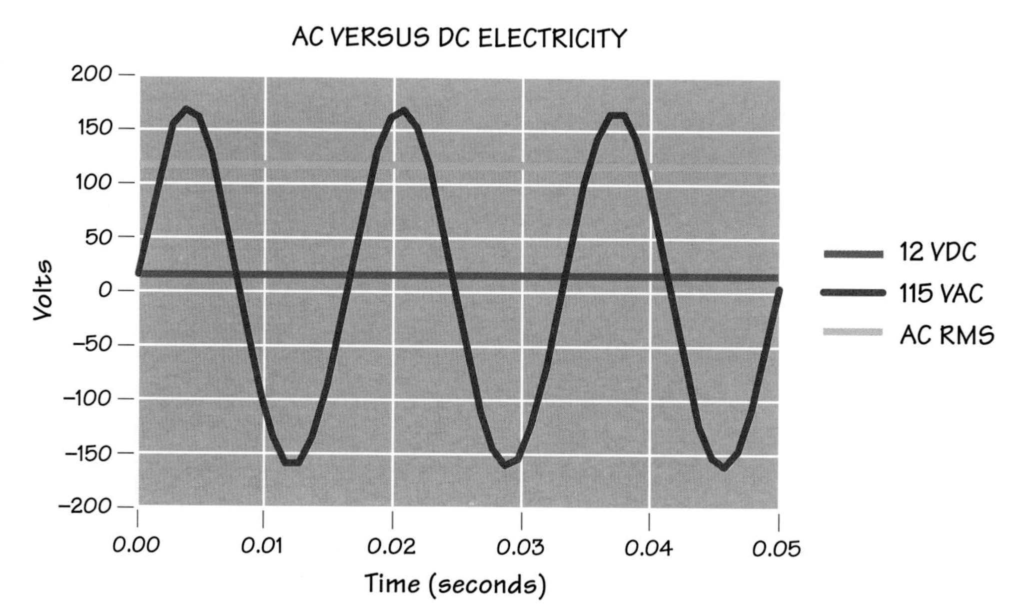

Figure 8.16: AC Versus DC Electricity

This graph compares the voltage outputs of two common electrical power sources: a 12-volt DC battery (12 VDC) and a 115-volt AC wall outlet (115 VAC). The dark blue horizontal line shows the constant 12 volts provided by the battery; it appears on the graph as a horizontal line because the voltage does not change from one moment to the next.

In contrast, the big red wiggly line shows the perpetually changing voltage provided by wall outlets like those common in U.S. homes and businesses. Although it is called "115 VAC," the actual voltage oscillates back and forth between about +165 V and -165 V. It is called 115 VAC because the power delivered to a resistive load by this alternating voltage is equal to the power that would be delivered to the load by a DC voltage source having a constant output of 115 V. This DC "equivalent" of an AC signal is called the root-mean-squared (RMS) voltage and is shown on the graph as an orange horizontal line labeled "AC RMS." For additional details, see the Tech Note: Quantifying AC Voltage.

TECH NOTE: QUANTIFYING AC VOLTAGE

How much voltage is provided by a wall outlet? It might surprise you to know that the voltage in a typical 115 VAC outlet in the United States actually varies between a low of about -165 volts and a high of about +165 volts, and it flips between these extremes exactly 60 times each second. These voltage values represent the peak excursions away from zero volts, but the oscillating signal spends most of each second somewhere in between these extremes.

Since most people care more about the average power delivered to their power tools and appliances than they do about the instantaneous voltage delivered, it makes sense to describe the voltage using a figure that can be plugged into the power-calculating formula (P = E x I). It turns out the meaningful way to get this average voltage is to use something called the **Root Mean Square (RMS)** method of averaging, which gives a result of 115 volts for the electrical power in a wall outlet. This method draws on the fact that the power delivered to a resistive load increases as the square of the voltage across the load. For example, tripling the voltage produces nine times the power.

RMS calculations basically square the value of the voltage at each instant to get an instantaneous power at each moment, then take the mean (average) of that power over one complete cycle of the sine wave, then take the square root of that average power to convert back to an equivalent DC voltage. Thus, the power available from a wall outlet with actual voltage oscillating between -165 and +165 volts is equivalent to the power available from a DC source providing a constant 115 volts. That's why the wall outlet is called a 115-volt outlet, even though it actually varies between about +/- 165 volts. (See Figure 8.16.)

Note that most volt meters and multimeters display RMS voltage when they measure an AC electrical signal. Less expensive models estimate the RMS voltage by measuring the peak voltages and then calculating the RMS voltage, based on the assumption that the signal is a sine wave. These meters will not report accurate values if the signal is not a perfect sine wave. More expensive meters generally offer "True RMS" readings, which accurately report the DC value that would give the same average power, even for waveforms that are not sine waves.

receivers, take DC power from their batteries and convert it to high-frequency AC oscillations used in radio communication circuits.

AC and DC sources of electrical power suitable for small underwater vehicles are discussed later in this chapter.

4.12. Series and Parallel Configurations

Whenever you have two or more batteries, lights, or other components in a circuit, they can be connected in different ways. Two particularly fundamental and important arrangements are given the names "series" and "parallel."

Components are said to be arranged in **series** when two or more components are connected end to end so that all the current flowing through one component must also flow through each of the other components. (See Figure 8.17.)

Although the amount of current flowing through two or more components in series will always be equal, the voltages across the different components does not need to be equal, and generally won't be (unless all the components are identical). The voltage

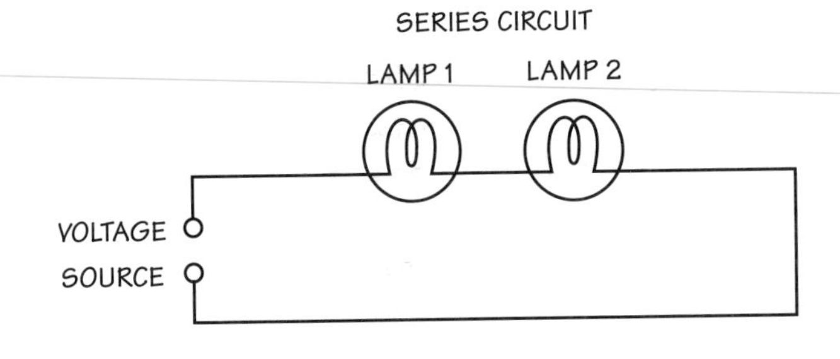

Figure 8.17: Components Arranged in Series

A series circuit connects devices—for example, light bulbs or batteries—end to end in a series, so that all electrical current flowing through one must also flow through the other.

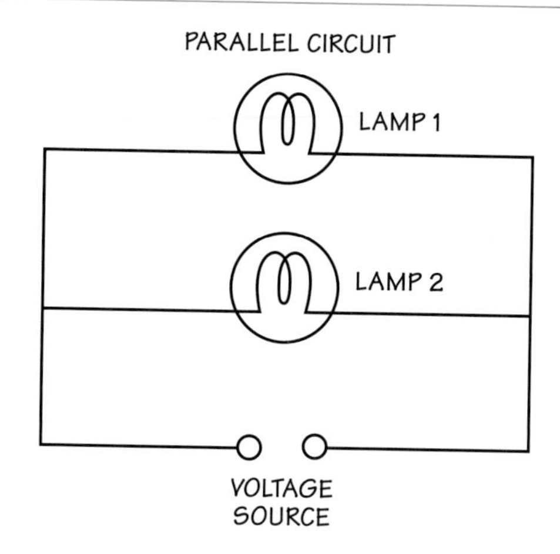

Figure 8.18: Components Arranged in Parallel

A parallel circuit arrangement connects two (or more) devices—for example, light bulbs or batteries—in a system, so that the current gets divided among them.

difference from one end of the series to the other will be equal to the sum of the voltage differences across each of the individual components.

The alternative is called a **parallel** arrangement (Figure 8.18). In this parallel arrangement, two or more components are exposed to exactly the same voltage; however, the current flowing into the parallel arrangement splits up, so that some fraction of the total current flows through each of the separate components. The split does not need to be equal, so the current through each component does not need to be equal, and usually it is not (unless all the components are identical).

5. How Much Electricity Does My Vehicle Need?

Now that you know some basic electricity concepts and vocabulary, it's time to start talking about the design of your electrical power system. The first step in this process is to develop a realistic sense of how much energy (and power) the vehicle will need to complete its missions successfully.

Remember that energy and power are not the same thing. Your vehicle needs plenty of both. The greatest demands for *power* in an underwater vehicle usually come from the propulsion system, which must work hard at times to propel the vehicle through the water or to maintain position against water currents. Often, the propulsion system is also the greatest consumer of *energy*. However, the propulsion system is rarely running at full throttle 100 percent of the time, so other systems that run continuously, such as lights used during a night dive or deep dive, may consume more total *energy* than the propulsion system during the course of a mission.

5.1. Power for Propulsion

Propulsion systems use power primarily to overcome energy lost to drag as the vehicle moves through the water. In the previous chapter, you learned two different ways—one theoretical, one empirical—to determine the drag on a vehicle at different speeds. You also learned that the thrust your propulsion system must produce to maintain a particular vehicle speed is equal in magnitude to the drag experienced by the vehicle while traveling at that speed. (See *Chapter 7: Moving and Maneuvering.*) Now it's time to look at the next step: equating your vehicle's measured or estimated thrust requirement to a power requirement.

Determining the minimum power *theoretically* needed to overcome drag at a given vehicle speed is straightforward, provided there are accurate numbers for the drag at that speed. For movement at a steady speed:

Equation 8.7

$$Power = Drag \times Speed$$

For example, if you towed a small vehicle at a steady speed of 1 m/s in a swimming pool, and it took 10 N of force to maintain that speed, then the minimum theoretical power needed to overcome the drag is 10 Nm/s = 10 W.

Unfortunately, this minimum value does not take into account any inefficiencies in propellers, motors, wiring, and so forth, so the power that must be supplied by a battery or other power source to move the vehicle at this speed will always be much higher than the calculated minimum. There is no simple way to predict precisely the power demand at the power source, because it depends on too many interdependent factors. The only thing you can bet on is that it will be greater than the minimum power value obtained by a drag-measuring experiment. In fact, it's not uncommon to have the power demand be more than 10 times greater!

However, it is possible to obtain a very rough estimate of the input power requirement using the following guideline, which is based on empirical data, rather than any theoretical calculations:

> For a small ROV or AUV propelled at speeds of less than 1 m/s by thrusters with electric motors and well-matched propellers, you can start by figuring that your thrusters will probably use 20–40 watts of electrical power per pound of thrust produced. There are professionally manufactured thrusters that beat this efficiency, but most thrusters designed and built by beginners do not.

If you are using a 12-volt battery as a power source, this translates into about 2 or 3 amps of electrical current per pound of thrust. So if your drag measurements in the pool indicate that you'll need 2 pounds of thrust for your vehicle, you'll probably want to plan for a couple of thrusters, each drawing something like 3 amps from a 12-volt battery. This sort of current draw is typical of a bilge pump motor with a small propeller on it.

Remember that no thruster will be on all the time, so your average power consumption will be less than the peak power consumption you would measure with the motor on. Thus, for example, if you run the thruster motors only half the time, your batteries will last twice as long as if you ran the motor continuously. This is particularly important in applications with limited battery power, such as vehicles powered from on-board batteries.

5.2. Other High-Power Systems

On a small remote-controlled or robotic vehicle, there are a few other potentially large consumers of power in addition to the propulsion system. These include bright lights (such as video lights), some cameras, and motors used for manipulator arms, water-sampling pumps, etc.

In most cases, these devices are being used for their intended purpose and draw power consistent with their specifications. For example, a 25-watt halogen light requires 25 watts of input power. Note that sometimes power is expressed indirectly as a combination of voltage and current. For example, a camera that requires 5 volts DC and draws 300 mA (milliamps) needs 5 V x 0.3 A = 1.5 W.

5.3. A Sample Power Budget

Table 8.2 shows the power budget for a real ROV, the *ROVing Otter* web-controlled ROV designed and built by students and faculty at California State University, Monterey Bay. The ROV is powered by an on-board battery and communicates with the surface via a standard Ethernet cable, much like those used to network computers in schools and office buildings.

It is worth noting that the low-power thrusters on this vehicle have proven too weak for most missions and will be upgraded to more powerful thrusters in the future. If you were designing a similar ROV, you'd probably want to plan for more powerful thrusters and correspondingly higher power and energy demands. That would improve performance, but would increase the peak power demands and drain the battery more quickly.

Table 8.2 assumes a 1.5-hour dive done at night (i.e., in the dark) and with very heavy thruster use (partly because these thrusters are so weak). It also assumes that most systems, except the thrusters and lights, are switched on half an hour prior to the dive (total ON time = 2 hours), to allow pre-dive testing of the web-based controls.

The values in the power column were determined by multiplying the voltage requirements of each device by the electrical current flowing through it.

Table 8.2: Sample Power Budget for a Small ROV

Device	Number	Power per device (maximum)	Total maximum power	Maximum ON time per dive	Total energy
Thruster motor	4	9 watts	36 watts	1.5 hrs	54 watt-hr
Video light	2	25 watts	50 watts	1.5 hrs (night dive)	75 watt-hr
Wireless Ethernet network switch	1	5 watts	5 watts	2 hrs	10 watt-hr
Network video camera	2	1.5 watts	3 watts	2 hrs	6 watt-hr
Camera tilt motor	1	9 watts	9 watts	negligible	negligible
Microcontroller (small computer)	2	0.5 watts	1 watt	2 hrs	2 watt-hr
Lasers	2	1 watt	2 watts	1.0 hr	2 watt-hr
Misc. sensors and other electronics	several	negligible	negligible	2 hours	negligible
TOTALS			**106 watts**		**149 watt-hr**

Any well-designed power system should be able to handle a "worst-case" scenario in which all systems are turned on at maximum-power at the same time. The "total maximum power" column in the table helps you keep track of this by recording the peak power required by each device on the ROV, then letting you add those together to get the maximum power demand theoretically possible for the entire vehicle. It may be unlikely that you'll have every thruster, light, camera, and other device pushed to its limit at exactly the same moment, but it's reassuring to know that you've designed a power system that could handle that situation without incident, if it ever arose. In this example, the table shows that you'd want a power system capable of delivering at least 106 watts.

Likewise, you want to make sure your power system stores and can deliver enough total energy to keep everything going through the end of the mission. After all, it's not good to run out of gas prematurely! The "total energy" column keeps track of this requirement and reveals that the energy supply for this vehicle should store at least 149 watt-hours of energy to complete this mission. (One watt-hour is 1/1000th of the kilowatt-hour unit of energy introduced in Section 2.1.4 of this chapter.)

This vehicle is powered by an on-board, sealed, lead-acid battery. Batteries will be discussed in much greater detail later in this chapter. For the purposes of this discussion, what you need to know about this battery is that it's specified as having a nominal energy capacity (when fully charged) of 144 watt-hours; that's just a hint shy of the 149 watt-hours recommended by the table. It's also specified as having a maximum power delivery capability of about 430 watts; that's roughly four times greater than the largest anticipated power demand.

At first it might seem that this battery is reasonably well matched to the energy and power requirements of this vehicle and its mission. However, actual dive experience suggests this is not quite so. It turns out there's some fine print in the battery specifications that can sneak up and bite you, if you're not careful. In this case, the 144 watt-hour energy capacity was specified for a 20-hour discharge rate. But this mission drains most of the energy from the battery in only 1.5 hours. The more rapid rate of discharge reduces the effective amount of energy available from the battery to less than 144 watt-hours. Fortunately, this reduction is somewhat offset by the fact that thruster use is never 100 percent during the mission. As a result, the battery usually does last for the full 1.5 hours of the mission. But there's another problem—the vehicle is noticeably less spunky after only an hour of night diving than it is at the beginning of the dive. That's because the maximum power output of a battery falls as it loses stored energy. The battery may have started with the ability to deliver 430 watts, but after losing a significant fraction of its original charge, it's no longer able to meet the power needs of the thrusters, so the vehicle begins to get sluggish.

Several important lessons can be gleaned from a careful study of this table and the associated performance of the real vehicle.

First, power-hungry devices are not necessarily energy-hungry. For example, the camera-tilt motor draws enough current to add significantly to the peak power demands in this vehicle, but is used so infrequently during dives that its impact on the total energy budget is negligible.

Second, low-power devices that are on all the time may add significantly to the total energy requirement without having much effect on the peak power requirement. This is particularly true if they are the only things that stay on for long periods of time. In this example, the camera, Ethernet network switch, and microcontroller collectively consume a steady 9 watts. That's enough to deplete the battery in about 15 hours, even if the high-powered thrusters and video lights were never used at all.

Third, it's not a good idea to cut things too closely. It looks on paper as though the 12 V, 12 Ah battery is about right for meeting the energy and power requirements for this night-dive mission. Practical experience with this vehicle reveals that the battery does meet the requirements, but just barely. The battery will power this vehicle for 1.5 hours at night without being completely dead, but performance is definitely reduced near the end of such a dive. For consistently good performance, it would be better to have a bigger battery.

Fourth, you can build in a bit of a safety margin by developing your energy and power-budget tables around worst-case scenarios, which are unlikely to occur in reality. In this particular case, the table was developed for a night dive, even though this ROV normally operates by daylight and uses its video lights sparingly, if at all. It also operates mostly in calm conditions, when even its wimpy thrusters do not have to work anywhere near the 100 percent capacity indicated in the table. Because of these realities, the *ROVing Otter* usually gets 2 to 3 hours or more of useful dive time under its normal operating conditions, even though the table predicts about 1.5 hours. It's always helpful to plan conservatively like this. After all, it's better to be pleasantly surprised when your vehicle can do a bit more than required than it is to discover that your vehicle can't complete its required mission.

6. Electric Power Sources for Small Vehicles

Most electrical devices used in homes, schools, and businesses are powered either by plugging them into a wall outlet or by using batteries. Both of these options can provide electric power for small underwater vehicles, though batteries are usually safer to use around water.

Wall outlets and batteries are both considered **voltage sources**, because they attempt to supply a well-controlled voltage (rather than a well-controlled current) to any appliance or other device connected to them. They rely on the electrical properties of the circuits in the device to limit the amount of current that flows from the voltage source in response to the applied voltage, just as a resistor limits the current flow through it, according to Ohm's Law. (See *Section 7.2: Fuses* and *Safety Note: Use Your Fuse* for a discussion of what can happen if the device fails to limit current to a safe level.)

Wall outlets provide an alternating (AC) voltage output, whereas batteries maintain a reasonably constant (DC) voltage output. A variety of common AC/DC power supplies and adapters can convert AC power from wall outlets into lower-voltage DC power for use with devices designed to run on DC voltages. The following sections discuss AC power, AC/DC adapters, and batteries in more detail.

6.1. AC Power

When you plug a power tool, kitchen appliance, stereo system, computer, or other device into a wall outlet, you are connecting it to a source of electrical power called the **mains**. This is basically the wiring infrastructure in a house or business that distributes electrical power from the power company to the various lights, outlets, and other circuitry in the building. In the United States, mains power is about 115 VAC oscillating at a frequency of 60 Hz. In other countries or in other settings (such as industrial settings or aboard an ocean-going ship), the voltage and/or frequency of the mains may be different, and the plugs may be different shapes to prevent accidental mix-ups.

AC power outlets provide a very convenient (though potentially dangerous) source of electrical power for many projects, because these outlets are abundant in and around most buildings. They can also be found on many boats, RVs, and other vehicles. If you don't happen to have an outlet exactly where you need it, you can use an extension cord to "move" electricity from a nearby outlet to your location. If there are no outlets within reach of an extension cord, you can even use a gasoline- or diesel-powered generator, or use solar panels, batteries, and an inverter to generate regular AC power wherever you need it, even in extremely remote places.

Most common U.S. outlets are capable of delivering about 15 A of current at 115 volts RMS. (See *Tech Note: Quantifying AC Voltage* earlier.) That translates into more than 1,700 watts of available power. This amount of power can be quite useful when you need it for electric heaters, power tools, or major appliances; however, it can also be dangerous and must be used with caution.

6.2. AC Power Safety

Never forget that the 115 VAC 60 Hz electricity in standard household outlets can and does kill people—roughly 100 people every year in the U.S. alone. Most of the victims were working on household wiring or electrical appliances at the time. The risk of electrocution is much greater in the presence of water, which is where you'll likely be working with the electronics inside your ROV or AUV, at least some of the time.

Given the widely recognized danger of mixing water and electricity, it may seem odd that commercial ROVs, both large and small, routinely rely on high voltages (often much higher than 115 VAC) for power. However, these vehicles and their control systems also incorporate advanced materials and design features for safe containment of high voltages in wet areas, even under "worst-case" operating conditions. These safety features are designed, built, and tested by professionals with years of experience. Moreover, these safety systems normally undergo rigorous inspections at regular intervals, and all personnel who work with these systems are trained to operate or repair them safely.

It is unrealistic to expect people without this advanced level of training and experience to produce equally effective and reliable safety systems. **Therefore, standard AC electrical power is *not recommended* for most underwater vehicle projects, particularly those being done by beginners or school groups.** There is simply too much risk of a fatal shock. Batteries, and to a limited extent, AC/DC power supplies and adapters (both discussed later in this chapter), provide a much safer alternative.

That said, you will still have occasion to use mains power. The power tools used to build most vehicles (in dry workshops), the battery chargers used to recharge vehicle batteries, and the TV monitors used to view an ROV's environment through its video camera are almost all designed to be powered from standard AC outlets. Therefore, you must know how to use this power source safely, even if you aren't going to use it to power your vehicle directly.

For starters, try to avoid using mains power near water. "Near water" includes damp concrete floors or pool decks, which conduct electricity quite well. For example, you can usually move batteries to a dry area away from water before trying to recharge them. If you need a TV monitor to see what your ROV's camera is seeing in the pool or under a dock, keep it far enough from the edge of the water that there is little or no danger of it or its power cords falling into the water or getting splashed.

If you must use mains power near a pool or other damp/wet area, always make absolutely sure that all electrical power supplied through an outlet (or extension cord or generator or inverter) to any device in an area near a body of water is coming through a **GFI** (**ground-fault interrupter**) outlet, GFI extension cord or other GFI protection device. (See the *Safety Note: A GFI Could Save Your Life.*)

Electricity is invisible, silent, and millions of times faster than your reflexes, so don't foolishly pretend you can dodge it. If you slip up, you may be injured or even killed before you even know you made a mistake. The electrical power from wall outlets can also start fires easily, and each year many people die in fires caused by faulty wiring.

To avoid becoming a statistic, exercise extreme care when using AC electrical power from an outlet or when repairing any device that normally plugs into an outlet. Always follow these basic safety precautions:

1. Never touch an electrical device that is plugged in if either you or the device is in contact with water.

2. If using an outlet-powered electrical device near water, make absolutely certain that all electrical power supplied to that device through an outlet (or generator or inverter) is coming through a GFI outlet, GFI extension cord, or other GFI-protected device. Press the "test" button on the GFI to make sure it's working properly before you trust it with your life.

3. If you are going to repair, adjust, test, or otherwise work on the "innards" of an electric device that is normally powered from a wall outlet, follow these additional safety precautions before you start this potentially lethal activity:

   a) Never work on any electrical system of any kind while it is plugged in. (Note: Turning off the device is NOT enough to protect you from electrocution if you are working on the internal circuitry of the device; it must be unplugged.)

   b) Always have a responsible person with you—preferably someone trained in CPR and first aid who knows how *not* to get electrocuted while rescuing someone who is being electrocuted.

SAFETY NOTE: A GFI COULD SAVE YOUR LIFE

You should always use a ground-fault interrupter (GFI)—also known as a ground-fault circuit interrupter (GFCI)—when using regular AC mains power from an outlet, extension cord, generator, inverter, or similar source anywhere near water.

Unlike a fuse or circuit breaker (both discussed later in this chapter), a GFI is *not* a form of current overload protection. It guards against a different type of electrical hazard—the one most commonly associated with electrocution in wet environments. This hazard is called a **ground fault.** It occurs when electrical current finds an "unauthorized" path to ground, possibly through a person's body.

So how does a GFI work? Normally, when a device is plugged into an outlet (whether a regular outlet or a GFI-protected outlet), *all* of the current flows from the outlet into the device through one prong of its plug, then returns to the same outlet through the plug's other prong. Thus, the amount of current flowing out through one side of the plug is *exactly* the same as the amount flowing in through the other side of the plug.

One of the most common ways that people get electrocuted, particularly around water, is that their body inadvertently becomes an alternative path for electric current seeking a shortcut to ground. Instead of returning to the outlet as it should, current takes a detour through the person's body, escaping from its usual return path to the outlet.

A GFI constantly compares the amount of current leaving the outlet with the amount of current coming back in. If the two aren't *exactly* equal, the GFI recognizes that some of the current must be taking an abnormal route and immediately shuts off all power to the outlet. A GFI can detect a ground fault condition and shut off power so quickly that sometimes people don't even know they were beginning to get a shock. GFIs are credited with saving many lives from electrocution. Now most building codes require their use in kitchens, bathrooms, locker rooms, swimming pool areas, and other areas near water.

Image courtesy of Dr. Steven W. Moore

Figure 8.19: A GFI Outlet

Many GFIs look and function like a regular outlet, except they usually have a couple of small pushbuttons and sometimes a small indicator light that are not found on other outlets.

One button is used to test the GFI. When pushed, it should "trip" the GFI, cutting off power to the outlet. The other button is used to reset the GFI and restore power. GFIs are also incorporated into some extension cords and may be built into some circuit breakers.

WARNING: A GFI does ***not*** protect against all possible electrocution hazards. Remember, a GFI works by looking for differences between the current flowing in and out of a circuit. It will not detect problems unless current fails to return to the outlet. For example, if you touch something connected to the "hot" wire from an outlet with one hand and simultaneously touch something else connected to the "neutral" wire with your other hand, you can get electrocuted without tripping the GFI. In this scenario, the GFI does not detect any difference between the outbound and inbound current, so it treats you like any other power-consuming "appliance" plugged into the outlet and does not shut off the power!

c) Work in a dry, uncluttered location.

d) Don't let your body become an electrical pathway to ground. Avoid working on circuits while leaning against water pipes, heating ducts, or other metal structures that might carry electrical current from your body to ground. Wear rubber-soled shoes and dry clothing to help insulate your skin from contact with current-carrying materials. Consider wearing rubber gloves.

e) Use non-metal or insulated tools whenever possible.

f) Double-check, then check again, to be sure the device is really unplugged before you reach in with hands or tools. Have a buddy repeat this check *independently* to make sure you didn't miss something. (For example, did you ensure the brown cord was unplugged without realizing that the device you're working on is the one with the white cord that is still plugged in?)

g) Remember that some circuit components can store dangerous amounts of electricity for a long time even after the device is unplugged. Before touching any wiring or other electrical components, be sure to short the contacts on any large capacitors and use a voltmeter or other reliable method to verify that no residual voltage is left in the circuit.

h) Note that many "old-timers" make it a habit to keep one hand always in their back pocket when working inside high-voltage electrical devices. This may be a bit inconvenient, but it greatly reduces the chance of electric current entering one hand and crossing directly through the heart muscle on its way to exiting through the other hand.

If you think all this safety stuff is excessive, remember the famous saying about airplane pilots: There are old pilots and there are bold pilots, but there are no old, bold pilots. A similar reality exists for those who work with high-voltage circuits.

SAFETY NOTE: AC POWER FROM GENERATORS OR INVERTERS

Portable generators, solar panels, and inverters are all great ways to obtain standard 115 VAC electrical power in remote locations, including boats at sea. However, electricity from these sources has all of the hazards normally associated with more conventional wall outlets, plus a few more.

For example, the safe use of portable gasoline- or diesel-powered generators requires more experience and caution than regular outlets do. In addition to the electrical shock hazards of outlets, these generators use fuels that are toxic and highly flammable. To make matters even worse, the fuel vapors are usually explosive. The electricity produced by the generator can create a spark that could ignite a fire or explosion in the presence of spilled fuel or fuel vapors.

Inverters work by taking electrical power from a battery (usually something like a 12 VDC car battery) and converting it into 115 VAC power. They are often used in boats or recreational vehicles to convert power from a car battery into standard AC power that can be used to run small appliances. They are also used to produce 115 VAC power in solar power systems. Solar panels produce DC electricity from sunlight. In a typical solar power system, this DC current is used to charge batteries during the day. An inverter then converts the stored battery power into AC power as needed, night or day. Anyone using an inverter for AC power must therefore be aware of electrical hazards, plus they must be aware of hazards associated with solar panels (some of which can produce dangerously high voltages in bright light) and batteries, which have their own host of potential hazards discussed later in this chapter.

SAFETY NOTE: AC/DC POWER SUPPLIES AND ADAPTERS

Many small electric devices, such as cell phones, laptop computers, battery chargers, and the like are powered through **AC-to-DC adapters**. These adapters, sometimes affectionately known as "wall warts," plug into a regular AC wall outlet at one end and have low-voltage DC electricity emerging through a little round plug at the other end of a wire.

It might seem like these adapters could provide a good option for converting the relatively high-voltage (and therefore dangerous) AC power in the mains into the low-voltage (and therefore safer) DC voltages used to propel small ROVs in water. If the DC wire is long enough, it seems the adapter could be plugged into an AC line in a dry place away from water with only the low-voltage DC wiring going out to the swimming pool or other area near water.

Unfortunately, this option is not quite as realistic as it might sound. In most cases, these adapters cannot produce anywhere near enough current to power an ROV, even a very small one. In those few adapters that can, the wires are not long enough to go very far, and if you try to lengthen them, you'll find you don't have much power left at the far end. (The reasons for this are described later in the chapter in *Section 7.4.1. Voltage Drop*.)

However, a few high-capacity "AC/DC power supplies" may have voltage and current capabilities sufficient to drive a small ROV at a modest distance. Examples include the power supplies designed for "quick-charging" dead car batteries and the 48 VDC, 1,000-watt switching power supplies used to provide power for ROVs during some of the MATE competitions.

WARNING: There is always a possibility that a short inside an adapter or power supply could accidentally connect the 115 VAC mains power directly to the (supposedly) low-voltage DC wires, so these adapters should be treated with the same respect afforded higher-voltage power sources and should be used only with a GFI-protected circuit when any part of the adapter (or someone handling it) will come into possible contact with water. (See *Safety Note: A GFI Could Save Your Life*.)

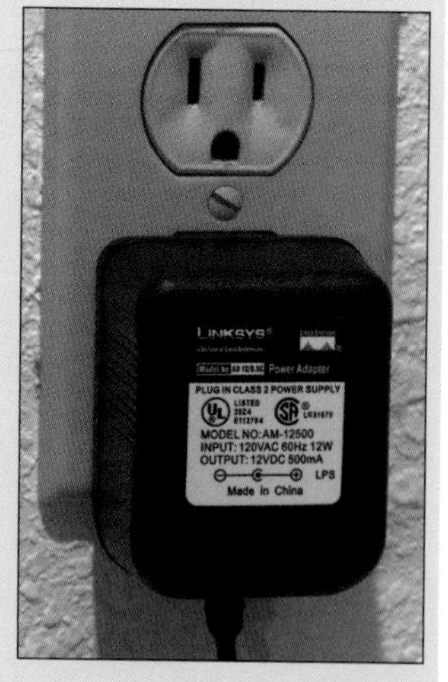

Image courtesy of Dr. Steven W. Moore

Figure 8.20: Wall Wart

AC power adapters, sometimes called wall warts, power bricks, or plug-in adapters, are a common and convenient way to convert relatively high-voltage AC power from a wall outlet into the lower DC voltages needed to power many of today's household electronic devices. Note that most are intended for indoor (i.e., dry) conditions only.

Figure 8.21: AC/DC Power Supply Options

The first photo shows a portable emergency automotive charger with built-in battery. The upper right photo shows a variable AC/DC power supply. The lower right photo is a 12 volt Mean Well AC/DC Switching Power Supply.

Note that some AC/DC switched power supplies can deliver over 1000 watts of power at low DC voltages. These can be used instead of batteries to power small ROVs in some cases. But because they plug into standard AC outlets, they should be treated with the same precautions as any other device powered off the AC mains, so use a GFI-protected circuit.

Image courtesy of Vickie Jensen

Image ourtesy of Randall Fox

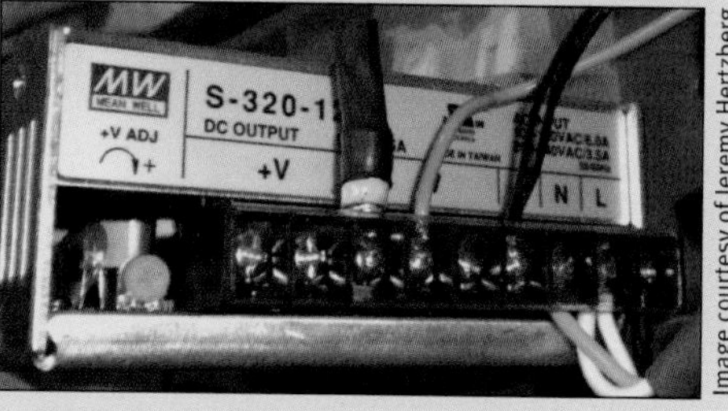

Image courtesy of Jeremy Hertzberg

6.3. An Introduction to Batteries

Low-voltage batteries provide an ideal, relatively safe source of electrical power for small underwater vehicles designed and built by students as school or club projects. These batteries are also excellent for vehicles constructed by adults who are less than 100 percent confident in their ability to work safely with high voltages around power.

On many AUVs, some small ROVs, and some manned submersibles, batteries power everything from thrusters and lights to communication and even life support systems. Even in larger vehicles that use diesel generators, nuclear reactors, or other power sources as their primary source, batteries play an important role as supplementary energy storage devices and sometimes as emergency backup power systems.

The advantages of batteries become clear when you match them up against the list of criteria for selecting a good power source (which was provided earlier in this chapter).

- Batteries suitable for small ROV/AUV projects are relatively easy to find and relatively affordable, particularly when you consider that many types are rechargeable and therefore reusable.
- They store a respectable amount of energy in a manageable weight and volume and can deliver plenty of power for typical ROV and AUV applications.
- Most (in particular, the non-flooded varieties) can be used upside-down, sideways, or in any other position without creating problems.
- As power sources go, they are very clean (not oily, greasy, or radioactive).
- It is tremendously easy to distribute power from batteries to devices that need the power, since you can do it simply by stringing wires; you don't need precisely aligned pulleys, belts, rotating shafts, etc.

Figure 8.22: On-Board Power

These student-built AUVs competing in the 2008 AUVSI and ONR International Autonomous Underwater Vehicle (AUV) Competition use on-board batteries for power.

Image courtesy of AUVSI Foundation, http://www.auvsi.org/competitions/water.cfm

- The energy density and power-to-weight ratios of batteries are not particularly good compared to those of fossil fuels and fossil fuel engines, but then this is a book about building underwater vehicles, not airplanes, so this is not such a big issue.
- Low-voltage batteries are much safer than most alternative power sources, though they do have a few hazards of their own that you should know about. (Battery hazards and safety are described below.)
- Finally, you can find plenty of good information on the web and elsewhere about how to choose, use, and care for batteries.

This is an exciting time in the world of battery development. There is high (and growing) consumer demand for battery-operated devices like cell phones, camcorders, laptop computers, hybrid and all-electric cars, radio-controlled airplanes, robots, and other gizmos. For these devices, consumers are demanding batteries that are smaller, lighter, more powerful, faster-charging, longer-lasting, more eco-friendly, and more affordable than in the past. Because of the huge sums of money to be made by those who can deliver what consumers want, a great deal of time and money is being invested in battery research right now, so battery technology is evolving at an unprecedented rate. That's great news for people designing underwater vehicles and countless other devices.

Unfortunately, it also means that the answer to the question, "What's the best type of battery to use for my project?" is a moving target. The best type of battery for your ROV project today may not be the best type of battery for your ROV project two years from now. Therefore, while this chapter discusses the pros and cons of some existing battery technologies, it also includes more general criteria you can use to compare battery types. Those criteria will remain useful for identifying the best battery types for your project, even as today's batteries become obsolete and newer, better ones take center stage.

6.3.1. Exactly What Is a Battery?

You may recall from *Chapter 4: Structure and Materials* that different kinds of metals have different affinities for electrons. If provided with a way to do so, electrons will flow from a less electronegative metal (one that doesn't hold on to electrons as tightly) to a more electronegative metal, leading to galvanic corrosion. Galvanic corrosion can have both good and bad consequences. The discussion in Chapter 4 focused on a *bad* consequence, namely that important structural or cosmetic metals often get eaten away by these corrosive reactions. The discussion of batteries in the present chapter will focus instead on a *good* consequence of galvanic corrosion, which is that energy can be extracted from the electrons and used to perform useful work as they move from one metal to the other. This is the electrochemical basis of how batteries work and provide power. In a sense, a battery is simply a container full of corrosion put to good use.

Technically, the term **battery** refers to a set of interconnected **galvanic cells**, where each galvanic cell consists of two electrodes (each made from a different type of metal) in contact with some corrosive liquid, paste, or gel that promotes a corrosion reaction. The galvanic reaction generates a voltage difference between the two electrodes, and this voltage can be used to drive electrons through a circuit as they try to move from the more negative electrode to the more positive electrode, thus powering the circuit.

Galvanic cell voltages range from less than 1 volt up to a maximum of about 3 volts, depending on the metals and corrosive chemicals used inside the cell. By linking more than one cell end to end in series, with the (-) electrode of one cell connected to the (-) electrode of the next, it's possible to create batteries with voltages higher than the voltage of individual cells. For example, a typical 12-volt car battery is actually made up of a series of six lead-acid galvanic cells, each with a voltage of just over 2 volts, connected in series.

Image courtesy of Dr. Steven W. Moore

Figure 8.23: Batteries

Batteries come in a tremendous variety of shapes, sizes, voltages, chemistries, energy capacities, and prices. This photo illustrates just a few fairly common ones on the smaller end of the size spectrum.

When it comes time to select batteries for your vehicle, research your options thoroughly so you get the optimal kind(s) available for your particular vehicle and mission, consistent of course with your overall project budget.

To avoid possible confusion, note that the term "battery" is commonly used both for true multi-cell batteries and for single galvanic cells. For example, the common 1.5-volt alkaline "dry cells" are more commonly called 1.5-volt alkaline "batteries," even though they contain only one cell and therefore aren't technically batteries at all. The remainder of this book follows the colloquial convention and uses the term "battery" to refer both to single galvanic cells and to batteries of more than one cell.

Not surprisingly, batteries are classified according to the metals and corrosion chemistry used in their galvanic cell(s). Some of these chemistries are designed for one-time use only; these are called **primary batteries**. The standard alkaline dry cells widely available in grocery stores, drug stores, and hardware stores are a familiar type of primary battery. The coin-shaped lithium batteries often used for calculators, digital wristwatches, and hearing aids are also primary batteries. Primary batteries cannot, in general, be recharged. In fact, trying to do so can be dangerous.

Other battery chemistries are designed to be rechargeable, so the batteries can be reused multiple times. Batteries with rechargeable chemistries are called **secondary batteries**. Secondary batteries are designed so that an external power source (the battery charger) can be used to force electrons through the system opposite the usual direction. This essentially runs the battery's internal corrosion reaction in reverse, more or less repairing the corroded metal inside and restoring the pre-corroded chemical condition. After being recharged in this way, the battery can be used to power circuits all over again. Popular secondary battery chemistries include lead-acid, Nickel-Cadmium (Ni-Cad), Nickel Metal Hydride (NiMH), and Lithium ion, or the newer Lithium polymer (Li-ion or Li-poly) batteries.

In addition to these common primary and secondary battery chemistries, there are dozens of other battery chemistries out there, but most of them are too specialized and/or too expensive and/or too experimental to be of much practical interest to the average gadget builder—at least for now.

Every battery (whether a single galvanic cell or a true battery) has two external terminals that provide a way to connect an electrical circuit to two different metal electrodes that are in contact with the highly corrosive solution, paste, or gel inside the battery. In some batteries, the terminals are just exposed metal surfaces on either end of the battery. In others, they are recessed inside some sort of socket or appear as metal tabs, bumps, bolts, or wires protruding from the battery.

If a complete electrical circuit is made so that electrons can flow from one terminal of the battery through the load (motor, light, or other device being powered) and back to the other terminal of the battery, then corrosion will proceed inside the battery and provide electrical power for the circuit. As the corrosion proceeds, the internal chemicals and/or metal electrodes will get used up, and the battery will eventually lose its ability to provide additional power. At that point, you have what's called a **dead battery**. If it's a secondary battery, you can recharge it. If not, the battery needs to be replaced, and the dead one needs to be disposed of properly. (See *Section 6.5.14. Ease of Acquisition and Disposal* and *Safety Note: Safe Battery Disposal.*)

Most batteries are available in several different sizes and shapes, and sometimes in different voltages made by connecting different numbers of cells together. For example, the familiar alkaline batteries sold in many stores routinely come in four cylindrical sizes (AAA, AA, C, and D), all with about 1.5-volt output, plus a rectangular 9-volt "transistor battery" form (a term deriving from early portable transistorized radios that required these higher voltage batteries). By comparison, lead-acid batteries usually come in larger rectangular packages and offer either 6 or 12 volts at their terminals.

6.4. Battery Safety

Batteries provide one of the safest ways to power a small electric vehicle. Nonetheless, all batteries, regardless of size, are potentially dangerous. Before you select batteries for your vehicle, review the hazards associated with each type you are considering. When you get your batteries, be sure to read all safety information that comes with the

SAFETY NOTE: CAR BATTERY CAUTIONS

For years, wet-cell lead-acid car batteries have been a popular source of power for hand-built underwater vehicles, so it's important to highlight the specific dangers associated with their use.

- Car batteries can easily deliver excessively high currents (way more than most small vehicles need), with more than enough power to melt metal tools, cause severe burns, or rapidly ignite a fire. In terms of power, they are unnecessarily dangerous overkill.
- The concentrated liquid sulfuric acid in these batteries is extremely caustic and can cause serious chemical burns to skin or eyes, and because it's in liquid form in car batteries, it can spill or splatter easily if the battery is tipped or dropped. (See *Safety Note: Safe Battery Transport.*)
- Hydrogen gas liberated from these batteries when they are severely discharged or when they are being charged can explode if ignited by a spark from the battery or any other source, and the explosion can splatter acid. Even if it's not ignited, the gas can build up to dangerous pressures whenever the batteries are used or charged in airtight containers, such as pressure canisters.
- Chemical reactions between the sulfuric acid in a car battery and the chloride ions in seawater can produce toxic chlorine gas.
- Finally, most car batteries are very heavy and require proper lifting technique to prevent injuries.

For all of these reasons, it is best to consider alternatives to car batteries. Select another, safer type of power source for hand-built underwater vehicles. If possible, use sealed lead-acid (SLA) batteries, which come in several varieties, including AGM (absorbed-glass mat) to provide similar battery chemistry in a safer package. (See Section 6.7.2 for more information on these types of batteries.)

Image courtesy of Jill Zande

Figure 8.24: Seeing is Believing

Car batteries are capable of delivering enough power to melt metal.

batteries and with any battery charger. Here's a quick summary of some of the potential hazards you need to think about when using batteries.

- **Fires**: Any battery powerful enough to propel an underwater vehicle is more than powerful enough to start a fire. This can happen, for example, if a short circuit causes a wire to get red-hot and ignite nearby flammable items. Even a tiny battery can make a spark big enough to ignite highly flammable materials or vapors. Some batteries (notably those containing lithium) can spontaneously burst into flames and burn intensely if they are charged incorrectly or if the lithium inside is ever exposed to oxygen. A lithium battery that may possibly have been damaged should be treated with extreme caution and stored away from flammable materials while awaiting proper disposal. Never try to disassemble any battery.
- **Heat-related burns**: Even if a short circuit doesn't actually start a fire, it can make wires, other components, or even the battery itself hot enough to burn your skin.
- **Explosions**: Many batteries can explode if overheated by incorrect recharging, shorting, or disposal in fire. Some, such as lead-acid batteries, give off hydrogen

SAFETY NOTE: SAFE BATTERY TRANSPORT

Special precautions should be taken when transporting any battery or set of batteries. There are two major dangers: 1) fire due to short-circuited battery terminals and 2) acid spills.

All batteries store energy, and most—even the little AAA-sized ones—are capable of delivering plenty of power to start a fire. Since batteries and other objects can (and usually do) shift around while being transported by cars, backpacks, or other means, it's important to make sure there's no way they can accidentally get shorted out when being moved. It's a good idea to keep batteries boxed separately from tools or other metal objects and to secure the batteries so that they can't roll around or fall over and so that metal objects can't slide or fall in among them.

As an added precaution on long or bumpy trips, you may want to tape over at least one of the terminals on each battery, so there's no way a complete circuit could form, even if something metal did fall across the battery terminals. Larger batteries with both terminals located on the same surface of the battery are particularly prone to shorting, since it's easy for a single metal object to contact both terminals at once. Never carry these batteries in a toolbox with metal tools like screwdrivers or wrenches. It's just way too easy for one of these tools to contact both terminals at the same time, thereby switching on an unsupervised "arc welder" in the back of your car! (This happens much more frequently than you might imagine, and the consequences can be disastrous.)

If friends or family help you pack the car, make sure they are aware that you are packing batteries and explain how to do so safely. The authors of this book know of at least one case where someone carefully packed the batteries far away from metal and even covered the terminals with some sweatshirts for added protection. A second person, who did not know the batteries were there, placed some wrenches on top of the sweatshirts. During the ensuing trip, the sweatshirts and wrenches shifted places slightly. One of the wrenches slid down and fell across the battery terminals. This created sparks and heated the wrench so much that it scorched one of the sweatshirt sleeves. Fortunately, somebody seated close to the batteries heard the zapping sound of the sparks, and noticed the situation and corrected it immediately. If it had gone undetected for even a few seconds longer, this unlikely scenario could have ignited the sweatshirts or other flammable objects nearby and created a serious car fire!

As discussed above, the hazard of spilled acid is a concern primarily with so called "wet" or "flooded" batteries, such as conventional car batteries, which contain strong acids or other corrosive electrolytes in liquid form. If these batteries tip over, they can spill their acid. Or acid can get sloshed out of them on a bumpy trip, even if they don't actually fall over. Battery acid is strong stuff that can burn skin and eat through metal. It is particularly dangerous (and illegal) to transport these batteries on commercial airliners, so don't even think about putting one in a suitcase for your next trip to an ROV contest.

or other explosive gases when being recharged or excessively discharged; this problem can be exacerbated if those gases are trapped and concentrated inside a pressure canister. Even if the gases don't ignite, their release inside a sealed pressure canister can generate pressures sufficient to turn the pressure canister into an explosion hazard.

- **Chemical burns**: Most batteries rely on strong acids or other caustic chemicals as part of their internal oxidation/reduction chemistry, and these chemicals can be dangerous.
- **Poisonous gases**: In some situations, batteries can give off toxic gases that can be harmful if inhaled.
- **Physical injuries**: Heavy batteries can fall off a table or slip out of your hands and break someone's foot. They can also cause back injuries if not lifted properly.

In addition to these general tips for safe battery use, see the related safety notes in this chapter.

6.5. Battery Performance Characteristics

Choosing the correct batteries for your ROV or AUV project is an important decision. Each battery chemistry, size, and package type has advantages as well as disadvantages. There is no single perfect answer, as the best choice always depends on the particular application.

As mentioned earlier, battery technology is evolving very quickly, so the best type of battery today may not be the best tomorrow. Therefore, rather than recommending specific batteries, this text starts with a list of general battery characteristics you can use to select from among available options now or in the future.

- voltage
- primary versus secondary batteries
- energy capacity
- energy density
- weight
- maximum power output
- discharge curves
- depth of discharge
- maximum charge rate
- temperature performance
- size and shape
- shelf life
- required maintenance
- ease of acquisition and disposal
- price

Most of the battery performance characteristics in this list and discussed in greater detail below are not detailed in typical consumer packaging. That packaging is not intended for a technical audience. Rather, it is designed to attract the general public with nebulous claims like "new," "more power," and "longer lasting." Anyone doing serious power system design work will need more detailed and precise quantitative information. For that, you will need to turn to industrial information sources written for engineers, who depend on accurate quantitative data about battery performance. You can find this information in a few places.

One of the best and easiest ways to find this data is to look on the website of a company that manufactures (not just sells) batteries. Panasonic, Energizer, and Duracell are a few examples. Many of these companies will include detailed technical information on their website, including **datasheets** on particular battery models.

Datasheets are summaries of the technical engineering details about various products, including batteries and other electronic components. Unfortunately, locating datasheets is easier said than done. To find them, try going to the home page of one of these major battery manufacturers. Ignore the glitzy marketing stuff about the latest consumer products and search instead for (usually small, subtle, all but hidden) links with names like "technical information," "industrial customers," or "OEM." (**OEM** stands for "original equipment manufacturer" and is a designation often used on parts that are designed to be built into other systems, rather than being finished, consumer-ready systems in their own right.) These links, if available, will usually lead you eventually to datasheets on each type of battery the company manufactures. In some cases, you can get this information from distributors that sell batteries, too, but usually not.

You may want to purchase only batteries for which such technical data is available from the manufacturer, because this detailed information can be very important in thoroughly understanding the power system performance of your vehicle. It's also a great way to learn more about batteries, how their performance is quantified, and how they behave under different conditions.

Most of these datasheets are written for engineers or other technical experts who already know their battery basics pretty well. To help you decipher datasheets and other technical documents you might uncover, this section of the chapter provides an introduction to most of the key performance

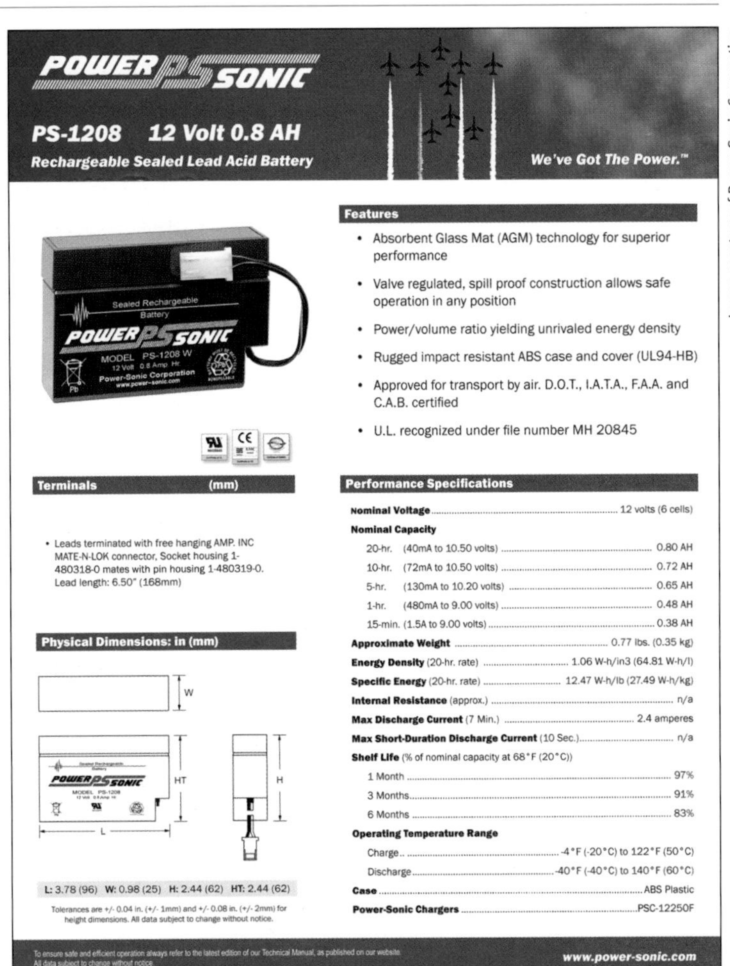

POWER PS SONIC

PS-1208 12 Volt 0.8 AH

Rechargeable Sealed Lead Acid Battery

We've Got The Power.™

Features

- Absorbent Glass Mat (AGM) technology for superior performance
- Valve regulated, spill proof construction allows safe operation in any position
- Power/volume ratio yielding unrivaled energy density
- Rugged impact resistant ABS case and cover (UL94-HB)
- Approved for transport by air. D.O.T., I.A.T.A., F.A.A. and C.A.B. certified
- U.L. recognized under file number MH 20845

Terminals (mm)

- Leads terminated with free hanging AMP. INC MATE-N-LOK connector, Socket housing 1-480318-0 mates with pin housing 1-480319-0. Lead length: 6.50" (168mm)

Physical Dimensions: in (mm)

L: 3.78 (96) W: 0.98 (25) H: 2.44 (62) HT: 2.44 (62)

Tolerances are +/- 0.04 in. (+/- 1mm) and +/- 0.08 in. (+/- 2mm) for height dimensions. All data subject to change without notice.

Performance Specifications

Nominal Voltage	12 volts (6 cells)
Nominal Capacity	
20-hr. (40mA to 10.50 volts)	0.80 AH
10-hr. (72mA to 10.50 volts)	0.72 AH
5-hr. (130mA to 10.20 volts)	0.65 AH
1-hr. (480mA to 9.00 volts)	0.48 AH
15-min. (1.5A to 9.00 volts)	0.38 AH
Approximate Weight	0.77 lbs. (0.35 kg)
Energy Density (20-hr. rate)	1.06 W-h/in3 (64.81 W-h/l)
Specific Energy (20-hr. rate)	12.47 W-h/lb (27.49 W-h/kg)
Internal Resistance (approx.)	n/a
Max Discharge Current (7 Min.)	2.4 amperes
Max Short-Duration Discharge Current (10 Sec.)	n/a
Shelf Life (% of nominal capacity at 68°F (20°C))	
1 Month	97%
3 Months	91%
6 Months	83%
Operating Temperature Range	
Charge	-4°F (-20°C) to 122°F (50°C)
Discharge	-40°F (-40°C) to 140°F (60°C)
Case	ABS Plastic
Power-Sonic Chargers	PSC-12250F

To ensure safe and efficient operation always refer to the latest edition of our Technical Manual, as published on our website. All data subject to change without notice.

www.power-sonic.com

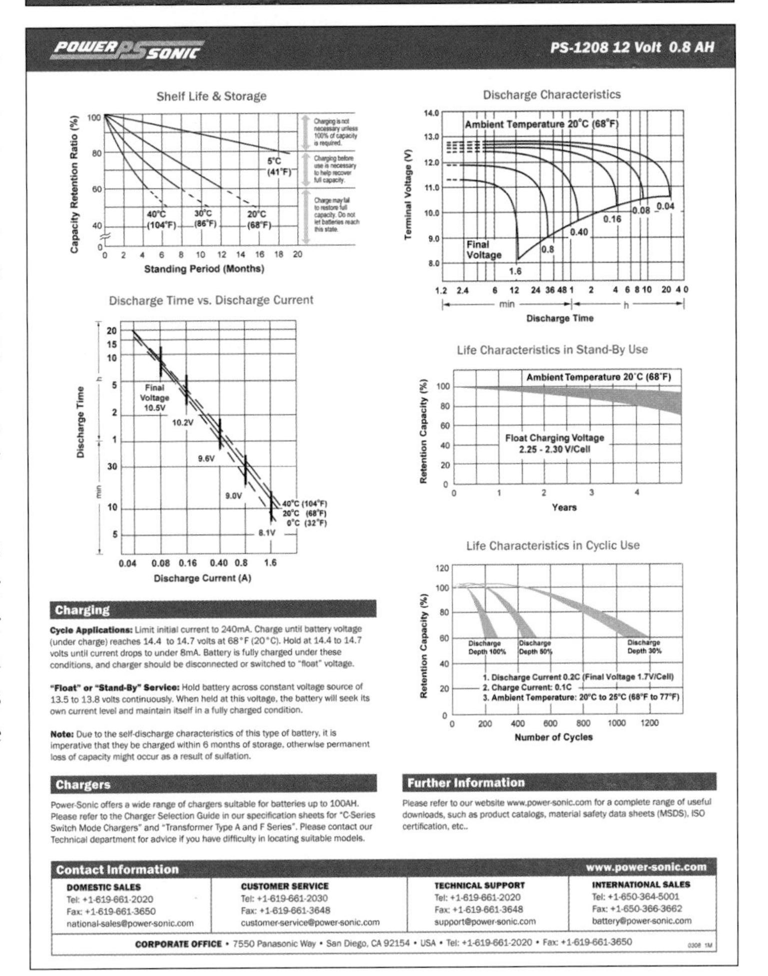

POWER PS SONIC PS-1208 12 Volt 0.8 AH

Charging

Cycle Applications: Limit initial current to 240mA. Charge until battery voltage (under charge) reaches 14.4 to 14.7 volts at 68°F (20°C). Hold at 14.4 to 14.7 volts until current drops to under 8mA. Battery is fully charged under these conditions, and charger should be disconnected or switched to "float" voltage.

"Float" or "Stand-By" Service: Hold battery across constant voltage source of 13.5 to 13.8 volts continuously. When held at this voltage, the battery will seek its own current level and maintain itself in a fully charged condition.

Note: Due to the self-discharge characteristics of this type of battery, it is imperative that they be charged within 6 months of storage, otherwise permanent loss of capacity might occur as a result of sulfation.

Chargers

Power-Sonic offers a wide range of chargers suitable for batteries up to 100AH. Please refer to the Charger Selection Guide in our specification sheets for "C-Series Switch Mode Chargers" and "Transformer Type A and F Series". Please contact our Technical department for advice if you have difficulty in locating suitable models.

Further Information

Please refer to our website www.power-sonic.com for a complete range of useful downloads, such as product catalogs, material safety data sheets (MSDS), ISO certification, etc..

Contact Information www.power-sonic.com

DOMESTIC SALES	CUSTOMER SERVICE	TECHNICAL SUPPORT	INTERNATIONAL SALES
Tel: +1-619-661-2020	Tel: +1-619-661-2030	Tel: +1-619-661-2020	Tel: +1-650-364-5001
Fax: +1-619-661-3650	Fax: +1-619-661-3648	Fax: +1-619-661-3648	Fax: +1-650-366-3662
national-sales@power-sonic.com	customer-service@power-sonic.com	support@power-sonic.com	battery@power-sonic.com

CORPORATE OFFICE • 7550 Panasonic Way • San Diego, CA 92154 • USA • Tel: +1-619-661-2020 • Fax: +1-619-661-3650

0308 1M

Image courtesy of Power-Sonic Corporation

Figure 8.25: Battery Datasheets

This is an example of a battery datasheet for a particular type of 12 volt sealed lead-acid battery. Note the detailed technical information.

characteristics used to describe and compare different types of batteries. Although batteries and their characteristics will evolve over time, these categories will likely remain relevant and useful for identifying the best batteries from among those available now and in the future. In addition to reading the material in this book, you may wish to consult other books or websites that provide general introductions to the technical aspects of batteries.

6.5.1. Voltage

The first characteristic to consider when selecting a battery is the battery voltage. Recall that voltage describes the electrical "pressure" with which the battery can push electrical current through a circuit. Higher-voltage batteries push harder than lower-voltage batteries.

Each electrical device on a vehicle will have a particular range of acceptable input voltages. In some cases, this may be a fairly narrow range (e.g., 4.5 to 5.5 VDC). In other cases, it may be a fairly wide range (e.g., 6 to 24 VDC). If you provide a device with less than its minimum-required voltage, the device may function poorly or not at all. If you provide more than the maximum-rated voltage, the device will likely become damaged or destroyed.

If you have more than one electric device on your vehicle, you may encounter a situation where you have non-intersecting voltage requirements. For example, one device might have a maximum of 3.3 VDC, while another has a minimum of 12 VDC. In situations like this, it may be easiest to provide different batteries for each device. Alternatively, you can get a battery that will supply the highest voltage needed, then use either a **voltage regulator** or a **DC-to-DC converter** to convert the higher voltage from the battery into lower voltages needed by some of the devices. (More information on voltage conversion is provided in Section 7.5.)

Common battery voltages (for fresh, fully charged batteries) include the following:

- The typical AAA, AA, C, and D batteries that abound in grocery stores, drug stores, gas stations, and hardware stores are generally single-cell batteries with a voltage of between 1.2 and 1.5 volts, depending on the chemistry involved.
- The little rectangular transistor batteries used for smoke detectors, some portable radios, and other devices are also common in stores and have a nominal voltage of about 9 volts (with some variation depending on chemistry).
- Car batteries, motorcycle batteries, and lead-acid gel cells typically have a voltage of 12 to 13.5 volts.
- There are also many specialty batteries—and collections of batteries called battery packs—that are used for hearing aids, remote-controlled racing cars and boats, laptop computers, cell phones, scientific instruments, and other specialized functions. These have quite a range of voltages, from a few volts to over 100 volts, depending on the need of the application.

Be aware that the descriptive voltages used for batteries are only approximate. For example, a fully charged "12-volt" lead-acid battery will actually provide about 13.6 volts initially, but gradually this amount will fall to, then below, 12 volts as the battery's energy is consumed.

Also be aware that most batteries are effectively dead long before their **no-load voltage** drops to zero. (The no-load voltage is the voltage measured across the battery terminals when the battery is not powering any circuit.) If you pull a lead-acid battery off a shelf, measure its no-load voltage, and discover that it is 10 volts, don't assume that the

TECH NOTE: ASKING THE RIGHT CURRENT AND VOLTAGE QUESTIONS

As you're evaluating batteries, make sure you return to your power budget (Section 5.3) to refresh your memory about voltage requirements and current requirements of your vehicle systems. Make sure the batteries you are considering can deliver sufficient voltage and power to all systems.

Remember that the maximum power a battery can deliver (in watts) is equal to its voltage (in volts) times its maximum current (in amps).

battery still contains $10/12^{th}$ of its maximum energy. In fact, a 10-volt lead-acid battery is essentially a dead battery. As soon as you hook this "10-volt" battery to any significant load, the voltage will plummet to a few volts or less. For more specific info, take a look at battery datasheets, as they often include graphs showing the relationship between charge state and no-load voltage; this info will help you understand how to estimate the percentage charge on a battery from the measured no-load voltage.

Finally, be aware that the voltage measured at a battery's terminals will depend on how much current is being drawn from the battery. As the load current is increased, the battery voltage at the terminals will drop. Electrical engineers actually think of a battery as an ideal voltage source in series with an internal resistance. As current increases, there is a voltage drop across the internal resistance (per Ohm's Law) that reduces the available voltage at the battery's terminals. This internal resistance grows as the battery loses charge, so the drop in voltage caused by current draw will become much more extreme as the battery becomes discharged. Most battery-testing devices take advantage of this fact by applying a load to the battery (i.e., drawing current from it), then measuring whether or not the battery voltage drops below some minimum acceptable value when it tries to supply that current. If the battery can deliver the current and maintain something close to its fully charged voltage, the battery is good. If the voltage drops way down when the load is applied, then the battery needs to be replaced or recharged.

6.5.2. Primary Versus Secondary Batteries

Recall that a primary battery is *not* designed to be recharged, whereas a secondary battery is. Primary batteries usually have slightly better performance in many ways than the corresponding secondary batteries. For example, primary batteries may be able to deliver more power and retain their charges longer while sitting on a shelf between uses.

While the choice of primary versus secondary battery can have a slight impact on performance, it can have a *huge* impact on cost, particularly over the lifetime of a project.

If you don't need the marginally better performance of primary batteries and can afford the initially higher cost of rechargeable batteries and a charger, then you can have a much lower cost per use in projects that go through batteries quickly. For example, a primary battery may cost only $2, whereas a similarly sized secondary may set the budget back $10, plus $25 for the charger ($35 dollars total). This sounds like the primary is a much better deal, but it may not be. For example, if your vehicle makes 100 dives and each dive drains one battery, the primaries will cost you $200 dollars (100 batteries at $2 each), whereas going with a single rechargeable battery and charger will cost you only $35, because the same battery and charger can be used over and over again for each of the 100 dives.

SAFETY NOTE: BATTERY-CHARGING SAFETY TIPS

- Never attempt to recharge a primary (i.e., non-rechargeable) battery, as this can cause an explosion.
- When recharging a secondary battery, be careful to use the right type of charger. In general, each type of battery chemistry and voltage requires its own unique type of charger. For example, NiMH and Ni-Cad batteries require different chargers, even for the same size and voltage of battery. In some cases, even minor variations in battery construction require different chargers. For example, sealed lead-acid batteries require lower charging currents than lead-acid batteries with liquid electrolyte, even though both types use the same basic galvanic corrosion reaction and produce the same voltage. Using a charger designed for a flooded 12-volt lead-acid car battery on a sealed 12-volt lead-acid battery can damage the sealed battery.
- Always use a manufacturer-approved charging unit with special high-performance batteries. In some cases, incorrect charging can create a serious and immediate risk of fire or explosion. This is particularly true with some of the high-performance battery packs, such as the Li-poly packs popular for use in R/C aircraft and other high-performance toy vehicles.
- Always charge batteries in a well-ventilated space, rather than inside a sealed container, such as an underwater pressure canister. Hydrogen or other gases released during charging can create a serious explosion hazard.

Image courtesy of Dr. Steven W. Moore

Figure 8.26: Charging a Battery

This picture shows a sealed lead-acid battery being charged by a battery charger designed specifically for this type of battery.

6.5.3. Energy Capacity

The total amount of useful energy stored in a battery is referred to as the **energy capacity** of the battery. Most battery manufacturers specify a battery's energy capacity indirectly, using units of **amp-hours** (Ah) on larger batteries, like car batteries, and **milliamp-hours** (mAh) on smaller batteries, like AA cells. One Ah = 1000 mAh.

At a basic level, the energy capacity expressed in amp-hours tells you how many amps (or milliamps) of current the battery can supply for one hour before the battery is dead. Roughly speaking, there is an inverse relationship between current and time. Thus, a 1 Ah battery should be able to provide 1 amp for 1 hour, or 2 amps for 1/2 hour or 0.1 amps for 10 hours. In reality, it's not quite that simple, but assume for the moment that it is.

Strictly speaking, multiplying amps by time does not give you an amount of energy. However, if you know the battery voltage, you can use the stated amp-hour capacity to calculate the amount of energy the battery can hold. Here's an example of how to do that: a 12-volt, 3.4 Ah battery should be able to maintain 12 volts while delivering 3.4 amps for a period of 1 hour. The two-step process below lets you calculate how much energy that represents (in joules):

Step 1: Multiply the voltage rating times the amp-hour rating to get the energy expressed in watt-hour, a valid unit of energy:

12 V x 3.4 Ah = 40.8 Wh

Step 2: (Optional: Use only if you want to convert to joules, which are watt-seconds):

40.8 watt-hours x 60 minutes/hour x 60 seconds/minute = 146,880 joules.

Now it's time to move a bit beyond the simplifying assumptions. The above calculations ignored a couple of details that you should not dismiss when actually designing your power system and selecting batteries for it. First, batteries deliver power more efficiently when discharged at a slow rate than when discharged rapidly. This means that the number of amp-hours you get out of a battery depends on how much current the motors and other systems draw from the battery. The fine print on the battery or datasheets will usually specify the conditions under which the Ah rating of the battery was determined. Commonly, this is a 20-hour discharge time for large batteries like lead-acid batteries and a one-hour discharge time for smaller batteries. Note that typical underwater vehicle operations may drain a battery more quickly than the discharge rate used by the manufacturer to determine the Ah rating of their battery. If this is the case for your vehicle, you can expect your battery capacity to be less than the advertised amp-hour rating.

A second factor ignored in the calculations above is that different manufacturers may have different definitions of what constitutes a dead battery. One manufacturer may consider a battery all used up when its voltage has dropped to 80 percent of its fully charged voltage, whereas another manufacturer of exactly the same type of battery may not consider it dead until its voltage has dropped to 50 percent of its fully charged voltage. Obviously, the second manufacturer could claim to get more amp-hours, even though there is essentially no difference between the batteries. Watch out for this when you try to compare battery specifications. Again, refer to the technical datasheets if you really want to know exactly what's going on.

6.5.4. Energy Density

The energy density of a battery is simply its energy capacity divided by the battery mass (**gravimetric energy density**) or volume (**volumetric energy density**). Gravimetric energy density is the one usually specified, with the units expressed as watt-hours per kilogram (Wh/kg) or equivalent. This parameter provides a good indication of how large and heavy a battery you'll need, to store all the energy your vehicle requires. For example, Lithium-ion batteries have relatively high energy density (e.g., 130 Wh/kg), whereas lead-acid batteries have relatively low energy density (e.g., 40 Wh/kg). If you need 160 Wh of capacity in a lead-acid battery, you'll need a battery with a mass of about 4 kilogram (=160/40), which weighs about 9 pounds.

6.5.5. Weight

Batteries, particularly large ones like car batteries, can be very heavy, and heavy batteries can be awkward and even dangerous to lift. If you need a lot of capacity, it may make more sense to get several small- or medium-sized batteries and connect them in parallel, rather than using one large battery. (See Section 6.6 for a discussion of series and parallel battery combinations.)

6.5.6. Maximum Power Output

The maximum power output, or **power capacity**, of a battery is the maximum power the battery is capable of delivering without damaging itself. Recall that energy and power are not the same thing. It's possible to have a battery that can store way more energy than you need, yet not be able to deliver that energy to your vehicle quickly enough. The result will be a vehicle that may run for a long time on one battery charge, but will also be sluggish and unresponsive. If a mission calls for a fast, spunky, high-performance vehicle, you may need to find a battery that can deliver more power.

Generally speaking, battery specifications do not state power capacities directly. Instead, they specify the maximum discharge *current* that the battery can deliver. This is sometimes called the **surge current** and is expressed in amps. This maximum current can also be expressed indirectly in terms of something called a **C-rate** (or C-rating). This is basically a conversion factor (with the strange units of "per hour") to convert the amp-hour rating of the battery into a maximum output current. For example, if a battery with a 10 Ah capacity rating has a discharge rating of 5C, then it can safely deliver up to 5 x 10 = 50 amps. To calculate the corresponding maximum power (in watts), multiply this maximum discharge current (in amps) by the battery voltage (in volts).

Keep in mind that running a battery near its maximum power output will severely reduce its effective energy capacity and shorten your run time.

6.5.7. Discharge Curves

All batteries lose voltage as they discharge. However, different battery chemistries and battery sizes behave differently in this regard. In the case of alkaline batteries, the battery voltage decreases steadily as the stored charge is used up. If you plot battery voltage against the percentage of full charge (or, alternatively, against the number of amp-hours delivered), you'll get a steadily sloping line.

In other cases, such as lithium batteries, the battery voltage remains nearly constant until the battery is almost dead, then drops precipitously just before the battery dies. Such batteries are said to have a flat discharge curve, because their graph looks like a horizontal line.

Which type of discharge pattern is better depends on what you're planning to do with the battery. For a flashlight, it might be better to have a battery that gradually decreases in voltage. That way the slowly dimming light will warn you that the battery is getting low while there's still time to hike back to camp. On the other hand, for a data recorder that might start giving you erroneous data below a certain voltage, the best choice might be a battery that will maintain a more or less constant voltage as long as possible, then simply quit. That way if the battery died, you'd have less data, but at least you'd know you could trust the data you'd gotten.

The discharge curve for a given battery will vary, depending on how quickly the battery is discharged. Therefore, one of these graphs will usually show a family of several curves, with each curve showing how the battery behaves under different rates of discharge. (Figure 8.27.) As described above with power output, these discharge rates are often expressed in terms of C-rates. Many older battery technologies had maximum discharge rates of about 1C, but newer ones can be much higher.

6.5.8. Depth of Discharge

Another characteristic that is important for secondary batteries used to power vehicles is the allowable **depth of discharge**. Some batteries, like conventional car batteries, are designed to deliver lots of power, but only for brief periods. These batteries are *not* designed to discharge very far before being recharged again. In a car, for example, the battery is always maintained in an almost- or completely charged state by the car's alternator. Once a car battery is allowed to discharge completely, it undergoes permanent damage that will shorten its useful lifetime.

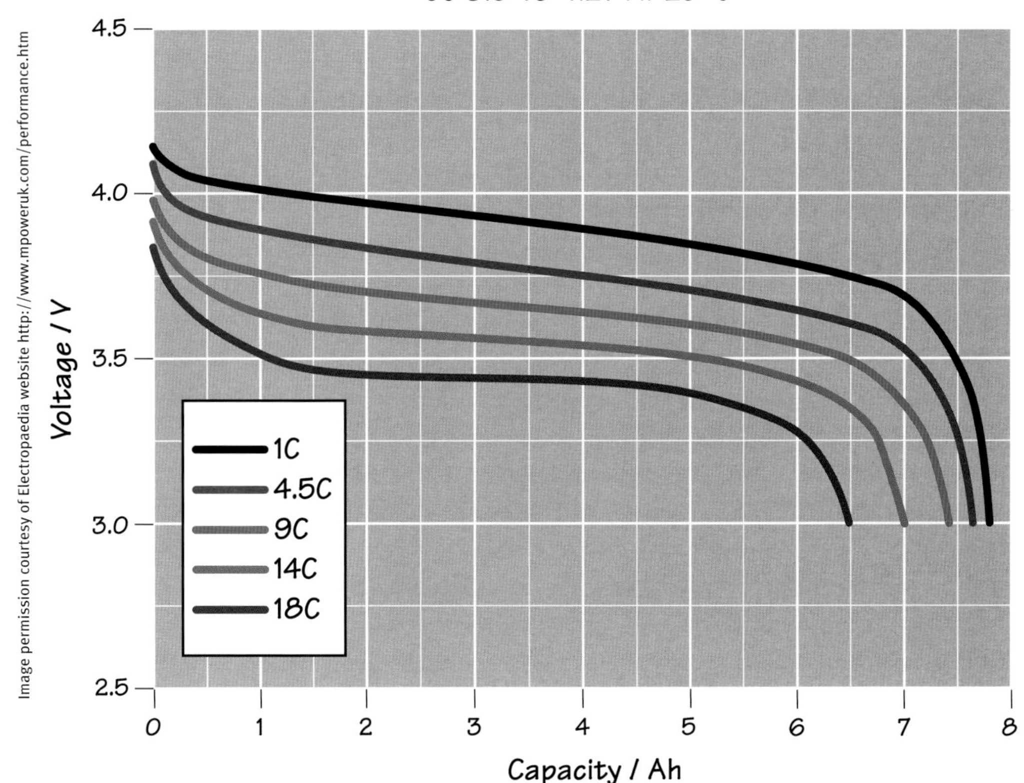

Image permission courtesy of Electropaedia website http://www.mpoweruk.com/performance.htm

Figure 8.27: Battery Discharge Curves

This sample battery discharge curve for a Lithium ion battery plots battery voltage as a function of the amp-hours delivered by the battery for several different rates of constant current (CC) discharge.

Note that under relatively light load (1C rate), the battery starts at about 4.3 volts. The voltage decreases steadily, and the battery finally dies a rapid death as it approaches the point where it has delivered almost 8 Ah. At higher rates (18C or about 18 x 8 = 144 A), the battery can muster only 3.7 volts to start and is effectively dead after delivering only about 6.5 Ah.

TECH NOTE: WHAT'S A MARINE BATTERY?

The term "marine battery" is a somewhat loosely defined category that may apply to wet batteries, AGM, gel cells, and others, depending on who's doing the defining. Generally speaking, marine batteries are like car batteries in that they are nominally 12-volt, lead-acid batteries, but then various manufacturers have "optimized" them for use in boats by having one or more of the following features:

- The lead plates inside may be more rugged, to withstand the pounding and vibration of a boat.
- Often they are sealed (utilizing gel or absorbent glass mat) to reduce the chance of spilling acid in a rocking boat.

Image courtesy of Scott Coté, Coté Consulting

Figure 8.28: Marine Batteries

- They may include greater deep-cycle capacity (using lower current for longer periods of time and discharging further before recharging is necessary), in contrast with the typical car battery (which is optimized for brief bursts of very high current to run the starter motor and which seldom discharges very far).
- Marine batteries are often designed so that wires can be connected securely with bolts through holes in the battery terminals rather than the posts of automobile batteries.
- They incorporate features to protect the battery from saltwater, diesel fuel, and other common marine contaminants. Some include built-in handles for easier, safer movement on unstable marine platforms.

Note that using marine batteries to deliver surface-supplied power still requires significant safety precautions.

Other batteries, called **deep-cycle batteries**, are specifically designed to handle repeated deep discharging and recharging cycles. Marine batteries are one example of a deep-cycle battery. Generally speaking, deep-cycle batteries are more appropriate for applications that rely on the batteries to deliver lots of energy between charges, including small underwater vehicles. You can find deep-cycle batteries by looking for batteries used in golf carts, electric wheelchairs, electric bicycles, and other electric vehicles. They are also used on boats, mobile homes, and other recreational vehicles, to provide power for inverters when the engine is not running.

6.5.9. Maximum Charge Rate

With secondary batteries, the time required to recharge the battery can sometimes be an important consideration, particularly if you need to recharge quickly between uses, like between ROV dives, for example. Charging rates, like discharging rates, are often expressed in terms of C values. For example, a 12 Ah battery that has a maximum charge rate of 0.2C or C/5 can be charged with a maximum current of 12/5 = 2.4 amps. The denominator in these C/X ratings give you a very rough estimate of the number of hours it will take to recharge the battery fully after a normal discharge cycle. In this example, C/5 would indicate a five-hour recharge time. If you provide more amps, you can potentially damage the battery. If you provide less, it won't hurt the battery, but recharging will take longer.

Some of the newer battery technologies have very fast charge rates of C/1 or even better, but most of the conventional battery chemistries have maximum recommended charge rates of C/5 or C/10.

6.5.10. Temperature Performance

Some batteries perform well over a wide range of temperatures, while others lose performance rapidly as temperature drops. Many are essentially useless near freezing. (Check a datasheet for this info.) So if you are building an underwater vehicle to explore a frozen-over pond or Arctic waters, this may be an important consideration. On the other hand, if you are building a vehicle to explore coral reefs off the Florida Keys, it's not going to be an issue.

6.5.11. Size and Shape

Battery size and shape may be issues to consider in your vehicle design, particularly if you plan to place the batteries on board the vehicle. As a general rule, bigger batteries can store more total energy and can deliver that energy at higher rates of power than smaller batteries having similar chemistry; however, bigger batteries are also harder to fit inside small underwater housings.

For reasons explained in *Chapter 5: Pressure Hulls and Canisters*, most electronics canisters are cylindrical in shape. Cylindrical battery shapes, such as the familiar C and D sized cells, can make efficient use of the space inside small cylindrical housings made from pieces of pipe. Unfortunately, larger capacity batteries, such as sealed lead-acid batteries, are usually much bigger and rectangular in shape, so they can require much larger cylinders to enclose them. This is especially true if the battery terminals are located near the corners of the battery, because the housing must then allow extra space for the connectors used to join wires to the battery.

The larger canister volumes, in turn, will require a heavier vehicle weight to maintain neutral buoyancy, since the larger canisters displace more water. A larger, heavier vehicle is more difficult to launch and recover safely. Moreover, the larger size generally creates more drag, reducing efficiency and undermining the benefits of having a larger battery in the first place. Keep in mind that placement of canisters containing heavy batteries will have a strong effect on your vehicle's flotation, trim, and stability for all the reasons discussed in *Chapter 4: Buoyancy, Stability, and Ballast.*

In general, if the batteries are going to reside on the vehicle, then pick a battery size and shape that allows you to get the required energy and power capacity into as small a cylinder (or set of cylinders) as possible. Also, be sure to think carefully about where you'll place those canisters on the vehicle frame and how you'll access them to charge or swap out batteries between dives. You'll be much happier if you've thought about these things *before* you buy expensive batteries.

6.5.12. Shelf Life

All batteries lose charge and voltage over time, even when they are not being used. This is sometimes called **self-discharge**. Some battery types, like alkaline cells, can retain most of their initial charge for years, whereas others, such as rechargeable NiMH cells, will go from being fully charged to nearly dead in just a few weeks or months. Again, the manufacturer datasheets are a good source for this information.

6.5.13. Required Maintenance

Most secondary batteries have recommended maintenance procedures, consisting of particular methods and schedules for discharging and recharging the batteries, limits on charging rates and voltages, and other constraints. Failure to comply with these maintenance recommendations can seriously reduce battery performance and longevity. In some cases, failure to follow proper charging procedures can even result in fire or explosions. Not surprisingly, the maintenance procedures are different for every type and size of battery. For example, Nickel-Cadmium batteries have a distinct **memory effect**, which means that if they are only partially discharged before recharging, the energy capacity thereafter is greatly reduced. And Li-poly battery packs can burst into balls of flame if charged with the wrong type of charger! Choose batteries with maintenance requirements you can realistically follow.

6.5.14. Ease of Acquisition and Disposal

Some batteries are very easy to find in any town. Others must be special-ordered from far away. In general, you'll save yourself headaches (and money) by settling on a vehicle design that does not require exotic, hard-to-find batteries.

Also, keep in mind that most batteries contain toxic chemicals and are categorized, officially, as hazardous waste. (See *Safety Note: Safe Battery Disposal.*) Fortunately, many communities offer free disposal of common battery types as part of a household hazardous waste program. Proper disposal of exotic battery chemistries may be more difficult.

6.5.15. Price

Battery prices vary dramatically, depending on the battery chemistry and the size of the battery. Not surprisingly, the batteries with the best performance under the widest range of conditions are also the most expensive.

Remember not to waste money on performance features you don't need. So, as with everything else in underwater vehicle design, it's important to first clarify your mission. Only then can you determine what sort of power will be needed and under what kinds of conditions. Once you know that, you can shop for the battery features that are really worth paying for.

SAFETY NOTE: SAFE BATTERY DISPOSAL

Primary batteries eventually run out of stored energy and become useless. Secondary batteries can be recharged numerous times, but even these rechargeables will eventually reach a point (usually after several hundred recharge cycles) when they will no longer hold a useful charge and must be replaced. This presents a problem, because many of the metals (e.g., lead, mercury, and cadmium) and other chemicals used in batteries are toxic to humans and the environment.

If batteries are simply tossed in the trash, they will most likely end up in a landfill, where rainwater percolating through the ground can absorb the toxic chemicals from rotting batteries and carry them into underground aquifers. From there, the contaminated water can make its way into wells or springs that supply drinking water for people and animals, as well as irrigation water used to grow the crops that people and farm animals eat. Disposing of batteries in a fire is also dangerous (they often explode) and makes the contamination problem even more serious by facilitating the rapid release of battery chemicals into the environment.

Most communities offer household hazardous waste programs that will accept unwanted batteries for free. Check with your local dump or recycling center for information about battery disposal programs in your area. Organizing a used-battery collection point in a convenient location at your school, office, or home can help immensely to facilitate proper battery disposal.

SAFETY NOTE: COMBINING BATTERIES SAFELY AND EFFECTIVELY

When combining batteries in series, remember that their voltages add together and that voltages above about 50 volts are considered dangerous, even in dry conditions. In wet conditions, it's a good idea to keep voltages less than about 15 volts. Don't defeat one of the main safety advantages of batteries (low voltage) by putting too many of them in series.

In addition to the electrocution risk, there are possible fire and explosion risks associated with improper battery combinations. For example, if you put two high-powered batteries of markedly unequal voltage or charge state in parallel, the one with greater voltage will try to charge the one of lesser voltage very quickly, because there will be little resistance to current flow, other than the internal resistance of the batteries. If this results in high enough currents, excessive heating could cause a fire or trigger a battery explosion.

Aside from these types of dangers, the main reason not to mix batteries is reduced performance. For example, current flow through batteries in series will be limited by the capabilities of the weakest battery.

For best performance and safety, always use fresh (or fully charged) batteries of the same size, type, age, voltage, and manufacturer when building series or parallel combinations of batteries.

6.6. Series and Parallel Battery Combinations

If you need higher voltages, more energy capacity, or greater peak power, you may be able to achieve these goals by combining two or more batteries in series and/or parallel combinations.

Connecting batteries in series will give you more voltage. For example, if you connect the (-) terminal of one 1.5 V battery to the (+) terminal of another 1.5 V battery, then the voltage at the (+) terminal of the first battery will be 3 volts above that at the (-) terminal of the second battery. Thus, the series combination will act as if it were a single 3 V battery. You could connect four 1.5 V batteries in series to get 6 volts. Connecting batteries in series like this can be a helpful trick if you have a motor that needs 6 volts supplied to it, but only have 1.5 V batteries available.

Note that while connecting batteries in series gives you more total energy in the form of higher voltage, it does not give you more amp-hour capacity. For example, if each of two 12 V batteries has a capacity of 5 Ah, and you connect them in series, you will end up with the equivalent of a 24 V battery, but it will still have only 5 Ah of battery capacity.

If you need greater current and/or longer run times (but not more voltage), you can connect batteries in parallel instead of in series. For example, if you take those same two 12 V, 5 Ah batteries and hook them up (+) to (+) and (-) to (-), then you'll get the equivalent of a 12 V, 10 Ah battery. It's the same voltage, but twice the capacity. This will give you the same 12 volts you had before, but with twice the run time and twice the peak power capability of the single battery.

If you need more voltage *and* more capacity, you can combine series and parallel arrangements of batteries to get the best of both worlds. (See Figure 8.29.)

When you combine batteries, you must do so carefully to avoid safety and performance problems. (See the *Safety Note: Combining Batteries Safely and Effectively.*)

Figure 8.29: Series and Parallel Combinations of Batteries

In this diagram, four batteries (A, B, C, and D) have their terminals connected by colored wires. Each battery by itself is 12 volts and has a capacity of 5 amp-hours. Batteries A and B are connected in series (i.e., one positive terminal to one negative terminal) by a blue wire, which effectively converts the remaining two terminals into the terminals of a single 24-volt battery (yellow); however, this 24-volt battery still has only a 5 amp-hour capacity. Likewise, C and D are joined in series by a second blue wire, making them into the equivalent of a 24-volt battery (lavender) with a 5 amp-hour capacity.

The green wires connect these two "24-volt" batteries in parallel (positive terminal to positive terminal, and negative terminal to negative terminal), doubling their effective capacity from 5 Ah to 10 Ah.

So a motor or other device connected from the red (+24 volt) wire to the black ground wire would be connected to the functional equivalent of a 24-volt, 10 Ah battery.

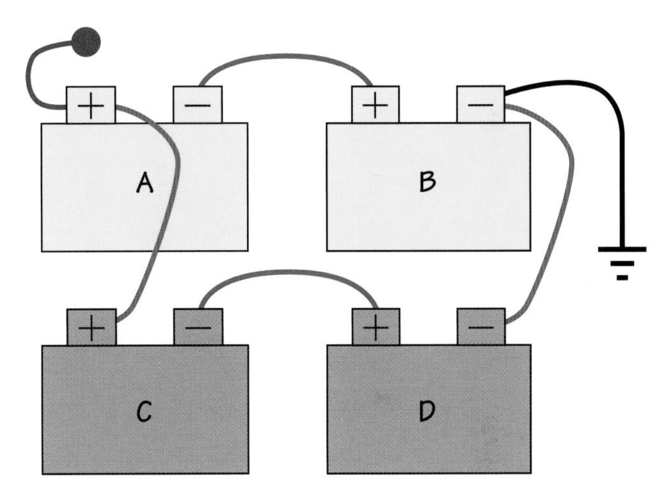

6.7. Contemporary Battery Choices

Among the generally affordable batteries available as of 2010, four or five types stand out as being particularly popular and/or promising for powering small underwater vehicles:

- Alkaline
- Sealed Lead-Acid (SLA, VRLA, and AGM)
- Nickel Metal Hydride (NiMH)
- Lithium ion (Li-ion) and the newer Lithium polymer (Li-poly)

These batteries are compared in Table 8.3 and described in more detail below.

Table 8.3: Comparison of Some Good Battery Chemistries for Small Vehicles

Name	Alkaline	Sealed Lead-acid	NiMH	Li-ion/Li-poly
Chemistry	Alkaline manganese dioxide	Lead-acid	Nickel Metal Hydride	Lithium-ion
Type	primary	secondary	secondary	secondary
Cell voltage	1.5	2	1.2	3
Common battery voltages	1.5 and 9	6 and 12	1.2	3 and 6
Peak load current	high	high	high	medium to high
Minimum time to recharge	N/A	8–16 hours	2–4 hours	2–4 hours
Recharge cycles (typical)	N/A	250	400	400
Overcharge tolerance	N/A	high	low	very low
Shelf life	years	months	days	weeks
Energy density (Wh/kg)	120	30–50	60–120	110–160
Additional comments	Widely available in every corner store. Low cost per battery, but generally not rechargeable.	Large, cheap, heavy, rectangular. Lead is toxic.	Quickly replacing Nickel-Cadmium (Ni-Cad), because they have higher energy density, less memory effect, and less environmental toxicity.	Serious fire and explosion hazard if damaged, improperly charged/discharged, or exposed to water. Properties listed here are average for this large and diverse group of related lithium battery chemistries.

6.7.1. Alkaline Cells

These ubiquitous primary cells are the most readily available batteries anywhere in the world. They are relatively inexpensive (for one-time use) and have good performance specifications (better than most secondary batteries.) They come in the familiar cylindrical AAA, AA, C, and D sizes, so they fit nicely inside small, cylindrical pressure housings. Their voltage is typically about 1.5 volts per cell.

Their main disadvantage is that most are not rechargeable (though there are some reusable versions being produced). Since most vehicles burn up battery energy pretty fast, the financial and environmental cost of constantly replacing alkaline batteries makes them much less popular than secondary batteries for most ROV or AUV projects.

6.7.2. Sealed Lead-Acid Batteries

Sealed lead-acid (SLA) batteries use the same galvanic cell chemistry as car batteries and other wet-cell, lead-acid storage batteries. However, instead of having liquid sulfuric acid sloshing around inside them and possibly spilling out, the caustic electrolyte has been rendered non-spillable in some way, making them safer to use. This may be done by turning the acid solution into a gel (e.g., gel cell batteries) or by absorbing it into a glass sponge (e.g., **absorbed glass mat** or **AGM** batteries). That means the batteries can be used upside-down, sideways, or in any other orientation—something not possible with conventional wet-cell batteries. Like regular wet-cell lead-acid batteries, all SLA batteries can give off hydrogen gas, particularly when being charged. That means they can't really be "sealed" in an absolute sense. Instead, they have valves to release any internal pressure. For this reason, SLA batteries are sometimes referred to as **valve-regulated lead-acid (VRLA)** batteries.

TECH NOTE: EXAMPLES OF BATTERY-POWERED UNDERWATER VEHICLES

- The experimental AUV *DepthX* (DEep Phreatic Thermal eXplorer) designed by Stone Aerospace uses Lithium polymer battery packs that allow it to operate for up to eight hours. It can run on the surface using wireless WiFi, utilize optical fiber for data uplink or control, or be programmed to operate autonomously.
- The autonomous underwater vehicle *SONIA 2007* was built by undergraduate students from École de Technologie Supérieure (ETS) to compete at the 10th International Autonomous Underwater Vehicle Competition sponsored by the Association for Unmanned Vehicle Systems Internation (AUVSI) and the Office of Naval Research (ONR). The AUV was powered by a Saft VH-D-9500 (NiMH) battery pack of 21 cells (7x3), 25.2 volt, custom-packed for an endurance of 120 minutes or better. The AUV employs six BDT150 thrusters.
- The *SeaEye B.O.S.S.* (Battery Operated Submersible System) was a compact observation hybrid ROV prototype. It employed a set of highly efficient, brushless DC thrusters powered by gel cell batteries in a detachable pod. Endurance was rated at from four to nine hours, with a duty cycle of 25 percent, depending on configuration.

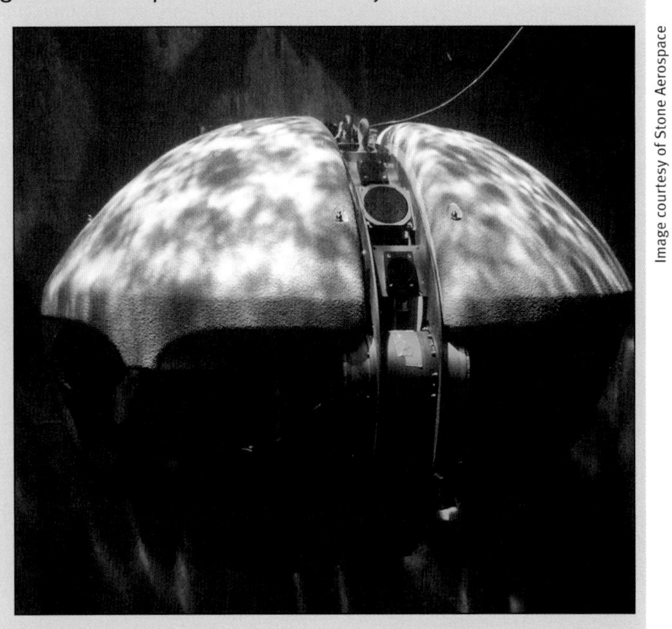

Image courtesy of Stone Aerospace

Figure 8.30: DepthX

In the case of gel cell batteries, the ions can diffuse through the gel (though more slowly than through water), so the battery chemistry still works, but has reduced power delivery capabilities. In the AGM battery, the electrolyte is still liquid, but it's absorbed into a sponge-like mat of glass fibers, much like water being absorbed into a damp sponge. These absorbed glass mat batteries are newer and generally have better performance characteristics than the gel cells, even rivaling that of their wet-cell ancestors. For this reason, AGMs are rapidly replacing the gelled varieties.

Image courtesy of Deepsea Power & Light

Figure 8.31: Submersible Power Systems

The SeaBattery™, manufactured by Deepsea Power & Light, is actually a submersible power system.

This rechargeable, lead-acid battery utilizes a non-liquid electrolyte, is sealed in a polyethylene case filled with inert oil, and is pressure-modified to perform in the high-pressure and low-temperature environment of the deep ocean. It is available in 12-, 24-, and 48- volts.

AGMs are presently a good choice for ROV and AUV batteries. They are widely available in 6-volt and 12-volt batteries, easily recharged, and cheap compared to other more exotic battery types, such as Lithium polymers. They come in rectangular shapes, which can make for some tricky mental gymnastics if you're trying to fit them into a cylindrical or spherical pressure can, but they can easily fit into oil-compensated rectangular cases, which can be placed outside the main electronics can. For an AUV or ROV with on-board batteries, this means battery weight can be partially offset by the added displacement of the battery case. Furthermore, having the battery case external to the vehicle makes it easy to change out a depleted pack and replace it with a fully charged spare. AGMs are available in amp-hour ratings ranging from about 1.2 Ah to as high as 50 Ah or more.

Always remember to charge and store SLA batteries in containers that will automatically vent any excess pressure.

6.7.3. Nickel Metal Hydride Cells (NiMH) and Nickel-Cadmium Cells (Ni-Cad) (or sometimes NiCd)

In the 1960s, scientists designed this battery cell so that its hydrogen chemistry is stored in a metal alloy, called a hydride, which can release hydrogen as required. Each

Figure 8.32: Lithium Batteries

Lithium-polymer batteries are stacked underneath the printed circuit board in this battery pod card cage from Cornell University's Triton *AUV. Notice the printed circuit board on top of the stack.*

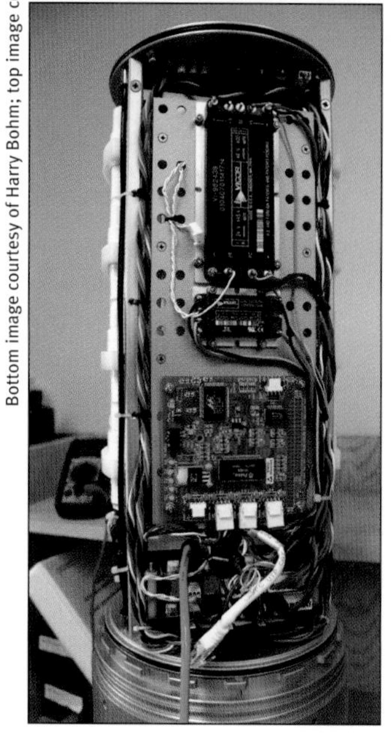

Bottom image courtesy of Harry Bohm; top image courtesy of Cornell University's Triton AUV Team

Figure 8.33: Control Circuitry for Lithium Polymer Battery Pack

This photograph of the GAVIA *AUV's Li-poly battery pod shows the complex control circuitry required to monitor safe charge and discharge rates.*

secondary cell produces 1.2 volts. **Nickel Metal Hydride** cells are rapidly replacing the once-popular **Nickel-Cadmium** cells, because they offer comparable or better performance in many respects without the **memory effect** problems that plague Ni-Cad batteries. The NiMH battery costs more and has half the service life of a Ni-Cad, but it offers approximately 30 percent more storage and discharge capacity (depending on the manufacturer). This means that NiMH cells can be used for running motors that pull currents of up to 30 amps. That said, Ni-Cads still offer greater peak power output capabilities and lower price, so they might be a better choice for some vehicles or missions. Like alkaline cells, both NiMH and Ni-Cad cells normally come in cylindrical AAA, AA, C, and D sizes, but each produces slightly less voltage (about 1.2 volts) than an alkaline cell. Charging some of these batteries requires a "smart charger" that properly discharges, then slowly recharges them.

Like other cells and batteries, these cells can be combined to produce stacks with different voltages and amp-hour capacities, so they can be used in vehicles requiring higher voltages for motors and other systems. These batteries, and more recently Lithium polymer batteries (described next), are commonly used as power sources for radio-controlled model cars and airplanes. The batteries come as stacks of cells that are conveniently shrink-wrapped in plastic, with the appropriate connector attached. The power requirements for these radio-controlled vehicles are often expressed in terms of the number of cells required per pack.

6.7.4. Lithium Batteries

This category of lithium batteries includes **Lithium ion (Li-ion)** and **Lithium polymer (Li-poly** or other abbreviations). (Sometimes you'll also see polymer lithium ion.) Lithium batteries are commonly used for laptop computers and mobile phones. They have one of the best energy-to-weight ratios of any battery, no memory effect, and a slow loss of charge when stored. These batteries are available either as a primary or secondary type. The rechargeable (secondary) versions are increasingly the battery of choice for small hand-launched AUVs, such as the Icelandic Hafmynd Corporation's AUV *GAVIA*.

Lithium polymer batteries are basically a new and improved version of the older Lithium ion ones. Both share similar basic chemistry, but improved construction techniques make Li-poly ones less expensive to manufacture, higher in energy, lighter, and more flexible in terms of shape and size. These last two characteristics allow Li-poly batteries to efficiently fill available space in the devices they power. In fact, one of the reasons mobile phones can be so small is because the Li-poly battery is manufactured in a thin rectangular shape.

The significant advantages of lithium batteries are offset by some distinct disadvantages, which make them a less than ideal choice for many projects. However, this battery type is evolving rapidly, so check out the most recent changes when making your selection. Here are some current disadvantages of lithium batteries you should know about. Most importantly, these batteries catch fire readily if damaged, if charged or discharged improperly, or if their innards are exposed to air or water. When they burn, they can do so rapidly and violently, launching themselves like a small rocket engine or simply bursting into a big ball of flame. In short, they can be extremely dangerous when used in unusual ways, like in a home-built ROV! Note that there's tremendous research and development going on with lithium batteries, so the information in the list below is subject to change.

- Lithium batteries begin to age as soon as manufactured, whereas other batteries age based only on the number of charge/discharge cycles. Thus, an older Li-poly battery will not last as long as a new one, solely because of its age. That's why you'll always want to look for the "manufactured on" date on any batteries you buy.

- The charge/discharge cycle for Li-ion and Li-poly batteries is lower than that of NiMH, non-spillable lead-acid, and Ni-Cad cells.
- Another annoying habit of lithium batteries is that they can suddenly fail with no warning as they get older or if their internal safety device unexpectedly deactivates the battery to prevent overheating and explosion.
- Note that lithium batteries should not be placed in oil-filled pressure housings, because they can absorb the oil into their cells, thereby upsetting their delicate internal chemistry.
- A Li-poly cell must never be discharged below a certain voltage, to avoid irreversible damage. Therefore, all systems involving lithium batteries should be equipped with a circuit that shuts down the system when the battery is discharged below the pre-defined threshold.
- The power densities are so high for a Lithium polymer cell that hooking them up in series and parallel, as you would a lead-acid or Ni-Cad battery, means there is a serious risk of fire and possible explosion if all the cells are not charged and discharged independently with a specialized control system or "smart charger" for each cell in the system.

Despite the fact that the technology is still emerging and has not been proven 100 percent safe, the model hobby world is now embracing the Lithium polymer battery and working to develop systems that can safely use and charge them. This promises to make them safer and easier to use in the near future.

7. Transmission and Distribution of Electrical Power

Once you have decided on a suitable battery or other electric power source for your vehicle, you need to figure out how to distribute the power from that source to all of the thrusters, cameras, lights, navigational sensors, and other devices that need it. Fortunately, power distribution is much easier with electrical systems than it is with mechanical or hydraulic systems. It's nearly as simple as running big enough wires to the right places. Still, there are some important issues and challenges to think about.

When planning the design of your power distribution system, start with the big picture and work your way down to the details. Figure 8.35 provides a great place to begin.

The electrical system for a small underwater vehicle generally consists of four major subsystems, plus the wires connecting them. Those major subsystems are: 1) a battery or other source of electric power, 2) a master fuse or circuit breaker located in the circuit very near the positive terminal of the battery, 3) some form of ON/OFF switch, and 4) the load, which consists of all of the equipment powered by the battery. The load is usually further subdivided into a number of electrical subsystems (A, B, C, etc.), such as thrusters, cameras, lights, navigational sensors, and payload tools. Each of these subsystems is often protected by its own fuse (not shown).

The preceding sections of this chapter discussed batteries. The remaining sections detail wires, fuses, and circuit breakers, and various methods for getting enough power at the right voltages distributed to each electric subsystem in the vehicle.

SAFETY NOTE: LITHIUM BATTERY CAUTION

All batteries can be dangerous if used improperly, but lithium batteries are particularly hazardous because of their unusual tendency to burst into flame. For some dramatic examples, you need only search the web for videos of "Li-poly explosion" or similar keywords. So be sure to read and follow all safety information if you use them.

TECH NOTE: INTRIGUING BATTERY ALTERNATIVES

Although batteries are probably the best source for low-voltage DC power for running ROVs and AUVs today, other technologies may be worth considering, particularly as these technologies mature in the years to come.

Fuel cells are an emerging technology that generates electricity by "burning" pure hydrogen with oxygen to make water in a controlled reaction. This bypasses the usual engine and generator steps used to convert more conventional fuels to electricity, so it is quite efficient (about 40 percent, compared to about 15 percent for diesel electric). The Norwegian AUV *Hugin* 3000 and 4500 rely on fuel-cell technology for successful seabed mapping, oil and gas industry surveys, hydrography, and marine research. Utilizing its semi fuel cell battery, the *Hugin* 3000 can travel at 4 knots for at least 60 hours with all sensors running. Though fuel-cell technology is still too experimental, too dangerous, and too expensive to be practical for small-scale, low-budget, home-built projects, it's probably worth keeping an eye on fuel-cell developments. Someday, improved fuel cells may become a good power source for small underwater vehicle projects.

The power sources for the family of *Hugin* AUVs are based on two different pressure-tolerant battery designs, thereby eliminating the need for pressure containers and thus the inherent risk for explosions due to battery hazards. The two types of battery technologies are Lithium polymer for shallower missions and aluminum oxygen semi fuel cell (ALHP FC) for deeper ops, with the latter producing electrical power from aluminum and oxygen.

In the spirit of the early diesel electric submarines, which recharged their batteries on the surface, then used them for power under water, **solar panels** could potentially be used to recharge an AUV's batteries while the vehicle was resting at the surface between dives. Enterprising school groups or others might want to explore the possibility of launching a battery-powered AUV with GPS navigation that would return to the surface for recharging whenever its batteries ran low on charge. In principle, such a vehicle could operate until the growth of barnacles or other marine organisms obstructed the solar panels or propellers and could potentially cover enormous distances to collect oceanographic data or perform other missions. With current technology, such a vehicle would need to spend a large portion of its time on the surface and might therefore be at the mercy of ocean currents, but the balance could shift in favor of more active exploration as solar technology improves in the future.

Image courtesy of Kongsberg Maritime Ltd.

Figure 8.34: **Hugin**

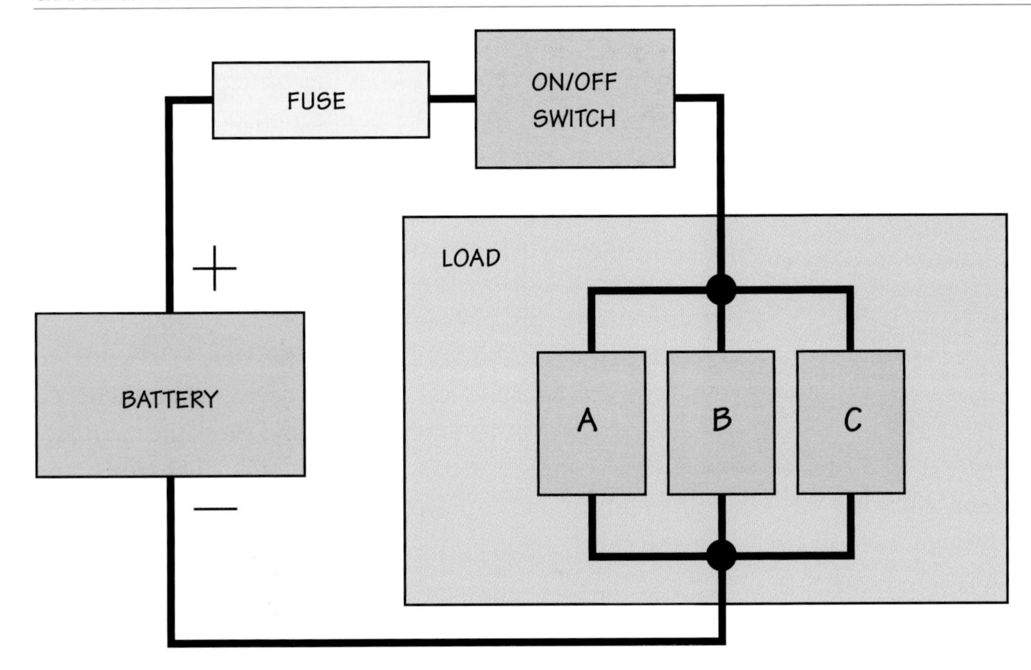

Figure 8.35: Overview of a Vehicle's Electrical System

7.1. Wires, Cables, and Connectors

Electrical power is usually transmitted from one place to another through wires, which may be bundled together into cables and connected to each other or to electrical devices by plugs or other connectors.

When it comes to electric circuits, the term **wire** is used to describe a long, slender, flexible strand of low-resistance material (usually copper) designed to carry electric current, with minimal loss of voltage, from one location to another. A **cable** refers to two or more wires bundled together. The tether on an ROV is essentially a specialized type of cable that not only contains a variety of conductors, but also, for larger vehicles, is made strong, abrasion-resistant, and neutrally buoyant.

Wires can vary in length, from shorter than one inch to many miles. Diameters can range from less than that of a human hair to up to several centimeters. Individual wires may have a solid conductor core or a stranded conductor core. A stranded core is manufactured by twisting many thin strands of metal together, giving the wire greater flexibility than it would have if made out of a single, thicker strand having equivalent cross-sectional area. (See Figure 8.36.)

Most wires are coated with a colored layer of flexible plastic or other non-conductive material called **insulation**. The insulation allows wires to cross one another or to be bundled together into a cable without creating a short circuit, and in high-voltage circuits, it protects people from shocks. The insulation is often color-coded to facilitate identification of particular wires in a circuit or in a cable containing many wires. Adopting and sticking to a particular color code in your own circuits can make them much easier to build, modify, and repair.

For example, in many low-voltage, battery-operated circuits, wires connected to the (+) terminal of the battery have red insulation, whereas wires connected to the (-) terminal of the battery have black insulation. In household wiring, green (or bare copper) wires are usually the safety (earth) ground, while white wires are usually the common (neutral) wires that return current to the power plant, and black wires are usually the so-called "hot" wires carrying the (dangerous) voltage.

Image courtesy of Dr. Steven W. Moore

Figure 8.36: Wires and Cables

Wires and cables come in many varieties. This photograph illustrates just a few of many common kinds. The rainbow-colored strap at the top is called ribbon cable, commonly used for digital signals in computers. The blue cable with four twisted pairs of wires emerging from it is called a Cat-5 cable, commonly used for computer networking. The brown wire is 14-gauge stranded wire, commonly used for power transmission. The yellow wire is 22-gauge solid conductor "hookup" wire. It's called this because it's often used to hook together the components in small hand-wired circuits. The tiny purple wire is 30-gauge wire, also used in small circuits where components are connected using a technique called "wire-wrapping" instead of soldering.

A **whip** (sometimes called a conductor whip) is simply one or more electrical wires in a waterproof casing that can be used to make connections outside pressure housings. Like other wires, whips come in various thicknesses.

A wide variety of **connectors** are available to join different wires or other parts of electrical circuits. An extension cord plug and wall outlet are a familiar pair of electrical connectors, as are the little stereo jacks used to connect headphones to audio devices, but there are hundreds of other kinds, too. Always check the manufacturer's specifications to make sure the wires and connectors you use can handle the amount of electrical current you expect to send through them.

Electrical connectors are an essential aspect of the circuitry that controls power to the vehicle's external components. Note that most electrical connectors are intended for dry use only. For underwater vehicles, you can use the dry connectors inside a dry pressure canister or an oil-filled one. But if the connector will come into contact with water, you will need to use a connector designed specifically for in-water use.

A handful of subsea connector companies, such as SeaCon, Teledyne Impulse, and Birns, make connectors suitable for small but deeper-diving vehicles. Research these companies and look at other similar projects to see which might have the best options to incorporate. The types of connectors required will become clear as you select the electrical components for your vehicle and draw a wiring schematic for its electrical distribution system for control, power, and sensors. (See *Tech Note: Electrical Connectors* in *Chapter 5: Pressure Hulls and Canisters* for additional information on electrical connectors.) (See Figure 8.37.)

Note that the whip lengths, splices, weight, size, orientation, and material of these subsea connectors often constrain where the pressure can and any external pressure-proofed components will be located on the framework, so it's important to research various options. Conversely, the canister's shape, size, and physical interior arrangement may affect the position of the penetrators on the can's endcap. Multiple cans will require more connectors to interconnect the various cans.

SAFETY NOTE: WIRE COLOR CODES

In many circuits, wires are color coded to identify their function. To the extent that these color codes are standardized, they serve to help reduce confusion and the likelihood of errors. Unfortunately, color coding is no guarantee of safety, particularly since the color codes differ among different sub-specialties. For example, in low-voltage DC circuits, it's quite common to use red for the 5-volt positive supply voltage and black for the ground. However, in standard household wiring in the U.S., ground is green or bare copper, and the black wire is actually the one carrying the dangerously high voltages! Thus, somebody familiar with the low-voltage standard could innocently assume the black wire in their house wiring was ground and get electrocuted.

Try to adhere to accepted color codes in your own circuit designs, but remember that since there are multiple standards and since people do make mistakes (meaning the standard may not be followed in all cases), you should never trust a color code in situations where a mistake could lead to electrocution or fire!

TECH NOTE: WIRE DIAMETER AND GAUGE NUMBERS

Wire diameters are often expressed in terms of gauge numbers. In the **American Wire Gauge (AWG)** standard, the larger the number, the thinner the wire. So a #2 AWG wire has a much larger diameter than a #30 AWG wire. The wiring used to carry power to outlets and light fixtures in most households or through a heavy-duty extension cord is usually about #14 AWG.

Table 8.4 lists some common wire gauges and some of their properties and uses. It also allows you to use Ohm's Law to calculate how much voltage drop you can expect per foot per amp of current. (A sample calculation is given in *Section 7.4.1. Voltage Drop.*)

Table 8.4: Some Standard AWG Wire Sizes and Properties

AWG	*Diameter of metal core (approximate inches)*	*Resistance (ohms per foot of length)*	*Maximum current* (Amperes)*	*Example of a typical use for this gauge wire*
#2	*0.258*	*0.00015*	*180*	*main power line into an apartment building*
#14	*0.064*	*0.001*	*55*	*typical electrical wiring inside house walls or inside a heavy-duty extension cord.*
#22	*0.025*	*0.017*	*7*	*hookup wire commonly used for hand-wired circuits*
#24	*0.020*	*0.026*	*3.5*	*one of the small wires within a phone jack or Ethernet cable*
#30	*0.010*	*0.103*	*0.9*	*very slender wires used in wire-wrapped circuits.*

** Note: Maximum current values are very situation-specific and should be viewed only as rough guidelines.*

7.2. Fuses

It would be nice if everything always went according to plan, but life just doesn't work that way. And it's fair to say that Murphy's Law ("Anything that can go wrong, will go wrong") appears to have a particular fondness for electric circuits. Fortunately **fuses** can help keep a small electrical problem from becoming a bigger headache. Fuses are conceptually simple devices designed to protect circuits and people from some of the hazards associated with short circuits, including fires and electrocution. They are common in battery-powered devices and AC appliances and should be included in every ROV or AUV project.

Image courtesy of Teledyne Impulse

Figure 8.37: Various Electrical Connectors for Underwater Use

Many circuit issues are just nuisance problems where the ROV or other gadget just doesn't work the way it's supposed to. But some of the problems are more serious and can present significant threats to the safety of people and property. One of the most common and potentially serious problems is a short circuit. A short can occur whenever a loose wire, dropped tool, flood of seawater, or other circumstance provides electricity with a shortcut around all or part of the regular load. Since the load normally limits the amount of current allowed to flow from the battery or other source, a short can cause a huge surge in current. This surge can quickly heat wires or other components red-hot and start a fire. A short

Images courtesy Dr. Steven W. Moore

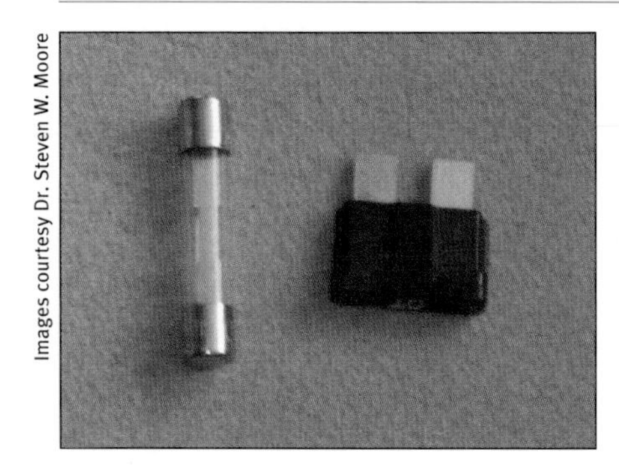

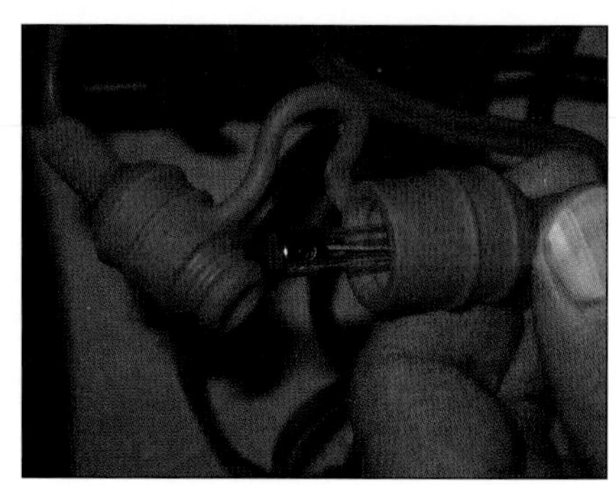

Figure 8.38: Fuses

The image on the left shows two common types of fuses—a cylindrical "cartridge" type and a flattened "blade" or "spade" type. Both types come in a wide range of different voltage and current ratings, and both types are appropriate for ROV or AUV use.

The middle image shows a cartridge fuse in one of many available types of sockets used to connect a fuse electrically with the circuit it is protecting.

The silhouetted image on the right shows a cartridge fuse that has "blown" in response to excessive current. You can see the gap where the thin section of metal wire inside the glass cylinder melted in two.

can also bring higher voltage parts of a circuit into direct electrical contact with parts that are normally at lower voltages, so in high-voltage circuits, a short can present a dangerous electrocution hazard.

7.2.1. How Fuses Work

A conventional fuse is, essentially, just a thin section of wire that's designed to be the "weakest link" in a circuit, but not so weak that it fails under normal operating conditions. A fuse is normally placed between the positive terminal of the battery and the rest of the circuit, so that all current leaving the battery must pass through the fuse. If a short develops, the excessive current flowing through the circuit will melt the fuse wire in two, cutting off current to the rest of the circuit before anything else in the circuit has enough time to overheat. When a fuse melts like this, it is said to have "blown." A blown fuse acts like a switch in the open (OFF) position; it cuts off power to the rest of the circuit, preventing further damage and reducing the likelihood of electric shock. Usually the fuse's thin section of wire is encased in a protective glass or plastic covering, so that the fuse wire itself won't damage anything when it gets hot and melts in two (Figure 8.38).

To be effective, the fuse must be able to withstand normal voltage and current levels in the circuit without failure, but fail more quickly than any other part of the circuit as soon a current overload condition occurs. Since the voltage and current requirements of different circuits are different, fuses come in a wide variety of voltage and current ratings.

One problem with a conventional fuse is that it's not reusable. Once it blows, it's dead and must be replaced, like a burned-out light bulb. Fortunately, fuses are meant to be replaced fairly easily; they usually plug or screw into fuse sockets. However, before a fuse is replaced, the cause of the excess current must be diagnosed and repaired, or the

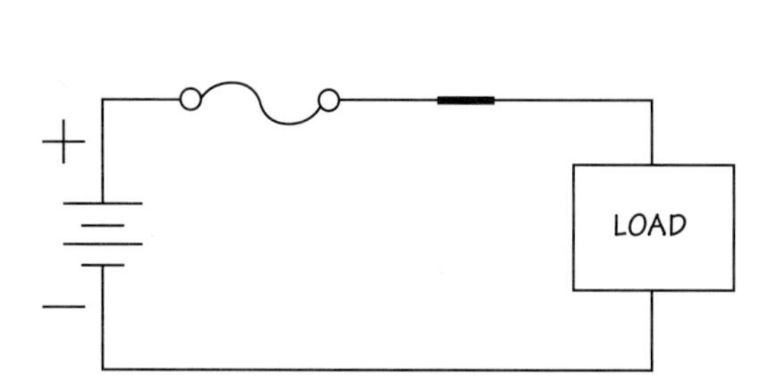

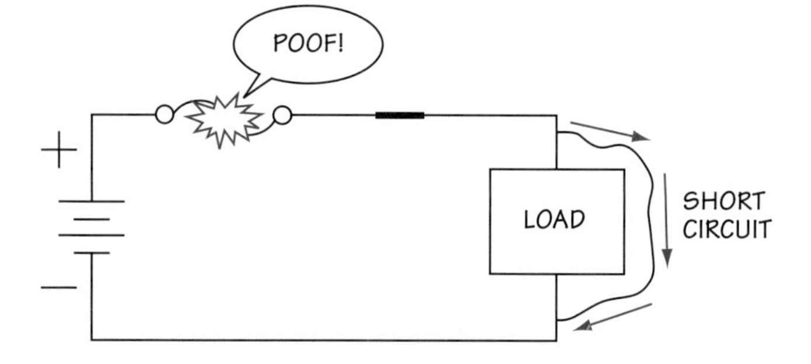

Figure 8.39: A Fuse-Protected Circuit

The stretched-out sideways "S" with the two small circles at either end is a common schematic symbol (in the U.S.) for a fuse. In this circuit, the fuse has been placed between the positive terminal of the battery and the power switch, so that all current leaving the battery must flow through the fuse. This is a good location for a fuse.

An unintended short circuit (red arrows) allows current to bypass the current limits normally imposed by the load, thereby increasing current levels and blowing the fuse. When the fuse blows (Poof!), it opens the circuit, stopping the excessive and unauthorized current flow and thereby reducing the risk of fire or other damage.

replacement fuse will blow, too. When you replace a fuse, make certain you are replacing it with a fuse having suitable voltage and current ratings.

7.2.2. Choosing the Right Fuse

When selecting a fuse to protect a circuit, you must make sure it will allow routine operation of the circuit, yet be the first thing to fail when something serious goes wrong. To select the right fuse, you must know the maximum voltages and currents to expect in your circuit when it's operating normally.

The voltage rating of the fuse needs to be higher than the expected voltage in your circuit, but beyond that, it doesn't really matter too much. It's fine to use a 12,000-volt fuse to protect a 12-volt circuit, except that you'll pay extra for a 12,000-volt fuse when you don't really need one.

The current rating, on the other hand, needs to be selected very carefully, because that's what the fuse is really using to protect you and your circuit. Make sure the current rating is somewhat (but not a lot) more than the expected maximum current under normal operating conditions. A short circuit will usually draw at least 10 times as much current as a properly functioning circuit (and sometimes thousands of times more current), so a fuse rated for 1.5 to 2 times the normal current should allow routine operation, yet blow reliably in the event of a short. For example, if you know your ROV will normally draw about 10 amps when all thrusters and other electrical systems are operating at full capacity, then a fuse rated for 15–20 amps would be a good choice. It's very important that your fuse be the "weakest link" in your circuit, so if your fuse is rated for 20 amps, you want to make sure the wiring in your circuit can handle at least 20 amps for longer than it would take the fuse to blow. Otherwise, you could end up with a bad situation in which the wires melt before the fuse does.

Fuses also come in "fast-blow" and "slow-blow" varieties. The fast ones provide extra protection, but are prone to irritating false alarms (blowing when there's not really a problem), particularly when used in motor circuits or other circuits that experience lots of brief current surges. In motor and relay circuits, it's often better to use slow-blow fuses, which tolerate these momentary overcurrents without blowing, but will blow if the overcurrent persists for more than a brief period.

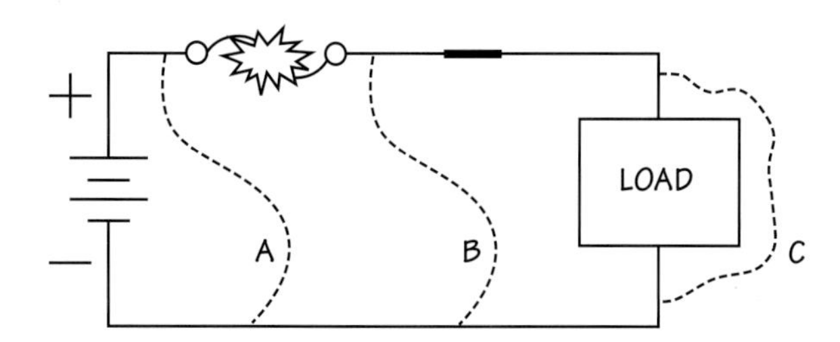

Figure 8.40: Proper Fuse Placement

For maximum protection in most circuits, a fuse should be placed as close as practicable to the positive terminal of the battery, even before the power switch, so that all current flowing out of the battery must go through the fuse before it can go anywhere else. For example, in this diagram, the fuse would provide protection against short circuits at locations B or C, which are farther from the battery than the fuse, but not from a short circuit at A, which is closer to the battery. The closer the fuse is to the battery, the greater the range of possible short-circuit situations it can guard against.

You may be wondering if it would work just as well to place the fuse close to the negative terminal of the battery. In terms of stopping excess current flow, yes, it would. With low-voltage circuits or circuits isolated from ground, it really doesn't make much difference. However, there are legitimate safety reasons for you to get in the habit of putting fuses close the positive battery terminal instead of the negative terminal, particularly if you think of the positive terminal as being analogous to the "hot" wire in AC mains wiring and the negative terminal as being analogous to the "neutral" mains wire. A blown fuse located on the return side of such a circuit would isolate the majority of the circuit from ground, while leaving that circuit connected to the "hot" high-voltage line. This would create a very dangerous situation for a person who touched the circuit while in contact with the earth ground.

SAFETY NOTE: USE YOUR FUSE

Fuses and related devices like circuit breakers are critical safety components in any circuits supplied with enough power to do damage when something goes wrong—and any energy source powerful enough to propel an ROV is also powerful enough to do some damage in the event of a short circuit.

It's always tempting to get your new circuits built and into their first test runs as quickly as possible, so it can also be temping to skip the fuse installation. After all, the fuse is just there as a safety precaution and shouldn't be needed if everything is working properly, right? Unfortunately, it's during those first tests that things are most likely to go seriously wrong. Furthermore, if your test is a success, you may be excited to move on to the next stage and forget about putting in the fuses. Therefore, instead of thinking of fuse installation as the last step in circuit construction, think about making it the first step. As soon as you get your new or recharged battery, hook it up to an appropriately sized fuse before you use the battery to power anything else.

Remember that if a fuse needs to be replaced (or reset) frequently, it suggests an underlying problem that needs to be diagnosed and fixed. Never ignore this type of problem. It could simply be an incorrectly sized fuse, or it could be a short or other serious problem somewhere in the circuitry.

Attempting to bypass a "problem" fuse by shorting across it with a paperclip or piece of tinfoil is a common but ill-advised practice that invites disaster. The fuse is there for a reason.

Figure 8.41: Short Circuit Disasters

Small disasters such as this "cooked" protoboard (upper right) quickly teach the importance of using fuses. Some lack-of-fuse disasters have greater heart-stopping consequences as shown in the two left photos of the fried innards of the Bot MATE-Tricks *ROV. Determined team members tackled an all-night wiring project and created simple ON/OFF controls for the ROV. The next day they went on to win the competition, as evidenced in the final photo.*

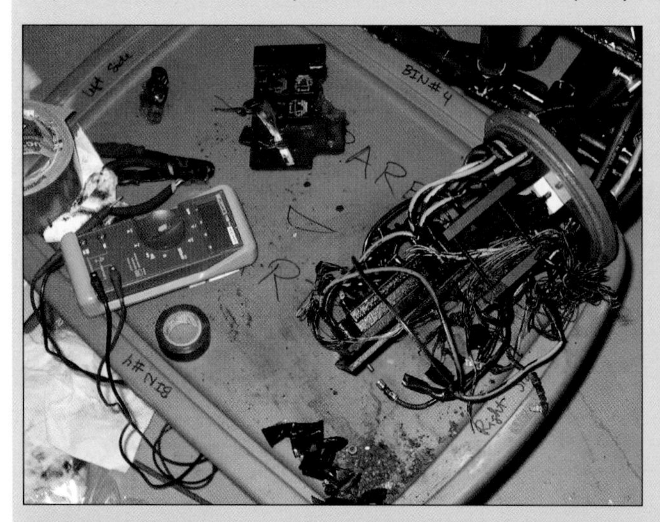

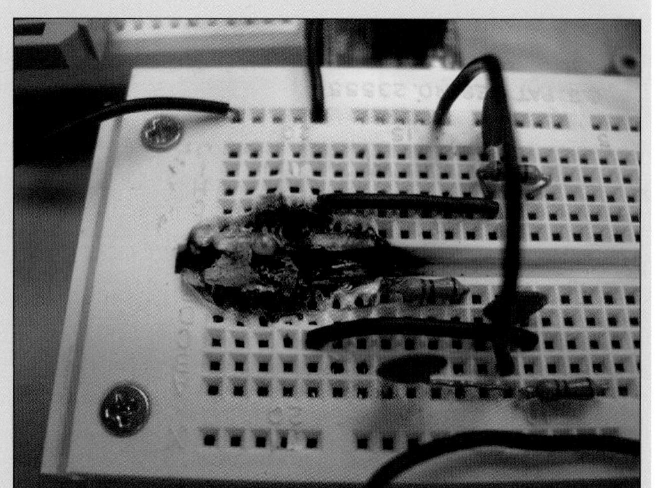

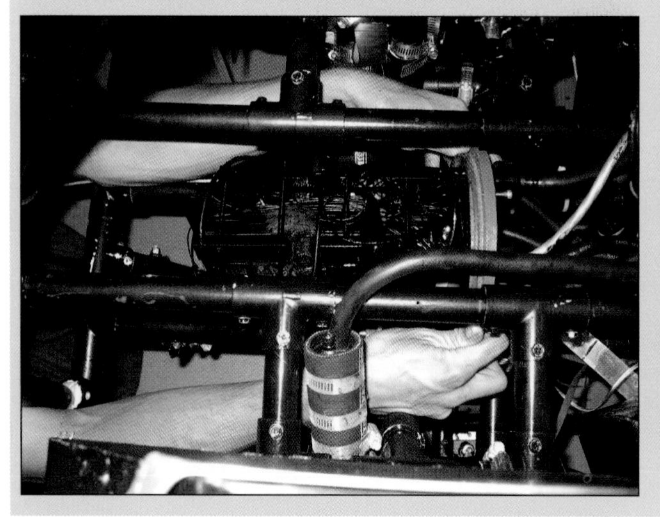

Images courtesy of Scott Coté, Coté Consulting

7.2.3. Circuit Breakers and Resettable (Auto-Reset) Fuses

Circuit breakers and **resettable fuses** are similar to conventional fuses in purpose; however, instead of "blowing" in response to excessive current, they get "tripped" and can be reset without having to be replaced. This difference means they can be reused over and over again. A resettable fuse works by increasing resistance dramatically when it gets hot. Excess current associated with a short circuit causes the resettable fuse to overheat. This in turn raises its electrical resistance enough to block most of the excess current and protect the circuit.

Image courtesy of Dr. Steven W. Moore

Figure 8.42: Circuit Breakers

This photograph of the front of a standard household "breaker box" shows 12 circuit breakers, each protecting a different circuit in the house. Most of these breakers are rated at 20 amps, but the two at lower left are rated at 30 amps each and joined together to provide special power for a larger appliance. The one that's third from the top on the left supplies power to bathroom outlets and includes a reset button for a ground-fault interrupter circuit (GFI).

For deeper-diving vehicles, on-board resettable fuses (also called auto-reset breakers) are convenient. When a current spike trips the auto-reset breaker, it resets after a short "cool-down" period, so there is no need to recover the vehicle and open it up to manually reset a breaker or replace a fuse. Note that if the electrical problem persists and the auto-reset breaker blows again, then you should pull the vehicle from the water and check for a problem in the system. In some cases, the use of an auto-reset fuse will allow an ROV or AUV to limp home under its own power, but the problem that caused the fuse to trip in the first place must then be diagnosed and repaired.

A circuit breaker that has tripped is essentially like a switch that has been flipped from the ON to OFF position automatically. When this happens, it simply needs to be reset. Depending on the type of circuit breaker, resetting is usually accomplished by flipping a switch in a particular way or by pushing a button. Circuit breakers are used commonly in household wiring and large industrial machinery, so for most small ROV/AUV projects, fuses are probably the better choice.

7.3. The Power Switch

Most electrical systems have a master switch located between the fuse and the rest of the circuitry to turn the entire electrical system ON or OFF. This can be important, since many cameras, sensors, and other systems draw current whenever they are connected to a power source. Unless there's a mechanism for cutting off the power, this will eventually deplete a battery and potentially damage it, even if the vehicle is just sitting in storage.

Strictly speaking, you don't actually need a formal power switch, since you can very effectively turn off the system by simply disconnecting the battery or unplugging the fuse, but a regular switch provides a more elegant and convenient solution. You can also get by without a master switch if your vehicle has only a few subsystems and if each of those subsystems has its own power switch. Turning off all of the subsystem switches at the same time is equivalent to turning off one master power switch.

The next chapter has an entire section devoted to switches and their use in controlling the flow of electricity through thrusters and other electrical systems, so they will not be discussed further here.

7.4. Transmitting Electrical Power over a Tether

Although AUVs, hybrid ROVs, and a few other underwater vehicles use on-board batteries for power, most ROVs—from the big commercial vehicles working in the ocean to tiny home-made vehicles plying the depths of a pool—rely on surface-supplied electrical power delivered to the vehicle through the umbilical or tether. This section of the chapter takes a closer look at what is required to deliver power effectively over a tether.

Getting surface-supplied electrical power to an ROV at depth requires a minimum of two wires. For example, if a battery on the surface is supplying power to the submerged vehicle, there must be one wire to carry current from the battery's positive terminal down to the vehicle and another one to carry the returning current back to the battery's negative terminal in order to complete the power circuit.

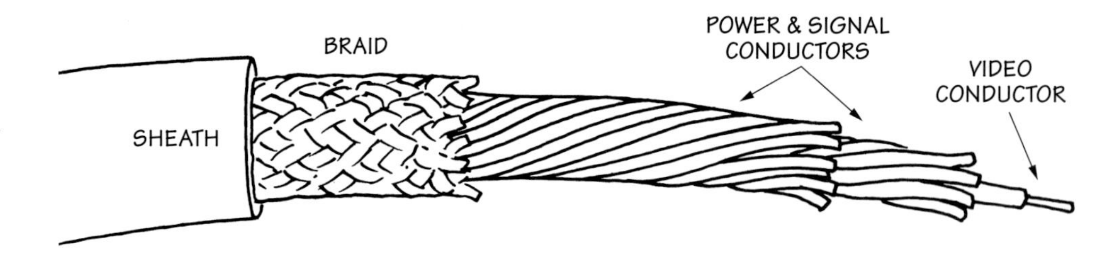

Figure 8.43: Complex Tether Cutaway View

7.4.1. Voltage Drop

Every wire, including the wires in an ROV tether, has some resistance in it. What this means is that some of the power you are trying to send down the tether to the ROV is being lost along the way, because it's heating the tether wire instead of powering the ROV. This power loss manifests itself in the form of a **voltage drop**, so the voltage available at the end of the tether is less than the voltage supplied at the top of the tether. This lost power reduces the overall efficiency of your vehicle's power system and undermines the performance of your vehicle, so you'll want to find some way to minimize the amount of wasted power.

Voltage drop brings up two important questions:

- How much power is being lost?
- How can you minimize the amount of this lost power?

Ohm's Law tells us that anytime current is flowing through the wires, some voltage will be lost as that current travels from one end of the wire to the other. This voltage drop (E) will be equal to the product of the current and the resistance:

Equation 8.8

$$E = I \times R$$

Based on this, you can substitute "*I x R*" in place of "E" inside the standard formula for power dissipation in a resistor (Section 4.10, Equation 8.6), to calculate the rate at which energy is lost from a wire due to resistive heating:

Equation 8.9

$$P = I \times E = I \times (I \times R) = I^2 \times R$$

Note that the current, I, is squared in this equation. That tells you that current has a really big effect on the amount of power wasted as heat. It also tells you, therefore, that a good way to reduce the wasted power is to reduce current in the wire. If you can cut the current in half, you will cut your power loss to $1/4^{th}$ of what it was before. Better yet, if you can reduce current to $1/10^{th}$ of what it was before, you can reduce power loss to $1/100^{th}$ of what it was before. The only way to reduce the current without also reducing total power reaching the ROV is to increase the voltage to compensate. This is why long-distance power transmission, such as that done through the high-tension power lines that criss-cross continents, is often done at very high voltages. It's simply more efficient.

Another approach to reducing the wasted power would be to reduce the resistance (R) in the wire. This could be done by making the wire shorter or by giving it greater cross-sectional area.

Unless you can somehow decrease the power requirements of your vehicle, this analysis leaves you with only four options:

1. You can shorten the tether.
2. You can increase the cross-sectional area of the tether wires to lower their resistance.
3. You can increase the voltage.
4. You can give up on transmitting power through the tether and place the batteries on board the vehicle instead.

Option #1 (shorter tether) is easy enough if your vehicle will be working in a swimming pool or from a boat, dock, or shore in a shallow area.

Option #2 (increased cross-sectional area) amounts to fattening the wires. This is easy if your tether is already fairly short and your vehicle is powerful enough to overcome the increased stiffness of a thicker tether. A heavy-duty extension cord, like those used for power tools, contains large-diameter conductors with low resistance and will often do the trick. However, if your tether needs to be longer than a few tens of meters and/or needs more than three or four wires in it, this option can get quite heavy, stiff, and expensive. Moreover, in strong currents, a thicker tether will add (often substantially) to the drag experienced by the vehicle. You could add more powerful thrusters to pull the fatter tether around, but that would require more power, leading you right back to the problem you started with. Note that one way to increase effective wire diameter, while maintaining some degree of flexibility, is to pigtail several smaller wires. (See the *Tech Note: Pigtailing Wires to Lower Resistance.*) If your motors or other systems require many amps of current, you will probably need to combine options #1 and #2.

Option #3 (higher voltage) is an attractive one, at least on paper, because the squared current in the equation means a modest increase in voltage can give you a big payoff in terms of reduced power loss. Public utility companies, which make their living by distributing vast quantities of electrical power over hundreds of miles of wire, have

TECH NOTE: PIGTAILING WIRES TO LOWER RESISTANCE

If you take two identical wires running in parallel and twist their exposed ends together (a process called "pigtailing"), you create the electrical equivalent of one wire with twice the cross-sectional area (and therefore half the electrical resistance). Pigtailing three or four wires together will reduce resistance even more. This can be a useful trick for reducing the resistance and associated voltage drops in your tether, particularly if you already happen to have some unused wires in the tether cable.

Even if you need to add new wires, pigtailing can be advantageous, because a bundle of pigtailed wires will be more flexible than a single thick wire having the same resistance as the bundle. For example, if you need to reduce the resistance of one of your tether wires to ¼ of what it is, you can replace that thin wire with a single wire that has twice the diameter (and therefore four times the cross-sectional area), or you can pigtail four of the thinner wires together. Both approaches will have the same effect on resistance and voltage drop, but the pigtailed version will be more flexible than the single-wire version.

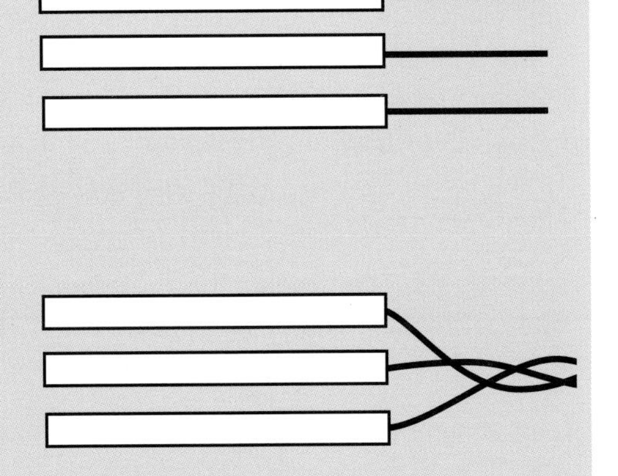

Figure 8.44: Pigtailing

noticed this. These companies don't want to waste power heating wires, and they don't want to waste money paying for miles of unnecessarily fat copper wires, either, so they use extremely high voltages (up to a million volts or more in some countries) to transmit power with minimal current. These companies typically use AC power instead of DC power, because AC power is much easier and cheaper to convert from one voltage to another. The same approach on a smaller scale is used to send power efficiently through a long tether to most work class ROVs. They, too, usually use high-voltage AC power, though the voltages they use are typically in the range of a few hundred to a few thousand volts. Unfortunately, high voltages can be deadly around water and require advanced techniques, materials, and training to ensure adequate safety, so option #3 is *not* a recommended solution for home- or school-built ROVs.

Option #4 (on-board batteries) is a viable choice for home- or school-built underwater vehicles, provided missions are short and easy enough that the vehicles do not require large quantities of stored energy. However, this option requires fairly sophisticated control electronics, because all switching of large electrical currents on the ROV must be done automatically on board the vehicle in response to encoded, low-power commands transmitted through the tether. Methods for doing this are described in *Chapter 9: Control and Navigation.*

Obviously, the solution you choose will depend on your mission and the resources you have available.

TECH NOTE: SAMPLE TETHER CALCULATIONS

Here's an example to illustrate how to do the calculations necessary to determine what wire diameter you might need in a tether. Suppose your project is for an ROV competition that calls for a vehicle with three thrusters, each made from one 12-volt, 3-amp motor. In addition, you are told that the only power source you'll be allowed to use is a 12-volt battery at the surface, and that your tether will need to be 18 meters (approx. 60 ft) long.

For starters, assume that you'll have a separate pair of tether wires supplying power to each motor, that your motors will be the only things powered by those wire pairs, and that each motor might need full power. If so, you know that each pair of wires needs to be able to carry up to 3 amps of current. Looking just at the maximum current ratings listed in Table 8.4, you might conclude that #24 wire, which can handle 3.5 amps, would be adequate.

That would be a mistake, because you must also consider the voltage drop in the tether. What would that voltage drop be using #24 AWG wire? To calculate the answer, use Ohm's Law: $E = I \times R$. Keep in mind that the electricity must make a round-trip journey through the tether (down and back up), so a 60-foot tether represents 120 feet (approx. 36 m) of #24 AWG wire through which the current must travel. Consulting the Resistance column of Table 8.4, you find that #24 wire would offer a resistance of 0.026 ohms/foot x 120 feet = 3.12 ohms. So running 3 amps through this diameter wire would result in a voltage drop of 3 amps x 3.12 ohms = approximately 9.4 volts. If you started with 12.0 volts at the surface, this would leave only about 12.0 - 9.4 = 2.6 volts at the bottom end of the tether—not nearly enough to run your 12-volt motors at full power!

Alternatively, if you used a thicker tether with #14 AWG wire, then your expected voltage drop would be only 0.36 volts (verify this number for practice with the calculations), so your motors would get almost all of the intended 12 volts.

If, for some reason, you needed a very thin, flexible tether and could not tolerate anything thicker than #24 AWG wires in it, then a 12-volt battery at the surface might not be able to deliver adequate power down to your ROV through the tether. In that case, the best option might be to place the batteries on board the ROV and use the tether only for very low-current command signals, rather than for power. The command signals won't lose significant voltage, because they aren't trying to send lots of current. As mentioned above, this approach would require some of the more sophisticated control schemes discussed in the next chapter, because all of the power switching would have to take place on board the ROV and be controlled remotely.

7.5. Accommodating Multiple Voltages

Even a simple ROV or AUV will usually have at least two or three separate electrical subsystems on board, and it's not uncommon for those subsystems to have different voltage requirements. For example, your vehicle might have thrusters with motors that require 12 volts, a camera that requires 10 volts, and navigational sensors that require 5 volts. This section of the chapter describes how you can accommodate such a range of required voltages.

One easy solution to this dilemma is to use self-contained subsystems that already include their own built-in battery systems for power. For example, you could make headlights for your ROV simply by strapping scuba-diving flashlights (with their own batteries) to the outside of your ROV.

A second solution is to provide centralized power for devices that normally run on batteries or use an AC-to-DC wall wart-style adapter. Odds are good that these devices can be modified to run from a centralized set of one or more low-voltage batteries. You will first need to determine what voltage each device requires and the maximum current it will draw. Then you'll have to provide a way to connect the required voltages and currents to the correct power supply wires inside the device. Remember that watts = volts x amps, so if a label on the device gives you any two of those values, it's easy to calculate the third one.

The simplest way of doing this is to provide different sets of batteries for each required voltage, then use wires to connect the devices in parallel to the appropriate batteries. This is actually good in terms of energy efficiency, but may be inconvenient in terms of physical space or weight. This option will also require that you keep buying (or recharging) several different kinds of batteries, a situation that could get tiresome and/or expensive.

An alternative approach, which uses fewer different types of batteries, is to use a single, relatively large-capacity battery whose voltage is appropriate for the highest-voltage device on the vehicle. Then use **voltage regulators** to create other lower voltages, as needed, from the power supplied by this single battery.

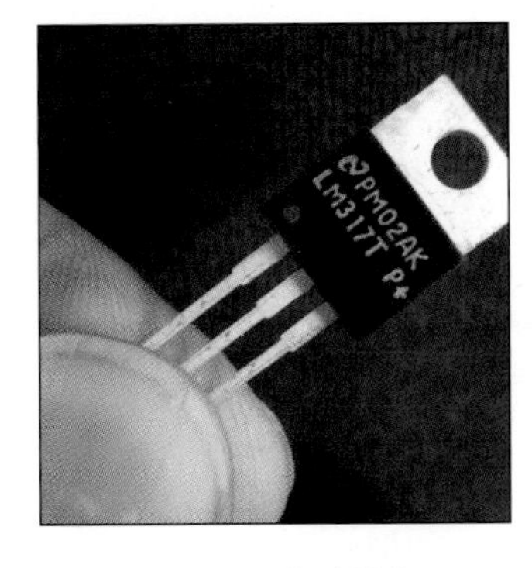

Figure 8.45: Typical Voltage Regulator

This is a voltage regulator in a three pin (wire) case called a TO-220 package. This package is used for some voltage regulators and many other components that can get hot. A piece of metal can be bolted to the hole in the top to act as a heat sink, drawing away excess heat and keeping the integrated circuit cooler.

Figure 8.46: Voltage Regulator and Heat Sink

This voltage regulator is shown next to a heat sink. The heat sink compound in the tube is used between the chip and the heat sink to improve heat transfer.

Figure 8.47: Voltage Regular and Heat Sink Installed

This voltage regulator has a heat sink attached to it and is installed on an ROV circuit board.

Images courtesy of Dr. Steven W. Moore

Image courtesy Dr. Steven W. Moore

Figure 8.48: DC-to-DC Converter

DC-to-DC converters or switching regulators, like regular (i.e., "linear") voltage regulators convert one DC voltage to another. However, there are a few significant differences.

Most DC-to-DC converters use a switched power scheme that makes them more efficient and allows them to run cooler than a typical linear voltage regulator. So they are often used in battery-operated circuits or where large voltage changes and/or large currents are involved. Some can also produce output voltages that are higher than the input voltages, which is something a linear voltage regulator cannot do.

If you need these features, consider a DC-to-DC converter. But if you don't, you'll probably find that a linear regulator is simpler and much less expensive to use. One other thing to consider is noise. The rapid internal pulse switching in a DC-to-DC converter can introduce electrical noise in a circuit, so a linear voltage regulator is usually a better choice for powering sensitive sensor circuits.

Voltage regulators are usually small transistor-like integrated circuits. They come in a wide variety of output voltages, with 3, 3.3, 5, 10, and 15 volts being particularly common. Some offer negative voltages. Most have at least three terminals: one for ground, one for the input voltage, and one for the output voltage. Most require additional electrical components, such as capacitors, so if you go this route be prepared to design, build, and solder some simple circuits. The additional components and directions for hooking them up will generally be specified in the voltage regulator's datasheet. This info is available from the manufacturer, usually as a pdf file that can be downloaded from the internet. Some of these datasheets are more user-friendly than others, so you may need to find a technicial or engineer to help you interpret some of the technical jargon and to help you build your first voltage regulator circuit.

Voltage regulators come in two broad categories: linear and switching. Linear regulators typically require fewer external components and often have better regulation performance, but they are less efficient than switching regulators, so they tend to get hot and waste more power, particularly in high-current situations with significant voltage drops between the input and output voltages. Switching regulators, on the other hand, are relatively efficient, but require more complex circuits. Some regulators of both types have adjustable output voltages; others have fixed output voltages.

Selecting an appropriate voltage regulator can be an involved process. In addition to the input voltage range and the output voltage(s) of the regulator, you should pay attention to power limits and "dropout." Dropout refers to the minimum allowable difference between the input voltage and the output voltage, and it usually depends on how much current your regulator needs to supply. For example, if a 5-volt regulator has a 2-volt dropout at the 1-amp current you need, then you'll need to supply that regulator with at least 7 volts (= 5-volt output + 2-volt dropout) for it to function properly. If you give it only 6 volts at the input, it will be able to give you only 4 volts at the output. Usually the more current, the larger the dropout. If you exceed the power limits, the regulator will overheat and/or shut off.

An alternative to the voltage regulator is the **DC-to-DC converter**, introduced along with voltage regulators in Section 6.5.1. These circuit assemblies tend to be more self-contained than voltage regulators; they can also produce output voltages that are higher than the input voltage supplied by the battery, but they also tend to be quite a bit more expensive.

7.6. Power Distribution Systems: Three Actual Examples

The following examples focus on three ROVs with very different power distribution systems. The first is a team-built competition vehicle that conforms to defined power specifications. The second ROV is an ocean-going, education-oriented vehicle where low voltage is an especially important safety consideration. The third is a commercial ROV where versatility and performance are key attributes, requiring a more complex power distribution system. Studying and comparing these various examples should give you a good idea of the many options available when designing a power distribution system to accomplish your own vehicle's mission.

7.6.1. Example #1: MATE ROV Competition

Each year, the Marine Advanced Technology Education (MATE) Center organizes annual student ROV-building competitions that draw thousands of participants from all over the United States, Canada, and other countries. These competitions generally take place in a swimming pool or test tank arena, usually involving a set of challenging underwater tasks modeled after real-world ROV missions.

For reasons of safety and fairness, MATE places stringent constraints on the power sources that may be used by the teams. In most cases, contestants are required to use power supplied from a poolside lead-acid battery equipped with a current-limiting fuse. The ROV design teams must power their ROVs through a tether, using energy from this battery. There are many possible ways to do this, and each year, teams come up with all sorts of ingenious solutions. One popular approach is illustrated in Figure 8.49 for a vehicle with three thrusters (two horizontal, one vertical) and one camera.

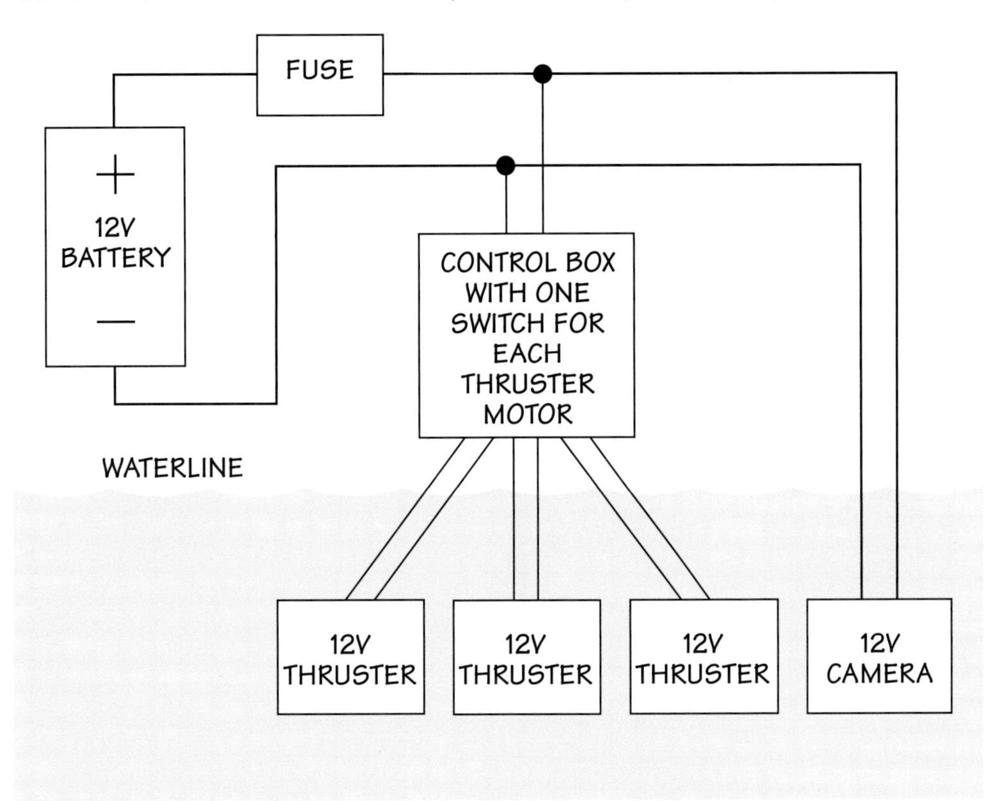

Figure 8.49: Typical Power Diagram for Competition ROV

7.6.2. Example #2: *ROVing Otter*

ROVing Otter is a web-controlled flying eyeball ROV designed for educational use in kelp forests in shallow ocean environments. This vehicle can be controlled remotely via the internet. In addition, it is designed for hands-on use and modification by local students. This aspect of its mission means that electrical safety in and around seawater in the presence of inexperienced students had to be a major factor in the design of this vehicle. Accordingly, one of the design constraints is that its maximum voltage is limited to 12 volts.

For reasons described earlier in this chapter, the 12-volt requirement, coupled with the long, slender, flexible tether (made from Cat-5e computer networking cable), required that the batteries be located on board the vehicle. Power is provided by a single 12 V, 12 Ah AGM sealed lead-acid battery.

One challenging aspect of the power system for this ROV is that the vehicle contains a number of different sub-circuits, with different voltage requirements. The thruster motors require 12 volts at up to 0.7 amps each (total 2.8 A for all four thrusters). The depth sensor used requires 10.0 volts for accurate depth readings. The network camera, network switch, microcontroller, and several other sub-circuits each require 5 volts, and the dual lasers used to measure objects in the camera view require 3.3 volts.

In addition, because of the logistics of deployment and post-dive delays in rinsing and drying the housing before it can be opened, the battery and other electronic components must be sealed inside the housing up to several hours prior to a dive and left in there for up to several hours afterward. Because the battery stores only enough energy to power the ROV for three to four hours, there needs to be some way to switch power on and off without opening the housing. This is no problem in ROVs where power is sent down the tether—you just avoid connecting power until it's needed. But with the power source on board and sealed inside, some system must be devised to allow remote ON/OFF control.

Figure 8.50 illustrates the power distribution system that solved the remote ON/OFF control problem for *ROVing Otter.* Current from the positive terminal of the battery is first routed through a fuse (to reduce the risk of fire and other damage in the event of a short), then to a **MOSFET switch.** This switch is a type of transistor used as an ON/OFF switch. Though located on the ROV, the MOSFET switch can be controlled from the surface by a tiny electrical signal sent down one of the four twisted pairs of wire in the Cat-5e tether. Thus, power from the ROV's on-board battery can be switched on or off, even while the vehicle is under water. (For more detailed information on MOSFETs, see Sections 5.2 and especially 5.12.1 in Chapter 9.)

Once past the MOSFET switch, the 12-volt power is routed directly to the thruster control circuitry and the video light control circuitry, as well as to two voltage regulator circuits: one that produces 10 V (to power the depth sensor) and one that produces 5 V (to power almost everything else that's left). A portion of the 5 V power is used to power a 3.3 V regulator for the lasers.

Figure 8.50: **ROVing Otter Power Distribution System *(major components in gray)***

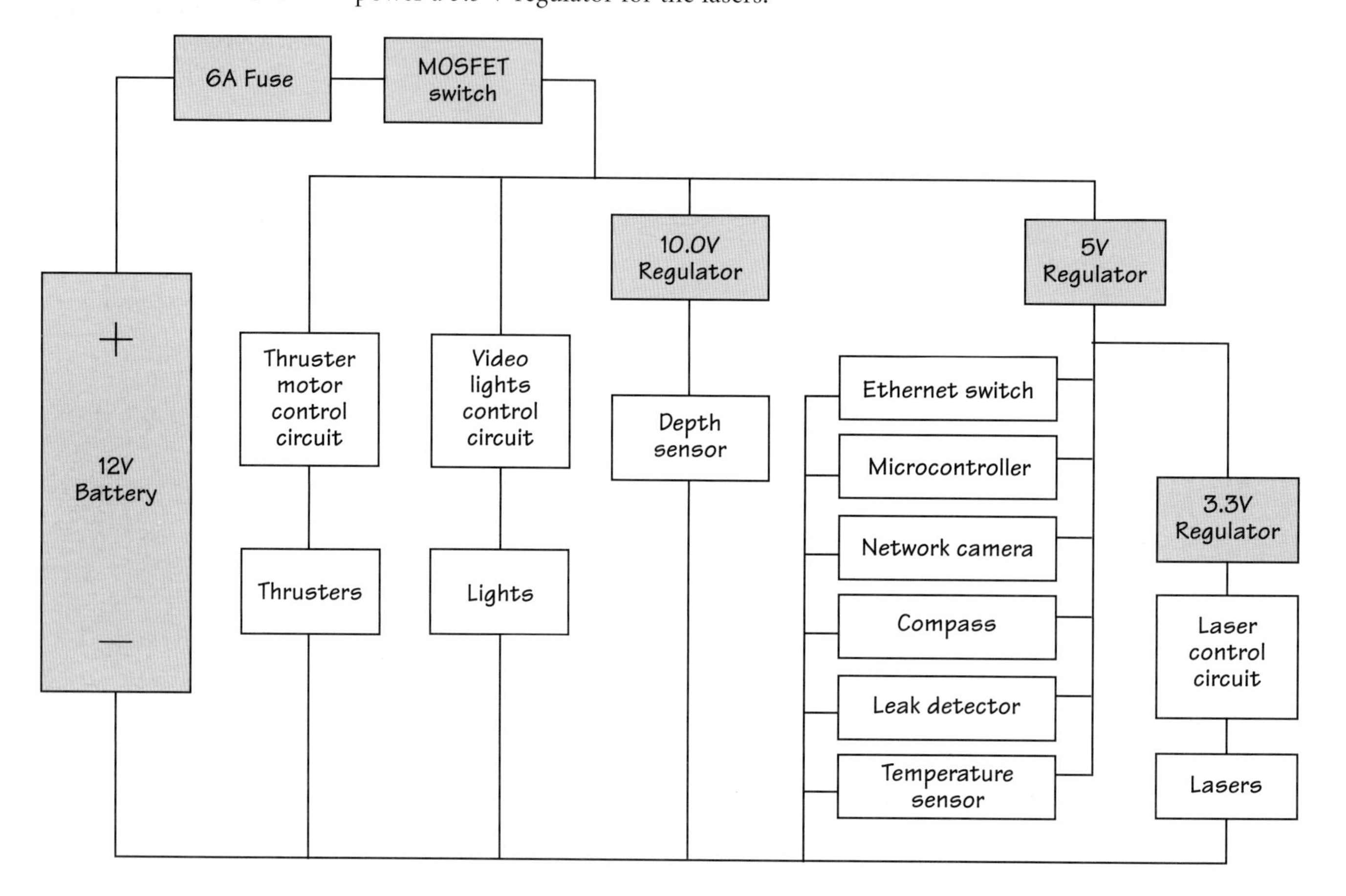

7.6.3. Example #3: A Commercial LCROV

Several companies manufacture small, portable, low-cost ROVs (LCROVs) intended for a wide variety of underwater inspection tasks and other jobs to depths of 100–300 meters (approx. 325–1,000 ft) or greater.

Though not nearly as powerful or deep diving as a full-featured, deep sea work class ROV, these LCROVs have an impressive list of features and can dive to surprising depths. For example, many have high-resolution color video cameras with video lights, optional grippers, and autopilots for depth and compass heading. In addition, most are small and light enough to be carried, launched, piloted, and retrieved safely and successfully by one person. While prices for LCROVs are still several thousand dollars for even the least expensive ones, they are considered low-cost, because their price tag is a small fraction of the cost of purchasing and operating a work class ROV.

A major difference between these commercial LCROVs and the previous two power distribution system examples is that most commercial units must be able to operate in deeper water for longer periods of time while doing work that requires much more power. On-board batteries are not a realistic option for delivering this amount of energy, so power must be delivered down the long tether. To avoid excessive losses from resistive heating in the tether wires, these vehicles often transmit the power using high voltages. (See Section 7.4 for a review of how this works.) At the top end of the tether on the surface, an AC transformer is used to "step up" the voltage to a higher

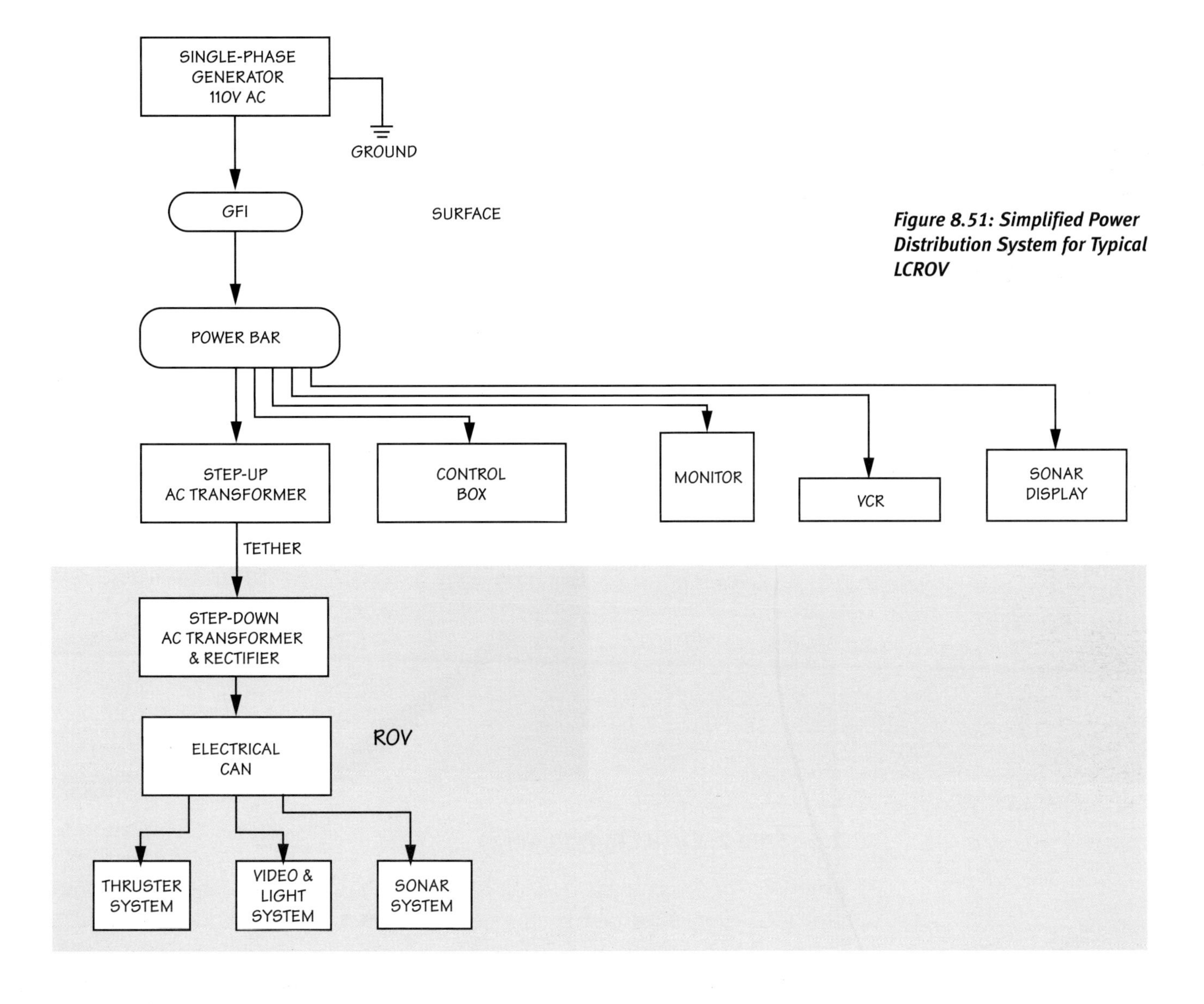

Figure 8.51: Simplified Power Distribution System for Typical LCROV

voltage. At the bottom end, on the ROV, a second transformer is used to knock the voltage back down. A rectifier is used to smooth out the ripples in the AC voltage, turning it into a set of nice, steady DC voltages that can be distributed to the equipment on the ROV. Figure 8.51 visually diagrams a power distribution system for a typical LCROV. Note that the wires for information flow (e.g., control signals, sonar data, camera images) are not shown.

SAFETY NOTE: AVOIDING THE ELECTROCUTION HAZARD

Commercially made LCROVs employ a number of good engineering practices to ensure that they do not present an electrocution hazard, in spite of the potentially dangerous voltages present in the tethers of many of them. This level of engineering is beyond the capabilities of most beginning and intermediate ROV builders, so stick with batteries and low voltages for home-built or school-built projects.

8. Steps in Circuit Design and Construction

8.1. Step 1: Circuit Design

Before you actually build a physical circuit, you'll want to work out a design on paper (or whiteboard or computer) that you think will do what you want it to do. For that professional look that will help others on your team read and understand your diagrams, practice using standard schematic symbols and other schematic conventions. You can find thousands of examples of schematics just by doing a web search for images using the key words "schematic diagram."

Although it may be tempting to skip this step, don't. The schematic diagram will help you in many ways. The process of generating a schematic forces you to think through all aspects of your basic circuit design. It also helps you communicate your design ideas to others for feedback, develop a shopping list for parts, determine how many wires you'll need in your tether, and so on. When you're all done, it serves as formal documentation of your circuit design.

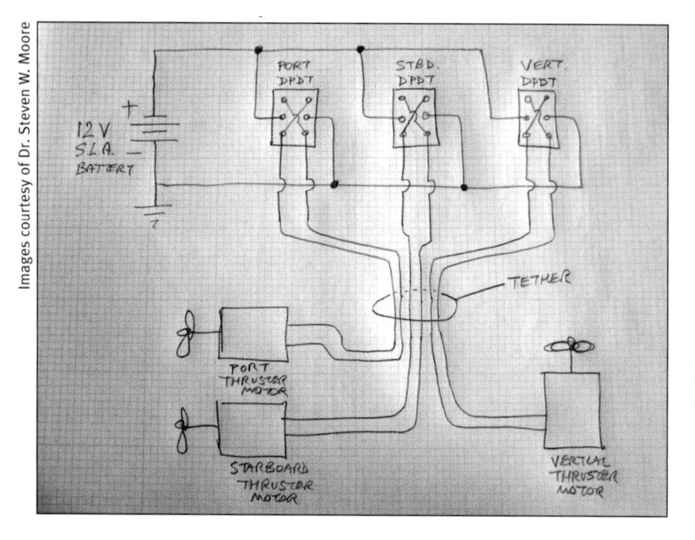

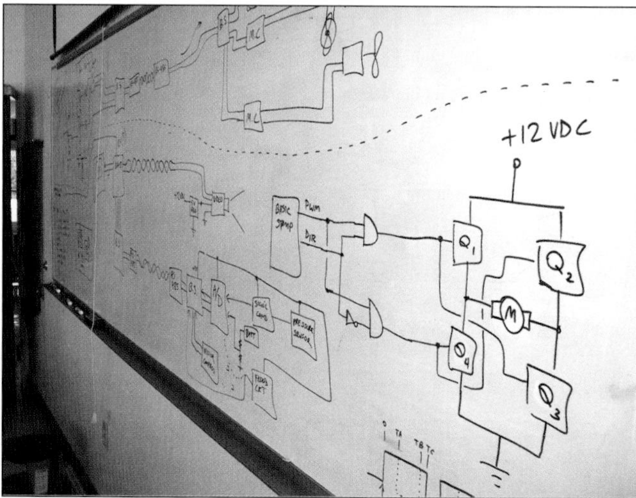

Images courtesy of Dr. Steven W. Moore

Figure 8.52: Schematic Diagrams

Every circuit should begin life as a schematic diagram showing all the components in your circuit and how they will be wired together.

8.2. Step 2: Circuit Prototyping

The next step in circuit design is to test your schematic ideas by hooking up temporary prototype circuits. You'll want to do this to make sure your circuit ideas work *before* you invest the much greater time and effort required to turn them into truly rugged and reliable circuits that can withstand the rigors of life at sea. Truth be told, circuit designs—even those by professionals—rarely work the first time. They almost always have hidden problems that don't surface until you actually build the circuit and try it out, and that's what makes prototyping so valuable!

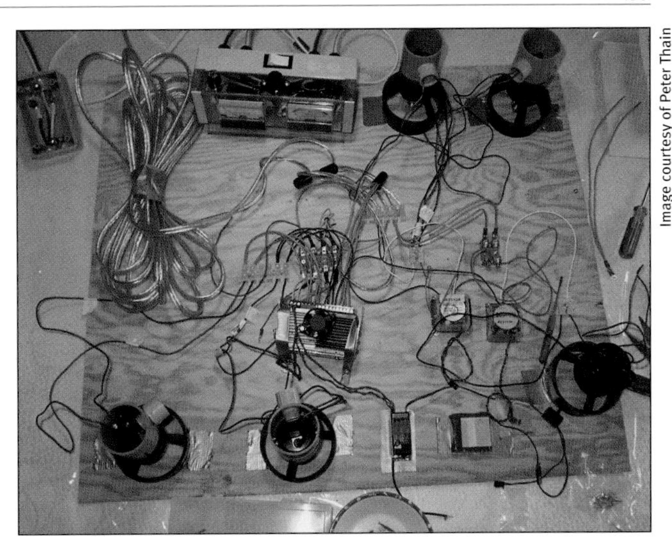

Image courtesy of Peter Thain

Figure 8.53: A Traditional "Breadboard"

A prototype circuit doesn't need to win any beauty contests, and it doesn't need to be rugged; it just needs to prove that your basic circuit design works. In this photo, some duct tape and a piece of plywood gets the job done by holding things in place long enough to verify correct circuit operation.

Breadboarding is a method of connecting the components of your circuit so you can see if the circuit actually does what you want it to. The simplest breadboard can be just a piece of plywood and duct tape (Figure 8.53). A more convenient solderless breadboard (Figure 8.55), sometimes called a "protoboard," allows you to hook up or modify circuit components simply by plugging them into the appropriate holes on the board. Either version allows you to create a working prototype of your circuit design.

There are lots of ways to make prototypes, and almost anything goes. However, any good prototype will exhibit the following features:

- It is comparatively quick, easy, and inexpensive to assemble.
- It allows easy access to all wires and components for testing and measurement.
- It is easy to change, so you can experiment with various modifications to your circuit until you get everything working properly.
- It is realistic enough to reveal any potential problems with your basic circuit design.

Images courtesy of Dr. Steven W. Moore

Figure 8.54: Different Test Leads

Short sections of wire with alligator clips (left) or their smaller cousins, called "minigrabber" or "micrograbber" clips (right), on either end are a great way to quickly interconnect a wide range of electronic components for prototype test-drives. These are for temporary prototype connections only, as they can easily come loose or touch and short things out. Be aware that most of these smaller test leads are not intended for high-current applications, so they can get hot and melt during some thruster tests if you leave the motors on too long.

Figure 8.55: Solderless Breadboard

For test-driving small, low-power circuits, including the "brains" of most control circuits, it's great to have access to one or more solderless breadboards. These are sometimes known as "protoboards." They are available at most places that sell tools for building and testing small electronic circuits. With these boards, you connect circuit components simply by plugging them into appropriate holes on the board. Modifying your circuit connections is a simple matter of plugging things into different sets of holes in the board.

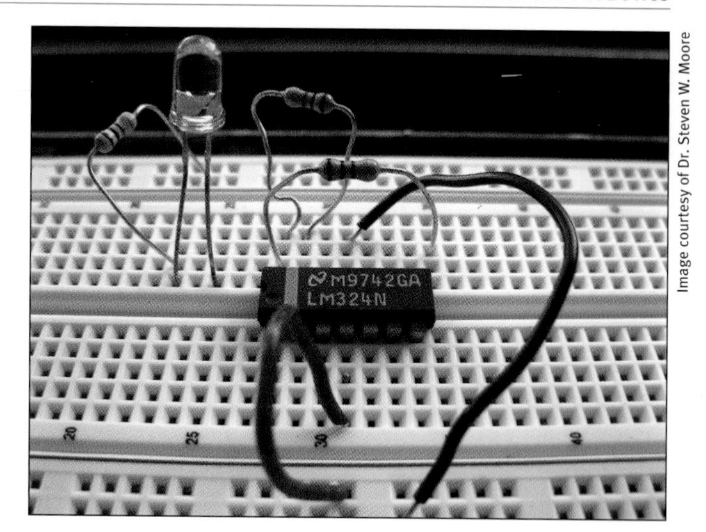

Image courtesy of Dr. Steven W. Moore

TECH NOTE: CIRCUIT TROUBLESHOOTING WITH TEST INSTRUMENTS

Most prototype circuits do not work properly when you first hook them up. That's just the way it is. And because electricity is invisible, it can be tricky to figure out why your circuit isn't working, unless you have some way to "see" what the electricity is doing (or not doing) in your circuit. Two test instruments are particularly helpful. The first is a **multimeter** (Figure 8.56), which is a simple and inexpensive instrument that allows you to measure voltage, current, resistance, and other electrical quantities in your circuit, thereby making it much easier to track down the source(s) of any problems.

For example, the voltage reading on your multimeter might reveal that battery voltage is not reaching the thruster, even though a visual inspection indicates that wires are connected properly. By tracing backwards through the wires with the multimeter until you find the place where you first lose the voltage, you can pinpoint the location of the broken wire or loose connection that caused the problem. See Section 9.4 of Chapter 12 for more about these helpful test instruments and how to use them.

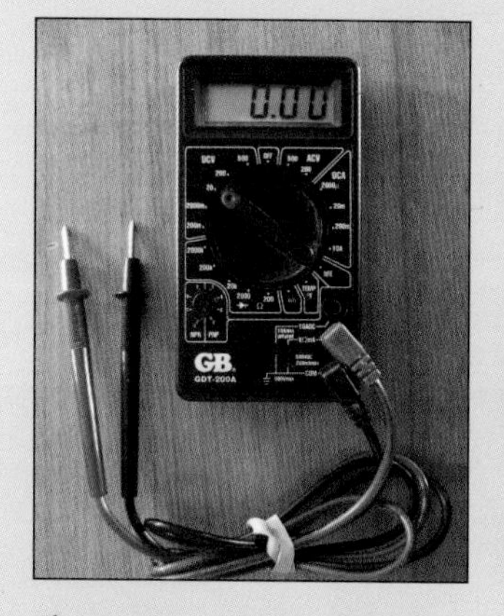

Figure 8.56: Multimeter

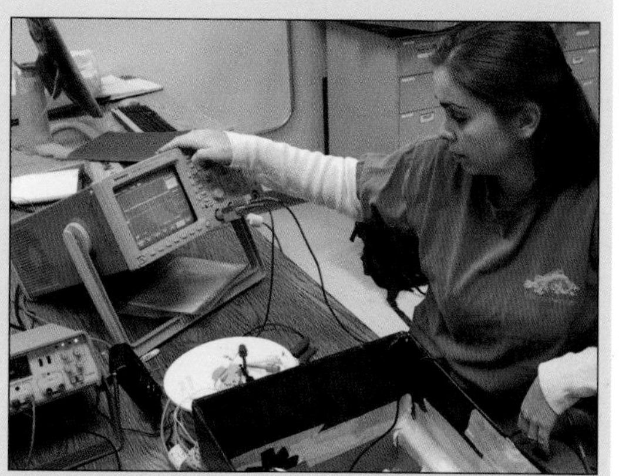

Figure 8.57: Oscilloscope

Images courtesy of Dr. Steven W. Moore

A multimeter is extremely helpful in debugging many simple circuit problems, but it cannot measure and display very rapidly changing electrical quantities, such as the voltage fluctuations in signals from a video camera or the digital control signals typical of automated control systems. To troubleshoot circuits that rely on these rapidly changing voltage signals, you need an instrument called an **oscilloscope** (Figure 8.57). These devices, affectionately known as "O-scopes," are usually much more expensive than a multimeter. Some cost thousands of dollars, but many schools have them for teaching physics or electronics and may make them available to students for school-related projects.

An oscilloscope measures voltages instant-by-instant and plots them in graphical form on a small screen. With the O-scope controls, you can adjust the time and voltage scales of the graph to view and analyze a wide variety of different signal types. Typical oscilloscopes are so fast they can resolve voltage changes lasting much less than 1 millionth of a second. Some newer models consist of a small box that plugs into a computer and uses the computer screen for display. Although these small units are usually not as fast or capable as the stand-alone versions, they are often less expensive and can be handy for troubleshooting ROV or AUV problems in the field or at competitions away from home.

For larger, heavier components and wires, a piece of plywood or other rigid surface works great as a base for your prototype circuit. Simply attach components to the board with screws, cable ties (through holes drilled in the board), or duct tape to hold them in place (see Figure 8.53), then make temporary wire connections between the components by twisting wires together, using alligator clip leads, or similar approaches. (See Figure 8.54.) For smaller circuit components, a solderless breadboard (see Figure 8.55) is often the best way to proceed.

Breadboarding your prototype circuit lets you work the kinks out of your circuit design and ensure that it works properly. This often takes awhile. Once everything is functioning, don't forget to update your schematics to reflect all changes you've made.

8.3. Step 3: Robust Circuit Construction

Image courtesy of Dr. Steven W. Moore

Figure 8.58: An Example of a Well-Constructed Circuit

This topside control circuit exhibits quality construction. All major components are grouped according to function and screwed securely to the baseboard. Wires are connected solidly to each component and to each other through the use of crimp-on terminals and terminal blocks. All tether wires are brought onto the board through one neat terminal block on the edge of the board.

The power lines for all four motor controllers come in through a fuse block that holds a separate fuse for each motor controller, and a central ***ground bus*** *is provided for ground connections. Excess wire is bundled neatly and tied with cable ties.*

As a result of this careful construction, this circuit could be transported in a simple box with little danger that wires will break or come loose, even if the box gets bumped or dropped.

Images courtesy of Dr. Steven W. Moore

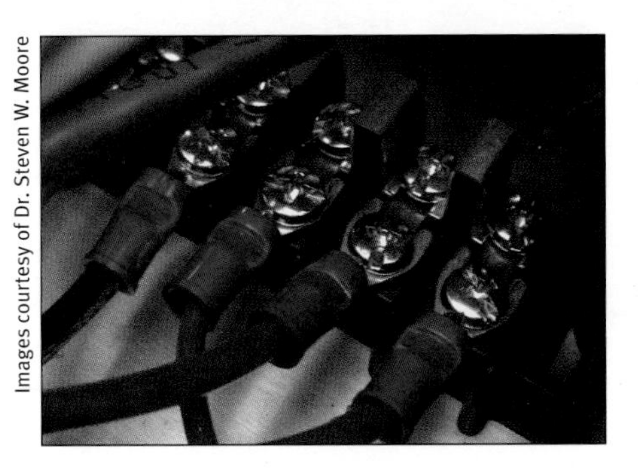

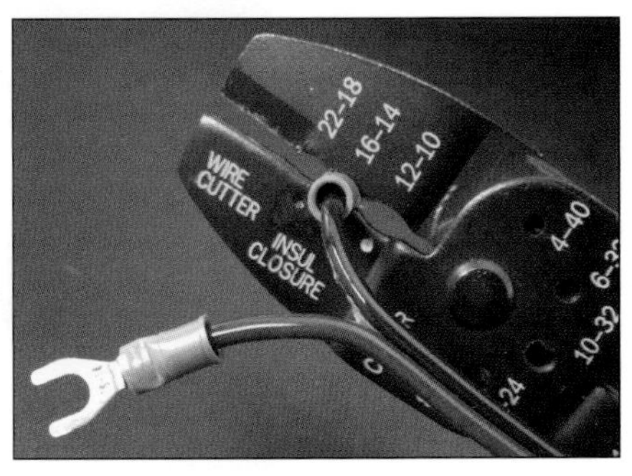

Figure 8.59: Using a Terminal Block

For larger-gauge wires, such as many thruster motor wires, solderless crimp-on terminals and terminal blocks provide an easy way to build reliable yet reversible circuit assemblies. For smaller-gauge wires, other techniques work better.

Once you've got a working prototype and have carefully recorded any changes in an updated schematic, then it's time to assemble your circuit in its final form. Of course, this final version must reproduce the electrical connections from your prototype. But it must do so in a way that can survive the inevitable bumps, vibrations, immersion, (after waterproofing) and other insults that will do their best to destroy your circuit during actual mission operations.

Image courtesy of Dr. Steven W. Moore

Figure 8.60: Wire Splice

Splices come in many forms. This one, like many, can be used to join two or three wires together simply by inserting the wires and then snapping the clip shut.

At this phase, one main goal is to anchor all components, wires, and connections securely in place so they cannot flop around or pop loose. Improperly anchored components act like loose cannons on the deck of an old battleship—they can do a lot of damage. They can bang around during transport over roads or in rough seas, snapping wires or shorting out other components. Even if all the heavy components are anchored, loose floppy wires can fail after a while due to metal fatigue caused by repeated flexing during periods of vibration. These wire breaks often occur inside the insulation where you can't see them, so they can be difficult and frustrating to track

Figure 8.61: Soldering

Soldering is a fun and useful skill. It's best learned from someone who has enough experience to show you how to make high-quality solder joints safely. Lead-free solders bearing the RoHS (Removal of Hazardous Substances) designation are now available and should be used when possible to reduce environmental toxicity.

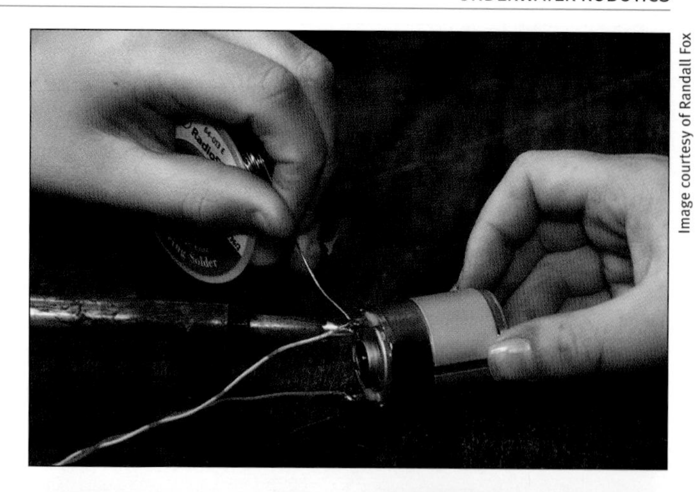

Image courtesy of Randall Fox

Image courtesy of Dr. Steven W. Moore

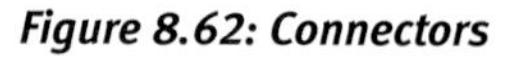

Figure 8.62: Connectors

There are literally hundreds of different kinds of connectors (sometimes these are called interconnects) designed for hooking wires together and for hooking wires to printed circuit boards. Some have screw terminals, some plug in, some snap together.

The best choice for any particular application depends on many factors. These include wire type and diameter, how many wires are involved, how frequently the connection needs to be disconnected (if ever), whether it will be exposed to water, etc. To get a sense for the variety, look under "interconnects" or "connectors" in an on-line or paper catalog from a large electronics parts distributor. These two photos show different types of connectors.

Image courtesy of Dr. Steven W. Moore

Figure 8.63: Point-to-Point Soldering

This photo shows point-to-point soldering used to join electric components. In this case, it's connecting parts mounted inside an aluminum chassis box.

Image courtesy of Dr. Steven W. Moore

Image courtesy of Dr. Steven W. Moore

Figure 8.64: Perfboard Circuit Construction

To join resistors, transistors, and other small components, you can use an insulated composite material with holes drilled in it (sometimes called Perfboard or Vectorboard). You just poke the wires from each component through the holes, bend them over to hold them in place, and solder the correct wires together on the back side of the board, as shown. It's not pretty, but it works. If the component leads aren't long enough, you can solder in additional sections of insulated wire.

Image courtesy of Dr. Steven W. Moore

Figure 8.65: Wire Wrap

This photo shows the back side of a wire-wrapped circuit board. The metal pins sticking up come from integrated circuit sockets on the opposite side of the board. Each pin has a square cross section, so as the stripped ends of the small (30 AWG) wires are wrapped about the pins using a special wire-wrap tool, the corners of the square pins bite into the wire. This results in good electrical and mechanical connections. If done properly, these connections are robust and do not require soldering. Best of all, they can be unwrapped easily and redone if you find a mistake after completing your board.

Image courtesy of Dr. Steven W. Moore

Figure 8.66: A Printed Circuit Board

In a printed circuit board, it's as if the metal wires connecting components are "printed" right onto the board. For an explanation of how they are really made, see Tech Note: Printed Circuit Boards.

Image courtesy of Randall Fox

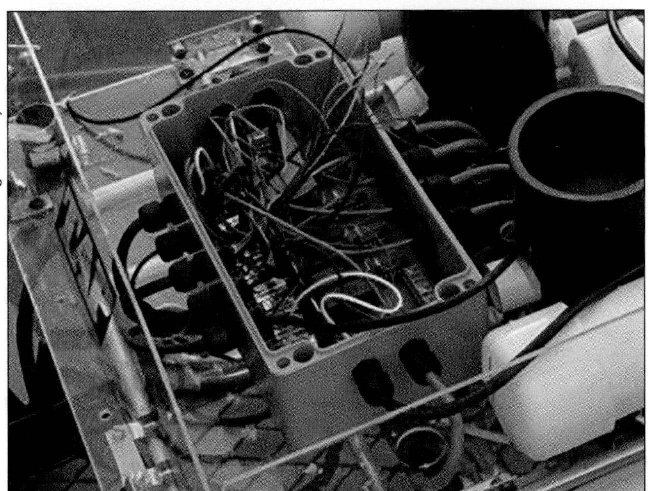

Figure 8.67: Waterproof Circuit Housing

This ROV has a watertight circuit box (lid removed in this photo) and uses watertight cable glands to pass wires through holes in the box without letting the water in. This approach is perfect for many shallow-water missions. Deeper missions would benefit from a cylindrical or spherical electronics housing, penetrators designed for higher pressure differentials, and some form of pressure compensation. In each case, it is important to make sure the circuit design is compatible with the housing.

TECH NOTE: PRINTED CIRCUIT BOARDS

A **printed circuit board**, or **PCB**, is the ingenious invention of the Austrian engineer Paul Eisler (1907–1995), who first made one in 1936 as part of a radio project. PCBs represent the ultimate in rugged circuit construction, so you should consider making one if you have very complex control circuits.

Most PCBs begin life as a thin, strong sheet of an insulated polymer/fiber composite (usually green in color), which is coated with an even thinner sheet of copper foil on one or both sides. To create an electrical circuit, copper is selectively removed from portions of the board. This leaves behind an intricate pattern of thin copper strips, or "traces," which function as electrical wires to interconnect resistors, transistors, integrated circuits (ICs), and other components soldered to the board. (See Figure 8.68.) Most circuit boards are designed on a computer that determines the best places to put the components and route the wires on the board.

Image courtesy of Dr. Steven W. Moore

Figure 8.68: Typical Printed Circuit Board

The dark lines on this printed circuit board are copper traces in silhouette.

To produce small numbers of prototype boards, computer-controlled milling machines are sometimes used to carve away the unwanted copper. However, for mass production of large numbers of identical boards, milling is too slow and expensive, so a process known as photolithography is used instead. First, the copper on each side of the board is coated with a special light-sensitive layer of plastic that becomes solvent-resistant when exposed to light. Next, light is shined on the board through a "mask" (basically a photographic negative of the circuit), so light reaches the plastic layer only where copper traces are supposed to be on the finished board. This toughens the plastic over the future trace locations. In the next step, a solvent is used to remove all the un-toughened plastic. Then the board is dipped into an acid bath that eats away any copper not protected by a layer of toughened plastic. Only the intended copper traces are left behind.

Although photolithography requires several steps, each of these steps is fairly easy to automate. More importantly, dozens (or hundreds) of boards can be processed in parallel in large vats of chemicals, allowing rapid mass production at relatively low cost per board. If you want to try photolithography, you can find kits for etching circuit boards at many hobby and electronics stores.

Some boards are designed for "through-hole" components. These components have metal pins that fit through tiny holes drilled into the board. However, most new boards are designed to include "surface-mount" components. The metal pins on these components stick out sideways and can be soldered to the surface of the board without

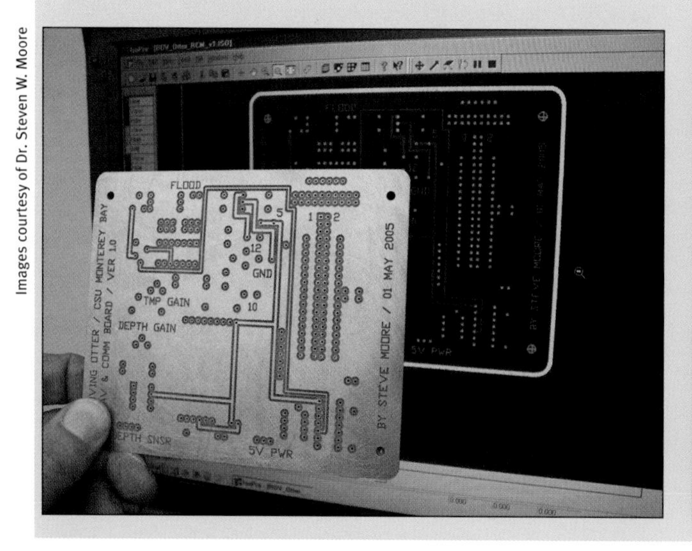

Images courtesy of Dr. Steven W. Moore

Figure 8.69: Circuit Boards from Milling Machines

Computer-controlled milling machines are sometimes used to make small numbers of printed circuit boards for prototyping.

requiring a hole. In general, surface-mount components are smaller than through-hole ones, so boards can be made smaller and lighter. But remember that smaller components can be tricky to solder by hand, and you'll still need holes to connect traces on opposite sides of the board.

If you'd like to try making a few of your own printed circuit boards, there are two reasonably affordable approaches for doing so. One is to get a do-it-yourself PCB photolithography kit from a hobby/electronics supplier. These use a simple form of photolithography with etching chemicals and home-made masks. These kits can be used at home or school under adult supervision. The other approach is to use a prototype PCB service, which you can find among the advertisements in hobby electronics magazines or on the web. (Try keywords "PCB prototype.") Typically, you download free software from these sites and use it to design your circuit board on your own computer, then you transmit your design via the internet to the PCB factory, along with some payment. A few days later, your completed boards arrive in the mail. It's fun, and you get professional-quality, custom PC boards.

down and fix. The other goal is to make sure your circuit construction lends itself to whatever method you will be using to waterproof and pressure-proof the underwater parts of your electronic circuits. As you work toward these goals of anchoring and water/pressure-proofing, keep in mind that you may want to keep some parts of your circuit easily accessible for routine testing and adjustment.

Large, heavy, topside components that don't need to fit inside a small waterproof cylinder on the vehicle can be securely screwed or bolted to a strong, rigid sheet of plywood, particleboard, or plastic. (See Figure 8.58.) Metal plates can also be used if you take care to avoid short circuits through the metal. In fact, you can use a metal plate as a **ground bus**, which is a central, low-resistance conductor to which the ground wires from all other components are connected. You will save yourself headaches if you invest the time to lay out your circuits in a well-organized fashion. It helps to have different portions of the board dedicated to different types of functions and wires running among them in neat bundles.

When it comes to interconnecting wires, the size of the wire often dictates the best method. For larger diameter wires (say, 16 AWG and larger) it's common to use crimp-on terminals and terminal blocks. (See Figure 8.59.) You can also make in-line splices (see Figure 8.60), but these generally are not as robust and versatile as terminals, since they add unsupported mass to a wire, increasing problems with wire flexing. If you do use them, make sure you anchor the splices securely so they don't flop around.

For smaller gauge wires, soldering (see Figure 8.61) and assorted connectors (see Figure 8.62) are the standard ways of connecting wires to each other and to various components.

If the components can be easily anchored to a chassis box or other support, you can solder wires directly from the terminals of one component to the terminals of other components. This is called **point-to-point soldering**. (See Figure 8.63.) Smaller components are usually soldered to perfboard (see Figure 8.64), plugged into wire-wrap sockets (see Figure 8.65), or soldered onto a printed circuit board (see Figure 8.66 and *Tech Note: Printed Circuit Boards*).

Whatever method you choose to use to assemble your circuits, make sure the one(s) used for any underwater parts of your circuit will fit inside a waterproof, pressure-proof housing and can be anchored securely inside that housing. (See Figure 8.67.)

9. Chapter Summary

This hefty chapter outlines the basics of energy (which is the amount of work that can be done) and power (which is how quickly that work can be done). Then it introduces the idea of efficiency, a particularly important concept in the world of underwater robotics. After introducing these fundamentals, the chapter presents a list of criteria that you can use to choose an appropriate power source for your vehicle and mission.

Electrical power turns out to be a logical choice for most of today's small ROV and AUV projects. Therefore, the remainder of this chapter focuses on understanding electrical power systems. While this textbook is not a primer for electricity, it does provide a general introduction to electricity and electric circuits, including vocabulary, concepts, and important safety tips. Once you have mastered these basics, the next section of the chapter shows you how to develop a power budget that will map out the electrical requirements of your vehicle's systems.

Once you have determined these electrical requirements, the next decision is how to supply them. Because many underwater vehicles rely on low-voltage batteries as a relatively safe, economical, and handy source of electrical power, this chapter then details battery basics: how they work, how they're rated, and how they differ. Understanding these options will give you the tools to evaluate various types of batteries and make the best choice for your particular project.

Finally, this chapter details how electrical power can be distributed from the power source you've chosen to your vehicle's thrusters, cameras, lights, navigation sensors, and other devices. This section of the chapter describes wires, cables, connectors, fuses, circuit breakers, and switches and explores the challenge of moving electrical power through a long tether. Next, there's a look at examples of different power distribution systems used in three ROVs. The chapter closes with a survey of various techniques you can use to assemble the real circuits you'll use in your vehicle.

When you've completed this chapter, you'll be ready to unleash the power your vehicle requires to accomplish its mission. But before you do, you will need to know how to control it. Not surprisingly, control is the focus of the next chapter.

Chapter 9

Control and Navigation

Chapter 9: **Control and Navigation**

Stories From Real Life: Human Torpedoes and Midget Subs

Chapter Outline

1. **Introduction**
2. **Control Systems**
   - 2.1. Open-Loop Versus Closed-Loop Control
   - 2.2. The Human Role in Vehicle Control Systems
3. **Navigation**
   - 3.1. Specifying Position or Location
   - 3.2. Navigational Instruments
4. **A Basic Control and Navigation System**
   - 4.1. An Overview of Electric Switches
   - 4.2. Using a Manual SPST Switch to Turn a Light ON or OFF
   - 4.3. Using a Manual DPDT Switch to Control Motor Direction
   - 4.4. Limitations of Manual Switch Control
   - 4.5. Adding a Basic Set of Navigational Sensors
5. **Advanced Control Options: Moving Beyond *SeaMATE***
   - 5.1. Do You Really Need Advanced Control?
   - 5.2. A Peek at the Possibilities
   - 5.3. Microcontrollers
   - 5.4. An Introduction to Electronic Signals and Communication
   - 5.5. Analog Data Transmission
   - 5.6. Digital Data Formats
   - 5.7. Digital Data Transmission
   - 5.8. Analog-to-Digital Conversion
   - 5.9. Signal Multiplexing
   - 5.10. Electronic Sensors and Sensor Circuits
   - 5.11. Data Display Options
   - 5.12. Using Transistors and Relays for Automated Power Control
   - 5.13. Motor Controllers
   - 5.14. Limiting Motor Travel
   - 5.15. Feedback Control Algorithms
6. **Chapter Summary**

Figure 9.1.cover: Piloting an ROV on an Offshore Oil Rig Mission

When an ROV is out of sight, working deep beneath the surface, control and navigation become major issues. A successful mission depends on accurate information about the ROV's position, effective remote control, and a skilled pilot.

In this photo, a pilot operates a VideoRay Deep Blue ROV on a natural gas rig in the middle of the Black Sea.

Image courtesy of Steve Van Meter, VideoRay LLC

Chapter Learning Outcomes

- Give examples of control systems used in ROVs and AUVs. Discuss the advantages and disadvantages of open-loop versus closed-loop control and of simple versus complicated control systems.
- Explain the purpose and function of the navigational instruments and motor control switches used on a simple ROV like *SeaMATE.*
- Explain what a microcontroller is and what role it can play in the control systems used for ROVs or AUVs.
- Explain how a microcontroller can get commands from a pilot through buttons, knobs, and joysticks or from various navigational sensors.
- Explain how a microcontroller can operate thruster motors, video lights, gripper arms, and other systems on a vehicle.
- Give examples of common control algorithms; list some possible causes and solutions for common control system malfunctions.

STORIES FROM REAL LIFE: Human Torpedoes and Midget Subs

A British Chariot*, piloted by its two divers wearing rebreather sets.*

During the years of World War I and World War II, most of the major naval powers concentrated on building fleets of sizable submarines. Working quietly, a few countries also added miniature versions to their arsenals. While many of these smaller vessels were imaginatively and enthusiastically designed, they were fraught with technical and financial difficulties. All relied on primitive controls, and all were difficult to navigate and stabilize. They were desperate craft for desperate times. Unfortunately, many were also unintentionally—or even intentionally—suicidal.

The very earliest submarines were all little vessels intended solely for warfare. Designer David Bushnell might be considered the father of midget submarine warfare, given the attack of his vessel, the *Turtle*, on HMS *Eagle* during the American War of Independence. Bushnell's tiny wooden craft was propelled by a pedaling pilot. It was armed with 666 N (150 lbs) of gunpowder and a clockwork fuse, all in a watertight package. When the *Turtle* came close to its intended target, the pilot was supposed to submerge and drill into the enemy hull with an auger, affix the charge, and then retreat a safe distance to await the explosion.

But first, the *Turtle* had to locate the enemy vessel. When submerged, the operator could rely only on a primitive depth gauge and a compass illuminated by foxfire (the phosphorescent light emitted by decayed wood), which required less air than a candle. While accounts of *Turtle*'s prowess vary significantly, there is little doubt it reached its target even though it failed to sink the ship. And word of the attack certainly caught the attention of combatants and inspired other submarine innovators. Over the ensuing decades, submarines slowly became bigger and better.

By the 1900s, the term "midget subs" came to specify vessels small enough to penetrate harbors defended by mines and submarine nets. Operators of these craft either cut through the protective mesh nets or hauled their vessels over or around them to get inside harbor waters and attack ships or facilities.

Around the globe, three basic types of miniature craft evolved:

1. human torpedoes (such as the Italian *Maiale* and British *Chariot*)
2. one-man submersibles (like the German *Biber* and Japanese *Kaiten*)
3. true midget submarines (such as the British *X-Craft* and Japanese *Ko-Hyoteki*)

The Italian *Mignatta* and *Maiale*

Early in 1915, Italy declared war on Austria-Hungary and entered World War I on the side of the Allies. The Italians initiated a new form of warfare when its navy began to use human torpedoes to attack the Austrian fleet, a tactic that didn't expose any Italian battleships in the process. The technique was fairly simple. In the case of the *Mignatta* of World War I, two men wearing specialized diving suits sat astride a machine built around a standard 35-centimeter (approx. 14-in) torpedo that was fitted with a propeller. Under cover of darkness, it was towed close to position, and the two operators sought out their enemy target, using visual reconnaissance. Once alongside, they attached two 1.6 kN (370 lbf) TNT charges to the ship with magnets—hence, the name *Mignatta* ("leech"). The *Mignatta* was never intended to submerge completely, but rather to be the key feature in stealth missions designed to attack enemy ships lying in defended harbors. Fortunately, the term "human torpedo" referred to its guidance system and controls, rather than the destruction of its target *and* operators, as the name seemed to suggest.

The Italian navy also had 11 actual midget submarines that were employed primarily on coastal defense patrols around Italy's harbors on the Adriatic Sea during World War I. This type of warfare, although not spectacular, was successful enough that it was continued during the inter-war period. In 1935, Italy established a military unit known as the Decima Mas that was dedicated to covert small-sub operations and that also developed an improved version of the human torpedo, called the *Maiale*.

This updated craft was propelled by an electric motor that delivered a top surface speed of 4.5 knots over a range of 6.4 km (approx. 4 mi). Operators sat behind a shield that deflected the strong current of water created by the torpedo's speed; the deflection created a quiet backwater, or eddy, for the operators and helped keep them from being dislodged by the force of the water. The forward man kept his eye on the target and used a lever to control both rudder and hydroplanes. In 1941, human torpedoes succeeded in penetrating Mediterranean harbor defenses and severely damaging HMS *Queen Elizabeth*, HMS *Valiant*, and a tanker. In response to the stealthy success of Decima Mas, the British developed their own version of the Italian human torpedo and dubbed it the *Chariot*.

The British *Chariot* and *X-Craft*

A British X-Craft *submarine at the surface, piloted by its captain. These vessels had no conning tower, and an open hatch could be swamped by even small waves.*

By 1945, all the major powers except the United States and Russia had some version of midget submarines in their navies. Britain was a formidable sea force and usually considered smaller, covert craft as the weapons of weaker powers. However, when Germany occupied Norway in 1940, Britain had to do something about the *Tirpitz*, Germany's 374-mN (42,000-ton) battleship, which now could lie in wait in a well-defended fjord. The best answer appeared to be the British human torpedo *Chariot*, which was developed specifically to take out the *Tirpitz*.

The *Chariot* was an imaginative design, but operators had serious difficulties delivering the torpedoes to their target sites, and in the end, the *Chariot* could not do the job. So Britain began to develop a versatile midget submarine called the *X-Craft* that could lay a magnetic mine in shallow, confined waters where conventional mine-laying methods wouldn't work. Production was highly secretive, with parts of the hulls built by two different companies and assembled by a third. *X-Craft* were designed to be towed by a mother sub to the vicinity of a target, then released to undertake their covert work. The control systems were miniaturized versions of those on a conventional sub. They were elementary, but hand-steering proved very effective, and the boats handled very well when they dived. In fact, when submerged, the *X-Craft* could be maneuvered astern or could move up and down while stopped, all necessary motions when attaching or laying a charge beneath a target.

Alas, the gyro compass, which was considered superior to the magnetic compass, gave endless trouble, as did other navigational aids. As well, conditions aboard the tiny craft were unbelievably squalid for the four crew members who might be aboard for 14 days. Despite these and other problems, the 15.5-meter (approx. 51-ft) *X-Craft* did succeed in severely crippling the German battleship *Tirpitz*. Later series of *X-Craft* were used for beach reconnaissance on the coast of Normandy, served as markers for the first landing craft on D-Day, and saw action in Norway and the Far East. One also sank the Japanese cruiser *Takao*.

The Japanese *Ko-Hyoteki* and *Kaiten*

As World War II approached, Japan anticipated a major sea battle with U.S. forces. But before that, their strategy was to whittle down the opposition, using improved torpedoes and midget subs. Under heavy secrecy, they began work on the *Ko-Hyoteki*, a 23.9-meter (78.4-ft) battery-driven submarine craft that had a conning tower amidships and carried two torpedoes. These midget craft could travel at 25 knots while submerged and had a range of 60 kilometers (approx. 37 mi); after releasing their torpedoes, the plan was that they would return to the mother ships. The design of the *Ko-Hyoteki* subs was fairly sophisticated, although they lacked internal venting for the torpedo tubes and had poor control when submerged. Furthermore, there was soon a shortage of experienced crews because of early war losses.

Several *Ko-Hyoteki* were launched against the U.S. fleet in Pearl Harbor and against Australian ships in Sydney Harbor. Technically, these midget subs were not suicide craft, although the crew's chances of surviving a mission were slim. Of the five *Ko-Hyoteki* used in the Pearl Harbor offensive, three were attacked and sunk by U.S. naval forces; the fate of the other two is unknown. Japanese claims that one of its midget subs sank the *Arizona* were bitterly disputed by naval aviators. Nonetheless, these small, stealthy underwater craft became the stuff of legends.

Later in World War II, as their situation became increasingly grim, the Japanese developed several versions of the *Kaiten*, a true one-man suicide submersible. These craft essentially inserted an operator into a very large torpedo and were similar in intention to the better-known kamikaze planes, whose pilots crashed their bomb-laden craft into enemy ships. Early designs may have allowed for the sub operator to escape before detonation, although in reality, this was probably not possible.

Prior to his one-way underwater mission, each young pilot wrote a farewell letter to his parents as part of a final ritual, then he squeezed into the midget sub, and it was sealed shut. Once near the target site, his craft was released from the deck of the parent submarine or surface ship. On the first part of its mission, each *Kaiten* traveled on a pre-set gyro course; then the sub surfaced, and its operator was expected to navigate the final stretch to the target, using visual reckoning and a crude periscope. However, given the craft's high rate of speed, the slightest mistake at the controls often caused the sub to miss its target. If the attack failed, the *Kaiten* was equipped with a mechanism for self-destruction. In fact, adding a human guidance system to what was essentially a torpedo never proved effective for these desperate suicide weapons.

The German *Neger* and *Seehund*

Germany developed its midget subs later than other countries did, so it took advantage of others' technical expertise and its own successful U-boat designs. However, German crews lacked leadership and training—traits that could not be borrowed. The first of the German weapons was the *Neger* (a larger model was called the *Marder*), a version of human torpedo in which the pilot relied on a wrist compass and a crude aiming device, including an aiming spike on the craft's nose. Unfortunately, the operator was positioned too low to see properly, and his shield was easily fouled by oil or debris. As well, the operator was frequently unable to release the torpedo.

As the war outlook dimmed, Germany produced a number of midget submarines that reflected more desperation than good naval design. The exception to this was the two-man *Seehund*, which was patterned after the recovered remains of two British *X-Craft*. The *Seehund* was armed with two torpedoes and capable of extended operations. An even smaller version, the *Hecht*, was armed with mines. The *Seehund* was a well-designed, effective weapon, but it damaged and destroyed relatively few ships, because it appeared so late in the war and because its crews were hampered by limited training and expertise.

The midget vessels of Italy, Britain, Japan, and Germany fall into several categories, according to their accomplishments. There were those that were practical and therefore successful—the *X-Craft*, *Seehund*, and *Maiale*. There were those that were enthusiastically designed but impractical—the *Chariot* and *Ko-Hyoteki*. And finally, there were those that were suicidal, either by accident or design—the *Neger*, *Kaiten*, and various derivatives.

All of these vessels relied on primitive controls. All were difficult to navigate, whether the operators utilized visual sighting or a compass. Most were hard to stabilize, and operators often had trouble regulating buoyancy. Steering mechanisms were complicated, and deadly battery exhaust was inadequately vented. Lacking adequate technical systems and often hampered by insufficient training, the operators of these small craft nonetheless struggled against tremendous odds to attempt their wartime objectives.

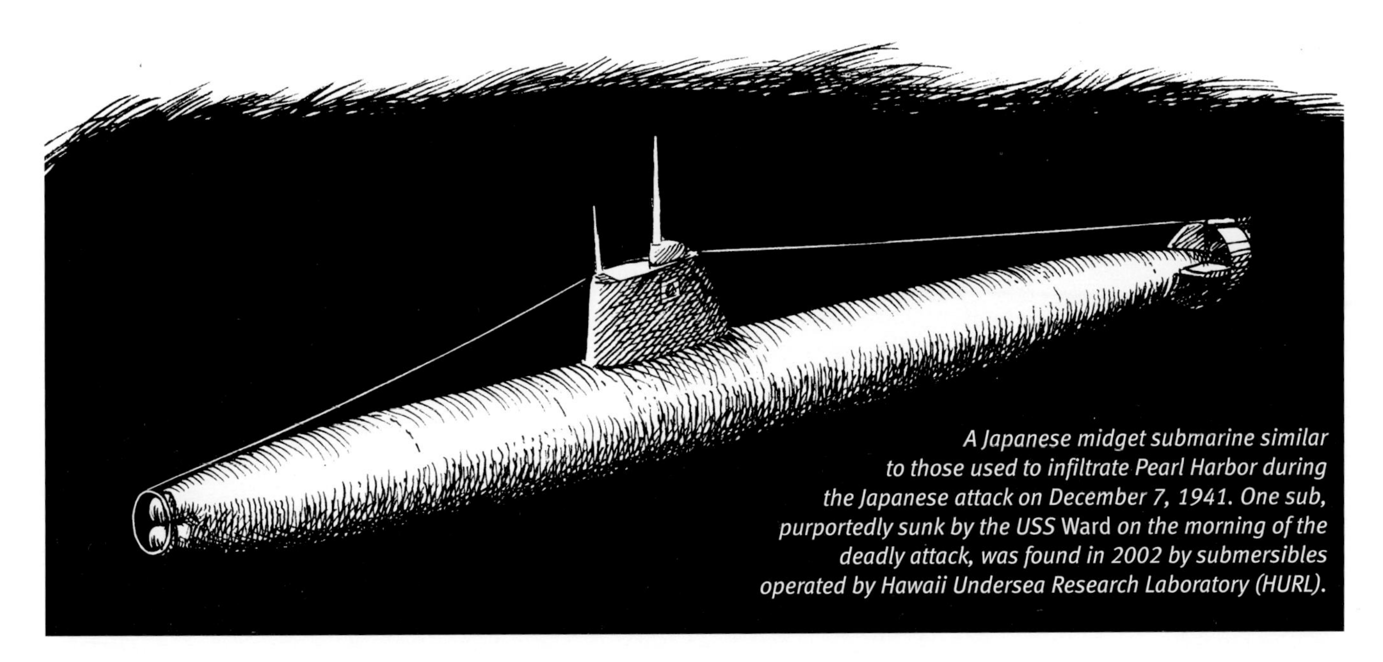

A Japanese midget submarine similar to those used to infiltrate Pearl Harbor during the Japanese attack on December 7, 1941. One sub, purportedly sunk by the USS Ward *on the morning of the deadly attack, was found in 2002 by submersibles operated by Hawaii Undersea Research Laboratory (HURL).*

1. Introduction

Imagine that you have been hired by geologists to recover a small but valuable seismometer deployed a year earlier on the seafloor. The instrument, which records earthquake activity, is anchored in about 1,000 meters of water many kilometers from shore. The geologists are hoping to analyze data recorded by the seismometer during a recent major quake and resulting tsunami. This information should help these scientists improve early-warning systems for coastal communities threatened by such devastating natural disasters. As the ROV pilot on a skilled team, you are excited about this opportunity to advance tsunami research. The ROV has just been secured to the ship's deck, and you are eagerly awaiting the moment of departure for this mission.

A few hours later, you and the ship are under way, headed out to sea. It's a foggy day, and within minutes, the coast has disappeared completely from view. Now you are surrounded on all sides by a vast, featureless expanse of gray ocean. Every direction looks the same. There are no roads or road signs to guide the ship to its destination. Even more disconcerting, there are no obvious landmarks to guide the ship safely back to port after the mission. Fortunately, getting the ship out and back safely is the captain's responsibility, not yours.

Image courtesy of NOAA

Figure 9.2: At the Controls

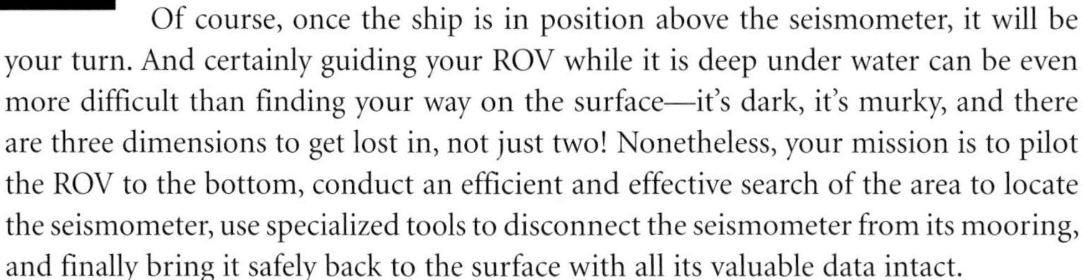

Of course, once the ship is in position above the seismometer, it will be your turn. And certainly guiding your ROV while it is deep under water can be even more difficult than finding your way on the surface—it's dark, it's murky, and there are three dimensions to get lost in, not just two! Nonetheless, your mission is to pilot the ROV to the bottom, conduct an efficient and effective search of the area to locate the seismometer, use specialized tools to disconnect the seismometer from its mooring, and finally bring it safely back to the surface with all its valuable data intact.

This mission is feasible because both the ship and the ROV are well equipped with effective tools for navigation and vehicle control. **Navigation** comprises the art, science, and technology of determining where your vehicle is now, comparing that location with its intended destination, and using that comparison to figure out what course the vehicle must follow next to reach its destination. **Control** refers to the regulation of vehicle speed and direction—as well as cameras, lights, tools, and other processes—so that your vehicle goes where it needs to go and does what it needs to do. Systems designed specifically to regulate such things are called **control systems**. This chapter is all about navigation and control. The control information is presented as two sections—introductory and advanced.

Obviously, most home-built ROVs and AUVs aren't going to be recovering seismometers in 1,000 meters of water, so their navigational and control capabilities don't need to be nearly so sophisticated; however, the success of every underwater vehicle mission, no matter how simple, depends on the coordinated control of at least a few vehicle systems. For example, every craft must be able to maneuver through the water, a task that usually requires coordinated control of at least three thrusters. And unless both the vehicle and its destination will be within plain sight of the pilot throughout the mission, some sort of navigation system will be required to tell the pilot where the vehicle is relative to its intended destination.

Rather than providing a comprehensive review of all control and navigation systems used on underwater vehicles, this chapter concentrates on how you can gather basic navigational data and control a set of electric thrusters to move a small underwater vehicle to its destination and back—in short, how to control its *position*. There are two great reasons for focusing on this goal. First, position control is the most universally needed form of control in underwater vehicles; everything from the simplest ROV operating in a wading pool to the most sophisticated AUV operating under the polar ice caps must be able to control where it's going. Second, effective position control encompasses just about every control system challenge you're likely to encounter in underwater vehicle design, from the use of navigational sensors to motor control strategies, so it provides a great learning framework. Once you know how to control vehicle position, you can easily adapt that knowledge to control just about any other system you can imagine adding to your vehicle.

This chapter begins with a general theoretical overview of control systems. This will provide a conceptual framework for exploring more practical aspects of control system design. Then the chapter looks at how control theory can be applied to the control of vehicle position. The middle of the chapter presents some straightforward methods you can use for navigation and thruster control in simple ROVs, like the *SeaMATE* ROV presented in Chapter 12. The final half of the chapter introduces more advanced methods for navigation and control suitable for use with AUVs and more sophisticated ROVs.

2. Control Systems

As the name suggests, a control system is a system that controls something. The term usually refers to machines or other equipment that have been designed and built by humans specifically for the purpose of regulating some particular quantity or process—the temperature of the air in a room or the speed of a motor, for example. However, the concept of control systems applies equally well to countless naturally occurring systems, too. For example, the physiological processes that regulate your body temperature and your heart rate are examples of naturally occurring control systems. Some control systems combine biological and human-made components. For instance, a person driving a car down a road is an integral part of a half-human/half-machine control system that regulates the speed and direction of the car.

Image courtesy of Randall Fox

Figure 9.3: A Simple ROV Control Box

Control systems engineered by people can be constructed in many different ways, made out of many different types of materials, and used to control many different kinds of things. (See *Tech Note: Toilet Control System.*) Most of the control systems used on modern ROVs and AUVs rely heavily on electronic components, even if they control something that is not electronic, such as vehicle depth. There are many reasons for the popularity of electronic control systems: 1) modularity (i.e., mix-and-match parts), 2) relatively low cost, 3) versatility, 4) small size, 5) ease of assembly, use, and maintenance, and 6) high reliability. Because electronic control systems have become so ubiquitous and important in underwater applications (and elsewhere), electronics is the primary focus of this chapter. However, control systems built out of mechanical, pneumatic (air-driven), or hydraulic (liquid-driven) components have also proven popular and effective in some underwater vehicle designs, so always keep an open mind when considering the design of control systems for your vehicle.

2.1. Open-Loop Versus Closed-Loop Control

Control systems come in two basic forms: **open-loop** and **closed-loop**. Both forms attempt to achieve some desired condition; however, the two forms differ in the way information flows from one part of the regulatory process to other parts. Figure 9.4 highlights the fundamental differences between the two types of control:

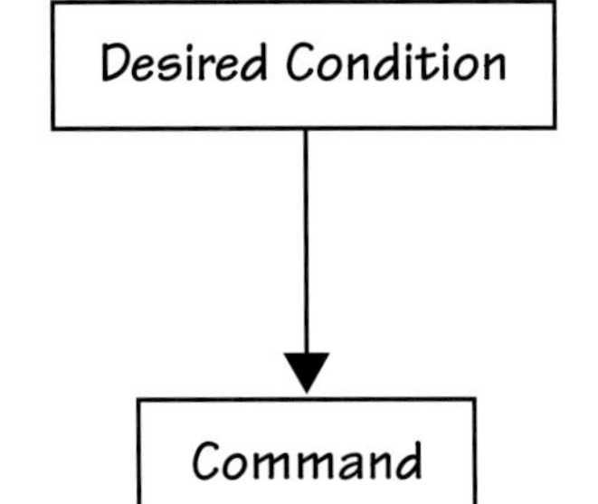

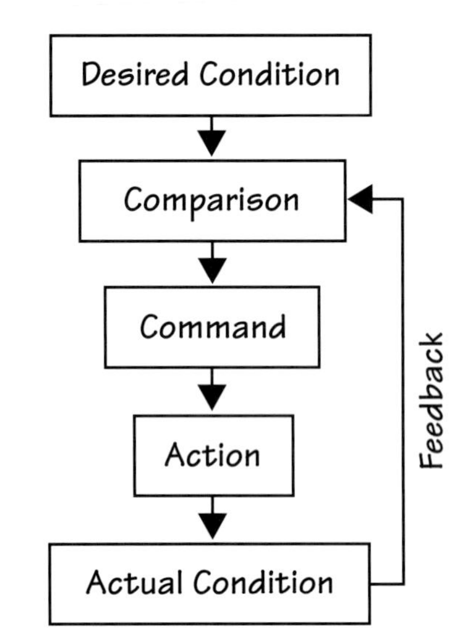

Figure 9.4: Comparison of Open-Loop and Closed-Loop Control Schemes

Open-loop control lacks the ability to compare actual conditions with desired conditions, because it does not have the information "feedback" pathway characteristic of all closed-loop control systems.

Open-loop control sends a command to produce some action in an attempt to make the actual condition match some desired condition. However, open-loop control *does not verify* that the action happened or that it produced the desired condition or result.

Closed-loop control, on the other hand, is not satisfied with *assuming* that the command produced the intended result. Instead, it monitors the actual condition and feeds that information back into the control process, where it can be compared to the desired condition to verify that the desired condition has actually been achieved. If the actual and desired conditions differ, it keeps adjusting the commands being sent until the actual condition does match the desired condition as closely as possible.

Open-loop control is simpler to implement, but closed-loop control usually results in a much closer match between the actual and desired conditions. The **feedback**, which creates the circular pathway or "loop" visible in the figure, is critical, because it allows information about the *present* actual condition to feed back around into the command process and thereby influence the *future* actual condition. This feature is so important that closed-loop control systems are often called **feedback control systems** or simply **feedback systems**. In fact, the term "control system," when used without the open-loop or closed-loop modifier, almost always refers to a closed-loop control system based on feedback.

TECH NOTE: TOILET CONTROL SYSTEM

Peek inside a household toilet tank (the tank, not the bowl) and you'll find a control system made out of plastic or metal floats, levers, and valves. This system is designed to control the amount of water stored in the tank between flushes. There needs to be enough for a good flush, but not enough to cause an overflow. When you push down on the flush lever, a big valve in the bottom of the tank opens, allowing water already stored in the tank to drain rapidly into the toilet bowl. This initiates the all-important flushing action. Inside the tank is a float connected to a small freshwater inlet valve. As the water level in the tank drops, the float drops with it, opening the valve. This starts refilling the tank as soon as the big outlet valve in the bottom of the tank has closed again, which it does automatically when the tank is nearly empty. As the water level in the tank slowly rises, the float rises with it, eventually shutting off the inlet valve just as the water reaches the correct level. This mechanism is an example of a closed-loop control system. After you read Section 2.1, see if you can explain what makes it a closed-loop system instead of an open-loop system.

A simple example can reinforce the difference between open-loop and closed-loop control. Imagine an automatic sprinkler system that waters plants according to a pre-programmed schedule, say 30 minutes every other day. This is an open-loop system. On average, this programmed schedule might provide about the right amount of water for the plants, but the system does nothing to confirm that. As a result, the plants might not get enough water during a hot, dry spell. And during a wet, stormy week, they'd probably get more water than they needed.

It's possible to turn this pre-programmed sprinkler system into a closed-loop system by adding a soil moisture sensor, then programming the system to water the plants when (and only when) the soil is too dry. An obvious advantage of this closed-loop system would be its ability to adjust the watering frequency to compensate for changes in rainfall. During a dry spell, it would water more often, and during rainy weather, it might not water at all. Every closed-loop control system offers this valuable ability to compensate for unpredictable, externally driven influences on the controlled condition. In fact, this is the main reason why closed-loop control systems were invented and why they are used so widely today.

HISTORIC HIGHLIGHT: WATT'S EARLY CLOSED-LOOP FLYBALL GOVERNOR

James Watt (the same guy the energy unit is named after) invented one of the most famous closed-loop control systems in order to control the speed of steam engines. Prior to Watt's invention, it was difficult to keep a complicated piece of machinery like a steam engine running at a constant speed, particularly if the load on the engine changed. Think of the example of a sawmill in those days, where a steam engine was turning a saw blade. When no log was being cut, there was very little load on the engine, so the engine tended to race too quickly. But as soon as the saw started to cut a log, the engine and saw blade slowed way down. If the power output was optimized to cut logs quickly, then when there was no log being cut, the engine would race so quickly that it could damage itself. But if the engine power was reduced to safe levels when no log was present, then it would barely manage to saw any log. This was a general problem with steam engines powering factory machinery at that time.

To solve problems like this, Watt rigged up an ingenious device called a "flyball governor," which consisted of a steam valve connected to a pair of heavy metal weights hanging from either side of a vertical rotating shaft driven by the engine. If the engine speed increased, the weights spun more quickly and got flung farther outward, away from the spinning shaft. As this happened, a mechanical linkage that connected the balls to the steam inlet valve reduced the amount of steam flowing into the engine, thereby preventing the engine speed from increasing beyond a certain level. On the other hand, if the load on the engine increased enough to slow the engine, the rotating shaft would start to spin more slowly, too, so the weights would fall back toward the shaft. This would open the steam valve, providing additional steam power to the engine. Thus, the engine was able to drive the added load without slowing down.

At first, Watt had trouble getting the system to work properly, but he carefully analyzed what was going wrong and made adjustments until his invention would automatically maintain a nearly constant engine speed, regardless of the load on the engine. Watt's flyball governor was so successful that it inspired the birth of modern control theory, a branch of mathematics and engineering that specifically focuses on improving the performance of closed-loop control systems.

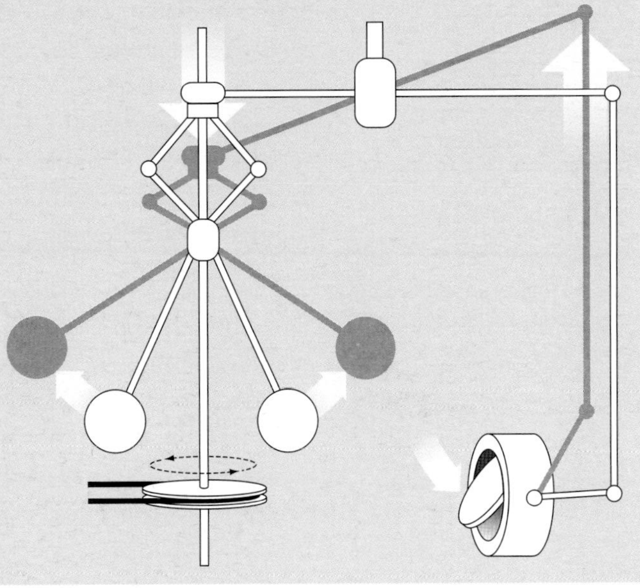

Figure 9.5: Flyball Governor

Another advantage of closed-loop control systems is their ability to compensate for minor imperfections in some parts of the control process itself. For example, our feedback-controlled watering system could (and would) compensate for slightly clogged water pipes by watering longer or more often, to make up for the reduced water flow rate. In many cases, feedback control can reduce overall system cost by allowing the use of less precise and less expensive parts in most parts of the system, yet achieve *better* performance than an open-loop system made from more expensive parts.

You will probably want to use closed-loop control systems rather than open-loop control systems in your underwater vehicle designs whenever you can. That's because closed-loop systems generally do a much better job of matching actual conditions to desired conditions, particularly in unfamiliar or unpredictable environments like the ocean.

Table 9.1: Examples of Control Capabilities for Different Vehicles in Different Operating Conditions

	Vehicle requires the ability to move or perform other actions	*Vehicle requires the ability to sense or measure the actual condition of the vehicle or payload*	*Vehicle requires the ability to compare actual to ideal conditions and to issue commands for appropriate actions*
Visible ROV	*Yes*	*No*	*No*
Hidden ROV	*Yes*	*Yes*	*No*
Hidden ROV w/ autopilot function(s)	*Yes*	*Yes*	*Only for autopilot functions*
AUV	*Yes*	*Yes*	*Yes*

2.2. The Human Role in Vehicle Control Systems

There is a reason that airplanes have pilots, trains have engineers, and cars have drivers. It's because humans are very good at performing the most difficult parts of the closed-loop control process that are needed to keep these vehicles moving quickly yet safely to wherever they are supposed to go. What parts of the closed-loop control system (shown again in Figure 9.6) are humans doing in these cases? First of all, humans define the *desired condition(s)*. It is the pilot or driver, not the vehicle, who decides where the vehicle should go and what route it should take to get there. Second, humans have eyes, ears, and other sensors for gathering detailed information about actual conditions, so they are also providing the *feedback*. Third, humans have sophisticated information processing systems (called brains), which they use to make the *comparison* between the actual conditions and the desired conditions, to determine what needs to happen next. Fourth, humans can operate steering wheels, brake pedals, and other mechanisms for issuing *commands* to the vehicle to modify the actual conditions (for example, a car's lane position or speed) any time the vehicle strays from the desired conditions. The only thing the vehicle does is to respond to the commands by doing the action (accelerate, turn, etc.), thereby directly influencing the actual conditions.

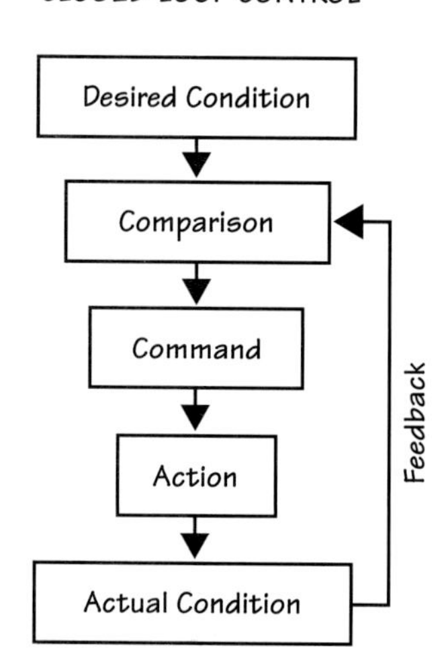

Figure 9.6: Closed-Loop Control

This illustration shows just the closed-loop portion of Figure 9.4.

As computers and electronic systems have become more sophisticated, it has become possible (though not necessarily easy or cost-effective) to replace many of these traditionally human functions with technology. For example, electronic sensors can provide feedback about actual conditions, and computers can perform the comparison between actual and desired conditions. Then the computer can issue commands to electronically controlled machines or other processes. Today, there are robotic vehicles in the air, on land, and under the sea, moving around entirely on their own and completing various jobs, all without any direct intervention from a human operator.

What all this means is that you have design options for control. As you begin to think about the design of your vehicle's control systems, you may want to spend some time carefully considering what role a human pilot will (or will not) play in the control of your vehicle.

In an underwater vehicle mission, the "desired condition" (or set of conditions) at each moment is almost always defined by a human (or a team of humans). For example, an ROV pilot might decide she wants the ROV to hover 2 meters above the seafloor. That altitude then becomes the desired condition, whether or not the ROV is actually hovering there yet. Even with an entirely autonomous vehicle, the desired conditions are usually specified ahead of time by the people in charge of the mission, then programmed into the vehicle's computer, so the AUV will know what it's supposed to do after being launched.

Look again at the bottom of Figure 9.6. Note that in the hovering ROV example the "action" and the "actual condition" are two things that happen under water, usually out of reach and out of sight of the human operator. For example, the action that changes the ROV's actual altitude above the seafloor might be rotation of a vertical thruster motor to turn a propeller. Other examples of actions that might take place to achieve various desired conditions could include the closing of a gripper jaw to grab a tool or the turning on of a light to illuminate a scene in front of the ROV's camera.

The three remaining processes—feedback, comparison, and command—offer places where you may (or may not) have leeway to choose whether you want to have the process handled by a human pilot or done electronically.

If it's possible to have a human pilot do all three of these things, then you have the option of making the vehicle's control system very simple. That's a good thing for beginning designers. For example, if the vehicle is a small ROV being operated in a shallow pool filled with clear water, where you can directly observe the vehicle's location and movements firsthand, then you can cover the feedback, the comparison, and the command processes (in addition to defining the desired conditions). The only part of the control system flowchart the ROV itself would need to provide is the ability to act on the pilot's commands. A simple set of thrusters on the vehicle operated by manual switches on the pilot's console would probably do the trick.

However, if your vehicle is an ROV being operated in deep or murky water, then you cannot directly observe where the vehicle is and what it's doing. That means you cannot provide the feedback. The ROV must therefore be equipped with sensors, such

SAFETY NOTE: CLOSED-LOOP SAFETY WARNING

This chapter provides just enough control system information to get you into trouble. Seriously. Closed-loop control systems are simple in concept, but in practice, they are notoriously difficult to implement in a reliable way. Often they appear to be working perfectly, but a very slight change in conditions can send them suddenly and unpredictably spinning wildly out of control. When this happens, the consequences can be dramatic and destructive.

Don't be afraid to experiment with closed-loop control systems in small, low-powered, non-critical projects, because watching your robotic gadget self-destruct can be both educational and entertaining. But NEVER trust life, limb, or valuable property to a closed-loop control system designed by amateurs. Always remember that the potential for disaster is there with any closed-loop system, even one that seems to be working properly.

TECH NOTE: LEARNING MORE ABOUT CONTROL SYSTEMS

If you want to get serious about learning good control system design in order to build truly reliable control systems, you'll need to combine some formal training in control theory with time spent working on real projects alongside experienced control system engineers. To get the theoretical training, plan on attending a college or university and majoring in some form of mechanical or electrical engineering. While in the program, take at least one course devoted entirely to control system design. These courses make heavy use of mathematics, so it helps to have completed courses in algebra, trigonometry, calculus, linear algebra, and differential equations beforehand.

Among the rewards for such efforts will be access to a wide variety of exciting and high-paying jobs in marine technology, robotics, and other engineering disciplines. Many entry-level jobs in these fields provide the opportunity to work alongside experienced control systems engineers and/or technicians who can teach you the practical side of control system design and troubleshooting.

as a camera, depth gauge, compass, or other instruments to measure those conditions and transmit that information to the surface for the pilot's use. Once you have that information from the sensors, you can proceed to make the comparisons between actual and desired conditions and can issue commands, as needed, to achieve the desired conditions.

If the vehicle is an AUV that will operate without human supervision throughout its entire mission, then it must take care of every aspect of closed-loop control all by itself. You will have no choice but to automate every part of the control loop for every control system on the vehicle. This is not easy to accomplish, but doing so can free vehicles to operate without tethers, so they can go places and do things a tethered vehicle cannot do. Therefore, designing and building a functional AUV can be very exciting and rewarding too.

It's worth noting that there are several ROV functions for which fully automated control is helpful, even when it is not strictly necessary. For example, a depth autopilot that can be switched on to hold the vehicle at a constant depth can free the pilot to concentrate on other matters during a challenging mission.

Table 9.1 highlights the tradeoffs between control system simplicity and vehicle capabilities. Note that if you choose to build a "visible ROV" (i.e., one you can directly observe in a swimming pool) as your first design project, you can concentrate on the action piece of the overall control loop without having to worry about engineering the sensory (feedback) or decision-making parts of the control system. As your knowledge and skills increase, you can move up to ROVs that work out of sight by adding sensors, then progress to ROVs with one or more autopilot features, such as depth or heading autopilots. Finally, you can try your hand at designing and building a completely autonomous vehicle.

3. Navigation

Above all else, a vehicle is a machine for transporting passengers, cargo, or other payloads from one place to another. In that respect, the most fundamental characteristic of any vehicle is its ability to move from its present location (wherever that is) to its intended destination. Viewed in the context of the closed-loop control system presented back in Figure 9.6, getting a vehicle to go where you want it to go is a matter of getting its actual position to match a particular desired position (the destination).

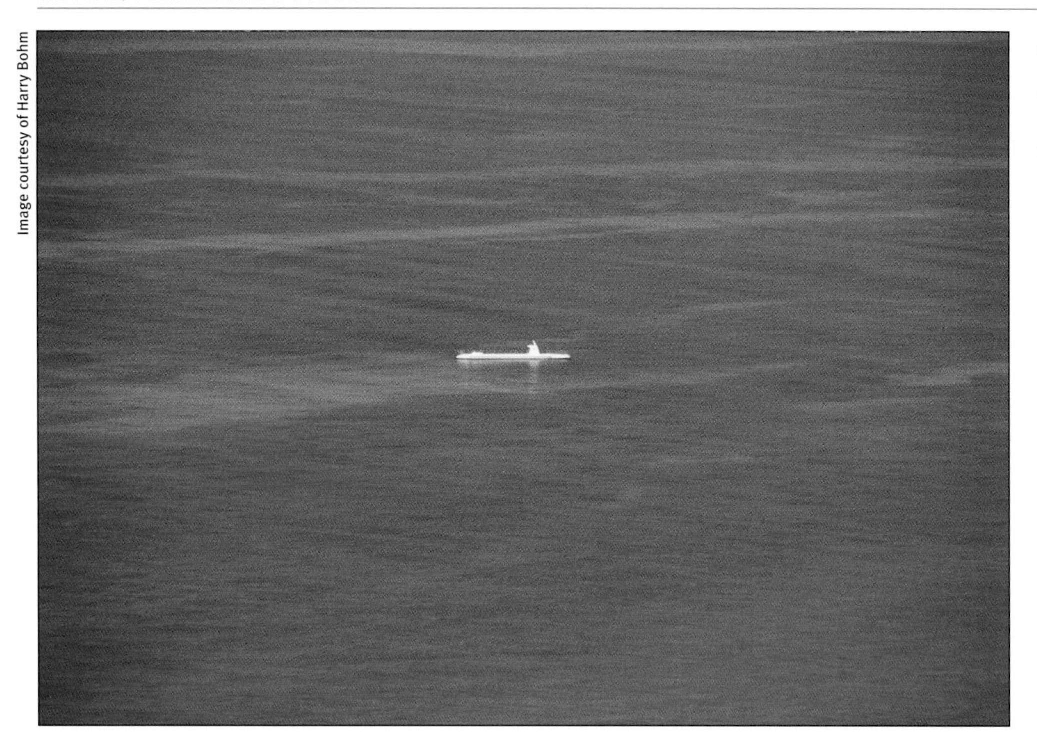

Figure 9.7: It's a Big Ocean!

Effective navigation is critical when you cannot see obvious landmarks.

Accomplishing this goal with a small electric ROV or AUV requires at least four distinct layers of control, nested one within another: 1) good control of vehicle *position*, which requires 2) good control of vehicle *movement* (i.e., speed and direction), which requires 3) good control of *thrust*, which requires 4) good control of the *electric currents* powering the thruster motors.

Effective closed-loop control of the first two layers in this hierarchy (position and movement) depends on accurate and timely feedback about the vehicle's actual position, speed, and direction.

Navigation comprises a suite of tools and techniques for providing this critical feedback information. This section of the chapter explores a variety of navigational techniques and instruments, with an emphasis on those most useful for low-budget ROV projects.

3.1. Specifying Position or Location

Effective navigation depends on clear, accurate, and precise information about the present position of a vehicle and the location of its intended destination. (For the purposes of this chapter, the word *location* will usually be used to refer to the whereabouts of something that does not move, like a town, whereas the word *position* will usually be used to describe the whereabouts of something that can move, like a ship or ROV.)

SAFETY NOTE: NAVIGATION AND SAFE BOATING

Navigation, both above and below water, is a huge and complicated subject unto itself, so this chapter can only ripple the surface. The simplified descriptions of navigation techniques provided in this book are *not* sufficient for guiding vessels carrying human passengers safely on large bodies of water. Many people have died from navigational errors while boating, so it's extremely important to know how to navigate safely (or to travel with someone who does) any time wind, water currents, changing weather, darkness, or distance may separate you from a clear view of familiar landmarks on shore.

On land, particularly in urban areas where street names, addresses, and familiar landmarks provide unambiguous reference points, specifying positions and locations is relatively easy: "I'll meet you at 3:00 pm on the corner of 4th Avenue and Miller Street, right in front of the bakery." However, finding a way to specify an exact position or location with similar precision on the open ocean, particularly under water, can be much more difficult. This section of the chapter describes some practical methods you can use to specify underwater locations in enough detail that you (or someone else) could guide a small ROV back to that same location to recover a valuable object you found (or left) at that location earlier.

TECH NOTE: ACCURACY VERSUS PRECISION

When it comes to specifying location (or any other quantity), the terms *accuracy* and *precision* are often used interchangeably, but this can lead to problems, because they are not the same thing. If a measurement or specification is accurate, it is correct. If a measurement or specification is precise, it is very specific or detailed. The best measurements are both accurate and precise; however, it is possible to be accurate without being precise or to be precise without being accurate. Stating that Tahiti is located in the Pacific Ocean is accurate, but not very precise, because the Pacific Ocean is a very big place. Good luck paddling to Tahiti if you don't have a more precise idea of where it is! On the other hand, stating that Tahiti is located in your uncle's house in New Jersey is much more precise, but not at all accurate.

If you use your ROV to deploy an instrument on the seafloor and want to return to recover that instrument later, your records of the instrument's location must be both accurate and precise. They must be accurate, because you don't want to waste your time looking in the wrong place. They must be precise, because limited visibility under water means that you won't be able to find the instrument in a reasonable amount of time, unless you are lucky enough to land practically on top of it.

Image courtesy of Harry Bohm

Figure 9.8: Distinctive Landmarks

3.1.1. Nearby Landmarks

The simplest method of specifying an underwater position or location, when it's available (which it usually isn't), is to use a distinctive **landmark** that is located immediately adjacent to the location of interest *and* can be found easily from a distance. For example, if you knew that your buddy's ROV team had accidentally dropped a tool while working just off the end of Fishing Pier #2 in Marinetown, your ROV team could probably recover the tool by searching for it right off the end of that exact pier. Unfortunately, with the exception of underwater sites located right next to obvious landmarks on shore or at the bottom of a buoy's mooring chain, this technique is rarely an option.

3.1.2. Dead Reckoning

One way of extending the usefulness of the landmark method to cover a somewhat wider area is to use a technique known as dead reckoning. Before the advent of modern navigational technologies, like GPS (described below), ancient mariners used dead reckoning with remarkable success to guide their vessels over vast ocean distances. This technique consists of starting at a known location, then heading away from that location in a particular direction at a particular speed for a particular amount of time. Since speed multiplied by time equals distance, this is (theoretically) equivalent to heading in a particular direction for a particular distance. For example, you might know that there's an especially interesting boulder that's home to a bunch of colorful fish about 50 meters due north of the end of the pier. If you know that your ROV travels 0.5 meters per second at maximum speed, you should be able to find the boulder by heading north from the end of the pier at full throttle for 100 seconds.

Though fairly simple in concept, dead reckoning has elements of open-loop control, because it assumes, but does not verify, that the vehicle is actually moving in the intended direction at the intended speed, and therefore ending up at the expected location. Even if you use closed-loop control to make sure the vehicle maintains a constant compass heading and thruster speed, currents or other influences can displace the vehicle from its presumed trajectory without any evidence that this is happening. (See *Tech Note: Dead Reckoning in Currents.*) In low visibility situations, these errors can cause you to miss your target entirely. These types of errors grow larger with distance, so dead reckoning works best on a small scale, particularly if you are looking for tiny objects in poor visibility. Given the huge distances involved, it's truly remarkable how well the ancient mariners were able to navigate across the oceans using dead reckoning, but keep in mind that they were usually looking for things the size of islands or continents, and they could often see these targets from many miles away. That means they could afford to be off course by a greater distance than you can when you're searching for a screwdriver in pea soup-like visibility with a little ROV.

One way to improve the long-range accuracy of dead reckoning is to break a long journey into a series of relatively short hops between stationary underwater landmarks. For example, if you know there's an old shopping cart littering the bottom of the harbor exactly halfway between the pier and that boulder with the fish, then you can first navigate to the shopping cart and from there, navigate to the boulder. If currents

TECH NOTE: DEAD RECKONING IN CURRENTS

Suppose you have used swimming pool tests to determined that your vehicle moves forward at 0.5 meters per second when at full throttle. Based on this calibration, you might reasonably expect that if you drive your vehicle at full throttle for one minute, it will move 30 meters (0.5 meters/second x 60 seconds = 30 meters). However, if you now try this in the ocean and are unaware that there's a current moving at 0.2 meters/second opposite your intended direction, your actual speed over the bottom would be reduced by this "headwind" to only 0.3 meters per second (0.5 m/s minus 0.2 m/s). So after one minute of travel, you would have moved only 18 meters (0.3 meters/second x 60 seconds = 18 meters), even though you thought you had moved 30.

The situation can be even more problematic with a side current. For example, suppose you use a compass to keep your vehicle pointed due north while operating at full forward throttle. In calm water, your vehicle would be headed north at 0.5 meters per second, as expected. However, if there was a current flowing at 0.2 meters/second from east to west, that current would be blowing you sideways off course, and you might not realize it. After one minute, you would have moved north by the expected 30 meters, but you would also have moved west by an unexpected 12 meters and could easily miss your intended target in murky water.

One strategy for dealing with this cross-current problem is a technique often used by scuba divers who face similar challenges of navigating with currents and limited visibility. Instead of watching the compass constantly while swimming, they start by using the compass to spot a distant object located along the intended heading. Once they start swimming toward that object, they focus on getting to the object, rather than paying attention to the compass. As they reach the object, they take another compass reading and look for other objects farther ahead along the same compass heading. By connecting the dots in this way, they avoid the problem of being blown off course by side currents.

TECH NOTE: THE LATITUDE AND LONGITUDE SYSTEM AND THE UNIVERSAL TRANSVERSE MERCATOR (UTM) SYSTEM

One universal way to describe location—one that does not rely on knowledge of local landmarks—is to specify the location in terms of **Latitude** and **Longitude** (sometimes abbreviated Lat/Long) coordinates. This is the method used by most modern ships and aircraft. In this system, a coordinate system is wrapped around the earth such that each and every point on the surface of the earth has a unique pair of coordinates. Latitude is measured in degrees north or south of the equator. Anything on the equator is given a latitude coordinate of zero degrees. Latitude is 90 degrees north at the north pole and 90 degrees south at the south pole. Longitude is specified in terms of the number of degrees east or west of a straight line drawn between the poles and running through the town of Greenwich, England. This line is known as the **Prime Meridian.**

There are a couple of standard formats for specifying Lat/Long coordinates.

- In one format, called decimal degrees (DD), fractional degrees are expressed as a decimal fraction.
- In another format, called degrees-minutes-seconds (DMS), fractional degrees are expressed as minutes and seconds, as if each degree were one hour. Occasionally decimal minutes are used instead of seconds; thus, 6'30" can be written as 6.5'.

It is common to use a superscript circle (°) to represent degrees, a single quote mark (') to denote minutes, and a double quote mark (") to denote seconds. Thus, 10 and a half degrees would be expressed as either 10.5° (DD) or as 10° 30' 00" (DMS).

Here are the Lat/Long coordinates for the Eiffel Tower in Paris, France, expressed in both formats:

DD: 48.8583 N Lat + 2.2945° E Long

DMS: 48° 51' 30" N Lat + 2° 17' 40"

The N and E in each designation refer to north and east, respectively, indicating that the Latitude value represents the number of degrees measured north of the equator, and the Longitude value represents the number of degrees measured east of the Prime Meridian.

Another popular coordinate method for specifying locations on planet Earth is the Universal Transverse Mercator (UTM) system. In this system, locations are specified in terms of the number of meters north or east of particular grid boundaries. Although there is not a simple relationship between the Lat/Long coordinate systems and the UTM coordinate system, both systems are used frequently enough that tools for converting between the two coordinate systems are readily available on the web and in most GPS receivers.

Here are the UTM coordinates for the Eiffel Tower:

31 U 448252 5411944

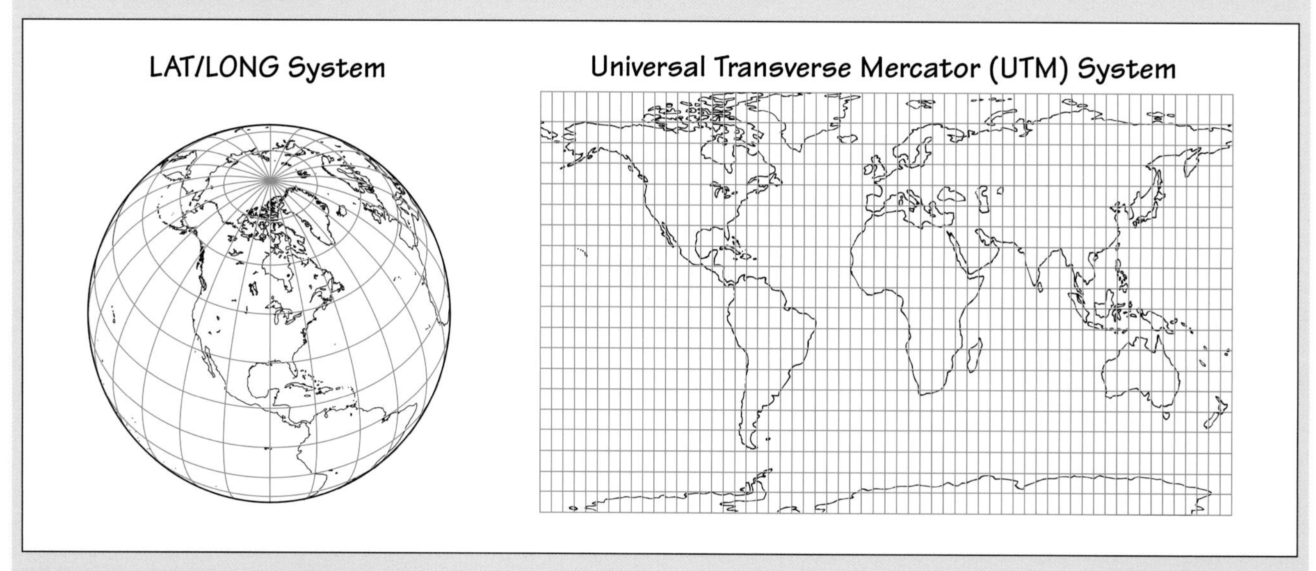

Figure 9.9: LAT/LONG and UTM Systems

have started to push you off course, you'll be able to detect and correct that problem when you notice the shopping cart off to one side instead of dead ahead. At that point, you can turn and head over to the shopping cart before continuing in the direction of the boulder. This puts you back on the intended course at the halfway point, which makes it much more likely the boulder will appear within your limits of visibility during the last leg of the trip.

3.1.3. Standardized Coordinate Systems and GPS

A number of standardized global coordinate systems have been established to facilitate navigation across the earth's entire surface, including areas like the oceans, where landmarks are not visible, and places far from home, where landmarks may be unfamiliar. The most well known of these is the **Latitude and Longitude (Lat/Long)** coordinate system. (See *Tech Note: The Latitude and Longitude System and the Universal Transverse Mercator (UTM) System.*) A closely related system is the **Universal Transverse Mercator (UTM)** coordinate system. These invisible coordinate grids provide a unique "address" for every point on the planet's surface.

The usefulness of these coordinate systems for the average citizen has increased dramatically in recent years with the availability of affordable **Global Positioning System (GPS)** receivers. These pocket-sized electronic devices receive coded radio signals from a collection of earth-orbiting navigational satellites and use those signals to calculate the receiver's exact position. They provide rapid, accurate, and precise position fixes, generally good to within a few meters, anywhere on the surface of the earth, any time of day or night. Some advanced models can pinpoint positions to within a centimeter (about ½ inch) and are now used in precise survey work.

Most GPS receivers can use successive position fixes to calculate and display the speed and direction of travel, along with present position. They can also estimate how long it will take to reach a destination given the current speed, and they can guide users directly to any specified location. Originally developed for military purposes, GPS receivers are now widely available for civilian use and serve as standard navigation equipment aboard ships, boats, and airplanes, as well as many cars and trucks. They are also popular with hikers and backpackers, who use them to navigate in the wilderness. Many models cost less than $100 (U.S.), and some are even being incorporated into cell phones, cameras, and other common portable devices. All of this makes GPS receivers extremely useful navigational tools for ROV and AUV projects.

Unfortunately, GPS is not directly useful for underwater navigation because GPS receivers require unobstructed line-of-sight views between the GPS receiver and the overhead satellites and because their radio signals do not penetrate water very well. However, GPS can assist with ROV and AUV operations, because it is an excellent tool for pinpointing specific locations on the surface of a body of water, like a lake or the ocean. Indeed, this is where GPS really shines and has proven most valuable—it provides location specificity that is as good as or better than most landmarks in a place where no landmarks exist!

If you own a GPS receiver you can bring with you on a boat or are launching your ROV from a boat already equipped with GPS, then this technology is probably your best option for positioning the boat directly over a desired dive location. It takes a little while to learn how to use a GPS unit properly, but it's well worth the effort.

Even though GPS does not work under water, AUVs are sometimes equipped with GPS receivers. This enables the AUV to verify its position, if it needs to, simply by coming to the surface.

TECH NOTE: USING LINES OF POSITION FOR TRIANGULATION

In geometry class, you may have learned that two straight, non-parallel lines intersect at exactly one point. This simple geometric principle is the basis of a navigational technique known as triangulation. (See Figure 9.10.) It can be used to describe precisely the location of your boat so that you (or someone else) can return to the exact same spot later; however, it works only if you can see landmarks (or other permanent, stationary objects) on shore from the vessel. To use triangulation, you must use landmarks to identify two different lines that each pass through the boat's position. Each of these lines is called a **line of position**, or **LOP**.

You can use either or both of two different methods to find and describe your LOPs. The first method is based on the geometric principle that two points uniquely determine a line. To use this method, find two landmarks on shore that line up when viewed from the boat's position. For example, you might notice that the vertical trunk of the tallest palm tree on the beach lines up exactly with the flagpole of a school located a few blocks inland. The tree and flagpole are two points that define a single LOP that passes through the boat's position. (Remember that you'll need to establish two LOPs for triangulation to work.) Your ability to relocate the exact same spot will be highest if the landmarks are far apart from each other, so one is way behind the other. (For example, it is better to use a tree on the beach and a radio tower on a distant mountaintop than to use two trees growing near each other on the beach.) Depending on how far you are from shore, a pair of binoculars may be helpful for seeing and lining up landmarks. Unfortunately, it's often difficult to find pairs of landmarks that line up perfectly with your boat's location, particularly if there are not a lot of distinctive landmarks present.

The second method of establishing an LOP relies on the geometric rule that a line is uniquely defined by a point and a direction through that point. To use this method, you take a compass bearing to a distinctive landmark on shore, such as a lighthouse. Using a compass to establish an LOP has the advantage that you need only one landmark per LOP, and that landmark doesn't have to line up with any other landmark. (Of course, you'll still need two LOPs.) However, this compass method is generally less precise than the two-point method, and you must be adept with a compass. For example, you must understand the differences between true north, magnetic north, and compass north, and you need to make sure you note which of those you were using at the time you recorded the direction. (See *Tech Note: Using a Magnetic Compass*.)

Triangulation works best if the two LOPs you choose are roughly at right angles (90 degrees) to each other. Whichever method(s) you select to establish your LOPs, remember to write down the information in sufficient detail so that someone else could reliably identify the same landmarks and the same compass bearings to arrive at the same LOPs and the same boat location. Although triangulation is used less often now that GPS has become widely available, it remains a valuable navigation technique.

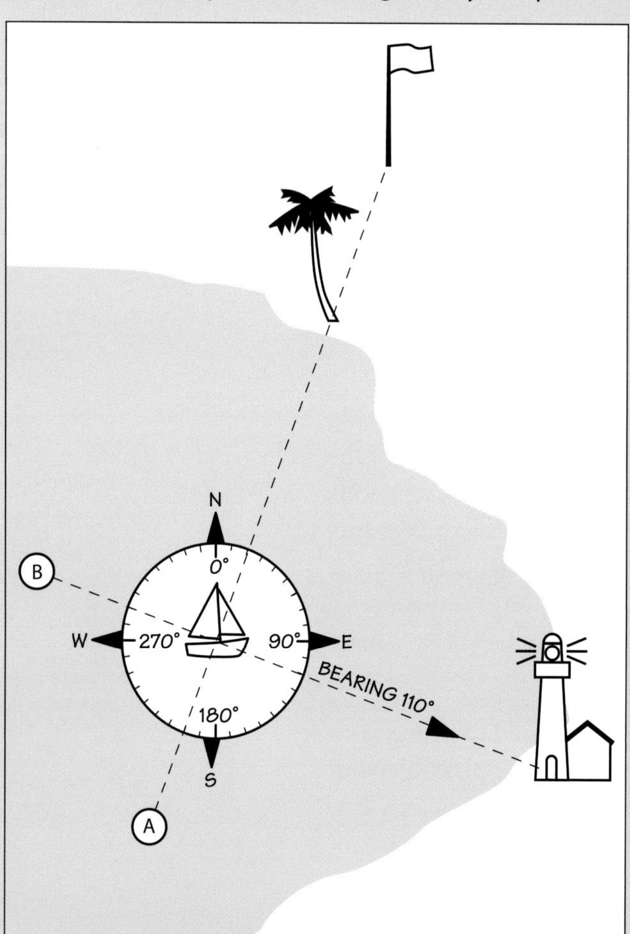

Figure 9.10: Triangulation

The intersection of two lines of position can be used to specify the position of a boat within sight of shore. In this illustration, one LOP (A) has been established by noting that a particular tree and flagpole line up with each other when viewed from the boat. The other LOP (B) has been established by measuring the compass heading toward another distinctive landmark (a lighthouse). In this illustration, the two LOPs are nearly perpendicular to each other, which is good, because it allows for more precise positioning of the boat.

3.1.4. Triangulation

If you don't have a GPS receiver, or if you forgot to install fresh batteries, there is an alternative method you can use to pinpoint positions or locations with GPS-like precision (sometimes better) near shore. It works as long as you are close enough to shore to see a variety of buoys, lighthouses, or other distinctive landmarks.

Triangulation consists of using pairs of landmarks, or landmarks coupled with direction readings (bearings) from a magnetic compass, to establish imaginary reference lines called **lines of position (LOP)**. Since two straight, intersecting lines cross at exactly one point, the vessel's position can be specified by identifying two intersecting LOPs that pass through that position. (For more details on how triangulation works, see *Tech Note: Using Lines of Position for Triangulation.*)

Traditionally, triangulation was used in combination with **nautical charts** to keep track of a vessel's position along a coastline or among islands whenever the vessel was within sight of land. Even with today's GPS-equipped boats, triangulation is used as a backup method of tracking vessel position—when safety is at stake, redundancy is a good thing, particularly since technology has a nasty way of failing at the least opportune moment.

When you don't have ready access to GPS, triangulation can provide a great way to record, and later return to, a surface location directly above some interesting underwater site. For example, if you accidentally dropped your keys overboard during a boat trip near shore and didn't have a GPS receiver available, you could use triangulation to record where you dropped the keys, then use it again to find the exact same location later.

3.2. Navigational Instruments

On dry land, you can easily reach most destinations of interest by following distinct highways, roads, or trails. The majority of these routes are well marked. If unsure which route to take, you can almost always find a detailed map that provides an overview of the network of available routes and their interconnections at whatever scale is most useful. In addition, it's usually possible to see for long distances, so you can navigate using distant features like mountain peaks or other landmarks to help establish your current location and direction of travel. You can often see your destination long before getting there, so you know how far away you are and which direction you need to travel to get there.

Alas, this is not so under water. Distinct routes are rare to non-existent, because there are no roads or trails. Landmarks are few and/or unfamiliar. Maps or charts of underwater features are scarce. Those that are available rarely provide the level of detail needed for precise navigation of ROVs or AUVs. Visibility is severely limited, too. The deeper parts of the ocean and many lakes or rivers are dark, both day and night. Even when light is present, the visibility in natural bodies of water is usually limited to a few meters or less, so there is no opportunity to use distant landmarks for visual navigation. In fact, it is quite possible to be staring straight at an important landmark or destination only 3 meters (approx. 10 ft) away and not be able to see it.

To make matters worse, natural bodies of water are often in motion. That means a vehicle can easily be carried off course by water currents unless a closed-loop position control system with access to accurate, real-time position information is being used to compensate for this kind of drift. For all these reasons and more, underwater navigation can be particularly challenging.

TECH NOTE: MARINE CHARTS

Mariners regularly have cause to shake their heads at Sunday sailors who take to the water with only a gas station roadmap for navigation. The fact is, a nautical chart differs significantly from an ordinary map. A chart represents the geographic features found under the water and along the coastline, whereas a map is primarily concerned with surface geographic features, particularly those on dry land.

The purpose of a chart is to facilitate the safe transit of vessels on the surface by presenting navigational information in an accurate, symbolic, and graphic format. In addition to indicating water depths and submerged hazards, marine charts provide identifying surface features such as headlands, rivers, harbors, and navigational buoys. Mariners also use charts to plot the position of a vessel, locate underwater objects such as shipwrecks, or mark a favorite fishing hole. Nautical charts are continually updated, so it's important to use the latest version and take note of updates found in the *Notices to Mariners* published periodically by government hydrological agencies such as NOAA.

Increasingly, printed paper charts are being replaced by electronic charts that are tied into the Global Positioning System. Some GPS receivers have electronic maps integrated into their displays.

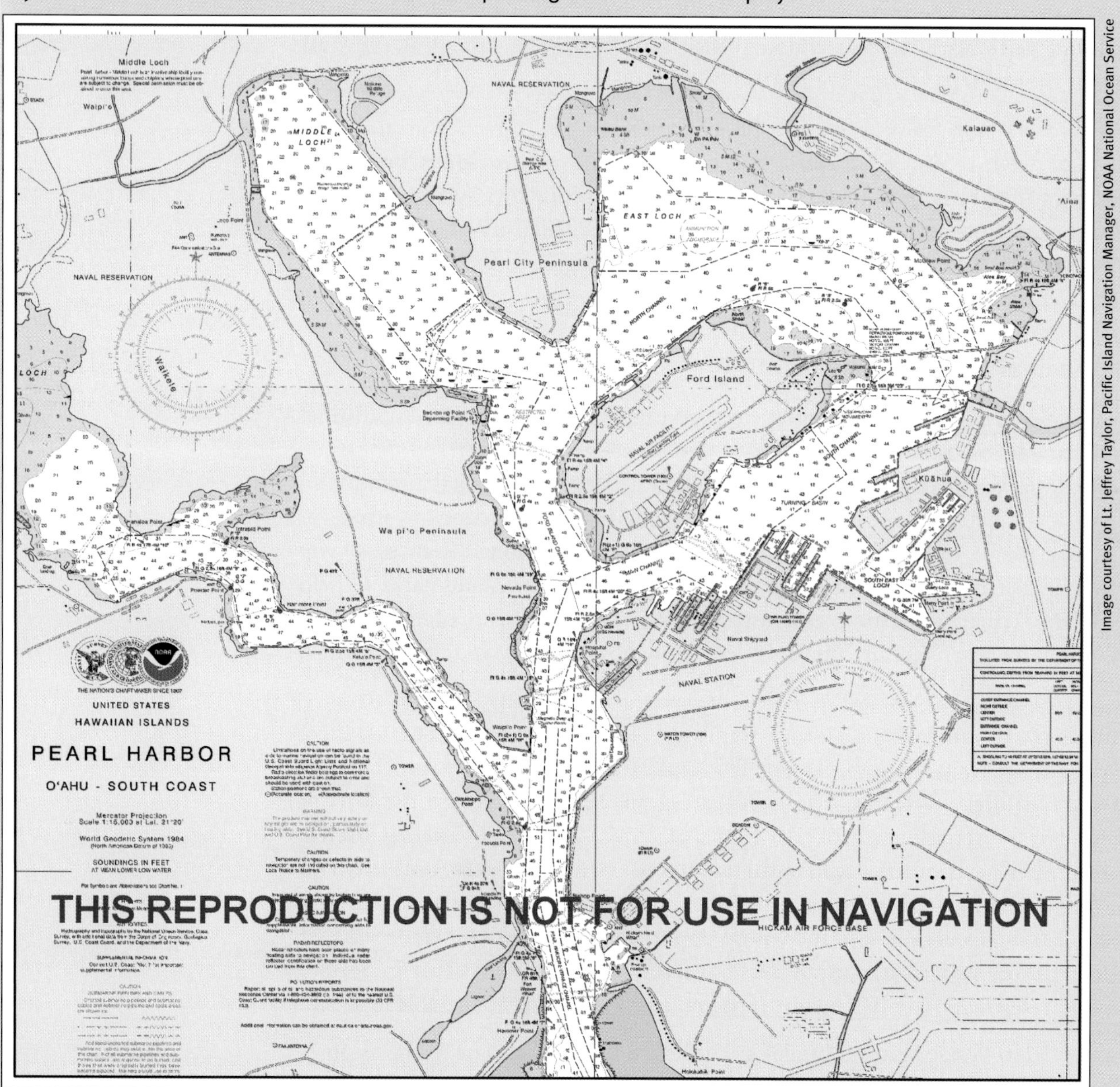

Image courtesy of Lt. Jeffrey Taylor, Pacific Island Navigation Manager, NOAA National Ocean Service

Figure 9.11: Marine Chart

This product contains nautical information for Pearl Harbor, reproduced from NOAA's National Ocean Service Chart 19366, 37th Edition, June 2007.

To deal with these difficulties, designers often rely on data from a suite of sophisticated sensors to provide accurate information about the position of the vehicle, the direction in which it is headed, and the presence of any obstacles or landmarks around it. These sensors may feed their information directly to a human pilot, who interprets the data and controls the vehicle remotely, or they may feed their information to one or more computers that perform automatic control of the vehicle's movements. In many cases, a combination of human and automated control is used. For example, a human pilot may control forward speed, while depth and heading are maintained by a feedback-controlled autopilot. (In addition to the introductory information on underwater sensors immediately below, see *Section 4.5. Adding a Basic Set of Navigational Sensors* for detailed descriptions about selecting and installing video cameras, compasses, depth sounders, and other sensors.)

3.2.1. Video Camera

Many people might not consider a video camera to be a navigational instrument; however, it can be incredibly useful for precise underwater navigation on small spatial scales in the range of meters to centimeters. This is especially true if there are recognizable and familiar subsea landmarks, or at least some distinctive navigational clues, like sand ripples, visible in the area. So unless your ROV is operating in a small swimming pool or body of water where the pilot has a clear view of it during the entire mission, the ROV will need to be equipped with a camera.

Image courtesy of Randall Fox

Figure 9.12: Navigating by Video Imagery

The true value of video cameras as navigational sensors lies in the fact that humans are fundamentally visual organisms. We are used to moving through the world using vision to provide most of our information about where we are, what obstacles we need to avoid, and what pathways we can take to get around those obstacles. Our human eyes and brains work together constantly to give us a detailed, moment-by-moment awareness of our surroundings. Putting a video camera on an ROV provides the pilot with a virtual underwater eyeball—that's why some small ROVs designed primarily for observation are affectionately known as "flying eyeballs." Video provides the pilot with a rich stream of detailed, real-time information in a format that a vision-oriented brain can readily interpret. In addition to showing an ROV pilot the vehicle's surroundings, a video camera provides a convenient way to relay information from other types of navigational instruments, such as a compass, to the pilot. (See also *Section 4.5.1: Choosing and Installing a Video Camera* later in this chapter.)

Although video cameras are probably the most common navigational tools found on ROVs, they are only rarely found on AUVs. That's because currently there is no effective way to transmit a video signal to the surface without a tether. And that's because radio signals don't travel well through water, and sonar signals don't offer enough bandwidth for video signals. The few AUVs that do use video cameras typically use them to record record data, storing these images on a computer hard drive or tape mechanism for viewing after the AUV has returned to the surface—rather than for real-time navigation. In a few AUVs, sophisticated image-processing algorithms are used with real-time video to enable a vehicle to follow jellyfish or other marine organisms for study, but this is not an easy thing to do.

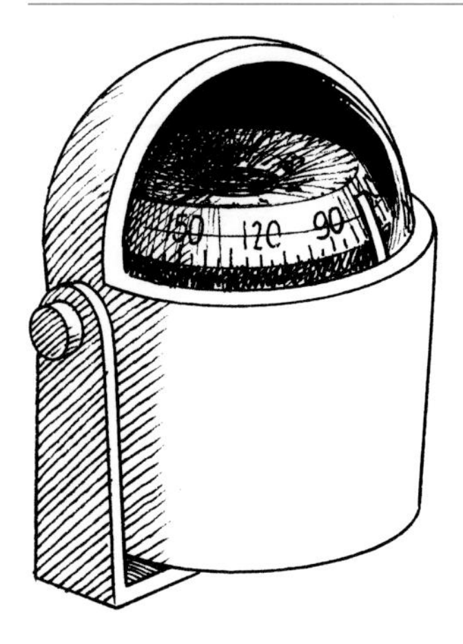

Figure 9.13: A Conventional Magnetic Compass

3.2.2. Compass

A conventional **magnetic compass** consists of a freely rotating magnetic needle contained in a protective enclosure of some sort. The needle spontaneously aligns itself with the earth's magnetic field, so that one end points (roughly) north and the other (roughly) south. (See *Tech Note: Using a Magnetic Compass* as well as *Chapter 11: Operations* for pragmatic tips on compass use with underwater craft.) A compass is one of the most fundamental, inexpensive navigational instruments you can put on an underwater craft. It's easy to mount one within view of your vehicle's camera, so the pilot can read it. A **gimbaled compass** is one that is specially mounted so that it remains level in spite of a vessel's rocking motion. Unlike GPS radio signals, magnetic fields penetrate water easily, so compasses work as well under water as they do on land. That's great news for ROV and AUV designers.

A **gyro compass** is a non-magnetic directional device that relies on a gyroscope that is sensitive to the earth's rotation and automatically aligns itself with true north.

TECH NOTE: USING A MAGNETIC COMPASS

A magnetic compass works by sensing the earth's natural magnetic field, which is oriented approximately (though not exactly) north-to-south. A simple mechanical compass consists of a magnetic needle balanced on a low-friction pivot point, so it can spin freely to align itself with the earth's magnetic field. The needle is usually surrounded by a ring inscribed with little marks for each of the 360 degrees in a circle. Using the compass needle for reference, if you rotate the ring until the zero degree mark is pointing north, then east will be at 90 degrees, south at 180 degrees, and west at 270 degrees. After orienting the ring properly, you can take a bearing on a landmark by looking across the center of the compass toward the landmark and noting which degree mark points at the landmark. Each compass is slightly different in its usage details, so consult a manual and practice.

When using a compass, be careful to distinguish among the following three different versions of north, because they affect how directions are interpreted and recorded.

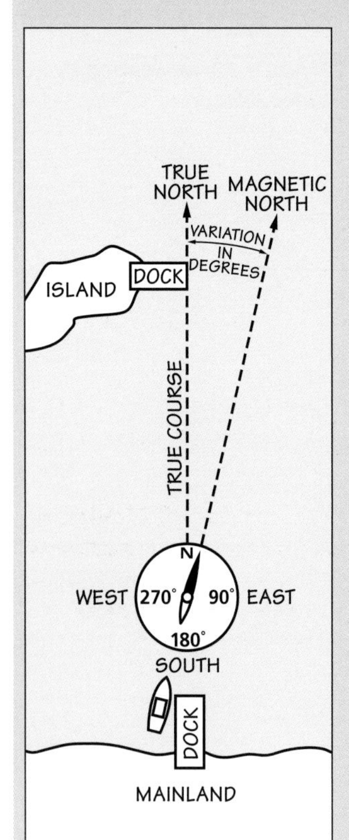

- **True north** is the direction toward the north pole, which is where the earth spins about its axis.
- **Magnetic north** is the direction toward the magnetic north pole, which is the place where the compass needle is expected to point. The magnetic north pole and the real north pole are not in the same place, so the direction to magnetic north normally differs from the direction to true north. This difference is known as the **declination** (or sometimes **variation**). Declination varies from location to location and even changes gradually from year to year. Check a recent nautical chart or topographic map near your operations area to get an up-to-date declination.
- **Compass north** is the direction in which the compass needle is pointing. Under ideal conditions this is the same as magnetic north, but it may differ if other magnets, pieces of iron or steel, or wires carrying electrical currents are nearby because these can create a local distortion in the earth's magnetic field.

Figure 9.14: The Compass Lies!

If you want to go from the mainland dock in Figure 9.14 to the island dock, you'll need to account for declination when determining which course to steer. The island is located due north from the dock. But if you just go in the direction the compass needle is pointing, you'll miss the island, because it's pointing to magnetic north, not true north.

An electronic version, such as a **fluxgate compass**, does not use needles and can't be read directly by a human, but it is easier to integrate into automated, computer-based electronic navigation systems. Techniques for doing this are described later in this chapter.

3.2.3. Depth Gauge

A depth gauge is another useful underwater navigational instrument. It tells you how deep your vehicle is in the water column and provides related navigational clues. For example, if you're searching for a reef that's known to be in 25 meters (approx. 82 ft) of water, and your ROV just found the muddy bottom at 15 meters (approx. 50 ft), then you know you probably need to head downslope to find the reef. In an area with a steeply sloped bottom, keeping an eye on depth while moving along the bottom can often provide lots of clues about where your vehicle is, particularly if you already know something about the bottom topography in the area.

3.2.4. Sonar

Advanced ROVs and AUVs usually navigate with the help of active sonar instrumentation, though the cost and complexity of most sonar equipment makes such use relatively rare on home-built ROVs. You'll recall that sonar works by sending sound pulses out through the water and timing how long it takes for echoes to return. Using the known speed of sound through water, the instruments can use the echo return time to calculate the distance to the object that reflected the sound.

Several different navigational sensors are based on sonar. One is called an **echosounder** (also referred to as an **altimeter** or **fathometer**). This system is used to measure the vertical distance between the vehicle and a hard surface, usually the seabed below, but sometimes ice or another hard surface overhead.

Sector scanning sonar is another type of sonar instrument that simplifies the work of piloting an ROV around the bottom in low-visibility situations. Sector scan sonar sweeps beams of sound horizontally in an arc or full circle, in order to scan for obstacles or other objects out in front, beside, or behind the ROV. In many ways, it's analogous to an underwater radar.

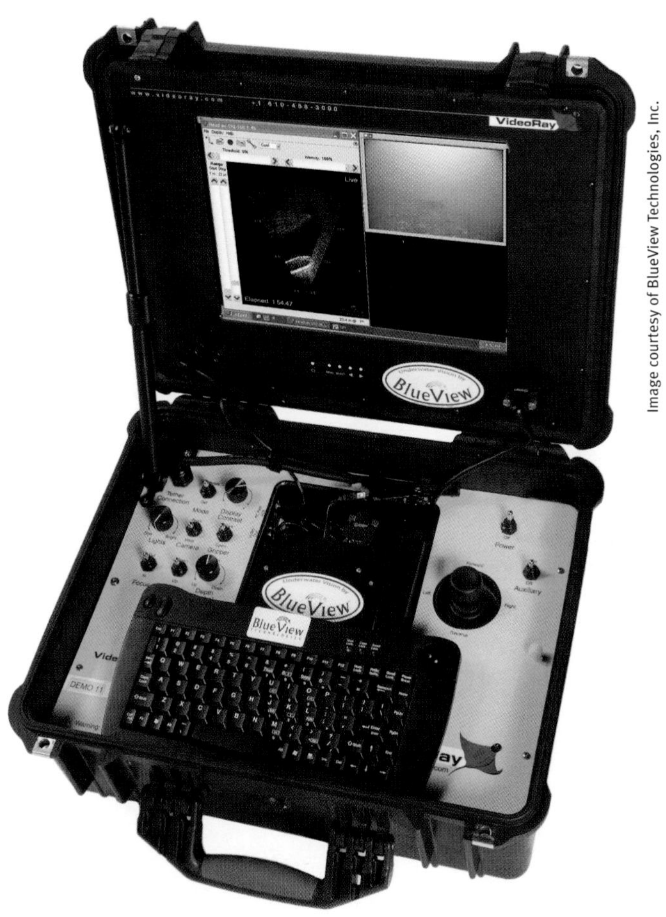

Image courtesy of BlueView Technologies, Inc.

Figure 9.15: Sonar for Navigation

The BlueView BV-250 Micro ROV Underwater Inspection® System is a conveniently packaged sonar navigation device.

Most sector scan sonars operate between a range of frequencies of 300 kHz to 1 MHz, with a nominal range of 200 meters (approx. 650 ft) for the lower frequency to approximately 50 meters (164 ft) for the high end. Some units are tunable (allowing the operator to select the best frequency for the application), while others are fixed. Micro-ROVs generally have very small but higher frequency scanning sonars. The image produced by the scan is displayed as a circular image which highlights "hard" or highly reflective objects against "soft" or less reflective objects in a grayscale or color-coded computer-generated graphic on a computer screen. This image also accurately indicates bearing and distance to objects. When used in conjunction with a compass and camera, scanning sonar provides the important navigation information needed to pilot an ROV in low-visibility waters.

Other types of sonar, called underwater **acoustic positioning systems**, are used for positioning an ROV, AUV, or towed equipment. These systems may include a seabed grid of deployed transponders where the ROV operates within the perimeter of the grid. Or there can be an omni-directional transmitter and receiver mounted on the hull of the support vessel that tracks the position of a transponder or responder mounted on the ROV. Whether it is a fixed bottom grid, or a shipborne system, each type has operational advantages and limitations. The system selection depends on the application and accuracy requirements of the program.

The acoustic tracking data comprises bearing angles and distances between the transponder(s) and the receiver. Using a dedicated computer, this acoustic positioning data is coupled with the ship's GPS position and true heading, thereby allowing the system operator to pinpoint precisely the underwater vehicle's position in terms of latitude, longitude, and depth.

Most sonar systems used on commercial robotic vehicles are way beyond the price range of the average low-budget vehicle. However, a number of hobby- and school-level ROV/AUV projects have experimented with using adaptations of inexpensive sonar depth sensors and "fish finder" sonar units for underwater navigation. So it may, in fact, be possible to create affordable sonar sensors for your project. (See *Chapter 10: Hydraulics and Payloads* and *Chapter 11: Operations* for information on other uses of sonar.)

3.2.5. GPS

As mentioned earlier, GPS is rarely used on ROVs, because it does not work beneath the surface; however, it is sometimes used on AUVs to give them a way to get a position fix when at the surface. On boats, GPS may be combined with sonar navigation to precisely locate an underwater vehicle in terms of latitude, longitude, and depth.

4. A Basic Control and Navigation System

Figure 9.16: Control Box Using Toggle Type Manual Switches

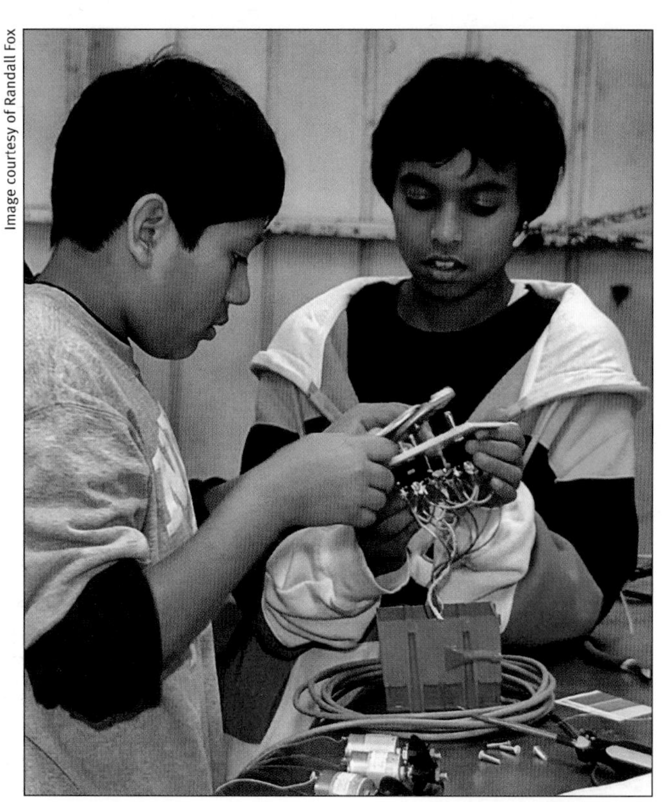

Image courtesy of Randall Fox

At this point in the chapter, here's where you're at in terms of control basics:

- You've learned some general principles about open- and closed-loop control systems.
- You've thought about how control of vehicle position involves a set of nested feedback loops that rely on control of vehicle movement (and ultimately on control of the electric current in thruster motors).
- You've been introduced to some navigational techniques and instrumentation useful for providing the position feedback needed for good control of vehicle position.

Now it's time to get down to the nuts and bolts of building a control system for your vehicle.

In smaller vehicles, the mechanical forces that propel and steer the vehicle are usually generated by electric thrusters, so this chapter focuses primarily on ways of controlling the electric currents that power thruster motors. It will also explore options for adding some basic navigational instruments to your vehicle—after all, knowing how to propel and steer your vehicle is of limited use if you don't know in which direction it needs to go.

Section 4 of this chapter details a relatively easy-to-build system for basic thruster control and simple navigation. Section 5 (the next major section of this chapter) covers more advanced control options, including some sophisticated and versatile motor control strategies, plus electronic sensors for navigation (and other purposes) that can be incorporated directly into computer-controlled systems for advanced ROVs or AUVs.

4.1. An Overview of Electric Switches

Control of most electronic devices, including thruster motors, lights, camera pan/tilt mechanisms, and various tools, is accomplished by using switches of one type or another to turn ON or OFF the flow of electrical current powering those devices. Most (but not all) of the switches commonly used for this purpose in underwater vehicles fall into one of four broad categories:

- manual switches
- relays
- magnetic switches
- transistor switches

Images courtesy of Dr. Steven W. Moore

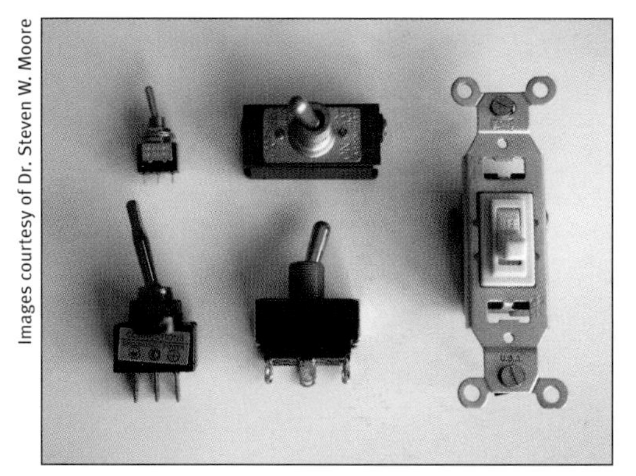

Figure 9.17: Common Toggle Switches

Toggle switches have a lever that you flip between either of two (sometimes three) positions.

Figure 9.18: Rotary Switches

Rotary switches usually have a rotating shaft with a knob on it, so you can rotate the dial to select connections to different circuits.

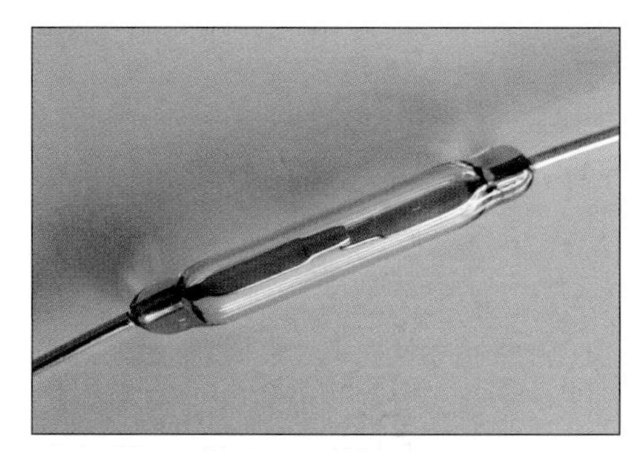

Figure 9.19: Magnetic Switches

Magnetic switches, like this magnetic reed switch, are opened and closed by nearby magnets. They can be an ideal way to transmit ON/OFF signals directly through the wall of a waterproof housing, reducing the number of places you need to drill holes to run wires through your housing.

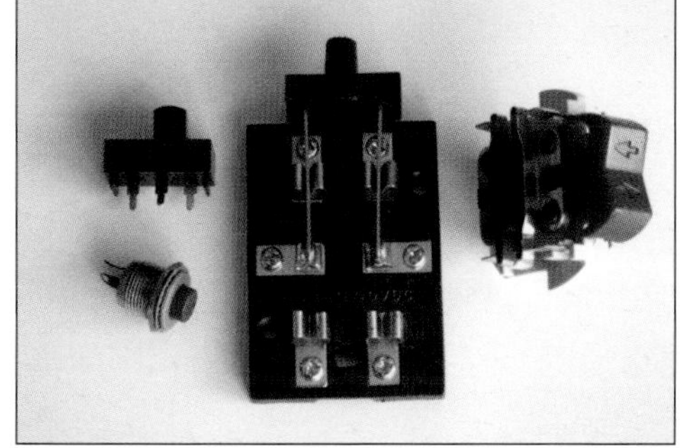

Figure 9.20: Other Common Switches

Other common switch types include include slider switches, push buttons, those funky-looking, old-fashioned knife switches popular in Frankenstein movies, and rocker switches.

TECH NOTE: EXTERNAL ON/OFF SWITCHES

Having an external waterproof ON/OFF switch for a vehicle with on-board batteries is a good expedient. It saves battery life when the ROV is not in use, because batteries will run down even if the vehicle is just on standby. When you are about to run a mission, the external switch makes it easy to turn on the ROV's power distribution system. Research the commercial options for such manual switches. For example, SeaCon manufactures a stainless steel–housed switch with a knob you turn to activate the power circuitry; it's expensive, but extremely robust. Another good option for an external ON/OFF switch would be some type of magnetic switch, like a reed switch, as mentioned in the text.

Manual switches are the most familiar of the four types of switches. They include the wall switches that turn room lights ON or OFF, the buttons that ring doorbells, and the keys on a computer keyboard or cellphone. They usually have a flippable lever (often called a "toggle"), a pushbutton, a rotating selector knob, or other mechanism operated by a mechanical force. That force most often comes from a human finger pushing the button or flipping the switch, but not always.

Magnetic switches, including **reed switches** and magnetic **proximity switches**, are simple variations on the manual switch theme, but are operated by movement of a nearby magnet instead of by an externally applied mechanical force. Inside the switch, the magnetic force moves parts that make or break electrical connections. Magnetic switches can be particularly useful in underwater systems, because they provide a way to control circuits inside a waterproof housing without having to drill a hole through the housing. Magnetic fields pass easily through glass, plastic, non-ferrous metals (e.g., aluminum), and some stainless steel alloys, so a magnet on the *outside* of the pressure housing can be used to operate a switch on the *inside* of the housing directly through the solid wall of the housing. The newer rare-earth magnets are particularly useful for this purpose because they are very strong.

Relays and **transistor switches** are both operated by electrical signals, rather than fingers or magnets. This makes them particularly valuable in systems that are controlled directly by computers or other electronic circuits, so they are the natural choice for most automated control systems.

This section of the chapter will focus primarily on manual switches. Relays and transistor switches will be covered in detail later in the advanced control section of this chapter.

4.1.1. Poles and Throws

Another way of classifying switches within these groups (particularly manual switches and relays) is based on their number of "poles" and "throws." Figure 9.21 lists several common pole-and-throw configurations, along with the corresponding schematic symbols for each. The symbols are inspired by the innards of some old-style rotary mechanical switches, though they apply equally well to other kinds of switches. The number of **poles** specifies how many separate circuits the switch can control simultaneously. This is usually one or two, but can be up to 20 or more. The number of **throws** specifies the number of active (ON) positions that can be selected for each circuit; the OFF position is not included in the number of throws.

Most switches have only one or two throws, but some offer many more. For example, a double pole double throw (DPDT) switch can control two separate circuits, and each circuit can have two possible ON positions in addition to the OFF position. These two ON positions might be "high" and "low" power or "forward" and "reverse."

SWITCH CONFIGURATION	SCHEMATIC
SPST (single-pole, single-throw)	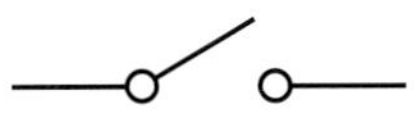
DPST (double-pole, single-throw)	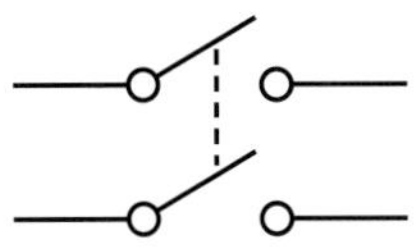
SPDT (single-pole, double-throw)	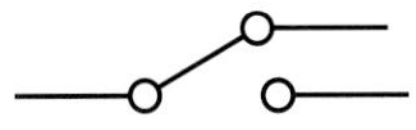
SPDT (single-pole, double-throw with central OFF position)	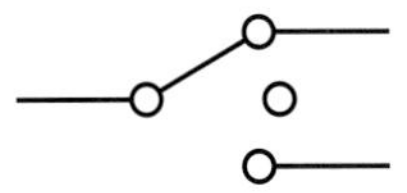
DPDT (double-pole, double-throw with central OFF position)	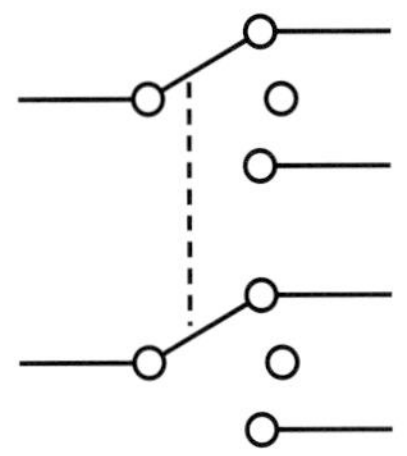

Figure 9.21: Abbreviated Names and Schematic Symbols for Common Switch Configurations

The slanted line(s) in each schematic symbol represents a movable electric conductor inside the switch that pivots about the dot on its left end to connect that dot electrically to one of the dots near its other end, thereby connecting the wires attached to those dots. Vertical dashed lines represent mechanical (not electrical) linkages between the switching mechanisms for each pole in a multi-pole switch. By convention, a single throw switch is almost always shown in its "open" or OFF position. The DPDT with a central OFF position is particularly useful for simple DC motor control circuits.

4.1.2. Momentary Switches

Many pushbuttons and toggle switches are spring-loaded, so they return to a particular position by default anytime nothing is actively pushing on them. In these spring-loaded switches, positions that are not the default position are termed "momentary," because the switch is in that position only momentarily (specifically, while someone or something is holding it there).

The "throw" terminals on momentary switches are often labeled with the abbreviations "NO" and "NC," which stand for "normally open" and "normally closed." For example, in an SPDT pushbutton switch, there will be three terminals. One is usually unlabeled and is the pole. Another, labeled NO, is not connected to the pole *until* the button is being pushed. The other, labeled NC, is always connected to the pole *except* when the button is being pushed.

In some three-position DPDT momentary switches, the central OFF position is the default, and each of the two ON positions is momentary. One common shorthand for this on the switch package or in the catalog is (ON)-OFF-(ON), where the parentheses indicate that the two ON positions are momentary. Be warned, however, that there are numerous other combinations out there to confuse the unwary. For example, you can find (ON)-OFF-ON switches, in which only one of the two ON positions is momentary. Also be aware that the abbreviation NC is often used in electrical schematic diagrams for "no connection" instead of "normally closed," so be careful not to confuse those two uses of the NC abbreviation. The context will usually make the right meaning obvious.

SAFETY NOTE: CHECK THE POWER SPECS

Before connecting any piece of equipment to a power source, check what voltage, or range of voltages, the device is designed to run on. On switches and some other components you should also check for maximum current ratings. Exceeding these limits may result in damaged equipment or fire.

4.1.3. Other Important Switch Characteristics

Another thing to be aware of when selecting switches is that some multiple throw switches are "make-before-break" and others are "break-before-make." You may wonder, "What's the difference?" Well, during a switching action, the make-before-break type makes the connection between the pole and the next throw terminal *before* it breaks the connection between the pole and the previous throw terminal. In some circuits, this can result in unintended short circuits between different parts of the circuit while the switch is changing positions. These brief short circuits can damage other components in the circuit and sometimes lead to nasty sparks, melted switches, or other problems. For most vehicle applications, you'll probably want to get break-before-make switches.

Finally, keep in mind that switches, like all electronic devices, can handle only a certain amount of voltage and current. The allowable limits are usually printed on the switch or on its packaging; if not, you may need to find a datasheet for the switch. Most common switches can easily handle the low voltages you are likely to be using with inexpensive ROVs or AUVs; however, many cannot handle the moderately high currents demanded by most thruster motors. Be sure to check the maximum voltage and current ratings for any switch you use to make sure it can handle at least as many volts and amps as you plan to use with it. (See *Safety Note: Check the Power Specs.*)

4.2. Using a Manual SPST Switch to Turn a Light ON or OFF

The best way to start learning about the control of electric systems is to start with something simple. Turning a light ON or OFF qualifies. And if you want to pilot your ROV at night or in deep, murky water (where it's dark even during the day), this lesson will have immediate practical value. If you have no such plans, consider this lesson a stepping stone to the next lesson, which covers forward and backward control of thruster motors.

A standard incandescent or halogen light bulb has two metal contacts or terminals (for electric power). It will light up whenever an appropriate voltage is placed across its terminals. (This assumes, of course, that the power source can also supply enough current to meet the light's demands.) While you could, in principle, use a screwdriver or soldering iron to connect and disconnect the power wires every time you wanted to turn the light ON or OFF, this would be extremely inconvenient. A switch provides a better way.

If you only need to turn something ON or OFF, then a simple SPST switch is usually a good choice. Recall that the SP (single-pole) means the switch is designed to control current flow in a single circuit, and the ST (single-throw) means that the switch has only one ON position. (It will also have an OFF position.)

Figure 9.22 shows how you can add an SPST switch to a light bulb circuit. The purpose of the switch is to introduce an easily reversible break, or gap, in the circuit. When the switch is in the "open" or OFF position shown, it's as if the wire is broken, so no electricity can flow around the circuit, and the light is OFF. When the switch is flipped to the "closed" or ON position, a complete circuit is formed, electric current can flow around the circuit, and the light is ON.

Note that it does not matter, functionally, where the switch is located in the circuit. As long as the switch can interrupt the flow of current around the loop, it will perform its function properly. (However, switch location may have safety implications. See *Safety Note: Where to Put the Switch in a Circuit.*)

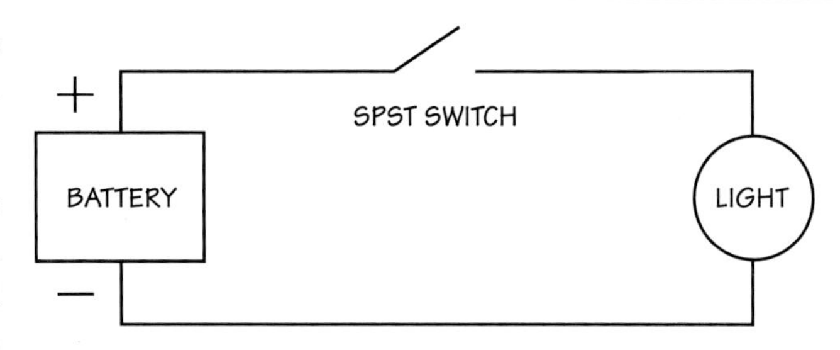

Figure 9.22: Using an SPST Switch to Control a Light Bulb

This schematic diagram of a simple light control circuit shows how an SPST switch can be connected to a battery and light bulb to control the flow of electrical current through the bulb.

This schematic shows that there is a wire coming from the positive terminal of the battery to one terminal of the SPST switch. (It does not matter which terminal.) It also shows a wire running from the other terminal of the switch to one terminal of the light bulb. (Again, it does not matter which terminal, at least for a conventional incandescent bulb.) Finally, there is a wire going from the remaining terminal of the light back to the negative terminal of the battery to complete the circuit.

4.3. Using a Manual DPDT Switch to Control Motor Direction

The previous section described how you can use an SPST switch to turn a light ON or OFF. The exact same technique can be used to turn a brushed DC motor ON or OFF, if that's all you want to do. However, when it comes to motors used in thrusters, grippers, and many other applications, you may also need to control motor *direction.* For example, you may need to run a motor in reverse to make a vehicle back up or to release a gripper's hold on some object.

The present discussion assumes that you are trying to control a brushed DC motor. These are the most common motors used for simple ROV projects, partly because they are the most common motors in bilge pumps. A small DC brushed motor will have either two separate wires coming out of it or it will have two terminals to which you can attach wires, usually by tightening a screw down on the end of a wire or by soldering a wire to the terminal. Turning this type of motor ON is a simple matter of connecting one of the two wires to the (+) terminal of a battery and the other to the (-) terminal of the battery (or other source of DC electrical power). This completes a circuit, allowing current to flow through the motor. Remember that the power source must have the correct voltage for that particular motor and be able to deliver enough current (amps) to power the motor properly. To turn the motor OFF, you simply disconnect it from the power source by disconnecting one of the wires.

Reversing the direction of a brushed DC motor is a simple matter of reversing the direction of current flow though the motor. This, in turn, is a simple matter of reversing the polarity of the voltage applied to the two motor terminals. (Note that most brushed DC motors are biased to run slightly better in one direction than the other; however, this asymmetry is not usually big enough to be bothersome in typical ROV applications.)

SAFETY NOTE: WHERE TO PUT THE SWITCH IN A CIRCUIT

For safety reasons, it's good to get in the habit of putting switches between the "hot" side of the power source and the device being powered. This is a better position than in the current return path, which is usually at or near zero volts relative to earth ground. (In a DC-powered circuit, the "hot" side is usually, but not always, the positive terminal of the battery.) If you put the switch on the return side instead, the device will be "hot" (i.e., at a voltage that is different than ground) when it's turned OFF. In low-voltage circuits, this is no big deal, but in high-voltage circuits, it presents a shock hazard, since a person touching the device could become a path for electrical current to flow to the electrical ground.

Later in this chapter, you'll see examples of transistor switches placed on the return side, which contradicts this safety advice. This is done for electrical convenience, based on the way these devices work, and is recommended here only for use with low-voltage circuits.

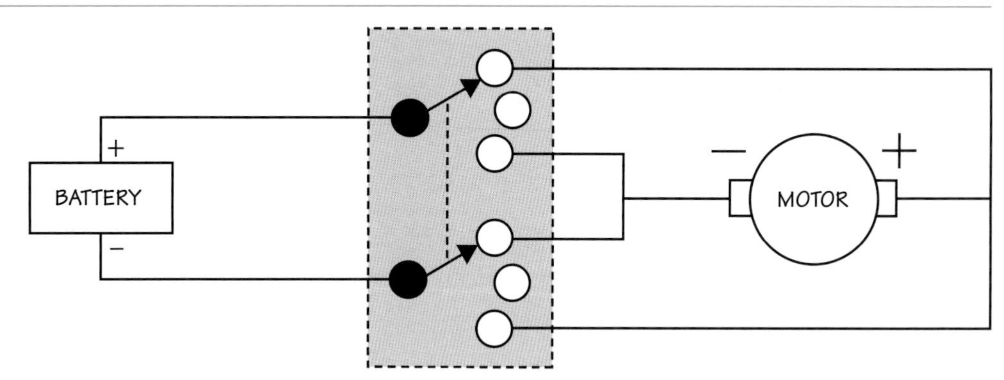

Figure 9.23: Using a DPDT Switch to Provide Directional Control

This schematic diagram illustrates how the terminals of a DPDT switch can be wired to the battery and motor terminals in order to provide directional control of a brushed DC motor. See text for full explanation.

To wire up a simple manual switch circuit for controlling motor direction, use a DPDT switch with a central OFF position. The momentary ones designated (ON)-OFF-(ON) work very well, because they automatically turn the motor OFF as soon as you let go of the switch. The wiring concept is illustrated in Figure 9.23.

In Figure 9.23, the large gray rectangle represents the DPDT switch, with its pole terminals shown as small black circles. The throw terminals and the OFF position are represented by small white circles. The battery has its positive (+) and negative (-) terminals labeled. The motor terminals have also been labeled (+) and (-) for reference; however, real DC motors often do not have them labeled—you just try it one way and reverse the wires if you want the motor to spin in the other direction. When the switch is flipped so that the arrows are pointed as shown, they connect the positive terminal of the battery to the positive terminal of the motor, and the negative terminal of the battery to the negative terminal of the motor. This will cause the motor to turn in the forward direction. But if the switch knob or toggle is moved so the arrows point to the lower white dots instead (a condition not shown), then the positive terminal of the battery will be connected to the negative terminal of the motor, and vice versa. Thus, changing the switch position reverses the polarity of the voltage applied to the motor and changes the direction of the motor's rotation.

Figure 9.24: Using a DPDT Switch for Forward and Reverse Control

This diagram shows how to wire a typical DPDT switch to provide forward and reverse control for a motor. The wire from A to F crosses but is not connected to the wire from C to D.

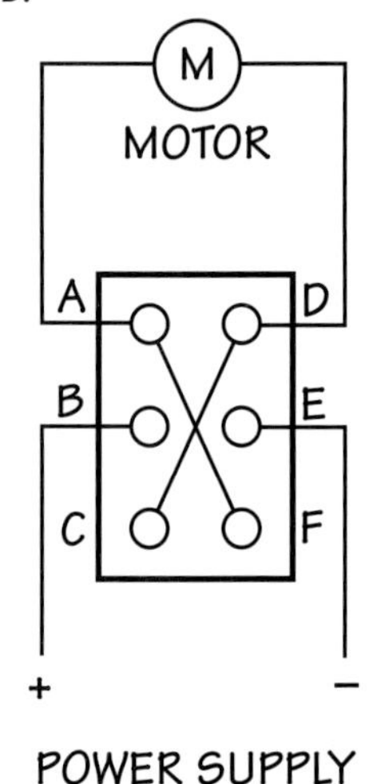

In practice, this arrangement can usually be achieved by wiring the six terminals on the back or bottom of a standard DPDT switch in an "X" pattern as shown in Figures 9.24 and 9.25. This is the switch configuration used for *SeaMATE* thruster control. Note that although this DPDT switch wiring allows bi-directional motor control, it does not offer speed control, dynamic braking, or other advanced motor control options. Those can be achieved through the use of transistor switches and motor controller circuits described later in this chapter.

Figure 9.25: Wiring a DPDT Switch for Small ROV Control

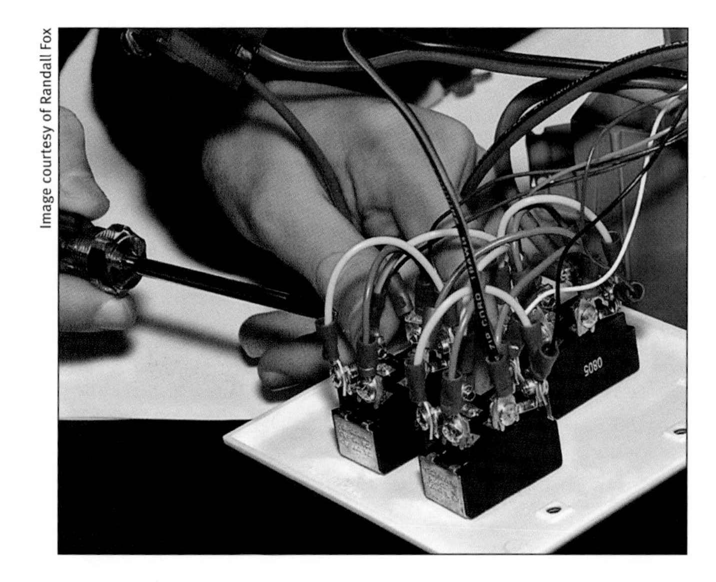

Image courtesy of Randall Fox

4.4. Limitations of Manual Switch Control

Using manual switches for direct control of electrical current has one major advantage over other more "sophisticated" approaches—it's simple. It also provides a way to learn some basic information about switches and switching of electrical current. You'll need to know that before working with any of the more advanced, electrically controlled switches like relays and transistors. However, controlling an underwater vehicle by supplying current to motors or other devices directly through manual switches does have some significant limitations. You should be aware of these issues before you make a final decision about how you will control your vehicle's motors and other devices:

- This technique is not suitable for controlling a large number of different devices through a single tether, because the large number of separate wires required to do this would require a fat, heavy, and unwieldy tether.
- The technique is not generally suitable for use with tethers more than 20 or 30 meters (approx. 66–100 ft) long, particularly in low-voltage applications, because of resistive power losses incurred when trying to run large amounts of current through long wires, as discussed in *Chapter 8: Power Systems.*
- This technique is not optimal for very precise control of vehicle movements or other precision control, because it does not allow for adjustments to motor speed, light intensity, and so forth.
- This technique cannot be used to control devices that require rapid, complex electrical pulse patterns for control. Such devices include brushless DC motors, stepper motors, and servos.
- The technique is not suitable for fully automated control systems, such as those found on AUVs and the autopilots of many advanced ROVs, because the computers or other electronic circuits coordinating the control cannot operate manual switches nearly as easily or effectively as they can operate transistor switches or relays.

For all these situations, the more advanced techniques covered later in this chapter are usually a better choice.

Figure 9.26: Cameras for Navigation and Recording

The pan-and-tilt unit on the work class Oceaneering Magnum 156 ROV shown here controls position of a number of cameras. The camera with the black plastic guard around it is a high definition video camera. The one with the red band is a low light black and white video camera, and the other is a normal color video camera.

The white square visible in the front of the camera for shooting stills is a desiccant pack to try and ensure that there will be no condensation to affect image quality. The water temperature on the seabed where this vehicle was working was very cold, nearly minus 1°C.

4.5. Adding a Basic Set of Navigational Sensors

The earlier *Section 3.2. Navigational Instruments* describes basic navigational sensors that are commonly used to ensure that underwater vehicles get where they need to go. This section gets into the nuts of bolts of what you need to know in order to easily equip your vehicle with a basic set of inexpensive and readily available sensors: a camera, a compass, depth sensor, etc. These sensors are ideal for many simple ROV tasks. However, if you want to get more complex in terms of monitoring these sensors electronically, you'll want to consult *Section 5.10. Electronic Sensors and Sensor Circuits.*

Image courtesy of Dr. Daniel Jones, SERPENT Project, National Oceanography Centre, Southampton

4.5.1. Choosing and Installing a Video Camera

The first sensor usually added to any ROV design is a video camera. Though internally much more complex than many other sensors, these versatile pieces of technological wizardry are nonetheless easy to use and affordable. For most small craft, a modest investment in video camera technology can return a wealth of information about where the ROV is and what it's doing, so in many cases it's the only sensor needed.

Camera Option Overview: There are many types of cameras to research. The main requirements are that they run on low-voltage DC, can see in low-light conditions, and have the highest resolution possible for your budget. (Note that a camera's light sensitivity is often expressed in "lux numbers," so look for one with a low lux rating.) Budget-minded ROV builders often use the tiny security cameras made for home or office surveillance. These are available from a large number of consumer electronics stores and on-line distributors, particularly those specializing in equipment for home and office security or spy gadgets. At present, these cameras come in both analog and digital models, though the growing popularity of digital models is slowly eclipsing the analog models. Analog models are generally the least expensive and easiest to use. They often deliver superior image quality in a given price range, too, so they're a good choice for first-time builders of camera-equipped ROVs. (See *Tech Note: What to Look for When Choosing a Camera.*)

As of 2009, analog camera prices are in the $50 (U.S.) range for basic black-and-white video cameras. Decent analog color cameras can be found for less than $100. Of course, there are much more expensive analog cameras, but the inexpensive ones provide adequate performance for most underwater missions. Besides, if the ROV floods and the camera drowns, a less expensive model may mean there's still some money left in the budget to replace it.

Compatible Video Format: The camera sending the image must use the same **video format** as the TV, computer monitor, or other equipment displaying or recording the image, or the system won't work. Unfortunately, there are a variety of video signal transmission formats in both analog and digital cameras, so you have to be careful to get compatible equipment. Most analog cameras available in the United States and

TECH NOTE: WHAT TO LOOK FOR WHEN CHOOSING A CAMERA

Whether you're checking out cameras in an electronics shop or ordering them from a catalog, here are some things to consider for your eyeball ROV:

- operating voltage
- how many amps of current the camera needs
- output signal format (analog versus digital, NTSC, PAL, Ethernet, etc.)
- field of view (FOV), also called angle of visual coverage (e.g., a 53° field of view is fairly narrow for underwater work)
- resolution (higher is generally better)
- ability to work in low-light conditions
- type of lens—you generally want a fairly wide-angle lens with an ability to focus up close. (Note that focus distance under water will be slightly different than in air.)
- color or black-and-white (Note: black-and-white cameras are cheaper and usually operate with lower light requirements.)
- shape of camera (e.g., a "lipstick" camera slides easily inside a pipe or other cylindrical housing)
- sound capability (e.g., some cameras have built-in microphones)
- cost
- whether the unit is in stock
- delivery time

Image courtesy of Dr. Steven W. Moore

Figure 9.27: Miniature Camera Mounted on ROV

Canada use something called the **NTSC format**, which has been a standard for decades. Other parts of the world use **PAL**, a different video signal transmission format. Having these two common yet incompatible formats in worldwide use makes trading videos between North America and other parts of the world difficult. **S-video** is closely related to NTSC, but is not an identical analog format. **High-definition television (HDTV)** cameras may require yet another format.

Evolving Video and Camera Technology: At the time of this writing, many digital cameras are **USB cameras** or **web cameras** designed to plug into a computer and be used for video conferencing. Others, called **IP cameras** or **network cameras**, use the Ethernet protocol and are designed to plug into standard Ethernet computer networks. If the IP cameras and the network to which they are connected are properly configured, the camera images can (optionally) be made accessible to viewers anywhere on the internet. While digital technology is opening up exciting possibilities for very long-distance remote control of ROVs via the internet, beginners will likely find that analog cameras are an easier place to start.

Waterproofing and Pressure-Proofing: Since most video cameras are not inherently waterproof, they must be housed in some waterproof and pressure-proof housing with a clear window or dome to shoot through. This can be the main ROV housing itself or a separate camera-only housing.

The main advantage to putting the camera inside the single primary canister is that you don't need to worry about waterproofing penetrations for electrical wires going between two separate housings. Of course, a camera mounted outside the primary pressure canister has some advantages, too. First, since there's no camera lens to accommodate, the main pressure canister doesn't need a viewport. Second, an external camera is likely to be in a smaller housing, offering you more choices as to where to position it for unobstructed viewing. One simple and effective (but non-reversible) method of waterproofing a camera so it can be mounted outside the main pressure can is to encapsulate, or "pot," the camera in epoxy. (See *Chapter 12: SeaMATE*, Section 8.2.1 for instructions on how to encapsulate a camera.)

Pan-and-Tilt Mountings: Mounting a camera on some sort of pan-and-tilt arrangement makes piloting easier, since you don't always have to spin the ROV around to view objects that are just out of view of a fixed, forward-looking lens. You can simply hold the ROV steady and move the camera. It's easiest to construct or purchase a pan-and-tilt mounting unit that will fit, along with your camera, inside a waterproof housing. However, if you choose to locate your camera outside the main pressure housing, you can still fabricate a pan-and-tilt component—but it's much more of a challenge, because the mechanism has to be pressure-proof and waterproof, corrosion-resistant, and able to operate by means of the vehicle's control system. It will also take at least four to six additional conductors that must be brought out from the pressure can via an electrical bulkhead connector. Research your options; most likely, you will have to modify mechanisms in order to make a suitable pan and tilt arrangement for your vehicle.

Figure 9.28: Student-Built Camera Tilt Mechanism

This mechanism, viewed here from below, only provides a tilt adjustment for the camera. To achieve a pan movement, you spin the ROV to the right or left with its thrusters.

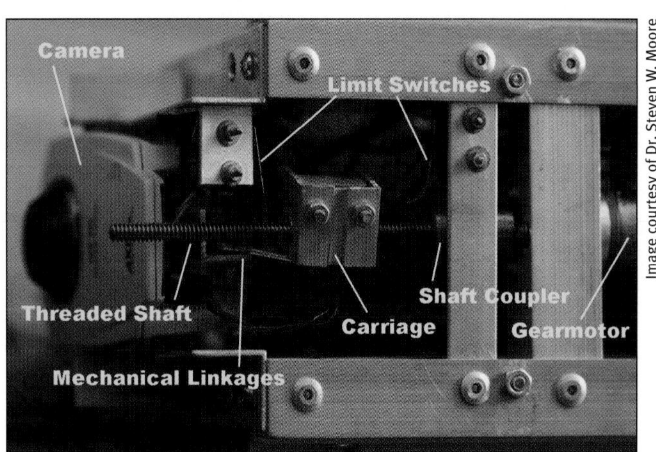

Image courtesy of Dr. Steven W. Moore

Figure 9.28 shows a camera tilt mechanism designed and built by students. The mechanism is viewed here from the underside of the vehicle after removal from the waterproof main pressure canister of the ROV. Camera tilt commands from the pilot energize the small electric

Image courtesy of Submertec Ltd.

Figure 9.29: Spyball, a Commercially Available Camera with Pan and Tilt System

gear motor, which is connected to a threaded shaft by a shaft coupling. As the motor turns, the shaft rotates at the modest rate of several revolutions per second. The small aluminum box labeled "carriage" contains a regular hex nut screwed onto the shaft and glued inside the carriage. As the threaded shaft rotates, the nut (and carriage) move along the length of the shaft, either forward or backward, depending on which way the motor is turning. Mechanical linkages connected between the carriage and the bottom of the freely-pivoting camera (hinged at its top end) translate the linear motion of the carriage into camera tilt. Limit switches cut power to the motor when the carriage runs into them, preventing the motor from damaging the mechanism or the motor by trying to travel too far. This is only one of many effective ways you could build a camera tilt mechanism. Standard R/C servo motors, such as those used to control rudder angle in radio-controlled model airplanes, provide another popular way of controlling the tilt angle of a camera.

Hookup and Testing: The simplified instructions provided here assume you are using a standard analog NTSC video camera. Procedures will differ for other types of cameras; follow the manufacturer's instructions. To test an analog NTSC camera, you must first supply it with appropriate power, as specified in the datasheet or instructions for the camera. If you're doing any custom wiring to bring power to the camera, be very careful not to mix up the (+) and (-) wires; doing so may kill the camera. Next, connect the camera's signal wires to the video input of a TV monitor, VCR, or computer with an NTSC-compatible analog video input. One connector commonly used for these signals is generally called an **RCA plug**. (See Figure 9.30.)

Image courtesy of Dr. Steven W. Moore

Figure 9.30: Hookup Basics

Coaxial cable and RCA connectors, such as the RCA plugs pictured here, are common ways of hooking up an underwater camera to transmit a video signal to the surface.

Presumably, you'll want to run the video signal from the ROV to the surface. The easiest and most cost-effective way to transmit the video signal is to use **coaxial cable** or some cable with twisted wire pairs in it as part of your tether. (Sections 5.5.1 and 5.5.2 provide more information about using coaxial cable and twisted pair wires.) There are wireless video cameras, but the radio signals they use to transmit the images won't penetrate water effectively, so they are of little use for underwater video transmission.

Anytime you send video over a long wire, the signal will degrade with distance. Analog signals can travel successfully through coaxial or twisted pair cable for distances of anywhere from less than 15 meters (approx. 50 ft) to more than 300 meters (approx. 990 ft), depending on the camera, type of cabling, quality of the receiving equipment, type of connectors used, and even the environment surrounding the cable. Note that for proper image transmission, you may need to terminate an analog video cable with a 75-ohm resistor connected between the signal wire and the electrical ground. Sometimes this termination resistor is built into the video equipment; other times you will have to add it. Check with the manufacturer, or just try it to see if the image quality improves.

With a USB digital camera, signals are typically limited to just a few meters. Network or IP cameras can go a bit farther—typically 100 meters (approx. 325 ft) through Cat-5 or Cat-5e computer networking cable to the nearest network hub or router. Well-designed fiber-optic systems can transmit video signals for a kilometer or more.

When the video signal finally gets to the surface, the next thing you need is some way to view it. For portable ROV operations, the tiny battery-powered TVs designed to allow sports fans to watch the broadcast of instant replays while sitting in the stadium

Images (left, center) courtesy of Steve Van Meter/VideoRay LLC

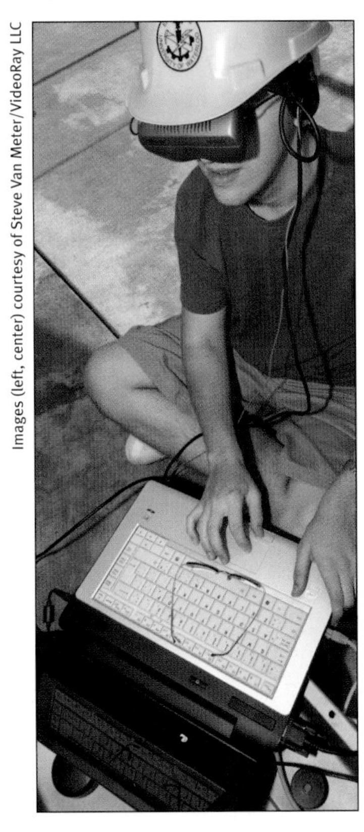

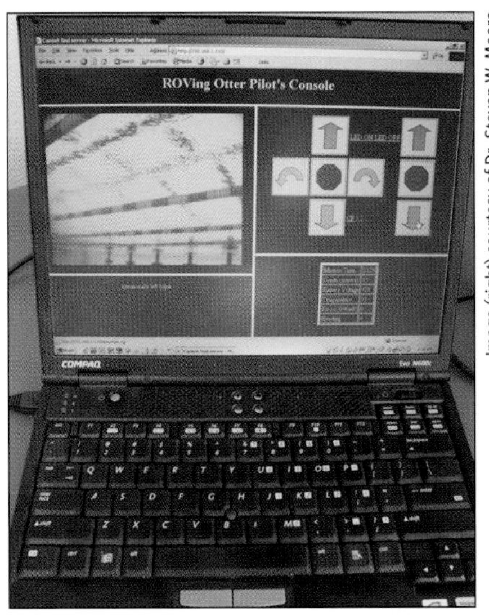

Image (right) courtesy of Dr. Steven W. Moore

Figure 9.31: Video Monitors

Modern video monitors can relay images to standard CRT screens, flatscreen monitors, or laptop computers, but viewing in a sunny environment still remains a challenge. Virtual reality goggles are another high-tech option for viewing video feedback.

bleachers are one option, but only if they have a place to plug in an external video signal. If you are using a regular TV plugged into a wall outlet near water, remember to use a GFI protected circuit! Some computers include an RCA jack for input of NTSC video signals. USB cameras are also fairly easy to set up and use with a portable computer, but remember that expensive computers and water don't mix well, and, as mentioned, USB cable lengths are quite limited. Network camera images can be viewed with a computer and web-browser software, but they can be more challenging to set up, particularly without some prior knowledge of IP addressing, router configurations, and other internet/ Ethernet technology requirements.

One final note about video camera selection: digital video technology is evolving very rapidly. In general, capabilities, convenience, and ease of use are going up, while camera size, power requirements, and price are going down, so be smart and check the web or other sources for the latest video developments when deciding what video technology to use for any ROV projects.

Lighting: Even cameras that can work in very low-light conditions generally require some underwater lights to illuminate the area in front of the camera. Note that these lights typically stay on all the time. As a result, they tend to consume a considerable amount of energy, sometimes even more than the thrusters. This is a big problem for hybrid ROVs and any AUVs with video recording, since they carry a limited supply of energy.

***Figure 9.32: Camera and Lights for* Nereus**

Woods Hole Oceanographic Institution equipped its HROV Nereus *with a battery of LED lights to illuminate missions to full ocean depth.*

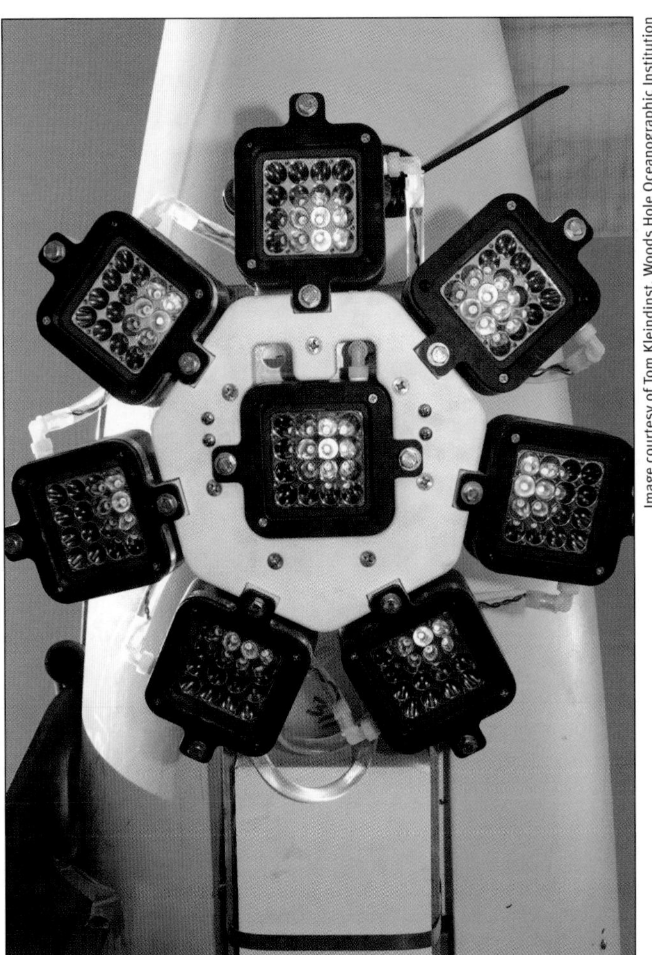

Image courtesy of Tom Kleindinst, Woods Hole Oceanographic Institution

Fortunately, advances in halogen and LED lamp technology have produced increasingly efficient lights that provide the required light intensity, using less power than conventional incandescent lighting uses. (See Figure 9.32.)

As noted in *Chapter 12: SeaMATE*, the simplest way to provide lighting for a small, deeper-diving ROV is to attach dive lights to the vehicle. Small, high-intensity LED or halogen dive lights are so bright and power-efficient that they often burn longer than one hour on a set of self-contained batteries. As a result, these lights would not be a burden on the onboard power distribution system, and there's no need to worry about waterproofing wire connections between the lights and the main electronics can. It would be a simple matter of changing both the ROV and lamp batteries at the same time, as part of the surface interval servicing. Pelican, Underwater Kinetics, Princeton Tec, and other dive light manufacturers make excellent dive lights that are rated to 100 meters (approx. 325 ft) or deeper. For all lighting options, you will have to fabricate an adjustable bracket to mount them on the frame.

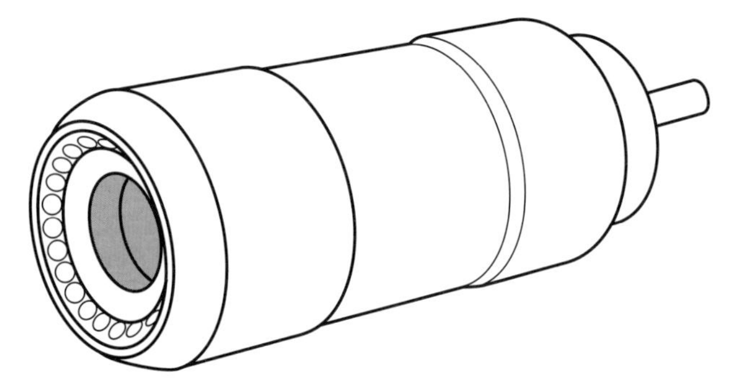

Figure 9.33: COTS Camera and Lights

DeepSea Power & Light's Mini-Sealite camera features built-in LEDs for illumination.

LEDs are now being integrated onto some video cameras. Typically, a ring of LEDs surrounds the lens of the camera, illuminating the area directly in front of the lens. (See Figure 9.33.) These can work well in clear water (e.g., a swimming pool), but are likely to produce unacceptable backscatter in cloudy or murky water. You may need to wire these LEDs into the vehicle power distribution system, and that may require some sort of voltage regulation. You could also make up your own array of encapsulated LEDs, attaching them at locations on the frame away from the camera to reduce the "whiteout" or "blizzard" effect caused by excessive backscatter of light from particles in the water.

Halogen lamps are not as power-efficient as LEDs (20–50 watts per lamp), but generally have higher lux, or lumen, ratings. You can get halogen lamps from any hardware store because they are used commonly in track lighting. They come complete with nice reflectors (either spot or flood), but usually need to be pressure-proofed by potting or placed in a diver's flashlight housing. However, instead of doing the latter, it is often better just to buy a halogen dive light with a wide-beam angle, remove the batteries, drill a hole in the housing, and run a water-blocked wire to the pressure can bulkhead connector to power the light.

4.5.2. Choosing and Installing a Compass and Depth Gauge

To navigate in places where the pilot can't observe the ROV directly, it really helps to supplement video information with data from a compass and depth gauge. The quickest and easiest way to do this is to mount a compass and depth gauge with easy-to-read displays in a location where their displays can be viewed through the camera without blocking too much of the camera's view. This simple approach is used to provide readouts from a compass and depth gauge for the *SeaMATE* ROV described in Chapter 12.

Inexpensive compasses, such as those spherical, oil-filled, self-leveling compasses made for use on car dashboards, will work fine as long as they are plastic on the outside (to resist corrosion), fluid-filled on the inside (to provide pressure compensation), and able to level themselves for more accurate reading. (See Figure 9.34.) An internally gimbaled (self-leveling), liquid-filled marine compass is most reliable, because it's less likely to stick, and the heading marker divisions are easy to see. Whatever style of compass you select, it should be easy to read on the video monitor.

Some types of scuba divers' compasses can work well, but most do not have built-in gimbals for self-leveling, and many are difficult to read accurately with a camera.

As for depth gauges, the older style divers' depth gauges with a needle that rotates to point to the depth (rather than the newer digital displays) can be easier to read in a video image. They may be less expensive, too, particularly if you can find a scuba diver who's upgrading to digital and would be willing to donate a used depth gauge to a good cause.

Unfortunately, this simple video approach does not lend itself well to the automated control of depth or heading necessary for AUVs or any ROV with a depth or heading autopilot feature. These automated systems are best implemented using depth sensors and compasses with electronic outputs. Techniques for automating these and other electronic sensors will be described along with other more advanced control options in the next section of this chapter.

Image courtesy of Dr. Steven W. Moore

Figure 9.34: An Automobile Dashboard Compass Adapted for Underwater Work

In this video image frame captured from ROVing Otter's *camera, you can see a fluid-filled magnetic compass ball mounted on a piece of PVC pipe being used to navigate across a rocky undersea reef near Monterey, California.*

The compass was originally intended for automobile dashboards and was purchased from an automotive supply store. It came with fluid inside to help damp out vibrations, but that fluid also provides convenient pressure compensation under water.

The green laser beams projecting into the distance are used to measure the size of objects in front of the ROV. The beams are 10 centimeters apart and precisely parallel, so they project a known size scale—two green dots 10 centimeters apart—on any object directly in front of the ROV. In this image, nothing blocks the beams, so they converge in the distance.)

The brown ruffly things growing on the rocks (note the one illuminated by the ROV's light to the right of the compass) are colonies of tiny marine animals known as bryozoans.

5. Advanced Control Options: Moving Beyond *SeaMATE*

The motor control system and navigational sensor system described in the previous sections work quite well for *SeaMATE* and similar ROVs, which are designed for simple missions emphasizing exploration of shallow ocean or lake environments. But what if you want to go beyond what *SeaMATE* offers? What if you need a longer, more flexible tether for deeper dives? You know you can't get it if you have to include a

Image courtesy of International Submarine Engineering Ltd.

Figure 9.35: A Typical Work Class ROV Control System

The advanced control required by heavy-duty work class ROVs operating hundreds of meters below the surface is well beyond the basic controls described in Section 4. Section 5 introduces concepts and techniques you need to understand before you can begin building advanced control systems.

separate pair of beefy, current-carrying wires for each thruster. What if you want joystick control of variable thruster speed for smooth, precise maneuvering? What if you want to move beyond ROVs altogether and step into the future by building your own AUV? This final section of the chapter provides the basic technical foundations needed to experiment with these possibilities and more.

One great thing about the more advanced control concepts, techniques, and terminology presented here is that they are universal; they'll work as well for terrestrial robots and automatic pet food dispensers as they will for underwater vehicles, so they can be employed in a wide range of fun and useful projects. As a result, the skills you develop working with these techniques can improve your chances of landing interesting work in any technology-related field.

A thorough description of all control techniques that might be useful on ROVs or AUVs is well beyond the scope of this book; however, this chapter does provide an overview of most of the key concepts and enough associated vocabulary to give you a solid foundation for learning more about advanced controls on your own through books, websites, conversations with retired engineers who may live next door, and other resources. This is where the learning starts to get especially fun and interesting, because advanced control techniques are one place where your own ingenuity and creativity can really shine through. It's where you can dream up new ways to customize a vehicle's capabilities, maybe even inventing something completely new!

5.1. Do You *Really* Need Advanced Control?

It's time for a reality check—do you really need (or even want) advanced control? While complex, high-tech control options sound exciting and have some obvious advantages in enabling your vehicle to do things it could not otherwise do, they also have their costs, both financially and in terms of the time, effort, and expertise needed to design, construct, and troubleshoot them. Therefore, it's important to decide early in the design spiral how much time, effort, and money you and your team want to invest in control systems. This section highlights some of the main advantages and disadvantages of opting for advanced automated control.

5.1.1. Disadvantages of High-Tech Control Systems

While it is tempting to assume that a technologically more advanced control system is always better, this is not normally the case. Here are some good reasons to avoid rushing prematurely into automated control:

- As control systems get more sophisticated, they tend to get much more expensive. There's almost no technical limit to how advanced (and therefore how expensive) a control system can get, so without careful monitoring, you can easily burn up all your budget on control electronics and not have enough left over to meet other critical mission needs.

- As control systems get more complicated, more things can go wrong. Remember, a simple control system that works is much more useful than a complicated one that doesn't.

- When automated feedback control systems fail, they often fail in dramatic and disastrous ways. In fact, a malfunctioning control system can be worse than no control system at all, because it can send a vehicle lurching wildly out of control, possibly destroying the vehicle, damaging something in its path, or even injuring someone nearby.

- You already have in your possession—at no additional charge—one of the most sophisticated control systems imaginable. It can quickly process phenomenal amounts of incoming data from a variety of finely tuned sensors. It can make comparisons between actual and desired conditions, make dozens of sophisticated decisions every second, and issue commands to initiate various vehicle actions. Better yet, it can continually monitor the effectiveness of its control and make adjustments to optimize control system performance over time. (In other words, it can learn and it gets better with practice.) That control system is, of course, your brain. It may not seem very high-tech to use a human instead of a computer for feedback control, but it can be much more effective. And when your mission is at stake, being effective is what matters most. Depending on the mission, you may not have the option of using human control (for example, an AUV will not have that option), but if you do, think twice before jumping into automated control.

5.1.2. Benefits of High-Tech Control Systems

In spite of the caveats above, if you can afford to splurge a bit on any particular aspect of your vehicle's design, then control system technology is probably a good place to do it. There are many advantages to advanced control if you have the time, money, expertise, and other resources needed to get it all working properly. Here are some of the main advantages:

- Fully automated control may be required for some types of missions. For example, if you are sending an AUV deep under the Arctic ice cap where a boat and tether can't follow, it will have to operate entirely on its own without a human pilot. In this case, if you can't equip your vehicle with fully automated control, you can't expect it to complete the mission.
- In some cases, the simple control strategies discussed in the previous section are not sufficient to meet the performance needs of the mission. For example, motor control with DPDT switches is fine if all you need is FORWARD-OFF-REVERSE control, but if you need precise control of thruster speed for careful maneuvering in tight quarters, you'll need a more advanced motor control scheme. Likewise, controlling the brightness of video lights or controlling the force with which a gripper jaw closes on a delicate sample will require more sophisticated control than can be done with the methods described so far.
- Advanced control strategies can make underwater tasks more intuitive for a human operator. For example, simple maneuvers like spinning to the right may require the coordinated action of two or more thrusters. A control system that translates intuitive joystick movements into corresponding thruster actions can make the vehicle much easier for the pilot to control. Likewise, reaching straight forward with a manipulator arm might require the coordinated action of three different manipulator joints. An appropriate control system could handle the joint angle details, so the human operator could focus on reaching forward rather than having to figure out the three different joint angles needed to produce a straight forward motion.
- In vehicles with many complex systems operating simultaneously, automating some simpler control tasks may free the human pilot to concentrate on the more difficult tasks. For example, an advanced work class ROV may need to control most or all of the following subsystems (and maybe more) simultaneously:

1) thrusters (commonly six or more)
2) manipulator arms/grippers (commonly one, sometimes two)
3) several cameras, each with tilt/pan/zoom/focus/iris control
4) video lights (on/off/brightness)
5) active ballast systems
6) lasers for object measurement and distance estimation
7) biological, chemical, or geological sampling equipment
8) welders, saws, wrenches, or other tools

- It would be difficult or impossible for a single pilot to control all of these at once without some help from automated control systems. While the vehicle you build may not have all of these subsystems, it could have many of them. Heck, it might include some control needs not normally found on a work class ROV, like control of a top-secret weapon designed to trounce your arch-rival during an underwater robot competition.
- Control system development is a fabulous learning experience. The skills you gain in designing, building, testing, and improving control systems will be useful skills that you can apply to a wide range of future projects and career opportunities. These include general problem-solving skills, information-gathering skills, and troubleshooting skills, as well as more specific abilities in areas like electronics, computer programming, and robotics.
- Control systems can grow with you. Start simple at a level that is challenging, yet comfortable. Then, as your experience and confidence grow, add more advanced control features to your projects.
- In competitions where vehicle size, weight, voltage, current, and other parameters are often constrained by the rules of the competition, precise vehicle control may offer one place where your team can gain an edge in the competition. (Just remember that time spent practicing with a simple open-loop control system, where a human brain closes the loop, may be more valuable than time spent building a fancy closed-loop control system!)
- Finally, humans like to control things, and that makes control system projects especially fun!

5.2. A Peek at the Possibilities

If you decide to go there, the world of advanced control will open up a huge range of options for providing your vehicles with custom functionality. The last half of this chapter is devoted to introducing you to a smorgasbord of different tools and techniques you can mix and match with each other (and with additional ideas you discover through books, websites, and conversations with other underwater robot enthusiasts) to give your vehicles a wide variety of advanced capabilities.

The ROV illustrated in Figure 9.36 features a pilot's control station, with a joystick for smooth, precise, and intuitive control of vehicle movement. It also has a variety of buttons and knobs for controlling things like the brightness of the video lights or switching on a depth-control autopilot.

This pilot station also includes a video screen that can display depth, battery voltage, leak warnings, and other valuable data superimposed on the live video image of fascinating marine life. (See *Section 5.11. Display Options* for more technical

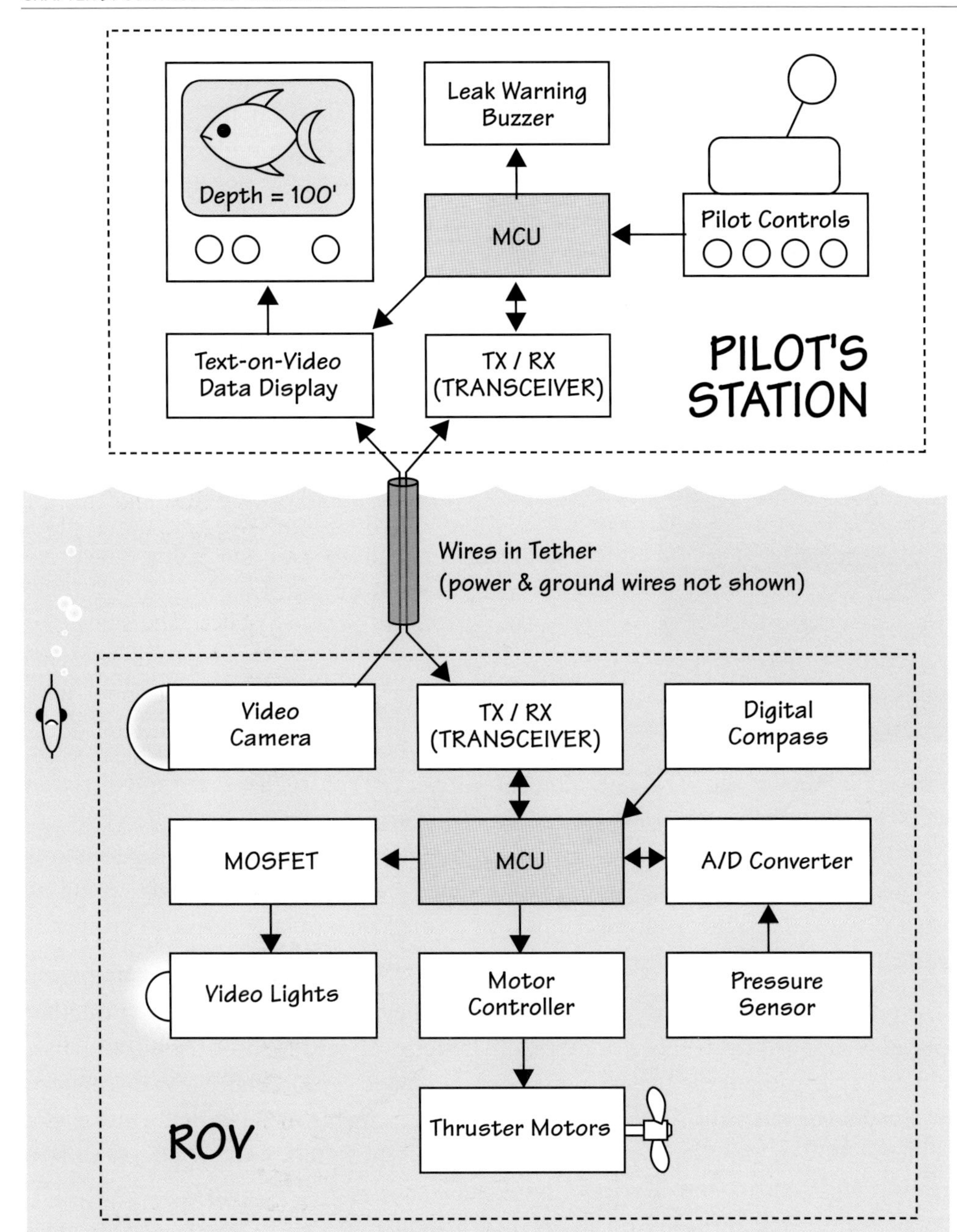

Figure 9.36:* SeaMATE *on Steroids

The hypothetical ROV represented in this diagram features a number of sensors, data display options, and advanced control capabilities that go beyond the simple navigation and motor control systems used in SeaMATE *(Chapter 12).*

Think of this figure as a roadmap to the rest of this chapter, which will focus in each section on a different part of this figure, beginning with the microcontrollers (MCUs) in the gray boxes.

information about text-on-video, also called video overlay.) Of course, the data display can be switched off, if you just want to go sightseeing. As a precaution, there's an audible warning buzzer that will sound anytime a leak is detected, even if the data display is turned off.

All of these control and data display features in the pilot's station are coordinated through a small computer known as a **microcontroller unit**, or **MCU** (the upper gray box in Figure 9.36). Special devices called transceivers (TX/RX) can be used to relay communication signals through the tether. This enables the MCU in the pilot's console to be used to relay information to a second MCU (the lower gray box in Figure 9.36) located aboard the ROV that is working beneath the water's surface. (TX is a standard abbreviation for transmitter and RX is standard for receiver; TX/RX indicates the two are combined into a transceiver.)

This second MCU serves as the main brain of the ROV and coordinates all of the underwater navigation and control operations. It receives information about the ROVs location from a variety of sensors, including a digital electronic compass and an analog pressure sensor used as a depth gauge. This microcontroller unit, like most

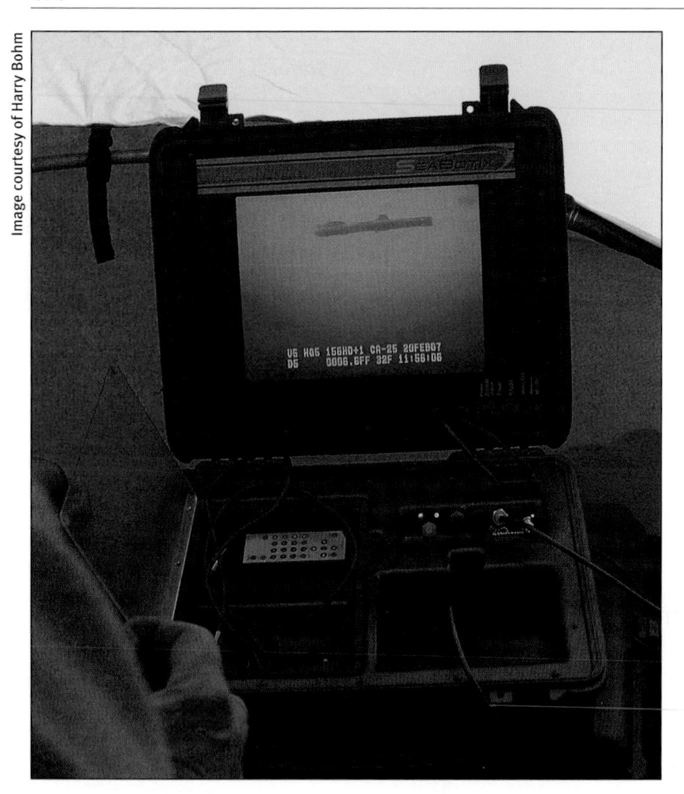

Image courtesy of Harry Bohm

Figure 9.37: Text-on-Video At Work

other computers, is designed to work with digital signals, so an analog-to-digital (A/D) converter is required to translate the analog signals from the pressure sensor into digital signals the MCU can understand. All of the navigational data is gathered by the ROVs MCU and sent up the tether to the surface MCU, so it can be added to the data displayed on the screen, whenever the pilot wants to see it.

In response to control commands sent down the tether from above, the ROV's microcontroller unit can turn the video lights ON or OFF and adjust their intensity. These bright lights require more electrical power than the MCU can supply by itself, so a type of transistor switch called a MOSFET amplifies the control signals to make them powerful enough to run the lights. (Refer to Section 5.12.1 for more information about these transistors.)

Likewise, weak but precise movement commands from the MCU are amplified by a motor controller circuit to give them enough oomph to drive several powerful thruster motors on the ROV. This motor controller can control the speed and direction of each thruster motor and also features built-in current-overload protection that will protect both the controller and the motors from damage if one of the propellers gets tangled in something and stalls a motor. Similar motor controller circuits could be used to control a camera tilt mechanism or a small manipulator arm and gripper, if you chose to add those features to this ROV at a later time.

The power system on this ROV is not shown in the diagram, so it's not clear whether power is supplied from above or carried in batteries aboard the ROV. You could do it either way. If you put the batteries on board and made your tether easy to disconnect, you could convert your ROV into a hybrid vehicle capable of functioning either as a tethered ROV or an independent AUV. Of course, you would need to add a few more navigational sensors and develop a more sophisticated MCU program, so the AUV could control the thrusters on its own in response to data from the navigational sensors.

The pages that follow will answer some of the questions you may be asking yourself about how you could add similar features to your own vehicles, including:

- What, exactly, is a microcontroller (MCU), and how is it used?
- How does the MCU communicate with other devices on the vehicle?
- How do the transceivers (RX/TX) send information over the tether?
- How can lights, lasers, grippers, and other devices be controlled?
- What is a motor controller and how does it work?
- What kinds of sensors can be added to a vehicle?
- How do I use an A/D converter, why is it needed, and how is it used?
- What options are available for displaying sensor data for a pilot?
- How can these devices and techniques be adapted to build an AUV?

5.3. Microcontrollers

If you really want to get serious about advanced control, the place to start is with microcontroller units (MCUs). These small integrated circuits are essentially tiny computers optimized for controlling machines. Think of them as little robotic brains. While not absolutely necessary for all forms of advanced control, they are by far the easiest and fastest way to make the leap into sophisticated vehicle control systems.

With microcontrollers, it's possible to give a vehicle the ability to monitor sensors, communicate with the pilot or other devices (including other MCUs), make decisions, and initiate actions—in a sense, providing the vehicle with a primitive form of intelligence. Microcontrollers can be used to add a depth control autopilot, send multiple complex commands over a single pair of wires (or optical fibers) in a long tether, and even control a completely self-contained AUV that will go out to sea on its own, complete a mission, then return home. Best of all, learning to use microcontrollers is really fun!

Although they function like little computers, microcontrollers are much smaller, much simpler, and much less expensive than a regular desktop or laptop computer. Many cost less than $5 (U.S.), and some even cost less than $1, though the most user-friendly ones are more expensive—typically about $50. They have no keyboard, monitor, or mouse, but they *are* programmable, and that's what makes them so versatile and so useful for creating customized control features.

Although you may not have heard of a microcontroller before, you and your family very likely own and use dozens of them every day! There's at least one microcontroller embedded inside almost every modern electronic gizmo—cell phones, TV remote controls, automobile anti-lock brakes, vending machines, kitchen appliances, digital wristwatches, and talking toys, to name just a few.

Because they are literally *embedded* inside these devices, they are included in an important class of control devices called **embedded controllers**. It would be difficult, if not impossible, to find a commercial ROV or AUV that did not use several microcontrollers or other types of embedded controllers to coordinate communication with its human operators, regulate motor speed, collect and transmit data from sensors, provide autopilot control for depth or heading, and perform countless other functions.

TECH NOTE: MICROPROCESSORS VERSUS MICROCONTROLLERS

Micro*processors* are sophisticated integrated circuits used as the brains in regular computers. They are optimized for extremely high-speed computation, raw number-crunching power, and versatility, so they can run an almost endless variety of software packages from word processing programs to graphics-intensive computer games. Typically, they are expensive, require a lot of power, and rely heavily on lots of additional hardware (motherboard, hard drive, memory chips, keyboard, mouse, monitor, etc.) to do their work.

Micro*controllers*, on the other hand, are optimized for small size, low cost, and low-power control of machines. They usually run only one simple program over and over again. A good example would be the microcontroller in a vending machine. It simply monitors the coin collector to make sure somebody has put the correct amount of money into the machine. It also pays attention to which candy bar button was pushed, then switches on the right motor for a few seconds to deliver the selected treat to the buyer. Of course, a regular computer could do this, too, but it would be a waste to use a computer worth hundreds of dollars to do this simple task, which can be done as effectively by a microcontroller costing only a couple of dollars. For most ROVs and AUVs, microcontrollers are the logical choice, although a few very advanced vehicles also include one or more microprocessors.

Image courtesy of Dr. Steven W. Moore

Figure 9.38: BASIC Stamp 2 (BS2) Microcontroller

The BASIC Stamp 2 (BS2) microcontroller, made by Parallax, Inc., is a small, learner-friendly, and programmable, electronic "brain" that can be used for simple ROV or AUV control functions.

Physically, most microcontrollers resemble any other integrated circuit or electronic chip in that they look like a small, rectangular bit of black or gray plastic with anywhere from 8 to 40 metal wires, called "pins," sticking out of them. A few others, like the model shown in Figure 9.38, look like tiny printed circuit boards covered with itty-bitty electronic components.

Most MCUs are optimized for low cost and functionality and require some specialized expertise to use properly, but a few are designed specifically for beginners and are learner-friendly. Though these beginner microcontrollers can be much more expensive than other MCUs, their ease of use is worth the extra money when you are first learning to use MCUs.

Examples of good MCUs for beginners include the BASIC Stamps by Parallax, Inc., the OOPic by Savage Innovations, the PICAXE by Revolution Education, Ltd. and the LEGO® MINDSTORMS or ROBOLAB kits, which include an easily programmable microcontroller or microprocessor inside the LEGO "brick". Once you have gained some experience with these education-oriented MCUs, you can shift to other microcontrollers that are smaller, faster, more powerful, and less expensive. The extensive PIC series of microcontrollers by Microchip and the AVR microcontrollers by Atmel are among the most popular of the MCUs in this more advanced category.

5.3.1. Programming a Microcontroller

Just like a regular computer, an MCU performs its job by following a set of instructions (called a **program**) that's been loaded into its memory. The process of creating these instructions and loading them into the MCU's memory is called "programming" the MCU. It is analogous to programming a computer.

The details of how you write and load a program vary somewhat, depending on the manufacturer and the type of microcontroller. Typically, you start by purchasing a programming kit. The kit includes a device called a "programmer," which is a circuit board (usually in a plastic box) that has a socket for the microcontroller, a cord to supply power, and a cable that connects the programmer to a regular laptop or desktop computer. There may also be a CD with programming software and tutorial. In some cases, the programmer is not needed—the MCU can be programmed after being inserted into its regular circuit (for example, inside your vehicle), as long as the circuit designers included a place to plug in the MCU programming cable from the computer. This is commonly called **in-circuit serial programming** or **ICSP**.

On the desktop or laptop computer, you would then open up the programming software provided by the MCU manufacturer (or a third party) to write/edit your MCU program and download it into the MCU through the programming cable. Many MCU manufacturers provide this software as a free download, to encourage purchase of their MCUs, but others charge a fee for it. Working within this software, you write the programs in whatever programming language is available for that microcontroller. Sometimes you need to use the manufacturer's special language. Other times you can use a general-purpose language, such as C, Java, or BASIC. Instructions for writing and loading programs into the MCU, plus sample programs to assist you in learning how to program the MCUs, are usually provided by the microcontroller manufacturer.

Each program enables the microcontroller to perform a few simple but useful tasks. (See *Tech Note: A Sample Microcontroller Program.*) For example, you can write a

TECH NOTE: A SAMPLE MICROCONTROLLER PROGRAM

The following is a section of a very short MCU program written for a BASIC Stamp 2 (BS2) microcontroller. This particular program is written in a programming language called PBASIC, which is the most common language used to program BASIC Stamp microcontrollers. The program instructs the BS2 to blink a small LED (connected to pin #3 of the MCU) repeatedly. This program might be a small portion of a larger program used to flash a warning light in the event of some malfunction. This program is only a few lines long. A regular program on a home-built ROV would normally be several dozen to several thousand lines long, depending on the complexity of the system.

```
MAIN:
HIGH 3
PAUSE 500
LOW 3
PAUSE 500
GO TO MAIN
```

The first line, "MAIN:" is just a label in the program that provides a recognizable destination for the "GO TO MAIN" statement found at the bottom of the program. The "HIGH 3" tells the MCU to raise the voltage on pin #3 to 5 volts. (HIGH = +5V, LOW = 0 V on a BASIC Stamp 2.) Since the LED light is wired to this pin, raising the voltage turns on the LED. The "PAUSE 500" tells the MCU to wait 500 milliseconds (which is one-half second) before proceeding to the next line of the program. This keeps the LED illuminated for 0.5 seconds. The "LOW 3" returns the voltage on pin #3 to zero, thereby turning off the LED. The second "PAUSE 500" keeps the light off for 0.5 seconds. The GO TO MAIN statement causes the program to return to the top and repeat over and over again instead of stopping each time it reaches the last instruction.

program that turns a microcontroller into a depth autopilot by telling it how to monitor readings from a depth sensor, compare those to a "desired depth" stored in memory, and control the speed and direction of vertical thrusters to make the actual depth match the desired depth. If the program works properly, the vehicle will dive down whenever the sensor says the depth is too shallow and will climb toward the surface whenever the sensor says the vehicle is too deep.

Some microcontrollers can be programmed only once. These are known (appropriately) as **One Time Programmable (OTP) microcontrollers**. Others (notably those with "flash" or **EEPROM** memory) can be programmed easily over and over again. Still others can be reprogrammed, but must have their memory erased for several minutes with a powerful UV light between programming sessions. For typical ROV or AUV projects, you'll want to use the flash memory versions, so you can quickly and easily change your programs to correct errors or to add new features without having to buy a new microcontroller or deal with UV erasing. Most of the beginner MCUs use flash memory, so they are easy to reprogram over and over again.

As mentioned, some of the newer microcontrollers offer "in-circuit" serial programming, which means that you don't have to unplug the microcontroller from your ROV circuit in order to reprogram it. This in-circuit programming is a very helpful feature for ROV/AUV development.

5.3.2. MCU Pin Functions

The different metal **pins** sticking out of a microcontroller have different functions. For the detailed function(s) of each pin on any particular MCU, you need to look at the technical datasheets for that model, because pin functions differ from one type of MCU to the next; however, some common patterns emerge. For example, on every

MCU, at least two pins are required to connect power to the microcontroller—one for connecting the positive wire from the power source and the other for connecting the ground wire. Two or three additional pins are needed for the connection made temporarily to a regular computer for programming the MCU. There is also usually one "reset" pin that can be used to restart the microcontroller in case of a programming glitch (much like rebooting a regular computer after a software crash). Most MCUs also have a couple of pins for connecting a crystal or other oscillator circuit that provides the timing pulses the MCU needs to run properly. On some beginner MCUs, the oscillator circuitry is already built in, so you don't need to worry about it.

Most of the remaining pins are available for general-purpose input and output of digital data. (The difference between digital and analog signals will be explained shortly.) These are known as **I/O pins** (for Input/Output). They are the pins through which the microcontroller interacts with its world. It can use them to gather data from sensors, communicate with other devices, or control other devices. With very few exceptions, I/O pins are allowed to be in one of only two possible voltage states. One of these states is designated as the low or "logic zero" state and is usually zero volts (relative to ground). The other is designated as the high or "logic one" state and is usually about +3 volts or +5 volts (relative to ground), depending on the type of microcontroller. The program you write for the microcontroller usually includes some instructions that tell the microcontroller which of its I/O pins to use for inputs and which to use for outputs.

When an I/O pin is configured as an input, the microcontroller can monitor the voltage present on that I/O pin. An input pin is generally connected to another circuit or device that will change the pin voltage in response to specific external conditions. That's how the microcontroller program can monitor signals from an external device. For example, if an I/O pin is connected to a pushbutton circuit that normally holds the I/O pin voltage high (e.g., at +5V) but pulls it low (e.g., down to zero volts) whenever the pushbutton is being pressed, then the microcontroller program can know whether or not the button is being pressed by checking to see whether that input pin is in a high or low state. This provides a way for you to talk to your microcontroller and tell it to do something. What it does in response to the pushbutton being pressed depends on what you have programmed it to do.

When an I/O pin is used as an output, the program can send signals out to control other devices by putting the pin in either the high or low state. This allows the microcontroller program to provide ON/OFF control (usually through a transistor or other power-amplifying circuit as described later in the chapter) to a light, relay, motor, or other device. The program can also use an output pin to send commands or data encoded as a series of pulses (periods of high output alternating with periods of low output, with particular timing). For example, a series of carefully timed pulses is one way that two microcontrollers can talk to one another or to other smart devices. Note that this provides a powerful technique for communicating over a tether—you can put one microcontroller in the pilot's control console at the top end of the tether and one on the ROV at the bottom end of the tether, and the two microcontrollers can relay information back and forth, using coded messages transmitted as series of pulses. This can greatly reduce the number of wires in a tether, because now you just need one fat pair of wires to send power to the ROV and one pair of skinny communication wires (one for pulses and one for reference ground), so that the MCU on the surface can tell the MCU on the ROV how to distribute the power delivered through the fat pair of wires. Details of how this can be done are provided later in this chapter.

5.3.3. MCU Limitations

Microcontrollers make wonderful mini-brains for ROVs, AUVs, and other robots, but they do have their limitations. For one thing, they are not nearly as fast or powerful as a regular computer when it comes to processing information, nor do they have anywhere near as much memory. In most ROVs, these limitations are not a problem; however, if you find that you need more processing speed or more memory, you may want to look into some of the stripped-down PCs known as **PC-104 boards**. These embedded PCs are made by many manufacturers. Just search the web for PC-104. They include the brains and some of the memory of a regular desktop computer, but save lots of space and power by eliminating the keyboard, mouse, monitor, and other non-essentials. Of course, it's possible to attach those peripherals temporarily to configure the system, load software, and program the computer, but then they are removed to conserve space and power. PC-104s run regular PC operating systems like Windows, Mac OS, or Linux, and can run regular PC software, too.

Microcontrollers have another significant limitation that's also shared by PC-104s. When it comes to electrical power output, these embedded systems are wimpy. That means if you want your microcontroller or embedded PC to control thruster motors, video lights, grippers, or any other device that requires significant electrical power, you will need to give it some help. Here's the problem: A typical microcontroller output pin can deliver a maximum of 3–5 volts and about 20 milliamps (mA) of current. PC pins are sometimes a little better, but only a little. This is enough power to communicate directly with most other electronic devices or to light a tiny LED indicator, but not much else. By way of comparison, a typical thruster made from a small bilge pump motor will require 12–24 volts and draw several thousand milliamps. Recall that electrical power is voltage times current, so a motor requires over 1,000 times as much power as the microcontroller pin can deliver! Fortunately, there are devices, including transistors, that can amplify low-powered MCU signals, giving them enough oomph to control motors and other power-hungry devices. Techniques for doing this are described later in this chapter.

5.4. An Introduction to Electronic Signals and Communication

To do its work, a microcontroller must be able to communicate with devices around it. For example, the microcontroller coordinating propulsion on an AUV may need to receive information from navigational sensors and send commands to other circuits that regulate thruster motor speed. Other types of electronic devices may need to communicate with each other, too. For example, a video camera needs to convey video images to a television or computer monitor for display.

For electronic devices to communicate effectively, two things must be present. First, there needs to be some medium—light, sound, radio waves, electricity, or whatever—capable of conveying information from one place to another. Second, there must be some agreed-upon language (called a **communication protocol**) that the devices can use to encode messages for transmission through that medium.

For example, if you wanted to send commands through an optical fiber to control thruster speed, you'd need to use light as the medium (since that's what can be sent over an optical fiber). You would also need to choose some characteristic of light (color, brightness, number of flashes, etc.) that you could change electronically in particular ways to encode different messages. For example, if you decided to use color, you could hook up some colored LEDs at the pilot's end of the fiber and some color-sensitive light detectors at the ROV end of the fiber. Then you could adopt a code in

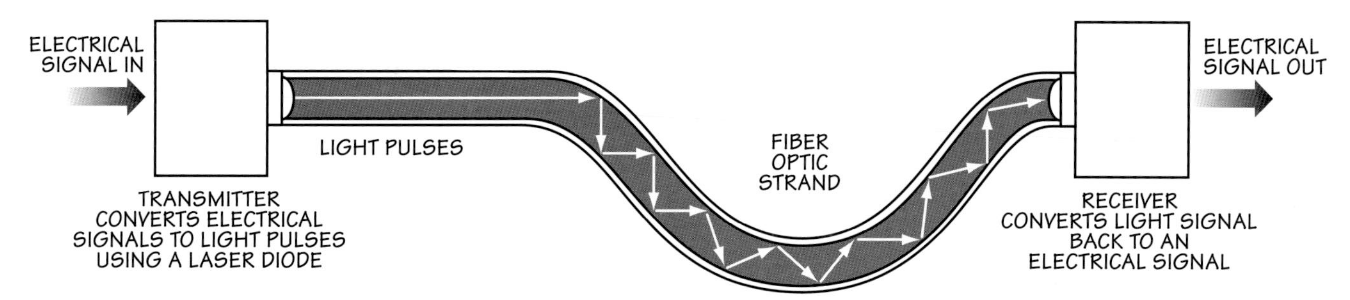

Figure 9.39: Fiber-Optic Signal Transmission

which purple light means "go forward," green light means "go backward," and blue light means "stop." As long as both the equipment sending the colors and the equipment receiving the colors knew how to interpret those colors, you would have an effective system for communicating commands to the thrusters.

You are always free to invent and use your own custom languages and codes, like the hypothetical colored light example above. However, it's usually much easier to use a pre-existing, standardized communication protocol, particularly one that's already in wide use. Doing so will allow you to take advantage of abundant, reliable, and relatively inexpensive hardware and software that already support those communication protocols. In the case of fiber-optic signals, for example, it's usually a pattern of light flashes, rather than colors, that are used to encode information.

This section of the chapter explores some of the most common standardized communication protocols used to send information between electronic devices. It begins with an introduction to analog and digital signals, then discusses analog and digital data transmission, and concludes with a description of analog-to-digital signal conversion.

5.4.1. Analog Versus Digital Signals

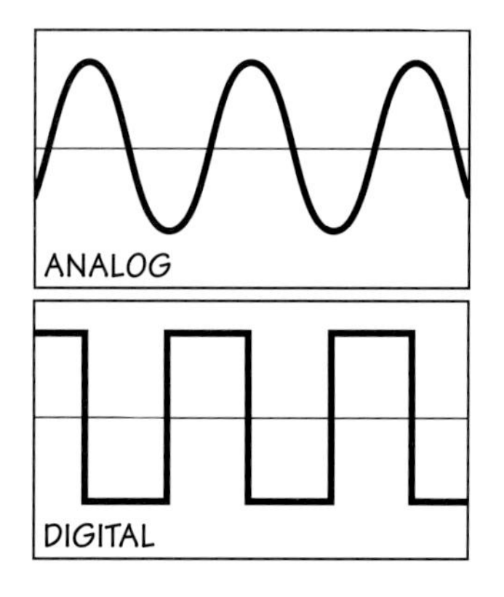

Figure 9.40: Analog Versus Digital Signals

Analog signals (top) vary smoothly and continuously, whereas digital signals (bottom) usually are restricted to one of only two values (high or low) at a time.

A **signal** is anything that changes with time and can be used to convey information. In most electrical systems, including those in ROVs and AUVs, signals usually take the form of fluctuating voltages or currents in metal wires. In optical fibers, the signals typically consist of changing light levels.

The signals that you will encounter as you work on underwater vehicles can be divided into two broad but important categories: analog and digital. An **analog signal** is one that can vary smoothly and take on an infinite number of values between its minimum and maximum values. A **digital signal**, on the other hand, is limited to a small number of distinct values—usually only two. Digital signals are by far the more common of the two in modern electronic communication, although analog signals still play an important role.

Many electronic sensors use analog signals to report their data. For example, an analog temperature sensor might generate 0.1 volts per degree Fahrenheit. If an output of 6.83 volts was measured from this sensor, and you knew the sensor generated 0.1 volts per degree F, then you would know the temperature was 68.3°F. Since temperature varies smoothly and continuously between different values, the output of this sensor would also vary smoothly and continuously. (See Figure 9.40.)

Digital signals are different. In an idealized digital world, there are no gradual transitions between values; only a small number of discrete values are allowed. In fact, the modern digital world is based on only two values. Symbolically, these values are represented by 1 and 0 (one and zero). In reality, they usually take the form of a particular voltage being present or absent on a wire, or of light being present or absent in an optical fiber. Electronic circuits lend themselves naturally to this two-value digital system, because they can so easily and naturally be switched between two distinct states: ON and OFF. That's one reason there are so many digital electronic devices today. An even bigger impetus for the growing popularity of digital communication formats is that digital signals are much less susceptible than analog signals to communication errors. For this reason, digital communication is generally preferred over analog communication whenever possible. (See *Tech Note: Digital Noise Immunity.*)

5.5. Analog Data Transmission

Although digital communication is preferred over analog communication for most routine data transmission needs, analog data transmission still plays an important role, particularly in the transmission of signals from analog sensors. These sensors tend to produce electrical signals (usually in the form of voltage, but sometimes charge, current, or resistance) that are proportional to the quantity being measured. When used to measure temperature, pressure, speed, direction, or other quantities that can vary smoothly and continuously, the output of these sensors is naturally an analog signal.

The simplest way of sending an analog signal over a wire is to use an analog voltage. A depth sensor, for example, might generate a voltage signal of 1 V for every 100 feet of depth, so if you measured 0.87 V coming from the sensor, then you would know your vehicle was 87 feet deep. Though simple, this method suffers from signal degradation with distance and is susceptible to electrical interference (noise) from other nearby electronics, radio broadcasts, and so on. A somewhat more robust method is to use current, rather than voltage. This is because current does not diminish with distance through a wire, as voltage does. (See *Section 5.5.3. Use the 4-20 mA Protocol.*)

Another common way of encoding analog signals is to use high-frequency oscillations, then modulate (i.e., adjust) either the amplitude (size) or frequency (cycles per second) of the oscillations in a way that represents the analog value you wish to send. This is how sound (an analog air-pressure signal) measured by a microphone is sent

TECH NOTE: DIGITAL NOISE IMMUNITY

In electrical circuits, the term "noise" refers not to sound, but to the inevitable random fluctuations that arise in electrical signals due to the random movement of electrons and other causes. Noise can be a big problem in communication systems because these random fluctuations can obscure the intended signal. Digital signals are much more resistant to corruption by noise than are analog signals.

To understand why this is so, consider the following example. If you receive a 4.9 V signal in an analog communication link, there's little choice but to assume that 4.9 V was sent, because any value is as valid as any other. But if you receive a 4.9 V signal in a digital link where the only two allowed values are 0.0 V and 5.0 V, you can be pretty confident that the 4.9 V signal received was originally sent as 5.0 V and that the 0.1 V difference is due to some random noise picked up along the way. You can therefore correct that error before you relay that signal to the next part of the system. After being passed from one receiver to the next many times, an analog signal can accumulate enough random errors to be very different from the original signal, whereas a digital signal, which has been stripped of these tiny errors at each step along the way, will arrive at its destination unscathed. (Techniques for reducing analog noise are discussed in Sections 5.5.1 – 5.5.4.)

In addition, digital signals lend themselves to error-checking routines that can detect missing "1's" or "0's" in the data stream. These methods are analogous to methods used with important multi-page documents, which will often specify how many pages are in the document, to make it easier to detect when a page is missing. For these reasons, digital communication links are much less prone to communication errors than are analog links.

over conventional AM (amplitude modulation) or FM (frequency modulation) radio broadcasts. For transmission through a tether, AM and FM signals can be transmitted through coaxial cable. (See Section 5.5.1.)

Because of an analog signal's high susceptibility to noise (i.e., signal degradation by contamination from other signals), it's critical to minimize the opportunity for noise to creep into analog data transmissions, particularly when those signals are traveling long distances (more than a few centimeters). The following sections cover various methods for protecting signals from noise.

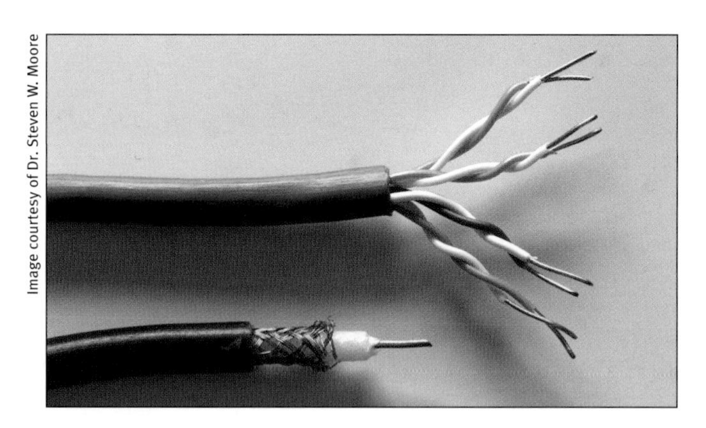

Image courtesy of Dr. Steven W. Moore

Figure 9.41: Cables for Analog Signals

These two cable types, and others, have been used successfully in the tethers of small ROVs to transmit analog and digital signals. The upper cable is Category-5 ("Cat-5") computer networking cable, which contains four twisted pairs of smaller wires. Though optimized to transmit digital signals between computer equipment, each twisted pair can also transmit analog signals, such as NTSC video.

The lower cable is RG-58 coaxial cable. It contains a central copper wire surrounded by insulating plastic (white), which in turn is surrounded by a layer of woven copper mesh and another layer of insulation (black). When grounded, the mesh acts as a shield to protect the signal on the core wire from external electrical interference. This makes it especially good for sending sensitive analog signals.

5.5.1. Use Coaxial Cable

If you wrap the wires carrying an analog signal in some foil, wire mesh, or other electrically conductive material connected to ground, you will partially shield the signals from external noise. This is the primary purpose of coaxial cable, which consists of one central wire surrounded by a (usually) grounded metallic mesh or foil that acts like a shield.

5.5.2. Use Twisted Pair Wire

Pairs of wires twisted around each other, when coupled with instrumentation amplifier chips or other devices that amplify and use the *difference* in voltage between the two wires, can be an effective means for transmitting analog signals. This works because both wires travel essentially the same physical path and are therefore exposed to the same noise-inducing electromagnetic fields. Since the noise picked up by the two wires is identical, it gets subtracted from itself when the difference between the signals on the two wires is taken at the other end of the wire. What's left is the original signal minus most of the noise.

5.5.3. Use the 4-20 mA Protocol

One of the most robust ways to send sensitive analog signals through electrically noisy environments is to transmit the signal using electrical current instead of voltage. This works because current signals are much less prone to interference than voltage signals are. One very common industry standard is to use current values between 4 and 20 milliamps to represent analog values. (Four mA is used instead of zero mA as a lower limit, in part so that a low sensor reading can be distinguished from a break in the wire.)

5.5.4. Avoid Getting Analog Data Wires Near Wires Carrying High Currents

Wires that carry large, rapidly changing electrical currents, such as those leading from batteries to thruster motors, can generate strong, fluctuating electric fields that can introduce noise into analog signal wires. So avoid bundling analog signal wires or placing sensors with tiny analog output signals alongside thrusters or current-carrying conductors.

5.6. Digital Data Formats

Transmission methods for digital data will be described shortly, but before doing that, it's important to look at some standard ways of encoding various things using patterns of only two digital values, represented by 1 and 0. After all, digital signals make perfect sense if all you need to send is a message that distinguishes two things: ON/OFF, or TRUE/FALSE, or YES/NO. But what if you want to send the number "7," the number

"3.14159," the letter "Q," or the color "periwinkle?" How about a movie or the haunting sound of a whale's call? How can these diverse things be sent using only 1 and 0? The trick is to agree on some standard codes for representing these other things as *patterns* of 1s and 0s.

5.6.1. Positive Integers

The most straightforward thing to encode digitally is a positive integer, such as the number 42. Unless otherwise specified, it's normal to assume that the "42" you see written on a page represents a base-10, or decimal, number. In other words, the 42 means (4 x 10) + (2 x 1). Similarly, the number 3,049 means (3 x 1,000) + (0 x 100) + (4 x 10) + (9 x 1). In the decimal number system, there are 10 digits to choose from (0, 1, 2, 3, 4, 5, 6, 7, 8, and 9). The right-most digit represents the "ones" column, and columns to the left represent increasing powers of 10 (1, 10, 100, etc.). **Binary** is similar to decimal, except that it's base-2 instead of base-10. That means there are two digits to choose from (1 and 0). It also means the right-most column is the "ones" column, and columns to the left represent increasing powers of two. That is, reading from right to left, each column represents a number twice the size of the preceding column (1, 2, 4, 8, 16, etc.) Thus, in the binary system, the number 101010 means (in decimal notation): (1 x 32) + (0 x 16) + (1 x 8) + (0 x 4) + (1 x 2) + (0 x 1) = 32 + 8 + 2 = 42 (decimal).

Each binary digit (1 or 0) is called a **bit**. By convention, bits are usually transmitted and processed in groups of eight bits at a time. Each of these groups of eight bits is called a **byte**. One byte can thus store binary numbers ranging from 00000000 to 11111111. These correspond to decimal values from 0 to 255.

Binary numbers are often written in a shorthand form known as **hexadecimal** or simply "**hex**." Hex is actually just base-16. Following the pattern you've already seen, base-16 has 16 digits, denoted 0, 1, 2, 3, 4, 5, 6, 7, 8, 9, A, B, C, D, E, F. (Lower-case letters can be used.) The right-most column is the "ones" column, and subsequent columns moving to the left are increasing powers of 16: 1, 16, 256, etc. It turns out that any group of four binary digits (sometimes called a "nibble") can be represented by a single hexadecimal digit, so a typical 8-bit byte can be represented by two hexadecimal digits. The table below gives some examples:

Table 9.2: Encoding Positive Integers

Decimal representation	*Binary representation*	*Hex representation*
0	0	0
16	10000	10
42	101010	2A
255	11111111	FF

5.6.2. Negative Integers

Negative integers are generally represented using a format known as "two's complement." While the format is a bit counterintuitive, it facilitates calculations involving both positive and negative numbers. Detailed explanations of two's complement are widely available on the web and in many technical books on computer fundamentals, so are not repeated here.

5.6.3. ASCII Characters

Many sensors, including GPS sensors and many other navigational sensors, transmit their information encoded as regular text characters, like the kind produced by a typewriter or computer. How is this possible if they can send only 1s and 0s? To facilitate the efficient storage and transmission of text in digital systems, a standard code known as **ASCII** (pronounced "ass-key") was developed. The ASCII code associates each of the printing characters on a typewriter (as well as most of the non-printing characters like tabs, spaces, new lines, and even the "bell" sound) with a particular integer between (decimal) 0 and 127. Some specialized characters, like the little "happy face" figure and a number of characters from other languages, are part of extended ASCII sets, using numbers in the range from 128 to 255, but these extended sets are less standardized. The standard ASCII table, as well as many of the extended tables, are widely available on the web or as appendices in many computer programming books.

Here's an example. To send the sentence "I love ROVs!" you would first send the ASCII code for a capital letter "I," which happens to be 01001001 (binary). Next, you would send the ASCII code for a blank space, which is 00100000. Then a lower case "l," and so on, until you end with the ASCII code for an exclamation point. The table below shows what this looks like in several different number formats

Table 9.3: ASCII Code for "I love ROVs!"

ASCII Character	Decimal value	Hex value	Binary value
I	73	49	01001001
(blank space)	32	20	00100000
l	108	6c	01101100
o	111	6f	01101111
v	118	76	01110110
e	101	65	01100101
(blank space)	32	20	00100000
R	82	52	01010010
O	79	4f	01001111
V	86	56	01010110
s	115	73	01110011
!	33	21	00100001

That sentence contains 12 characters (including the blank spaces and punctuation mark). If the (12 x 8 = 96) 1s and 0s corresponding to the 12 binary bytes shown in the right-most column (plus a few extra bits required to synchronize timing) were sent to most computer monitors, LCD screens, or other text displays, then the sentence "I love ROVs!" would appear on the display.

5.6.4. Floating Point Numbers

Numbers such as 3.14159 can be stored in digital formats, too. There are many ways of doing this, but one simple method is simply to store two integer numbers—the one to the left of the decimal point and the one to the right of that point. Another is to store the sequence of ASCII characters used to write the number, including the ASCII code for period (the decimal point). A particularly common approach involves converting the number to scientific notation, then storing the mantissa (the fractional part of a logarithm to the base-10) and exponent (the symbol indicating the number of times a base number is used as a factor) as integer values in binary.

5.6.5. Images and Sounds

Even colors, images, sounds, and movies can be stored and transmitted using binary data. In all cases, the secret is to adopt a standard format for the data, so that the patterns of 1s and 0s can be interpreted correctly. For example, colors are usually stored as three bytes of data. It turns out that you can make just about any color on a computer monitor by mixing red, green, and blue light in different proportions. So to specify a color, you just specify the intensity (on a 1-byte scale of 0 to 255) for each of these three colors in order (R, G, B). Thus, for example, the hexadecimal number 0000FF represents the color blue. Images are created simply by specifying the color of each of the thousands or millions of pixels that make up the image. Although fairly simple in concept, this can require millions of 1s and 0s per image. Movies require even more data, because they are a series of still images shown at a rate of many images per second. Sound can likewise be transmitted as a series of 1s and 0s. There are several different standard ways of doing this, some of which are used by digital sound media, such as CDs and DVDs.

Anytime you want to send information over a tether or other communication link using digital means, you can make up your own protocols and data formats if you want to. However, it's almost always easiest to stick with existing standards.

5.7. Digital Data Transmission

Once you've determined an appropriate pattern of 1s and 0s, you still need some physical way to transmit those 1s and 0s over wires, optical fibers, radio waves, or other media between devices. Many different standardized options exist for doing this. Ethernet, USB, and Wi-Fi are some of the more familiar ones. Associated with each of these common marketing names is one or more precise technical specifications approved by the Institute for Electrical and Electronics Engineers (IEEE) and given some reference number. For example, the Wi-Fi standards used for wireless Ethernet communications in computer networks are known among engineers and network technicians as the IEEE 802.11 series of standards.

In most cases, the standards include specifications about both the physical connections (what type of cables or wires, what type of connectors, etc.), as well as the nature of the signals transmitted (what voltages, currents, or light intensities represent a 1 and which represent a 0, how many bits per second are allowed, and so on). It's relatively easy to convert from electrical pulses to optical or radio signals, using appropriate optical or radio transmitters and receivers, so the discussion here will focus on transmission of 1s and 0s encoded as electrical pulses in metal wires. For example, you could send a 1 down a wire by raising the voltage of the wire to +5 V and send a 0 by dropping it back down to 0 V.

5.7.1. Serial and Parallel Transmission of Digital Data

When you send digital data over wires, you sometimes have the option of sending it as serial data or as parallel data.

- **Parallel data transmission** protocols send several bits at the same time, using a different wire for each bit. It's common for parallel systems to use eight data wires to send one byte at a time, though some systems use 16, 32, 64, or other numbers of wires to send larger numbers of bits at one time. Parallel data transmission is used primarily where extremely high data transmission speeds are needed and distances are short enough that wire cost is not an issue. For example, parallel data transmission is the normal way of connecting a computer's main processor chip with the memory chips located on the same printed circuit board.
- **Serial data transmission** protocols send data one bit at a time. To send a byte of data, a serial protocol sends eight bits in sequence, one after another, down the

same wire (or optical fiber or whatever). To send 10,000 bytes, you just need to send 80,000 bits, one after the other. (It's a tad more complicated than this, but this simple explanation captures the basic idea.) Serial communication is used over longer distances (more than a few centimeters) and is common in cables connecting physically separate devices, like two computers in adjacent rooms. Serial data transmission is the obvious choice for use in ROV tethers.

In addition to the wires that carry the data, both serial and parallel methods may include one or more wires serving support functions. For example, whenever the signal is sent as a voltage signal (e.g., 0 volts to represent a logical 0 and +5 volts to represent a logical 1), there must usually be an extra wire called the **signal ground** that is used as a reference against which these voltages are measured. There may also be extra wires used to coordinate the timing and rate of data transfer.

5.7.2. Synchronous Versus Asynchronous Data Transmission

Timing is critical in digital data transmission, particularly when a series of 1s or 0s follow each other down a wire. Otherwise, there's no way to tell whether a single +5 V pulse on the data line corresponds to just a single bit (e.g., 1) or a sequence of identical bits (e.g., 11111111) all bunched together. Two fundamentally different schemes are used to control the timing of data transmission. One is called synchronous, and the other is called asynchronous. Both can be used with either serial or parallel data transmission formats.

- In **synchronous data transmission**, one or more wires *in addition to* the data wires and the ground wire are used to send a timing signal (usually a series of pulses) that coordinates the timing of data transmission on the data wire(s). This timing signal is usually called the **clock signal**, and the wire carrying it is usually labeled CLK on schematic diagrams. The signal on the CLK wire is usually controlled by the sending device. Each **clock pulse** tells the receiving device that a new bit of data is on the data line and ready to be read. This prevents the receiving device from accidentally missing a bit, accidentally reading the same bit twice, or trying to read data right during a transition from one bit to the next, all of which could result in errors.

- In **asynchronous data transmission**, there is no clock signal or clock wire. Instead, both the sending and receiving devices keep a reasonably accurate timer running, and they synchronize their clocks frequently, so they can coordinate the timing of when bits are sent and received. The clocks are resynchronized by resetting them at the exact instant that a scheduled change occurs from a 1 to a 0 or vice versa. This happens about once every 10 bits or so in most cases, so the clocks on either side do not need to be terribly accurate. If one clock thought there were 24 hours in a day and the other thought there were 22 hours, that would be close enough for error-free transmission.

5.7.3. Common Serial Data Transmission Protocols

There are a number of serial data transmission protocols in wide use, and it's good for any aspiring ROV or AUV builder to be familiar with at least a few of the most common ones. Once you are, it's easy to pick up others as needed. Here's a list of some good ones to learn early on:

- **RS-232**: For communication with digital devices over short to moderately long distances (30 meters—approx. 100 ft—or more), the classic favorite for over two decades has been an asynchronous serial communication protocol known as **RS-232**. This is the communication protocol associated with the venerable serial port found on most computers in the 1990s. (But it is different than the newer USB serial ports and cables.) These older serial cables normally sport either 9- or 25-pin "D" style connectors (and 9 or 25 wires in the cable), but for simple data transmission, only three of the wires are necessary: ground, transmit, and receive (usually denoted GND, TX, and RX, respectively). The details of this protocol are widely available on the web and in many electronics, computer, and robotics books, so they won't be repeated here.

If there is one protocol you should study and learn, RS-232 is probably it. Although RS-232 is slow and low-tech by present-day standards (most computers no longer even offer an RS-232 serial port), it is a relatively simple and low-cost alternative that continues to serve a useful purpose for many interested in robotics, ROVs, AUVs, and computer control of various devices. In fact, most microcontrollers, many sensors, many motor controllers, and many other peripherals still use (or at least offer) an RS-232 communication interface or something closely related to it. Moreover, materials that explain how the newer protocols work frequently do so by comparing and contrasting the newer protocol with RS-232, so it's an advantage to understand the basics of RS-232. If necessary, you can buy converters to go between RS-232 and USB or Ethernet.

One warning: The formal RS-232 specification involves voltages not used by most modern microcontrollers. To connect a microcontroller to a true RS-232 device (such as some computers), you may need to use a "232 driver" chip, such as a MAX232, to make the appropriate voltage conversions. Some systems require this conversion, and others work fine without it. If in doubt, or if you need reliable communication, it's better to use it.

- **RS-485**: In many ways, RS-485 is an improved version of RS-232. Like the original version, it is an asynchronous serial communication protocol that uses two signal wires and (usually) one ground wire; however, RS-485 can send data more reliably over longer distances (and is therefore useful with longer ROV tethers). It also allows more than two devices to share a single set of wires, provided those devices coordinate their transmissions in some way so that the devices aren't trying to talk at the same time. You can buy RS-485 transceiver chips that will convert a microcontroller's RS-232 or similar signals into RS-485 signals.

- **I2C and SPI**: For communication with nearby digital sensors, other MCUs, and other digital devices (including some memory chips), it's common to see any of several 2-, 3-, or 4-wire communication schemes involving synchronous serial data transmission. Popular among these are the I^2C and SPI standards. Both were developed by private companies for use with their microcontroller products, but have caught on and become popular enough that they are widely used by many manufacturers. The details of these standards, like others, can be found on the web or in books.

- **Ethernet**: This is not a simple protocol to master, but it is currently the most commonly used protocol for transmitting data over computer networks and the internet, so many relatively inexpensive video surveillance cameras, sensors, and other devices use it. Numerous books and websites explain it, if you are interested in learning about the details. If you learn enough about this protocol and some related subjects like TCP/IP, configuring network routers, web browsers, and embedded Ethernet controllers, you can control robots and ROVs remotely over computer networks, possibly even over the internet! Ethernet protocol is mentioned here to encourage interested individuals to learn more. To do so, search on keywords and titles like "TCP/IP," "computer networking," "Ethernet," and "Embedded Ethernet Control."

- **USB**: This is a fast serial communication standard. As of this writing, USB is a commonly used standard for interconnecting computer peripherals. As with Ethernet, the details are not easy to learn; however, there are plenty of commercially available adapters and other devices out there that can handle the details for you. For example, you can buy serial-to-USB adapters that will take a standard RS-232 input and translate it into a USB output, or vice versa.
- **New Stuff**: Computer networks and telecommunication networks are increasing in speed, complexity, and capability all the time. While it's not possible to predict the future of tech developments, it does look bright for those with technical interests and skills. No doubt, even better communication protocols for ROVs and AUVs are just around the corner. Keep up with the latest developments by surfing the web, reading books, and subscribing to publications devoted to fun electronic projects.

5.8. Analog-to-Digital Conversion

As mentioned earlier, microcontrollers, regular computers, and many other electronic devices work primarily or exclusively with digital signals, because these signals are less prone to corruption by noise and because they are easy to process, distribute, and display using all these digital electronic devices. However, the natural world is largely analog, and sensors that measure that world often generate analog signals, so it's important to be able to convert those analog signals into a digital format. This process is called **analog-to-digital (A/D) conversion**.

Since digital signals are less sensitive to corruption by noise than analog signals, the sooner you do this conversion, the less you have to worry about noise messing up your data. For this reason, many sensors do A/D conversion internally, thereby becoming digital sensors. If your sensors have an analog output and you need to send the data a long distance (e.g., up a tether), you should ideally convert the sensor's analog signals to digital signals within a few centimeters (inches) of the sensor.

To go from analog to digital, you use a specialized type of integrated circuit chip known as an **analog-to-digital converter** (also known as an **A/D converter**, **A/D** or **ADC**). There are hundreds or thousands of different brands and models of these converters available through major electronics parts distributors such as Digi-Key or Jameco. Many microcontrollers also include internal A/D converters connected to some of their I/O pins. In these cases, the pins can be configured in software to be used as a regular digital I/O pin or be configured as an analog input.

Most A/D converters accept analog voltages as their inputs, though some accept currents such as the 4–20 mA industrial standard mentioned above. An A/D converter will have a standard input range, such as 0 to +5 volts or –10 to +10 volts. Your sensor output type and range should match (or nearly match) that of your A/D converter for greatest benefit. It is often necessary to use an **operational amplifier (op-amp)** or **instrumentation amplifier** (both are types of integrated circuits) to amplify small sensor output signals and make them big enough to cover most of the input range for a typical A/D converter. Op-amp circuits are beyond the scope of this book, but these useful circuits for processing analog signals are covered in most introductory electronics books, and you can find good examples with explanations on the web.

To assign a digital number to an analog signal coming in, the A/D converter first divides its input range into a certain number of equal sized "bins." The bins are numbered sequentially, and the A/D converter reports which bin the analog value falls into. The number of bins is described in terms of the number of bits needed to specify all the bin numbers. For example, an 8-bit converter divides its input range into 256 bins, which is 2 raised to the 8th power. A 10-bit converter uses 1,024 bins. A 12-bit converter uses 4,096 bins. A 16-bit converter uses 65,536 bins. The more bits or bins, the finer the resolution in input voltage. (See *Tech Note: A/D Conversion Example* for a sample A/D converter calculation.)

TECH NOTE: A/D CONVERSION EXAMPLE

Question: what is the digital output of an 8-bit A/D converter with a 0 to +5 V input range when it's receiving an input signal of exactly 3.7000 volts? Answer: An 8-bit converter divides its input range into 256 equal bins. The first bin is number 0, and the last is number 255. In this case, each bin is worth 5 V/256 = 0.01953 V, or 19.53 millivolts (mV). Now, 3.7 V divided by 0.01953 V = 189.45, so the A/D converter would report a digital value of 189 (A/D converters report only integer values, not decimal fractions).

As long as the microcontroller or other device receiving this digital data knows that each bin represents 19.53 mV, it can calculate back to a value that is very nearly the original analog value. In this case, a microcontroller receiving the number 189 and knowing the conversion factor (19.53 mV per count) would calculate a voltage of (0.01953 V/count x 189 counts = 3.6912 V, an error of much less than 1 percent. For most routine measurement purposes, this is close enough to the real value. If greater precision were needed, an A/D converter with greater resolution, such as a 10-, 12-, or 16-bit converter, could be used.

5.9. Signal Multiplexing

In smaller ROVs with short tethers, it's feasible to run a separate pair of wires down the tether for each thruster, light, camera, or other electric device. In this scheme, control of the devices at the other end is a simple matter of supplying power (or not) to the pair of wires associated with the device you want to control. This is the approach used in *SeaMATE*. However, the added weight, stiffness, and expense of all those wires make this approach impractical for complex vehicles having dozens of electrical devices controlled through very long tethers. In general, these vehicles use a single pair of power conductors, plus a small number of thin electrical wires or fiber-optics to handle the control signals. This is made possible through a technique called **signal multiplexing**, or just **multiplexing**. In this method, different signals are encoded in such a way that they can all be sent together down one pair of wires (or an optical fiber), then separated again at the other end.

There are lots of different ways of doing this. Most can be used with either analog or digital signals. One common method is called **frequency division multiplexing**. This is how TV stations, cell phones, police radios, and hundreds of other radio signals can all share the same "airwaves" without drowning each other out. Basically, each signal is broadcast on a different frequency (or small range of frequencies), and the receiving equipment "listens" to just one frequency at a time. For example, when you change radio stations, you are telling your radio to listen to a signal coming in on a different frequency. Frequency division multiplexing is commonly used in radio controllers for R/C model cars and airplanes where, for example, one frequency might carry the signal that tells the vehicle how fast to go, and another might carry the signal telling it whether to turn right or left.

Time-division multiplexing is another common method of multiplexing. In this form of multiplexing, signals take turns traveling over the wires (or airwaves). For example, you might send a series of three commands down to an ROV, one after another, then pause for a specified amount of time to allow your ROV to transmit vehicle depth and heading information back up to the surface over the same pair of wires.

If you have a good grasp of electronics and have some experience working with microcontrollers, you can experiment with implementing some standard multiplexing schemes or even invent your own. Alternatively, you can also buy a wide variety of commercial **multiplexers** (often abbreviated **MUX**) and adapt them for ROV use. *Section 5.10.3 Adding Joystick Control* and *Tech Note: Adapted R/C Controllers* in this chapter describe how you can use the MUX already built into a commercial R/C hobby controller to send multiple command signals more or less simultaneously to your ROV over a single pair of wires.

5.10. Electronic Sensors and Sensor Circuits

Now that you've been introduced to microcontrollers and the fundamentals of electronic communication, it's time to start thinking seriously about how you can use microcontrollers to enhance the capabilities of your vehicle. Basically, this boils down to three things:

1. You can learn how to provide *inputs* to your microcontroller, so that it can receive pilot commands and/or receive information from various types of navigational or other sensors.
2. You can discover how to connect *outputs* from your microcontroller to motors, lights, and other devices on the vehicle, so it can control those devices.
3. You can program your microcontroller to make it the "brains" of your vehicle's control system; if you're successful at this, the commands it sends to the motors and other devices in response to data from the sensors will help the "actual conditions" match "desired conditions."

This section of the chapter takes a look at the first of these three things by exploring ways you can provide sensory input to your microcontroller. In many cases, that input will come from buttons, knobs, joysticks, or other "sensors" operated by a pilot to send commands to the vehicle. In other cases, that input may come from on-board sensors of many different types.

5.10.1. Connecting a Pushbutton to Your Microcontroller

The simplest form of input you can provide to your microcontroller is a single bit (1 or 0) of data from a pushbutton or other manual switch. Though simple, it's also very useful. It provides a straightforward way for a pilot to issue some types of commands to the microcontroller. Such commands could include things like asking the microcontroller to turn on a light or activate a depth autopilot mode. It also provides a way for some on-board systems to provide status information to the microcontroller. For example, a button might be mounted in the path of a camera tilt mechanism to tell the microcontroller when the camera has reached the full limit of its allowable movement.

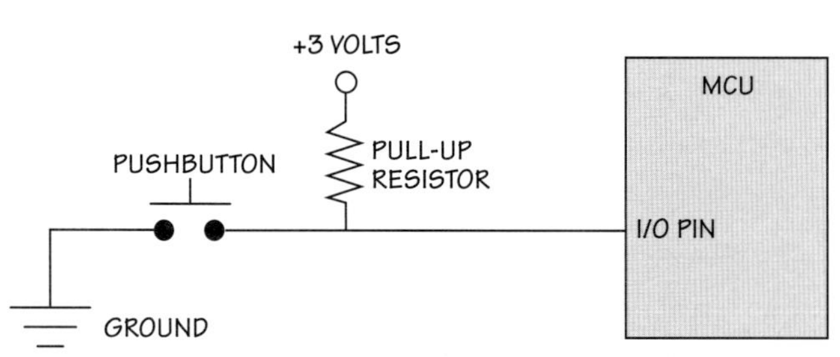

Figure 9.42: Using a Pushbutton with a Microcontroller

This simple circuit provides an easy way for a microcontroller to monitor the status of a pushbutton or other manual switch.

One of the MCU's I/O pins is first configured (usually through instructions in its program) to be an input pin. This pin is then connected through a resistor (usually 1,000 to 10,000 ohms is about right) to a voltage that the MCU will interpret as a "high" or logic 1 value. This resistor is called a "pull-up" resistor, because it pulls the I/O pin voltage up to the source voltage anytime the button is not pushed.

When the button is pushed, the I/O pin gets connected directly to ground. This forces the input to 0 V, because the +3 V coming through the resistor can't compete with the direct short to ground. As soon as the button is released, the short to ground disappears, and the pull-up resistor immediately pulls the I/O pin voltage back up to +3 V.

The trick here is to devise a simple switch circuit that changes the voltage present on one of the microcontroller's I/O pins, depending on the status of a switch or button. For example, if you were using a microcontroller that interpreted +3 V as a logical 1, and 0 V as a logical 0, you could rig up a circuit that would put +3 V on a wire connected to that I/O pin whenever the button wasn't pushed, but would pull that voltage down to 0 V whenever the button was pushed. The circuit shown in Figure 9.42 would do this. Different buttons could be connected to different I/O pins in this way, so the microcontroller would know which button had been pushed (i.e., which command had been issued) by paying attention to which I/O pin changed from a logical 1 to 0.

5.10.2. Using a Potentiometer for Adjustable Input

The little pushbutton circuit just described is a great way to tell your MCU that you want it to turn a light ON or OFF, but what if you want to adjust the light's brightness? How do you tell your microcontroller that you want to dim the light intensity to only 40 percent of full power? How do you tell a manipulator arm joint to rotate to particular angle? One way to do these things is to use two microcontrollers. One on the surface can monitor the position of a rotating knob or slider control, so it can send that information in the form of digital signals to a second microcontroller located under water on the ROV.

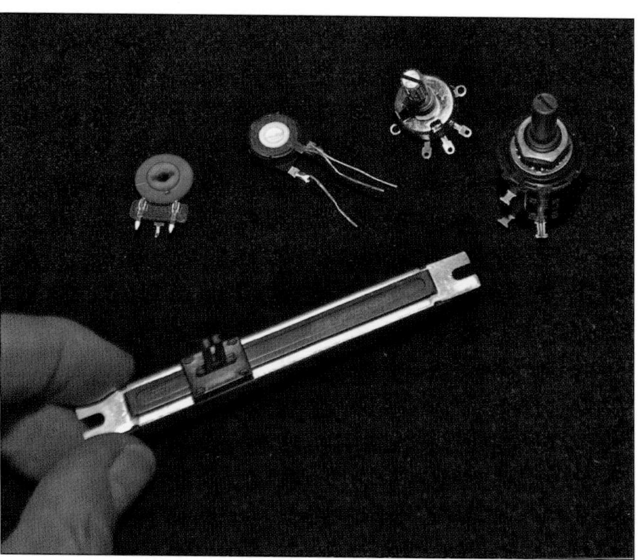

Image courtesy of Harry Bohm

Figure 9.43: Potentiometer Assortment

Potentiometers and rheostats come in a wide variety of sizes and styles. This photo illustrates some common types used for small control systems. One type is a trimmer potentiometer, or "trimpot", which is designed to be set once with a screwdriver to fine tune some part of a circuit, then left alone.

A rotary potentiometer ("pot"), has a shaft that is usually attached to a knob for frequent adjustment of motor speed, light brightness, stereo volume or other parameters. A third type is a sliding linear potentiometer which is common. For an explanation of how potentiometers and rheostats work, see Tech Note: How a Potentiometer Works.

In general, these knob and slider controls involve an adjustable resistor called a **potentiometer**. Each potentiometer is a three-terminal device that can be wired up in different ways, depending on what you want to do with it. Inside most potentiometers is a long, coiled piece of resistive wire. Each end of the wire is connected to a terminal on the outside of the potentiometer. The third terminal is connected to a movable wire that sweeps along the coiled resistive wire as the potentiometer knob is rotated (or as the slider slides). (See Figure 9.44 and Te*ch Note: How a Potentiometer Works.*)

One wiring option is a **variable resistor**. In this configuration, which uses only the center sweeper terminal and one end terminal, the knob adjusts the resistance between those two terminals. The popular BS2 microcontroller manual provides instructions for building a simple circuit and using built-in PBASIC commands to read the resistance (and hence the knob position) of a potentiometer by measuring how long it takes to discharge a capacitor through that resistance.

A potentiometer can also be wired up in a configuration called a **voltage divider**. In this configuration, one end terminal is hooked to a supply voltage (e.g., +5 V), and the other is hooked to ground (0 V). The middle, sweeper terminal will have an output voltage that varies between 0 and +5 V, depending on the position of the knob. This output voltage can be fed to an A/D converter connected to an MCU, so the MCU can measure the voltage and from that, determine the knob or slider position.

5.10.3. Adding Joystick Control

A joystick, like those used with many interactive video games or with radio-controlled (R/C) model airplanes and cars, can provide a wonderfully intuitive way to control vehicle movements with great precision. The simplest joysticks provide detailed information about the angular position of the stick in two dimensions—usually denoted X and Y. More sophisticated joysticks may include twisting action and multiple buttons or sliders for additional control.

At its core, the X-Y feature of a joystick is usually made possible through two potentiometers hidden inside the joystick mechanism. As the stick is moved forward and backward, one of the potentiometers is rotated. As the stick moves left and right, the other is rotated. Most often, these potentiometers are already wired up as voltage dividers, so the X and Y positions of the stick are represented by voltages present on wires coming from the central sweeper terminals of the potentiometers.

In older joysticks designed for computer games, these analog voltages were delivered by wire to A/D converters on special game boards inside the computer. Newer joysticks for computer games, including dedicated game boxes like PlayStation and Xbox, usually process the voltages with internal A/D converters and microcontrollers, transmitting their information digitally to the computer through a standard USB interface, wireless interface, or proprietary custom interface. Joysticks intended for radio control functions generally use A/D converters and microcontrollers to encode the stick positions as precisely timed radio pulses broadcast to radio receivers on the remote-controlled vehicle. Any of these joystick types can be used to control an ROV, but the old-style analog joysticks and the R/C joysticks are usually the easiest to incorporate into home-made ROV designs. (See Figure 9.45.)

To use one of the old-style computer game joysticks, open it up to discover which wires supply power to each potentiometer and which ones carry the voltage outputs from the potentiometers. Then you can use your own A/D converters and microcontrollers to read the joystick's X and Y axis information. (This is not trivial, but it's definitely doable, and it's a great learning experience!)

TECH NOTE: HOW A POTENTIOMETER WORKS

This big potentiometer is like a rotary potentiometer on steroids. It's made to handle larger currents than its smaller counterparts can. Best of all, it's innards are exposed, so you can see how a potentiometer works. The rust-colored paint coats most of a long, resistive wire coiled around a doughnut-shaped piece of ceramic, but there's a section of each turn exposed (where the person's thumb is touching the coiled wire).

This particular wire has a total resistance of 100 ohms and can handle up to 1.73 amps, as indicated on the side of the potentiometer. The ends of the coiled wire are accessible via two metal terminals: one (call it "A") is visible under the "100 OHM" label, and the other (call it "C") is barely visible under the "1.73 AMP" label. A third terminal (call it "B"), clearly visible between the first two, connects to a central shaft and from there over a spring-loaded bridge-like arm that makes electrical contact with the coiled wire. When you turn the big black knob, the bridge rotates, and the contact point slides along the coiled wire.

In this photograph, the middle terminal is touching the coiled wire about 2/3 of the way around from the A terminal to the B terminal, so the resistance between A and B is about 2/3 of 100 ohms, or about 67 ohms. The resistance between B and C is about 1/3 of 100 ohms, or about 33 ohms. If you use just terminals A and B (or just C and B), the potentiometer can be used as a rheostat (adjustable resistor) with an adjustable range of 0 to 100 ohms. If you use all three terminals, you can do fancier things like create a voltage divider. In this case, if you hook 12 volts to terminal A and ground C, then you can "dial in" any voltage you want between 0 and 12 volts on terminal B. Right now, there would be about 4 volts on B, because the bridge is 1/3 of the way around from the 0 volt terminal to the 12 volt terminal, but rotating the knob would change that.

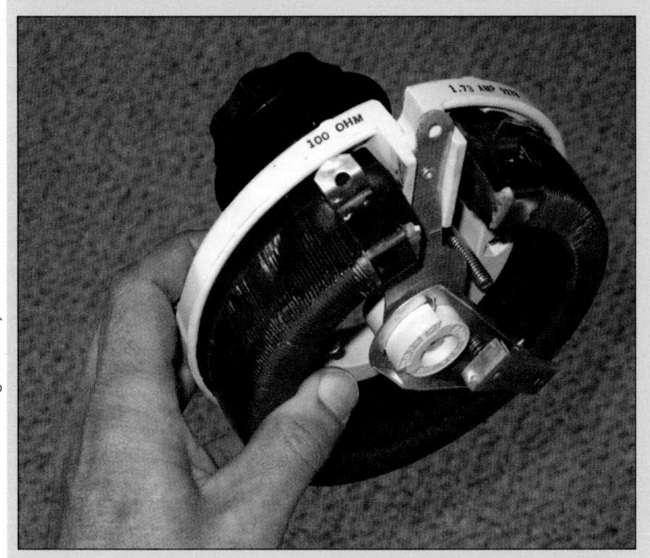

Image courtesy of Dr. Steven W. Moore and Dr. Fred Watson

Figure 9.44: A Large Potentiometer

Image courtesy of Harry Bohm

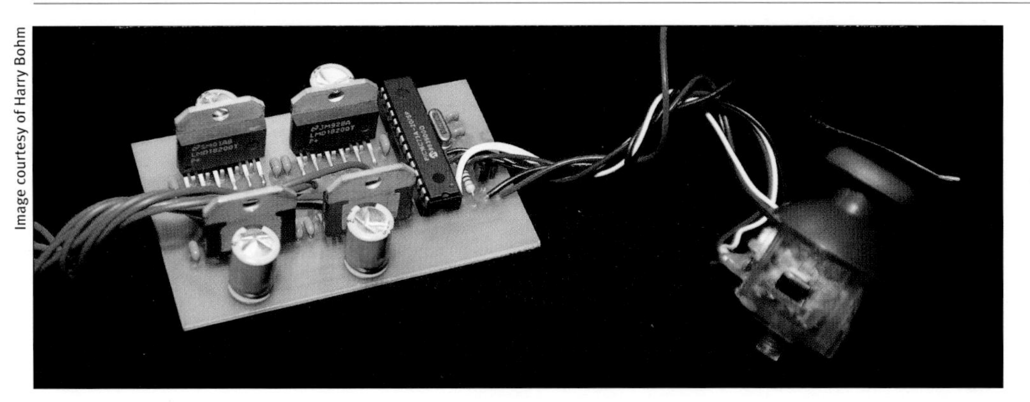

Figure 9.45: Motor Controller Option

This example of a student-built motor controller uses a video game thumb joystick.

If you want to grab the digital data directly from the newer computer and game box joysticks, you should search the web for information about how to do this, since it varies with each make and model of joystick. These joystick outputs are generally optimized for use with high-performance desktop computers running standard operating systems and having complex joystick driver software, so they are not necessarily easy to interface with a microcontroller. If you're using a PC-104 computer inside your vehicle, that may be a different story. Of course, you can also hack into most of these newer joysticks and go straight to the potentiometers, bypassing the elaborate electronics inside. In fact, you can sometimes find on-line instructions for hacking these joysticks and using them to control terrestrial robots. These same techniques would work well for controlling underwater ROVs.

With R/C joysticks, you can follow the same idea of opening them up to get at the potentiometer outputs inside, or you can leave them intact and simply reroute their radio signals through a coaxial cable in a tether. To do this, you need to take the internal wire that normally hooks to the antenna and connect it instead to the central conductor in the coaxial cable. You also need to hook the outer conductor (the cable shield) to ground. At the other end, you attach the shield to the receiver's ground and connect the central conductor to the place where the receiving antenna normally connects to the rest of the circuitry.

These R/C controllers transmit several different channels of data to control different things in the vehicle. For example, in a model airplane, one channel might control the throttle, another the rudder, and another the ailerons. A compatible receiver will know how to decode these transmissions. Normally, these receivers translate the coded signals into voltage pulses used to control servo motors inside the vehicle. The details of these servo control pulses are described later in this chapter, in Section 5.13.2 which discusses motor control.

If you want to control servos so they adjust something like camera tilt on your ROV, you can directly connect the receiver output to compatible servo motors. If you want to use it to control thruster motor speed and direction, you may be able to find motor controller circuits that are already compatible with those signals, or you may need to use a microcontroller to "translate" those signals into other commands appropriate for thruster motor control. See *Tech Note: Adapted R/C Controllers* for some specific suggestions.

Admittedly, this is a rather broad-brush overview of how to incorporate joystick control in your vehicles. But this is one place where advanced control gets complicated enough that this book cannot realistically cover all the details, especially since the details keep changing quickly as new styles of joystick evolve. Instead, the goal of this section is to make you aware of some of the promising possibilities and point you in the right direction to learn more on your own. Keep in mind that none of this is

TECH NOTE: ADAPTED R/C CONTROLLERS

Radio-control (R/C) controllers have been adapted to operate a number of small ROVs. For example, *Mini-ROVer*, one of the first LCROVs developed by Chris Nicholson, used a 7-channel R/C unit. It encoded the analog control voltages from the joystick, queued them in the multiplexer circuit, and sent them down to the ROV on an FM carrier wave. Instead of traveling through the radio antenna, the signal was modified to pass through a coaxial cable in the tether. The result was an elegant way to control a small ROV.

Images courtesy of Randall Fox (left) and Harry Bohm (right)

Figure 9.46: Topside and Interior Adaptation of RC Controls for Underwater Vehicles

The left hand photo shows a standard handheld remote control unit that is used normally for model airplanes. In this case, it's been adapted for an ROV.

The photo on the right shows circuitry that interprets commands from the RC controller and uses them to operate the thrusters.

These R/C controller systems consist of a hand-held transmitter unit, which houses the controls and the multiplexer unit, plus a compatible receiver unit, which resides on the vehicle and includes a demultiplexer to separate the control signals and route them to the appropriate motors or other controlled devices. You can find these R/C controller systems at many hobby stores that sell radio-controlled airplanes, boats, tanks, or other vehicles. This means that you can use a relatively thin and flexible coaxial cable (such as size RG-174) in your tether to connect the transmitter and receiver directly, instead of relying on the usual radio signals, which won't go far through water. The R/C transmitter/receiver package comes in many different flavors, which you'll want to research. Currently, the main choices are between AM or FM analog or digital types, such as the proprietary Futaba PCM1024 digital multiplexing scheme. Digital R/C controllers are usually programmable.

The number of R/C transmitter channels varies by the model—as low as two to more than 14. A single channel is able to control one actuator on the robot. Thus, a deeper diving hybrid with four thrusters would need a transmitter and corresponding receiver with a minimum of four channels. However, seven- and eight-channel units are common; you'd be smart to get one of these, since it would easily accommodate future expansions of any ROV subsystems, such as a manipulator or a camera tilt mechanism.

The device that makes R/C adaptable to low-cost ROVs is the solid-state motor controller, sometimes called an **electronic speed controller**, or **ESC**, in model hobby circles. Vantech, Castle, and dozens of other companies make ones that have been used successfully in underwater vehicles. Hobbyists have been pushing the speed barrier for model racing cars, which require high rates of power delivered in short time periods. Typically, the electric motors that power these cars pull currents in excess of 15 amps and may peak at 30 and 50 amps for short periods of time. R/C manufacturers have met these demands, producing small, high-capacity electronic motor controllers for this purpose. This is good news for underwater robot designers, since thrusters that pull up to 30 amps at 26 volts can be controlled by these heavy-duty R/C motor controllers. One warning: airplanes don't fly very well backward, so ESCs designed for controlling model airplane motors usually do *not* offer a reverse direction. Since reverse is essential for most ROV maneuvering, make sure you get an ESC that is not only compatible with your R/C unit, but also has the ability to control multiple motors in both forward and reverse directions. You might, for example, look for R/C controllers designed specifically for twin-screw model boats or twin-track tanks that have propulsive systems similar to those of ROVs. This is called a dual function mode, and it allows you to control both right and left horizontal thrusters with one joystick. A number of companies make R/C digital control systems that can be configured to operate in this dual mode, so you'll want to check out these options.

particularly easy to do, but none of it is impossible, either. In fact, lots of people do it with great success. Be resourceful about finding out what others have done well and be creative in customizing what they've done to fit your particular mission objectives.

5.10.4. Leak Detection Sensor (also called Water Intrusion Sensor)

Switch your attention now from sensors that provide input primarily to an MCU in the pilot's console to sensors that provide input *directly* to an MCU on the ROV or AUV itself. The first example covered here is a leak detection sensor, commonly called a water alarm. Nobody ever plans for their vehicle to leak, but sometimes it happens. If you have an early warning system to alert you that water is getting into the pressure can, you may be able to return the ROV or AUV to the surface in time to save most or all of the internal electronics from a soggy death.

There are a couple of ways to approach this. The simplest is to use seawater to short two wires mounted next to each other (but not touching) in the bottom of the canister. If this completes a circuit with a battery, resistor, and LED, then the LED will light up. This LED can be placed in front of the camera to provide a visual warning that water is getting in.

A somewhat more sophisticated version uses a transistor to amplify the signal, allowing the sensor wires to be in the tether for a very bright LED warning, or even a

TECH NOTE: A TRANSISTOR-BASED LEAK DETECTOR CIRCUIT

This circuit uses an inexpensive and widely available 2N3904 NPN transistor as a switch to control a warning light or buzzer. Like other transistors, this one has three wires connecting it to the rest of the circuit. The top one of these three wires in the diagram (called the collector) is wired to a light or buzzer. The one with the little arrow (called the emitter) is connected to the negative terminal of the battery.

The remaining transistor wire is called the base. If the base is at zero volts, the transistor switch is OFF, and no current can flow from the transistor's collector to the emitter, so the warning light or buzzer is OFF. The two wires shown ending in dots are stripped bare of insulation at their ends and mounted near each other—but not touching—in the lowest part of the ROV, at the spot where water would first accumulate in the event of a leak. One of these wires is connected directly to the positive terminal of the battery. The other is connected to the base of the transistor and also (through a 1 megaohm resistor) to the negative terminal of the battery.

In the absence of water, there is no flow of electrical current between the two stripped wires, because air is a good insulator. Therefore, the 1 megaohm resistor discharges any non-zero voltage on the wire connected to the base of the transistor, ensuring that the transistor (and the alarm) is turned OFF.

However, if water leaks in, the water forms a relatively low-resistance electrical connection between the two bare wires. This creates a short circuit that connects the base of the transistor to the positive side of the battery, turning the transistor switch ON and allowing current to flow through the warning light or buzzer. The main circuit, including the transistor, resistor, and warning device, can be located on the control box at the surface, with a pair of thin sensor wires in the tether, so the pilot will know immediately if there is a leak, even if the camera system fails.

Before each dive, the warning system can be tested easily by placing a small piece of metal across the two wires to produce a temporary short.

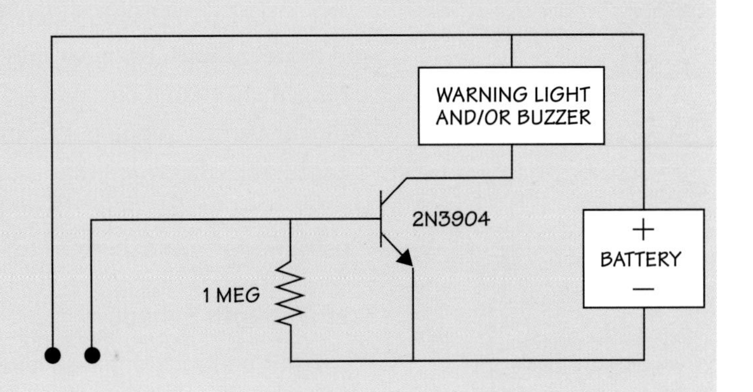

Figure 9.47: Schematic Diagram of Leak Detector Circuit

buzzer, on the surface. A circuit for doing this is shown in Figure 9.45. The figure caption explains how it works, but you may find it easier to understand this circuit after learning more about transistors later in this chapter.

5.10.5. Battery Voltage Sensor

Cars have gas gauges to tell you when your car is about to run out of gas. For the same reason, it's helpful to have a battery voltage sensor to help you monitor how much energy is left in your batteries. This is particularly important in an AUV or hybrid ROV, where you are limited to the energy stored on board so you need to know when it's time to turn around and come home.

For an ROV equipped with a camera, you can purchase electronic voltage meter kits that illuminate a row of LED lights at various voltage levels (all lights showing = maximum voltage; one or two showing = low voltage). These kits can be adapted for ROV use by placing the LEDs in the camera's lower or upper field of view, thus enabling the pilot to easily observe battery voltage. Alternatively, you can build more advanced circuitry from scratch. Generally, this consists of a pair of resistors used as a voltage divider to scale the battery voltage to a range that matches the input of an A/D converter connected to a microcontroller. This MCU can be set up to transmit voltage information to a topside computer for pilot use or used directly to control AUV activities. Some advanced digital R/C controllers offer battery voltage monitoring and on-screen display capability, which could probably be adapted easily for ROV use.

5.10.6. Compass

As noted earlier in this chapter, a compass is one of the most basic navigational tools and is nearly essential for keeping track of direction in poor-visibility underwater environments. It can also help you keep track of the total number of rotations of the ROV, so that the tether doesn't inadvertently get twisted during a dive. You already know that the easiest way to get a compass heading for an ROV is to stick a waterproof and pressure-proof compass within view of the video camera. This easy method works well if a human pilot is available to watch the video and control the heading. However, if the compass data must be converted to an electrical signal for digital display on a computer screen, or so that a computer or microcontroller can monitor the heading and steer the vehicle automatically, then you'll need to install an electronic compass, such as a fluxgate compass, instead.

Quite a number of electronic compasses are available. One of the simplest for ROV use has four signal lines, one for each of the cardinal directions. If the compass is oriented to the north, the voltage on the north signal line goes high (logical 0 becomes logical 1), and this can be detected if that signal line is connected to one of the I/O pins of a microcontroller. Likewise for east, south, and west. Four intermediate directions are also detectable (e.g., NE is identified by both the N and E lines going high). Other more sophisticated, electronic compasses can provide greater resolution and more options, but generally these require a more complex hardware and software arrangement with a computer or microcontroller. Most use a synchronous serial data protocol through which the microcontroller requests a heading, then receives the digital results. Be aware that most compasses will give you an incorrect reading if they are not held perfectly horizontal, and only a few gimbaled models offer self-leveling capabilities or some form of tilt compensation.

5.10.7. Depth Sensor

As mentioned earlier, a depth sensor can be a very helpful addition to an underwater vehicle. Placing a diver's depth gauge in front of the camera works fine if you want the pilot to monitor and control depth, but not if you want automated depth control for a depth autopilot or AUV navigation. In these cases, you'll need to integrate an electronic depth sensor into your vehicle. This presents some serious hardware and software design issues, so this task is best for someone with prior electronics and programming experience who is eager to take on a new challenge. Don't try this for your first electronics project.

The standard way of electronically measuring depth is to use an electronic *pressure* sensor to measure the ambient water pressure, then convert the pressure reading to depth, using the knowledge that one additional atmosphere of pressure results from every 10 meters (approx. 33 feet) of increasing depth in seawater (approx. 34 feet in freshwater). Most of these pressure sensors work by having a thin, delicate metal membrane that bulges slightly whenever a pressure differential is applied across the membrane. Tiny strain gauges, which change electrical resistance as they stretch, are glued to this membrane and connected into a circuit known as a **Wheatstone Bridge** (Figure 9.48). This useful circuit is exquisitely sensitive to slight changes in the resistance of the strain gauges and hence to pressure-induced changes in the membrane shape. A Wheatstone bridge circuit has two wires coming in, which are used to supply power to the bridge, and two wires coming out, which are used to sense the output signal. The pressure signal is read by measuring the (usually very small) difference in voltage between the two output wires. An instrumentation amplifier chip amplifies the tiny signal to make it large enough to work with comfortably. Then the amplified signal is fed into an analog-to-digital (A/D) converter, which changes it to a digital format that a microcontroller can read on its I/O pins.

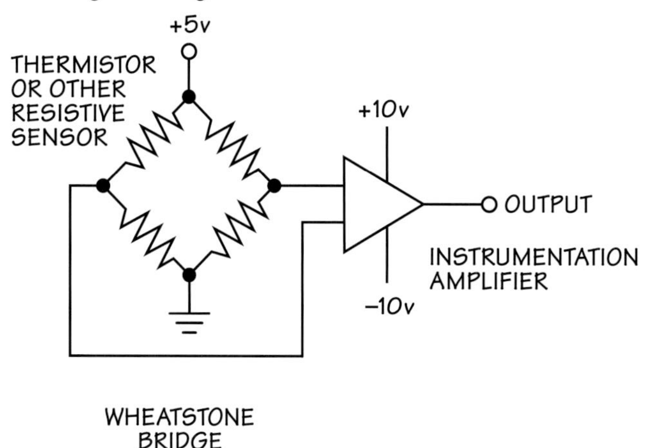

Figure 9.48: Wheatstone Bridge Circuit

A wheatstone bridge consists of four resistors arranged as shown in this diagram. One (or more) of the resistors changes value slightly in response to some measured quantity (temperature, strain, light levels, etc.). This causes a tiny imbalance in the voltages at the inputs to the instrumentation amplifier, which amplifies the difference and outputs the amplified signal.

Unless you have experience in the design of analog amplifier circuits for very small signals (less than 1 mV), you'll probably want to buy a pressure sensor that has an amplifier circuit already built into it. These usually cost a bit more, but the extra cost is generally worth it.

Pay careful attention to the detailed specifications before purchasing a pressure sensor. Thousands of models exist, but only a few are appropriate for a diving vehicle. The most common ones can only be used in dry air (not water) and have pressure ranges too low or too high to be useful for a typical underwater vehicle. Make sure that any part of the sensor that will come into contact with water is made of stainless steel or some other material that can tolerate a corrosive water environment. Most pressure sensors designed for measuring high water pressure are shaped like a bolt and are designed to screw into a threaded hole drilled through a wall separating the high- and low-pressure areas—this would be the wall of your ROVs pressure canister. Some pressure sensors come with a built in O-ring designed to seal tightly against leaks when the sensor is screwed snugly into a flat, smooth, clean surface, such as the endcap on a housing.

Your sensor must, of course, be able to withstand the range of pressures anticipated (plus a bit more as a safety margin), yet provide the depth resolution you need. Most models come in several different pressure ranges. It's usually best to pick one with a normal working range that is between 120 and 200 percent of the expected maximum depth. Of course, you may need to convert between atmospheres, p.s.i., Pascals, or feet of water to translate dive depth needs into the appropriate sensor pressure range. (Revisit Chapter 3, Chapter 4, and *Appendix III: Hydrostatic Pressure.*)

Many manufacturers allow you to specify a reference pressure option for their sensor. Basically, this identifies what's on the side of the membrane opposite the pressure being measured. Here are the usual options:

- **Absolute**: In some cases, the reference pressure is provided by a small vacuum-pumped chamber with zero pressure. These gauges typically provide readings of absolute pressure (i.e., they'll read 1 atmosphere of pressure at sea level).
- **Gauge**: In other cases, it's a sealed chamber with air at 1 atmosphere pressure. These will read gauge pressure (i.e., 0 atmospheres at sea level).
- **Relative**: The third and most common option simply has the low-pressure side of the membrane vented, so it is exposed to the air inside an ROV, for example. This type normally reads gauge pressure if the inside of the ROV is at 1 atmosphere, but not otherwise.

Generally, the vacuum-pumped absolute pressure gauges seem to be the easiest to work with (because they always give you a positive pressure reading, which you can subtract in software to zero your depth reading at the surface). Unfortunately, they are also a bit more expensive and harder to find.

One final thing to consider: the Wheatstone bridge in a pressure sensor needs an "excitation" source. That's basically just a power source for the bridge. Most require a common DC voltage, like 5, 10, or 12 VDC. These are easy to use. But a few require AC voltages or particular currents (instead of particular voltages), so check carefully to make sure the required excitation is one you can supply easily—ideally, a DC voltage already supplied to other devices in your vehicle.

5.10.8. Global Positioning System (GPS)

GPS can be tremendously useful for positioning a boat over an ROV dive site, and if this is what you want to use it for, then it's easiest to use a boat's built-in GPS receiver or carry a hand-held GPS unit to mark the boat's position. However, if you want to give GPS awareness to an AUV, so it can find out where it is each time it surfaces, then you'll want to get an OEM (original equipment manufacturer) version of a GPS receiver and connect it to a microcontroller. OEM devices are designed to be built into other systems. They lack the stylish case and user-friendly buttons found on a regular GPS unit, but they provide electrical contacts and (usually) a datasheet with instructions telling you how to interface them with a computer or microcontroller—something the regular consumer versions lack. These OEM GPS receivers are available on-line from several GPS manufacturers, as well as from a number of other sites, including electronic component distributors and companies that sell sensors and other parts for robot projects. These robotics-parts vendors can be particularly helpful, since they've already selected for GPS units that are easy to integrate into robots, and that's basically what an AUV is.

Image courtesy of Parallax Inc.

Figure 9.49: A GPS Receiver (OEM Version)

This small GPS receiver from Parallax, Inc. is one of several that are easy to interface with a microcontroller. It can be used to provide an AUV with the ability to determine its precise location at sea whenever the AUV is at the surface.

When configured properly by the microcontroller, the GPS receiver will output a stream of GPS data in a standardized ASCII character format known as **NMEA format** (for **National Marine Electronics Association**), which the microcontroller can receive and interpret. This approach requires knowledge of ASCII and the NMEA formats (which can be found on the web). It also requires considerable skill in programming microcontrollers and integrating them with sensors, because the GPS unit will send a lot more information than you need, and it will be in a format that can't easily be used directly for navigational computations. You'll need to write a program that's smart enough to sift through all the information, pick out the useful parts, and translate those ASCII character strings into numbers you can work with in navigational calculations.

5.10.9. Sonar

An OEM echosounder altimeter can be installed, using techniques similar to those required to install an OEM GPS unit, provided the unit outputs data in NMEA format. Installing more sophisticated sonar systems on a vehicle is a very challenging project. In fact, it's recommended only for those with advanced skills in hardware and software interfacing. Searching the web for projects in which people have hacked inexpensive sonar fish finders might be another way to get some good ideas for how to incorporate affordable sonar into your vehicle designs.

More advanced sonar systems, like sector scanning sonar, could also be added to your vehicle, but these system are generally expensive and even more complicated to incorporate in home-made control circuitry.

5.10.10. Other Sensor Options

Many additional types of sensors not yet discussed could be useful for underwater vehicle projects. Here are some examples:

- Three-axis accelerometers and other inertial motion sensors can be used to monitor vehicle movements and assist with navigation or dynamic vehicle stabilization.
- Touch sensors can be used to monitor the strength of a manipulator arm's grip on an object or to tell if the vehicle has bumped into something.
- Temperature sensors, chemical sensors, and others can collect oceanographic data or information about the water quality in a local pond.

As your knowledge and experience with sensors and feedback control systems grows, you will no doubt come up with your own creative ideas for using a variety of sensors and incorporating the information from some of these sensors into the feedback control systems on your vehicles.

5.11. Data Display Options

For closed-loop control where the pilot, rather than a microcontroller or computer, is closing the loop by monitoring conditions and making the control decisions, it's essential that the pilot be able to receive and interpret information from the ROV's sensors. There are lots of techniques for doing this. Here are four common ones:

1) The easiest method is the one described for *SeaMATE* in Chapter 12—simply position sensors with visual displays in front of the ROVs video camera, so their data displays are visible in the video image that the pilot sees.

2) **On-Screen Display (OSD)** or **Text-on-Video**: This useful feature displays information from various ROV sensors (e.g., compass heading, depth, water alarms, and power levels) directly on the pilot's monitor. However, an on-screen display feature like this generally needs to be integrated with a more advanced control system. If you are using analog video and electronic sensors with digital outputs interfaced to a microcontroller, then a text-on-video circuit, such as the BOB series (e.g., Bob IV) by Decade Engineering, can superimpose data from those sensors directly on top of the video image (refer back to Figure 9.37). With this method, it's possible to control when and where each piece of

Figure 9.50: BOB-4 for Text-on-Video

The BOB series of circuit boards from Decade Engineering follows coded instructions from a microcontroller to overlay text information onto a video image. If the microcontroller is connected to appropriate sensors, this board can be used to add real-time depth, heading, battery voltage, or other data to the live image coming from a small ROV's video camera.

Image courtesy of Dr. Steven W. Moore

sensor data is displayed on the screen. It's even possible to change the data display options in the middle of a dive. For example, you might want to see lots of navigational data displayed on the screen during routine piloting, but be able to turn off the navigation data for an unobscured view when something really cool, like a big shark, shows up on screen. You can't do that with a diver's gauge strapped in front of a camera—if it's blocking the view, it's blocking the view, period. With the electronic data display, you could also choose to have a warning message flash on the screen when a leak detection sensor reports water coming in.

3) Another option is a dedicated LCD display for text and/or graphic display of data on the pilot's console. You will need a microcontroller to control the display of information on the LCD panel. As usual, examples and instructions for interfacing LCD displays with microcontrollers can be found on the web.

4) Finally, if your ROV carries an embedded Ethernet controller with the ability to serve web pages, you could have a web page displaying sensor data, possibly alongside the video image from an IP camera. This is not easy to do, but it's a fun challenge for advanced students and has the distinct advantage of allowing remote monitoring of vehicle status from anywhere on the network. Potentially, this could include anywhere on the internet, and therefore, anywhere in the world.

5.12. Using Transistors and Relays for Automated Power Control

Recall that one of the major limitations of microcontrollers is that their I/O pins cannot put out enough power to run anything much bigger than a tiny LED. What's the point of being able to gather and process information from a pilot or a set of navigational sensors if you can't use that information to control high-power devices like thrusters and video lights? In other words, what good is a "smart" AUV if it can't move?

Fortunately, it is possible to control power-hungry machines from a wimpy MCU. The secret is to give your MCU's I/O pins some muscle by routing their control commands through fabulous little devices known as **transistors**. Normally, these transistors can provide all the power amplification you need, but if you find you need even more, you can use a transistor to drive a relay, as explained in Section 5.12.4.

5.12.1. Transistors

Transistors are three-terminal semiconductor components that function as amplifiers or switches by taking a small electrical input signal and using it to control a much larger electrical current.

When the input signal falls within a fairly narrow range of values known as the transistor's "linear range," the transistor acts like a proportional amplifier; the output current is a scaled-up version of the input signal. If you double the input signal, it doubles the output. Transistors operating in their linear range are used in stereo systems at concerts to convert tiny electrical signals from microphones into the beefy currents needed to drive massive woofer speakers. Learning to use transistors in their linear range is fun and useful, but also relatively difficult and will not be covered in this book.

However, when driven by input signals that are either too low or too high to fall within the narrow linear input range, the transistor acts a lot like a simple SPST switch. If the input signal is too small to reach the linear range, the transistor blocks current flow between its other two terminals, acting very much like an SPST switch in the OFF position. The transistor is said to be in "cutoff." On the other hand, when the input signal exceeds the linear range, the transistor goes into what's called its "saturation" state. In this state, it allows electrical currents—even large currents—to flow freely between its other two terminals. Thus, it acts like an SPST switch in the ON position.

NPN PNP

Figure 9.51: Bipolar Junction Transistors (BJTs)

There are two basic types of BJTs: NPN and PNP, each with its own schematic symbol. They differ in the polarity of the voltage that needs to be applied between their collector and emitter. They also differ in the direction of current flow, through the base, required to turn them on.

In an NPN transistor, the collector (C) voltage must be higher than the emitter (E) voltage, and adequate current flow into the base (B) turns it ON.

In a PNP transistor, the emitter voltage must be higher than the collector voltage, and adequate current flow out of the base turns it ON.

The I/O pins on an MCU can deliver enough power to switch between the cutoff and saturation states of many transistors. In other words, MCUs can use transistors to help them switch large electric currents ON and OFF, thereby allowing these smart, but weak, little mini-computers to control high-power devices.

Transistors come in literally thousands of different models, but most can be divided into two major categories: **Bipolar Junction Transistors (BJTs)** and **Field Effect Transistors (FETs)**. BJTs have three terminals called the base, collector, and emitter (abbreviated B, C, E). If a voltage is placed across the collector and emitter, then a small current flowing in or out of the base terminal can be used to control a much larger current (often 100 times larger or more) flowing between the collector and emitter. (Remember, voltage does not flow; current does, as explained in *Chapter 8: Power Systems.*) If no current is flowing through the base terminal, then no current flows between the collector and emitter, and the transistor acts like an SPST switch that is turned OFF. If enough current flows through the base terminal to saturate the transistor (i.e., to move it past its linear range to where the current flow between collector and emitter is the maximum possible), then the effective resistance between the collector and emitter drops to a very low value and the transistor acts like an SPST switch that is ON.

FETs are similar to BJTs, except that the three terminals are called the gate, drain, and source (G, D, S), and it's a *voltage* (rather than a current) applied to the gate that determines the effective resistance (and hence the ease of current flow) between the drain and source. One family of FETs is particularly good at quickly switching large currents ON and OFF. These are called **Metal-Oxide-Semiconductor Field Effect Transistors**, or more commonly just **MOSFETs**, and they are an excellent choice for connecting MCUs to ROV motors, lights, and other high-current devices. (MOSFETs are introduced in Section 5.2; see also *Tech Note: How Motor Controllers Work* later in this chapter.)

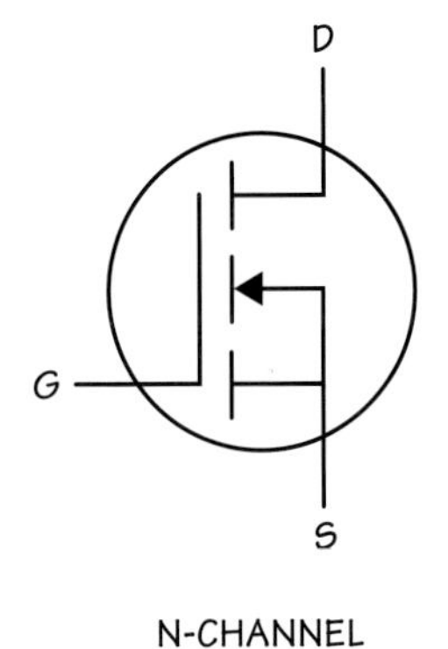

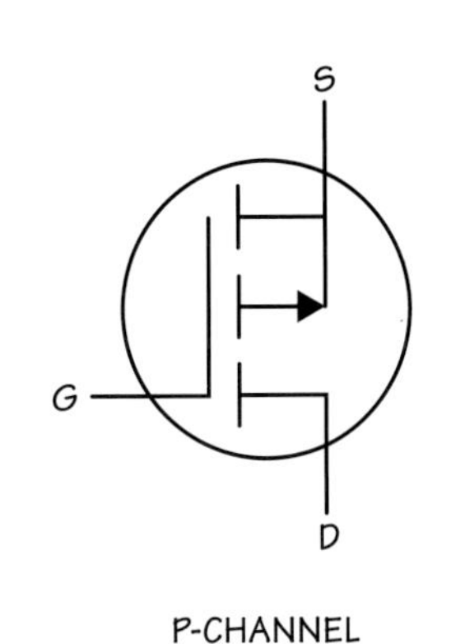

Figure 9.52: N-Channel and P-Channel MOSFETS

Metal-Oxide Semiconductor Field Effect Transistors (MOSFETs) come in both N-channel and P-channel varieties. These are roughly analogous to the NPN and PNP types in BJTs. The letters G, D, and S stand for gate, drain, and source, respectively. Note the different arrow directions and the different positions of D and S in the schematic diagrams for the N- and P-channel devices.

The diagrams shown in this figure represent enhancement mode MOSFETs, because a signal applied to their gate enhances the flow of current through the transistor. N- and P-channel MOSFETs also come in depletion mode varieties (not pictured here). In depletion mode MOSFETs, a signal applied to the gate decreases the flow of current through the transistor.

TECH NOTE: HOW TO KILL YOUR MOSFET

There are three easy ways to kill or wound a MOSFET. First, like many other semiconductor electronic components, MOSFETs are extremely sensitive to static electricity and can be damaged or destroyed simply by contact with your finger if you don't do it right! After being zapped in this way, they won't look damaged. In fact, they'll look perfectly normal, but they just won't work properly. Believe it or not, simply walking across a carpeted room can generate thousands of volts of static electricity on your body, and it takes only about 7 volts to damage the gate of most MOSFETs.

Before removing a MOSFET from its protective static-safe bags or other antistatic packaging, make sure you are touching both the bag and the circuit into which you will plug the MOSFET. (Remember to disconnect any power sources before touching a circuit!) Maintain physical contact with all these things while removing the MOSFET and plugging it into the circuit. This will neutralize any voltage differences between you, the MOSFET, and the circuit. To further safeguard against MOSFET damage from static electricity, avoid working on carpeting or on plastic surfaces (wood or paper is usually ok; special antistatic foam is even better). Also avoid working on electronics in clothing made from synthetic fibers, including polypro and fleece, because these fibers generate high-voltage static electricity quite easily.

A second way to kill your MOSFET is through inductive kick, as explained in *Section 5.12.5*. Devices with coils of wire wrapped around iron cores, including motors and relays, are inductive. That means it's difficult to suddenly change the amount of electrical current flowing through them. When the electrical current flowing through inductive devices is suddenly stopped, a brief but high-voltage spike is produced. This voltage spike can zap your MOSFET. Some MOSFETS include built-in diodes to protect against this hazard by providing an escape route for the current to short out the high-voltage spike. While it is possible to design diode circuits to protect against inductive kick, it's easier to buy MOSFETs that already have protective diodes built into them.

The third way to kill your MOSFET is by running too much current through it. This is particularly problematic while the MOSFET is operating in its linear range (i.e., turned only partway on). This can cause it to overheat, potentially melting down the circuit, burning fingers, or even starting a fire, while also damaging the MOSFET.

When using a MOSFET as a switch, make sure the input signal you feed to its gate is turning the MOSFET completely ON or completely OFF, not partway in between. Look for special MOSFETS designed to switch completely ON in response to the "logic level" signals from a microcontroller.

When shopping for MOSFETs, you'll have lots of choices. Make sure the ones you choose can comfortably handle the maximum currents you expect to be controlling with them. Also, make sure they are designed specifically for logic level inputs (most are not); otherwise, they may not switch ON completely. If they don't do that, they'll fail to deliver full power to the device they're controlling, and they'll also get excessively hot, potentially melting parts of your circuit. You can get them in lots of different physical packages, too. The easiest for most beginners to use are the TO-220 or TO-220AB packages. Before you handle a MOSFET, be aware that they are very easily damaged by static electricity and require special handling procedures. (See *Tech Note: How to Kill Your MOSFET.*)

5.12.2. Using a MOSFET to Turn ON a Video Light

Figure 9.53 illustrates a simple circuit that allows a 5 V microcontroller with maximum I/O pin currents of 20 milliamps to control a bright, 12 V, 24 W video light that draws 2 A (2,000 milliamps) of current. This simple circuit can be used to control many high-power devices, including motors (but first read the warning about inductive kick in Section 5.12.5).

Figure 9.54 summarizes the three approaches discussed so far for switching a motor ON or OFF at the far end of a tether when the battery is located aboard the ROV.

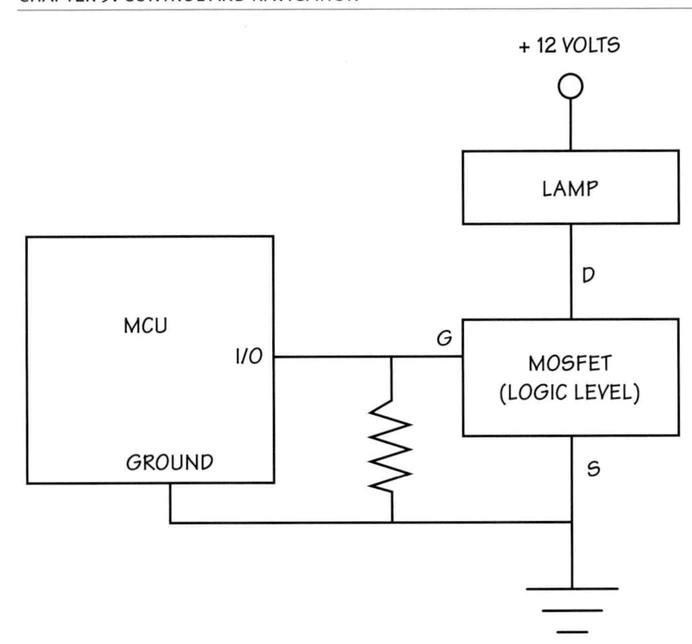

Figure 9.53: Using a MOSFET to Turn ON a Light

This circuit shows how an N-channel enhancement mode MOSFET with logic level gate can enable an MCU to operate a device that requires more voltage and more power than the MCU can deliver. When the MCU raises its output to +5 volts, the MOSFET's gate (G) also goes to +5 volts. Since this is a logic level MOSFET specifically designed to reach saturation with a typical 3–5 volt "logic level" input from an MCU, this puts the MOSFET into saturation, effectively shorting its drain (D) to its source (S). This places 12 volts across the lamp, turning the lamp ON.

When the MCU lowers its output to 0 volts, the transistor shuts OFF, effectively disconnecting the lamp from ground and blocking current flow through the lamp. The lamp goes off. The resistor shown in the circuit is not strictly necessary, but helps drain any residual charge from the gate to ensure that the MOSFET is completely OFF anytime the MCU is not actively trying to turn it ON.

TECH NOTE: CONTROLLING A MOTOR FROM THE SURFACE

This figure provides a comparison of three ways of using a manual switch on a pilot's control console to control an ROV motor that is supplied with power from an onboard battery. This is for illustrative purposes only and does not include any way to control the direction of the motor.

SWITCH A: a manual surface switch directly controls large motor current sent through the tether. This method wastes lots of power by heating the tether and reduces power available to the motor.

SWITCH B: a manual surface switch controls low current through the tether to operate an electromagnetic coil on a relay, which in turn controls larger motor current directly. This is much more efficient, but still wastes some power in the tether and may leave you without enough to control the relay reliably.

SWITCH C: a manual surface switch controls a voltage signal, using essentially no current through the tether, to operate a MOSFET transistor switch that directly controls motor current. This method is so efficient it can be done with very thin wires, allowing for a light, flexible tether.

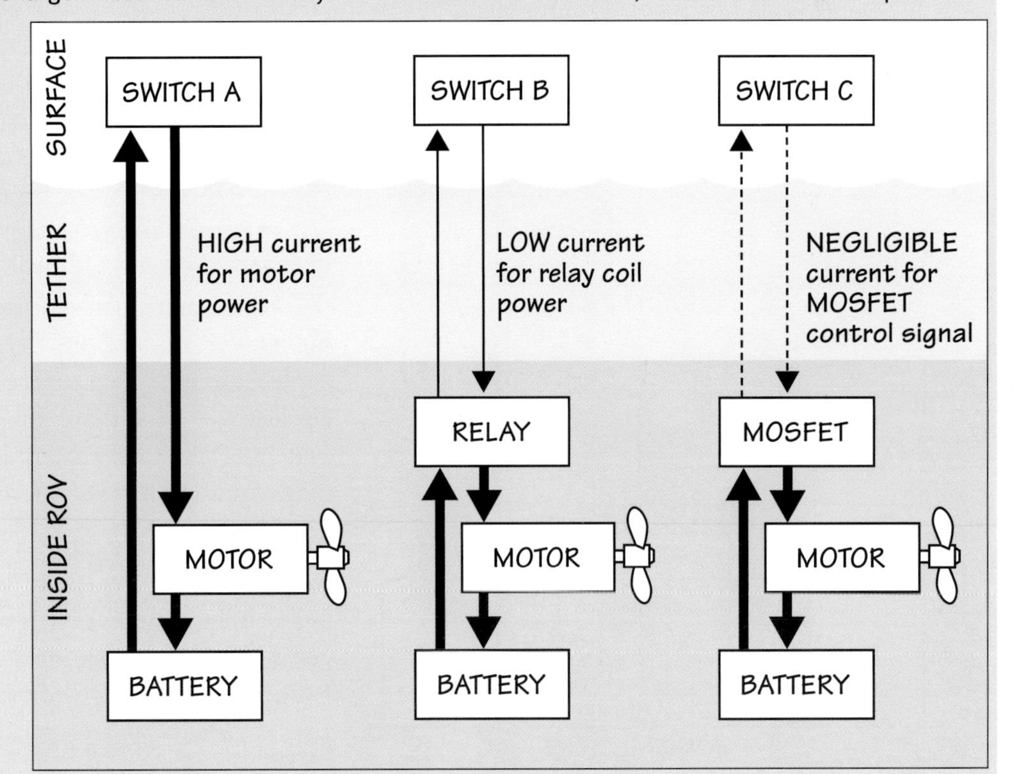

Figure 9.54: Three Ways of Controlling an ROV Motor from the Surface

5.12.3. Pulse Width Modulation

If you ever want to control the brightness of a light or regulate the speed of a motor, you'll need to find some way to adjust the power flowing through the device. Since you have learned about potentiometers or rheostats (both are types of adjustable resistors), you may be tempted to control motor speed (or light brightness or other such things) by using a potentiometer to regulate the voltage supplied to the device. At first, this may seem like a good idea—and it will often work, but this approach has significant drawbacks and should be avoided, particularly in controlling high-power devices. For example, although reducing the voltage to a DC motor will slow the motor, it will also reduce the torque the motor can produce, effectively making it weaker than it needs to be. Worse still, resistors and rheostats work by siphoning off and throwing away (as heat) the energy that would otherwise have gone to the motor. At best, this is inefficient; at worst, it can cause enough heating in the potentiometer or rheostat to destroy it.

There's a better way. Microcontrollers can change the state of their output pins from high to low and back to high again extremely quickly—thousands of times per second, or even faster. Moreover, they can easily and precisely control the fraction of time this train of pulses spends on high versus low. Amazingly, MOSFETs can keep pace as they switch power ON and OFF, following the pulse train produced by the MCU. Thus, microcontrollers can control the *average* power delivered to a device in addition to simply controlling whether it's ON or OFF. The technique is known as **Pulse Width Modulation**, or **PWM**. The fraction of time the pulse train is in the active (ON) state is called the **duty cycle**. In general, the duty cycle can range all the way from 0 to 100 percent. Thus, for example, you could deliver 20 percent of full power to a motor or lamp by generating a pulse train that spent only 20 percent of its time in the active state. This would have the MOSFET turned on only 20 percent of the time.

Figure 9.55: Sample PWM Signals

The fraction of time a pulse train spends in the active (usually high) state is called the duty cycle. This figure shows three different pulse trains. All have the same frequency, but they differ in their duty cycles.

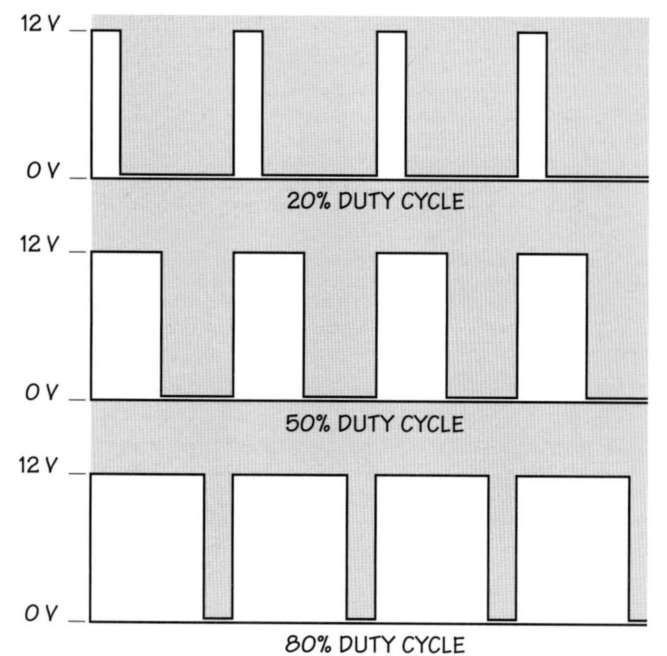

PWM control systems can be made much more efficient than any of the common alternatives that limit power by siphoning off energy as heat. Recall that electrical power dissipated in a device equals the voltage across the device multiplied by the current through it. When the transistors used in PWM are fully ON, they have very little resistance, so essentially a full voltage drop occurs across the load (i.e., the light bulb, the motor, or whatever), rather than across the transistors. Therefore, the voltage drop across the transistor is very low, which makes the power dissipated by the transistor while in the ON state very low, too. When the transistors are OFF, the voltage across the transistors is higher, but there's zero current flow, so no power dissipation. Thus, when PWM is used, the vast majority of the power goes directly to the intended load, without being wasted as heat in the control circuit. (That said, even the best PWM control circuits are less than 100 percent efficient. Those switching very high current loads, such as large motors, may require heat sinks or even fans for cooling.)

Another advantage of PWM, particularly for motor control, is that a typical brushed DC motor will retain nearly full torque, even at a very slow speed, because it's getting its full voltage during those ON pulses. Thus, PWM offers speed control without sacrificing torque.

5.12.4. Relays

Some of today's MOSFETs, even the ones routinely available from distributors of electronic components, can handle over 100 amps of continuous current under normal circumstances. That's more than enough for thruster motors or any other devices typically found on a small ROV or AUV. However, in the unlikely event that you can't find a MOSFET beefy enough to handle your vehicle's needs, you can resort to relays for really high power requirements.

Relays predate transistors by decades and are quite simple in their basic concept. They are essentially manual switches operated by electromagnets, rather than human fingers. A coil of wire wrapped around a ferrous (iron or steel) metal core creates a temporary magnet inside the relay any time electric current flows through the coil. The magnetic field produced while the coil is energized controls the switching action by physically attracting a metal plate attached to some spring-loaded contacts. When power to the coil is cut, the spring pulls the contacts back to their resting positions.

Like manual switches, relays come in SPST, SPDT, DPDT, and other combinations. In general, a relay will have two more terminals than the corresponding type of manual switch. The two extra terminals are connected to the two ends of the electromagnet coil and are used to operate the relay. For example, a typical DPDT relay has eight terminals: six for the DPDT switch part, plus two extras for the coil.

The switch contacts can be made very robust, so they can handle huge voltages and currents. Thus, a modest current used to activate the electromagnet can be used to control much larger currents flowing through the relay's contacts. As with any manual switch, you must make sure the contacts in the relay you use are rated for the maximum voltage and current they need to control, including motor stall currents, which can be much larger than the usual operating currents. Similarly, you must ensure that the voltage and current required to operate the coil are within the capabilities of whatever device is powering the coil to control the relay. Although the coil voltages and currents are typically much less than the maximum voltages and currents a relay can control, those coil currents may still be more than a microcontroller can handle. Therefore, signals from a microcontroller usually need a little boost from a transistor before they can control a relay.

Unfortunately, relays do not operate quickly enough to be useful for PWM control. Moreover, their contacts would wear out quickly if subjected to such rapid cycling.

Figure 9.56: Anatomy and Function of a DPDT Relay

This is an older-style, double-pole, double-throw (DPST) relay. It offers one normally open and one normally closed contact on each side of the relay. When the electromagnet is not *energized, contacts on the thin copper strips extending from the spring-loaded armature touch the upper terminals on each side of the relay, forming electrical connections between them. Meanwhile, there are tiny gaps between the copper strip and the lower terminal, preventing electrical current flow between the copper strips and the lower terminals. (The blue arrows show one of the gaps.)*

However, whenever the electromagnet is energized, the roles of the upper and lower terminals reverse; the armature and the copper strips get pulled down toward the electromagnet, breaking the electrical connection at the upper terminals and forming an electrical connection on the lower ones.

Since the copper strips and terminals can support much larger currents than those required to energize the electromagnet coil, a relatively weak circuit used to power the coil can switch ON or OFF a much larger and more powerful circuit wired to the copper strips and terminals.

TECH NOTE: RELAY OPTION DETAIL

Figure 9.57 shows in a bit more detail how the relay option (Switch B) in Figure 9.54 works.

In this diagram, a small SPST switch (upper left) located on the pilot's surface control console controls a small current flowing through tether wires (thin lines) to operate the electromagnet coil inside a relay (gray box) on the ROV. When (and only when) the pilot's switch is closed, the relay will become energized and create a magnetic field that will pull closed (dashed arrow) the sturdy SPST switch contacts inside the relay. This permits large currents to flow locally from the battery to the motor through short, thick wires (heavy black lines).

A MOSFET switch used in place of the manual pilot's switch and placed aboard the ROV would allow an on-board microcontroller to operate this motor, which might be useful in an AUV. Note that these particular relay and MOSFET circuits can only turn the motor ON or OFF; they cannot provide direction or speed control.

Figure 9.57: Using a Relay for More Efficient Motor Control through a Long Tether

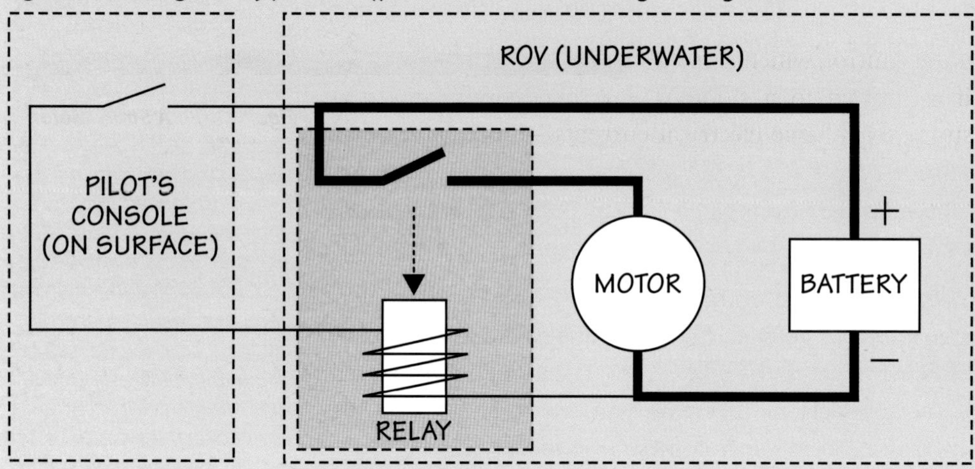

5.12.5. Inductive Kick

Warning: Using a transistor to drive a relay, motor, solenoid, or any other inductive device (one with a coil of wire around a piece of iron or steel) must be done carefully to avoid damaging the transistor and nearby circuitry. These coils can produce a huge voltage spike of up to a several thousand volts each time current through the coil is suddenly stopped, even if the original voltage sent to the coil was only a few volts. This phenomenon is known as **inductive kick**. It can destroy a transistor, as well as any MCU or other device connected to the transistor. It is prudent to use a diode (a semiconductor device that acts as a one-way valve for electrical current), as shown in Figure 9.58, to protect against inductive kick anytime a transistor is being used to control a relay, motor, solenoid, or other inductive device. (See below for more information about diodes.)

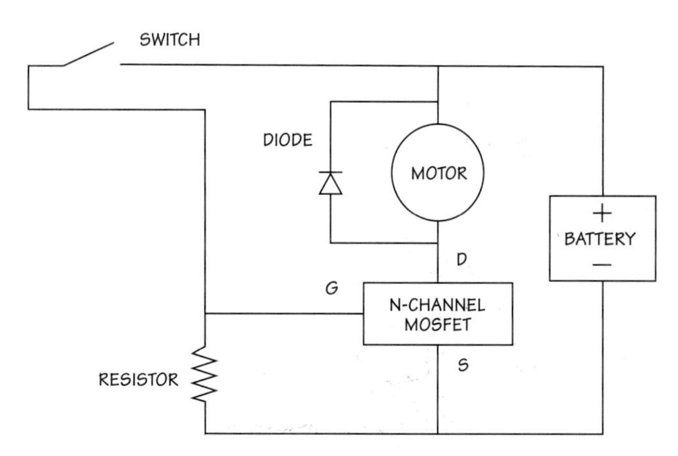

Figure 9.58: Using a Diode to Protect a MOSFET from Inductive Kick

A MOSFET or other transistor used to switch an inductive load, such as a relay coil, motor, or solenoid, can be destroyed by inductive kick. In this circuit, a diode (triangle symbol with bar across tip) has been placed in parallel with the inductive device (in this case a motor), to provide an alternative current path for current that is suddenly blocked from flowing through the MOSFET. The orientation of the diode is important.

5.13. Motor Controllers

Motor controllers, sometimes called **electronic speed controllers** (**ESCs**), are specialized circuits designed specifically to control the speed and direction of electric motors. They provide a great example of how transistors and (in some cases) relays can improve ROV performance. When building an AUV or adding autopilot features to a more advanced ROV, or even when simply trying to avoid large tether currents by putting motor control in the hands of an MCU on board the ROV, it's necessary to have some way for a microcontroller or computer to control the motors.

Motors present an especially difficult control challenge. Whereas lights usually require only ON/OFF control and (rarely) brightness control, motors commonly demand ON/OFF control, FORWARD/BACKWARD control, speed or acceleration control, torque control, and in some cases, active braking control, which stops a motor more quickly than if it coasted to a stop. In addition, motors can draw surprisingly large electrical currents—much larger than their normal operating currents—right when they first start up and whenever they stall (i.e., get stuck and can't turn). Finally, when motors stop or change directions, they often produce big voltage spikes (the inductive kick described earlier) that can quickly destroy poorly designed electronic control circuits.

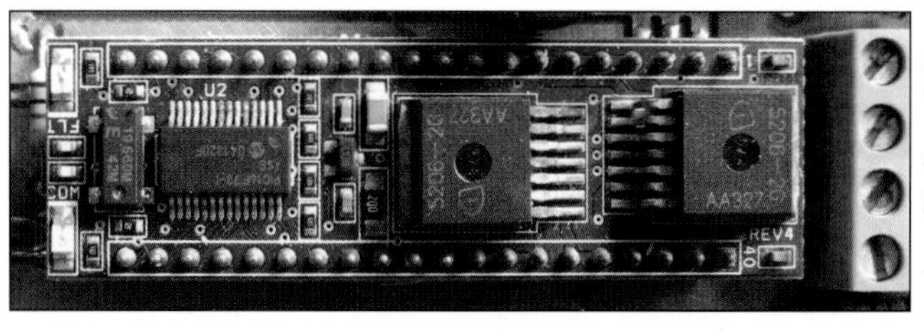

Image courtesy of Dr. Steven W. Moore

Figure 9.59: A Small Motor Controller

This small but versatile motor controller (called a Motor Mind C) is just one of hundreds of types available. This one fits into a socket on a printed circuit board and is designed to provide independent control of two small (less than 2 Amp) DC electric motors at once.

Larger motor controllers are often stand-alone, may include fans or other cooling mechanisms, and are designed to be bolted onto a support surface.

The little Motor Mind C in this photo would be suitable for use in a very small ROV or AUV operated in a pool. However, you'd probably want something capable of handling larger motor currents for any vehicle that is larger, needs to move quickly, or will be operating in areas with significant water movement, such as rivers or the ocean.

A motor controller is a circuit specifically designed to grapple with these challenges. A good motor controller will take simple, low-power commands for motor speed and direction and translate those commands into the actual patterns of larger electrical currents needed to produce the desired behavior from the motor. While it is possible (but not easy) to build your own motor controllers out of individual transistors and other components, most people start by purchasing a commercially available motor controller.

A typical brushed DC motor controller has several places for connecting wires:

- There are two beefy wires coming from the battery (or other DC electric power source) to the motor controller.
- Two additional thick wires (positive and negative) deliver power from the motor controller to the motor.
- Other (usually smaller) wires bring commands to the motor controller.
- There may be additional wires to supply separate power for the motor controller circuitry, particularly if that circuitry requires a different voltage than the motors.

Some motor controllers have places to connect a heat sink, which is a piece of metal used to draw excess heat away from the controller.

5.13.1. Choosing a Motor Controller for a Brushed DC Motor

Motor control is big business, with thousands of industries needing to control motor speed for all sorts of critical tasks—transportation, conveyor belts, positioning systems, and many more. To support this enormous demand, many companies specialize entirely in motor control technologies, both electronic hardware and software, so there is no shortage of motor controller options. In fact, there are so many motor controllers on the market that it may feel overwhelming to sift through them all. It works to narrow the search by going to distributors specializing in parts for small- to medium-sized robotic projects; generally, they have already sifted through the masses of motor controllers and picked out ones with good specifications for a range of motor voltages and sizes similar to those commonly used in small ROV and AUV projects.

Keep in mind that different types of motors require different types of motor controllers. For example, one designed for brushed DC motors generally won't work for controlling a brushless DC motor. (These motor types and their control were introduced in *Chapter 7: Moving and Maneuvering* and are described in more detail in Section 5.13.2.) Furthermore, those designed for controlling brushed DC motors come in a variety of different voltage and current ratings, and some can control more than one motor at a time. In general, those that can handle more current and/or more motors are bigger and more expensive. Special features, such as built-in overload protection, the ability to apply active braking, or the ability to receive commands in more than one standard format, also cost extra.

When choosing a brushed DC motor controller, consider these factors:

- How many motors can it control at once? Sometimes getting a motor controller that can control more than one of your thrusters or other motors at a time can save money and space.
- Are the minimum and maximum voltages it can supply compatible with your motor requirements?
- What's the maximum current available for each motor? Can it handle the stall current for the motors selected? If so, for how long?
- Does it have built-in systems to protect it from overheating, excessive current (as when a motor stalls), accidentally reversed battery polarity, and other common problems? These features may cost a few extra bucks but they're less expensive that replacing a whole motor controller every time something goes wrong.
- If the motor controller has automatic shutoff to protect it from overheating, can it be reset remotely after the motor controller has cooled off, or will your ROV be dead in the water?
- Will any required extras, such as heat sinks or other cooling accessories, be added (at extra cost)? Will these added parts fit and function inside an enclosed pressure can?
- What type of input signals can the motor controller accept as commands? Some use an analog signal from a potentiometer. Others use standard R/C car or plane control signals. Still others are designed to receive ASCII coded instructions over a serial data link. Some accept all three formats, providing good flexibility.
- Does the controller provide a way to monitor its status? Some have LED lights to warn of overheating or other problems. Some can send coded status messages back to a computer over a serial port connection.
- How big is the controller physically, and what shape is it? Will it fit, and can it be securely mounted inside a compact pressure canister?
- Is it provided with easy, yet reliable, ways to connect battery power, motor wires, and control signals?

TECH NOTE: HOW MOTOR CONTROLLERS WORK

Even if you don't plan to build your own motor controller, knowing something about their internal workings can help you make more informed decisions when you go to purchase a commercially available one.

To control motor *speed*, most motor controllers rely on a built-in microcontroller and MOSFETs for pulse width modulation (PWM) of the currents flowing to the motors. To control motor *direction*, a typical motor controller for brushed DC motors will use a DPDT relay or its transistor-based equivalent, called an "H-bridge" (described below). The DPDT relay controls direction in exactly the same way a manual DPDT switch can be used to control direction (as covered earlier in the chapter), except it's controlled electronically, rather than by a manual switch.

Controlling motor direction with transistors is a bit trickier. It's not possible to just run out and buy a DPDT transistor. By their very nature, transistors used as switches act like SPST switches. Each transistor can control only one circuit at a time, and there are only two choices of position—ON or OFF. Nonetheless, it is possible (indeed, very common) to use transistors to control motor direction. The trick is to arrange four transistors in a special circuit configuration known as an **H-bridge**. It's called an H-bridge because the schematic diagram for the circuit resembles a capital letter "H" as you can see in Figure 9.60.

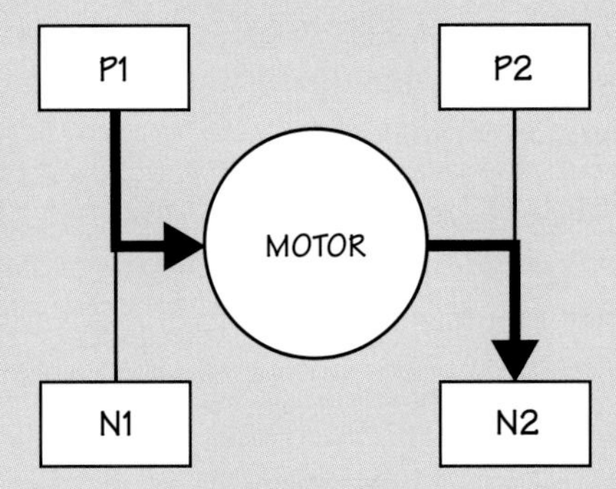

Figure 9.60: Simplified H-Bridge

Four transistors arranged in an "H" as shown in this diagram can control motor direction.

Although Bipolar Junction Transistors (BJTs) can be used, most modern H-bridges designed for motor control use MOSFETs, because of their more efficient power-handling capabilities. In Figure 9.60, two P-channel MOSFETs (P1 and P2) and two N-channel MOSFETs (N1 and N2) are used to control the direction of current flow through a motor. Both of the P-channel MOSFETs are connected to the positive terminal of the battery (connections not shown), and both of the N-channel MOSFETs are connected to the negative terminal of the battery (again, connections not shown). If P1 and N2 are both switched ON while P2 and N1 are OFF, then the current path highlighted in bold is open, and current flows down through P1, through the motor from left to right, then out through N2. If, on the other hand, P2 and N1 are ON while P1 and N2 are OFF, then current flows in the opposite direction (right to left) through the motor, so the motor rotates in the opposite direction.

Interestingly, the H-bridge provides a way to apply **dynamic braking** to a DC brushed motor. Of course, simply shutting off the supply of power to a DC motor will allow it to coast gradually to a stop. However, actively shorting one terminal of the motor to the other while the motor is spinning will cause it to actively resist its own rotation—it will screech (literally) to a sudden halt. With an H-bridge, this braking action can be activated easily by turning N1 and N2 ON simultaneously (while P1 and P2 are OFF), or turning P1 and P2 simultaneously ON (while N1 and N2 are OFF). (See *Safety Note: H-Bridge Safety*.)

Image courtesy of Dr. Steven W. Moore

Figure 9.61: LMD 18200 H-Bridge Chip

This popular integrated circuit, or IC, is basically an H-bridge in a chip. The model shown (LMD 18200) can handle the heavy lifting of modest motor currents (up to about 3 amps), but it does not have much in the way of built-in smarts, and it needs to be soldered into a circuit board for connections to power, motors, etc. If you want speed control, it will be up to you to provide a microcontroller and to write the program needed to generate the required pulse-width modulated speed control signals, direction signals, braking signals, etc.

SAFETY NOTE: H-BRIDGE SAFETY

There are dangers in using this H-Bridge/MOSFET scheme, particularly if you try to build the H-bridge and its controller yourself. For example, you must take great care to make sure that P1 and N1 are never switched ON at the same time. If they are, a direct short will be created across the battery terminals. Engineers jokingly refer to this H-bridge condition as the "fuse test." Good motor control systems are designed to minimize the chance of this and other disasters. They always include a fuse to protect against fires or other serious damage, should something not go according to plan. You, too, should always include a fuse in series with the battery or other power supply. The amp rating on the fuse should be no more than about twice the maximum current that the vehicle is expected to draw under normal operations (i.e., with all thrusters operating at 100 percent power in water with all lights and other devices turned ON.)

5.13.2. Controlling Other Types of Motors

Up to this point, the discussion of motors and motor control has focused primarily on brushed DC motors, because these are the ones most commonly used in home-made ROVs and AUVs. But there are at least three other categories of motors that may be of use to designers of small underwater vehicles: (1) brushless DC motors, (2) stepper motors, and (3) servos.

1. **Brushless DC motors** are used in the same ways as brushed DC motors. (Section 5 of *Chapter 7: Moving and Maneuvering* discusses various motor types.) Brushless motors offer a number of performance advantages over brushed motors, but at substantial additional cost. Among the performance advantages for underwater vehicle use are:

- better reliability and longevity
- greater efficiency (particularly in low-load situations)
- better heat dissipation
- less electrical noise
- better operation in oil-compensated underwater housings

Although the brushless motors themselves are typically a bit more expensive than similarly powered brushed motors, the real added expense arises because each brushless motor *must* be run with a sophisticated electronic motor controller made specifically for that particular type of brushless motor. These motors will not rotate if connected directly to DC power.

In spite of their higher cost, small, lightweight brushless motors are popular for driving the propellers on remote-controlled model airplanes. And many ROVs and AUVs rely on brushless motors, because they have somewhat better performance and work better than brushed motors in oil-compensated housings.

2. **Stepper motors** are motors that take tiny rotational steps of only a few degrees at a time in response to electrical pulses sent by a special stepper motor control circuit. These motors can be rotated a precise number of degrees and held there by sending the appropriate number of properly sequenced pulses. They are often used to drive worm screws or other threaded shafts in devices like printers and plotters, where extremely precise, repeatable positioning ability is more important than speed or other motor parameters.

If driving a threaded shaft, each full 360° rotation of the motor rotates the shaft one full revolution and therefore moves something driven by the shaft threads by the distance separating adjacent threads. Thus, it is possible to get repeatable positioning to within better than 1/1000 of an inch! In an ROV, this type of motor could be useful for precise positioning of a camera tilt mechanism or for sucking a precise volume of water into a syringe for water sample testing. You can control stepper motors with commercial stepper controllers, or you can build your own from microcontrollers and small power amplifiers called stepper motor drivers.

3. **Servo motors** (or **servos**, for short) are motors that are generally (but not always) set up to rotate less than 360° total to a specified angle, then hold that position. They are popular for controlling the leg joint angles on small robots or adjusting the rudder angle and other control surface angles on remote-controlled model airplanes. Like stepper motors, they could be used to control a camera tilt angle, manipulator arm grippers, and other such devices. Servo position is usually controlled by a series of electrical pulses, where the width of the pulse ranges between 1.0 and 2.0 milliseconds. At 1.0 ms, the servo is turned all the way in one direction, at 2.0 ms, it's all the way in the opposite direction, and at 1.5 ms, it's halfway in between. Some servos use these pulse width codes to control speed and direction, rather than position.

Image courtesy of Dr. Steven W. Moore

Figure 9.62: Typical Radio-Controlled Servo Motor

If you are experienced with electronics and want to try building controllers for any of these types of motors, you will find plenty of examples and instructions on the web or in electronics books and magazines. If you're more inclined to buy one or get it donated, be aware that the control circuits and software for each type of servo motor can be very distinct from the other types. Be sure to get a controller that's compatible with the type of motor you're planning to use.

5.14. Limiting Motor Travel

In some cases, you may want to limit how far a motor can turn in each direction in order to prevent damage that might occur if the motor were inadvertently told (or simply allowed) to go too far in one direction. For example, if a strong motor attempted to keep closing gripper jaws after they were already closed completely, something could break. Likewise, if a camera tilt mechanism tilted too far, it could run the camera lens into something and damage the lens, the tilt mechanism, or whatever the lens ran into.

A common way of protecting motor-controlled systems from this type of damage is to use **limit switches** in combination with **diodes**. Limit switches are usually just manual pushbutton switches mounted at a location where some part of the mechanism moved by the motor will bump into them and activate the switch. The switches can be wired so that they cut power to the motor if the motor tries to go too far and bumps into the switch. Sometimes little levers or other mechanisms are added, to make it easier for the motor movements to operate the switch reliably. Limit switches work best with gear motors or other mechanisms where the motor has been geared way down, so things move slowly, even when the motor is spinning quickly. Otherwise, cutting power may not stop the motion quickly enough.

Diodes are semiconductor devices that act as one-way valves for electrical current. Each limit switch is wired so that it is normally closed (i.e., it normally conducts electricity), but opens when the motor has reached its authorized limit of travel in one particular direction. Diodes are placed across each switch and oriented to allow the motor to be driven away from an opened switch. (Otherwise, you'd never be able to reverse the motor direction once it had hit a limit switch.)

The schematic diagram in Figure 9.63 illustrates how limit switches and diodes can be used to protect a motorized mechanism from going too far.

If you are using microcontrollers for motor control, it's possible to limit motor motion through the use of regular limit switches that talk to the microcontroller instead of cutting power directly, essentially letting the microcontroller know there's a problem, so it can tell the motor to stop. You can also use optical or magnetic sensors that tell these "smart" circuits when to stop turning a motor in a particular direction. These "information only" sensors may be easier to use than traditional mechanical limit switches; however, they don't directly stop the motor. They only put in a polite request. If something goes wrong with the complex "smart" circuitry, that request may be ignored with unhappy consequences, so there's a lot to be said for the good old-fashioned mechanical limit switches with diodes that literally cut power to the motor, rather than simply asking it to stop.

TECH NOTE: LIMIT SWITCHES

This schematic diagram shows how diodes and limit switches can be used to limit how far a motor can move in each direction. A battery (top) provides power to the motor through a DPDT switch, which can be used to control the direction of motor movement by reversing the polarity of the voltage applied to the motor. At the moment, it is connecting the positive voltage to the right side of the diagram, and this is causing the gear motor to rotate in a direction that moves the heavy black bar slowly toward the right. The diodes are at the far right and left sides of the diagram, shown as a triangle with a small bar on its tip. A diode acts as a one-way valve for electric current. In this case, the diodes are oriented to allow current to flow up through them, but not down.

Connected in parallel with each diode is an SPST switch used as a limit switch. At the moment shown in the diagram, the motor has pushed the heavy black bar far enough to the right that the bar has opened the limit switch on the right side of the figure, cutting off the clockwise flow of current that was causing the motor to push the bar to the right. (The diode on the right will not allow current to flow down through it, so there's no way for the current to bypass the open switch in that direction.) If the DPDT switch is reversed, so that the left side of the motor gets connected to the (+) voltage, current will be able to flow down through the left limit switch, through the motor, and back up through the diode on the right side, bypassing the open switch in that direction and allowing the motor to back the bar away from the right-most limit switch.

The motor can continue to move the bar to the left until the bar runs into the left-most limit switch, at which point counterclockwise flow of current around the circuit will be blocked, and further movement to the left will be impossible. However, the right-most limit switch will be closed again, so it will be possible to move the bar back toward the right. Note that any limit switches and diodes used must be able to handle the normal motor currents without overheating.

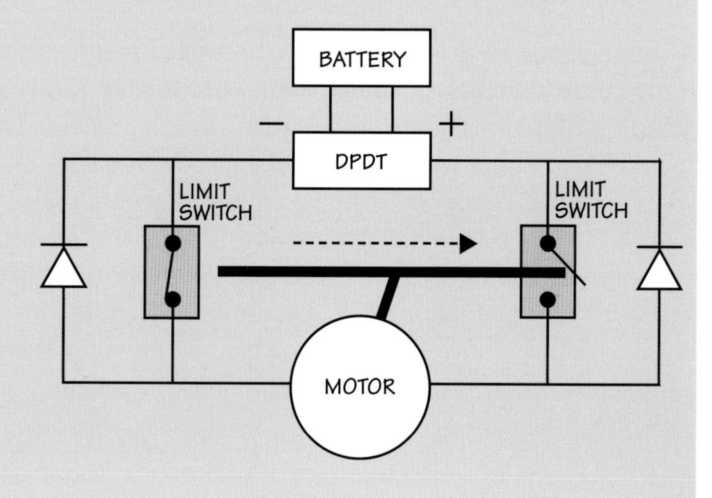

Figure 9.63: Using Diodes and Limit Switches to Control Motor Movement

5.15. Feedback Control Algorithms

An **algorithm** is a set of steps or rules for doing something. Control algorithms are computer programs or other sets of instructions that determine what commands get issued to try to make the actual conditions match the desired conditions.

In control theory jargon, the difference between the actual and desired conditions is called the "error signal" or simply the "error." There are a number of fairly standard algorithms for translating an error signal into a command that will cause an action to reduce the size of the error (i.e., move the actual condition closer to the desired condition). These algorithms range from very simple to very complicated. All of them can be implemented in hardware or software, but software is easiest to adjust later, if necessary. A thorough introduction to control algorithms and how to write the programs or build the hardware that implements those algorithms is well beyond the scope of this book, but here's a quick summary of the basic idea behind some of the popular ones. This can be used as a starting point for anyone who wants to do a little research to learn more about using these (or other) control algorithms in their own underwater projects.

- **ON/OFF** or "**Bang-Bang**" Control: This is the simplest of the standard control algorithms. Most home heating systems work this way—if it's too cold, the furnace turns ON. If not, it turns OFF. There is no option to regulate how strongly the heater turns ON. It's simply ON or OFF. This type of control, shown in Figure 9.64, is the simplest to implement, but is too limited in its performance for many applications.

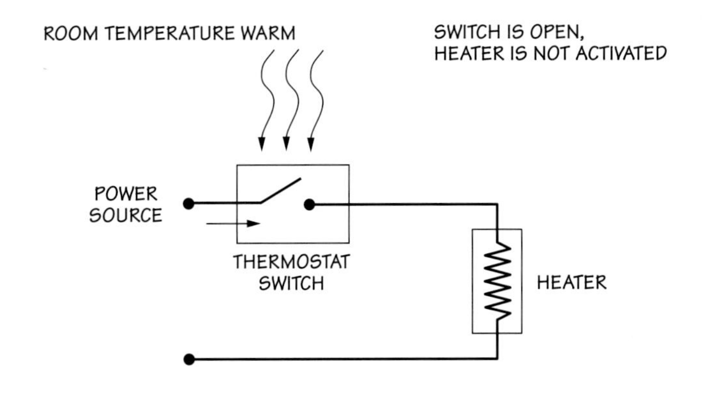

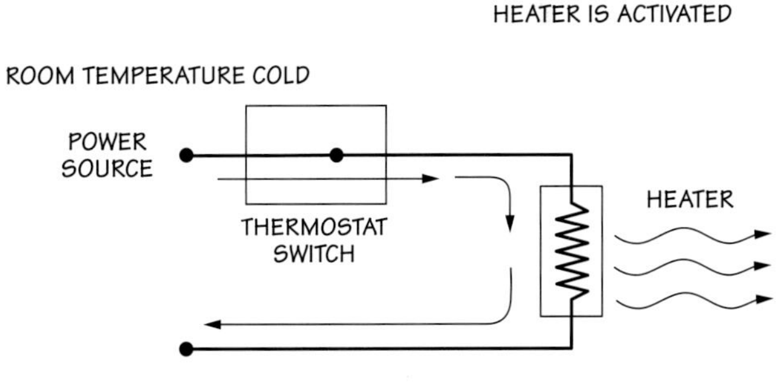

Figure 9.64: Simple Thermostat Circuit

- **Proportional ("P") Control**: This is one notch more sophisticated than ON/OFF control. It adjusts the *strength* of the corrective action to match the size of the error. If the desired and actual results are very different, the control system will attempt a large correction to bring them back into agreement quickly, but if they are only slightly different, then a much gentler correction will be imposed. In general, this allows more rapid corrections to large errors while reducing the risk of overshooting the desired result. Unfortunately, proportional control has two problems that limit the usefulness of this simple strategy. First, in the presence of an external force that's trying to push the system away from the desired value, P control will always leave a slight, steady error. You can reduce the error by increasing the strength of the corrective action, but only up to a point. If you go too far, a second problem will rear its ugly head: the corrections will overshoot the desired result and lead to sometimes violent oscillations.

- **PID Control**: The industry standard for motor control and many other control applications is something called PID (pronounced pee-eye-dee) control. PID stands for Proportional + Integral + Derivative. (These terms are explained below). PID provides improved performance over simple P control. In PID control, the signal going to the motor (or other thing being controlled) is derived from the sum of three different control signals:

1) The first of these (the "P" part) is proportional to the error between the actual and desired results. It's identical to the P control discussed above. Its role is to make the strength of the corrective action appropriate for the size of the error.

2) The second signal (the integral or "I" part) is proportional to the accumulation of error over time. It's called integral, because that is mathematical terminology (from calculus) for adding up lots of little amounts of something to get the total. The integral part grows gradually stronger the longer an error persists, and helps force any small, otherwise persistent errors down to zero.

3) The third signal (the derivative or "D" part) is proportional to the rate of change of error. (In calculus terminology, "derivative" relates to the rate at which something is changing.) This part puts on the brakes when corrections are happening quickly, and this helps prevent the control system from overshooting the desired final value. Its presence allows the strength of the P term to be increased beyond the point that would otherwise result in oscillations. This permits faster corrections.

Sample programs for performing PID control are widely available on the web in a variety of different computer languages, including some microcontroller programming languages.

- **Adaptive Control**: The latest advances in control system design actually incorporate features that actively monitor the effectiveness of the control algorithm being used and, if necessary, adjust that strategy to improve performance as much as possible. Incorporating adaptive control strategies in your underwater vehicle requires an advanced understanding of control system performance plus advanced programming skills, but it's an exciting field for anyone who wants to pursue an active interest (and possibly a career) in control systems.

5.15.1. Troubleshooting Abnormal Control System Behavior

As mentioned near the beginning of this chapter, closed-loop control systems are simple in basic concept, but in practice they are notoriously difficult to implement successfully. Sometimes they don't seem to do anything. Sometimes they seem to work, but not very well. Sometimes they just act silly. And sometimes they fail in dramatic (and occasionally dangerous) ways, often without warning.

This final section of the chapter summarizes a few of the most common and perplexing problems encountered by people putting together closed-loop control systems. Potential malfunctions of a depth control autopilot operating in a swimming pool are used to illustrate the rather bizarre behaviors normally associated with each type of problem. Of course, these wacky behaviors are not limited to depth autopilots—corresponding behaviors, with corresponding causes and solutions, can show up in *any* closed loop control system.

Sample Bizarre Behavior #1: Instead of hovering at the desired depth, the vehicle races all the way to the surface or all the way to the bottom and stays there.

There are at least four possible causes of this problem:

1. The first possible cause is not really a control system problem at all. It may simply be that the vehicle is ballasted improperly (i.e., it's too dense or too buoyant), and the control system isn't powerful enough to overcome the vehicle's natural tendency to sink or float. To avoid this problem, shut off the thrusters and make sure the vehicle is close to neutrally buoyant.

2. The second possible cause is that the depth sensor information is not being received or processed properly. Perhaps the sensor is damaged, or wires from the sensor have come loose, or there's a problem in the microcontroller program (software) such that the sensor readings are not being updated or used properly.

3. The third possibility is that signal polarity has accidentally been reversed somewhere in the control loop, so the vehicle drives upward whenever it's supposed to drive downward, and vice versa. The cause can be as simple as a pair of crossed motor wires or as subtle as an improperly placed minus sign somewhere inside the control software.

4. The fourth possible explanation is that the control system is actually working fine, except that it thinks the desired depth is somewhere above the surface of the water or below the bottom of the pool. Check carefully to make sure a realistic depth has been specified and that this specification is interpreted properly throughout the entire control loop. For example, if the person piloting the vehicle in an 8-foot deep swimming pool entered a desired depth of "6," thinking "6 feet," but the control system was programmed (by someone else) to interpret the input value as "6 *meters*" (which is about 20 feet), the ROV would hit the bottom of the 8-foot deep pool before reaching its desired depth of 6 meters. Tracking down this type of error can be tricky, because there are lots of places in the control system hardware and software where a number gets converted from one form to another (e.g., from typed value to voltage on a wire). Somewhere in the feedback loop, the conversion factor may be wrong.

Sample Bizarre Behavior #2: The control system exhibits a sluggish or incomplete response.

Sometimes a feedback control system will attempt to drive the vehicle in an appropriate direction, but it will do so more slowly than desired or will do so incompletely (perhaps going only 80 percent of the way to the desired depth).

1. One possible explanation is simply that the thrusters are too weak to achieve the desired value effectively.

2. A second possibility, especially in the case of proportional control, is that the proportionality constant (the "gain") is set too low. Increasing the gain, that is, increasing the strength and speed of corrective action taken by the control system, can help, if it's possible to do so. But at some point, this approach may lead to another, potentially more serious behavior, as discussed next.

Sample Bizarre Behavior #3: The control system drives the vehicle to more or less the correct depth, but instead of sitting still calmly, the vehicle oscillates up and down like a bouncing ball.

Even a properly wired control system with perfectly functional sensors and actuators can have serious problems. When control systems regulate mechanical systems, such as wheels or propellers, a common symptom of this type of problem is oscillations or vibrations. These vibrations often start small, but in powerful systems, can grow large enough to literally rip the vehicle apart!

Here's what's going wrong: In any closed-loop control system, it takes some time for the control signals to propagate around the control loop. This time is known as the "loop delay." If the loop delay is long enough to allow the control system to drive the system past its desired condition before the motors or other actuators get the message to stop, then a series of alternating over-corrections will result. Using the 6-foot dive depth as an example, the feedback-controlled depth autopilot might get the signal that the vehicle is at 4 feet when it's supposed to be at 6 feet, so it would switch on the down thrusters. If there was a substantial delay in the feedback loop, so that when the sensors were reporting arrival at the 6-foot level, the vehicle was already at 7 feet, then the "down" thrusters wouldn't shut off until 7 feet. But a few moments later, the sensor data would catch up, and the system would realize it's too deep and turn on the "up" thrusters. Then the problem would repeat itself in the opposite direction as soon as the vehicle control system learned (belatedly) that it was above the 6-foot depth.

If the loop delay isn't too bad, these oscillations will gradually diminish in size and finally fade away completely. If the delay is a little longer, the oscillations may continue indefinitely at about the same size. If the delay is longer still, the oscillations can actually begin to grow in size, so that each over-correction is bigger than the last one. In such a case, the size of the oscillations will keep growing unless something stops it—like crashing into the bottom of the pool. Occasionally nothing stops it until the vehicle rips itself apart.

To cure this type of problem, you have two possible courses of action. The best is to try to reduce the delay in the feedback loop, if possible. The next best is to reduce the gain (i.e., the strength of the corrective action) of the control system, to slow its response to errors. This will make the whole system more sluggish, but may prevent it from overshooting the desired value, thereby preventing uncontrolled oscillations.

6. Chapter Summary

This chapter uses the themes of navigation and vehicle position control to introduce you to the sometimes wild and challenging (but imminently useful) world of underwater vehicle control. It provides a general theoretical overview of control systems in which you are introduced to the all-important concept of feedback. Feedback improves control system performance by letting a control system know whether or not it has finished doing what it was supposed to do. If not, the feedback tells it how it needs to change the actual condition to make it match the desired condition. Some control systems rely heavily on humans to perform many of the control functions, whereas others are fully automated and require no human control.

When the goal of control is to reach a particular destination, then you are trying to control vehicle position. This in turn requires control of vehicle movement. Navigation is a suite of tools and techniques for figuring out where a vessel is, how quickly it's moving, and which way it's headed. This navigational information provides the feedback needed to determine an optimal course to the vehicle's destination. The navigation portion of the chapter highlights several navigational instruments useful in ROV and AUV operations, including the underwater video camera, magnetic compass, depth gauge, sonar, and GPS receiver. It also describes some useful navigational techniques, including the use of landmarks, triangulation, dead reckoning, and GPS coordinates.

Following the general introductory information in the first sections, the chapter looks at the practical side of building a simple control system for navigating and propelling a vehicle. This involves a set of basic navigational instruments (video camera, diver's compass, and diver's depth gauge), plus DPDT switches used to control thruster motor direction (but not speed).

For those of you eager to earn your techie stripes, the chapter concludes with a lengthy section that introduces a variety of more advanced techniques for vehicle control and navigation. Microcontrollers (MCUs), essentially small programmable computers, are the brains behind most of this advanced control. After introducing microcontrollers and the ways they can communicate with other devices (including a variety of digital data transmission protocols), the chapter presents techniques for providing the microcontrollers with information from the pilot and from electronic sensors. Additionally, this advanced section discusses how the weak control signals produced by an MCU can be amplified by transistors, relays, and motor controllers to drive motors, lights, and other power-hungry systems on the vehicle.

Finally, the chapter concludes with a brief look at some common algorithms for controlling vehicle behavior and offers suggestions on how to recognize and repair common control system malfunctions.

Now that you can control your vehicle, it's time to do something with it. Chapter 10 will give you some ideas for the payloads and tools that will help you do just that.

Chapter 10

Hydraulics and Payloads

Chapter 10: **Hydraulics and Payloads**

Stories From Real Life: Lethbridge's Diving Barrel

Chapter Outline

1. **Introduction**
2. **Hydraulic Mechanisms**
    - 2.1. How Hydraulic Systems Work
    - 2.2. Force Amplification
    - 2.3. Hydraulic Versus Electrical Power Delivery
    - 2.4. How to Build Your Own Hydraulic System
    - 2.5. The Pneumatic System Option
3. **Manipulators**
    - 3.1. Components of a Manipulator System
    - 3.2. Home-Built Manipulators—A Case Study
4. **Underwater Tasks and Tools**
    - 4.1. Tow Sleds
    - 4.2. Tools
5. **Considerations When Designing Payloads**
    - 5.1. Mass, Buoyancy, and Stability
    - 5.2. Lifting
    - 5.3. Reaction Forces
    - 5.4. System Interference
6. **Examples of Payload Options for Simple ROVs**
    - 6.1. Gripper Variations
7. **Chapter Summary**

Chapter Learning Outcomes

- Describe what a payload is and why it's important for an underwater robotic vehicle. Why are multiple and/or interchangeable payloads particularly useful?
- Explain the basic components of a standard hydraulic system and how these systems can be used to transfer force, motion, and power from a prime mover to an actuator.
- Contrast a single-function manipulator with a more complex multi-function version.
- Describe options for building a single- or double-function manipulator for a simple ROV like *SeaMATE*.
- Describe different types of payload tools carried by various commercial underwater robotic vehicles.

***Figure 10.1.cover:* Johnson-Sea-Link**

Sporting a full array of sophisticated instruments, cameras, and payload tools, the Johnson-Sea-Link *submersible descends on another mission for science.*

Image courtesy of Scott Olson, Harbor Branch Oceanographic Institution

STORIES FROM REAL LIFE: Lethbridge's Diving Barrel

Image courtesy of Naval Undersea Museum

From earliest times, humans have endeavored to work under water. The challenge was often daunting, sometimes deadly. Today, subsea vehicles utilize a variety of payload tools to accomplish their specific mission objectives. Robotic craft offer a safer alternative, allowing the operator to remain safely at the surface. From there a pilot can direct tools that mimic human arms and hands in their ability to conduct repairs, collect specimens, or retrieve objects, but with greatly magnified force.

Other instruments serve as the craft's underwater eyes and ears, recording what humans cannot easily see or hear. Sophisticated sensors also log and analyze an astounding range of data from the depths, creating a greater understanding of this silent world.

All of these tools, instruments, and cameras enable underwater vehicles to perform the work that pays the bills—hence, the appropriate term "payloads." As Lethbridge's inventive diving barrel pointed out, the key to working under water successfully is matching the task, the vehicle, and the payload.

Time and again, the seafloor has offered treasures to those who could find ways to reach them and bring them back to the surface. Sometimes this bounty was a natural harvest—sponges, pearls, seafood, or more recently, oil and gas. Other times, it was the debris of a tragic occurrence—centuries of cargo-laden ships wrecked in storms, or nuclear subs felled by technological or maintenance faults. Throughout history, the quest has been to devise ways to venture safely under water and perform the work required. It is this last challenge that involves the design, construction, and utilization of effective tools—the payload.

Certainly, early breath-hold divers readied themselves with a collecting net and knife, as well as a rock to aid their descent. But attempts to develop a combination of man and machine that would extend working time on the bottom really began in the early eighteenth century. In 1715, an Englishman named John Lethbridge became intrigued by the possibility of building some sort of device that would enable him to recover items from sunken wrecks. Looking about his farm for suitable materials, he spied a large barrel, or hogshead, that was big enough to hold his body. The easiest way to find out whether he could breathe inside it was to crawl in, bung the barrel tightly closed, and see what happened.

Lethbridge managed half an hour "without communication of air." But would the same hold true under water? Lethbridge dug a trench, flooded it with water, and sealed himself inside the submerged hogshead. He was delighted to report that it seemed even easier to stay under water!

Lethbridge's next step was to hire a barrel maker to build an actual diving machine. It consisted of oak panels 6 feet (approx. 2 m) long, reinforced with iron hoops. The shape was a truncated cone 3 feet (approx. 1 m) wide at the head end, tapering to 2 ½ feet (approx. 75 cm) at the foot. Once the diver had entered at the top end, he slipped his arms out two waterproofed holes that were essentially well-greased leather sleeves that were then tied tightly around his biceps. A top cover was bolted in place, then the machine was heavily ballasted with iron castings and suspended from a surface vessel. On the wreck site, the diver could look out a piece of glass 4 inches (approx. 10 cm) in diameter and 1 ¼ inches (approx. 3.5 cm) thick that was inset into one side. Once his target was spotted, the diver could employ the payload—his arms and simple tools—to salvage those treasures.

Lethbridge reported being able to stay submerged for 34 minutes before signaling to be pulled up. And rather than emerging from the chamber, he simply had air pumped in through forward and aft openings. Then the device was lowered again, and he went back to work. In case of emergency, Lethbridge had only to release the ballast from inside, and his diving machine resurfaced under its own buoyancy. Lethbridge's invention was effective only to depths of 30–60 ft (approx. 9–18 m), although he did achieve 72 feet (approx. 22 m) "with great difficulty." However, even at these modest depths, his patented diving machine worked! Documents indicate that he enjoyed great success in the treasure-salvage business for a brief period.

Similarly, in 1728, Captain Jacob Rowe's patented diving engine was used to recover treasure from the *Adelaar*. In both cases, a single diver encased in a chamber used his hands or simple tools to connect hooks or cables onto valuables that could then be hoisted to the surface. Not surprisingly, strong, dexterous human arms, wrists, and hands remain the prototype for the sophisticated mechanical manipulators and "grabbers" so commonplace on today's manned and unmanned underwater vehicles.

Like these early salvage devices, the non-military manned submersibles that followed all involved some type of payload for observation and usually some device for salvage or collection of specimens. For example, in 1894, Simon Lake launched his *Argonaut Jr.*, which featured an airlock, or bottom hatch, that could be opened to facilitate the recovery of clams, oysters, or rocks.

In 1897, an Italian, Count Piatti Dal Pozzo, constructed a heavy spherical "submarine" called *La France* that was equipped with working arms. When suspended from an overhead barge that was anchored over a sunken wreck, crew inside the iron ball could look out the portholes to a site illuminated by electric lights and use the working claws to handle tools and conduct salvage tasks. Understandably, Pozzo's invention was dubbed "the Underwater Worker."

Other submersibles of the early 1900s also boasted workable salvage arms. For example, the Italian inventor Cavaliere Pino launched his egg-shaped *Submarine Worker* that was 3 meters (approx. 10 ft) in diameter. It was reputedly designed to allow a crew of two to work at 90 meters (approx. 300 feet). A priest at the cathedral in Tunis designed the *Bou-Korn* to assist Greek sponge fishermen. His 5-meter (approx. 16-ft) lemon-shaped hull had three propellers—one at the stern and one on each side—as well as maneuverable arms. The *Bou-Korn* was used successfully for many years.

While Beebe's bathysphere, launched in 1930, set important scientific records and precedents by venturing into significantly greater depths than ever before, its payload was limited to two humans who had no way to interact with their watery surroundings. Bolted securely inside the stout sphere and dangling off a cable below a barge in the British West Indies, William Beebe and Otis Barton could only peer out of the viewports to observe marine life at previously unseen depths. Essentially, Beebe's bathysphere was a precursor to modern eyeball ROVs. Other observation-only craft, such as *Trieste*, quickly gave way to submersibles—and eventually robots—

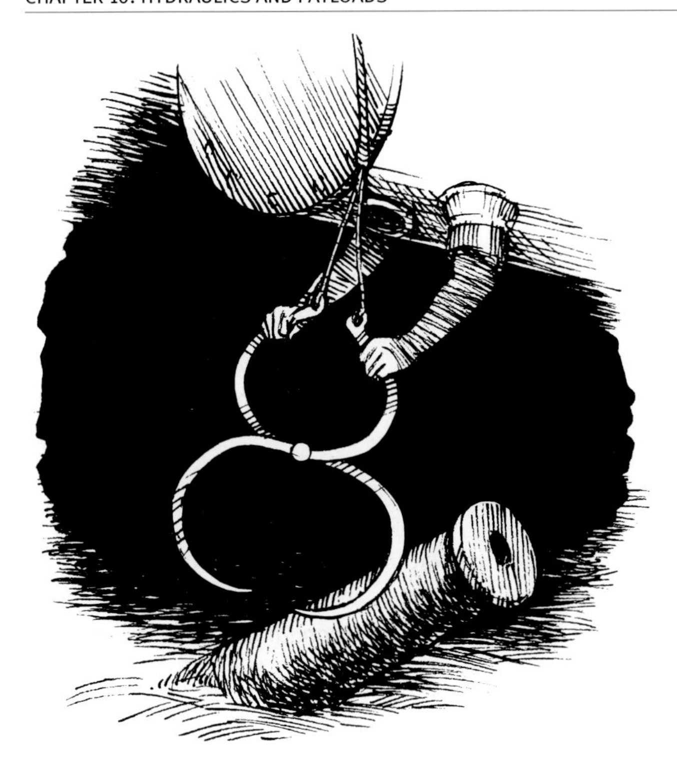

equipped with one, then two or more mechanical arms, as well a wide array of sensors.

Certainly, observation—whether by eye, camera, sonar, or laser devices—is still a key aspect of any subsea mission. But equally important are payloads that perform underwater work. Many of these tools are increasingly mission-specific and technologically sophisticated—they range from various oil-field "hot stab" tools to creative student-built devices to accomplish competition tasks. Payloads often depend on an integrated effort; for example, locating a pipeline leak, then repairing it. Some tools require significant force multiplication, generally provided through hydraulics, for cutting, moving, and lifting. Some require less—usually, scientific tools need to be quieter, gentler, and less obtrusive in order to observe and recover specimens in their natural environment.

Like designing underwater vehicles, designing and building effective payloads can require knowledge of new materials, miniaturization, robotic engineering, control theory, and computerized data collection. Oh, yes, and imagination! However, the objective has not changed since Lethbridge was first sealed in his makeshift diving chamber—to safely accomplish work under water.

Revisiting Lethbridge's Diving Barrel

Over the years, there has been some suggestion that John Lethbridge, like many other inventors, may have exaggerated his exploits—that a diver could never have worked with his arms exposed at the ambient pressure of 18 meters (approx. 60 ft), as Lethbridge claimed to have done. In the 1980s, Robert Stenuit, a record-setting diver himself, set out to prove (or disprove) that Lethbridge's diving machine could perform as claimed.

Stenuit arranged for a replica of the oak diving barrel to be built, using only materials and techniques that would have been available in the early 1700s. He used a blueprint of the invention that was based on a set of plans and sketches drawn back then by a French industrial spy, evidently a common practice even in Lethbridge's day!

When Stenuit first tested the replica, it became readily apparent that the leather sleeves, as drawn, leaked badly. Stenuit surmised that someone had deliberately misled the spy, feeding him false information about the critical components of Lethbridge's patented diving machine. Stenuit then tried several alternate sleeve configurations, coming up with a strap-and-buckle arrangement that sealed the water out and allowed him to work under water, although with some discomfort.

Robert Stenuit successfully tested Lethbridge's invention to 9 meters (approx. 30 feet), the maximum depth of his diving test tank, and said he believed he could have gone to much greater depths. Historical records suggest that John Lethbridge utilized his diving machine for several successful salvage operations in the 1700s. More than 200 years later, a fellow diver proved that this inventive Englishman understood the value of a workable payload and certainly earned his place in subsea history.

1. Introduction

Figure 10.2: Mission-Specific Payloads

Each of these student-built ROVs includes an innovative, mission-specific tool as its payload. Each of the tools was designed to answer the challenge of a specific competition scenario.

Most commercial trucks, planes, boats, and even mules earn their keep by transporting passengers and/or cargo from one place to another in exchange for some form of payment. Each load of people or goods that brings in revenue could be considered a **payload**.

Submersibles, ROVs, and AUVs also carry payloads, though of a different sort. Larger ROVs are often designed to carry interchangeable **tool trays** that bolt onto the vehicle's frame or slide in and out like drawers in a dresser.

Images courtesy of the 2008 MATE International ROV Competition Dive Support Team

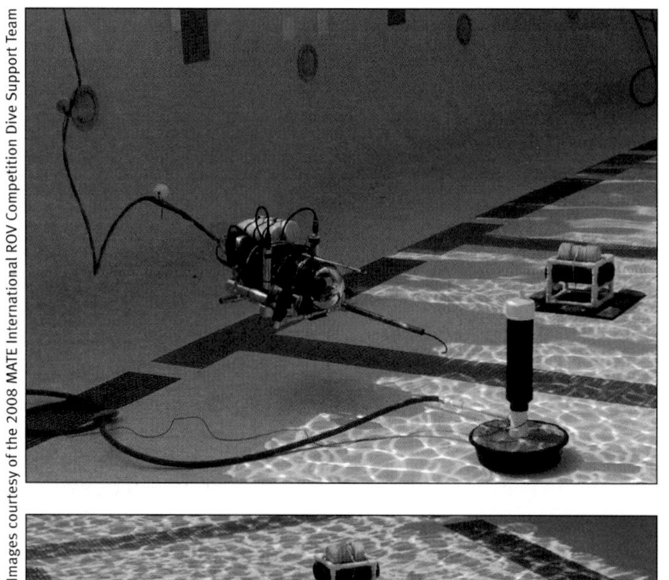

For these vehicles, those tools and tool trays are the payload because the work they accomplish is what makes the mission profitable. The main robotic vehicle serves to transport the tool tray to and from the worksite and usually doubles as the rigid platform that supports the tools, so they can do their work. Usually, the vehicle also supplies electrical and/or hydraulic power to whatever tools or data-gathering instruments are included in the tray. The flexibility of tool trays allows a single ROV to be configured quickly and easily for a multitude of different functions. So by swapping one tray for another between missions, the same ROV could theoretically be welding damaged pipeline one day with one set of tools, scrubbing barnacles off an underwater drill rig the next day with a different set of tools, and collecting seafloor sediment cores for a science experiment the day after that.

Image courtesy of Saab Seaeye Ltd.

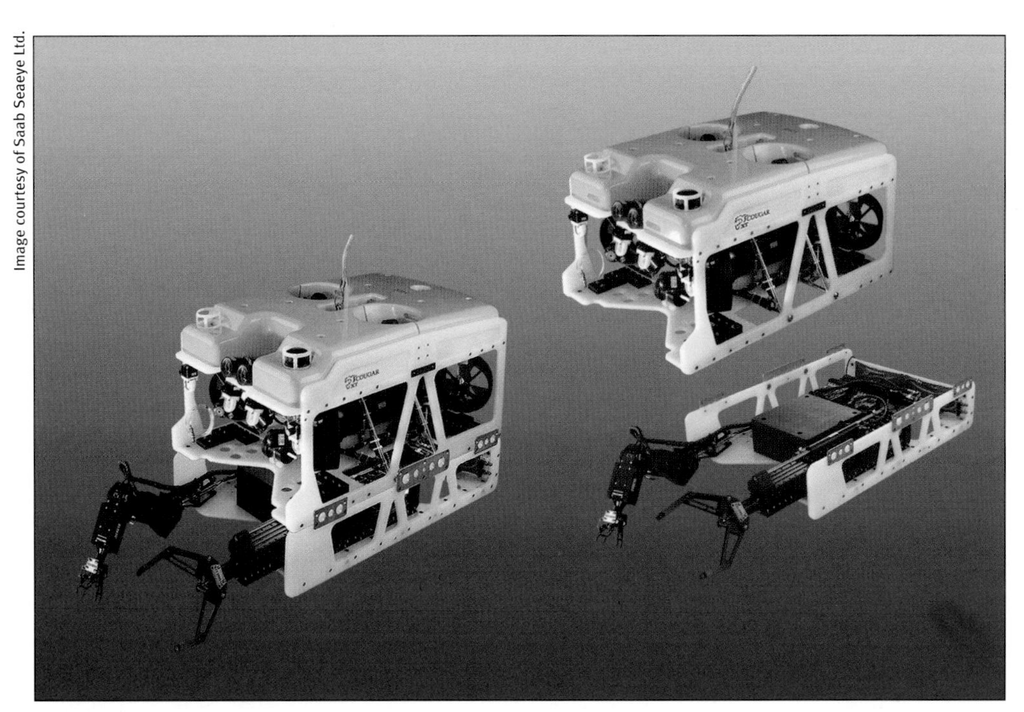

Figure 10.3: Tools on Demand

The Seaeye Cougar can function in observation mode or can easily switch to other mission tasks with the additional of a tool tray/ manipulator combo.

Tool trays are interchangeable units that are typically mounted underneath or in the forward section of an ROV or submersible. Tool trays can be designed to carry sample bottles, dredgers, corers, trenching or jetting pumps, fiber-optic spools, etc.

Using interchangeable tool trays allows the vehicle to be reconfigured easily and quickly for a variety of tasks, thereby saving time and keeping costs low. It also means that the work class ROV is a versatile platform able to perform a multitude of tasks, depending on the tool trays selected.

Technicians generally assemble the various tools onto different sleds at the shoreside facility; these modular units are then loaded aboard the support ship, where they are ready for fast mounting on the vehicle at sea, as required.

Commercial ROVs are often tasked with jobs that require monstrous forces, such as severing heavy cables or digging deep trenches in the seafloor. Although electric tools and electric power can be used for these operations, it is more common on large commercial ROVs to use hydraulic tools and hydraulic power systems to perform such Herculean tasks. Simplified versions of these hydraulic systems can be useful for powering payload tools on small ROVs, so this chapter begins with an introduction to hydraulic systems.

On larger commercial and scientific ROVs, standard subsystems such as cameras, navigational sonar, and manipulator arms are usually considered to be essential parts of the vehicle itself, rather than payloads. However, any specialized tool or instrument may be considered a payload if it is not part of the base configuration of the robot/ ROV/AUV. For example, manipulator arms may be viewed as a payload on very small ROVs and AUVs, since they they are optional tools that can be attached or removed to reconfigure the vehicle for a particular mission. This chapter adopts a very generalized definition of payload that includes almost any tools or instruments beyond the frame, tether, thrusters, basic power system, and basic navigational instruments of the vehicle. In particular, it treats manipulator arms and grippers as payloads, and devotes significant space near the end of the chapter to exploring ways you can build and power manipulator arms for your own vehicle.

Before diving in to design your own tools and manipulator arms, it's worth noting that payload tools can be extremely simple and inexpensive, such as the pick-up probe used on the *SeaMATE* ROV, or as sophisticated and expensive as a hi-tech manipulator. In many instances, advanced "manips" are considered complete robotic systems in their own right. In fact, designing the payload tools can be as challenging as designing the robotic platform that carries them.

2. Hydraulic Mechanisms

If you've ever watched a bulldozer, backhoe, or excavator (Figure 10.4) moving tons of dirt and rock around on a construction site, you've witnessed the awesome power capabilities of hydraulic machines. The same technology that enables these engineering marvels to move mountains on dry land can be put to work just as effectively under water. For example, many work class ROVs use hydraulic systems to power strong thrusters, manipulator arms, and a wide variety of saws, cutters, and other tools. Hydraulic systems are used mostly with large, heavy machinery, but there are also times when a simplified hydraulic system can prove useful on a small home- or school-built ROV.

2.1. How Hydraulic Systems Work

Often people think a hydraulic system is a source of power, but this is not quite accurate. Rather, a hydraulic system is a means of *distributing* power from one place to another. It does this by using liquid pressure to transmit *force* and *motion*. The energy for a hydraulic system typically comes from electricity or hydrocarbon fuels, such as diesel fuel. This energy is first transformed into kinetic energy by a **prime mover**. The prime mover is usually just an internal combustion engine or an electric motor connected by a drive shaft to a **hydraulic pump** (Figure 10.5 and 10.8). As the shaft rotates, the pump pressurizes and circulates a liquid (usually a specialized oil known as **hydraulic fluid**) through pipes, hoses, or tubes—referred to collectively as "hydraulic lines" or just "lines." The pressurized liquid in the lines flows through one or more **hydraulic actuators,** such as rotary motors or linear rams. (A linear hydraulic

Image courtesy of Al Harvey, The Slide Farm

Figure 10.4: Hydraulic Excavator

Each of the four hydraulic pistons visible on this excavator can generate many tons of force, yet each also offers precision control to dig, lift, move, and dump heavy loads of dirt and rocks safely.

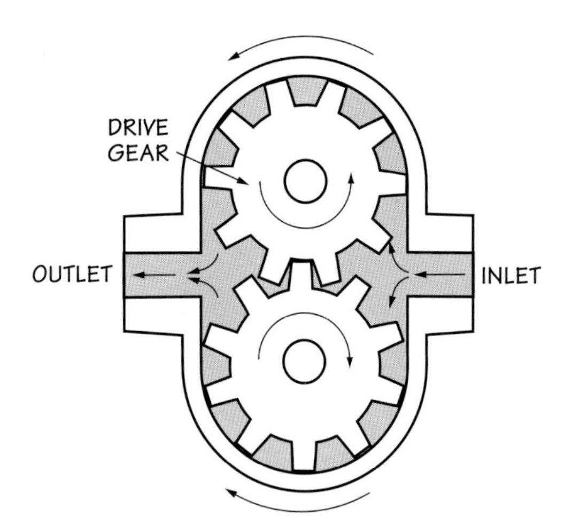

Figure 10.5: Gear Pump

In a gear pump (one of several common types of hydraulic pumps), one gear is turned by a diesel engine, electric motor, or other prime mover. A second gear meshes with, and is turned by, the first. The spaces between the gear teeth transport liquid along the inner walls of the pump from the inlet to the outlet.

ram is basically a piston in a cylinder.) These actuators move whatever they are connected to when the fluid pressure presses on pistons, gears, or other mechanisms inside the actuator. For example, Figure 10.7 shows a commercial ROV manipulator arm that uses double-acting cylinders (a type of hydraulic actuator that uses fluid pressure to push *and* to pull) for control of joint angles. As hydraulic fluid passes through an actuator, it loses a lot of its pressure, because the energy in the fluid has been extracted and converted into useful work by the actuator. The now low-pressure hydraulic fluid is then returned to a reservoir and cycled through the pump again.

Hydraulic force transfer, which is ultimately the basis of the **hydraulic power transfer** that makes these systems so useful, is based on something called Pascal's Law, which says that fluid confined in a container or any closed system of pipes or tubing can transmit force more or less instantaneously from one part of the system to every other

HISTORIC HIGHLIGHT: PASCAL'S LAW AND EARLY HYDRAULICS

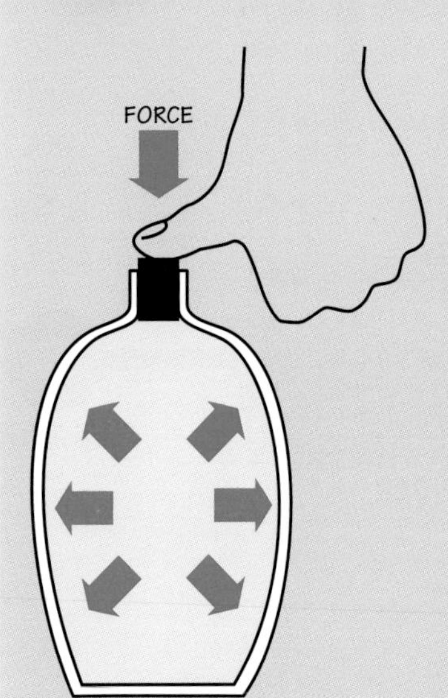

In the seventeenth century, the French scientist Blaise Pascal (1623–62) discovered the principle that a fluid confined in a container such as a pipe or piston and subjected to an outside force could instantaneously transmit that force to the other end of the container. His understanding, called **Pascal's Law**, states:

Pressure applied to a confined fluid is transmitted undiminished in all directions and acts with equal force on equal areas and at right angles to them.

This law forms the basis of hydraulic systems and pneumatic systems too. (See *Section 2.5. The Pneumatic System Option.*) With hydraulics, if you apply a force to a fluid at one end of a system of pipes or tubing, the fluid can apply a force to something else at the other end of the system, thereby transferring the force. With pneumatic systems, it's a pressurized gas that does the force transfer.

Joseph Bramah (1748–1814), a British mechanic and inventor, conceived and built the first hydraulic press, using Pascal's hydrostatic principle. His invention galvanized mechanical technology, becoming the foundation of hydraulic machinery and today's modern machine tools. The device created liquid pressure by use of a hand-pumped piston, just like that used in a modern hydraulic jack. Later, this invention was improved by connecting an engine to the pump.

Figure 10.6: Hydrostatic Pressure

When a plastic bottle is filled with water, capped with a cork, and the cork pushed in, the walls of the bottle bulge outward because of increased internal pressure. This illustrates Pascal's Law that fluid can transmit force.

Figure 10.7: Magnum-5 Mini Manipulator

The Magnum-5 Mini has a lifting capacity of 150 pounds and is typical of a 5-function manipulator suitable for light duty ROVs. ISE also offers regular and mini arms with 1000-pound lifting capacities.

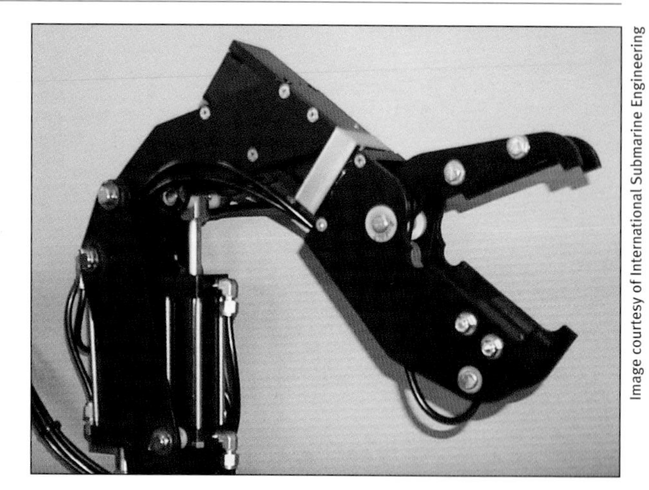

Image courtesy of International Submarine Engineering

Figure 10.8: Hydraulic System with Double-Acting Cylinder

This pair of diagrams illustrates how a hydraulic system can use a double-acting cylinder to either push or pull on something.

One diagram shows the valve set to push; the other shows it set to pull. In both diagrams, red represents pressurized fluid, blue represents unpressurized fluid, and the small grey arrows adjacent to the pipes or hoses show the direction of fluid flow. The energy used to move the piston in the cylinder comes from a hydraulic pump, which sucks unpressurized fluid from a collecting reservoir and pumps it under pressure into an accumulator. The accumulator stores energy from the pump and helps to maintain steady fluid pressure in the system. This is like a battery which stores energy from a charger and helps maintain steady voltage in a circuit.

When the valve is rotated into the "push" position, one of two diagonal holes drilled through it routes the pressurized fluid to the left side of the piston, forcing the piston outward and pushing on whatever is connected to it. For example, it might be used to open the door of a sampling box on an ROV.

(continued opposite)

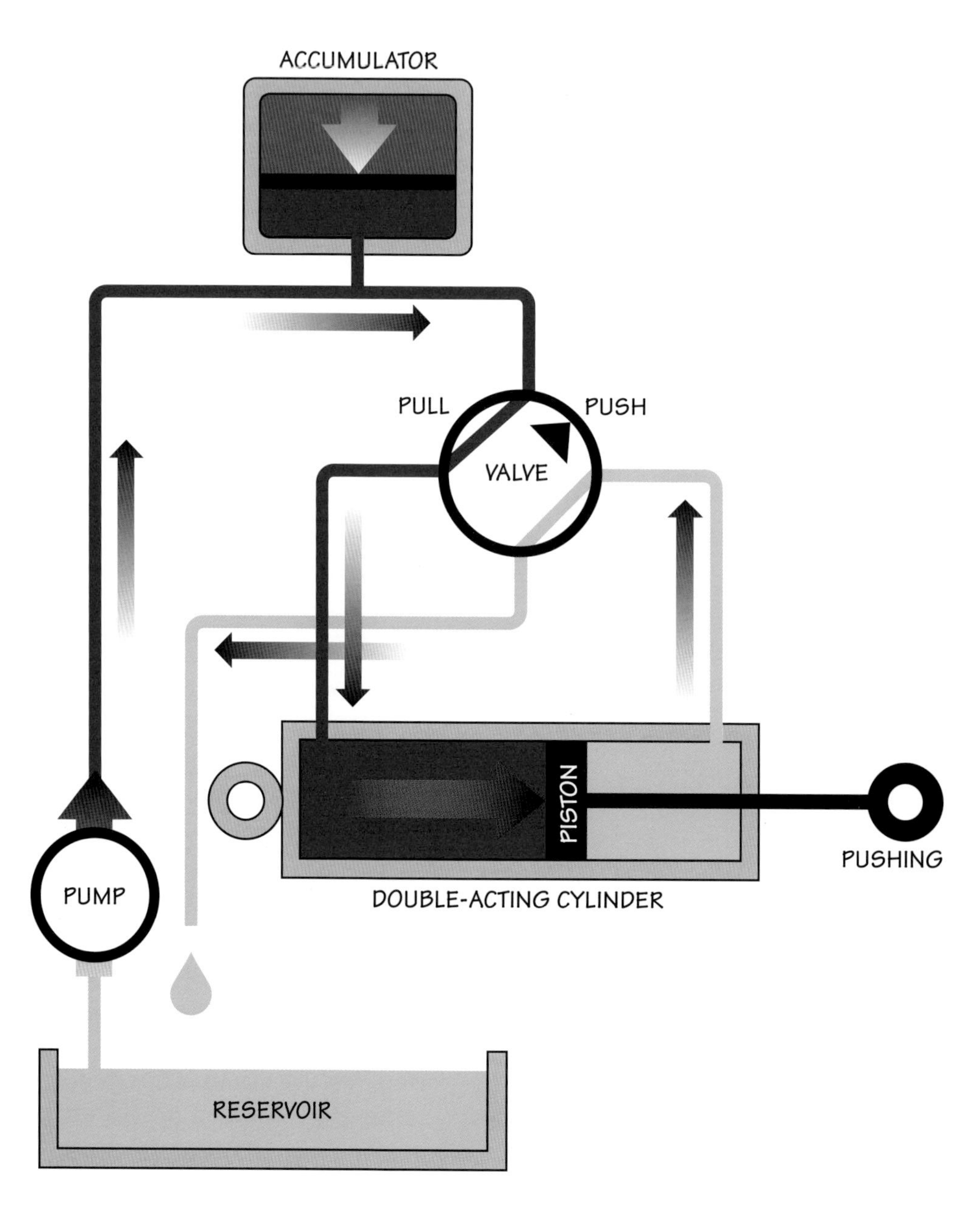

TECH NOTE: HYDRAULIC POWER SYSTEMS

The type of hydraulic system discussed in this chapter is sometimes referred to as a **power hydraulic system** rather than a **hydraulic power system**. This alternate terminology emphasizes the fact that these systems are capable of delivering mechanical power from the prime mover to the actuators and whatever's attached to them. Mechanical energy, often called "work," can be quantified as a force multiplied by distance, so lifting a car a certain distance or forcefully compressing a sheet of paper on a printing press both require a combination of force and distance. Therefore, they both require energy. If you want to accomplish these tasks in a limited amount of time, then that energy must be delivered at a certain rate. Recall that power is the rate at which energy is delivered (or converted between forms), so these systems that move as they push or pull things are quite appropriately described by some as *power* hydraulic systems. Regardless of which terminology you use, the quality that makes hydraulic systems so popular is that they are capable of delivering surprisingly high power, given the small size and weight of the lines and actuators required.

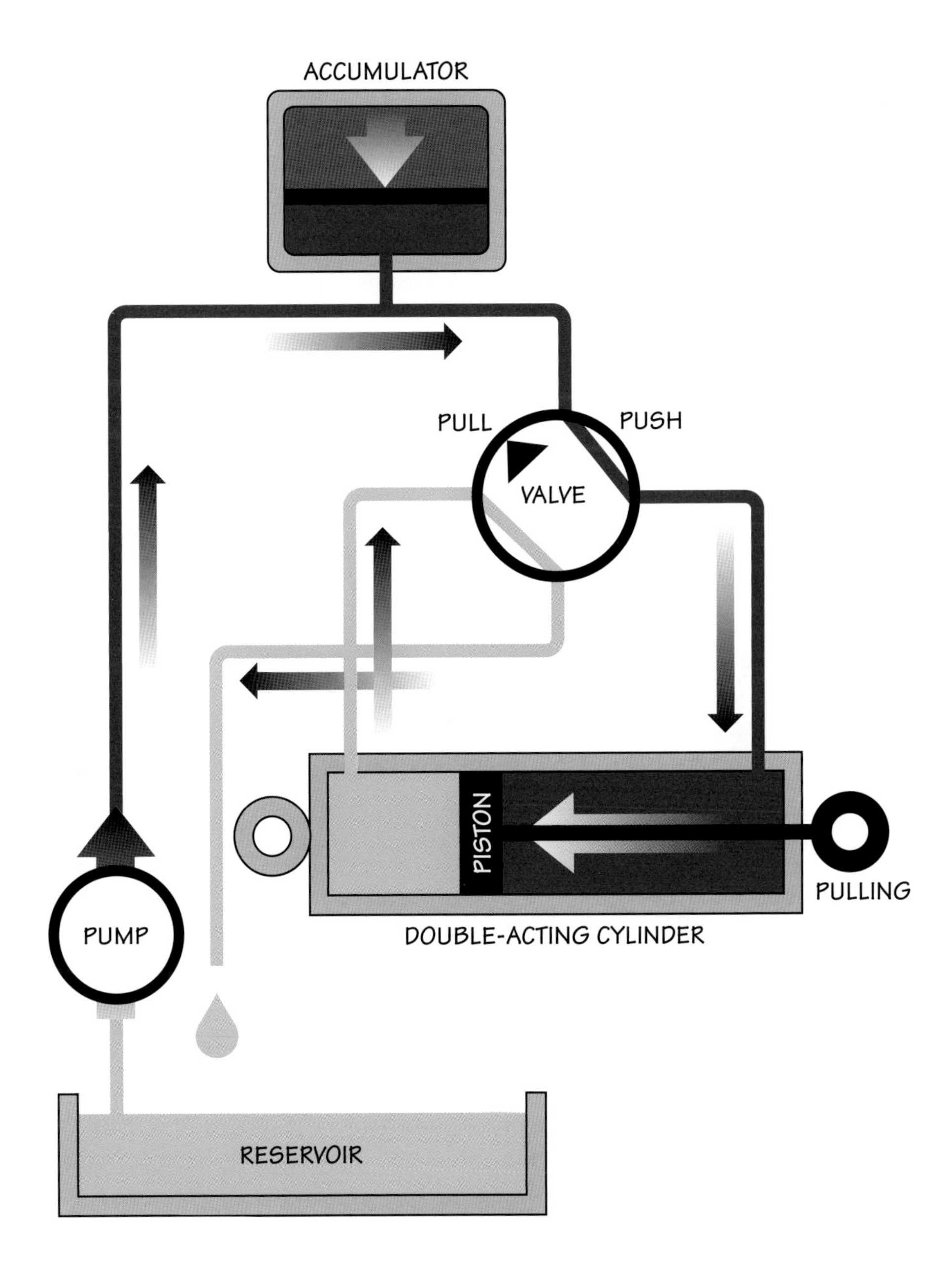

(continued from p. 538)

As the piston moves to the right, fluid on the right side of the piston is squeezed out of the cylinder through a second diagonal hole in the valve and drips into the reservoir.

The second diagram shows the valve rotated into the "pull" position. Now the two diagonal holes drilled through it reverse roles, and pressurized fluid is now routed to the right side of the cylinder. This causes the piston to move to the left, pulling on whatever is connected to it.

In the sampling box example, this would close the door. If the valve is rotated to a position halfway between its "push" and "pull" settings, neither of the diagonal holes in the valve will connect with the lines to and from the piston, and fluid will be unable to move in or out of the cylinder. This will effectively lock the piston in place.

One advantage of hydraulic systems is that they can hold a constant position like this without requiring any further input of power. With the valve in the central "locked" position, the pump could be turned off, and the piston would still hold its position, even if it was supporting a heavy load.

part. (See *Historic Highlight: Pascal's Law and Early Hydraulics.*) This can be demonstrated easily if you fill a plastic soda bottle with water clear to the top, leaving no airspace. Now empty out just enough water to get a cork inside the bottle neck. Push hard. The cork will go in until it reaches the water, then it will stop. If you continue pushing hard, the sides of the container will begin to bulge outward as in Figure 10.6. This demonstrates Pascal's Law—that fluid conducts or transmits the force on the cork to the bottom and sides of the container.

In many (but not all) hydraulic systems, a special chamber called an **accumulator** is used to store pressurized fluid from the pump. The accumulator can temporarily deliver stored fluid to actuators at rates exceeding the maximum output of the hydraulic pump. This allows for bursts of high-power output without requiring a large and expensive high-power pump. Pressure in the accumulator is maintained by pressing on the fluid with a heavily weighted piston, a spring-loaded piston, or pressurized gas on the other side of a flexible bladder or diaphragm located inside the accumulator. Note that while the accumulator can increase the instantaneous power beyond what the pump can deliver, it cannot increase the average power delivered. If you need more power on average, you'll need a bigger pump.

Figure 10.8 shows the components of a very simple hydraulic system that might be used to operate a double-acting cylinder, such as those visible on the excavator in Figure 10.4 and the ROV manipulator in Figure 10.7. This common cylinder-and-piston configuration creates a linear ram that can both push and pull, depending on the position of a valve; however, it does not offer very fine control of movement or force. Variable-rate pumps and adjustable flow valves provide more control in more advanced hydraulic systems.

Note that most hydraulic systems rely on fluid pressure within *closed* networks of pipes, hoses, or other conduits to generate forces and movements by pushing against movable surfaces. Although some fluid moves from one place to another as it transmits power, it is the *pressure* of the fluid, rather than the inertia or the weight of the fluid, that does the real work. This is in contrast to something like a water wheel, which is an open (unpressurized) system that uses the weight and/or momentum of flowing liquid to turn the wheel and perform work. Note that there are also some *open* hydraulic systems used on ROVs or other undersea vehicles. These are not closed systems. They don't need to be because they use seawater instead of oil as their hydraulic fluid. Since the vehicle is surrounded by seawater, the pumps in these systems can suck the water directly from the ocean and feed it under pressure to cylinders or other hydraulic actuators. When the water eventually exits the actuator, it can simply be ejected right back into the ocean, rather than being returned to a reservoir, as it would be in a completely closed system.

2.2. Force Amplification

If you get a flat in the tire on your car and need to change it, you can use a jack to lift the automobile's weight off the tire so you can change it. The jack provides a form of force amplification that allows you to lift something much heavier than you could lift unassisted. Hydraulic systems excel at force amplification. That's why they are so good at generating very impressive forces to move boulders, split wood, and perform countless other tasks that require huge forces. (Indeed, many car jacks are hydraulic.)

To understand how hydraulic force amplification works, it's helpful to start with a more familiar, non-hydraulic example: the pry bar. If you've ever needed to lift a heavy refrigerator or desk, or move a big rock, you may have used a long, stiff, metal rod

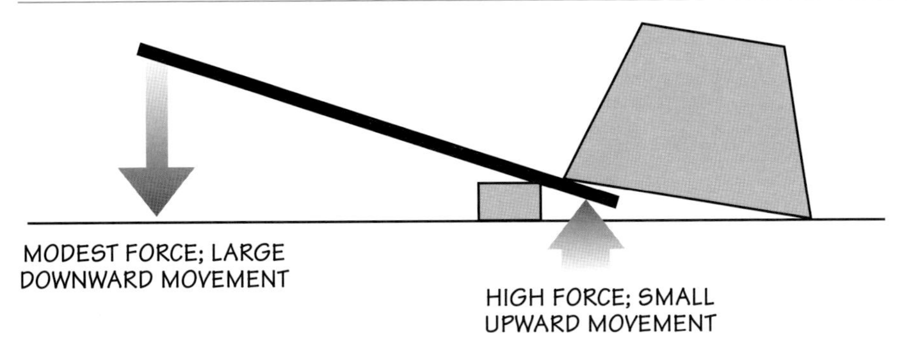

Figure 10.9: Using a Pry Bar to Amplify Force

A pry bar amplifies force at the expense of distance and speed.

called a pry bar to do it. You just wedge one end of the bar under the object you want to lift, slip a block of wood under the rod close to the heavy object, and push down on the raised end of the bar (Figure 10.9). Like the car jack, the pry bar allows you to lift something much heavier than you could lift without it. Unfortunately, the laws of physics won't let you get something for nothing, so this extra force comes at a price: you must sacrifice distance and speed. The pry bar effectively trades distance for force by converting a long-distance downward movement with modest force at your end of the bar into a short-distance upward movement with much higher force at the end under the heavy object.

A similar sort of force amplification can be achieved in a hydraulic system by using liquid pressure and pistons instead of a pry bar. Recall from *Chapter 5: Pressure Hulls and Canisters* that the force exerted on a surface by a pressurized fluid acts perpendicular to that surface and is equal in magnitude to the fluid pressure multiplied by the area of the surface:

Equation 10.1

$$F = P \times A$$

This means there are two ways to increase the lifting force that such a pressure can produce. One way is to increase the pressure of the fluid applied to the piston in the cylinder. (This is analogous to pushing harder on the pry bar.) The other is to increase the diameter of the piston, so that more surface area is exposed to the available fluid pressure. (This is analogous to using a longer pry bar for more leverage.) Both methods are used commonly in hydraulic systems.

1. Increasing piston diameter is a straightforward method. Just make an actuator with a larger-diameter cylinder and fit it with a larger-diameter piston.
2. Increasing pressure requires a pump that can produce high-pressure fluid flow. To generate high pressures, hydraulic pump designs take advantage of the fact that this equation can be rearranged to solve for pressure:

Equation 10.2

$$P = \frac{F}{A}$$

This means there are two ways to generate high pressure at the pump outlet. One is to push harder on the fluid (for example, by using a stronger prime mover to operate the pump). The other is to push on the fluid, using something that has a *smaller* surface area, such as a tiny piston or a pair of small gears in a gear pump. (Remember, dividing a force by a smaller number is the same as multiplying it by a larger number.)

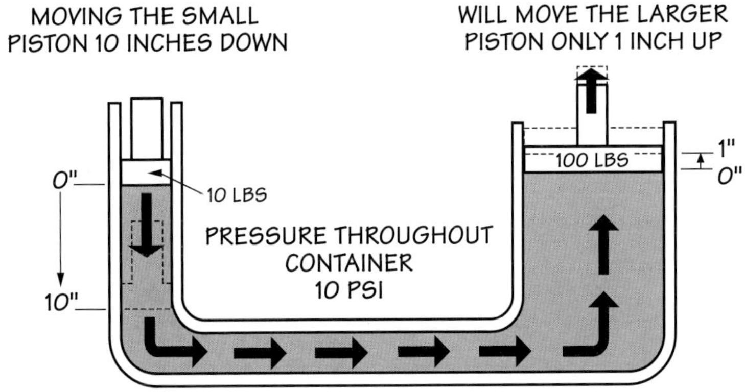

Figure 10.10: Using a Hydraulic System to Amplify Force

In this hydraulic system, a small force is used to move a large weight. The input force of 10 pounds on a 1 square-inch piston produces a pressure of 10 psi throughout the liquid in the container. This pressure will support 1 pound for every square inch of surface area, so a 10 square-inch piston can support a weight of 100 pounds.

If the 10-lb weight is pushed down 10 inches, it will force 10 cubic inches (10 inches x 1 sq inch = 10 cubic inches) toward the larger weight, moving that 100-lb weight up 1 inch (10 cubic inches/10 sq inches = 1 inch).

MOVING THE SMALL PISTON 10 INCHES DOWN
WILL MOVE THE LARGER PISTON ONLY 1 INCH UP
0"
10"
10 LBS
PRESSURE THROUGHOUT CONTAINER 10 PSI
100 LBS
1"
0"

In summary, to produce a large force, you can use a strong prime mover to press with high force on a small piston to generate a very high pump output pressure. This very high pressure can then be applied to a large actuator piston to produce a truly enormous force. In large industrial hydraulic systems, it's not uncommon to have pumps capable of producing pressures of several thousand psi and to have those pressures act on pistons that have dozens of square inches of area to produce forces of a hundred tons or more!

Of course, the Law of Conservation of Energy says there's no such thing as a free lunch—in exchange for this greater force, you get reduced distance and speed, just as you do with a longer lever arm on a pry bar. The pump will have to pump a lot of fluid to move that large piston very far.

2.3. Hydraulic Versus Electrical Power Delivery

In *Chapter 8: Power Systems*, you learned that electricity can be used to transmit power from a source (such as a battery) to one or more devices (such as electric motors) that use the power to perform useful work. In some respects, hydraulic mechanisms are simply an alternative way to distribute power to various parts of a vehicle. It's just that hydraulic systems use liquid flowing through lines or hoses, instead of electric charges flowing through wires, to transmit the power. In fact, hydraulic systems are remarkably similar to electrical systems in many respects, and you can actually use what you already know about electrical power distribution to help you understand hydraulic systems better (Table 10.1).

Table 10.1: Parallel Examples of Electrical and Hydraulic Power Delivery

In an Electrical System	*In a Hydraulic System*
A battery generates a voltage, which is basically an electrical "pressure."	*A hydraulic pump generates pressure in a fluid.*
The voltage pushes a flow of electric charges through wires.	*The pressure pushes a flow of liquid through tubes and hoses.*
The voltage-driven flow of current through an electric motor or other electric device can be used to perform useful work.	*The pressure-driven flow of liquid through a hydraulic piston or other hydraulic mechanism can be used to perform useful work.*
The flow of electric current can be controlled with switches.	*The flow of hydraulic fluid can be controlled with valves.*

These similarities raise some important questions: What advantages does each offer? When would you use one over the other? Is there ever a reason to use both at the same time?

Electrical systems are generally more versatile, because they can be used to power an enormous variety of different things: lights, heaters, communication devices, motors, and so on. Hydraulic systems are comparatively limited in their range of applications, being used almost exclusively for powering the movement of mechanical systems. Electrical power is much easier to transmit over long distances, including the length of a long ROV tether. Electrical systems are also much cleaner and generally easier to install and maintain.

However, whenever very large forces, torques, or instantaneous power outputs are required, hydraulic actuators (hydraulic pistons for linear motion and hydraulic motors for rotary motion) may be preferred. They are generally much smaller, lighter, simpler, and more rugged than electrical counterparts having similar forced and power capabilities.

In work class ROVs, it's common to use (and even combine) both electrical and hydraulic systems. Electricity is used to deliver power down the tether and to operate navigational sensors, lights, cameras, and many other devices on the vehicle, but it's also used to operate an electric motor that serves as a prime mover to run a hydraulic pump. This pump in turn provides hydraulic power for thrusters (Figure 10.11), manipulators, and payload tools. In underwater applications, hydraulic systems offer the added advantage that they are inherently pressure-resistant and less susceptible to damage by seawater than electric systems. As with electrical systems, the actuator can be located some distance from the prime mover.

Figure 10.11: Hydraulic Thrusters

These SMD hydraulic thrusters are made up of a hydraulic motor, matched propeller, and duct. Hydraulic fluid flowing under pressure through the motor spins the propeller and produces thrust for the ROV.

The cost/benefit ratio generally favors hydraulic systems only in cases where the need for high forces and power justify the high initial cost of installing a hydraulic system and the ongoing costs associated with routine maintenance, which can be substantial. This is rarely the case with small vehicles, so most of them rely exclusively on electricity. That said, you don't have to go for a full-blown hydraulic system. You may find that a few simple concepts borrowed from larger hydraulic machines can be put to good use easily and inexpensively in your small vehicle designs. Indeed, hydraulic systems have been used successfully on many small ROVs, particularly for control of grippers and manipulator arms, so it's worth learning a little bit about how they work and how you can build a simple hydraulic system for your own vehicle.

2.4. How to Build Your Own Hydraulic System

For small home-built ROVs or AUVs, a full-fledged hydraulic system—even a small one—is not usually cost-effective. However, some of the elements of more complex hydraulic systems can be incorporated into small, shallow diving ROVs with great success.

For example, many ROV teams have used pairs of plastic syringes (without needles) to transmit fluid pressure and motion from "pump" syringes (used as the controls) to "actuator" syringes via aquarium tubing running alongside the tether. (See Figure 10.13 for an example of this concept.)

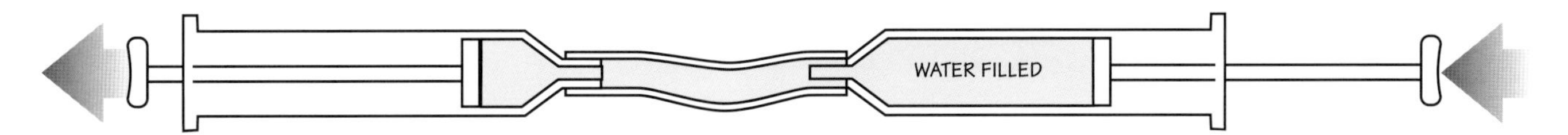

Figure 10.12: Simple Hydraulic Syringe System

You can use syringes and tubing to transfer fluid in a basic hydraulic system.

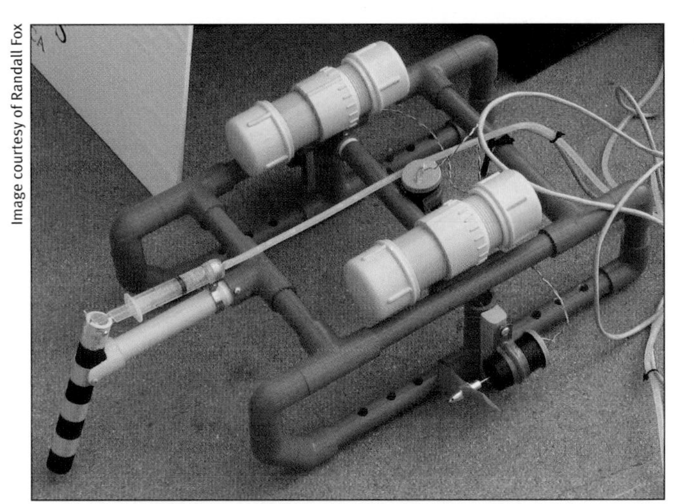

Image courtesy of Randall Fox

Figure 10.13: ROV With Syringe Hydraulics

A basic hydraulic system made out of syringes can be adapted to create simple, surface-controlled manipulators on an ROV.

Image courtesy of Steve Van Meter, VideoRay LLC

Figure 10.14: Basic Hydraulics and Grippers

The dual grippers on this student-built ROV are hydraulically activated.

Note that hydraulics for large commercial applications typically use hydraulic fluid (a specialized oil) in their lines. However, water works just fine for small-scale hydraulic systems made with plastic and rubber components. Best of all, water is easier, cheaper, less messy, and more environmentally friendly.

Recall that voltage is lost when electric current flows through long wires. In much the same manner, fluid pressure is lost when fluid flows through long tubes. Therefore, this type of system works well only with fairly short lengths of tubing. And just as electrical power losses can be reduced by using thicker wires and higher voltages, hydraulic power losses can be reduced by using larger-diameter tubing and higher pressures. Of course, both of these approaches have disadvantages, too. Larger tubes are more expensive and stiffer (making the tether bundle stiffer), and excessive pressures are likely to pop the tubing right off the end of the syringes!

2.5. The Pneumatic System Option

Pneumatic systems are much like hydraulic systems, except that they use pressurized gases, such as air or nitrogen, rather than pressurized liquids, to transfer the forces and motions. Thus, most of what you have just learned about hydraulic systems is applicable to pneumatic systems. However, one important difference between the two systems is that pneumatic systems exhibit a characteristic springiness, called "compliance," whereas hydraulic systems tend to hold any given position much more rigidly. The extra springiness can be a problem in some cases, because it makes precise

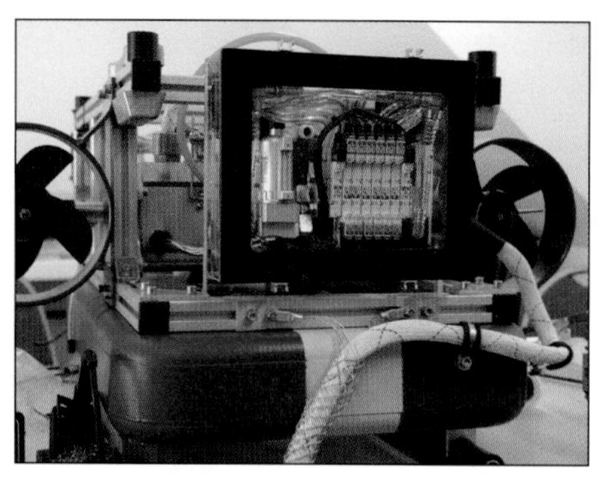

Images courtesy of Lee Thieman, SeaTech 4-H Club

Figure 10.15: Electronic Control of Hydraulics Using a Solenoid Valve

Chapter 9 emphasizes how electronics can be used to control the movement of vehicles or manipulators through the use of electric motors; however, electronics can also be used to control hydraulic or pneumatic systems by activating solenoid valves. This ROV, built by the Sea-Tech 4-H Club for the 2009 MATE International ROV Competition, is a good example. The team focused on refining arm utility, incorporating seven functions in each one, controlled by a relatively new miniature 4-way, 5-port, closed center pneumatic solenoid valve design supplied by TPC Mechatronics. This valve allows for discrete positioning of a double-acting pneumatic actuator. The compact design fit an eight-station assembly on a 6 inch long manifold.

Each manipulator was controlled by a relatively new miniature electronically controlled pneumatic solenoid valve supplied by TPC Mechatronics. It's visible in the plastic case mounted on the rear of the ROV, as shown in the right-hand photo.

SAFETY NOTE: PNEUMATIC EXPLOSION HAZARD

Pneumatic power systems are ideal for many small, low-pressure applications, such as toys, because they are easy to build and there is no risk of a leaky mess. However, large, high-pressure applications almost always use liquids (hydraulics) instead. One major reason has to do with safety. Gases compress relatively easily, so a large, high-pressure pneumatic system can store plenty of energy to unleash a violent explosion if something ruptures.

By comparison, oil and other liquids are incompressible so they can be used to transmit force and energy at very high pressures without storing energy. This means there is little risk of explosion. So use the pneumatic approach only for very low pressure control systems.

position control more difficult; however, in other situations, it can be an advantage. For example, human musculoskeletal systems are compliant, so pneumatic systems often do a better job than hydraulic systems of reproducing realistic human movements and forces in humanoid robots. Another important difference has to do with safety, so be sure to read *Safety Note: Pneumatic Explosion Hazard.*

3. Manipulators

Thousands of times in a day, you use your arms and hands to accomplish work—to shove pizza into your mouth, floss your teeth, search your backpack for missing homework, scratch the dog, lift a chair, repair a broken bike chain. Without question, they're the most versatile "payload tool" on the human body. Not surprisingly, the most common payload tool used on both manned and unmanned underwater vehicles is a robotic arm, known as a **manipulator**. This device is either electrically or hydraulically powered and mimics the motion of a human arm. Manipulators are remotely controlled devices in their own right and may be as complex to design and build as the underwater vehicle itself. While used extensively in shore-based manufacturing operations, nuclear power plants, and excavating equipment, manipulators also have been adapted successfully to a variety of subsea uses, especially manned submersibles and ROVs. Some subsea craft carry only one robotic arm, but others are equipped with two, each of which may have different functionality.

Manipulators perform several critical tasks:

- They provide a remotely located "hand," effectively extending the reach of the pilot, who is either inside a pressure hull or stationed on a surface support platform.

Image courtesy of Kraft Telerobotics

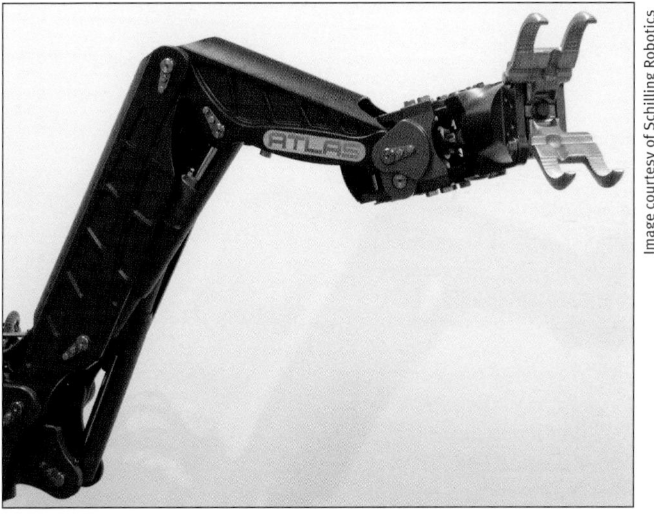

Image courtesy of Schilling Robotics

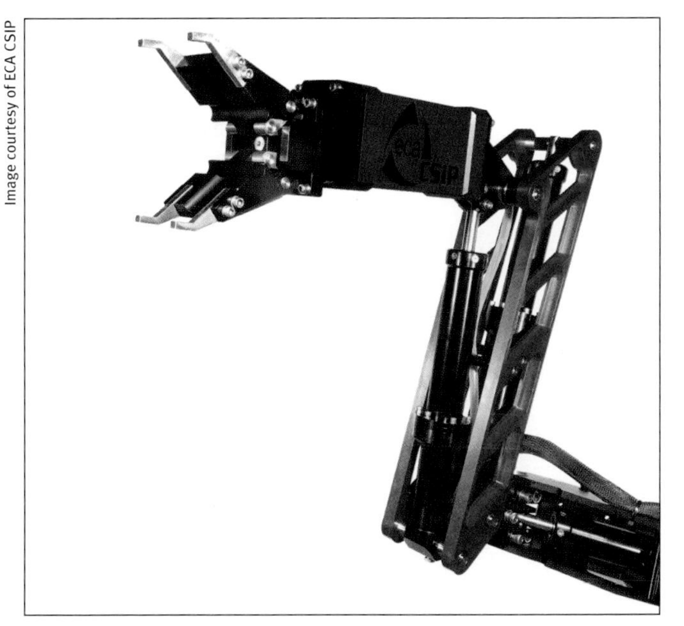

Image courtesy of ECA CSIP

Figure 10.16: Hydraulically Controlled ROV Manipulators

A hydraulic ROV manipulator combines linear actuators (hydraulic cylinders) and rotary actuators (hydraulic motors).

The Kraft Predator force feedback arm (top photo) is used for various underwater tasks including scientific work, deep ocean archaeology, and the excavation of ancient ship wrecks. The Schilling Robotics Atlas manipulator arm (bottom photo) is a new class of manipulator specifically suited for high-performing vehicles under the most demanding conditions.

Figure 10.17: Electrically Controlled ROV Manipulator

This all-electric manipulator by CSIP, a UK subsidiary of ECA, features linear actuators powered by brushless electric motors instead of hydraulics.

- They can significantly increase, or magnify, the pilot's strength to move and lift objects.
- They can perform tasks in hazardous environments (e.g., deep water, nuclear reactors, etc.) that cannot be done safely by means of direct intervention.

Of course, no robotic arm is without its limitations:

- Compared to a human arm and hand, a manipulator has limited dexterity. (That said, note that some robotic wrist joints can actually rotate 360° or more, which is greater than the human wrist's rotational capacity.)
- Most manipulators lack a sense of touch, although some advanced manipulator systems do include tactile sensors for precision gripping. Sensors can also provide the pilot with **force feedback**, thereby giving a sense of the intensity of the grip. (See below for a more detailed description of force feedback.)
- Currently, the size, complexity, weight, and cost of a commercial robotic arm limit the types of vehicles that can carry one.

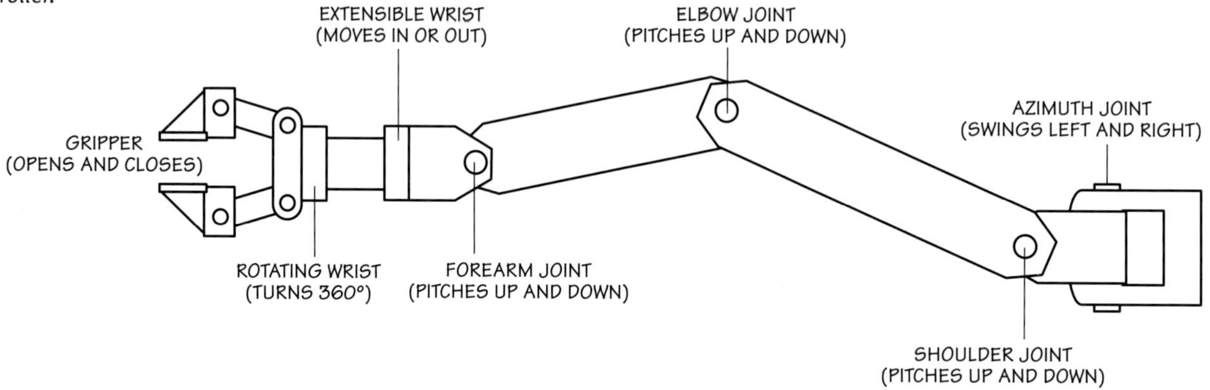

Figure 10.18: Simplified Diagram of a Robotic Arm

The manipulator is sometimes referred to as the "slave" arm. The pilot control interface and associated control electronics is the corresponding "master controller."

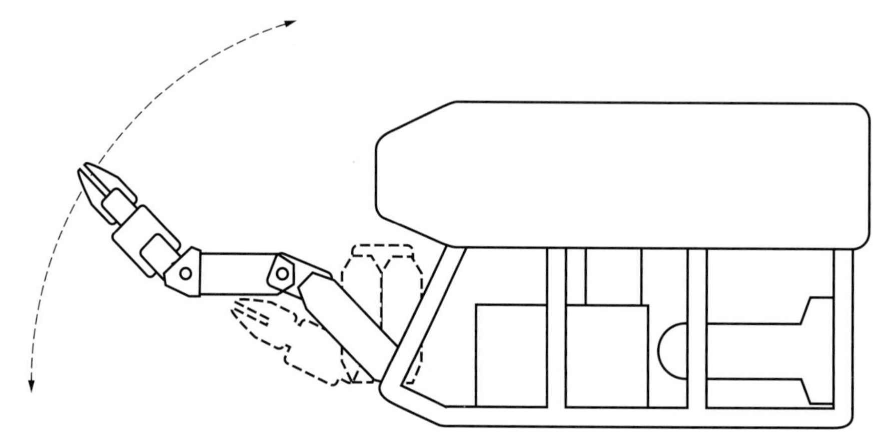

Figure 10.19: Manipulator Mounted on ROV

A manipulator system is a robotic system in its own right. In effect, it's a robot mounted on another robot (the ROV).

This simplified diagram shows a robotic arm mounted on an ROV and ready for work. The heavy dashed lines show the arm in its retracted position. The lighter dashed lines show its range of motion.

3.1. Components of a Manipulator System

Much of what you've learned about controlling robotic vehicles also applies to controlling their manipulators. Both the remotely operated vehicle and the remotely operated manipulator depend on surface control and a way to transfer those control signals to the robot in order to complete underwater tasks. Both also rely on visual or other closed-loop feedback to provide information to the pilot.

Manipulators can be either hydraulically or electrically powered. Both types share similar features. Hydraulically activated manipulators are the logical choice for commercial work when great force is required. For lighter-duty applications, electrically activated manipulators work well. Although less powerful, electric arms have several advantages: they are generally cheaper to produce, they utilize a simpler control system, and they require less maintenance. Electric systems are also quieter, which is an advantage in some tasks. Generally, smaller ROVs carry electric arms that can be useful for sampling or attaching a light line to objects.

3.1.1. Master Controller and Slave Controller

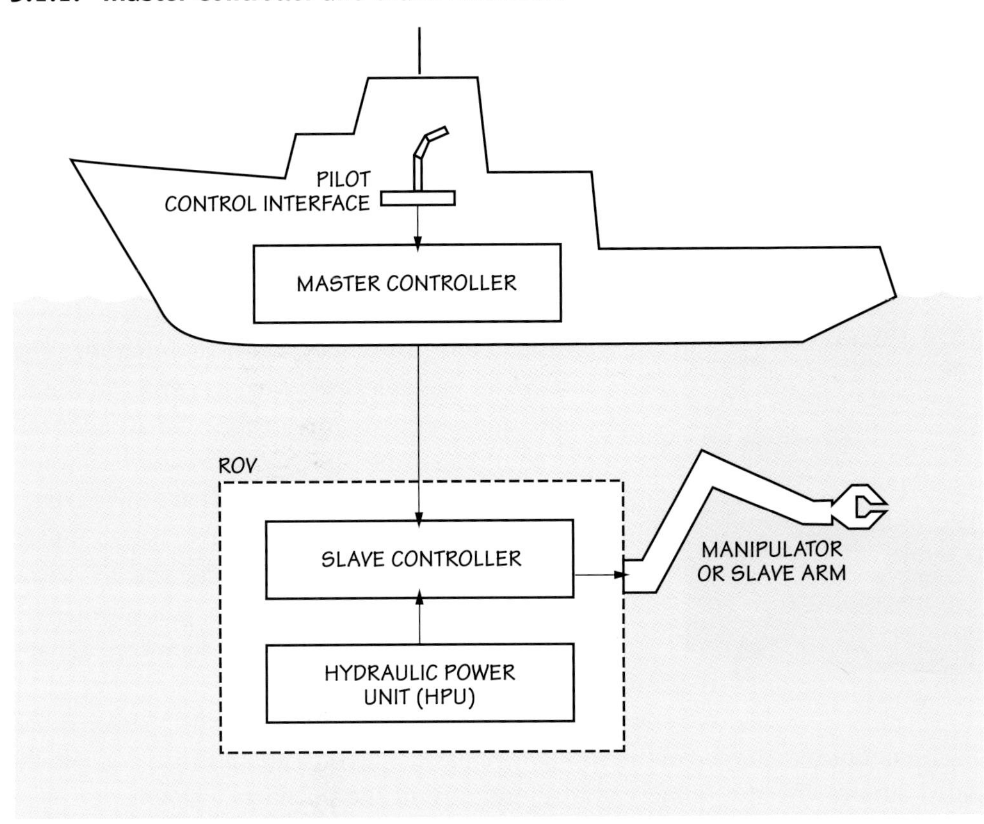

Figure 10.20: Simplified Diagram of a Manipulator System

Hydraulic manipulators generally feature the following components: master controller, hydraulic power unit (HPU), slave controller or interface unit, hydraulic cylinders or other actuators, and an end effector (gripper or other tool attached to the end of the arm).

There are different ways a pilot can communicate instructions to a manipulator arm. In most systems, he uses either a joystick or scale model of the manipulator (Figure 10.21). Each joint in this scale model has a servo pot (potentiometer) attached to it. These pots continuously measure the angle of each joint and digitally encode them. Regardless of the details, these user-machine interfaces are part of a "**master controller**" system which translates pilot commands into coded signals sent down the tether.

At the bottom end of the tether on the ROV there is a "**slave controller**" which decodes the commands coming down the tether and uses them to operate the manipulator arm. The term "slave" refers to the fact that the manipulator follows exactly the motion commanded by the master controller. If the master twists or bends, so does the manipulator.

If the arm uses hydraulic actuators, power is provided by a **hydraulic power unit**, or **HPU** (power + accumulator), to generate pressurized hydraulic fluid. The slave controller uses the decoded signals to operate electrically controlled hydraulic valves that regulate the flow of hydraulic fluid to the arm's actuators. If the manipulator uses electrically driven actuators, power is delivered through the tether or by an on-board battery. The slave controller uses motor control circuits to route electrical current to the arm's actuators.

3.1.2. The Arm Itself

Your arms may be the most versatile of all your body bits because of their amazing range of motion. Not surprisingly, engineers have designed most manipulators so that

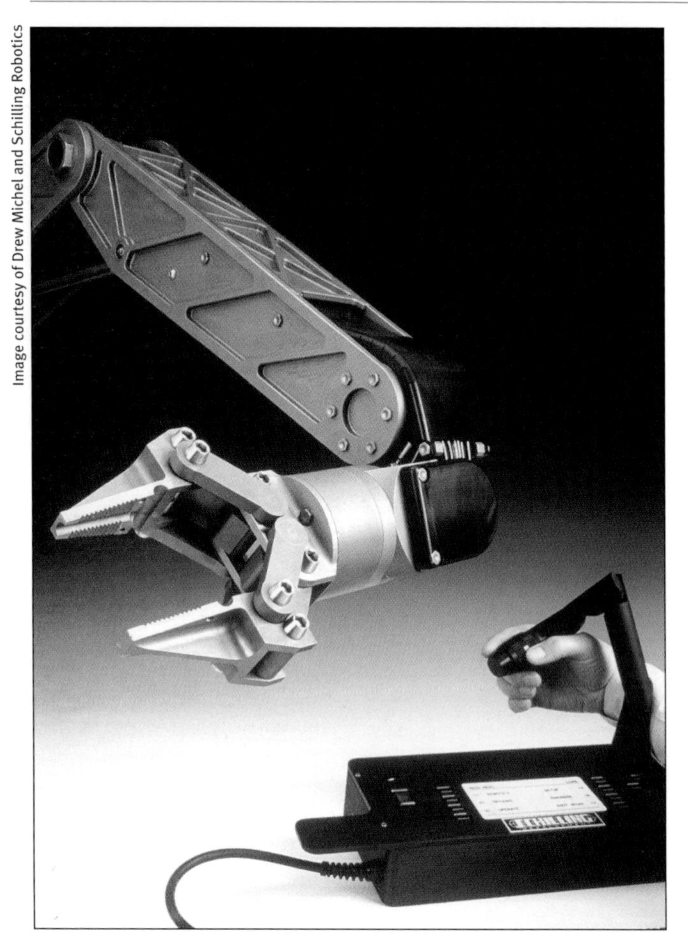

Image courtesy of Drew Michel and Schilling Robotics

Figure 10.21: Controller Unit

In the Schilling Control Interface Unit pictured, the arm-like appendage attached to the top of the master controller is a scaled model of the actual robotic arm with sensors built into the joints.

These sensors send control signals via the master controller to the slave manipulator, which mimics the joint angles and therefore the overall motions produced by the interface unit. There are also other styles of interface units that are panel- mounted away from the master controller.

they resemble the human arm's mechanical function. Like a human arm, manips have joints that connect the various portions of the mechanical arm. Each of these joints rotates or extends (or both). This joint movement is produced by actuators that are either cylinder-enclosed pistons (rams) or rotary motors. One actuator produces one range of motion—for a cylinder, that means extending out or pushing back in, and for a rotary motor, that means turning counterclockwise or clockwise. These cylinders and rotary motors work in concert to produce the complex extension and rotational movements of the manipulator.

Each distinct type of movement (bending or rotating at a joint) that an arm can make is referred to as a **degree of freedom** (**DOF**) or sometimes as a **function**. The human arm, for example, has seven degrees of freedom: The whole upper arm (humerus bone) rotating at the shoulder joint can move up/down, forward/backward, or spin on the long axis of the bone. (That's three DOF.) In addition, the elbow can flex/extend. (That's a fourth DOF.) The forearm can twist, enabling you to orient your palm so it faces up or down. (That's a fifth DOF.) And your wrist can flex up/down or right/left. (That's two more DOF, for a total of seven—more that most robotic arms.) On a robotic arm, the gripper or other device mounted at the end of the arm (where the hand is on a human arm) is generally called an **end effector**. Some manufacturers count the opening/closing action of a gripper as an additional DOF; others do not.

In theory, a six-DOF arm can move a gripper into any position (X, Y, Z) and any orientation (roll, pitch, and yaw). However, the human arm's "redundant" seventh DOF is useful for collision avoidance by allowing more than one way to achieve a given hand location and orientation. You can demonstrate this by picking up a glass

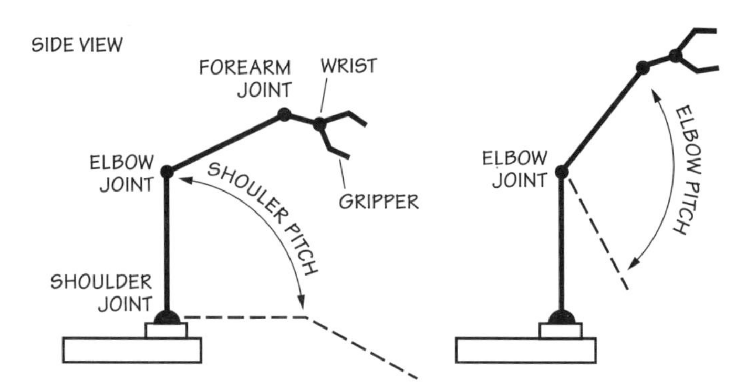

Figure 10.22: Range of Motion for a Manipulator

The motion of a manipulator resembles that of a swing desk lamp.

Shoulder joint (2 DOF): Rotate left or right about vertical axis + pitch up or down
Elbow joint (1 DOF): Pitch up or down
Forearm joint (1 DOF): Side to side motion
Wrist (1 DOF): Rotate about its long axis
Gripper (1 DOF): Open or close in linear motion

of water from the table with your elbow either up or down. Simple robot arms have fewer DOF; the most sophisticated robot arms have seven DOF.

Almost all work class ROVs and commercial deep diving manned submersibles are equipped with two arms, or **dual manipulators**. As with humans, two arms work better than one. Unlike human arms, which have the same joints, the manipulator on the left is often a four-function arm, while the one on the right is a six-function arm. The left manip is used for heavier work and/or for anchoring the vehicle in place. The more dexterous right arm is used for tasks requiring finesse and complicated positioning. The two manipulators can operate simultaneously, just as you use two arms and hands to control objects.

Image courtesy of VideoRay LLC

Figure 10.23: ROV with Single-Function Arm

A single-function manipulator has a fixed arm length, but the gripper at the end can open or close linearly; that is, it has two opposing jaws that do not curl or rotate.

In this photo, the VideoRay Pro3 GTO recovers a mock IED (Improvised Explosive Device) with the U.S. Navy. The ROV completes its mission using only a basic single-function manipulator.

3.1.3. End Effectors

If you wanted to switch from writing a report to cutting down a tree, you'd put down your pencil and pick up a saw. With a manipulator, it's slightly different. You'd actually unscrew your hand (with the pencil) and attach a saw to your wrist. In either case, you've simply traded one end effector tool for another. **End effector** is a general term that describes the device attached to the wrist of the robotic arm—it's the business end of the manipulator. The other aspects of manipulator movement (the shoulder, elbow, forearm, and wrist) are all designed to get the end effector in a position to do the actual work. Sometimes special-purpose end effectors must be designed for a particular job, but several common ones are readily available:

- **Gripper**: This is the most common end effector and is simply a vise-like mechanism or grabber. For an electric gripper, an electric motor drives a worm screw that opens or closes the jaws. For a hydraulic gripper, a hydraulic ram opens and closes the jaws. Rotation of the gripper is made by the manipulator's wrist and is not usually integral within the gripper mechanism itself. Grippers are sometimes classified as "precision grip" or "power grip." The human hand can do both; it can thread a needle or hold a hammer. Most robot grippers can do only a power grip. A few sophisticated ones can do both.
- **Mechanical cutter**: A gripper can be outfitted with sharp-edged, hardened-steel cutting blades. Instead of grabbing an object, the gripper surrounds a cable or line and severs it. If a cutter has to cut very large cable, a specially designed cutter that can exert more pressure than a modified gripper is installed on the arm. Another option is to use a rotary cutter.
- **Scrubber**: ROVs are used routinely to clean the underwater structures of oil-rig platforms; they do

Figure 10.24: ROV with Multi-Function Manipulator

Larger ROVs have one or more manipulators with multiple DOF. The five-function arm is the workhorse of the work class ROV. It's usually hydraulic and is used for heavier tasks.

In this photo from the 2004 Mountains in the Sea expedition, the Institute for Exploration (IFE) ROV Hercules *uses a 7-function Predator manipulator arm to recover a scientific experiment deployed the previous year.*

Image courtesy of OceanExplorer NOAA

Image courtesy of Drew Michel and Schilling Robotics

Figure 10.25: Various End Effectors

much of this work using a rotary scrub brush mounted on a mechanical arm. This type of end effector can scrub away the marine growth that fouls these structures and clean the supports so that structural inspections can proceed. Sometimes a high-pressure water jet can also be mounted on a manipulator arm to clean steel supports.

- **Arc welding cutter**: Ordinary hydraulic cutters may have difficulty cutting thick steel, so an electric arc welder can be attached as an end effector on one mechanical arm. A second arm positions the welding rod correctly in relation to the work, thus ensuring good clean cuts.

- **Specimen collector**: On science expeditions, manipulator arms can be outfitted with specimen collectors. These are usually special low-power suction dredges that can slurp up floating biological specimens such as jellyfish or more stationary bottom samples. The specimens are then piped to collecting bottles located on a sampling work tool tray, called a carousel. For larger specimens, vehicles such as MBARI's *Ventana* also carry gallon-size clear plastic jars with motorized top and bottom lids that can shut rapidly to trap creatures inside.

3.1.4. Feedback Control

Because manipulator tasks usually require precision control, they are almost always controlled through closed-loop feedback systems. This means that information about what the arm is doing must be fed back to the pilot.

Visual images from a video camera are the simplest and most universal form of feedback. Usually, a camera mounted on the body of the ROV provides an overview of the arm's action. In addition, some arms have a camera mounted on or near the manipulator's wrist, to provide close inspection or monitoring of end effector activity.

Some advanced manipulators provide additional types of feedback. For example, grippers may include force feedback sensors that can measure how hard the jaws are gripping. This information can be relayed back up the tether to the master controller. The replicator arm on the controller can be equipped with little motors that provide resistance to the pilot's hand, based on the force feedback, thereby giving the pilot a physical sense of how much pressure is being applied to the object being grabbed. Similar force feedback can be used in all of the manipulator joints. In such a case, if the gripper is lifting something heavy, force feedback also imparts some of this "weighty" feel back to the pilot.

The advantage of force feedback is that it increases the amount of information the pilot has when trying to grasp items and lift them from the bottom. Since strong hydraulic arms are force multipliers, it's easy to apply too much pressure and accidentally crush delicate objects, so force feedback is especially important when gathering delicate or crucial objects, such as artifacts from shipwrecks or flight recorder boxes from aircraft disaster sites.

TECH NOTE: UNDERWATER ROBOTIC HANDS

Any doctor, biologist, or inventor will tell you that the human hand is amazingly complex. In fact, you've probably heard that your "opposable thumb," which allows you to grasp, is what makes humans more adept than "primitive" animals. If it were that simple, crabs would rule the world! More precisely, the unique ability of the human hand is linked largely to the ability of the human thumb to "index." That means your thumb can directly oppose each of your individual fingertips, centering itself if more than a single finger is employed, as in grasping. Moreover, your hand does all of this reflexively, without conscious thought—a remarkable feat!

Traditionally, **Atmospheric Diving Suits (ADS)** have had to rely on a two-jawed manipulator in place of the human hand. This manipulator operates much like a pair of pliers—squeezing the handles inside the ADS causes a set of jaws to close on an object outside of the suit. (To understand the difficulty of executing an underwater task that requires significant manual dexterity, try attempting the same task on the surface using two pairs of pliers.) With practice and good eye/hand coordination skills, the operator in an ADS can perform simple tasks. However, to complete complex tasks, the diver inside the Atmospheric Diving Suit needs the ability to use his or her hands in the same manner as a scuba or helmet diver does. Until now, this has not been possible.

The development of an ADS manipulator, or end-effector, that could match or nearly match the dexterity of a gloved hand would require that the external "fingers" not only mimic the exact movements of the inside "master" hand, but also provide full, 100 percent reflexive index-ability of the external thumb, in concert with the other digits employed. In addition, the outside "slave hand" should also provide directly proportional sensory feedback of pressure, weight, etc., to the inside master hand (yours!).

Previous attempts to build such a device have achieved reasonably close mechanical matches to the geometry and motion of the individual fingers, but the designers and engineers were not able to provide a true, rotating, index-able thumb to oppose those fingers. At this writing, an electronically controlled version is under development for use on ROVs and deep submersibles. And there is also considerable interest and discussion with the national space agencies of several countries on the use of this "Prehensor" as a possible alternative to the conventional space-suit gloves.

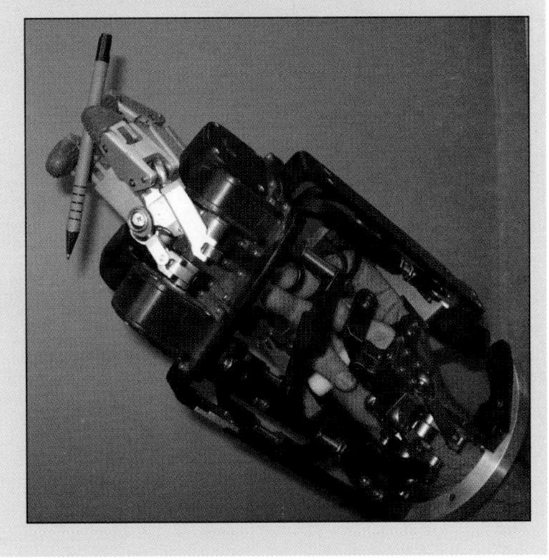

Image courtesy of Vickie Jensen and Nuytco Research Ltd.

Figure 10.26: The Prehensor Prototype

Nuytco Research developed the unique capabilities of its Prehensor for the Nuytco ADS Exosuit. *But the company also expects the system could replace existing simple jaw-style ADS manipulators once beta testing is completed and production begins.*

3.2. Home-Built Manipulators—A Case Study

Equipped with only a camera and a basic probe, simple eyeball ROVs like *SeaMATE* in Chapter 12 can certainly get you into some fun missions. But it won't be long before you run up against some of the limitations of such a basic design, and you'll find yourself dreaming up ways to give your vehicle more capabilities. This will lure you into the next design project, and the next (and the next).

For example, sooner or later (probably sooner), that camera eyeball on your ROV is going to spot something you just can't resist picking up. In most cases, a simple probe won't be up to the task, so you'll want to start thinking about how to design a more sophisticated manipulator or gripper.

Any number of people have attempted the challenge of designing and building some sort of retrieval tool for their robot—and soon an even more sophisticated arm. This kind of design challenge inevitably involves a trial and error process. You start with an idea, try an early prototype, see which things don't work, generate new solutions and ideas, try a second prototype, etc. It's a learning process that seems to be equal parts frustration and elation. But such a gradual evolution process, coupled with your own resourcefulness, often results in a successful gripper or manipulator that is not only workable, but often a far cry from your original idea. Just keep in mind that successful improvements are usually based on failures. So if you're not trying, you're not learning!

In the section that follows, one person walks you through his adventures of designing and building a manip arm for his ROVs. Connecticut-based ROV builder Steve Thone describes himself as an "ordinary guy" who loves building underwater robots "for fun." In Section 3.2., Thone shares his thoughts, design ideas, construction experiences and his photographs of various manipulator projects for his *SEAFOX* ROV.

3.2.1. Thone's Story: Driven by Curiosity

"I got started building ROVs as a result of fishing. I had a fish finder on my boat that would show the depth and contour of the bottom of the lake, but I wanted to see the actual structure of the bottom and where the fish were hiding. One day, while watching the Discovery Channel, I saw a show about remotely operated vehicles and started thinking, 'That's what I need.'"

Figure 10.27: Prototype Design and Model

Thone says: "I drew my idea and made a simple prototype using balsa wood, just to see if the idea was feasible. This prototype was never even hooked up to a motor, but it seemed like it should work ok."

All images courtesy of Steve Thone

"Back in the early '90s, I couldn't find hardly any information about building ROVs, so I fooled around with different designs for about a year. There was a lot of trial and error and a lot of failed attempts. I almost gave up a few times, but persistence finally paid off. Finally I found the website www.ROV.net and saw they were running a design competition. That inspired me, so I actually finished my first ROV, *Stinger*. And it won! Since then, I've built a number of ROVs; more recently, I've been working on the challenge of designing and building manipulators for my vehicles."

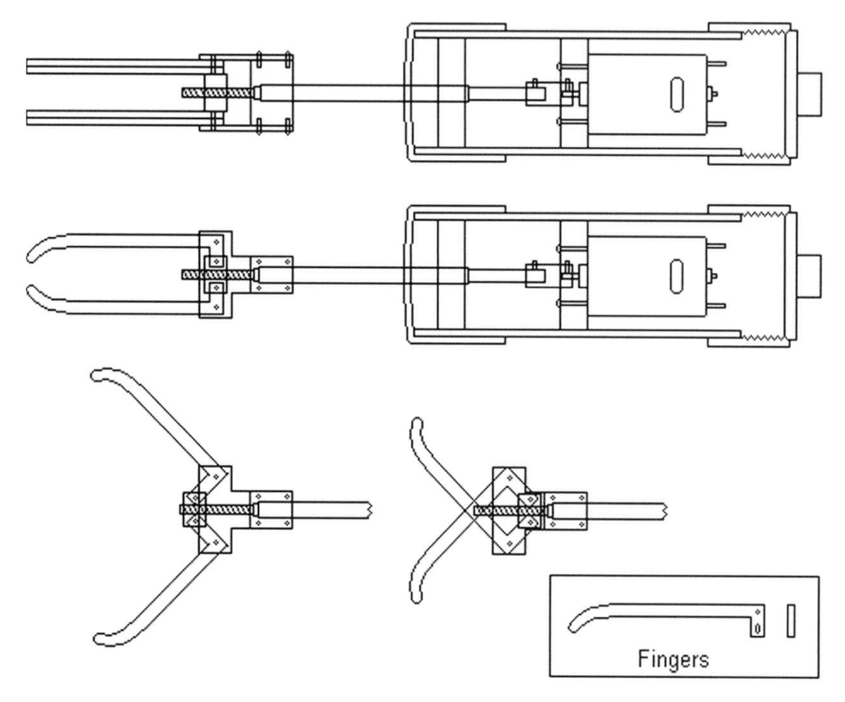

3.2.2. Thone's Story: Early Prototypes

"These first two mockups were my initial attempts at designing a manipulator before I had any real tools. Although they never went farther than the prototype builds pictured here, I could see that they would work fine for a simple manipulator."

3.2.3. Thone's Story: My First Real Manipulator Attempt

"The second ROV I built was based on the *SEAFOX* plans in the nifty book *Build Your Own Underwater Robot and Other Wet Projects*. When I got it up and running, the challenge was to actually add some sort of an electrically powered manipulator. To keep it simple, I started with a single-function one with just jaws that opened and closed. What I came up with uses a geared motor to operate. The fingers are opened and closed with a separate set of gears. The gripper assembly was hacked from an old Radio Shack Armatron toy, which is pictured here. A manipulator like this could work on any basic ROV like *SEAFOX* or *SeaMATE*."

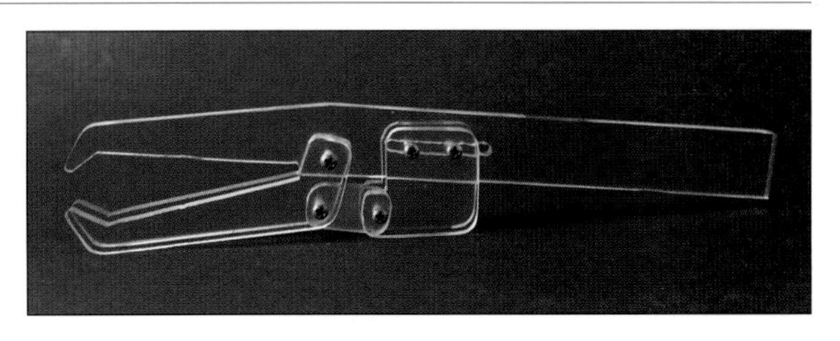

Figure 10.28: Second Prototype

Thone says: "I cut my second prototype out of Plexiglas. Both of these ideas work on the principle of a geared motor turning a threaded shaft or rod. As the rod turns, it pushes a block back and forth. The gripper fingers are attached to this block and have a pivot point, which makes them open and close when the block moves. It's a pretty basic design. Doing these two mockups made me realize that the design was feasible."

"It's just a basic claw with an open-and-close function, which is accomplished by means of a motorized planetary gearbox. I bought one from MicroMark that has 14 different gearbox combinations, which can provide a complete output speed range from 13.8 to 6,560 rpms. The gearbox is sealed in the same type of housing I designed for the thrusters on my original *Stinger* ROV, but this one has an actual seal on the shaft. In fact, I might try using one of these for a thruster motor in the future."

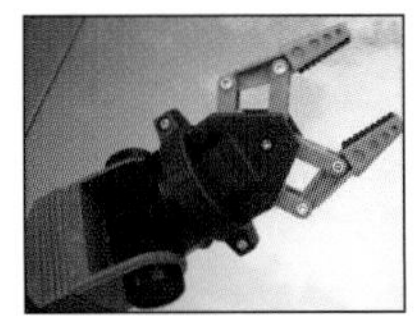

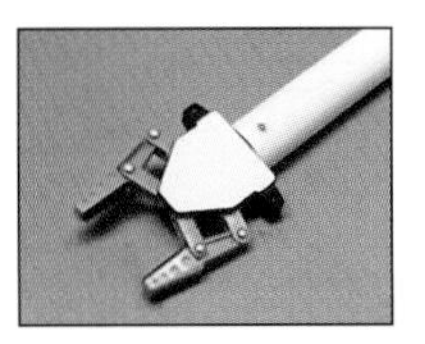

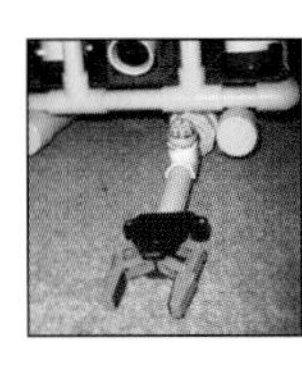

Figure 10.29: Robotic Toy Adapted to Basic Claw

"Unfortunately, the first time I took the ROV out with the manipulator, I forgot to put Teflon tape on the threads of the plug to the housing, so it filled with water and caused all kinds of buoyancy problems. Then I forgot to dry the motor out; it rusted up solid and had to be replaced. Live and learn, I guess."

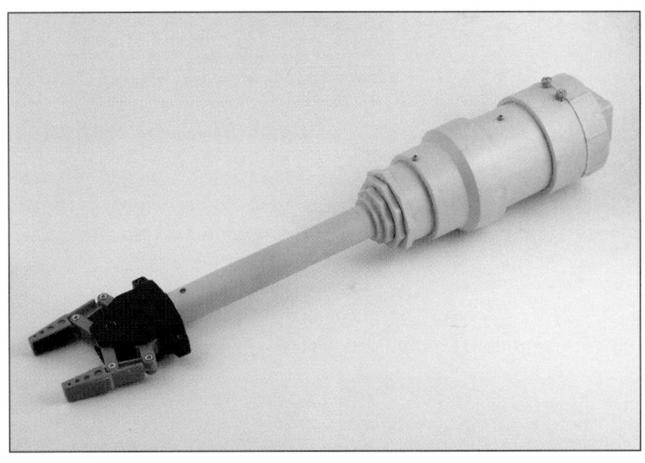

Figure 10.30: Completed Manipulator and Housing

Thone says: "This is the completed manipulator and housing for the manipulator. It can be attached or removed from the ROV in minutes with the use of zip ties. Power is hooked to the two bolts running through the housing; you can see them at the end of the housing."

"While this manipulator is a fun payload tool and I learned a lot building it, the arm is not very practical, since my *SEAFOX* ROV was not very powerful to begin with. Most smaller ROVs have a standard fixed manipulator with only the ability to open and close the gripper fingers—just like the one I'd finished building. These manipulators are usually mounted parallel to, or at the bottom of, the ROV and attached either to one of the skids or between them. This is how I had mine mounted originally, and this arrangement works fine for gripping anything that is off the bottom. However, if you're trying to retrieve something small that's located directly on the seabed, there's a problem getting the manipulator into position low enough to grip the item. It's a task that can be almost impossible if the bottom is not perfectly flat. And even if the bottom is flat, there's always the possibility of kicking up silt, because the ROV tends to bounce along the bottom when trying to get into position for a retrieval."

"Since I wanted to be able to retrieve small objects from the bottom without running into those problems, I came up with the idea of adding the ability to rotate the manipulator arm up or down during the dive, depending on the condition required. I figured this would allow me to reach down and retrieve the object while allowing the ROV to remain slightly off the bottom. My plan for rotating a manipulator arm may seem complex, but I actually came up with the idea in order to make the gripper easier to use on a basic ROV."

Figure 10.31: Common PVC Pipes and Fittings

3.2.4. Thone's Story: Moving to a Multi-Function Arm

"The next manipulator I designed for my retrofitted *SEAFOX* worked on the same principle as the first prototypes. I already knew that adding new functions isn't always easy, depending on the control method you're using. In my case, it would have required adding more wires to the tether. However, I didn't want to add any more bulk there, so I figured I could get around that problem if I just combined the functions I already had. My original *SEAFOX* ROV had thruster control, of course, and I had built a standard manipulator, detailed above, as well as a tilting mechanism for the camera. I solved the problem by simply moving the tilt function to the new manipulator arm, then mounting the camera on the arm itself. That way I was able to achieve the results I was looking for without adding any more wires or motors."

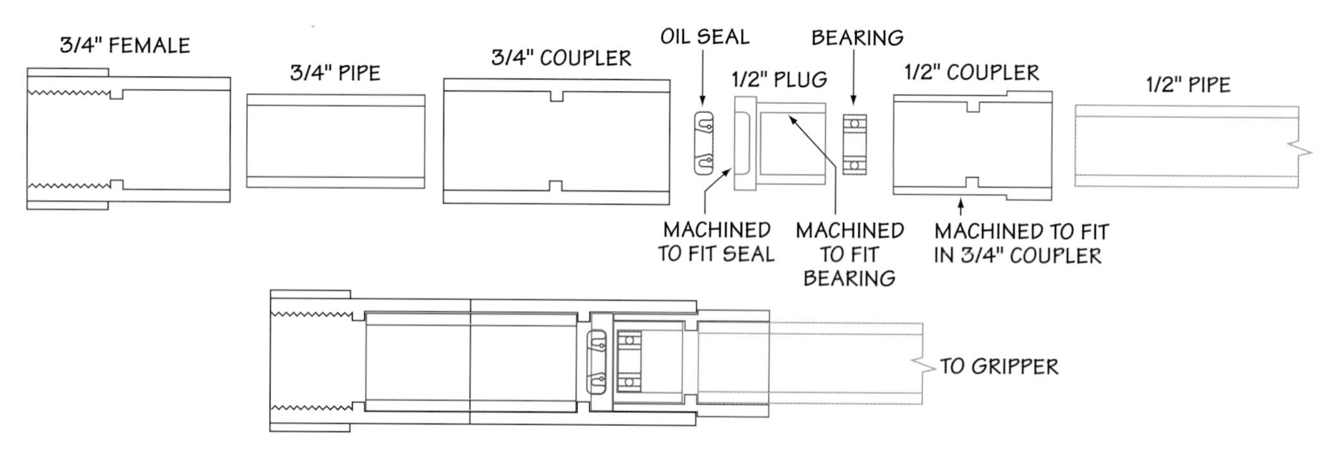

Figure 10.32: PVC Diagram

Thone says: "The new manipulator arm was made from common PVC pipe and fittings, as detailed in the diagram. Most of the parts where machined to fit together so as to give the whole ROV and manipulator a smoother look, but this arm could probably be made using only simple hand tools and a drill press."

10.33: Completed Manipulator Details

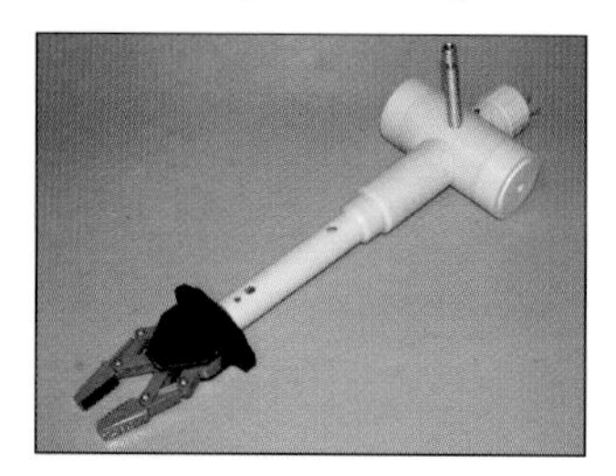

"The aluminum rod sticking up in the middle of the completed arm assembly is the camera mount."

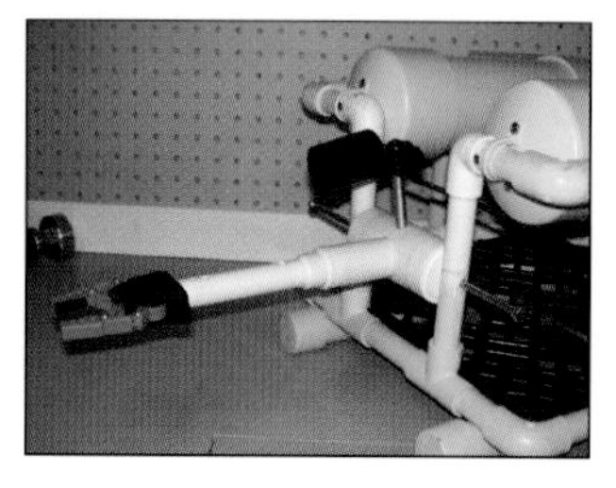

"A few 1/4-20 bolts hold the manipulator in place for a quick mock-up."

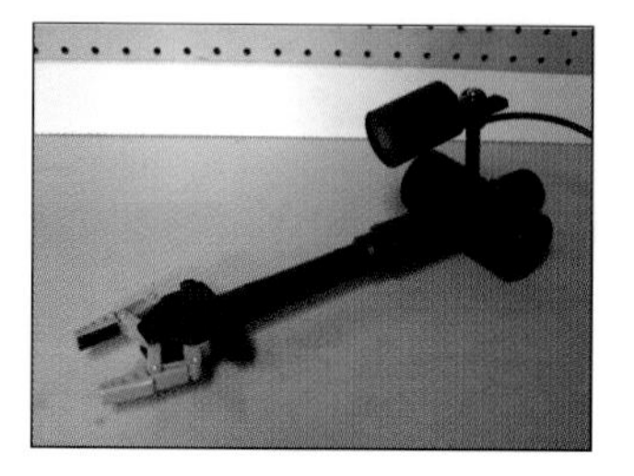

"Here the finished assembly has been painted and the camera has been mounted to the manipulator arm with an aluminum rod machined to fit this particular camera's original mount. I also added an auxiliary light, which tilts along with the camera and manipulator."

"In my updated version, I re-used the gripper mechanism from my original SEAFOX *manipulator, which I'd cannibalized from an old Armatron toy. It uses a simple geared motor to turn a shaft that opens and closes the fingers via a worm gear setup in the gripper. I put the geared motor in the manipulator arm, with the shaft exiting the housing through a motorcycle oil seal, to keep the housing waterproof."*

10.34: Geared Motor for Gripper Rod

Thone says: "This is the geared motor that turns the rod on the gripper. I removed the gear so that now a simple shaft and adapter connect to the rod on the gripper."

"An important thing to remember is that manipulators are only as strong as the ROV they are on. For example, if I were to try and pick up too large an object, the whole ROV would become unstable and uncontrollable, because the load in the manipulator would throw off the balance of the ROV. You can, however, still recover larger objects with a smaller ROV. One way is to use the manipulator to take down a separate hook and line so that the hook is held in the gripper and the line is manually fed out of the boat during the dive. Position the ROV so the gripper can place the hook on the object being retrieved, then release the hook from the gripper and pull the line in by hand from the boat to retrieve the object."

Figure 10.35: Making a Plug for Sealing Motor Housing

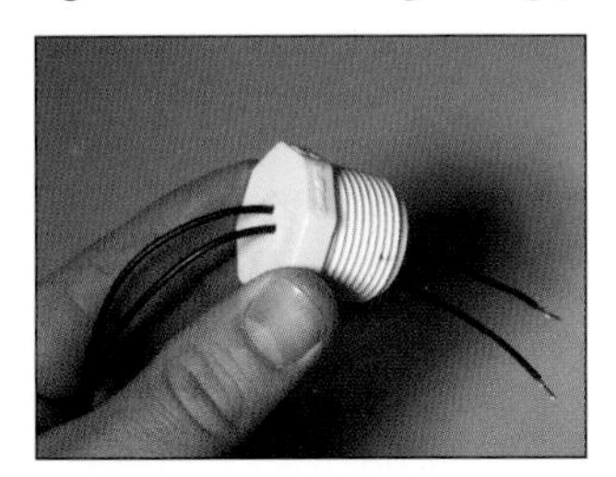

Thone says: "A simple PVC plug seals up the motor housing on the arm. To make this, I started by drilling two small holes for the wires to pass through."

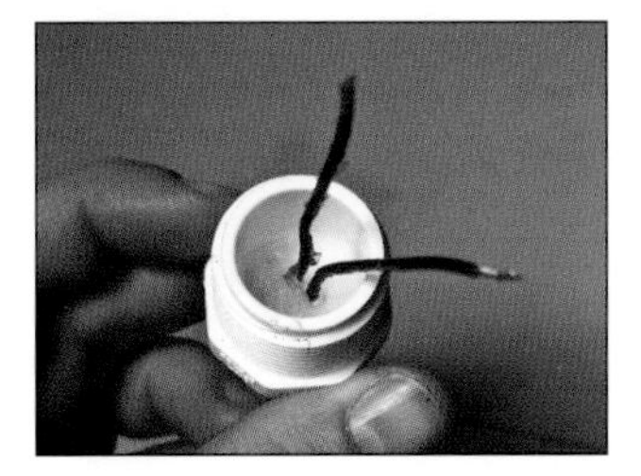

"Next, I added a little silicone in the bottom of the plug. The silicone gave the wires a bit of strain relief."

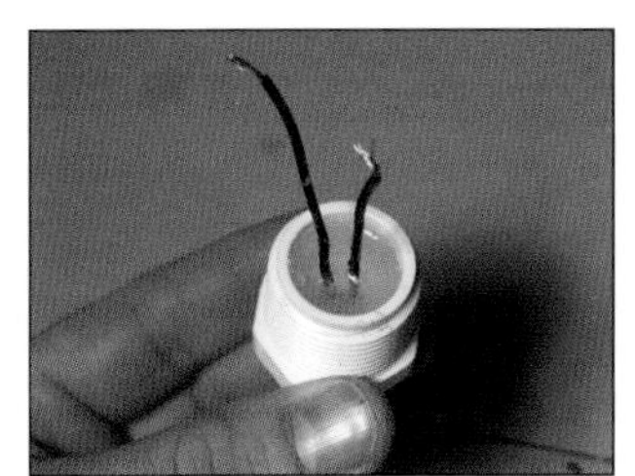

"The final step was to fill the rest of the plug with potting epoxy."

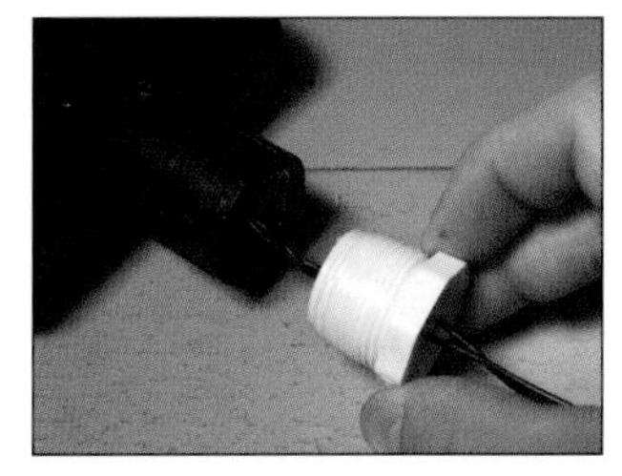

"Then I soldered the wires to the motor. By pre-twisting the plug and wires a few turns, it meant the plug could be screwed in without the wires binding up inside the housing."

Figure 10.36: Gear Assembly for Arm Pivot

Thone says: "This is the gear and shaft that attaches to the pivot point of the arm mount."

"The next set of gears is attached to the PVC frame."

"Then they are secured to the frame with a 6/32 screw, which still allows the gears to rotate freely."

"This is the mount for the geared motor that will rotate the whole arm assembly."

"Here are two views of the mounted motor."

Figure 10.37: Completed Manipulator

Thone says: "Here is the completed manipulator and camera setup. The rotation of the entire arm is accomplished via an Airpax 12 VDC gearhead motor and four gears (pictured above). I used this gear setup to achieve the torque needed to lift the arm and to slow the tilt speed down. In the beginning, I left the motor for the tilt exposed to the water, but in the near future, I will probably have to construct a waterproof housing for it because it has already shown signs of failure after the first few times of use."

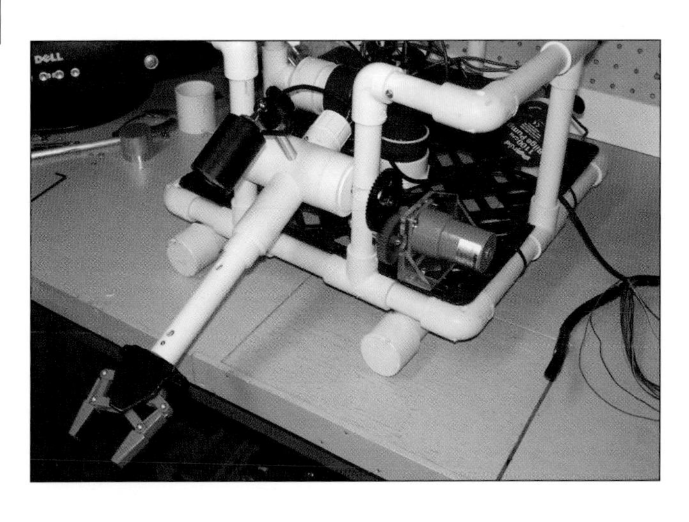

Figure 10.38: ***Manual Rotation***

Thone says: "This photo shows how the manipulator can be rotated manually. Here the grippers are in the vertical position, but they can be rotated to any angle, depending on the requirements of the dive. The only downside is that the ROV has to surface in order to manually change the angle of the gripper. So my next challenge will be to add another motor that will change the angle on the fly."

"In theory, you should be able to control the opening and closing of the gripper's fingers as well as rotate the entire wrist assembly with just one motor. When the fingers are fully opened, the whole wrist assembly would rotate in a counterclockwise direction; when fully closed, it would rotate in the other direction."

3.2.5. Thone's Story: Sharing the Learning

Steve Thone's case study is typical of the trial and error process all inventors go through. And like many of them, he is happy to share his ideas via his website (www.homebuiltrovs.com). He says, "It's always great to hear from some inspired kid in school, but I have to laugh, because they assume I'm some kind of engineer or something and want to know where I went to school. The truth is I'm just your everyday, average guy without any technical background. I never went to college or tech school. I just build underwater robots because they're a fun challenge!"

4. Underwater Tasks and Tools

A carpenter needs a hammer and saw for working with wood, an electrician relies on a multimeter for measuring voltage, and a plumber requires a pipe wrench for tightening pipes. Similarly, every underwater task requires a specific set of payload tools in order to complete the mission. While the tools themselves vary (see below), so do the ways in which they can be carried or mounted, as detailed here:

- as a separately towed craft, called a **tow sled**
- as part of an interchangeable **tool tray**, or **tool sled**, that generally fits underneath or forward in an ROV or submersible
- as part of the permanent structure of an underwater craft
- as an end effector on one or more of the vehicle's robotic arms

This section concentrates on underwater tools, with tow sleds and tool trays described first, followed by a sampling of the large variety of tools used by undersea craft to accomplish their missions.

4.1. Tow Sleds

As introduced in Chapter 1, a tow sled is a towed underwater platform loaded with an assortment of payload tools—essentially an undersea Swiss Army knife. Propulsion is supplied by the towing vessel, which pulls the equipment in a survey or search grid at a specified depth. Traditionally, tow sleds have been used for two basic types of missions—visual survey and/or sonar mapping. While they have often made headlines for their role in helping to discover sunken ships, tow sleds are also important for commercial surveys, mapping potential oil/gas fields, and for military purposes. With

HISTORIC HIGHLIGHT: THE DEVELOPMENT OF TOW SLEDS

Image courtesy of Woods Hole Oceanographic Institution

Figure 10.39:* ANGUS—*Early Tow Sled

Tow sleds such as the *ANGUS* (for Acoustically Navigated Geological Underwater Survey) represented an early technological deep water exploration advance in the 1970s. Aptly nicknamed "dope on a rope," the heavy-duty steel sled originally carried downward-looking, black-and-white film cameras that clicked away while being towed over the area to be explored. The result was a photographic mosaic consisting of thousands of images, but no one could view the results until the sled had been hauled back on board, the cameras emptied, and the film developed. Also, there was no way to pinpoint the location of any discovery on the ocean bottom.

Driven by his quest to discover the *Titanic*, marine archaeologist Bob Ballard envisioned that the next generation of deep towed vehicles would provide real-time visual presence on the ocean floor. *Argo* was typical of this new unmanned craft, armed with video cameras, as well as two sonar systems—a forward-looking one to detect obstacles in its path and a side scanning sonar for investigating the bottom. This collection of equipment was linked to a surface ship by means of a long fiber-optic cable. Furthermore, *Argo* became home to a smaller robot named *Jason* (originally *Jason, Jr.*), which could utilize its own independent propulsion to gather specimens or take cameras into riskier places. In fact, *Argo* did find and record the tragic shipwreck in 1985. The veteran tow sled *ANGUS* was even part of the action, when close-up runs were too risky for *Argo*. These tow sleds and others like them would go on to discover and explore hundreds of shipwrecks around the world.

a tow sled, success in locating specific underwater targets, such as shipwrecks, depends on its being pulled in carefully overlapped passes, a laborious survey grid technique referred to as "mowing the lawn." (See *Chapter 11: Operations* for more on search grids.)

Sometimes tow sleds are considered to be a separate class of underwater vehicle, but not so in this text. Certainly, they come in all sizes and configurations. The most basic carries only a single sensor, such as a camera or side scan sonar. Note that when a side scan unit is pulled as a separate unit, it's commonly referred to as a **towfish**. Both sonar and camera devices can be incorporated into a modest tow sled. At the other end of the size and complexity spectrum is a massive, elongated framework boasting an array of light-sensing devices, specialized sonar, manipulator arms, supplementary thrusters, or even small ROVs deployable "on the fly."

Although tow sleds have been reliable workhorses for many decades, their operation can be compromised by bad surface weather and tether drag. With the continued development of AUVs, however, many of the search and survey jobs traditionally assigned to tow sleds (and ROVs) are being taken over by tetherless underwater vehicles that are not affected by surface weather and that require less expensive (i.e., usually smaller) support ships for launch and recovery.

4.2. Tools

The range of tasks that underwater robots can handle is huge—as long as they have the right tools. Cameras are perhaps the most universal of all underwater tools, providing actual images of underwater terrain, structures, and objects. However, when

HISTORIC HIGHLIGHT: DSRV—THE REAL MISSION

During the Cold War years, the real mission statement of many deep sea vehicles was the covert collection of information. But often, this military objective was disguised by another, more innocuous one. Indeed, if even half the U.S. deep water spy stories are true, they would certainly rival any of James Bond's exploits. The success of these stealth maneuvers almost always depended on highly specialized vehicles and payloads.

A case in point is the development of the Deep Submergence Rescue Vehicles, or DSRVs, which looked as benign as a city bus. In theory, the U.S. Navy's DSRV-1 (also known as *Mystic*) and DSRV-2 (*Avalon*) were built to rescue crew from sunken submarines, hence the vehicle's ability to mate to another submarine hatch. But in reality, they were designed to operate as deep as 1,981 meters (approx. 6,500 ft), greater than any sub's crush depth. In fact, their main job was to collect information about Soviet deep sea activities. To this end, they had to be extremely stealthy—so the DSRVs required no surface support ship and could, in fact, hitch a ride atop a submerged submarine for easy, unobserved transport anywhere in the watery world.

To accomplish their mission, these two DSRVs were equipped with powerful sensors for scanning the seabed and with the latest in navigation equipment, computers, and displays; they also boasted elaborate manipulators, winches, and claws that enabled those aboard to clear debris from sunken foreign subs or to tap into subsea cables to intercept Soviet communications. All of this came at a cost, of course. When *Mystic* and *Avalon* entered the Cold War fracas in 1971 and 1972, the total bill for the two vehicles had escalated from $3 million each to $110 million apiece! However, many felt the cost was justified because of the richness of war paraphernalia scattered on the seafloor. Indeed, one Russian sub that went down in 1986 yielded two nuclear reactors, two nuclear torpedoes, and 16 long-range missiles with nuclear armaments—a virtual treasure trove for the United States to pick apart and analyze, all compliments of the DSRVs and their sophisticated payloads.

Images courtesy of Jay Henson and Wikipedia Commons

Figure 10.40: DSRV-2 **Avalon**

The Avalon *being launched from a surface support ship for training operations.*

Figure 10.41: A DSRV Attached to a French Submarine

Although it's not a common practice, occasionally smaller underwater vehicles are designed to piggyback on aircraft or submarines. This allows the "passenger vehicle" to conserve fuel and speeds delivery to the mission site.

equipped with highly specialized, mission-specific tools, underwater vehicles can also lay and repair communication cables, handle routine inspections and repair for the oil and gas industry, monitor hazardous areas such as nuclear power plants, help to construct underwater structures, find and recover delicate archaeological artifacts, collect scientific specimens, etc. The list of tools in the following sections gives you some sense of their wide variety and assignments, but it is only a fraction of the specialized equipment available. Keep in mind that many of these tools can be mounted on interchangeable tools trays that can be easily and quickly swapped out, allowing the ROV to perform multiple functions with a minimum of time lost at the surface. (See Figure 10.3 at the beginning of this chapter.)

4.2.1. Cameras

Single or multiple cameras are extremely important tools and among the most basic on any underwater vehicle. They're used for navigation, locating items, performing inspections, and monitoring the work that other tools are doing. In short, everything that needs to be seen underwater is visualized with some type of visual sensor. In the 1940s, only basic still and movie cameras were available for underwater use. Today, there are also highly sophisticated video and digital cameras, as well as those that can function in low-light conditions.

Image courtesy of Harry Bohm

Figure 10.42: Small But Mighty Effective ROV

Chris Roper preps for an easy hand launch of a SeaBotix LBV-150. The micro-ROV comes equipped with a high resolution color camera as its primary payload but can be customized with additional tools such as a 3-jaw grabber, sonar, a tracking system, and external lighting.

Cameras play a dual role. They are often crucial for navigation (see *Chapter 9: Control and Navigation*), but they also are basic to most payloads. This payload role was certainly the case with the *MIR* submersibles that carried high-definition IMAX format cameras down to the *Titanic* in order to obtain the footage necessary for the film *Titanica*. These specialized cameras were payload-specific, with a mission that was totally different from that of standard submersible imaging gear used for navigation or general observation purposes.

Image courtesy of Ian MacDonald, Texas A&M, and NOAA

Figure 10.43: Lighting and Cameras on the ROV Global Explorer

Diagram of the lighting and cameras on the Global Explorer ROV, designed and developed by Deep Sea Systems Inc.

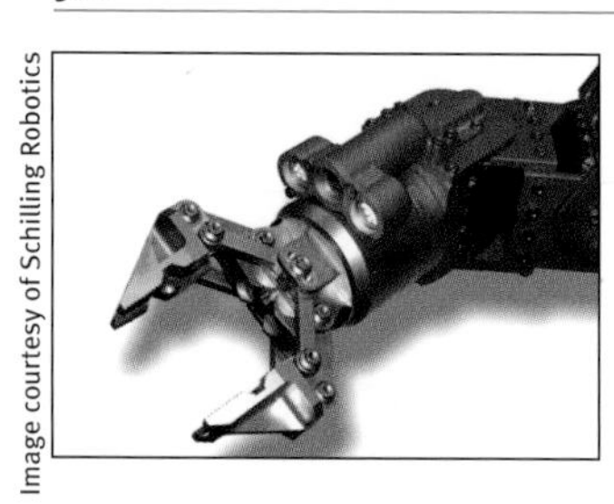

Image courtesy of Schilling Robotics

Figure 10.44: Integrated Camera and Lights

Schilling's Titan manipulator can include an integrated high-resolution wrist camera with two LED lights, an integrated titanium housing, and internal cabling.

Image courtesy of Dr. Daniel Jones, SERPENT Project, National Oceanography Centre, Southampton

Figure 10.45: Targeting the Work

Look closely and you will see that one of the manipulators on this Oceaneering Magnum 155 ROV can light and photograph its intended worksite at the Rosebank North Field, Faroe-Shetland Channel, UK.

On some missions, an ROV may carry as many as eight different cameras as part of its scientific payload, including a rear-facing camera to facilitate tether management. Of course, most cameras for deep-water work require specialized lighting—a further addition to the payload. When MBARI's *Ventana* is deployed in the waters of Monterey Bay, it's often carrying these different types of cameras:

- The main camera captures broadcast-quality images.
- A macro unit can zoom in on very small marine creatures.
- Another camera is located higher up for overviews.
- Another tiny light and camera, next to the specimen-collecting mechanisms, allows scientists to view their catch.

4.2.2. Scanning Lasers

The **scanning laser** may become the camera of the future. This invention sends out a single blue-green laser beam to scan objects in a very precise and fast motion. The energy of a laser beam hitting and being reflected from an object is measured in much the same way as in an active sonar system, but the resulting image is high in resolution. Compared to sonars, scanning lasers are limited in range; however, they can scan greater distances than camera systems. Better yet, they attain acceptable resolution and accurate distance information even in low-visibility water because of intensive digital image-enhancement capabilities in the processor. As scanning lasers improve, they may replace conventional cameras, but current versions are extremely expensive.

4.2.3. Cleaning, Inspection, and Testing Tools

Ships' hulls and underwater structures such as pilings, oil-well platform legs, pipelines, and concrete structures, must be inspected periodically for flaws, weakness, and corrosion. The first step in this inspection process is generally cleaning, since marine growth and corrosion often obscure the surface of these underwater structures. A variety of tools and techniques are used for cleaning. Some scrub away unwanted materials with rotating brushes; others use high pressure water jets, cavitation bubbles, suction, or even intense sound. These tools are tailored for the size, surface, and structural configuration of what they are cleaning. Once the surface is clear, cameras can conduct the next step, a visual assessment. Even then, some flaws may still not be visible so other specialized tools may be employed to test structural soundness.

Ships' hulls are routinely cleaned, inspected, and re-conditioned when vessels are hauled out of the water for routine maintenance in a shipyard. But the cleaning, inspecting, and testing procedures for underwater structures are typically conducted by ROVs equipped with the appropriate tools in their payloads.

4.2.4. Oil Rig Underwater Manifold Tools

Many of the critical manifolds (sets of valves) and well heads on offshore oil rigs are located on the seabed. They regulate the flow of oil to the pipelines carrying the crude to the refinery or to a riser for loading an oil tanker. Inspecting, maintaining, and repairing these underwater manifolds and well heads is all part of an extremely specialized field of ROV operations, requiring mission-specific tools and techniques developed by the oil industry. Other specialized equipment relating to work on the oil and gas industry's underwater rigs and well heads are commonly called **hot stab tools**, to designate a special design of hydraulic connector. These come in various configurations and generally relate to transferring fluid (typically hydraulic fluid) to another tool or piece of subsea equipment. For example, a hot stab can be connected to an ROV's hydraulic pressure unit (HPU) in order to transfer hydraulic fluid from the ROV and activate subsea valves.

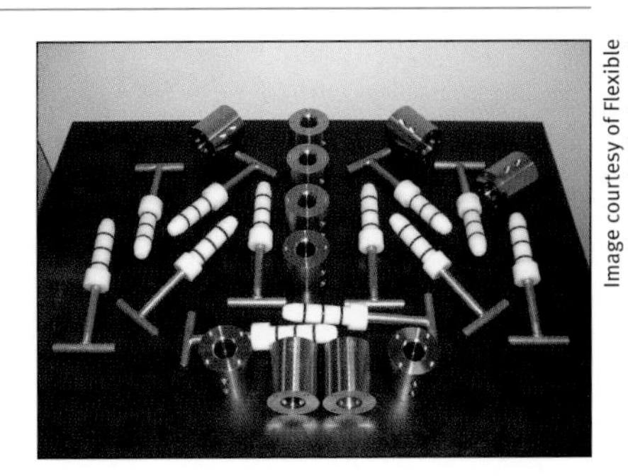

Image courtesy of Flexible Engineered Solutions Ltd.

Figure 10.46: Assortment of Hot Stab Tools

Hot stab tools are special hydraulic connectors used in servicing undersea oil rigs (Figure 10.47). This assortment is typical of the industry standard, making it easy for ROVs to service manifolds and perform maintenance on these deep water structures.

Image courtesy of GRI Simulations Inc.

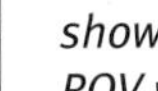

Figure 10.47: Hot Stab Training Exercise

This image is part of a training simulator for ROV pilots. It shows an Ultra Heavy Duty ROV with Rigmaster and T-4 manipulators operating a hot stab in a Chevron Petronius Project pipeline repair system simulation.

4.2.5. Cutters, Drillers, and Corers

Hydraulic cutters are one of the most useful types of tools for a work class ROV. Most of these cutters are used to sever large-diameter cable or cut some types of metal sheet or bar. Sometimes for light-duty jobs, cutter blades are inserted in the gripper. Cutters are basic to underwater construction and maintenance work, but they can also be useful in salvage operations where debris may have fouled the object that needs to be lifted.

Hydraulic core drill tools are highly specialized and mission specific. Some are designed to attach to oil rig structures in order to drill holes in thick steel. Other core drill tools are used to bore and recover geological samples from underwater rock or from the seabed. The largest, most heavy-duty drill tools are those used for establishing underwater oil/gas wells.

4.2.6. Gradiometers, Magnetometers, and Metal Detectors

Gradiometers, magnetometers and metal detectors are all magnetometry instruments that can detect ferrous metals under water; not surprisingly, they're used for some types of geological surveys and archaeological searches and for locating buried cables.

Figure 10.48: Magnetometer for Detecting Magnetic Fields of Ferrous Objects

The Geometrics G-882 Magnetometer in this photo was used during the hunt for the Alligator*, a Civil War-era vessel lost in 1863 during a fierce storm.*

This high-tech underwater search was conducted by the National Oceanic and Atmospheric Administration (NOAA), with support from the Office of Naval Research (ONR).

Magnetic gradiometer sensors are commonly mounted on AUVs and ROVs to detect and measure changes in the magnetic field as the vehicle traverses the seabottom on a systematic grid pattern. A magnetometer, or "mag," measures magnetic fields generated by submerged ferrous objects such as ship hulls, engines, cannon, deck gear, and cargoes. Typically, mags are mounted on a towfish and are used to cover large areas. Author Clive Cussler used a magnetometer, along with painstaking research and other search techniques, to locate the remains of more than 60 shipwrecks, including the Confederate submarine *Hunley*.

Metal detectors do the same job as magnetometers, but to do so, they must be positioned very close to the bottom. The detector consists of a copper-wound magnetic coil that senses ferrous metals on the seafloor or even buried beneath the mud. As the disc-shaped coil sensor is swept over the bottom, any metal nearby disturbs its magnetic field, creating a signal that is sent to electronic circuitry, which reads out the magnitude of the signal—the larger the metal mass, the higher the magnitude of the signal. Metal detectors and magnetometers are seldom mounted on ROVs, since most underwater craft have a lot of metal components themselves that would trigger false readings. They won't work very well in swimming pools, either, since any concrete pool is laced with metal re-bar.

Figure 10.49: ROPOS ROV Hooking Up to Cable-Laying System

This photo shows crew lowering the ROPOS (Remotely Operated Platform for Ocean Sciences) ROV onto a remotely operated cable laying system (ROCLS). This specialized tool attachment allows for the deployment of smaller diameter cables that cannot be laid by a cable ship.

4.2.7. Trenching Tools and Cable and Pipeline Detectors

Despite the proliferation of communications satellites, undersea communications cables remain an important transmission medium for the telecommunications industry. Today, ROVs and specialized trenching robots are used to lay and maintain a variety of fiber-optic, copper-wire and other communication cables that crisscross the ocean floor. They also lay pipelines and power cables. Often these pipes or cables need to be buried in the seafloor to protect them from damage that can be inflicted by anchors or by fishing boats dragging massive trawl nets. One of the ways to bury a line is to use a specialized marine vehicle or a trenching/jetting tool mounted onto an ROV or submersible to actually cut a trench in the soft mud bottom by means of a high-pressure water jet. Some vehicles are designed to do this trenching operation as new cable is spooled out. These machines often feature a "depressor," which basically makes sure the cable is pushed down into the trench created by the water jet before the muddy slurry settles back into the trench. This improves the chances of getting the cable seated in the bottom of the trench.

However, there are many occasions when previously laid sections of pipe or cable must be uncovered for repair or improvements. Depending on the work required, ROVs carry jetting tools to uncover buried cables, lift them for repairs, and then rebury them. Cable vehicles and tools must be able to deal with all types of seabed—steep grades, solid rock or coral, holes, and even extremely soft soil.

Of course, before any type of repair or maintenance work can happen, the pipe or cable must first be located—another task for an underwater platform outfitted with the right tool, such as a gradiometer (see Section 4.2.6) or sub-bottom sonar profiler (Section 4.2.11).

4.2.8. Marine Science Sensors

ROVs, AUVs, and submersibles often carry marine science sensors as payloads. They can provide data on temperature, salinity, turbidity, conductivity, pressure, chlorophyll, pH, dissolved oxygen, current speed, etc. One of the basic electronic sensing instruments in oceanography is the **CTD profiler** which combines three sensors in one instrument—a temperature thermistor, a conductance cell, and a pressure sensor. This data is digitally stored, then downloaded to a computer, where it is processed into graphs and data tables. From this information, the CTD profiler can calculate water density and sound speed precisely at any depth.

Often scientific tools get created as the need for the information arises. For example, the MBARI ROV *Ventana*'s need for an underwater odometer led to the development of a low-speed flow meter that calculates linear distance traveled.

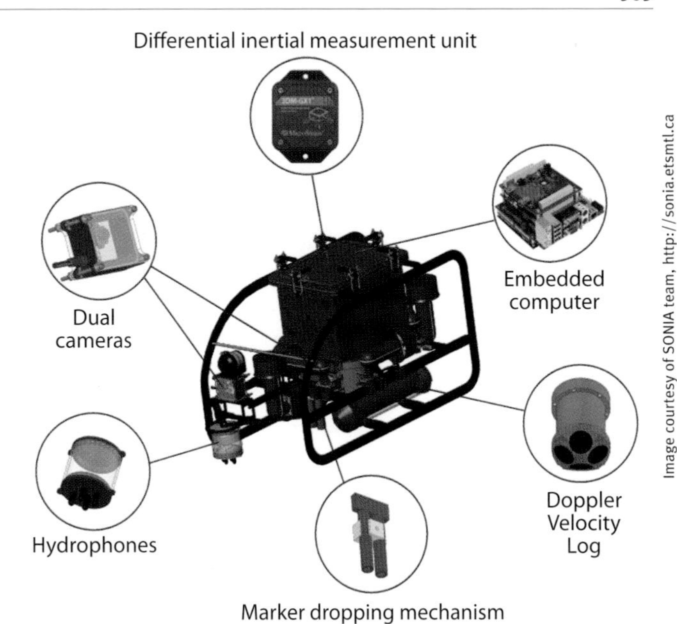

Image courtesy of SONIA team, http://sonia.etsmtl.ca

Figure 10.50: SONIA Sensors

SONIA 2008 *is an autonomous underwater vehicle built by undergraduate students from the École de Technologie Supérieure (ETS).*

The AUV's basic functions are autonomous inspection, exploration, recovery, marking, and mapping. Its broad collection of high-level tools provides facilities for easy reconfiguration of mission tasks and on-line and offline telemetry data access.

The AUV is composed of a main hull, which houses the onboard computer and the electronic mainframe, six thrusters, and a set of sensors.

TECH NOTE: HOT STUFF!

A **thermistor** is a passive sensor that measures temperature. Ocean water temperature is an important variable for oceanographers, because it affects fish stocks, plankton growth, sound speed, and weather patterns. Knowing the temperature profile of a column of water gives information on the speed of sound, thermoclines (abrupt change in temperature at a specific depth), and water density.

Image courtesy of Woods Hole Oceanographic Institution

Occasionally, temperature sampling can be more challenging than expected. The first time scientists aboard *Alvin* attempted to push a temperature probe into the inky-black smoke spewing out of strange seafloor structures, they decided the readout had to be wrong. It wasn't until they got back to the surface and could actually examine the thermometer that they realized the unaccountably high readings were probably correct. The tip of the thermometer, which was made of polyvinyl chloride material rated to 180° C (356° F), had melted! In fact, later dives with a recalibrated thermometer recorded temperatures hot enough to melt lead.

Figure 10.51: A High-Temperature Thermistor

A high-temperature probe mounted on the submersible Alvin *is inserted into a hydrothermal vent to measure the water temperature.*

TECH NOTE: THE SAME DOPPLER ANIMAL DOES DIFFERENT TRICKS

The **Doppler shift** is a change in the apparent frequency (pitch) of a sound (or other wave), caused by relative movement between the sound source and the listener. You've heard an example of this Doppler shift on land, when a car goes speeding past you and the whine of its engine is slightly higher in pitch as the vehicle approaches and slightly lower once it speeds past. ("Zeeeeeeeeoooooоommm.") This occurs because the sound waves are bunched up ahead of the speeding car and spread out behind it, due to the motion of the vehicle. The same principle can be used under water to measure the relative motion of sound sources and objects that reflect the sound.

An **Acoustic Doppler Current Profiler (ADCP)** measures currents—or more accurately, water velocities (speed and direction)—that run past a stationary ADCP placed in the water column. In this case, the sound bounces off tiny bubbles or other particles suspended in the water. If the ADCP is stationary, any water movements—even wave heights—can be measured with this device. The ADCP can also be used on boats or underwater vehicles to measure current in relation to vehicle movement.

The **Doppler Velocity Log (DVL)** is used in **Acoustic Doppler Navigation** to guide ships and underwater vehicles. It measures the Doppler shift of (typically four) acoustic signals transmitted diagonally toward the bottom. By interpreting measured frequency shifts in the echoes coming from the bottom, a very accurate measure of vehicle speed and direction over the bottom can be calculated moment by moment. This, in turn, can be used to update the vehicle's position in relation to its starting position. DVL navigation information is used in conjunction with other navigation sensors, such as pitch and roll sensors, compasses, depth sensors, altimeters, and vehicle propeller rpms (speed).

AUVs such as Hafmynd Corporation's *GAVIA* (see Figure 10.52) have both an ADCP and DVL working together. The ADCP is located on the top half of the sensor module, with the DVL in the lower half, pointed toward the bottom. By using both ADCP and DVL, the vehicle can navigate and simultaneously determine water column currents even while moving. DVLs and ADCPs are tremendously useful acoustic devices. On ROVs such as the submarine rescue vehicle built by OceanWorks, the DVL aids in the careful navigation necessary in order to make a successful mating of the transfer skirt with a submarine's escape hatch.

Image courtesy of Harry Bohm

Figure 10.52:* GAVIA *AUV Outfitted with Multiple Sensors

This GAVIA *AUV carries a SeaBird CTD. It looks like a ray gun and is mounted in a black syntactic foam block to offset its weight.*

In addition, GAVIA *carries a turbidity sensor, ADCP, DVL, and side scan sonar as standard payload tools. The AUV has also been outfitted with Geoswath sonar for under-ice Arctic Ocean work.*

TECH NOTE: *TOW-YO*

Tow-Yo is an especially interesting example of a towed vehicle equipped with an array of water sampling sensors. Invented by NOAA scientist Ed Baker, this device is like a towed laboratory that gathers data on water temperatures, chemicals, and particles as it moves through the water. Except *Tow-Yo* doesn't move along at any even depth; instead, it's like a yo-yo, doing 300-meter (approx. 990-ft) gyrations as it is pulled through the water. This unique sawtooth pattern allows it to sample water and collect data at a variety of depths and locations. With additional scientific equipment, *Tow-Yo* has been able to sample minute amounts of seawater at a range of depths and immediately analyze them for metal content.

4.2.9. Side Scan Sonar

Side scan sonar is like shining an acoustic flashlight sideways and interpreting the highlights and shadows to infer size and shape of objects or terrain on the bottom (Figure 10.53). Such a "sound picture" is the result of computer processing of returned sonar echoes from transducers mounted on a ship's hull or on either side of a torpedo-shaped towfish that is pulled behind and below a vessel, near the seafloor.

Specialized side scan sonars are used for bottom surveys that map the contours of the ocean floor. Other types facilitate finding lost objects or surveying routes for communication cables or pipelines. Increasingly since the early 1990s, side scan sonar devices have been mounted on AUVs, which provide a very stable platform for undersea surveys. Furthermore, conventional survey methods that involve ships towing sonar sensors in deep ocean work are expensive procedures, so sonar-mounted AUVs have become commercially competitive.

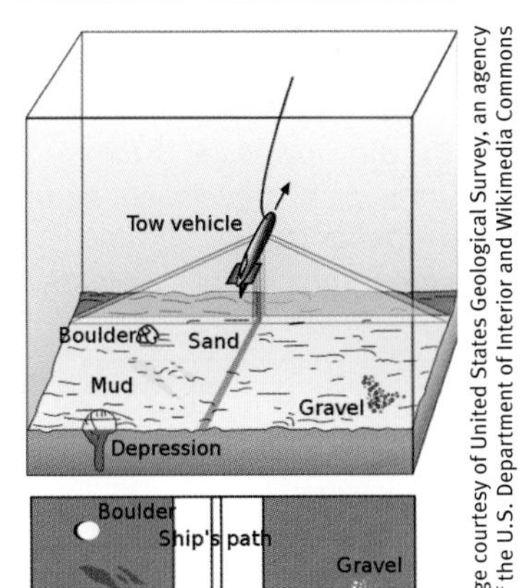

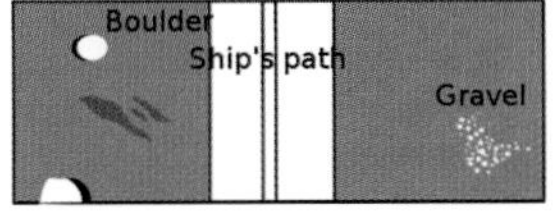

Image courtesy of United States Geological Survey, an agency of the U.S. Department of Interior and Wikimedia Commons

Figure 10.53: Diagram of how Side Scan Sonar Works

Interpreting sonar images requires skill and attention to detail. In the gray side scan image above, the black shadow on the far left-hand side of the boulder shows that this object was sticking up from the sea floor, whereas the shadow on the right-hand side of the other roundish "object" shows that it was actually a depression.

Image courtesy of Edgetech Marine

Figure 10.54: EdgeTech 4200 PM Side Scan Sonar Towfish

The side scan sonar towfish is towed under water behind a vessel. It contains the acoustic transmitters and receivers needed to generate side scan images of the seafloor.

Image courtesy of Edgetech Marine

Figure 10.55: Side Scan Shipwreck Image

As demonstrated by this image of a sunken ship, Side Scan information can be quite detailed. It was generated from side scan echo data collected by the EdgeTech towfish shown in Figure 10.54.

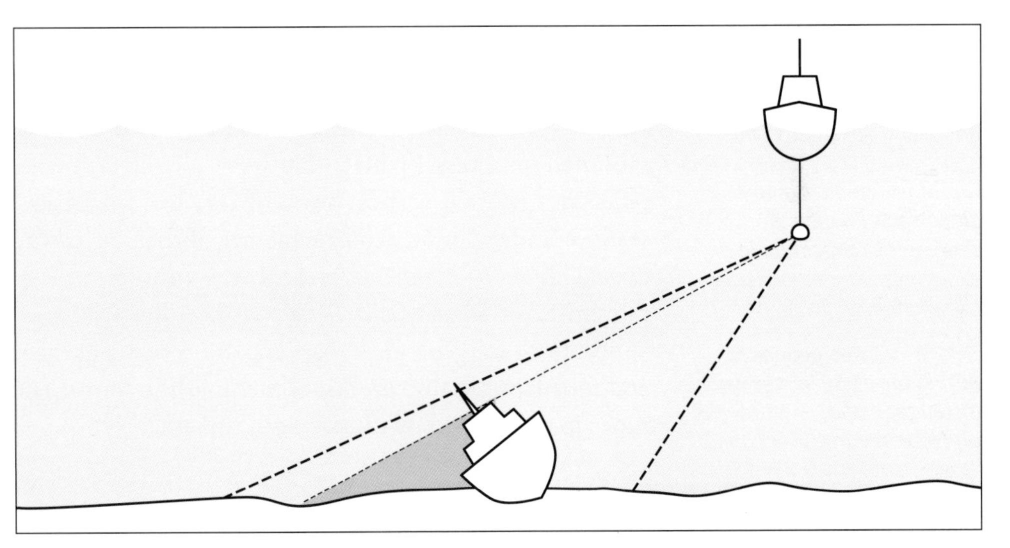

Figure 10.56: Acoustic Shadow of a Shipwreck

A target protruding from the bottom will create an acoustic shadow. This shadow is what side scan and sector scan operators look for when searching for lost objects under the water. Generally, the longer the shadow, the higher off the bottom an object is.

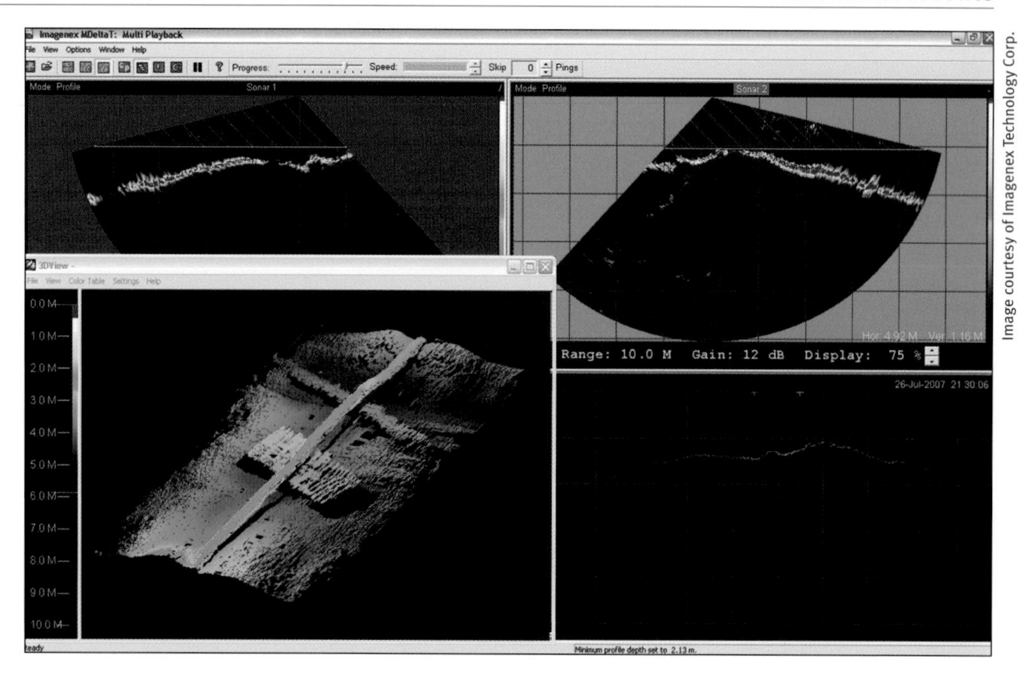

Image courtesy of Imagenex Technology Corp.

Figure 10.57: Multibeam Sonar Surveys

The pipeline image in the lower left was generated by an Imagenex multibeam sonar unit. It shows depth contours for a section of pipe and the flexible concrete "mattress" (underneath) that is used for protection, support, and/or stabilization.

4.2.10. Multibeam Sonar

Multibeam sonar generally emits sound waves from underneath a ship's hull, though it is increasingly being used on AUVs. Its fan-shaped coverage results in a swath of depth soundings over a large area that can be used to map bathymetry (depth contours) in detail. High resolution multibeam images can provide much of the same information as side scan sonar while giving accurate depth information as well. (See multibeam sonar images in Figure 10.57 as well as Figure 3.18 in Chapter 3 which shows the depth contours of giant underwater sand waves.)

4.2.11. Sub-Bottom Profilers

Another sonar variation is the sub-bottom profiler (SBP). This acoustic tool is usually mounted on a vessel with sound waves directed down into the seafloor sediments. It relies on a narrow, vertical beam of very low frequency sound waves to penetrate the seafloor and create an acoustic picture of rock or sediment layers and to reveal objects buried there. If a pipe is present, the profiler will show that information on the monitor. Not surprisingly then, sub-bottom profilers have long been used in ROV applications for geological mapping, for looking at the depth of pipe and cable burial, and for mapping shipwrecks. Sometimes the sensor is used independently and other times in conjunction with side scan imaging.

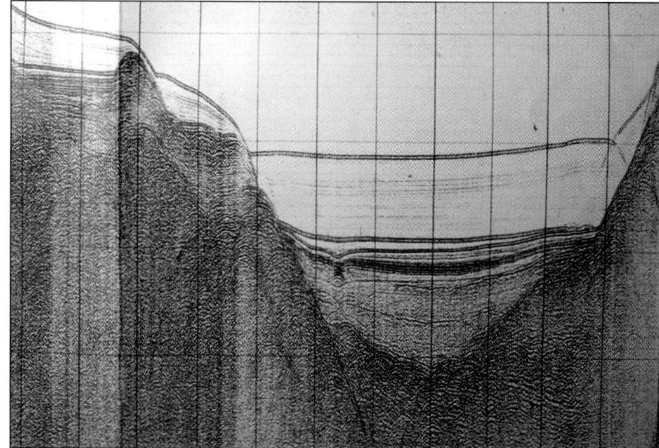

Images courtesy of Mark Atherton

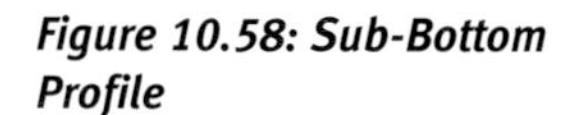

Figure 10.58: Sub-Bottom Profile

The photo on the left shows an oceanographic printer and a SBP record being generated for a shallow water engineering study. The record shows the seabed and sediment layers overlaying bedrock.

The sub-bottom profile on the right shows that when conditions allow, the same low power SBP is capable of penetrating 100 meters or more. Each of the dominant horizontal lines represents approximately 75 meters. This record is a good example showing delination of sediment stratification in excess of 100 meters below the seabed.

This record was generated during the 1989 survey for the installation of the natural gas pipeline to Vancouver Island.

4.2.12. Tools for Underwater Mine Detection and Torpedo Recovery

Finding military targets under water requires many of the tools already listed: high-resolution and low-light cameras, sector scan, side scan, and multibeam sonar, and sub-bottom profilers. In addition, the robot's payload may include special manipulators for cutting cables and neutralizing mines. Other mission-specific tools actually destroy the mines. For example, the GEC-Marconi's wire-guided ROV *Archerfish* carries a small warhead that can be detonated when the expendable ROV is close to the mine, thus eliminating the threat completely.

Some research has focused on small Autonomous Legged Underwater Vehicles (ALUVs) that are built to locate and neutralize (or blow up) mines in the hazardous surf zone. Tetherless vehicles such as *Ariel* are essentially crab-like walking robots that seek out and locate mines, using various sensors and tools, such as touch, magnetic gradiometers, metal detectors, ultrasonics, etc. Built to be dispensable, these highly sophisticated robots also have the ability to communicate with others of their kind and stand by their discovery until told to destroy the mine with their onboard explosive charges.

High-tech torpedoes are very expensive devices, so recovering and reusing them after they have been fired in the test-firing ranges makes good financial sense. To do so, the military has developed specialized underwater robots to handle torpedo recovery. These craft carry sensor payloads similar to those that detect mines; however, since their mission is to bring spent torpedoes back to the surface, these vehicles also have powerful robotic claws to grab onto the torpedoes.

5. Considerations When Designing Payloads

Designing a payload tool is similar to designing an underwater vehicle—the emphasis must be primarily on the mission tasks and how to accomplish them. Keep in mind any mission-specific constraints. For example, the pressure squeeze at deeper depths can impact joint action in a manipulator as well as seals for waterproofing. Of course, if you have really deep pockets you could buy a ready-made manipulator for deep water.

When designing a payload, remember that each decision often affects some other part of the vehicle, so it's important to keep the whole system in mind as you work on this important subsystem. (See *Chapter 2: Design Toolkit.*) Some payload tools are extremely small so may not have a significant impact on other subsystems. For example, *SeaMATE* uses only a camera and a straight stick, or probe, attached to its frame. Whereas a larger, more complex gripper will affect vehicle buoyancy, power, control systems, etc.

The following chapter sections will examine some of the fundamental issues you'll need to consider when designing a payload system and evaluating how it will impact other vehicle systems.

5.1. Mass, Buoyancy, and Stability

Anything put on an underwater vehicle will add mass. As you know by now, that can affect a number of things, including vehicle attitude (yaw, pitch, roll), buoyancy, and stability. For example, manipulators are typically positioned in the forward part of the structure. Since these tools are usually negatively buoyant, they add an enormous amount of "weight" in the front end, putting the vehicle off balance. This lack of trim must be compensated by adding more flotation directly above the manipulator.

Images courtesy of Harry Bohm

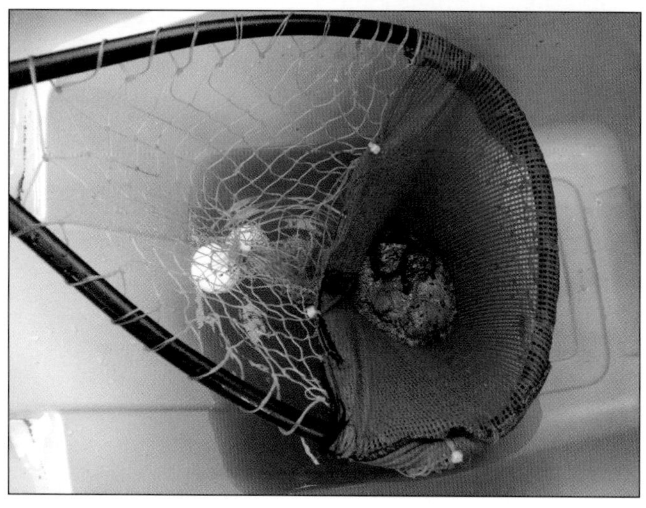

Figure 10.59: Keep it Simple

Keep it simple—that's good advice for a payload tool, particularly an improvised one. In this photo a simple dip net attached to the LBV *ROV enables retrieval of geological samples from a lake bottom.*

Another consideration: when a manipulator is extended outward, it levers the front of the vehicle down. As a result, it may not be possible to use the manipulator in mid-water. In order to conduct work, such a vehicle might have to be sitting hard on the bottom or anchored some way in order to operate this type of payload tool. The mass and buoyancy of a manipulator also means the robotic arm should be positioned lower down on the frame, to increase stability. If placed higher up, then it will make the vehicle even less stable. (See *Chapter 4: Buoyancy, Stability, and Ballast.*)

In very sophisticated and powerful ROVs, thruster action can maintain the dynamic stability of an ROV while grasping and lifting objects by providing compensatory torques. Thus, when a robotic arm is extended with a load, a forward thruster pushing upward may compensate for a droop in attitude. Unfortunately, this level of control is not a realistic option without a fairly advanced control system, so you'll need to be aware of this potential problem when adding or subtracting payloads.

5.2. Lifting

Some payload tools are designed to lift objects. The problem, of course, is that adding any significant load to the vehicle can dramatically disturb its attitude and stability. Making a heavy lift to the surface often requires additional flotation (such as soft ballast tanks) or thrust (from multiple vertical thrusters), to offset the imbalance caused by a heavy object.

5.3. Reaction Forces

Any time an underwater vehicle pushes or pulls on something, blasts a jet of water to dig a trench, or tries to scrub or rotate anything, then Newton's 3rd law of motion comes into play. It says that any force applied by the vehicle will be met by an equal and opposite force applied to the vehicle. For example, because an ROV is "light weight" in the water, any strong jet of water discharged from the end of a hose to dig a trench will also act like a thruster, pushing the ROV in the opposite direction. Therefore some sort of compensating mechanism or strategy is needed in order to counteract this force. Here are some basic strategies to do this:

- Anchor the ROV to a structure or have it sit hard on the bottom.
- Counter the reaction with the vehicle's thrusters (locating the vehicle's vertical thruster(s) above the line of action of the jetting arms).
- Make the vehicle as heavy as possible (but still light enough that the thrusters can maneuver the system).
- Design the jets so that they have a much lower net impact on the vertical force on the vehicle (e.g., having some portion of the water jetting upward at the same time as downward).
- Use a variable buoyancy system to achieve the same goals as in the second bulleted item above (i.e., heavy during trenching ops, light during flight).

5.4. System Interference

Some payloads, especially scientific measuring devices, are very sensitive and can be adversely affected by interactions with other components or devices on the vehicle. For example, sonar or camera imaging could be blocked by framework struts, so these payload tools need to be located in a position clear of all obstructions. The camera's view can also be obscured by thruster wash that stirs up silt. A temperature sensor needs to be located free of any turbulence from the thrusters in order to get accurate measurements. Magnetic sensors need to be far from large chunks of steel and current-carrying wires. So always be aware of interference between the payload and the rest of the vehicle systems; this will help you position them on the underwater vehicle for greatest effectiveness.

6. Examples of Payload Options for Simple ROVs

Chapter 12: SeaMATE provides basic instructions for adding a very simple pick-up probe to a shallow diving ROV. However, more sophisticated missions often require more complex payloads, as evidenced in the three payload examples that follow.

6.1. Gripper Variations

In the summer of 2007, teams from around the world met at the sixth annual MATE International ROV Competition, held at the Marine Institute of Memorial University in St. John's, Newfoundland. In conjunction with the International Polar Year (IPY), the 2007 competition theme focused on Earth's polar regions and their impact on global climate, the environment, natural ecosystems, and human societies. Student teams were challenged to design and build vehicles to deal with the same work situations and environmental conditions faced by scientists and engineers working in polar environments—ocean observing, scientific research, and offshore industry operations in frigid waters.

The ice tank at the Institute of Ocean Technology (IOT) provided a very real sense of extreme conditions, with teams having to navigate their ROVs under the ice sheet to collect simulated "algae" samples (ping pong balls) and deploy an acoustic sensor. In the IOT's tow tank, student ROVs heaved and rolled in waves while preparing a simulated wellhead for oil production. The flume tank challenged students to pilot their tethered craft in a current and recover a severed anchor off an ocean-observing buoy. Not surprisingly, student teams came up with any number of creative payload variations in order to tackle these polar missions.

For instance, the three teams cited below devised three very different ideas for a gripper. These examples highlight the importance of "what if" thinking and creativity. You can stretch your own ideas about payload variations by talking with other designers, reading technical journals, participating in local, regional, or national competitions, and browsing student reports.

6.1.1. Carbonear Collegiate (Carbonear, Newfoundland)

At the 2007 international competition, Carbonear Collegiate won the "Sharkpedo Award" for innovation, originality, and "thinking outside the box" for its gripper design. The team's *Red October* ROV used its front-mounted gripper to perform a variety of mission tasks, but it was the construction of this gripper that made it unique and award-winning.

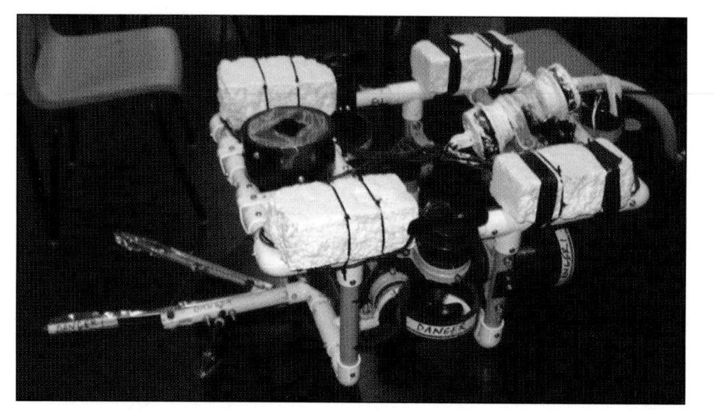

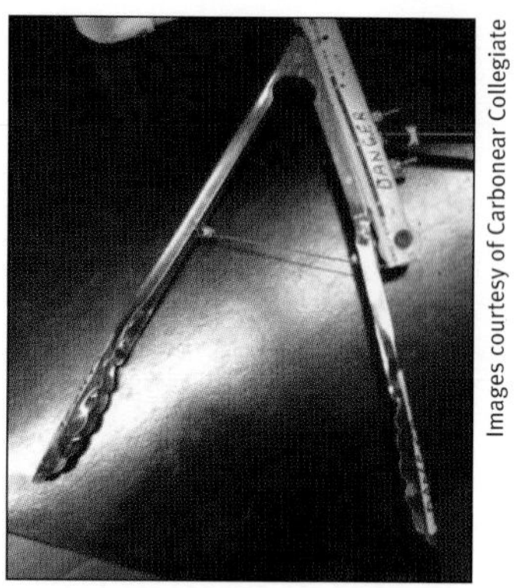

Images courtesy of Carbonear Collegiate

***Figure 10.60: (left) Carbonear Collegiate's ROV* Red October**

Figure 10.61: (right) A Most Basic Gripper for Carbonear Collegiate's ROV

The gripper consisted of a simple set of stainless steel salad tongs (costing a whopping $1), actuated by a bilge pump motor that was mounted to the bottom of the *Red October* ROV. When activated by a switch, the motor retracted a cable attached to one side of the tongs, thereby closing the gripper. Closure speed could be adjusted by a variable resistor. For ease of movement, the cable ran over a pulley, just visible in Figure 10.61. Both cable and pulley were salvaged from an old photocopier found in the school. In every aspect, Carbonear Collegiate's gripper was a clear case of innovative thinking and adaptation—and at low cost!

6.1.2. Flower Mound High School (Flower Mound, Texas)

Students from Flower Mound High School designed a claw for their *Flo-Mo* ROV that could easily handle the range of ice-tank tasks, from collecting samples and deploying a sensor device to grabbing a benthic jellyfish. The tow tank mission involved positioning a hot stab tool into a simulated well head, a task that required the claw to hold the tool at an angle of 45° from the horizontal claw arm. To accomplish this, the team had to modify the claw blades (not shown). They also made a special arm/hook assembly for the wellhead cap removal and gasket insertion.

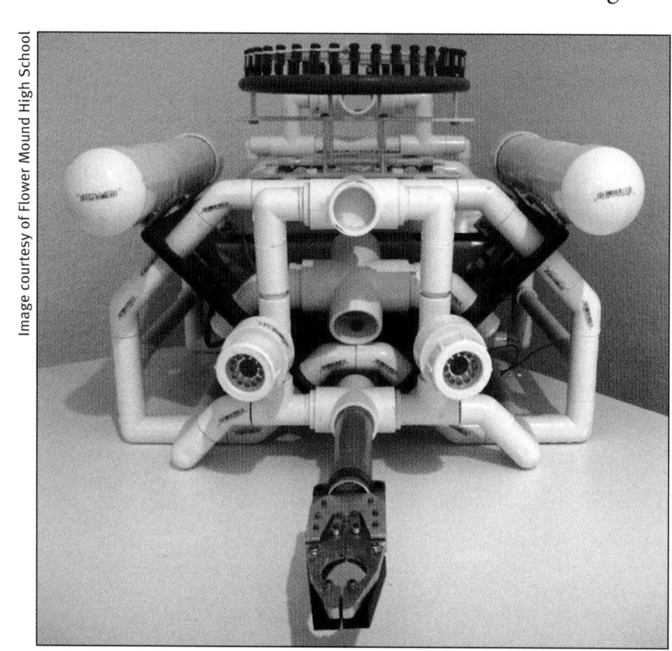

Image courtesy of Flower Mound High School

***Figure 10.62: Flower Mound High School's ROV* Flo-Mo 1**

The Flower Mound team designed the claw parts using CAD software, then fabricated them with shop tools. The original claw design utilized an electric motor with planetary gears and a linear drive screw, mounted internally in an aluminum tube. They chose this form of drive mechanism because the clamping force from the motor torque was significant.

In the process of designing and fabricating their claw, the team encountered three particular challenges:

1. waterproofing the geared electric motor
2. setting travel limits for the drive screw
3. finding available parts for bushings and seals to fit within the aluminum tube

They resolved the first and third issues successfully, but did not have time to work out the problem of travel limit control before participating in the Texas regional competition. Nevertheless, they won!

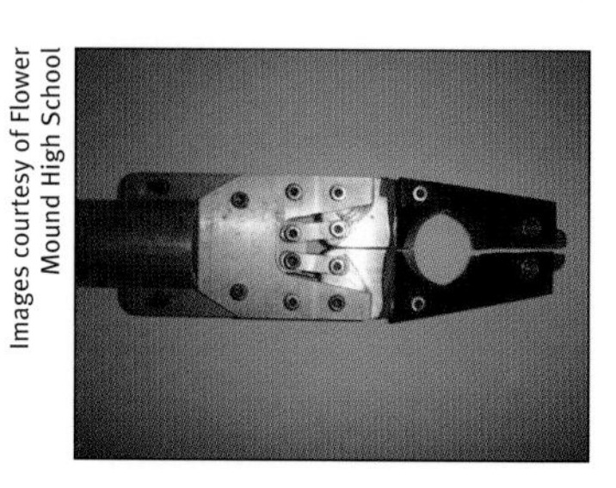

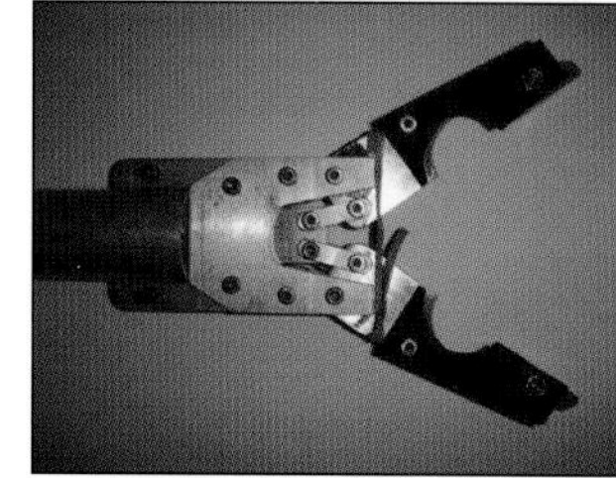

Images courtesy of Flower Mound High School

Figure 10.63: Claw Mechanism Closed and Open

As they prepared to move on to the international event, the team decided to go back to an earlier idea—using an air-actuated, piston-driven claw. They chose a single-action piston with spring return, which enabled the claw to open when pressure was applied and to close by means of the spring return. The only problem with this power system was a relatively weak clamping force because of the spring return. Fortunately, clamping force was not a critical component of any of the 2007 mission tasks. The gripper worked, and the team took third place at the international competition. With an eye to future challenges, students plan to improve the clamping force, using a double-action piston and solenoid valve switch.

6.1.3. Long Beach City College (Long Beach, California)

At the start of the design process, the Long Beach City College (LBCC) student team all agreed that a gripper was definitely essential for their ROV. However, rather than engineer an all-purpose manipulator, the team decided to design and build one to handle the very specific 2007 competition missions. Their gripper was loosely based on the "Jaws of Life" hydraulic tool used by rescue crews to cut open mangled vehicles and extricate anyone trapped inside.

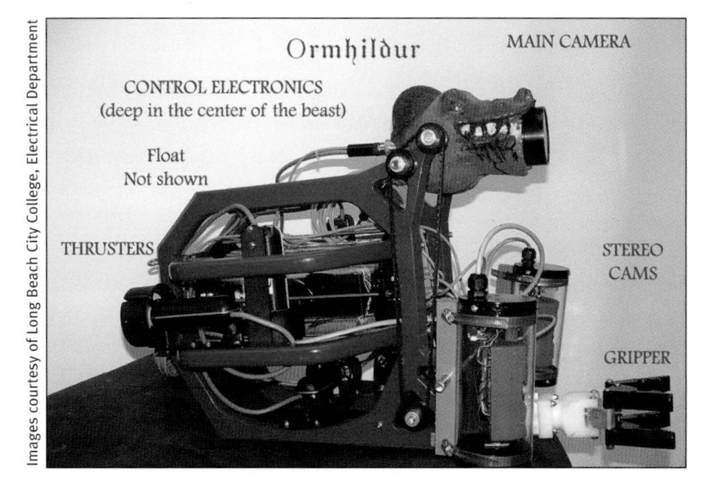

Images courtesy of Long Beach City College, Electrical Department

***Figure 10.64: Long Beach City College's ROV* Ormhildur**

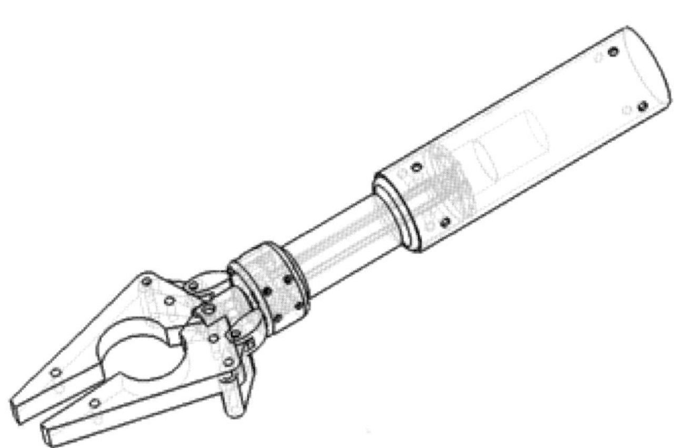

Figure 10.65: SolidWorks Rendering of Initial Electrically Operated Gripper

Image courtesy of Long Beach City College, Electrical Department

Figure 10.66: Close-Up of Gripper Jaws

First, the team refined the idea using SolidWorks software. Then they fabricated the tool from the same material as their ROV's frame. Its jaws were cut so as to provide a gripping surface in front of a circular opening. This opening provided clearance behind the jagged gripping surface for longer objects. The opening and closing motions were controlled by a simple actuator driving a ram back and forth. Originally, this linear movement was controlled electrically, but later, the team opted for using hydraulics, activated by a lever on the surface.

In the end, LBCC's gripper arm featured three settings: 90° straight off the front, 45° degrees down from the front, and straight down. Most importantly, it allowed their *Ormhildur* ROV to complete a multitude of tasks in a timely manner without having to change payload tools for each task. While the team didn't place in the top three at the international event, the students' positive attitude and encouragement of other competitors earned them the team spirit award.

Three different ideas, three different designs, three different grippers. And all of these payload variations completed their mission tasks successfully!

7. Chapter Summary

ROVs and AUVs rely on a variety of payloads to accomplish their missions. Many of these vehicles, particularly the larger ones, utilize payload tools that are powered by hydraulics. Therefore, this chapter opens with a discussion about hydraulic systems and how they work. Hydraulic systems can be used to transfer mechanical power, apply forces, or control position of various devices. Whether working on land or under water, the basic parts of any hydraulic system are a prime mover (electric motor or internal combustion engine), hydraulic pump, lines that carry the pressurized fluid, and one or more actuators, which are the working end of the system. While most commercial ROVs feature complex and expensive hydraulic systems, some simple adaptations can be crafted for home-built vehicles.

The second half of this chapter focuses on payloads—those tools or other devices added to an underwater craft that enable it to complete its specific mission tasks. They are called payloads because the work they do (after the vehicle carries them to the work site) is what really "pays" for the mission by producing revenue or some other tangible achievement. Payload tools must suit the mission and the type of underwater vehicle carrying them. For example, the majority of AUVs carry remote sensing devices, primarily for scientific and survey missions. Work class ROVs are typically outfitted with a manipulator (usually two on heavy duty models) and other tools specific to the task at hand.

Manipulators are extremely versatile tools, ranging from a simple one-function (i.e., one DOF) arm to the almost human-like movements of a sophisticated seven-function manipulator equipped with force feedback. This versatility is further enhanced by the variety of specialized end effectors for manipulators.

Many payload tools are interchangeable, thereby adding functionality and economy. The tools themselves are extremely diverse and highly specialized, ranging from versatile manipulators and various types of cameras, to sonar systems, scientific instrumentation, hydraulic cutters and welders, and even mine detection devices. Some payload assortments are grouped onto modular tool trays that can be easily swapped out.

This chapter concludes with several real-life examples of how school teams have created and implemented grippers for their underwater robots. Their innovative designs should provide inspiration and intriguing options for your own payload ideas.

Chapter 11

Operations

Chapter 11: **Operations**

Stories From Real Life: *Alvin* Ops—Submersible, Team, and Ships

Chapter Outline

1. **Introduction**
2. **Operations Overview**
    2.1. Search and Recovery Ops
    2.2. Other Types of Ops
3. **Operational Phases**
    3.1 Mission Considerations and Risk Assessment
    3.2 Ops Planning
    3.3 Staging and Mobilization
    3.4 On-Site Activities
    3.5. Demobilization
4. **Chapter Summary**

Chapter Learning Outcomes

- Describe the basic concepts and considerations in managing an underwater operation.
- Detail the steps involved in organizing and executing a shallow water underwater search and recovery operation using a small ROV.
- Explain how to conduct a search pattern and describe the types of equipment that might be involved in such an operation.

Figure 11.1.cover: Ship Hull Inspection with a Micro-ROV

In addition to an appropriate vehicle equipped with the right sensors, mission success required effective planning, personnel, and execution.

This chapter cover photo shows a VideoRay Pro 3 GTO, equipped with a BlueView P-900 High Definition imaging sonar, deployed on a hull inspection operation in Jamaica. The mission for possible contraband attached to the ship's hull turned up two canisters containing several hundred pounds of illegal drugs.

Image courtesy of Steve Van Meter, VideoRay LLC

STORIES FROM REAL LIFE: *Alvin* Ops—Submersible, Team, and Ships

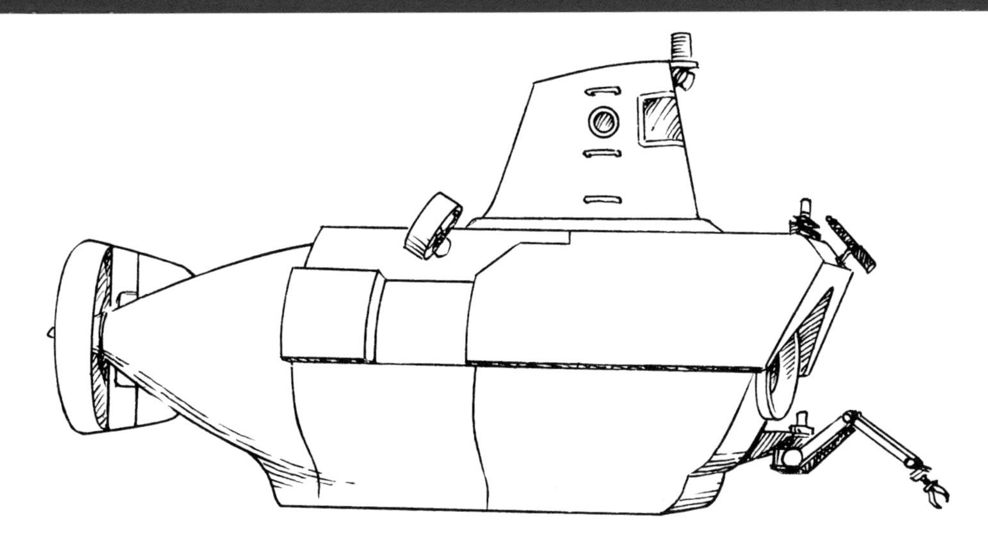

The tough little submersible Alvin *has been front-page news for more than four decades. Headline expeditions include the first-ever manned dive to hydrothermal vents, the discovery of cold-water hydrocarbon seeps, the first submersible to visit the remains of the* Titanic, *and a close-up study of the Mid-Atlantic Ridge through participation in Project FAMOUS (for French-American Mid-Ocean Undersea Study).*

The success of more than 4,000 dives for this subsea workhorse is testimony to meticulous ops planning and amazing personnel—captains, crew, technicians, scientists, and pilots. Another factor has been Alvin's *often unheralded support ships over the years: the unconventional* Lulu, *the* Atlantis II, *and now* Atlantis.

Alvin—the Idea

Before the mid-1950s, oceanographers seldom ventured beneath the surface. Scuba was still fairly new, so most scientists interested in the ocean depths explored them from the deck of a ship, by towing nets and examining the catch. Or they dangled cameras into the inky waters and sent down instruments to obtain samples and temperature data. Then in 1956, a conference of deep sea scientists discussed a radically better idea—conducting research from manned submersibles. Participants included Bob Dietz and Jacques Piccard, scientists involved with the *Trieste* project, and Allyn Vine, a young geophysicist working at Woods Hole Oceanographic Institution (WHOI), who had spent time on submarines during World War II. Vine proposed a research craft that had significant viewports and could be driven like a submarine but could dive much deeper.

Biologist Andreas Rechnitzer and *Trieste* pilots Don Walsh and Larry Shumaker agreed with Vine's idea. Together they came up with the design of a much smaller craft that had bigger windows, easy maneuverability, and the capacity to dive to 900–1,500 meters (approx. 2,950–4,925 ft). Unlike *Trieste*, it would not use gasoline for buoyancy and would not cost $1,000 per dive. Harold "Bud" Froehlich, a General Mills engineer who had already made a mechanical arm for *Trieste*, drew up a sketch for a deep diving vehicle 2.4 meters wide by 5.5 meters long (approx. 8 ft x 18 ft). There would be room inside a steel passenger sphere to accommodate two occupants, who could look out portholes in front and control a mechanical arm. The craft would hover, spin, and dive to 1,800 meters (approx. 5,900 ft), giving oceanographers access to nearly half the water volume of the ocean. Froehlich called his design *Seapup*.

Unfortunately, the U.S. Navy's Deep Submergence Program declared it was a waste of time to explore any deeper than 750 meters (approx. 2,500 ft). Rechnitzer quit in disgust. Fortunately, Swede Momsen, the Chief of Undersea Warfare at the Office of Naval Research (ONR), liked Froehlich's design. So did the Deep Submergence Group. This small collection of scientists, engineers, and technicians from WHOI, included Bill Rainnie, hired as "alternate" pilot, and Earl Hays, full-time manager of the project and chairman of the new Applied Oceanography Department. The Office of Naval Research and Woods Hole were already well into negotiations with the head of Reynolds Metals Company to build and rent an aluminum submarine called *Aluminaut*, but the talks regarding this very different kind of deep diving vessel were not going well. Maybe *Seapup* could serve as an interim explorer.

Momsen finagled some funding from the Navy, and in 1962, Bud Froehlich and General Mills got the nod to build the radically different underwater vehicle for $498,500. Even before the contract signing, the Deep Submergence Group began calling the sub *Alvin*, after Al Vine. The submersible, a combination of spacecraft and submarine, would work under water during the day and recharge its golf-cart batteries at night, so it needed to be small and light enough to be portable. It would be equipped with skids, so it could land on the seafloor; the rest of the time, a pilot would "fly" it. *Alvin* was finally launched in 1964 and went to work for Woods Hole Oceanographic Institution, because no other marine lab wanted it.

Support Ships

A submersible without a support ship is like the proverbial fish out of water. However, so much of the effort, design, and budget for *Alvin* had gone directly into the vehicle itself, there were few resources left for a support vessel. The next best scenario was to find or create a stripped-down version that would make do until funds became available for a "real" mother ship. WHOI engineers decided that a towed catamaran vessel would provide ample space for launch and recovery of *Alvin*. Two large mothballed pontoons, originally built for minesweeping, were available for the skeletal start of a catamaran. Fortunately, the *Alvin* team knew just the man who could make a mother ship out of a bunch of junk and surplus parts.

Big Dan Clark was known as the Paul Bunyan of Woods Hole. He scrawled his bid on a cedar shingle and won the contract. His marine construction gang began work in mid-winter in Cape Cod. The twin pontoons were connected fore and aft by two steel arches. One pontoon would hold machinery; the other—soon nicknamed the "Tube of Doom"—would serve as living quarters. *Alvin* would rest mid-deck and be lifted in and out of the water by a winch. When an engineer discovered four huge, rusty diesel outboard motors at another shipyard, the planned mother ship was modified to become a self-propelled vessel. Officially, the 2,660 kN (299-ton) vessel was *Deep Submergence Research Vehicle Tender No. 1*, but after a shipyard visit from Al Vine's mother, the support ship became her namesake—*Lulu*.

The Team

In March 1965, *Alvin,* her support ship and crew members were ready to go to work. Initially, some 16 to 19 men called *Lulu* home: eight or nine were sailors who crewed the mother ship, the others were technicians—mechanics, electricians, and electronics people who were core members of the Alvin Group, as the Deep Submergence Group had begun to call itself. In the beginning, *Alvin* had only two pilots, Bill Rainnie and "Mac" McCamis. William "Skip" Marquet, an instrumentation engineer, was considered the outfit's high-level brains. George Broderson became crew chief, a job that entailed being bosun, honcho, safety officer, teacher, peacemaker, and all-around man in charge. He signed the checklist during launch and recovery, which meant ensuring that each individual had done his job. He also double-checked everything, building in a margin of safety in case any operation turned hairy.

Broderson knew that communication is key to running successful missions. In the water, communication between surface ship and submersible was mainly by means of underwater telephone—when it worked. Launch and recovery ops were physically demanding. Often the order was for all hands on deck, from galley staff to members of the scientific party, in order to keep the submersible from bashing into the mother ship. Even when *Alvin* was safely back on deck, the job wasn't over. The sub still had to be cleaned, checked, and recharged. As well, repairs had to be made to whatever technical problems had cropped up. Even small jobs were important, like replacing the carbon dioxide scrubber each morning and sponging out the bucketful of water that collected in the passenger sphere from condensation and sweat. None of it was glamorous work; all of it was essential.

Making Headlines

Early in 1966, the Alvin Group had just about completed the submersible's first annual overhaul when an unexpected phone call changed their plans. *Alvin* and her Woods Hole crew were suddenly heading to Spain, to assist the Perry submersible *Cubmarine* and the larger *Aluminaut* in searching for a 70-kiloton hydrogen bomb that had been dropped into the ocean after a refueling collision between a B-52 bomber and a KC-135 jet tanker over Spain's Mediterranean coast. Crude underwater navigation and official directives made the difficult search conditions even more treacherous. But on its nineteenth dive, *Alvin* finally located the deadly bomb, still wrapped in a huge, billowing parachute. The unmanned robot *CURV* made the recovery. Submersibles and underwater robots were suddenly headline news!

Alvin and crew came back from Spain and rejoined *Lulu*. The makeshift mother ship had undergone a much-needed refit, but things still did not necessarily go smoothly during 1967's first full season of diving for science. The ongoing Vietnam War meant that funding hopes for a new mother ship were dim, so there was nothing to do but continue to work while refining tools, equipment, and techniques wherever possible.

The constant push of the work schedule and the equally constant mechanical strain of work at sea finally took their toll. With only one more expedition to go in the 1968 season, the ship's cables snapped just as *Alvin* was being launched. The submersible's occupants just barely managed to scramble out before *Alvin* disappeared into the sea. Gone.

Seven months later, the submersible was finally recovered and brought back to Woods Hole. Work began on the crippled sub, but the funding picture was bleak and the job extensive; *Alvin* did not return to work in 1970.

In 1971 *Alvin* was finally ready to go back to work, but contracts were scarce. With no guarantee of Navy funding, even the future of the Alvin Group was uncertain. The only solution seemed to be to begin charging scientists to use the submersible. A young Bob Ballard joined WHOI; his assignment was to find funding sources and missions for *Alvin*. Ballard was full of ideas, although some people just shook their heads when he talked of needing a truly exciting feat that would make the submersible a household word—something like finding the *Titanic*. Such a feat seemed only an implausible dream.

Financing for the submersible's 1971 season involved a hash of funders, barters, and IOUs, as Ballard worked to fill out a paying dive schedule. When the Alvin Group finally secured backing for a new titanium passenger sphere that would enable the sub to double its diving depth to 3,600 meters (approx. 12,000 ft), *Alvin* underwent a major refit at Woods Hole during the winter and spring of 1972–73. Although the submersible's operating budget never seemed to grow much from year to year, now scientific institutions everywhere were eager to test the submersible's new features. It seemed as though every dive resulted in discoveries, including hydrothermal vents in the Eastern Pacific. In the 1970s, *Alvin* was one of the submersibles used by French and U.S. scientists for mapping, sampling, and photographing the Mid-Atlantic Ridge. When *Alvin* made her thousandth dive in February 1980, she was by far the most experienced deep diving craft in the world; for the first time, lack of funding was no longer such a serious threat.

At the end of the 1982 season, a task force decided it was time for *Alvin*'s mother ship, *Lulu*, to retire. She steamed home to a large welcoming crowd, having given 18 years of service. With still more funding cutbacks on the horizon, the only viable replacement option seemed to be *Atlantis II*. The ship was already due for a major overhaul, so a stern-mounted A-frame was added to the work order. At 64 meters (approx. 210 ft), *Atlantis II* was double *Lulu*'s length; the vessel also carried 50 berths and could stay at sea for a month—twice as long as *Lulu*. Things were looking up.

In 1984, *Alvin* discovered cold-water hydrocarbon seeps in the Gulf of Mexico. And in 1985, Bob Ballard did what he said he would do—locate the wreckage of the *Titanic*! Ballard knew that diving any manned submersible into the shipwreck was out of the question, so an unmanned robot, dubbed *Jason Junior*, was sent out and controlled from *Alvin*. Despite doubts about its performance, this new ROV did manage to penetrate the famous wreck and record pictures for the entire waiting world.

At the turn of the millennium, *Alvin* was finally paired with another mother ship, this time one equal to the submersible's capabilities. Built in 1997 for the U.S. Navy, the 83.5-meter (approx. 274-ft) *Atlantis* is one of the most sophisticated research vessels afloat, boasting 344.66 square meters (approx. 3,710 sq. ft) of laboratory space and a 60-day endurance capability. It's the kind of support vessel that can assist *Alvin* in continuing its phenomenal underwater activities.

While many of *Alvin*'s exploits are legendary, all are based on careful ops planning. Insiders also know that the submersible's remarkable accomplishments and decades-long track record have been enabled by her support vessels and most certainly the dedication, skill, and flexibility of her pilots, scientists, technicians, divers, captains, and crew.

Figure 11.2: A Highly Complex Recovery Operation

Once the Confederate submarine was discovered by the National Underwater Marine Agency in 1995, raising the H.L. Hunley *required a carefully planned and orchestrated operation with several surface support platforms and underwater divers.*

This photo shows the moment when the vintage sub, cradled in a steel truss, is about to break the surface of the water in Charleston Harbor. The submarine was then placed on a barge and brought to the Warren Lasch Conservation Center for subsequent evaluation and restoration.

The complex recovery plan evolved only after a realistic evaluation of the craft's structural integrity. Besides the normal ops variables of weather, tides, and unforeseen accidents, this mission also had to deal with extremely low visibility, high currents, and jellyfish! Since the successful recovery, the remains of Hunley's *crew have been given a proper burial. And subsequent analysis of the sub's construction has resulted in increased awareness of its structural complexity and sophistication for a human-powered craft built a century and a half earlier!*

Image by cgphoto.com, courtesy of Friends of the *Hunley*

1. Introduction

The previous chapters have concentrated on the physical principles involved in designing the underwater vehicle and payload equipment necessary to carry out a mission. This chapter focuses on what happens once that vehicle is ready to go. This is the underwater **operations**, or "**ops**," phase of your vehicle's mission.

The goal of any underwater operation is to successfully complete a mission. In the relatively simple recovery mission detailed in Chapter 2, students had to design and build an ROV that could retrieve bozonium canisters off the bottom of Monterey Harbor. But there were actually two parts to this mission—in addition to creating the ROV, the students also needed to come up with an operations plan for conducting the search and recovery operation. This second part of their mission involved a procedure for getting the group and gear down to the docks, a strategy for searching the harbor bottom, a method of snagging the cylinders and returning them to the surface, and an exit plan for moving gear and crew back to the shop for debriefing. All of those steps were essential parts of the students' ops phase. Similarly, a deep ocean campaign, such as the search for the *Titanic*, required specialized underwater vehicles, of course. But successful completion of the mission also relied heavily on organizing a team of highly trained personnel, ocean-going vessels equipped with high-tech machinery, the necessary equipment and supplies, and a comprehensive plan.

2. Operations Overview

Any underwater vehicle operation, whether simple or highly complicated, consists of the same basic elements:

- a practical plan of action
- advance preparation of the equipment, gear, and people required at the ops site
- effective search and/or recovery strategies
- a team with the training and staying power (sometimes working in extremely adverse conditions) to run the mission to its conclusion

TECH NOTE: BASIC SKILL SET FOR OPS

An understanding of basic navigation and boatmanship skills is advisable in order to safely conduct any mission, particularly those in open water. Mastering the following skills is helpful:

Knots: Know how to tie four types of knots: a bowline, clove hitch, sheet bend, figure-eight on a bight (or figure-eight loop).

Lines: Be able to coil a line or tether in various patterns so that it doesn't kink or tangle and so that it pays out easily.

Boats: Be able to throw a line and tie up a boat. Also practice handling a dinghy, rowboat, or canoe efficiently and safely.

Anchoring: Be able to effectively anchor a small boat to run ROV search ops.

Tides: Learn how to read a tide table (if working at sea).

Radio: Know how to use a hand-held walkie-talkie.

Navigation: Know how to read and use a marine chart, as well as how to operate a GPS.

First Aid: Know basic first aid.

Water Safety: Know water safety skills, such as how to don a life jacket (technically called a personal flotation device, or PFD) and swim with one. Be familiar with the dangers and treatment of hypothermia, and how to conduct a man-overboard rescue.

There are many good references on boatmanship and water safety; vintage ones such as *Boatmen's Guide to Light Salvage* by George H. Reid are worth tracking down and extremely reasonable in price.

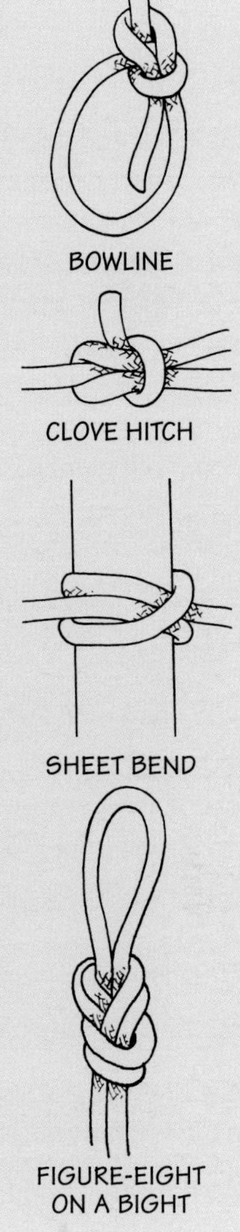

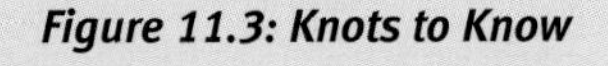

Figure 11.3: Knots to Know

These are the essential elements of a mission, whether millions of dollars are riding on an ROV operation to find and repair a pipeline leak at 900 meters, or whether low-budget student-built robots are being put to the test in an ROV competition. It's important to know that running a successful and safe operation involves different skill sets than designing or building an underwater craft. Often, the key to success is a team of people with diverse, flexible expertise, the ability to multi-task and "think on the fly," and the determination to see the mission through to completion.

2.1. Search and Recovery Ops

As an example of an underwater operation, this chapter focuses on the procedures for running a basic underwater search and recovery operation using a small robotic vehicle. To do so, the text incorporates a mission scenario based on a true-life event. (See *Tech Note: Mission Scenario.*) It takes place in shallow water (10 m/approx. 33 ft). The search and recovery op uses a small ROV similar to the *SeaMATE* ROV in Chapter 12 to find and recover a radio-controlled model boat, dubbed the MV *Zande* (MV is standard marine notation for Motorized Vessel) that has been lost in a small body of water called Conway's Pond. You can follow along with this adventure in many of the Tech Notes throughout this chapter. While this is a relatively simple operation, the techniques parallel those used for more challenging commercial, scientific, and military operations done at sea.

SAFETY NOTE: ALL-AROUND SAFETY

The issue of safety includes both safety in the workplace and safety at sea. Safe working practices at sea include adequate training, proper gear, equipment in good working order, clear communication, practiced emergency procedures, basic seamanship skills, first aid knowledge, and common sense. Solid leadership skills are particularly essential in the area of safety, but safety is everyone's responsibility, so the more team members with first aid certification and marine emergency training, the better. Just remember to discuss safety issues, equipment, and mandatory safety briefings during the planning stage, as well as on the actual mission.

Safety Checklist for a Dock-Based Operation:

- Include a first aid kit in the expedition packing.
- Remind people to bring/wear sunscreen, sunglasses, and non-skid, closed-toe footwear.
- Make sure drivers of all cars or other (land) vehicles are properly licensed and insured.
- Do not overload land vehicles or boats.
- Provide personal flotation devices (PFDs) for dockside crew and everyone in a watercraft.
- Make sure there is some type of flotation rescue equipment on hand at the dock.
- Bring a pike pole to extend your reach when recovering the ROV or other items that may fall into the water.
- Don't crowd the work area on the dock.
- Provide GFI-equipped extension cords if using shore power at the dock.
- When transporting batteries, use protective carrying cases.
- Think about any other situations that may become hazardous (i.e., bee stings, lightning storms, slippery docks) and plan appropriate strategies.
- Check school/workplace requirements for insurance and secure signed waivers of liability before setting off, if required.

TECH NOTE: MISSION SCENARIO

Flash!! A wooden radio-controlled tugboat model sank today in Conway's Pond. The 1/16th scale replica of the 90-foot (approx. 27 m) heritage steam tug MV (Motor Vessel) Zande *is complete with a scaled, hand-crafted steam engine. It is believed a cooling water intake hose for the steam engine burst, causing water to flood the hull. The model tug now lies somewhere in the bird sanctuary pond. Your mission is to find and recover the MV* Zande *with a* SeaMATE *type of ROV.*

This mission scenario is based on a real event. A high school science teacher attending the MATE Summer Institute for Faculty Development was building a *SeaMATE* type of ROV as part of the course. (See www.marinetech.org for more information on Summer Institutes.) She confided to the course instructor that she had been hoping to use it for a very special purpose—to find and recover a radio-controlled model boat her father had lost in the small, shallow lake near his home. Her father had spent many hours building his model, only to have it sink on its maiden voyage. The course instructor worked with her to come up with a plan for a search and recovery operation using her ROV and a rowboat.

***Figure 11.4: The Missing MV* Zande**

TECH NOTE: THE FUNDAMENTAL PRINCIPLE OF MARITIME SALVAGE

A salvage operation is one type of recovery in which a person or group (called the salvor) is attempting to retrieve something of value and lay claim to it. You may be surprised to discover that finding something underwater and recovering it doesn't automatically mean it's yours. Maritime law is complex on this issue, but the main principle underlying salvage law is that a salvor does not have a right to bring up just anything that has been found on the bottom. The law considers that sunken objects still belong to the person or company who owned them when they were lost. Thus, the salvor must get permission to salvage anything of value, be it historic, commercial, or private; otherwise, the recovery operation can be construed as theft. While salvage is definitely not a "finders keepers" scenario, the salvor is generally entitled to some compensation for the recovery effort and expense.

Note that the operations discussion in this chapter assumes that the model tugboat lost in the Conway Pond scenario is of no historical or commercial value, and that those searching for the small tug have permission from the owner to find and recover the object.

TECH NOTE: PLATFORMS

The subsea industry uses the term **platform** to describe any fixed, flying, or floating surface that can support a vehicle, its crew, and the machinery used to launch and recover the vehicle. The same platform may also be used to lift objects off the bottom that the vehicle has found or rigged for a recovery.

A platform doesn't necessarily have to be mobile. So a beach, dock, or oil rig can all be considered platforms. Huge drill ships and even rowboats can both serve as mobile platforms for on-site operations; it just depends on the scale of the operation and the mission. For example, an ROV inspecting the pilings of a dock may function just fine from a pier. However, repairing a pipeline at a depth of 300 meters (approx. 990 ft) and some 160 kilometers (approx. 100 mi) offshore will definitely require the services of a seagoing mobile platform, such as a support ship with dynamic positioning capability. (See *Section 3.2.2. Assessing Platform Options.*)

Image courtesy of unknown photographer.

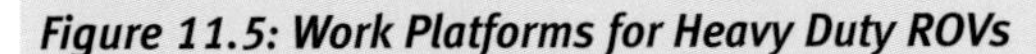

Figure 11.5: Work Platforms for Heavy Duty ROVs

Semi-submersible drill rigs like the Deepwater Nautilus, *shown here in hurricane seas with the support ship* Laney Chouest, *are common work platforms for heavy duty ROVs.*

By now you know that underwater vehicles are involved in a wide variety of missions. That said, the type of ops that most commonly make newspaper headlines are search and recovery missions.

Smaller search operations often rely on an ROV equipped with a camera or simple sensor. More advanced search ops may use ROVs, AUVs or tow sleds loaded with additional sensors such as complex sonar arrays, magnetometers, etc. These sensors are most effective when coupled with carefully executed search patterns to deal with complicating issues such as the depth of the search and the murkiness of the water.

2.2. Other Types of Ops

Drilling support and maintenance of offshore oil rigs is the primary mission for the majority of work class ROVs. Offshore construction projects also utilize work class ROVs for oil production field installation, maintenance, repair, and decommissioning. As well, these vehicles are used for technical and safety support in the construction of piers, bridge pilings, pipeline installation (inspection and repair), cable laying (inspection and repair), tunneling, and tunnel inspection and repair. Complex projects generally rely on commercial divers and robotic underwater technologies, topside cranes, vessels, construction gangs, and a whole host of specialized equipment. Of course, there are also numerous military operations, as well as ongoing scientific missions. This chapter will concentrate on a search and recovery mission utilizing a small *SeaMATE* ROV.

3. Operational Phases

Ops can be pretty daunting affairs, since they involve a number of unknowns, not the least of which are weather and sea conditions. It's important for the crew to feel that the operation is well planned and that they are as prepared as they possibly can be for any eventuality. To meet that expectation, it is helpful to break the mission into these basic phases:

1. Mission Considerations and Risk Assessment
2. Ops Planning
   - Researching the Target and Site
   - Assessing Platform Options
   - Developing a Detailed Plan
   - Preparing to Implement the Plan
3. Mobilization
   - Staging and Transporting Equipment
4. On-Site Activities
   - Handling the Unexpected
   - Navigating and Piloting Techniques for ROVs
   - Establishing Search Grids
   - Positioning for Recovery
   - Recovering the Target
5. Demobilization
   - Heading for the Barn
   - Debriefing
   - Writing Reports

These five phases apply to mini-operations, such as that detailed in the model tugboat search below, as well as to larger deep ocean campaigns. Of course, this methodology may need to be modified to fit your specific circumstances, but it serves as a useful starting point.

3.1. Mission Considerations and Risk Assessment

Before starting to plan the details of any operation, it's essential to assess the situation, including opportunities and risks, that could affect the success of the mission. For a search and recover operation it makes good sense to find out as much as possible about the circumstances surrounding the loss of the object in question. This information will tell you where it is most likely to be found, will hint at the resources needed, and can raise the odds of completing the mission successfully. This information is also important in assessing the risks associated with recovering the object. These include a consideration of possible danger to people and equipment, the costs involved, and the amount of time it will likely take to complete the mission.

Generally, the degree of risk you are willing to shoulder is dependent on the training, experience, and money you can bring to running the operation. For example, the military can take greater risks to salvage a vessel carrying secret munitions, because they have the experience, crews, and equipment to deal with a highly hazardous and potentially dangerous environment. It's ok—indeed smart—to opt out of a mission you feel is dangerously beyond your skills, equipment, or experience.

If you decide to go ahead with the mission, risk assessment helps in deciding which type of equipment you'll need for the operation. For example, in the mission scenario at Conway's Pond, an ROV is only one tool that could be used to locate and recover the model tug. You might think scuba diving is a viable alternative, but there are significant safety and liability issues, as well as costs. The ROV is probably the cheaper option and one that is definitely safer.

Despite the limitations of small, inexpensive, shallow-diving craft, you can run surprisingly effective search and recovery missions in shallow, calm waters. Murkiness may challenge the pilot, but that can be offset by using a methodical search grid. A small rowboat or canoe can act as a surface platform. And if the target is relatively light, you can use the ROV to attach a line so it can be hauled up by hand.

Budget is always an important consideration, but the investment for something the scale of the MV *Zande* mission is minimal. Above and beyond the cost of the ROV, you'll need to pay for gas and possibly rental fees for a car to get to the site and/or a small boat. Other required gear such as equipment to measure depth, some rope, and other small bits of hardware easily falls within a $100 budget. So even if you don't find the model, there's not much to lose by trying. The risk is certainly worth taking.

3.2. Ops Planning

A small, basic mission may involve only a few people; larger ops will have many crew members, some with specialized skills. Regardless of size, somebody has to function as the ops supervisor. In the MV *Zande* search, that's you.

Once you've assessed the risk of the mission and determined that it's feasible, given your expertise, equipment, and budget, then it's time to begin planning. Any underwater search and recovery operation, whether simple or complex, starts with developing a detailed plan based on your strategy for finding and subsequently recovering the lost object. Once you've settled on that strategy, then your plan orchestrates how personnel and equipment will be acquired, transported, and deployed to the mission site. The plan must also include a realistic budget and a method to track costs as the operation is run. Any plan should be flexible, to allow for unexpected contingencies.

The two most basic operations management tools to help in planning any operation are: (1) making lists and (2) figuring out schedules. Keep in mind that budget is always an overall consideration.

TECH NOTE: RESEARCH AND MORE RESEARCH

In his non-fiction book *The Sea Hunters*, Clive Cussler details the basic techniques that he and his volunteers have used to locate more than 60 historic shipwrecks. "Research is the key," he states emphatically. "You can never do enough research. This is so vital, I'll repeat it. *You can never do enough research.*"

Investigation and study can either improve the odds of finding a ship, thereby making the search feasible, or tell you it's hopeless, Cussler explains.

When looking for shipwrecks in rivers or harbors, it's important to investigate historic changes in a coastline, early salvage accounts, correspondence, and newspaper articles, as well as subsequent dredging or building activities that might affect where a ship ended up.

He adds that there are very few wrecks that have been found without exhaustive background work being done first. "Sure you can get lucky, but don't bet your bank account on it."

3.2.1. Researching the Target and Site

The goal of researching a mission scenario is to establish a probable search area, then narrow that area down even more. Some of this research will need to be done right away (e.g., while witness memories are still fresh). Other research may need to wait until later (e.g., when you can visit the site).

It's worth it to do a good job with this research step (asking questions, gathering data, etc.) because narrowing down the probable search area can save you considerable time and money as well as significantly increase your chances of a successful mission.

Even in ponds and small lakes, something the size of a model tug can be elusive; the murky water and silt make the search considerably more difficult. And with limited tools for the job (basically a canoe and a home-built ROV), such a small tug could go undiscovered quite easily. So start your research by asking and answering these basic questions:

1. When did it happen?
2. Where did it happen?
3. How deep is it?
4. What's the visibility?
5. What's the bottom like?
6. Who saw it happen?
7. What *exactly* are you looking for?

These questions will naturally expand to become more specific and detailed as you delve deeper into the event. For example, your initial research to find that model tug in Conway's Pond might result in the following information:

1. When Did It Happen?

The rule in underwater search is that the longer an object is under water, the harder it is to find, *and* the harder you'll have to work to get the information needed to find it. For a model tug in a pond, this might not seem like a big issue, especially if the event occurred less than a month or two before. But as time goes by, things can change.

One of those changes is in people's memories. If you spoke with eyewitnesses to the sinking within an hour or even a few days, you'd get a pretty good description of what they saw and perhaps even of where they thought the tug had gone down. But memory becomes less reliable over time, so the accuracy of the recollection has a higher likelihood of error.

Other changes may be caused by environmental and human events. For example, presume a year has passed, and during that time, the pond froze. If the tug was in a shallow part of the lake, it could have sustained ice damage. Or the pond might have silted up and buried the tug if the region experienced heavy rains or flooding. Or someone else may have attempted the salvage without you knowing about it, and the tug may no longer be in the pond. Even if they were unsuccessful at recovering the tug, they may have dragged it away from its original position on the bottom. And if many years have passed since the initial sinking, then the wood and machinery on the tug would probably have deteriorated to the point that the model boat would not be feasible to repair, even if salvaged.

The passage of time can also change the value of a lost object. Usually, the longer something lies on the bottom (especially the hull and machinery of a vessel), the less commercial value it has. On the other hand, precious artifacts, such as gold, silver, and jewels on a shipwreck, can be worth a lot, regardless of age. In fact, salvaged Roman coins or Spanish pieces of eight may have an historic value that greatly exceeds the current market value of the metal. Even non-precious artifacts from old shipwrecks can increase in value. When nautical archeologist Peter Throckmorton discovered a Bronze Age ship off Cape Gelidonya, Turkey, the wreck turned out to be over 3,000 years old. In his book *The Sea Remembers*, Throckmorton explains that the ship was only a common Mycenaean trading vessel with no gold or jewels. However, it turned out to be a treasure trove of copper ingots. Even more intriguing, the wreck divulged priceless artifacts that revealed what life was like during the ancient Trojan Wars.

So ultimately is the model tug sunk in Conway's Pond worth salvaging? Mostly, it means a lot to the builder who lost it, reflecting long hours of labor. There's also the value of the hull and the custom-built steam engine, which, with a little restoration, might be quite serviceable. Of course, the electronics would have to be replaced.

For those conducting the search, there's the challenge of finding the lost model and the satisfaction of recovering it. So attempting this search and recovery mission sounds feasible and worthwhile, assuming it's only been a few weeks since the MV *Zande* sank and that it has not been disturbed in any significant way.

2. Where Did It Happen?

The location of the sinking is generally the biggest factor to consider when making an underwater ops plan. Obviously, if something goes down in the mid-Pacific, the underwater operation will be considerably more complex than if it occurs in Conway's Pond. Once you sort out this aspect of the ops plan, other factors, such as equipment, personnel, vehicle type, and budget, naturally fall into line.

Risk evaluation is also highly dependent on location. For example, the risk from environmental factors (e.g., wind, currents, distance, waves, cold, etc.) is much lower in Conway's Pond than if the model were lost in a larger body of water.

TECH NOTE: "IT OUGHT TO BE RIGHT THERE!"

"It went down right there . . . Shouldn't be too hard to find." These famous last words are often spoken by those who don't understand the difficulties of searching a body of water. In fact, many factors can put a sunken object somewhere else rather than "right where it should be." Here are some of them:

Current: A current running at the time of a sinking can carry a foundering object far from the point of its demise, especially if it sinks slowly. Even when the object is on the bottom, a strong current can shift its position or eventually cover it with silt or debris.

Wind: If driven by wind, a vessel taking on water can drive a long way from its last reported position before it actually sinks.

Veering: Often a sinking object doesn't just fall straight down. Instead, it veers away from the spot where it disappeared on the surface. You can see this effect if you take a penny and toss it into large jar of water. Watch how it wobbles and moves on its way to the bottom. The degree of veer is dictated by the shape, aspect, and buoyancy of the object. For example, a flat, light sheet of aluminum will wobble erratically if dropped flat into the water, but it will sink straight down if the aspect is edge down toward the bottom. Only heavy, compact objects, such as cannonballs, round rocks (flat rocks wobble), or even a set of car keys, will sink more or less straight down. And, of course, an object that is only slightly negatively buoyant will take longer to fall to the bottom than one that is more negatively buoyant, so it may be subjected to more external forces that can cause it to veer off.

TECH NOTE: ASSESSING THE CONWAY'S POND LOCATION

For this scenario, presume the pond is in an urban nature park within the limits of a city where you live. Nature trails surround the pond, and a parking lot is on the south shore. It's a short walk of 30 meters (approx. 100 ft) from the parking lot to a small wooden dock that sits about 30 cm (approx. 1 ft) above the water and extends out for 9 meters (approx. 30 ft). The pond itself is small, surrounded by trees and marsh, so there is no danger of high waves, current, or long distances to transit. So all you would need is a van or truck that can carry three people and a canoe, plus other gear. Fortunately, the cost for all this is well within your modest budget. While a map of the pond would be helpful, you could get by with the park guide that has the pond and trails marked on it.

Potentially Useful Information About Conway's Pond

- Pond is located 5 km (approx. 3 mi) from city center on Sullivan Drive and inside Clough Nature Park
- Pond is roughly ovoid in shape
- Length: 183 meters (approx. 600 ft)
- Width: 61 meters (approx. 200 ft)
- Maximum depth of water: 6 meters (approx. 20 ft)
- Visibility: estimated visibility in bright sunlight is 2 meter (approx. 6 ft)
- Bottom: mud and silt, tree branches, bulrushes, lily pads around edges of pond
- Parking lot 30 meters (approx. 100 ft) from south end of pond
- 120 VAC outlet available at nature center
- Pressurized water (garden hose faucet) available on picnic grounds near dock
- Footpath goes right around pond's perimeter
- One footbridge crosses out-flowing stream on east shore, with clear view of pond from midway on bridge
- One 6-meter (approx. 20-ft) high bird watching observation tower on north shore of pond
- A small island for nesting birds is located near the western shore
- Pond is surrounded by marshland
- Canoe launching dock on south end of pond 30 meters (approx. 100 ft) from parking lot

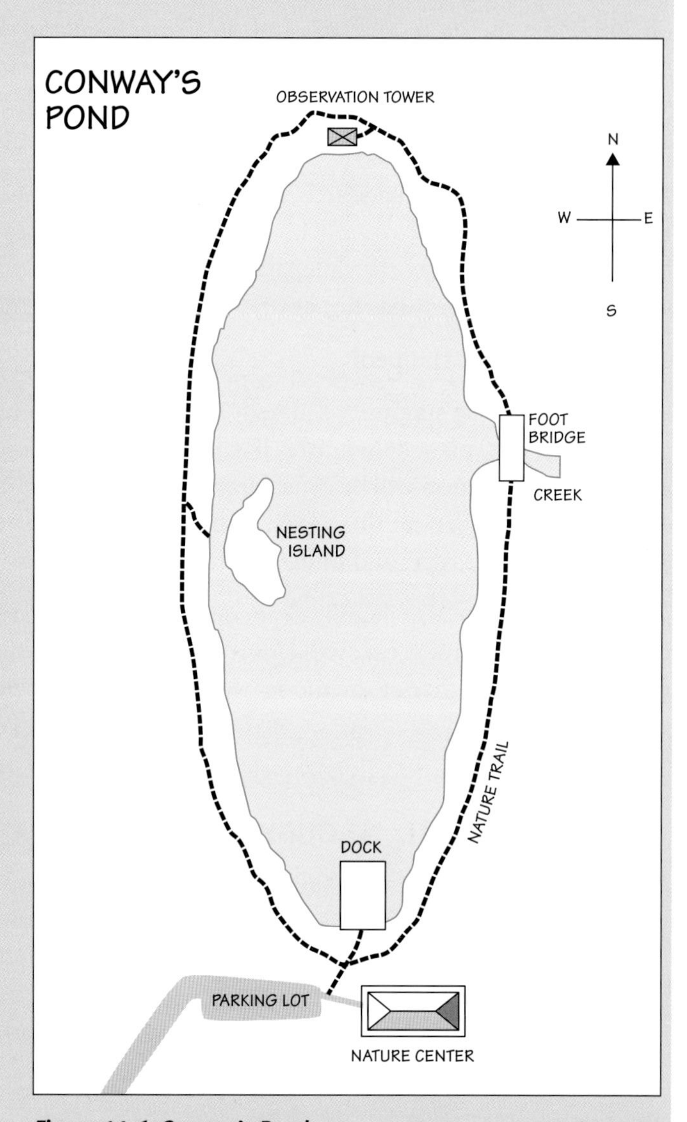

Figure 11.6: Conway's Pond

3. How Deep Is It?

Depth is a huge factor in any underwater operation. Fortunately, working in 6 meters of water is a whole lot simpler than working at 60 meters or 600 meters. In the shallower depth, you need only a small boat, some sturdy line, the ability to tie a good knot, and a strong arm to haul the vehicle in and out of the boat. But to go a lot deeper, you'll need long cables, winches, and possibly a crane to get equipment over the side and back on board again.

Another factor inexorably linked to depth is water pressure. You already know the importance of having vehicles and gear that can withstand the ambient pressure found at the depth the mission is to take place. But depth also impacts lifting gear and platforms. So, the general rule of thumb is the deeper you go, the more mass the salvage equipment must have.

Even with a relatively simple mission such as that in Conway's Pond, it's important to know how deep the water is. Typically, a pond will be shallower closer to shore and deeper in the middle sections, but you need to know for sure. Fortunately, getting accurate soundings, or depth measurements, is easy in a shallow body of water. You'll need a marked line (or a fiberglass tape measure) with a weight and a canoe or rowboat to get out on the pond. Simply lower the weighted line to the bottom and read off the depth on the marked line at the water's surface. For this scenario, assume a maximum depth of about 6 meters (approx. 20 ft). That immediately lets you know that it's feasible to use a *SeaMATE* type of ROV with a depth rating of 10 meters (approx. 30 ft) to conduct the salvage.

A more technologically advanced way to measure depths in Conway's Pond is to use a portable fish finder as a depth sounder. These devices have become very common and relatively inexpensive. Any depth sounder is basically a simple type of active sonar that points toward the bottom. It sends sound pulses down into the water where they bounce off objects and return to the sonar instrument. The return time for the echo is used to calculate the distance to objects below. Fish finders are optimized to show echoes from fish, but they also do a good job of showing how deep the water is, because the

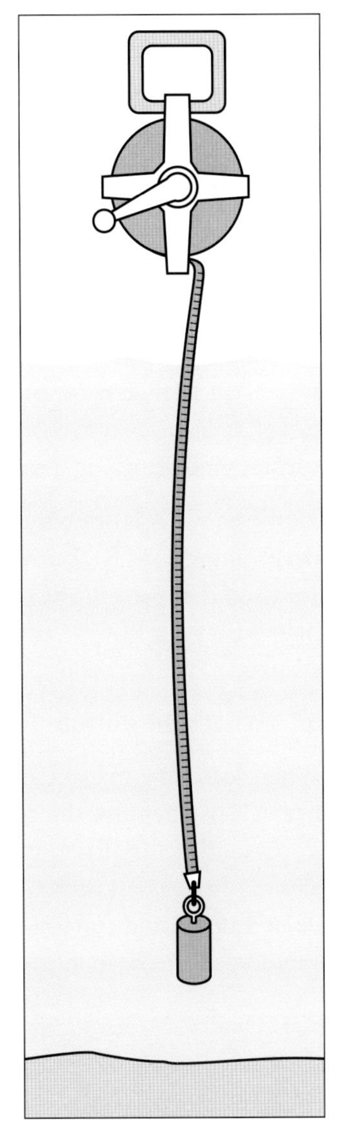

Figure 11.7: Home-Made Sounding Line

You can make a sounding line from a 30-meter fiberglass measuring tape that you can buy at a hardware store. Add a modest weight of a pound or so to the clip at the tape's beginning. Lower the weighted line to the bottom and record the measurement on the tape at the water level. To retrieve, simply reel in the tape with its built-in handle or pull it up hand over hand.

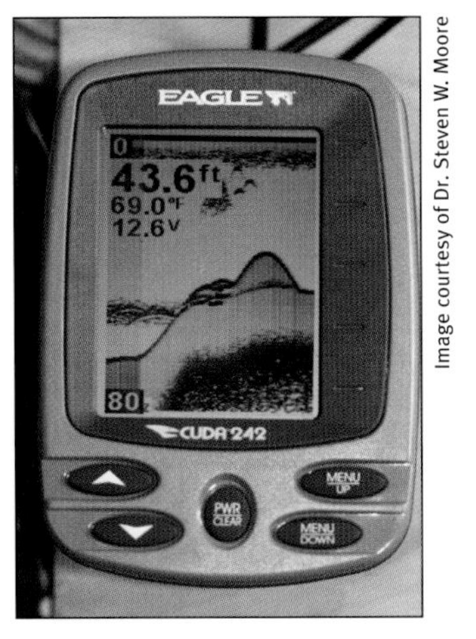

Image courtesy of Dr. Steven W. Moore

Figure 11.8: Fish Finder as Depth Sounder

Portable depth sounders or fish finders are suitable for small boats like canoes and skiffs.

HISTORIC HIGHLIGHT: ORIGIN OF THE FATHOM

A **fathom** is a nautical unit of depth measurement equal to 6 feet (approx. 1.8 m) that is still used on some marine charts today to indicate soundings. But why use such an odd unit to measure depth? The word fathom is derived from the Dutch word "vadem" and the Germanic word "faden." In the original sense of the word, it described someone who embraced with their arms outstretched. In the early European age of sail, seamen would measure lengths of line by taking hold of it and pulling it out as they stretched their arms wide apart. The length between outstretched arms is close to 6 feet.

This method of measurement was fast and convenient when taking soundings. In earlier times, a lead line was commonly used to measure water depths. If the ship was stationary, the sailor would simply lower a length of line with a lead weight attached to the end. If the ship was moving, he'd throw the line out ahead of the ship and wait until the line was vertical before taking the measurement. Either way, the sailor noted the mark where the surface of the water met the line. Then he hauled it back in, stretching his arms wide with the line between his hands. Hence, one "embrace," or fathom, became the standard unit of depth. The number of times the sailor's arms were outstretched equaled the water depth in fathoms. This basic technique is still useful for mariners, even today.

bottom usually produces a strong echo. The water depth is often displayed by the sonar unit as a numerical value, a graph, or both (Figure 11.8). Running a boat equipped with a depth sounder or fish finder around the pond in a regular pattern will result in a fairly good idea of how deep various parts of the pond actually are, and you can transfer the soundings to your own map/drawing. (See Section 3.4.2 for effective search pattern descriptions.)

4. What's the Visibility?

It's pretty obvious that the clearer the water, the easier it is to see what's down there. In fact, you wouldn't even need an ROV to search for the tug in Conway's Pond if the water had 10 meters (approx. 30 ft) visibility; in that case you could probably spot the target by using a face mask or by looking through a glass-bottomed bucket held at the surface while running a canoe around the pond.

However, water is often murky, and Conway's pond is no exception. It's about 6 meters deep (approx. 20 ft), but the visibility is only about 2 meters (approx. 6 ft). In other words, if the tug is in the deeper parts of the pond, you won't be able to see it without help from your ROV's camera or a drop camera (a weighted underwater video camera suspended from a cable).

5. What's the Bottom Like?

When searching for sunken objects, there's a good reason to ask "What's the bottom like?" That's because the composition of the bottom and the type of terrain are both important factors to consider when choosing the search strategy. This information allows you to assess difficulties in locating your target (e.g., boulders, canyons, or other objects that could conceal the target from view) as well as entanglement hazards (e.g., submerged tree branches, debris, or fishing nets).

TECH NOTE: TYPES OF BOTTOM TERRAIN

Flat Bottom: This is the most desirable terrain for a search. Because the bottom is flat, anything that has sunk onto this type of ground should protrude above it and be fairly easy to spot with cameras or sonar. However, very soft mud is easy to stir up, reducing visibility, and can sometimes bury the object you are looking for. Lack of landmarks or features on a flat bottom can make it difficult for visual navigation of a vehicle.

Rises and Depressions in Flat Bottom: A flat bottom is not always perfectly flat—often, there are small rises and depressions. These can occur naturally or they can be man-made, such as those found in harbors from channel dredging or ships dropping anchor. These irregularities in the bottom may hide a search target from view. This is a big problem with side scan sonar but can also be an issue for visual surveys with a camera. (See Figure 11.9 and *Tech Note: Piloting Tips for Sector Scan Sonar Navigation* in Section 3.4.2.)

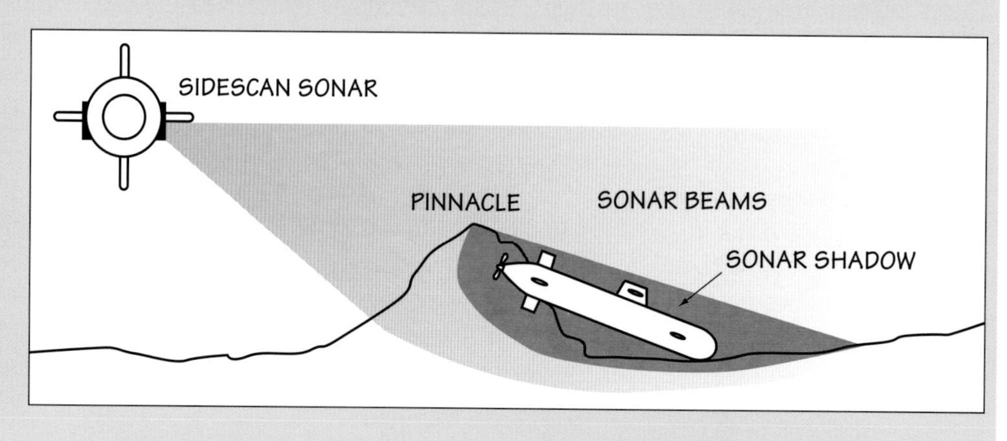

Figure 11.9: Depression Hiding Object from Camera View or Side Scan Search

Mounds: These are outcrops of rock or accretions from underwater springs that have built up layers of sediment over time. Mounds can vary in height, rising up to several meters off a flat bottom. One advantage of mounds is that they can provide a landmark for navigation. But a disadvantage is that they may hide a target from view, just as rises and depressions can.

Rubble and Boulder Fields: Rubble and boulders vary in size and are often found on sloped ground. Riprap is the term for angular rock intentionally placed around shorelines to prevent erosion; it also serves as the base for man-made structures such as piers, railway beds, docks, or jetties. Generally, ROVs and AUVs have no problem navigating around such a bottom. And there is no silting up due to thruster wash since most of the time there are no appreciable amounts of mud or sand. However, small items are difficult to detect since the object could be wedged down between the stones. For metal objects such as keys, coins, or jewelry, a metal detector might be the best sensor to use. Depending on visibility, a camera is also effective. Note that sonar can give false returns and clutter, especially when shooting up or down slope. Angular rock is especially prone to acoustic bounce, or "multipathing" of the return acoustic signals, making it hard to distinguish the object.

Slopes, Ledges, and Cliffs: Many lake basins, coastlines, and fjords have steep underwater slopes, cliffs, and ledges along their shores as a result of the geologic processes. This is generally challenging underwater terrain for conducting acoustic searches because of interference from multiple echoes (multipathing). However, searching slopes with tethered camera-equipped vehicles is easier. Drop cameras also work well. The slope terrain provides a natural orientation (up or down slope) that is a useful aid in navigating. Slopes also have more recognizable features that can be used as reference points. Be aware, however, that slopes collect a lot of natural and unnatural debris, such as fallen trees, polyethylene line, wire rope, monofilament fishing line, shopping carts, and other scrap metal, all of which can foul a towed or independently powered vehicle.

Outcrops, Pinnacles, and Reefs: These are large, rocky bedrock protuberances that emerge dramatically from the seabottom. They can form the foundation of reefs and shoals and also, depending on depth, can be partially or fully exposed above the surface during low water or tides. A typical pinnacle is surrounded by a sand or mud bottom that rises in a series of ledges and cliffs. In limestone or lava flows, there may even be caverns in its structure. As with mounds, these structures often rise suddenly from the seabed, so anticipating and avoiding them can be particularly challenging on towed surveys (Figure 11.10). These structures can also hide objects. Using an ROV to work a reef or pinnacle is relatively easy, although natural growth such as kelp and seaweed can foul a vehicle. Camera use is again dependent on the visibility of the area. Tides and wind can create currents around outcrops, pinnacles, and reefs, thereby affecting the handling of a vehicle. Waves are a real problem, with ocean reefs particularly subject to wave swells and surges.

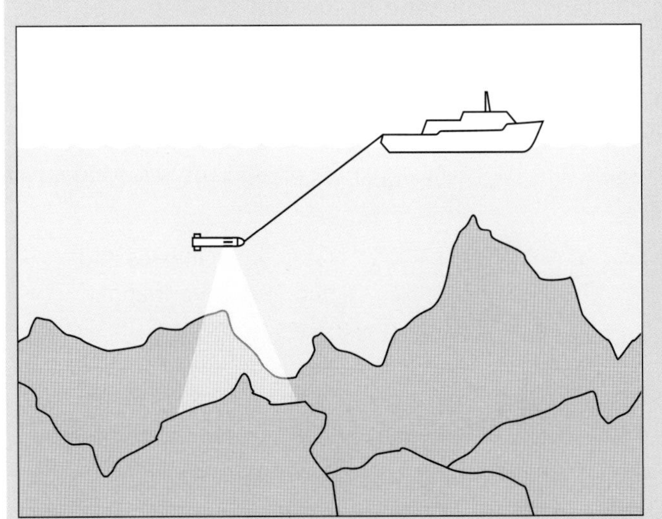

Figure 11.10: Sudden Terrain Changes

Underwater topography sometimes changes dramatically, presenting many challenges to underwater vehicles and their pilots.

Caverns and Caves: These large underground chambers are formed by natural geological processes. An ROV's tether is one of the biggest drawbacks of unmanned cave exploration, since twists, turns, and obstacles can easily entangle the machine. In clear water, visual sensing is not an issue, but if the water is silted, then determining the way through a cave is nearly impossible. Sonar is also problematic, since a transducer surrounded by rock walls in close proximity causes the acoustic beams to multipath, resulting in a confused incoming signal. However, the technology of cave exploration using hybrid vehicles and sensing systems is improving.

6. Who Saw It Happen?

Eyewitnesses often give the best clues as to where and how an object went down. If there is even minor agreement in people's accounts as to location of a sinking event, there's a greater chance of finding the object than if none of the stories match up.

Just remember that bad information is often worse than no information, because it can send you off on a wild goose chase that will cost time, money, and perhaps your reputation. So when investigating the loss of a shipwreck or even a set of keys dropped off a dock, it's a good habit to question the validity of the data you are gathering. If someone says, "The keys are right there," see if anyone else saw the accident or can confirm the first person's observation. If there's more than one witness, ensure their accounts match up, or check which parts are similar. If someone says there's a cargo of gold on board a wreck, verify it by looking for the ship's manifest to see the listing of cargo taken on board at its port of departure, or at least check historical accounts to see if such a cargo was likely for ships traveling that route. Verification and corroboration of the data will get you looking in the right places.

7. What *Exactly* Are You Looking For?

The size, weight, and material composition of a sunken object will dictate the search equipment and/or method to use and at what range the object will likely be detected. This information will also help you predict area coverage and the time the search will take, so you will need to find out the following:

- Object's size and shape
- Material composition
- Mass to raise

Object's Size and Shape: Size impacts the type of equipment necessary for the search and recovery operation. Similarly, the material(s) the object is made of can be a factor in determining what search equipment or method to use and at what range it will likely be detected. So this is vital information to get *before* beginning your search.

Also, the level of difficulty in locating an object under water is directly related to its size and shape. Generally, the smaller the object, the harder it is to find, especially if the search area is big. If you doubt this, try covering your eyes and throwing a quarter

TECH NOTE: MODEL TUG SPECS

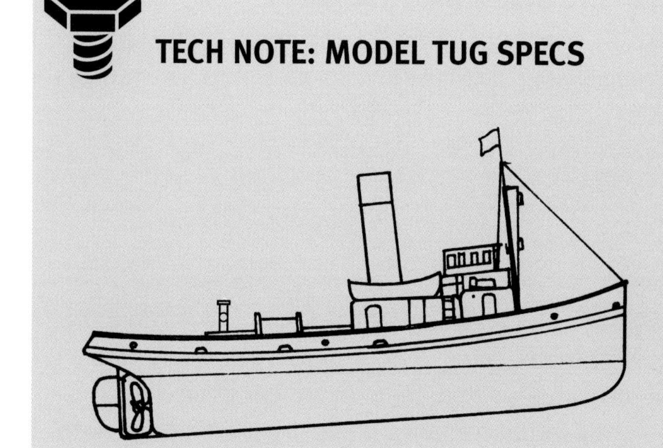

***Figure 11.11: MV* Zande**

Length: 1.15 m (approx. 3.75 ft)

Breadth: 38 cm (approx. 1.25 ft)

Height of hull (top of wheelhouse to keel): 45 cm (approx. 1.5 ft)

Weight (in air): 155.5N (approx. 35 lb)

Hull: fiberglass

Upper houseworks: wood

Engine: steam

Ballast: lead poured in keel

Colors: black hull, beige and white houseworks

TECH NOTE: WITNESS ACCOUNTS OF MODEL TUG SINKING

The Conway's Pond scenario has four eyewitnesses, and here are their accounts of the sinking:

Eyewitness #1: Owner of tug. Location: On dock, piloting tug by radio control during sinking. Eye height above water: 2 meters (approx. 6 ft). Account: Tug was moving in a westerly direction when it began to lie low in the water; then sank. The tug's position was in a line with the observation tower and the dock, about 90 meters (approx. 300 ft) from the dock.

Eyewitness #2: Bird watcher on observation tower. Eye height above water: 6 meters (approx. 20 ft). Account: Viewed event with binoculars. Claims the tug sank about 75 meters (approx. 245 ft) away from tower, in line with the dock.

Eyewitness #3: 12-year-old boy on dock. Eye height above water: 1.5 meters (approx. 5 ft). Account: Tug went down about 60 meters (approx. 200 ft) from the dock, but slightly to the west of the tower-dock line.

Eyewitness #4: Adult on east trail bridge. Eye height above water: 2 meters (approx. 6 ft). Account: Tug went down in line with a bird-nesting island (located in western part of pond) and the bridge. Estimated distance: Unknown.

Notice that these witnesses all reported the same event, but with some differences. The estimated distances from the dock vary. Note that typically, the lower a person is in relation to the water, the harder it is to accurately estimate distance. Also, people have different ways of guessing distance, so their figures will vary. One good thing is that witnesses #1 and #2 agree that the tug went down in line with the tower and the dock. Witness #3 says it was not on the line, but close. The clincher comes from witness #4, who gives a bearing on the tug as between two different landmarks, the bridge and a nesting island. This is great, since there's now a cross-bearing that transects the tower-dock line. Theoretically, the tug should be where the two lines of position cross. Now all that remains is to note these positions on a map or chart of Conway's Pond. (See *Operations Task #1* below.)

into the shallow end of a swimming pool, then searching for it without a mask. Chances are you'll spend a lot of time groping around the pool bottom. However, if you toss something larger, such as a plastic milk crate, into the pool, the odds are greater that you'll find the crate sooner than you'll find the quarter. So in the same-size search area, size matters. If you look at the missing tug's relative size compared to the area of Conway's Pond, you'll see that your search is challenging—comparable to finding the small, flat coin in the pool.

The milk crate was easier to find for two important reasons—first, it was larger in size, and second, its shape stuck up higher off the bottom than the coin. This means you could use a faster, wider sweeping motion with your arms to search for it, whereas the coin required keeping your hand flat against the pool bottom, thereby forcing you to go more slowly. So an object's height off the bottom is a huge factor in searches. This is especially true when doing acoustic side scan sonar searches. The higher the object, the more likely it will cast a nice clean acoustic shadow behind it. The rule is that any object that is low or covered over will be more challenging to locate, regardless of whether you are using visual, acoustic, or even dragline techniques. Note also that shape can be an important factor for retrieval. It would be much easier to snag and lift the milk carton with a big hook than to try and pick up a quarter the same way.

Material Composition: Knowing the physical composition of a target object is important. For one thing, this information can affect the ease with which the object can be spotted visually. For example, a dark object may show up well against white sand, but not against dark rock. The object's materials can also affect the search options--for example an iron object might be found relatively easily with a magnetometer or metal detector, even when buried or lost in very murky water. Depending on the conditions, some materials may rot or corrode away with time. Finally, smooth or slippery materials can make recovery more challenging, whereas rough materials or surfaces may be easier to grab and pick up.

Mass to Raise: The heavier the object to recover, the more difficult it is to get it to the surface. Recovering the coin in the pool is simple—just clutch it in your hand and bring it up. But the missing model tug is a bit heavier and could be as much as 6 meters (approx. 20 ft) down. Since it is more massive than the *SeaMATE* ROV can probably recover, the robot will need to attach some sort of line to the tug, so someone in the canoe can bring it to the surface. Certainly, this operation is more complex than retrieving a coin from the pool bottom by hand. The ROV has to be fitted with a means to get a line around the model or to snag it for retrieval. The pilot has to be skilled in piloting the craft to do this job. And the canoe also has to be stabilized, so it doesn't tip when lifting the tug off the bottom to the surface.

3.2.2. Assessing Platform Options

Once you have an idea of probable location for your search operation, you can start thinking about how you might find the target and recover it. Generally this involves an assessment of which might be the most suitable platform to use. As explained earlier, the term "platform" refers to the primary work site that will facilitate getting machinery, personnel, and underwater vehicles in and out of the water for your ops. A support platform can either be fixed (e.g., a pier or drill platform) or mobile (e.g., a rowboat or oceanographic ship). Choosing and deploying the right support platform is crucial to the success of an operation, regardless of its size. (See the *Tech Note: Platforms* earlier in this chapter.)

Most undersea operations are supported by surface vessels. The choice of which type of floating mobile support platform depends on several factors:

- the working depth of the water, which impacts tether length and the need for appropriate winches and cable drums
- the size of the subsea vehicle
- the number of support crew
- the nature of the work to be done
- the necessity of additional equipment to support the operation
- the anticipated sea conditions and duration of the operation
- the location of the operations
- the distance from shore-supported logistics
- the availability and appropriateness of various platforms for the task
- the budget

Image courtesy of Harry Bohm

Figure 11.12: Catamaran-Style Canoes as Support Platforms

Floating support platforms can range in size from a canoe or rigid-hulled craft to a large offshore support vessel outfitted with specialized heavy-duty deck machinery and scientific technical equipment, all operated by specially trained personnel. These specialized support ships can be either "ships of opportunity" (i.e., a system temporarily installed on an available vessel) or a dedicated support platform (i.e., a ship with a permanent installation for an underwater system). This can impact the cost of ship time considerably, especially since the day rate for a major support vessel can easily run into tens of thousands of dollars.

At the other end of the size spectrum, small, inexpensive, humble craft have surprising capabilities. For example, Simon Fraser University's Underwater Research Lab routinely uses canoes for AUV ops in small Canadian lakes. For added stability and greater carrying capacity, two canoes get lashed together like a catamaran (Figure 11.12). In good weather, snowstorms, and even on the ice, these canoes have acted as fully adequate support vessels. Regardless of its size, any support platform should be selected with safety foremost in mind.

Common Support Platforms for Underwater Ops

Shore-Based Work: Using the shore as a support system provides two primary advantages—it is solid, and it doesn't move. Depending on the site, it's even possible to bring in cranes to launch and recover vehicles or set up accommodations, labs, and workshops. The suitability of the shore as a platform is very much dependent on topography, location, accessibility, and sea state.

Image courtesy of Harry Bohm

Figure 11.13: Hand Launching From Shore

Docks: A dock offers several advantages as a support platform, primarily providing access to deeper water from the shore and often allowing watercraft to moor alongside. A crane or ramp positioned on the dock can further facilitate the launch and recovery. A dock can also provide a transition structure for loading a subsea craft and its supplies from a truck or rail car. A small, lightweight ROV like *SeaMATE* is easily launched and operated from a dock. The major disadvantage of a dock is that it can provide access only to waters near the shoreline. Also, docks are affected by changes in waves, currents, and tidal heights, which can complicate loading, launching, and recovery operations if the dock is exposed or experiences big changes in tide at its location.

Image courtesy of Harry Bohm

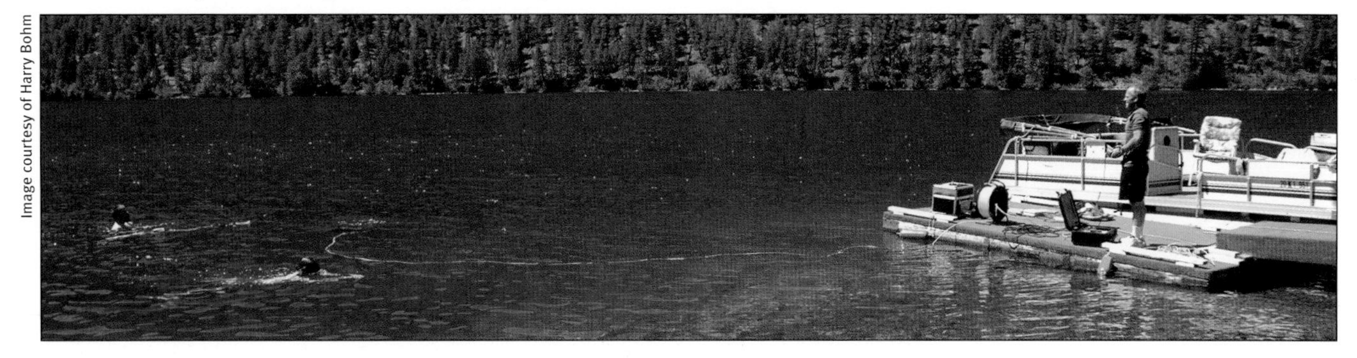

Figure 11.14: Using a Dock as an Ops Platform

Dams, Bridges, and Causeways: Dams, bridges, and causeways can often provide a solid structure right over or adjacent to a worksite. Underwater tasks that are typically done from such platforms include anode replacement, inspections, construction, repair, and recovery of lost objects or persons.

Image courtesy of Transocean

Offshore Oil and Gas Platforms: Offshore oil and gas platforms are man-made islands constructed of steel, concrete, and and rock rubble (called riprap). Their job is to support oil and gas drilling/production, to withstand extreme weather conditions, and to provide a secure and comfortable environment for workers. To accomplish that, these self-contained factories on the ocean are equipped with crew quarters, heli-pads, cranes, fuel storage, generators, and the machinery used in the actual oil production or extraction. Some are located close to shore, but others may be positioned in deep water, hundreds of miles from shore. The two basic types of platform are floating and fixed. Floating platforms are used in depths over 450 meters (approx. 1,500 ft). All offshore platforms rely on divers, ROVs, or AUVs as essential components of their operations.

***Figure 11.15: The Semi-Submersible Drill Rig* Polar Pioneer**

The Polar Pioneer *is a semi-submersible drilling unit, capable of operating in harsh environments and water depths up to 500 meters (1,650 ft).*

Small Craft: Most shallow water ops can be conducted from very small boats, including outboard speedboats, rowboats, sailboats, small cruisers, and rigid-hulled inflatable (RHI) boats. The inflatable has become the small workboat of choice for much underwater work. It is an extremely fast, stable, seaworthy vessel. Being rubber-sided, it can push up against a submersible without causing damage. Its low freeboard means the boat is ideal for deploying and recovering divers. In calm waters, inflatables can even be used for side scan sonar work and missions with small ROVs or AUVs.

Images courtesy of Dr. Steven W. Moore (left), Harry Bohm (middle), and Kongsberg Maritime (right)

Figure 11.16: Typical Small Craft as Ops Platforms

Figure 11.17: Launch and Recovery Transport (LRT) Barge

The Hawaii Undersea Research Lab (HURL) operates its Pisces *submersible in open seas when studying the undersea volcano Lohihi. For years this submersible barge design allowed them to safely launch and recover the submersible in swells as high as 3.5 meters (approx. 12 ft).*

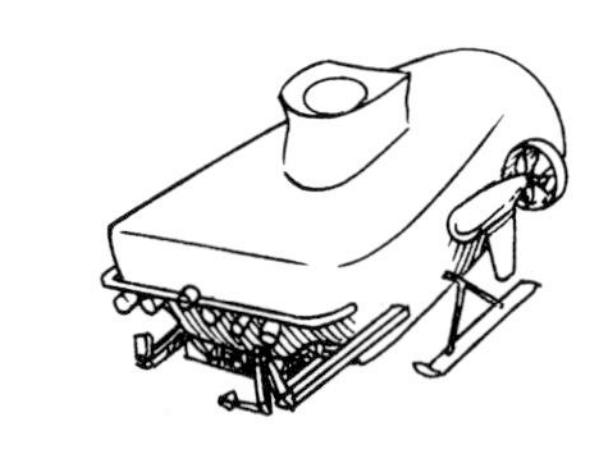

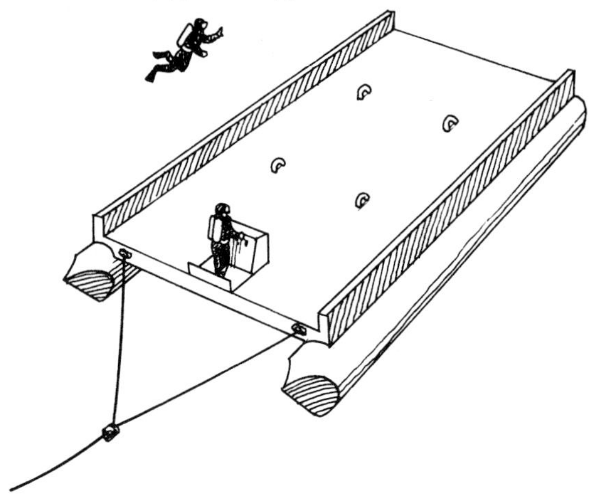

Commercial Fishing Boats and Tugs: Commercial fish boats and tugs are extremely useful vessels to use in undersea operations, because they are versatile and seaworthy. Depending on their size, some vessels have accommodations for crew, some sort of lifting rig, and good fuel endurance. Most are limited in how far offshore they can work, and normally they have no way to anchor in deep water and no dynamic positioning capabilities. Still, they are excellent craft for work in coastal areas.

Barges as Support Platforms: There are many classes of barges, from simple deck barges and ramp barges to large, heavy lift capacity derrick barges. Some barges are self-propelled, but most are towed by tugs. The advantage of a barge is that it can be loaded with enormous amounts of equipment and cranes. A barge can also be equipped with winches to anchor it on site.

Occasionally, specially built submersible barges are put to work in the subsea industry. The Hawaii Undersea Research Lab (HURL) built one called the LRT (Launch and Recovery Transport) that actually went under water to launch and recover submersibles. A cross between a barge, a dry dock, and a wet submersible, the vessel was used to transport HURL's Pisces submersible. To launch the vehicle, divers would open a series of vent valves in the transport barge to flood the ballast tanks. Once the barge was under water and hovering at depth, the divers released the submersible, and the barge resurfaced when compressed air was blown into its ballast tanks. This same procedure was used to recover the submersible.

Drill Ships: Essentially, drill ships are mobile deep water drilling rigs that search for oil, gas, or other resources in very deep water. They use dynamic positioning to stay on site, rather than anchors. These ships have a unique construction to facilitate launch and recovery ops: a hole or "moonpool" is situated in the center of the ship. This moonpool provides open access to the water surface. During drilling operations, the opening is used for lowering the drill pipe, but it has the additional advantage of providing convenient launch and recovery access for ROVs.

Image courtesy of Transocean

***Figure 11.18: The Drill Ship* Discoverer Enterprise**

The Discoverer Enterprise *is the first ultra-deepwater drillship to use Transocean's patented dual-activity drilling technology that saves time by using two drilling systems in a single derrick.*

Oceanographic Ships: Worldwide, there are large, highly specialized ships capable of supporting a wide variety of missions. Companies such as Saipem America charter these ships on a long-term basis for deepwater construction and drill support work with particular ROVs such as Sonsub's 200-hp *Innovator.* Other ships support scientific exploration. In the United States, there is a unique research fleet of more than 25 oceanographic research vessels called the UNOLS fleet. UNOLS stands for the University-National Oceanographic Laboratory System, a consortium of some 57 academic institutions that participate in federally funded oceanographic research. These institutions either operate or use vessels in the fleet. Established in 1971, UNOLS has amassed the largest, most capable fleet of oceanographic research vessels in the world.

Images courtesy of Saipem America Inc.

***Figure 11.19: Saipem America* Chloe Candies *and Sonsub* Innovator**

Submarines and Submersibles: Small undersea craft can even be deployed from submarines and submersibles. Submarines can deploy AUVs and ROVs, as well as torpedoes. Submersibles have also been used to launch ROVs. Perhaps the most famous example of ROV deployment from a submersible was *Jason Jr.*, a small tethered robot launched from *Alvin*, which was able to "fly" its cameras inside the wreck of the *Titanic* and bring back never-before-seen footage of the ill-fated ocean liner. James Cameron also used the *Mir* submersibles to deploy small ROVs to take footage for his motion picture *Titanic* and documentary *Ghosts of the Abyss*. Some submersibles, notably the U.S. Navy's stealthy DSRV fleet (*Mystic* and *Avalon*), were capable of hitching a secret ride on submarines that could take them to their mission sites. This piggyback technique was pioneered by the Japanese during WWII, when their subs carried mini-subs called *Kaiten*.

Unusual Support Designs: SWATH (for Small Waterplane Area Twin Hull) vessels have a twin-hull design, somewhat like a catamaran. But unlike catamarans, these hulls generally stay completely submerged, thereby reducing hull drag and providing good stability. MBARI's SWATH vessel *Western Flyer* supports operations of their deep-diving ROV *Doc Ricketts*. The design is like two submerged tubes with thin legs extending out of the water to carry a big flat deck or platform—somewhat like an oil rig standing on two submarines. It has a moonpool with doors that accordion up and out of the way to expose the water. When closed, the doors form a garage floor for vehicle maintenance.

Figure 11.20: SWATH

This photograph of Monterey Bay Aquarium Research Institute's flagship research vessel Western Flyer *shows the twin hulls that give the SWATH vessel extra stability at sea. MBARI's deep-diving ROV* Doc Ricketts *is launched from a moonpool between the two hulls.*

FLIP (for Floating Instrument Platform) is another extremely unusual support ship design. The 108-meter (approx. 354-ft) long spoon-shaped buoy was built by the U.S. Navy in 1962 and is now operated by the Scripps Institution of Oceanography. The submarine-like hull holds a series of ballast tanks that can be strategically flooded until a portion of the top end of the vessel rises out of the water and the bulk of the structure hangs below the surface, thus providing an extremely stable work platform. *FLIP* has no propulsive power, so must be towed to the site.

Figure 11.21: FLIP

Scripps Institution of Oceanography's FLoating Instrument Platform, *or* FLIP, *conducts sea trials off San Diego in May 2009.*

3.2.3. Developing a Detailed Plan

Once you've completed your basic research, the next step in planning an operation is to sort through your research data and begin to join together all the pieces of the puzzle into a detailed operations plan. Of course, the exact details of any ops plan will depend heavily on the unique character of the mission and what you've learned through your research about the site and the target of the search. Often this plan is best expressed as a series of tangible tasks that you can check off as you complete them.

1. Generate a map showing the search area and the probable location of the target.
2. List all possible search options and identify those that are most promising.
3. Figure out the equipment and personnel needed to conduct such a search.
4. Figure out how to mark the area for navigation.
5. Decide what search grids or techniques to use.
6. Figure out how to mark the target when you find it.
7. Decide how to position the boat or other platform chosen for recovery.
8. List the tools and techniques necessary for recovery.
9. Think about what kind of follow-up will be needed to put closure on the mission.

For an example of an ops plan formulated as a task list, see the *Tech Note: Conway's Pond Ops Tasks.*

No matter how carefully you lay out an operations plan, remember that unexpected things will always happen. So take some time to imagine worst-case scenarios and ensure that your plan includes the time, equipment, and other resources to deal with them. For example, in the case of technical malfunctions you'll want to bring tools for repairs in the field, "how-to" manuals for technical equipment, replacement parts, and spare fuses, batteries, and weights. For unexpected scheduling issues, think about how the ops plan might have to change in case of rain, storms, etc. Don't forget to consider medical emergencies such as a team member who gets the flu and can't make the trip, a crew member who gets hurt on the job, or someone who develops a severe allergic reaction. "Expect the unexpected" is always a good rule to keep in mind; it's one that will help you handle emergencies with a minimum of additional stress.

TECH NOTE: CONWAY'S POND OPS TASKS

Here are the ops tasks synthesized from the information about the lost model tug scenario:

Operations Tasks: Search and Recovery of MV *Zande* in Conway's Pond

1. Produce a map of the pond and plot lines of position, based on eyewitness accounts. Plot a "most probable" primary search area for the tug. Plot secondary search areas, in case tug is not found in the primary area.
2. Although it's unlikely that the tug drifted into shallow water, first paddle the canoe around the shore of the pond to see if this might have happened. Include the area by the bridge where the creek flows out and near the island. If the tug is still not found, begin ops for the underwater search.
3. Set up range markers (Figure 11.22) to help the canoe navigate the search area.
4. Run a search grid to locate the tug.
5. When found, mark the sunken tug's location.
6. Anchor over the tug, positioning the canoe and ROV for recovery phase.
7. Use the ROV to get a lift line attached to the tug.
8. Raise the tug and secure it.
9. Demobilize.

TECH NOTE: CONWAY'S POND OPERATIONS TASK #1—MAPPING THE SEARCH AREA

Find a suitable map or chart of Conway's Pond and plot the most probable position of the lost object, based on the eyewitness accounts. This simple scenario doesn't require any nautical charting or navigation techniques; you can get pretty accurate Lines of Position (LOP) just by using the landmarks on the shore. Start by plotting the LOP given by witness #1; do the same for the rest of the witnesses. The most probable position of the model tug is where the lines intersect. Admittedly, there's some discrepancy in the position given by witness #3, but the other three LOPs agree. So you can safely assume this intersect would be a good place to start your search. If you cannot find the tug in the vicinity of the primary search area, then use the position given by witness #3 as the focal point for your secondary search area. (Review *Tech Note: Using Lines of Position for Triangulation* in Chapter 9, if necessary.)

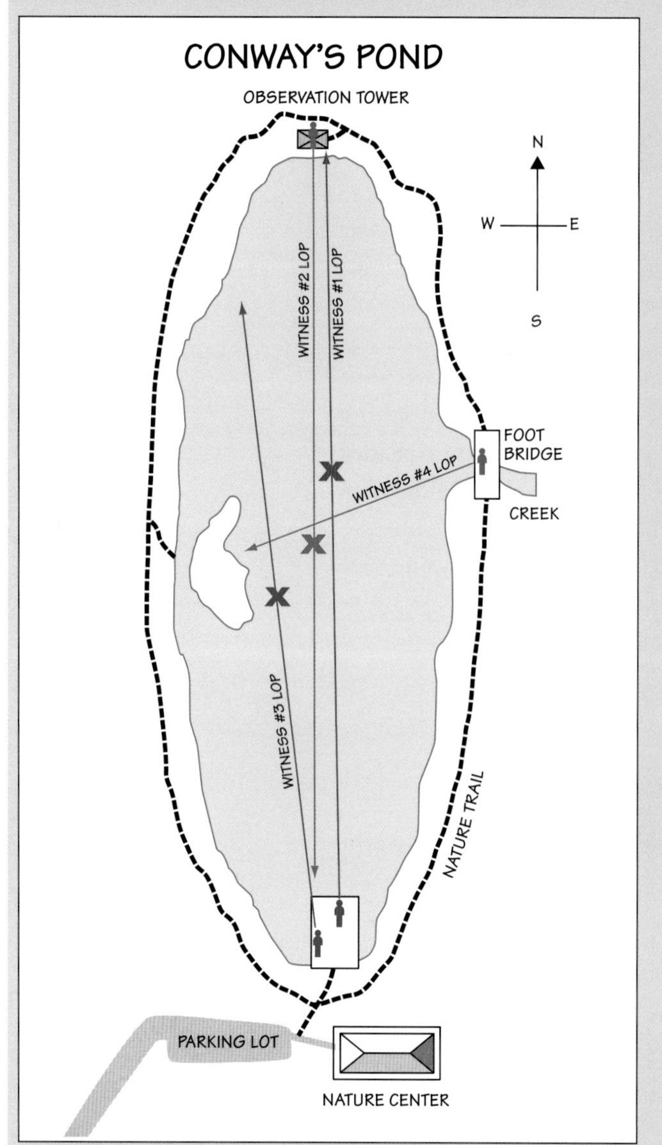

Figure 11.22: Conway's Pond with LOPs for Four Witnesses

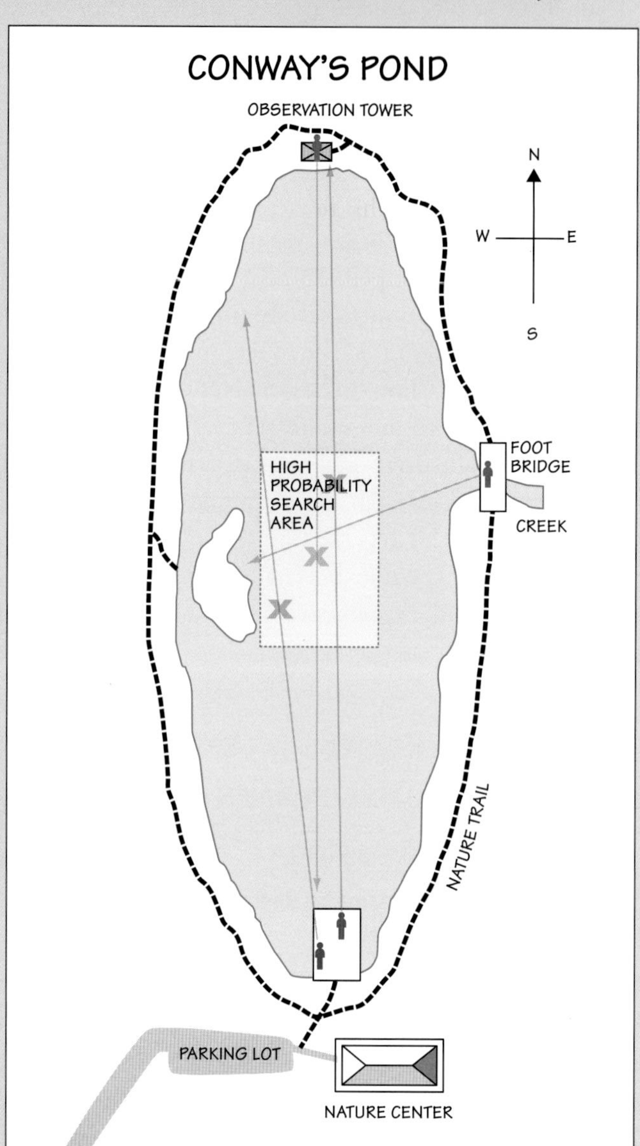

Figure 11.23: Probable Search Area

3.2.4. Preparing to Implement the Plan

Making Lists: Once you've got your general ops plan figured out, it's time to look at each of the requirements you've decided on and begin making lists of the specifics needed to complete each task.

As pointed out in *Chapter 2: Design Toolkit*, a complete list is a most important tool, whether you're designing an underwater robot or organizing the logistics of any operation. Get a small notebook to carry with you at all times. Date the page you are working on. As you think of details you need to take care of or things to bring, write them down before they are forgotten. Compiling these lists may seem tedious, but on-site search and recovery operations can easily get sidetracked by a crucial, missing item. Thinking ahead and paying attention to details are basic to any successful operation. And don't forget to plan for the unexpected!

When planning any mission, it would be ideal to visit the ops site beforehand to get as much actual information about the situation as possible. Often this isn't feasible, so one option is to use the technique of visualization to help you plan. Concentrate on imagining the circumstances at the time of loss—weather, sea conditions, and any other known circumstances. Then run different search scenarios through your mind. Visualize an underwater vehicle traveling along and how it could deal with weather, obstacles, silt, currents, or anything that might impede the mission. You'll also want to compare notes with those who are familiar with the area. Then write down your ideas and list any gear to help deal with various scenarios. Later, you can organize your notes as steps or tasks to be completed, as in this suggested guide:

- **Transportation**: This relates to anything required to move people and supplies to the worksite and back again. In larger ops, this may involve ships, boats, cars or trucks, planes, helicopters, as well as loading/unloading arrangements at the pier. Smaller ops may only require the family car. Remember transport materials, such as strapping and pallets, padding/packing materials, and containers of various sizes, that may be required. Transportation paperwork, such as parking fees, tickets, passports and/or visas, freight arrangements, bills of lading, or customs documents, may also be necessary. Increasing emphasis on security and customs vigilance may necessitate special documentation, declarations, and inspection of gear, especially if you're transporting techie-looking devices with wires and batteries. Plan for contingencies, in case some of the gear is disallowed. Note that batteries are considered as dangerous goods for aircraft conveyance and must be packaged according to the regulations governing transportation of dangerous goods.
- **Communications**: This involves any means needed to communicate with team members and mission supporters (companies, bosses, officials, transportation crews, dockworkers, Coast Guard, weather agencies, etc.) before, during, and after the mission. So think about the various types of communications required (e.g., weather briefings, progress reports, scheduled meetings, emergency calls, etc.), then plan the necessary equipment/methods (radios, telephones, memos, emails, computer communications, satellite communications, pagers, etc.).
- **Crewing**: List all the people, their job titles on your expedition, contact numbers, and next of kin (in case of emergencies). Include medical information if at sea for extended time periods. If certification of competency is required for particular personnel, that paperwork must be checked and filed.
- **Power**: This includes all the equipment required to supply the required power for the underwater vehicle and peripheral equipment. Larger ops may require fuel and generators, which must be installed and grounded according to applicable electrical codes. Make sure there are extra batteries for cameras, flashlights, GPS, portable recording equipment, self-powered scientific sensors like a CTD sensor, as well as "just in case" backup supplies and equipment.
- **Tools**: List and organize all the tools required to diagnose and carry out field adjustments and repairs. Double up on the ones that routinely go "swimming," such as side cutters, pliers, and wrenches.

TECH NOTE: BUDGET HEMORRHAGE

Financial risk is an all too common hazard of any underwater operation. During a 2003 expedition to the Black Sea, Bob Ballard endeavored to excavate ancient deepwater shipwrecks. But that activity came to a crashing stop when his group was denied permission by Turkish government officials to proceed to the operations site, thereby forcing the Woods Hole Oceanographic Institution research ship *Knorr* to sit idle at a pier in Sniop, Turkey.

Things were not looking good. The bureaucratic red tape secured the ship to the dock with more strength than any mooring line could and was costing the expedition $40,000 per day! In a May 2004 *National Geographic* article, Ballard reported, "We're hemorrhaging money."

If the delay had gone much longer, the operation would have had to be cancelled. Fortunately, local authorities who were sympathetic to his plight made the ship's crew "temporary" citizens of Sniop, which nullified the restraining paperwork; two days later, they were allowed to go to sea and proceed with their mission.

Figure 11.24: General Safety Gear

- **Food and Crew Comfort Supplies**: Besides main meals, the crew will need snacks, coffee/tea, and beverages. Be aware of special food requirements of crew and observers, such as vegetarian and diabetic diets. Remember things like hand cream, degreaser soap, laundry soap, dishwashing liquids, cutlery, sunscreen, etc.
- **Special Environmental Clothing**: This does not include personal clothing, but refers instead to items that are expensive and normally not owned by individuals but that are necessary to protect protect crew members so they can do their job at sea. Such items might include foul-weather gear, diving suits, exposure suits, specialized work gloves, etc.
- **Safety Gear**: This category includes special items like PFDs (life jackets), first aid kits, fire extinguishers, GFIs, emergency blankets, stretchers, and the like. Make sure any vessel being chartered has all the requisite Coast Guard safety inspections and equipment, including a qualified crew, if required. Most importantly, ensure that there is an up-to-date crew roster, listing all home addresses and nearest relative, in case someone needs to be notified of a medical emergency.
- **Underwater Vehicle and Associated Equipment**: This is all the gear and special tools needed to get the vehicle into the water, carry out maintenance, and complete the mission.
- **Other Equipment**: You need to plan for all the gear and equipment needed to make the ops happen in normal as well as abnormal circumstances. Such equipment can include cranes, airlifts, air bags, inflatable boats, shackles, etc. Often, different groups in a larger operation (e.g., divers, scientists, and cooks) may require specialty items; it is usually best if they organize this gear themselves then submit the list to the operations supervisor.
- **Budgeting**: A convenient, satisfactory method for keeping track of expenses is absolutely essential.

As you organize your overall list, remember that there will still be items to add and remove. As the list gets longer, devise sub-categories and record specific details. Remember that for any list to be effective, it needs to be detailed, even down to the number, size, and type of screws you might need.

Scheduling: Once you have begun a comprehensive list of what the mission operation will require, then you naturally begin to figure out the scheduling necessary to bring together the equipment, vehicle, people, and ships or boats necessary to run the mission. In some cases, you will already have the experience to know how much lead time will be required for various aspects of your mission plan. Other times, you'll have to research this information. In the end, scheduling is a vital time management tool that provides a framework around which to organize the operation—it includes knowing when to have the crew and gear down to the dock, estimating the time of departure (ETD), figuring transit time to the ops site, and demobilizing after the mission is completed.

TECH NOTE: OPS IMPLEMENTATION LIST FOR CONWAY'S POND

Here's a sample listing of the gear necessary for the Conway's Pond search and recovery scenario:

Transportation:

- Van or truck to move gear and canoe to lake; fill gas tank beforehand
- One canoe with electric motor and three paddles (Note that the electric motor would make running search patterns easier.)
- Plastic totes for packing gear

Communications:

- Fully charged mobile phone(s) and address book and/or or telephone directory
- One set of hand-held walkie-talkies (in case cell phones don't have reception).

Crewing:

- Three people: two in the canoe, with one running the motor (or paddling) and one handling the tether and piloting the ROV; one supervisor on shore to track position of canoe on the pond, set up range markers, and coordinate on-site activities. Use walkie-talkies with adequate range or cell phones (make sure they're charged up) to communicate with each other.
- List of contact information for team

Power:

- 12-volt, deep cycle battery to power electric motor, ROV, and portable depth sounder
- Charger for 12-volt, deep cycle battery
- Battery spares and chargers for flashlight, walkie-talkies, handheld GPS and all battery-powered devices
- Extension cord and power strip (120 VAC power outlet available at parking lot)

Tools:

- Tool box with basic hand tools, including hammer, screwdrivers, pliers, wrenches, and saws
- Battery-powered drill and drill bits
- Soldering iron and solder
- Multimeter
- Electrical tape, masking tape, and various fasteners (including 50 cable ties)
- Scrap pieces of lumber and plywood
- Rags and paper towels

Food and Crew Comfort Supplies:

- Bag lunches and snacks
- Hot/cold beverages, including water

Special Environmental Items:

- Rain gear
- Sunscreen, hats, and sunglasses

Underwater Vehicle and Associated Equipment:

- ROV, tether, control box, and associated cabling to connect to 12-volt battery
- TV monitor or laptop for viewing camera images
- Extra fuses

(Continued on next page)

- Payload tool(s)—some kind of probe that can take a lift line down to the tug, 3 large fish hooks
- Extra weights to orient the ROV to function as a towed drop camera during the search procedure—these will be removed for recovery phase when the ROV is used to attach line to tug
- ROV compass
- Dark-colored tarp to cover ROV pilot and electronics in the canoe in case of brilliant sunshine or rain
- Spare ROV lights

Other Search Equipment:

- 30 meters of ¼" polypropylene line
- 15 meters of ¼" polypropylene line marked off with electrical tape in 3-meter increments, to use for measuring search area as per LOPs on chart
- 6 meters of lightweight chain
- 4 shackles to match chain links
- Map of pond and park trails
- Two pairs of binoculars
- Two handheld magnetic compasses
- Handheld GPS unit (optional)
- Portable depth sounder and transducer pole (or home-made sounding line)
- Clamps for attaching depth sounder to canoe
- Clipboard with clear plastic cover sheet, waterproof paper notebook, and pencils
- Ruler, protractor, and drawing compass; graph paper
- Four range-mark rods made up at shop
- 15-meter waterproof tape measure
- Roll of strong twine, flagging tape
- 3 meters of 14- or 16-gauge solid mild steel or copper wire, bendable by hand and pliers
- Four buoys with 10 meters of line and weights
- Two canoe anchors (Use ¼" poly rope and chain to attach anchors.)

Organizing the Team: Once you've laid out your overall plan, it's time to figure out the people who will make it happen. Generally you want people who can multitask and "think on the fly," particularly when unexpected situations arise that require creative solutions. Often they need to be flexible and able to function in more than one role, if required.

Job responsibilities and titles vary with the scale of the operation and the management structure of whoever is in charge of running it. For example, larger commercial or scientific ops will have a separate mission leader who is responsible for selecting appropriate team members. However, with smaller-scale operations such as a class or club participating in a competition, a teacher often fills the role of organizing and coordinating the team. Regardless of the size of the team, members need to know which equipment or phase of the operation they will be responsible for and what those responsibilities include. At some point, paperwork also becomes important, whether it's ensuring that crew members have proper technical certifications and passports for international jobs or parental permission for school teams.

Another key person for any size mission is the ops supervisor. For large jobs this is often a separate position while for smaller, informal groups, the team captain often functions as the ops supervisor. Regardless of the size of the mission, the ops supervisor needs to be an experienced, organized, detail-oriented leader who's concerned for the well-being and safety of the team on site.

TECH NOTE: THE TEAM FOR CONWAY'S POND

The team breakdown for Conway's Pond is similar to that of the three-person team described in this section, but the jobs differ a bit. The pilot for the ROV has to handle both the tether and pilot the ROV. That's because the boat operator will be too busy keeping a proper heading, running a search grid using the ranges, compass, or GPS, and following directions from the shore-side supervisor. In this scenario, the supervisor would best be positioned on shore, ideally on top of the bird-watching tower, so as to have a bird's-eye view of the pond. From there, the supervisor can use the ranges and compass bearings to track the canoe, keeping it inside the search area, and can relay this information to the boat operator via radio. The supervisor should also document what areas have or have not been searched.

Even a small lake- or sea-going operation operates best with at least three people on the team—a pilot, a supervisor, and a tether handler. The pilot is in charge of operating and piloting the ROV, two jobs that require someone with good concentration and spatial sense. Even so, piloting is always a team effort, and effective communication between the pilot and other team members is critical to a safe and successful mission. Contrary to popular belief, the pilot is *not* in charge of the underwater operation—the supervisor is. That's because the supervisor actually has a better overall view of what is going on, while the pilot is generally preoccupied with the details of operating the robot.

The supervisor's job is to develop a game plan for the mission while working in consultation with the team. Being in charge involves making log entries concerning the mission, time in and out of the water, depth, navigation, etc. In short, the supervisor's job is to oversee mission operations and to piece together a picture of where the ROV has been, is, and will be. On large jobs, the supervisor must also be aware of the ship's position, changing weather conditions, deck crew deployment, and winch operation. Wind and wave heights dictate the decision to launch an underwater vehicle and, more importantly, affect the decision as to when and how to recover a vehicle safely. These conditions can change dramatically during the course of a single dive. The ability of the support platform to handle rough weather, the capacity of the lifting rigs, and the experience of the crew are also factors in this equation. Above all else, safety and minimizing risk are always on the supervisor's mind.

The third member of the ROV team, the tether manager, maintains and deploys the tether. With small craft, this may mean attaching floats to the tether, keeping the tether clear of obstructions near the surface (e.g., propellers), bringing the tether in and out on instruction from the pilot, and making sure the tether is either reeled properly on the winch drum or coiled in a figure eight on the support platform. A good person working the tether enables the pilot to accomplish tasks efficiently and effectively.

3.3. Staging and Mobilization

Packing, or "staging," is the time when all the planning starts to feel real. For small-scale ops with two to ten people, the best approach is to have each person organize whatever gear is necessary for their part of the ops. Then gather all items in a central place, packing as much as possible into plastic or metal totes (Figure 11.25). Totes make loading and offloading much easier; they also protect contents from rain, spray, hard knocks, and loss. For gear that doesn't fit into a box, make sure the items are secured, to make handling as easy as possible. Label everything with an identifying code or a listing of contents. Then put a list of the contents of each box either on a master list or on a waterproof sheet placed inside the tote. Enter each container's identifier on the

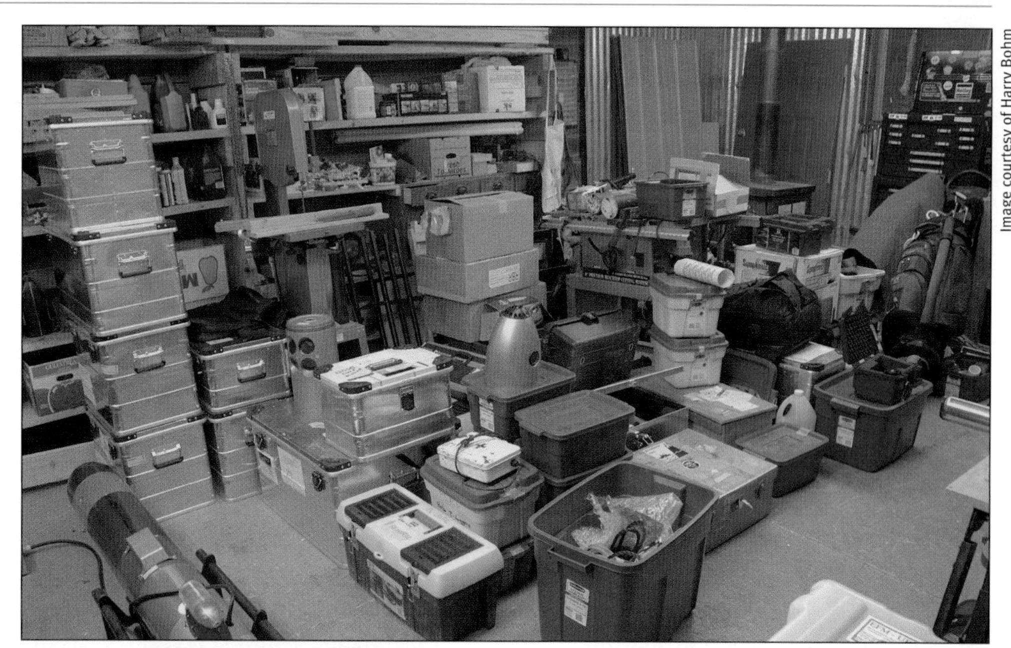

Image courtesy of Harry Bohm

Figure 11.25: Staging Gear and Equipment

Staging a field operation involves organizing gear and stowing them in bins suitable for transport to the site. Here plastic bins commonly used for household storage and tougher watertight equipment cases are used for the organization, storage, and transport of vehicle modules and associated gear.

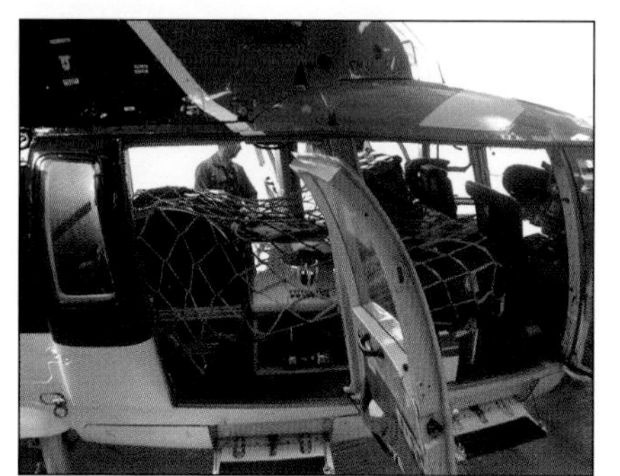

Left and right images courtesy of Harry Bohm; center image courtesy of Remote Presence, Ireland

Figure 11.26: Transport Options

Transporting underwater vehicles and equipment can be as simple as commandeering the family vehicle (left) or as elaborate as hiring a helicopter (center) for a remote delivery. However, the larger the size of the equipment to be transported, as in the photo on the right of OceanWorks' submarine rescue vehicle, the greater the complexity of the move.

master list. Use duct tape to secure boxes and any lids that are not latched. Then stash the totes and other containers in a pile in one place; that way, when it's loading time, you know that all the gear that has to go is in one spot. When it's time to get everything into transport vehicles, use your list and check off items as they are loaded. At the end of the ops, check off each item again as it comes back to the dock or home site.

3.4. On-Site Activities

Once you've arrived at the site and unloaded all the gear, it's tempting to get the ROV in the water and get started with the search. But there are some important tasks to complete first, like setting up navigational markers. However, even before you do that it would make a lot of sense to take your canoe and use your own eyeballs to search everywhere you can easily see the bottom, just in case the tugboat has drifted into some place where you can spot it and retrieve it by hand. (See *Tech Note: Conway's Pond Operations Task #2*).

TECH NOTE: CONWAY'S POND OPERATIONS TASK #2—CHECK THE SHALLOWS

Even before you set up any navigation aids for your search area, use the canoe to paddle around the perimeter of the pond and search for the model tug in the shallows, just in case it drifted off unseen when partially submerged. This option is unlikely, but checking the shallows is easy and eliminates this possibility. One particular area of focus would be around the bridge, where a creek runs out of the pond, creating a current that could tend to carry floating objects in this direction, albeit very slowly. Another might be the shallows near the east side of the island, based on account of Witness #3. If you don't see any evidence of the tug, then proceed to task #3.

Typically, the mission consists of two phases, once all gear, equipment, and personnel get to the ops site. First, there's the search phase activity; once the object is found, the recovery phase gets under way. A search can be conducted in many variations, with the choice depending on location, seabed conditions, station-keeping ability of the support platform, weather, sea state, the size of the search area, target size, the type of sensor gear being used (sensor coverage), and the vehicles that can carry them. Aside from physically launching and recovering vehicles from the platform, the two major ops activities are setting up an underwater navigation system and following a search pattern (grid). All search work is for naught unless you can pinpoint where the underwater vehicle is, or where the surface vessel is, at any given point on the grid. The accuracy of this data is what will yield the highest success rate in locating and, afterward, relocating your target.

TECH NOTE: DEALING WITH MURPHY

Murphy's Law states that if something can go wrong, it will. So it is important that any ops plan be flexible enough to allow for contingencies. These unforeseen situations often require quick, creative thinking. While each team member will bring unique solutions and strategies to critical circumstances, here are some general pearls of wisdom based on firsthand experience:

- Stay calm and focused when things go wrong. This gives your team a sense that you are in control of the situation and that solutions can be found.
- Be patient. It takes time to get things done at sea or even in a lake or pond, and there's always a learning curve at the beginning of an op, even for experienced hands. Take your time, but work steadily.
- Be persistent. Operations can be extremely tedious, with hours, sometimes days, of staring at monitors. Add to this the impact of nasty weather, a touch of seasickness, a dwindling budget, and frustrating setbacks. These are normal for ops. Persistence is the best asset to bring to a mission.
- Pace yourself and the crew. On long ops, it's important to get adequate rest, so establish a watch rotation, if necessary. A tired crew is an inefficient one and makes a lot more mistakes. Save the all-nighters for real emergencies.
- Don't get discouraged. There is always a solution.
- Stay positive. This attitude is infectious. Besides, placing blame on others or making excuses only makes matters worse.
- Get as much ops experience as you can. Going to sea with an experienced crew is the best way to learn about common types of problems and solutions in underwater operations.

3.4.1. Setting Up Navigational Aids

Once you've ruled out any easy options, it's time to get serious about implementing the search. But first you need to determine where to look. This involves taking the proposed search area map you made and transferring that information to the physical ops site so you are definitely working inside the area marked on your map. There are a number of ways to do this, including the use of range markers or GPS to establish lines of position (LOPs). (LOPs and GPS are both introduced in the navigation section of *Chapter 9: Control and Navigation.*)

You may recall that one way to establish a LOP is to line up two objects on shore. These objects are sometimes called **range markers**. If you're real lucky, you'll be able to use pre-existing objects on shore as range markers (Figure 11.29). If not, you can create your own from pieces of PVC pipe or other suitable materials (Figure 11.27).

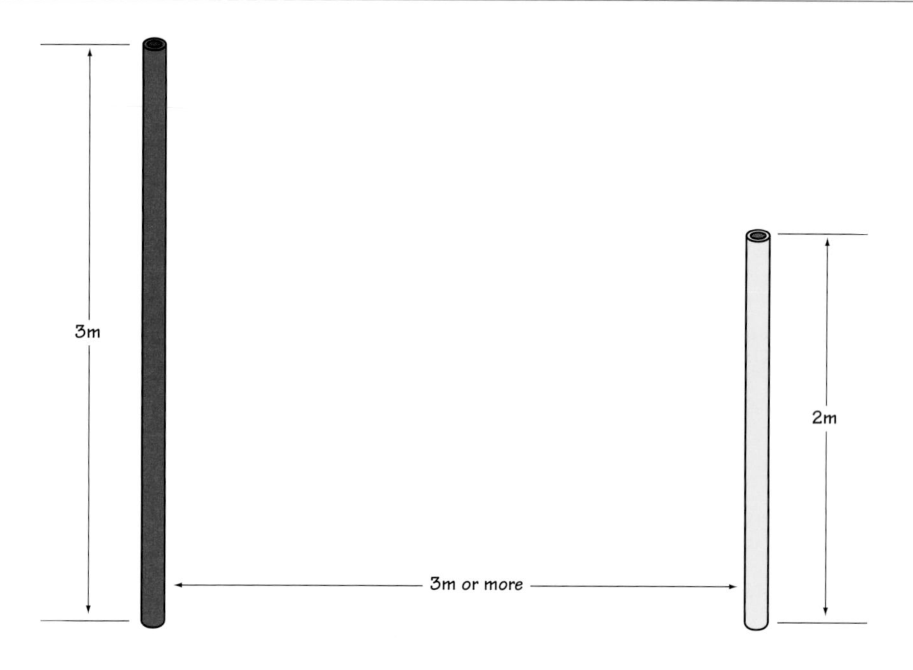

Figure 11.27: Setting Up Range Markers

You can set up your own pairs of range markers on shore to help you determine the lines of position that will mark your search area. Range poles can be made from 1-inch PVC pipe, with the shorter, front pipe that's closer to the water being 2 meters tall, and the taller, rear one that's farther away from the water being about 3 meters or more. (The heights are approximate, depending on circumstances.) The distance separating the 2 poles should be no less than 3 meters (approx. 10 ft).

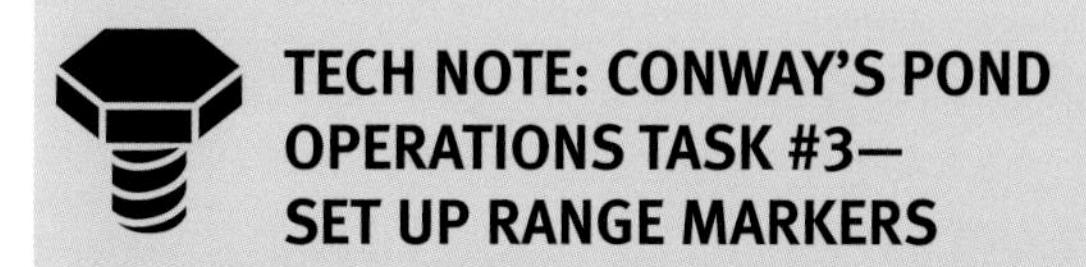

TECH NOTE: CONWAY'S POND OPERATIONS TASK #3— SET UP RANGE MARKERS

Range markers refer to existing or man-made objects on shore that help you establish lines of position. In turn, these LOPS will enable you to establish and mark the search area on the water. In the Conway's Pond example, range markers help you recreate the lines of position provided by the witnesses. For this task, you will want to set up two ranges—one is on the dock that aligns the dock and the observation tower, as given by the LOPs from witnesses #1 and #2; the second is set up on the bridge, aligned to the bridge-island LOP, as observed by witness #4. That done, you can paddle out from the dock, keeping the dock-tower range in line. When the bridge-island range lines up, stop and drop a buoy. This will mark the center of your primary search area. After you drop the buoy, send the ROV down the line for a quick visual check of the immediate area, just in case the model is right there. If it isn't, then pull up the ROV and begin a systematic search.

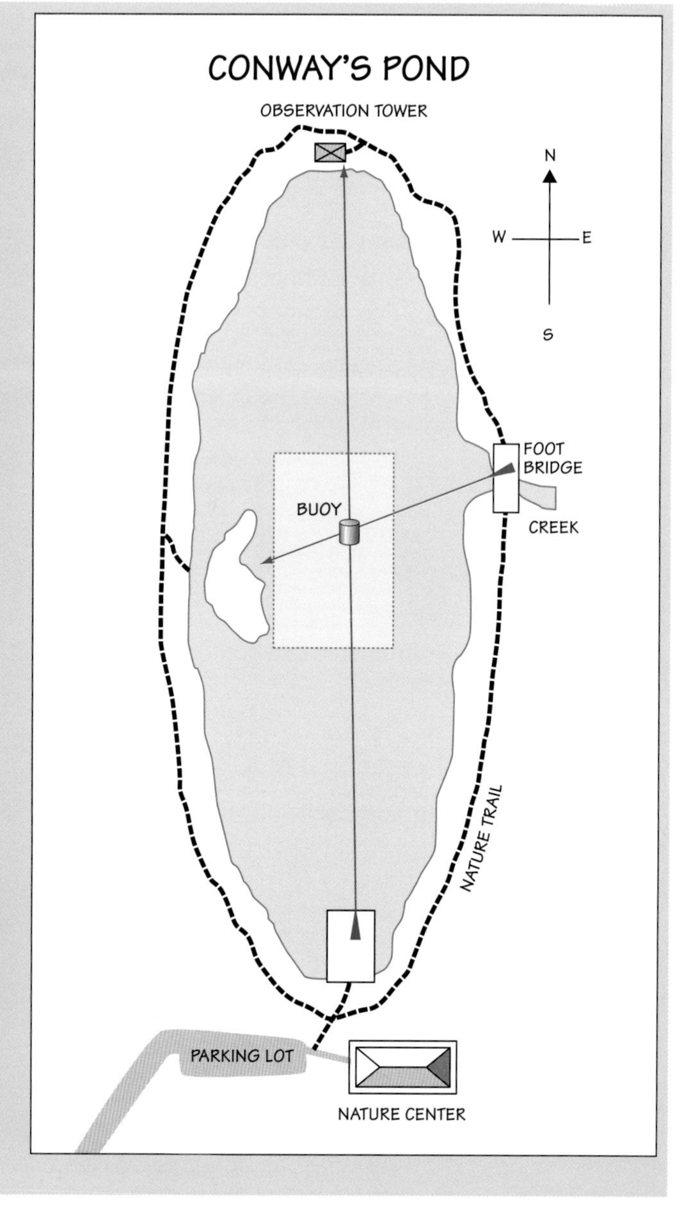

Figure 11.28: Buoy Marking Primary Search Area

Once you've set up range markers on the dock and bridge, next paddle out until the two sets of ranges are each in line. That is the intersection point of the two bearings. Drop a buoy to mark the spot, as shown in this figure; this will make it easier to run your search grid around the intersection point. Alternatively, you can use a GPS instead of ranges to fix the location of your search grid, but in many ways, ranges are much easier to use and more accurate for small-area searches.

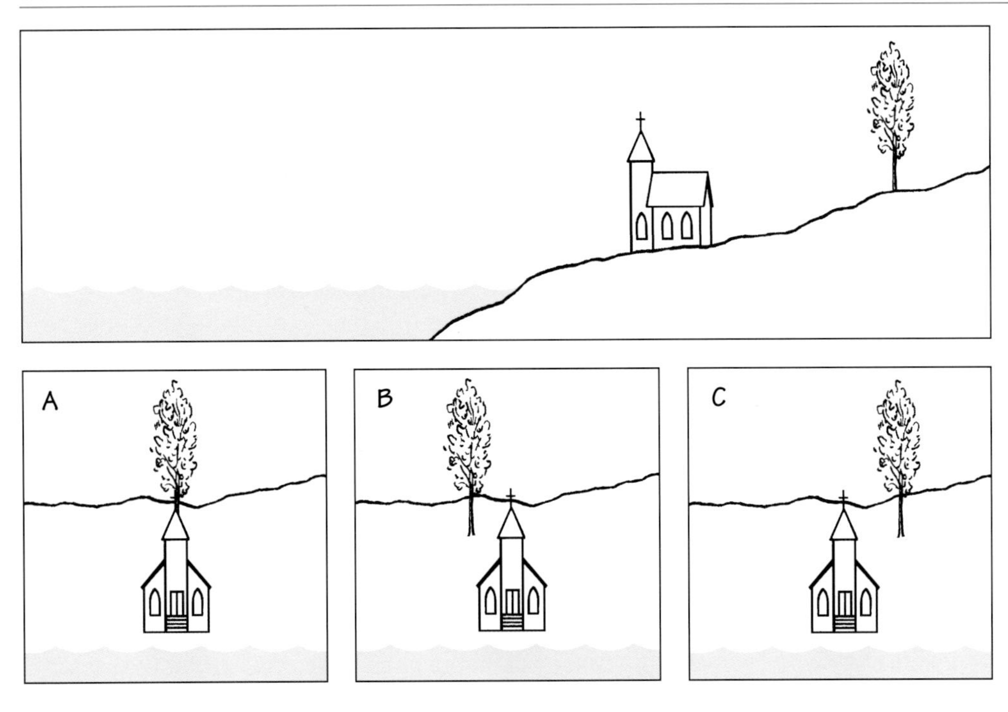

Figure 11.29: Using Range Markers

Choose your range markers so that the taller one is behind the foremost one. If so, you'll be accurately on the range line when the two range markers line up, as in A. If the shorter range marker appears to the right of the taller marker, as in B, then you're to the left of the line. If the shorter tower appears to be to the left of the taller range marker, as in C, then you're to the right of the range line.

3.4.2. Conducting the Search

As the name implies, a search and recovery mission consists of two basic activities. First, there's the search phase. Then when the object is found, the recovery phase gets underway.

You're finally ready to begin the search phase once you've marked the search area, confirmed that the ROV is working properly, and put your boat(s) in the water. A search can be conducted with many variations, depending on location, weather and sea state, the size of the search area, target size, seabed conditions, the ability of the support platform to maintain a constant position, the type of sensor gear being employed, and the vehicles carrying them. Once these issues have been addressed, there is still one outstanding key element in any successful search—technique. This relates to both the pilot's skill with the ROV and to selection of an appropriate, methodical search grid. Both of these issues are detailed below.

Small ROV Navigation and Piloting Techniques: As soon as you're ready to begin looking for the missing target, attention shifts to the person running the ROV. Admittedly, navigating and piloting an ROV is a bit of a black art. But the more you do it, the more you learn, and the more you get a feel for the operation of your vehicle. This is true whether you're piloting a hand-built eyeball ROV or a much larger work class robot.

More complex ROVs are often equipped with sonar, a turn counter, auto-heading, auto-depth, and perhaps an acoustic positioning system. All this additional information makes the navigation task easier. Any small ROV, like *SeaMATE*, is actually more challenging to pilot, especially if you are trying to work and/or navigate in low-visibility situations. Without variable thruster control or auto-depth and -heading functions, you have to get a feel for the vehicle characteristics in order to make it hover or run in a straight line for any distance. Logging hours of practice time is the only way to hone these skills. Below are some of the skills good pilots work to develop and improve.

1. **Spatial Awareness**: The pilot must develop an enhanced sense of spatial awareness from limited artificial imaging sensors. First of all, the images from the ROV's camera are two-dimensional, so they provide very little depth perception. Field of

TECH NOTE: USING A GPS OR A HANDHELD BEARING COMPASS IN SEARCHES

Handheld GPS units are helpful for identifying your approximate position, but their limited resolution (typically plus or minus several meters) makes them less than ideal for running small, camera-based search missions. It may be better to use range markers and buoys to follow your search grid patterns.

However, if you want to practice using your GPS, here are the basic steps for setting up a search area for Conway's Pond with a GPS unit that has a magnetic fluxgate compass feature. (Note: If your GPS has no compass built into the unit, you can always use a standard magnetic compass to take the stationary bearings).

Step 1: Walk out to the end of the dock, turn on the GPS, and let it acquire its satellites. Stand still for at least five minutes to get the most accurate fix possible, then note the position in your notebook. Set it as waypoint "A" on the GPS. This is now your land-based reference point for the search grid. Also note how much accuracy you are getting from the GPS unit—3 meters, 6 meters, 10 meters, etc., remembering that the higher the numbers, the lower your accuracy will be. Anything from 3 to 10 meters is acceptable, but if the numbers are higher than this, you should resort to the range-marker method detailed in Ops Task #3, or use a surveyor's transit to plot positions on the pond.

Step 2: Look for the observation tower at the opposite side of the pond and take a bearing on it with the GPS. Note this bearing. This will be your course to steer when you take the canoe out on the pond to set the buoy at the center of the search area.

Step 3: Now get into the canoe and travel out on the dock-tower bearing. Note the distance traveled. When you get to the location deduced from the intersecting bearings given by witnesses #1, 3, and 4, drop a buoy. This marks the center of your search area. Set a waypoint for this buoy on the GPS and label it as "1."

Step 4: From the center buoy, take a bearing on the bridge and run the canoe in that direction for 30 meters (approx. 100 ft), using the waypoint you set at your center-of-search origin. Drop a buoy. This buoy marks one of the outer edges of the search area. Set this as waypoint "2." Then travel back to the central buoy, take a bearing on the island, travel 30 meters in that direction, and drop another buoy. Set this as waypoint "3." Head back for the center buoy and take a bearing on the tower. Now travel out 30 meters in that direction, drop a buoy, and set this as waypoint "4." Finally, head back to the center buoy and take a bearing on the dock. Travel 30 meters in this direction, drop a buoy, and set this as waypoint "5." You've now delineated a square search area with a center buoy and are ready to run your search grid.

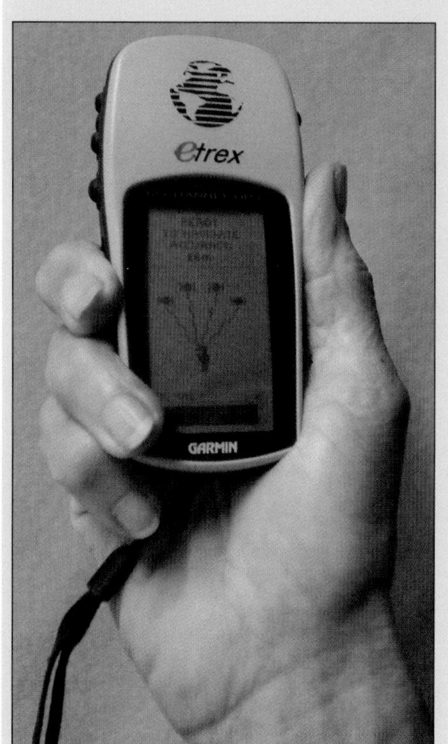

Figure 11.30 (left): Portable GPS

A handheld GPS unit can be used for small boat navigation and running search and recovery ops. Just make sure to bring along extra batteries.

Figure 11.31 (right): Handheld Bearing Compass

Another useful device in search and recovery ops is a handheld bearing compass. It can be used in conjunction with a GPS that does not have a compass feature or as the main tool for determining your bearings in navigating around a search grid.

Images courtesy of Harry Bohm

view is also limited, so a pilot cannot rely on peripheral vision to provide perspective. Low visibility in the water also makes accurate perception difficult.

Besides developing an awareness of what's around the ROV, a pilot must also have a good feel for the actual shape and size of the vehicle, visualizing how long, tall, and wide it is. This is particularly important, since the pilot cannot physically see behind or beside the vehicle. In the case of an ROV that's examining a shipwreck, imagine that the pilot spots a hole in the hull and decides to take the ROV inside for a photographic reconnaissance of the wreck. Using the joystick controls, the pilot nudges the ROV into the opening, but it gets stuck. If the pilot is familiar enough with the vehicle's configuration, he or she may know that craft may have hung up on the skid just out of sight of the camera. That kind of knowledge helps in getting the ROV out of the problem and may even allow the pilot to try another, more successful, angle of entry. Or a good pilot may decide not to enter that wreck's opening even before the ROV gets hung up, having estimated that the size of the entryway is too tight for the ROV to enter safely.

2. **Tether Position**: A good pilot is always aware of the position and length of tether in the water, since the risk of entanglement or entrapment is of constant concern. Clear communication between pilot and tether handler is crucial. Complex ROVs generally have several cameras, including one that looks back and up toward the tether. Others have a rotation counter, so the pilot is aware of making a complete turn. Rotating the tether more than 360° is asking for trouble, since it is easy to stress the cable and make a bight that can hang up the robot. Then there's the sea story of the pilot who accidentally tied a knot in the tether by running a 360° circle, then accidentally piloting the ROV through the middle. This is one maneuver you don't want to have to undo at 600 meters (approx. 2,000 ft).

3. **Handling Characteristics of the Vehicle**: Every underwater vehicle has a character of its own, so the pilot has to learn the distinctiveness of each vehicle. While it might be an overstatement to attribute a personality to an underwater craft, each vehicle responds to the controls in a unique way. Some of these characteristics are:

   - *Stopping distance:* How quickly will the vehicle stop in response to reverse thrust?
   - *Turning bias:* How much bias or skewing occurs when the vehicle is turned counterclockwise (CCW) or clockwise (CW)? ROVs, especially the smaller ones, do not make a perfect circle when they turn. Usually, tether drag is the major cause for the circle becoming skewed or elliptical in shape. So the pilot has to compensate for this turning bias when maneuvering.
   - *Power systems:* Sometimes running multiple thrusters and lights will affect the amount of power available for maneuvering. This may be particularly true of ROVs such as *SeaMATE*, where power distribution and loads have not been designed with a power controller and regulator. In such cases, there may be power drops to the thrusters if more than one is being operated at a time. These power drops will affect turning and stopping rates.
   - *Delays in the command-to-action response:* With practice, a pilot will develop a feel for the inherent delay between a command and the ROV's action response. For example, there may be a noticeable delay between the time when the joystick is pushed forward and the actual motion is executed. Inertia, power levels, and computer or control-circuit timing delays may all contribute to this lag.

- *Tether drag:* The pilot can sense if the tether is biasing ROV movements by the visual clues on the compass and camera. Tether drag is particularly pronounced in high-current areas, and such situations can hinder effective movements of the ROV. A weighty tether can adversely influence a small ROV; Figure 11.32 shows how adding a float can reduce tether weight. Figure 11.34 illustrates how tether length can also impact tether drag—too little cable increases tension and raises the vehicle; too much cable increases drag and pulls the vehicle backward.

Figure 11.32: Using Floats to Reduce Tether Weight

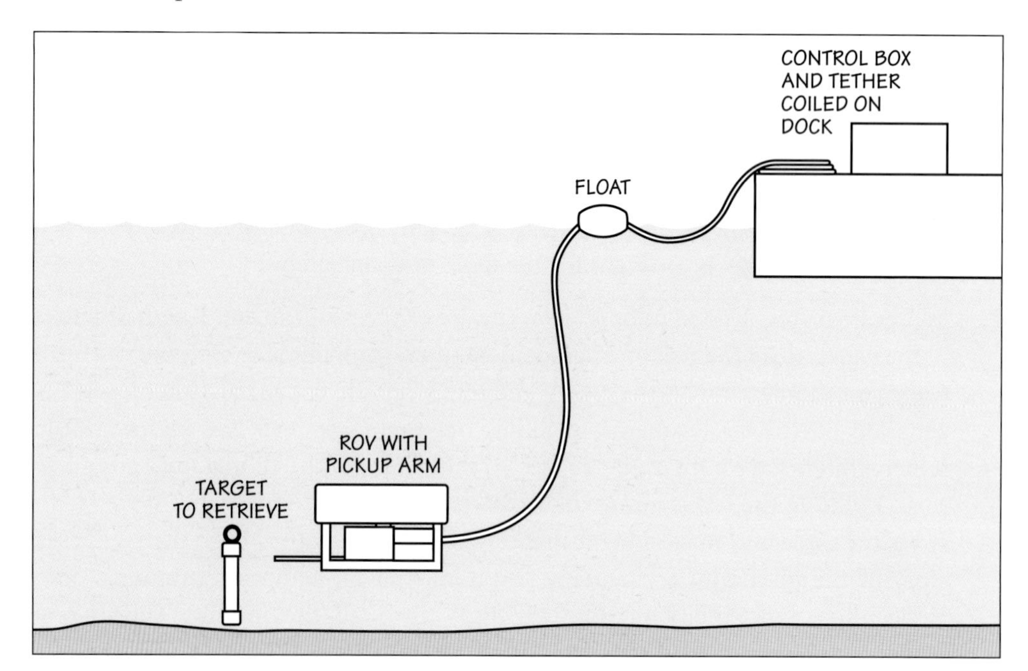

Figure 11.33: Effects of Tether Length

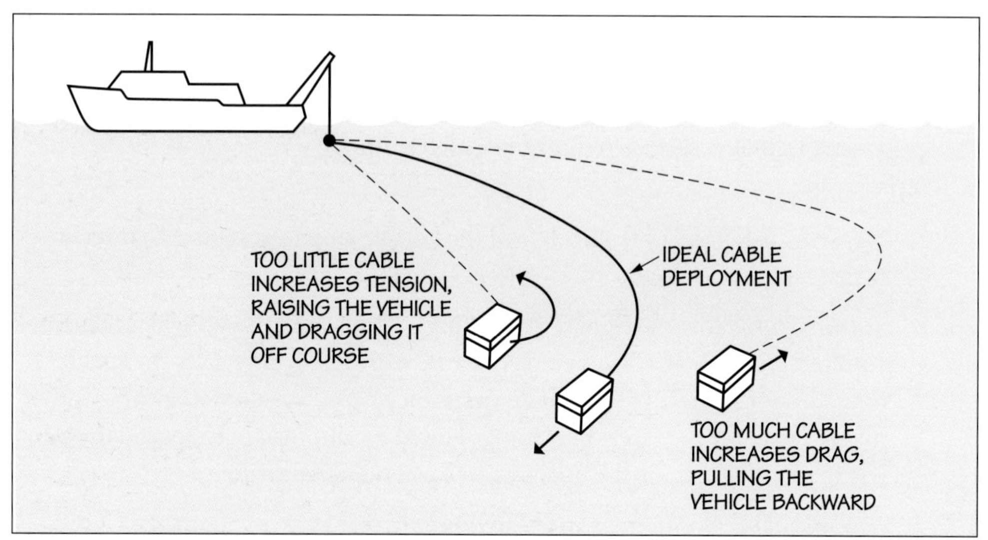

4. **Piloting Tips**: A pre-dive vehicle checklist is not really necessary for a simple *SeaMATE* operation, but it is good practice to devise and run an informal checklist just for practice, since it's mandatory in most commercial and scientific ROV ops. Once you've completed the checklist, it's time to put the ROV in the water utilizing the vehicle's launch and recovery system. The dive commences once the vehicle is detached from its launch equipment and is floating at the surface. Here's a procedure many ROV pilots use for diving a small LCROV without a cage or garage:

   - Before diving, note the compass heading you want the ROV to travel along when it reaches the bottom. Turn the ROV so it faces in that direction.

- Now dive the vehicle, maintaining that heading. Your goal is to descend straight down without moving forward. Generally, the ROV has greater directional accuracy if it descends to the bottom (or to working depth) first, then moves ahead. As you make this descent, watch the particles in the water as they move past the camera lens. In low-visibility situations or in mid-water, ROV pilots use the relative motion of these particles on the screen to help them see if they are descending or ascending. Since there are no other visual reference points in mid-water, even in clear conditions, the easiest way a

SAFETY NOTE: REAL-LIFE SAFETY CHECKS

Science is the driving force of *Alvin*'s missions, but safety is also a significant concern for those on the mother ship, for the swimmers in the water, and for the crew in the submersible. This Safety Note is based on November 16, 2007 log entries by *Alvin*'s long-time chief pilot, BLee Williams.

Before *Alvin* even gets wet each morning, the "Team Atlantis" crew is required to go through a 17-page pre-dive checklist to ensure that the submersible and the A-frame crane that lifts it into and out of the water are safe. As an extra precaution, just before sealing the hatch, the pilot goes over all 17 pages and verifies that each critical item has been checked.

But that's not the end of the safety checks. As soon as *Alvin* is launched, the pilot ensures that the circuits involved with dropping the sub's dive weights (for ease in resurfacing) are functioning; he also checks the circuits that trigger the explosive bolts, which would dump various pieces of equipment in an emergency. Once the pilot has verified these circuits, then he signals the swimmers to remove the ropes supporting the scientific gear mounted on the front of the submersible. Then the connection attaching the "science basket" to the submersible is verified.

Next, the pilot begins a series of checks on *Alvin*'s electrical systems. He confirms that the leak detector is not reporting water in any of the 25 different areas it monitors. He also ensures that the appropriate tracking frequency is set in the Nav (navigation) Box and verifies that all life support systems are functioning properly. Then he checks in with the pilot monitoring the dive on board the mother ship *Atlantis* via the underwater telephone (or UQC).

Once this series of checks is deemed satisfactory (or "sat"), then the pilot instructs the swimmers to "turn on the light, open the valve." This refers to *Alvin*'s Submarine Identification Light, another safety tracking device should the sub surface at night or in poor weather. Opening the valve refers to readying the ballast tanks so the pilot will be able to admit water into them when the sub is ready to begin its dive.

Only now does the pilot radio the ship, checking each item yet again as he recounts its status. "*Atlantis*, *Alvin*; the hatch is shut, the valve is open, the light is on, oxygen's on and the scrubber is scrubbing, tracking is on 12.5, the UQC check was sat, leaks, dumps, and grounds are normal. Request a launch altitude and permission to dive when the swimmers are clear." On board the mother ship *Atlantis*, the surface controller listens to the pilot's report and initials each entry on the Post-Launch/Pre-Dive Checklist. He has already verified water depth (2,500 meters—approx. 1.5 mi) for the day. Now, if all conditions for making a dive are within the operating envelope, the Surface Controller replies, "*Alvin*, *Atlantis*; your launch altitude is 2,500 meters; you are clear to dive when the swimmers are clear."

All that remains is for the swimmers to verify that they are back in the support boat *Avon*. Once this radio verification is received, the pilot responds, "Roger *Avon*, this is *Alvin* diving . . . " Finally, he vents the ballast tanks, and *Alvin* begins yet another descent to explore the deep ocean.

Image courtesy of Woods Hole Oceanographic Institution

Figure 11.34:* Alvin *with Swimmers

pilot can assess whether the ROV is progressing up or down is by this relative motion. Changes in depth gauge readings provide additional info. Once the ROV is on the bottom, you may have to wait for the silt to clear. Then you can re-orient the vehicle to the desired heading and start your search pattern.

ROVs are notorious for not being able to travel in a straight line. Larger ROVs get around this by using autopilot mode. Since micro ROVs generally don't have one, try this procedure for traveling in a straight line along the bottom:

- Once the ROV is on the bottom, orient it towards the desired compass heading.
- Watching the compass, pick up a bottom landmark just at the edge of the video image on the screen that is in the direction of the heading. Lift the ROV slightly off the bottom by reducing the vertical thruster.

TECH NOTE: SECTOR SCAN SONAR NAVIGATION

Sector scanning sonar really simplifies piloting an ROV in low-visibility situations. In fact, it is often the only viable method for navigating under water. The procedure for piloting an ROV equipped with sector scanning sonar is similar to the technique of traveling in a straight line along the bottom, except that the sonar picture provides the target to aim for. Here's a review of the procedure:

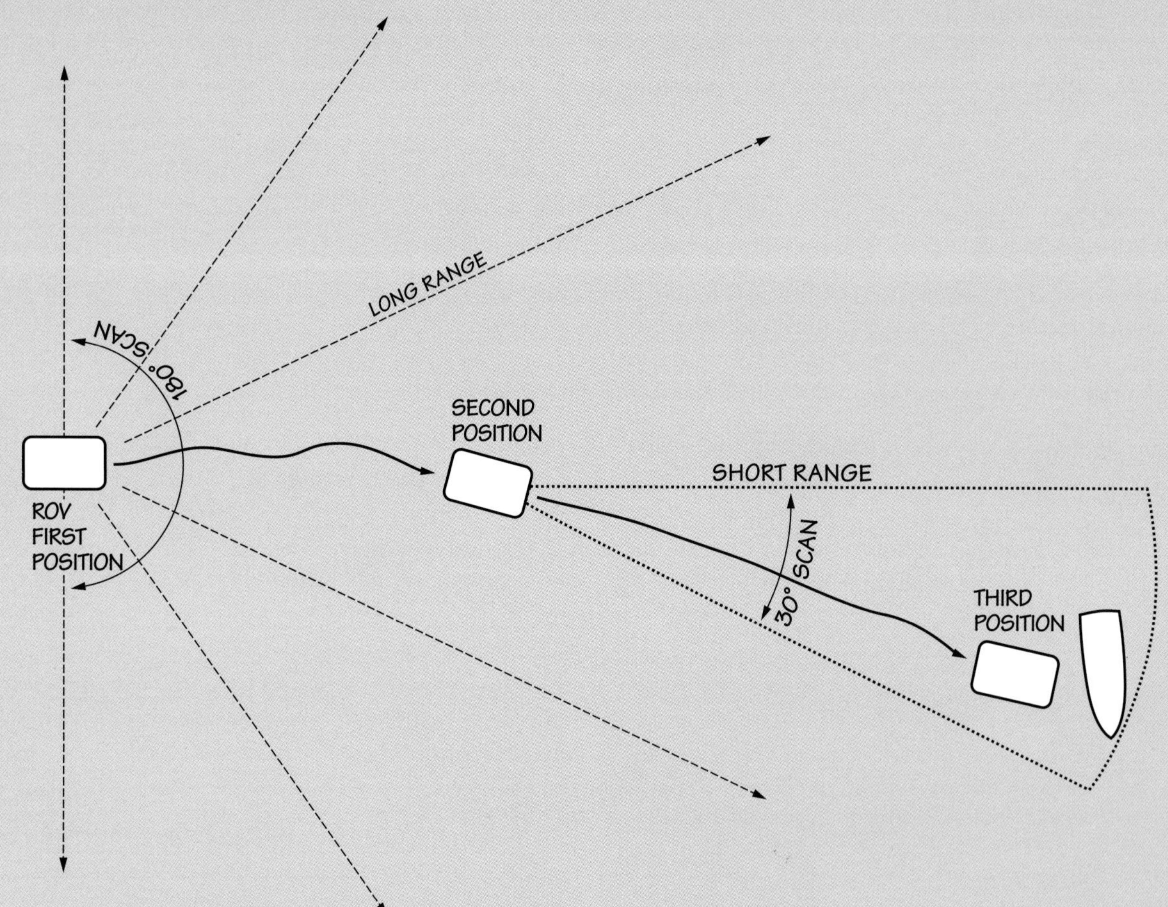

Figure 11.35: Target Location with Sector Scanning Sonar

In the first position shown, the ROV uses a wide angle, long-range scan to locate the target, take a compass bearing, and follow it toward the target. In the second position, the ROV uses a narrower, short-range scan to image the target, set a near bearing, and follow it closer to the target. The third position utilizes a very short range scan; in fact, the ROV may even be close enough that its camera can actually see the target.

- Drive ahead to the object. Once you reach it, apply down-thrust and settle on the bottom. To avoid stirring up silt, use only enough vertical thruster power to stop the ROV.
- Now check the heading and correct your course by spinning the vehicle slowly until it is pointing in the desired direction of travel. Pick another spot at the limit of visibility and move to it. Repeat as needed to reach your destination.

In murky water, the pilot has to rely on the compass a lot more to keep the ROV traveling in a straight line. For best results, keep the duration of each run short, and correct your compass heading often. Sometimes, the particles in the water can give you some clue as to whether you are moving off course. If the relative motion of the particles advances to the left or right of the screen, then the vehicle is turning. If it's going on a straight course, the particles will seem to come right at you.

Running the vehicle through cloudy silt makes for fatigued and wandering eyes. And if you watch only the compass, it's easy to wander in a circle—a phenomenon called "chasing the compass." This is particularly true with quick-turning ROVs, since it takes a while for a magnetic compass to catch up to the actual movement the vehicle has already completed.

Common Search Grids: No matter how skilled the pilot is, a haphazard search or one that relies entirely on hunches runs the risk of missing the area where the target is hiding. A better approach is to use a search grid. A search grid is basically just a very systematic exploration of the bottom terrain, using a sensor (or a group of them) that is either mounted on the underwater vehicle or the surface vessel. This search procedure tracks a specified area, or grid, so that you can confidently determine if the object you are looking for is or is not there. To date, a search grid is the most effective technique, when coupled with available sensor technology, to find things that have been lost. Fortunately, several good search grid options are listed below. You'll want to evaluate which makes the most sense for the type of sensor(s) you're using and for your particular ops scenario.

1. **Mowing-the-Lawn Search Grid**: Just as a lawnmower cuts a series of parallel swaths through the grass, this aptly named pattern consists of a series of long, straight runs, called "legs," by a vessel or underwater vehicle. The sensor is either towed or mounted directly on the craft as it makes a transit along a specified heading for the length of the search area. At the end of each leg, the vessel or underwater vehicle makes a U-turn and proceeds down another leg that is parallel to the first one, as shown in Figure. 11.36. The goal is to run all of the legs systematically, until the entire search area has been covered. This basic search grid pattern works well with most sensors.

   The distance between legs on the grid should be based on the range of the sensor(s) you are using for the search. For example, if you are slowly towing a downward-looking drop camera that can see 1 meter (about 3 feet) to either side of center, then it scans a swath 2 meters wide. Theoretically, then, the distance separating adjacent legs in the grid could be 2 meters apart. However, that is theory; reality is quite different. Wind, current, waves, and even

Figure 11.36: The Mowing-the-Lawn Search Grid

A search vehicle running a "mowing-the-lawn" search grid travels back and forth along a series of parallel paths, called "legs."

However, no vessel (especially a towed vessel, like the camera sled in the drawing) travels in a perfectly straight line; therefore, the searched areas on neighboring legs should overlap to ensure that there are no missed sections or "gaps" in the search coverage.

If there are weather or current issues that make it harder to travel in a straight line, the overlap area should be increased even further to avoid gaps.

Note that side scan sonar may require greater overlap. See Tech Note: Additional Overlap Considerations for Sonar Searches.

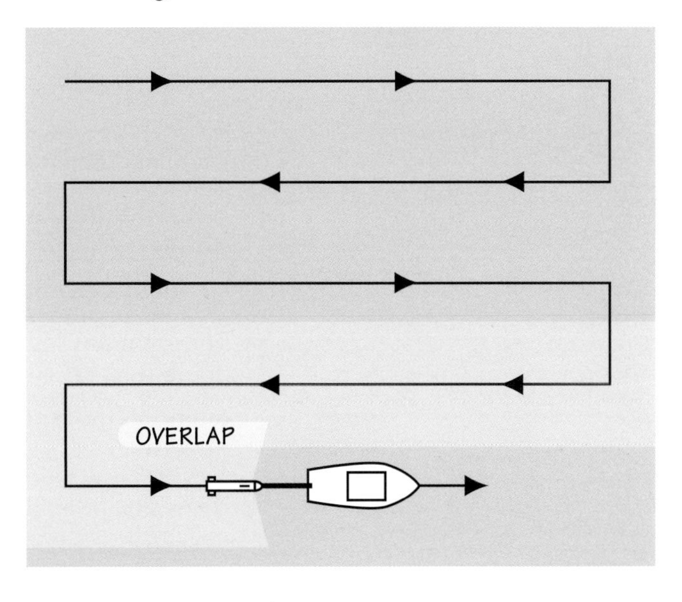

sloppy steering will force a search craft to deviate from its intended straight-line course while on a leg and will necessitate frequent course corrections. In other words the “straight” legs will actually be slightly wiggly lines, and the distance separating them will vary.

To compensate for this, you’ll want to decrease the spacing between the legs, so that on the next pass, the camera (or other sensor) view overlaps the pervious search area enough that no gaps are formed, even with the navigation wiggles.

The general rule of thumb is that the overlap should be about 25 percent of the side-to-side range of the sensor. Hence, in the 2-meter wide camera view example, the overlap between legs should be 0.5 meters (25% of 2 meters). This means that distance separating the centers of the two legs should be 1.5 meters, not 2 meters. This overlap percentage should be increased if sea conditions are affecting the quality of the sensor returns or are causing dramatic course corrections.

Most search sensors used for searches function best when towed in an even, steady manner over a flat bottom. Problems in running a mowing-the-lawn search grid not only occur with side-to-side sensor displacement, but also with up-and-down motion caused by bottom topography or rough seas. Heavy seas and irregular terrain can force you to increase and decrease the camera altitude continually. Even if the terrain changes aren’t enough to make you alter the camera depth, changes in distance between the camera and the bottom can change the width of the observed swath. The shorter the distance between camera and bottom, the narrower the observed swath. These fluctuations in swath width can cause blind spots or distortions in the sensor information.

Here are some practical tips to consider when running a mowing-the-lawn search grid in a small area like Conway’s Pond.

- When working the grid with a drop camera arrangement, keep in mind the amount of overlap you’ll need. Cameras will cover only a small area per pass, especially in limited visibility.
- Pay special attention to the course you are steering and try not to deviate from the track.
- Turns can be problematic; your goal is to make a smooth, even turn.
- Use the buoys and range markers (or GPS) as an aid to stay on the correct leg of the grid. Note that the precision of GPS is suited to large, sonar-based searches.
- When you do locate the object, use a buoy to mark the spot. If using a GPS, set the waypoint. As a backup, take note of the position relative to the other landmarks and buoys.
- Watch the drop camera towing speed. If you go too quickly, the camera will rise off the bottom. If you do want to go faster, either add more weight or let out more tether. Just remember, there are limits to speed.
- The optimum sensor speed limit for a camera is determined by visibility and the altitude off the bottom. The greater the visibility, the higher the camera can be off the bottom, and the faster you can travel.
- The size of the object you are searching for also dictates the speed of towing. If it is small, you’ll need to get closer to resolve the object and hence need to tow more slowly.

TECH NOTE: CONWAY'S POND OPERATIONS TASK #4— RUN A SEARCH GRID

Once the search area is outlined, ask yourself what type of search grid would be most effective to run. Several factors will influence this decision: the nature of the bottom, water visibility, the camera's field of view, and the length of the ROV tether.

In this scenario, the underwater terrain is primarily a soft mud bottom that slopes gently to maximum depth near the middle of the pond. There's some debris, such as branches and fallen trees, which could present a snagging hazard to an ROV being towed through the pond. This is a challenge, but untangling from debris like this is not impossible. Often, backtracking the boat along the path just covered will free up the vehicle. As well, branches that have been in the water awhile are weakened by decomposition and may snap easily if the snagged ROV is retrieved by pulling it up by its tether.

The muddy bottom also poses a real problem. Running an ROV around on the bottom with thrusters going full out will definitely stir up silt, thereby clouding the camera's view, so it's best to run your search without touching the bottom. The length of tether also limits search options. If the tether is 10 meters long and the search depth is 6 meters, you've got only an extra 4 meters of slack to travel along the bottom, and that's not very much for conducting a search.

The solution to the debris, silt, and tether-length problems may be to rig the ROV to function like a drop camera during the search phase, towing it around in a systematic pattern. To do this, weight the ROV so it hangs pretty much straight down, with the camera pointing at the bottom, as in Figure 11.37. Then use the canoe to tow the ROV very slowly in a search-grid pattern, keeping it about 1–1.5 meters off the bottom. This technique works even better in higher visibility situations, since keeping the ROV higher off the bottom generates a larger field of view. However in 2-meter visibility, 1.5 meters is about the maximum altitude. Fortunately, that is enough for this search area, particularly with such a good cross-bearing from the LOPs. Remember, you cannot travel quickly, or the hydrodynamics of moving water acting on the vehicle will cause the ROV to lift off the bottom. Using the heaviest weights possible will help keep the ROV down and may allow you to increase the towing speed slightly, without generating significant lift. (See *Common Search Grids* in Section 3.4.2.)

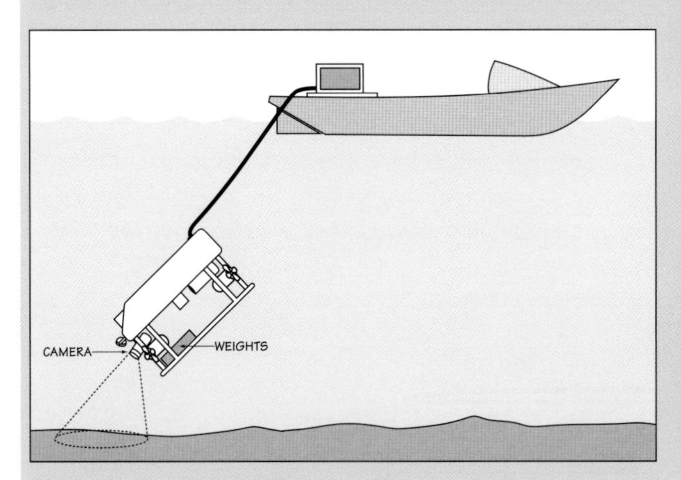

Figure 11.37: Deploying the ROV as a Drop Camera

Adding a couple of pounds of lead weight to the front end of a small ROV should allow it to function like a drop camera. Note that this technique will not work well if you're towing faster than .5m/sec, since the tether and ROV will create drag forces at this speed that will easily overcome any weights you've added, and the camera will no longer point at the bottom.

2. **Arc (Circular) Pattern from a Dock or Anchored Platform**: Mowing-the-lawn is only one of several effective search pattern grids. It is probably the most common search pattern used by large vessels searching vast areas with towed camera sleds, side scan sonar, or multibeam sonar searches. However, these vessels are commonly equipped with advanced navigational systems for precise above-water and below-water navigation.

   A very simple ROV that does not have a sophisticated navigation system may have better luck with a couple of other search patterns: an arc grid and a straight-run grid. These are particularly effective for a SeaMATE-type ROV operating from a stationary platform, such as a dock or a securely anchored boat. Both of these search patterns incorporate the initial rigging setup detailed in the *Tech Note: Rigging Setup for a Shallow Water Search Pattern.*

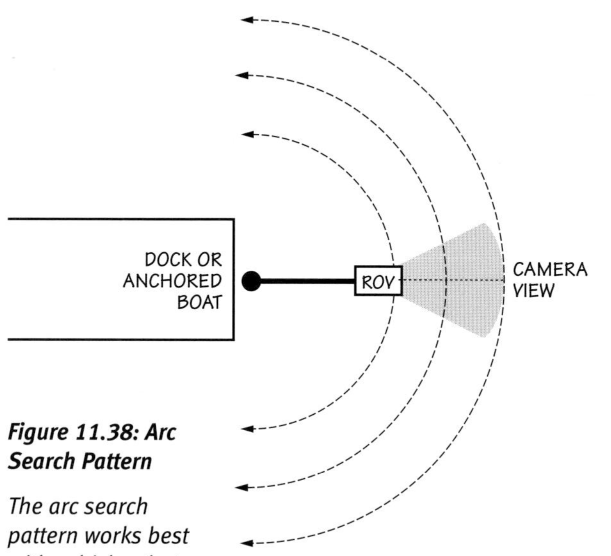

Figure 11.38: Arc Search Pattern

The arc search pattern works best with vehicles that can pull straight ahead and sideways ("crabwalk") at the same time. This keeps tension on the tether while the ROV moves sideways along the arc. It may also be helpful to mount the ROV's camera sideways so it's looking along the arc instead of across it.

To run arc search grid, begin by rigging the ROV with an eyebolt on a weighted block, as shown in Figure 11.41.

- Pilot the ROV in a series of arcs of ever-increasing radii from the eyebolt block, Begin by holding the tether and driving the vehicle over the bottom in an arc. At the end of the arc, let out more tether so that the next arc is further out (i.e., has a larger radius). Increase each arc by the increment of your choice. Visibility determines the amount of tether you let out, so if visibility conditions are really bad, use smaller increments. And always make sure there's some overlap with the area you just covered in the previous arc.

- Hopefully, you will have spotted the target before you've reached the end of your tether length. If not, pull the ROV back and move the eyebolt and weighted block to another locating, making sure to overlap with the area you just covered to ensure you don't miss the target. Start another arc search grid.

- Sketch a small map of all the areas you have covered to you know what has been searched and what has not.

- Note that as you conduct the arc search pattern, the ROV tether may actually snag or sweep the object. This is a good thing if the target object is heavily weighted on the bottom, because as you pull back on the tether, the ROV will

TECH NOTE: RIGGING SETUP FOR A SHALLOW WATER SEARCH PATTERN

A very simple ROV that does not have a sophisticated navigation system requires accurate positioning for the search grid to be effective. The rigging setup shown in Figure 11.41 and detailed here helps maintain accurate positioning for a small ROV, regardless of whether you're running an arc pattern or a straight-run pattern. The key is to operate the ROV from a stationary platform (e.g., a dock or securely anchored boat) in order to execute an accurate shallow water search.

- Run the ROV tether through an eyebolt that is attached to a weight.
- Drop the weight to the bottom directly below the platform you will be working from.
- Measure and note the water depth.
- Take up the slack in the tether until the ROV is tight against the eyebolt.
- Mark the tether with a piece of electrical tape just where it touches the surface of the water.
- Work back from this point, marking the tether above the waterline with electrical tape at intervals of 30 cm (approx. 1 ft).
- Now you're ready to begin piloting the ROV in a series of arcs (arc pattern) or a series of radial lines (straight-run pattern).

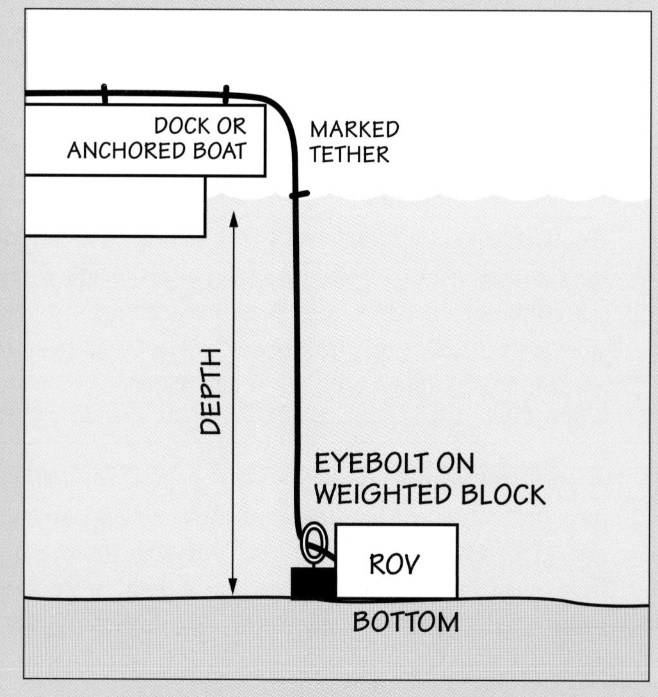

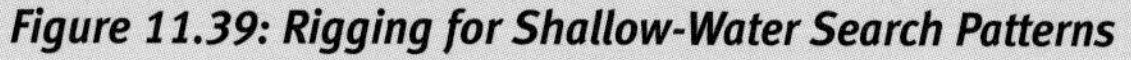

Figure 11.39: Rigging for Shallow-Water Search Patterns

be dragged back to the object. However, if the object is light, the tether may drag it away and complicate the search.

The arc search grid works best with an ROV that can crabwalk (i.e., travel sideways while looking straight ahead). However, you can still use an arc grid with a simpler 3-thruster ROV, such as *SeaMATE*, by attaching the tether to the side of the vehicle and then driving it forward around the arcs.

3. **Straight-Run (Radial) Search from a Dock or Anchored Platform**: When underwater visibility is good, you may be able to conduct a straight-line search. Note that ROVs have difficulty moving in a straight line, so it pays to practice doing this. Here's the technique for a systematic linear search radiating out from a dock or anchored platform:

- Rig the ROV to a weighted eyebolt block, as shown in Figure 11.39.
- Run the ROV away from the eyebolt on a compass heading to the perimeter of the search area. Then retrieve the ROV over the same course simply by pulling in on the tether.
- Choose another compass heading, perhaps 5-10° from the first leg, ensuring there will be overlap out near the periphery. Make another run out, then retrieve the ROV back to the dock. Repeat this until you have covered the search area or found the object.
- If you still haven't spotted the target, move the eyebolt and weight to a new location and try again.

It's important to note that with this type of search pattern, the change in course heading is determined by the visibility and by the size of the object you are searching for. For a small object such as a model tugboat, a 5-10 degree course change could mean you'd easily miss the object. That's because every degree of change separates the center of each leg by 1.74 meters at 100 meters distance. So a 5° change in course heading equates to 8.5 meters and double that for a 10° degree course change.

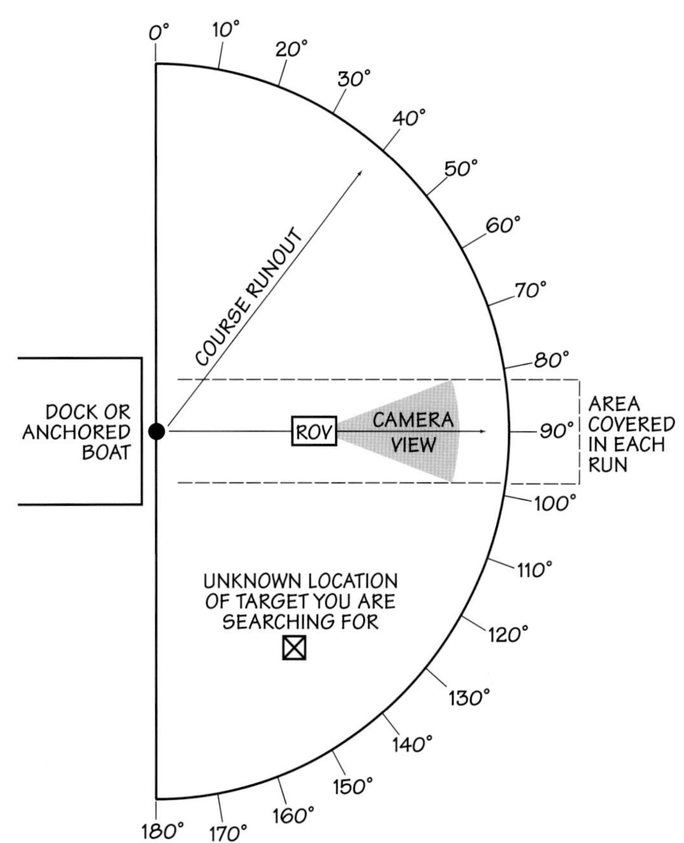

Figure 11.40: Straight-Run Search

TECH NOTE: SEARCH GRIDS AND PULLING ON THE TETHER

Note that shallow water ROVs taking part in competitions in clear water conditions generally do not require the use of search patterns. The rules may also not allow pulling on the tether, to encourage students to get additional piloting experience.

However, when building a simple ROV such as *SeaMATE*, the use of a tether strain relief ensures that you can lift the lightweight ROV without damage to its tether. This allows you to retrieve the ROV by pulling on the tether, a technique that is particularly effective when navigating in murky water. Pulling back on the tether is a fast retrieval method and also allows the pilot to know exactly which course the ROV is returning on when using a weighted eyebolt block.

TECH NOTE: SENSOR LIMITATIONS

Every type of sensor has limitations in what it can detect. Some of these limitations are obvious. For example, most pressure sensors do a predictably lousy job of measuring light intensity. But other limitations are more subtle and more likely to get you into trouble.

Consider limitations in sonar resolution. **Resolution** refers to the smallest details that can be distinguished in any image, whether from sonar, a light-based camera, or any other image-forming device. Sonar is a fabulous tool for scanning large areas of the bottom to find objects lost in deep, murky water—a task for which a traditional light-based camera is nearly useless. But sonar images typically have much lower resolution than photographic images, so if you're not careful (and not experienced at reading sonar returns) you can misinterpret a sonar image and think an object wasn't there, when in fact it was.

Resolution problems can't be solved just by enlarging the image; beyond some point, all you get is a bigger blur. Therefore it's important to understand exactly what resolution is and to know what factors affect it. The two yellow-toned pictures in Figure 11.41 are side scan sonar images of a submerged bicycle. (The inset one is a magnified portion of the larger image.) The other inset is a photograph of a different bicycle set up on pavement (in air, not water) to create optical shadows resembling the acoustic shadows visible in the sonar image. The yardstick provides a size scale. Note that the camera image reveals more detail than the sonar image—you can even make out spokes on the wheels. And it's not just because the bicycle in the camera image is bigger. If you magnified the sonar image until the bicycle in it appeared exactly the same size as in the photograph, you still could not see the spokes. That level of detail simply isn't present in the original sonar image. Additional magnification would just give you a bigger blur where the spokes should be.

With sonar, many things can be done to improve resolution, but each has its drawbacks. For example, you can move the sensor closer to the sea floor, but that reduces the area covered on each pass, so you need to spend more time at sea to complete the survey. Or you can increase the frequency of the sound used, but higher frequency sounds don't travel as far through water as lower frequency sounds do. It's important to be aware of these and related sensor issues, so you can optimize your search without risk of missing something you're looking for.

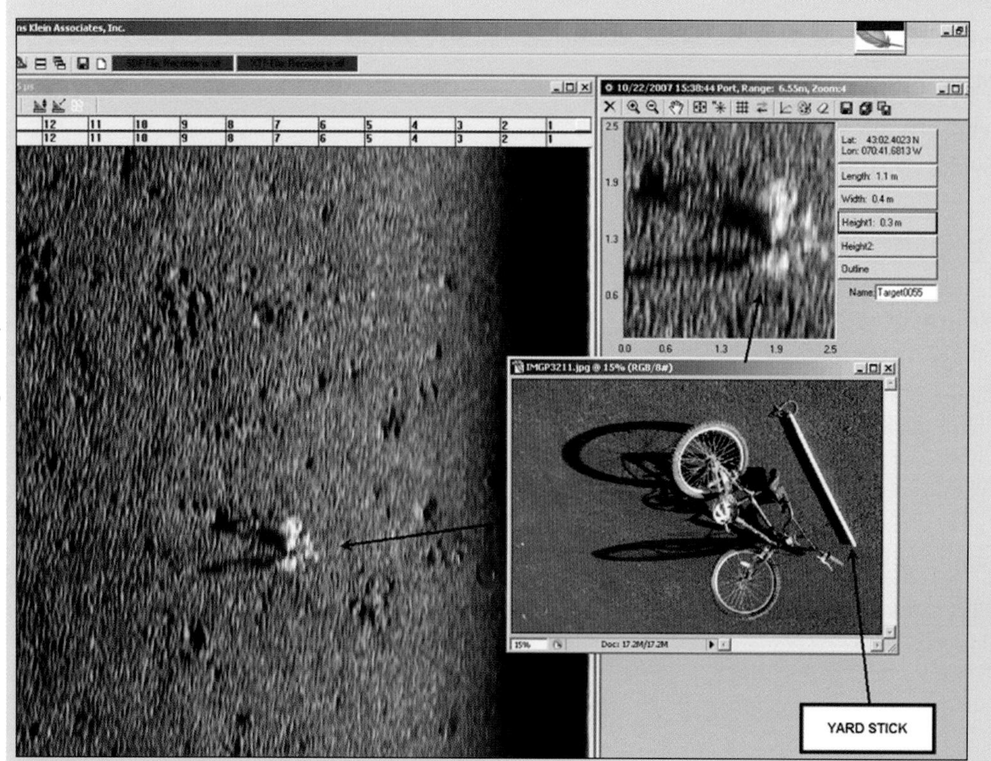

Image courtesy of L-3 Communications Klein Associates, Inc.

Surveys done with video cameras on ROVs or tow sleds also have limitations. If you are too far away from the bottom, the murky water blocks your view. If you move closer, you need to slow down or things on the screen whiz by so fast that you can't correctly identify them.

Figure 11.41: Image Resolution

TECH NOTE: MAKING THE BEST OF SENSOR LIMITATIONS

Often, it is best to have alternative strategies to mitigate a sensor's shortcomings. One of these is to figure out the characteristics of the object that can help make it easier to find. For example, presume you have been hired to find a small anchor and chain that a yacht lost as it tried to anchor in 7.5 meters (approx. 24 ft) of water. You have a *SeaMATE* type of eyeball ROV on hand, so your sensor is going to be the camera. Luckily, the bottom is sandy, and there's a visibility of 15 meters (approx. 50 ft).

What characteristics of the anchor and its chain will help in locating it? And what exactly was the situation when it was lost? These are fundamental questions you always ask before conducting a search. For example, if the yacht was in the act of anchoring, it would have been backing up so the anchor would dig into the sand, so it is likely that the whole length of the chain would have been stretched out when the unsecured end ran off the deck. So the chain would provide a linear target to look for on the bottom, and a linear target is much easier to find than a pile of rope heaped on top of the anchor.

Knowing this, you question the skipper of the yacht further and ascertain the most likely place to set down a buoy to mark the starting point of your search grid. Then configuring the ROV as a drop cam, you can run a series of long, lazy "S" turns downwind and/or in the direction of the current flow from where the anchor had initially been dropped. If that position is based on reliable information, you should easily run across the anchor rode in a few passes. Bob Ballard used this very technique to find the *Titanic* with deep sea towed cameras. He knew the ship had a linear debris field, so he calculated that what he had to do was come across that field with the camera, then decide whether to go left or right, and follow it to the ship. And that is exactly what happened.

TECH NOTE: ADDITIONAL OVERLAP CONSIDERATIONS FOR SONAR SEARCHES

This bird's-eye view shows a boat pulling a towfish equipped with a side scan sonar transducer in a search for a sunken wreck.

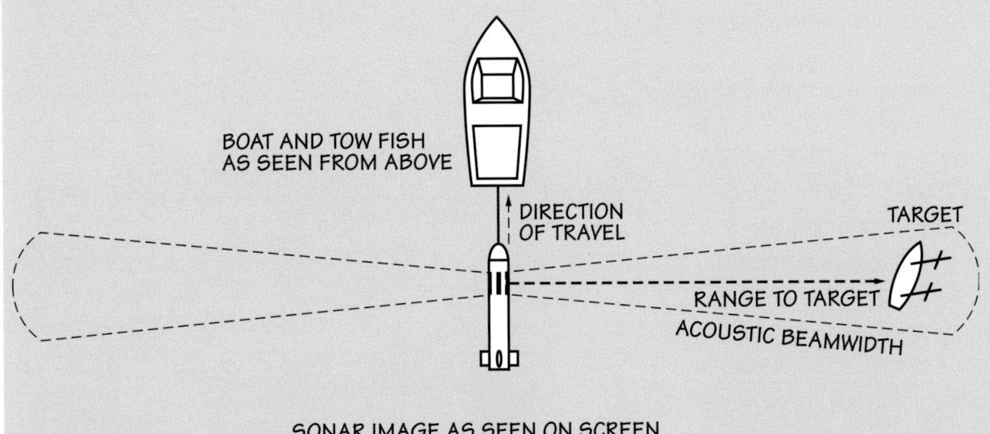

Note that in the sonar image, the shaded section directly below the towfish is blank, because the acoustic beams do not image this part of the bottom. As a result, it is important to run a grid pattern with more than 50 percent overlap so that the search beam from one leg scans the blind spot from the previous leg. Note that the overlap for the camera search shown in Figure 11.36 would not be nearly enough.

Figure 11.42: Side Scan Blind Spot

TECH NOTE: ALTERNATIVE SEARCH TECHNIQUES

There are certainly other search strategies and techniques besides using an ROV. However, if you opt to use alternatives such as towing a grapple hook or a dragline arrangement, you will still need to use a systematic search grid. In fact, it's even more important to be methodical and orderly because you *are* operating blind without a camera.

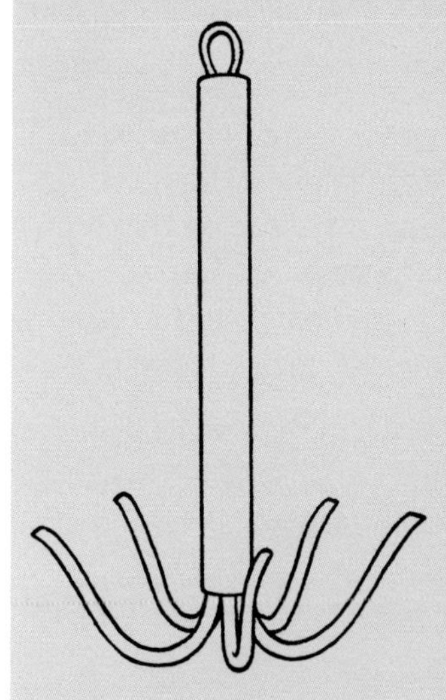

The Grappling Alternative: A traditional but effective search technique for bigger, heavier items involves towing a grapple hook behind a boat in a search grid. A grapple is a modified stock anchor with three to six steel tines instead of the normal two; it's connected by a short length of chain to a long length of line. A boat pulls the grapple, which is dragged along the bottom. When it hooks something, the boat is slowed or stopped. Depending on the size and weight of the snagged object, it is either hauled up to the surface, or a diver or ROV is sent down the line to see what it is. Of course, a grapple is not foolproof, since it can easily snag rocks, branches, or other debris. However, given the right bottom terrain and a target object like an outboard motor or an anchor rode with enough protrusions on it for the grapple to hook onto, this can be an effective search technique for large objects. Note that it does not guarantee 100 percent bottom coverage.

Figure 11.43: Grapple Hook

The Dragline Alternative: The dragline method is another low-tech but effective method for conducting a bottom search. It requires two powered boats (two canoes with electric motors would work for the Conway's Pond scenario) and a long cable or weighted line with a length of chain attached. The line is attached to the two boats that drag it in concert over the bottom along a search grid. If an object is snagged, the ROV can be sent down the line to identify it. Using a dragline has several advantages—it's cheap to configure, and you can cover a huge amount of ground in the search area quickly. The biggest drawback with this system is the number of non-target objects that can be snagged on the bottom. If fact, this would be a frustrating exercise if the bottom had too many trees, branches, or boulders. But if the bottom is relatively clear, then a dragline is a good technique to consider. Admittedly, it stirs up a lot of silt, but this is a minor problem, because sending the ROV down the dragline means it should be able to follow the line right to the object. Once there, the silt will settle if you wait just a bit.

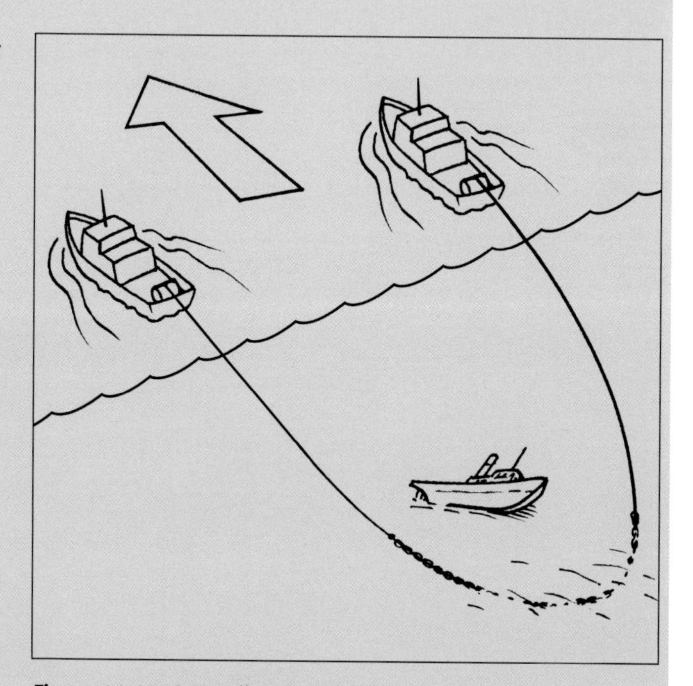

Figure 11.44: A Dragline Search

TECH NOTE: CONWAY'S POND OPERATIONS TASK #5— MARKING THE SUNKEN TUG'S LOCATION

Once the tug is found, mark the spot using a float attached to a length of line and weight. If using a GPS, set a waypoint.

Marking the Location: When your search locates the target, it's important to mark the spot with some sort of marker buoy, unless you plan an immediate recovery. A weight, line, and buoy easily serve as a good marker device. Don't waste time trying to tie the marker line onto the target—just get it in close and make sure the weight is heavy enough that the buoy won't get dragged away by wind or wave action. Be sure to give yourself backup coordinates by taking sightings and/or GPS coordinates to mark the spot.

3.4.3. Positioning for Recovery

Regardless of the size and type of mobile support platform used, the big challenge is keeping it stationary and in position for recovery of the object. Weather and currents can push the vessel away from the target area. The vessel may also move up and down with the waves, an action that adds further complexity to the recovery efforts. The deeper the water, the more difficult it is to maintain position over the site and to provide a support system capable of completing the recovery portion of the mission.

The most common methods of maintaining a position over an op site are:

1. anchoring
2. live boating
3. dynamic positioning

1. **Anchoring**: Simply put, anchoring involves securing a heavy weight to the end of a line, tying the other end to the craft, and then throwing the anchor weight overboard. This very basic arrangement—often as simple as a concrete-filled coffee can and rope combination—still works for many a rowboat. In fact, for the Conway's Pond mission scenario, any small anchor is probably a good option for anchoring the canoe when it comes time to make the lift.

   For centuries, sailors relied on this basic weight-and-line anchoring technique to hold a small ship. But as watercraft evolved into larger vessels, mariners found that a heavy weight alone did not have sufficient holding power, particularly in bad weather. So other methods were devised, and the modern anchor evolved (Figures 11.45 and 11.46). No longer just a heavy weight on a line, contemporary anchors are designed to hold ground by digging into the bottom. An anchor does this by burying its flukes into the mud at the bottom, somewhat like a plow digs into the earth, and the weight of the mud does the holding against the ship. A modern anchor is joined to the anchor line by a length of chain, which serves to keep the shank lowered, thereby maximizing the resistance of the flukes against being pulled out by the vessel. This chain weights the shank to a near horizontal position in order to get a good holding angle.

   For a 6- to 18-meter (approx. 20- to 60-ft) boat, the most common anchoring system consists of a stout line, or anchor rode, shackled to a length of chain secured to the anchor. The end of the anchor rode is secured to the boat. Larger vessels will use only chain in their anchoring systems and, of course, heavier anchors. Other types of rode may include steel wire rope or combinations of chain and wire rope.

   The anchoring process consists of dropping or "letting go" the anchor and paying out enough length of anchor rode so as to equal about three times the depth of water the vessel is in. Then the rode is "tied off," or secured, either around a bit or

Figure 11.45: Basic Anchor Anatomy

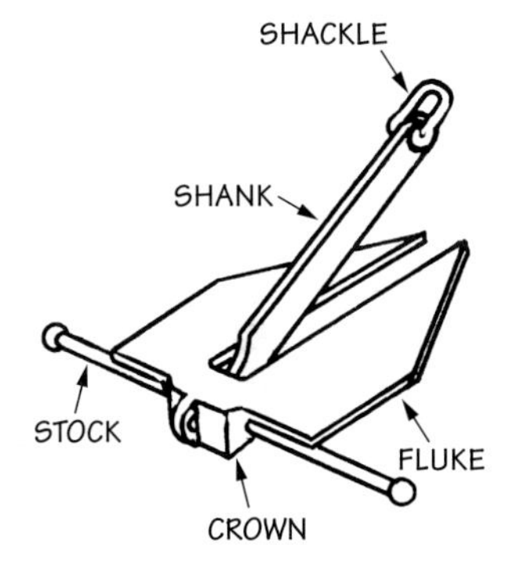

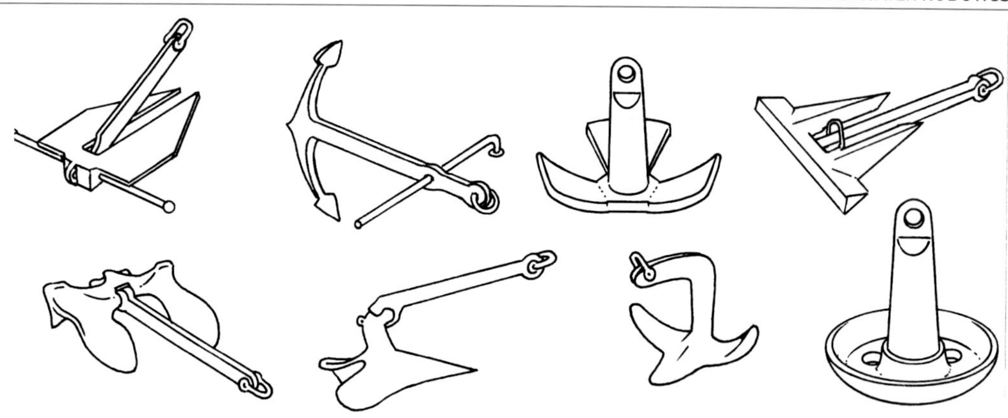

Figure 11.46: Types of Anchors

cleat on the deck or to the brake set on a winch or windlass. The term *scope* refers to the amount of line let out when anchoring. In fact, it is a term that describes the ratio between the depth of water the vessel is anchoring in and the amount of anchor line let out. The more scope, the better the holding power, because the rode angle is increased and thus is not pulling up on the anchor. Too little scope, and the holding power of the anchor is reduced.

The general rule of thumb for scope is three times the length of anchor rode to the depth. Thus, in 30 meters (approx. 100 ft) of water, you should have about 100 meters (approx. 330 ft) of line out. In bad weather, it is advisable to increase the scope, because swells can lift the boat, thus increasing the depth and shortening the scope. The technique for hauling the anchor back on board depends on the weight of the anchor and size of the vessel. For a small skiff or canoe, you just haul it in by hand. A yacht typically uses a small electrically driven anchor winch. Large ships and barges use a powerful winch to haul in the chain and anchor.

Figure 11.47: How An Anchor Works

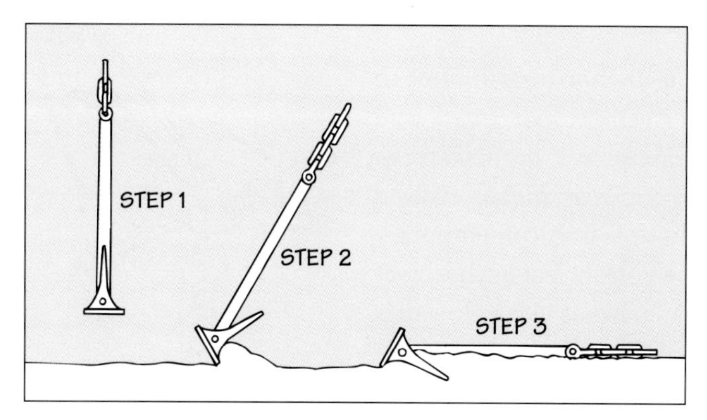

The depth limit for anchoring most craft is about 60 meters (approx. 200 ft), but salvage vessels, supply tugs, or barges can be specially rigged to anchor in 300 meters (approx. 990 ft) of depth. However, this depth requires at least 900 meters (approx. 2,950 ft) of anchor rode, which in turn requires massive winches and winch drums. Often, two or more anchors are deployed, since multi-point anchoring allows the vessel to sit right over an op site with minimal movement. Using more than

Figure 11.48: Proper Anchor Scope

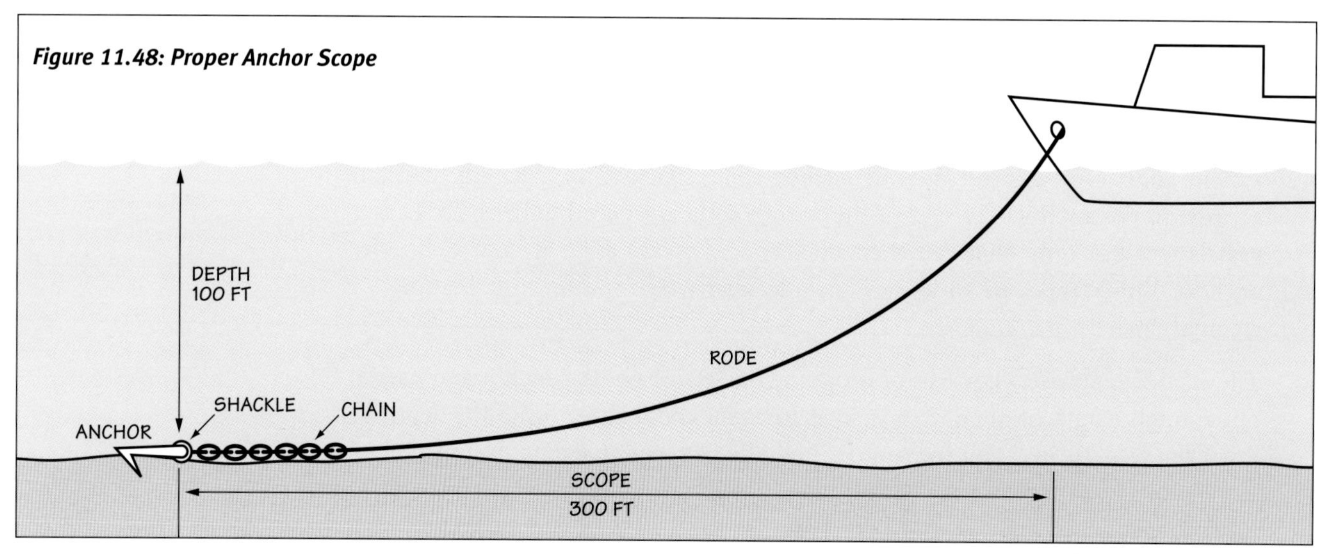

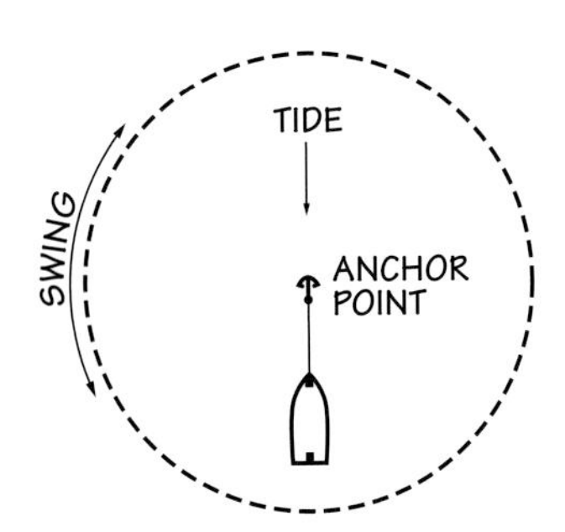

Figure 11.49: Single-Point Anchoring

The simplest anchoring technique is the single-point mooring. The vessel swings off the anchor rope or chain so that it always faces into the wind and waves. This reduces the frontal area showing and makes the craft more seaworthy in heavy weather.

However, swinging with the tide and wind is actually a disadvantage of single-point mooring for underwater operations, since the ROV's tether may become entangled in the vessel's propellers and rudders. This can occur even in calm conditions with only a slight breeze and moderate tide.

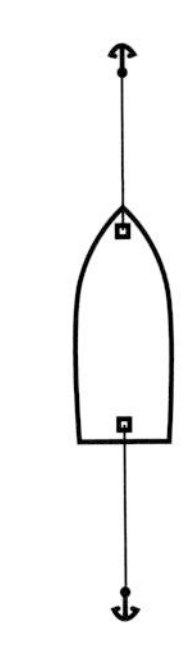

Figure 11.50: Two-Point Anchoring

A two-point (bow and stern) anchoring system keeps a vessel in position without drifting. Because the boat or ship cannot swing, it needs to be anchored facing into the weather and tide. Where there are tidal changes, the vessel's bow-to-stern alignment should always be parallel to the tidal stream. If the vessel is broadside to such a water flow, the pressure on the hull may cause the ship to drag anchor.

Any vessel utilizing a two-point moor still has a tendency to move out of position laterally. Two-point mooring is relatively easy and allows the vessel to take up and shift position over the bottom within the scope of the two anchors.

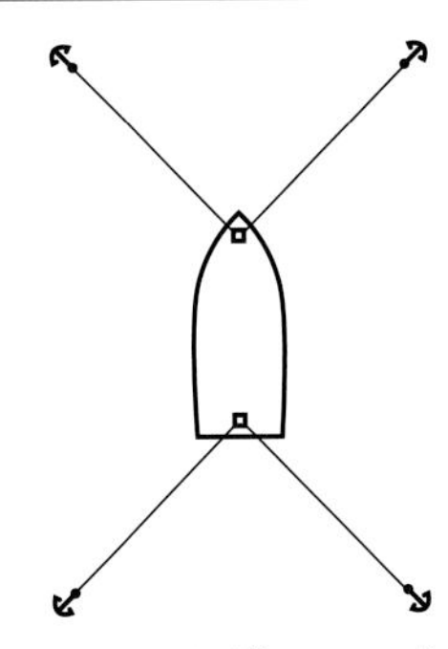

Figure 11.51: Three- and Four-Point Anchoring

Setting three or four anchors provides a more secure situation. The advantage is that the ship can be shifted in four directions while at anchor, simply by hauling in or easing out on the different anchor lines.

This is the optimum anchor arrangement for underwater work, since it gives the support ship the best chance of maintaining position. The only drawback for open-water work occurs if the weather worsens. Multiple anchors constrain the ship from swinging into the weather, so in foul weather, the vessel must let go the anchors and ride on a single mooring for the duration of the storm.

TECH NOTE: CONWAY'S POND OPERATIONS TASK #6—POSITIONING FOR RECOVERY

For this recovery phase, you'll need to take the weights off the ROV, so it's no longer functioning just as a drop camera. Once the ROV is reconfigured and you've checked its buoyancy, you can begin the recovery phase. Anchor the canoe over the model, setting two anchors, one fore and one aft, if optimal. Once anchored, you may elect to remove the marker float if it impedes the ROV's maneuvering.

one anchor also allows controlled placement of a barge or vessel over a site, by means of hauling in and letting out on different winches. Figures 11.49, 50, and 51 illustrate the most common anchoring options for underwater vehicle operations.

2. **Live Boating**: Anchoring is usually an effective option for smaller vessels working with underwater vehicles, especially ROVs. But some ROV recovery missions call for a "live boat" alternative in which the boat's engines operate continuously to maintain a stationary position over the ops site. Live boating works best in calm waters, with minimal tidal current and within site of fixed visual cues from the shore or from navigation markers. It requires intense concentration from the skipper to make the constant adjustments of throttle and rudder to keep the boat on station. A momentary lapse can quickly put the vessel off the mark.

The biggest drawback of this technique is the danger to the underwater vehicle—the support boat's propellers are constantly engaged, and the tether of the ROV may get tangled in them. It is best to avoid live boating in situations of high swells, strong tides, or wind, where the boat can get pushed off site, dragging an ROV into obstructions on the seafloor and entangling it. The advantage of not having to anchor a vessel while engaged in underwater operations is balanced by the difficulty in manually maintaining position and the danger of damaging the tether with the prop.

3. **Dynamic Positioning System (DPS):** Basically, dynamic positioning is automated live boating. This technique was first employed in 1961 by the drill ship *CUSS I*, which was equipped with four giant outboard motors that worked together to maintain position continually. Now, ships are specially built for dynamic positioning with two main props aft and additional bow thrusters forward and aft. The system gets its navigation information from GPS, radar, motion detectors, and acoustic pingers set on the bottom. This information is input into a dedicated computer programmed with the op-site coordinates. These coordinates are then turned into a control signal that operates the props and bow thrusters. If the ship moves off position, the appropriate thrusters rev up and move the vessel back on station. Under ideal conditions, a ship using dynamic positioning can maintain a coordinate of +/-1 meter (approx. 3 ft). This allows the deck crew to confidently launch, position, and recover submersibles, ROVs, AUVs, salvage gear, or oceanographic instrumentation. Of course, maintaining position always relies on the integrity of the navigational input and can be impacted by external forces such as wind, tide, and wave heights.

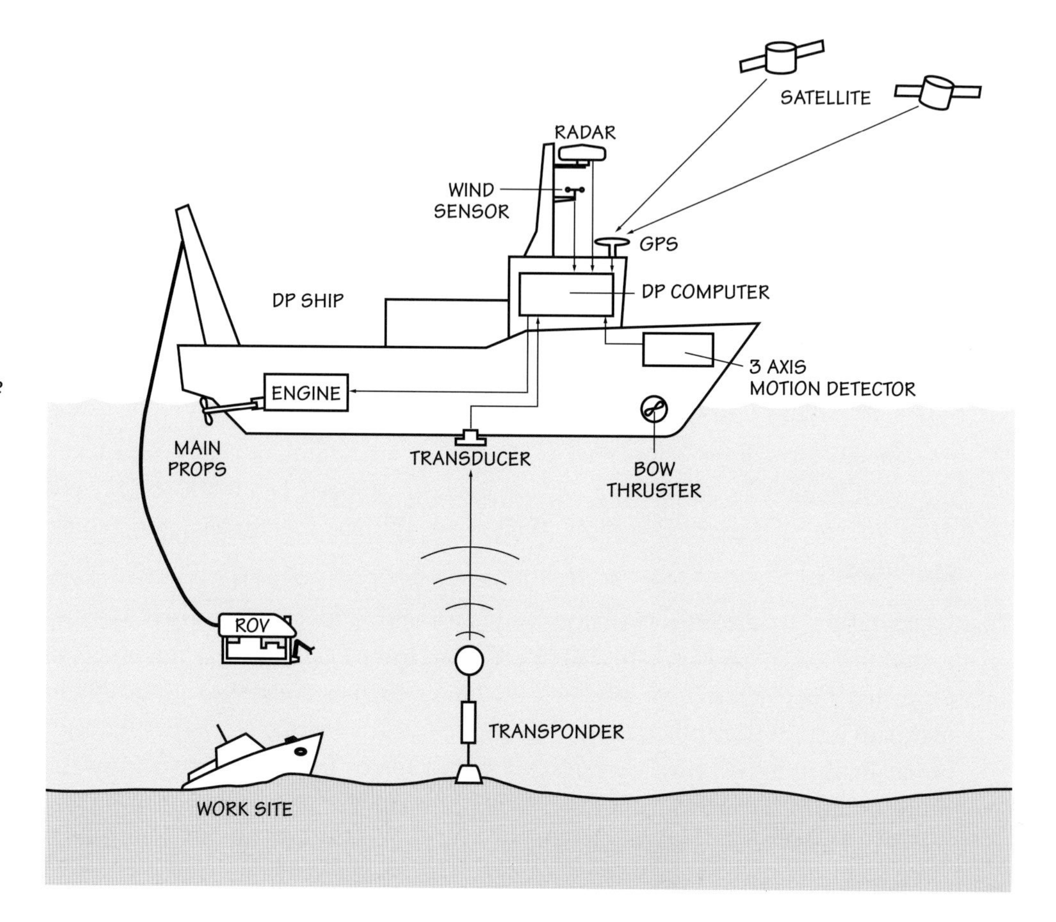

Figure 11.52: Dynamic Positioning System

This simplified diagram of a dynamic positioning system depicts how the vessel keeps stationary over a point on the bottom.

The ship's sophisticated computer-controlled navigation system takes accurate GPS positions and inputs that data into its dynamic positioning thruster control system, which in turn directs bow and stern thrusters and/or the main propellers to counteract the effects of wind, waves, and current.

3.4.4. Making the Recovery

Certainly a successful recovery often depends on utilizing the right technology and equipment, but it's also worth it to study the heritage of professional divers, salvors, and ocean engineers whose experience and valuable tricks of the salvage trade have evolved over many years.

In the real world of modern unmanned salvage operations, the ROV is still the only robotic vehicle that can be used for recovery ops. AUVs and towed vehicles are still essentially survey vehicles; they're very good for mapping or running a search grid, but not great for the heavy-duty, high-endurance work that a recovery op entails. The ROV gives the greatest bang for buck in terms of lifting capacity and underwater endurance. Its tether gives the pilot real-time control and observation capability, as well as providing almost unlimited endurance. Furthermore, a work class ROV can carry and use heavy-duty payload tools, such as manipulators and cutting gear, as well as high-definition imaging and detection systems.

Admittedly, many recovery strategies are specific to each situation, and each job brings its own set of circumstances and problems to overcome. But it is possible to describe the basic sequence common to all recovery ops, once the support platform is positioned over the target object. They are as follows:

1. choosing the lift mechanism
2. rigging for the lift
3. making the lift
4. securing the recovered object

Keep in mind that it's easy to get caught up in planning fancy, complex lift ops. The smart thing is to go for the safest, most straightforward recovery option once the target has been located and marked.

Choosing the Lift Mechanism: Once the object has been located, the next challenge is figuring out which tool(s) or mechanism(s) will get it to the surface. (See *Tech Note: Lifting Options.*) Stationary cranes and deck winches are most commonly used for lifting heavy objects off the bottom or for moving gear and vehicles around the deck or over the side. Lift bags or barrels are also handy recovery devices; they use air to displace water and produce a buoyant force that functions somewhat like soft buoyancy tanks in submarines. Other in-the-water lifting gear includes large floats or barges that can be modified to serve as lifts by harnessing tidal flux or by ballasting and de-ballasting them with water. ROVs may work in conjunction with any of these mechanisms, helping to rig the object for its lift. Choosing the right mechanism and the right size of mechanism for the job is not only important for ensuring a successful recovery, but it can be a big safety factor for team members.

SAFETY NOTE: SAFE WORKING LOADS AND BOOM ANGLES

Regardless of the type of crane chosen for the job, each has a safe working load (SWL) limit—this is the weight it can lift safely when the boom is extended in the horizontal (or near horizontal) position. For example, if the boom has SWL 10T painted on it, then 10 tons is the maximum weight the boom can lift at a 90° angle to the mast (the weakest position for any crane). As the boom is raised, the load-lifting capacity increases. The general safety rule is never to exceed the SWL of a boom unless the load limits for various other angles have been inscribed on the boom.

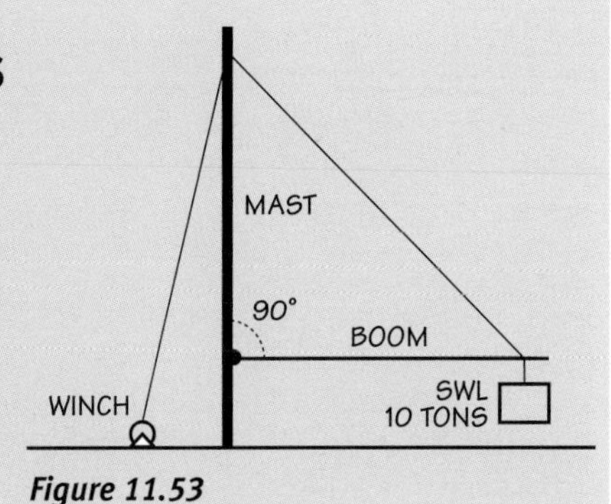

Figure 11.53

TECH NOTE: LIFTING OPTIONS

All recovery ops require lifting gear of some sort. Some of this equipment, such as winches and cranes, may be situated on the recovery platform itself; others, such as ROVs or lift bags, may work in the water. Figures 11.54-57 show some of the more common types of lifting gear found on or deployed from support platforms.

Image courtesy of the Institute of Nautical Archaeology

Recovery with a Lift Bag and Barrel: Lift bags and barrels rely on compressed air to displace a volume of water and hence exert a buoyant force on the object to which they are attached. This same principle is used to submerge and surface submarines. The volume of air expansion in the lift bag is regulated by a dump valve. The barrel is less sophisticated, with only a hole cut in its bottom, so as the air expands during the ascent, the excess can escape through the hole. This maintains an equilibrium in the volume of air inside the barrel, resulting in a steady ascent rate. A somewhat different lifting approach takes advantage of tidal flux. For this method, the hookup between barrel or barge and object is made at low tide; as the tide rises, the barrel/barge rises, thereby lifting the object off the bottom. A small boat or barrels rafted together can utilize tidal change to recover items or move anchor blocks.

Figure 11.54: Lift Bags (or Buckets!)

Lift bags are inflatable sacks, often used to raise heavy objects to the surface. In this photo, fragile ship's timbers from an 11th century Byzantine shipwreck excavation are raised to the surface on a lift box, with buoyancy ingeniously provided by three air-filled buckets instead of lift bags.

Recovery with a Barge: If positioned over a sunken object and made fast to it by means of lines, a barge can also serve as a lifting device. First, the barge is ballasted with water to lower it; then the slack lines are tightened and the water pumped out of the barge, causing it to rise. Thus, the barge acts a giant float to lift the object. Such barge lifts are the only way large vessels, such as the Russian submarine *Kursk*, can be brought up and transported to shallower waters or dry docks after they are recovered.

Lifting with Davits, Winches, Cranes, and A-Frames: A small, single mechanical arm with a hand-operated or powered winch, called a **davit**, provides more lifting capacity than the hand-over-hand retrieval of an object. It consists of a rigid boom that can swing horizontally and a hand-cranked or small powered winch used in conjunction with a block (pulley) to lift loads of less than 2.2 kN (approx. 500 lbs).

Image courtesy of Drew Michel

Figure 11.55: Hydraulic Crane

Whether mounted on board or ashore, a **derrick crane**'s long boom reach and swing enhances its versatility, enabling it to handle many tasks. The most common model is the single-swing boom. A sailboat mast and boom can be similarly modified to swing an ROV over the side or even to bring a man overboard back on deck. Any simple mast and boom arrangement will exert the greatest amount of pull when the line or cable is run through a set of blocks (pulleys) that multiply the amount of force that can be transferred to the lifting wires.

A **winch** generally refers to a motorized spool used to reel in a rope, cable, tether, or net. Often a winch is used in conjunction with a crane

Image courtesy of Harry Bohm

Figure 11.56: Crane Launch of an OceanWorks Submarine Rescue Vehicle

or davit to lift heavy objects. In the context of larger underwater robots, a winch is used to reel in (or out) and store the long umbilical or tether of an ROV.

A **heave-compensated crane** is a special rig designed to launch and retrieve underwater craft at sea. Although the design varies widely from crane to crane, each has a special mechanism that compensates for the rise and fall of waves. Active heave-compensation means that the crane is controlled by a computer that senses the motion of the crane and then triggers actuators that compensate for any heave motion. Passive compensation refers to a mechanically controlled compensation system with no computer or sensors involved.

Image courtesy of Woods Hole Oceanographic Institution

An **A-Frame crane** can lift extremely heavy loads and is often heave-compensated. However, unlike the derrick boom, the A-frame does not swing horizontally. It is fixed to the bow or stern of a barge or ship, and the operator extends it by changing the angle of the A-frame legs. The vertical load lift is done with a winch cable, pulley, and hook. In deep water recovery work, large heave-compensated A-frames are used routinely in conjunction with deck winches to recover sunken objects, as well as submersibles, and lift them onto the deck.

Figure 11.57: A-Frame Crane

The Atlantis *mother ship is equipped with a large heave-compensated A-frame crane, for launching and retrieving* Alvin.

Rigging for the Lift: The term *rigging* has a number of definitions, depending on usage. In marine circles, rigging is defined as the general name for ropes, chains, and wires that hold masts and associated cross pieces for rigging sails in place. But in salvage, rigging has a slightly different, though related, meaning. It refers to the system of ropes, straps, cables, or other lifting hardware (including shackles, blocks, etc.) used to support and control a load that is being raised by a crane or winch. Rigging can be simple or complex, depending on how heavy the object is and how deep it is; not surprisingly, heavier and deeper is the most complex.

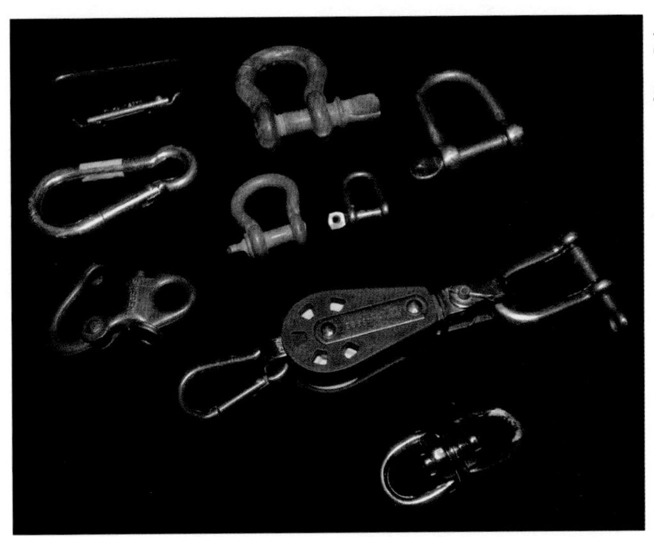

Image courtesy of Harry Bohm

Figure 11.58: Various Types of Shackles

Shackles come in a wide variety of sizes, weights, and metals. The shackles pictured here are suitable for small salvage operations.

For an ROV to facilitate any recovery operation, it must have the capability to rig up a submerged object for lifting. That might mean adapting existing rigging hardware to fit the ROV's manipulator or modifying the manipulator or other payload tool(s) to fit the rigging. Success at this requires innovative thinking, research (checking out publications, books, and other similar jobs), some trial-and-error work, and a single-minded determination to find the best solution. For example, there are a variety of ways you can adapt rigging to *SeaMATE*'s frame and probe in order to pass a line around a lifting point on a sunken boat.

Lifting and Securing the Object: It is important to identify a solid place, or lifting point, on the target object where you can attach the lines to make the lift. (See *Tech Note: Lifting the Model Tug.*) The lifting point must be strong enough to resist the forces acting on it as it is being lifted. Double check to make sure that your rigging will work with the attachment point. The location of the lifting point is also important, since objects vary with regard to their center of gravity. For example, a sunken motorboat has most of its mass aft, while the bow is "light" by comparison. So if you use the bow cleat as your lifting point, you know the boat will come up with the motor

TECH NOTE: CONWAY'S POND OPERATIONS TASK #7—ATTACHING A LIFT LINE

There are a number of possible techniques for using the ROV to bring down a line and attach it to the tug. But first, survey the tug to see how it lies on the bottom, and identify possible strong points on the model for attaching a line. Then decide on a technique. For example, you might be able to snag the tug with a large fish hook on a weighted line. The method you choose will determine how you'll need to rig the ROV. The robot's camera will help you guide the line to the tug. If you are repeatedly unsuccessful attaching a line with one method, try another.

TECH NOTE: LIFTING THE MODEL TUG

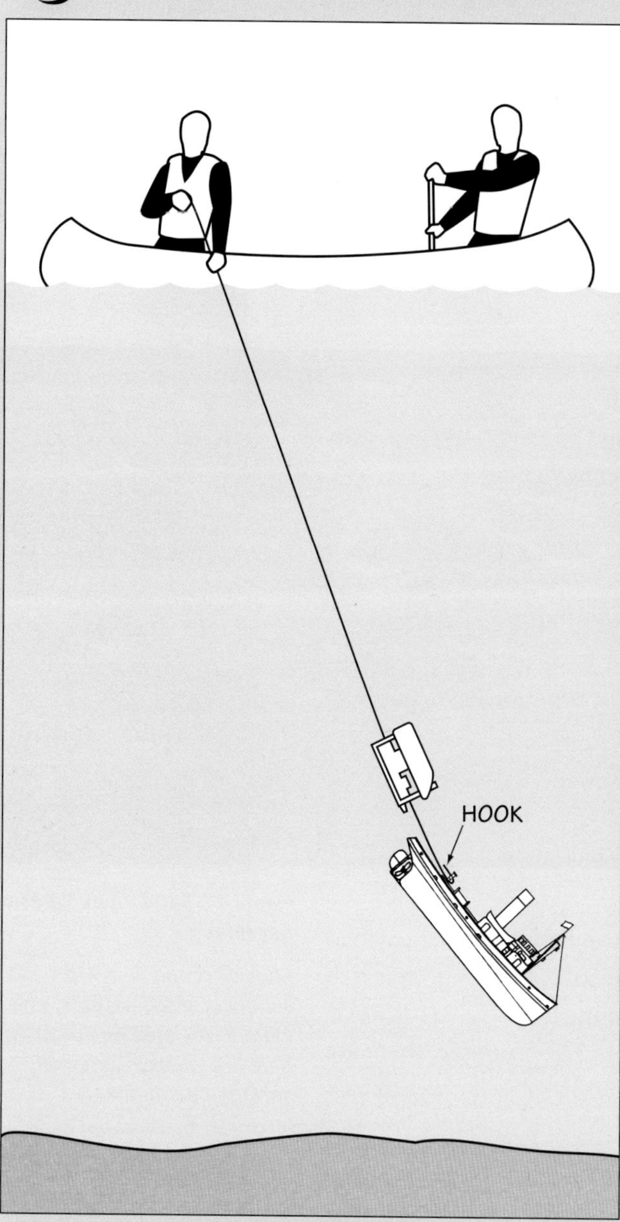

Figure 11.59: Recovering the MV **Zande**

In Conway's Pond, where the waters are almost always calm, there are no weather conditions short of a severe summer storm or full-blown blizzard that would prevent making a lift of the model tug.

Before making the lift, you'll want to use the ROV's camera to scope out the best lifting points on the model tug—hopefully there are several. They might include a railing, or a rudder post at the stern, if there is one. It's also important to see how the tug sits on the bottom. Is it upright? Is the bow or stern raised? Is there any debris on top? You will want to evaluate any factors that might impede the lift. Once that's done, you can begin rigging for lifting the MV *Zande* from the bottom of Conway's Pond.

In the Conway's Pond scenario, rigging the *SeaMATE* ROV so as to hook onto the MV *Zande* would probably take a lot of effort and time. One option would be sending a strong swimmer down the marker buoy line to pick up the model tug or to attach a line to it.

However, given the safety and liability issues associated with putting a swimmer in the water, it's probably wisest to use the ROV for your recovery option, even though it may take a little longer. The best option with an ROV is to use it to attach a recovery line to the tug. You can do this by fashioning and attaching some sort of hook and line to the ROV frame. Then you can fly the ROV down to the tug and either snag the railing or hook onto some other strong point on the model. Once the hook connection is secure, simply haul in on the tether, pulling the tug and ROV up together.

Once the model tug is at the surface, the easiest and simplest method of securing it is just to lift it carefully aboard the canoe. Then you can paddle back to the dock, where you may want to rinse it gently with fresh water before turning it over to the delighted owner. Now you can celebrate a successful mission and get ready for demobilization!

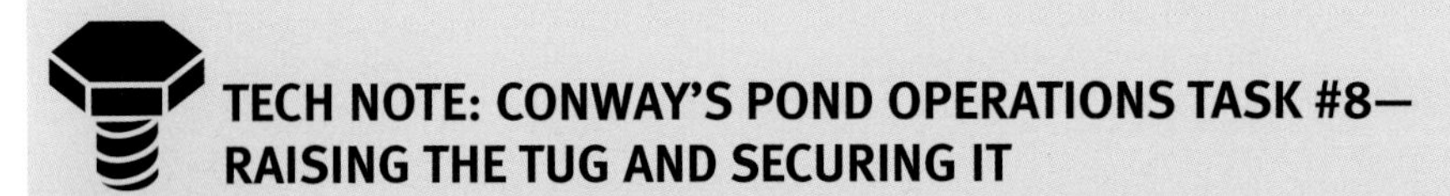

TECH NOTE: CONWAY'S POND OPERATIONS TASK #8— RAISING THE TUG AND SECURING IT

Once you've hooked the tug securely, ensure that the canoe is stable, so that one of you can safely reach over the side to retrieve the ROV, then haul up the tug. Pull slowly and evenly, so the tug doesn't jerk off the line. You may want to have a dip net ready. Once the tug is in the canoe, pull up all the buoy markers and head for shore.

dangling down. That's not an issue for the underwater aspect of the lift, but you might encounter problems at the surface when trying to secure the boat for towing or when pumping the water out of it if the heavy aft end is hanging 10 meters below the bow. If you can find the center of balance and lift from two or more points, then the object will come up more or less on an even keel.

When the rigging is securely connected to the object, it's time to make the lift. One of the most important aspects of the lift is to make sure the sea state is calm enough, since a pitching, heaving deck in a growing sea makes for dangerous work. Also, the increased forces acting on the rigging can snap the line or shackle, and the object of all your efforts can be lost once again. The sea state in which you can lift depends a lot on the support platform, depth, the type and quality of the lifting gear, and of course, the mass of the object being raised. As a rule of thumb, the bigger the mass, the calmer the waters have to be.

Getting the object to the surface seems like the end of the recovery battle. But it's not over yet. You still have to secure it for transport back to a shoreside dock or repair facility. This last step may seem pretty unimportant, but it can be crucial—just ask one of the teams from the 2002 MATE International ROV Competition. They had just about swept up all the objects they needed to recover from the bottom of the pool in order to win the competition. Victory seemed theirs. Then, as their ROV ascended to the poolside with the treasures, the claw holding them opened prematurely and dropped the entire horde back to the bottom—all their hard work was gone in a split second. Another more serious event occurred in 1968, as the mother ship *Lulu* prepared to launch *Alvin* for the final dive of the season. A cable suddenly snapped during launch procedure, sending the submersible to the bottom—fortunately minus crew members. Seven months later, *Alvin* was finally recovered.

These two examples illustrate what every mariner knows—the sea is full of surprises, so never figure the recovery is over until the object is recovered and secured.

One way to secure an object after a lift is to bring it on board the support platform. Another is to tie it off alongside. In the case of salvaged boats or ships, sometimes they can be brought to the surface and pumped out enough to regain their buoyancy, allowing them to be towed to port. In such situations, it is wise to mount a pump on the deck of the salvaged vessel and station a crew member nearby, in case the hull begins to leak again while being towed. Salvaged vessels such as the *Ehime Maru* and *Kursk* were never raised to the surface—instead, they were secured underneath their support platforms for easier towing to shallow waters.

3.5. Demobilization

Celebrating a successful recovery operation is always important, but there's still work to do. Demobilization includes getting back to shore, handing off the object,

TECH NOTE: CONWAY'S POND OPERATIONS TASK #9—DEMOBILIZING

Once ashore, wash off the tug, ROV, buoy markers, and other gear with freshwater. Assess and note the condition of the recovered tug. Contact the owner and make arrangements to pick up the model tug. Pack up all gear and any litter you may have generated. Write down any ideas for improvements to the ROV or suggestions for more effective recovery techniques.

accounting for equipment, materials, and personnel, washing and restowing gear, maintaining equipment, organizing debriefing sessions, and writing reports.

Heading for the Barn: This common phrase is used to describe getting everyone and everything back to home base and getting the gear cleaned and stowed away. It involves checking things off those lists you made earlier, to ensure that everything is accounted for. This is also a good time to write down things that have gone missing or been damaged and to note items you wished you'd had. Carefully wash off all gear.

In the case of underwater vehicles, there are always routine maintenance procedures that must be conducted after a prescribed number of operational hours and following every job. Technicians need to record these procedures and note any repairs, additional maintenance, or adaptations conducted on site. They also need to log suggestions for work that should take place once the vehicle is back at home base. On big jobs, other team members may need to conduct similar demobilization procedures for their own specialized equipment.

Debriefing: Debriefing is the process of discussing and recording what went right, what went wrong, and what could have been improved. These sessions can be important planning and problem-solving sessions for the next day's work or for future ops. Debriefing is most effective if it takes place as soon as possible after the mission (or a segment of it) is completed. On big operations, specialized groups (such as deck crew or divers) may have their own debriefing sessions with team leaders, who then report the results of those sessions to the project manager.

Writing Reports: Report writing seldom generates enthusiasm on anyone's part, but it is true that no job is done until the paperwork is completed. Often, there are a variety of reports to do, from budgetary accounting to performance evaluation of the underwater vehicle and writing up mission results. Before starting to write, be sure you are clear on the purpose of the report and its intended readership. When you do start to organize your thoughts, you really see the wisdom of taking notes as you go because you've got the raw data handy.

A good report should include performance data, accurate details and observations from various team members, and the supervisor's overall perspective on the project. As in the debriefing sessions, note what went right, what wasn't effective or simply didn't work, and suggestions for improving the operation. Make sure copies are distributed and filed for easy reference in the future. In commercial work, these reports are required by the client and may be used to assess if the work was done to contract performance requirements. A poorly written report can result in confusion, delay, or even non-payment of contractual fees. A thorough report becomes an extremely valuable document for planning the next mission and possibly for securing the next contract. (For a review of report writing, see *Evaluating Ops and Writing a Report* in *Chapter 2: Design Toolkit.*)

4. Chapter Summary

The preceding technical chapters have focused on how to design and build your vehicle's various subsystems. This chapter is different. It focuses on a completed vehicle that is ready to take on the challenge of an underwater mission—specifically a search and recovery operation.

The chapter outlines a series of steps that apply to all types of operations. Initially this sequence involves assessing the risks, researching the options and gathering information, then deciding on a practical plan. In the case of search and recovery ops, that plan will include defining the search area, then selecting appropriate cameras, sonar, or other sensors to locate the missing object.

Once the major planning decisions are made, the next step is mobilization—assembling and transporting the necessary crew, equipment, and gear. When on site, it's important to define and mark the search area and then choose the most appropriate search grid for the search phase.

When the target of the search is located, the ops scenario shifts to inspection/ assessment and recovery considerations. This involves figuring out how to make the lift and selecting appropriate equipment and gear to do so. If the lift is successful, then the object must be secured for transport.

Last but not least, there are the essential steps of demobilization: getting everyone and everything back to the dock, stowing and returning gear, holding debriefing sessions, and writing final reports. The successful completion of all of these phases, not just the recovery of the object in question, is true cause for celebrating a mission accomplished!

Chapter 12

SeaMATE

Chapter 12: *SeaMATE*

Stories From Real Life: Jules Verne—Fact, Fiction, and Inspiration

Chapter Outline

Figure 12.1.cover: Roll Up Your Sleeves and Dive In!

In classrooms around the globe, students are learning the science and excitement of designing and building robotic vehicles like the SeaMATE *ROV detailed in this chapter. Once they master these basics and gain hands-on experience, they can explore more advanced craft like the ROV pictured here or AUVs.*

Image courtesy of Steve Van Meter, VideoRay LLC

Chapter Learning Outcomes

- Build a simple shallow diving ROV using hardware-store technology.
- Conduct an underwater mission.
- Identify key issues involved in designing a deeper-diving vehicle.
- Research and evaluate system options and components for an ROV capable of going to 100 meters (approx. 325 ft).

STORIES FROM REAL LIFE: Jules Verne—Fact, Fiction, and Inspiration

When Twenty Thousand Leagues Under the Sea *was first published in 1869, it was hailed as a prophetic glimpse into the future. Readers around the world were captivated, even inspired, by Jules Verne's amazing tale of an electric-powered submarine, battles with monstrous squid, and abundant harvests from undersea gardens. These exploits may have seemed like bizarre projections, but Verne based them on actual scientific knowledge of his day.*

Our own recent discoveries of strange life forms in the harsh environment of the ocean are no less amazing. And, like Verne, we now contemplate the possibility of imaginative vehicles that may unlock the secrets of water-based life—only this time, on the moons of other planets. You, too, may find inspiration from Verne's writings—or even from textbooks like this one!

As a young man working in the Paris stock market, Jules Verne discovered he could supplement his earnings by writing short articles on scientific and historical topics. He spent long hours in the library, researching reference books, scientific journals, and newspapers. This gave him the idea for an innovative type of novel that would blend scientific fact and fiction. Teaming up with a feisty publisher, Verne began churning out a series of adventure articles that ended up as amazing novels.

Because his books often cited famous academic authors and their works (along with wholly fictitious ones) and even included lengthy scientific passages, many readers presumed Jules Verne was a scientist, an inventor, or even a geographer. In fact, he was a good researcher and an innovative writer.

Jules Verne is considered by many to be the "father of science fiction." Most readers thought novels such as *Five Weeks in a Balloon, Journey to the Center of the Earth, From the Earth to the Moon, Twenty Thousand Leagues Under the Sea*, and *Mysterious Island* were wildly imaginative, futuristic works. Even today, readers of *Twenty Thousand Leagues Under the Sea* see the novel as a look into the future of subsea exploration and technology, but in fact, the oceanographic world that Verne portrayed was really an inventive understanding of the scientific knowledge and experiments of his time.

Verne's books were set in the mid-1800s, when ordinary people were just beginning to come to grips with the possibility of mass travel. It was a period of rapid industrialization, with railways linking countries and ships linking continents. Travel under water and even in the air was a tantalizing possibility. Submarines had already been built and tested by the late 1700s. Verne may have even named his fictional sub *Nautilus* after Robert Fulton's prototype built in Paris in 1800.

The author's tale of sophisticated submarine travel certainly captivated the imagination of Simon Lake and reportedly encouraged the innovative American youngster to dream of a world of underwater vehicles that had yet to exist. It is perhaps no coincidence that when Lake eventually designed and constructed his own submersible in 1895, the *Argonaut Jr.* had an underwater diver lockout system, just as Captain Nemo's *Nautilus* had. Furthermore, both undersea craft were ostensibly designed for underwater exploration, rather than war.

During his lifetime, Jules Verne produced some 60 novels, many of which served as inspiration to generations of scientists, engineers, inventors, and explorers. In their 1997 *Scientific American* article, "Jules Verne, Misunderstood Visionary," Arthur B. Evans and Ron Miller cite a number of notables who admit to being influenced by Verne's scientific visions; these include William Beebe (creator and pilot of the first bathysphere), Admiral Richard Byrd (pioneer explorer of Antarctica), Yuri Gagarin (first human to fly in space), and Neil Armstrong (first astronaut to walk on the moon).

What does the excitement of an action-packed, historic underwater thriller have to do with the future of subsea exploration today? Just as Verne's readers were contemplating the possibilities of travel to distant places for the first time in the mid-1800s, so we are taking on the challenge of exploration in previously inaccessible regions. However, now we have the technological capability of sending robotic devices to inhospitable or totally hazardous environments, where they can safely see, sense, and sample the harsh surroundings for us.

Lake Vostok

One such challenge is to penetrate the thick layer of ice covering Lake Vostok, the largest of some 76 subglacial lakes in Antarctica discovered in 1996 by British and Russian scientists. The discovery has been hailed as the biggest geographic feature discovered on Earth in the twentieth century. Not surprisingly, scientists have been eager to discover what microbial life might be locked in the remote lake's pristine waters. By 1998, Soviet scientists had already drilled some 3.5 kilometers (approx. 2.2 miles) into the ice—even so, they were still some 120 meters (approx. 400 ft) above the lake's surface. Ice crystals taken from this depth have already proven a goldmine to biologists, who have discovered microorganisms and rod-shaped bacteria. The drilling was stopped, however, so as not to contaminate the water with the kerosene used to keep the boreholes open.

Reaching Lake Vostok's long-sealed frigid waters to see what organisms may exist in this exceptionally difficult environment requires innovative equipment and methods. For example, a remotely operated ice probe, or **cryobot**, might melt its way down through the thick surface ice. Once at the lake, it could then release a separate robot, or **hydrobot**, that would sample water and lake-bottom sediment, as well as take photographs. This information would then be transmitted back to the surface by radio or cable; perhaps designers might even be able to find a way to retrieve actual samples.

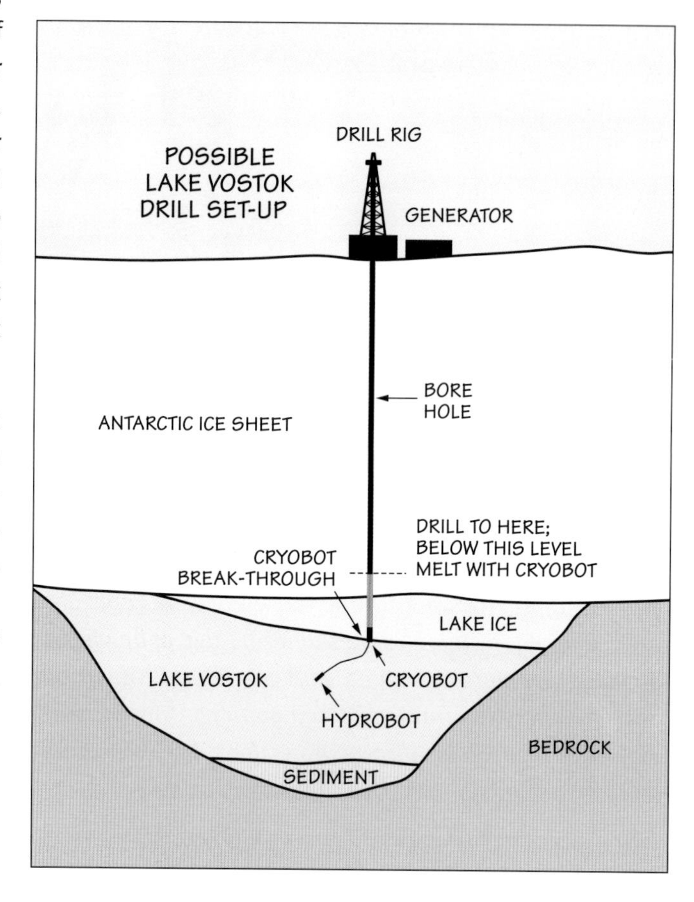

Certainly getting a probe down to Lake Vostok is a huge challenge, but the real test will be to obtain data without contaminating its sealed waters with surface organisms. After penetrating much of the ice shield, how does such a cryobot sterilize itself before entering the pristine water of this subglacial lake? Some scientists have suggested using a powerful ultraviolet light or a hydrogen peroxide spritz as sterilizing agents. Whatever technique is chosen, it seems wise to try it out first. Some scientists suggest targeting a smaller subglacial lake first; others recommend sending a cryobot through the ice shelf and into the Ross Sea off Antarctica, where any contamination from a failed sterilization procedure wouldn't be quite so catastrophic.

Jupiter's Europa

Mastering the technology to sample the ice-locked waters of Lake Vostok may be the first step in a somewhat similar but far more complex mission to see whether or not life might exist elsewhere in our solar system. In 1995, the space probe *Galileo* sent back intriguing data about Europa, one of the four moons orbiting the planet Jupiter. This particular young orb appears to have an ocean some 100 kilometers (approx. 62 miles) deep, covered by a rind of ice. Since Europa appears to be seismically active, scientists speculate the possibility of deep sea vents similar to those on our own planet. This brings up an astounding question: if life forms began in Earth's primordial ocean, might they not also be found in Europa's depths?

The first step in unlocking that riddle would be to launch a mission to confirm the existence of a subsurface ocean that to date has only been suggested by photographic data. Then the next logical mission would be to land and examine Europa's surface, utilizing remote robotic technology to sample the icy surface. Ultimately, it might be possible to design a cryobot, carried by a space capsule to a pre-selected site where it would melt through the icy crust and release a hydrobot that would reveal the secrets of Europa's deep ocean.

Both of these projects present some formidable hurdles. First, there's the difficulty of even accessing the sites. Then, there's the technological challenge of designing probes that can penetrate hundreds of meters, if not kilometers, of ice to reach the previously untouched water. There's also the matter of securing samples and either transmitting that data or retrieving actual materials. Such advanced technological maneuvers would, it is hoped, be carried out without contaminating the pristine bodies of water with microbes from the surface, the intermediate ice, or the probes themselves. All these challenges depend on obtaining obtaining sufficient funding and putting together a cooperative, multinational scientific investigation team.

If any of these tasks seem impossible, think back on the historic sagas of subsea pioneers who triumphed over insurmountable odds in their quests to discover and explore the unknown—and seemingly inaccessible—underwater world. Remember that throughout the last century, scientists debated whether anything could live in the harsh environment of the ocean depths. Now we know that the answer to that query is a resounding "yes!" Not only is there an amazing array of previously unknown life forms existing in the ocean, but some of them are nourished in ways previously thought impossible.

Perhaps the parallel question of the new century is whether life of any type—whether complex or simple—exists elsewhere in our solar system. Jules Verne would approve of a futuristic look at underwater exploration on far distant moons such as Europa. It's a near-fictional proposal based on cutting-edge scientific knowledge and experiments. Odds are that inventive minds will answer the question of life on other planets, always remembering that we may be in for some surprises!

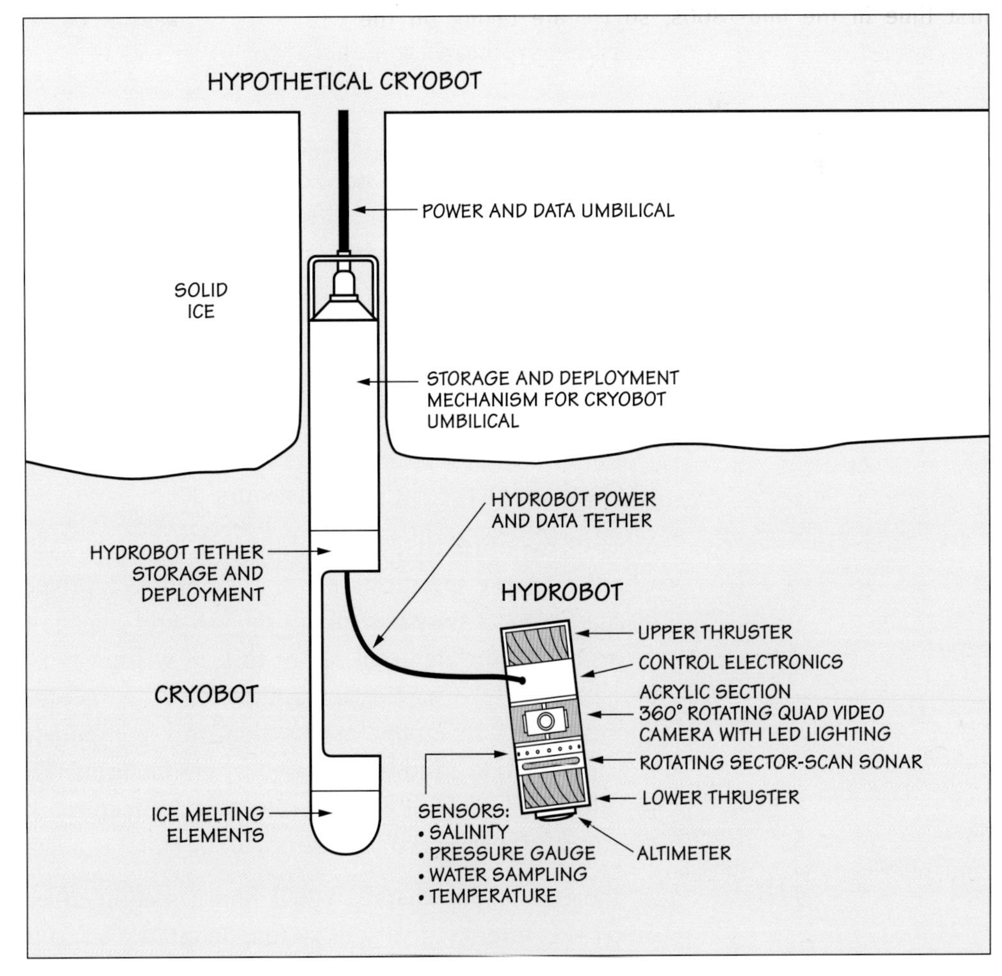

1. Introduction

Image courtesy of Massachusetts Institute of Technology Sea Grant College Program and students from Gloucester High School

Figure 12.2: First-Time ROVers

Students are now building ROVs in classrooms around the world. This hands-on learning fosters an interest in science and future education, teaches problem-solving skills, encourages interaction with mentors, and instills pride and confidence.

Image courtesy of Randall Fox

There are two excellent ways to learn about underwater vehicles—the first is to read everything you can about them. The other way is to actually build one—that's where the practical learning and fun really begin.

This chapter presents information and instructions for building an ROV called *SeaMATE.* This underwater robot project is designed to integrate the theoretical information you've been studying in this book with the construction of an actual shallow diving ROV capable of performing simple search and recovery operations in freshwater or saltwater to a depth of 10 meters (approx. 33 ft).

This final chapter is designed so that you can coordinate a section of the *SeaMATE* project with the corresponding technical chapter that details relevant information. For example, you might want to build the framework while studying *Chapter 4: Structure and Materials.* Alternatively, you may choose to construct the ROV independent of the textbook chapters—the choice is yours.

Most of the other chapters in this book have emphasized metric units. However, in Chapter 12, the specifications for the *SeaMATE* project are given in imperial units, because most of the parts sold in the U.S. and Canada are sized using this system. All prices quoted are applicable at the time of writing and are in U.S. dollars. When parts are specified by a particular manufacturer's name, this does not denote a product endorsement—it simply means that this component was readily available, was reasonably priced, and worked well when the *SeaMATE* prototype was built. Feel free to substitute comparable components manufactured by other companies, provided they meet similar specifications or can be suitably adapted.

At the end of this chapter, you'll find a section called *What's Next? Going Beyond SeaMATE.* It looks at various system adaptations and possibilities you might explore that would take your thinking and learning to the next level. That could consist of upgrading a specific vehicle system, such as building a more technically sophisticated payload tool, or it might involve a rethink of the entire vehicle concept, such as designing a deeper-diving ROV capable of working to 100 meters, or even an experimental AUV or a hybrid vehicle. The *What's Next?* section is a wrap-up of all the information you've learned throughout this book. It offers some guidance and encouragement for you to systematically explore the next level of challenges and solutions in the world of underwater robotics. Enjoy!

1.1. What Is *SeaMATE*?

SeaMATE is a small, robust, surface-powered ROV constructed from materials readily available from hardware stores, electronics shops, and hobby stores. Although the working depth limit of this ROV is only 33 feet, it follows the same physical principles that govern the design of larger, more complex vehicles. And it is definitely a fun project!

1.2. What Can It Do?

The *SeaMATE* underwater robot can do the following:

- operate to 33 feet in freshwater or saltwater
- maneuver forward or backward, turn right or left, spin right or left, and move up or down
- navigate under water using a compass (optional)
- use its small video camera and light for inspection missions, such as looking at the hull and prop(s) of a boat
- locate and recover small objects from shallow depths, e.g., keys lost off a dock
- be modified for use in ROV competitions or science fairs
- function as a learning aid for science classes and technical subjects
- serve as an introductory project, so you can gain the skills and knowledge necessary for building more advanced underwater vehicles

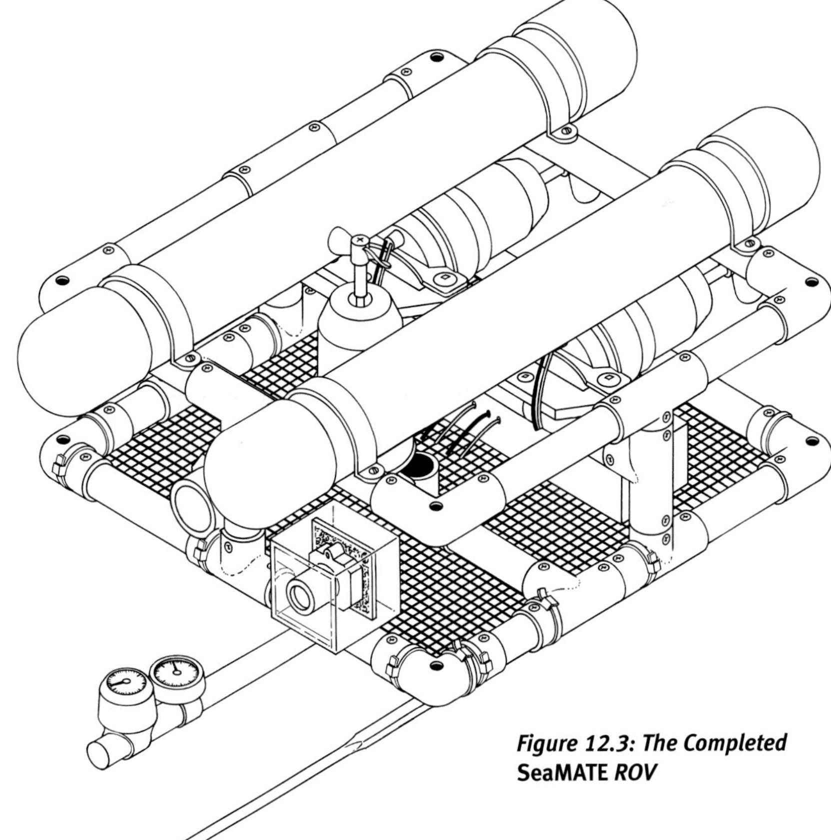

Figure 12.3: The Completed* SeaMATE *ROV

1.3. What Can't It Do?

SeaMATE cannot:

- operate reliably in depths greater than 33 feet without modifications
- carry payloads and/or instrumentation weighing more than a pound in water without additional flotation
- navigate in a perfectly straight line

- travel more quickly than 1.6 feet per second (approx. 0.5 m/s) or operate effectively in currents faster than a few inches per second

The *SeaMATE* project offers pragmatic advice and plans for building an unmanned, shallow diving underwater vehicle. As your experience and mission requirements grow, you may want to move beyond these suggestions. That's when you might want to dip into the section *What's Next? Going Beyond SeaMATE* at the end of this chapter.

2. Building the *SeaMATE* ROV

2.1. Budget

If you purchase all parts and materials, the *SeaMATE* underwater robot will probably cost no more than $200. Note that this estimate includes a video camera and electrical components, but does *not* include a power source, depth gauge, or compass. Nor does it include any tools or test instruments. Budget costs for component parts and tools/ test instruments can be reduced by scrounging materials, asking for donations, and by shopping carefully.

2.2. Skills

SeaMATE is a relatively simple ROV to build, but it requires some rudimentary skills in order to tackle this project safely and successfully. You should be able (or willing to learn) to:

- sketch simple plans on graph paper
- read simple construction sketches
- cut and fit PVC pipe and fittings together
- follow simple wiring diagrams (schematics)
- build simple circuits (strip, splice, and solder wires, use heat shrink tubing and cable ties, run conductors neatly, etc.)
- measure voltage, current, and resistance with a multimeter
- drill and tap a hole for a threaded fitting or fastener

SeaMATE expands upon the ROV projects found in the book *Build Your Own Underwater Robot and Other Wet Projects* by Harry Bohm and Vickie Jensen. That book has basic tips on cutting PVC pipe, soldering, etc. If you need help with other skills, any library will have books on basic electricity and tool use. But the best way to learn is to ask for instruction from someone who has already mastered these skills.

2.3. Workspace

Before starting on any ROV project, it is important to set up a place to work. Some people will already have access to a workshop space. Students may be able to use a school's industrial training shop, an electronics classroom, or a science lab. Even if you have no dedicated workspace available, it's relatively easy to put together a mini underwater robotics workshop in which to build and test *SeaMATE* in a garage, basement, kitchen, bedroom, or regular classroom.

An underwater robotics workspace should include the following elements:

- a workbench with access to power

TECH NOTE: THE VALUE OF GUIDED HANDS-ON LEARNING

Image courtesy of Randall Fox

Figure 12.4: Taking on the Challenge

When Frank Barrows took on the challenge of teaching MATE's first ever ROV class at Monterey Peninsula College in 2000, he stepped up to the plate and made the transition from teaching automotive technology to subsea robotics. "It's no different than building any engine," he would remind students, "except it has to be able to work under water!"

Chapter 12: SeaMATE makes the transition from theory to hands-on learning. This is where it all becomes real! Students and teachers agree that nothing equals these "do it yourself" understandings.

However, one key factor can elevate this experience to an even higher level—and that's an expert, innovative mentor who knows how to motivate students. Frank Barrows did that for his classes at Monterey Peninsula College when he tasked them with the mission of recovering deadly "bozonium" canisters off the bottom of Monterey Harbor. (See Chapter 2 to review those mission details.)

Dr. Tom Consi is another inventive teacher who added a novel twist to his ROV mission. While at MIT's Department of Ocean Engineering, Consi challenged introductory students to build their own PVC ROV. This hands-on learning was interspersed with targeted mini-lectures on a variety of ocean engineering subjects, such as pressure housings, hydrodynamics, and motors. When the fleet of ROVs was completed, the class headed to the New England Aquarium for sea trials off the dock in Boston Harbor. But like any resourceful teacher, Consi had "prepped" the site ahead of time with suitable bait. So when his students attached tasty squid to their ROV frames with cable ties and lowered the vehicles into the water for a look around, they encountered an "attack of the killer crabs," a vivid battle captured on their ROV cameras. Not surprisingly, these students were keen to pursue further underwater studies!

- a set of basic tools and test equipment (e.g., a drill, heat gun, multimeter, etc.)
- a small water tank, aptly named a test tank, for testing your vehicle and its parts (Figure 12.1 and 12.5). Note that when it comes time for sea trials, you'll need access to a pool or similar site for testing your ROV and practicing piloting skills.)

2.3.1. Workbench

A small table or desk that's at least 2 feet x 4 feet can function as a workbench. A chair or stool is a handy addition. If there is no wall outlet nearby, use a power strip (also called a power bar) extension cord to bring power to the bench. Mount a swing-arm lamp to one side of the workbench for extra illumination. Adding a small shelf along the back edge of the worktable will serve as a convenient place for small-parts bins and for a toolbox that holds your hand tools. Include space for a soldering iron, a power supply, a multimeter, and a small fixed or portable bench vise. Below the table add a small set of drawers or shelves to hold all your larger tools, such as an electric drill. Equally handy is a larger shelf or some stackable rubber containers with lids for your robot, its parts, and the numerous odds and ends needed for such projects.

2.3.2. Test Tank

When building an underwater robot, it's important to have some sort of water tank to test various project components. But don't worry—all that's really needed initially for the *SeaMATE* ROV is a large, clean garbage can filled with water. This can be placed outside with a cover, or if you have a basement or garage location for your workspace, it can go beside the workbench. Keep sponges, rags, or paper towels handy for wiping up spills. It's not smart to put a test tank in a kitchen, bedroom, or classroom unless the floor is of a water-resistant material. In a pinch, it's possible to use a bathtub as a test tank, although it may not be deep enough for some of the testing required. Of course, more advanced projects may require a larger test tank or pool. And once *SeaMATE* is completed, you'll want to conduct sea trials in a larger pool.

Image courtesy of Peter Thain

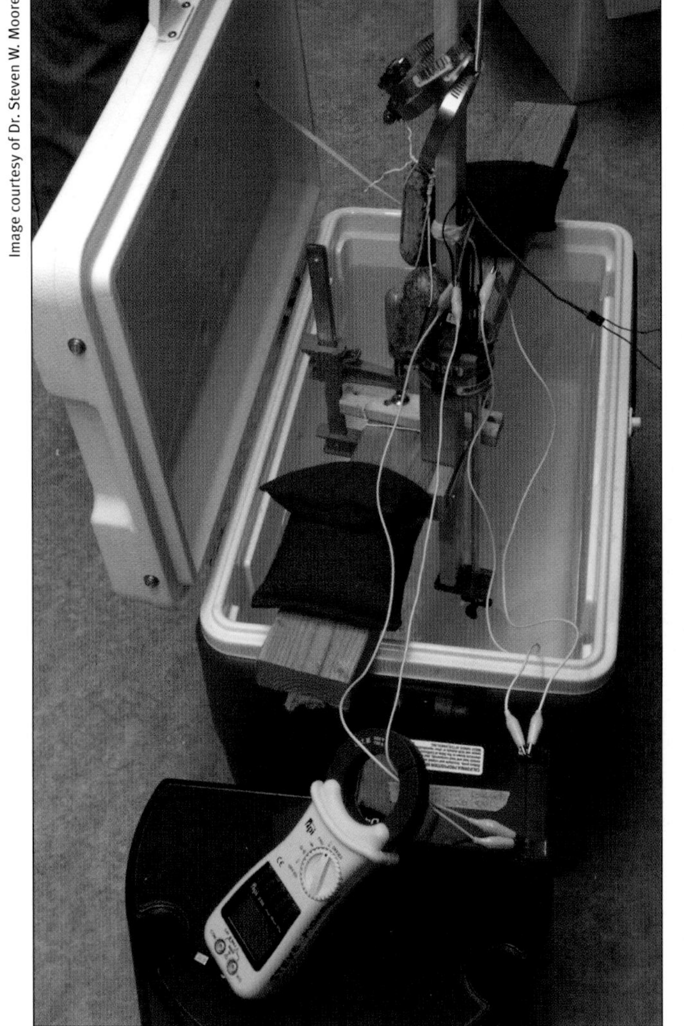

Image courtesy of Dr. Steven W. Moore

Image courtesy of David Warner, Rancho Santa Fe School

Figure 12.5: Test Tank Options

Among the many test tank options are livestock tanks (plastic and metal), inflatable pools, and even a food cooler.

SAFETY NOTE: WORKSPACE SAFETY

It is always wise to keep safety in mind, even in a small workplace. If you get into the habit of thinking about safety now, that mindset will carry over to future projects and jobs. Here are the safety basics for the *SeaMATE* project:

Images courtesy of Harry Bohm

- Keep your workbench tidy. Messes lead to accidents.
- If you're working around water (for example, next to a test tank), make sure all of your power comes through a GFI-protected outlet. (For an explanation of GFIs, see *Chapter 3: Working in Water* and *Chapter 8: Power Systems*.)
- Make sure all your power tools get turned off after a work session. It's easy to do this if you run all your workbench power tools from a power strip (or power bar) and make sure to turn it off or unplug it when you leave.
- Unplug and stow power tools after you have finished any work session.
- Make sure you have enough illumination in the workspace.
- Roll up sleeves and tuck in loose clothing when working. If you have long hair, pull it back out of harm's way. Pay attention to jewelry that could be a snagging hazard with power tools or that could burn you if the jewelry shorts out a battery while you're working with hot or rotating power tools.
- Use safety glasses when drilling, cutting materials, and soldering.
- Use protective rubber gloves when handling solvents and glues.
- Read all warning labels and ventilate the workspace when using solvents or glues. Use recommended respirators or dust masks when indicated by the product or activity.
- If in doubt as to how to use a power tool, ask for help.
- Use power saws and other tools that have the potential for serious injury only when others who can render assistance are nearby.

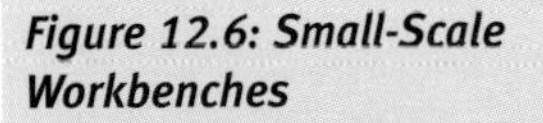

Figure 12.6: Small-Scale Workbenches

These small workbenches are suitable for building a "hardware-store technology" type of ROV.

2.3.3. Cart

A lightweight ROV can be carried easily by hand, but if you're working with larger vehicles and tethers, you'll want to keep your eyes out for some kind of sturdy cart for hauling robots, tethers, and other gear (Figure 12.7).

Figure 12.7: Lots of Transport Options

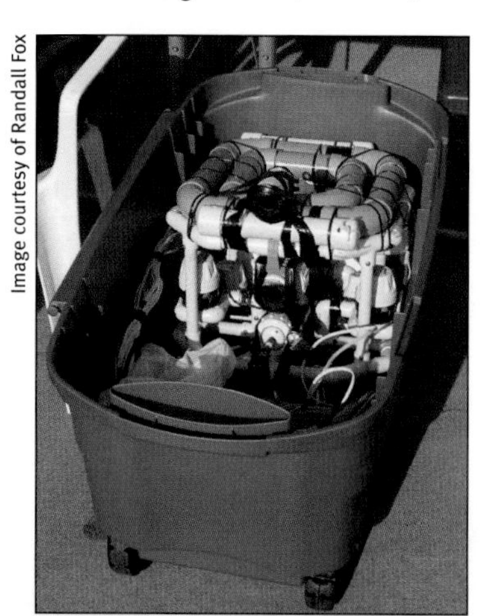

Image courtesy of Randall Fox

Image courtesy of Harry Bohm

Image courtesy of Randall Fox

Image courtesy of Scott Coté, Coté Consulting

SAFETY NOTE: ELECTRICITY AND WATER DO NOT MIX

Whenever you put an undersea vehicle into water, there's the potential for getting an electrically conductive medium (the water) in contact with electrical circuitry. This isn't *supposed* to occur with a properly designed and built vehicle, but water has a nasty habit of finding its way into places where it shouldn't be. For that reason, it is always good practice to follow these general safety guidelines when operating electrical devices around or in the water.

- Any time you're plugging into an outlet for power, make sure you reduce your risk of serious electric shock by using a GFI outlet.
- Fuse all AC and DC circuits.
- Dry your hands and workbench of any water before energizing any electrical devices.
- Disconnect the power from any circuit before removing protective coverings. This includes removing circuitry from pressure cans on any underwater vehicle.
- This textbook recommends you avoid the use of high-voltage circuitry entirely, but if you do use them on an underwater vehicle, make sure you mark any high-voltage sources with distinctive warning labels.
- Protect all exposed electrical circuitry from spray or rain.

Image courtesy of Harry Bohm

Figure 12.8: Basic Tools

Common tools include a soldering iron, multimeter, 12-volt DC power supply, heat gun, needle-nose pliers, side cutters, small screwdrivers, and a vise, among others. The bench is well lit with a swing-arm lamp. A couple of power strips (power bars) make it easy to plug in power supplies and tools.

2.4. Basic Tools, Devices, and Supplies

Building any undersea vehicle such as *SeaMATE* requires a selection of basic workshop tools, as well as devices such as electronic test equipment and a power supply. If you're not familiar with using them, you'll need an experienced person to show you how to do so safely and effectively. The list that follows is not comprehensive, but it will get you started. Add tools and devices as your own needs and budget dictate.

The minimum requirement is an assortment of saws, drills, files, screwdrivers, wrenches, and pliers. Access to a few power tools is great, but not absolutely necessary. Power tools will definitely make fabrication easier and generally improve the quality of the finished product. However, all power tools designed for cutting or drilling can be hazardous if not used properly, so make sure you've received appropriate training in how to use a power tool safely, and remember, never work alone when using potentially dangerous tools.

2.4.1. Basic Tools and Test Equipment

Screwdrivers: You'll want an assortment of various-sized screwdrivers, including different sizes of flat-bladed, Phillips (with the star-shaped end) or Robertsons (the square end commonly used in Canada). Get a Phillips #1 and #2 or a Robertson #1, #2, and #3.

Pliers: A variety of pliers is necessary for a range of jobs. These include regular pliers, a small needle-nose pliers, a wire stripper, small diagonal wire cutter.

Saws and cutters: You'll need a hack saw or back saw. A woodworking saw will be ok, but other types with finer teeth that can handle metal and plastic are much more effective. For cutting PVC pipe, you can use a back saw and miter box, a PVC pipe cutter, or a plumber's tubing cutter (Figure 12.17).

Drill and drill bits: A light-duty, variable-speed, battery-powered drill (sometimes called a "drill-driver" because it can be used as a screwdriver, too) with a standard 3/8th-inch) drill chuck is ideal. There's no need to buy expensive drill bits since most of the drilling for this project is in plastic rather than metal. But you will need a set of drill bits ranging in size from 1/16th to 3/8th inch. (Note that imperial sizes are commonly used in the U.S. and Canada; metric sizes are used in most other countries.)

Soldering equipment: A basic soldering set-up includes a soldering iron, holder, rosin core solder, and a solder-wiping sponge. There are two basic types of soldering irons—portable and bench-type. The portable irons are the least expensive. If you do buy one of these, purchase a higher wattage one and get a stand for it. The stand makes it easy to use the iron and prevents accidents (e.g., burning the table, burning your hand, or having the iron roll off the bench). The bench-type soldering iron integrates a transformer and holder into one unit, with the iron placed in the holder. Professionals prefer the more expensive bench soldering irons because they have better heat regulation. Regardless of the type of soldering iron you buy, also get a solder-wiping sponge and a roll of rosin core solder. Note that lead-free solder is available and certainly is environmentally kinder and less toxic for the user. Solder only in a well ventilated workspace.

Figure 12.9: Learning with a Mentor

Working with an experienced mentor is an excellent way to learn how to solder effectively and safely.

Images courtesy of Randall Fox

Figure 12.10: Soldering Close-Up

When soldering, a clamp or small vise holds the piece steady and helps avoid burned fingers.

Taps and dies: For cutting screw threads inside a hole you use a "tap," and for putting threads on the outside of a shaft you use a "die." You can buy small "tap and die" sets that include an assortment of different taps and dies along with handles to turn them. *SeaMATE* uses a 1/8th inch NPT (National Pipe Tapered) pipe tap, which is included in some sets. If not, you can buy it separately. Be careful not to confuse NPT (tapered) tap with the NPS (National Pipe Straight) tap.

Measuring and drawing tools: You'll want both a ruler (a see-through type is especially good) and a measuring tape, as well as a calculator. Basic drawing tools consist of pencils (including colored) and eraser, a drawing compass, T-square, straightedge, protractor, shape template, and graph paper. On a computer, you'll want some type of CAD drawing program. Machinist layout tools include calipers, micrometer, and Vernier scale.

Power strip (also called a **power bar**): This is a type of extension cord that has multiple outlets at one end. Mount the power strip in a convenient spot on the bench so you can easily plug and unplug devices without having to duck under the table.

Figure 12.11: Electric Glue Gun in Action

Multimeter: This handy device measures voltage, current, resistance, and continuity. There are many types of multimeters; the smallest and least expensive (around $20 to $30 dollars) is quite adequate, but if you get one with a digital display rather than a moving needle, it will be easier to read. It's also wise to choose one that can read up to 10 amps, though these may be more expensive. Make sure to read and save the multimeter manual; alternatively, you can find general instructions on the web. If you're going to be measuring larger currents, you may want to invest in a clamp meter or current shunt.

Heat gun: This device resembles a hair dryer but puts out a very hot stream of air and is useful for many operations in building an ROV. The most common use is for shrinking electrical insulation tubes ("heat shrink tubing") to effectively insulate bare electric conductors. An inexpensive heat gun from a paint, hardware, or hobby store is all that's needed. Just be careful when using a heat gun—it's very easy to burn and melt things with this tool.

Electric glue gun and glue sticks: The electric glue gun is another indispensable project tool used to make fast glue jobs. Again, beware of the tip; it gets really hot and can cause a nasty burn. A tin or aluminum pie plate makes a safe spot to lay the glue gun when in use, or buy a glue gun that comes with a stand.

Fixed or portable bench vise: A bench vise holds small parts securely when soldering, drilling, cutting, or filing. It's practically indispensable, since it functions as your third hand. Portable vices clamp onto the table edge or just stand on their own pedestal. Fixed vises are fastened down to the bench top with screws or bolts.

Old electric frying pan: Melting wax for potting thrusters, etc., can be a dangerous procedure if not done properly. An old electric frying pan is one of the safest methods to heat water until hot enough to melt wax, so look for a used one in a second-hand store. Never attempt to melt wax in a microwave oven. Don't try to melt it on a stove, even in a double-boiler, because wax spilled onto a hot burner can combust easily.

SAFETY NOTE: HEAT HAZARD

Soldering irons, electric glue guns, and hot-air heat guns are high-heat sources that can cause severe burns. Avoid touching the tip of a soldering iron or glue gun or putting your hand in front of the hot-air stream—it's hot enough to blister paint! When not in use, put your soldering iron in its holder and lay the hot glue gun in a metal tray or aluminum pie plate. Read the manufacturer's instructions and precautions that come with these tools. Always remember to unplug these devices when you're done with them.

2.4.2. Optional Tools and Test Equipment

These tools are not necessary for constructing the *SeaMATE* ROV, but they are useful to have, particularly if you are planning on building a deeper-diving ROV. Be sure to observe all safety precautions when using these tools.

Power saw: A power saw is nice for rapid and precise cutting of metal and plastic materials. Different types of saws are suited to different jobs. For example, a table saw works well for flat sheets of plastic, metal, or wood. A chop saw or band saw works better for cutting long bars or pipes. An electric jigsaw does a good job of cutting curves and making irregular cuts in plastic, metal, and wood sheets. Different types of materials may also require specialty blades. For example, a regular blade for sawing wood may chip or shatter acrylic plastic.

Drill press: This machine is necessary to accurately drill brackets and structural parts on a deeper-diving ROV. A standard benchtop model is fine and will allow for more precise and safer work than a hand-held drill.

Metal lathe: A small metal lathe with a 4-inch (approx. 10-cm) swing allows an experienced operator to machine endcaps and O-ring grooves for very small pressure cans. (One example is TAIG Tools' Micro Lathe II for about $175; note that you'll probably want or need several accessories, which can quickly increase the price.) Alternatively, you can find a machine shop that will fabricate these parts for you, but that's usually costly, too.

Oscilloscope: This piece of test equipment is used to measure and display rapidly-changing voltage signals in complex circuits, such as electronic control systems. You won't need one for a basic *SeaMATE*-type ROV, but you will need access to an oscilloscope for troubleshooting more advanced circuits. Good oscilloscopes are pricey (typically $1000 or more), but if you're willing to live without certain features you can find less expensive ones, including ones that plug into your computer and use its screen for display. Many schools and colleges use them for teaching physics or electronics classes, so you may be able to access one there.

SAFETY NOTE: SAFE USE OF ELECTRIC POWER TOOLS

- Note that any type of power saw or power drill can be a very dangerous tool if misused. So learn how to operate a power tool safely and correctly. When using any potentially dangerous tool, don't work alone.
- Always read and follow the manufacturer's safety guidelines that come with any new power tool.
- If you are at all unsure about how to use the tool, find someone who is knowledgeable and can teach its proper use. Another alternative is to find someone who can assist by fabricating the parts you need.

SAFETY NOTE: LEAD-ACID BATTERY WARNING

WARNING: Each year in the United States, OSHA lists some 80,000 reported injuries associated with car batteries and similar wet lead-acid batteries. So review the safety precautions covered in *Chapter 8: Power Sources* before using these kinds of batteries.

2.4.3. Power Sources

The *SeaMATE* project requires 12 volts DC. There are two common ways to get this. One is to use a 12-volt battery; the other is to use a power supply that coverts 120-volt AC household current to 12 volts DC. (Battery and power supply options are discussed in greater detail in *Chapter 8: Power Systems.*)

Batteries: A 12-volt sealed lead-acid (SLA) battery (with a minimum 6 amp-hour capacity) is ideal. These batteries are small, do not require a protective case, and can be placed right on top of the workbench. The common car battery or "wet" lead-acid battery is not recommended unless you take special precautions to handle and charge it safely. (See *Safety Note: Lead-Acid Battery Warning.*) Regardless of what type of battery you choose, make sure to attach an in-line fuse.

Power supply: Be advised that most inexpensive power supplies cannot deliver enough current to operate thruster motors. Make sure the power supply you select can deliver at least 10 amps. Those that can tend to be expensive. For this reason, a battery and battery charger may be a better choice.

TECH NOTE: USING AN IN-LINE FUSE

Whenever you are using battery power, be sure to attach an in-line fuse between your battery and your ROV. The current rating of the fuse must be greater than the anticipated current through the circuit under normal operating conditions, but less than the amount of current it would take to overheat wires or other parts of the circuit.

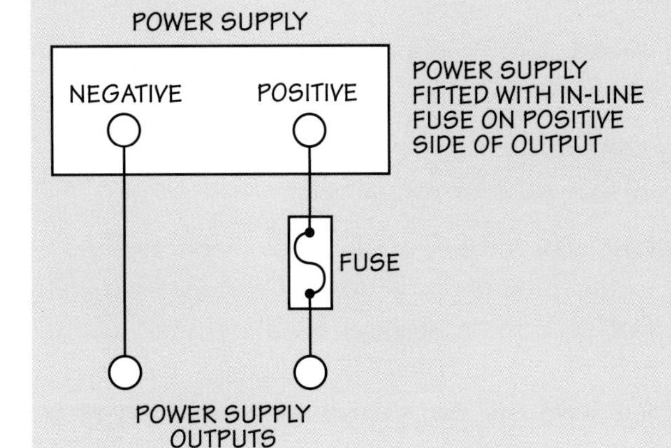

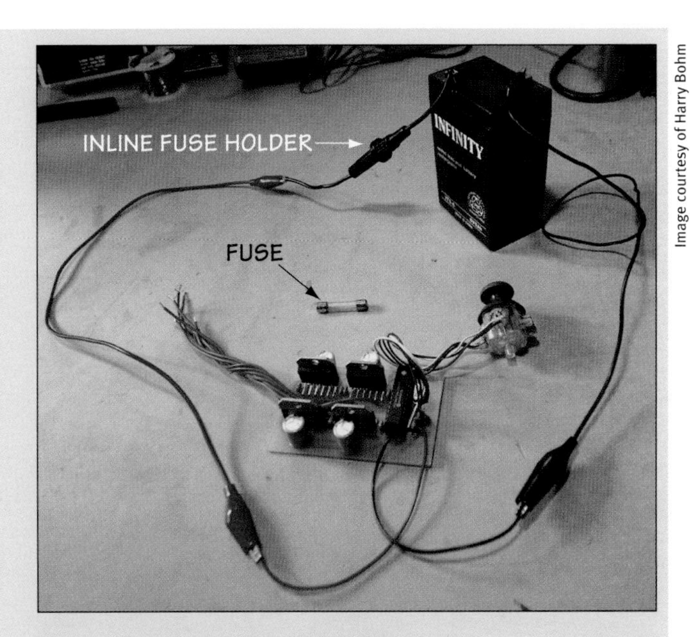

Image courtesy of Harry Bohm

Figure 12.12: In-Line Fuse Schematic and Example

In this photo, the battery features an in-line fuse wired into the positive terminal and hooked up to a motor controller for testing. This type of fuse holder uses common glass tube-style fuses (also called cartridge fuses) found in in most automotive shops.

2.4.4. Essential Supplies

Tape: a roll each of black electrical tape, duct tape, and masking tape

50–100 pieces of 1/8-inch x 12-inch cable ties: Sizes vary from manufacturer to manufacturer, so find a size that is close to this specification.

Sandpaper: a couple of sheets of fine and medium grit

Heavy-duty scissors capable of cutting plastic screen

Old ceramic cup or mug for melting wax

Isopropyl alcohol (70%)

Plywood scraps: 1 or 2 pieces, 4 inches x 4 inches x 3/4th inch, to place under your work when drilling to prevent the drill from damaging any work surface.

Lint-free tissues or cloth for cleaning camera ports

Marking and note-taking supplies: permanent marker pen with medium-point tip, pencils, and notepad

General cleanup: paper towels, rags, and newspapers, plus a mild, all-purpose spray cleaner for countertops

2.4.5. Safety Gear

Safety glasses: These are essential to protect your eyes, so it's smart to wear them whenever using power tools, soldering, pouring caustic liquids, or working around

Image courtesy of Vickie Jensen

Figure 12.13: Safety Gear to the Max!

Wearing safety gear is always a good idea. In this photo, Scott Coté is the safety stylin' guy and emcee at an underwater robot competition.

"wet" lead-acid batteries. Chips can easily hit your eye when you're cutting or drilling metal or plastic materials. It's a good safety habit to wear safety glasses whenever you're around a workbench, no matter what procedure is under way.

Rubber, latex, or nitrile gloves: These protect your hands from glue, chemicals, or anything that could harm the skin. Note: some chemicals will eat through certain types of protective gloves, so read the product recommendations for a safe type of glove material. Also note that some people have latex allergies.

Ear protectors: Many power tools put out a high level of potentially ear-damaging noise. Use earplugs or headphone-style protectors if you, or someone nearby, is using noisy tools.

First aid kit: Get a small basic first aid kit and keep it near your work area.

Fire extinguisher: Every workspace should have a small kitchen-style ABC fire extinguisher.

Ventilation fan and dust masks: A few chemical products used in this project can be toxic if not used correctly. Always read the manufacturer's instructions on storage, handling, and use. It's important to have lots of fresh air moving around the workbench when using paints, glues, and cements, so open the windows and use a small fan to help create air movement. To avoid breathing excess dust created by cutting plastics and wood with power tools, wear a filter mask.

Oven mitts: Use these when handling hot materials, such as hot melted wax.

3. Sourcing Materials

3.1. Shopping for Parts and Materials

While it's not quite time to start building, it *is* time to start buying! This section of Chapter 12 lists all the parts and supplies needed to construct *SeaMATE.* But before running off to the store, take a little time and skim through all the project sections. This will provide an overall picture of what to expect and what supplies you might need (or already have). You may also come up with questions. Write them down and get answers *before* spending any money.

The local hardware store or home improvement store is a good place to start when sourcing structural materials, tools, and devices for any underwater vehicle project. However, unless your local store features an "underwater vehicle supplies" aisle, it will probably be necessary to use your imagination and creativity, inventing underwater vehicle parts out of whatever is available. For example, sheets of polycarbonate or acrylic plastic that might be used for a camera viewport will likely be found in the window repair section. Extruded aluminum, a possible framing material, may be found in the bathroom section, because it's often used for the trim around shower doors.

Make it a point to wander up and down *all* the aisles of the store, taking mental notes on the whereabouts of anything that looks promising for underwater vehicle construction. Better yet, make some notes in your notebook. Plumbing and outdoor areas can be particularly productive (that's where you'll find the PVC and ABS plastic pipe and fittings), because many items in these sections are designed to be compatible with water. Don't forget to look for items at hobby shops and electronics stores, to get ideas before plunking down any major cash.

For specialized stuff like sensors, check out dive shops or automotive supply places. Secondhand stores sometimes yield real treasures, such as geared toys that can be cannibalized for mechanical arms. If you live in a rural or remote area, you may have to do most of your shopping on-line. With a computer, internet access, and a credit card, it's easy to order exactly what you need at any of hundreds of on-line distributors and enjoy the benefit of having materials delivered directly to your doorstep. *Appendix V: How to Find Parts* provides suggestions for researching sources for parts and materials.

Regardless of how you do your shopping, jot down the various options in a notebook, including what they cost and where to get them. If you are ordering materials, remember to find out if things are in stock and to include taxes and shipping expenses in your budget calculations. Taking time to do this kind of research will provide a variety of ideas, teach you a lot about options and availability, *and* probably save you some money.

When the parts list in this chapter specifies a particular material or piece of equipment, know that this is the component that was readily available at the time of publication, that was reasonably priced, and that worked well. However, it may not be available in your area or on the day you need it. Feel free to substitute, modify, or adapt. Just remember that any significant change will likely impact other components or subsystems.

There are two approaches to gathering materials. One is to buy all the parts at the beginning, before starting any construction. The other is to procure parts as you go along. Each approach has its advantages.

Similarly, you can choose to build *SeaMATE* in any order that works for you. For example, if you want to start with the video camera or the control box and build the frame last, that shouldn't be a problem. However, it is a good idea to read through all the helpful Tech Notes and Safety Notes that are in every *SeaMATE* project section first—they're excellent advice (often based on lessons learned the hard way) that could save you time and trouble.

TECH NOTE: SUBSTITUTING PARTS

If you cannot obtain the parts in this list, it's perfectly ok to use substitutes or modify the design of *SeaMATE* to reflect availability of parts in your area.

Just be aware that substituting parts might affect other components in the vehicle's subsystems. For example, changing the diameter of the floats will require a different size of pipe clamps for attaching them to the frame. Or adding extra thrusters will change the number of wires in the tether.

The point is that all underwater vehicle systems are integrated, so when you change one component, check to see how it might affect the performance or capability of other parts.

SAFETY NOTE: MATERIAL SAFETY DATA SHEETS

When you are working with glues, paints, lubricants, solvents, and other potentially harmful substances, you should always review and keep on file a copy of the **Material Safety Data Sheet**, or **MSDS**, for that substance.

An MSDS is a short document designed to inform workers and emergency personnel about potential safety issues associated with a substance; it also provides guidelines for safe storage, use, and disposal of the substance. Most MSDSs are a few pages long. Regardless of length, each MSDS includes a basic description of the substance and its physical properties, as well as an accounting of potential hazards associated with the substance, including its degree of flammability, any tendency to explode, short-term toxic effects of exposure (e.g., chemical burns, allergic reactions, etc.), and long-term toxic effects (e.g., cancer, nerve damage, etc.).

MSDSs are available for free from the product manufacturer or on-line. Many schools, colleges, and universities also keep a wide selection of MSDSs on file to cover all the substances used in labs and workshops. To find out more, just search for "MSDS" on the web.

3.2. *SeaMATE* Parts List

The complete parts list for *SeaMATE* is given below for easy reference. Specific subsystem parts are also repeated under the appropriate section in the chapter.

3.2.1. Miscellaneous Hardware and Supplies

Various fasteners and supplies are used throughout the project. They're not specified in the subsystems parts lists, but should be part of the general supplies of any workbench.

- 100 #6 x ½-inch pan- or round-head stainless steel sheet metal screws (Note that this number of screws is in excess of what you'll actually need for the project, but it is usually much cheaper to buy in lots of 100 or by the box, rather than purchasing individual screws. Also, Phillips or Robertson screws are easier to use than flat-slotted screws, since your screwdriver won't slip off as easily.)
- 100 #6 stainless steel washers (As noted above, you won't need this many, but purchasing by the box saves money.)
- one small can of PVC cement
- 50–100 pieces of 1/8-inch x 12-inch cable ties (You'll have extras and will save money buying them in lots of 100 or by the package.)
- three packs of toilet bowl ring sealant wax
- one box of toothpicks

TECH NOTE: SCHEDULE 40, 80, AND 120 PIPE AND FITTINGS

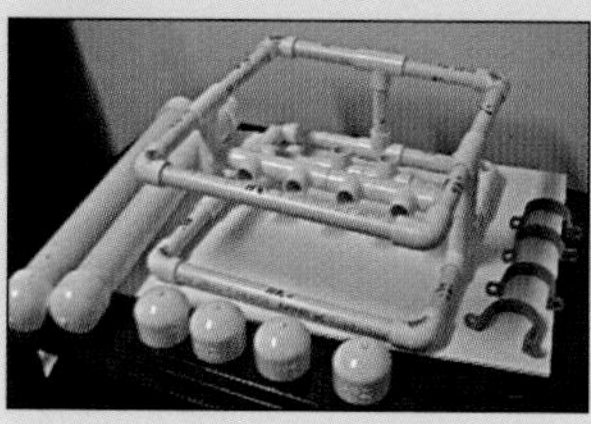

Image courtesy of Harry Bohm

***Figure 12.14: PVC Materials for* SeaMATE**

SeaMATE uses PVC pipe for its structure, since this plastic pipe is so widely available and can handle the ROV's depth rating. *Chapter 4: Structure and Materials* provides a good deal of information about PVC pipe, sizes, and fittings (see *Tech Note: PVC Pipe Grades* in *Section 6.2. Plastics* of that chapter), as well as other materials that can be used for ROV structure.

Remember that PVC pipe diameter designations, like "½-inch," are standardized but are not actually the size they claim they are! If the text says to get some ½-inch pipe, it's referring to the stuff that's classified, labeled, and sold as "½-inch" pipe, even though your tape measure will try to convince you it's something else.

3.2.2. Structural Subsystem Parts

- 10 feet of ½-inch Schedule 40 PVC pipe. Alternatively, you can use Class 200 PVC pipe, which is also ½-inch pipe.
- eight ½-inch Schedule 40 PVC elbow fittings
- sixteen ½-inch Schedule 40 PVC tee fittings
- one 16-inch x 16-inch square of ¼-inch mesh plastic screen material to serve as the "floor" of your ROV. This plastic hardware cloth should be available at your local hardware store, but you can also improvise.

3.2.3. Ballast Subsystem Parts

- one 36-inch (or longer) piece of 1½-inch Schedule 40 or Class 160 PVC pipe
- four 1½-inch Schedule 40 PVC endcap fittings
- four 1½-inch plastic PVC pipe clamps

- approximately 2 pounds of lead weights or equivalent: Scuba weights or deep sea fishing weights work well for this purpose. Try to get several smaller weights rather than just one large one, since you won't know the exact quantity of weights needed for ballast until sea trials. Weights that have some sort of slot or loop will be easier to attach securely to your ROV.

3.2.4. Tether, Junction Box, and Power Distribution Subsystems Parts

- 200 feet of #22 AWG (American Wire Gauge) speaker wire
- 50 feet of RG-174 coaxial cable (for camera video signal)
- one plastic project box about 4 inches wide x 6 inches long x 2 inches high: A smaller box will work, but it becomes difficult to do any wiring inside it.

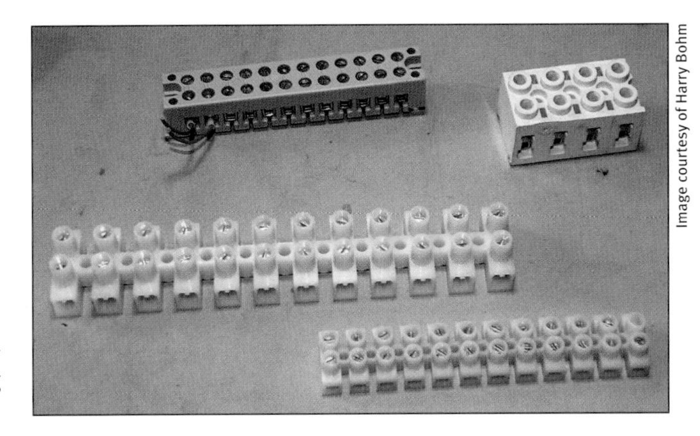

Image courtesy of Harry Bohm

Figure 12.15: Barrier (Terminal) Strip Options

Various styles of barrier strips are available. Note that barrier or terminal strips are referred to as connector strips, or chocolate blocks, in the U.K.

- one 12-terminal miniature barrier strip #22 AWG gauge (also referred to as terminal strip)
- five RCA plugs: Omit these if you are wiring the tether directly to the control box.
- 50 feet of polypropylene hollow braid rope, 3/8 inch in diameter (optional)
- one 3/8-inch fid: A fid is designed for splicing hollow braid line, but in this project, it is used for threading the tether conductors inside the hollow braid rope. Remember to buy a fid at a marine supply shop when you purchase your polypropylene rope. (An easy alternative is to use a chopstick in place of a fid.)

3.2.5. Propulsion Subsystem Parts

- three Johnson Pump Company bilge pump cartridges part no. 32-1550C - 550 gph (gallons per hour): This is the replaceable motor cartridge part of a 12-volt bilge pump made by the Johnson Pump Company. Note that for this project, you need buy only the cartridge portion and not the complete bilge pump.
- three #6 32 x ¾-inch stainless steel bolts
- three #6 x 1-inch threaded standoffs
- three Dumas propellers (model #3003) 1½-inch diameter 2-bladed nylon prop

3.2.6. Sensor/Navigation Subsystem Parts

Underwater Camera Parts

- one CCD (charge-coupled device) video camera, such as the PC 166XS sold by Supercircuits: Note that many tiny, inexpensive video cameras are available—check out various websites, catalogs, and stores in your area. Make absolutely sure that any camera you get is compatible with the 12 VDC battery power used in *SeaMATE.* Many of the smaller video cameras are designed for use with 5 volt DC supplies or low-voltage AC power; those models generally will not work with a 12 VDC supply and may be permanently damaged by connection to such a supply.
- a cable running between camera and junction box for camera power and video signal: Note that this cable may come with the camera or you may have to make it or extend it yourself.

- one transparent plastic box that will hold the camera: This clear box needs to be at least 1½ inches wide by 1½ inches long by 1½ inches deep in order to fit most small video cameras. No lid is necessary. The part of the box that the camera will look through needs to be optically clear with no seams, ripples, lines, etc.
- one video monitor
- 1 small tube of silicone sealant (5-minute epoxy is a good alternative): Get the smallest amount you can buy, since you'll be using only a dab. Silicone is used to secure the camera lens to the bottom of the transparent box. Use a toothpick or sliver of wood for an applicator.
- one 8-oz epoxy or polyurethane compound kit for potting: Get compound with a longer curing time (1-2 hours). This is available at hobby shops, hardware stores, fiberglass supply outlets or via the internet. Toilet bowl wax is another option.
- one cable strain relief: for camera cable to junction box
- 1 foot of 1/8th-inch heat shrink tubing (if splicing the video cable to the camera) and a heat gun tool for heat source

Underwater Lighting

No specific light construction is offered here, since there are many low-cost and effective options for making underwater lights. Have fun researching, designing, and fabricating your lighting system. The *SeaMATE* junction box has spare terminals that can supply 12 volts DC for lights, if you wish to tap into it. Figure 12.24 shows how to wire this up. (*Chapter 9: Control and Navigation* provides information on lighting options.)

Optional Navigational Sensors

These sensors are optional, but make piloting your ROV easier in water with low visibility. They would not be necessary in a swimming pool.

- one diver's depth gauge
- one automotive dashboard compass: Note that the metal rebar used in the concrete walls of most swimming pools can confuse compass readings.
- material to construct a bracket to hold a depth gauge and compass in view of your ROV's camera.

3.2.7. Control Subsystem Parts

- three small DPDT (double pole, double throw) switches with center OFF Make sure they're rated for at least 5 amps.
- one small SPST (single pole, single throw) switch for light and camera power Make sure it's rated for at least 5 amps.
- 10 feet of #16 AWG lamp cord or equivalent for 12-volt power cord
- one cable strain relief for 12-volt power cord
- 20 feet of #22 AWG red hook-up wire
- 20 feet of #22 AWG black hook-up wire
- one large pair of alligator clips for the end of the 12-volt DC power cord. An alternative connector can be substituted to match the connectors on whatever power supply you use.

- one plastic project box approx. 6 inches x 4 inches x 2 inches
- one miniature barrier (terminal) strip #22 AWG gauge
- four RCA bulkhead jacks: Omit if wiring the tether directly to the control box.
- one fuse holder: Note that there are many types, but any basic style of fuse holder will work as long as it fits inside the dimensions of your control box housing.
- two 8-amp fuses to fit fuse holder: Keep one as a spare.

3.2.8. Payload Subsystem Parts

ROV payload tools and manipulators can be nearly as complex and costly as the robot itself. Because of the numerous payload tool options possible, we suggest you start with an extremely simple pick-up probe. Once you've mastered retrieval of simple objects, you can then think about more complex manipulators and payloads. (See *Chapter 10: Hydraulics and Payloads* for more elaborate options.)

- one 6- to 12-inch plastic or metal rod or wooden chopstick
- tape (or bracket) to securely attach the probe to the frame

3.2.9. Parts for Potting the Junction Box

- one package of toilet bowl wax
- electric frying pan and ceramic cup for safely melting wax
- hot glue gun

4. Constructing the Frame

The *SeaMATE* project begins with building the frame. This activity parallels the technical information in *Chapter 4: Structure and Materials.* However, you can start fabricating the ROV with any subsystem you like. The design spiral will eventually bring you around to all the other sections. So it's perfectly fine to begin by fabricating other subsystems and work on the frame when it suits your own project schedule. Note that this particular frame shape is the one designed by *SeaMATE*'s inventor, Harry Bohm, but it can be easily modified to suit your specific mission goals.

As mentioned in *Chapter 4: Structure and Materials*, PVC pipe (from ½-inch to about 2-inch diameter, depending on the size of the vehicle) is the common choice for a simple underwater vehicle frame. PVC is cheap, readily available, relatively lightweight, and very easy to cut and drill. (See *Tech Notes: Measuring Pipe Lengths Accurately* and *Cutting PVC Pipe.*) PVC pipe also comes with a wide variety of fittings that make it easy to assemble sections of pipe into differently shaped frames—often you can do this without any specialized tools.

If you want to glue the frame sections together for added strength and rigidity, special PVC primers and glues (you'll want to use both) are usually located in the same hardware store aisle as the PVC pipe. On the other hand, if you want the frame to come apart easily for transport or to reuse the parts later, it's easy to drill and screw through the fittings and pipes. (This is the technique selected for *SeaMATE.*) Some builders just press-fit the sections of pipe into the fittings, but this is not recommended, since it is possible the pipes might wiggle out of their fittings over time.

TECH NOTE: WATERTIGHT VERSUS FREE-FLOODING PIPE FRAMES

It is possible to make *SeaMATE*'s frame watertight by using pipe solvent cement to seal all the joints. Doing so would give an extra measure of buoyancy to the vehicle because of the increase in displacement.

However, from a practical point of view, this method has certain disadvantages:

- It is harder to square up the frame.
- Construction time is increased.
- Sufficient ventilation is necessary for the gluing procedure.
- You have to work quickly before the solvent cement bonds. (Seconds count!)
- You cannot reuse parts if you make a mistake.
- You cannot drill any holes in the frame to fasten components; otherwise, water can leak into the tubing.
- The frame now has a collapse depth limit, although it should easily withstand the pressure at 66 feet (approx. 20 m).
- The frame may leak anyway, and then you have a real problem, because the water accumulating in the frame will make for unpredictable buoyancy and cause stability problems due to the free surface effect (see *Chapter 6: Buoyancy, Stability, and Ballast.*)

In general, the advantage in buoyancy gained from sealing the frame is outweighed by its disadvantages.

4.1. Parts

To begin the frame, gather the following parts:

Structural Subsystem Parts

- 10 feet of ½-inch Schedule 40 PVC pipe. Alternatively, you can use Class 200 PVC pipe, which is also ½-inch pipe.
- eight ½-inch Schedule 40 PVC elbow fittings
- sixteen ½-inch Schedule 40 PVC tee fittings
- one 16-inch x 16-inch square of ¼-inch sturdy plastic mesh screen material to serve as the "floor" of your ROV. This plastic mesh (sometimes called hardware cloth) should be available at your local hardware store, but you can also improvise. Depending on the size of your frame, you may not need this much mesh, but you can always cut it to fit.

4.2. Procedure

1. Spread out all the PVC fittings on your workbench. Sort out the tees from the elbows.
2. Start laying out the fittings used in the frame in the same configuration as shown in Figure 12.18. **Do not cut any pipe yet!** Before making any cuts, read and heed the *Tech Note: Measuring Pipe Lengths Accurately.*
3. Measure and cut all the very short PVC pipe lengths that join abutting fittings. Use a felt marker or pencil to write the length on each piece. Note that these short, stubby pipe sections are not visible in Figure 12.18 because they are completely hidden inside some of the tees.

TECH NOTE: MEASURING PIPE LENGTHS ACCURATELY

Achieving a frame that is truly squared can be a bit more complicated than just cutting pieces of pipe all the same size. That's because these pipe sections are fitted inside PVC fittings, each of which may—or may not—be uniform, since individual manufacturers have not standardized the dimensions of their fittings. That means, for example, that a tee, elbow, or coupling from Company A may not have the same socket depth as the ones made by Company B. A variation as little as 1/8 inch is significant for the lengths of the pipe cut for this project and subsequently for the squaring of the frame.

So to be truly accurate and get the squared frame that you want, determine the length of pipe between any two fittings by using the following procedure:

1. Firmly insert a length of pipe into the socket until it cannot go any further. You may have to lightly tap it or wiggle it in a bit.
2. Make a mark on the pipe where it meets the end of the socket.
3. Remove the pipe.
4. Measure the distance from the end of the pipe that was inserted in the fitting to the mark. Record this socket depth measurement with a marker pen on the pipe and in your journal. Use this marked pipe to check the socket depth for all the fittings. The same socket depth should be the same if all fittings are from the same manufacturer. If you're using fittings from a variety of manufacturers, measure each fitting and write socket depth on the outside of each fitting.
5. Next, check the required distance needed between the branch ends of the two fittings. Then calculate for the total length of pipe to cut, using this formula:

   length of pipe to cut = length between branch ends + socket depth A + socket depth B

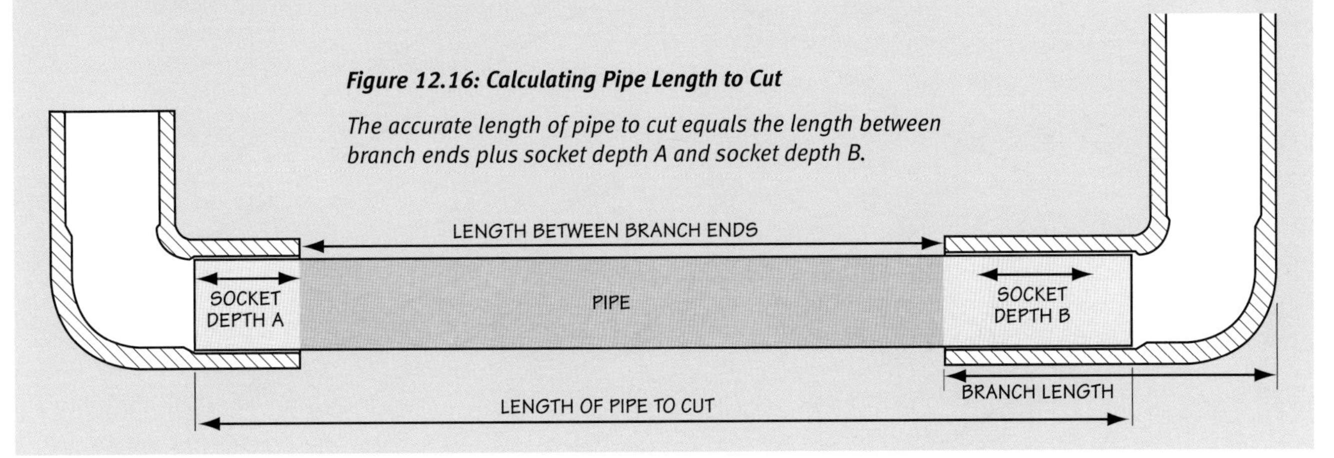

Figure 12.16: Calculating Pipe Length to Cut

The accurate length of pipe to cut equals the length between branch ends plus socket depth A and socket depth B.

4. Use the short pipe sections to connect all the fittings that directly touch each other, which includes most of the tees, as shown in Figure 12.18. Assemble all the fittings that connect to each other, using Figure 12.18 as a guide.
5. Take one of the pieces you just assembled in the previous step or start with a fitting in one corner of the frame. Measure and cut the next pipe length you will need to join your starter piece to the next fitting, as shown in Figure 12.18. Remember to make sure to account for any differences in branch lengths of your fittings. Then assemble the pipe length you just cut. For now, just lightly tap the pieces together.
6. Continue to cut one pipe length and fit it as you work your way along the frame illustration. Do NOT cut all the pieces ahead of time. If you cut one piece at a time and assemble as you go, the chances of error are smaller and easily corrected.

TECH NOTE: CUTTING PVC PIPE

Special PVC pipe-cutters work a bit like modified scissors to make cutting smaller pieces of pipe very quick and easy. However, be careful with these, as they ratchet and can't easily be released if you catch your finger in them.

PVC pipe can also be cut with a plumber's tubing cutter, but be aware that an inexpensive one tends to raise a bump on the edge of the cut pipe—this prevents the pipe from fully inserting into any fitting.

Another alternative is to cut the pipe with a back saw and miter box. However, the burr created by this method will need to be sanded smooth.

Figure 12.17: Options for Cutting PVC

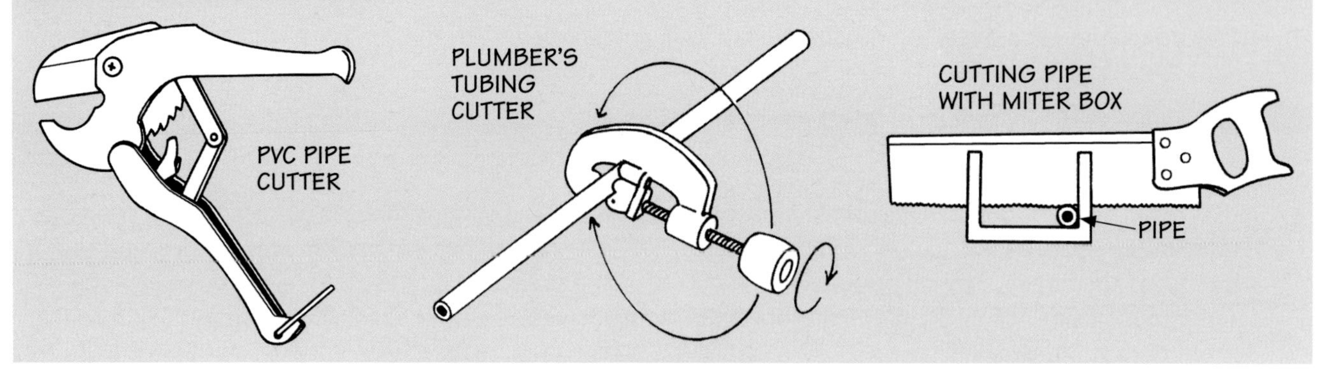

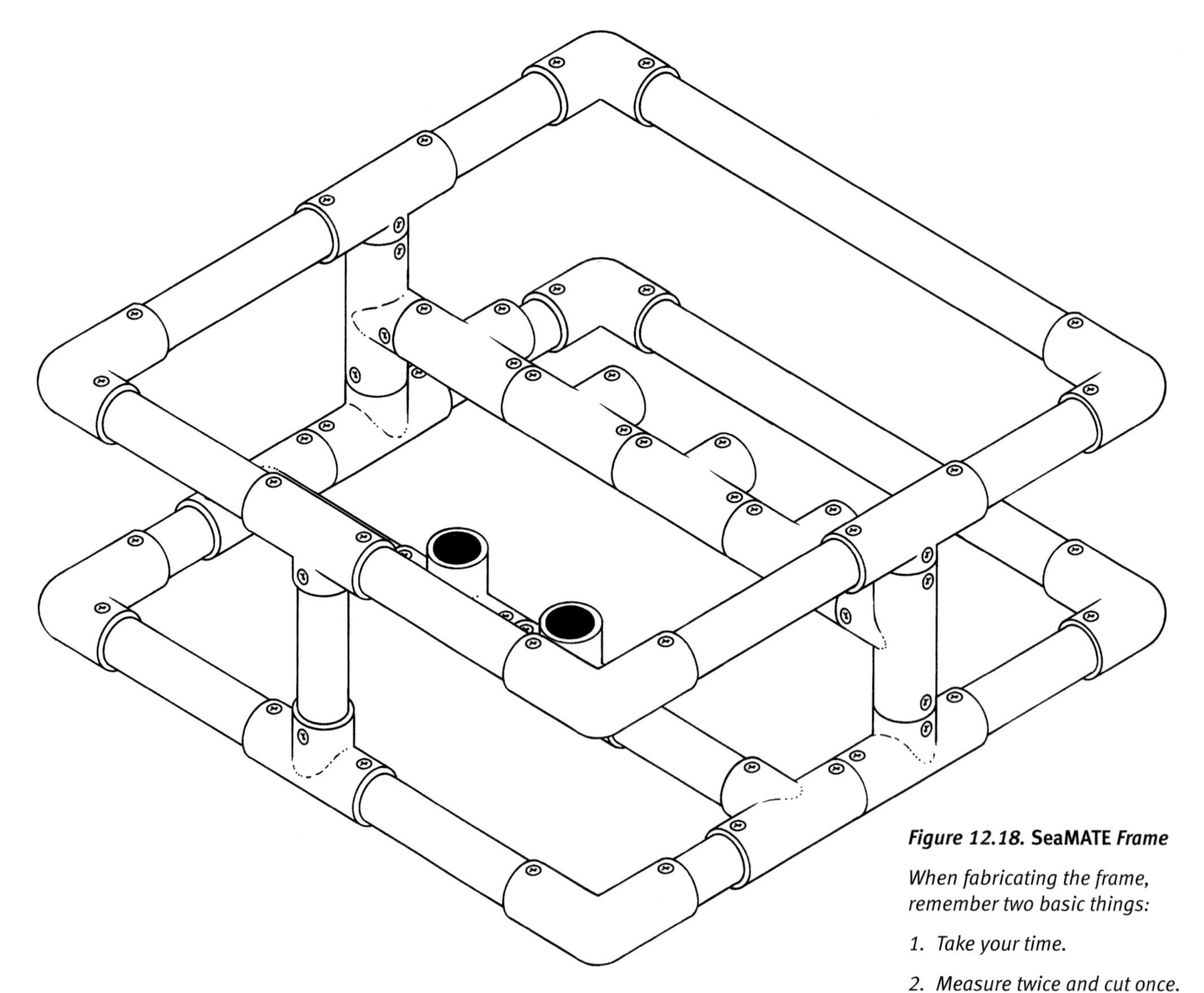

Figure 12.18.* SeaMATE *Frame

When fabricating the frame, remember two basic things:

1. *Take your time.*
2. *Measure twice and cut once.*

Use Figure 12.18 as your guide. When all the pipe lengths and fittings are assembled, visually align the fittings to make sure the frame is square. Once you are satisfied, lightly tap the fittings and pipe together to set the pipe tightly inside the sockets of the fittings.

7. Once the frame is aligned, the next step is to fasten the joints of all the fittings with the #6 x ½-inch screws. Use a 1/8th-inch drill bit to drill a pilot hole where the pipe is inserted into the fitting. Then screw a #6 x ½-inch fastener into this pilot hole. Do this for the entire frame. Use only one fastener for each joint.

8. Drill ¼-inch holes all the way through the corners of the frame, as shown in Figure 12.19, for a total of eight holes. This will allow free flooding of the frame structure when the ROV is in the water. (See *Tech Note: Wet Frame Physics.*)

9. Finally, measure and cut the plastic screening with scissors so that it fits the bottom of the ROV frame. Attach it to the frame with cable ties, as needed. Cut flush the protruding tips of the cable ties and trim off any excess screening with scissors. This plastic screening (or any similar sturdy mesh) will provide a handy platform for carrying ballast weights. (See Figure 12.20.)

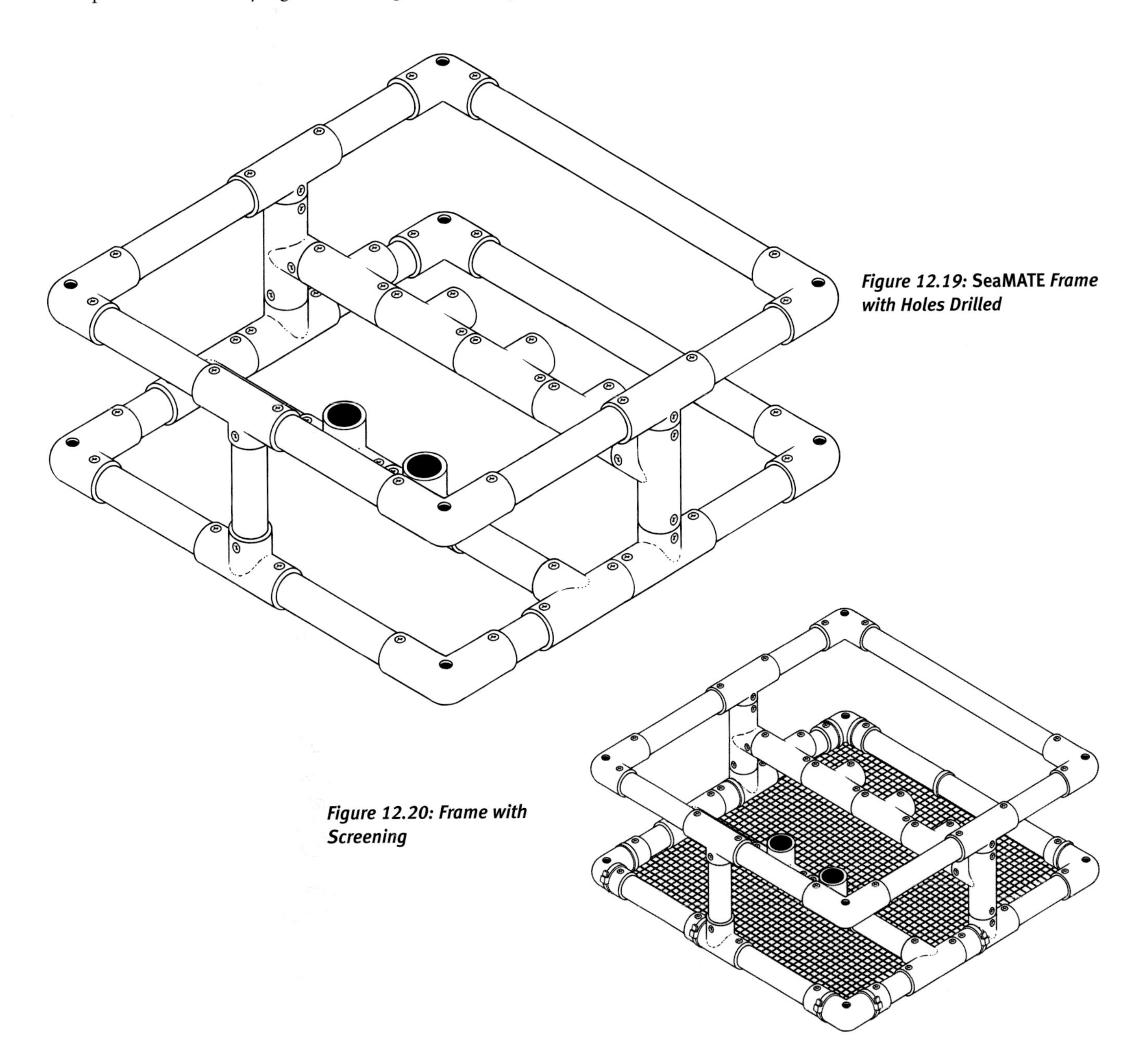

Figure 12.19:* SeaMATE *Frame with Holes Drilled

Figure 12.20: Frame with Screening

TECH NOTE: WET FRAME PHYSICS

SeaMATE's frame resists water pressure by allowing liquid inside the PVC pipes (a technique called free-flooding). The holes drilled in the frame allow water to flood the inner frame, thereby equalizing interior and exterior pressure, so there's no pressure differential. In other words, the water pressure acts evenly on both the inside and outside surfaces of the PVC pipe walls. This means that only the actual plastic itself has to resist water pressure, and plastic is essentially an incompressible solid, so that's easy to do. In fact, *SeaMATE*'s frame could theoretically resist the incredible seawater pressure at full ocean depth (35,000 ft). (See *Chapter 5: Pressure Hulls and Canisters*.)

To make the frame free-flooding, the PVC sections are not glued together using solvent cement, as is normally done in plumbing installations. Instead, the pipe and fittings are pushed together dry (called a dry fit), then secured with small screws. Once the frame is assembled, additional ¼-inch holes are drilled through the pipe fittings (Figure 12.19). The plans specify that one hole in each corner is drilled through on the topside of the upper frame and one hole in each corner is drilled through in the lower side of the bottom frame, for a total of eight holes. The lower holes allow the water to flood into the frame, while the upper holes allow any air inside the pipes to be completely purged. If only the holes at the bottom of the frame were drilled, the trapped air would prevent the water from completely flooding inside the pipes. It's the same principle that keeps air inside an inverted jar that is being pushed under water. The water rises only so far inside the jar, compressing the air until it equals the water pressure.

Air trapped inside the frame also makes it hard to ballast the ROV for diving. ROVs normally use the principle of fixed or static displacement, which means the volume of fluid displaced by the ROV at the surface is the same as when it's working under water. In a static displacement system, the ROV is weighted (ballasted), so when it is placed in the water, it just barely floats or is just slightly positive (near to neutral) in buoyancy. Thus, the ROV's vertical thruster can easily push the vehicle down or up in the water column. Once the vehicle is ballasted at the surface to this buoyancy state, no adjustments are normally made, regardless of what depth the vehicle is designed to dive to.

However, if air is trapped inside a free-flooding frame, the air will compress as the ROV dives, reducing the vehicle's displacement and making it negatively buoyant. In fact, it may become so negatively buoyant or "heavy" that the vertical thruster may not be able to make it hover or rise again. In that case, you will have to haul it back up by its tether. This is why it is very important to purge all the air in free-flooding frames. Drilling top and bottom holes in the frame is a very simple way to achieve this.

5. Constructing the Ballast Subsystem

SeaMATE's ballast subsystem is a static one. Once the ROV's buoyancy and attitude are adjusted at the surface, no other adjustments are made after the vehicle submerges. The vertical thruster is the only method for actively controlling the vehicle's depth.

The following section lists the parts you'll need to gather in order to construct the *SeaMATE* ballast system.

5.1. Parts

- one 36-inch (or longer) piece 1½-inch Schedule 40 or Class 160 PVC pipe
- four 1½-inch Schedule 40 PVC endcap fittings
- four 1½-inch plastic PVC pipe clamps
- approximately 2 pounds of lead weights or equivalent: Try to get several smaller weights rather than just one large one, since you won't know the exact quantity of weights you'll need for ballast until sea trials. Weights that have some sort of slot or loop will be easier to attach securely to your ROV.

5.2. Procedure

1. Measure, mark, and cut two 15-inch long pieces from the 39-inch long 1½-inch Schedule 40 or Class 160 PVC pipe. Lightly sandpaper away any burrs on the cut edges. You can choose to make the floats longer, but do not make them shorter or they won't provide enough buoyancy.
2. Cover the workbench with newspaper. Make sure your workspace is well ventilated. Put on a pair of protective gloves before opening the can of PVC cement. Then carefully glue one endcap to each end of the two 15-inch pipe lengths. This will give you two buoyancy tubes as pictured in Figure 12.22. Silicone sealant is another alternative to PVC cement for gluing the endcaps to the pipe. Just make sure to read and follow the manufacturer's handling instructions before using the product.
3. Set the floats aside to cure (harden).
4. After the floats have hardened, take one of the tubes and a pipe clamp and find a good position for it on the frame. Mark the location of the holes you will be drilling. Check Figure 12.22 for position. Do this for all four pipe clamps.
5. Using a 1/8-inch drill bit, drill pilot holes at the locations you have marked for the clamps.
6. Align one float on the frame, placing the pipe clamps on the float to line up with the pilot holes you've drilled. Use the #6 x ½-inch stainless steel sheet metal screws to fasten the float to the frame, as shown in Figure 12.22. Repeat this for the second float.

TECH NOTE: HOW FLOTATION TUBES RESIST PRESSURE

SeaMATE's PVC flotation tubes provide a buoyant force. They are made of 1½-inch diameter Class 160 PVC plastic pipe, 15 inches long, and two PVC endcaps. When these endcaps are glued on, they provide a watertight seal, so the tube contains air at atmospheric pressure. Because the air inside the pipe provides very little counter-pressure to the ambient water pressure outside, the pipe itself has to provide most of the pressure-resistant strength. *SeaMATE*'s PVC flotation tubes easily resist the pressure found at 33 feet for several reasons: their cylindrical shape, the plastic material itself, and the thickness of the pipe wall. This means that for its weight, cost, and ease of construction, the Schedule 40 or Class 160 PVC pipe is a good choice for small, shallow-water ROVs.

An alternative to PVC floats are plastic flotation spheres (sometimes called trawl floats). Shapewise, these spheres are by far the best shape for resisting water pressure. If you compare the pressure resistance of the 4-inch diameter plastic trawl floats with that of the PVC cylinders, the spherical floats can withstand a gauge pressure of 135 psi at 300 feet, whereas the Class 160 tubes might not even withstand 65 psi gauge pressure at 150 feet before failing. Obviously, spheres have a superior ability to resist pressure.

So why do the *SeaMATE* plans call for cylinders? The answer is easy: supply, cost, and drag. PVC pipe is cheaper and readily available in hardware stores, whereas the spheres are more expensive and found only in commercial fishing supply outlets located in coastal communities. Furthermore, the cylindrical shapes moving endwise through the water offer less drag than a sphere of comparable volume. Since other parts of *SeaMATE* (i.e., the thrusters) can withstand only the pressure normally found at 33 feet, having the stronger spheres isn't really necessary. If you go on to construct a deeper-diving vehicle, remember these spheres as a good buoyancy option.

Figure 12.21: Nalkon Trawl Float

Although you've bought an assortment of small weights as part of the overall ballast subsystem, wait to add them when *SeaMATE* is undergoing sea trials.

Note: If necessary, the floats can be removed from the frame to facilitate further construction of the power, propulsion, sensor/navigation, control, and payload subsystems. The floats and clamps are easily reattached afterward.

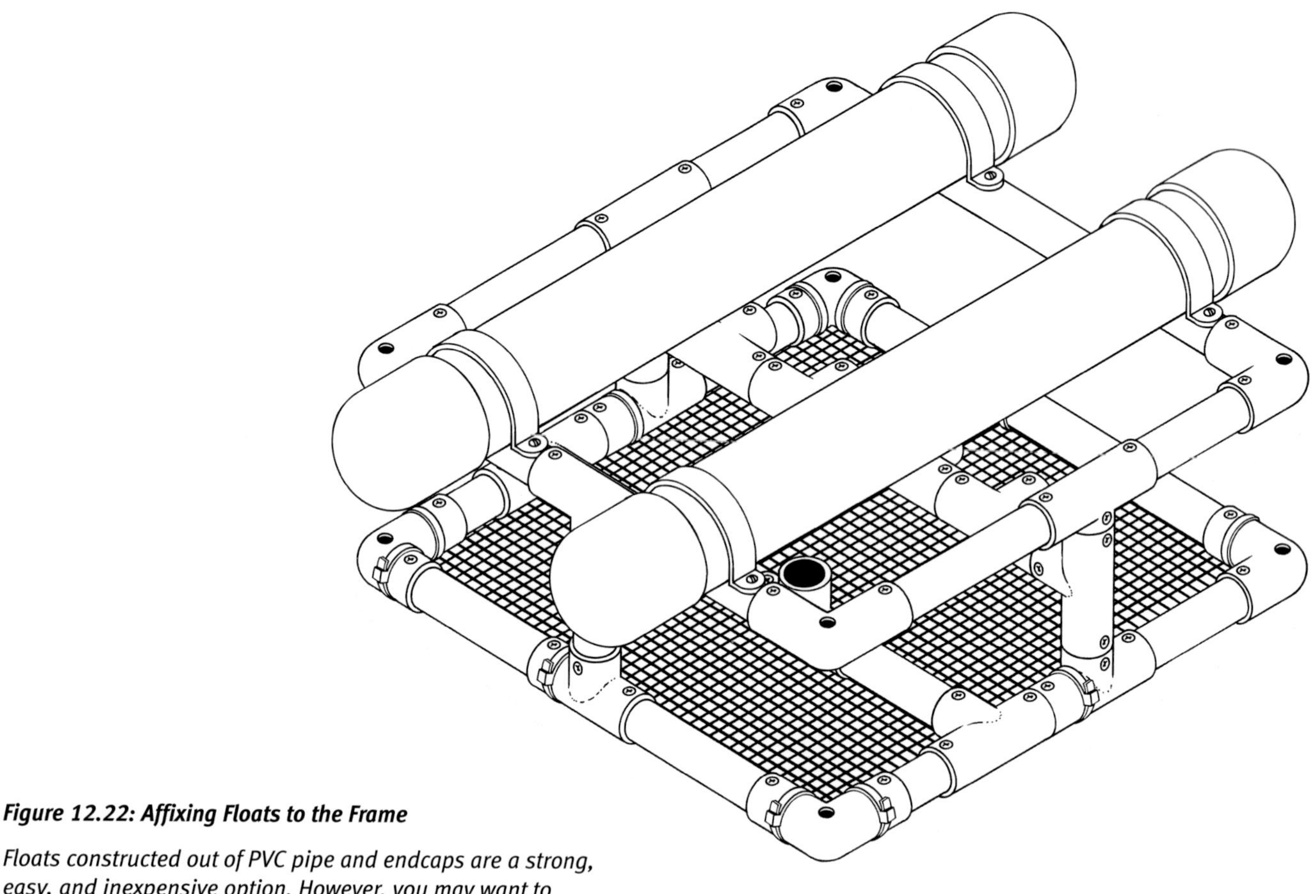

Figure 12.22: Affixing Floats to the Frame

Floats constructed out of PVC pipe and endcaps are a strong, easy, and inexpensive option. However, you may want to experiment with other ideas and sizes. Just be aware that depth significantly impacts some types of flotation (e.g., pipe insulation or plastic soda bottles) more than others.

SAFETY NOTE: DON'T HESITATE. VENTILATE!

PVC cement fumes can be toxic and flammable, so use this product only in a well-ventilated room. Before opening the can, read and follow all advisories and instructions on the container, so you know what precautions to take.

SAFETY NOTE: LEAD HANDLING HAZARD

Lead is a toxic metal, so after you've handled any lead weights, be sure to wash your hands well before eating. An option is to wear latex gloves. It's also possible to paint (or dip) lead weights with a liquid rubber coating so you don't have to handle the lead directly. Consider using alternative weights, such as steel nuts and bolts. Just remember to coat steel ones to prevent rusting.

TECH NOTE: STYROFOAM FLOTATION

Styrofoam is the trade name of polystyrene foam insulation, first introduced in the U.S. in 1954. The term in now in common generic use. Foamed polystyrene has thousands of tiny air spaces in its structure. These air spaces are *open cell*, which means they are not totally enclosed spaces. Because styrofoam is inexpensive and readily available, it is tempting to use it for flotation for ROV projects. However, you should know that over time, water pressure will cause liquid to infiltrate the open cell air spaces, thereby saturating the foam and causing it to lose some flotation capacity.

Styrofoam also compresses when subjected to water pressures, resulting in a further loss of displacement. In other words, styrofoam loses buoyancy (gets "heavier") for two reasons:

1. The air has been replaced by water, which is denser.
2. Its air space displacement has been decreased.

An additional disadvantage of this type of foam is that it is quite soft and easily damaged. Thus, closed cell–type foams are better than styrofoam for underwater robot projects. (See *Chapter 4: Structure and Materials*.)

TECH NOTE: PIPE INSULATION USED AS FLOTATION

Pipe insulation is another common open cell flotation material used in homebuilt ROVs. It's readily available from any hardware store, is tubular in shape, and is designed to fit over ½-inch, ¾-inch, or 1-inch diameter (Schedule 40/ Class 160) copper or plastic plumbing pipe. There's a slit down one side, which makes it easy to install over the pipe. It can be cut to length easily with scissors or a knife. Once in place, it can be secured to the pipe with either electrical or duct tape. Pipe insulation is convenient for ROVs that do not go deeper than 6 feet (approx. 2 m). Like styrofoam, pipe insulation will begin to compress significantly and lose its buoyancy, so it is not recommended for any mission depth greater than 3 meters (approx. 10 ft).

6. Constructing the Tether and Junction Box (Electrical Termination Can)

Before getting into the details of how to build the tether and junction box, it's helpful to understand the layout of the overall electrical system of which they are a part (Figure 12.23).

Unlike a work-class ROV, which usually gets its power from a generator, *SeaMATE* uses a simple 12 VDC battery for power. A gel cell or other type of sealed lead-acid battery is recommended. Alternative 12 VDC sources, such as a car battery or AC power supply with an output of 12 VDC and at least 8-10 amps, could be substituted but would require extra safety precautions. (Review relevant safety notes in *Chapter 8: Power Systems.*)

Inside the ROV's control box, electrical power from the battery is distributed through a terminal strip to four switches, three that control the thrusters and one that turns the camera and light ON and OFF. There is one pair of wires going to each switch. The four pairs of wires connect to four RCA jacks (bulkhead-mounted RCA sockets). The RCA sockets are where the tether wires will be connected. Alternatively, you can skip the RCA sockets and plugs and connect the tether wires directly into the control box. Note that if you do it this way, the tether cannot be disconnected. (See *Section 11.1. Potting the Junction Box* in this chapter.)

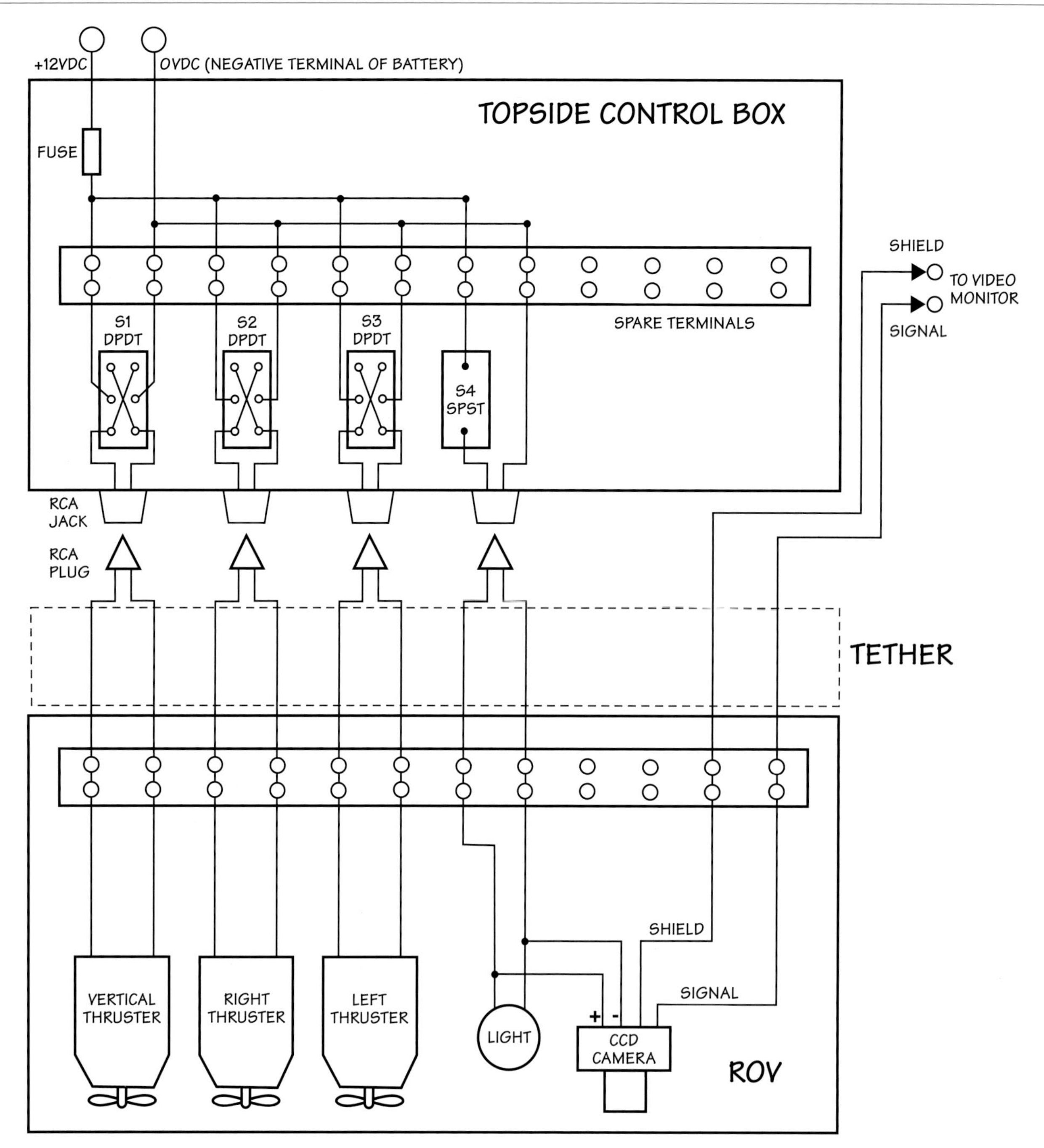

Figure 12.23:* SeaMATE *Wiring Diagram

From the control box, the tether carries energy to the ROV where the junction box (electrical termination canister) is located. This termination can is like an underwater electrical junction box. Note that the terms "junction box" and "electrical termination canister" are used interchangeably in this project. On *SeaMATE*, all the conductors from the tether terminate in this junction box. Inside the canister, the tether conductors are connected to a screw terminal, or barrier strip, that serves as a convenient way to organize and route the various conductors going to the thrusters, camera, and light. Of course, the challenge is to waterproof all those electrical connections in the termination can.

SeaMATE's tether consists of four pairs of #22 speaker wire and one length of RG-174 coaxial cable. Three pairs of the speaker wire run power down to each respective thruster (vertical, left, and right). The fourth pair of speaker wire powers the light and the camera. (Some video cameras come supplied with an extra-long cable that has both power and video signal conductors inside. You can use this video cable instead of the RG-174 and camera/light conductor described in these instructions.) The fifth pair of wires in the tether is inside a thin, flexible coaxial cable, type RG-174. It carries the video signal directly from the video camera up to the video monitor. This coaxial cable does not pass through the control box. Although this cable may look like one wire, you'll recall that a coaxial cable contains two conductors—the core and the shield—so it's really two wires.

At the top or "dry" end of the tether, each pair of wires is terminated with an RCA plug. The three thruster connections and the camera/light power connections plug into the control box RCA jacks. The video signal RCA plug connects directly into the video monitor's VIDEO IN RCA jack. At the other end of the ROV tether, also called the "wet end," the cable terminates at the ROV junction box where the electricity gets distributed to the thrusters, camera, and light.

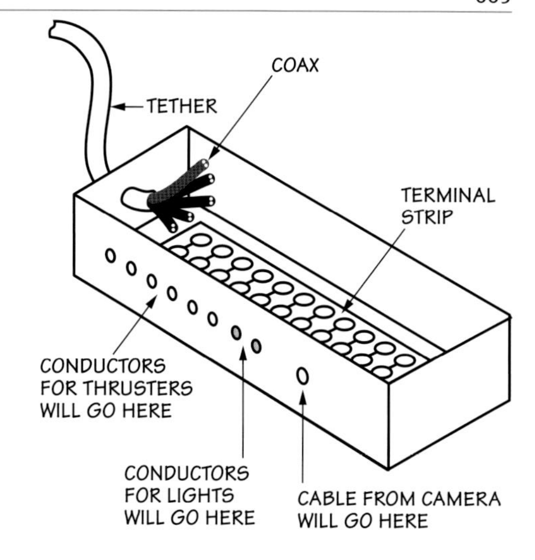

Figure 12.24: Junction Box

TECH NOTE: RCA CONNECTORS

These connectors are typically used for audio/video purposes, such as VCRs, camcorders, and home stereo systems. There's a standard color code used in audio and video applications to delineate the function of the RCA connector (developed by the **R**adio **C**orporation of **A**merica, hence the name). Red (right channel) and black or white (left channel) are for audio. Yellow is used exclusively for video input/output. Note that the socket (also called a jack) is the female connector, and the plug is the male connector.

This project adapts an RCA connector for *SeaMATE*'s tether connection to the control box for several reasons—it's inexpensive, reliable, and readily available. Also, it can easily handle the low voltage and currents used to power the thrusters and camera/lights. Note: Do not use an RCA connector for high-voltage or high-current applications, since it is not rated for that kind of usage.

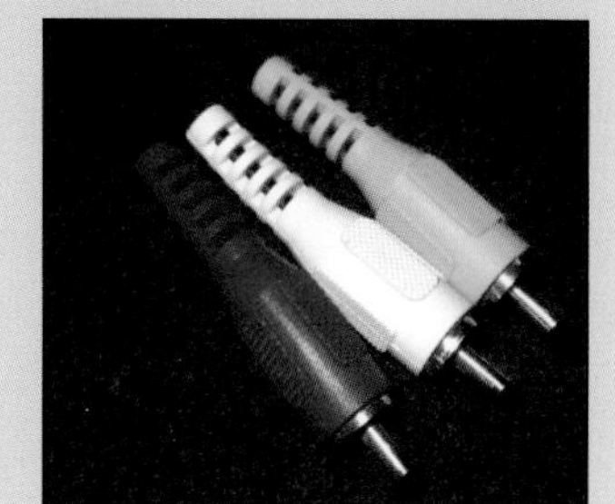

Image courtesy of Harry Bohm

Figure 12.25: Typical RCA Connectors

SeaMATE's tether meets most of the following requirements:

- It offers moderate flexibility.
- It confines the conductors into as small as possible tether cross-section width.
- The tether is long enough to reach 33 feet operating depth.
- It is strong enough to lift the ROV in and out of the water.
- The tether resists abrasions and impacts.
- It can be laid out without tangling for use and coiled for storage.
- It has conductors capable of carrying adequate power to the ROV (i.e., minimal voltage drop).
- The tether can have floats attached to make it neutrally buoyant or slightly positive and to keep it from dragging on the bottom.
- It can be unplugged from the control box for storage and maintenance.
- It is neat and tidy.

6.1. Parts

Gather the parts necessary to construct the tether and junction box. Note that the battery will not be needed until it's time to conduct electrical systems testing.

- 200 feet of #22 AWG (American Wire Gauge) speaker wire: Use this to make the four pairs of conductors that will carry electricity to the thrusters and camera/ lights via the tether.
- 50 feet of RG-174 coaxial cable: Use this for the camera video signal.
- one plastic project box about 4 inches wide x 6 inches long x 2 inches high: A smaller box will work, but it becomes difficult to do any wiring inside it.
- one 12-terminal miniature barrier (terminal) strip #22 AWG gauge
- five RCA plugs
- 50 feet of polypropylene hollow braid rope, 3/8 inch in diameter (optional)
- one 3/8-inch fid: A fid is designed for splicing hollow braid line, but in this project, it helps you thread the tether conductors inside the hollow braid rope. Remember to buy a fid at a marine supply shop when you purchase your polypropylene rope. (You can always use a chopstick in place of a fid, but it won't be as easy.)

6.1.1. Two Options for Constructing the Tether

There are two ways to construct *SeaMATE*'s 40-foot tether, and you should decide which method you want to use before you cut any speaker wire. The simplest way is to bundle all the wires and tape them together every 18 inches with black electrical tape or duct tape. If you opt for this method, then the polypropylene hollow braid rope and fid specified above in the parts list are not necessary.

With a bit more effort, you can make a compact, robust umbilical that looks quite professional. (The procedure is described below.) By carefully taping all the conductor wires to a fid (a stiff stick or tool somewhat like a chopstick or metal rod), you can pass them through the inside of a hollow braid rope. The rope sheathing provides protection and keeps the wires neatly confined, yet retains a good measure of flexibility.

Before attempting the fid method of tether protection, practice running conductors through the rope, using the surplus speaker wire from your 200-foot supply. There should be enough extra to cut four pieces that are each 10 feet in length. Don't bother using any RG-174 for this practice session, since it is really thin wire, and omitting it for practice won't matter. To develop your technique, follow the numbered instructions given below for sheathing the tether, only use the shorter 10-foot wires and a shorter 10-foot piece of hollow braid rope. Start the actual tether construction only after you have successfully mastered the practice version of the tether.

TECH NOTE: TETHER SHEATHING OPTIONS

A high-tech alternative to using hollow braid rope for sheathing a tether is a product called Techflex. It is made in the same way that hollow braid rope is, but the strands are more flexible and the product comes in a variety of colors and sizes. It is designed specifically to sheath wiring runs on motorcycles, custom cars, and boats where a high-tech look is desired. Note that it can also be used in ROV construction to bundle wiring runs inside pressure cans or on the framework. Check sources on-line for this and similar product options.

6.2. Procedure for Constructing the Tether

1. Cut four 40-foot lengths of speaker wire to serve as conductors (also called conductor pairs). Note that each conductor pair consists of two strands paired together: one wire is copper-colored and the other is silver-colored. Copper is designated as negative and silver is positive. The four lengths of conductors will serve as follows:

   a) 1 pair #22 speaker wire for vertical thruster

   b) 1 pair #22 speaker wire for left horizontal thruster

   c) 1 pair #22 speaker wire for right horizontal thruster

   d) 1 pair #22 speaker wire for the light and camera

2. Cut the coaxial cable (RG-174 or equivalent) to a 50-foot length. Note that the additional length topside is needed for connecting this wire to the monitor.

3. Now cut a 36-foot length of polypropylene hollow braid rope. Use a heat gun to melt the strands on each end to prevent them from unraveling, taking care not to melt the end shut. **Caution: The plastic can get very hot, and melted poly rope can easily burn your hands.**

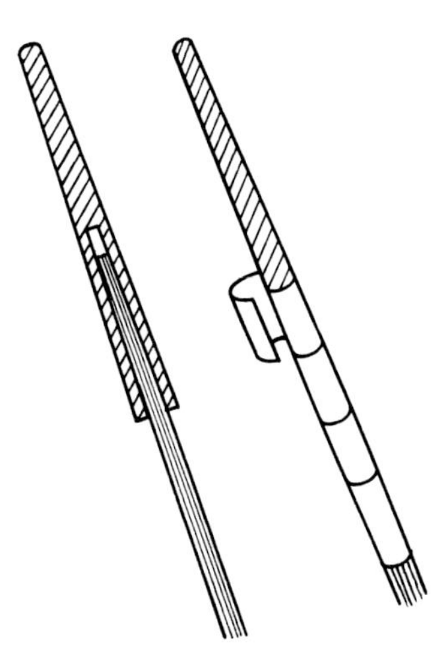

Figure 12.26: Hollow Fid with Conductors Inserted

4. Label each end of each conductor pair. Devise your own code or abbreviations, perhaps using different colored markers or masking tape, to label which length of speaker wire is for the vertical thruster, which is for the left horizontal thruster, etc. This is an important step, because once the wires are inside the rope sheathing, you cannot distinguish which wire is which.

5. Clear a long working area on the floor and straighten out the four pairs of 40-foot length speaker wires and one 50-foot RG-174 coaxial cable length, for a total of five pairs. If you opt to use an all-in-one video power cable instead of using the RG-174, simply omit the RG-174. You will still need the fourth speaker wire pair to power the light. The integrated camera cable can be either taped to the outside of the tether or run inside the sheath. The choice is up to you.

6. Line up the conductors at one end. If using a wooden chopstick as your fid, blunt one end and smooth it round with sandpaper, so it can pass through the hollow braid without snagging. Note that a fid designed for splicing hollow braid line is already smoothed. It's also hollow, so taping the group of conductors to it is much easier.

7. Now attach one end of each of the conductor pairs to the fid, using black electrical tape (Figure 12.26). You may have to stagger the conductors so that they're not a large lump. Try to make your wrapping as small and smooth as possible, since it's extremely difficult to work a lump of cable through the rope, even with the fid. The trick is to keep the bundled conductors the same diameter as the fid. This won't be possible with a smaller diameter chopstick, but it should work as long as your taping is smooth and tight.

8. Double-check that the conductors are securely attached to the fid with the tape. Pulling the conductors through the rope creates a fair bit of tension, and you don't want the conductors to be torn away from the fid while inside the rope.

9. Starting about 4 inches from the end of the hollow braid rope, insert the fid through the braid. Note that you don't start sheathing at the very end of the rope, because it's just too difficult to do so.

10. Work the fid and conductors through the sheath by bunching up the hollow braid rope and then pushing the fid forward. Do only about 5 or 6 feet at a time, then work the fid out of the braid. (Hint: Exit through the space between the braids rather than through the fibers of each braid.)

11. Now pull the wires through until only 36 inches of conductors and 9 feet of RG-174 are left outside of the rope from the initial spot where you entered the rope braid. You need this amount outside the sheath so the conductors can be connected to the control box and video monitor. Note that the extra long length of coax cable allows it to reach a video monitor that might just be positioned farther away than the control box. Later on, if you find these lengths are too long, just trim them off to the desired length. Remember, it is always easier to cut off excess wire than to lengthen it! As a precaution, wrap a few turns of black electrical tape around the top, or "dry," end to hold the connector wires in place and to prevent any more from being accidentally pulled through the sheath.

12. Once you've pulled the wires through the first section of rope sheath, re-enter the rope at the same place you just exited (Figure 12.27). Continue the bunch-and-push procedure used before for another 5 or 6 feet. Then exit the rope sheath with the fid and pull the connector wires until they are taut inside the second section of rope. Now re-enter the rope at the same place with the fid for another bunch-and-push session. Repeat this procedure until the full length of the conductor bundle has been pulled through to the the bottom, or "wet," end of the rope. There should be around 12 inches of speaker wire and coax sticking out of the wet end of the rope.

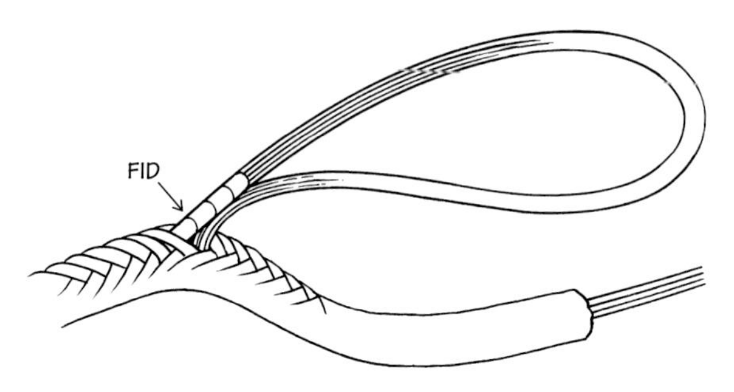

Figure 12.27: Threading the Fid through the Rope

13. Wrap some black electrical tape tightly around the wet end of the rope to secure the bundle.

14. Now go back to the top end of the tether and get ready to solder one of the four RCA plugs to each conductor pair that goes to the control box. Note that the center pin in the RCA plug is generally positive and the outer ring is generally negative. Label the plugs so you know which plug will connect to which jack on the control box. The coverings for RCA connectors come in black, red, white, green, and yellow. Remember to slide the color coded coverings up the wire before soldering the wires to the terminals. Afterward, pull the coverings down and tighten them back on. You can use red for right thruster, green for left thruster, white for the vertical thruster, and black for the camera/light power socket. Remember to reserve yellow for the video coaxial cable, since yellow is the standard color code for video signal. **Caution: The yellow-jacketed connector should be plugged only into the video monitor and never into a power jack on the control box. If it is plugged into a 12-volt RCA jack, the electrical surge will likely destroy the CCD camera!** To avoid this possibly disastrous mix-up, many builders use different types of connectors for the camera, making it impossible to accidentally connect the camera to the thruster power.

Figure 12.28: Basic Soldering Info

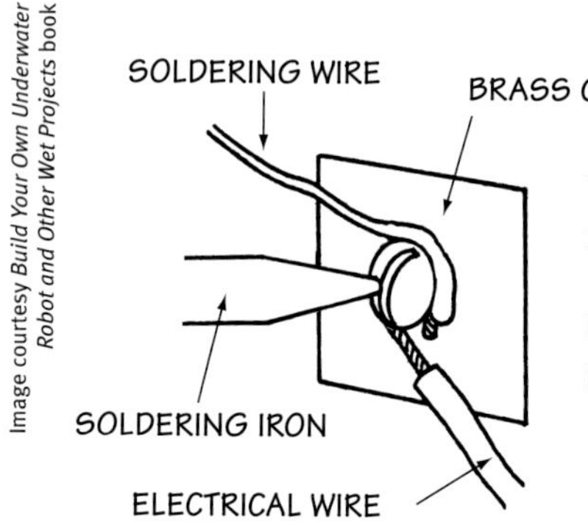

Image courtesy *Build Your Own Underwater Robot and Other Wet Projects* book

TECH NOTE: CONTINUITY TESTING THE TETHER

Continuity refers to the conductivity of a material—in this case, wire. If you forgot to label the ends of the conductors of your tether before pulling them through the rope, you can use a multimeter to determine which wire is which by using the following procedure for continuity testing:

1. Set up the multimeter dial to measure continuity. (Check the multimeter manual or the internet for instructions if you're not sure how to do this.)
2. Strip a ¼-inch length of insulation off each speaker wire in the tether on the wet end. Then strip off the same amount of insulation from the control box end of each speaker wire, unless you have already soldered on the RCA plugs on the control box end. There's no need to strip the RG-174, since it is obvious from both ends which wire is the coax cable.
3. Connect the red lead from the multimeter to one conductor on the tether's wet end.
4. Now touch the black lead from the multimeter to each bare wire end on the control box end. Do this until you get a beep from the meter. The beep means there's continuity. Label that wire at both ends. For example, you might use abbreviations like: right thruster +, right thruster -, left thruster +, left thruster -, vert +, vert - , light/cam +, light/cam-. Or you might prefer to color code the ends. The choice is up to you. Just be sure to label them!
5. Continue this procedure for all the wires until all wires are identified and labeled.

6.3. Procedure for Constructing the Junction Box

The junction box, or electrical termination can, is used to make a waterproof transition from the wet end of the tether to the connections for the thrusters, camera, and light. It is similar in function to the electrical pressure canister on an industrial ROV. Here's how to construct it:

1. Take the project box and begin transforming it into your junction box by drilling a hole slightly smaller in diameter than the tether's outer diameter, as shown in Figure 12.29. Do *not* make it larger than the tether's diameter. It needs to be a snug fit. If it's too small, it can always be reamed out to the right size.
2. Drill six holes that are exactly the same diameter as the wires that will come from each of the three bilge pump thrusters. (See Figure 12.29 for location of these holes.) Again, if you're not sure about hole size, drill the holes slightly smaller and ream them out to get a snug fit. If you have decided that you want lights that are powered by the vehicle supply, now is also a good time to drill two more holes for the extra pair of wires that will be used for a light. Note that if you've substituted some other type of lighting that does not require its own power source, you can omit drilling these extra two holes.

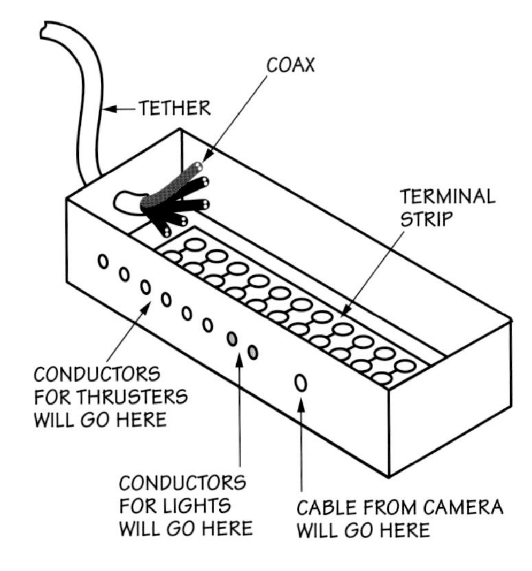

Figure 12.29: Junction Box

This illustration is the same as Figure 12.24, repeated here for your convenience.

3. Normally, you would drill all the holes in your junction box at the same time, as shown in Figure 11.29; however, *SeaMATE* presents several lighting and camera options in Section 8, so do not drill holes for the lighting wires or for the video camera cable until you have made those decisions. You can drill the required holes later when you begin working on the camera and waterproofing in Section 8.2. (See Section 8.2.2. *Procedure for Wiring the Camera* for optional camera installation and cabling.)

4. Secure the barrier (terminal) strip inside the box, using hot glue, epoxy, or a couple of #6 x ½-inch screws. Check Figure 12.30 for optimal position.
5. Drill four sets of 3/16-inch holes in the bottom of the junction box as shown. These will be used to pass cable ties through the junction box, to secure it to the frame.
6. Carefully wiggle the wet end of the tether through the hole into the junction box. The 12 inches of wire should all come through, as well as ½ inch of rope sheathing, which should be inside the hole in the junction box.
7. Secure the tether using a cable strain relief which functions to prevent the tether wires from pulling out of the box, as shown in Figure 12.30. Alternatively, you can snug two cable ties onto the tether on either side of the box wall.
8. Carefully maneuver the junction box/tether assembly through the center of the upper framework. Secure the junction box to the ROV, using cable ties.
9. Do *not* trim the wires on the wet end or wire up the junction box just yet. Other things must be done before this can happen.

Figure 12.30: Junction Box Wiring

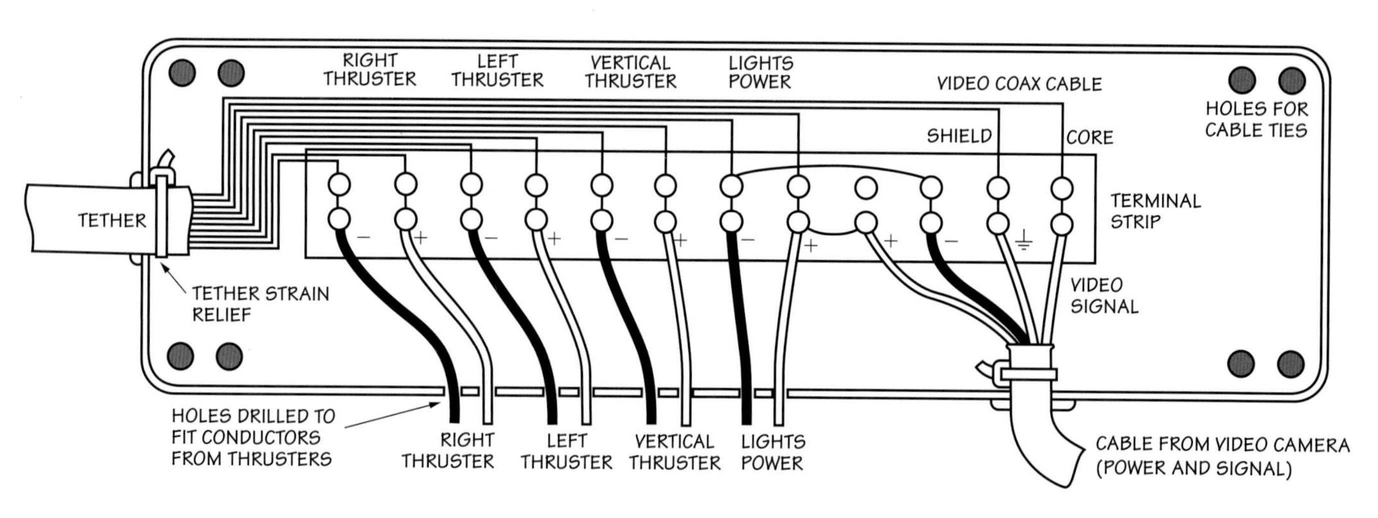

7. Constructing the Thrusters

The next step in constructing *SeaMATE* is to assemble the thrusters and mount them in the framework.

SeaMATE uses three thrusters to produce motion. Two of the thrusters are mounted horizontally, one on the port side and one on the starboard side, approximately 6-8 inches (15-20 centimeters) apart. As described in *Chapter 7: Moving and Maneuvering*, this pair of horizontal thrusters can generate quite a variety of useful motions in the horizontal plane even with *SeaMATE's* simple control system, which lacks thruster speed control. These horizontal motions include: straight forward, straight reverse, forward while turning right, forward while turning left, reverse while turning right, reverse while turning left, and spinning either right or left in one spot without moving forward or backward. The remaining thruster is mounted vertically to control up and down movement of the vehicle.

SeaMATE's thrusters are made from standard 550 gph (gallons per hour) bilge pump cartridges manufactured by the Johnson Pump Company. They are waterproof and fully submersible. Each bilge pump cartridge has a plastic housing that surrounds an internal 12-volt electric motor. This housing is factory-sealed throughout, except for two penetrations: one for the motor's shaft and one for the wires. Both use seals (rubber-like inserts) to prevent water from getting in.

Although bilge pumps were never designed for use as ROV thrusters, they have a surprising capability to withstand water pressure up to 33 feet (approx. 10 m). With a small 1½-inch, two-bladed nylon propeller (Dumas model #3003) attached to the motor shaft, they develop around 1 pound (4.4N) of thrust each at 12.5 volts (measured at the motor), with an average current draw of around 3 amps. These bilge pumps are low-cost and suitable for other shallow-water robot projects.

TECH NOTE: PROPELLER MATCHING

If you cannot find the Dumas props specified, another good option for the RC model boat props are the "3/16" Drive Dog Props found at the hobby supplier www.hobby-lobby.com. They are very inexpensive and work much better than cropped airplane propellers.

Whichever props you select, you'll have to get resourceful and test alternatives. (See *Chapter 7: Moving and Maneuvering* for a discussion of thruster/prop matching.) However, for testing alternative props to pair with the *SeaMATE* bilge pump thrusters, the thruster test jig in Figure 12.31 will work well, because it is designed to keep the bilge pump cartridge submerged, so it will not burn out. (There are many design options for thruster test jigs; note that the variation in Chapter 7 has the motor housing mounted out of the water, so be careful not to run it for more than a few seconds at a time when using it with bilge pumps, or they may overheat.)

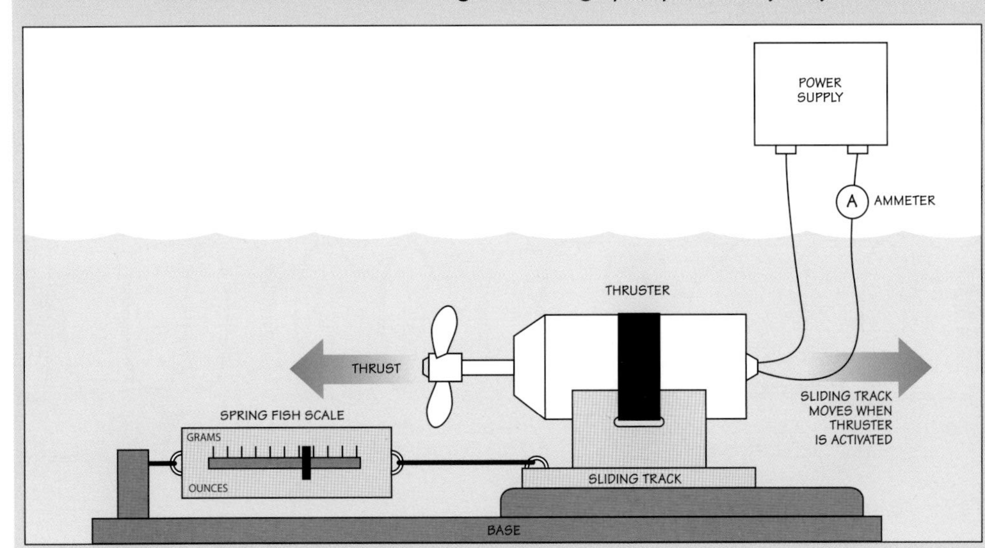

Figure 12.31: Thruster Test Jig

7.1. Parts

You'll need the following parts to build the thrusters:

- three Johnson Pump Company bilge pump cartridges part no. 32-1550C - 550 gph (gallons per hour): This is the replaceable motor cartridge part of a 12-volt bilge pump made by the Johnson Pump Company. Note that for this project you need buy only the cartridge portion and not the complete bilge pump.
- three #6 32 x ¾-inch stainless steel bolts
- three #6 x 1-inch threaded nylon standoffs: Get these at a commercial electronic supply shop or on-line distributor, not at a consumer electronics store.
- three Dumas propellers model #3003 1½-inch diameter 2-bladed nylon prop

TECH NOTE: BILGE PUMP SEALS

Figure 12.32: U-Cup Seal

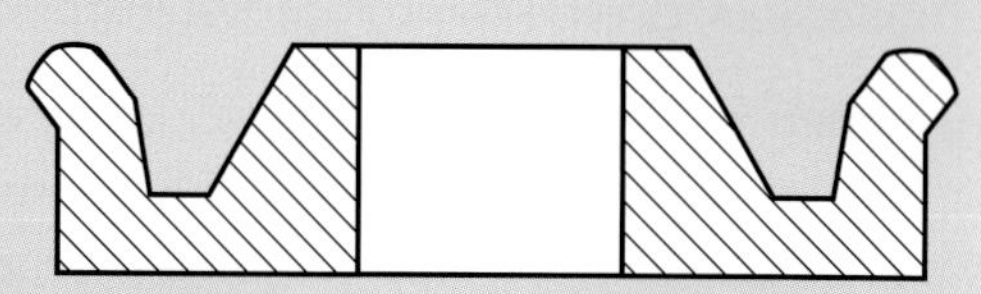

The motor shaft in each bilge pump uses a **U-cup seal** that allows it to rotate, yet keeps water from entering the housing. This component is designed to take advantage of water pressure to make a watertight seal against the shaft.

This type of seal works well as a rotary shaft seal in shallower depths, but has two major disadvantages for deeper water. The first problem is that the greater the water pressure, the more squeeze is put on the shaft. That squeeze creates more friction for the motor to overcome. As the motor goes deeper, at some point its ability to overcome the additional friction is compromised, and the motor slows and/or stalls. The second problem is that they're not designed for high differential pressures, so can fail at greater depths.

The second penetration in the bilge pump housing is for the power wires. The seal for this opening also uses water pressure to make the rubbery material press tightly against the housing and wires. This method of sealing wires is fine for up to 33 feet (approx. 10 m), but the weakness of the design is that at increasing pressures, the seal can be forced out of its seat and pushed right into the housing. This is the major cause of leakage in these types of thrusters.

TECH NOTE: SUBSTITUTING BILGE PUMPS AND PROPS

If Johnson bilge pump cartridges are not available, you can substitute a different brand, such as the Rule 500. However, these will require some adaptation, such as constructing a mounting bracket. The Rule 500 pump does not have the winged mounting tabs, as the Johnson bilge pump cartridges do.

You may encounter another related issue: the specified props in this project may not be available. If substituting a prop, you will have to fabricate another type of shaft attachment and conduct propeller matching tests to ensure the prop you have chosen is well matched to the bilge pump motor. (See the discussion of propeller test results in *Chapter 7: Moving and Maneuvering*.)

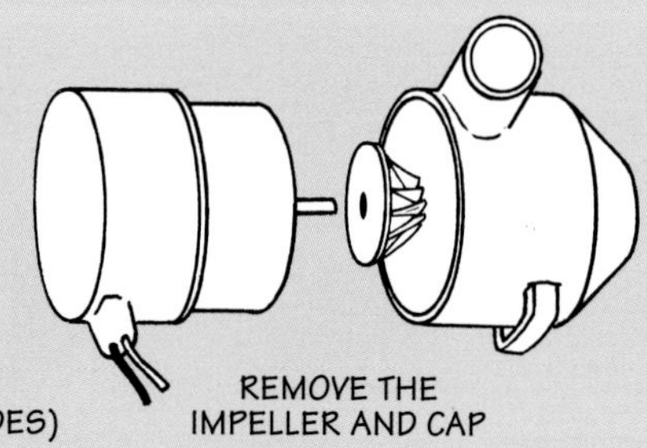

Figure 12.33: Adapting a Rule 500 Bilge Pump Cartridge

Images courtesy of *Build Your Own Underwater Robot and Other Wet Projects*

7.2. Procedure

1. Take a Johnson bilge pump cartridge and remove the impeller from the shaft.

2. Take a 1-inch nylon standoff and test to see if it fits snugly on the end of the bilge pump shaft. A standoff can be used as an adapter to connect the propeller to the motor shaft. Doing so also moves the prop farther from the motor housing for more efficient water flow.

3. Attach the prop to the end of the standoff, using a #6 x ½-inch stainless steel sheet metal screw. Make sure the prop is mounted so that a clockwise rotation creates a forward push. CAUTION: Gently tighten the screw; do not overtighten, as the prop hub (called a "boss") may split.

4. Attach the other two propellers to the two remaining cartridges.

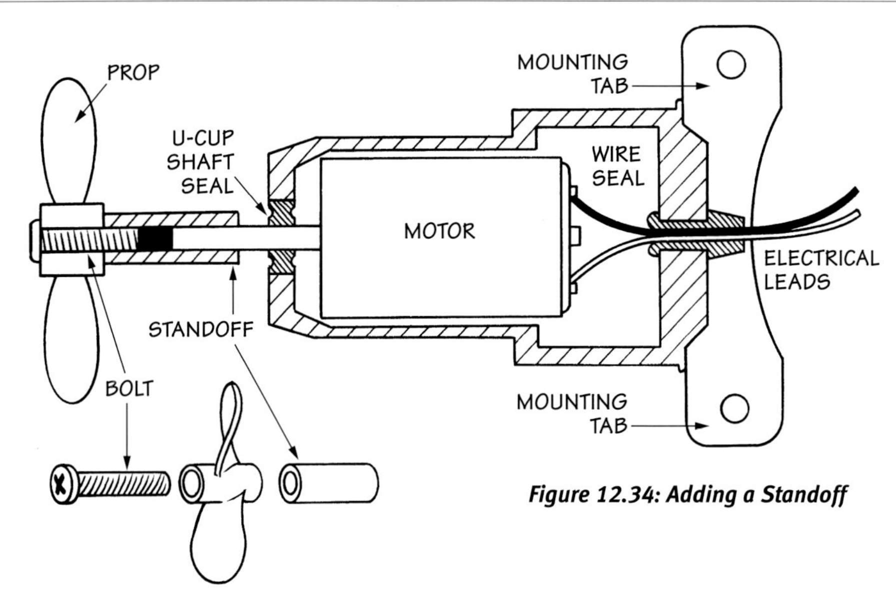

Figure 12.34: Adding a Standoff

TECH NOTE: BILGE PUMP SHAFT SEAL DAMAGE

You can easily damage a bilge pump shaft seal by running the motor at high speed for longer than 15 seconds when it is not in the water. The shaft seal works well in water because the liquid acts as a lubricant and cooling agent. In air, however, the friction of the rapidly rotating shaft overheats the rubbery material that is in contact with the shaft, destroying its ability to seal effectively. In most cases, the damage goes unnoticed until you put the ROV in the water and the thruster housing floods. Unfortunately, there is no way to easily repair this damage, short of replacing the bilge pump, because the pump is not designed for easy replacement of a defective shaft seal.

5. Choose two of the thrusters to mount horizontally. Note how in Figure 12.35 you can see how they are positioned on the frame and how the mounting flanges (tabs) on each of the horizontal thrusters will connect to the PVC tees by screws.
6. On each of the mounting tabs, use a marking pen or pencil to mark the position of a hole to be drilled for the mounting screw.
7. Using a 3/16-inch drill bit, bore a hole at each mark on the mounting tabs.
8. Now place the first horizontal thruster in position against the horizontal tee support. Align it so that it is level and at right angles to the frame. Mark the tees through the holes in the mounting tab. Then do this for the second horizontal thruster.
9. Use a 1/8-inch drill bit to drill the horizontal mounting tees at each of the marks you just made.
10. Using two #6 x ½-inch sheet metal screws with two #6 washers, attach the horizontal thruster.
11. Use this same method to attach the vertical thruster to its pair of tees, although the tees and attachment position are hidden from view in Figure 12.35. The correct position for the vertical thruster should be evident on your framework, but you may want to refer back to Figure 12.18 for a picture.

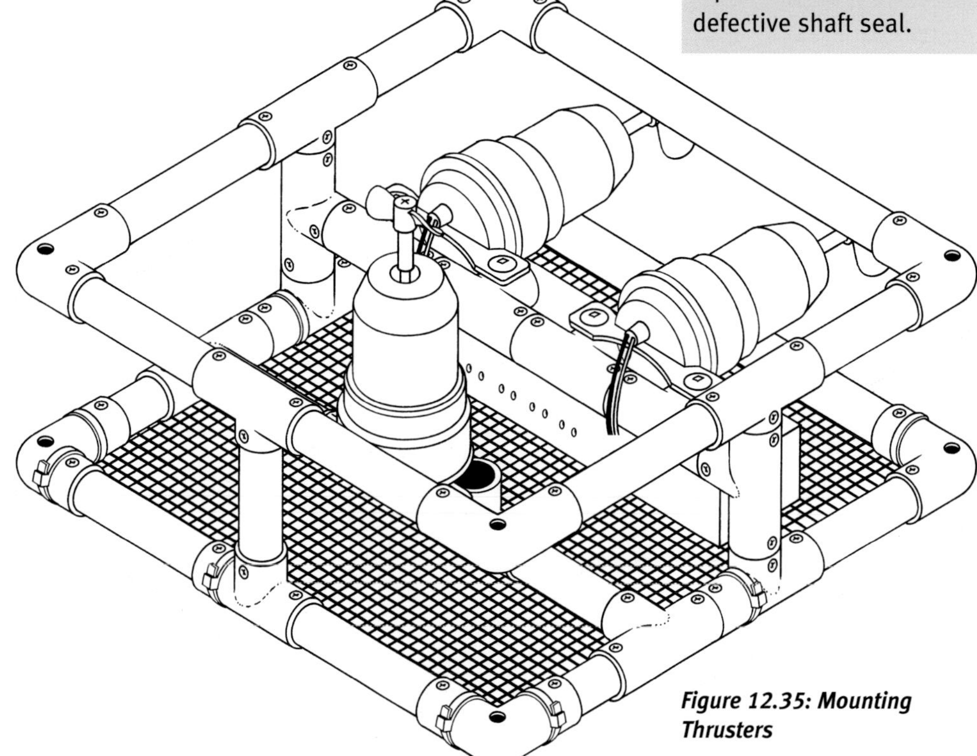

Figure 12.35: Mounting Thrusters

TECH NOTE: BOARD CAMERAS

Many types of video cameras can be used for a project like *SeaMATE*, but a small, inexpensive **board camera** similar to the one pictured here is ideal and is the type indicated in the *SeaMATE* instructions. A board camera is a bare-bones camera (usually video) made by attaching a small lens (glass or pinhole) to a small printed circuit board that includes a **CCD** or **CMOS** electronic image sensor (located behind the lens). It also includes some additional electronics to convert electrical signals from the image sensor into a standardized video output format.

Because of their small size and low power requirements, board cameras are popular for covert surveillance. You can find dozens of models available for modest prices through on-line distributors specializing in video security and surveillance equipment. Just search for "board camera" on the web.

Pay special attention to the input voltage requirements and the video output format of any camera you are considering. You want to be sure to match the camera's voltage requirements with those of the power supply you are using. Note that some smaller video cameras can be used with 5 volt DC supplies or low-voltage AC power; however, those models are not suitable for the 12 VDC power source in the *SeaMATE* project and can be permanently damaged by it. As for video output format options, NTSC is currently the dominant format in the United States, but many other countries use the PAL format. (Both are discussed in *Chapter 9: Control and Navigation*). These venerable analog standards may fade from dominance as technology moves toward higher-resolution television based on digital signals. Just remember that the output format of the camera must match the input format of whatever video monitor, television set, or video recorder you will use with the camera.

A board camera may come with a pre-attached cable, a detachable cable, or no cable at all. Cameras that don't come with a cable will have some place to attach a cable, like the orange-colored Molex-style connector on the camera shown here. If you are planning to pot the camera for underwater use, it might make most sense to solder your tether wires directly to the cable connector pins on the camera's board prior to potting. That's probably easier than trying to waterproof a separate cable and cable connectors or to splice tether wires to an existing cable.

Most board cameras will have four pins or wires for their connection. One pair supplies power (+V and ground) to the camera circuitry. The other carries the video output (signal and ground). You may need to check the camera datasheet or look for markings on the board to figure out which pin is which. Be careful not to mix them up, or you may fry the camera.

Note that the pinhole models are usually less expensive than otherwise-similar glass lens models. However, for underwater viewing, it might be worth the slightly extra cost for a glass lens, because this type generally offers better image quality and better low-light performance.

Image courtesy of Dr. Steven W. Moore

Figure 12.36: A Typical Board Camera

Many models of small, reasonably priced board cameras are available from on-line distributors and will work just fine for basic, shallow-diving robots once they are waterproofed.

Most board cameras have a wide-angle lens that naturally provides a large depth of field. (In other words, most objects will be more or less in focus all the time.) But if you want very precise focus at a particular distance, especially up close, you might want to get a camera with an adjustable focus lens and experiment to find the focus setting that works best for your vehicle. Note that the focus distance under water will be slightly different than in air. You won't be able to change the focus after you pot the camera, so you'll need to figure out some way to test and adjust underwater focus before potting, without getting the camera wet. For example, you could try focusing on something inside a fish tank with the camera pressed up against the dry side of the aquarium glass.

8. Adding an Underwater Video Camera, Lighting, and Sensors

Now it's time to add navigational sensors to your ROV. These can include a compass and depth gauge, but the best place to start is with a video camera.

8.1. CCD Video Cameras

SeaMATE's main image sensor is a CCD board camera. (See *Tech Note: Board Cameras.*) These board cameras are basically stripped-down video cameras, as shown in Figure 12.36. **CCD** stands for **charge-coupled device**—basically an electronic chip that detects light images and sends them up a special video cable to a video monitor. The alternative to CCD technology is a **CMOS chip**. Traditionally CMOS sensors have had lower sensitivity, lower resolution, and lower quality than CCDs, but they are improving and generally offer lower cost and great battery life, so you may want to conduct further research.

There are two common types of CCD cameras—standard monochrome (black-and-white) and color. Typically, they are around 1 inch wide x 1 inch long x ¾ inch high and weigh very little. The most inexpensive CCDs take black-and-white images. Color cameras have almost the same dimensions as the monochrome ones, but are slightly more expensive. Some models even have very tiny microphones built into their circuit boards so you can hear sounds that are made by the things seen with the camera. (See *Tech Note: What to Look for When Choosing a Camera* in *Chapter 9: Control and Navigation.*)

All camera models have some way to focus the image on to the sensor. There are three common means of doing this: the pinhole, the fixed focus optic lens, and the adjustable focus optic lens. The *Tech Note: Board Cameras* details some of the differences in these three types, but you will also want to research manufacturers' catalogues and websites before deciding which type of camera would best suit your project needs. Even the least expensive monochrome camera will be suitable for the *SeaMATE* ROV.

A CCD video camera must first be waterproofed and pressure-proofed, then mounted on *SeaMATE*. Fortunately, there are very simple and inexpensive ways to do this.

8.2. Pressure-Proofing a Camera

You'll often read about "waterproofing" a piece of equipment—the more inclusive term used here is "pressure-proofing," since the component has to contend with both issues: pressure and liquid. The traditional way of pressure-proofing a camera is to construct a tough housing around it. Such a camera housing generally has a cylindrical shell made of plastics or metals, with one penetration opening and seal for the wires located at the back or side of the housing. At the front of the housing is a transparent viewport (Lexan, acrylic, sapphire, or glass) for the camera lens, fitted with a rubber seal. This viewport can be flat or hemispherical, although the hemispherical shape can typically resist higher water pressures and offers better image quality with less distortion. Note that used underwater camera housings or underwater flashlight housings, if you can find them, make excellent pressure canisters—and they already come with the requisite O-ring seals and a built-in window.

Another method of pressure-proofing a camera is simply to entomb it within a solid block of plastic, rubber, or wax. This involves pouring a potting compound (usually liquid epoxy, polyurethane or melted wax) into a container that houses the camera.

Image courtesy of Randall Fox

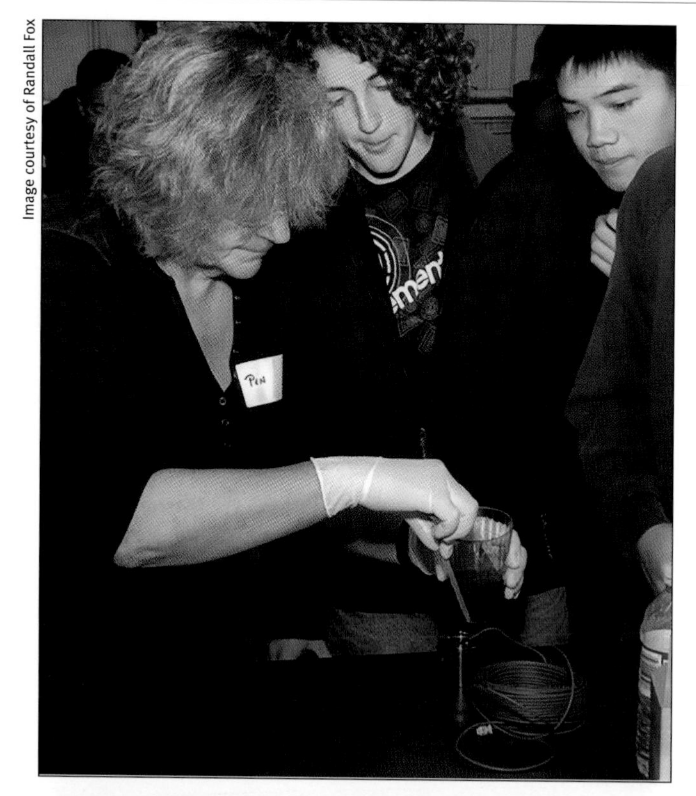

Image courtesy of Randall Fox

Image courtesy of Dr. Susan Crawford, Naval Undersea Museum Foundation

Figure 12.37: Potting Set-Up with Epoxy Potting Compound

The liquid compound then hardens (cures), sealing the camera and its connecting wires inside the container. A potted camera can easily withstand water pressure to 130 feet (approx. 40 m). Remember, however, that the camera cannot be removed if it has been encapsulated with a polymer. The advantage of wax is that it's possible to remove and reuse or repair the camera.

The simple potting technique explained in the instructions below is based on suggestions made by Nuytco Research Ltd. and VideoRay LLC. It should allow you to successfully waterproof a camera to depths of 100 feet (approx. 30 m). This set of instructions uses polyurethane as the potting compound, but you can opt to use wax instead.

8.2.1. Camera Parts and Materials

To pot your camera, you'll need the items listed below.

- one CCD video camera, such as the model PC 166XS sold by Supercircuits: Note that there are many tiny, inexpensive video cameras available—check out various websites, catalogs, and stores in your area. Make absolutely sure that any camera you get is compatible with the 12 VDC battery power used in *SeaMATE*. Many of the smaller video cameras are designed for use with 5 volt DC supplies or low-voltage AC power; those models generally will not work with a 12 VDC supply and may even be permanently damaged by connection to such a supply.
- one transparent plastic box, big enough to hold the camera: A clear polystyrene jar (Figure 12.37) works well, too. This clear container needs to be at least 1½ inches wide by 1½ inches long by 1½ inches deep in order to fit most small cameras. No lid is necessary. The part of the box that the camera will look through needs to be optically clear with no seams, ripples, lines, etc.
- one video monitor compatible with camera video output format
- one small tube of silicone sealant: (5-minute epoxy is a good, faster alternative.) Get the smallest amount you can buy, since you'll be using only a dab. Silicone is used to secure the camera lens to the bottom of the transparent box. Use a toothpick or sliver of wood for an applicator.
- one 8-oz epoxy or polyurethane compound kit for potting: Get compound with a longer curing time (1-2 hours). This is available at hobby shops, hardware stores, fiberglass supply outlets or via the internet. Toilet bowl wax is another option.

- a cable running between camera and junction box for camera power and video signal: Note that this cable may come with the camera or you may have to make it or extend it yourself.
- one cable strain relief to grip and protect the cable where it enters the junction box
- 1 foot each of 1/8th-inch heat shrink tubing and 1/4th-inch heat shrink tubing

SAFETY NOTE: POTTING

Epoxy resins can cause allergic reactions in some people so use latex gloves to protect your hands. Work only in a well-ventilated room. If you spread newspapers over the work area, you'll protect the surface and clean-up is easy. Dispose of unused compound only after it has hardened.

8.2.2. Procedure for Wiring the Camera

Before proceeding, read and make sure you comprehend any instructions supplied with the camera. Note that any procedures such as soldering wires to the camera board, encapsulating the camera, or any other modifications generally void the manufacturer's warranty. In particular, remember that you will be supplying power to the camera by wiring it into the ROV's battery or other power source. **Be extremely careful that the voltage you supply is compatible with the camera you are using, and avoid accidentally mixing up the wires when making splices or hooking into the junction box, or you will likely destroy the camera.**

Depending on the type of camera you've selected, it may or may not come with its own cable and connector. If it does come with its own cable, that cable may be short (too short to reach your junction box), or quite long (long enough to run all the way to the surface in your tether), or somewhere in between.

There are lots of different ways to connect wires from your camera to the junction box. Just remember that anything you're going to do at the camera end of the wires must be done before you pot the camera; once entombed in plastic, the wires will be unreachable and changes will be nearly impossible. The diagrams in Figure 12.39 illustrate some options to consider, depending on the details of the camera, wires, and cables you have available.

In each of the figure diagrams below, the big gray square represents the clear plastic box filled with epoxy, polyurethane, or other potting compound. The dark gray T-shaped object represents a side view of the board camera with its lens glued to the floor of the box. The thick black line represents a cable going to the junction box or directly to the surface in the tether. The thinner

Figure 12.38: Wiring and Potting Overview

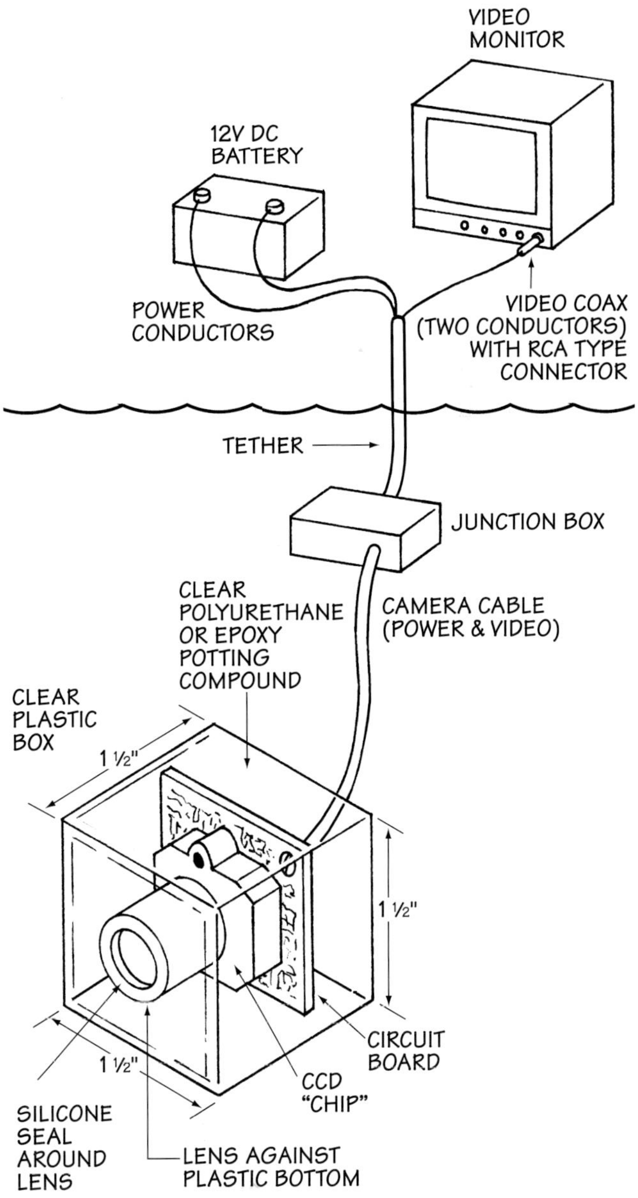

Image courtesy *Build Your Own Underwater Robot and Other Wet Projects*

black line represents a short segment of cable. Small white rectangles represent standard pluggable connectors, which are not inherently waterproof. A white circle represents a waterproof electrical connector or a waterproofed splice. (See *Tech Note: Waterproof Wire Splices.*)

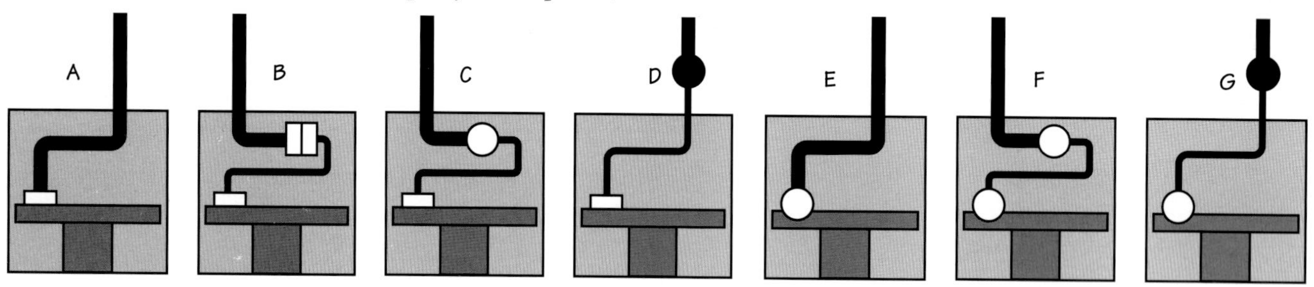

Figure 12.39: Some Options for Wiring Up Your Camera

Option A is easiest and works great if your camera cable is long enough to reach the surface, or at least to reach your junction box.

Option B is next easiest, if you have the appropriate cables and connectors.

Option C is probably the most commonly used, because it's a good choice when the camera comes with a plug-in cable that's too short to reach the junction box.

Option D is also acceptable, but requires a waterproof splice.

Option E is an alternative if the camera does not come with any cable at all. You just use another suitable cable with the right number of wires and solder each wire to its corresponding pins or other connection points on the camera board.

Option F is like E, but is helpful when the main cable is too stiff or bulky to permit easy soldering without risk of damage to the delicate camera board. This option allows you to solder smaller, more flexible wires directly to the board, then splice those smaller wires to the main cable a little farther from the camera.

Option G is simply an alternative version of F, but with a waterproof splice made outside the camera encapsulation box.

Note that in all cases, any non-waterproofed connectors, splices, or solder joints are encapsulated to protect them from water. Also note that every cable has a bend and travels some distance through the potting compound before emerging into the water. This is essential to provide strain relief for a robust waterproof seal around the cable.

Also note that in every case where you are cutting and soldering or splicing wires, you must be *extremely* careful not to mix up which wire is which. For example, if you inadvertently swap the (+) and (-) power wires or route (+) power to the video output, you will probably fry your camera. Remember if you are using coaxial cable for the video signal that the main video output signal should be connected to the center conductor and the video ground should be connected to the coax cable shield. Remember that as soon as you flip the ON switch, the camera will be connected directly to 12-volts from the battery, so make sure the camera model you are using is compatible with a 12 VDC power source.

Lastly, be sure to trace and label all wires before you cut and solder them. When you think you are done, double-check everything visually *before* applying power. Then apply power and make sure you can see a video image from the camera on a monitor to verify that everything is working properly. The time to identify and correct any wiring problems is *before* potting the camera, not after.

Likewise, if you have selected a camera with an adjustable focus optic lens, this is also the time to adjust the focus. (See *Tech Note: Board Cameras* for instructions on how to do this.) At the risk of sounding repetitious, test your camera and its wiring and adjust the focus (if you have this option) *before* encapsulating the camera.

TECH NOTE: WATERPROOF WIRE SPLICES

If your tether wires ever get damaged, or if the wires from your bilge pumps aren't quite long enough to reach your junction box, you may need to splice in an extra section of wire. This can be tricky if the splice will be under water, where you not only need to make a good mechanical and electrical connection, but you also need to do it in a way that won't short out when the new connection gets wet. The method outlined in this Tech Note is a good general-purpose method for making waterproof wire splices for these kinds of situations.

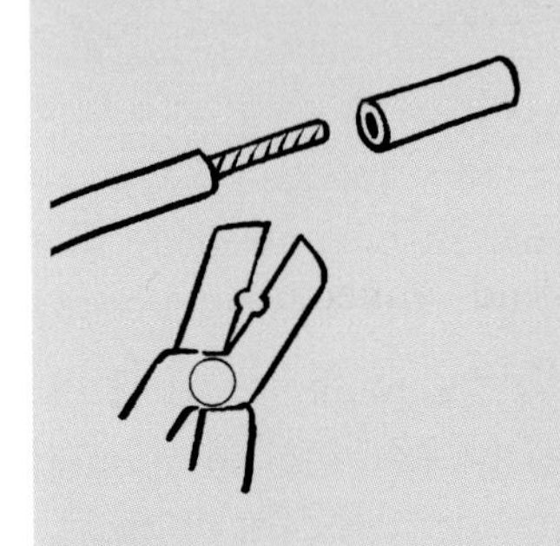

The epoxy step in Figure 12.40 is the part that makes the splice waterproof, so if your join won't be exposed to water, you can leave out the epoxy step and simply make a splice that's suitable for dry or otherwise protected conditions. For example, if your splices will reside inside an air-filled or oil-filled pressure canister, they don't require the epoxy step. Likewise, if you need to splice tether wires onto wires coming from your camera, you may be able to create a waterproof splice simply by making the join close to the camera. That way you encapsulate the splice along with the camera when you pot the camera with compound, as shown in Figure 12.42.

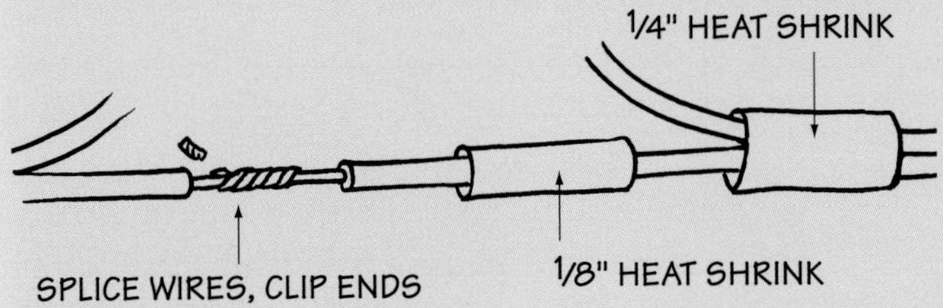

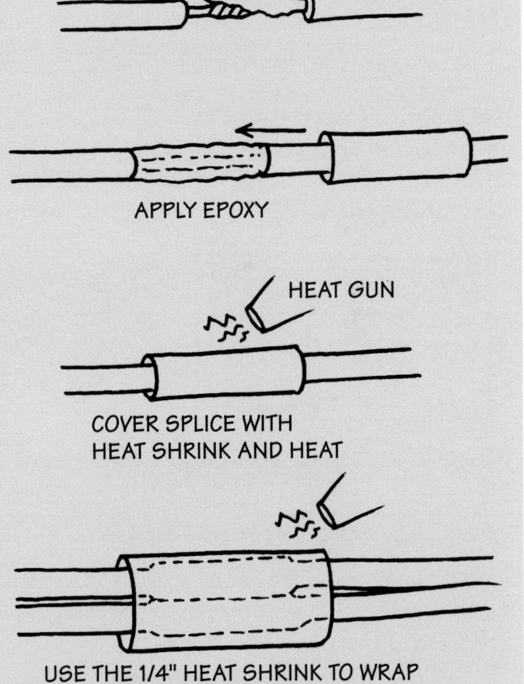

Figure 12.40: Steps for Waterproofing a Wire Splice

To make a waterproof splice, follow these directions:

1. *Strip both wires.*
2. *Cut a piece of heat shrink tubing longer than the splice and put it on one of the conductors.*
3. *Twist the two stripped ends together so they lie flat.*
4. *Solder them.*
5. *Mix 5-minute epoxy and liberally cover the splice, overlapping onto the insulation.*
6. *Now slide the heat shrink tubing over the splice and shrink it, using a heat gun.*
7. *Use the ¼-inch heat shrink to wrap the two individually wrapped splices. Wipe off any epoxy drips and let the epoxy cure.*

Images courtesy *Build Your Own Programmable LEGO Submersible.*

8.2.3 Procedure for Potting the Camera

Once you've completed all the wiring preparation, it's finally time to pot your camera. Read the instructions first, assemble the necessary materials, and prep your work site before beginning the encapsulation process.

1. Practice positioning the camera in the box, with the lens against a flat, blemish-free part of the transparent bottom. Figure out where it fits best, making sure that it won't move or fall over in that position. Check that the video and power wire

splice fits easily within the plastic box dimensions, since they will be potted with the camera unit.

2. Take the camera out of the box and spread a fine bead of silicone on the flat ring around the camera lens, as illustrated. A toothpick can be handy for applying only a small amount. Do not get any on the lens. Note: If there is too much silicone on the ring, it can ooze out and smear the lens as you proceed to the next step.

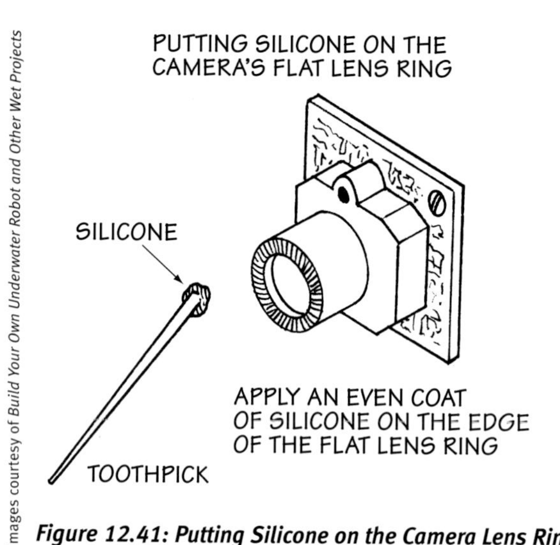

Figure 12.41: Putting Silicone on the Camera Lens Ring

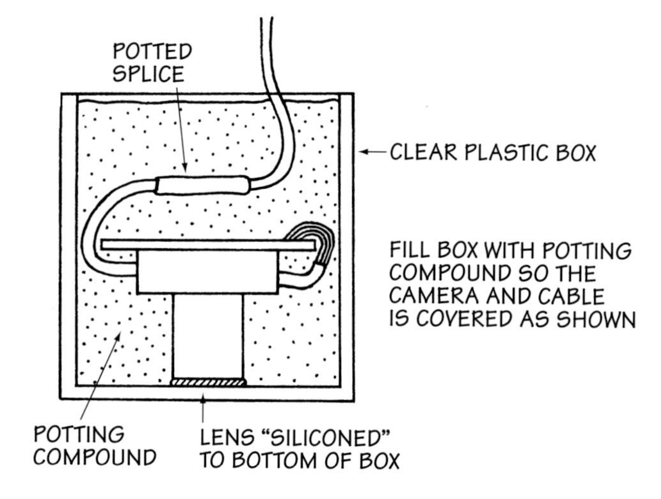

Figure 12.42: Potting the Camera and Wires

This illustration shows the box filled with potting compound so that the camera and cable are covered. (See also Figure 12.37.)

3. Carefully place the camera back into the box, seating the lens against the transparent bottom. Take care not to move it around, as that will smear the silicone.

4. Wait until the 5-minute silicone sealant cures. After it has set and the camera is fixed to the plastic box, tuck the wire splices of the video and power leads so they are inside the transparent box, ensuring that they will be completely covered by the potting compound when it is poured.

5. Read the directions on the potting compound package and prep your workspace for the pour.

6. Mix up the potting compound (epoxy or polyurethane) only after the silicone sealant has cured. If you select epoxy as your potting compound, get a large enough can or jar of epoxy to fill the transparent box to the brim. Choose a type with a curing time of 1-2 hours. Alternatively, you can melt a quantity of toilet bowl wax and use that instead of the polyurethane or epoxy compound. Caution: Use appropriate protective gloves to protect your hands. Use oven mitts if handling hot wax.

7. Pour the potting compound into the clear box, completely covering the camera. Make sure the wires and connectors (or spliced video cable) are also covered. Note: Leave the potted box to cure at room temperature. The instructions for the potting compound should tell you how long curing will take. Do not apply extra heat, hoping for a shorter curing time. If you're using wax, it takes at least 30 minutes for this quantity to harden, but to be on the safe side, leave it for one hour without disturbance.

8.3. Mounting the Camera and Prepping the Junction Box

Devise a bracket to mount the camera to the front of the ROV frame where it will have an unobstructed view (Figure 12.43). Ideally, this bracket should be adjustable, so the camera can be easily positioned for the best viewing.

Once you've figured out where to mount your camera on the frame, it's much easier to figure out exactly how much wiring you'll need to connect the camera to the junction box (including at least 6 inches for hookups inside the box). When you have established the necessary length, you can cut off any excess cable. Now drill a hole in the junction box to accommodate the diameter of the 4-conductor camera cable coming from your camera. (See Figure 12.29 and 12.30 for suggested position.) Make sure there are at least 6 inches of cable inside the box. Don't wire it to the barrier/terminal strip yet. That will happen in Section 9.3.1. Use a cable strain relief to prevent the camera conductor cable from pulling out of the box.

If you are going to use lights that draw power from the junction box, drill two small holes in the junction box to accommodate a #22 AWG speaker wire pair if you haven't done so already. Then cut a length of speaker wire and push it through the holes you just drilled in the junction box. This will supply 12 volts DC to the lights. Make sure there are at least 6 inches of wire inside the box, but don't hook it to the barrier/terminal strip just yet.

Image courtesy of Randall Fox

Image courtesy of Dr. Steven W. Moore

Figure 12.43: Mounting a Camera

No matter how fancy your craft is, you want to ensure that the mounting apparatus for its camera allows a clear field of view for the mission. As your technological experience and confidence grows, you may want to devise some way so you can adjust the camera mount.

8.4. Underwater Lighting Options

As noted in Chapter 9, lights can be very useful—often essential—for navigating and for illuminating objects that you want to see with the underwater camera. You'll note that no specific lights or lighting parts are given for *SeaMATE*. That's because there are a variety of lighting options to choose from. So you are encouraged to consider the choices given below and to explore the lighting systems used by other similar shallow-diving ROV projects before making your decision. Note that there is an option for two wires into the junction box layout that can power your lights.

Option 1: Many modern video cameras have low-light capabilities, so an additional light may not be necessary in clear, shallow water in daylight conditions.

Option 2: Consider using white LED lights. An advantage of LEDs is that they use very low amounts of power, yet put out a good white light. They are now very common in automotive taillights, flashlights, and dive lights. You can modify these to illuminate the area in front of the camera.

Option 3: Many commercial cameras now integrate LED lights around the lens (Figure 12.44). Product options are developing rapidly, so look on-line to see what is currently available. There are some interesting and low-cost kits out there now (less than $100) that include an LED illuminated camera and a color monitor.

Option 4: The simplest way to provide lighting is to attach dive lights to the ROV. Small high-intensity LED or halogen dive lights are so bright and power-efficient that it is possible they'll burn longer than one hour on a set of self-contained batteries. Pelican, Underwater Kinetics, Princeton Tec, and other dive light manufacturers make excellent dive lights that are rated to 100 meters (approx. 325 ft) or deeper.

No particular light components are specified here for *SeaMATE*, but the extra pair of wires in *SeaMATE*'s tether assumes that you will power the light from the surface power supply; the instructions that follow are based on that presumption. However, there are many low-cost and effective options, so research, design, and fabricate the underwater lighting option that will work best for you, given your camera selection. Note that you may need to adjust the instructions to drill and wire up the junction box accordingly. Most of these lighting options require that you come up with some sort of adjustable bracket to mount the light(s) on the frame.

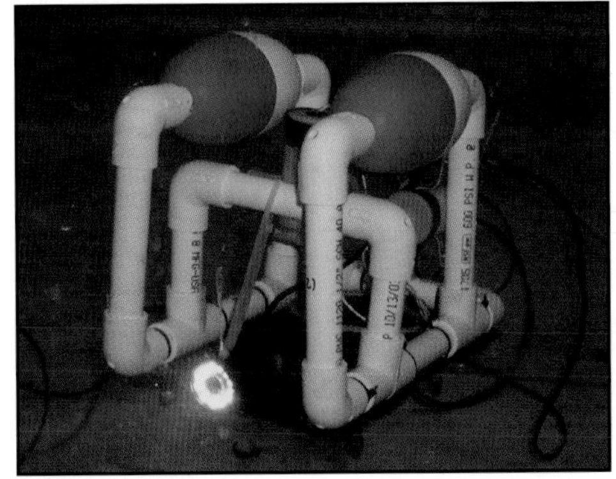

Image courtesy of Massachusetts Institute of Technology, Sea Grant College Program

Figure 12.44: LED Lights

Using a camera with LED lights around the lens is only one of many lighting options.

8.5 Adding Other Navigation Sensors

Other sensors, such as a compass and a depth gauge, can also provide valuable underwater navigational data for your ROV. Look for a small, fluid-filled compass, such as those used on car dashboards, that is mounted in clear plastic housing. Check out diving shops for an inexpensive depth gauge. Some vehicle designers build in flood-detection warning circuits that sound an alarm when a leak occurs. (*Chapter 9: Control and Navigation* describes how to do this and provides more information on sensor options.)

8.5.1. Parts

- one divers' depth gauge
- one automotive dashboard compass
- material to construct a bracket to attach a depth gauge and compass to the ROV

8.5.2. Procedure

1. Use tape, plastic cable ties, or screws to mount the sensors onto any type of mounting bracket that fits your robot's design and the devices you've selected. It is helpful if the bracket has some way to adjust the sensor so that it can be easily seen by the camera without blocking too much of its view.
2. Since the wiring has not been completed at this stage, it will be difficult to adjust the position of the compass and depth gauge in front of the camera. This work should be done during the sea trials stage in the test tank or pool. (Remember that the rebar in concrete pool walls can affect compass readings.)

9. Fabricating the Control Box, Wiring the Junction Box, and Testing

SeaMATE employs one of the simplest types of control systems for running an ROV—three DPDT (double pole double throw) switches. These control the operation of the thrusters and are hardwired—that is, one pair of speaker wires is directly wired to one thruster, with no intervening electronic control circuits or multiplexing. So when you flip the switch, the circuit closes and the thruster turns. This type of control scheme is called **bang-bang control**. That means that when the switch is activated, the motor turns full ON or OFF—bang on, bang off. There is no option for incremental change in the rate of the rotation, commonly known as speed control.

However, the switch is able to reverse the spin of the motor, since it can be wired to reverse the polarity of the voltage applied to the motor. And when you change the polarity, a brushed DC motor rotates the other way. If you run each thruster in different directions for different durations and also use combinations of REVERSE and FORWARD, you can achieve all the basic movements needed to maneuver an ROV in calm water. (Review *Chapter 7: Moving and Maneuvering* and *Chapter 9: Control and Navigation* for simple or more complex control options.)

This section of the *SeaMATE* project actually involves three major construction steps:

- drilling and wiring the control box
- wiring the junction box
- testing the circuits for the whole vehicle

But before you can complete these steps, you'll need to gather the necessary parts.

9.1. Parts

- three small DPDT (double pole, double throw) switches with center OFF: make sure they're rated for at least 5 amps.
- one small SPST (single pole, single throw) switch for light and camera power: make sure it's rated for at least 5 amps.
- 10 feet of #16 AWG lamp cord or equivalent for 12-volt power cord
- one cable strain relief for 12-volt power cord
- 20 feet of #22 AWG red hookup wire
- 20 feet of #22 AWG black hookup wire
- one large pair of alligator clips for the end of the 12-volt DC power cord: Note that an alternative connector can be substituted to match the connectors on whatever power supply you use.
- one plastic project box approximately 6 inches x 4 inches x 2 inches
- one miniature barrier (terminal) strip #22 AWG gauge
- four RCA bulkhead jacks: Omit if you're wiring the tether directly to control box.
- one fuse holder: There are many types, but any basic style of fuse holder will work as long as it fits inside the dimensions of your control box housing.
- two 8-amp fuses to fit fuse holder: Keep one as a spare.

9.2. Procedure

These instructions describe how to build a very simple control box.

1. Take a project box, figure out and mark the placement of all the holes you'll be drilling for:

   - the three DPDT thruster switches
   - the SPST light switch
   - the 12-volt power cord
   - the four RCA bulkhead jacks
   - the fuse holder

   Figure 12.45 shows a suggested layout for the type of project specified here. Remember that you will also be mounting a 12-post miniature terminal strip inside the box, so make sure the placement of that terminal strip will not interfere with the position of the switches and the RCA jacks that extend into the box. A little extra room will also facilitate connecting the wires.

2. To mount the three thruster switches and the single lamp switch, choose a drill bit that matches the stem diameter of the switch, then drill the holes marked for the switches. Next, select a smaller bit that fits the width of the locator prong and drill the four locator holes for the switches. As shown in Figure 12.47, these holes prevent the switches from spinning loose.

3. Mount the four switches in their respective holes to test fit them.

4. The *SeaMATE* plans call for using RCA jacks in the control box, so that you can unplug your tether for easier storage and transport. To do this drill four holes for the jacks in roughly the locations indicated in Figure 12.45. You'll need to look at the threaded part of the RCA jack to determine what size drill bit to use. You also have the option of skipping the RCA jacks, if you're willing to wire your tether permanently to the control box. This simplifies construction, but makes it harder

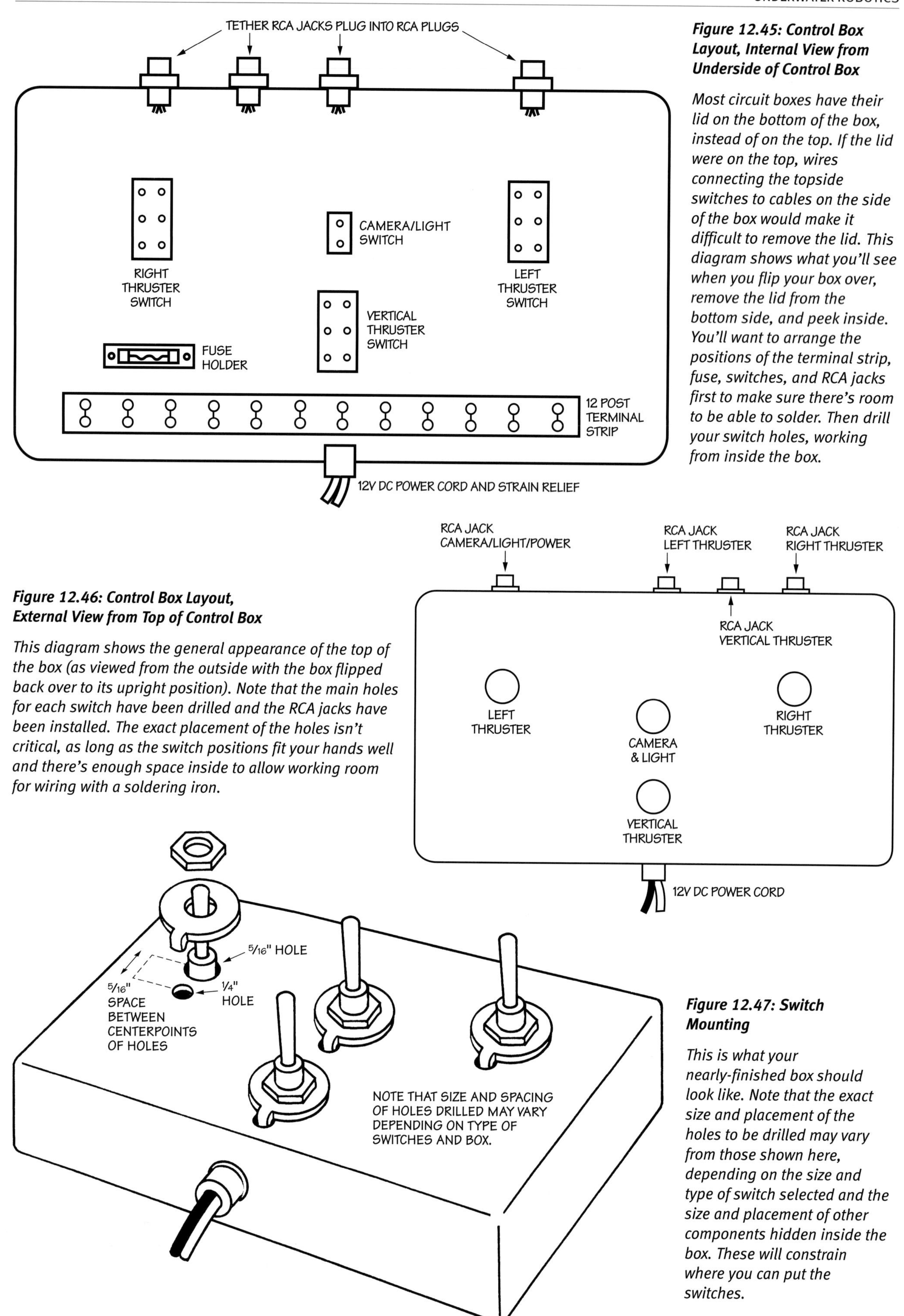

Figure 12.45: Control Box Layout, Internal View from Underside of Control Box

Most circuit boxes have their lid on the bottom of the box, instead of on the top. If the lid were on the top, wires connecting the topside switches to cables on the side of the box would make it difficult to remove the lid. This diagram shows what you'll see when you flip your box over, remove the lid from the bottom side, and peek inside. You'll want to arrange the positions of the terminal strip, fuse, switches, and RCA jacks first to make sure there's room to be able to solder. Then drill your switch holes, working from inside the box.

Figure 12.46: Control Box Layout, External View from Top of Control Box

This diagram shows the general appearance of the top of the box (as viewed from the outside with the box flipped back over to its upright position). Note that the main holes for each switch have been drilled and the RCA jacks have been installed. The exact placement of the holes isn't critical, as long as the switch positions fit your hands well and there's enough space inside to allow working room for wiring with a soldering iron.

Figure 12.47: Switch Mounting

This is what your nearly-finished box should look like. Note that the exact size and placement of the holes to be drilled may vary from those shown here, depending on the size and type of switch selected and the size and placement of other components hidden inside the box. These will constrain where you can put the switches.

to pack, unpack, and store the tether during ops. If you decide to go without the RCA jacks, don't drill the four RCA jack holes in the control box. Instead, replace them with a single hole barely big enough to accommodate all four pairs of power-carrying tether wires. Then bring those wires inside the box through the hole. Make sure you pull in enough wire to reach the switches, plus a bit more to give you some wiggle room. Use a cable tie, cable strain relief, or other mechanism to keep the tether from getting yanked out of the box, which would put strain on the solder joints and possibly break them loose. Then use wire strippers to strip some insulation off the ends of the wires and solder each wire to its correct switch terminal. Consult the schematic diagram (Figure 12.48) to help you determine which wire goes where.

5. Next, drill a hole for the strain relief that will hold the power cord. There are many types of strain relief devices, so measure the type you've chosen to determine the size of drill bit to use. Drill a hole and install the cable and strain relief so that there are at least 8 inches of slack cable inside the control box.

6. Drill holes to mount the 12-post terminal strip inside the box.

7. Determine placement of the fuse holder and mount it inside the box (Figure 12.48). Depending on the type of fuse holder you select, this may require screws, cable ties, or glue. Alternatively, you could use an in-line fuse holder, which attaches outside the control box on the cord supplying the power from the battery. Don't insert the fuse yet.

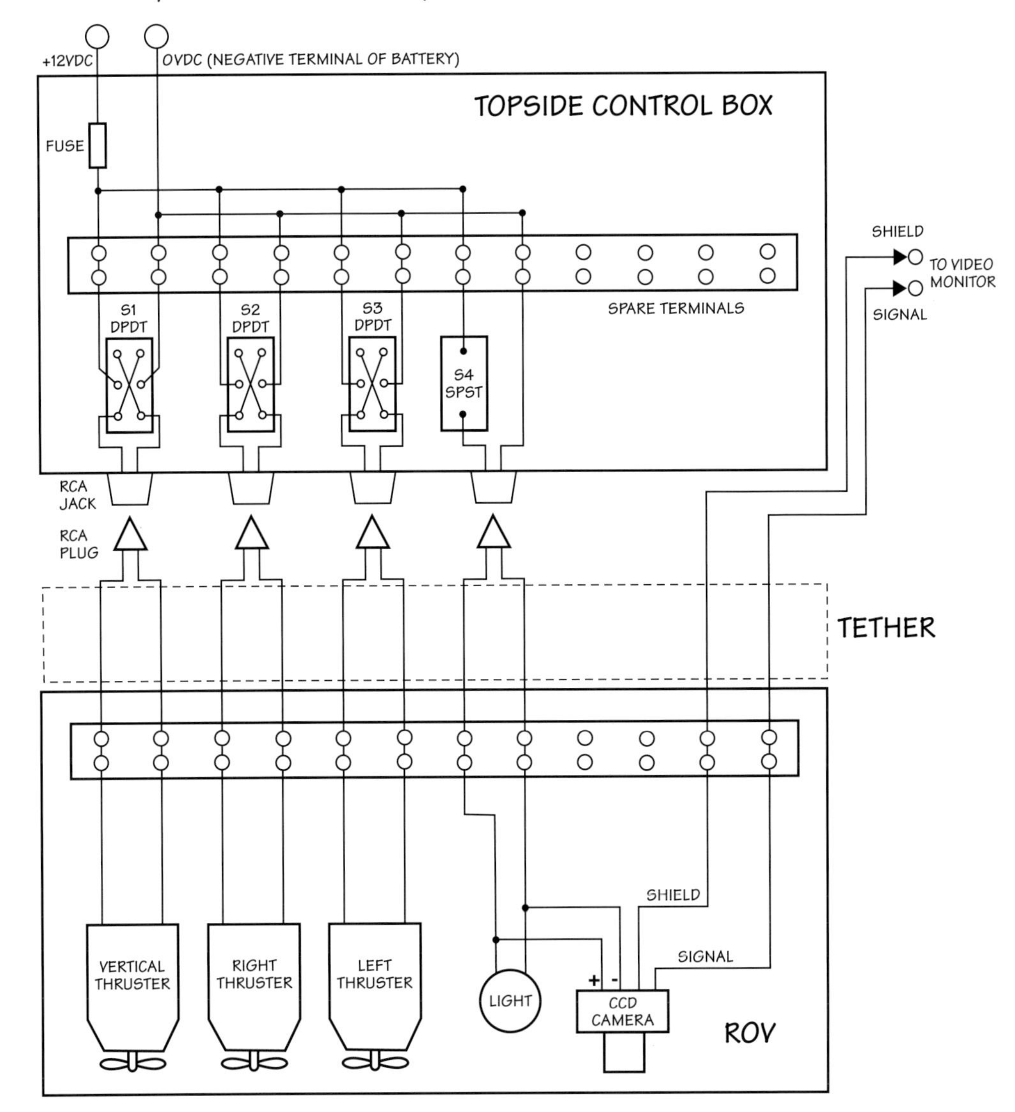

Figure 12.48: Schematic

This image is repeated for your convenience.

TECH NOTE: TIPS FOR CONSTRUCTING THE CONTROL BOX

When making the control box, place the controls so they are easy to reach and so the pilot's hand movements will be made in a logical order. For example, the right switch controlling the right thrusters is located on the right side of control box and operated by the right hand. The left switch controlling the left thruster is located on the left side of the control box and operated by the left hand. The vertical switch can be located in the center of the box and operated by either hand. Make sure the distances between the switches are such that the pilot can reach them comfortably.

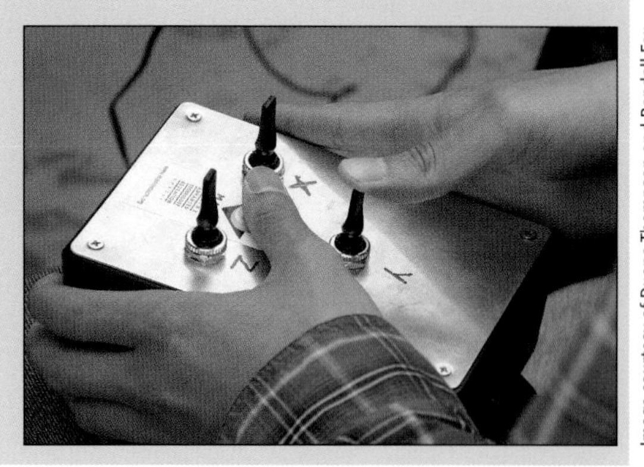

Figure 12.49: Convenient Switch Position

Image courtesy of Duane Thompson and Randall Fox

8. Follow the schematic and solder the wiring inside the control box. This includes several different steps with quite a few wires, so carefully follow the schematic illustration in Figure 12.48 as you progress through each step. You will use the red and black wires for most of the internal wiring. First, wire up the "X" on each DPDT switch to make thruster direction reversible. Then wire the power cord—the negative (ground) wire will go directly to the terminal strip, and the positive wire will go to the fuse. Use a red hook-up wire to connect the other side of the fuse to the terminal strip. Next use red (positive) or black (ground) wire, as appropriate, to distribute positive voltage and ground to each of the four switches via the terminal strip. Use red and black wires to connect RCA jacks to the switches, putting red wire on the center conductor. (Remember, however, that this wire may act as positive voltage or ground, depending on the switch position.) Don't forget to hook up the last wires for camera and light power, too. It's helpful if you check off or color/highlight each wire on the schematic as you complete it to make sure you don't forget anything.
9. Double-check your wiring.
10. Solder the two alligator clips to the external end of the 12-volt power cord. The clips make it easy to attach the power cord to a battery or other power supply. If you have other power line connectors, such as banana plugs, solder those on instead of the alligator clips.
11. Finally, go through all of the wiring connections again, tracing them to triple-check that you have not made any mistakes.

9.3. Wiring the Junction Box

9.3.1. Procedure

1. Take the junction box that you worked on earlier and run all the thruster wires into it through the appropriate holes you've already drilled. (See *Section 6: Constructing the Tether and Junction Box.*)
2. Working with one conductor pair at a time, trim the wires to 4–6 inches in length. Do not make them any shorter, since you'll need some slack in the wires. If you leave them overly long, the wire takes up too much space in the junction box. Do not trim and cut all the wires at the same time, because it's easy to lose track of which wire is which.

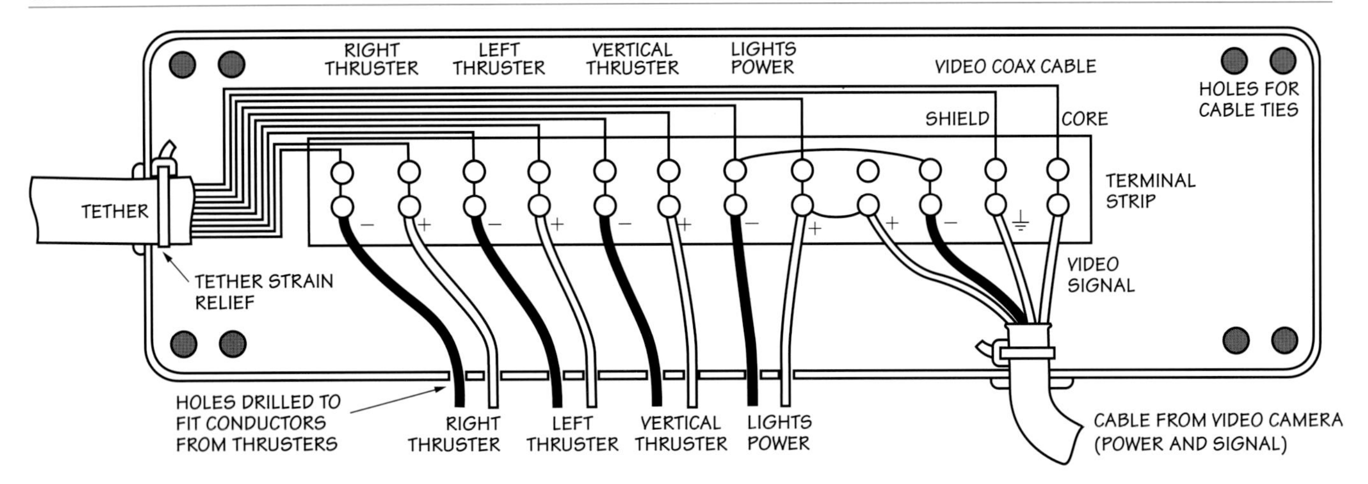

Figure 12.50: Junction Box Wiring

This wiring illustration is included here again for your convenience.

3. Strip the insulation off the end of one conductor pair and connect the wires to the barrier/terminal strip, according to the lower half of the Figure 12.50 schematic diagram. Check off each wire you connect against this schematic.
4. Continue this procedure for every wire in the junction box—thrusters, camera, and optional lighting, if you choose to install it.
5. Once the wiring is completed, pull back any slack wires and neaten them. Make sure no wires are extending above the top edge of the box.
6. Once again, physically trace and double- and triple-check all the wiring on the vehicle. Correct any mistakes. At this point, DO NOT pour any wax into the junction box. The system has yet to be systematically tested with the power ON.

9.4. Testing for Wiring, Thrusters, Camera, and Light(s)

If you encounter problems at any time in these tests, **stop immediately**. Before going on, read the *Tech Note: Troubleshooting Electric Circuits* and diagnose the problem. Then correct it. Remember that blowing a fuse is a sure sign of a shorted circuit or current overload.

9.4.1. Wiring Test

Once the wiring is done and circuits have been visually checked to ensure they are correct, it's time to energize the vehicle, one circuit at a time. Here's a suggested procedure:

1. Make sure all the switches on the control box are in the center, or OFF, position.
2. Lift off the cover plate of the control box if it is not already open and locate the fuse holder. If there is a fuse already in the holder, remove it. Do *not* connect the tether. Leave the lid open.
3. Connect the power cord to the appropriate power supply terminals; red typically for positive, black for negative.
4. Use a multimeter to measure the voltage at the power supply source. It should be between +12 volts and +13.5 volts DC.
5. Use a multimeter to measure the voltage at the fuse holder in relation to ground. If the voltage is the same as the power source voltage, unplug the power cord and install the fuse. Turn on the power and measure the voltage again. If the voltage is not the same, there's a wiring problem somewhere, which you'll need to resolve before moving on.

6. If all the voltages are ok, flip all the thruster switches to the FORWARD position. Turn the camera/light switch ON.
7. Set your multimeter to read DC volts and measure the voltage at the inside of the vertical, left, and right thruster RCA jacks as shown and described in Figure 12.52. They should all read +12 to +13.5 volts DC.
8. Measure the voltage at the video power RCA socket. It should read +12 to +13.5 volts DC.
9. Flip all the thruster switches so they are in the REVERSE position. Turn the camera/light switch OFF.

TECH NOTE: MULTIMETERS

An inexpensive multimeter, similar to the digital model pictured here, is perfect for working on *SeaMATE* and many other robotic projects. It's an essential piece of test equipment because it lets you "see" otherwise invisible electrical quantities.

Hundreds of different models are available. Like almost all of them, the one pictured here can measure and display DC voltage, AC voltage, current (amps), and resistance (ohms). As well, it can check for electrical continuity between two points in a circuit (indicated by an audible beep). Like many meters, it also offers a few less-standardized, and generally less important, features—in this case, a transistor tester, diode checker, and thermometer.

When shopping for a multimeter, look for models with a maximum current rating of 10 amps or more. Many max out at about 200 mA (0.2 amps), which is not enough for measuring the 3-4 amps typical of bilge pump motors. The ability to measure and display capacitance (not offered on the model pictured here) is another helpful feature, since it's notoriously difficult to read capacitor values directly off the component.

Whatever meter you use, be sure to read the instruction manual thoroughly and/or have a knowledgeable person teach you how to use it properly. You'll be glad you did. Once you become comfortable using a multimeter, it will become your best friend for troubleshooting circuits that aren't working properly.

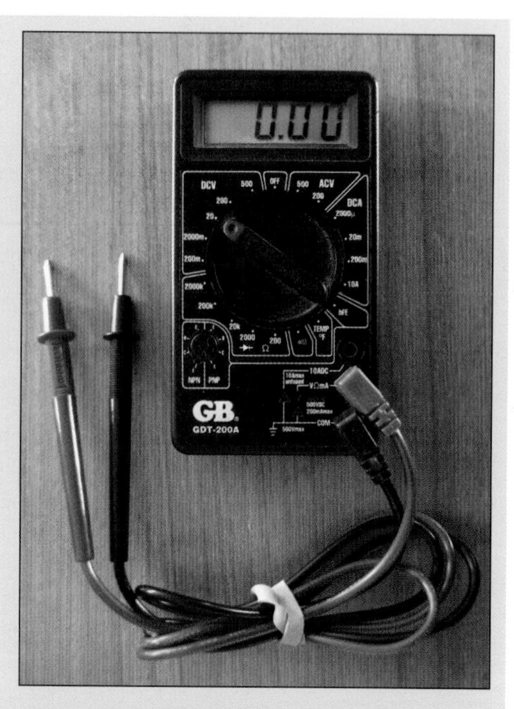

Image courtesy of Dr. Steven W. Moore

Figure 12.51: Multimeter

See Tech Note: Circuit Troubleshooting with Test Instruments *in Chapter 8, Section 8.2. for a bit more information about multimeter use.*

Image courtesy of Dr. Steven W. Moore

Figure 12.52: Measuring RCA Jack Voltage

Before connecting your thrusters, lights, or camera to your control box, you should check to make sure the correct voltages are present at each of the four jacks for each position of the control box switches.

To measure the voltage on one of the RCA jacks, set your multimeter to measure DC volts. (Most meters offer several DC voltage ranges to choose from; the 20 VDC range setting should be about right.) Then touch the positive (red) probe to the metal (not the colored plastic) inside the jack while holding the negative (black) multimeter probe against the metal on the outside of the jack, as shown in this test of an RCA jack on a printed circuit board.

It may help to have one person read the meter display while another person concentrates on holding the metal probe tips in place, so they don't slip and short something out.

10. Measure the voltage at the RCA sockets. All the thruster sockets should read –12 to –13.5 volts DC, except the video power RCA socket, which should be zero, since the switch is OFF.
11. Disconnect the battery.

9.4.2. Thruster Tests

1. Flip all thruster switches to the OFF position. Flip the camera/light switch OFF.
2. Connect the tether RCA plugs to the corresponding RCA jacks on the control box; left, right, vertical, and camera/light. Caution: The yellow RCA plug from the tether is for video out to the monitor. **Never connect the yellow RCA plug to a 12-volt power jack. If 12 volts is sent down the RG-174, it will almost certainly destroy the camera.**
3. Connect and turn on the power supply.
4. The next step is to test your thrusters. **Remember not to run them in air for more than 15 seconds** or you may overheat and damage the shaft seal. (See *Tech Note: Bilge Pump Shaft Seal Damage.*) To begin the test, flip the left thruster switch to the FORWARD position. The left thruster should turn ON and rotate the prop in a clockwise rotation. You should feel a bit of wind come off the prop if you put your hand close to it. If it does not run in the correct direction, check to see that the polarity of the power cord plugs is connected to the battery correctly. Red is positive, black is negative. If the battery connections are correct, then swap the motor wires at the terminal strip in the junction box. This should make the motor run in the correct direction.
5. Now flip the left thruster switch to the REVERSE position. The prop should spin in a counterclockwise direction. Remember not to run the thruster more than 15 seconds.
6. Follow the same procedure for the right thruster (for less than 15 seconds).
7. Follow the same procedure for the vertical thruster.
8. Next, run the right and left thrusters together.
9. Run all the thrusters together (for less than 15 seconds).

If you have completed all of these steps without any problems, you have successfully wired the thrusters.

9.4.3. Camera Test

Next you want to set up for the video camera test.

1. Plug the yellow RCA VIDEO OUT plug from the tether into the VIDEO IN RCA jack on the monitor. Turn ON the monitor.
2. Flip the camera switch on the control panel to the ON position. An image should appear on the monitor screen.
3. If you've powered the optional lights through this circuit, they should also turn ON and OFF when you move the camera switch. The junction box wiring diagram will show you how the light is wired to this circuit.

This completes the electrical tests. If everything worked as expected, you can turn the power off and ready to pot (encapsulate) the junction box. If you have had problems, then diagnose them by reading *Tech Note: Troubleshooting Electric Circuits.* Resolve the issue before proceeding.

TECH NOTE: TROUBLESHOOTING ELECTRIC CIRCUITS

The ability to troubleshoot (i.e., diagnose and repair) electrical problems and other technical malfunctions is a valuable skill. It's also one that's easy to learn, if you approach it the right way. The idea of troubleshooting was introduced back in *Chapter 2: Design Toolkit*, but here it's explored in more depth using four suggestions that will help you pinpoint the source of trouble quickly and reliably.

1. **Understand the normal behavior of the circuit.** You can't spot something that's not normal if you don't know what "normal" looks like. When attempting to troubleshoot a circuit, you should have a good sense of what the circuit is supposed to do and how it's supposed to work. Then you can look for places in the circuit where the electricity is not doing what it's supposed to be doing. Since electricity is invisible, test instruments like multimeters and oscilloscopes, which allow you to measure and display electrical quantities in real time, can be extremely helpful in tracking down subtle electrical problems.

2. **Be scientific in your approach.** A scientist figures things out by a) asking questions, b) proposing possible answers (hypotheses) for each question, c) making distinct predictions that follow logically from each hypothesis, and d) eliminating any hypotheses when the associated predictions don't match reality. Through a process of elimination, incorrect hypotheses are weeded out, leaving the scientist with the best possible answer to the question. It may sound mysterious, but science is really just the formal application of common sense. You can apply this same common-sense approach to figure out the reason an electrical circuit isn't working properly.

   For example, if your thruster isn't spinning when you turn it on, make a list of possible explanations (hypotheses): dead battery, disconnected battery, blown fuse, broken wires, fried switch, burned out motor, jammed propeller. Then, for each hypothesis, make some logical predictions: If the battery is dead, then the lights shouldn't work either and swapping in a fresh battery ought to fix the problem. If the wire to the switch is broken, a multimeter ought to reveal that voltage at one of the wire isn't getting to the opposite end and temporarily bypassing the wire with a second one ought to make the thruster work. And so on.

3. **Check the easy things first.** Some predictions are very easy to check ("Are the wires connected to the battery?"). Others are more difficult and time consuming ("Can I get the vertical thruster to work by rewiring it to the horizontal controls?"). Experienced people usually check the quick and easy things first, even when they suspect the problem lies elsewhere. Often the correct explanation is something very simple (e.g., somebody forgot to turn on the power). In such cases, the quick check saves a lot of time and effort. If the quick tests don't reveal the problem, at least you know the extra effort you put toward testing the other hypotheses isn't wasted.

   In addition to a simple visual check of the wire connections, another quick and easy technique is to gently wiggle wires, one at a time, while the power is on. If the problem comes and goes intermittently while you are wiggling a wire, then that wire probably has a break or unreliable connection somewhere. Just be careful that you don't wiggle things so much that you break or loosen new wires in the process, and never attempt this on a live high-voltage circuit.

4. **Divide and conquer!** Imagine that you have a very complicated circuit consisting of 1,024 components hooked together end-to-end in a long chain, and even though you're putting power into the front end, you're not getting the electrical behavior you expect out of the other end. Where is the problem located? You could start at the front, systematically checking to make sure that power is getting to the first component, then move on to make sure the appropriate power or signal is getting to the second component, and so on down the line until you find the problem. That method would take you 1,024 tests to check the entire circuit. Of course, the problem is as likely to be in the first half of the circuit as in the second half, so you would probably find it long before checking all 1,024 components. Still, it could take a while. In fact, a little math can show you that it would take, on average, 512 tests to locate the problem using this technique.

   But what if you took a different approach? Instead of starting with the first component, start by checking to see if you have power (or other expected electrical signal) at the midway point in the circuit, between parts #512 and #513. If you detect an abnormality there, you know the problem is located somewhere in the first half of the circuit, between part #1 and part #512. Otherwise, the problem must be located in the second half of the circuit, between part #513 and #1024. Either way, you've just identified 512 parts you probably do not need to bother checking!

Suppose you've determined that the problem is between #1 and #512. Now divide that chunk in half again by checking for an appropriate electrical signal between part #256 and #257. The answer will allow you to eliminate another 256 components from the list of likely suspects. Thus, in only two tests, you've eliminated 512 + 256 = 768 possible suspects! It turns out that if you cut 1,024 in half again and again like this only ten times, you'll have narrowed the problem down to a single component, because 1,024 divided in half ten times equals one.

The take-home message here is that you can pinpoint a problem much more efficiently if you divide and conquer, looking for opportunities to eliminate big fractions of the circuit from consideration. In the original thruster example above, you might use a multimeter to check for proper voltages at the output of the thruster power switch. If the voltage goes on and off as expected when you flip the switch, that's a strong indication that the problem lies in the wires going to the thruster or the thruster itself, rather than in the battery, the switch, or the wires between those two. At that point, you don't need to waste time checking the battery, the switch, or the connecting wires because you know the problem is farther downstream.

10. Constructing Payload Tools

In order to use *SeaMATE* to retrieve small objects off the harbor bottom, you'll need to figure out just what the ROV might be capable of recovering, given its limitations of thrust, vision, and maneuverability. In other words, in order to fashion an effective payload tool, you have to know the specs of the object to be recovered as well as the capabilities (hence, limitations) of your vehicle.

10.1. Pick-Up Probe

If the object to be recovered has a convenient attachment point, such as the ring on the bozonium canisters, all you may need for a payload tool is a basic pick-up probe made from a rod, a length of coat hanger, or even a chopstick. Of course, you'll have to fasten the probe to the front of the ROV in view of the camera. Once you get the hang of retrieval, it's fun to construct other objects that the probe can recover and to practice doing so.

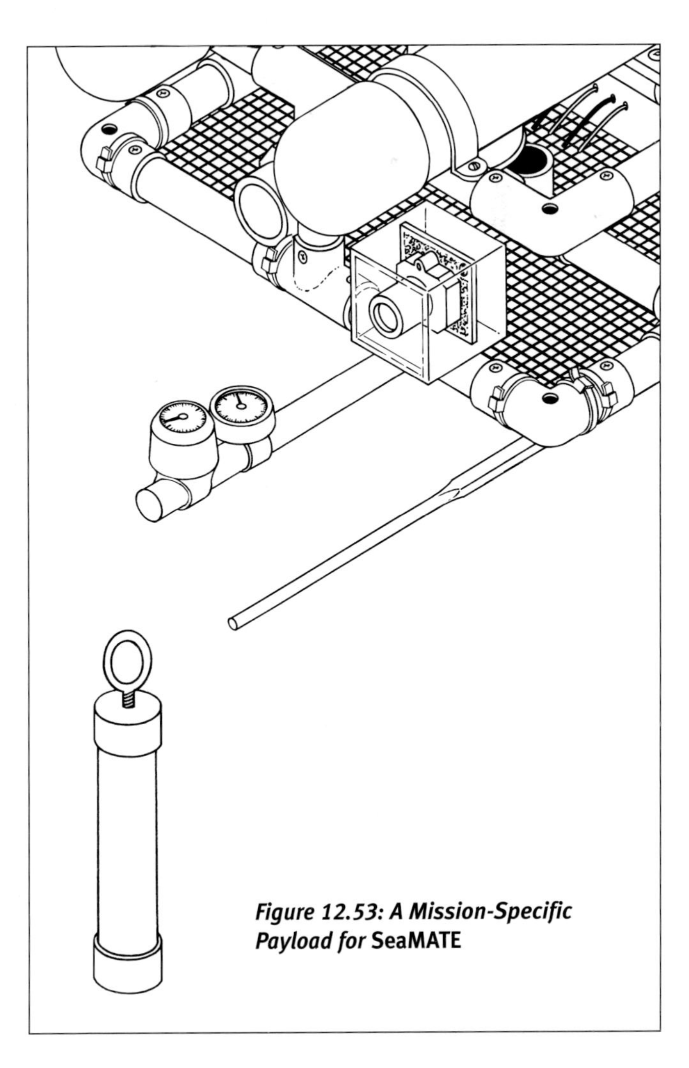

***Figure 12.53: A Mission-Specific Payload for* SeaMATE**

10.1.1. Parts

- one 6- to 12-inch probe, fashioned from a plastic or metal rod or wooden chopstick
- tape (or bracket) to attach the probe securely to the frame

10.1.2. Procedure

Attach the probe to the front of the ROV with cable ties or strong waterproof tape so that it's in the camera's view range. You need to ensure the probe has a strong attachment point to the ROV. If you opt to use tape, you may find that duct tape works better than electrical tape for underwater missions. For even greater strength, you may decide to fashion some sort of bracket for attaching the probe. Experiment with positioning the probe so as to have the best view of the tool, the compass, depth sensor, and the object being recovered.

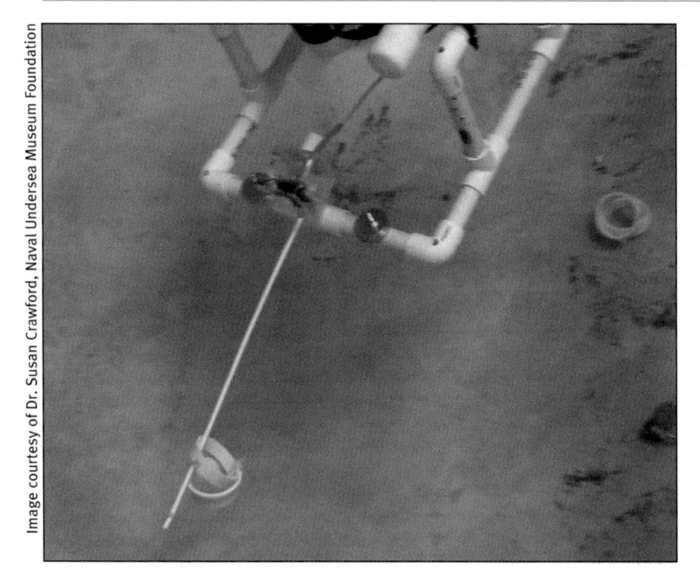

Image courtesy of Dr. Susan Crawford, Naval Undersea Museum Foundation

Figure 12.54: Simple and Effective

You can pick up a great deal of piloting skill using a simple probe to "spear" or capture targets.

Image courtesy of Randall Fox

Figure 12.55: Other Attachments

A cluster of sturdy, plastic knives created a cage that let this ROV bring back samples of simulated marine growth in a competition.

10.2. Other Payload Options

Depending on what you want your ROV to retrieve, you can expand the concept of a simple pickup probe in any number of ways. For example, you can double the number of probes, fixing one to each side of the lower front frame, then attach a piece of mesh between them. This easily creates a recovery basket, as shown in Figure 12.56.

Another approach is to replace a probe with some other object. For example, a strong, lightweight Lexan fork, commonly used for camping, might be used to recover a lost key ring, clothing, or any object that can be snagged by the fork's tines. Or you could attach a very small dip net or child's sand shovel to retrieve samples of the bottom. Experiment with modifying a long-handled barbeque fork or try adapting a large fishhook for grabbing cloth or wooden objects. Just be careful not to snag logs or heavy objects on the bottom that could trap your ROV.

These are just a few of the most basic payload tool ideas for *SeaMATE*. You'll come up with plenty of other simple possibilities when you put your imagination to work. Other examples of somewhat more complex tools for shallow diving ROVs are illustrated in *Chapter 10: Hydraulics and Payloads.*

Figure 12.56: Expanding the Simple Probe Idea

Look closely and you'll see that this ROV has mesh attached between the two probes on its frame. This easy adaptation allowed the ROV to accomplish its retrieval mission.

Image courtesy of Randall Fox

11. Preparing for Sea Trials and Ops

At this point, *SeaMATE* is almost completely fabricated and the circuits have been tested. All that's left is to pressure-proof (i.e., pot) the junction box and make various sensor and tether adjustments before sea trials can be performed.

One waterproofing technique, commonly known as **potting**, or **encapsulation**, has already been discussed in *Chapter 5: Pressure Hulls and Canisters.* This technique is based on the fact that solids are far less compressible under pressure than gas is. By surrounding the wiring with a waterproof solid, it is possible to effectively waterproof and pressure-proof the wiring. Wax is commonly used as the material for this process, since it's waterproof and can easily change state (e.g., from solid to liquid to solid) within a very small range of temperature—it liquefies at just below the boiling point of water and solidifies quickly at room temperature.

Encapsulating *SeaMATE*'s electrical termination can simply involve melting wax and pouring it into the canister until completely full. When the wax cools into a solid state, the termination can is both water- and pressure-proofed. If you need to make repairs to the termination can, the wax can be removed at any time by scooping it out or by remelting it with a heat gun and pouring it out. Once the repair is completed, the wires can be resealed by pouring more melted wax back into the junction box.

Materials other than wax can be used for encapsulating the wiring. Polymers (plastics) such as acrylic, epoxy, or polyurethane can be mixed and then poured into the termination can. The advantage of polymers is that they solidify much harder than wax does. However, once they are poured into the can, there is no way to remove the material and effect repairs at a future time. Polymers are also much more expensive and require special preparation and ventilation procedures. Other pressure-proofing techniques, such as oil compensation and air compensation, are discussed in detail in *Chapter 5: Pressure Hulls and Canisters.*

11.1. Potting the Junction Box

Potting, or encapsulating, the junction box involves pouring hot liquid wax. When the wax hardens, it forms a watertight seal over all the wires; since the wax becomes a solid, it can resist a considerable amount of water pressure. Toilet bowl wax seems to have the best qualities to seal the wires. As noted previously, this wax becomes very sticky when solid. Paraffin wax is not recommended, because it does not have the same sticky qualities and does not adhere well to the wire insulation.

SAFETY NOTE: COMBUSTIBLE WAX

Wax melted in a cup or mug placed in boiling water is not likely to reach a temperature greater than the boiling point of water which is 212° F (100° C). This is much lower than the flashpoint of wax so it can't easily catch fire. However, wax melted directly over a burner may combust. Never melt wax in a microwave oven. When handling the hot cup of melted wax, use oven mitts to avoid getting burned.

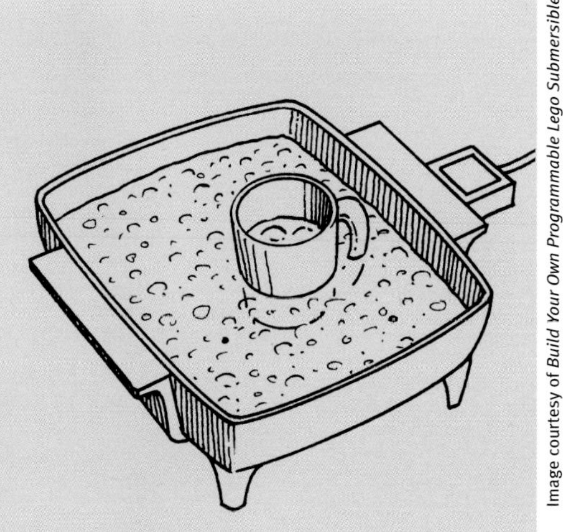

Image courtesy of *Build Your Own Programmable Lego Submersible*

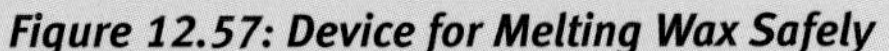

Figure 12.57: Device for Melting Wax Safely

An old electric frying pan is one of the safest methods for melting wax for waterproofing thrusters and cameras.

TECH NOTE: POTTING BASICS

Suitable Waxes for Potting: Not all waxes are suitable for potting (encapsulating) wiring that is exposed to water. For example, although paraffin wax is easily available, it is not a good choice, because it does not bond well to wire insulation in its solid state. This means that under pressure, water will be forced between tiny gaps between the wire and wax, allowing moisture to leak inside the wire. Use either beeswax or the more economical toilet bowl ring sealant wax available at any plumbing store. In their solid state, these types of wax are very sticky and make a tight bond to the wires.

De-Waxing Potted Parts: It is easy to remove a wax-encapsulated part from its waterproof housing. However, the part will be coated with a sticky wax residue that needs to be removed. Use 70 percent isopropyl alcohol as a solvent to de-wax the part, applying with a toothbrush, Q-tip, or cloth as needed. You may want to wear latex gloves and work in a well-ventilated area when applying isopropyl alcohol. Remember that 70 percent isopropyl alcohol is very flammable, so be sure to keep it away from any heat source. Read and follow the safety guidelines printed on the bottle.

11.1.1. Parts and Tools

- one package of toilet bowl wax
- hot glue gun
- electric frying pan and ceramic cup for melting wax safely

11.1.2. Procedure

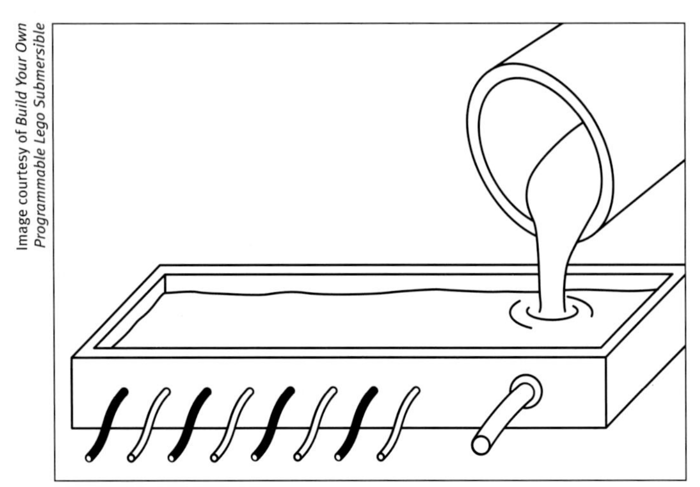

Image courtesy of *Build Your Own Programmable Lego Submersible*

Figure 12.58: Potting Technique

1. Use a hot glue gun to seal every opening in the junction box with hot glue, including the holes where the conductors come through.
2. Half-fill the electric frying pan with water and heat it to simmer.
3. Cut up some wax and place it in a cup. Place the cup in the simmering water.
4. When the wax is completely melted, pour it carefully into the junction box. Continue to pour until the box is filled to the brim and all conductors have been covered with wax.
5. Do not disturb the box until the wax has solidified. Then screw the box lid on.
6. If you need to remove the wax to make a repair or change the wiring, just scoop out as much of the wax as needed or carefully reheat the wax with a heat gun on LOW temperature setting. When the repair is completed, melt more wax and re-fill the box.

11.2. Finishing the Fabrication

Once you've completed potting the junction box, you've also essentially completed fabricating *SeaMATE*. Congratulations! Your finished ROV should look like Figure 12.59. All that remains now is a series of adjustments that are part of prepping the vehicle for sea trials.

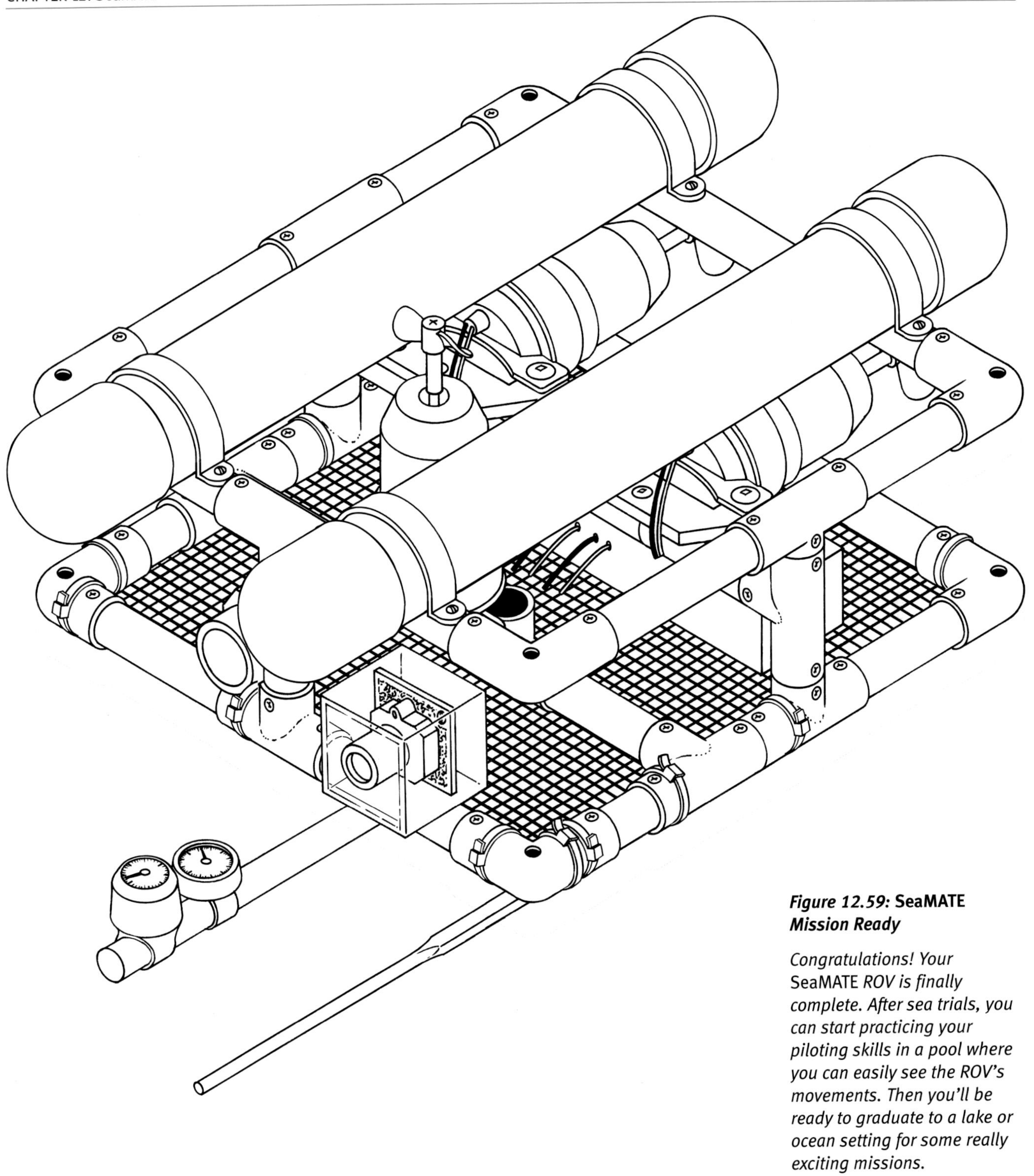

Figure 12.59:* SeaMATE *Mission Ready

Congratulations! Your SeaMATE *ROV is finally complete. After sea trials, you can start practicing your piloting skills in a pool where you can easily see the ROV's movements. Then you'll be ready to graduate to a lake or ocean setting for some really exciting missions.*

11.3. Getting Ready for Sea Trials

Sea trials for *SeaMATE* include these steps:

- securing the tether to the ROV frame
- adjusting the ballast and vehicle trim
- reducing the adverse effects of the tether on the ROV
- adjusting camera view angle
- adjusting the light angle to reduce backscatter (marine "snow")

11.3.1. Securing the Tether

Use a cable tie, string, or tape to secure the tether to the frame of the vehicle. Experiment to see the best point for doing this—you want the vehicle to be able to move with the least amount of tether interference. The goal here is to make sure that a tug on the tether doesn't yank directly on the junction box or its connections.

TECH NOTE: WHAT TO DO IF YOUR VEHICLE FLOODS

Leaks are a common occurrence, particularly during the testing stage of any new design. To minimize damage to the vehicle, it's important to notice if your vehicle begins acting a little heavier than usual and to know what to do if your vehicle gets water inside.

If your vehicle floods, act immediately, especially if the vehicle is in saltwater:

- Disconnect any external electrical power *if you can do so safely.*
- Open the vehicle and pour out the water.
- If there are internal batteries, disconnect them immediately, because applied voltages greatly accelerate the corrosion process.
- Rinse all affected parts under tapwater as soon as possible—seconds count!
- After rinsing the affected components, gently blot them dry with cloth or paper towels.
- Complete the drying process by warming the affected circuits in sunshine, under a hair dryer (on low setting and with care to avoid melting anything), or in an oven set to its lowest setting.

The good news is that a very quick response to a leak can often save many electronic components. (See *Tech Note: What to Do If Your Vehicle Floods* in *Chapter 5: Pressure Hulls and Canisters.*)

11.3.2. Adjusting Ballast and Trim

It's finally time to put *SeaMATE* in the water! But before you can see how your ROV performs, it's important to get it properly ballasted and trimmed.

1. Begin by re-attaching the two floats to *SeaMATE* if they have been removed. Do not add any ballast weights yet.

2. Now place *SeaMATE* in at least 18 inches of water. (Before you do this, you might want to read *Tech Note: What to Do if Your Vehicle Floods.*) Shake and tip the frame to dislodge any trapped air bubbles.

3. The vehicle should float high out of the water. Now add ballast weights to the bottom mesh until the ROV just barely floats at the waterline. If *SeaMATE* tips down on one end or to one side, redistribute the weights to adjust trim. The ROV should rest evenly or slightly up in the front and down in the stern. Once you're satisfied with the vehicle's buoyancy and trim, you can attach the weights to the mesh, using cable ties or some other method, so they won't shift or fall off during ops, launch, or recovery. If *SeaMATE* does not float, you'll have to increase buoyancy by increasing the size of the PVC floats or by adding flotation of one type or another. (See *Chapter 6: Buoyancy, Stability, and Ballast* if you encounter any significant problems.)

4. Turn on the vertical thruster. *SeaMATE* should descend. If it doesn't, try reversing the vertical thrust direction. If the ROV still doesn't descend, you probably need to add more weight to your vehicle.

SAFETY NOTE: GFI REMINDER

Don't forget to plug your monitor into a GFI-protected outlet or similar power device when working near water.

TECH NOTE: REMOVING TRAPPED AIR

Sometimes even a free-flooding frame can still contain trapped air when launched. To dislodge any air, shake and tip the frame until no more bubbles escape from the vent holes. If you find the process still not effective, you might consider drilling more holes in the frame or increasing the diameter of the holes already drilled.

11.3.3. Reducing Adverse Tether Effects

The amount of tether in the water affects any ROV's attitude, buoyancy, and freedom to maneuver. These techniques can reduce the effects the tether has on vehicle operation:

- Use only the amount of tether length necessary to allow the ROV to reach its working depth.
- Put a float on the tether to the depth that *SeaMATE* is operating.
- Add small floats along the length of the tether.
- Use good tether management and handling from surface. (See *Tech Note: Handling System for Small ROV Tethers.*)

11.3.4. Adjusting Camera View Angle

Adjust the camera viewing angle so that it not only sees the pick-up probe and the compass/depth sounder sensors in the lower portion of the screen, but so that it still has a large portion of the field of view in order to image what is in front of the vehicle. Getting this adjustment right really makes the pilot's job easier.

TECH NOTE: HANDLING SYSTEM FOR SMALL ROV TETHERS

It's important to think about tether handling before the actual mission. Small ROVs do not normally use winches that keep the tether neatly wrapped around a drum, so the tether is usually laid in a figure eight on the deck or dock. Utilizing a figure eight prevents the tether from kinking or tangling. In addition to using this pattern, it's important to ensure that all crew and team members know never to step on the tether. Doing so can grind bits of dirt in through the outer sheath, possibly causing a leak or damaging the conductors.

As a length of tether is passed out into the water, it adds weight to the vehicle. Ideally, the tether should be near neutral buoyancy, but this is not always possible to achieve. So small spherical pressure-resistant floats can be tied onto the tether to give it lift. However, it is important not to put too many on the cable so that it becomes overly buoyant to the point where it actually lifts the ROV. Getting the right balance is a matter of experience, gained by trial and error.

Always designate and utilize someone as tether tender. Pulling the tether in and out of the water can seem like a boring job, but it is crucial to the safe, efficient operation of any ROV. If too much cable is allowed to pass into the water, the tether may get tangled, trapping the ROV on the bottom. Too little cable, and the ROV will be pulled around by its tether, unable to maneuver effectively.

11.3.5. Adjusting Light Angle to Reduce Backscatter

Mount the light onto the frame, placing it so that the subject in the camera's field of view is illuminated at angles that minimize the backscatter, or marine "snow," caused by particles in the water (Figure 12.60). This is best done by placing the light off to the side and slightly behind the camera housing. Position it as far off to the side (and/or above or below) as you can, to further reduce backscatter.

Figure 12.60: Optimum Light Placement

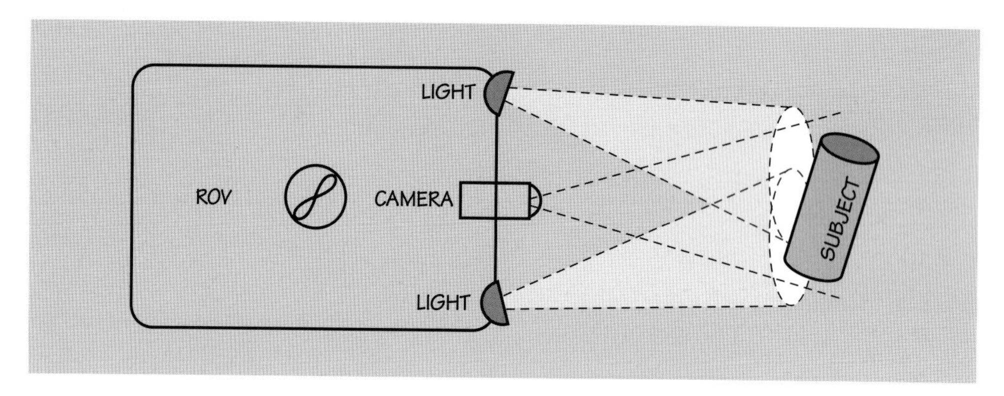

11.3.6. Adjusting Video Monitor

Once you're satisfied with your camera and lighting adjustments and the vehicle is in the water, you're ready to "see" what's out there. But if you're working in bright light conditions, it may be necessary to create some sort of shield or covering in order to see the images on your monitor. This can be as simple as a blanket or coat over top of the pilot and the monitor, or you may opt for a slightly more structured tube, tent, or rubber tote (Figure 12.61).

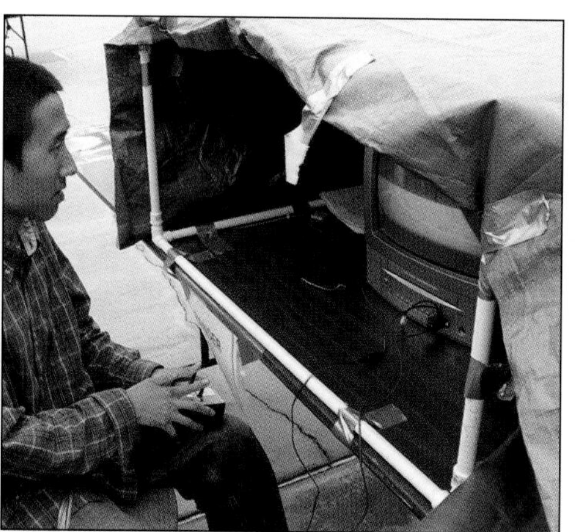

Images (left and middle) courtesy of Randall Fox

Image courtesy of Dr. Steven W. Moore

Figure 12.61: Video Monitoring in Bright Sunlight

11.4. Piloting Practice

By manipulating the control switches in a number of combinations, you can make *SeaMATE* travel in the following directions:

- up/down
- forward/reverse
- turn right forward
- turn left forward
- turn right reverse
- turn left reverse
- spin right
- spin left

You'll want to practice these various combinations until they become automatic. You'll also want to become proficient in sending *SeaMATE* out on a straight line course, as well as having the ROV descend to the bottom without traveling forward. Once you feel comfortable with these basic maneuvers, try moving or retrieving an object with the pick-up probe. (See *Chapter 11: Operations* for information regarding navigation techniques and search grid patterns when piloting basic ROVs under water.)

With practice, you'll also become familiar with the vehicle's response time, the effect of the tether on vehicle operations, and the impact of various sea conditions on operations. Remember, practice is the best way to become proficient at piloting *SeaMATE.*

Images courtesy of Randall Fox

Figure 12.62: Now the Piloting Challenge Begins!

Practice and improve your piloting skills before *undertaking your mission.*

Images courtesy of Gail Drake, Battlefield High School, NVCC

Figure 12.63: Obstacle Courses

Teams from 11 Virginia high schools participating in the 2009 Williams County Schools Robotics Competition worked at perfecting their piloting skills by running obstacle courses such as this one.

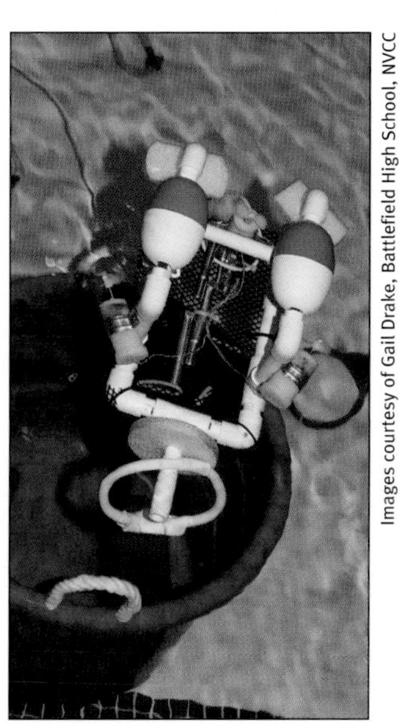

Images courtesy of Gail Drake, Battlefield High School, NVCC

Figure 12.64: Relay Races

The Prince William County Schools Robotics Competition includes timed races, in which ROVs spear and deposit as many rings as possible into a collection bucket. Other events include a game where ROVs must work together to earn points.

12. What's Next? Going Beyond *SeaMATE*

Image courtesy of Dr. Susan Crawford, Naval Undersea Museum Foundation

Figure 12.65: Have Some Fun!

Once you understand the basics of building a simple ROV like SeaMATE, *it will turn you loose to consider any number of fun challenges, such as this "underwater bug."*

SeaMATE may mark the final chapter of this textbook, but don't think it needs to be the end of your adventure in underwater vehicle technology. Far from it—*SeaMATE* is just the beginning! If you've absorbed most of what this book has to offer and gained some first-hand experience constructing your own shallow water ROV, then you have already demonstrated the initiative, know-how, teamwork skills, and creativity that will allow you to move far beyond *SeaMATE*. In fact, the possibilities are nearly endless.

This chapter concludes with a look ahead to several possible options, any of which would greatly expand your underwater technology adventure. You could modify or upgrade a specific system on *SeaMATE*, you could go deeper, you could design a hybrid vehicle, you could do away with the tether completely, or you might even become part of a team designing a vehicle to search for life on other planets.

12.1. Modifying *SeaMATE*

When you first start piloting *SeaMATE* (or its equivalent), you'll enjoy a rapid increase in the variety and complexity of tasks you can perform with it, because your skills will improve with practice. Sooner or later, though, your advancing skills will run up against hard limits imposed by the capabilities of the vehicle itself, and you'll find yourself wishing it could do more. Being frustrated by such limitations can be a great thing when it inspires you to explore ways to push beyond those limitations. One obvious way to do so is to upgrade *SeaMATE*, giving it new or improved capabilities.

Some options you might want to consider include:

- substituting more powerful thrusters for improved speed and maneuverability
- upgrading the payload tool from a simple probe to a gripper or maybe even a multi-function manipulator arm
- installing a fancier camera and lighting system for better video quality
- improving the control system by adding variable speed control for the thrusters, a depth/heading autopilot, or other features
- adding a smooth fairing to improve efficiency and reduce the risk of entanglement

This is by no means an exhaustive list. Let your unique mission opportunities and creativity define your direction. The point is that there are lots of ways you could expand the capabilities of *SeaMATE* or any similar vehicle.

Just remember that any change you make in one system will invariably ripple through many or all of the other systems, as described in *Chapter 2: Design Toolkit*. For example, you won't be able to add a fancy manipulator arm without also modifying the distribution of ballast and flotation on the frame, reworking the power distribution system, and upgrading the control system to include controls for the arm.

12.2. Going Deeper

SeaMATE is a shallow diving vehicle limited to the same depths that can be explored easily by snorkeling or recreational scuba diving. It's a great vehicle for learning the basics of underwater vehicle design, but the main advantage of most ROVs is that they allow exploration and work at depths *beyond* those where scuba divers can safely go. Therefore, a logical next step is to create a vehicle that can dive deeper than the 40-meter depth considered the safe limit for most recreational diving. An ROV that could explore to 100 meters would be well within the technical capabilities of any group of underwater technology enthusiasts who have studied this book; yet, such a craft would be capable of working at depths unreachable by most divers. Imagine using something you built with your own hands to explore a seafloor or lakebed that no other human being has ever seen before!

If you decide to pursue this sensible, yet ambitious goal of going deeper, you will need to overcome several major challenges:

- **Longer tether**: To allow enough slack in the tether to maneuver comfortably at 100 meters depth, particularly in the presence of any currents, you will probably need a tether much longer than 100 meters in length. The longer tether will mean more drag and consequently require more powerful thrusters. Increased electrical resistance in the longer tether, coupled with this demand for more thruster power, will probably force you to start with higher voltages at the surface or to use onboard batteries (essentially creating a hybrid vehicle, as detailed below). Both options require a redesigned power system. And the higher-voltage option carries with it significant safety concerns, which mandate special precautions. For a review of some of these problems and solutions, see *Chapter 8: Power Systems.* A significantly longer tether also means you'll need a beefier winch and a more complex launch and retrieval system.

- **Increased water pressure**: At 100 meters, the pressure is 10 atmospheres gauge, or about 150 psi. That's more than enough to promote serious leaks or to crush a casually constructed pressure canister. If you do get a leak in a canister, the internal pressure can increase enough to turn the canister into a dangerous pipe bomb when it returns to the surface, so oil compensation or strong canisters with pressure-relief valves become critical. In addition, many forms of flotation that work just fine in a swimming pool will compress significantly and lose much of their buoyancy at 100 meters depth. For a review of these problems and related solutions, see *Chapter 5: Pressure Hulls and Canisters* and *Chapter 6: Buoyancy, Stability, and Ballast.*

- **No direct observation**: Even in clear water, you will not be able to see your vehicle working 100 meters below. You must therefore rely entirely on cameras and other navigational instruments to give you the remote "situational awareness" needed to determine your location and accomplish your mission without getting trapped or entangled in obstacles. At a minimum, you'll need a video camera, compass, and depth gauge capable of performing at the greater mission depth. Additional cameras looking in other directions, an altimeter, and scanning sonar are other options you may want to consider, though each adds considerably to the cost and complexity of the vehicle. These issues and solutions are covered in *Chapter 9: Control and Navigation.*

- **Low light levels**: Even in clear, tropical waters, things start getting pretty dark at 100 meters, and in murky water, it can be pitch black that far beneath the surface.

Your camera will therefore need some bright lights for illumination. Adding these lights will impact the frame design, electrical power system, and control systems. If you are depending on your camera to observe an external compass and depth gauge, you'll need to make sure they're illuminated with enough light that they can be easily read in the camera's image. Lighting issues and techniques are covered in *Chapter 9: Control and Navigation.*

- **No rescue**: At this depth, you are beyond safe scuba diving limits, so there's no easy way for anyone to rescue your vehicle if it does get trapped or entangled 100 meters down. You don't want to lose a vehicle you've spent so much time, energy, and money building, so it's worth doing whatever you can to avoid getting trapped. Adding a rear-facing camera, a smooth, snag-resistant fairing, and a tether sturdy enough to tug on *hard* without damage are some of the options you might want to consider.

You have three basic options for overcoming these deeper-diving challenges:

1. Modify *SeaMATE*, so it can dive deeper.
2. Buy a commercially available LCROV capable of 100-meter dives.
3. Start from scratch to plan and build a new ROV designed from the beginning to handle 100-meter depths.

The first option—modifying *SeaMATE*—may seem appealing, particularly if you've already got a finished or nearly finished *SeaMATE* sitting on your workbench. But consider what's really involved. If you add a longer tether, you've got to switch to on-board batteries or high voltages, as mentioned above. That means completely redesigning and rebuilding the power supply system. It also means making a new pressure canister to house the batteries (if you go with the battery option) or the transformers and associated electronics (if you go with the higher-voltage option). Then there are the thrusters. Bilge pump motors are unlikely to stay watertight at 100 meters and probably wouldn't have enough thrust to drag that long tether around, anyway, so you'd need to replace those with bigger ones that have pressure compensation or much better seals. The new electronics canister and the bigger thrusters would mandate changes in the ballast and flotation and would require a complete redesign of the frame. The list goes on and on. And each upgrade would boomerang back around and force you to make other upgrades. Pretty soon, what you'd likely end up with is something resembling a motley collection of patches used to patch earlier patches—a lot of work for a not so very satisfactory solution.

The second option—purchasing a commercially manufactured LCROV capable of diving to at least 100 meters—is certainly a faster and easier solution; however, note that LC (which, if you recall, stands for "Low Cost") is a relative term. Very few groups (school groups or otherwise) can afford the thousands of dollars required to purchase one of these ready-made vehicles. Besides, buying an ROV just isn't the same as creating your own. It bypasses the fun, the learning, the camaraderie, and the incredible sense of accomplishment that come with working as a team to design, build, and pilot a successful home-grown vehicle.

If the first and second options leave you feeling high and dry, don't despair! The third option—designing and building your own ROV from scratch—is likely to be a viable option for most readers of this book. By using the design methodology described in *Chapter 2: Design Toolkit,* coupled with your now greater experience and the advanced material sprinkled throughout the technical chapters, you can take your ideas for a deeper-diving vehicle and transform them into a concept design in which every vehicle system works seamlessly with every other system to directly support this new mission of exploration 100 meters beneath the surface.

As you know by now, the beauty of running through a concept design exercise is that you get a pretty good idea of what will be involved in building your own craft (money, time, vehicle capability, etc.), without having to commit to actually fabricating it. This helps you make an informed choice—to buy, to build, or to postpone the idea as being unrealistic for now, given your time and budget constraints. As you work on your concept design, remember that liberal use of commercial off-the-shelf components can give you the convenience, time-savings, and reliability of professionally engineered components, while retaining the flexibility, creativity, cost savings, and sense of accomplishment that come with designing and assembling most of the vehicle yourself.

Image courtesy of Scott Fraser, Long Beach City College Electrical Technology Dept.

Figure 12.66: Check Mate

Building and operating a ROV is much like playing a chess match; you have to constantly consider all the variables because leaving one out opens you up to an attack!

This ROV chess challenge involved two existing ROVs and five students at Long Beach City College. In attempting to play underwater chess, they discovered that a giant board with magnetic bases on the chess pieces was a necessity. So were excellent piloting skills and patience!

12.3. Using Multiple Vehicles

The decision to go with multiple vehicles usually depends on the specific mission statement. Even so, with the exception of really complicated oil-field jobs, most times there's seldom a need for more than two vehicles. Furthermore, multiple vehicle fleets work best with AUVs, so building a fleet of ROVs probably isn't the best use of time or money. Good reasons for building multiple vehicles include the following:

- **Rescue and observation**: Building *two* similar ROVs, each capable of rescuing (or at least observing) the other one is a great backup plan. The two ROVs can dive together and keep each other out of harm's way. It's what scuba divers call "the buddy system." Manned vehicle systems often run with two craft for safety or have an ROV on standby.

  An added bonus of building multiple vehicles is that each ROV can capture cool pictures or video of the other one working in the depths. The Russian submersibles *Mir 1* and *Mir 2* are familiar examples of underwater craft that observe and film each other on missions to the *Titanic.*

- **Mission tasks**: Sometimes a mission seems to suggest the option of using more than one vehicle to do the required work. For example, the third annual MATE International ROV Competition scenario was sited on an imaginary underwater reef located near NOAA's Florida Keys National Marine Sanctuary. Based on a preliminary survey, the missions for Ranger class vehicles involved collecting specimens, identifying a wreck, finding the source of methane gas, exploring the caverns of that area, and recovering a towfish that had been lost during the initial survey. Two teams responded to this challenge by building multiple ROVs, each of which

Figure 12.67: Two ROVs May Be Twice as Good!

In this photo, the Cambridge Rindge and Latin School team celebrated the performance skills of both their Snoopy *and* Woodstock *ROVs at the 2004 MATE International ROV Competition. The Monterey Peninsula College team won the Thinking Out of the Box Award for their dual ROVs accompanied by a radio-controlled surface vehicle.*

Image courtesy of Harry Bohm

was optimized to address a particular aspect of the overall mission, rather than attempting one vehicle capable of all missions.

12.4. Going Tetherless

Another very different direction you can probe, if you want to go beyond *SeaMATE*, is to venture into the exciting world of autonomous vehicles. Few things are more fun and rewarding, or more ambitious, than creating a true underwater robot that can strike out on its own without any help from a human pilot, accomplish its mission, and return home safely.

Image courtesy of MIT *ORCA* AUV Project

Figure 12.68: Taking on the AUV Challenge

The MIT team built its ORCA VI *AUV for the 2005 International Autonomous Underwater Vehicle Competition.*

It's best to start with a shallow-diving AUV in a small, calm body of freshwater, like a swimming pool. As your skills and confidence increase through experience, you can push the envelope farther by going into larger, murkier, or deeper bodies of water. Keep in mind that a deep-diving AUV will encounter most of the same issues as a deep-diving ROV *plus* a host of challenges unique to autonomous robots. Therefore, don't mix autonomous operation with great depth until you are ready for a true test of your abilities and are willing to suffer the possible consequences of failing that test—including permanent loss of your vehicle!

Here are some of the major issues you will face if you choose to go down the AUV path:

- **Self-contained power**: Power can be supplied to an ROV from the surface via its tether. As long as the surface power source remains available, the ROV can keep moving. And if the power source fails, a small ROV can be reeled in, like a fish on a line, and easily recovered. Not so with an autonomous vehicle. All the vehicle's energy supply must be stored on board (or extracted in some very clever way from the environment). For this reason, supplying adequate AUV power for the duration of a mission is extremely difficult. AUV designers must pay extra close attention to issue of power storage and vehicle efficiency. So matching props to motors, maintaining the slowest speeds compatible with the mission, and adding streamlined fairings all become critically important. Creative ideas, like the passive propulsion of sawtooth gliders, can greatly extend the range of AUVs.

- **Navigation**: Without a doubt, the single greatest challenge facing an autonomous underwater vehicle is accurate navigation. The vehicle must be able to tell where it is and which way it's headed, so it can reach all of its destinations successfully, especially that final one: home. With a typical small ROV on a fairly short tether, it's easy to know approximately where the ROV is—it's near the boat or dock from which it's being operated. Couple that information with live feedback from a video camera, and it's comparatively easy for an experienced pilot familiar with the area to tell more or less precisely where the vehicle is and what it's doing during its mission. Bringing an ROV home can be a simple matter of turning around and following the tether back to the surface. In extreme cases, a small ROV can literally be "reeled in" by its tether.

  On the other hand, an AUV has no tether to tie it to a particular geographic locality. In a large body of water with currents, like the ocean, it could end up almost anywhere. Furthermore, underwater video images, which provide such a rich source of detailed information to a human observer, are essentially useless for AUV navigation. That's because typical video images are too complex for the computer systems on most AUVs to interpret. Many outstanding scientists and engineers are working on this problem right now, but it may be years before machine vision becomes sophisticated enough, simple enough, and inexpensive enough to be of much use for home-made AUV navigation.

Fortunately, there are some fairly easy and affordable navigation alternatives. GPS is great for telling a vehicle precisely where it is on the surface, and many modern GPS units are affordable and easy to interface with micro-controllers—the problem is that GPS works *only* on the surface. While under water, the vehicle must rely on other sensors, such as compasses, depth gauges, and/or inertial guidance systems. In some situations, it may be possible to deploy an array of acoustic beacons, so your AUV can use their acoustic signals to determine its position. A number of navigation and control options are discussed in detail in *Chapter 9: Control and Navigation.*

None of these approaches is easy, but all of them are possible when the information provided in this book is supplemented with review of product literature, web research, and conversations with others who have tried similar things. Whatever approach(es) you choose, be sure to thoroughly test your vehicle's navigation system in a swimming pool or other confined body of water before turning the vehicle loose in a place where it might get lost and be difficult to recover. (Remember however that magnetic compass data can be skewed by the rebar used to reinforce the bottom and sides of swimming pools.)

- **Decision-making**: When an ROV is in the water, most or all of the decisions are made by the pilot. The ROV needs only to respond to simple pilot commands: Turn right. Go up. Close manipulator jaw. An AUV, on the other hand, is responsible for making all of its own decisions throughout the mission (at least while it's under water and out of communication range). These decisions are based on the mission goals and information provided by the AUV's sensors. If the vehicle makes a poor decision, it could botch the mission or disappear forever.

  This level of internal "smarts" requires one or more micro-controllers or even a full-blown computer, plus some pretty advanced programming. This hardware/software combination must be able to monitor and interpret information from all the vehicle's sensors and have precise, coordinated control over all of the vehicle's functions. This advanced material is introduced in the second half of *Chapter 9: Control and Navigation*, but it takes experience programming real vehicles in the real world and seeing how they handle a wide variety of situations before you can expect to get really good at it.

- **Unanticipated events or situations**: Good decision-making becomes all the more critical when the AUV gets pushed off course by current, runs into some unexpected obstacle, or encounters another situation that forces a change in plans or actions. The person programming the vehicle must anticipate as many of these situations as possible and provide enough sensors and enough "intelligence" to recognize these situations and respond with appropriate actions. Closed-loop control is critical for dealing with these unexpected situations and is usually needed even in perfectly routine situations which still involve some

Images courtesy of San Diego iBotics Student Engineering Society

Figure 12.69:* The Stingray *AUV

The 2008 Stingray *is a fully autonomous vehicle constructed by the San Diego iBotics Student Engineering Society for entry into AUVSI & ONR's 11th International Autonomous Underwater Vehicle Competition. The carbon-fiber shell mimics the form of a stingray and provides very low drag.*

degree of drift or other uncertainty. In fact, there is no way that open-loop "dead reckoning" would suffice for any but the most trivial of missions in a bathtub. Alas, good closed-loop control is not easy to accomplish. But then again, nobody said building a successful AUV would be easy. To review information on closed-loop control, including its advantages and special challenges, consult *Chapter 9: Control and Navigation.*

- **Vehicle loss**: It's a big ocean out there. If you launch your vehicle at sea (or even in a large lake) and something goes wrong with its power or navigation, there's a very good chance you'll never see it again. That would be a big loss so you'll want to take special steps to improve its likelihood of recovery. For starters, make sure it defaults to positive buoyancy, so it will float back to the surface if it loses its ability to propel itself. Equip it with bright colors, flashing lights, a radio transmitter, satellite transceiver, or other mechanism (all independently powered, in case the main power source has failed) that will help you locate the vehicle on the surface if it fails to come home. Oh, and don't forget to put your name and phone number or email address in waterproof writing on both the outside and inside of the AUV, in case a friendly fisherman or beachcomber finds your lost vehicle someday. That's not likely, but it does happen.

12.5. Designing a Hybrid

The biggest roadblock for a small, deeper diving ROV is its long, unwieldy tether. One solution is to reduce the bulk of the tether by switching to on-board battery power, as described earlier. This certainly helps, but you still have some of the limitations of a tether. Another option (described in the previous section) is to get rid of the tether entirely, by making an AUV.

***Figure 12.70:* PURL I**

PURL I *was a testbed AUV/ROV to study the feasibility of a small hybrid underwater vehicle.*

Image courtesy of Simon Fraser University Underwater Research Lab

A hybrid vehicle offers the best of both worlds. It gets all its power from an on-board source, so it can operate independently as an AUV. However, it can also use a thin, communications-only tether for direct, real-time pilot control (including live video) without suffering the same level of drag caused by a thicker, power-carrying tether.

Here are the basic considerations for building hybrid ROV/AUV vehicles:

- **Power**: You will need to research on-board power options. Be aware that your choice will have a major impact on the speed your vehicle can travel and the length of time it will be able to work under water before it runs out of energy. *Chapter 8: Power Systems* describes power options, safety considerations, and reminds you that battery technology is rapidly evolving.
- **Control**: The control systems for hybrid vehicles are inherently more complex than those on *SeaMATE.* Since the tether is not carrying power directly to the motors, there must be some way to encode and decode commands and other information (usually digital) sent through the tether. And if the tether is not being used (AUV mode), then the vehicle must have enough "smarts" to operate completely on its own. (See the second half of *Chapter 9: Control and Navigation* for more advanced control options.)

- **Launch and Recovery (LARS)**: Because smaller, lighter tethers are easier to manage, you have more options for launching and recovering hybrid vehicles. If the vehicle itself is not very heavy, one or two people can easily handle LARS from a small boat or dock.
- **Testing**: Hybrid vehicles can offer a less risky introduction to the world of AUVs. The first few times you test the AUV mode on your vehicle, it's wise to leave the tether attached so you can switch back to pilot control or troubleshoot code and system status on the fly if you run into trouble. Disconnect the tether only after you've worked out all the bugs. A wise intermediate step is to test your vehicle's AUV mode in the confined environment like a pool or small pond before turning it loose in a larger body of water where you might not be able to make a recovery if it goes missing.

12.6. Exploring Other Worlds

There is mounting evidence that Earth is not the only object in our solar system, or beyond it, capable of supporting life. As mentioned earlier, at least one of Jupiter's moons, Europa, is thought to have a large liquid water ocean beneath its frozen exterior. Recent successes with robotic vehicles exploring the dry surface of Mars and automated probes photographing the surfaces of other planets and moons have made it clear that exploration of other worlds by autonomous robots is no longer just a pipe dream. What wonders—living or otherwise—wait to be discovered by those with the knowledge and skills needed to build robotic space probes?

Admittedly, designing and building a robot capable of reaching, exploring, and reporting back from the undersea world of Europa or another world is not something that's going to be done by a couple of people working out of their garage as a weekend project. Like the other successful missions to distant worlds, it will require lots of preparation, lots of money, and lots of creative people working together to design and build something totally new, totally amazing.

Whether you realize it or not, you are now started on your way to being a contributing member of such a team. The material presented in this book has emphasized basic principles of physics, chemistry, and engineering over cut-and-dried recipes. These principles apply just as well on Mars or Europa as they do here on Earth. Sure, there are some differences, as noted in Chapter 3, but only in the specific numbers, not the basic concepts.

Images courtesy of Stone Aerospace

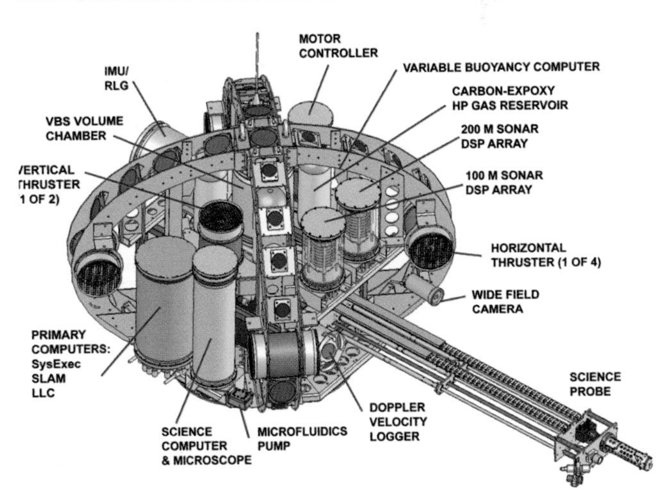

Figure 12.71: Future Exploration

Stone Aerospace developed the AUV DepthX *to test the ability of a mobile robotic platform to autonomously explore and map in 3D where no external navigation aiding is possible and in unknown environments.*

The AUV is part of NASA's planetary robotic efforts to explore the sub-surface ocean of Europa, the 4th moon of Jupiter. The extended science probe shown in the diagram retracts flush with the vehicle during normal transit and exploration.

For example, pressure will increase with depth at a different rate on Europa than it will here on Earth because of the different gravitational accelerations on the two celestial bodies. However, if you understand how gravity and a fluid's mass density interact to produce weight, and if you understand how the weight of a column of water causes pressure to increase with depth, then it's a straightforward jump to figure out how quickly the pressure changes on Europa, all you need to do is look up the gravitational acceleration for that moon and crunch a few numbers in your calculator.

In case you do have a yearning to explore oceans on distant planets or moons, here are some issues you may want to consider:

- **Different gravitational acceleration**: Different planetary compositions and different planetary sizes mean different gravitational accelerations. These directly affect the presence and composition of any atmosphere or ocean and change the rate at which pressure changes with depth in any bodies of liquid that might be present.
- **Different temperatures**: The temperature at the surface of different celestial bodies can be quite different than on Earth. Accordingly, the phase (solid, liquid, or gas) of many materials will be different. For example, on Earth the oceans are made of liquid water, and methane is a gas. On a much colder planet, all the water might be frozen solid, and the "oceans" might be made of liquid methane. These temperature extremes would also have profound implications for the materials used to build vehicles. A structural material that is solid here on Earth might melt on a much hotter planet, and a soft, pliable rubber gasket used as a seal here might freeze and shatter on a much colder planet.
- **Different chemical environment**: The chemical composition of the atmospheres, oceans, and soils of other planets and moons differ from those of Earth. The same basic elements and molecules are often present, but in different amounts and/or in different physical phases. Because of these chemical differences, materials that work well in Earth's oceans might corrode quickly or even dissolve away in the oceans of other worlds. Conversely, materials that are not suitable for use here may be ideal there.
- **Limited power options**: Obviously, even our nearest celestial neighbor is very far away. A tether or extension cord for power would be impractical, to say the least. And you can't just zoom back to Earth every now and then to change batteries, either. Solar power and nuclear power might be viable options, but both would take some careful planning and engineering.
- **Limited communications**: Radio waves travel quite well through the vacuum of space, and they have already been used to control robots on Mars, so we know they are a viable option—at least for communicating with something on the surface of another world when Earth is in the sky overhead. However, if the mission operations are on the backside of another planet or moon, or if they are conducted under water, then communication gets more complicated because some form of relay station is required. And remember that even though radio waves travel at the speed of light, it takes some time to traverse the vast distances of space.

  This means that communication delays must be factored into any control or communication scheme. Because of these delays, any vehicle used on another world must be at least partly autonomous. For example, when Earth and Mars are at opposite ends of their orbits, it can take over 20 minutes for a camera image transmitted by one of the *Mars Rovers* to reach Earth, and another 20 minutes for a response from NASA personnel to reach the *Rover*. In other words, these robots need enough sensors and on-board "intelligence" to recognize a cliff or other hazard and respond appropriately during the 40+ minutes it can take for NASA to see the problem and yell "Stop!"

- **Less predictable conditions and events**: We know disconcertingly little about our own oceans and continue to be surprised by underwater volcanic eruptions, landslides, jellyfish population explosions, and other things that interfere with the operation of undersea vehicles. We know far less about oceans on other worlds. Any robot serving as the first explorer in an alien sea is certain to encounter situations that its designers could not have anticipated, at least not in any detail. Such a robot will therefore need to be extremely good at quickly assessing its environment, identifying probable threats and opportunities, and making wise decisions about what to do (or not to do) next—all without any immediate help from its human parents. Providing these robots with this kind of artificial intelligence is perhaps the greatest challenge facing the designers and builders of these pioneering machines.

Such extra-terrestrial exploration efforts will require teams of experts from a variety of different fields—astronomy, physics, chemistry, geology, biology, mathematics, electrical engineering, mechanical engineering, computer programming, and others—to understand these challenges and to develop all the clever solutions needed to overcome them. One of the great things about getting involved in ROV and AUV design at an early stage in your life is that it introduces you to many of these subjects and gives you the foundation to make meaningful contributions to any team working to solve one or more of these related challenges.

12.7. Knowledge, Experience, and Dreams

Regardless of which challenge you take on next, at this point you have several distinct advantages:

1. You now know the important scientific principles involved in building successful underwater craft.
2. You've read stories of what happens when inventors before you have ignored or been unaware of these basic physical and chemical principles.
3. You have access to new technologies, information, and materials that were not available to those early inventors.
4. You have gained valuable hands-on experience through building projects like *SeaMATE*.
5. You know the importance of daring to dream.

Figure 12.72

13. Chapter Summary

By following the step-by-step instructions of this chapter, you'll end up with your own version of the shallow diving *SeaMATE* ROV. But the real learning involves understanding how and why this ROV is constructed to accommodate the physics and chemistry of the watery environment. In other words, the project is about applying what you've learned from the previous textbook chapters in order to fabricate a vehicle that can carry out an actual underwater mission.

Building a shallow diving ROV includes many more skills than simply putting the components together. You need to be able to source materials and equipment for the project, choose where to get components, see if they will fit within your budgetary constraints, and decide on feasible substitutions, if required. You need to be able to analyze the mission in order to select appropriate sensors and construct a pick-up tool. Once the vehicle nears completion, you have to be able to troubleshoot circuits, then trim, ballast, and make other adjustments as part of sea trials.

Learning to pilot an ROV is another important skill—and piloting a well-built vehicle is especially exciting. After all, the point of building any ROV, including *SeaMATE,* is ultimately to be able to conduct a successful underwater mission. Your successes—and your frustrations—will serve you well as you contemplate your next vehicle.

Although Chapter 12 is the final chapter in this textbook, in many ways the *SeaMATE* project is just a beginning. It will undoubtedly get you thinking about other modifications, options, and missions. Not surprisingly then, the last section of this chapter is called *What's Next? Going Beyond SeaMATE.* It presents a number of ideas—upgrading a single system on *SeaMATE,* using multiple vehicles, figuring out a tetherless AUV, and designing a hybrid vehicle. The choice and the challenge are up to you. As the authors of this textbook, we eagerly await the results of your dreams and efforts!

THE MATE TEXTBOOK TEAM

Dr. Steven Moore, Vickie Jensen, Harry Bohm, and Jill Zande (Missing is illustrator Nola Johnston)

Dr. Steven W. Moore, co-author

Steve is passionate about his family, marine life, robotic gadgets, and education. By age twelve he was building underwater camera housings to photograph reef fish and designing electronic burglar alarms to keep siblings out of a top-secret laboratory in his bedroom closet. A few years later he completed a zoology degree at the University of California (UC) Davis, then went on to complete a Ph.D. in bioengineering from UC Berkeley and UC San Francisco. He is presently a full-time, gadget-building zoologist and professor at California State University Monterey Bay (CSUMB), where he teaches courses in marine biology, physics, nature photography, and electronics/robotics.

Harry Bohm, co-author

As a teenager, Harry became fascinated with diving, designing underwater vehicles, and exploration. His diverse background includes working as an Outward Bound instructor, voyaging on traditional sailing vessels, operating towboats, conducting salvage operations, managing Simon Fraser University's Underwater Research Lab, and working as an educator and technical consultant. He conceived of the *Sea Perch* ROV project that is now in wide use in many science classes and clubs across the United States. Part of each year he volunteers and photographs in Thailand and India; when in Vancouver he works on underwater robotics projects.

Vickie Jensen, co-author and editor

Vickie thrives on challenge, keeping several careers—as writer, editor, lecturer, photographer, and book distributor on the go at once. She earned her maritime waterwings as editor of *Westcoast Mariner Magazine*, traveling aboard tugs, tankers, fishing boats, dredges, or charter craft each month. She and Harry Bohm co-authored *Build Your Own Underwater Robot* and *Build Your Own Programmable Lego Submersible.* The MATE textbook is Vickie's 10th non-fiction book and harkens back to her childhood dream of having magic glasses to view the wonders of the underwater world. She lives, writes, and operates her book publishing/distribution company Westcoast Words in Vancouver, Canada, and Puerto Vallarta, Mexico.

Jill Zande, textbook project coordinator

Jill spent childhood summers exploring the creek behind her home in northwestern Pennsylvania. Between that and the local swimming pool, her parents had a hard time keeping her out of the water. She discovered the world of underwater vehicles during her Master's research when she traveled to the bottom of the Gulf of Mexico in a submersible. So when it came to developing the MATE's ROV competition program (and serving as the textbook project coordinator), she dove in. To her amazement and delight, the competition has grown into a network of events that are held around the world and inspire thousands of students and teachers to get wet each year.

Nola Johnston, illustrator and layout artist

Nola prefers staying on top of the water, rather than venturing under it (although in the past she has had involuntary submersible experiences due to an inclination for whitewater canoeing). She specializes in educational and interpretive design and illustration, with a particular focus on natural and cultural history. She also teaches design courses for the Emily Carr University of Art and Design and the BC Institute of Technology in Vancouver, Canada. In recent research projects, Nola has looked into the use of virtual worlds in the field of education. This is the third book on underwater robotics that she has worked on.

Marine Advanced Technology Education (MATE) Center

The Marine Advanced Technology Education Center is one of more than 30 Advanced Technological Education Centers established with funding from the National Science Foundation. Headquartered at Monterey Peninsula College, in Monterey, California, MATE is a national partnership of community colleges, research institutions, professional societies, government organizations, and marine industries working to improve marine technical education and meet the needs of the ocean workforce. MATE's student ROV competitions, its summer institutes for faculty development, and this textbook are all examples of products and services developed by the Center to fulfill its mission.

APPENDIX I : Dimensions, Units, and Conversion Factors

To avoid mistakes in communicating information about size, weight, speed or other quantities in technical fields, it's essential to attach meaningful units to every number. Three *miles* is not the same as 3 *inches*, though both units refer to the same dimension (length), and it is therefore possible to convert from miles to inches or vice versa. However, three *miles* is definitely not the same as three *seconds*, because these don't even refer to the same dimension. Miles are units of *length*, whereas seconds are units of *time*. Since they have different dimension, it's not possible to convert between them.

This appendix provides information about dimensions and units frequently encountered when designing, building, or operating small underwater vehicles. Please note that some of these units (for example, "newtons," a unit of force) may not

Table AI.1: Fundamental Dimensions & Some Associated Units

Dimension	SI (Metric) Base Units	Other Common Metric Units	Common Imperial Units
Length (L)	meter (m)	millimeter (mm) centimeter (cm) kilometer (km)	inch (in) foot (ft) yard (yd) statute mile (mi) nautical mile (nmi)
Mass (M)	kilogram (kg)	gram (g)	pound mass (lbm) slug
Time (T)	second (s)	millisecond (ms) minute (min) hour (hr) day (d) year (yr)	same as metric
Electric Current	ampere (A)	milliamp (mA)	same as metric
Temperature	kelvin (K)	Celsius (°C); also known as centigrade	Fahrenheit (°F)
Countable Quantity	mole (mol)	—	dozen
Luminous Intensity	candela (cd)	—	same as metric

Table AI.2: Common Derived Dimensions & Some Associated Units

Dimension (Parentheses show relationship to fundamental dimensions)	Common Units used to measure this Dimension*	
	Metric	Imperial
Area (L^2)	square meters (m^2) square centimeters (cm^2) square kilometers (km^2)	square inches (sq. in., in^2) square feet (sq. ft., ft^2) square yards (sq. yd., yd^2) square miles (mi^2)
Volume (L^3)	cubic meters (m^3) liters (l) cubic centimeters (cc or cm^3) milliliters (ml)	cubic feet (ft^3) gallons (gal)
Speed (LT^{-1})	meters per second (m/s, ms^{-1}) kilometer per hour (km/h)	feet per second (ft/s) miles per hour (mph) knots (kn; kts)
Acceleration (LT^{-2})	meters per second per second (m/s^2, ms^{-2})	feet per second per second (ft/s^2)
Force (MLT^{-2})	newtons (N) dynes (g cm s^{-2}) tonnes (t)	pounds force (lbf) ounces (oz) tons (t)
Pressure ($ML^{-1}T^{-2}$)	pascals (Pa) newtons per square meter (N/m^2) millimeters of mercury (mmHg) bars (bar) atmospheres (atm)	pounds per square inch (psi) atmospheres (atm)
Torque (ML^2T^{-2})	newton-meters (Nm)	foot-pounds (ft-lbf)
Electromotive force ($ML^2A^{-1}T^3$)	volts (V)	same as metric

** This table includes only those units commonly used in design or construction of small ROVs and AUVs; there are many others.*

be familiar to all readers. Other units may be familiar but often are used incorrectly. For example, "kilograms" are units of mass, but they are often used informally to quantify weight (which has dimensions of force, not mass). When working in technical fields, it's important to use accurate terminology. This includes paying careful attention to the dimensions and units of all quantities used.

Part 1: Dimensions and Units

The seven dimensions in the left column of Table AI.1 are frequently regarded as fundamental; all other dimensions (including those in Table AI.2) can be derived from simple combinations of these seven fundamental dimensions.

In the International System of Units (SI), a standardized version of the metric system, each of the fundamental dimensions has an associated "base unit," as well as other units that can be used with the dimension. For example, the "meter" is the base unit of length, but centimeters and kilometers are also used frequently. The tables emphasize metric and imperial units commonly encountered in work with small underwater vehicles.

Each metric unit can be multiplied by standard factors simply by affixing an appropriate prefix to the name of the unit. For example, the prefix "kilo" means "multiply by 1000", so a "kilometer" is 1000 meters. Table AI.3 summarizes most or all of the prefixes you'll need for underwater vehicle work.

Part 2: Metric Prefixes

In the metric system, standard prefixes may be affixed to the front of any metric unit to multiply it by a specified power of ten. Once you learn to use these prefixes, you'll find they make the metric system much easier to use than the imperial system. Each standard prefix has a corresponding standard abbreviation, which may be added to the unit abbreviation. Important: all unit and prefix abbreviations are case-sensitive; capitalization matters. For example, a millimeter (mm) is 1/1000 of a meter whereas a megameter (Mm) is 1 million meters. Thus one Mm is 1 *billion* times bigger than a mm, so be careful!

Table AI.3: Metric Prefixes

*Prefix**	*Symbol*	*means multiply by...*
Tera	*T*	10^{12} = 1,000,000,000,000
Giga	*G*	10^{9} = 1,000,000,000
Mega	*M*	10^{6} = 1,000,000
kilo	*K*	10^{3} = 1,000
(no prefix)		1
*centi **	*c*	10^{-2} = 0.01
milli	*m*	10^{-3} = 0.001
micro	*μ (Greek "mu")*	10^{-6} = 0.000001
nano	*n*	10^{-9} = 0.000000001
pico	*p*	10^{-12} = 0.000000000001

** This table includes only those prefixes commonly used in design or construction of ROVs and AUVs; somewhat more extensive lists can be found easily on the web.*

*** Note that "centi" (1/100) is an exception to the general rule of increasing or decreasing by a factor of 1000.*

Part 3: Conversion Factors

Conversion calculators are widely available on the web. However, you can also use the conversion factors below and a hand-held calculator for most routine conversions, particularly if you extend them by using the table technique described in Part 4 of this appendix.

Note that the following list of conversion factors is organized by dimension.

Also note, the ≅ symbol means, "is approximately equal to." Conversion factors given with the "equals" sign (=) are exact.

Length

1 in = 1000 mils = 2.54 cm
1 ft = 12 in
1 yd = 3 ft
1 nmi = 1852 m ≅1.1508 mi
1 m ≅ 1.094 yd
1 fathom = 6 feet
1 mi = 5280 ft ≅ 1609 m

Mass

1 kg ≅ 2.2046 lbm
1 slug ≅ 32.174 lbm

Time

1 min = 60 s
1 hr = 60 min = 3600 s
1 day = 24 hr = 1,440 min = 86,400 s

Temperature

Use the following formulas to convert from temperatures expressed in one set of temperature units to temperatures expressed in another set of units:

K = °C + 273.15
°C = (°F – 32) × (5/9)
°F = (°C × 9/5) + 32

Area

1 ft^2 = 144 in^2
1 in^2 = 6.4516 cm^2
1 m^2 = 10,000 cm^2 ≅ 1,550 in^2
1 cm^2 = 100 mm^2

Volume

1 ft^3 = 1728 in^3 ≅ 7.48 US gal ≅ 28.317 liters
1 US gal ≅ 3.784 liters
1 liter = 1000 cc = 1000 ml = 1000 cm^3
1 m^3 = 1000 liters

Speed:

1 mph ≅ 1.609 kph
1 kn = 1.151 mph = 1.852 kph = 0.5144 ms^{-1}
1 ms^{-1} ≅ 3.281 ft/s ≅ 2.237 mph

Acceleration

1 m/s^2 ≅ 3.281 ft/s^2

Weight

(see Force)

Force (including weight*)

1 lbf ≅ 4.45 N
1 gf (gram force) = weight of 1 gram of mass on Earth ≅ 0.0098 N

* *Reminder: Weight is a force. Weight and mass are not the same thing. Weight is the force of attraction created between two masses (for example, you and planet Earth) due to gravity.*

Pressure

1.00 atm ≅ 14.7 psi ≅ 101 kPa ≅ 1.01 bar
1.00 atm = 760 torr (1 torr = 1 mmHg)

Torque

1 lbf-ft = 192 ozf-in ≅ 1.356 N-m

Energy

1 cal ≅ 4.19 J (Joules)

Note: the dietary Calorie (denoted by a capital "C") is 1000 cal or 4,190 J

Power

1 Watt = 1 J/s
1 HP ≅ 746 Watt

Part 4: Using Unit Cancellation Tables to Derive Other Conversion Factors

Sometimes it's tricky to determine whether you need to multiply or divide when doing conversions. Unit cancellation tables are a useful trick that makes it easy to do even complicated, multi-step conversions between different units without making mistakes.

If you have a unit conversion question phrased in the form, "How many X are in Y?" then start by putting Y (with its units) in the upper left cell of a two-row table. Add one or more columns to the right of that to hold conversion factors as needed.

Each column in the table represents one conversion factor expressed as a fraction (numerator in the top row, denominator in the bottom). Arrange conversion factors so that all units *except* the ones you want in your final answer cancel. In other words, make sure that each time an unwanted unit appears in the top row of one column, that same unit appears in the bottom row of a different column. Put a line through such units to cross them out. The units you want in your final answer must appear in the top row. One exception is in the case of derived units like miles per hour, where "miles" would need to appear in the top row and anything after the "per" (in this case "hour") would need to appear in the bottom row. Once your units are all taken care of, focus on the numbers instead of the units; multiply all the top numbers and divide by all the bottom numbers to get the final numeric answer.

Here are some examples:

How many centimeters in 1 inch?

- You can easily find the answer to this question in the list of conversion factors without using a table, but this simple example will show you how to set up a table that can be used to solve more complicated conversions later.
- Since the question takes the form "How many (somethings) are in 1 inch?" start by writing 1 inch in the upper left cell.
- Then look in the conversion factor list above for a conversion factor that relates inches and centimeters. You find the conversion factor 1 inch = 2.54 cm, so you add a column to your table and enter that conversion factor in the table with "cm" on the top and "in" on the bottom, because "cm" is what you want for your final answer and you need "in" in the bottom row to cancel the "in" in the top row of the first column.
- Note that after you've cancelled all units that appear in both the top and bottom rows, only the desired "cm" units remain in the top row. That means you're ready to multiply all the numbers in the top row and divide by all the numbers in the bottom row to get your final answer.

1 ~~in~~	2.54 cm
	1 ~~in~~

Answer = [(1 x 2.54) / (1)] = 2.54 cm

How many centimeters in 7.5 inches?

- This one is very much like the first one, but now we have 7.5 inches instead of 1 inch in the upper left table cell.

7.5 ~~in~~	2.54 cm
	1 ~~in~~

Answer = [(7.5 x 2.54) / (1)] = 19.05 cm

How many inches in 1 centimeter?

- This is also similar to the first example, but with one important difference. Here the question is reversed, so instead of putting 1 inch in the upper left box, we put 1 centimeter there. Another difference is that we must now put the 2.54 in the *lower* box, so the centimeters will cancel, leaving inches as the surviving unit for our final answer.
- Note that since 2.54 now appears in the bottom row, we end up dividing by 2.54 rather than multiplying.

1 ~~cm~~	1 in
	2.54 ~~cm~~

Answer = [(1 x 1) / (2.54)] ≅ 0.3937 in

How many yards in 10 nautical miles?

- The previous examples were just warm up exercises. Now it's time to experience the real power of these unit cancellation tables.
- Note that there is no direct conversion given in the list above for converting nautical miles to yards. No problem. With the table we can combine other conversion factors to create the one we need.
- As before, start by putting the 10 nautical miles in the upper left box of the table. Now look for conversion factors involving nautical miles per hour that will allow

you to convert that unit into some other unit, such as regular miles. The first two columns essentially answer the question, "How many miles in 10 nautical miles?" Next we need to convert that to feet. Then convert that result to yards.

- It sounds tricky, but it's easy. In each case, just throw in a conversion factor that gets you closer to the units you want. For each new conversion factor, add a column and make sure you put the conversion number in the correct row, so the units you don't want cancel and the units you do want are left over in the right row.

- Note especially that in order to cancel the "ft" in the top row, we needed to flip the feet-per-yard conversion factor, so the 3 is in the bottom row.

10 ~~nmi~~	1.1508 ~~mi~~	5280 ~~ft~~	1 yd
	1 ~~nmi~~	1 ~~mi~~	3 ~~ft~~

Answer = [(10 x 1.508 x 5280 x 1) / (1 x 3)] ≅ 20,254 yds

How many feet per second are you traveling when moving at 60 miles per hour?

- This table approach works just as well for derived units, like speed, that are formed from combinations of fundamental units, in this case length and time.

- The table and calculation below convert 60 mph into an equivalent number of ft/s. See if you can follow the table and the calculations through step-by-step and explain how they work to arrive at the correct answer. If you can, you understand this technique well and can apply it yourself whenever you need to convert units.

60 ~~mi~~	5280 ft	1 ~~hr~~
1 ~~hr~~	1 ~~mi~~	3600 s

Answer = [(60 x 5280 x 1) / (1 x 1 x 3600)] = 88 ft/s

Part 5: Scientific Notation

Very large or very small numbers that would normally contain lots of zeros are more easily and concisely represented using a special shorthand known as scientific notation. In this format, numbers are multiplied by ten raised to some power (exponent). The exponent tells you how far (and in which direction) to move the decimal point. If the exponent on the ten is a positive number, the decimal place moves to the right by that number of places. If the exponent on the ten is negative, the decimal point moves to the left by that number of places. The number that gets multiplied by a power of ten is usually written so that it contains exactly one digit to the left of the decimal place.

Here are some examples:

Number of objects in 1 dozen = 12 = 1.2×10^{1}

Speed of light in a vacuum ≅ 300,000,000 m/s = 3×10^{8} m/s

Mass of the earth ≅
5,980,000,000,000,000,000,000,000 kg = 5.98×10^{24} kg

Number of meters in 1 millimeter = 0.001 = 1×10^{-3}

Diameter of human hair = 0.000050 m = 5.0×10^{-5} m

Mass of one electron ≅
0.000000000000000000000000000000911 kg = 9.11×10^{-31} kg

APPENDIX II: Useful Constants, Formulas, and Equations

Constants

π (Pi) ≅ 3.14159
(circle's circumference divided by its diamater)

G ≅ 6.67 x 10^{-11} Nm^2/kg^2
(universal gravitational constant)

g ≅ 9.8 m/s^2 ≅ 32 ft/s^2
(free-fall acceleration due to gravity at earth's surface)

For additional constants related specifically to the properties of water or other materials, see Appendix IV.

IMPORTANT NOTE:

Most of the formulas and equations below are provided in a generic algebraic form that can be used with any consistent set of units, whether SI, or imperial, or something else entirely. *However, the results you get will only be meaningful if the units on both sides of the equals sign match.*

For example, the familiar formula d = rt (distance = rate x time) will give you a correct answer for distance (in miles) if you multiply speed (in miles per hour) by time (in hours). However, it will NOT give you a correct answer (in miles) if you multiply speed (in miles per hour) by time (in seconds), because the seconds and hours won't cancel. Likewise, the formula will give you a correct answer for distance (in meters) if you multiply speed (in meters per second) by time (in seconds), but not if you multiply meters per second by hours.

Note that sometimes units may be consistent even when they don't at first appear to be. For example, the Pascal (Pa) is a unit of pressure defined as 1 newton per square meter (N/m^2), so if you had "Pa" as your units on one side of the equals sign and "N/m^2" on the other, you would be fine; those units are basically the same thing written in two different ways. They are consistent units.

Algebraic Notation

The equations presented in this appendix use standard algebraic notation:

- Letters (including Greek letters) represent variables, which are numbers that can take on different values when applied to different situations. For example, H could be used to represent the height of an object, and two different sized objects might have different values for H.
- Two numbers (including variables) placed next to each are multiplied. Thus 2H means twice the value of H.
- A superscript number, like n, immediately following another number, like R, means R raised to the nth power. Thus R^3 means R cubed, which is R multiplied by R multiplied by R.
- By default, multiplication and division are always performed before addition and subtraction; however, this order can be overridden by parentheses. The parentheses enclose operations that should be performed first. For example, if a = 2 and b = 3, then 2a+b = 7, whereas 2(a+b) = 10.

Some Useful Circle and Triangle Facts and Formulas

1 full circle = 360 degrees

A "right angle" = 90 degrees

Pythagorean Theorem:

For a right triangle with sides a and b and hypoteneuse c,
$a^2 + b^2 = c^2$

The sum of interior angles of any triangle = 180 degrees

Newton's Laws

Newton's *First Law of Motion* says, basically, that every object will maintain its speed and direction unless and until some net (i.e., unbalanced) force acts on it to change its speed and/or direction. This first law is embedded within the second law (see below), since setting the acceleration equal to zero in the second law is the same as saying "no change in speed or direction."

Newton's *Second Law of Motion* describes the relationship among the net force (F_{net}) acting on an object, the object's mass (m), and the object's acceleration (a).

$$F_{net} = ma$$

(The net force and the acceleration are always in the same direction.)

Netwon's *Third Law of Motion* basically says that when one object pushes or pulls on another (either through direct contact or through a non-contact force such as gravity), both objects experience exactly the same sized force, but in opposite directions.

Newton's *Law of Universal Gravitation* is another one of his famous laws. Here F representes the force of gravitational attraction between two spherical objects; m1 and m2 are the masses of the two objects; G is the universal gravity constant (approximately 6.67 x 10^{-11} $m^3\ kg^{-1}\ s^{-2}$); and d is the distance separating the centers of the two objects.

$$F = \frac{m_1 m_2 G}{d^2}$$

Mass and Weight

To convert mass (m) to weight (W) on earth's surface, use Newton's second law, but with weight (W) as the force, and g = 9.8 m/s^2 as the acceleration.

$$W = mg$$

(The preceding formula can also be used for calculating weight on other planets or moons, provided you subsitute the appropriate acceleration in place of g.)

Density and Specific Gravity

The density (ρ) of an object or material sample is equal to its mass (m) divided by its volume (V).

$$\rho = \frac{m}{V}$$

The specific gravity (SG) of a material is equal to its density divided by the density of pure water.

$$SG_{material} = \frac{\rho_{material}}{\rho_{water}}$$

The specific gravity is greater than one for materials that are denser than water and less than one for materials that are less dense than water.

Hydrostatic Fluid Pressure

The following very general formula can be used to calculate the change in pressure (Δp) associated with a specific change in depth (Δh) in any *incompressible* fluid having density (ρ_{fluid}) subjected to any *constant* acceleration (a). For example, this formula can be used to calculate pressure changes under an ocean of liquid water on Europa (one of Jupiter's moons) or even under the sea of liquid methane thought to exist on Titan (one of Saturn's moons).

$$\Delta p = \rho_{fluid} a \Delta h$$

In the specific case of water on or near earth's surface, where the freefall acceleration due to gravity is $g = 9.8\ m/s^2$, this becomes:

$$\Delta p = \rho_{H_2O} g \Delta h$$

For additional information about calculating hydrostatic pressure on earth, including tables of pressure as a function of depth in both freshwater and seawater, please refer to *Appendix III: Hydrostatic Pressure.*

Absolute and Gauge Pressure

Under water, the total pressure (called the absolute pressure) is the sum of two pressures: the pressure of the air pressing down on the surface of the water, which is one atmosphere (1 atm) of pressure, and the additional pressure (P_{gauge}) caused by the weight of the water pressing down from above.

$$P_{absolute} = P_{gauge} + 1\ atm$$

Note that tire pressure gauges and most other pressure gauges display *gauge* pressure, so they read 0 in air at sea level, even though they are really exposed to 1 atmosphere of pressure there. However, for calculating changes in gas volume associated with changes in depth, you should always use the *absolute* pressure.

Hydrostatic Force

When hydrostatic pressure acts on a surface, it applies a force perpendicular to the surface. Hydrostatic forces can be huge and are one of greatest threats to underwater vehicles. The magnitude of the net hydrostatic force pushing inward (F_{in}) on a section of pressure hull is equal to the pressure difference between inside and outside ($p_{out} - p_{in}$) multiplied by the surface area (A) exposed to the pressure difference.

$$F_{in} = (p_{out} - p_{in})A$$

Some Useful Geometry Formulas

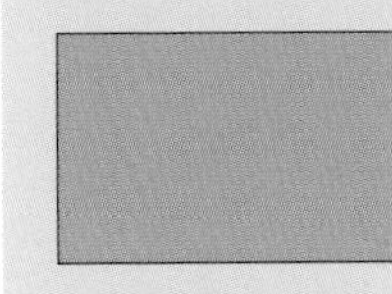

Rectangle or Square
(Length = L, Height = H)

Area = LH
Perimeter = 2(L+H)

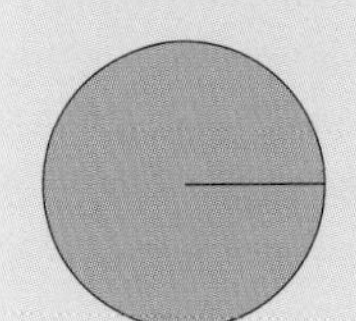

Circle
(Radius = R)

Area = πR^2
Circumference = $2\pi R$

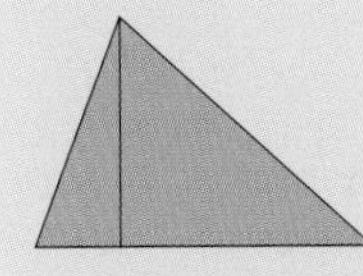

Triangle
(Base = B, Height = H)

Area = $\frac{1}{2}BH$

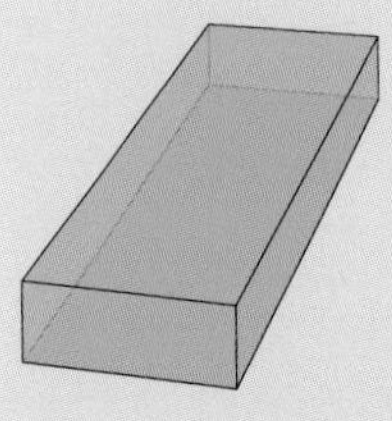

Rectangular Block or Cube
(Length=L, Width=W, Height=H)

Area = 2(LW + LH + WH)
Volume = LWH

Note that for a cube L=W=H, so...
Area = $6L^2$
Volume = L^3

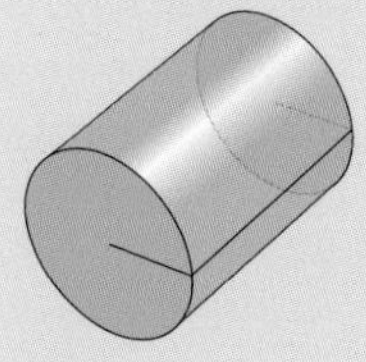

Cylinder
(Length = L, Radius = R)

Area of circle at each end = πR^2
Area without ends = $2\pi RL$
Total Area = $2\pi R^2 + 2\pi RL$
Volume = $\pi R^2 L$

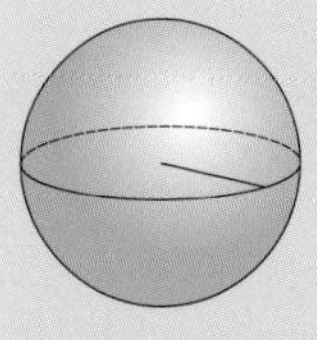

Sphere
(Radius = R)

Area = $4\pi R^2$
Volume = $\frac{4}{3}\pi R^3$

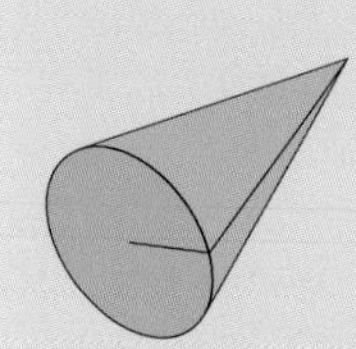

Cone
(Height = H, Radius of base = R)

Area of circular base = πR^2
Area without base = πRH
Total area = $\pi R^2 + \pi RH$
Volume = $\frac{1}{3}\pi R^2 H$

In most pressure hulls, pressure housings, and pressure canisters, the internal pressure (p_{in}) is about 1 atmosphere, so the pressure difference ($p_{out}-p_{in}$) is simply the gauge pressure of the surrounding water.

Buoyant Force

The buoyant force (B) created when an object displaces a volume of water (V) is equal to the weight of the displaced water. (That's Archimedes' Principle.) The weight of the displaced water is the displaced volume of water (V) multiplied by the density (mass per volume, not weight per volume) of the water (ρ) multiplied by the acceleration of gravity ($g = 9.8$ m/s^2).

$$B = V\rho g$$

Effective Weight

When an object is partially or completely immersed in water, some or all of its weight (W) is offset by the buoyant force (B), resulting in a reduced *apparent* weight. This apparent weight is called the "effective weight" (E), or sometimes the "wet weight," "submerged weight," or "in-water weight." If the effective weight is a negative number, the object floats.

$$E = W - B$$

Drag

When an object moves through water (or water flows past an object), relative motion between the object and the fluid is resisted by a friction-like force known as drag. Drag (D) for ROVs and objects of similar size and speed depends on the shape of the object (which influences the drag coefficient, C_d), the density of the fluid (ρ), the projected area (A) (See Figure 7.6 in *Chapter 7: Moving and Maneuvering* for an explanation of projected area.), and the relative speed (v) between object and fluid as follows:

$$D = \tfrac{1}{2} C_d \rho A v^2$$

In practice, the complex geometry of most small ROVs (and the corresponding uncertainty in C_d and A) makes it easier to measure drag empirically than to calculate it.

Behavior of Gasses under Pressure

Air, nitrogen, helium, and other gasses commonly associated with underwater technology applications follow closely the predictions of the Ideal Gas Law, which is explained in detail in most introductory college chemistry textbooks. This law says that the pressure (p), volume (V), number of gas molecules measured in moles (n), universal gas constant (R), and absolute temperature (T) are related as follows:

$$pV = nRT$$

From a pragmatic standpoint, the quantities on the right side of this equation can be treated as approximately constant in many small, shallow-diving ROV and AUV applications, so this equation reduces to PV = constant, or...

$$p \alpha \frac{1}{V}$$

(The little "α" symbol means "is proportional to.")

In other words, for a fixed quantity of gas that doesn't undergo major temperature changes, pressure and volume are inversely proportional to each other. So if one doubles, the other gets cut in half. If one triples, the other gets reduced to 1/3, and so on. This has important implications for diving physiology and for compressible flotation systems, including soft ballast systems.

Torque

The torque or "twisting force" applied by a wrench or similar mechanism is equal to the length of the lever arm (the wrench handle) multiplied by the portion of the force directed perpendicular to that lever arm.

$$\tau = F_{\perp} L$$

Power and Efficiency

Power is the rate at which energy is flowing or being converted.

$$P = E/t$$

The power (P) thrusters must provide to move a vehicle through the water at constant speed is the product of the speed (v) and the drag (D) at that speed.

$$P = Dv$$

Efficiency is the fraction (commonly expressed as a percent) of total energy put into a process that emerges in the form you consider useful. If both the numerator and denominator are divided by time, an equivalent expression based on power instead of energy is obtained:

$$Efficiency = 100\% \; x \left[\frac{Energy\ Output\ (desired\ form)}{Energy\ Input\ (total)}\right]$$

$$= 100\% \; x \left[\frac{Power\ Output\ (desired\ form)}{Power\ Input\ (total)}\right]$$

Electricity

Ohm's Law defines the electrical resistance (R) of a wire, resistor, or other resistive electrical device as the ratio between the voltage (E) across the device and the electrical current (I) running through it.

$$R = E/I$$

(Note that this equation works well for DC circuits, in which voltages and currents are constant, but for AC circuits, in which voltages and currents fluctuate, you may need to use a more generalized version of resistance called impedance. Check an introductory textbook on electronics for details.)

The power dissipated (i.e. released) by a resistor, wire, or other resistive load in a circuit is:

$$P = IE = I^2R$$

APPENDIX III: Hydrostatic Pressure

As a vehicle dives deeper under water, the hydrostatic pressure surrounding it increases. This pressure can crush a vehicle or compress flotation and alter its buoyancy, so hydrostatic pressure must be accounted for in vehicle design. The definitions, equations, and tables below may be used to estimate the hydrostatic pressure in water at various depths.

Definition of Pressure

Pressure = force per unit area

Common Units for Pressure

(To convert between these units, see Appendix I or the next section)

Atmospheres (atm)

Bar

Kilopascals (kPa)

Pounds of force per square inch (psi)

Torr (= mmHg)

Typical Atmospheric (air) Pressure near Sea Level

1.00 atm ≈ 101 kPa ≈ 1.01 bar ≈ 14.7 psi ≈ 760 torr

Calculating Hydrostatic Pressure under Water

In seawater, pressure increases by 1 atm for every 10.0 meters (about 33 feet) of depth. Therefore, when working in seawater, the gauge pressure (expressed in atmospheres) is approximately the depth, d, (in meters) divided by 10 m/atm.

$$P_{gauge} \approx \frac{d}{10}$$

This simple formula is generally adequate for estimating pressures in freshwater, too, though it underestimates the actual pressure by 2 to 3%. If you need an answer that's accurate to within 1% or so, use 10.3 m/atm (about 34 feet/atm) for freshwater instead of 10 m/atm.

Remember that absolute pressure is equal to the gauge pressure + 1 atmosphere. (The extra atmosphere of pressure comes from the weight of all the air above sea level.)

$$P_{absolute} \approx P_{gauge} + 1 \text{ atmosphere}$$

For example, at a depth of 25 meters, the gauge pressure is 25/10 = 2.5 atmospheres, and the absolute pressure is (2.5 + 1 = 3.5) atmospheres.

When calculating hydrostatic pressure-related forces on rigid hulls or canisters you must use the *difference* in pressure between the outside and the inside of the container. Therefore, if the canister is filled with air (or other gasses) at approximately one atmosphere pressure (absolute), you should use the *gauge* pressure of the water in your calculations. For example, a rigid canister sealed at the surface with air trapped inside, then taken to a depth of 25 meters, will have a pressure difference across it of 3.5 atmospheres absolute on the outside minus 1 atmosphere absolute on the inside for a differential

Table AIII.1: FRESHWATER

Depth		Gauge Pressure		
feet	meters	atm	psi	kPa
0	0	0.00	0.00	0
1		0.03	0.43	3
2		0.06	0.87	6
3		0.09	1.30	9
	1	0.10	1.42	10
4		0.12	1.73	12
5		0.15	2.17	15
6		0.18	2.60	18
	2	0.19	2.84	20
7		0.21	3.03	21
8		0.24	3.47	24
9		0.27	3.90	27
	3	0.29	4.27	29
10		0.30	4.33	30
12		0.36	5.20	36
	4	0.39	5.69	39
15		0.44	6.50	45
	5	0.49	7.11	49
	6	0.58	8.53	59
20		0.59	8.67	60
	7	0.68	10.0	69
25		0.74	10.8	75
	8	0.78	11.4	78
	9	0.87	12.8	88
30		0.89	13.0	90
	10	0.97	14.2	98
	12	1.16	17.1	118
40		1.18	17.3	119
	15	1.46	21.3	147
50		1.48	21.7	149
60		1.78	26.0	179
	20	1.94	28.4	196
70		2.07	30.3	209
80		2.37	34.7	239
90		2.66	39.0	269
	30	2.91	42.7	294
100		2.96	43.3	299
	40	3.88	56.9	392
150		4.44	65.0	448
	50	4.85	71.1	490
200		5.92	86.7	597
	100	9.71	142	980
500		14.8	217	1490
	200	19.4	284	1960
1000		29.6	433	2990
	500	48.5	711	4900
2000		59.2	867	5970
	1000	97.1	1420	9810
5000		148	2170	14900
5390*	1640*	159	2330	16100

** This is the approximate depth of the deepest known freshwater body in the world, Russia's Lake Baikal. It is over a mile deep.*

pressure of 2.5 atmospheres. This is simply the gauge pressure at that depth. For additional information, see *Chapter 5: Pressure Hulls and Canisters.*

On the other hand, when calculating volume changes in gasses that can be compressed by the water pressure, you should use the absolute pressure. So, for example, if an open-bottomed diving bell is taken from the surface to 25 meters depth, the pressure acting on the air inside will go from 1 atmosphere absolute at the surface to 3.5 atmospheres absolute (at depth), so the air volume inside the bell will have been compressed to 1/3.5, or about 29% of its surface volume. For additional information, see *Chapter 6: Buoyancy, Stability, and Ballast.*

Calculating Pressures in Other Fluids or on Other Planets

To calculate pressures in fluids other than water or in fluids subjected to a gravitational acceleration different than that on the earth, see the more general hydrostatic pressure formula provided in Appendix II.

Hydrostatic Pressure Tables (for Earth!)

The tables in this appendix list the approximate hydrostatic pressure in freshwater (first table) and seawater (second table) at various depths on Earth. They assume a water density of 1,000 kg/m^3 and 1025 kg/m^3, respectively. Water density varies slightly with temperature, water purity, and other factors, so all values in the tables have been rounded to three significant figures. (Pay careful attention to the location of the decimal points.) This precision is more than adequate for most small ROV/AUV purposes. For depths not included in the tables, use the formulas provided above, or (for slightly greater accuracy) interpolate between table values.

Table AIII.2: SEAWATER

Depth		Gauge Pressure		
feet	meters	atm	psi	kPa
0	0	0.00	0.00	0.00
1		0.03	0.44	3.06
2		0.06	0.89	6.12
3		0.09	1.33	9.19
	1	0.10	1.46	10.0
4		0.12	1.78	12.2
5		0.15	2.22	15.3
6		0.18	2.67	18.4
	2	0.20	2.92	20.1
7		0.21	3.11	21.4
8		0.24	3.55	24.5
9		0.27	4.00	27.6
	3	0.30	4.37	30.1
10		0.30	4.44	30.6
12		0.36	5.33	36.7
	4	0.40	5.83	40.2
15		0.45	6.66	45.9
	5	0.50	7.29	50.2
	6	0.60	8.75	60.3
20		0.61	8.89	61.2
	7	0.70	10.2	70.3
25		0.76	11.1	76.5
	8	0.80	11.7	80.4
	9	0.90	13.1	90.4
30		0.91	13.3	91.9
	10	1.00	14.6	100
	12	1.19	17.5	121
40		1.21	17.8	122
	15	1.49	21.9	151
50		1.52	22.2	153
60		1.82	26.7	184
	20	1.99	29.2	201
70		2.12	31.1	214
80		2.43	35.5	245
90		2.73	40.0	276
	30	2.99	43.7	301
100		3.03	44.4	306
	40	3.98	58.3	402
150		4.55	66.6	459
	50	4.98	72.9	502
200		6.07	88.9	612
	100	9.95	146	1010
500		15.2	222	1530
	200	19.9	292	2010
1000		30.3	444	3060
	500	49.8	729	5030
2000		60.7	889	6120
	1000	99.5	1460	10100
5000		152	2220	15300
	2000	199	2920	20200
10000		303	4440	30600
	5000	498	7290	50200
20000		607	8890	61200
	10000	995	14600	100000
35800**	10900**	1100	16000	111000

*** This is the approximate depth of the deepest known spot in the ocean, the Challenger Deep. It is located in the Mariana Trench in the western Pacific and is nearly 7 miles deep!*

APPENDIX IV: Material Properties

This appendix compiles some helpful information about the physical/engineering properties of water as well as a number of useful materials for building small, inexpensive ROVs or AUVs. Most of this information has already been presented elsewhere in this book, but here it is all gathered in one convenient location.

If you seek information beyond what is provided in this appendix (either other materials and/or other properties not listed here), you may be able to find what you are looking for in one of these helpful sources:

- For information about the most important physical and chemical properties of different elements (materials composed of a single type of atom), look for a chart called the **Periodic Table**. This famous table features prominently in almost every high school and college chemistry textbook. You can also find many renditions of it on the web—just do a search for images with the keywords "periodic table."
- The **CRC Handbook of Chemistry and Physics**, published for the last several decades in a series of regularly updated editions, is a widely-used and widely-cited source for detailed physical and chemical properties of a tremendous range of materials. It's a huge book with tiny print, so it may take a while to find what you're looking for, but it's probably in there somewhere! Many high school and college libraries have it.
- Many **manufacturers** and **distributors** of metals, plastics, and other building materials make information about the material properties of their products and related materials available on their websites or in their catalogs.
- As always, be sure to ask **experienced people** for their advice on which materials to use as well as why to use them, where to get more information, and where to obtain them.

PART 1: Properties of Water

Water is special. Here are some fun and interesting facts about it:

- Approximately 71% of earth's surface is covered with water.
- Water is the only substance that occurs naturally in significant quantities on earth in all three of the following states of matter: solid, liquid, and gas.
- Water is one of very few substances that *expand* upon freezing, so its solid form (ice) floats on its liquid form (water).
- The water molecule (H_2O) has an asymmetric charge distribution which gives it a tendency to form hydrogen bonds with other water molecules and many other substances. This property is behind water's surface tension and its ability to dissolve an unusually large variety of other substances.
- Water has an extremely high heat capacity (ability to absorb heat energy without changing temperature very much) and a high heat of vaporization (amount of heat energy that must be added to boil away a certain amount of water once boiling temperature is reached). Both of these properties combined with the huge size of the ocean make water a powerful factor in stabilizing Earth's climate.

Approximate Physical Properties of Water

Note: The low-precision values given in the table below are adequate for routine calculations related to the design and operation of small, non-critical ROVs. For more precise values, which are dependent on exact temperature, pressure, salinity and other factors, or for additional properties not listed here, consult credible web pages or a scholarly text reference, such as *Air and Water: The Biology and Physics of Life's Media* by Mark Denny (1995, Princeton University Press).

Table AIV.1: Some Approximate Properties of Water

Water Property	*Quantity*		*Units*
	Freshwater	*Seawater*	
Density	*1000*	*1025*	*kg/m³*
Salinity	*0*	*35*	*‰ (parts per thousand)*
Electrical conductivity (Note: actual value is strongly dependent on salinity)	*0.005 (highly variable)*	*4.8*	*S/m (siemens per meter)*
Speed of sound (Note: actual value is strongly dependent on temperature)	*1450*	*1500*	*m/s*

PART 2: Properties of Structural Materials

Table AIV.2: Approximate Properties of Some Metals and Plastics Used in Underwater Vehicle Structures

Material	Strength (Ultimate tensile strength in MPa)	Stiffness (Elastic modulus in GPa)	Density (kg/m³)	Corrosion Resistance	Cost Comparison (Approximate retail price for a solid round rod, 1-inch diameter, 36-inch length as of March 2009, in U.S. dollars)	
					Per rod	Per kg
Carbon steel (medium)	500	200	7860	poor	$20	$6
Stainless steel (316)	550	195	8000	good to excellent	$50	$13
Aluminum (6061)	310	70	2700	good	$15	$12
Titanium (Ti-6Al-4V)	1000	110	4430	excellent	$500	$244
Naval brass (485)	430	100	8440	fair to excellent	$75	$19
Bronze (316, tempered)	450	115	8860	good to excellent	$125	$30
PVC (extruded)	46	1.9	1360	excellent	$6	$10
ABS (molded)	40	2.2	1050	excellent	$17	$35
Polycarbonate (extruded)	67	2.4	1200	excellent	$26	$47
Acetal (cast)	60	3.1	1420	excellent	$15	$22
Acrylic (cast)	80	3.3	1200	excellent	$14	$25

Note: These are approximate, averaged values intended for general comparison purposes only. The physical properties for any given piece of material may vary considerably from those listed, depending on exact alloy composition, temperature, environmental exposure, and other factors. Do not rely on these generic numbers for precise or critical design decisions. Likewise, prices among vendors are affected by the quantity of material purchased, and they fluctuate with market conditions.

Sources: www.onlinemetals.com, www.matweb.com, and www.mcmaster.com, accessed March 2009.

Table AIV.3: Galvanic Compatibility Chart

BASE METAL	FASTENER: Aluminum	Carbon Steel	Silicon Bronze	316 Stainless Steel
Aluminum	Neutral	Compatible[1]	Not Compatible	Not Compatible[2]
Steel and Cast Iron	Not Compatible	Neutral	Compatible	Compatible
Copper	Not Compatible	Not Compatible	Compatible	Compatible
316 Stainless Steel	Not Compatible	Not Compatible	Not Compatible	Neutral[3]

Source: abbreviated and excerpted from a chart supplied by the International Nickel Company, Inc. (INCO)

[1] *This combination is generally compatible, but some enlargement of the bolt hole may occur over time.*

[2] *Base metal tends to be cathodic because its larger surface area increases the electrical potential of the metal.*

[3] *There may be some corrosion under the head of the bolt over time.*

APPENDIX V: How to Find Parts

Introduction

Any ROV or AUV project will require parts, not to mention raw materials like metal rods or plastic pipe from which to fabricate custom parts. This appendix offers tips and suggestions for finding what you need.

For most groups doing the types of projects described in this book, budgets are limited to no more than a few hundred or at most a few thousand dollars. Therefore, the emphasis here is on sources for low-cost parts and materials.

This appendix is divided into several sections: 1) The Challenge, 2) General Sources of Information about Parts, 3) Free Parts and Materials, 4) Local Shopping, 5) On-Line Shopping, 6) A Sampling of Major On-Line Distributors, and 7) Suggested Sources Organized by Vehicle Subsystem, which includes a table of keywords for on-line searches.

1. The Challenge

Finding parts for an underwater vehicle project is fundamentally more challenging than most other "shopping" activities. If you want to buy some clothes, you go to the clothing store. If you want to buy some food, you go to the grocery store or supermarket. And so on. But unless you live in a very unusual community, you can't just run down to the local ROV Parts Market to get all the parts you need for your underwater robot.

In addition to a lack of one-stop-shops to supply your parts, you'll encounter other obstacles. In particular, many of the parts you seek will not be advertised in the local paper, on prime time TV, or in any other consumer advertisements. That's because the typical consumer buys and uses kitchen appliances, power tools, or other machines, but does not design or build them. Because of this, you will need to expand your search beyond mainstream consumer advertising, and you will probably need to repurpose typical consumer items in creative ways.

Remember to keep an open mind and think broadly and creatively about what you can use for parts and where you can find them. This appendix should serve as a useful starting point, but don't limit yourself to sources suggested here.

2. General Sources of Information about Parts

Before running out to get parts, you'll want to consult as many of the following sources as are readily available to you for advice and information about what to get and where to get it.

- **Internet Web Searches:** You can always try searching the web for products. Almost every company has a web presence these days. The only downside is that there is so much on the web that your search can be a bit like looking for a needle in a haystack. The additional parts information sources listed below can help you focus your web searches to make them more productive by guiding you to particularly helpful websites (as well as other sources of parts and information). Near the end of this appendix, you'll also find a list of keywords you can use in web searches to help you locate items of interest for budget-conscious ROV/AUV builders.

- **Experienced People:** There's no better source of highly relevant, up-to-date, parts information than talking to experienced people who are working on (or have recently worked on) projects similar to yours. These people can provide you with essential advice on which parts to get and which to avoid. ROV/AUV competitions, robotics clubs, and similar events or organizations tend to concentrate large numbers of these people in one place at one time and provide extremely fertile venues for gathering ideas about what parts to buy, where to get them, how to put them all together, and how to test the final product. And don't forget to consult friends, relatives, or neighbors who have worked in technical, engineering, or scientific careers or hobbies. They can also be a rich source of information about parts for projects.

- **Technology-Oriented Hobby Magazines:** Another excellent source of information about parts for small underwater ROV and AUV projects is the many magazines geared toward readers interested in robotics, radio-controlled vehicles, or electronics. (A few examples of specific magazines in this category are listed in Appendix VI.) These magazines usually feature articles about projects in which the authors describe the projects they have done and list the parts they used to complete them. They are also chock-full of advertisements from companies that make parts useful for that readership, and therefore probably useful for you, too!

- **ThomasNet** (http://www.thomasnet.com): The venerable Thomas Register, which for decades came in the form of a multi-volume book that linked industrial purchasers with industrial suppliers, is now available and searchable on-line. ThomasNet summarizes products and services offered by each of a huge number of US and Canadian industrial parts and supply companies (including OEM parts manufacturers), and they are organized conveniently by the type of product or service the company provides. If you need a brushless DC motor, just search ThomasNet, and you'll get a list of dozens of companies that make and sell them. The basic facts about each company include the company website and other contact information, so you can call or e-mail them or just visit their website to get more information about their products. This resource is definitely geared toward the big industrial players rather than the small-time ROV builder, but their versatile search engine allows you to find companies that manufacture the kinds of parts or materials you need. Even if some of those companies don't sell small quantities to individuals, most will be happy to direct you to a distributor who sells their products in your area.

- **Web Pages for College or High School Courses:** Some instructors who teach engineering courses focusing on ROVs, AUVs, Robots, or Mechatronics post web pages with links to dozens or hundreds of companies that sell parts useful for student-built projects. Sometimes these are grouped by type of product or annotated with helpful information about each company. These pages are not always easy to find, but they are valuable if you can locate them.
- **Appendix VI**: Note that many of the general sources of information summarized in Appendix VI are also great sources for information about parts.

3. Free Parts and Materials

- **Salvaged Junk**: One man's junk is another man's treasure. You may be fortunate enough to have access to an ample supply of scrap pieces of plastic, metal, or other building materials that are still big enough and in good enough condition to be useful. Likewise, you may have the opportunity to extract functional parts or materials from the abandoned carcasses of earlier projects, broken toys, or defunct appliances. Keep in mind that it may take considerable time and effort to extract these parts in functional form, and they may not be the exact size or shape you need to fit other parts of your vehicle, so you may need to spend additional time modifying the parts. You must weigh those disadvantages carefully. Ultimately, it may be better to buy, so you can get exactly what you want and spend your limited time on assembling your vehicle rather than disassembling junk.
- **Gifts**: Birthdays, graduations, and other gift-giving occasions may provide an opportunity to request parts or supplies (or tools) for your projects. Some people will want to surprise you, but others may welcome a specific request, so they can be sure they're getting you something you really want and can use. For these latter folks, plan ahead and let friends and family members know what you need well in advance. Be as specific as you can, right down to the model number, quantity, and price. Also provide the phone number or web site of at least one vendor that sells what you're asking for, since most of your potential gift-givers won't be experienced at shopping for ROV/AUV parts.
- **Donations**: You may be able to get monetary donations from friends, relatives, or local organizations who want to support your project. These folks may also be happy to give you perfectly functional motors, propellers, lights, or other items they no longer use or need. If you have a well-organized school or club team that will be participating in a regional contest or other high-visibility event, you may also be able get donations of parts, materials, or tools from companies that make or sell those items. Such donations are usually given with the understanding that you will help promote the company by letting your audience know about the company (or companies) that helped your team. You can do this by putting stickers with the company logos on your vehicle, thanking them in person and on your team website, and verbally thanking them in public when you give your speech at the awards ceremony.

4. Local Shopping

Although you won't find an ROV/AUV parts store *per se*, several local stores are likely to carry consumer items that can be repurposed as underwater vehicle parts.

- **Hardware Stores/Home Improvement Stores**: These are excellent sources for PVC pipe or aluminum angles to use in frame construction; miscellaneous nuts, bolts, screws, washers, and other fittings; electrical wiring; screen to keep algae out of props; and countless other items that may be useful for building ROVs. Local hardware stores are also an excellent source for basic tools.
- **Plumbing Stores or Irrigation Supply Stores**: Both are useful for underwater vehicle parts, because they carry lots of products designed for use in or with water. These include pipes, gaskets, and seals.
- **Boating/Fishing/Outdoor Sports Stores**: Any of these may carry trolling motors, propellers, marine batteries, waterproof compasses, floats, lead weights, water-resistant switches, and many other items useful for ROVs.
- **Electronics Stores**: Sometimes there may be all or part of the store devoted to basic electronic parts and tools, such as wires, cables, switches, integrated circuits, soldering irons, etc.
- **Toy/Hobby Stores**: Look for those that sell radio controlled or other remotely controlled toys that can be dissected to extract useful components for control systems.
- **Art/Craft Stores**: These tend to have a wide variety of not-too-expensive items. A few of these, including graph paper, waterproof paints, expanded polystyrene forms, casting resins, and simple tools like hot melt glue guns, might prove useful for design, construction, or decoration of underwater vehicles.
- **Kitchen Suppliers**: Such stores often carry water-tight plastic containers that are useful for very low-pressure housings or potted electronics, salad tongs that can be modified into a simple gripper, and other items that a little ingenuity and creativity can transform from kitchen implements into underwater vehicle components.
- **Appliance Parts Dealers**: Those that sell replacement parts to repair household appliances can be a great source for parts. However, these stores usually have their stock stored behind the counter and organized by brand and part number, so you usually need to know the make and model of an appliance that the part normally goes in, as well as the specific part number, before you can shop there effectively.
- **Automotive Parts Shops**: These stores carry sealed lead acid batteries, wires, switches, fluid-filled dashboard compasses, and many other parts that can be useful in ROV construction.

- **Plastics Suppliers**: Here you can find acrylic sheets (usually in a wide variety of colors), clear tubing, fiberglass resins, casting resins, plastic hemispheres (which can be used for camera ports), plastic boxes, plastic glues (cements) and many other useful materials, if you are lucky enough to have a large store in your area.

- **Dive Shops**: sell diver's depth gauges, underwater compasses, lead diving weights, and a variety of other seawater-proof and pressure-proof items that may prove useful.

5. On-Line Shopping

When it comes to shopping for parts on line, it will help if you recognize that your shopping needs are less like those of a typical consumer and more like those of a small company manufacturing a complex mechanical consumer product, such as a blender, washing machine, or airplane. Companies that make the parts for these kinds of machines are known as **Original Equipment Manufacturers,** and the parts used to make the machines are known as **OEM** parts. You won't find OEM parts advertised in the newspaper or on TV. The companies that make and sell OEM parts to OEM companies usually advertise through channels that are largely hidden from the mainstream consumer. Fortunately, the web is one channel shared by both consumer and OEM advertising, so you can find OEM parts sellers on the web if you look in the right places. One trick is to include the keyword "OEM" in some of your web searches or to look for the tiny, non-obvious links with labels like "OEM," "technical," "industrial customers," etc. on company websites.

It also helps to understand the difference between a manufacturer and a distributor, because shopping for OEM parts can lead to dead ends and frustration if you don't understand the difference. **Manufacturers** are companies that actually *make* the products, whether they are tools, equipment, materials, or supplies. Parts manufacturers typically specialize in a small range of related products and produce them in enormous quantities. Although a few parts manufacturers will sell small quantities directly to individuals, most sell only to other companies and only in huge quantities. For example, a semiconductor manufacturer making microcontrollers might have a minimum order quantity of 10,000 units, and their typical "customer" might be something like a cell phone company that builds the microcontrollers into their line of consumer cell phones. Unless you needed 10,000 microcontrollers for your ROV, you probably would not be able to buy directly from this manufacturer.

On the other hand, **distributors** are companies that make a living by buying large orders from manufacturers at bulk-order prices, then reselling the items in small quantities to individuals or small companies who are willing to pay more per-item in exchange for not having to buy huge quantities. For example, a distributor of parts for robotics enthusiasts might buy 10,000 microcontrollers from the manufacturer above for $1 each (a total of $10,000), then turn around and sell them to you and other individuals for $3 each, so you can use them in your robot projects. You get a great deal, because it costs you only $3 instead of $10,000 to get one microcontroller, and the distributor makes a profit because they will gross $30,000 on their original $10,000 investment (a $20,000 profit) by the time the sell all 10,000 microcontrollers. Another advantage of distributors is that they will often carry groups of related parts that fit together, even if those parts are made by different manufacturers, so you can do one-stop shopping.

Although you'll probably buy most of your parts through distributors, it's still good to know about the manufacturers, because a manufacturer's web site often provides more detailed technical specifications about each part than a distributor's site does.

See Section 7 below for suggested keywords you can use to begin your parts search for each vehicle subsystem.

6. A Sampling of Major On-Line Distributors

If you're looking for a place to start your comparison-shopping, try checking out the major on-line distributors listed here. All have been in business for decades. All have truly gargantuan, yet accessible inventories. And all have well-earned reputations for user-friendly websites, prompt delivery, and good customer service. If you can forego one or more of these benefits, you may be able to find specific items for lower prices from other sources, but these are excellent places to get a sense of what's available and to do some one-stop shopping.

Industrial Tools, Parts, Supplies, and Materials

- McMaster-Carr (http://www.mcmaster.com)
- Grainger (http://www.grainger.com)

Electronic Tools, Components, and Supplies

- Digi-Key (http://www.digikey.com)
- Mouser (http://www.grainger.com)
- Newark Electronics (http://www.newark.com)
- All Electronics Corp. (http://www.allelectronics.com)
- Jameco Electronics (http://www.jameco.com)

7. Suggested Sources Organized by Vehicle Subsystem

To help you get started, Table AV.1 (on the next page) summarizes suggested sources for parts organized by vehicle subsystem. Keep in mind that these are just suggestions for where to start your search; they are by no means an exhaustive list of all the possibilities.

Table AV.1: Suggested Sources and Keywords

Vehicle Subsystem	***Local Shopping***	***Suggested Keywords for On-Line Searches (Hint: search for singular not plural terms)***
Frame (PVC pipe & fittings, aluminum angles or other aluminum extrusions)	*PVC pipe & fittings from hardware store or irrigation supply. Nuts, bolts, screws or other fasteners from hardware store. Metal parts from hardware store, metal supply, or welding supply.*	*PVC pipe, PVC fitting, ABS, metal, plastic, aluminum extrusion, tubing, fastener, hardware*
Pressure Canisters and/or Encapsulation	*Larger diameter PVC pipe from hardware store or irrigation supply. Underwater flashlight housings from dive shops. Thick PVC, acrylic, or polycarbonate sheets for endcaps, and clear acrylic domes for camera ports from plastic stores. Potting compounds from plastic stores, larger electronics stores.*	*PVC pipe, metal tubing, dive light, potting compound, epoxy, subsea housing, waterproof case*
Fairings	*Plastic stores or hardware stores for fiberglass and epoxy resins. Plastic stores for sheets of polystyrene or other thermoplastic you can use for do-it-yourself vacuum forming.*	*Fiberglass, epoxy resin, DIY vacuum forming, thermoforming*
Static Buoyancy Control	*Fishing/boating stores for floats and weights. Dive shops for weights. Hardware stores for PVC pipe, foam pipe insulation.*	*Fishing float, trawl net float, pipe insulation, fishing weight, dive weight*
Propulsion/Thrusters	*Boating supply stores for bilge pump motors, trolling motors, large plastic propellers. Hobby stores with model boats/planes for small plastic props. For brushless motors to use in oil-compensated housings, try hobby stores that sell R/C cars or planes.*	*bilge pump, bilge pump cartridge, trolling motor, small DC motor, BLDC motor, planetary gearmotor, robot motor, plastic propeller, model boat propeller, model plane propeller, ROV thruster*
Power & Power Distribution	*Automotive/motorcycle stores for sealed lead acid batteries. Hobby stores that sell radio-controlled vehicles for high performance battery packs. Automotive, hardware, car stereo, electrical, or electronics supply stores for wire.*	*battery, SLA battery, battery pack, wire, stereo wire, extension cord, GFI, GFI extension cord, DC-DC converter, voltage regulator, switching converter*
Basic Control	*Automotive supply, marine supply for DPDT toggle switches. See above for wire sources.*	*DPDT toggle switch, wire, stereo wire, cable, plastic boxes*
Advanced Control	*Local electronics store (if it has good selection of microcontrollers and other components); otherwise, on-line is your best bet.*	*robotics, robot, microcontroller, MCU, radio control, model airplane, ethernet cable, embedded system*
Cameras	*Home/business security stores or consumer electronics stores.*	*security camera, video surveillance, CCTV system, board camera, PCB camera*
Other Sensors	*Dive shops for depth gauges. Dive shops, automotive/boating/kayaking stores for waterproof compasses. Electronics parts distributors for electronic sensors.*	*sensor, depth gauge, compass, electronic sensor*
Manipulators & Grippers	*Simple grippers can be made from modified salad tongs (kitchen store) or pliers (hardware store). Simple hydraulic systems can be made from syringes and plastic tubing available at some pet/aquarium stores, hardware stores, craft stores.*	*plastic syringe, plastic tubing, pneumatic robot, pneumatic fitting, hydraulic robot, hydraulic fitting*

APPENDIX VI: How to Find Additional Information

This weighty book covers a lot of material, but it's just the tip of the iceberg when it comes to the ocean of information that exists about underwater vehicles, underwater robotics, and related technologies. It would be impossible to list individually every source for information related to this interdisciplinary subject. Instead, Appendix VI provides more generalized suggestions on where and how you can mine this vast sea of knowledge as efficiently as possible to locate the specific information you need.

1. World Wide Web

The world wide web on the internet is a phenomenally rich source of searchable, on-line information about an endless variety of subjects, including underwater vehicle technologies. It is also a great place to find out about more traditional sources of information including paper publications, professional organizations, events, and contests.

The most basic way to find information on the web is to use a keyword search to locate some pages of interest, then follow hyperlinks on those pages to find other related information. For example, you could get started by searching on the keywords "ROV," "AUV," "underwater robot," or many of the words in the glossary and index of this book.

But a simple keyword search is just the beginning. To get the most out of what's available on the web, it's worth investing some time to learn how to use the advanced search options (like "exact phrase" searches and Boolean searches) and organizational tools (like hierarchically arranged bookmarks) available on most web browsers.

Boolean searches are an especially powerful search technique that is often overlooked because this option is usually hidden under the "advanced search" features. Depending on your browser, it may not be labeled "Boolean," but you can recognize this search option by its use of standard Boolean logic operators like AND, OR, NOT, and AND NOT (all usually capitalized) or key word categories like "contains all of these words" but "none of these words." For example, you can use a Boolean search to find information about electrical relays for your control system without having to sift through lots of sports pages about relay races. Or you could do a Boolean search to find pages that contain the names of at least two different robotics competitions you know—a clever way to find out the names of competitions you *don't* know by locating *lists* of robotics competitions. With practice, you'll quickly discover how useful Boolean searches can be.

Don't forget to bookmark the goldmines. Every now and then you'll come across a page that has exactly what you've been looking for or one that has links to dozens or even hundreds of other very useful sites. Use your browser's "bookmark" feature to record links to these pages, so you can easily find them again later. As your bookmark list grows, take advantage of browsers that allow you to organize those bookmarks into folders or other hierarchical schemes.

2. Books and Magazines (Periodicals)

- **Technical Books and Manuals**: There are quite a number of technical or semi-technical books on the subjects of ROVs, AUVs/UUVs, submarines, submersibles, and other underwater technologies. Those that deal specifically with underwater technologies are geared toward an experienced professional audience, including undersea operations managers, ROV pilots, and naval engineers. As a result, they can be a bit hard to find, challenging to understand, and expensive, but you may be able to locate one or more of them in larger libraries. However, there are considerably more extensive, accessible, and affordable books for robotics enthusiasts. Most (but not all) of these deal exclusively with terrestrial robotics, but there is so much overlap between terrestrial robotics and subsea robotics in areas such as control systems, gearing, etc. that these robotics books can be very useful resources for the subsea vehicle designer.

- **Tales of Adventure and Discovery**: There are many exciting books about undersea exploration, discovery, or adventure. While some are more factual in their technical, historical, and/or biographical details than others, all of them can be excellent sources of inspiration.

- **Trade Journals**: Most industries, including various subsea industries, publish magazine-like periodicals called trade journals that are designed to keep professionals in those industries abreast of the latest developments, technologies, and services. These magazines usually feature a mix of interesting articles and advertisements for products and services used by that industry. Although they tend to be a bit too specialized for general reading, trade journals provide a glimpse into what's happening in the industry, including what kinds of jobs are available, the equipment and training required, where the action is happening, and who the leading companies are in any particular specialty. Examples of trade journals related to the subsea industry include *Sea Technology*, *Marine Technology Reporter*, and *Ocean News and Technology*.

- **Tech-Oriented Hobby Magazines**: There are quite a number of magazines geared toward readers who are interested in technology-oriented hobbies, including robotics, remote controlled cars/planes/boats, electronics, general gadget design/construction, and the like. Examples include *Robot Magazine*, *SERVO Magazine*, *MAKE Magazine*, *RC Driver*, *Nuts and Volts*, and *Circuit Cellar*. Some are written for the beginner, others for those with more experience. In addition to having very interesting and relevant articles, including occasional articles about ROV and AUV design, these publications are chock-full of advertisements for products and services of great interest to underwater vehicle designers.

3. Television

- Television stations that cater to audiences interested in history, science, technology, exploration, and the like occasionally run documentaries featuring underwater technologies. These programs can be quite interesting, informative, and inspirational.

4. Organizations and Events

- **Educational Organizations**: Many schools, summer camp programs, and other educational organizations (including the Marine Advanced Technology Education Center that produced this book) have come to recognize the tremendous educational value of using underwater vehicle projects as a fun way to learn about science, technology, project management, and teamwork. As well, these hands-on activities and mentor opportunities encourage further education possibilities and career options. Depending on the size and the specific educational goals of the organization, they may provide one or more of the following sources of information: 1) student or teacher workshops where you can learn how to build simple underwater vehicles, 2) websites with links to useful information about underwater robots and related topics, 3) local, regional, national, or international ROV or AUV competitions, 4) internships or other experiences that connect students directly with the subsea industry, 5) underwater robotics curriculum materials and support for teachers, and 6) ROV plans or even kits for schools, plus fabrication advice. In addition to regular schools and educational programs, there are highly-specialized training programs operated by the subsea industry that prepare people for careers as commercial divers, ROV pilots, and other subsea industry careers.

- **Professional Societies**: Like people in any career field, those who work in subsea technology industries have created professional societies to help them network with each other and with vendors who sell the equipment, parts, and supplies they need. Many of these societies, such as the *Marine Technology Society* (*MTS*), include strong educational outreach components that provide access to information about careers, technologies, and other opportunities in the field. Don't forget to check out societies in related fields, like the *Institute of Electrical and Electronics Engineers (IEEE)* and the *American Society of Mechanical Engineers (ASME)*, both of which are enormous societies with specialized sub-branches, including some specifically related to underwater design and/or control systems.

- **Clubs**: Robot clubs or similar organizations provide a place where like-minded people get together from time to time to share ideas and have fun building things. The members of these clubs can be a great source for information and camaraderie. Ask around to find out if there are any clubs in your area that have a focus on robotics, electronics, radio-controlled planes, model submarines, or any other theme that might overlap with underwater vehicle technology. Also there are a number of on-line robotics clubs, including some that specialize in underwater robots.

- **Competitions and Contests**: Going to an ROV, AUV, or robot competition can be a great way to meet lots of people, exchange information, and see examples of creative solutions to the types of challenges you might be facing with your underwater vehicle designs. In addition to meeting other contestants, you can usually find contest judges, industry representatives, or other experts who have years of experience and are willing to answer questions.

GLOSSARY

Glossary Notes

This glossary provides the definitions of words as they are used in this underwater robotics textbook. Many of these words have other definitions in other contexts.

Generally most glossary words are boldface the first time they appear in the text, but occasionally major terms are emboldened a second time in a later chapter or chapter section that focuses on that concept. Some glossary terms come from chapter or section headings or refer to a general topic.

Glossary terms used within the definition of other glossary terms are italicized the first time they appear in a definition.

The glossary does not include the names of surface and underwater craft or the names of organizations, academic institutions, and companies.

A

A-frame. A special type of crane with a shape somewhat reminiscent of a capital letter "A" that is fitted on the bow or stern of some ships and can lift extremely heavy loads. A-frames are often *heave-compensated.*

A-to-D converter. See *analog-to-digital converter.*

ABS. See *acrylonitrile-butadine-styrene.*

Absolute pressure. *Pressure* measured relative to a vacuum, which has zero pressure. See also *gauge pressure.*

Absorbed glass mat (AGM) battery. A type of *valve-regulated lead-acid battery* in which the electrolyte is rendered non-spillable by absorbing it into a *glass* sponge.

Absorption. In physics, reduction in the *strength* or intensity of *sound* or light (or other type of propagating *energy*) as a result of a conversion of energy into another form (usually heat) in the medium through which the sound or light is traveling.

AC. See *alternating current.*

AC motor. A type of electric motor powered with *alternating current.*

AC-to-DC adapter. A familiar device, sometimes called a "*wall wart,*" that plugs into a wall outlet and converts the *AC* electrical *power* in the wall into *DC* power at lower, safer *voltage* to power or recharge a small electric device, such as a cell phone, laptop computer, or *digital camera.*

Accumulator. In *hydraulic systems* a mechanism whereby the *pressure* of *hydraulic fluid* lifts a *weight,* compresses a spring, or compresses a gas to store *energy.* Useful for smoothing out uneven pressure pulses from the *prime mover* and, more importantly, for delivering bursts of high *power,* which allows for a lower power, less expensive *prime mover.*

Acetal. A type of *plastic* known for its low friction surface.

Acoustic Doppler current profiler (ADCP). A tool that oceanographers use to measure the speed and direction of water *currents* simultaneously at different *depths* in the *water column.*

Acoustic Doppler navigation. A method of *navigation* for ships and subsea vehicles that relies on detection of the *Doppler shift* occurring in each of (typically) four transmitted acoustic *signals* when those signals bounce off features on the bottom. Allows precise measurement of the vessel's speed and direction over the bottom in a way that is largely unaffected by winds and water *currents.*

Acoustic positioning system. A *navigational* system that relies on *sonar* to precisely locate underwater objects, such as an *ROV,* equipped with an acoustic transponder. An array of *hydrophones* is used to measure the distance from the transponder to each hydrophone in the array, thereby allowing calculation of its position relative to the array. Often used to determine the precise location of an ROV relative to a surface support ship, then combined with shipboard *GPS* (which can tell precisely where the ship is, but does not work under water) to determine the precise latitude, longitude, and *depth* coordinates for the ROV.

Acrylic. A type of *plastic* known for its excellent optical clarity.

Acrylonitrile-butadine-styrene (ABS). A type of *thermoplastic* used frequently to make *ABS pipe, ABS pipe* fittings, and a variety of other things.

Active ballast system. See *dynamic ballast system.*

Active sonar. Any *sonar* system that acts as its own *sound* source. For example, many sonar systems send out "pings" or other sounds, then listen for returning echoes. Using the known speed of sound in water, these systems can calculate the distance to any object that reflects the sound by measuring how long it takes the echo to return from that object. See also *sonar, passive sonar.*

A/D converter. See *analog-to-digital converter.*

Adaptive control. A type of self-adjusting *control system algorithm* that monitors the effectiveness of its present *control* strategy and makes changes in the *control algorithm* as needed to optimize performance.

ADC. See *analog-to-digital converter.*

ADCP. See *acoustic Doppler current profiler.*

Air compensation. A form of *gas compensation* in which compressed air is used as the pressurized gas inside the *canister* or *housing.*

Air-compensated housing. An undersea *canister* or *housing* protected from *hydrostatic pressure* and leaks by *air compensation.*

Airlock. Usually a small chamber with pressure-proof doors used to move people or supplies between areas of differing *pressure.* For example, *submarines* may be equipped with a floodable airlock that allows divers to enter or leave the sub under water without flooding or pressurizing the whole interior of the sub.

Algorithm. A set of step-by-step procedures for accomplishing some task.

Alkaline battery. A *battery* composed of one or more *alkaline cells.*

Alkaline cell. A *galvanic cell,* popularly called an "*alkaline battery,*" that relies on an electrochemical reaction between *zinc* and manganese dioxide in the presence of a potassium hydroxide electrolyte to make electricity. Has a typical cell *voltage* of about 1.5 *volts* and popularly available in the familiar cylindrical AAA, AA, C, and D sizes. Among the most available and affordable batteries worldwide. Generally not *rechargeable,* but there are some exceptions.

Alloy. A homogeneous mixture of two of two or more elemental *metals* (and sometimes other non-metallic elements such as carbon) used to obtain a new metal with improved *strength*, corrosion-resistance, machinability, or other useful properties.

Alternating current (AC). A form of electrical *signal* used for transmitting *power* or information and characterized by having a *voltage* that changes rapidly in time (typically oscillating back and forth dozens to billions of times per second). AC signals with a frequency of 50 or 60 cycles per second are commonly used to transmit electrical power over long distances and are the usual form of power available from standard wall outlets.

Altimeter. See *echosounder*.

Aluminum. An elemental *metal* used most commonly in aluminum *alloys*. Some of these alloys have been used extensively for *pressure hulls* and *canisters*. Though stiff, many aluminum alloys are easy to cut and drill, even with hand tools.

Ambient pressure. The *pressure* of the surrounding medium (usually air or water) at a particular altitude or *depth*. Example: if a container is left open, its internal pressure will equal the ambient pressure.

American Wire Gauge (AWG). A standard numerical scale used to describe the diameter of electrical *wires*. The larger the number, the thinner the wire. So a #2 AWG wire has a much larger diameter than a #30 AWG wire. The diameter associated with each AWG number can be looked up on the web or in many electrical texts.

Amp. See *ampere*.

Ampere (A). A standard unit of *electric current* used in both the *metric* and *imperial systems*. Commonly called an "*amp*." One ampere is defined as a current of one *coulomb* per second.

Amp-hour (Ah). A unit of electric *charge* equal to the amount of charge delivered by a steady *current* of 1 *amp* flowing for 1 hour. Frequently used as an indirect measure of the amount of *energy* stored in a *battery*. See also *energy capacity*.

Analog ground. A *signal ground* used for very small, delicate *voltage signals*, such as those produced by many *analog sensors*. Similar to any other *reference ground* except that great pains are usually taken in the design of the *circuit* to ensure that the analog ground is protected from all likely sources of interference, so its voltage remains very *stable* at all times. Compare *digital ground*. See also *ground* for other types of electrical ground.

Analog sensor. A *sensor* with an output *signal* that can vary smoothly and continuously within its range of possible values.

Analog signal. A *signal* that can vary smoothly and continuously, essentially taking on an infinite number of possible values, between its minimum and maximum possible values.

Analog-to-digital (A/D) conversion. The process of converting information from an analog signal to a digital signal.

Analog-to-digital converter. An *integrated circuit* that performs *analog-to-digital conversion*. Typically accepts an analog *voltage* or *current* as an input, converts it internally to a digital form, and outputs a digital value corresponding to the analog input value. Also known as an *A-to-D converter*, *A/D converter*, or *ADC*.

Analog video format. Any *video format* that relies on analog data storage or transmission. Common examples include HDTV, DVD, Mini-DV, MPEG-4, MPEG-2, and many others.

Anode. The *wire*, electrode, or other *conductor* on a polarized electrical device through which *electric current* flows into the device. This is the negative terminal on a typical *battery* while it is supplying electrical *power* to other devices, and the positive terminal on a typical device powered from the battery. However, the roles can reverse, as when the positive terminal of a battery becomes the anode while the battery is charging. See also *cathode* and *electric current*.

Anodization. An electrochemical process used to coat some *metals* (mainly *aluminum*) with a hard, protective, oxide layer that functions like super-hard paint to protect the surface from scratches and further corrosion.

Aphotic zone. Deeper layers in a water body, where there is not sufficient light to support significant *photosynthetic* activity.

Aqua-lung. See *scuba gear*.

Arc search pattern. A *search grid* that consists of following a series of concentric arcs of progressively larger radii. Also known as a *circular search pattern*.

Archimedes' Principle. A useful observation (made by Archimedes) that the *magnitude* of the *buoyant force* acting upward on a completely or partially submerged object is equal to the *weight* of the water displaced by the object.

***Argo* float.** A free-drifting, *autonomous*, *CTD*, that records measurements of seawater *salinity* and *temperature* at various *depths*, then, at regular intervals comes to the surface and transmits that data along with its position to a relay satellite.

ARV. See *Hybrid underwater vehicle*.

ASCII. A standardized code in which all conventional keyboard characters (including blank characters like spaces, tabs, and new lines) are equated with a binary number between 0 and 127 (decimal). Extended ASCII codes, somewhat less standardized, include binary codes for other symbols like happy faces, hearts, etc. Allows text to be stored and transmitted digitally.

ASDIC. See *sonar*.

Assembly drawings. Detailed drawings that show how various parts of the vehicle are put together to create the finished vehicle.

Asynchronous data transmission. A form of *serial data transmission* in which independent clocks, one on the sending device and one on the receiving device, are used to coordinate transmission of digital data between the devices.

Atm. A common abbreviation for the unit of *pressure* known as one *atmosphere*.

Atmosphere. A layer of gas surrounding a planet. Also, a unit of *pressure* equal to the *force* per area normally exerted by the *weight* of Earth's atmosphere on objects at sea level; this unit of pressure is often abbreviated "*atm*" as is approximately equal to 101,000 *newtons* per square *meter*, or 14.7 *pounds per square inch*.

Atmospheric Diving Suit (ADS). A rigid-walled, one-person diving suit with a near 1-atmosphere internal *pressure* and articulated joints that allows a person to work at great *depths* (up to 600 m) without having to decompress after a dive. Some are equipped with propulsion systems and essentially function as "wearable" HOVs. Most ADSs are used for military or commercial *missions*.

Attenuation. The fading or weakening of a waveform, such as *sound*, light, radio *signals*, electrical signals, or water *waves*, as the distance from the source increases. Attenuation can be caused by a variety of factors.

Autonomous. Self-directed or self-controlled. Automatic. Not requiring a human *pilot* or operator.

Autonomous Underwater Vehicle (AUV). A tetherless, *unmanned vehicle* that operates as an independent robot with no physical link to the surface and no direct *pilot* control.

AUV. See *Autonomous Underwater Vehicle.*

B

Backscatter. A smattering of white specks or streaks on an underwater image that occurs when bright light from strobes or video lights is reflected directly back into the *camera* when it reflects off tiny marine plankton, suspended sediment, or other particles in the water.

Ballast. *Weight(s)* placed low on a vessel to decrease its *buoyancy* and/or lower the *center of gravity* to increase *stability.*

Ballast system. A system of *weights,* floats, or other components that collectively *control* an underwater vehicle's *buoyancy* and *trim.*

Bang-bang control. The simplest of the common *control algorithms.* Also known as ON/OFF control, because it can turn things on or off, but cannot otherwise regulate the speed or strength of the *control signals.* Compare *proportional control* and *PID control.*

Bar. A *metric* unit for *pressure* defined as 100,000 *Pascals.*

Bar stock. Long rods of solid material that may be round, square, rectangular, or hexagonal in cross section.

Barometer. A device used to forecast weather changes by measuring changes in the *absolute pressure* of the earth's atmosphere.

Barometric pressure. The *pressure* of the *atmosphere* at a given place and time, which is generally close to, but not exactly equal to, the standard unit of pressure known as one *atmosphere.*

Bathyscaph (also **bathyscaphe**). An independent diving vessel with its own system for regulating *buoyancy.* The most famous example is *Trieste,* which in 1960 set a record by descending nearly 11,000 *meters* to the deepest place in the ocean.

Bathysphere. A spherical non-maneuverable chamber housing a small crew, suspended by a *cable* beneath a ship in order to observe deepsea life.

Battery. Technically, a set of interconnected *galvanic cells* used as a source of electrical *power.* Also used in everyday speech to refer to single galvanic cells, such as 1.5 volt AA, C, and D sized cells, commonly used to power toys, flashlights, and other portable electric devices.

BDC motor. See *brushed direct current motor.*

BG. The distance separating the *center of buoyancy* from the *center of gravity.*

BJT. See *biplolar junction transistor.*

Bilge pump. A small electric pump used to remove water from the bilge, a low spot inside a boat where water collects and must be removed periodically. Bilge pumps can be modified to make excellent *thruster*s for small, shallow-water *ROV* and *AUV* projects.

Binary. In computer science and related fields, including robotics, a reference to the base-2 number system commonly used to represent numbers and a wide variety of other things in computers. Unlike the more familiar base-10 number system, which has ten digits (0 through 9) and columns representing powers of ten (1, 10, 100, etc.), the binary system has only two digits (0 and 1) and columns representing powers of two (1, 2, 4, 8, 16, etc.). Compare *octal* and *hexadecimal.*

Biodiversity. A quantitative measure of the total number and/or relative abundance of different species in an area. Several different mathematical formulas are used to quantify biodiversity, and there is debate about which are most relevant or meaningful in any particular habitat or ecosystem.

Biofouling. Growth of attached marine or aquatic life, such as algae, barnacles, or sponges on boat hulls, underwater cameras ports, or other human-made surfaces that is thick enough to interfere with the normal operation of those surfaces.

Bipolar junction transistor (BJT). The original type of *transistor.* Relatively inexpensive and widely used; however, being replaced by newer transistor technologies, such as *field effect transistors* in most other applications.

Bit. A single *binary* digit, which can have either of two possible values, usually represented as either zero or one.

Blade. Many definitions, but in this text used to refer to a *propeller blade.*

Blade back. The relatively flat surface of a *propeller blade* located on the opposite side of the *blade* from the *blade face.*

Blade face. The relatively flat surface of a *propeller blade* you would see if you were following behind a vehicle propelled forward by the *propeller.* Normally the high-*pressure* side of the propeller blade.

Blade root. The point where a *propeller blade* attaches to the *propeller hub.*

BLDC motor. See *brushless direct current motor.*

Block. A rectangular chunk of solid material, usually used as a starting point for machining parts.

Board camera. A board *camera* is basically the innards of an electronic camera (often a video camera) made by attaching a small lens (*glass* or pinhole) to a small *printed circuit board* that includes a *CCD image sensor* or a *CMOS image sensor* located behind the lens. It also includes some additional electronics to convert electrical *signals* from the image *sensor* into a standardized video or still image output format.

Bollard pull. See *bollard thrust.*

Bollard thrust. The maximum pulling (or pushing) *force* a *thruster* or vessel can produce when tied to a strong, stationary anchor point, or otherwise not moving. Also known as *bollard pull.* An important specification for tug boats, *ROVs,* and other slow-moving vehicles that must maneuver against *tether drag* or other heavy loads.

Boss. See *penetrator boss* and *flange.*

Bottom-crawling ROV. A type of *ROV* that crawls along the bottom, typically by means of caterpillar treads. Most are large and heavily weighted. *Power,* launch and retrieval normally require support from a ship, though a few can crawl into the water from shore.

Boyle's Law. A statement that gas volume is inversely proportional to absolute pressure, provided that the number of gas molecules and the temperature remain constant. This is one of several special cases of the more general *Ideal Gas Law.*

Brainstorming. A method of generating a long and fairly complete list of possibilities (for later evaluation) by having a group of people share ideas in a supportive, non-critical way that allows each idea presented to inspire a stream of other related ideas. For maximum benefit, evaluation and critique of ideas must take place after, not during, the brainstorming session.

Brass. A hard, strong, corrosion-resistant *alloy* made by combining copper and *zinc*. See also *naval brass*.

Breadboard. In electronics, a surface upon which to build and test a *protoype circuit*. May be as simple as a piece of plywood or as complex as a commercially available circuit prototyping station.

Breadboarding. The process of assembling a *prototype circuit* for testing. See also *breadboard*.

Breaking wave. See *surf*.

British Thermal Units (BTU). An *imperial* unit of *energy*. One BTU = about 1055 *joules*.

Bronze. A hard, strong, corrosion-resistant *alloy* made by combining copper and (usually) tin.

Brushed direct current (BDC) motor. A type of *DC motor* that relies on a mechanical *commutator*. The name derives from the fact that these commutators have spring-loaded, electrically-conductive pads or contacts, called "brushes," which transfer *electrical current* from the non-moving *stator* to the spinning *rotor*.

Brushed motor controller. A *motor controller* designed for use with *brushed DC motors*.

Brushless direct current (BLDC) motor. A type of *DC motor* that relies on a solid-state electrical *commutator*, which may be located inside the motor but is frequently located outside the motor in a separate circuit. A *BLDC* lacks the "brushes" found in the commutator of a typical *brushed DC motor*.

Brushless motor controller. A *motor controller* designed for use with *brushless DC motors*.

Bulkhead. A sturdy wall or partition in a ship, *submarine*, or airplane to provide structural support and/or separate the space. On ships and submarines, they are often designed to be very strong and are equipped with water-tight doors and *penetrators*, so that sections of the vessel can be sealed off from the others in the event of a serious hull breach or other flooding.

Bulkhead penetrator. A *penetrator* designed to seal a hole in a *bulkhead* or other flat wall. Commonly used to pass electrical *power* or *signals* through the *endcap* of a *pressure canister*.

Buoyancy. The overall tendency of an object to float in a fluid. Buoyancy is a *force* that depends on both the *weight* of the object and on the *buoyant force* acting on the object. It is exactly the opposite of a quantity called the *effective weight*.

Buoyant force. The upward *force* exerted by a *fluid* on an object that is partially or completely immersed in it. Affects the *buoyancy* of the object by partially or completely offsetting the object's *weight*.

Byte. A set of eight *bits*.

C

Cable. In an electrical context, two or more *wires* bundled together for transmission of electrical *power* or information.

Cable-controlled Undersea Recovery Vehicle (CURV). One of the earliest types of *ROVs*. Developed by the US Navy in the 1950's to recover valuable military ordnance (weapons) lost at sea.

Caisson disease. See *decompression sickness*.

Calorie (Spelled with capital "C"). A *metric* unit of *energy* equal to one *kilocalorie*, or 1000 calories. Same as the dietary *Calorie* often used on food labels.

calorie (Spelled with lower-case "c"). A *metric* unit for *energy* defined as the amount of heat energy it takes to raise the *temperature* of one gram of pure liquid water by one degree centigrade. One calorie is equivalent to about 4.184 *joules*. Note that 1 Calorie = 1000 calories, so you must be extremely careful about capitalization of this word.

Camera. A device for recording light-based images. See also *board camera, CCD camera, CCD image sensor, CCD video camera, IP camera*, and *web camera*.

Camera port. A *viewport* designed specifically for a camera.

Can. See *pressure canister*.

Canister. See *pressure canister*.

Capacitor. An electronic component that stores *energy* in the form of an *electric field*.

Capture frame. A type of *LARS* in which the vehicle is securely held in an elevator-like system that moves along a set of vertical tracks or rails. Constrains the ROV's movements so that it cannot damage itself or the launch *platform* during the transition through surface *waves* and swell. Also called a *cursor system*.

Card cage. A rack or similar frame-like structure that supports *printed circuit boards* or other electronic components inside a *pressure canister* and can be slid out of the *canister* to simplify accessing and servicing of the circuits.

Casting. A shape formed by pouring molten *metal*, polymer resins, or other liquids into a mold, then allowing the liquid to harden.

Caterpillar drive. See *magnetohydrodynamic drive*.

Cathode. The *wire*, electrode, or other *conductor* on a polarized electrical device through which *electrical current* flows out of the device. See also *anode*.

Cathodic protection. A method of protecting *metal* from *corrosion* by making sure the *metal* you want to protect is the *cathode* in any corrosive electrochemical reactions that might be taking place. There are various ways to do this, including *sacrificial anodes* and externally imposed electrical *voltages*.

CB. See *center of buoyancy*.

CCD camera. A camera, either still or video, that uses a *CCD image sensor* to convert a light image into electrical signals, which can be recorded or transmitted.

CCD image sensor. A type of electronic image *sensor* commonly used in video cameras and digital still cameras. Consists of an array of light-sensitive pixels that accumulate *charge* at a rate dependent on light intensity, then transfer that charge via a charge coupled device (hence "CCD") to circuitry that processes the information into a standard still image or video frame format. Compare *CMOS image sensor*.

CCD video camera. See *CCD camera*.

Center of buoyancy (CB). A single point in or near each partially or completely submerged object that is effectively the *center of gravity* of the *mass* of *liquid* displaced by the object. Useful as a concept, because many physics and engineering problems involving the *buoyant forces* acting on complex, three-dimensional objects (like *ROVs*) can be solved more easily by assuming that all of the buoyant forces acting on the individual parts of the object can be treated as if they all acted together at the *center of buoyancy*.

Center of gravity (CG). A single point in or near each object that is effectively the average location of that object's *mass*. Useful as a concept, because many physics and engineering problems involving the movement of complex, three-dimensional objects (like *ROVs*) can be solved more easily by treating all of the object's mass and *weight* as if it were concentrated at the object's *center of gravity*, rather than being distributed (as it really is) in a complex way among the various parts of the object. Also called the *center of mass*.

Center of mass. See *center of gravity*.

Ceramics. A group of materials that is made by using intense heat to partially melt and fuse clay particles or similar non-organic, non-metalic particles. Includes ancient forms such as pottery, and newer forms which are stronger and less brittle and have been used successfully in some of the most demanding *ROV* and *AUV* applications to date, including *pressure canisters* used at the deepest part of the ocean.

CG. See *center of gravity*.

Charge. A fundamental property of matter assigned to certain subatomic particles by humans as a way of helping to explain a group of observed phenomenon loosely classified under the heading "electricity." Charge comes in two forms, denoted positive (+), which is associated with the protons in an atom's nucleus, and negative (-), which is associated with electrons that orbit the atom's nucleus. See also *ion*, *voltage* and *electric current*.

Chassis ground. An electrical connection to the *metal* box or "chassis" surrounding an electrical *circuit*, as in a power tool or appliance. Normally, for safety reasons, the chassis ground is connected to an *earth ground*. See also *ground* for other types of electrical ground.

Chemosynthesis. A biochemical process whereby *energy* is extracted from inorganic molecules by certain microorganisms and used to make biologically important molecules and grow new cells. Ecologically analogous to *photosynthesis*, but possible even in places where there is no light. See also *chemosynthetic community*.

Chemosynthetic community. A rare type of biological community that derives its *energy* for life from *chemosynthesis* by special microbes, rather than from *photosynthesis* by plants or algae. The most famous of these communities are associated with deep sea *hydrothermal vents*. See also *chemosynthesis*.

Chip. In electronics, an informal term for an "*integrated circuit*." See *integrated circuit*.

Circuit. A closed loop (or part of it) through which *electric current* can flow. Typically includes a *voltage* source (such as a battery), a *load* (such as a light bulb or motor), and *switches* (to turn the load OFF and ON).

Circuit breaker. Similar to a conventional *fuse* in purpose; however, it is more like a *switch* that automatically flips to the OFF position in the event of a *current* overload. Once the *short circuit* is remedied, the circuit breaker can be switched back to the ON position without needing to be replaced. See also *fuse* and *resettable fuse*.

Circular search pattern. See *arc search pattern*.

Clearing the ears. See *Vansalva maneuver*.

Clock pulse. One pulse in a chain of pulses used as a *clock signal*.

Clock signal. A series of *clock pulses* used to coordinate the timing of data transmission (or other events) between different parts of an electrical *circuit*, or between separate electrical devices that need to coordinate their activities.

Closed-loop control. A form of *control* in which the *control system* receives *feedback* telling it the actual condition or value of the quantity being controlled. The *control system* can use this information to adjust its control efforts for a better match between the actual condition and the desired condition. Because the feedback provides the ability to detect and correct errors, *closed-loop control* is generally more effective than *open-loop control*, though it is more complicated to implement.

Closed-loop control system. A *control system* that uses *feedback* for *closed-loop control*. Also known as a *feedback control system*.

CMOS chip. An *integrated circuit* manufactured using complementary *metal* oxide silicon technology.

CMOS image sensor. A type of electronic image *sensor* commonly used in video cameras and digital still cameras. Consists of an array of light-sensitive pixels that accumulate *charge* at a rate dependent on light intensity, then process that information using *semiconductor* circuitry built into each pixel to create an image. The name (CMOS) derives from the type of semiconductor manufacturing process involved, which is called complementary *metal* oxide silicon (CMOS). Compare *CCD image sensor*.

Coaxial cable. A type of 2-*conductor* electrical *cable* designed for the transmission of radio-frequency or other high-frequency electrical *signals* (but also used for other purposes). Consists of a central *metal* conductor surrounded by a special *insulating* layer, which in turn is surrounded by a sheath of braided metal or foil used as electrical shielding to isolate the signal on the central conductor from things outside the *cable*. The shield is usually covered by its own layer of *insulation*. Comes in various sizes.

Coil. In a *relay*, the *electromagnet* used to move the electrical contacts.

Cold War. A period in world history lasting from the 1950's to the 1990's characterized by deep mistrust and constant preparation for war (but no actual fighting) between dominant world powers of the time, notably the United States (USA) and the Soviet Union (USSR). The Cold War motivated the development of many undersea technologies, including nuclear *submarines*, *sonar*, *ROVs*, and other sophisticated *navigational* technologies like *GPS*.

Commercial-off-the-shelf (COTS) parts. Parts or systems you can buy already made through conventional retail stores or on-line vendors, rather than having to make them yourself. Use of COTS parts simplifies and expedites the construction process and sometimes reduces the overall project cost.

Communication protocol. In electronics, an agreed-upon language or code that electronic devices can use to encode messages for transmission through some medium (e.g., light, *sound*, radio *waves*, electricity).

Commutator. A mechanism in (or associated with) a *DC motor* that switches the direction of *electrical current* through the motor's *electromagnets* at precisely the right times to produce rotation of the motor. See also *brushed direct current (BDC) motor* and *brushless direct current (BLDC) motor*.

Compass. A navigational device used to determine direction. For various types, see *fluxgate compass*, *gimbaled compass*, *gyro compass*, *magnetic compass*.

Compass north. The direction in which the *compass* needle (or equivalent) is actually pointing. This is ideally the same as *magnetic north*, but it may differ if pieces of *iron*, *steel*, or other ferrous *metals*, other magnets, or *electrical currents* (which generate magnetic fields) are nearby.

Composites. A class of materials consisting of strands or particles of one solid substance embedded inside another solid substance, providing added *strength* and toughness.

Compressibility. The ease with which an object or substance can be squeezed into a smaller volume.

Concept design. The end result of the planning phase in designing an underwater vehicle. Such designs include detailed drawings, *functional specs*, and often a model of the vehicle.

Concept drawing. During the final stages of the design phase, a fairly specific and realistic sketch of the proposed vehicle, indicating its anticipated size, shape, and materials, as well as the arrangement and type(s) of cameras, lights, *thrusters*, *payloads*, etc. Once approved, it will form the basis for more detailed *fabrication drawings*.

Conductivity, Temperature, and Depth (CTD) sensor. A commonly used oceanographic instrument that records the water's *electrical conductivity* and *temperature* at various *depths*. *Pressure sensor* readings are usually used to determine the depths at which each of the other readings is taken. The temperature and *conductivity* readings can be combined to calculate the *salinity*. Thus, *CTD* data can be used to calculate and plot how temperature and salinity vary with depth. Also known as a *CTD profiler*.

Conductivity. See *electrical conductivity* and *thermal conductivity*.

Conductor. In an electrical context, a material or substance, such as *metal*, that has high *conductivity* (i.e., it offers low *resistance*) and therefore conducts electricity easily. Also an object, such as a *wire*, made from such a material that is used to carry electricity from one location to another.

Conductor whip. An electrical *wire* in a waterproof casing so it can be used to make connections outside *pressure housings*.

Conical port. A thick *glass* or *plastic viewport* resembling a cone in shape and designed for use in deep diving, where the slanted sides of the port help support the tremendous *forces* pressing inward upon the viewport and help to seal the port against leaks.

Connector. See *electrical connector*.

Constraints. In the vehicle design process, factors such as a limited budget, lack of access to tools, the laws of physics, or other conditions that limit the design options you can realistically consider.

Control. In the context of underwater "*navigation* and *control*," the word "control" refers to the regulation of vehicle speed, direction, cameras, and other systems to accomplish the *mission*.

Control algorithm. An *algorithm* used by a *control system* to accomplish its *control* task.

Control room. The room, often portable, from which a work class *ROV* or other large ROV is controlled. The control room includes a *pilot's* console, monitors, and other equipment and is often built inside a modified shipping container or trailer for easier transport between ships or other work *platforms*. Also known as a *control van* or *control shack*.

Control shack. See *control room*.

Control system. A *system* designed and used to *control* some machine or process. The generic term "control system" usually refers to a *closed-loop control system*, though the term technically includes *open-loop control systems* as well.

Control van. See *control room*.

Corrosion. An electrochemical process, usually accelerated by moisture, in which chemical reactions gradually eat away a material, typically *metal*. Also the discoloration or structural damage resulting from that process.

COTS. See *commercial-off-the-shelf parts*.

Cotter pin. A small, inexpensive, *metal* rod or *pin*, typically doubled back on itself, that is designed to be inserted through a hole drilled sideways through a *shaft*. Used to keep something such as a wheel *hub* or *propeller hub* from sliding along the shaft or rotating relative to the shaft.

Coulomb. A standard quantity of electrical *charge*. Equal to the amount of charge present on approximately 6.25×10^{18} electrons.

Couple. In physics and engineering, a pair of equally strong *forces* that act in opposite directions along different, but parallel lines to produce *torque*, but no net *force*.

C-rate. A common way of specifying the rate at which a *battery* is being charged or discharged. Defines "C" as the number of *amps* that would drain the battery in one hour, according to the stated *amp-hour energy* capacity rating of the battery, then expresses *charge* or discharge *currents* as a multiple or fraction of that C. For example, if a battery has an energy capacity rating of 10 Ah, then C = 10 amps, so a current of 10 amps flowing out of the battery would be described as a 1C discharge rate. A current of 30 amps flowing out of the same battery would be considered a 3C discharge rate, and a current of 2 amps into the battery would be a C/5 charging rate. See also *C-rating*.

C-rating. A *battery* performance parameter that specifies, in terms of a *C-rate*, the maximum *current* the battery can safely deliver for an extended period of time.

Crazing. A network of tiny cracks that forms in *acrylic plastic* and some other materials over time.

Cryobot. Generally, a descriptive term that could be applied to any robot designed to operate in extremely cold conditions. More specifically, the name used for robotic probes designed to melt their way down through thick ice to reach liquid water in lakes or polar seas below. See also *hydrobot*.

CTD. See *conductivity, temperature, and depth sensor*.

CTD profiler. See *conductivity, temperature, and depth sensor*.

Current. Two definitions are used commonly in this text: 1) Water current: A more-or-less steady flow of water moving in a particular direction. 2) *Electrical current*: A flow of electrically *charged* particles. See also *ampere*.

Current source. An electric *power* source that attempts to provide a predictable *electric current*, while allowing the *load* to determine the *voltage* required to make that amount of *current* flow. (Current sources are not common in day-to-day experience, except that they are often hidden within specialized *circuits*.) Compare *voltage source*.

Cursor system. See *capture frame*.

D

Datasheet. A document, ranging in size from one page to hundreds (usually just a few) that summarize the technical specifications and typical uses of a technical product, such as an electrical *circuit* component.

Davit. A small, crane-like mechanism that can reach over the side of a dock or vessel to lift moderately heavy objects in and out of the water. Commonly consists of a rigid boom that can swing in azimuth and a hand-cranked or small powered *winch* used in conjunction with a block (set of pulleys) to lift loads of less than 2.2 kN (500 lbs).

DC. See *direct current*.

DC motor. A type of electric motor powered with *direct current*.

DC-to-DC converter. An alternative to a *voltage regulator*, these *circuit* assemblies usually include a *switching regulator* and associated support circuitry. As a result, they are easier to use, but they cost more than typical voltage regulators. Unlike most voltage regulators, some DC-to-DC converters can produce higher output *voltages* than the input voltage supplied to them by the *battery*.

DCS. See *decompression sickness*.

Dead battery. A *battery* that has exhausted most or all of the *energy* stored in it and is no longer able to deliver the *voltage* and/or *current* needed. If a *secondary battery*, it can usually be recharged and reused. If a *primary battery*, it needs to be disposed of properly.

Dead reckoning. A form of *navigation* that consists of starting at a known location, then heading away from that location in a particular direction at a particular speed for a particular amount of time. Simple, but prone to error, especially in the presence of cross winds or *currents*.

Decimal. In computer science and related fields, including robotics, a reference to the base-10 number system, which is familiar and easy for most humans to use, but which is not particularly well suited to digital computing. Compare *binary*, *octal*, and *hexadecimal*.

Deck cable. A *cable* running between the *control room* and the *winch* aboard the ship or other *platform* from which an *ROV* is being operated. This cable contains all the conductors and optical fibers that go to the ROV via the *umbilical/tether*. It may also contain additional conductors for voice and video communication between the *winch operator* and the control room. See also *umbilical, tether*.

Declination. The difference between *true north* and *magnetic north*. Varies with location on Earth. Usually specified in degrees and sometimes called *variation*.

Decompression sickness (DCS). A potentially life-threatening medical condition that occurs when small gas bubbles form in the tissues of a person as a result of a rapid drop in external *pressure*. In diving, this can occur when a diver who has been breathing pressurized air at *depth* ascends too quickly. Its effects range from a mild rash or joint pain to severe paralysis and death, depending on the size, number, and location of the bubbles. Also known as "*the bends*" or "*caisson disease*."

Decompression stop. A period of time during which a diver returning to the surface from a deep dive stops for a specified number of minutes at a particular *depth* to avoid *decompression sickness*. These stops allow time for dissolved gasses to diffuse out of the diver's tissues without forming dangerous bubbles. Deep dives often require a number of decompression stops for several minutes, each at progressively shallower depths.

Deep-cycle battery. A *secondary* (rechargeable) *battery* that is designed to withstand a large *depth of discharge* between chargings without damage.

Degrees of freedom (DOF). In a mechanical *system*, such as a robotic arm, the number of independent hinged joints, rotations, linear extensions, or other separate opportunities for controlled movement. Example: in an *ROV manipulator*, wrist rotation would be one DOF, rotation about the elbow joint would be another DOF, etc. In manipulator arms, the number of DOFs is sometimes described as the number of *functions*.

Demobilization. The final phase of an underwater *operation*. Includes returning to shore, cleaning, inventorying, organizing, and storing all equipment and supplies; discussing what worked well, what didn't, and what to do next time for an even more successful *mission*; and delivering any recovered items, data, or reports to the client.

Density. The amount of *mass* per unit volume of a substance or object. Density is a more useful measure of "heaviness" than *weight* when working in water, because density provides a direct indication of whether an object (or substance) will float or sink. Example: A 1000 *pound* log floats while a 1 pound rock sinks; this happens even though the log is much heavier than the rock, because the log has lower density.

Depth. Distance beneath the surface (of a body of water).

Depth gauge. A device carried by a diver or underwater vehicle for measuring and displaying the *depth* of that diver or vehicle. Usually works by measuring the *hydrostatic pressure*, but displaying that *pressure* in terms of an equivalent depth; therefore, accurate depth readings require that the gauge has been calibrated for the *density* of the water in which it is being used.

Depth of discharge. The extent to which a *battery* has been discharged, commonly expressed as a fraction of its total *energy* capacity.

Depth of field (DOF). In *camera* systems, the range of distances, near to far, that is in sharp focus.

Depth sensor. A *sensor* for measuring and reporting *depth*. Most depth sensors are actually *pressure sensors* that have simply had their output calibrated to display the depth at which the measured *pressure* is found.

Derrick crane. A type of crane consisting lifting ropes and pulleys hanging from the end of a long boom that can rotate about a vertical axis to lift and swing a load.

Design methodology. A series of design stages and procedures to plan, define, design, build, test, and implement the specific systems in designing and building an *ROV* or *AUV*.

Design spiral. One helpful strategy for dealing with the challenge of designing a complex system (such as an *ROV* or *AUV*) in which a design change in any one *subsystem* tends to impact the design of many (or all) of the other *subsystems*.

Design trade-offs. Compromises that must be made during the design process, because optimizing one feature inevitably has a negative impact on other features.

Dial pressure gauge. A familiar style of *pressure gauge* with a round dial and a needle that rotates to point to a number or mark showing the *pressure*. Commonly used for measuring the pressure of air in tires and of liquids or gasses in *pipes* or tanks.

Digital camera. Any *camera* that records or transmits its image or video data in a *digital* format.

Digital ground. A *signal ground* used for digital *circuits*. Normally identified as digital ground only in circuits that also contain a separate *analog ground*. In such circuits, the distinction is important, because digital circuits tend to produce lots of *voltage* spikes on their *ground* and *power lines*. Although these spikes don't usually bother the digital circuitry, they can create large errors in the measurement of tiny *analog signals*.

Digital sensor. A *sensor* with an output that is constrained to take on a finite number of discrete values within its range of possible values.

Digital signal. A *signal* that is limited to a small number of distinct values—usually only two—at any one instant.

Digital video camera. Any video *camera* that stores or transmits its video image in one or more *digital video formats*.

Digital video format. Any *video format* that relies on digital data storage or transmission. Common examples include *NTSC, PAL, S-video*, and many others.

Diode. A two-terminal *semiconductor* device often used as a one-way gate or valve for *electric current* flow in electric *circuits*. See also *light-emitting diode*.

Direct current (DC). A form of electrical *signal* used primarily for short-distance *power* transmission within small *circuits* and characterized by having a *voltage* (and *current*) that remains more-or-less constant in time. Most *battery*-powered circuits are examples of circuits powered by direct current.

Discharge curve. A graphical representation of how a particular type of *battery's voltage* drops as *energy* is drained from the *battery*. The vertical axis of such a graph usually shows the voltage at the battery terminals. The horizontal axis is variously expressed in units of time, Ah or mAh delivered, percentage of total capacity discharged, or other measures of how "used up" the battery is.

Displacement. The volume of water displaced (i.e., pushed out of the way) by a boat or vehicle when placed in water. Also, the *weight* of that amount of water.

Dive planes. Wing-like projections on a *submarine* or other fast-moving underwater craft that can be angled up or down, like a horizontal rudder, to help the vehicle dive or surface.

Diving bell. One of the earliest, reliable means of providing humans with a supply of air to breathe under water. Consists of a heavy church bell or similarly shaped enclosure lowered by rope or chain from a ship or dock. Swimmers could do salvage work on the bottom, periodically sticking their head up inside the bell to take a breath from the pocket of air trapped inside.

DIY. Acronym for "do-it-yourself," which refers to jobs or projects you do yourself at home or school, rather than calling in an experienced professional to do them or purchasing ready-made components.

DOF. Abbreviation for two very different quantities relevant to subsea vehicles: *degree of freedom* and *depth of field*.

Dome port. A *viewport* consisting of a portion of a clear *glass* or *plastic* sphere. This shape has good *pressure* resistance and minimizes optical distortion caused by *refraction*.

Doppler shift. In acoustics, a change in the apparent frequency (*pitch*) of a *sound* source that occurs when an object making or reflecting the sound is moving toward or away from the listener. See also *acoustic Doppler navigation, Doppler velocity log*, and *acoustic Doppler current profiler*.

Double-pole double-throw (DPDT) switch. A six-terminal *switch* with two electrically separate, but mechanically linked switches inside. Each of these switches-within-a-switch typically has one OFF position located between two different ON positions. Used commonly to provide directional *control* for *thrusters* in simple *ROV control systems*. See also *pole* and *throw*.

DPDT switch. See *double-pole double-throw switch*.

Drag. A *force* produced by, and always opposing, movement of an object through a *fluid* (and/or movement of fluid past an object). Composed of two components: *pressure drag* and *viscous drag*, though one typically dominates in any particular application. Under normal circumstances most of an underwater vehicle's propulsive *thrust* is used to overcome drag.

Drag coefficient. A number used in calculating drag that captures the contribution of object shape to the amount of drag. Streamlined objects have lower drag coefficients than boxy or parachute-shaped objects.

Drop weights. Detachable *weights* used as *ballast* in *submersibles*. May be used to pull the vehicle gently to the bottom, then dropped for the return to the surface as a *power* saving strategy, or may be used only in emergencies to gain immediate *positive buoyancy*.

Dual manipulators. *Manipulators* that are paired, two per vehicle, so they can assist each other much like the left and right hand of a human can work together to unscrew the lid on a jar or perform other tasks.

Ducted propeller. A *propeller* mounted inside a short section of open cylinder, which is used to protect the propeller from damage, protect people from injury, and/or to increase the *thrust* and *efficiency* of the propeller.

Ductile materials. Materials like putty or copper which, once deformed, will stay deformed. Compare *elastic materials*.

Duty cycle. In a series of electric pulses, such as those used in *pulse width modulation*, the fraction of time spent in the ON (or active) state. Example: something with a duty cycle of 80% fluctuates rapidly between ON and OFF, but spends 80% of its time ON.

DVL. See *Doppler velocity log*.

Dynamic ballast system. A *ballast system* that uses compressed gasses, pistons, or other active methods to make changes in the *displacement* or *density* of a vehicle during a dive, thereby changing the vehicle's *buoyancy* and helping to drive it upward or downward through the water. Also known as an *active ballast system*.

Dynamic braking. A method of slowing and stopping a motor by extracting *energy* much faster than if it were simply allowed to coast to a stop. A *brushed DC motor* can be dynamically braked by disconnecting it from *power*, then *shorting* together its leads while the motor is still spinning. This type of dynamic braking is included as a feature in many *motor controllers*.

Dynamic seal. Any type of seal made between parts that move in relation to each other, such as between a *pressure housing* and a rotating *motor shaft* or a sliding piston shaft. *U-cup seals* are an example of a type of seal often used for dynamic seals. Compare *static seal*.

E

Earth ground. A type of *ground* in an electric *circuit* that is electrically connected via a low *resistance* pathway directly to planet Earth. Commonly used in combination with *circuit breakers, fuses*, or similar devices as part of an electrical safety system designed to reduce the risk of fires and electrocution in the event of a *short circuit*. See also *ground* for other types of electrical ground.

Earthing. The process of connecting part of a *circuit* to *earth ground*.

Echosounder altimeter. See *echosounder*.

Echosounder or **Echo sounder**. A device that uses *sonar* to measure the distance to the seafloor or lake bottom. Used on boats to measure water *depth*, but also used on *ROVs* and *AUVs* to measure the altitude of the vehicle above the bottom.

Eddy. A rotating or spinning *mass* of water.

EEPROM. Acronym for Electrically-Erasable, Programmable, Read-Only Memory. A type of digital memory that retains its data even if *power* is lost, yet allows that data to be erased and overwritten as needed. Also known as *flash memory*.

Effective weight. The "apparent" *weight* of an object immersed in a fluid, which is less than its actual weight by an amount equal to the *buoyant force* pushing upward on that object. The effective weight may be positive, zero, or negative depending on the *density* of the object and the fluid. An object with a negative effective weight floats. Effective weight is also known as *wet-weight*, *submerged weight*, and *in-water weight*, and is exactly the opposite of *buoyancy*.

Efficiency. The amount of energy or power you get out of a machine or process in the form you want (heat, light, mechanical work, etc.) expressed as a fraction of the total energy or power you put into the machine or process.

Elastic materials. Materials such as *rubber* or spring *steel*, which, once deformed, will bounce back to their original shape. Compare *ductile materials*.

Electric current. A flow of electrically *charged* particles. *Current* is defined as flowing in the direction from higher to lower *voltages* within a *circuit*, and therefore corresponds to the direction of movement of positively charged particles, such as sodium *ions* in seawater, and opposite the direction of movement of negatively charged particles, such as the electrons flowing through a *metal wire*. See also *direct current* and *alternating current*.

Electric field. A characteristic of the space surrounding particles or other objects having electric *charge*. Can be used to describe the *strength* and direction of the *force* acting on a charged particle located near other charged particles. In particular, if a positive and negative charge are separated from one another, then a third charge placed in between the first two will be attracted to the one with opposite charge and repelled from the one with identical charge. The strength of an electric field between two points is characterized by the *voltage*.

Electrical conductivity. A measure of how easily *electrical current* can flow through a substance. See *conductivity*.

Electrical connector. A plug or socket used to interconnect *wires*, electrical *cables*, *printed circuit boards*, or other parts of electric *circuits*. Specialized *subsea connectors* are required to keep underwater electrical connections from *shorting* out when wet.

Electrically-erasable, programmable, read-only memory. See *EEPROM*.

Electrolysis. See *electrolytic corrosion*.

Electrolytic corrosion. A *corrosion* process fundamentally similar to *galvanic corrosion*, but accelerated by the application (intentionally or unintentionally) of an external *voltage*.

Electromagnet. A type of magnet that can be turned on or off using electricity. Usually made by wrapping a long *wire* many times around an iron core. Activated by passing *electric current* through the wire. The *windings* in an electric motor and the *coil* in a *relay* are examples of *electromagnets*.

Electronic pressure sensor. A *pressure sensor* with an output in the form of an electronic *signal*, which is designed to be connected to a computer or other electronic system rather than read directly by a human being.

Electronic speed control (ESC). A type of *motor controller*, often used in radio controlled model cars and airplanes. All *ESCs* can *control* the speed of a motor, but some may not be able to control its direction.

Electronics can. See *electronics canister*.

Electronics canister. A *pressure canister* designed specifically to protect electronic circuits from exposure to water at *depth*. Also known as an *electronics can*.

Embedded controller. A *microcontroller* (or much less commonly, a *microprocessor* or computer) built into a machine or device to give it some information processing ability without requiring connection to an external computer. Widely used (much more common than computers), but rarely noticed, because they are "embedded" inside other systems, like kitchen appliances, cars, cell phones, electronic toys, etc. Often used to *control ROVs*, *AUVs*, and other robotic vehicles.

Emergent property. A new characteristic or capability that results from the interactions among the parts of a *system* and does not exist without those interactions.

Encapsulation. See *potting*.

End effector. The *gripper*, tool, or other device attached at the end of a *manipulator* arm or other robotic arm. Performs the work after the arm has moved it into the correct position and orientation.

Endcap. The cap or lid used to close and seal the end of a cylindrical *pressure canister*.

Energy. A very important quantity in physics and engineering that essentially provides a measure of the amount of work that a system is capable of doing. Energy can exist in many different forms including (but not limited to): heat, light, chemical fuels, mechanical movement, electrical energy, and even *mass*. The *SI* unit of energy is the *joule*. Compare *power*, which is not the same as energy.

Energy capacity. A *battery* performance specification that gives the total amount of useful *energy* the battery can store. Sometimes expressed in legitimate energy units, such as *joules* or *watt-hours*, but more often expressed indirectly in terms of *amp-hours (Ah)*, a measure of *charge* storage capacity. For true energy units, multiply amp-hours by battery *voltage* to get watt-hours.

Energy density. A *battery* performance specification that indicates how much *energy* can be stored in a battery of a particular size. Useful in determining how large a battery you need for your application. There are two common versions: *gravimetric energy density* and *volumetric energy density*.

Entanglement. A threat to underwater vehicles caused by becoming caught in kelp or seagrass, branches or roots of submerged trees, abandoned nets, or other debris.

Epoxy. A type of *plastic* material sold as a pair of two separate liquids (a resin and a hardener), which are later mixed together as needed to form a glue or *potting compound*. Upon mixing, the hardener facilitates a chemical reaction that cross-links the molecules in the resin and turns the liquid mixture into solid *plastic*.

Ethernet communication protocol. An extremely popular serial *communication protocol* used to transmit information over computer networks. Most computers, network printers, *IP cameras*, other such devices connect to each other and to the Internet using the Ethernet communication protocol, including its standard cables and other hardware.

Euphotic zone. The upper layer in a water body, where light penetrates far enough to support significant *photosynthetic* activity.

Europa. One of Jupiter's moons. Surface features are consistent with the possibility that the moon may have a deep, ice-covered ocean of liquid which could, in principle, support extraterrestrial life. As a result, there are a number of proposals pending to send space probes with *AUVs* to explore the ocean beneath Europa's icy surface.

Eustachian tube. A small passageway leading from the back of the throat to the middle ear. Important in *pressure* equalization across the eardrum. See *Vansalva maneuver.*

Extrusion. A long piece of usually straight material manufactured by forcing melted *aluminum*, *thermoplastic*, or similar substances through a specially shaped opening to produce a long rod-like *structure* with a cross-sectional shape determined by the shape of the opening.

Eyeball ROV. See *Observation class ROV.*

F

Fabrication. The actual construction phase of vehicle development, which follows after completion of the detail design and *procurement* of parts and materials.

Fabrication drawings. Detailed drawings that show how each custom part of the vehicle is to be constructed.

Face seal. A type of *endcap* seal. Also called a *flange seal* because it employs a flat *flange*, glued or welded to the end of the *pressure can*. An *O-ring* or *rubber gasket* is then sandwiched between this flange and the *endcap*, which is usually held in place with retaining bolts and water *pressure*.

Fairing. A skin or shell that covers all or part of the *frame* of many (but not all) vehicles to give them a smoother, more streamlined shape for better speed and *energy efficiency*.

Fathometer. A particular trademarked type of *echosounder*.

FEA. See *finite element analysis*.

Feedback. In the context of *control systems*, feedback refers to the return of information (about the effectiveness of the *control* efforts) back to the part of the control system that adjusts those efforts to achieve the best possible control. For example in a control system designed to regulate the speed of a motor, feedback information about the actual motor speed as measured by a speed *sensor* might be returned to the *control circuit*, so that control circuit could fine-tune the motor speed to match the desired value. It's called feedback, because this information is "fed back" into the decision-making process.

Feedback control system. See *closed-loop control system*.

FET. Abbreviation for *field effect transistor*.

Fiber-optic communication cable. A *cable* containing specially prepared *glass* fibers (known as "fiber-optics" or "*optical fibers*") through which pulses of light can be transmitted over distances of several kilometers to send information from one end of the cable to the other. A typical *ROV umbilical* or *tether* will include copper *wires* for *power* transmission and optical fibers for data transmission. Hybrid vehicles operating in ROV mode do not need copper wires in the tether, since they have on-board power and may use an extremely thin and flexible cable containing only optical fibers.

Field effect transistor (FET). A type of *transistor* that requires very little input *current* to *control* the output current. Popular in *battery*-operated devices. See also *metal oxide semiconductor field effect transistor.*

Fill port. A small opening in a *pressure canister* designed for filling the *canister* with nitrogen, oil, or other liquids/gasses.

Finite Element Analysis. An advanced computer simulation technique that represents a (usually solid) object as if it were composed of a large number of tiny pieces, then analyzes mechanical forces, heat flow, or other interactions between neighboring pieces to determine how the forces will deform the object or the heat flow will change the temperature of various regions of the object. Generally used to identify the portion of the object that is under the greatest thermal or mechanical stress and is therefore most likely to fail, so that the object can be reinforced or cooled in appropriate places to prevent failure.

Fixed-bladed prop. *Props* that have a constant *pitch* that cannot be changed, so in order to change vessel speed, the rotational speed of the *propeller shaft* must be changed; to reverse direction, the propeller's direction of rotation must be reversed. Compare *variable-pitch prop.*

Flange. A circular piece of thick *plate*, welded around a *hatch* or other hull opening to structurally reinforce the area around the hole and sometimes to provide a smooth surface for sealing. Essentially a large *boss*.

Flange seal. See *face seal.*

Flash memory. See *EEPROM.*

Flat port. A simple type of *viewport* made from a thick, flat *plate* of *glass* or *plastic*. See also *conical port* and *dome port.*

Fluid. A liquid or gas.

Fluid density. The density of a fluid.

Fluid pressure. The *pressure* exerted by a *fluid*.

Fluxgate compass. A type of electronic *compass* that uses *electric currents* flowing through two or more coils of *wire*, instead of a traditional magnetized compass needle, to measure the direction of the earth's magnetic field. Comparatively easy to interface with electronic systems for digital display or transmission of compass data, or for use in automated *navigation* systems.

Flying eyeball. See *Observation class ROV.*

Foot-pound (ft-lb). An imperial unit used for both *energy* and *torque*. As a unit of energy, one ft-lb is the amount of energy it takes to move something a distance of one foot while applying a force of one pound in the direction of motion. This is equivalent to about 1.356 *joules* of energy. As a unit of torque, one ft-lb is the torque generated when a force of one pound is applied at a perpendicular distance of one foot from a pivot point. In this context, it is equivalent to 1.356 newton-meters of torque.

Force. In physics and engineering, a physical push or pull that can be measured and described completely in terms of two attributes: 1) a *magnitude* (*i.e.*, how strong the push or pull is) and 2) a direction. See also *newton (N)* and *pound force (lbf)*.

Force feedback. A method of providing a sense of touch to the human operator of a robotic *gripper*, *manipulator* arm, or similar mechanism, so the operator can "feel" how hard she is gripping, pushing, or pulling on something with the remote robotic arm. Requires that *force sensors* on the gripper somehow regulate the amount of mechanical *resistance* in the operator's controls.

Fractional horsepower motor. An electric motor with a maximum *power* output of less than one *horsepower* (746 *watts*). This is the category of motor used most commonly on small *ROVs* and *AUVs*.

Frame. In an underwater vehicle's *structure*, the frame (also known as the *framework*) provides the primary skeleton of the vehicle, usually made of interconnected beams, struts, *plates*, or other load-bearing members. Defines vehicle's overall shape and provides mechanical support and attachment points for other vehicle components.

Framework. See *frame*.

Free surface effect. A potentially dangerous loss of vessel trim or stability that can occur if heavy weights or liquids shift position when the vessel leans or tips in rough seas.

Frequency division multiplexing. A *signal multiplexing* strategy wherein each signal is transmitted on a different frequency. The receiver monitors each frequency separately to sort out the signals. Example: different radio stations can share the airwaves by each using a different band of frequencies.

Frontal area. Used in theoretical calculations of drag, frontal area is the projected area of an object presented to flow. If you take a photograph of a vehicle from directly in front of it (based on the direction its moving relative to the water), blow that photo up to life-size, draw an outline around the vehicle in the photo, and measure the surface area inside that outline, that area is the vehicle's frontal area.

Fuel cell. An emerging technology that generates electricity by "burning" pure hydrogen with oxygen to make water in a controlled reaction.

Function. Many definitions. For the one used as a *manipulator* arm specification, see *degrees of freedom*.

Functional specifications. During the design process, a "boiled down" summary of the anticipated performance requirements for your vehicle. After construction and testing, a similar summary of the actual measured performance capabilities of your vehicle.

Fuse. A device designed to protect *circuits* and people from some of the hazards associated with *short circuits*, including fires and electrocution. Typically consists of a thin *metal wire* or strip designed to melt in a safe, controlled manner, at a predetermined *current* level. If *power* for the circuit is routed through the fuse, excessive current caused by a short circuit will cause the fuse to melt, thereby cutting off electrical power to the circuit. See also *circuit breaker* and *resettable fuse*.

G

Galvanic cell. The basic building block of a *battery*. Typically consists of two electrodes made of different *metals* with some corrosive liquid, paste, or gel between them to promote *galvanic corrosion*. The more electronegative electrode (i.e., the one made from the metal least willing to give up electrons) will try to steal electrons from the less electronegative one, resulting in a *voltage* difference between the electrodes. If a complete electrical *circuit* is made so that electrons can flow from one electrode of the cell, through the *load* (motor, light, or other device being powered), and back to the other electrode of the cell, then *corrosion* will proceed inside the battery and provide electrical *power* for the external load.

Galvanic corrosion. A spontaneous form of *corrosion* that occurs when two dissimilar *metals* touch (or otherwise become electrically connected) in the presence of moisture. This usually causes corrosion of one of the two metals but not the other.

Garage tether management system. One type of *tether management system*. See the *tether management system* entry for details.

Gas compensation. The practice of controlling the amount of air, nitrogen, or other pressurized gas inside an undersea *canister* or *housing* so that the internal gas *pressure* closely matches the *ambient pressure* of water outside the canister at all times. This eliminates any significant *pressure differential*, so it reduces the likelihood and severity of leaks and protects the canister from being crushed by *hydrostatic pressure*; however, it requires a complex system of compressed gas cylinders and pressure regulating valves. It also carries the risk of possible explosion, if the valve system fails to release internal pressure properly during ascent. *Oil compensation* is a simpler and safer alternative that can be used in most cases.

Gas-compensated housing. An undersea *canister* or *housing* protected from *hydrostatic pressure* and leaks by *gas compensation*.

Gasket. A usually thin, flat layer of *rubber* or other material compressed tightly between two flat surfaces to form a *pressure-proof* seal that prevents water or other fluids from leaking through the gap between the surfaces.

Gauge pressure. Formally, *pressure* measured relative to a standard 1 *atm pressure* reference. In practice, pressure measurements made with a gauge that uses *ambient pressure* as the reference pressure. On land, this ambient pressure is usually, but not always, about 1 *atmosphere* of pressure. See also *absolute pressure*.

Gear motor. An electric motor of any type with a set of gears attached to the *motor shaft*. The gears, often contained in a *gearbox*, are usually used to produce greater *torque* (twisting *force*) and/or lower rotational speed of the output shaft.

Gear ratio. An important feature of a *gearbox* or similar set of gears that quantifies how many revolutions the *motor shaft* (or other rotational *energy* input) must complete to produce one full revolution of the *shaft* coming out of the gearbox. Example: A gear ratio of 4:1 could be used to spin a *propeller* at ¼ the speed of the motor shaft, but with about 4 times as much *torque*.

Gearbox. A set of interacting gears contained within some supporting box. Usually connected to a motor or engine to increase *torque* at the expense of rotational speed.

Gel cell. A type of *valve-regulated lead-acid battery* in which the electrolyte is rendered non-spillable by converting it to a gel instead of a liquid.

GFCI. See *ground-fault interrupter*.

GFI. See *ground-fault interrupter*.

Gimballed compass. A *compass* that is specially mounted so that it remains level even on a rocking boat, banking airplane, or other moving vessel.

Glass. A material made by melting sand or other inorganic, non-metallic substances, then allowing the melt to cool and solidify without forming crystals. A familiar material for windows and lenses. Also used in *glass-reinforced plastics*, *syntactic foam*, and other products useful for subsea vehicle design.

Glass sphere. A type of *pressure housing* made by joining two thick, strong, hemispherical *glass* domes to produce a hollow, *glass*, pressure-resistant sphere with enough space inside for electronics or other vehicle systems that need to be kept dry.

Glass-reinforced plastic (GRP). A *composite* material made by embedding glass fibers in *epoxy* or similar *plastics*.

Glider. In undersea technology, an extremely *energy*-efficient type of *AUV* that uses wing-like *structures* to travel great distances by gliding, much as an airplane glider can cover vast distances without an engine. However, unlike an airplane glider, undersea gliders can switch from gliding downhill to gliding uphill by changing their *buoyancy* and can use this trick to cover vast distances in a vertical ziz-zag, or sawtooth, pattern.

Global Positioning System (GPS). A global *navigation* system originally developed for military use, but now widely used for both military and civilian navigational purposes. Consists of a system of 24 earth-orbiting *GPS satellites* that transmit *navigation signals* to portable *GPS receivers* on or near earth's surface. *GPS* positions are routinely accurate to within a few *meters* and with special techniques can be accurate to within less than 1 cm. GPS is frequently used for navigational purposes on land, on water, and in the air, as well as for survey work.

GPS. See *global positioning system*.

GPS receiver. A usually portable, electronic, *navigational* device that reports its precise position anywhere on (or above) the earth's surface based on *signals* it receives from earth-orbiting *GPS satellites*. Can be used for personal, vehicle, or robot navigation. See also *global positioning system*.

GPS satellite. One of about 24 earth-orbiting satellites that provide highly-precise *navigational signals* to *GPS receivers* on or near the earth's surface as part of the *global positioning system*.

Gradiometer. An instrument used to measure the rate of change in some physical quantity over distance. Gradiometers that measure the rate of change in magnetic fields by comparing *magnetometer* readings from different locations are sometimes used to detect undersea pipelines, *cables*, or shipwrecks.

Grams force (gf). The *weight* of a one-gram *mass* in Earth's *gravity*, a useful unit for figuring weight and *displacement* during small vehicle projects.

Gravimetric energy density. A *battery's energy capacity* divided by its *mass*. See also *energy density* and compare *volumetric energy density*.

Gravitational acceleration (g). The *weight* of an object divided by its *mass*. The value obtained is independent of the size or mass of the object, but does depend on the local *strength* of *gravity*. On Earth, the value of gravitational acceleration is approximately 9.8 m/s^2 (32 $feet/s^2$). This corresponds to the acceleration with which the object's weight would cause it to fall toward Earth in a vacuum.

Gravitational force. A *force* associated with *gravity*.

Gravity. An observed property of our universe whereby a *force* of attraction exists between any two objects that have *mass*. This *gravitational force* is quantified by *Newton's Law of Universal Gravitation* and is usually too small to notice unless at least one of the objects is the size of a planet or moon. The gravitational attraction between Earth and an object located on or near the surface of Earth causes the force commonly called the object's *weight*. Thus, *gravity* is also indirectly the cause of *hydrostatic pressure* and *buoyancy*, both of which are caused by the weight of water.

Gripper. A pincer-like pair of mechanical jaws that forms a simple, yet versatile tool. Can be mounted directly on the front of an *ROV*, or mounted on the end of a robotic *manipulator* arm, as an *end effector*.

Ground. In an electrical *circuit*, generally the return path for *electrical current* to the *power source* and/or a reference point against which to measure *voltages* present elsewhere in the circuit. See also *reference ground*, *signal ground*, *analog ground*, *digital ground*, *chassis ground*, and *earth ground*.

Ground bus. A large, conveniently located, and often conspicuous ground *wire* or other grounded conductor to which other parts of *circuit* that need to be grounded can be connected easily.

Ground fault. A potentially fatal situation that occurs when *electrical current* finds an "unauthorized" path to *ground*, possibly through a person's body. To protect against the danger of ground faults, you should always use a *GFI* when working with outlets, extension cords, *inverters*, or other high *voltage* electricity sources near water.

Ground-fault circuit interrupter (GFCI). See *ground-fault interrupter*.

Ground-fault interrupter (GFI). A device used in *series* with a standard *AC* electrical *power source*, such as an electrical outlet, to shut off *power* automatically if the *current* returning to the device does not exactly match the current leaving the device, because that inequality suggests a potentially dangerous *ground fault* situation.

Grounding. The process of connecting part of a *circuit* to *ground*.

GRP. See *glass-reinforced plastic*.

Gulf Stream current. A major, powerful, and swift *current* that originates in the Gulf of Mexico, then flows north along the eastern coastlines of the United States and Newfoundland before crossing the Atlantic.

Gyre. A very large *eddy*, typically on the scale of an entire ocean basin.

Gyro compass. A *navigational* instrument that relies on interactions between the earth's rotation and the spin of a motorized gyroscope to align the gyroscope's axis of rotation with the earth's north/south axis. Unlike a conventional *magnetic compass*, a gyro compass finds true north, rather than *magnetic north* and is not sensitive to distortions in the earth's local magnetic field. Gyro compasses are widely used on ships, *submarines*, and deep-diving *submersibles*.

H

H-bridge. An electronic *circuit*, typically consisting of four interconnected *transistor switches*, that can be used to reverse the direction of *current* flow through a *load*. Commonly used in *motor controllers* to reverse the direction of motor rotation.

Halocline. A layer of water in which *salinity* changes rapidly with *depth*.

Handedness (of a propeller). Refers to whether a *propeller* is right handed or left handed. See *right handed propeller* and *left handed propeller* for explanation.

Hard ballast tank. A rigid-walled, pressure-proof chamber used to control buoyancy in a submarine or submersible by changing the vessel's weight, rather than its displacement. Built to resist pressure at depths greater than the operating depth of the vessel, they are usually filled or emptied of water by a high-pressure pump. Compare *soft ballast tank*.

Hard-hat diving helmet. An underwater diving helmet made of copper, *brass*, or *bronze*. The hard hat was sealed to a canvas suit and had an air hose that supplied air to the diver from a surface pump. Used from the mid 1800s until the 1960s, when *scuba* and other new technologies made hard-hat diving obsolete.

Hatch. An opening through which people and gear pass in and out of a *pressure hull*.

HDTV. One of several standard *digital video formats*.

Heat capacity. The amount of heat *energy* that must be added to one unit quantity of substance to increase its *temperature* by one unit of temperature. Example: The heat capacity of water is 1 *calorie* per cubic centimeter per degree centigrade, because it takes 1 calorie of heat energy to increase the temperature of 1 cubic centimeter of water by 1 degree centigrade.

Heat sink. Something cool that draws significant quantities of heat away from something warmer. Often used as a way to keep a heat-generating machine or process from overheating. Examples include *metal* fins attached to electric *circuit* components, the coolant and radiator in a car, and ocean water used to absorb heat from *thrusters* and underwater video lights.

Heave-compensated. This term describes a crane, *A-frame*, or *winch* used in ship-based launch and recovery *operations* that is able to move in a way that compensates for the up-and-down motion (heave) of the ship in large ocean swell. Heave compensated systems help to decouple the ship's motion from the *umbilical/tether*, so that this motion does not yank up and down on the *ROV*.

Hex. In computer science and related fields, usually short for *hexadecimal*.

Hexadecimal. In computer science and related fields, including robotics, a reference to the base-16 number system. Compare *binary*, *octal*, and *decimal*.

Hg. The chemical symbol for mercury. *Barometric pressure*, used for weather forecasting, is often expressed in millimeters of mercury (mm Hg) or inches of mercury (in. Hg), based on how far the barometric pressure could lift a column of mercury.

Horsepower (Hp). An *imperial* unit for *power*, originally based on the amount of *power* that a strong horse could produce; one horsepower is the amount of mechanical power required to lift 33,000 *pounds* one foot in one minute. One Hp = 746W.

Hot stab. Any of several kinds of tools designed to make or use connections between pressurized subsea *hydraulic systems*. Example: a hot stab tool may be used to transfer *hydraulic fluid* under *pressure* from an *ROV's hydraulic power unit* to hydraulically controlled equipment on the seafloor.

Housing. See *pressure canister*.

HOV. See *manned vehicle*.

HPU. See *hydraulic power unit*.

HROV. See *hybrid underwater vehicle*.

Hub. On a *propeller*, the central solid disc or cone that fits over the end of the *propeller shaft* and supports the *propeller blades*, transferring rotational *energy* from the *shaft* to the *blades*, and transferring *thrust forces* from the blades back to the shaft to propel the vehicle.

Hubless propeller. An innovative style of *propeller* in which the *propeller blades* point inward from the inside of a rotating cylinder rather than pointing outward from a *hub*. Based on an unusual *BLDC motor* design in which the rotating cylinder is essentially the *rotor* in a BLDC motor. Unlike a regular propeller, a hubless propeller does not easily get tangled in ropes, kelp, or nets.

Human-occupied vehicle (HOV). See *manned vehicle*.

Hybrid Underwater Vehicle (HUV). This term usually refers to a vehicle that can be configured to operate either as a *tethered ROV* with direct *pilot control* or as an untethered *AUV* that operates without a human pilot. Alternative acronyms include *ARV* (AUV+ROV) and *HROV* (hybrid ROV). These terms occasionally refer to other unusual modifications or combinations of ROVs and AUVs, such as groups of AUVs launched and recovered by an ROV.

Hydraulic. See *hydraulic system*.

Hydraulic actuator. Any mechanism, such as a *hydraulic cylinder* or hydraulic motor, that converts *power* transmitted to it (by the flow of pressurized hydraulic fluid) into mechanical work by moving while applying *force* or *torque*.

Hydraulic cylinder. A common type of *hydraulic actuator* consisting of a moveable piston inside a cylinder. Pressurized *fluid* flow delivered to one side of the piston moves the piston, which can be linked to other mechanisms to move them with tremendous *force*.

Hydraulic fluid. A *fluid* used to transmit *power* in a *hydraulic system*. Usually a form of oil.

Hydraulic power system. See *hydraulic system*.

Hydraulic power transfer. Transmission of mechanical power from one location to another through the movement of pressurized liquid in a *hydraulic system*. Based on *Pascal's Law*.

Hydraulic power unit (HPU). A complete *system* for providing a steady, pressurized flow of hydraulic fluid, including a *prime mover*, *accumulator*, cooling system, and attachment points for hydraulic *lines*.

Hydraulic pump. A pump used to move and pressurize *hydraulic fluid* in a *hydraulic system*.

Hydraulic system. A system of pumps, valves, tubing, pistons, or other components for distributing *power* from one place to another by moving liquid under *pressure* to transmit force and motion at useful speeds. Also known as a *hydraulic power system* or sometimes a *power hydraulic system*.

Hydroacoustics. The study of *sound* is called acoustics, and the sub-specialty that examines sound propagation through water and its potential practical applications, including *sonar*, is called hydroacoustics. See also *sonar*.

Hydrobot. A descriptive term that has been applied to a variety robots and remotely controlled vehicles designed to operate in water, including exploratory *ROVs* or *AUVs* designed to be released into an ice-covered body of water by another robot. See *cryobot*.

Hydrodynamic pressure. *Pressure* caused by movement of water. Compare *hydrostatic pressure*.

Hydrophone. Similar to a microphone, but optimized for picking up underwater *sounds*.

Hydrostatic pressure. *Pressure* applied by (non-moving) water, usually as a consequence of *depth*, and the *weight* of the water pressing down from above.

Hydrothermal vent. A fissure in a planet's surface from which geothermally heated water is emitted, either on land or undersea.

I

I/O pin. Standard abbreviation for "input/output" *pins* on a *microcontroller* or other digital *integrated circuit.* These pins are usually used to send or receive digital data from other devices.

I²C communication protocol. A popular synchronous serial communication protocol used for communication between *microcontrollers* and other *integrated circuits.* It requires only two *wires* (plus *ground*), yet can support communication between multiple devices sharing those two wires. See *Serial data transmission protocols.*

IC. See *integrated circuit.*

ICSP. See *in-circuit serial programming.*

Ideal Gas Law. An algebraic equation (PV = nRT), important in chemistry and physics, that summarizes how ideal (and most real) gasses respond to changes in pressure, volume, number of gas molecules present, and/or absolute temperature.

Impeller. Similar to a *propeller,* but contained within a conduit and optimized to accelerate water through an opening to create a powerful jet of water.

Impeller-driven water jet. A type of propulsive system that uses an *impeller* to eject a jet of high-speed water through a nozzle.

Imperial system. A system of measurement units, including feet, *pounds,* and gallons, that originated in Britain and was used extensively (with minor variations) in many countries around the world for decades. Also known as the British or English system of units. Now replaced in most countries by the *metric Systèm International d'Unités (SI).* The United States is one of the very few countries still making extensive use of imperial units for commerce, industry, and routine household measurement tasks.

In-circuit serial programming (ICSP). A feature of some *microcontrollers* that allows them to be programmed (or re-programmed) after they have already been *soldered* into a circuit.

Inductive kick. A potentially-damaging *voltage* spike caused when *current* flowing through an inductive device, such as a motor *winding* or *relay coil,* is suddenly turned off. Can destroy *control circuits* for these devices, if those control circuits are not properly designed to handle the kick.

Inductor. An electronic component used to store *energy* in the form of an *electric current.*

Industrial revolution. An important period in western history, lasting from about 1760 to 1870. Characterized by the development of new inventions and processes that greatly accelerated the development of other new machines, including underwater vehicles. Pivotal changes in this period included the development of steam engines that replaced human and animal muscle *power* for many tasks, and mass production methods that allowed the manufacture of large numbers of nearly identical, interchangeable parts, and centralization of production in factories.

Inertia. See *mass.*

In-kind contributions. Donations of parts or services, as opposed to money.

Instrumentation amplifier. An analog *integrated circuit* that is designed to amplify, by a precisely controlled amount, the *voltage* difference present between its two inputs. Used commonly in *sensor* circuits and other instrumentation. Not the same thing as an *operational amplifier,* though their *schematic* symbols look identical.

Insulation. On an electrical *wire* or other electrical component, a layer of flexible *plastic* or other non-conductive material, which allows the wires to cross one another or to be bundled together into a *cable* without creating a *short circuit.* On high *voltage* systems, it helps protect people from shocks.

Insulator. In an electrical context, a material or substance, such as *glass* and most *plastics,* that has a low *conductivity* (i.e., high *resistance* to *electric current*). Also, an object made from such material that is used to block or prevent the flow of electricity.

Integrated circuit (IC). A complex, solid-state (i.e., no moving parts), miniaturized electronic *circuit* that may contain thousands or even millions of *transistors* and other components, all manufactured together on a single tiny piece of *semiconductor* material by using patterned chemical etching and other methods. Usually embedded in a small block of gray *epoxy* with *metal* legs (called *pins*) sticking out, so the IC can be *soldered* into a *printed circuit board.* Also known as a "*chip.*" Familiar examples include computer *microprocessors.*

Integrated pressure sensor. A type of *electronic pressure sensor* based on one or more *integrated circuits.*

Intervention. In subsea jargon, the term intervention refers to physical interaction with objects on the sea floor, such as conducting installation, repairs, or adjustments on undersea valves, pipelines, or telecommunication *cables.*

Inverter. A device which takes *DC* electrical *power* from a *battery* and converts it into higher *voltage* (e.g., 115 VAC) power suitable for devices that normally plug into a wall outlet.

In-water weight. See *effective weight.*

Ion. An atom or small molecule in which the number of protons does not equal the number of electrons, thereby resulting in a non-zero net *charge* on the particle.

IP camera. A family of *digital video cameras* designed to be connected (either wirelessly or by Ethernet *cable*) into a standard Ethernet computer network. Communicates via standard Ethernet packets and standard network protocols. Used especially for web-based monitoring of remote locations, because the images can be viewed from anywhere in the world that has internet access, provided the *camera* network is connected to the internet and all the servers and network routers between the viewer and the camera are configured to allow it viewing of the images. Also known as a *network camera.* See also *web cam.*

J

Jam jar seal. A type of *endcap* seal that works very much like the lid on a typical jar of jam which has threads that allow the endcap to be screwed down onto the *canister* and in the process compress a *gasket* or *O-ring* to form a water-tight seal.

Joule (J). The standard *SI* (*metric*) unit of *energy*--the amount of energy required to move an object one *meter* using one *newton* of *force* applied in the same direction as the motion. Same as *newton-meters (Nm)*; see also *foot pounds (ft-lbs).*

K

Key. When used in the context of propellers, a piece of *metal* that fits into a *keyway* to transfer *torque* from a *propeller shaft* to a large *propeller.*

Keyway. A set of one or more lengthwise slots cut into the *shaft* and *hub* of larger propellers and designed to fit a rectangular piece of *metal*, called a *key*, which acts to transfer *torque* from the *propeller shaft* to the *propeller*.

Kilocalorie. A *metric* unit of *energy* equal to 1000 *calories* (lower case "c"). Also equal to 1 *Calorie* (capital "C").

Kilogram (kg). The *SI* unit for *mass*.

Kilowatt (kW). A *metric* unit of *power* which equals 1000 *watts*. See *watt*.

Kilowatt-hour (kWh). A *metric* unit of *energy*. One kWh = 3.6 million *joules*. A smaller version of this unit is a *watt-hour (Wh)*, which equals 1/1000 kWh. See also *watt*, *watt-hour*, and *kilowatt*.

Kort nozzle. A type of *ducted propeller* that uses a specially shaped duct around the *propeller* to improve *thrust* and *efficiency*.

L

Landmark. A distinctive, conspicuous, stationary feature of the landscape, such as a building or mountain, that is useful for *navigation*.

LARS. See *launch and recovery system*.

Latitude and Longitude coordinate system. A standardized global coordinate system which allows every location on earth to be specified in terms of Lat/Long coordinates (Latitude measured in degrees N or S of the Equator; Longitude in terms of degrees E or W of the *Prime Meridian*).

Launch and recovery system (LARS). A system used to move mid- to large-sized *ROVs* or similar vehicles safely in or out of the water, even in moderately rough seas. A *LARS* typically includes a crane or *A-frame* for lifting the heavy vehicle off the deck during launch and returning it to the deck upon recovery, a *winch* to roll up the long *umbilical/tether*, and a *slip ring assembly* to keep the *cables* from twisting as the winch rotates.

Law of Conservation of Energy. One of the most important laws in physics, confirmed by a large number of very careful measurements: the total amount of *energy* present before a particular process or event is the same as the total amount of energy present during and after that process or event. Energy can change forms, but it is neither created nor destroyed.

LCROV. See *Low-cost ROV*.

Lead-acid battery. A type of *battery* made from cells that have lead electrodes and use sulfuric acid as the electrolyte. The car battery is a familiar example. See also *valve-regulated lead-acid battery*.

Leading edge. The edge of a *propeller blade* that cleaves or bites into the water (when the *prop* is spinning in its normal forward direction). Usually straighter than the *trailing edge*.

LED. See *light-emitting diode*.

Left handed (LH) propeller. A *prop* that pushes a vessel forward when rotating counter-clockwise as viewed from behind the vessel. Compare *right handed propeller*.

Lever arm. See *moment arm*.

LH propeller. See *left handed propeller*.

Lift. A *force* that acts at right angles to the direction of vehicle motion through the water (or water motion past the vehicle). Usually suggesting an upward force, *lift* can also act sideways or downwards.

Light emitting diode (LED). A type of *diode* that emits light when conducting *current*.

Limit switch. A *switch*, placed in the path of an object whose motion is controlled by a motor, that is used to turn off the motor when the object has gone as far as it is supposed to go. Used to prevent damage or injury that might result if the object were moved too far. Example: limit switches are used commonly in robotic arms to limit how far each joint can move.

Line. A rope, especially one used aboard a ship or boat. Also a section of *pipe* or *tubing* used to conduct *hydraulic fluid* in a *hydraulic system* or gas in a *pneumatic system*.

Line of position (LOP). A straight line (either conceptual or actually drawn on a *nautical chart*) used in *navigation* to determine a vessel's position by means of *triangulation*. Commonly determined by lining up a pair of *range markers* or other *landmarks*, or by taking a compass bearing to (or from) a single landmark.

Linear regulator. A type of *voltage regulator* that uses analog circuitry to *control* its output *voltage*. Typically less expensive than *switching regulators*, require fewer external components to work properly, and often have better regulation performance, but also tend to waste a larger fraction of *power*, which makes them less ideal for many *battery*-powered systems and can cause overheating problems.

Literature review. A thorough search for written information related to the topic of interest.

Lithium battery. This term refers generally to *lithium ion batteries*, *lithium polymer batteries*, or other *batteries* based on electrochemical reactions involving the element lithium.

Lithium ion (Li-ion) battery. Any of a broad range of *batteries*, both primary and secondary, based on electrochemical reactions involving the element lithium. These batteries are generally noted for having good *energy density*, but require special handling and charging procedures.

Lithium polymer (LiPo or Li-poly) battery. A relatively new class of lithium-based *secondary battery* chemistries popular for a variety of portable applications, including remotely controlled vehicles, because of their very high *energy densities*, high *surge current* capabilities, and ability to be manufactured in arbitrary shapes to take advantage of limited available space in small, handheld devices. Require extra safety precautions to avoid fire/explosion.

Load. In electrical systems, the process(es) or device(s) being powered by the *battery* or other electrical *power source*. Example: in a flashlight *circuit*, the light bulb is the *load*.

Long-shore current. A *current* moving parallel to shore near the shoreline.

Low-Cost ROV (LCROV). Easily portable, smaller all-electric vehicles that cost less than $50,000 and typically operate at less than 300 *meters* (1,000 ft). There are also *ROVs* available for $10,000 or less called *VLCROVs* (*Very Low-Cost ROVs*).

M

Magnetic compass. A traditional *navigational* aid with a freely spinning magnetic needle that tends to align itself with the earth's magnetic field. The needle points roughly north from most places in the world and can be used as a directional reference.

Magnetic coupling. A mechanism that relies on interactions between magnets on opposite sides of a *pressure housing* wall to transfer rotational *energy* through the wall. Can be used, for example, to transfer energy from a motor inside a *thruster* housing to a *propeller* outside, without requiring a rotating *shaft* to pass through the housing wall. This greatly reduces the probability of a leak.

Magnetic north. The direction toward the magnetic north pole, which is the place where a *compass* needle is expected to point (but see also *compass north*). The magnetic north pole and the real (true) north pole are not in the same place, so the direction to magnetic north normally differs from the direction to *true north*. See also *declination*.

Magnetic switch. An electrical *switch* operated by movement of a nearby magnet. Useful under water, because these switches can be located inside a waterproof *housing* and operated from outside by passing a magnet along the outside of the housing.

Magnetohydrodynamic drive (MHD). Also known as *caterpillar drive* or *M-drive*, MHD uses magnetic fields and *electric current* to push seawater past the vehicle, thereby generating *thrust*.

Magnetometer. An instrument used to measure magnetic fields. Can be used as a *metal detector* to find any kinds of metal that cause a local distortion in the earth's magnetic field.

Magnitude. Size. Used commonly in physics and engineering to refer to the numerical value of a *force*, velocity, or other *vector* quantity, without specifying the direction of that vector.

Magnus effect drive. A propulsion system that uses rotating smooth cylinders to generate a propulsive *pressure differential*.

Mains. The wiring infrastructure in a house or business that distributes electrical *power* from the power company to the various *circuits* and components in the building. In North America, mains power is usually about 115 VAC oscillating at a frequency of 60 Hz.

Manip. Common abbreviation for *manipulator*.

Manipulator. A robotic arm. Many large *ROVs* and some smaller ones are equipped with one or more of these versatile devices.

Manned vehicle. A vehicle, such as a *submarine* or *submersible*, that carries a human crew and sometimes human passengers. For historical reasons, these *human-occupied vehicles* (*HOVs*) are still known as "manned vehicles," though women often serve as crew and/or observers in today's world.

Manometer. A simple U-shaped, clear, *glass* or *plastic* tube partly filled with colored water, *mercury*, or some other *fluid* and used to determine the *pressure differential* between the two ends of the tube by measuring how far the pressure differential moves the top of the fluid on one side of the U above the top of the fluid on the other side of the U.

Manual switch. An electric *switch* operated by a person's fingers, a machine cog, or some other manual means. (As opposed to being operated by an optical, magnetic, or electrical *signal*.)

Marine snow. A constant "rain" of bits of fish waste, mucous, dead *phytoplankton*, and other organic debris that transports nutrients from the *euphotic zone* into darker, less productive layers of water below. This ecologically important source of carbon and other nutrients in the deep sea is called "snow" because *backscatter* causes it to look like snow falling at night when viewed with *ROV* lights and cameras in the ocean's dark depths.

Mass. A measure of an object's intrinsic resistance to acceleration (i.e., to changes in its speed or direction of motion). Also known as *inertia*. Regarded as a good measure of the amount of matter in an object, particularly for work in space or on other planets/moons, because, unlike *weight*, it is not dependent on the *strength* of the local gravitational field. See also *kilogram (kg)* and *pound mass (lbm)*.

Master controller. The portion of a *servo system* that translates *pilot* hand movements or other commands into electronic *control signals* sent to the *slave controller*.

Material Safety Data Sheet (MSDS). A sheet that summarizes, for each chemical or mixture of chemicals sold by a company, how to use, store, and dispose of the substance safely. It also summarizes the basic physical and chemical properties of the substance. Generally available from the manufacturers, distributors, or on-line sources for free.

MCU. See *microcontroller*.

M-drive. See *magnetohydrodynamic drive*.

Memory effect. A permanent reduction in the amount of *charge* (and therefore *energy*) a *secondary battery* can store after it has been recharged one or more times without being fully discharged prior to each charging. This phenomenon is particularly pronounced in *nickel cadmium batteries*, but also appears to a lesser extent in some other types of batteries.

Metal. Any of a large number of materials made primarily from atomic elements that conduct heat and electricity well due to the presence of loosely held electrons in their atoms. Also *alloys* made from mixtures of these elemental metals. See also *steel, aluminum, titanium, brass* and *bronze*.

Metal detector. An instrument used to detect the presence of buried or otherwise hidden metal, usually by detecting changes in the resonant frequency of an inductive *circuit* caused by the presence of nearby *conductors*.

Metal oxide semiconductor field effect transistor (MOSFET). A type of *field effect transistor* known for its ability to handle high *currents*, among other attributes. Frequently used as a *transistor switch* in the *H-bridge* of *motor controller* circuits.

Meter. The base *SI* (*metric*) unit for length. Roughly 39.37 inches long.

Metric system. A standardized system of measurement units, most of which are related to each other by simple multiples of ten and/ or measurable properties of pure water. See also *Systèm International d'Unités (SI)*.

MHD. See *magnetohydrodynamic drive*.

Microcontroller. A small *integrated circuit* that is essentially a tiny computer-on-a-*chip* optimized for controlling machines. Can be used to give a vehicle the ability to monitor *sensors*, communicate with a *pilot* or other devices (including other *microcontrollers*), make decisions, and initiate actions--in a sense, providing the vehicle with a primitive form of intelligence. Also known as a *microcontroller unit* or *MCU*. Compare *microprocessor*.

Microcontroller unit (MCU). See *microcontroller*.

Microprocessor. An *integrated circuit* designed for processing information quickly and used as the "brain" in a typical computer.

Micro-ROV. See *Observation class ROV*.

Milliamp-hour (mAh). 1/1000th of one *amp-hour*.

Mission. The goal or work one hopes to accomplish during a dive or a set of dives.

Mission statement. In the vehicle design process, a concise statement summarizing, in a sentence or two, what the proposed vehicle must do to complete its *mission.*

Mission tasks. In the vehicle design process, a detailed list of step-by-step actions or tasks that the vehicle must complete, in sequence, to accomplish the goal(s) specified in the *mission statement.*

Mixed layer. A layer of water, near the surface of very deep bodies of water, where mixing due to the action of wind and *waves* or other factors tends to break down *thermoclines, pycnoclines, haloclines,* and other distinct layers, creating a relatively homogeneous layer of water. Its thickness varies enormously with location, season, weather, and other factors.

Mixed-gas diving. Diving while using breathing gas mixtures that differ from air in the types and amounts of each gas present. Carefully designed mixtures are used to reduce or eliminate some of the problems associated with breathing compressed air at great *depths,* including *nitrogen narcosis, decompression sickness,* and oxygen toxicity.

MKS system. See *Systèm International d'Unités (SI).*

Mobilization. The phase in an *operation* when gear and personnel are transported from the *staging* area to the *operations* site.

Mocking up. The process of building a *prototype.* Same as *prototyping.*

Moment arm. A distance used in calculating *torque.* Specifically, the shortest distance between the pivot point (axis of rotation) and the line of action of an applied *force.* Same as *lever arm.*

Momentary switch. An electric *switch* that is spring-loaded, returning automatically to its original position once the finger *force* (or other activating agent) is removed. For example, a standard doorbell button.

Moonpool. An opening in some ship's decks through which an *ROV* can be launched and recovered.

MOSFET. See *metal oxide semiconductor field effect transistor.*

MOSFET switch. A *MOSFET* used as an electrically controlled *switch.*

Motor controller. A *circuit* designed to *control* the speed (and sometimes direction, angular position, *torque,* or other aspects) of one or more electric motors. Typically consists of a *microcontroller* that regulates the activity of one or more *H-bridges.*

Motor Shaft. A *shaft,* usually extending from the *rotor* portion of a motor, used to transmit rotational *torque* and *energy* produced by the motor, to a wheel, *propeller* or other rotational mechanism.

Mowing-the-lawn search grid. A *search grid* in which the vessel conducting the search moves back and forth along parallel lines, proceeding in a straight line for a specified distance, then making a hairpin turn and returning along another straight line parallel to the first, then making another hairpin turn to start a third line parallel to the second, and so on. Reminiscent of the pattern commonly used by a person mowing a lawn.

MSDS. See *material safety data sheet.*

Multibeam sonar. A *sonar* technology used to map the seafloor. Similar to an *echosounder* in basic concept, but with the equivalent of multiple *sound* beams arranged into a fan-shaped array to survey a wide swath of bottom beneath the survey ship or *AUV.* Computers process the returning echo data to generate a map of seafloor *depth* at each point. High resolution *multibeam* systems can detect ship wrecks and other objects as localized decreases in bottom depth.

Multimeter. A small and fairly inexpensive, but versatile, electronic test and measurement instrument that is very helpful for diagnosing and *troubleshooting* problems in simple electrical circuits. Most offer the ability to measure and display *AC voltage, DC* voltage, *resistance,* and *electrical current,* and additional features. Any person or team working on an underwater vehicle's *power* or *control systems* will want access to a *multimeter.*

Multiplexing. See *signal multiplexing.*

N

Nautical chart. A chart presenting information designed to facilitate safe and accurate *navigation* on water. Uses standard symbols to indicate the position of important landmarks, buoys, submerged reefs or other hazards, and other useful information.

Naval brass. A *brass alloy* specifically formulated for high corrosion-resistance in seawater.

Navigation. The art, science, and technology of determining where you or your vehicle are now, comparing that location with the intended destination, and using that comparison to figure out which direction you or your vehicle need to go next to reach your destination.

Negative buoyancy. A condition in which an object's *density* is greater than that of the *fluid* in which it is immersed, so the object sinks.

Network camera. See *IP camera.*

Networking. This word has many definitions, but in the context of this book it refers to the process of consciously establishing an ever-growing number of mutually beneficial relationships with other people, usually by having people you already know introduce you to people you don't yet know who may have access to information, resources, or additional contacts that would be helpful to you.

Neutral buoyancy. A condition in which an object's *density* is equal to that of the *fluid* in which it is immersed, so the object neither floats nor sinks.

Newton (N). The *SI* unit of *force.* Defined as the amount of net force required to accelerate a 1 kg *mass* at the rate of 1 m/s^2.

Newton's Law of Gravity. See *Newton's Law of Universal Gravitation.*

Newton's Law of Universal Gravitation. An equation, developed by Sir Isaac Newton and published in 1687, stating that the *gravitational force* of attraction between two *masses* is equal to the product of the masses divided by the square of the distance between them. (Also known as *Newton's Law of Gravity*).

Newton's Laws of Motion. Three mathematical equations formulated by Sir Isaac Newton that describe accurately and completely how the motion of all objects from the size of tiny specks to entire planets and even galaxies is controlled by the mechanical *forces* pushing or pulling on those objects. These "laws" form the core of any mathematical analysis of underwater vehicle motion.

Newton-meter (Nm). A *metric* unit of *energy,* precisely equivalent to *joules* (1 Nm = 1 J). See also *foot-pounds (ft-lbs).* An *SI* (metric) unit used for both *energy* and *torque.* As a unit of energy, one Nm is the amount of energy it takes to move something a distance of one meter while applying a force of one newton in the direction of motion. As a unit of torque, one Nm is the torque generated when a force of one newton is applied at a perpendicular distance of one meter from a pivot point. Compare *foot-pound.*

Nickel Metal Hydride (NiMH) battery. A type of *secondary* (rechargeable) *battery* with properties, including the ability to deliver high *surge currents,* that make them popular for powering vehicles (among other things). Similar in many ways to *nickel-cadmium batteries,* but with reduced *memory effect* problems.

Nickel-Cadmium (NiCad, nicad, or NiCd) battery. A type of *secondary* (rechargeable) *battery* with properties, including the ability to deliver high *surge currents,* that make them popular for powering vehicles (among other things). However, their popularity is limited in part by a pronounced *memory effect* and the more recent development of *nickel metal hydride batteries,* and *lithium batteries,* which offer similar or improved performance in many respects, but without the same memory effect problems.

Nitrogen narcosis. A potentially dangerous, mind-altering effect, similar to that caused by narcotic drugs, experienced while breathing air or other nitrogen-containing gas mixtures at elevated *pressures.* For divers breathing air, nitrogen narcosis typically begins at depths of about 30 *meters* (100 feet) and increases in severity with increasing *depth.* It can lead to poor decisions with fatal consequences for the diver and/or would-be rescuers.

Nitrogen purging. A technique for removing most of the water vapor and oxygen from the interior of a *pressure canister* by filling it with nitrogen instead of air. This is done to prevent the formation of water condensation (which can fog *camera* ports) and to reduce or eliminate *corrosion* processes inside the *canister.*

NMEA format. A standardized *ASCII* data communication format used commonly for communication among different types of *navigational* equipment, such as *GPS* units, *echosounders,* and other equipment used aboard marine vessels. NMEA stands for the National Marine Electronics Association.

No-load voltage. The *voltage* present at the terminals of a *battery* or other electric *power supply* when there is nothing drawing power from it.

NTSC. One of several standard *analog video formats.*

O

Observation class ROV. A usually small *ROV* equipped with a *camera* and some navigational *sensors,* but usually no other tools or payloads. Designed primarily as a way to *pilot* a video camera to a particular underwater location to visually search, inspect, or explore an area. Also known as an *eyeball ROV, flying eyeball,* or *micro-ROV.*

Ocean observing system (OOS). A coordinated suite of vehicles and *sensors* including land-, sea-, undersea-, air-, and satellite-based technologies for collecting, analyzing, storing, and displaying a wide range of biological and physical oceanographic data over a region of the ocean in near real-time. These systems, in which *AUVs* and *ROVs* commonly play a major role, provide a holistic view of ocean processes not possible from measurements made at one point in time or space by a single vehicle or sensor and are becoming an important tool in understanding ocean processes.

Octal. In computer science and related fields, including robotics, a reference to the base-8 number system. Compare *binary, decimal,* and *hexadecimal.*

OEM. Acronym for *original equipment manufacturer.*

Ohm (Ω). A unit of electrical *resistance* numerically equal to the number of *volts* required to produce a *current* of 1 *ampere.* See *Ohm's Law* for more information.

Ohm's Law. (R = E/I). Probably the most useful of all equations in electronics, Ohm's Law defines the resistance (R) of an object (such as a length of wire) as the voltage (E) across the object divided by the amount of current (I) flowing through it in response to that voltage.

Oil compensation. The common practice of filling an undersea *canister* or *housing* with oil (or similar liquid), which is virtually incompressible, to protect the canister from leaking or being crushed by *hydrostatic pressure* at *depth.* See also *gas compensation.*

Oil-compensated housing. An undersea *canister* or *housing* protected from *hydrostatic pressure* and leaks by *oil compensation.*

ON/OFF control. See *bang-bang control.*

One time programmable (OTP) microcontroller. A type of *microcontroller* that can only be programmed once. Best avoided by beginners or others who are likely to want to fix mistakes in their programs or to upgrade their programs to add more features.

OOS. See *ocean observing system.*

Open circuit. A gap somewhere in a *circuit,* preventing *current* from flowing around the loop. May be intentional (e.g., a *switch*) or unintentional (e.g., a broken *wire*).

Open-loop control. A form of *control* in which there is no *feedback.* Compare *closed-loop control.*

Operational amplifier. An analog *integrated circuit* that can be wired up in quite a variety of different ways to perform a number of useful functions. These include a variety of standard mathematical operations, like addition, subtraction, multiplication, and division. Widely used in *sensor* circuits, analog *control circuits,* and other applications. Not the same thing as an instrumentation amplifier, though their *schematic* symbols look identical.

Operations ("Ops"). In the context of subsea vehicle use, the activities conducted with an already completed, functioning vehicle to complete a *mission.* Does not include all the work that goes into designing, building, and testing the vehicle, but does include the planning and execution of all the logistics related to vehicle transport to/from the dive site, planning *search grids* or other underwater activities, *navigation,* crew support, safety, and many other aspects of conducting a successful mission.

Ops. See *operations.*

Optical fiber. See *fiber-optic communication cable.*

Original equipment manufacturer. There are conflicting definitions for this term, but in this text and many websites, *OEM* refers broadly to parts that are not generally sold directly to consumers but rather are included inside other products (such as cars, kitchen appliances, etc.) that consumers buy. In the automotive industry, OEM refers more specifically to a replacement part manufactured by the same company that made the original part used in the car.

O-ring. A donut-shaped piece of *rubber* or other elastic material, commonly sandwiched in a groove between two smooth surfaces to create a water-tight seal. See also *static seal, dynamic seal.*

Oscilloscope. A type of electronic test and measurement instrument used to display rapidly varying *voltage signals* in a graphical form.

Useful for analyzing and *troubleshooting* complex electrical systems, such as *microcontroller*-based *control systems* or communication systems.

OTP. See *one time programmable microcontroller.*

Oxidation. The process of removing electrons from atoms. Oxidation of *metal* atoms causes metal *corrosion.*

P

Pa. Abbreviation for the *Pascal*, a unit of *pressure.*

PAL. One of several standard *analog video formats.*

Parallel. Electrical components are said to be arranged in *parallel* when their ends are connected in such a way that they are exposed to the same *voltage*, but the *current* flowing to the group of components can split up into separate smaller currents that flow through each component, then recombine on the other side. Thus, the components need not have the same amount of current flowing through them. See also *parallel data transmission.*

Parallel data transmission. A form of digital data transmission in which more than one *bit* of data (typically 8, 16, 32, or more) is sent simultaneously. Generally faster than *serial data transmission*, but requires more *wires* and is reliable only for short distance communication. See also *parallel* for another use of that term in electronics.

Pascal (Pa). An *SI* (*metric*) unit for *pressure* defined as 1 *newton* per square *meter.*

Pascal's Law. A law of physics that forms the basis for *power* transfer in *hydraulic systems.* It states: "The *pressure* applied to a confined *fluid* is transmitted undiminished in all directions and acts with equal *force* on equal areas and at right angles to them."

Passive sonar. Any sonar system that listens for *sounds* produced by, or reflected from, other sources, but does *not* generate sounds of its own. In contrast to *active sonar* systems, a passive sonar system must rely on ambient sound to function; however, it does not produce sounds that might give away its location or disturb marine life, so it is preferred for covert military *operations* and some forms of biological research. See also *sonar, active sonar.*

Payload. The set of tools, instruments, collecting devices, or other equipment carried by an underwater vehicle in support of its *mission*, thereby allowing it to produce revenue or some tangible achievement. Generally removable and distinct from the basic structural, propulsion, navigation, and other systems that comprise the vehicle itself.

PC-104 board. Essentially a stripped down, desktop PC computer (no keyboard, mouse, monitor, etc.) Often used as *embedded controllers* in complex *AUVs* or other machines that need more information processing speed and *power* than a *microcontroller* or two can provide.

Penetrator. A specialized part or set of parts that facilitates leak-proof passage of electric *wires, hydraulic lines, pipes*, or other such things through a hole drilled in a *bulkhead, pressure hull*, or *pressure canister.* The *penetrator* basically fills and seals the space between the wires (or whatever) and the edges of the hole, so water can't leak past.

Penetrator boss. A thickened, reinforced portion of a *pressure hull, pressure housing*, or *bulkhead* used to strengthen the area around a hole made for a *penetrator.*

Performance requirements. In the vehicle design process, a list of specific capabilities the vehicle must have in order to complete all of the *mission tasks*, given the *depths, currents, temperatures*, obstacles, and other realities of the underwater environment in which it will be working.

Periscope. An optical system traditionally involving mirrors and telescope-like lenses inside a long, thin vertical tube that allows a *submarine* crew to peek surreptitiously above the surface of the water while the submarine is still submerged and hidden from enemy view. Modern periscopes often substitute video cameras and *optical fiber* links in place of the mirrors and lenses.

pH. A unit of measurement used to quantify the acidity or alkalinity of water.

Photosynthesis. A biochemical process whereby plants, algae, and some other organisms use *energy* from light and carbon-dioxide from air or water to build organic molecules like sugar that can be used as a source of energy and carbon atoms by these and other life forms. Ecologically critical as the primary source of energy, accessible carbon, and oxygen for almost all life on Earth. Photosynthesis is not possible in the dark waters of the deep ocean.

Phytoplankton. *Photosynthetic* algae that are particularly abundant in nutrient-rich waters at relatively shallow depths, where sunlight can reach them. A common source of *turbidity* in the *euphotic zone.*

PID control. PID is a very popular *control algorithm* used in many industrial and other *control systems.* It usually provides better performance over *proportional (P) control* and greatly improved performance over *bang-bang control.* In PID control, the *signal* going to the motor (or other thing being controlled) is derived from the sum of three different *control* signals (designated P, I, and D), each of which solves a different problem associated with simpler control systems.

Pigtailing. Twisting two or more *wires* together in *parallel* to create the electrical equivalent of one wire with a larger cross section. A useful trick if you need a thick wire and only have thin wires available, or if you want a more flexible version of a thick wire.

Pilot. During an *ROV mission*, the pilot is the ROV operator who "flies" the vehicle. He or she usually doubles as the electronics technician and is responsible for maintaining the *control systems*, navigational *sensors*, and other electronic equipment. See also *supervisor, winch operator.*

Pin. In electronics, one of the (usually several to many) short stiff *wires* commonly emerging from an *integrated circuit.* Each pin normally gets *soldered* to a different wire in a *printed circuit board* to form electrical connections between the *IC* and other components on the board. Each pin typically has a different purpose. The function of each pin is specified in the datasheet for that particular *IC.* See also *I/O pin.*

Pipe. A rigid, hollow cylinder with a long length compared to its diameter. Pipes are among the most useful building materials for underwater vehicle frames and *pressure canisters.*

Piston seal. A specially-machined *endcap* seal that fits down inside a *pressure can*, like a piston in a cylinder, with an *O-ring* sandwiched between the "piston" and the inside of the can.

Pitch. Two definitions of pitch are used commonly in this book: 1) The forward or backward angle through which a vessel, such as a boat or airplane, has tipped away from its normal upright position. 2) The theoretical distance a *propeller* would move forward through the air or water after exactly one full revolution, based on the angle of the *blades*, if there was no *slip.*

Planetary gear motor. A motor with an attached *gearbox* that contains gears arranged internally in what is called a "planetary" configuration. This particular arrangement has one or more layers, each with a central gear that drives several little gears revolving around it (like planets). Planetary gear motors have an output *shaft* lined up with the *motor shaft*, and other features useful in many applications, including *thruster* design.

Plastics. Organic polymers consisting of interlocking chains of carbon-based molecules. Because of their low cost, easy availability, excellent *corrosion* resistance, and ease of *fabrication*, plastics are also the dominant structural materials in most small, low-budget underwater vehicles. Common plastics used in the construction of small underwater vehicles include *PVC, ABS, polycarbonate,* and *acrylic.*

Plate. A flat layer of raw material, such as *metal* or *plastic*, commonly used as a starting point for machining parts.

Platform. In the context of subsea *operations*, any fixed, flying, or floating surface that can support a vehicle, its crew, and the machinery used to launch and recover the vehicle. The platform may also be used to salvage a lost object off the bottom, after the object has been found and rigged for recovery.

Pneumatic system. Similar to a *hydraulic system*, but uses the *pressure* of compressed air or other gases, rather than liquid pressure, to transfer mechanical *power.*

Point-to-point soldering. A *circuit* construction technique in which electrical connections are made by hand-*soldering wires* between individual components.

Pole. In electric *switch* specifications, the number of "poles" refers to the number of separate *circuits* that can be controlled by the switch. Example: a 2-pole switch can *control* two separate circuits simultaneously.

Polycarbonate. A type of *plastic* known for its toughness and impact resistance.

Polyurethane. A *rubbery* polymer used for a variety of purposes, including mold-making and the *potting* of electronic circuits.

Polyvinyl chloride (PVC). A type of *thermoplastic* used frequently to make PVC *pipe*, PVC *pipe* fittings, and many other things.

Positive buoyancy. A condition in which an object's *density* is less than that of the *fluid* in which it is immersed, so the object floats.

Potentiometer. In electronics, a three-terminal electrical component in which two terminals are at opposite ends of a long, usually coiled, resistive *wire* and the third terminal is connected to a slider that moves along the wire, usually in response to the turning of a knob. Depending on how the three terminals are wired into a circuit, the potentiometer can function as a *rheostat* (adjustable *resistor*) or as a *voltage divider.*

Potting. A method of waterproofing non-moving components, such as electrical circuits, by embedding them in a solid material (typically *epoxy* or *polyurethane*), thereby often eliminating the need for any *housing* whatsoever. Also known as *encapsulation.*

Potting compound. A liquid *epoxy, polyurethane,* or other material that hardens into a solid or *rubbery* waterproof substance and can be used to encapsulate electronic circuits or other non-moving parts to protect then from exposure to water.

Pound (lb). An *imperial* unit usually used for *force*, but occasionally *mass*, and therefore somewhat ambiguous. Better to use *pound mass (lbm)* or *pound force (lbf)* to be clear about the meaning. See also *slug.*

Pound force (lbf). An *imperial* unit for *force*, roughly equivalent to 4.448 *newtons.* See also *pound (lb).*

Pound mass (lbm). An *imperial* unit for *mass*, roughly equivalent to 0.454 *kilograms.* See also *pound (lb).*

Pounds per square inch (psi). A commonly used *imperial* unit for *pressure* or *stress.*

Power. The rate at which *energy* is being used or converted between forms. Defined as energy per unit time. Essentially a measure of how fast work can be done. Compare *energy*, which is not the same as power.

Power capacity. When used as a *battery* performance specification, the maximum *power* a battery is capable of delivering without damaging itself.

Power distribution unit (PDU). Some *ROVs*, especially work class vehicles, employ a PDU to distribute high *voltage power* from the generator to the various subsystems, often through transformers, which lower the voltage.

Power hydraulic system. See *hydraulic system.*

Power source. Something that delivers mechanical power, electrical power, or some other form of *power* (a rate of energy flow) in a form that can be used easily to perform various tasks. Example: A *battery* is a power source because it provides electrical power that can be used to operate an electric motor or other electric device.

Practical salinity unit (psu). A standardized, dimensionless, unit for reporting the *salinity* of a water sample based on the sample's *electrical conductivity.* Established as part of UNESCO's international Practical Salinity Scale of 1978, these units are widely used and roughly comparable in numerical value to salinity values obtained by more conventional (but more difficult and time consuming) methods used to measure the parts per thousand *mass* of salt dissolved in the water.

Pressure. In the context of this book, pressure is a way of describing how hard a *fluid* (liquid or gas) presses against an object in contact with the fluid. Formally defined as the *force* applied by the fluid to a small portion of the object's surface divided by the area of that portion. This force is always directed perpendicular to the surface, and it increases with *depth*, since it is caused by the *weight* of the fluid above. *Pressure* represents a form of compressive *stress.*

Pressure canister. A usually small, waterproof, pressure-proof container, typically cylindrical in form, used to protect cameras, lights, *control* electronics, or other equipment from water and water *pressure.* Many *ROV* and *AUV* systems are protected inside pressure canisters. Also called *canister, can, housing,* or *pressure housing.* See also *pressure hull.*

Pressure differential. A difference in *pressure* from one location to another, as in the pressure differential between the interior and exterior of a *pressure canister* at *depth.* An important concept and quantity in subsea design, because it is pressure differentials, rather than pressure alone, that creates the pressure-related *forces* that sometimes damage or crush vehicles.

Pressure drag. That portion of total *drag* caused by the *inertia* of the *fluid.* The dominant drag component for most *ROVs* and *AUVs.*

Pressure gauge. A device used to display measurements from a *pressure sensor*, which is often integrated into the gauge. The familiar "tire gauge" is an example of a pressure gauge. See also *dial pressure gauge* and *integrated pressure sensor.*

Pressure housing. See *pressure canister.*

Pressure hull. A large, strong, waterproof, pressure-proof hull or shell surrounding people and equipment on a *submarine* or *submersible* to protect them from water and extreme water pressure. Normally cylindrical or spherical in shape to resist collapse under pressure. See also *pressure canister.*

Pressure hull penetrator. A *penetrator* designed for, or being used on, a *pressure hull.*

Pressure relief valve. A safety mechanism designed to vent excessive *pressure* from a *pressure canister* or other container automatically.

Pressure sensor. A device for measuring *pressure.*

Primary battery. A non-rechargeable *battery.*

Prime Meridian. A straight line drawn between the poles and running through the town of Greenwich, England. Longitude is specified in terms of the number of degrees east or west of this line.

Prime mover. In a *hydraulic system*, the diesel engine, electric motor, or similar device used to drive the *hydraulic pump*(s), thereby converting other forms of *power* into hydraulic power.

Printed circuit board. A thin, rigid, sheet of resin-like material to which electronic components, such as *integrated circuits*, are attached and interconnected by flat *metal wires* (called "traces") adhering to the surface of the board. *Circuit boards* are a robust method for constructing circuits and they lend themselves to mass production. They are extremely common in computers, cell phones, and other electronic devices.

Procurement. The process of purchasing or obtaining donations of needed parts, materials, and supplies. Typically occurs between the design and *fabrication* stages, though it may also happen at various times during fabrication.

Product research. A process whereby information about available products you might use in your project is gathered and evaluated using as many approaches as are available, including web searches, reading of product brochures, phone calls with sales representatives, conversations with others who have used the products, etc.

Program. In the context of computers or robotics, including *ROVs* and *AUVs*, a *program* is usually a set of instructions entered into a computer, *microprocessor*, or *microcontroller*, that tell it what to do. As a verb, the word *program* means to write such a set of instructions and/or enter those instructions into a computer, microprocessor, or microcontroller.

Project report. A well-organized summary of a *mission* or assignment including an explanation of the mission or purpose of the project, methods used, a discussion of achievements and challenges during the dive(s), and overall conclusions including recommendations for future work.

Prop. In this text, shortened terminology for *propeller.*

Prop torque. See *propeller torque.*

Propeller. A familiar mechanical device rotated, often at high speed, by an engine or motor to provide *thrust* in air or water.

Propeller blade. One of the several twisted fins or foils extending outwards from the central *hub* of the *prop.* The *blades* push against the water when the *propeller* rotates, thereby producing *thrust.*

Propeller shaft. A *shaft* used to transmit *torque* and rotational *energy* to a *propeller.*

Propeller torque. The *torque* applied to a vessel by a rotating *propeller* as an "equal and opposite" reaction to the torque applied by the vessel to the propeller. Prop torque can cause problems like unwanted *roll* in vehicles that are not sufficiently *stable.*

Proportional (P) control. A *control algorithm* intermediate in complexity and performance between *bang-bang control* and *PID control.* P control adjusts the *strength* of the corrective action to match the size of the error.

Propwalk. A slight sideways *force* created by a spinning *prop.* Most noticeable in high-speed boats and usually not a problem in *ROVs* and *AUVs.* It can be eliminated by using two *propellers* in *parallel*, with a *right-handed (RH) prop* on one side and a *left-handed (LH) prop* on the other.

Prototype. A relatively quick, easy, and inexpensive mock-up, model, or draft version of a proposed part, *circuit*, or machine. Used to determine whether or not the basic design idea will work and to identify any needed or recommended changes before investing the time, money, and effort to build a final, robust version. Especially important for parts that are extremely difficult, expensive, or time-consuming to build (since you don't want to figure out after the fact that you wasted all that time, *energy*, or money). Also important when you plan to make many copies of a part, circuit, or mechanism, such as a *thruster* (since you want to make sure that one works before you build many).

Prototyping. The process of building a *prototype.* Same as *mocking up.*

Proximity switch. A *switch* used to detect the close approach or presence of another object. Used frequently in factory automation, security systems, and robotics. Depending on the design of the switch, may use optical, magnetic, acoustic, capacitive, or inductive *signals* to detect the presence of a nearby object.

psi. See *pounds per square inch.*

psia. A common *imperial* unit for *absolute pressure* measured in *psi.*

psig. A common *imperial* unit for *gauge pressure* measured in *psi.*

Pulse width modulation (PWM). A digital method of producing or approximating features of an *analog signal.* Can be used, for example, as a way for digital circuits to *control* the brightness of a light or to regulate the speed of a motor. Entails very rapidly switching a *circuit* ON or OFF (often hundreds or thousands of times per second) while adjusting the average *power* delivered by controlling the *duty cycle* of the series of pulses.

PVC. See *polyvinyl chloride.*

Pycnocline. A layer of water in which *density* changes rapidly with *depth.*

R

Radial search pattern. See *straight-run search pattern.*

RCA connector. A popular style of two-*conductor electrical connector* originally introduced in the 1940's by Radio Corporation of America. Consists of an *RCA plug*, which plugs into an *RCA jack.* The jack is also known as an *RCA socket.* Each half has a central *signal*-carrying *metal conductor* separated by a color-coded ring of *plastic* from a metal ring, which is normally connected to *ground.* Used primarily for audio and video signals, but can also used for other purposes.

RCA plug. The "plug" half of an *RCA connector.* See *RCA connector* for details.

RCA socket. The "socket" half of an *RCA connector*. See *RCA connector.*

RCV. See *Remotely operated vehicle.*

Rebreather. A type of breathing equipment in which exhaled air is treated to replenish oxygen and remove carbon-dioxide, then reused. Originally invented for rescuing workers trapped in flooded mines, later adapted for subsea use. Still used for specialized diving situations, but largely replaced by simpler, safer *scuba gear* for routine, shallow-water diving.

Rechargeable battery. See *secondary battery.*

Recovery operation. An *operation* in which the primary goal is to retrieve an object off the bottom. Usually conducted after a *search operation* has already located the object of interest.

Reed switch. A type of *magnetic switch* in which a flattened *metal wire*, superficially resembling the reed of a wind instrument, rests in contact with another flattened wire that can be moved by a nearby magnet to open or close the *switch*. Usually enclosed in hollow *glass* or *plastic*.

Reference ground. A point in an electrical *circuit* defined as having zero *volts* and used as a reference against which to measure *voltages* in other parts of the circuit, much as sea level is often defined as an altitude of zero and then other altitudes are measured relative to sea level. Often, but not always, the place in a *DC* circuit where the voltage is lowest. See also *ground* for other types of electrical ground.

Refraction. Bending of light rays, ocean *waves*, and similar wave-related phenomenon as the waves cross into areas where their speed increases or decreases. Example: A *camera* lens can refract (bend) light rays to focus an image because light waves travel more slowly through *glass* than they do through air.

Relay. An electric *switch* similar in many ways to a conventional *manual switch*, but controlled electrically through an *electromagnet* rather than by direct mechanical *force* applied to a button or toggle. Normally used as a way of allowing relatively small *electrical currents* to *control* much larger ones.

Remotely Operated Vehicle (ROV). An unmanned robotic vehicle operated by remote control. The term usually refers to a vehicle used under water, but controlled from the surface. The *pilot* directs the robot to perform underwater work, communicating with the vehicle by means of a *cable* (*tether*), which transmits electrical *power* to the robot and allows data, video, and other information to be sent back to the pilot. Early versions were sometimes called "remotely controlled vehicles" (RCV), but that was also the brand name of an early vehicle, so to avoid confusion the term ROV was adopted for general use.

Resettable fuse. Similar to a conventional *fuse* in purpose; however, excess *current* associated with a *short circuit* causes a resettable fuse to overheat. This in turn raises its electrical *resistance* enough to block most of the excess current and protect the circuit. Once the resettable fuse cools, it will again allow current flow. See also *fuse* and *circuit breaker.*

Resistance (R). The electrical analog of friction. Whenever *electrical current* flows through a *wire*, light bulb, or any other object, it encounters resistance, which converts some of the *energy* of the flowing electricity into heat and causes a drop in *voltage* from one end of the object to the other. See also *Ohm.*

Resistor. A two-terminal electronic component with a roughly constant value of *resistance.*

Resolution. The size of the smallest details that can be distinguished in an image, particularly a photograph or *sonar* image.

Reynolds number (Re). A dimensionless number expressing the ratio (and therefore the relative importance) of *inertial forces* (for example, *pressure drag*) to viscous forces (for example, *viscous drag*) in *fluid* flow situations.

RH propeller. See *right handed propeller.*

Rheostat. An adjustable *resistor.* See also *potentiometer.*

Rigging. In maritime usage, rigging is the general name for all the ropes, chains, and other items that hold masts, beams (spars and yards), and sails in place. In salvage work, rigging refers to the gear needed to lift a submerged object.

Right handed (RH) propeller. A *prop* that pushes a vessel forward when rotating clockwise as viewed from behind the vessel. Compare *left handed propeller.*

Righting torque. A torque that counteracts tipping (pitch or roll) of a vessel and so tends to return the vessel to its normal upright position. A righting torque thus contributes to vessel *stability.*

Rip-tide. A *current* that flows away from shore into deeper water. Not a true *tide.* Usually a return route for large volumes of water carried ashore by breaking *waves* or a place where two long shore currents flowing in opposite directions meet and head out to sea.

Risk Assessment. An evaluation of the risks involved in attempting a particular underwater *operation.* Used to determine whether or not to try conducting the operation.

Roll. The angle through which a vessel, such as a boat or airplane, has leaned to the left or right of its normal upright position.

Root Mean Square (RMS). A method of calculating the *power* in an *AC signal.* Gives a result that is roughly equivalent to the value of a *DC* signal having the same power.

Rotor. The rotating part of an electric motor, usually the inside.

ROV. See *Remotely Operated Vehicle.*

RS-232. An old, *asynchronous, serial data communication protocol.* No longer used much in new equipment, but once used very widely for computer peripherals and other purposes. Though slow by modern day standards, RS-232 is still popular as a relatively simple, inexpensive, data communication protocol for *DIY* projects, including home-made *ROVs* and *AUVs.*

RS-485. An *asynchronous, serial data communication protocol* developed to address some problems with RS-232, an earlier protocol. Reliable over longer distances than RS-232 and also allows for multiple devices to share one set of communication *wires.*

Rubber. A generally soft, elastic material frequently used to make *O-rings, gaskets,* and other types of *seals.*

S

Sacrificial anode. A piece of *zinc* or other *metal* with relatively low affinity for electrons attached to a ship hull, pipeline, or other valuable *structure* made from a more electronegative metal, such as *steel,* to protect that valuable structure from *corrosion.* The sacrificial anode works by giving up its electrons to the structural metal to keep that metal from being oxidized. In the process, the sacrificial anode oxidizes and gradually corrodes away, so it must be inspected and replaced periodically to ensure continued protection of the valuable structure. See also *cathodic protection* and *zinc*.

Safety factor. The maximum estimated *strength* or capacity of a *structure* or *system* divided by the anticipated strength or capacity needed under normal use. Example: an *ROV* engineered to function down to a *depth* of 1000 *meters* before being crushed, but actually operated at a maximum depth of 500 meters, would be operating with a safety factor of 2.

Salinity. A measure of the overall concentration of salts dissolved in water. Usually expressed in parts per thousand by *mass* (e.g., grams of dissolved salt per 1000 grams of seawater) or by *practical salinity units*, which are numerically comparable to parts per thousand, but are measured using *electrical conductivity*. Example: 1000 grams of seawater containing 35 grams of salt has a salinity of 35 parts per thousand.

Salinity profile. A graph or other representation of how *salinity* varies with *depth* in a particular *water column* at a particular time. *CTD sensors* are often used to obtain salinity profiles.

Saturation diving. A diving technique in which the gasses dissolved in a diver's tissues are allowed to come to equilibrium with the partial *pressure* of the gasses in the diver's air supply or other breathing gas mixture. Once the diver's tissues are saturated in this way, the diver can continue to work at that *depth* indefinitely without requiring additional decompression time.

Scanning laser. A relatively new tool for subsea *navigation* and mapping that sends out a single blue-green laser beam to scan objects in a very precise and fast motion.

Schematic. A technical drawing or diagram used to show how various electronic components within a *circuit* are connected together, using standardized symbols to show the conductive pathway. Typically emphasizes the flow of *electric current* and information through the circuit, rather than the physical layout of the circuit components. Also known as a *schematic diagram*.

Scuba. Acronym for Self-Contained Underwater Breathing Apparatus. Based on the "aqua-lung" system originally developed by Jacques Couteau and Emil Gagnan shortly after WWII, scuba is now widely used for recreational diving and shallow work-related diving. See *scuba gear*.

Scuba gear. A set of diving equipment consisting of a pressurized air tank, hoses, mouthpiece, and *pressure* regulating valves that deliver air from a cylinder ("scuba tank") to the diver's lungs at just the right pressure to breath under water. Unlike a *rebreather*, conventional *open-circuit* scuba does not recycle the exhaled breathing gasses; it just releases them into the water.

Sea trial. An in-water test conducted after *fabrication* is complete, but before attempting the real *mission*, to evaluate the reliability and endurance of the vehicle under conditions similar to the real mission conditions. Sea trial results can be used to make any needed adjustments or other modifications prior to attempting the real mission.

Seal screw. A specially designed screw in a *pressure canister* or *endcap* that can be loosened to let internal *pressure* come to equilibrium with outside pressure in a safe, controlled manner before trying to open the *canister*. The seal screw must be re-tightened to make the canister watertight for a dive.

Sealed lead-acid (SLA) battery. See *valve-regulated lead-acid battery*.

Search grid. A methodical search pattern used for systematic exploration of the bottom, particularly in an attempt to locate a missing object (or confidently confirm that the object is not there), but sometimes for mapping an area. Common search grids include the *mowing-the-lawn search grid*, the *straight-run search pattern*, and the *arc grid search pattern*.

Search operation. An *operation* in which the primary goal is to look for and locate something under water.

Secchi depth. See *Secchi disk* for explanation.

Secchi disk. A simple device for measuring *visibility* in water. Consists of a round white disk with a bold black and white pattern, usually lowered into the water on a rope until it reaches the *depth* where the black and white pattern it just barely visible. This depth is known as the *Secchi depth* and provides a convenient, standardized, measure of visibility.

Secondary battery. A *battery* that is designed to be discharged and recharged repeatedly. Also known as a *rechargeable battery*.

Sector scanning sonar. A *sonar*-based navigational system frequently used for *ROV navigation*. Akin to traditional aircraft radar, which scans (usually horizontally) through an arc, completing a partial or complete circle, and displays the echo returns on a circular screen. The screen is interpreted as if it were an aerial view or map of the surroundings, with the vehicle located in the center of the circle.

Self-Contained Underwater Breathing Aparatus. See *scuba*, *scuba gear*.

Self-discharge. A process observed in all *batteries* whereby they gradually lose *charge*, even when not being used to *power* a circuit. The rate of self-discharge depends on the type of *battery*, *temperature*, and other conditions.

Semiconductor. A material which has electrical properties that fall between those of *insulators* and *conductors*, and which can have its properties modified through the addition of tiny amounts of impurity to make it more or less conductive. Also, a *circuit* device, such as a *diode*, *transistor*, or *integrated circuit*, made from such materials.

Sensor. A device that converts a physical or chemical quantity, such as *temperature*, *salinity*, or *pressure*, into a *signal* that can be used, recorded, displayed, or interpreted easily by a machine or person. Many modern sensors convert the sensed quantity into an analog or digital electrical signal.

Serial data transmission. A form of digital data transmission in which *bits* of data are sent one after the other. Generally slower than *parallel data transmission*, but requires fewer *wires* and can be used reliably over longer distances.

Series. In electrical *circuits*, components are said to be arranged in series when two or more components are connected end-to-end so that all the *current* flowing through one component must also flow through each of the other components.

Servo. Shortened term for *servo motor*. Also an adjective used to describe parts of a *servo system*.

Servo motor. A type of motor designed to rotate its *shaft* through an angle that mimics or copies the angle of rotation of a knob or other *control* device, thereby allowing those input movements to be reproduced at a remote location and/or with amplified *torque*.

Servo pot. See *servo potentiometer*.

Servo potentiometer. A *potentiometer* used to provide knob, *shaft*, joint, or other angle information for a *servo system*.

Servo system. A type of *control system* designed to make a machine mimic the movements of a knob, lever, or other command input device moved by a human operator. For example, a servo motor rotating through an angle of 42 degrees clockwise in response to a human operator rotating a *control* knob 42 degrees clockwise.

Shaft. Many definitions, but it this text generally used for a long, cylindrical *rod* or *bar* that rotates to transmit mechanical *energy* from a motor or engine to a wheel, *propeller*, or other rotating mechanism.

Shaft coupler. A device used to connect one rotating *shaft* to another. Some types include the ability to accommodate slight misalignment between the shafts.

Sheet. Similar to a *plate*, but sometimes thinner and more flexible.

Short. In the context of electrical *circuits*: Noun = Abbreviated term for a *short circuit*. Verb = to create or cause a short circuit.

Short circuit. An error condition in which *electric current* uses an unauthorized "shortcut" to bypass some or all of the usual *load* that it's supposed to flow through. This can result in dangerously high current levels capable of melting *wires* and starting fires. It can also create an electrocution hazard.

SI. See *Systèm International d'Unités (SI)*.

Side scan sonar. A technology for seafloor mapping and object detection that uses *sonar* to scan an area of the seafloor on either side of a *towfish*. Computer processing of the returning echoes generates images in which objects and other seafloor features are made visible by the acoustic highlights and shadows they create.

Signal. Something that can convey information from one place to another in the way it changes (or doesn't change) over time.

Signal ground. A *reference ground* used specifically for measuring the time-varying *voltages* associated with information-carrying *analog* or *digital signals*. See also *ground* for other types of electrical ground.

Signal multiplexing (MUX). Sending several different signals or messages, more or less simultaneously, over the same *wire*, *optical fiber*, or other communication channel. Useful when it is impractical or impossible to dedicate a separate wire or other communication line to each message. See also *frequency division multiplexing* and *time division multiplexing*.

Single-pole single-throw (SPST) switch. A simple *ON/OFF switch*. It has two terminals which are electrically connected when the switch is ON and disconnected when the switch is OFF. See also *pole* and *throw*.

Skin friction drag. See *viscous drag*.

Slave controller. The portion of a *servo system* that interprets electronic *control signals* from the *master controller* and uses them to *control* the motors, *hydraulic* valves, or other devices that actually produce the machine movements.

Slip. The difference between the actual forward progress of a *propeller* through air or water compared with that predicted by the propeller's *pitch* alone. Slip occurs because air and water give under *pressure*, much as sand gives underfoot when you climb a steep sand dune, making your actual elevation gain less than it would be if you had taken the same number of steps on a solid stairway.

Slip ring assembly. A mechanism that transfers electrical *power* and data (and sometimes optical data) between a rotating object and a non-rotating object. In *ROV* applications, it is used on the *winch* to transfer power and data between the non-rotating deck *cable* and the rotating winch drum upon which the *umbilical* or *tether* are stored. Without the slip ring assembly, the cables would get twisted and damaged as the winch drum turned.

Slug. An archaic, rarely used *imperial* unit for *mass*. Roughly equal to 14.59 *kilograms*. Also a very cute, squishy, slimy, little mollusk related to a snail, but without a shell.

Socket depth. In this book, the distance to which a *PVC pipe* can be inserted into a pipe fitting.

Soft ballast tank. A type of ballast chamber used to control buoyancy in a submarine or submersible by changing the vessel's displacement, rather than its weight. Soft ballast tanks are open to the surrounding water and ambient pressure through openings on their undersides. Pressurized air can be admitted to the top of the chamber to force water out the bottom, or vented to allow water back in. Compare *hard ballast tank*.

Solar panel. A usually flat frame supporting a layer of material that directly converts sunlight into electric *power*.

Solder. A special *metal alloy* that melts at *temperatures* much lower than the temperatures required to melt most other metals. Used during *soldering* to form *electrical connections* between *circuit* components.

Soldering. A technique for making permanent *electrical connections* between *wires* and electrical components by joining the wires from those components with *solder*. A *soldering iron* is used to heat the wires past the melting point of solder, then solder is melted onto the wires and allowed to cool until it hardens, forming a good electrical connection between the wires.

Soldering iron. A hand-held tool with a *metal* tip that gets very hot and is used to provide localized heat that melts solder for soldering electrical components together.

Solderless breadboard. A commercially available type of *breadboard* that facilitates rapid assembly and modification of *prototype* electronic *circuits*. The surface of the board is covered with anywhere from dozens to thousands of small holes. Many sets of holes are electrically interconnected inside the board, so connections between circuit components can be made simply by plugging the components into the right sets of holes. Modifying connections in the circuit is as easy as moving a component to a different set of holes.

Sonar. A technology that uses *sound* for navigation, object detection, and communication, usually under water, where sound penetrates better than light. Originally developed to detect *submarines* and known as *ASDIC* (for "Allied Submarine Detection Investigation Committee"). Later standardized as sonar (an acronym for "SOund Navigation And Ranging") and used for a much wider array of purposes from seafloor mapping and *ROV* navigation to fish finding. See also *active sonar* and *passive sonar*.

SOSUS. See *sound surveillance system*.

Sound. The propagation of mechanical vibrations through a solid, liquid, or gas.

Sound Surveillance System (SOSUS). A network of global bottom-mounted hydrophones arrayed across the North Atlantic to monitor Soviet *submarines*. Deployed in 1961 and later expanded.

Specific gravity. A measure of *density* calculated by dividing the density of a given object or substance by the density of pure water. Substances with a specific gravity greater than one are denser than water, and so will sink in it, whereas those with a specific gravity less than one are less dense than water and will float in it.

Spherical port. See *dome port.*

SPI communication protocol. A popular synchronous serial communication protocol used for communication between *microcontrollers* and other *integrated circuits.* It requires four *wires* (plus *ground*) and supports communication between multiple devices sharing the same wires. See *Serial data transmission protocols.*

SPST switch. See *single-pole single-throw switch.*

Stability. With regard to *pitch* and *roll* in undersea vehicles, stability refers to the ability of the vehicle to return to an upright position (*i.e.* to right itself) whenever it is tipped or flipped over. A vehicle having this ability is said to be *stable.*

Stable. Exhibiting *stability.*

Staging. The phase in an *operation* when you actually locate, organize, and pack all the gear for transport to the work site.

Stainless steel. A corrosion-resistant *steel alloy* made by combining iron with chromium (and sometimes other materials).

Static ballast system. A simple *ballast system* composed of fixed *weights* and floats, usually designed to make a vehicle about neutrally buoyant, so it hangs effortlessly in the water.

Static seal. Any type of seal made between parts that do not move in relation to each other. *O-rings* and *gaskets* are frequently used for static seals. Compare *dynamic seal.*

Stator. The stationary part of an electric *motor*, usually the outside.

Steel. A general term referring to a large group of different *metal alloys* in which iron is the dominant elemental metal. Additives including carbon, manganese, and silicon, cause the different alloys to differ in their hardness, *stiffness, corrosion* resistance, and other properties.

Stepper motor. A type of *DC motor* designed specifically for precise, repeatable position control. Rather than rotating smoothly, a stepper motor makes discrete steps of a few degrees or less for each electronic pulse received from a special *control* circuit. Used commonly in machines like computer printers that must move to exactly the right location on a page.

Stiffness. A measure of how much *force* it takes to bend or deform a *structure* by a certain amount.

Stiffness-to-weight ratio. A numeric value expressing how much *stiffness* you can expect to get out of a certain amount (*weight*) of a particular structural material.

Straight-run search pattern. A search pattern that consists of a series of straight lines running radially outward from a common starting point and separated by a fixed angle. Also known as a *radial search pattern.*

Strain. In mechanical engineering, a measure of deformation (in response to *stress*) usually expressed as a change in length, thickness, or other dimension divided by the original length, thickness, etc.

Strength. A measure of how much *force* a *structure* or part can withstand before breaking or suffering other permanent damage.

Strength-to-weight ratio. A numeric value expressing how much *strength* you can expect to get out of a certain amount (*weight*) of a particular structural material.

Stress. In mechanical engineering, the *force* per unit area applied to a *structure* or material.

Structurally reliant ROV. An *ROV* that is usually attached permanently or semi-permanently to an underwater *structure*, such as an oil rig, and moved along it by means of pulleys, *cables*, or other mechanisms for cleaning, inspecting, testing, and maintenance work.

Structure. In an underwater vehicle, the *system* that provides physical support, protection from *pressure*, attachment points, and possibly streamlining for other parts of the vehicle. Typically consists of the *frame, pressure hull* (or *pressure canisters*), and *fairing.*

Sub-bottom profiler. A *sonar* instrument that uses *sound waves* to penetrate the surface of the seafloor, returning echos from objects hidden under sediment. Used for mapping, geology, cable/ pipeline maintenance, archaeology, and treasure hunting.

Submarine. An *HOV* designed to move and operate under water, usually without the support of a surface ship. Most are used for military purposes, though a few are used for scientific research or undersea tourism. Submarines can be quite large (>150 m in length) and capable of carrying over 100 crew members and/or passengers.

Submerged weight. See *effective weight.*

Submersible. An *HOV*, generally smaller than a submarine and usually fitted with specialized *sensors*, robotic arms, and other gear to facilitate exploration and scientific research. These vehicles are capable of carrying a small crew and one or a few scientists and generally require the services of a support vessel.

Subsea connector. See *underwater connector.*

Subsystem. A *system* that functions as one of the parts within a larger system.

Supersystem. A *system* composed of other systems.

Supervisor. In large *ROV operations*, this is the person leading the ROV team, which usually includes at least two other people: a *pilot* and *winch operator*. The supervisor must be able to see the "big picture" and often has experience in a variety of subspecialties, including commercial diving, electronics, and *hydraulics*. See also *pilot, winch operator.*

Surf. When open ocean *swell* approaches shallow water, the lower portions slow down due to friction with the bottom, causing the upper parts of the swell to outrun the lower parts. As a result, the *swell* piles up and forms surf, sometimes called a *breaking wave.*

Surge. A form of water movement found near the bottom in shallow coastal areas where swell on the surface gets transformed into linear back-and-forth water motion along the seafloor.

Surge current. A brief, but large rush of electrical *current* in a *circuit.* As a *battery* specification: the maximum current a battery can safely deliver for a brief period of time without damage. Usually expressed in *amps*, but can also be expressed indirectly in terms of something called a *C-rate* (or *C-rating*), which is basically a conversion to convert the *amp-hour* rating of the battery into a maximum output current.

S-video. One of several standard *analog video formats.*

Swell. A form of water movement in which the *waves* spread out into a smooth undulating form that that causes seagoing ships to slowly rise and fall at sea.

Switch. A mechanism for creating a reversible *insulating* gap in a *circuit*. Can be used to block or allow the flow of *electric current* through the *circuit*, thereby by turning a *load* ON or OFF.

Switching regulator. A type of *voltage regulator* that adjusts digital pulses and averages them over time to *control* output *voltage*. Generally more *energy*-efficient than *linear regulators*, so often used in high *power* applications, but require more complex support *circuits* and generate more electrical "noise," which can interfere with *analog sensors*.

Synchronous data transmission. A form of *serial data transmission* in which a shared timing *signal*, such as a *clock pulse* on a separate *wire* connected between the sending and receiving device, is used to coordinate the transfer of digital data. The clock pulse tells the receiving device when the next *bit* of valid data from the sending device is available on the data wire(s).

Syntactic foam. A stiff, nearly incompressible flotation material made by embedding very tiny (almost microscopic), hollow, *glass* spheres in *epoxy*. While not as buoyant near the surface as some more familiar forms of foam, it maintains most of its *buoyancy* all the way down to full ocean *depth* and has become the standard flotation material used on *work class ROVs* and other deep-diving vehicles with *static ballast systems*.

System. A collection of parts (and/or processes) that work together to do (or be) something the individual parts cannot do (or be) by themselves.

Systèm International d'Unités (SI). A precisely-defined, highly standardized, subset of the *metric system* and now the most widely used system of measurement in the world. (Adopted by all but 3 of nearly 200 countries.) Characterized by a small number of distinct "base units" including the *meter*, *kilogram*, and second; this system is also known as the *MKS system*.

T

Taps and dies. Hardened *metal* cutting tools for adding screw threads to holes or *rods*. Taps are used to cut threads on the inside of a hole, and dies are used to cut threads on the outside of a cylindrical rod. Often sold in tap and die "sets" that include an assortment of standard thread sizes.

Teflon. A hard, slippery, *plastic*-like substance used for some types of bearings and other applications where low-friction is critical.

Telepresence. A technology-mediated sense of being physically present in a distant location. Usually created when a human *pilot* operates a remotely controlled vehicle in real time from afar while receiving live video images (sometimes 3-dimensional) from cameras located on the vehicle. *ROVs* provide a form of underwater telepresence.

Temperature. Technically, a measure of the average kinetic *energy* of molecules in a substance. Temperature is important in a wide range of physical, chemical, and biological processes.

Tether. On a small *ROV* working under water, this is the *cable* which carries electrical *power* and commands from the surface to the ROV and returns video *signals*, *sensor* readings, and other data from the ROV to the *pilot*. On larger ROVs, the term may refer only to the relatively short, thin cable connecting the vehicle to its underwater *garage* or *cage*. See also *umbilical*, *deck cable*.

Tether Management System (TMS). A mechanical system for efficiently handling *tether* deployment and recovery. The TMS may be located on the ship or oil *platform* deck for *ROVs* in which a single *cable* functions as both *umbilical* and tether, or it may be located under water between the end of the umbilical and the beginning of the tether when the two cables are distinct. When located at the end of the umbilical, it is usually housed in one of two basic arrangements: (1) A *garage*, sometimes called a *cage* or *launcher*, which is a frame, dangling deep under water on the end of the umbilical, that the ROV is flown out of (for launch) or into (for recovery), or (2) a *tophat* assembly, which has the ROV temporarily attached beneath a frame that houses the motor-driven tether reel and associated *slip ring assembly*.

Tethered free-swimming ROV. An *ROV* that is *tethered* to a ship or other *platform* on the surface but is otherwise free to move around in three-dimensions, propelled by its *thrusters*.

The bends. See *decompression sickness*.

The Doppler Velocity Log (DVL). An automatic log of *acoustic Doppler navigation* readings which provides a cumulative record of vessel movements, allowing precise determination of a vessel's position relative to its starting point.

Thermal conductivity. A measure of a material's ability to conduct heat.

Thermistor. A type of *temperature sensor* based on a *resistor* whose *resistance* changes with temperature in a predictable way. An electronic *circuit* can measure the temperature by measuring the electrical resistance of the thermistor.

Thermocline. A region of water in which the *temperature* changes quickly with *depth*, such as where the relatively warm water of the upper *mixed layer* in the ocean meets the cold, *unmixed layer* of water below.

Thermoplastic. A group of different *plastics* that soften and can be molded or extruded when heated. Examples: *PVC*, *ABS*.

Throw. In electric *switch* specifications, the number of possible ON positions offered by the switch. Example: a triple-throw switch used in a 3-speed appliance could have positions for high, medium, and low (in addition to OFF).

Thrust. An actively generated, *energy*-requiring, propulsive *force* used specifically to cause or *control* vehicle movement.

Thruster. A device for generating a propulsive *force* when there's nothing solid to push against. On *ROVs* and *AUVs*, a thruster typically consists of a *propeller* attached to the rotating *shaft* of an electric or *hydraulic* motor.

Thruster test jig. A mechanism which you can design and build at home or school to measure the *thrust* produced by a *thruster*. Useful while experimenting with different combinations of motors and *propellers* to find the best combination for your underwater vehicle.

Tidal current. A *current* that flows in response to changing *tides*.

Tides. Changes in the height of the ocean surface (and inlets such as bays, harbors, and estuaries connected to the ocean) caused by interactions between the gravitational attraction of the sun and moon, the *mass* of the ocean water, the rotation of the earth, and undersea topography. See also *tidal current*.

Time-division multiplexing. A *signal multiplexing* technique wherein *signals* (usually digital signals) are broken into short segments, which then take turns traveling over the *wires*, *optical fibers*, airwaves, or other channel, before being reassembled correctly at the other end. Example: A video sent over the internet is a huge file that will be broken into small packets of data, sent along with other packets containing *bits* of e-mail, web pages, and other internet traffic, then reassembled at the other end as the packets arrive and are separated from the non-video packets.

Titanium. An elemental *metal* most commonly used in several titanium *alloys*, which offer excellent resistance to saltwater *corrosion*, a very high *strength-to-weight ratio*, and high resistance to fracture.

TMS. See *tether management system.*

Tool sled. Essentially a toolbox carried by an *ROV* (usually a large work class or scientific ROV) to provide it with the specialized suite of tools, instruments, or other *payload* items it needs for a particular *mission*. Tool sleds containing different payloads can be swapped between dives, allowing a single ROV to perform an enormous range of different tasks.

Tool tray. A modular drawer or similar supporting *structure* pre-loaded with a set of tools, containers, instruments or other gear needed for a particular *mission*, then mounted on an *ROV* or *submersible* as part of its *payload*. Different tool trays can be prepared in advance with different sets of gear, then swapped quickly when a vehicle returns to the surface, allowing rapid re-configuration of the vehicle for the next dive.

Tophat tether management system. One type of *tether management system*. See the *tether management system* entry for details.

Torpedo. A type of undersea weapon. The term torpedo originally referred to floating explosive devices, which today would be called mines. Later the term was applied to explosives mounted at the end of a long pole, or spar, on the front of ships or *submarines*, which would ram an enemy vessel. Today, the term usually refers to unmanned, self-propelled, missile-like underwater vehicles carrying explosive warheads. Modern torpedos can be launched from submarines, ships, or airplanes, are usually driven by *propellers*, and have sophisticated navigational systems to help them find their target.

Torque. Noun: Informally, a twisting *force*, such as that applied by a wrench to a bolt. The *magnitude* of a torque is equal to the force multiplied by the *moment arm*. Verb: To exert a twisting force. See also *foot-pound* and *newton-meter*.

Tow sled. A platform, towed under water behind a ship or boat that is equipped with cameras and other *sensors* or instrumentation to collect information about the *water column* or the lakebed/seafloor. Sometimes equipped with *dive planes* to adjust *depth* or altitude, but otherwise powered only by the ship or boat towing it.

Towfish. A *side-scan sonar* instrument, shaped like a *torpedo*, that is towed under water behind a survey ship or boat to conduct a side-scan sonar survey of the seafloor or lake bottom.

Trailing edge. The rear edge of a *propeller blade* (when the *prop* is spinning in its normal forward direction). Usually rounder than the *leading edge* of the *blade*.

Transistor switch. A transistor used as an electrically-controlled *SPST switch*.

Transistor. A type of *semiconductor* device commonly having three *wires* and the useful property that a small electrical *signal* on one of the wires can be used to regulate the amount of *current* flowing between the other two wires. Can be used as an amplifier or as an electrically-controlled *SPST switch*. Invented in the 1950's and revolutionized electronics. Now forms the main building block of most *integrated* circuits, including computer *microprocessors*. Comes in two basic forms: *bipolar junction transistors (BJTs)* and *field effect transistors (FETs)*.

Triangulation. A method of determining position based on finding the intersection of two straight lines, which are often called *lines of position.*

Trim. As a verb, the word trim means to adjust a vessel's orientation (specifically, its pitch and roll) so that it is perfectly upright or otherwise oriented correctly. In the case of an undersea vehicle, the term may also include adjusting the vehicle's buoyancy along with pitch and roll. As a noun, the word trim refers to the physical orientation of a vessel, especially the pitch and roll angles relative to the vessel's normal or upright orientation. In the case of an undersea vehicle, this concept of trim may include the vehicle's buoyancy in addition to its pitch and roll.

Trim tank. One tank in a network of two or more tanks that are interconnected and arranged inside a *submarine*, *submersible*, or aircraft, in such a way that *liquid* can be pumped from one tank to another to shift the *center of gravity* of the vehicle. This allows easy, controlled, adjustment of vehicle *trim* while underway.

Trimming a vehicle. The process of adjusting a vehicle's *trim*.

Trolling motor. A type of *battery*-powered *thruster* used commonly to propel a small boat slowly and quietly for sport fishing. Can be adapted for use as an *ROV* or *AUV* thruster.

Troubleshoot. Using a systematic, efficient, and effective approach to diagnose and correct a problem in a part or system that isn't working properly. Also troubleshooting.

True north. The direction toward the north pole, which is where the earth spins about its axis.

Tsunami. A rare, but sudden, localized shift in sea level, usually as the result of a large earthquake, that can generate potentially dangerous *currents* or *waves* (called "tidal waves"), which sweep inland over low-lying coastal areas with little or no warning, pushing deadly walls of churning debris in their paths.

Tubing. A long, flexible, hollow cylinder, like a garden hose, usually made from plastic or *rubber* and used as a conduit through which to move fluid.

Turbid. Having high *turbidity*.

Turbidity. Cloudiness in water caused by the presence of suspended particles, such as silt or tiny plankton.

Turbidity current. A downward *current* caused by the rapid sinking of dense, *turbid*, particulate-laden water, such as might occur near the mouth of a river carrying very muddy water into the ocean. Turbidity currents are thought to play a role in the scouring of some undersea canyons.

U

U-boat. Abbreviation for *unterseeboot.*

U-cup seal. A type of *shaft* seal, often made from *rubber* or *plastic*, that is suitable for sealing rotating *motor shafts* or linear piston shafts against modest *pressure differentials*. Commonly used, for example, to seal *bilge pump* motor shafts.

Umbilical. A heavy-duty, armored electrical *cable* stored on the ship's *winch* drum and attached to a large *ROV's tether*, serving as the lift wire in launch and recovery of the vehicle. See also *tether*, *tether management system (TMS)*.

Undersea habitat. Subsea working and living quarters, anchored to the sea floor, where divers can live and work under water without coming to the surface for many days at a time. Also a place where a particular organism or group of marine organisms lives because of the presence of physical and biological conditions suitable for those organisms. Examples: a coral reef habitat or a kelp forest habitat.

Underwater connector. An *electrical connector* designed to be used under water. (Note that most can be plugged in or unplugged only while the connector is dry.)

Underwater habitat. See *undersea habitat.*

Universal Transverse Mercator (UTM) coordinate system. A standardized global coordinate system in which locations are specified in terms of the number of *meters* north or east of particular grid boundaries called Mercator lines, which are part of a particular rectangular grid superimposed on an ellipsoidal projection of the Earth's surface.

Unmanned undersea system (UUS). See *unmanned vehicle.*

Unmanned undersea vehicle (UUV). See *unmanned vehicle.*

Unmanned vehicle. A vehicle, such as an *ROV* or *AUV*, that operates without human occupants. Underwater, unmanned vehicles are also known by the acronyms *UUV* (*unmanned underwater vehicle*) and *UUS* (*unmanned undersea system*). See also *manned vehicle.*

Unterseeboot. German word for a *submarine* (lit. "undersea boat").

USB camera. A family of *digital video cameras* designed to plug into the *USB* port of a computer and allow video or still images to be recorded on the computer, e-mailed, or posted on the web. Popular for informal video conferencing. See also *web cam.*

USB communication protocol. A popular protocol for connecting computer keyboards, mice, monitors, printers, *USB cameras*, external hard drives, and other devices to computers.

UUS. See *unmanned vehicle.*

UUV. See *unmanned vehicle.*

V

Valve-regulated lead-acid (VRLA) battery. A type of *lead-acid battery* based on the same chemistry used in car batteries, but with the dangerous acid electrolyte rendered non-spillable and with the normal air vents replaced by safety pressure relief valves. These batteries are cleaner, safer, and easier to use than conventional flooded lead-acid batteries, yet preserve many of the performance characteristics of the flooded variety. Also known as a *sealed lead-acid battery.* See also *absorbent glass mat battery* and *gel cell.*

Vansalva maneuver. A technique for equalizing the *pressure* (i.e., eliminating the *pressure differential*) across the human eardrum to prevent pain and possible injury while descending during a dive. Consists of closing the mouth, pinching the nose, and blowing gently to squeeze air through the *Eustachian tube* into the middle ear. Also known as *clearing the ears.*

Variable resistor. See *potentiometer.*

Variable-pitch prop. Propellers designed so that the *propeller blades* can be rotated to change the *pitch* of the *propeller* while the *prop* is rotating. This allows the speed and sometimes the direction of the vessel to be changed without having to change the speed at which the *propeller shaft* is rotating.

Variation. See *declination.*

Vector. A quantity, such as a *force* or velocity, which has both a *magnitude* and a direction associated with it.

Vectored thrusters. *Thrusters* that are used in pairs with their *thrust* directed along non-parallel lines to create a wide range of possible vehicle movements.

Video format. The specific protocol (set of rules) used by a particular piece of video equipment to encode, transmit, record, or display video information. There are many standard video formats in common use. Each one specifies, for example, whether the *signal* is analog or digital, the dimensions of the image in terms of horizontal and vertical lines or pixels, and many other details. In general, camera, video recording system, and playback/display system must all use the same format. See also *digital video format* and *analog video format.*

Viewport. A thick, strong, *pressure*-proof window installed in a *pressure hull* or *pressure canister* to allow viewing of the subsea world during a dive.

Viscosity. A measure of the resistance of a *fluid* to rapid deformation. Informally, the "thickness" or "goopiness" of a fluid. Example: syrup has higher viscosity than water.

Viscous drag. That portion of total *drag* caused by the *viscosity* of the fluid. Also known as *skin friction drag.*

Visibility. The greatest distance at which objects can be observed visually through water at a particular place, time, and direction. Visibility is typically limited by *turbidity* and available light levels.

VLCROV. See *Low-cost ROV.*

Voltage (V, sometimes E or e). A unit of electro-motive *force* used in both the *metric* and *imperial systems* of measurement. You can think of voltage as a measure of electrical "pressure" that can push *electrical current* through *wires.*

Voltage divider. A pair of *resistors* wired together to provide a *voltage* that's a known fraction of another voltage in the circuit. The fraction is set by the ratio of the *resistance* values of the two *resistors.*

Voltage drop. The decrease in *voltage* between one point in a *circuit* and another point as a result of *energy* lost (usually as heat) when *electrical current* flows between those points.

Voltage regulator. An electrical device designed to maintain a constant *voltage* level for a *circuit*, in spite of possible fluctuations in the supply voltage. See also *linear regulator* and *switching regulator.*

Voltage source. An electric power source, such as a *battery* or a wall outlet that attempts to provide a predictable *voltage* while allowing the *load* to determine the *electrical current* that will flow in response to that voltage. Compare *current source.*

Volumetric energy density. A *battery's energy capacity* divided by its *volume.* See also *energy density* and compare *gravimetric energy density.*

W

Wall Wart. Nickname for an *AC-to-DC adapter* that plugs into a wall outlet.

Water column. A useful concept, rather than a physical reality, that consists of an imaginary column-shaped "tube" of water extending vertically from the surface to the floor of an ocean or lake. The concept is useful in measuring and describing the patterns with which *temperature*, *salinity*, and other water properties vary with *depth*.

Watt (W). A *metric* unit for measuring, quantifying and expressing *power*. One watt = 1 *joule*/second. One *kilowatt (kW)* is 1000 watts.

Watt-hour (Wh). A *metric* unit of *energy* equal to 3600 *joules* or 1/1000 kWh. See also *joules* and *kilowatt-hours*.

Waves. A form of water movement usually produced when wind blows over the surface of water forming ripples and then pushing them along, adding energy and amplifying their size. Also wave form patterns occurring in sound, energy, light, radio, electricity.

Web cam. A video camera, usually a *USB camera* or *IP camera* connected to the internet, whose video stream (or at least occasional "snapshot" frames) can be viewed over the internet via a standard web browser.

Web camera. See *web cam*.

Weight. The *force* of attraction between two objects as a result of *gravity*, especially the downward force experienced by an object pulled by gravity toward the center of a planet, moon, or other large celestial body.

Weight statement table. A chart, table, or spreadsheet used to keep track of the *weight* and *buoyancy* contributed to the vehicle by each of its parts. This is used to predict early in the design process whether the vehicle will float or sink, so that appropriate adjustments to its buoyancy can be made at the design phase, before parts are purchased and before construction begins.

Wet weight. See *effective weight*.

Wheatstone bridge. A simple, yet exquisitely sensitive, *circuit* configuration used commonly in precision *sensors*.

Whip. See *conductor whip*.

Winch operator. During large *ROV operations*, one team member (the winch operator) typically controls the *winch* for launch and recovery and performs *tether* management during the dive. This person typically doubles as the mechanical technician who maintains all the *hydraulics* and other mechanical systems.

Winch. A motorized spool or drum used to "reel in" and neatly store long *cables*. In large *ROV operations*, powerful winches with huge spools are used to manage as much as several kilometers of heavy *umbilical*/*tether*.

Windings. The coils of *wire* inside an electric motor, which convert *electric current* into magnetic fields that cause the motor to rotate.

Wire. A conductive pathway, usually made of a long, thin, *metal* (usually copper) core surrounded by a layer of *insulating plastic*, used to route electricity to particular locations.

Work class ROV. A type of *ROV* that is large, powerful, and versatile. These highly-maneuverable vehicles can be equipped with a variety of tools and *sensors* to perform numerous underwater tasks. Commonly used for heavy commercial and industrial tasks, such as those associated with the undersea oil and gas industry.

Y

Yaw. The angle through which a vessel, such as a boat or airplane, has turned right or left away from its previous course or other specified direction.

Z

Zero BG vehicle. A vehicle designed to have BG equal to zero, so it has no preferred upright orientation. Such a vehicle can work equally well sideways, upside down, or in any other orientation, so it is versatile, but such vehicles can also be very difficult to control.

Zinc. 1) A type of *metal* frequently used to make *sacrificial anodes*. 2) A *sacrificial anode* made from zinc metal. See also *cathodic protection*.

INDEX

Notes: Page references to figures are in *italics.* AUV stands for autonomous underwater vehicles; ROV, for remotely operated vehicles.

A

B

C

R

S

T

U

V

W

X

Y

Z